4th International Conference on Intelligent Circuits and Systems

Journal of Physics: Conference Series Volume 2327

Online
7-9 April 2022

ISBN: 978-1-7138-6046-4
ISSN: 1742-6588

Printed from e-media with permission by:

Curran Associates, Inc.
57 Morehouse Lane
Red Hook, NY 12571

Some format issues inherent in the e-media version may also appear in this print version.

Institute of Physics
Dirac House, Temple Back
Bristol BS1 6BE UK

Phone: 44 1 17 929 7481
Fax: 44 1 17 920 0979

techtracking@iop.org

Additional copies of this publication are available from:

Curran Associates, Inc.
57 Morehouse Lane
Red Hook, NY 12571 USA
Phone: 845-758-0400
Fax: 845-758-2633
Email: curran@proceedings.com
Web: www.proceedings.com

TABLE OF CONTENTS

Journal of Physics: Conference Series **2327** (2022) 011001 doi:10.1088/1742-6596/2327/1/011001

Preface

The ICICS 2022 (International Conference on Intelligent Circuits and Systems '2022) is the 4th series of the conference which is organised biannually since 2016. ICICS 2022 was organised (Virtually) by School of Electronics and Electrical Engineering of Lovely Professional University, Punjab, India on 08th -09th April 2022.

The 4th International Conference on Intelligent Circuits and Systems '2022 was organized by the Organizing Committee ICICS 2022, initially scheduled offline on 08th -09th April 2022 at the Lovely Professional University, Punjab, India. Considering the criticalities induced by the spread of COVID-19, difficulties in travelling, social distancing, and strict rules to follow, the Organizing Committee made the difficult decision to organize the 4th International Conference on Intelligent Circuits and Systems '2022 completely online on 08th -09th April 2022, using the Zoom platform, to give the opportunity to the authors to introduce their valuable papers in 11 online parallel technical sessions. A total of 107 research papers were provisionally accepted and presented, out of which 78 research papers are finally accepted for publication in the 4th International Conference on Intelligent Circuits and Systems '2022.

The conference has highlighted particular areas of promising technological development, and supported the unifying themes like Emerging Technologies for sustainable development in Electrical Engineering, Intelligent Circuits, Intelligent Systems and Health care sector.

The 4th ICICS 2022 program scheduled three Keynote lectures given by Prof (Dr.) Joshua R Smith, Zeutschel Professor, Allen School of CSE and Department of ECE, University of Washington, USA, Prof (Dr.) Om P. Malik, Professor Emeritus University of Calgary, Canada and Dr. C. S. Ozveren Electrical Power Engineering Consultant at Oz Sustainable Energy Solutions Dundee, Scotland, U.K., while each author confined his talk in 15 minutes, including questions and answers. The participants were from many countries like UGANDA, Malaysia, USA, CANADA, UAE, and 17 different states of India, with a total number of 107 researchers connected to the 4th ICICS 2022 online event.

We would like to acknowledge all of those who supported ICICS 2022. The help and contribution of each individual and institution was instrumental in the success of the conference. In particular, we would like to thank the organizing committee for its valuable inputs in shaping the conference program and reviewing the submitted papers. We would also like to thank conference alerts.co.in for promoting the conference.

List of Committees, Co-chairs, International Advisory Committee, Local Organising Committee and Contact details for the declaration of these titles are available in this pdf.

Journal of Physics: Conference Series **2327** (2022) 011001 doi:10.1088/1742-6596/2327/1/011001

COMMITTEE MEMBERS

ORGANIZING COMMITTEE

Chief Patron

Shri Ashok Mittal, Chancellor, Lovely Professional University, Punjab, India.

Patrons

Mrs. Rashmi Mittal, Pro-Chancellor, Lovely Professional University, Punjab, India.

Er. H.R Singla, Director General, Lovely Professional University, Punjab, India.

Conference General Chair

Prof. Lovi Raj Gupta, Pro Vice chancellor, Lovely Professional University, Punjab, India

Prof. Dinesh Kumar Singh, Executive Director, All India Institute of Medical Science, Bathinda, Punjab, India

Dr. Monica Gulati, Dean, Division of Research and Development

Conference Chair

Prof. Bhupinder Verma, Dean, School of Electronics and Electrical Engineering

Conference Co-Chair

Dr. Gaurav Sethi, Additional Dean, School of Electronics and Electrical Engineering

Publication Chair

Dr. Amit Kumar Singh, Associate Professor, School of Electronics and Electrical Engineering

Dr. Akhil Gupta, Associate Professor, School of Electronics and Electrical Engineering

Conference Organizer

Dr. Amit Kumar Singh, Organizing Secretary, Associate Professor, School of Electronics and Electrical Engineering

Dr. Akhil Gupta, Organizing Secretary, Associate Professor, School of Electronics and Electrical Engineering

Conference Co-Organizer

Dr. Megha, Joint-Organizing Secretary, Associate Professor, School of Electronics and Electrical Engineering

Dr. Raam Dheep, Joint-Organizing Secretary, Assistant Professor, School of Electronics and Electrical Engineering

TECHNICAL PROGRAM COMMITTEE

International Advisory Board Members

Dr. Ramesh Bansal, University of Sharjah

Dr. Vinod kumar Shukla, Amity University Dubai, UAE

Dr. Chetna Nagpal, Abu Dhabi Vocational Education and Training Institute (ADVETI)

Dr. Antonio Sergio Bezerra Sombra, Universidade Federal do Ceara

Dr. JoyDev Ghosh, Tomsk Univerity, Russia

Dr. Huseyin Haci, North East University, Nikosia

Dr. Jako Zoltan, Broad Bit Energy Technology

Dr. Swarup Bhunia, Case Western Reserve University, USA

Dr. Tejinder Singh, University of Waterloo

Dr. Ateet Dutta, UNAM Mexico

Dr. Harpreet Singh, Wayne State University

Dr. Joshua R Smith, University of Washington

Dr. O.P. Malik, University of Calgary

Dr. C S Ozveren, Oz Sustainable Energy Solutions

Dr. Patrick D Cerna, Haramaya University

Dr. Bamidele Adebisi, Manchester Metropolitan University

Dr. Otsuki Tomoaki, Keio University

Dr. Marco Di Renzo, Paris Saclay University

Dr. Ning Zhang, University of Windsor

National Advisory Board Members

Dr. Ghanshyam Singh, MNIT Jaipur

Dr. Akhilesh Pathak, AIIMS Bathinda

Dr. Y.R. Sood, NIT Hamirpur

Dr. Deepak Garg, Bennett University, Greater Noida.

Dr. Davinder Singh Saini, Chandigarh College of Engineering

Dr. Prerna Gaur, NSIT, Delhi

Dr. Bhaskar Bhattacharya, IILM Gurgaon

Dr. Promod Kumar Singh, Sharda University

Dr. B.S. Saini, NIT Jalandhar

Dr. Dilbag Singh, NIT Jalandhar

Dr. Rajesh Khanna, Thapar University

Dr. Kulbeer Singh, Thapar University

Dr. Neelu Jain, PEC

Dr. Rajoo Pandey, NIT Kurukshetra

Dr. R.K.Agarwal, NIT Kurukshetra

Dr. Samarjeet Ghosh, Thapar University

Dr. Suhas G Sapate, ADCET

Dr. Anil Kumar, Chandigarh College of Engineering

Dr. Sunil Dhingra, USIC Kurukshetra

Dr. Dharmendra, IIT Roorkee

Dr. Arun Khosla, NIT Jalandhar

Dr. Manoj Duhan, D Crust , Muruthal

Dr. Bhasker Gupta, Chandigarh College of Engineering

Dr. Neeraj Kumar, Thapar University

Dr. Manik Lal Das, DAIICT

Dr. Prasanjeet Majumdar, DAIICT

Dr. K.S.Nangla, NIT Jalandhar

Dr. Harsh Kumar Verma, NIT Jalandhar

Dr. Sunil Singla, Thapar University

Dr. Vikas Mittal, NIT Kurukshetra

Dr. Partho Preetam Roy, IIT Roorkee

Dr. Chirag Modi, NIT Goa

Dr. Meenakshi Sood, NITTR

Dr. Shruti Jain, JP Universoty of Infotech

Dr. R K Jha, IIIT, Jabalpur

Dr. Dheeraj Khatod, Indian Institute of Technology Roorkee, India

Dr. Chinmoy Saha, Indian Institute of Space Science and Technology

Dr. Raghvendra Kumar Chaudhary, IIT Dhanbad

Dr. Manmeet Kaur, AIIMS bathind

Dr. Kavin Khatri, AIIMS bathinda

Dr. Ankita, AIIMS bathinda

Dr. Pankaj Dahiya, DTU, New Delhi

Dr. V S K V Harish, NSUT, Delhi

Dr. Neeraj Gupta, National Institute of Technology Srinagar Jammu & Kashmir, India

Dr. Farhad Ilahi Bakhsh, National Institute of Technology Srinagar, Jammu & Kashmir, India

Dr. Rajkumar Viral, Amity University Uttar Pradesh, Noida, India

Dr. Suresh Limkar, AISSMS IOIT Pune, India

Dr. Tushar Kanti Bera, NIT Durgapur

Dr. Jyoteesh Malhotra, GNDU

Dr. Sanjay Jain, Thapar University

Dr. Balwinder Raj, NITTR Chandigarh

Dr. Velamuri Suresh, SYMBIOSIS International University

Dr Jai Shanker Prasad Kesari, GL Bajaj ITM

LOCAL COMMITTEE

Technical Committee

Dr. Kamal Kumar Sharma, Professor, School of Electronics and Electrical Engineering

Dr. Shamik Chatterjee, Assistant Professor, School of Electronics and Electrical Engineering

Finance Committee

Mr. R K Sharma, Assistant Professor, School of Electronics and Electrical Engineering

Dr. Raam Dheep, Assistant Professor, School of Electronics and Electrical Engineering

Mr. Shashi Pal, Administrative Officer, School of Electronics and Electrical Engineering

Coordination Committee

Ms. Kanu Sharma, Assistant Professor, School of Electronics and Electrical Engineering

Dr. Megha, Associate Professor, School of Electronics and Electrical Engineering

Publication Committee

Dr. Amit Kumar Singh, Associate Professor, School of Electronics and Electrical Engineering

Dr. Akhil Gupta, Associate Professor, School of Electronics and Electrical Engineering

Journal of Physics: Conference Series **2327** (2022) 011002 doi:10.1088/1742-6596/2327/1/011002

Peer Review Statement

All papers published in this volume have been reviewed through processes administered by the Editors. Reviews were conducted by expert referees to the professional and scientific standards expected of a proceedings journal published by IOP Publishing.

Type of peer review: Double anonymous

Conference submission management system: Morressier

Number of submissions received: 280

Number of submissions sent for review: 280

Number of submissions accepted: 78

Acceptance Rate (Submissions Accepted / Submissions Received × 100): 27.8 %

Average number of reviews per paper: 2.00

Total number of reviewers involved: 14

Contact person for queries:

Name: Akhil Gupta

Affiliation: Lovely Professional University

Email: akhil.20239@lpu.co.in

Performance Evaluation of Conventional Inverters Driven PMSM Drive using Microcontroller

[1]Rakesh G. Shriwastava, [2]Nilesh C. Ghuge,[3] D.D.Palande, [4]Ashwini Tidke

[1]Professor, Electrical Engineering Department, MCOERC, Nashik (M.S.) - India
[2]Associate Professor, Mechanical Engineering Department, MCOERC, Nashik (M.S.), India
[3]Associate Professor, Mechanical Engineering Department, MCOERC, Nashik (M.S.), India
[4]Assistant Professor, Electrical Engineering Department, MCOERC, Nashik (M.S.) ,India
rakesh.shriwastava@matoshri.edu.in

Abstract. This paper focuses on Performance evaluation of conventional inverters driven PMSM drive using microcontroller. The purpose of this paper is to decrease the ripples in torque of Micro controllers based PMSM drive. The innovative method consists of conventional inverters, switched-mode power supply (SMPS), PMSM motor and Microcontroller. It is used to maximize fundamental component of torque also. The both the results of the suggested Micro controller are compared on the basis of torque and speed and it proves that reduces the torque ripple and improve the dynamic response of the system in comparative analysis. This paper organized introduction, mathematical model of the PMSM, proposed Microcontroller technique, hardware and simulation results and conclusion in different sections.

Keywords: Switched-mode power supply (SMPS), conventional inverters, Microcontroller, Permanent magnet synchronous motors (PMSM), EPROM, Clock generator.

1. Introduction

In industrial application, Electric motors (EMs) and generators are the primary workhorses. The generators convert energy transformation to charge the batteries for motor operation. Required torque developed by motors for drive the wheels. The permanent magnet, switched reluctance, and induction motors are used in industrial application [1-4]. This paper explains Simulation and Hardware Implementation of VSI -Fed PMSM drives for automotive application. For the speed &torque control,VSI fed PMSM drive is one of the method used. The VSI fed PMSM drive system is designed, simulated and implemented using Microcontroller. A hardware setup is designed and implemented based on a Four-pole 0.25 HP PMSM.The control hardware consists of, Main power circuit, Control circuit, Isolator and driver circuit and PMSM [5-11]. The fixed dc voltage is obtained from three phase rectifier circuit and used shunt capacitor for filter purpose. The output of three phase rectifier circuit is given to MOSFET bridge inverters in power circuit. [12]. The function of control circuit is to turn the MOSFET by using Gating pulses Isolator and driver circuit is used for proper isolation by using opt isolator [25-26]. The steps involved in any drive system are design, building its hardware and its testing. Once the system is design, depending on designed values of various components, the system hardware is developed. Once the assembly of hardware is completed then it is tested by carrying out various experiments on the developed system to test whether the system is working satisfactory or not. If it is not working properly then the error must be removed and the system must be updated and modified. The drive developed here is operating variable frequency mode. The theme of the paper is Simulation and Hardware analysis of torque minimization of PMSM drive for automotive application.

2. Mathematical Model of PMSM

a) PMSM Modelling in rotor reference frame

There are two stator windings in the dq reference frame and field winding is absent in rotor of PMSM. The direct-axis winding is along with the axis of magnetic pole. If PMSM running in anti-clockwise direction, at the speed of 'ωr',

The induced voltage in the direct-axis winding:

$$u_d = R_d i_d + \frac{d\lambda_d}{dt} - \omega_r \lambda_q \quad (1)$$

Where, 'id' and *'Rd'* are called direct-axis stator current and resistance respectively

The induced voltage in the quadrature-axis winding:

$$u_q = R_q i_q + \frac{d\lambda_q}{dt} - \omega_r \lambda_d \quad (2)$$

Where, 'Rq' and '*iq*' are called the quadrature -axis resistance and current of stator.

$$\lambda_d = L_d i_d + \lambda_m \quad (3)$$

λd = flux-linkage in the direct-axis of stator in webers,λm is the PM rotor flux

$$\lambda_q = L_q i_q \quad (4)$$

λq = flux-linkage in the quadrature-axis stator (Wb)

In this case of quadrature-axis, there are no magnets so *λm* is absent.

Considering round rotor PMSM, we have

$$L_d = L_q \quad (5)$$

The PMSM torque equation is given by:

$$T_e = \frac{3}{2}\frac{p}{2}(\lambda_d i_q - \lambda_q i_d) \quad (6)$$

Substituting for *'λd'* and *'λq'* in the torque equation of PMSM,

$$T_e = \frac{3}{2}\frac{p}{2}[(\lambda_d i_d + \lambda_m) i_q - L_q i_q i_d] \quad (7)$$

$$T_e = \frac{3}{2}\frac{p}{2}[(L_d - L_q) i_d i_q + \lambda_m i_q] \quad (8)$$

The two components of torque developed are:

$$Reluctance\ torque = \frac{3}{2}\frac{p}{2}(L_d - L_q) i_d i_q \quad (9)$$

$$field\ torque = \frac{3}{2}\frac{p}{2}\lambda_m i_q \quad (10)$$

$$T_e = \frac{3}{2}\frac{p}{2}\lambda_m i_q \quad (11)$$

In a round rotor PMSM the electromagnetic torque present is the field torque present due to the PM flux linkage, *λm*. For a chosen PMSM, the PM rotor flux-linkage *(λm)* and the number of poles (*p*) is constant. Hence, for the round-rotor PMSM, the electromagnetic torque equation is

$$T_e = Kt\, i_q \quad (12)$$

Where, Kt = Torque constant

$$K_t = \frac{3}{2}\frac{p}{2}\lambda_m \quad (13)$$

Therefore, electro-magnetic torque is given by

$$T_e = T_l + B\omega_m + J\frac{d\omega_m}{dt} \quad (14)$$

ωm = rotor's mechanical speed, *ωr* = rotor's electrical speed

3. Simulation Result

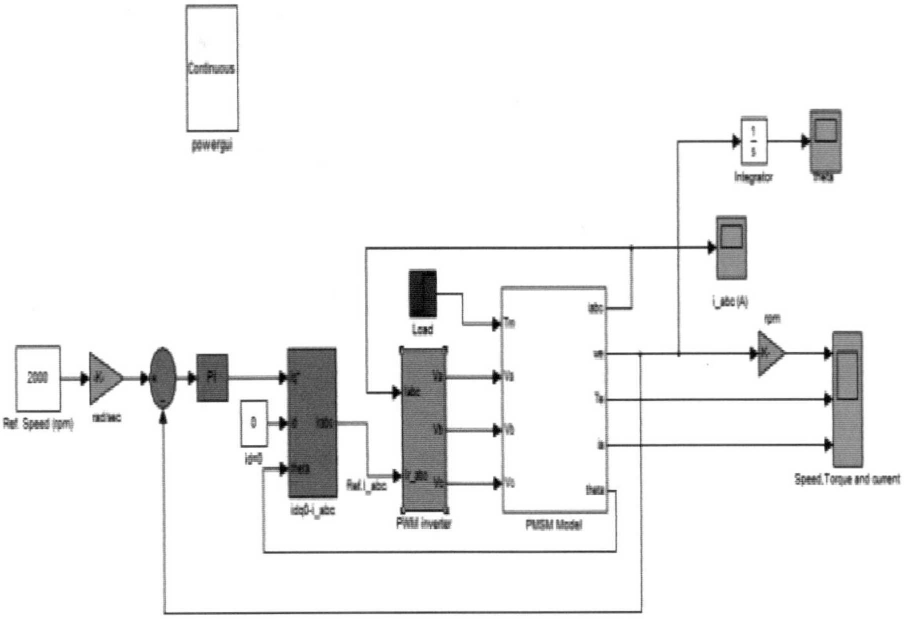

Figure 1.Simulation diagram

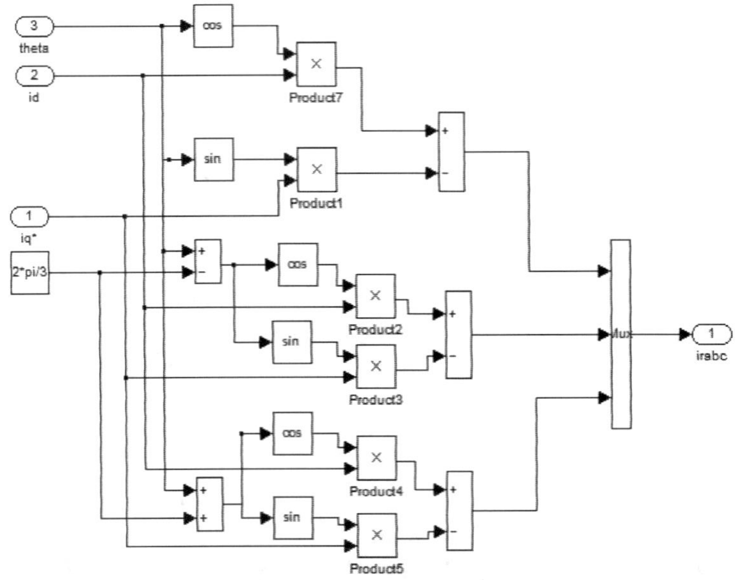

Figure 2.D-axis &Q-axis Currents model

Journal of Physics: Conference Series **2327** (2022) 012001 doi:10.1088/1742-6596/2327/1/012001

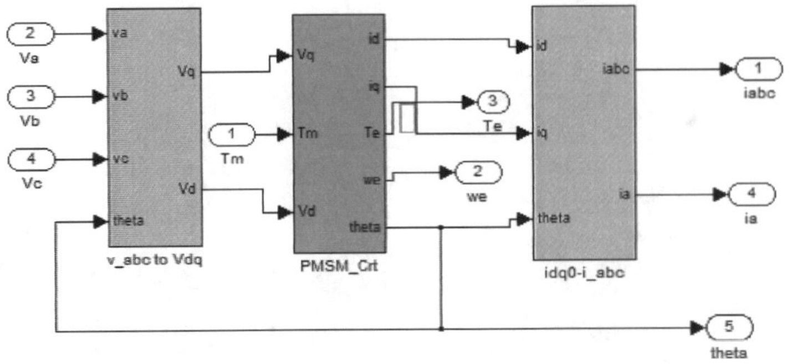

Figure3.PM Synchronous Motor model

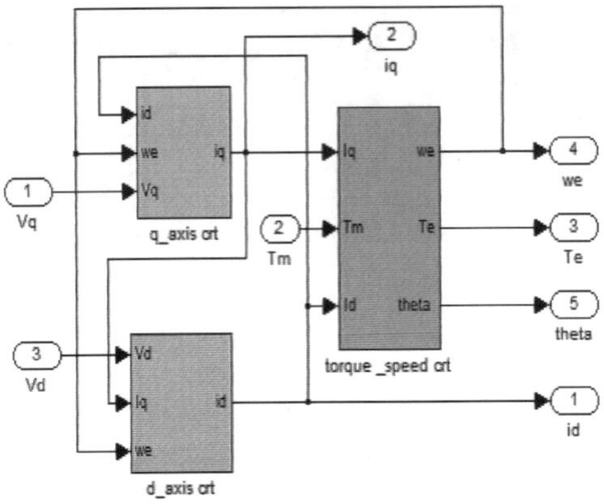

Figure4. PMSM Characteristics diagram

The simulation response of Voltage source inverter-fed PMSM drives at different speed range is shown in figure 5 to figure 9.

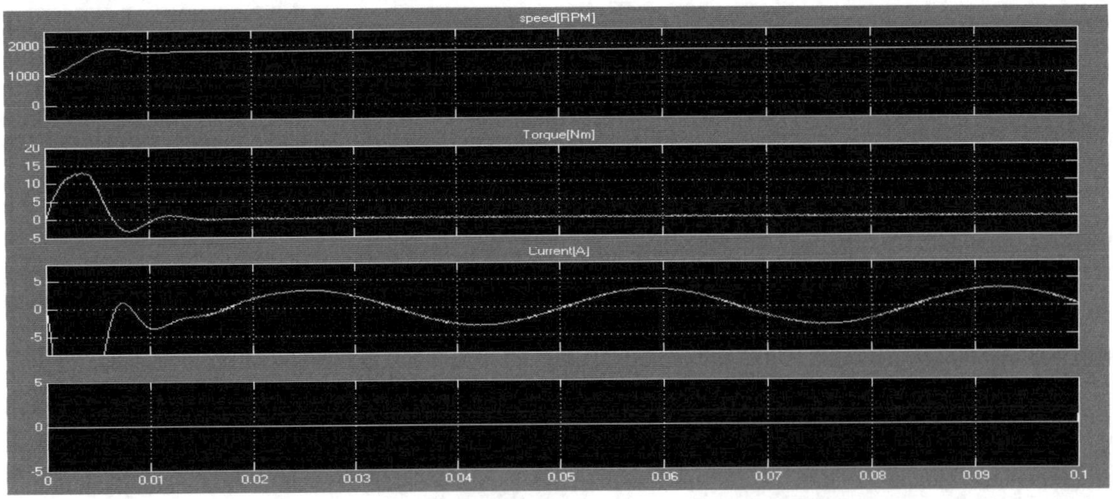

Figure 5.Speed, torque and Current response at 1000 to 1800 rpm

Journal of Physics: Conference Series **2327** (2022) 012001 doi:10.1088/1742-6596/2327/1/012001

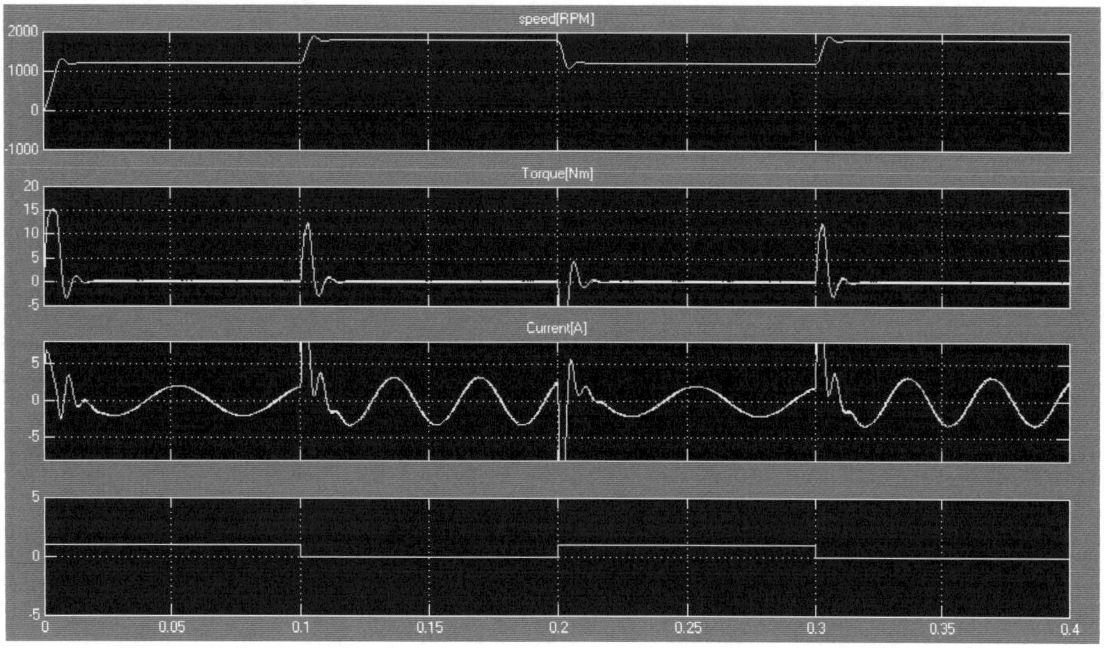

Figure 6.Speed ,torque and Current response at 1200 to 1800 rpm

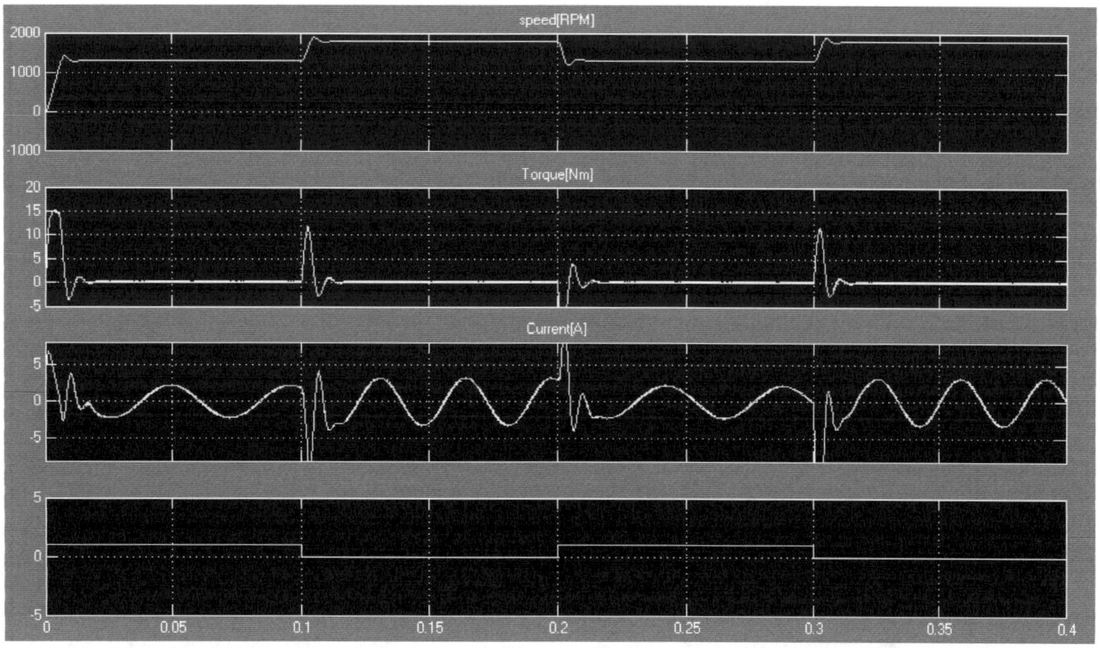

Figure 7.Speed ,torque and Current response at 1400 to 1800 rpm

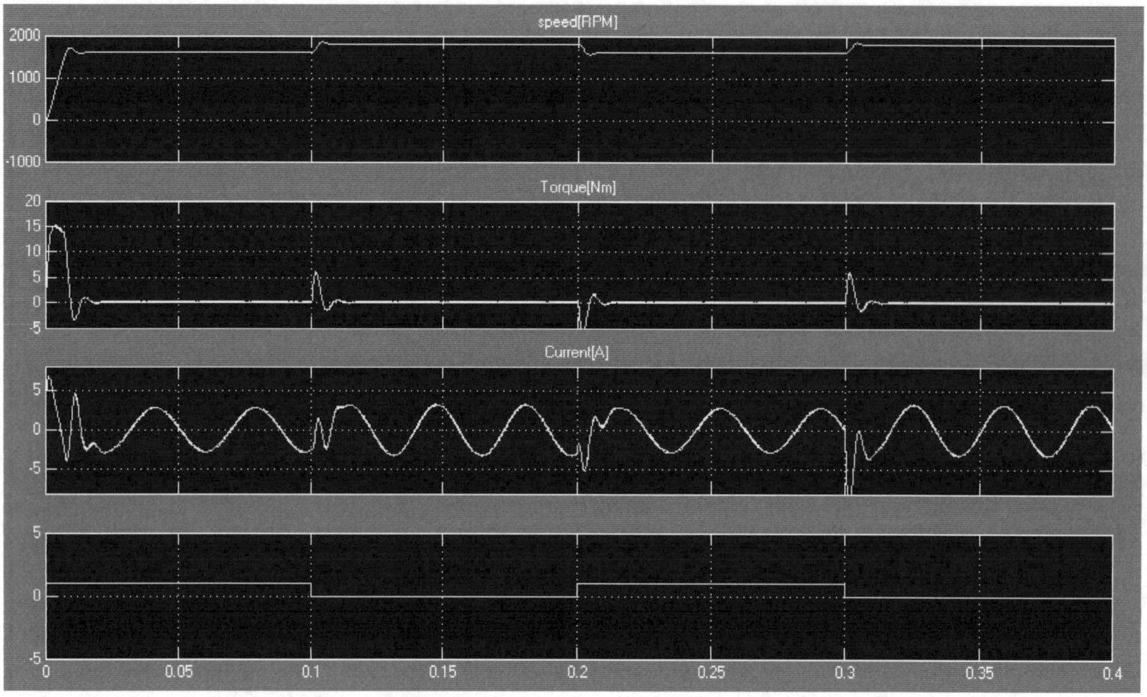

Figure 8.Speed ,torque and Current response at 1600 to 1800 rpm

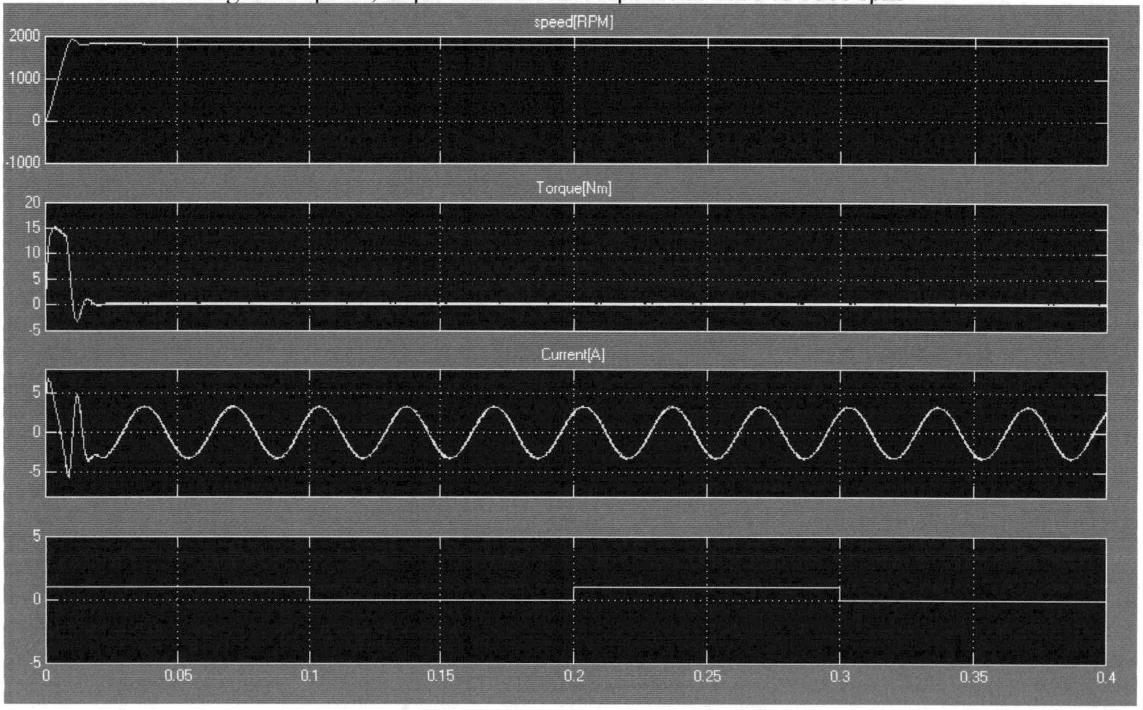

Figure 9.Speed, torque and Current response at 1800 rpm

4. Hardware Result Analysis

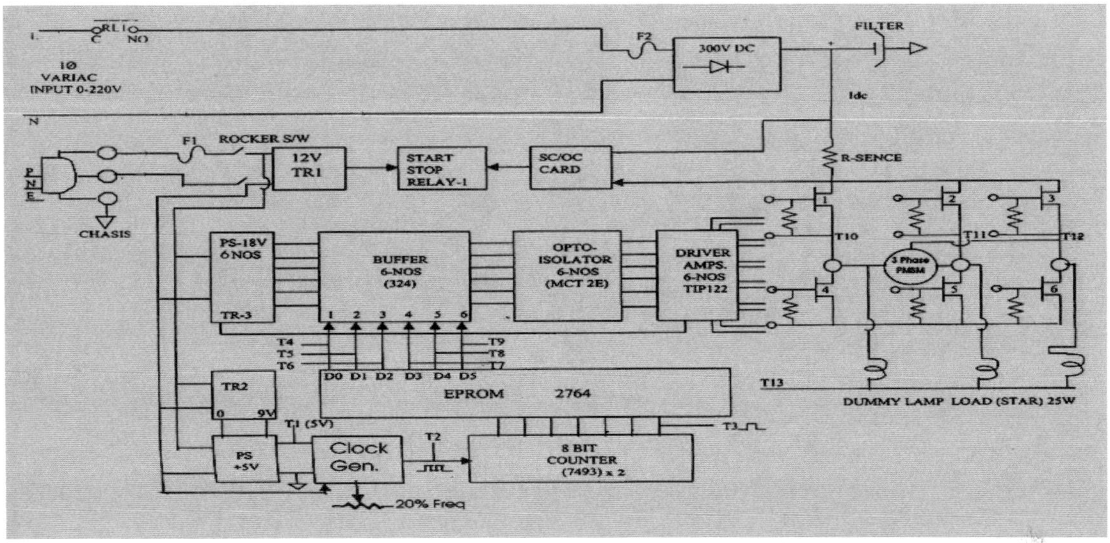

Figure 10.Block diagram of conventional inverters driven PMSM

Initially the drive operation was worked on dummy load (3 phase star connected lamps) by connecting the motor to the drive. Various outputs are taken out from the control card on the drive. Firstly the frequency knob on the drive was kept at its minimum position. The converter card was then switched on. The power supply status was indicated by all the LEDs inside the unit. By using CRO probes, the waveforms at various test points were observed. While observing the waveforms on CRO, care was taken to use the unearthed CRO only. The basic timer IC 555 output frequency was observed at the test point. This frequency can be changed by changing the position of frequency knob. The pulses for various gate drives were observed at the test points. The output frequency was observed at the test point. The minimum and maximum frequency was noted and tabulated in the result table. The frequency knob position was slowly changed along with the set speed pot. Slowly the lamps were turned on as the soft start in the result table. The output line and phase voltage waveforms for the three phases were observed on the CRO at the corresponding test points. After observing the satisfactory performance of the control card then the motor was connected to the drive. The dummy lamp load was now removed and the motor was connected in the output connector. Thus the motor runs at variable speed depending on the frequency. After testing the various test points the dummy star connected load is removed and across the output of the bridge inverter PMSM is connected. Then gradually increasing the load on the shaft of the motor for different values of frequency speed of the motor is measured.

Table No.1: Motor speed Variation by frequency

Sr.No.	Time (m.s.)	Frequency (Hz)	Actual Speed (rpm)	Measured Speed (rpm)	Voltage (Volts)
1	30	33.4	995	1005	260
2	25	40.1	1250	1219	265
3	22	45.5	1462	1370	265
4	20	50.2	1530	1518	265
5	18	55.6	1765	1672	265
6	17	59.2	1785	1795	265

Journal of Physics: Conference Series **2327** (2022) 012001 doi:10.1088/1742-6596/2327/1/012001

Table No.2: Motor speed Variation by load at 33.4

Sr. No.	Load (gm)	Frequency (Hz)	Actual Speed (rpm)	Expected Speed (rpm)	Torque (N-m)	Output Power (W)
1	500	33.4	1110	995	0.15715	16.59
2	1000	33.4	1110	995	0.2743	35.28
3	1500	33.4	1110	995	0.4814	48.99
4	2000	33.4	1110	995	0.6186	65.45
5	2500	33.4	1110	995	0.8357	79.98
6	3000	33.4	1110	995	0.9829	99.49

Table No.3: Motor speed Variation by load at 50.2Hz

Sr. No.	Load (gm)	Frequency (Hz)	Actual Speed (rpm)	Expected Speed (rpm)	Torque (N-m)	Output Power (W)
1	500	50.2	1500	1540	0.16715	25.89
2	1000	50.2	1500	1540	0.3043	49.96
3	1500	50.2	1500	1540	0.4814	76.45
4	2000	50.2	1500	1540	0.6886	99.79
5	2500	50.2	1500	1540	0.7857	125.51
6	3000	50.2	1500	1540	0.9229	148.78

Table No.4: Motor speed Variation by load at 59.2 Hz

Sr. No.	Load (gm)	Frequency (Hz)	Actual Speed (rpm)	Expected Speed (rpm)	Torque (N-m)	Output Power (W)
1	500	59.2	1700	1770	0.1571	29.64
2	1000	59.2	1700	1770	0.3243	58.16
3	1500	59.2	1700	1770	0.4914	89.73
4	2000	59.2	1700	1770	0.6588	115.33
5	2500	59.2	1700	1770	0.7957	142.90
6	3000	59.2	1700	1770	0.9229	172.49

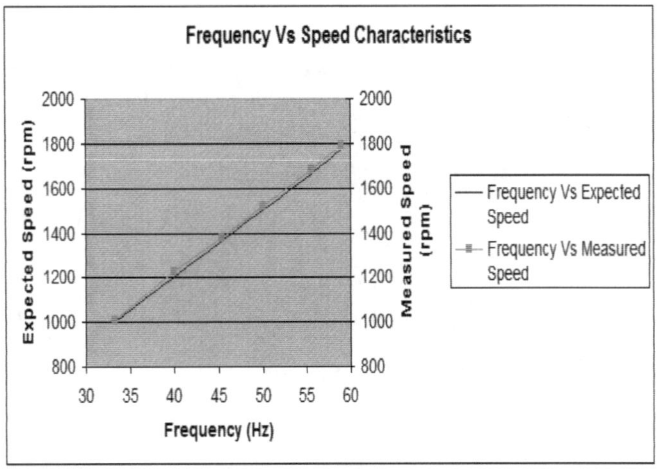

Figure 11.Speed- Frequency characteristics

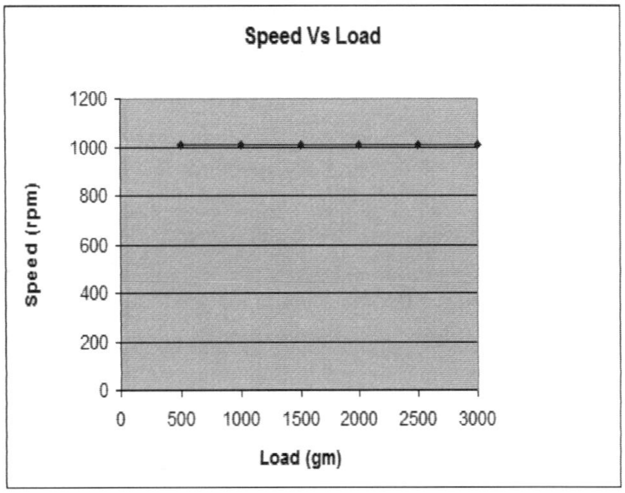

Figure 12.Load -Speed Characteristics at 33.3 Hz

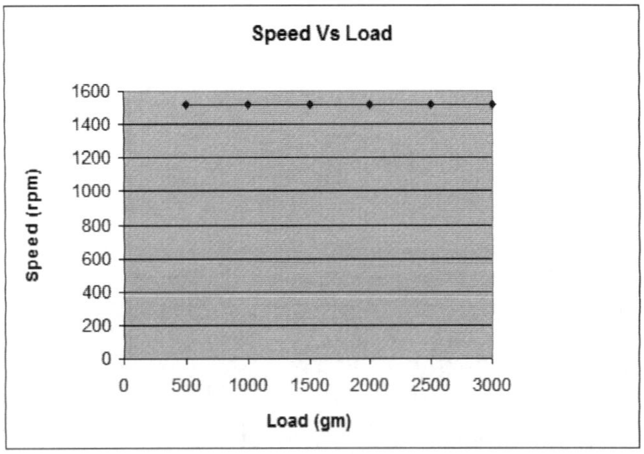

Figure 13.Load -Speed Characteristics at 50 Hz

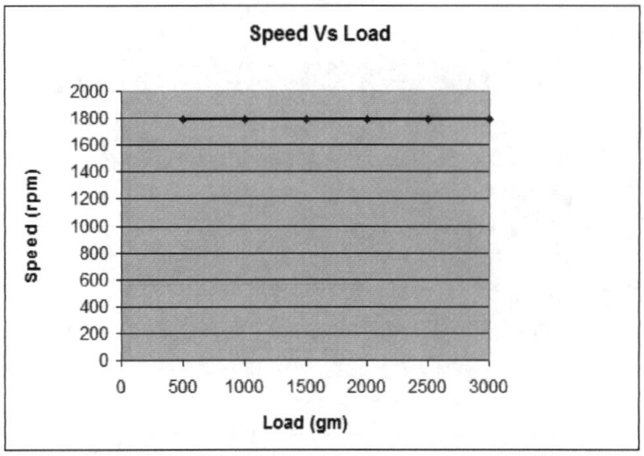

Figure 14. Load -Speed Characteristics at 59 Hz

Conclusion

This paper focus on simulation and hardware analysis of conventional inverters driven PMSM drive using microcontroller. A detailed Simulink model of torque minimization of PMSM drive has being developed. It is observed that, at constant frequency, speed remains constant irrespective of load. The motor runs at synchronous speed. Speed also gets changed accordingly the inverter frequency .The overall motor performance can be very well judged from the performance characteristic shown. Finally we conclude that simulation results and hardware implementation shall give better performance analysis. Hence it is used in automotive application.

https://orcid.org/0000-0001-8196-8299
https://orcid.org/ 0000-0002-0544-2006
https://orcid.org/ 0000-0002-1322-4076
https://orcid.org/0000-0001-6689-6880

References

[1] B.K.Bose, Modern Power Electronics an AC Drives (Prentice-Hall, 2002).
[2] W. P. Jung, et al., "Improvement of control characteristics of interior permanent-magnet synchronous motor for electrical vehicle,". IEEE Transactions on Industry Application, vol. 37, pp.1754-1760, Nov. 2001.
[3] L. Zhong, et al., "A direct torque controller for permanent magnet synchronous motor drives," IEEE Transactions on EnergyConversion, vol. 14, pp. 637-642, Sep. 1999.
[4] M. F. Rahman, et al., "A direct torque-controlled interior permanent magnet synchronous motor drives incorporating field weakening," IEEE Transactions on Industry Application, vol. 34, pp. 1246-1253, Nov./Dec. 1998.
[5] G. Buja, et al., "Direct torque control of PWM inverter-fed AC motors-a survey," IEEE Transactions on Industrial Electronics, vol. 51, pp. 744-757, Aug. 2004.

[6] D. Casadei, et al., "FOC and DTC: two viable schemes for induction motors torque control," IEEE Transactions on Power Electronics, vol. 17, pp. 779-787, Sep. 2002.
[7] Y. H. Li et al., "Simulation study on the effect of voltage Vector on torque in direct torque control system of permanent magnet synchronous motor," in Proc.2007 IEEE Industrial Electronics and applications Conf., pp. 1521-1525, 2007.
[8] Y. H. Li, et al., "A novel switching table to suppress unreasonable torque ripple for the PMSM DTC drives," in Proc.2008 IEEE Electrical Machines and Systems Conf., pp. 972-977, 2008.
[9] Y. H. Li, et al., "The control of stator Flux and torque in the surface permanent magnet synchronous motor direct torquecontrol system," in Proc.2009 IEEE Industrial Electronics and applications Conf., pp. 2004-2009, 2009.
[10] Y. H. Li, et al., "A Simplified Voltage Vector Selection Strategy for the Permanent Magnet Synchronous Motor DirectTorque Control Drive with Low Torque Ripple and Fixed Switching Frequency," in Proc.2010 IEEE Electrical Machines and Systems Conf., pp. 674-679, 2010.
[11] Y. H. Li, et al., "The Comparison of Control Strategies for the interior PMSM drive used in the Electric Vehicle," in Proc.the 25th World Electric Vehicle Symposium and Exposition, paper 4C-7, 2010.
[12] F. Heydari, A. Sheikholeslami, K. G. Firouzjah amd S. Lesan. Predictive Field-Oriented Control of PMSM with Space VectorModulation Technique Front. Electr. Electron. Eng. China(2010).
[13] M. S. Merzoug and F. Naceri,Comparison of Field-Oriented Control and Direct Torque Control for Permanent Magnet Synchronous Motor (PMSM) World Academy of Science, Engineering and Technology 45(2008).
[14] Jorge Zambada, Microchip Corporation, Sensorless Field Oriented Control of PMSM Motors. Microchip Technology Inc., 2007.
[15] Dr.R.G.Shriwastava, Dr. M. B. Daigavane,Dr. N.B.Wagh, "Comparative Analysis of Three Phase 2-Level VSI with 3-Level and 5-Level DCMLI Using CB-SVM",Published in IEEE International Conference on Electrical, Electronics, Computers, Communication, Mechanical and Computing (EECCMC-2018),at Priyadarshini Engineering College, Chettiyappanur, Vaniyambadi - 635751, Vellore District, Tamil Nadu, India. Conference records no 978-1-5386-4304-4/18/$31.00 ©2018 IEEE.
[16] Dr.R.G.Shriwastava,Dr. Nandkumar Wagh,Dr.Santaji K.Shinde "Implementation of DTC-Controlled PMSM Driven By A Matrix Converter", published in International Journal of Recent Technology and Engineering (IJRTE) Volume-8 Issue-5, January 2020,PageNo:4420-4424,ISSN:2277-3878,Retrieval Number: E6641018520/2020©BEIESP DOI:10.35940/ijrte.E6641.018520,.
[17] Dr.R.G.Shriwastava, Mr. D.R.Bhise,Dr. Mr. Pravin Nagrale "Comparative Analysis of FOC Based Three Level DCMLI Driven PMSM Drive", Published in IEEE Explore ICITAET 2019 Conference record no ISBN -978-1-7281-1901-4 Page no-26-31
[18] Dr.Rakesh G. Shriwastava, Mohan P. Thakre,Manoj D. Patil, Digamber R.Bhise " Performance Analysis of CB-SVM Three-Phase Five-Level DCMLI Fed PMSM Drive", published in Turkish Journal of Computer and Mathematics Education Vol.12 No.12 (2021), 887-89

A novel design of hybrid EV's based on flux additive DC–DC converter with super capacitor

Gondu Vykunta Rao [1*], Madhuri Bayya[2], Aruna Bharathi M [2], Baratam Murali [3]

[1]Birla Institute of Technology & Science, BITS Pilani, Secunderabad, Telangana, 500078.
[2]Geethanjali College of Engineering and Technology, Affi to JNTH, Cheeryal, Keesara (M), Medchal (Dist.)-501301.
[3]Sri Venkateshwara Engineering & Technology (SVCET) Affi to JNTU, NH16, Chennai - Srikakulam Hwy, Etcherla, Andhra Pradesh.

E-mail: vykunta.rao@wilp.bits-pilani.ac.in

Abstract. Now-a-days hybrid electric are employed with different dc-dc converters. Here in this paper, a flux additive dc-dc converter is put forward and analysed. The collation with conventional strategies we attain an enhanced output gain. The input sources like photovoltaic and fuel cell while they both are considered for power supply as primary sources and the photovoltaics are implemented for charging the BSS. The application of photovoltaic improves the efficiency of the BSS. The photovoltaic cells voltage is used to fed the converter. The implementation of Super Capacitor improves the output gain. The proposed converter supplies the demand to load side by operating to power resources simultaneously. The power supervision is analyses and then it is employed in control methodology. A simulation model is implemented for analyzation and to prove the efficiency of proposed converter.

1. Introduction

Greenhouse gases extract up the radiation from sun and discharge it into atmosphere which provokes the worsening of surface temperature of earth which causes the effect of global warming. Most of conventional vehicles emits these greenhouse gases and there is need for replacing these vehicles with electrical vehicles (EVs) which are clean sources of energy whose emissions are nil. The consideration of hybrid electrical vehicles (HEVs) is having greater efficiency when compared to conventional vehicles. EVs are mainly dependent on BSS as they carry BSS within them as power source [1]. Despite of having a few hiccups like more charging periods, the EV's are highly developed for ameliorating the advantages and the efficiency of them with implementing the bidirectional flow to facilitate the V2G technology in them. We are very well aware of solar powered EVs and their relocation and sizing in the distribution system is complicated and impractical currently [2]. So, we use PVs as the power source to charge the battery. Here fuel cell (FC) is the dominant power supplier. FC has many assets like low effusions, producing clean energy, higher operating efficiency, high current density potential [3]. With these assets there comes some liabilities like heavy costs and substandard transient performance.

Electrical vehicles are classified into many types and among them some vehicles are modified with using BSS with FC and Solar PVs. HEVs are those modified vehicles and they are having pros like fuel-wise economy, more flexibility with operating strategy, overcoming of FC cold start, minimal cost per unit power [4]. System here we are implementing is powered by FC and battery via PV source. Here in this paper, we use the V2G technology for converter. A flux additive DC-DC converter for PV/FC/battery is the proposed model in this paper. The converter has high advantage as it is having bidirectional power flow and also one need to note that more the number of elements, less is the converter efficiency. In general, the PV cost is primarily higher and after we start to gain advantages from the PV panel.

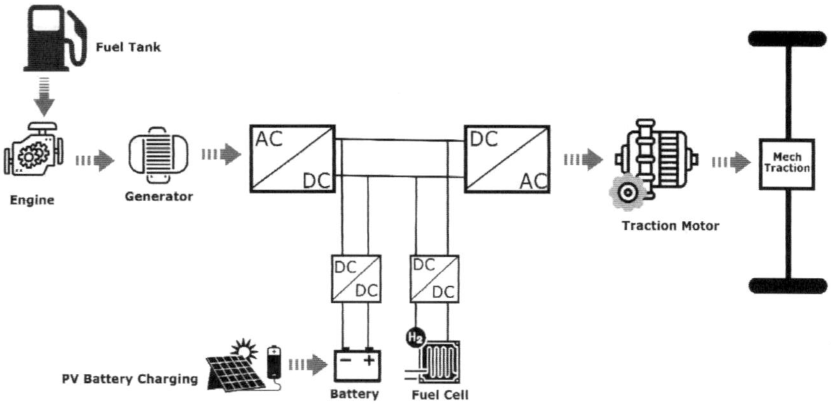

Figure 1. Schematic diagram representing basic multi-input DC-DC converter

We use MPPT algorithm to pull out maximum power from the PV panel [5]. Here the converter used is flux additive dc-dc converter which has a transformer circuit with two primary windings and one secondary winding in it. Transformer contains the duo coils which are electrically isolated with following the faradays laws for electromagnetic induction. The magnetic flux is created with the above law by voltage and current in the primary of the transformer which induces the emf in secondary coil of the transformer. Since the absence of moving shafts or rotors in the electrical vehicle to supply energy and so one can say that there are no frictional and winding losses in electrical vehicles. A global overview for current status of EV is discussed with state of art with its contributions in advanced engineering in transportation and grid balancing. Usage of multiple power sources make the system more reliable, flexible and efficient. Dropping out one of the power sources effects the output voltage. So, it is necessary to maintain the multi-input to the converter to ameliorate the performance characteristics. In general, the rechargeable cells/batteries act as power sources in common for EVs. For depicting higher performance levels compared with ICEs, the EVs operate with multi-input sources like batteries, FC, PV, super capacitors, ultra-capacitors etc., The basic configuration is depicted in Fig. 1. This figure portrays about the fuel tank connected to ICE and the generator connection is given which gives mechanical energy derived from electrical energy. In DC-DC converters one come across the super-capacitors acting as input sources similar to that of battery sources. In the transformer present in converter one can find coils isolated which facilitates the magnetic power transfer. Fluxes operation decides the type of converter. If the fluxes are in same direction and they get summed up to give additive effect of flux. If the fluxes are in opposite direction the get differentiated to give less resultant flux. In this paper flux additive converter is considered which operates on the principle of additivity of flux.

Here flux addition takes place with implementing multiple inputs for regulating the output voltages. The main objective in this paper is to adopt the PV, FC, battery to supply the EV and to improve the output voltage and performance of proposed converter in presence of super capacitor.

2. Proposed converter topology

The circuit of proposed flux additive dc-dc converter is depicted in Fig. 2. It primarily contains buck-boost type of converter with a super capacitor in it pictured in Fig. 3 and the BSS for storage of electric energy. The proposed converter mainly has four components i.e., the PV-B stage input circuit, the FC stage input circuit, the output stage circuit and the transformer. The PV-B stage input circuit contains primary winding of TPV-B with nPV-B turns in it and produces the flux φPV-B. The FC stage input circuit contains primary winding of TFC with nFC turns in it and produces the flux φFC. The output stage input circuit contains secondary winding of TT with nT turns in it and produces the flux φO. The transformer is a static electrical device which shifts electrical power from one circuit to another magnetically. The faraday's law of electromagnetic induction is the principle involved in the working of transformer.

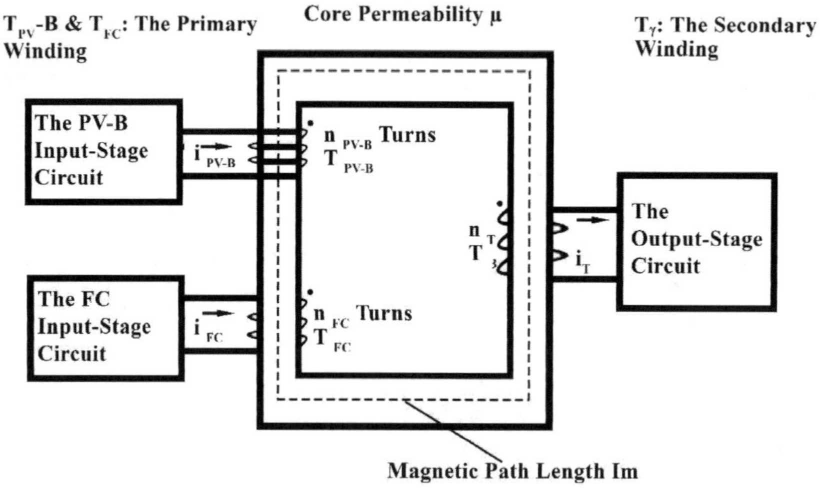

Figure 2. Block representation of Flux additive DC-DC converter

The converter proposed suits the requirement of the hybrid electric vehicles. The functioning of the converter with multi sources is depicted the section allocated for power management and MPPT algorithm part. The VPV-B and VFC are the independent sources of electrical energy whose characteristics influence the output. The inductances of PV-B and FC are given by L1 and L2 which acts as filters respectively. As L1 and L2 being in series with the sources make them act as current sources. D1, D2, D3, D4, D5, D6, D7, D8 are the diodes that control the modes. The capacitor C1 is used to improve the output gain and output voltage filter is given as the output capacitor Co. The converter operates with almost continuous conducting mode to supply the smooth current output with minimal ripple amount. The presumptions are made for simpler investigation of the proposed model and are given as:

- The choke inductance L1 and L2 are having greater inductances as the flow of current to be maintained constant.

- The idealized transformer is considered in order to ignore the leakage inductance and magnetized current.
- All the switches and diodes should be treated as ideal.
- Output Capacitor used as filter should have greater value to maintain output voltage constant [7].

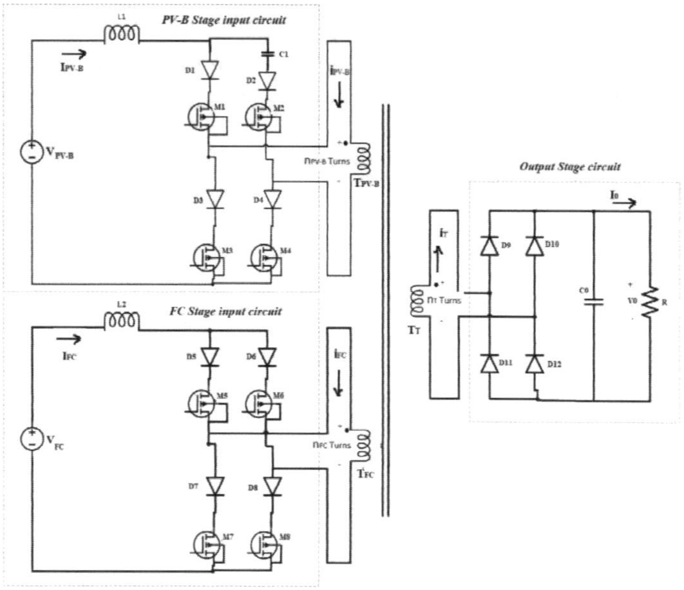

Figure 3. Circuit of Flux Additive DC-DC converter

The flux additive DC-DC converter having twelve different operating modes out of which we depict only mode1, mode 3, mode 7, mode 9 for result analysis. The timing diagram of gate pulses for 8 MOSFETs varying from t1-t6 are given in Fig. 4.

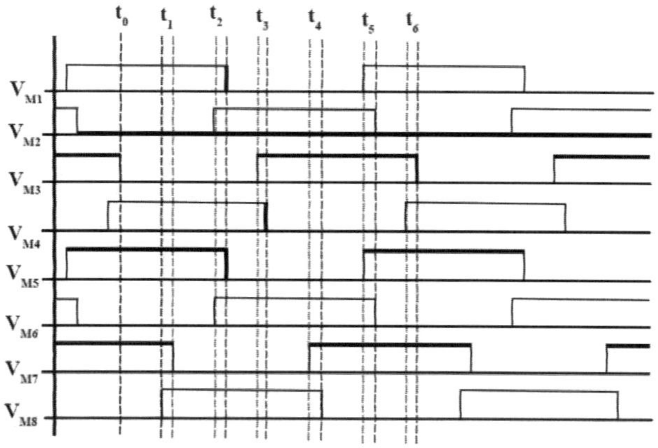

Figure 4. Gate signals of the MOSFETS

3. Operating Modes

Initially before mode 1 the input currents of transformer IPV-B and IFC are in freewheeling mode and the mode is operated through D1-M1-D3-M3 and D5-M5-D7-M7 respectively. The output stage diodes are in reverse bias and the output stage is inactive due to zero power transfer in between the primary and secondary coils.

3.1. Mode 1 Operation (t_0<=t<t_1)

Mode 1 operates when M3 is off at t0 duration. The Fig. 5 depicts the diagonal MOSFETs work in PV-B input stage and the transfer of power from input to the output circuit is shown mode 1 where the primary current IPV-B flows via VPV-B-L1-D1-M1-T1-D4-M4-VPV-B.

The second stage input current flow is given by VFC-L2-D5-M5-D7-M7-VFC. It shows that the second stage in free-wheeling state which implies that there is no power transferring in between primary and secondary circuit.

The flux φPV-B generated due to the current IPV-B induces the emf on secondary output winding circuit. D9 and D12 are in conducting stage due to emf induced by primary coil of transformer.

3.2. Mode 3 Operation (t2<=t<t3)

Mode 3 operates with turning on the M7 so that the current flow IPV-B takes place via VPV-B-L1-D1-M1-T1-D4-M4-VPV-B and IFC takes the path VFC-L2-D5-M5-T2-D8-M8-VFC. Now it can be said that both the input circuits in primary end are supply the power to secondary end of the transformer. Hence it can be said that the power is delivered to the load both individually and simultaneously.

The flux produced in the circuit adds up and the output is improved when compared with mode. The output circuit operation remains unchanged like in mode 1 and mode 2. The mode 3 operation is depicted in figure 6.

3.3. Mode 7 and Mode 9 Operation

The mode 7 starts when t=t6 and since the M4 turned off at t6, with that the mode 1 repeats itself but with the opposite polarities of voltage and current and the same follows in mode 9 while it is symmetrical to mode 3 with the opposite polarities of voltage and currents.

Both mode 7 and mode 9 are depicted with their operations in figure 7 and figure 8 respectively. We find remaining modes 10,11,12 symmetric to mode 4,5,6 respectively with opposite polarities. Hence these opposite polarities of voltages and current justifies the bidirectional power flow in the proposed converter.

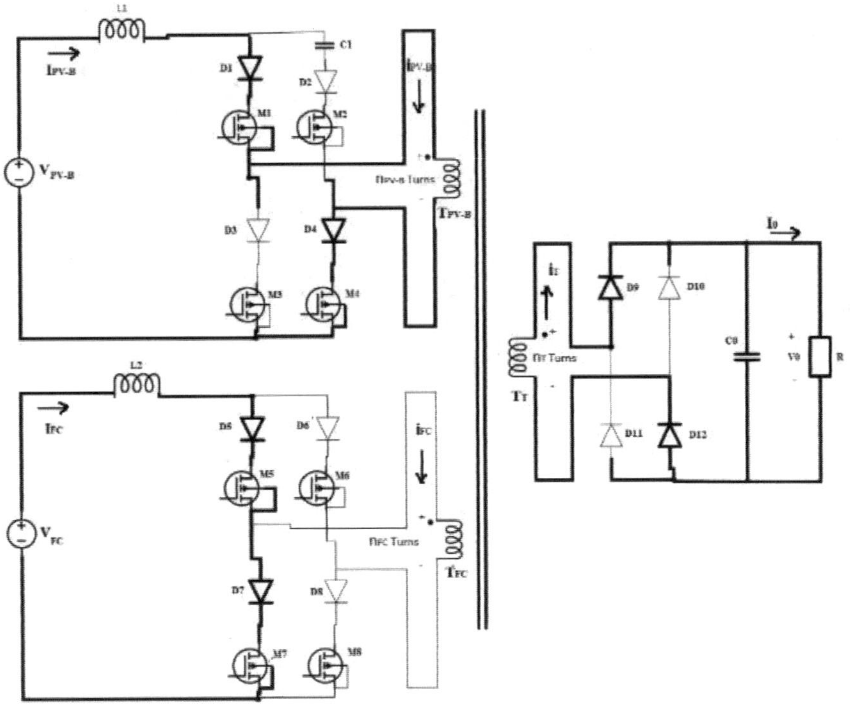

Figure 5. Mode 1&3 operation

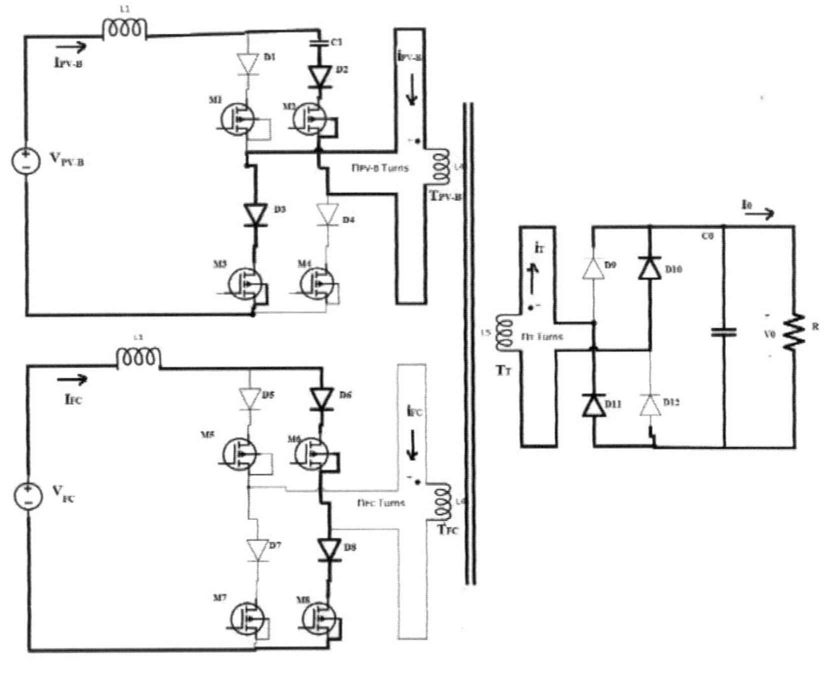

Figure 6. Mode 7&9 operation

4. Power Management and MPPT Algorithm

The multi-input structure of HEV is shown in Fig. 1 in which the FC and battery charged by PV are the main sources to supply power. The roof top PV panel is employed to improve the efficiency of HEV by reducing the fuel usage. The battery charging is done by PV, FC where in order to track maximum power and extract it to run HEV, one can implement the system in optimal region. The criteria for optimality are given by

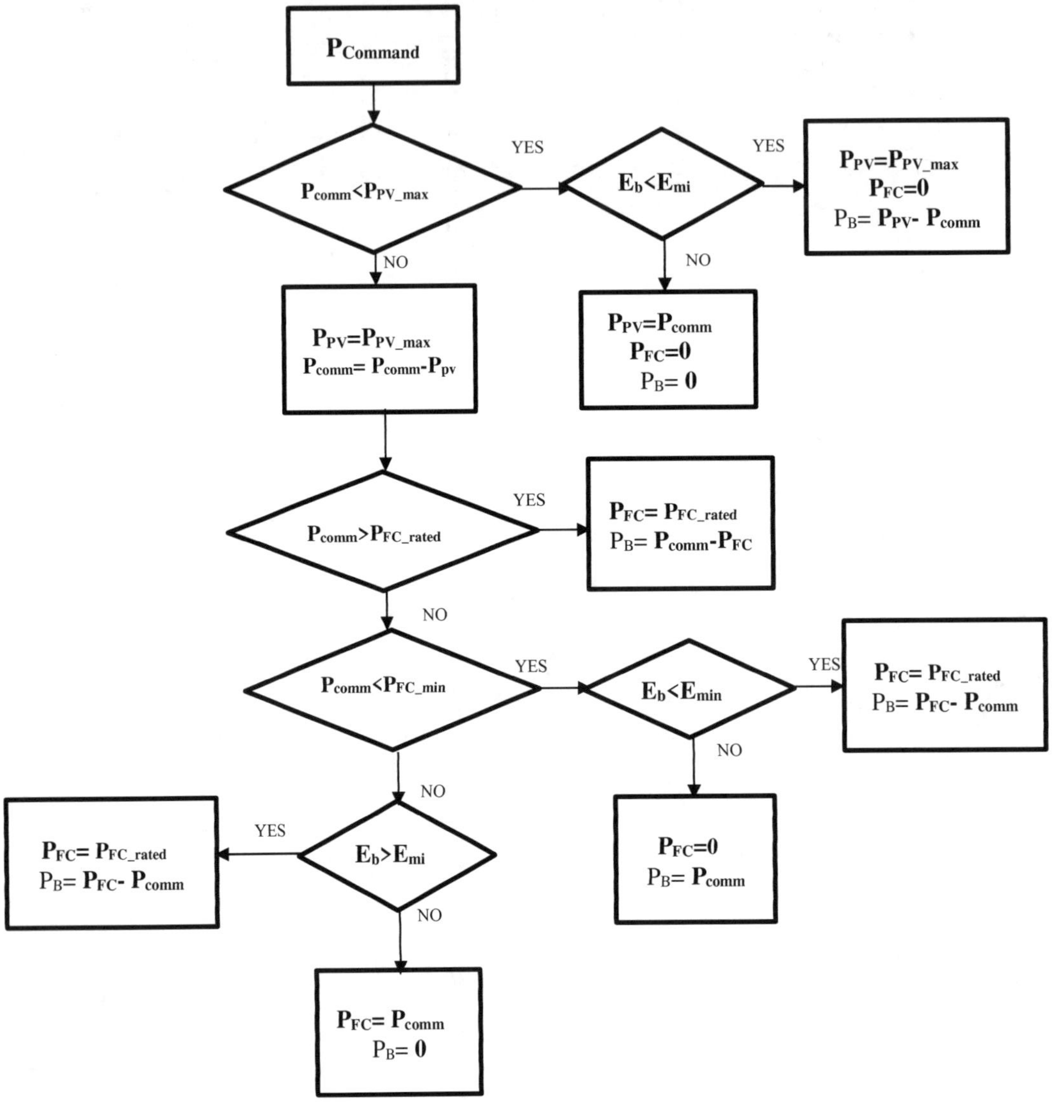

Figure 7. Flow chart representation of MPPT

- Electrical motor provides the power demanded every time.

- PV panel and FC are to be operated optimally
- The battery energy should always be in limits of optimality
- The power is managed with the following strategy described below and the flow chart depicted in Fig. 9 gives overview of the MPPT algorithm.
- Using the control signal we track the acceleration and brake and we attain the command power demanded by the operator of HEV.
- If the commanded power is lower than the maximum PV power and battery energy level is lower than the minimal energy level, PV will operate at maximum power point and the extra energy is used to store in the battery.
- If the commanded power is lower than the maximum PV power but battery energy level is greater than minimal energy level, then PV will produce commanded power and battery stays in off condition. In the duo the FC stays in off condition.
- If commanded power is greater than the maximum PV power, PV panel will work at maximum power point and new commanded power is given by

$$P_{comm}^{new} = P_{comm} - P_{PV}$$

- New commanded power when collated with FCs rated power, the new commanded power is more than the rated power of FC gives FC working at its rated power and the remaining power demand is met by battery.

5. Results & Case Studies

The accomplishment of the converter which is proposed is studied by the 80W prototype simulation is carried out and the circuits are framed and run in MATLAB with a switching frequency of 50KHz which is depicted in four phases for observing output operation. The proposed converter is applied for different applications like industrial, smart houses, EVs, DG technologies etc., The electrical sources considered for producing the magnetic flux via converter are PV panels, FC and battery. By neglecting the transient behavior of the sources of power we can place the DC power supplies for attaining the required results. The input sources are set to 80V and 50V. the lithium-ion battery with rating of 12V 7Ah is considered for the storing of energy in the battery. Lithium-ion batteries are utilized in a large scale as they are having best performance in portable gadgets in electronic stuff when compared to others. Because of higher switching frequencies we opt for iron cores in inductors. The L1 and L2 are given as 470 and 200 μH respectively. The capacitor rating is 470 μF for all capacitors. The resistance is given as 72Ω. The results for different modes are obtained i.e., with usage of HEV, PV, battery to identify the operation. Operation modes are explained in previous section.

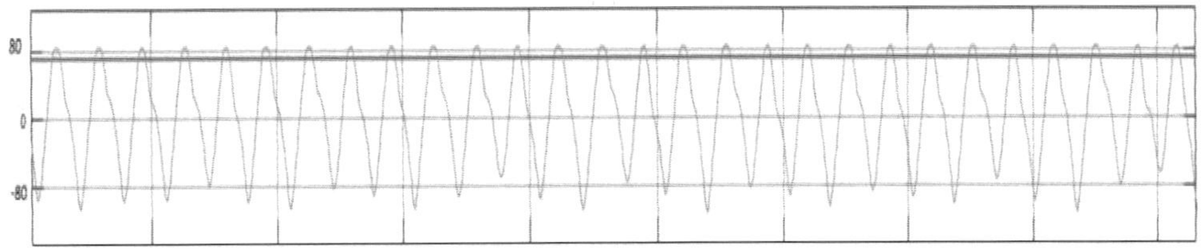

Figure 8. Transformer TPV-B voltage

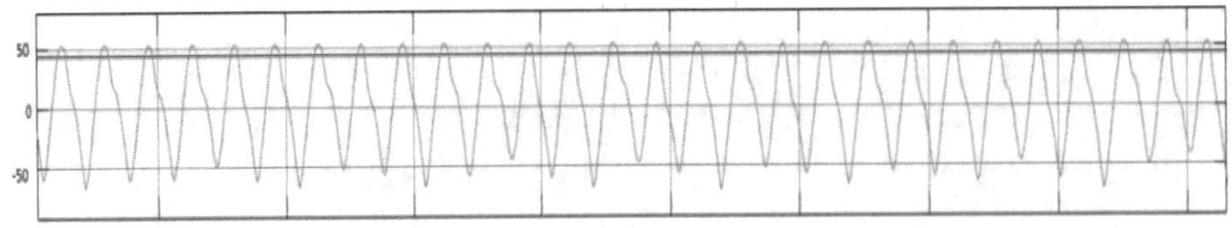

Figure 9. Transformer TFC voltage

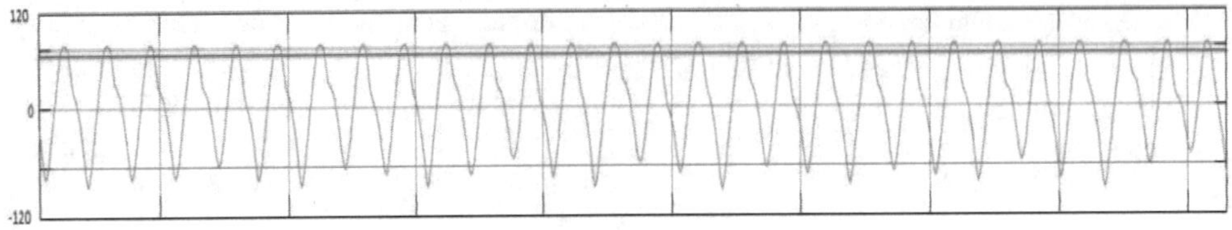

Figure 10. Transformer TT voltage

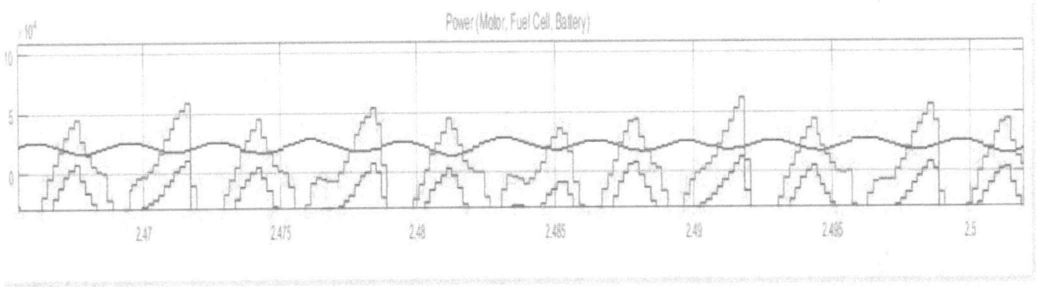

Figure 11. Power from motor, fuel cell and battery

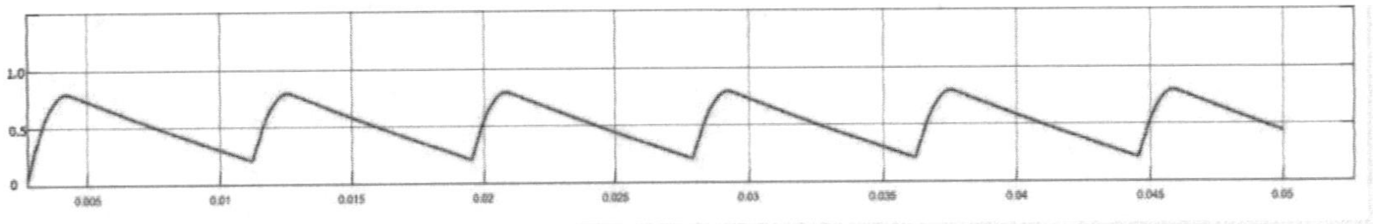

Figure 12. Voltage ripple waveform

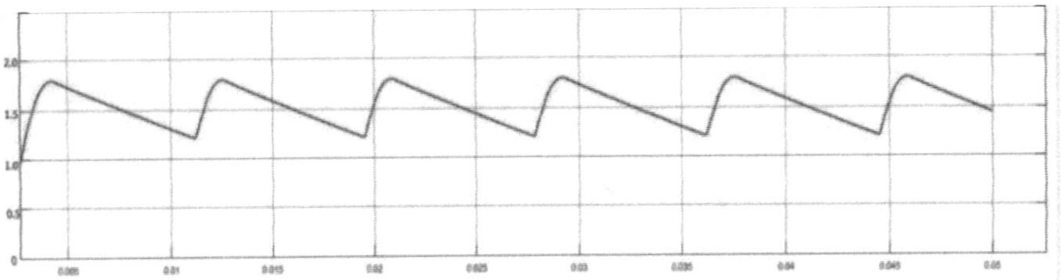

Figure 13. Current ripple waveform

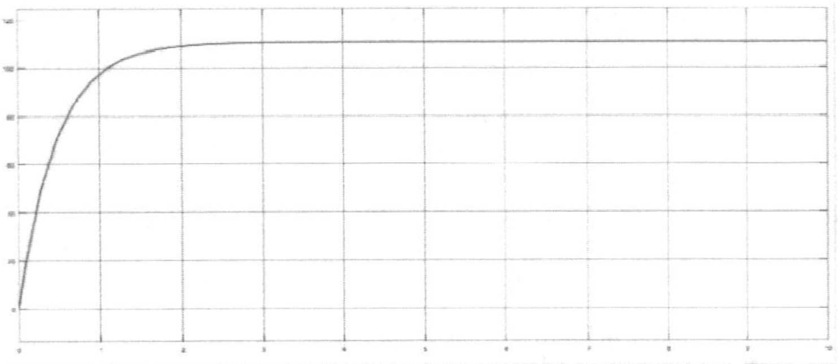

Figure 14. Output Voltage of Mode 3

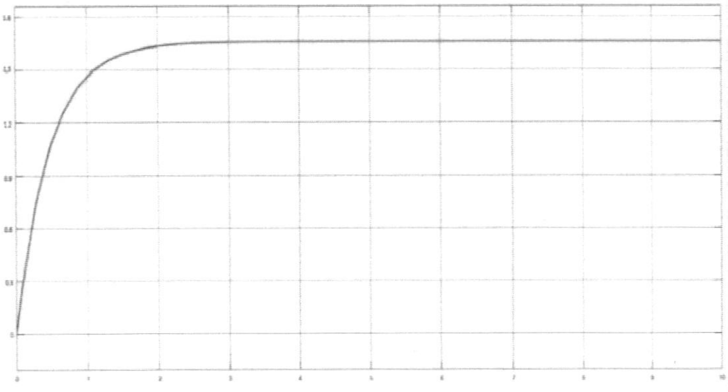

Figure 15. Output Current of Mode 3

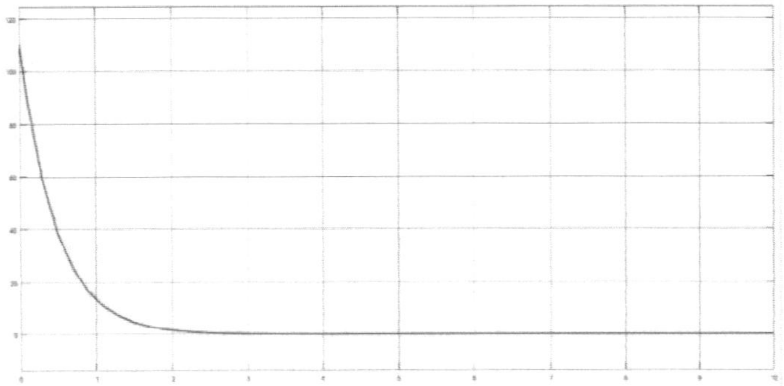

Figure 16. Output Voltage of Mode 9

The results attained depicts the efficiency of proposed strategy. Fig. 10-12 represents the transformer winding voltages. Hence one can estimate the how the converter operates in different modes based on gate pulses of MOSFETs. The overshoot in inductor current can be accepted. Fig.13 depicts the motor, battery, FC powers. The current in transformer is little bit low. The voltage ripple is 1% and the current ripple is 1.78% which showcase the efficiency of our proposed model shown in Fig. 14 and 15 respectively. Along with that output voltage and output current of mode 3 are given in Fig. 16 and Fig. 17 respectively. Also output voltage and output current of mode 9 are given in Fig. 18 and Fig. 19 respectively. The obtained output voltage is 110.56V and output current is 1.52A which is improved when compared to multi-input DC-DC converter.

6. Conclusion

In this paper, a flux additive DC-DC converter is proposed and its modes are observed keenly for their operation. The above stated operating modes gives the effectiveness of the converter and its operation with single source and simultaneously both sources and bidirectional power flow through mode 7. We can observe the operating modes from 1-6 operating in motor mode and the mode7-12 operates with regenerative brake mode. So, from this we use this converter for HEV application because of this bidirectional power flow. The MPPT algorithm is accomplished, and power management is used to implement the controller's orders. However, the converter's performance will be improved by using power management and the MPPT technique. Furthermore, an experimental design of the system converted is built in MATLAB, and the results are recorded and shown. The converter's analysis and performance are demonstrated by the results.

7. References

[1] A. Ostadi and M. Kazerani, 2014 Optimal sizing of the battery unit in a plug-in electric vehicle, *IEEE Trans. Veh. Technol,* vol. 63, no. 7, pp. 3077–3084.

[2] P. Mulhall, S. M. Lukic, S. G. Wirashingha, Y.-J. Lee, and A. Emadi, 2010 Solar-assisted electric auto rickshaw three-wheeler, *IEEE Trans. Veh.Technol.,* vol. 59, no. 5, pp. 2298—2307.

[3] H. J. Chiu and L. W. Lin, 2006 A bidirectional dc-dc converter for fuel cell electric vehicle driving system, *IEEE Trans. Power Electron.,* vol. 21, no. 4, pp. 950–958.

[4] T. Markel, M. Zolot, K. B. Wipke, and A. A. Pesaran, 2003 Energy storage requirements for hybrid fuel cell vehicles," *presented at the Adv. Autom. Battery Conf.,* Nice, France.

[5] M. A. G. de Brito, L. Galotto, L. P. Sampaio, G. de Azevedo e Melo, and C. A. Canesin, 2013 Evaluation of the main MPPT techniques for photovoltaic applications, *IEEE Trans. Ind. Electron.,* vol. 60, no. 3, pp. 1156–1167.

[6] M. Chen, Y.C. Liu, F.Y. Wu, and T.F. Wu, 2001 Multi- Input DC-DC Converter based on the Flux Additivity, *36th Annual Industry Applications Conference,* vol. 3, pp. 1866 –1873.

[7] Gowri shankara C.K, Abhishek J Gowda, S G Srivani, 2016 Improvements on Spacecraft Power regulators, *Biennial International Conference on Power and Energy Systems: Towards Sustainable Energy (PESTSE).*

[8] O. Hegazy, R. Barrero, J. Van Mierlo, P. Lataire, N. Omar, and T. Coosemans, 2013 An advanced power electronics interface for electric vehicles applications, *IEEE Trans. Power Electron.,* vol. 28, no. 12, pp.1-14.

[9] Y. M. Chen, A. Q. Huang, and X. Yu, 2013, A high step-up three-port DC-DC converter for stand-alone PV/battery power systems, *IEEE Trans. Power Electron.,* vol. 28, no. 11, pp. 5049–5062.

[10] A. Khaligh and Z. Li, 2010, Battery, ultracapacitor, fuel cell, and hybrid energy storage systems for electric, hybrid electric, fuel cell, and plug-in hybrid electric vehicles: State of the art, *IEEE Trans. Veh. Technol.,* vol. 59, no. 6, pp. 2806–2814.

[11] Krishna P. Yalamanchili and Mehdi Ferdowsi, 2005, Review of Multiple Input DC-DC Converters for Electric and Hybrid Vehicles Multi-Input DC/DC Converter Based on the Flux Additivity, *Power Electronics and Motor Drives Laboratory*, pp.7803-9280.

[12] C. Chan, 2007 The state of the art of electric, hybrid, and fuel cell vehicles, *Proceedings of the IEEE,* vol. 95, pp. 704–718.

Frequency Monitoring Analysis based Power Quality Operational Data Management System

K.SaiHimajaChowdary[1] , Vijaykumar Mantri[2], M.Neelakantappa[3]

[1]Amazon Development Center India Private Ltd, Bangalore

[2,3]Dept. of Information Technology, B V Raju Institute of Technology, Narsapur

Email of Corresponding Author: neelakantappa.m@bvrit.ac.in

Abstract. In this work, a low-cost measurement solution (named PyFreqDAQ) has been developed to monitor the frequency of the electrical network, which is regarded a quality parameter of the electrical supply, and at the same time provide a more precise result thanks to the usage of the Allan variance. Software (Grafana) is available for free and can be used to track the frequency's change over time in order to better understand how it is maintained within certain parameters. Two distinct techniques of calculating the frequency have been used to show the device's power consumption. For starters, we'll follow the guidelines established by IEC 61000-4-30 The Allan variance-based approach, on the other hand, uses an algorithm that offers a more precise frequency measurement and a GPS-traceable reference value. With the Allan variance-based method, one can expect improved precision and accuracy, as well as faster calculations and a wider range of observations. A significant advantage over other applications (e.g. labVIEWTM) is the flexibility demonstrated by Python while programming the data capture card and interacting with the database Because Python is a more powerful programming language, it allows for operations and calculations to be carried out using simple codes, while also allowing for minimal computational load and consuming considerably less working memory. To date, there hasn't been any comparative research done between Python and LabVIEW™.

Keywords: Big data; data acquisition; frequency stability; Grafana-based lab; power quality; GPS reference; frequency measurement

1. INTRODUCTION

Both the utility and the customer end of the supply chain are harmed by low-quality electricity. Focus on the quality of electricity that is being delivered to loads is critical. Please continue reading to learn more about the reasons of poor power quality, how to measure it, and how to improve it.

As an alternative, "power quality" refers to a grid's ability to supply electricity to customers and the ability of equipment to absorb the power it receives. The term "good power quality" refers to a device's ability to produce a sinusoidal waveform at its rated voltage and frequency. There must be a distinction between static and dynamic characteristics of a power supply system to determine its quality (Power Quality, PQ). Since electrical machines rely heavily on frequency stability, this is one of the most important parameters. Special mention requires those that operate in synchronized mode, for which the deviations that occur with respect to their nominal value required for correct operation, are inadmissible [1].

A power system's efficiency and cost can be significantly impacted by the quality of the power it receives. Low-quality power means that a system's performance will suffer and it won't be able to operate correctly. Customers are more conscious of power quality than ever before, thus many governments have changed their regulations so that electric utilities must ensure that the power quality meets the requirements they set. Modern equipment is also more susceptible to power quality fluctuations. Power quality is a growing concern for manufacturers, utilities, and consumers alike.

As the characteristic parameters of the network can oscillate within a permitted range without being inappropriate for the equipment, it is essential to monitor whether they fall within these limits. However, in practice, the European frequency (50 Hz) is fairly unstable due to the fact that the supplied signal is not exactly sinusoidal, with permissible limitations of 50 Hz for systems with synchronized connections and 50 Hz in systems that are disconnected. [2, 3].

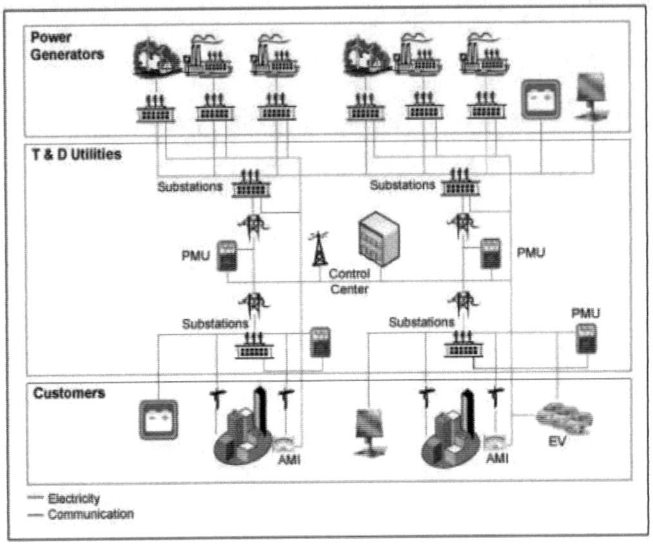

Figure 1: PMU stands for point-of-use meters, which are installed across the network.

Figure 1 depicts the whole electrical network, complete with power meters [5] installed at strategic locations. Voltage magnitude and phase angle are typically measured by most PMUs, assuming that the frequency remains constant [7]. In addition, the total cost of procuring, installing, and commissioning a PMU ranges from USD 40,000 to $180,000, depending on the size of the project. PMU's [10] indirect measurements specifically ignore this first notion. These operations are essentially based on the formulation of a mathematical problem whose purpose is to fit the measurements to a sinusoidal curve. When the sinusoidal curve is not optimal, the PMU is unable to perform the correct correction. In this approach, the poorer the correction is and the lower the quality of the phasor generated, the less sinusoidal the voltage supply signal. New tactics are needed to supplement existing ones based on PMUs because of the widespread use of distributed and random generation and nonlinear loads in the modern grid .

GPS receiver-data acquisition (DAQ) system [12] binomial is used to build the instrument hardware, which is powered by a Python-based software application. The first has a precision of 1 ppm and delivers a traceable time stamp through a 1 pulse per second (PPS) signal. While large software schematics and programming codes may cause delays, the proposed PQ system can collect frequency values every 10 seconds [2] with an unprecedented temporal resolution of more than 20 milliseconds [16,17]..

2. CAUSES OF POOR POWER QUALITY

There is a need to determine the elements that contribute to a power system's low power quality. Uncertain occurrences, utility, consumer, and manufacturer are all possible causes.

2.1 Uncertain Events :
Random events like faults, resonance, and lightning surges produce the majority of power quality issues. Disruptions in the electric utility are common.

2.2 Utility:
Poor power quality can be traced back to the utility in three different ways.

Generation end: Factors like expansion, maintenance planning, outages, and load shifting exacerbate power quality issues at the generating end.

Transmission end: Wind, voltage changes, lightning, incorrect voltage regulation devices, and other factors can cause transmission lines to lose power quality.

Distribution end: Voltage dips, interruptions, transients, spikes, transformer energization, etc. all contribute to poor power quality in the distribution system.

2.3 Consumer:
The consumer is largely responsible for poor power quality. Customers' non-linear loads are to blame for power system harmonics. As voltage increases, the impedance of an electrical load changes. Even with sinusoidal voltage, the non-linear load's fluctuating impedance results in non-sinusoidal current draw. It is possible that the power system and the loads it serves can be harmed by harmonic currents in non-sinusoidal currents interfering with the system's impedance and causing voltage distortion.

2.4 Manufacturer :
There are two ways in which manufacturers might be connected to power quality issues:

Standards: Using a device with poor power quality can be caused by a lack of industry standards for anything from installation to testing to certification to purchasing to sale.

Equipment sensitivity: Equipment's sensitivity might produce power quality concerns if the electrical environment is unsuitable due to its high sensitivity.

3. COMMON POWER QUALITY ISSUES AND PARAMETERS

3.1 Transients:
Pulses with high intensity that last for a short period of time in a sinusoidal waveform are known as transients. In either case, the source of transients might be either internal or external to the facility, meaning that they can come from anywhere. Lightning, wind, transformer switching, and other natural phenomena are examples of external influences. Internal sources include system faults, load switching, and arcing. In order to avoid equipment damage, such as dielectric failure, fracture, insulation flashover, and overload, these waveform distortions must be eliminated. A lack of quality is a result of this.

3.2 Voltage Variations:
Voltage fluctuation is a word used to describe how a system's nominal voltage can differ from the actual value. Voltage fluctuations can be caused by a variety of things, including equipment failure, control dysfunction, or the functioning of fuses or circuit breakers. Load shifting or activating heavy loads can cause voltage sag or dips when used with large motors, single lines to ground faults, and ground faults. Variations in voltage are caused by voltages that are either too low or too high. When the system is overloaded, undervoltages occur, and overvoltages occur when the system is outfitted with lighter loads than the utility's voltage level, respectively.

3.3 Unbalanced Voltages:
The voltages in a three-phase system are said to be "unbalanced" if there is a large or small phase difference between the two phases. Blown fuses in any of the three phases, unequal load distribution, and a lack of overhead transmission line transposition are the most common causes of voltage imbalance in power networks. Poor power quality can be the result of such uneven voltages harming or damaging electrical equipment.

3.4 Flickers:
The visual sensation is unstable due to the quick changes in the load currents caused by the constant voltage variations of the supplied electricity. Human eyes are harmed by the quick and obvious shift in brightness of a light bulb. The most prevalent causes of flickering include sudden load fluctuations, motor drives, arc furnaces, welding machines, and so on. As a result, the flickering raises doubts about the quality of the power source.

3.5 Distorted Waveforms:
Distortion of a waveform occurs when it departs from the steady state sinusoidal waveform. Electric noise and DC offset are a few of the types of distortion that might occur. Switching devices, inductor leakage, and other inductor loads can create DC offsets in an AC system, which are referred to as DC offsets. Due to overheating, an equipment's lifespan may be reduced as a result of a DC offset. When a sinewave has a frequency equal to or greater than the fundamental frequency, it is a harmonic waveform. Harmonics in the power system are mostly caused by non-linear loads, switching devices, and the like, which can lead to the failure of control devices, losses in equipment, and the like. An additional waveform distortion is the addition of unwanted electric signals to the voltage or current waveforms of the power system. Corona effect, faulty power system connections, and electronic gadgets are among the most common causes of electric noise. They all have a negative effect on power quality and must be addressed.

3.6 Total Harmonic Distortion (THD):
Total Harmonic Distortion is a term used to describe the distortion of waveform harmonics (THD). An increase in THD has a negative impact on power quality. The power quality will be lower if the system has higher harmonic distortion, and the reverse is also true. RMS harmonic content to fundamental is the ratio of THD:

$$\mathbf{THD} = \frac{\sqrt{\sum_{n=2}^{\infty} V_{n-rms}^{2}}}{V_{fund-rms}}$$

where $V_{n\text{-}rms}$ is the RMS voltage of nth harmonic in the signal and $V_{fund\text{-}rms}$ is the RMS voltage of the fundamental frequency.

3.7 Power Factor:
There is a direct correlation between the two, namely, power quality and power factor. Closer to a power factor of 1 suggests excellent power quality. It's no coincidence that as power factor drops below one, so do costs and quality of power.

3.8 Varying Frequency:
Frequency variations refer to changes in the magnitude of the frequency (50 or 60 Hz). Because of an imbalance between generation and demand, a deviation from this fundamental value can occur in the power system. Frequency changes can also be caused by problems with the transmission system. All

electrical equipment are designed to operate at their rated frequency, and when this number changes, it can have a detrimental effect on them

3.9 Temporary interruptions:
Electric supply to consumers can also be interrupted or cut off i.e., total loss of power occurs, and voltage becomes zero, for short or long periods of time in a localized area. This interruption might be planned by your utility or can also be a result of the following causes:

- **Short term power interruptions** usually occur because of insulation failure, lightning and insulator flashovers etc.
- **Long term interruptions** can be caused by equipment failure in the power system network, human errors, unfavorable weather conditions, poor coordination of protection devices etc.

3.10 Voltage notching:
Normal operation of power electronic devices (rectifiers, SCRs, etc.) causes current to switch between phases, resulting in a notch in the quality of the output voltage. In the electrical system, the abrupt change in voltage due to the notch excite the natural frequencies. This results in an increase in the number of non-characteristic harmonics in the voltage of the system. Due to these high frequency harmonics created by notching, sensitive logic and communication electronic circuits in the facility can be damaged, radio interference might occur.

4. PROPOSED METHODOLOGY

4.1 System Description and Arrangement :
From here onwards, the proposed distributed instrument will be described. The hardware consists essentially of a chassis and a data acquisition card (DAQ) from the manufacturer NI™, which receives, on the one hand, the voltage input (50 Hz, 220 Vrms) from the electrical network and, on the other, the pulsed signal (1 PPS) from the GPS receiver (Symmetricon) disciplined clock. The pulses of the latter are used as traceable references for time domain stamping and as references to calculate the frequency according to the Allan variance (Figure 2). By using Ethernet, the acquisition card sends its data over to the computer. They are then processed, and the results of the processing are entered into the database afterward. Python-coded software handles all of this for us. In order to monitor the electrical network, these data can be used to create a graph of the frequency's change over time.

Data is transferred to online software by the c-DAQ module, which collects the signals.

The monitoring solution developed according to this strategy allows for obtaining data from multiple measurement sources, and putting them in common in a database, which is available on the display. With this, since a large amount of signal data from different points can be observed and analyzed (compared), the maintenance of the network can be carried out using spatial and temporal compression strategies, which offer a simplified and truthful statistical vision of the huge amount of initial data (big data).

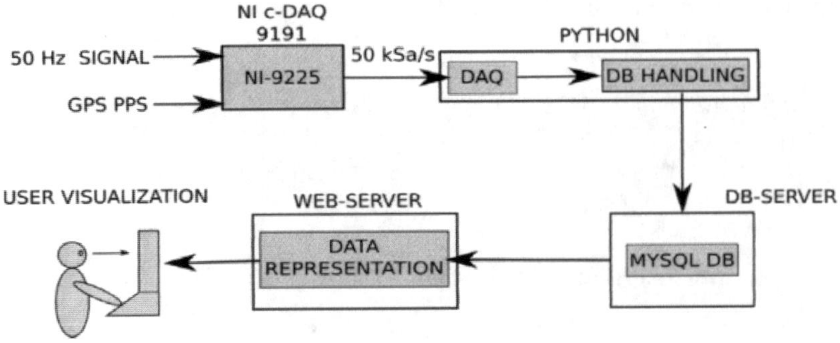

Figure 2. System block diagram.

In order to obtain an overview of the measurement solution, Figure 2 shows a block diagram, which indicates the direction of the signal flow and the interactions between the different elements that make up the system.

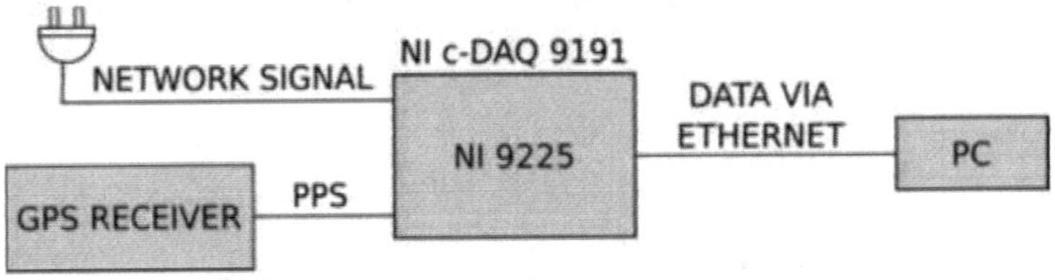

Figure 3. Scheme of the measurement chain.

The DAQ card (NI 9225) inside the chassis is used to collect network and GPS-PPS signal data (c-DAQ-9191). It then goes through the data and sends it over Ethernet to the PC.

To measure PyFreqDAQ, users need a GPS receiver for synchronisation, a network plug for signal acquisition, and an DAQ card to receive and send both signals to thier PC. Figure 3 depicts this in full depth.

4.2 Frequency performance indicators:

1) Rate Of Change Of Frequency (ROCOF): In the event of an imbalance between active power generation and demand, this metric is used to measure the frequency gradient [9]. When a generation is lost, the frequency immediately begins to deviate from the rated value.

For qualitative evaluation of frequency performance, ROCOF can be computed in two ways: first, using an approximation derived and given by (2), and second, by computing the slope of the frequency decrease in a fixed time window (0.5s).

$$\mathbf{RoCoF} = \frac{\mathbf{df}}{\mathbf{dt}} \qquad (1)$$

$$\mathbf{RoCoF} = \frac{\mathbf{f \times \Delta P}}{\mathbf{2(E_{sys} - E_{lost})}} \qquad (2)$$

where ΔP is the MWs lost (i.e power deficit), f is the system frequency, Esys is the system kinetic energy in MWs, Elost is the kinetic energy lost in MWs. For frequency measurement and protection, the ROCOF's relevance is based on the equipment's ability to acquire data quickly enough to detect changes in frequency.

4.3 Transient performance indicators :
The stability margin is based on the power angle: Percentage value of maximum angular deviation between electrical system's synchronous machines.
This indicator is defined as follows:.

$$\mathbf{Margin} = \frac{\mathbf{360^0 - \delta max}}{\mathbf{360^0 + \delta max}} \times \mathbf{100} \qquad (3)$$

where δmax is the maximum angle separation of any two generators of the system at the same time in the post fault response.

5. RESULTS AND DISCUSSION

5.1 PyFreqDAQ Operation and Frequency Calculation :

Inputs to the NI 9225 card include a voltage line signal from a wall plug and a GPS-PPS signal for an absolute time reference (sampling frequency: 50 kS/s). Once the PC software has received the raw data via Ethernet, it can analyse it.

The "DAQ" project, the first of the two, continuously reads data from the NI 9225. There are two methods of carrying out a calculation. Using the PPS, the theoretical 50 Hz frequency is calculated to have a theoretical cycle time of 20 ms.

Using a Schmitt trigger, a voltage can be controlled (it returns a square waveform with the same frequency as the input signal). An excessively high output is produced when the input voltage is higher than the predetermined upper threshold; an excessively low output is produced when the input voltage is lower than the predetermined lower threshold. Regardless of whether the input voltage is above or below the two thresholds, the output is unchanged (Figure 4). Once this square waveform has been fragmented into its cycles, calculations can be performed.

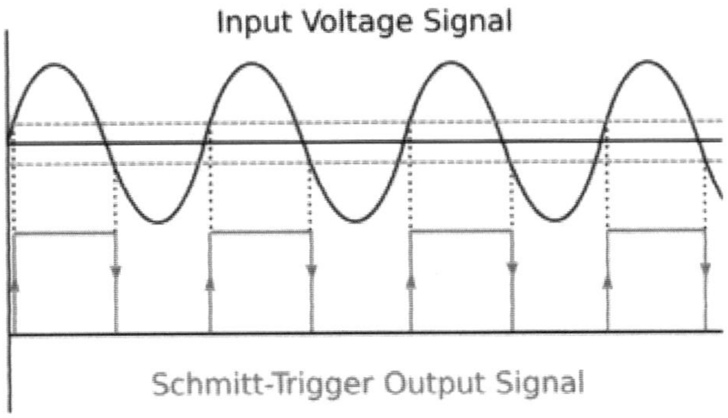

Figure 4. Schmitt-trigger effect produced with the input voltage signal

Here, we have a black sinusoid representing the input voltage, and a blue square for what we get as an output. In order to keep things simple, scales and units are omitted.

Journal of Physics: Conference Series **2327** (2022) 012003 doi:10.1088/1742-6596/2327/1/012003

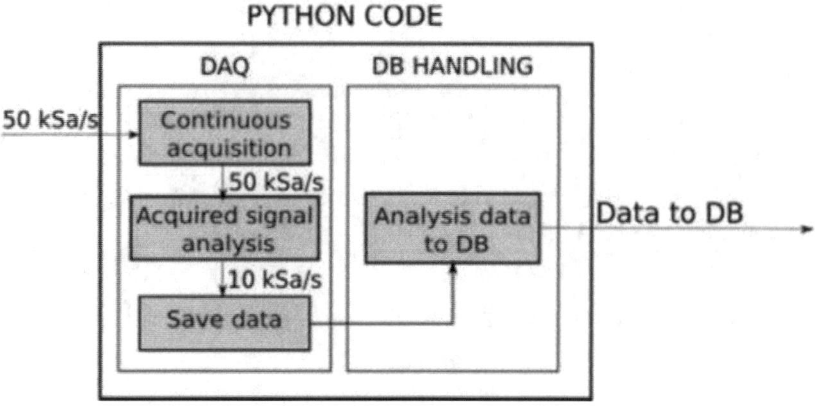

Figure 5. Python code operation with the two projects

To acquire and transfer data to the second project, the first project, DAQ, registers data at a rate of 50 kSa/s. database management and storage are handled by the DB Handling project.

Data processing, calculations, and sending information to the database are handled by two separate Python software projects. Receiving data from the NI 9225, calculating frequency, and sending the results to the second are the responsibilities of the first. It is the second project's responsibility to receive data from the DAQ and organise it within the database that is referred to as "DAQ." "DB handling" refers to this section. Figure 5 depicts the projects' workflow in a simplified form.

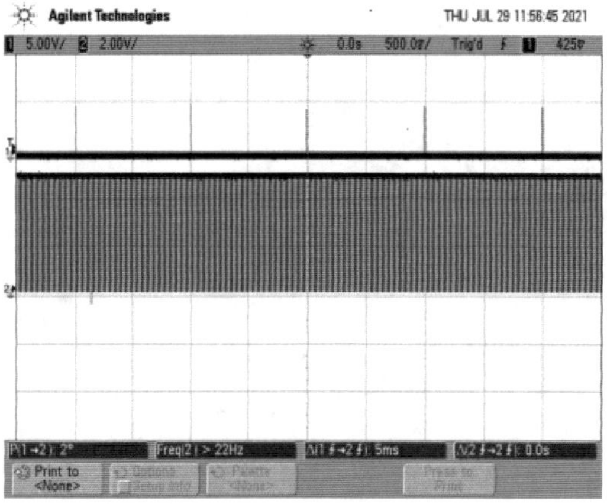

Figure 6. Schmitt-trigger output signal and PPS pulses

Five PPS pulses can be seen at the top while the total of the cycles are monitored in the signal below.

The next two illustrations, Figures 6 and 7, show the real PPS signal from the GPS receiver and Schmitt-trigger signals, at different sampling rates, from the lab oscilloscope

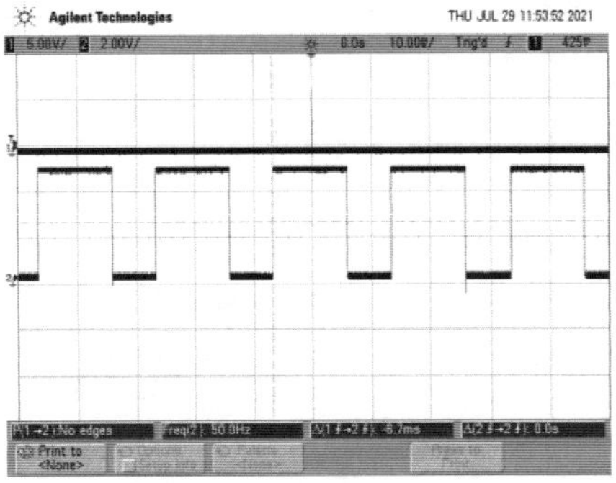

Figure 7. Schmitt-trigger application result for the input voltage signal. In this sample only 1 PPS is seen, but the square signal from the Smith-trigger is clearly observed.

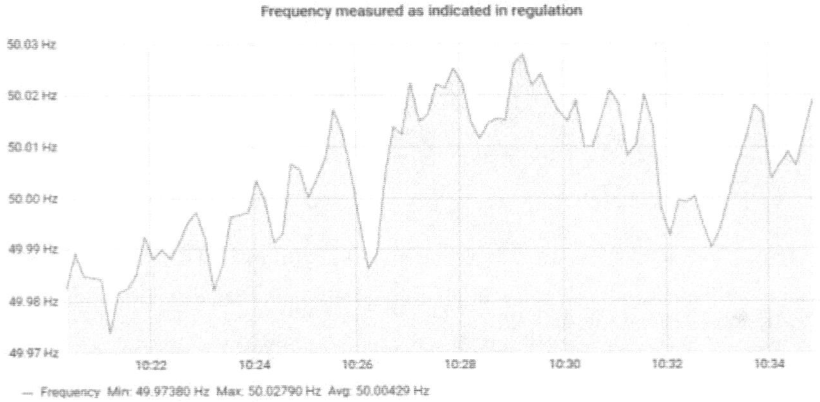

Figure 8. Evolution of the measured frequency (vs. time) following the norm, with values each 10 s, over 15 min

Frequency cannot exceed the limits set by UNE-EN 50160:2011 and must be measured using 10-s periods, according to this international standard. Figure 8 depicts the 15-minute evolution of frequency as shown by the norm.

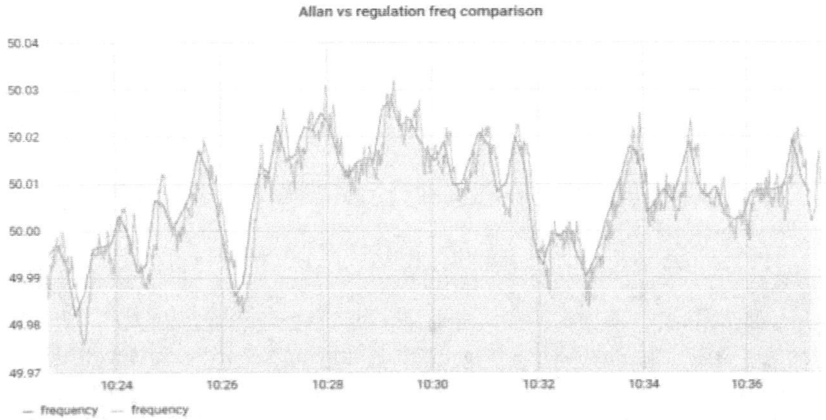

Figure 9. Frequency timeline following the regulation (green line) and following the algorithm using Allan Variance (orange line)

The following Grafana dashboard compares the evolution of the two frequencies in order to examine the differences between the two methods of frequency calculation. A time interval of 15 min has been chosen, as can be observed in Figure 9.

6. CONCLUSION

In this work, a low-cost measurement solution (called PyFreqDAQ) has been proposed to monitor the frequency of the electrical network, considered as a quality parameter of the electrical supply, and at the same time offer a more precise result thanks to the use of the Allan variance. The hardware is managed by Python, and open-source software (Grafana) is available that allows for monitoring the evolution of the frequency, in order to study the limits between which it is maintained. This measurement solution is original in that it does not use high-level visual programming software for data acquisition and data management (e.g., labVIEW™), thus lightening the computational load, an issue that could be the subject of future work. The use of Python to program the data acquisition card and manipulate the database has shown much greater flexibility than other applications (e.g., labVIEW™). In effect, Python is a more powerful programming language, and consequently, it allows operations and calculations to be carried out using simple codes, while allowing for a low computational load and consuming much less working memory. In future works, a Python vs. labVIEW™ comparative could be studied, which would be unique research, to date.

References

[1] Phadke, Arun G., and Tianshu Bi. "Phasor measurement units, WAMS, and their applications in protection and control of power systems." Journal of Modern Power Systems and Clean Energy 6, no. 4 (2018): 619-629.

[2] Gholami, Mostafa, Mohammad J. Sanjari, Mostafa Safari, Mahdi Akbari, and Mohammadreza R. Kamali. "Static security assessment of power systems: A review." International Transactions on Electrical Energy Systems 30, no. 9 (2020): e12432.

[3] Alimi, Oyeniyi Akeem, Khmaies Ouahada, and Adnan M. Abu-Mahfouz. "A review of machine learning approaches to power system security and stability." IEEE Access 8 (2020): 113512-113531.

[4] Phadke AG, Thorp JS (2008) Synchronized phasor measurements and their applications, 2nd edn. Springer, New York

[5] China Satellite Navigation System Management Office (2013) Development report for Beidou Satellite Navigation System (version 2.2). Beijing

[6] Sunitha R, Kumar SK, Mathew AT. Online static security assessment module using artificial neural networks.IEEE Trans Power Syst.2013;28(4):4328-4335.

[7] Sekhar P, Mohanty S. An online power system static security assessment module using multi-layer perceptron and radial basis function network. Int J Electr Power Energy Syst. 2016;76:165-173.

[8] Mitra P, Vittal V, Keel B, Mistry J. A systematic approach ton-1-1 analysis for power system security assessment. IEEE Power Energy Technol Syst J. 2016;3(2):71-80.

[9] Swarup K, Corthis PB. Power system static security assessment using self-organizing neural network. J Indian Inst Sci. 2013;86(4):327.

[10] Srivani J, Swarup K. Power system static security assessment and evaluation using external system equivalents.Int J Electr Power EnergySyst. 2008;30(2):83-92.

[11] Buraimoh E, Ariyo F, Davidson I. Power system static security enhancement through interline power flow controller, Proceedings of the2018 IEEE PES/IAS Power Africa Conference, 2018, pp. 330-335. IEEE.

[12] Li Y, Li Y, Sun Y. Online static security assessment of power systems based on LASSO algorithm. Appl Sci. 2018;8(9):1442.

[13] Dhandhia A, Pandya V. Binary classification of The static security of the power system using support vector machine, Proceedings of the International Conference on Thermal Engineering (ICTEA), Vol. 2019, 2019.3
[14] Jmii H, Meddeb A, Abbes M, Chebbi S. An Intelligent Combination Method for Static Security Assessment, Proceedings of the 2019International Conference on Advanced Systems and Emergent Technologies (IC_ASET), 2019, pp. 290-294. IEEE.
[15] "Raccoon Mountain." https://www.tva.com/energy/our-power-system/hydroelectric/raccoon-mountain
[16] "Grid Protection Alliance." https://www.gridprotectionalliance.org/.
[17] "Hitachi-ABB Power Grids." https://www.hitachiabb-powergrids.com/offering/p roduct-and-system/high-voltage-switchgear-and-breakers/monitoring-and-cont rolled-switching/circuit-breaker-sentinel.
[18] "GE Grid Solutions." https://www.gegridsolutions.com/md/catalog/cbwatch3. htm.
[19] "PingThings PredictiveGrid." https://www.pingthings.io/.
[20] "OSIsoft PI Server." https://www.osisoft.com/pi-system/pi-core/pi-server.
[21] Axpo, "Electricity market facts and figures: The Swiss electricity market,"; 2020

Journal of Physics: Conference Series **2327** (2022) 012004 doi:10.1088/1742-6596/2327/1/012004

Novel Micro flow Sensor for Air Purge Method to Monitor the State of Charge of Lead Acid Battery

Yashwant Adhav[1], Dayaram Sonawane[2] and Chetankumar Patil[3]

1 Cummins College of Engineering for Women, Pune-411052, India
2 Department of Instrumentation and Control Engineering, College of Engineering, Pune-411005, India
3 Department of Instrumentation and Control Engineering, College of Engineering, Pune-411005, India

E-mail: yashwant.adhav@cumminscollege.in, dns.instru@coep.ac.in, cyp.instru@coep.ac.in

Abstract. The foremost aim of the present research study is to measure specific gravity of lead-acid batteries and further know batterys state-of-charge (SoC). In this article, air purge method is used to monitor the status of a battery; back differential pressure is measured using MPX4006 DP MEMS pressure sensor. In air purge method, it is essential to adjust the rate of bubbles in acid. This can be achieved by controlling the compressors speed or controlling the air flow by controllers. When air purge tubes are inserted through a batterys cap, it is difficult to observe bubbles produced in the acid as the battery is nontransparent. If the bubble rate is too slow, the system lowers the output signal and hence stops working. During the calibration of the instrument, air flow measurement through the pipe is of prime importance. In this method, FS1012 MEMS gas flow sensor is used to monitor the discharged air flow. FS1012 is based on the thermo-transfer principle. Compared to the resistive technique, this flow sensor consists of a thermopile for air flow sensing whose signal-noise ratio is exceptional. The output of the sensor is exponential; signal conditioning circuit has been designed to get linear output 0 to 3.3 V standard range. The amplified output of signal conditioning circuit is monitored using TM4C123GXL arm controller. FS1020 is first time proposed to monitor the air flow of a small air purge probe used to monitor the state of charge of a lead-acid battery. FS1020 ensures an ideal flow of the air bubbler system, which improves the performance of the battery charge monitoring system. Conventional flow sensors for example, rotameter cannot be used for very low flow measurement; they are costlier as well as bulky. The minimum flow required to produce sufficient bubbles in acid is 0.50 liter per minute (LPM) which is measured by FS 1012.

1. Introduction

The present research study mainly attempts to design a real-time measurement system for a specific gravity of acid of a lead-acid battery using an air purge tube that can be inserted in a small quantity of acid (the novelty is already patented [1]). Specific gravity directly indicates the percent of the charge in a battery. There are several methods to determine the state of the charge (SoC) of a battery, however, among them, the method based on specific gravity measurement techniques is more accurate, precise and valid [2]. At present, lead-acid batteries are still the most frequently utilized energy storage systems (ESS) [3][5]. These batteries, having a competitive price and high capacity, can transform energy into electrical energy. ESS can be driven from several batteries, namely fuelcell batteries, Li-ion batteries and lead-acid batteries.

Lead-acid batteries too have a long history of use and have sophisticated technology. Lead-acid battery is extensively utilized such as photovoltaic (PV) wind power generation systems, uninterruptible power systems (UPSs), the automotive field and aerospace [6][8] because of its precise and stable voltage a broad range of operating temperatures characteristics and low costs.

Specific gravity measurement technique with air purge method compulsory requires an air flow measurement system to ensure proper bubble rate in a lead-acid battery. It should not be too fast or too slow for proper functioning of an air purge tube. If the rate of bubble, emerging from the end of air tubes, is too low, backpressure is not developed and hence sensor fails to detect the signal. As the air purge tube is immersed in the acid inside the battery due to non-transparent plastic, it is difficult to observe whether bubbles are emerging out or not. Therefore, measuring the flow of air through a tube is of a prime importance. In the previous state of the art, no research has been carried out on the measurement of air flow inside the SoC measuring air purge tube. The other challenge is to measure a very low flow rate, as it is a flow rate that produces only bubbles.

Various types of flow sensors and transducers have been developed and fabricated during the last few decades [9]. The flow as evaluated is considered to be compensated by the systems temperature change, chemical composition pressure variation and change in velocity [10]. Mostly, flow is calculated by measuring volume, mass, fluid velocity [11]. Also, bulk micromachining and surface micromachining form the emerging technology that is used to fabricate MEMS sensors for a variety of applications such as medical, sports exercise, safety and weather monitoring departments [12]. Microfluidic flow sensors are miniaturized devices having high resolution and are the cost-effective alternative to existing sensors [13]. MEMS flow sensors are utilized in different industrial applications like autonomous underwater vehicles, mobility and object tracking [14], flow estimation in medical applications surgical procedure, diagnostic equipment, therapy areas [15]. Several types of fluid flow sensing techniques have been preferred in industry, such as the rotameter, turbine flow meter, and electromagnetic flow meter, hot air anemometer, however, they are bulky in size and are costly too. Due to the features such as inexpensiveness, easy to handle, the flow sensor based on thermo calorific principle proposed for air flow measurement in air purge SG measurement method.

The conventional flow sensors have limitations of size, cost and accuracy. FS1012 MEMS flow sensor is proposed to measure the air flow rate of the air purge system. FS1012 MEMS mass flow sensor works on thermo-transfer principle. FS1012 MEMS flow sensor is used to measure air flow discharged by a compressor in the acid of the battery. This method is vibration and shock-resistant, needs low power, high sensitivity, and the response time is very fast i.e., less than 5 seconds. This sensor has an added advantage like its output is in millivolt and its operating temperature ranges between 0 C and +85C. Further, it measures flow in the range of 0 L.P.M to 2 L.P.M.

This paper consists of three main sections. In Section 1, the design of the MEMS flow sensor-based air purge system has been presented. Section 2 highlights the Proteus simulation and signal conditioning circuit hardware design of the MEMS flow sensor. Section 3 analyses the result obtained with adequate discussion. Section 4 offers the conclusion and provides the future scope of research.

2. FS-1012 MEMS Flow Sensor

2.1. Working Principle of FS-1012

The air mass flow rate can be calculated from the specific heat equation:

$$Q = H/C_P(T_2 - T_1) \tag{1}$$

As shown in Eq. (1), H is heat transferred, Q is mass flow rate of the air, C_p is specific heat of the gas and T_1 is temperature of air noted by thermopile1 and T_2 is temperature of air sensed by thermopile 2. As shown in Figure 1, thermal MEMS sensor consists of three essential element heater and two thermopiles. Heater is placed at the centre between two thermopiles. When air flows through the pipe, it is first heated by the heaters. The temperature of the air is measured using a thermopile. The thermopile is symmetrically placed to both sides of the heater upstream and downstream of air flow. When no air flows through the pipe, the temperature shown by thermopile-1 and thermopile-2 is the same. The symmetrical temperature profile appears at non-zero flow conditions. For the lowest flow rate, the temperature shown by thermopile 1 is very high. As the flow rate starts to increase due to the laminar flow of air, the temperature is carried by air molecules from the inlet to the outlet and hence the temperature of thermopile-1 starts to drop. The temperature profile starts to deviate from the inlet to the outlet as the flow of the air starts to increase. Thermopile-1 and thermopile-2 are thermally isolated so that only the rate of temperature variation is the effect of flow variation.

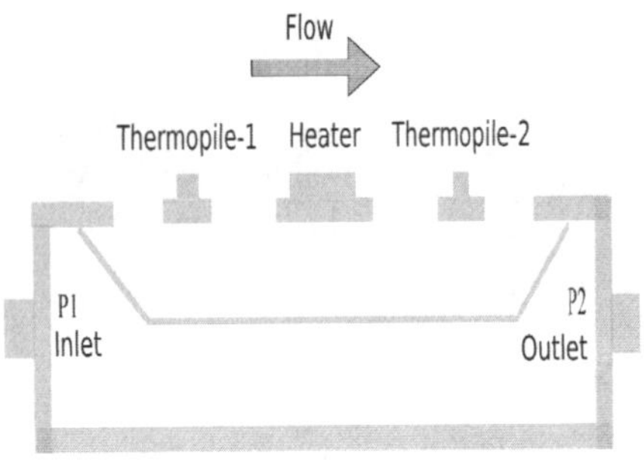

Figure 1. Internal structure MEMS flow sensor shows heater and two thermopiles at inlet port P1 and outlet port P2.

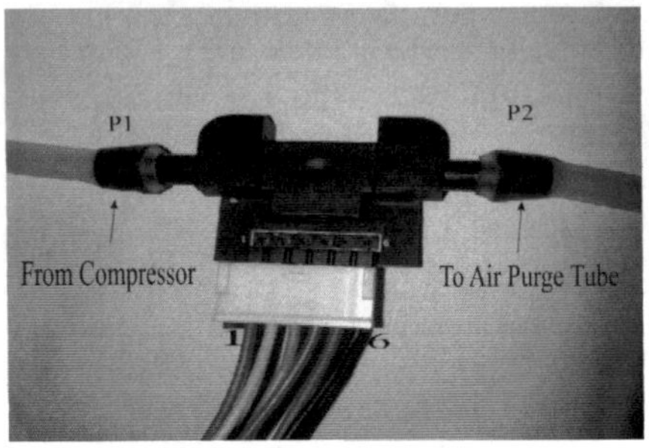

Figure 2. FS-1012 MEMS Sensor for air flow measurement having an air inlet and outlet port.

As shown in Figure 2, the pin connection diagram is mentioned further: pin-1 output of thermopile-1, the pin-2 ground of thermopile-1, pin-3 supply voltage of heater, pin-4 ground for the heater, pin-5 ground of thermopile-2, pin-6 output of thermopile-2, pin-5 and pin-6 are generally used for the liquid fluid. Single-ended thermopile output is used for signal amplification.

2.2. MEMS Flow Measurement System

As shown in Figure 3, the MEMS-based flow measurement system consists of an aquarium pump, FS-1020 MEMS flow sensor, operational amplifier -OPO7 and LM-324, TM4C123GXL ARM controller. Initially, during calibration, it is essential to adjust a proper air flow through the air purge system. It is also essential to monitor a perfectly adjusted flow throughout the working operation of the system. FS1020 MEMS flow sensor is connected between the air compressor and the air purge tubes. Air flow through the air purge tubes and FS1012 is controlled using a control valve. This control valve also protects the sensor. As shown in Figure 3, the output of the MEMS sensor is amplified using an OP-07 operational amplifier. In the next stage, the output is linearized and finally filtered out. TM4C123GXL ARM controller is used for averaging the samples, as well as for monitoring data with high speed and high resolution.

2.3. Signal Conditioning of FS-1020 using Proteus Simulator and Hardware

As shown in Figure 4, MEMS flow sensor signal conditioning has been designed using OP-07 and LM-324 and RC filter. The circuit is simulated with Proteus software and later, this is implemented on the breadboard as shown in Figure 5.

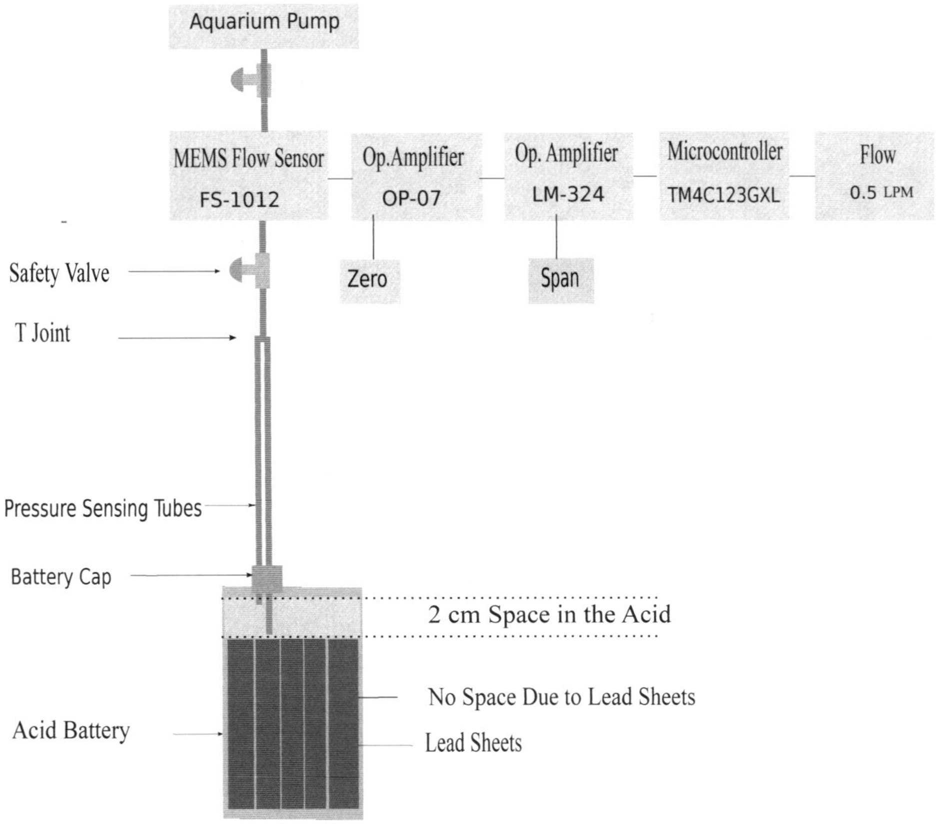

Figure 3. Air flow measurement system with FS-1012 MEMS sensor for SG monitoring system of lead-acid battery with op amplifier and TMC123GXL microcontroller.

As shown in Figure 5, MEMS flow sensor shows the output of thermopile-1 which is 126 mV at 0 LPM and 110 mV at 2 LPM. The difference between the output of the minimum flow rate and the maximum flow rate is 16 mV. As shown in Figure 5, 16 mV signal is amplified in such a way that it produces 3.3 Volt. TM4C123GXL ARM controller is used for the monitoring of the flow parameter. To get maximum sensitivity, the output signal of the MEMS flow sensor is amplified from 0 Volt to 3.3 Volt. At the initial stage, 126 mV is adjusted by pot, P1 is a simulated output of the sensor. 110 mV voltage is adjusted by pot P2, so the difference of the voltage provided to the operational amplifier (OP-07) is 16 mV. The gain of the operational amplifier (OP-07) is set 100 times. Therefore, 16 mV voltage is amplified up to 1.6 volts. The output of the sensor is amplified 100 times at this first stage. In the second stage, non-inverting amplifier is used to double the voltage. In the second stage, the output of OP-07 is amplified using span compensation (P4 pot) to get 3.3 V. In the third stage inverting differential amplifier is used to get linear output. Due to the differential output of this stage for the same input, this third stage produces 0 volt. At the third stage, as 0 LPM output is desired, constant output of 3.3 V is adjusted using pot P3 and provided to the non-inverting stage of the amplifier. In the fourth stage, low pass filter is used to get a noise-free output. For 0-volt input signal, TM4C123GXL arm controller displays 0 LPM flow of the air. This output is obtained by making zero flow of the air purge tube.

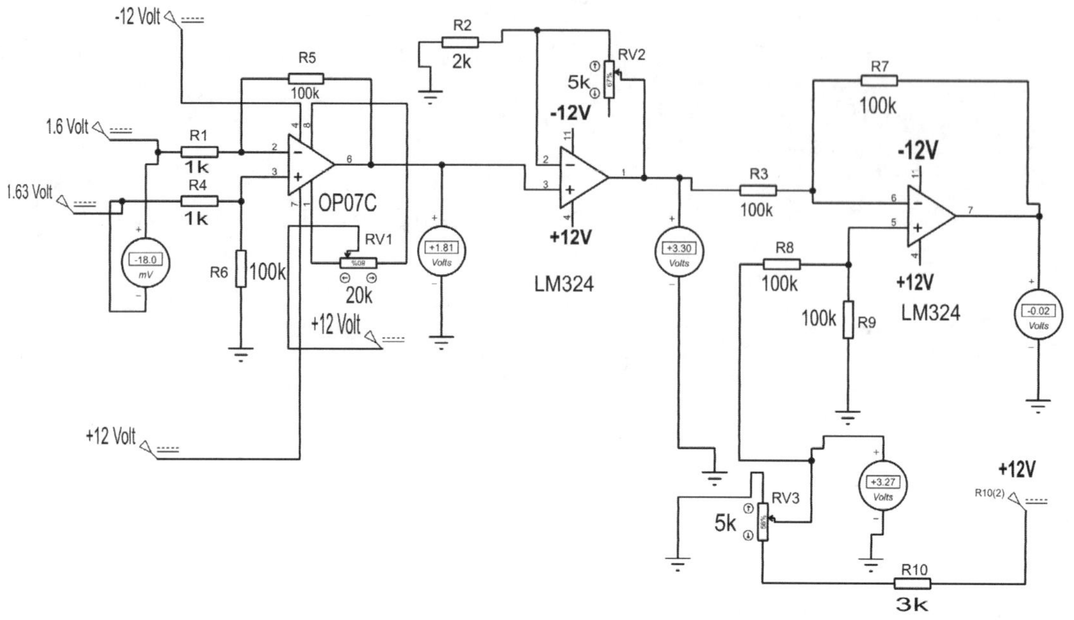

Figure 4. MEMS flow sensor signal conditioning using OP-07, LM-324 and RC filter using Proteus software.

2.4. TM4C123GXL ARM Cortex-M4F Microcontroller

As shown in Figure 6, TM4C123GXL uses ARM Cortex-M4F technology in microcontrollers. The ARM Cortex-M processor has excellent features like high-performance, low cost, minimum memory implementation, less pin count, and low power consumption during the performance. However, it has excellent computational performance as well as an exceptional response to interrupts.

The algorithm of the proposed system is shown in Figure 7. The air flow is sensed by the FS-1020 flow sensor then the signal is read using the analog read command. Further, the signal has been displayed as an air flow in L.P.M.

2.5. Calibration

This proposed sensor system is designed in such a way that it can be calibrated and utilized for air flow measurement of different lead-acid batteries. The calibration procedure is mentioned as per further steps. Air flow can be changed by using compressor controlled potentiometer. When no air flows, zero compensation knob is adjusted so that the display device shows 0 LPM Then, the higher flow is allowed to pass and thereafter the span compensation knob is adjusted until the display shows 2 LPM. This procedure is repeated 4-5 times to set the proposed system to measure flow from 0 LPM to 2 LPM. Calibraton is an important aspect to get more accurate

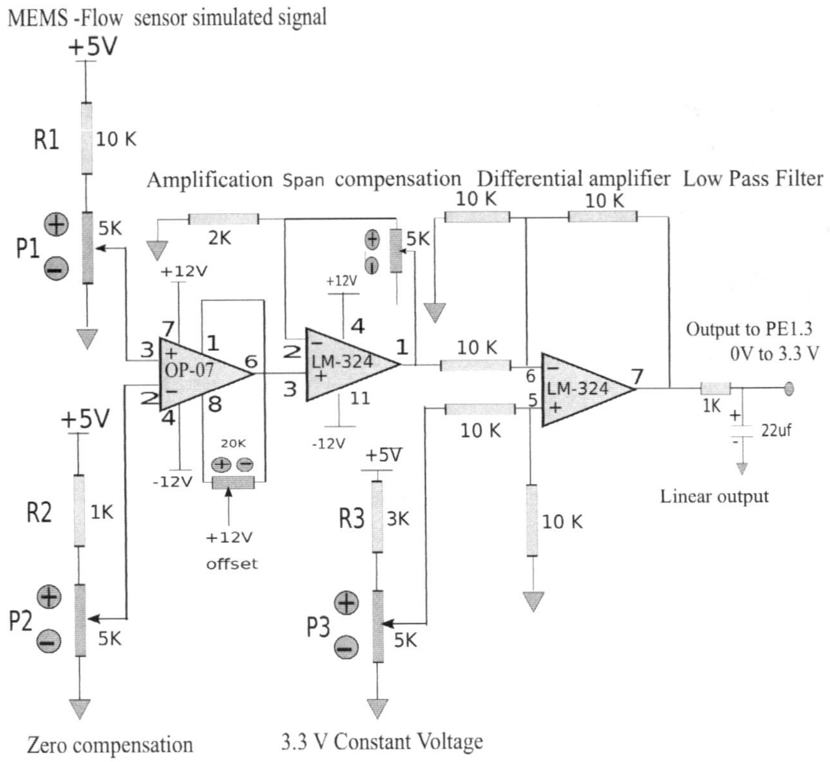

Figure 5. MEMS flow sensor signal conditioning using OP-07 and LM-324 and RC filter.

result and should be done periodically

3. Results and Discussion

As shown in Table 1, when maximum air flow is allowed through the air purge system, the output of thermopile-1 starts to decrease. For maximum flow, thermopile-1 produces a 108 V signal. Due to the same voltage at amplifier input at the first stage, it produces 0 V. The output of the sensor is amplified 100 times at the first stage. As this is zero volts, the amplified output of the second stage is also 0 V. The output obtained at the third stage is as desired i.e. 3.3 V output. A constant output of 3.3 V is already fixed and is provided to the non-inverting stage of the amplifier. As a result, we get 3.3 V output at differential amplifier in the third stage. In the fourth stage, low pass filter is used to get a noise-free output. At the final stage, the output obtained is 3.3 V, which is to be expressed as 2 LPM using TM4C123GXL.

Table 1. Maximum air flow

Sensor o/p (mV)	OP-07(mV)	amplifier(mV)	differential stage(V)	filtered output V	Display(LPM)
108	0	0	0	3.3	2
108 (fix)	-	-	3.3 (fix)	-	-

As shown in Table 2, when the air flow is not allowed through the air purge system, the output of thermopile-1 starts to increase. For no flow condition, thermopile-1 produces a 126 mV signal. Due to the fixed voltage at the amplifier input at the first stage, it produces a

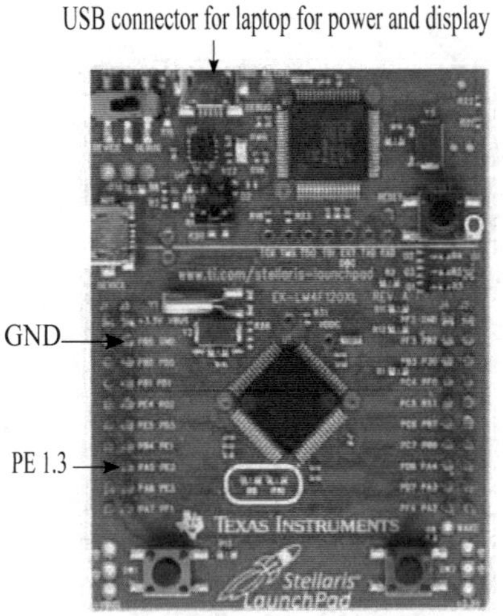

Figure 6. TM4C123GXL ARM Cortex-M4F microcontroller PE 1.3 for analog input and a USB connector for the laptop to display flow parameters.

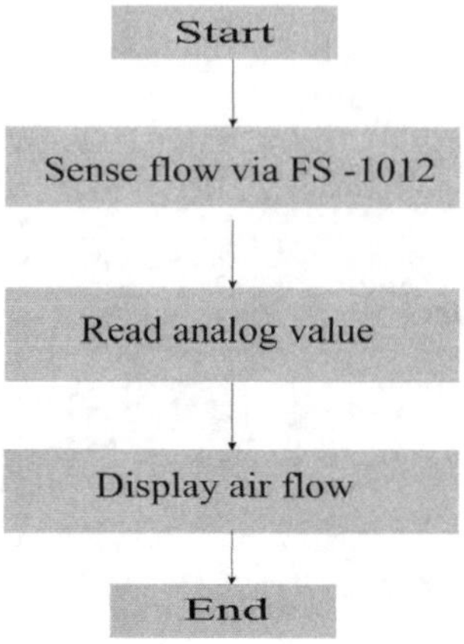

Figure 7. Flow chart to measure the flow of air inside air purge tube.

differential voltage of 16 mV. The output of the sensor is amplified 100 times at the first stage. As this is 16 millivolts, the amplified output of the second stage is also 1.6 V. The output obtained at the third stage is as desired i.e. 0 V output. A constant output of 3.3 V is already fixed and provided to the non-inverting stage of the amplifier. As a result, at last 0 V output is seen at the differential amplifier in the third stage. In the fourth stage, low pass filter is used to get a noise-free output. At the final stage, the output obtained is 0 V which is to be expressed as 0 LPM using TM4C123GXL.

Table 2. No air flow

Sensor o/p (mV)	OP-07(mV)	amplifier(mV)	differential stage(V)	filtered output V	Display(LPM)
126	16	1.6	3.3	0	O
108 (fix)	-	-	3.3 (fix)	-	-

During the experiment, it is observed that to get bubbles at the bottom of the pipe, the output shown by a thermopile-1 is 122.8 mV, and it is 0.500 LPM as shown by TM4C123GXL. For a proper functioning of the air purge system, it is essential to adjust the flow rate of 0.500 LPM. As shown in Table 1, it is observed that if the air flow is maximum, the lowest output is shown by the sensor.

As shown in Figure 8, for no air flow condition, the sensor output is 126 mV and for 2 LPM, the sensor output is 108 mV. From the sensor output, the flow of the air is determined. From the graph shown in Figure 8, sensor output is directly proportional to air flow. Two-point calibration method is used to characterize this micro flow sensor.

As shown in Table 3, flow condition is simulated using proteus for no flow condition. The sensor output is maximum that is 126 mV which is conditioned as 0 V using signal conditioning circuit. For maximum flow condition, the sensor output is minimum, that is, 108 mV which is presented as 3.26 V.

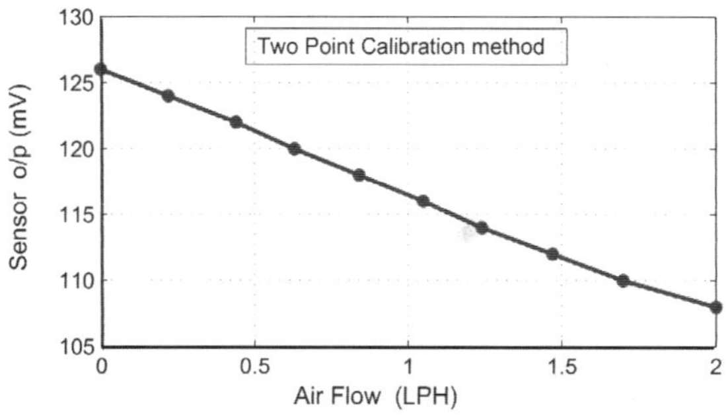

Figure 8. Sensor output is directly proportional to the air flow.

Table 3. Simulation result using proteus software

Sr. No.	Sensor o/p (mV)	Amplified o/p (mV)
1	108	3.26
2	110	2.90
3	112	2.53
4	114	2.17
5	116	1.80
6	118	1.44
7	120	1.07
8	122	0.71
9	124	0.34
10	126	0

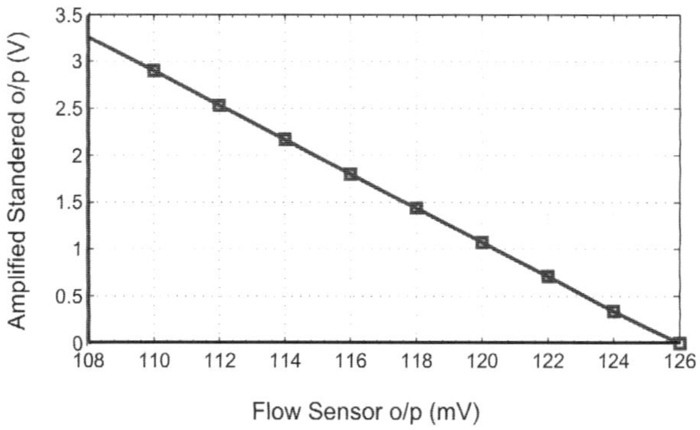

Figure 9. Sensor output range is linearly amplified in standard range 0 to 3.3 V.

As shown in Figure 9 and as per the results obtained by the proteus simulator, for no air flow condition, the sensor output is 126 mV and for 2 L.P.M, the sensor output is 108 mV. If the mico flow sensor voltage increased from 108 mV to 126 mV, the output obtained varies linearly from 3.26 V to 0 V.

4. Conclusion

In this applied research study, the practical problem of low air flow measurement is addressed. The proposed solution consists of sensor selection, design of the circuit, microcontroller program development and display to get near-time benefits to know the SoC of the battery. For the first time, a MEMS flow sensor is introduced to know the air bubble flow rate of the lead-acid battery. It is designed to cut down manpower and reduce hardware systems and get high accuracy. The air flow rate is 0.50 LPM is noted with this instrument and it is difficult to measure with a conventional flow sensor like rotameter. Mostly square root extractor IC is required in the circuit to measure flow rate but here it is eliminated from this flow signal conditioning circuit. This measurement system ensures the bubble rate inside the battery. This is a real-time measurement system. FS-1020 MEMS flow sensor is used to measure the flow of discharged air in the non-transparent lead-acid battery. If the flow of the air is maintained at 0.50 LPM, the system works efficiently. In the future, the flow sensor can be directly used to measure the SoC of the battery instead of the MEMS pressure sensor as the flow is directly proportional to differential pressure. The Same sensor with relay can be utilized to know air flow from compressor towards acid battery as well as from the flow, differential pressure can be monitored as a SOC of the battery. This research offers a low-cost air flow measurement solution for a lead-acid battery as well as different industrial applications.

Acknowledgments

The authors thank Instrumentation and Control Department, College of Engineering, Pune (C.O.E.P) for the financial support provided in patent filing related to SG measurement of lead acid battery and further process in the research study. (The Patent Office Journal No. 15/2020, Indian Patent)

References

[1] C Y Patil and Y G Adhav 2020 *The Patent Office J.* **15** 16896
[2] C Tang and J Lin 2020 Online Autonomous Specific Gravity Measurement Strategy for Lead-Acid Batteries *IEEE Sensors J.* **20** 1980
[3] S Kim, M Kwon, and S Choi 2018 Operation and Control Strategy of a New Hybrid ESS- UPS System *IEEE Trans. on Power Electronics* **33** 4746
[4] M Lei, Z Yang, Y Wang, H Xu, L Meng, J C. Vasquez and J M Guerrero 2018 MPC-Based ESS Control Method for PV Power Smoothing Applications *IEEE Trans. on Power Electronics* **33** 2136
[5] A Clerici, E Tironi, and F Castelli-Dezza 2018 Multiport Converters and ESS on 3-kV DC Railway Lines: Case Study for Braking Energy Savings *IEEE Trans. on Industry Applications* **54** 2740
[6] B Vulturescu, S Butterbach, and C Forgez 2014 Experimental Considerations on the Battery Lifetime of a Hybrid Power Source Made of Ultracapacitors and Lead-Acid Batteries *IEEE J. of Emerging and Selected Topics in Power Electronics* **2** 701
[7] M Greenleaf, O Dalchand, H Li and J P Zheng 2015 A Temperature Dependent Study of Sealed Lead-Acid Batteries Using Physical Equivalent Circuit Modeling With Impedance Spectra Derived High Current/Power Correction *IEEE Trans. on Sustainable Energy* **6** 380
[8] P B L Neto, O R Saavedra, and L A de Souza Ribeiro 2018 A Dual-Battery Storage Bank Configuration for Isolated Microgrids Based on Renewable Sources *IEEE Trans. on Sustainable Energy* **4** 1618
[9] H M Choi, B R Yoon, C G Kim and Y M Choi 2011 Evaluation of flowmeters for heat metering *Flow measurement and instrumentation* **22** 475
[10] G de Graaf, A Abarca, M Ghaderi and R F Wolffenbuttel 2015 A MEMS Flow Compensated Thermal Conductivity Detector for Gas Sensing *Eurosensors* **120** 1265
[11] B Tian, H Li, H Yang, D Song, X Bai and Y Zhao 2018 A MEMS SOI-based piezoresistive fluid flow sensor *Review of scientific instruments* **89** 025001
[12] C M Ho, Y C Tai 1998 Micro-electro-mechanical-systems (MEMS) and fluid flows *Annual review of fluid mechanics* **30** 579

[13] R Bogue 2013 Recent developments in MEMS sensors: a review of applications, markets and technologies *Sensor review* **33** 300
[14] I V J H K, B C Claus and J C Kinsey 2018 , A navigation solution using a MEMS IMU, Model-Based Dead-Reckoning, and One-Way-Travel-Time acoustic range measurements for autonomous underwater vehicles *IEEE Journal of Oceanic Engineering* 1
[15] S Silvestri, E Schena 2012 Micromachined flow sensors in biomedical applications *Micromachines* **3** 225

Performance evaluation of thermoelectric generator employing SEPIC converter and incremental conductance maximum power point tracking for electric vehicles

N. Jaswanth* [1], **G. Raam Dheep**$ [2]
*ORCID: 0000-0003-4434-7295, $ORCID : 0000-0001-8470-2437
School of Electronics and Electrical Engineering, Lovely Professional University, Punjab, India
[1]jaswanth055@gmail.com, [2]raam.25227@lpu.co.in

Abstract:

This article introduced a unique MPPT controller for thermoelectric generators. Our method uses a modified incremental conductance (IC) algorithm to track MPP efficiently. The proposed method is simple to install, fast-tracking, and stable. Traditionally, MPPT trades off steady-state and transition-state properties. Various designs flows are explored. The study's common parameters are fast convergence, efficiency, and low oscillations. Use a proposed strategy that effectively handles these difficulties. The proposed methodology improves TEG MPPT efficiency and convergence times.

Keywords: Maximum Power Point Tracking (MPPT), Thermo Electric Generator (TEG), Incremental conductance (IC), Single Ended Primary Inductor (SEPIC) Converter, MATLAB-Simulink.

1. Introduction

Recently, the rising energy consumption has accelerated. Living standards improvements increase energy demand, causing outages and rising fossil fuel prices, prompting many utilities and customers to find and utilise more reliable energy sources. Solar, wind, thermal, water flow, ocean energy, and fuel cells are all examples of renewable energy. There is almost 75% wastage in transport[1]-[3] and industry. Thus, waste heat recovery can contribute a significant quantity of energy to everyday needs, especially transportation. Thanks to modern technology, With the help of a thermoelectric generator, you may salvage wasted heat (TEG), which converts heat into electricity[4]-[5]. Technologies that do not require moving parts or the production of environmentally hazardous waste provide low-cost electricity and environmentally friendly technology [6].

The technology of TE devices has advanced significantly in recent years, and as a result, their efficiencies are growing dramatically, particularly as a result of advancements in TE materials and geometrical device design. [7],[8]. Although TEG has poor efficiency, there is still much room for improvement in this area. The TEG system's low efficiency is a serious problem, and work is being done to increase the devices' performance. To maximise the TEG device's capacity, it is important to extract the maximum electricity. There are several ways in which power conditioning methods can be used to ensure that the greatest power is extracted from a TEG system, such as TE material improvement and geometrical enhancement. Using a SEPIC converter and impedance matching are components of the power conditioning process. The impedance matching process balances the TEG system's internal and external load resistance. As a rule of thumb, if you want maximum power output from a TEG module, the ideal electrical load should be equal to the TEG module's internal resistance [9]. Thus, the maximum power point tracking (MPPT) SEPIC converter is utilised to ensure a constant voltage and the highest possible power output [10]. Some of the difficulties encountered in the extraction of MPP are as follows: The overshooting problem, setbacks caused by delayed convergence periods, a constant scaling factor, and the high cost of implementation are all concerns commonly associated with standard derivative MPPT algorithms.

To overcome the issues of TEG, the current work presents a novel Incremental conductance MPPT approach. The new control strategy divides the operating region into two halves depending on how close

Journal of Physics: Conference Series **2327** (2022) 012005 doi:10.1088/1742-6596/2327/1/012005

the searched MPP is to the current location. The step size of the new control strategy alternates between two fixed values. The proposed technique has several advantages over other approaches because it is not dependent on population optimisation and does not require scaling factors. It also requires a basic algorithm with minimal design changes, making it very adaptable to other methods.

2. System Configuration

The SEPIC converter has a changeable dc voltage setting that keeps the output consistent. Although the SEPIC is very similar to the buck-boost converter, unlike a standard buck-boost converter, it has the advantage of using a series capacitor to transport energy from the input to the output. The TEG includes SEPIC converters, incremental conductance MPPT control, and load. All of the TEG's components are depicted in Fig 1.

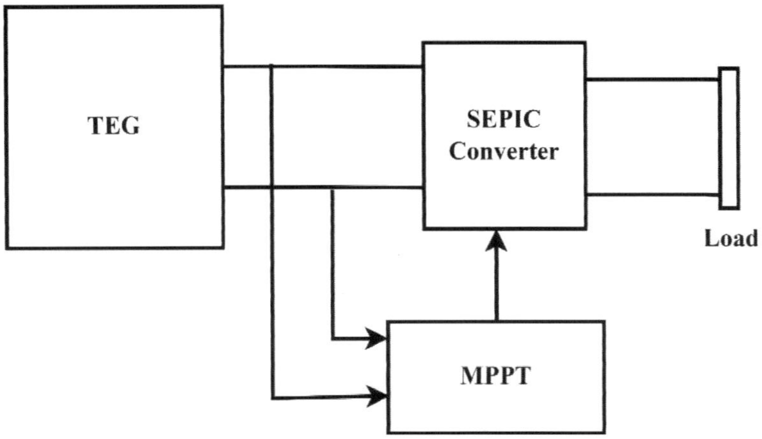

Figure. 1. General block diagram of TEG system with MPPT

2.1. Switching Converter Analysis

In connection to a TEG, a conversion converter is always required. The use of a no-return diode is even required when connecting the battery directly to a TEG. Most likely, the output voltage must be adjusted for the planned use. The most widely used SEPIC converter is shown in Fig. 2 is connected to the load through the SEPIC converter. SEPIC topology provides excellent MPPT efficiency and minimal ripples of current. TEG voltage can be precisely controlled, and energy is transferred from the TEG to the load regardless of the temperature.

Because of two inductors, the SEPIC switching power supply accepts a wider range of inductors. It has the advantage of allowing lower ripple currents. As long as the current flow is positive through the inductor L_1 at the current input, a boost converter appears to be a SEPIC converter. The capacitor CP has been utilised, allowing this converter to eliminate input-output isolation and, thus, an advantage. The condenser Cp protects from short circuits or overload. Because SEPIC power supplies cannot provide outputs with voltages as low as zero volts, SEPIC power supplies have the disadvantage of setting the output voltage as low as possible.

2.2. Incremental Conductance MPPT Algorithm

Thermal energy is becoming increasingly popular as a source of electrical power. However, the efficiency of conversion is still quite poor. The TEG system (MPPT) provides high-efficiency power, and the TEG system with MPPT in MATLAB/Simulink is depicted in Fig.4. Because of its simplicity, speed, and effectiveness, it is widely used, the incremental conductance approach is frequently considered an MPPT methodology. The conductance decreases and the incremental conductance rises on the right side of the MPP. The MPP is reached when the TEG power derivative by voltage (dP/dV) is zero. In this strategy, the fundamental equations are as follows:

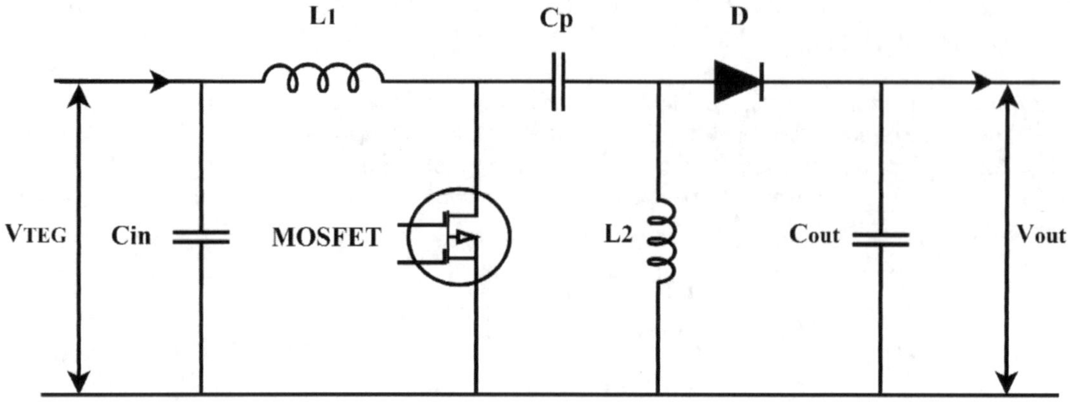

Figure. 2. Circuit diagram of SEPIC converter

$$\frac{dP}{dV} = 0 \quad (1)$$

Eq. (1) can be rewritten as:

$$\frac{dP}{dV} = \frac{d(IV)}{dV} = 1 + V\frac{(dI)}{dV} = 0 \quad (2)$$

At MPP: $\frac{dI}{dV} = -\frac{I}{V}$ (3)

At left of MPP: $\frac{dI}{dV} > -\frac{I}{V}$ (4)

At right of MPP: $\frac{dI}{dV} < -\frac{I}{V}$ (5)

Figure 3 depicts a flowchart of the incremental conductance approach.

The traditional MPPT technique relies on maintaining a uniform step size to achieve good performance. A slow convergence around MPP characterizes TEG power oscillations; however, they cannot function under rapidly changing weather circumstances. With a bigger step size, it is possible to accomplish faster tracking while maintaining steady-state tracking, but this is impossible to achieve with a larger step size. These conundrums necessitate a large number of slower contributions. This has led to numerous contributions with various step sizes, and great progress has been made in this area. The system should automatically adjust step size to establish a suitable balance of dynamics and oscillations. Simpler, faster, and less oscillatory are some of the advantages of the IC MPPT method proposed in this study. The variable step size IC MPPT method is shown in Fig. 3.

Here is the approach with a variable step size that has been offered:

$$D(K) = D\,(K-1)\ \pm N * \left|\frac{dP}{(dV-dI)}\right| \quad (6)$$

In this case, D(k) represents the converter's current duty cycle and D(k-1) its previous duty cycle (k-1). N is the scaling factor used to determine the step size while sampling.

Journal of Physics: Conference Series **2327** (2022) 012005 doi:10.1088/1742-6596/2327/1/012005

dI, dV, and dP are the TEG output current, voltage, and power derivates. With the use of formulas:

$$dI(K) = I(K) - I(K-1) \quad (7)$$

$$dV(K) = V(K) - V(K-1) \quad (8)$$

$$dP(K) = P(K) - P(K-1) \quad (9)$$

Where I(k), V(k), and P(k) are the output current, voltage, and power of the array at the time of the measurement (k)

For a given time instant, the TEG array's output current (k-1), voltage (k-1), and power are denoted by these values: (k-1)

The step size can be varied. The MPPT approach is dynamic (large step size in the far left and far right of MPP, however very small step size near the MPP and equal to zero at MPP).

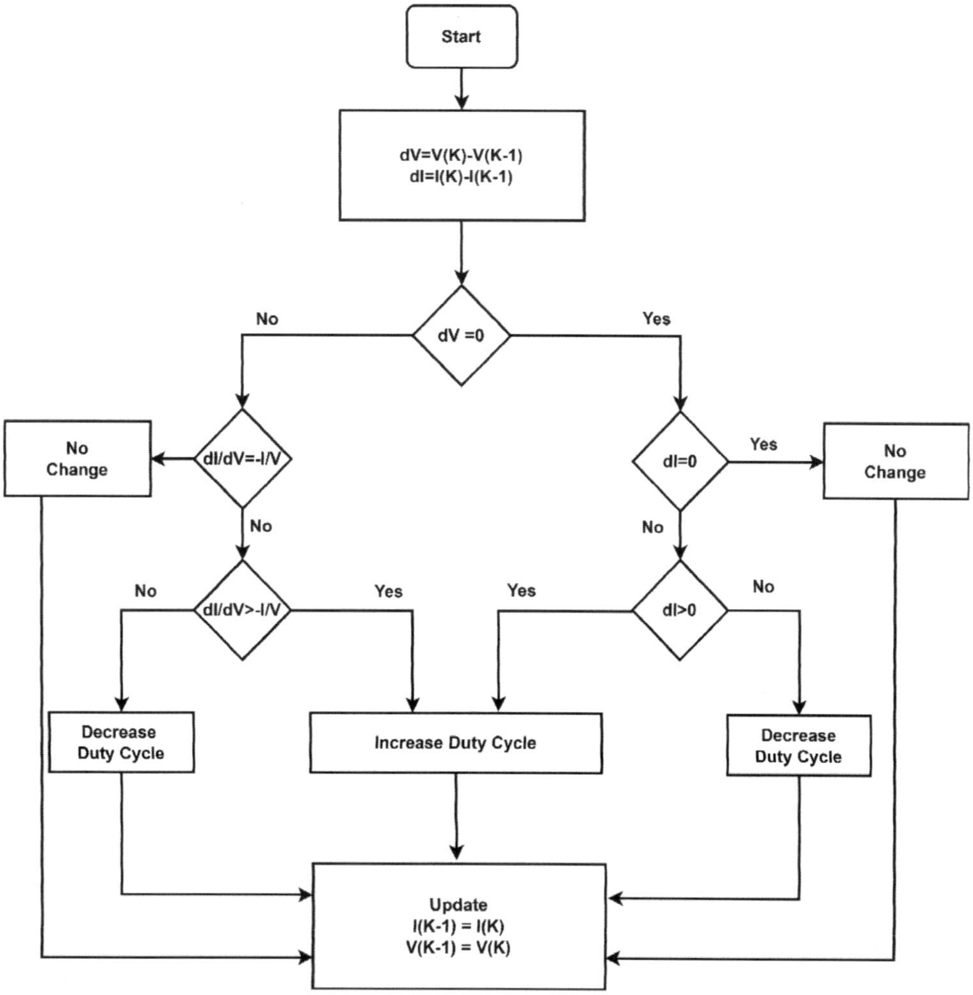

Figure 3: IC approach flowchart with direct control

Journal of Physics: Conference Series **2327** (2022) 012005 doi:10.1088/1742-6596/2327/1/012005

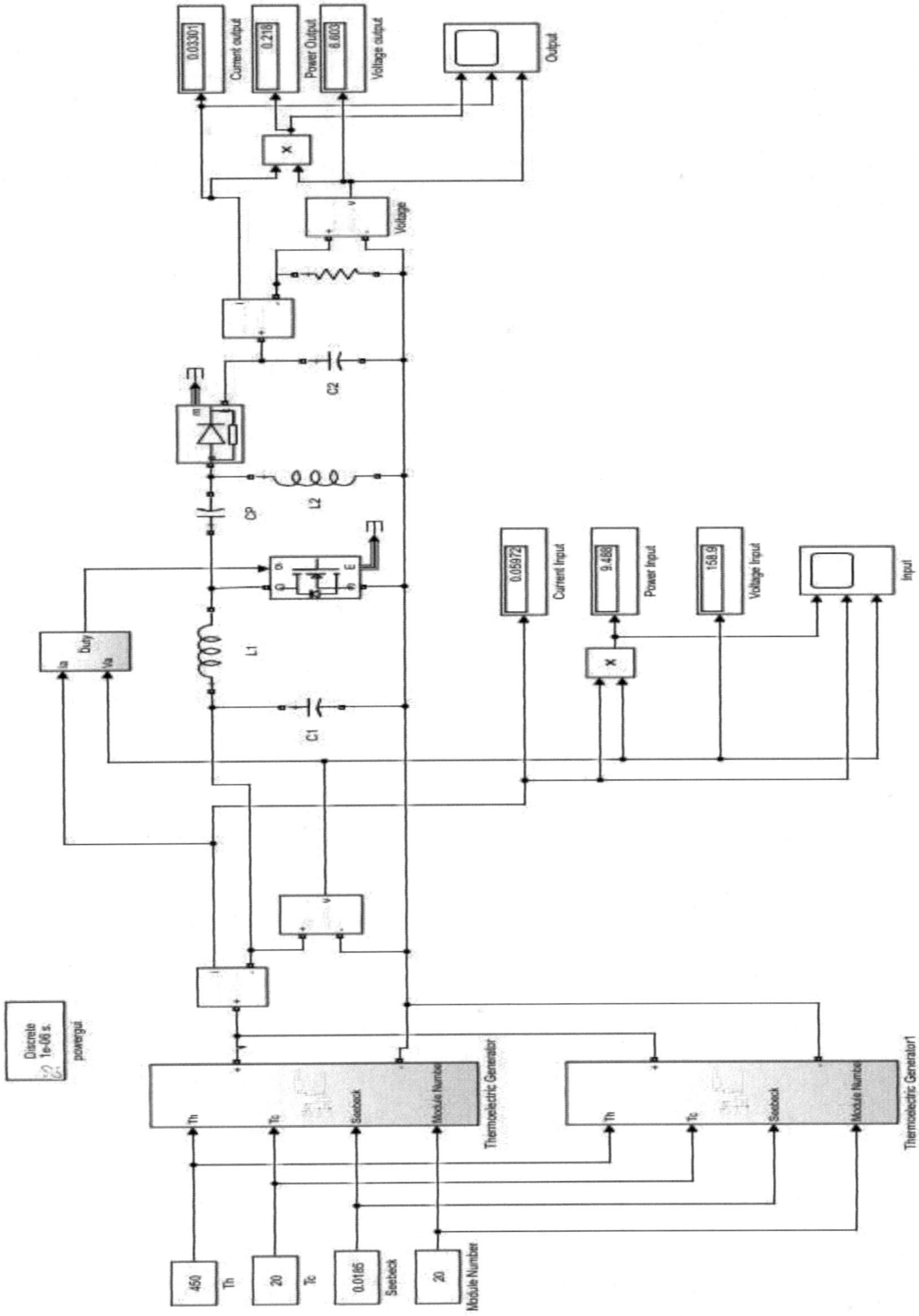

Figure.4. MATLAB/ Simulink circuit for proposed MPPT for TEG

3. RESULTS AND DISCUSSION

Figure 4 depicts the circuit model created using MATLAB/SIMULINK modeling software.

The following reactants were generated between 200 and 450 degrees Celsius: Temperatures ranging from 200 to 450 degrees Celsius can be used to evaluate the efficiency of the conversion system (C). The cold side temperature can be set to 50 degrees celsius. The TEG modules are connected in series/parallel and modulated using an algorithm to produce this simulation. Electricity is generated from thermal energy by a thermoelectric generator using an incremental conductance MPpt with a SEPIC converter.

The output voltage vs load resistance is shown graphically in Figure 5 at various hot system sides. Fig. 5 shows that the output voltage rose gradually when the load resistance was modified at different temperatures. Figure 5 depicts the output voltage of 52 volts and the hot side temperature of 450 ^{0}C with an impedance of 80 ohms.

Hot side temperature has an impact on load resistance Vs output power, as shown in Figure 6. When the load resistance was varied, the output power rose until it reached its maximum value. Power output at 450 ^{0}C may be shown in figure 6 when the output resistance is 80 ohms. According to the findings, at 80 ohms and 450 ^{0}C, the maximum output power is 35.5 watts. In all circumstances, the output power increases with temperature. Temperature differences have been blamed for the rise in the hot-side temperature.

When we examine Fig.7, we can see how varying hot side temperatures affect the Output current based on load resistance. Convertor output voltage drops in proportion to increased converter load, and, as a result, higher converter temperatures are also shown. Thus, an increased output current is recorded at 450 ^{0}C on the hot side (Th).

Voltage, current, power, and load resistance are plotted in Fig. 8. Input voltage is also raised by converter load, making it obvious that the voltage rises by however much voltage are detected between zero resistance and 80 Ω. It was found that the proposed energy gathering method was more efficient than standard methods.

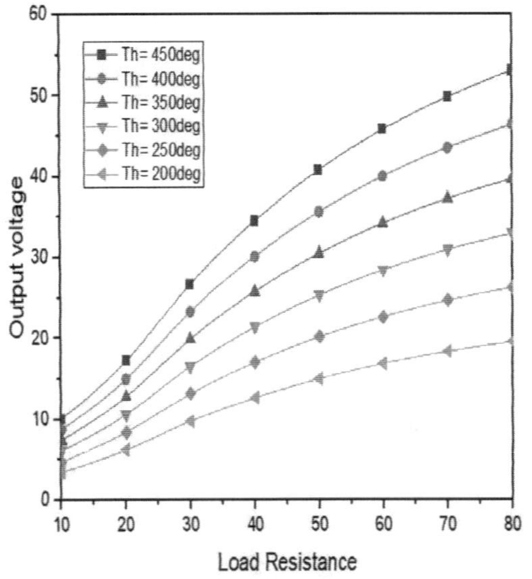

(a)

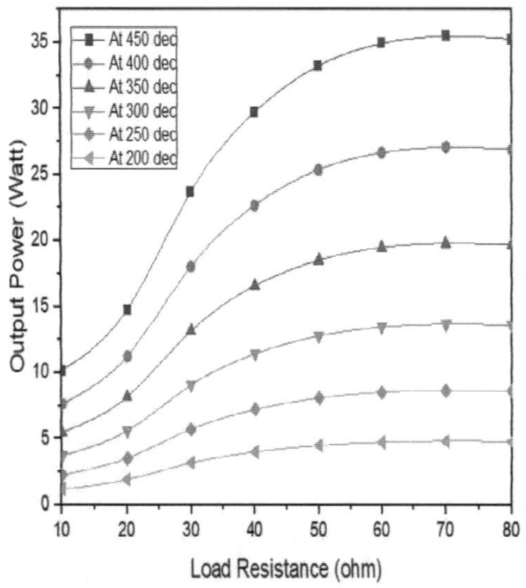

(b)

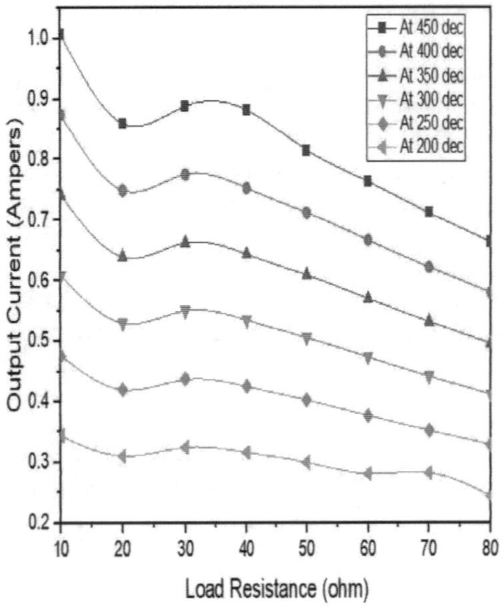

(c)

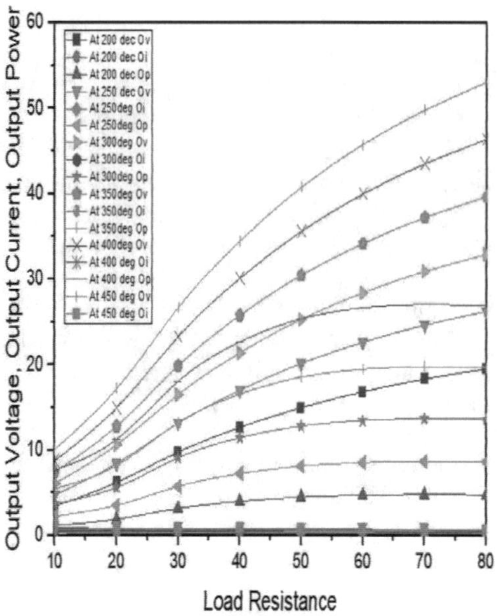

(d)

Fig. 5. At different hot side temperatures (a) Output voltage Vs load resistance, (b)Output power Vs load resistance, (c) Output current Vs load resistance at different, (d). The output voltage, output current, output power, Vs load

4. CONCLUSION

A modified Incremental Conductance MPPT for a thermoelectric generator is presented in this work using Matlab/Simulink. The TEG system for simulation includes the TEG model, the SEPIC converter, and the modified Incremental conductance MPPT controller. This investigation showed MPPT results depending on temperature and load resistance. The results support the proposed MPPT controller's higher performance. Considering the results, we can say that the suggested MPPT controller outperforms previous approaches. The rapidity of response, high output voltage, large current and low oscillations are advantages of this design. In light of the findings, the Incremental conductance MPP with SEPIC converter for TEG is acceptable for E-Vehicle applications.

REFERENCES:

[1] Hamiton M. Recent advances in energy harvesting technology and techniques. IECON 2012- 38th Annual conference on IEEE industrial electronics society, Montreal; 2012.

[2] Vullers R, Schaijk R, Doms I, Hoof CV, Mertrns R, Micropower energy harvesting. Solid-State Electron 2009; 53: 684-93.

[3] Wang C, Hung C. Effect of irreversibilities on the performance of thermoelectric generator investigated using energy analysis. J Therm Sci Technol 2013; 8:1-14.

[4] Ramli MAM, Twaha S, AI-Hamouz Z. Analysing the potential and progress of distributed generation application in Saudi Arabia: the case of solar and wind resources. Renew Sustain Energy Rev 2017(70): 287-97.

[5] Zheng XF, Liu CX, Yan YY, Wang Q. A review of thermoelectric research- recent developments and potentials for sustainable and renewable energy applications. Renew Sustain Energy Rev 2014; 32:486-503.

[6] Seetawan T, Singsoog K, Srichai S. Thermoelectric energy conversion of $Ca_3Co_4O_9$/n-$CaMnO_3$ module. The 6th International Conference on Applied Energy- ICAE2014,0:20144, p:2-5.
[7] Liu W, Jie Q, Kim Hs, Ren Z. Current progress and future challenges in thermoelectric power generation from materials to devices. Acta Mater 2015;87(155): 357-76.
[8] Liu X, Deng YD, Li Z, Su CQ. Performance analysis of a waste heat recovery thermoelectric generation system for automotive application. Energy Conver Manage 2015; 90:121-7.
[9] Rumeli MF, Tan L, Date A, Singh B, Akbarzadeh A. Simultaneous power generation and heat recovery using a heat pipe assisted thermoelectric generator system. Energy Conver Manage 2015;91:110-9.
[10] Mamur H, Ahiska R. Application of a DC-DC boost converter with maximum power point tracking for low power thermoelectric generators. Energy Conver Manage 2015; 97:265-72.

Journal of Physics: Conference Series **2327** (2022) 012006 doi:10.1088/1742-6596/2327/1/012006

Investigation of increasing non-linear load impact on power quality in the distribution system

Navin Maury[1], Amit Kumar Singh[2]

[1,2]School of electronics and electrical engineering
Lovely professional university
Jalandhar Punjab, India
navinkumar122015@gmail.com, amit.20267@lpu.co.in

Abstract. Now a day's power quality problem daily increases due to the use of non-linear load in all sectors. So in this paper, we work on power quality improvement help of DVR (dynamic voltage restorer). Power quality issues are present time most thinkful because all sectors use power electronics devices and automation (PLC). In any sector without the use of power electronics devices and non-linear load create more problems because now a day all sectors use embedded and automation systems. The power quality main issue are sag, swells, unbalance, fault, and harmonic in this paper we control and reduce the sag in a fixed distribution system with the help of DVR with a PID controller. PID controller connected in series with DVR which is the controller the voltage sag and its frequency. This combination improves the power quality in the distribution system we can see the improvement by a simulation result shown in the paper we use 11kv subsystem and three-phase step-down transformer which converts 11kv to 440v. This project improves the power quality issue on the distribution side.

1. Introduction

The power quality issue in the distribution system is more compared to generation and transmission line systems because in generation and transmission line systems non-linear load is less used compared to the distribution system. This non-linear load is the main cause of power quality this non-linear load produces the sag, swells, harmonic noise, flicker, etc. So we work only on power quality improvement on the distribution side. For power quality improvement, more technology is used but we use DVR (dynamic voltage restorer) because DVR injects voltage in series, and with the help of the controller it controls the voltage and frequency so we can not only improve power quality we can control the voltage and frequency according to our use. In this project, we use a DVR with a series-connected PID controller. PID controller consists of PWM generator, two-level converter, three-phase transformer 12 terminal, abc to dq0 controller, dq0 to abc controller. This all function controls the voltage fluctuation and frequency. In the project take a fixed area of power distribution system where the main supply is 11kv connected with a three-phase step-down transformer which converts 11kv supply into 440v for the small consumer. In the distribution system connect a fault and some RLC load this fault switching time 0.016 to 0.083 more explain in the simulation section.

2. Power quality

Power quality is a simple word the ability to function properly of electrical devices in an electric system. Power quality shows the efficiency of an electric system that is how much it consumes the power and how much it gives the output. When we talk about power quality there are several areas considered like the continuity of service, variation in voltage magnitude, transient voltages, and currents. Good power quality can be defined as a study supply voltage in other words the prescribed range, steady ac frequency close to the rated value, and smooth voltage curve waveform.

2.1. Power quality issues

Power quality presents the ability of electric equipment to consume the energy being supplied to it. The power quality issues n number of like harmonic, sag, swell, fault, unbalance voltage, voltage instability etc.

2.1.1. Harmonic. Harmonics are sinusoidal voltage or current having frequencies that are integer multiple of the frequency at which the supply system is designed to operate known as the fundamental frequency. when any distortion in pure sinusoidal voltage and current frequency is called harmonic. The first harmonic is also called 1st harmonic and the first three harmonics 100Hz is called 2nd harmonic. 150 Hz 3rd harmonic, 200 Hz is called 4th harmonic and any addition of waves with these frequencies is periodic at 50 Hz.

2.1.2. Voltage sag. The voltage sag is a short-duration (0.1 and 0.9pu) decrease in the RMS voltage and current power frequency for a duration of 0.5 to 1 cycles. Due to voltage sag always creating the dipping problem in the distribution system and system is not properly work system failure chances also increases.

2.1.3. voltage swell. The voltage swell is increase in the voltage in a short duration (1.1pu to 1.8pu) and current power frequency duration from 0.5 to 1 minute. Due to voltage swell system is taking over-voltage and system failure chances increases.

2.1.4 Fault. Fault in distribution system presents the abnormal condition that involves electrical failure of distribution system equipment. A fault in the circuit is the disturbance, which interfaces the normal operation. when fault accrues in any system then current suddenly increases in the system.

3. DVR

The dynamic voltage restorer (DVR) is a series voltage booster or voltage injection device which injects the energy in form of voltage in circuit. Dynamic voltage restorer connects with series in the system at the load side. Dynamic voltage restorer is a power electronics device because it makes many electronics devices like filter, battery, series VSI, controller. These power electronic devices use to protect the sensitive loads and various disturbances in the power supply.

The main work of DVR is supplying a voltage of desired magnitude and frequency for the system. DVR restores the voltage and this passes the supply when the source voltage is unbalanced then it supplies the desired voltage for the required voltage.

In a DVR voltage source converter typically one or more convert connected in series to provide the required voltage rating.

3.1 DVR control action

3.1.1. when supply voltage normal condition then DVR is operating standby mode means zero voltage injection in the system.

3.1.2. when occurs the voltage sag and swell in the system then DVR needs to inject the voltage in three single-phase voltages in a very short duration.

3.1.3 when fault downstream of DVR then converts working and bypassed temporarily using thyristor switch for protection of DVR to overcurrents.

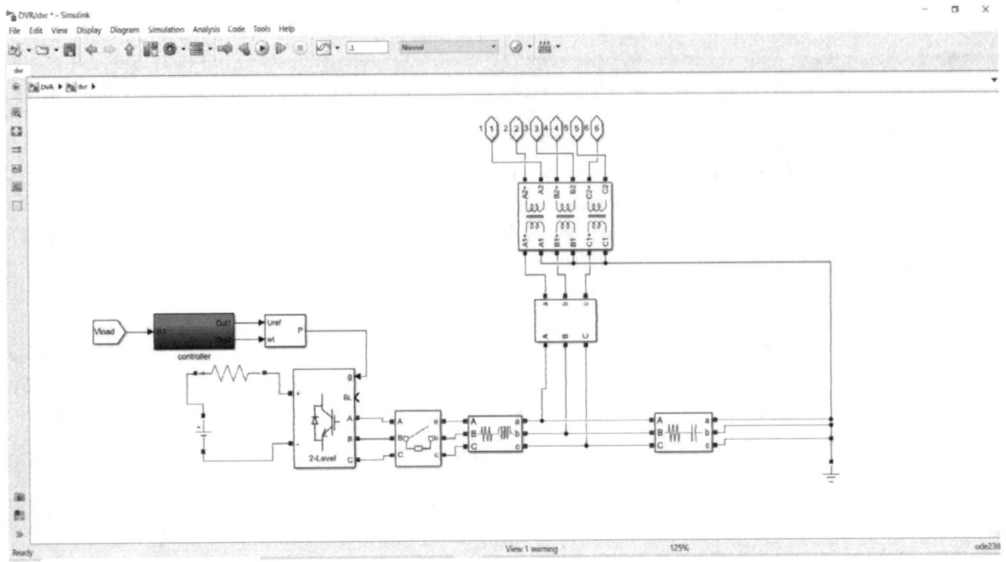

Figure 1 – DVR

3.2 DVR Devices

The above DVR figure consists of many power electronics devices following –

3.2.1. Controller The controller controls the voltage magnitude and frequency when the system is under voltage sag and swell conditions

3.2.2. PWM generator(2-Level) The PWM generator generates the pulses for carrier pulse width modulation convert use of two-level technology. The PWM can control the switching devices like GTO, IGBT.

3.2.3. Two-level converter. The two-level converter controls the signal which comes in the PWM generator it controls the signal by six firing pulses.

3.2.4. Three-phase breaker. The three-phase breaker when the circuit does close or open always connects in series between the input and output.

3.2.5. The three-phase transformer 12 terminal. The three-phase transformer 12 terminal transformer implements three single phases two winding linear transformers. The three-phase transformer 12 terminal is used when we do not know circuits are star or delta connected.

4. PID controller

PID (Proportional, Integral, Derivative) controller supplies the continuous various output within a control and loop feedback. PID controller removes error signal and increases the frequency efficiency. In PID controller integral and derivative remove the deviation error and manage repaid process moment. The PID controller compares the output signal the generate the desired signal.

PID controller connects with DVR in series which controls the system signal and takes the desired signal for system balancing.

Journal of Physics: Conference Series **2327** (2022) 012006 doi:10.1088/1742-6596/2327/1/012006

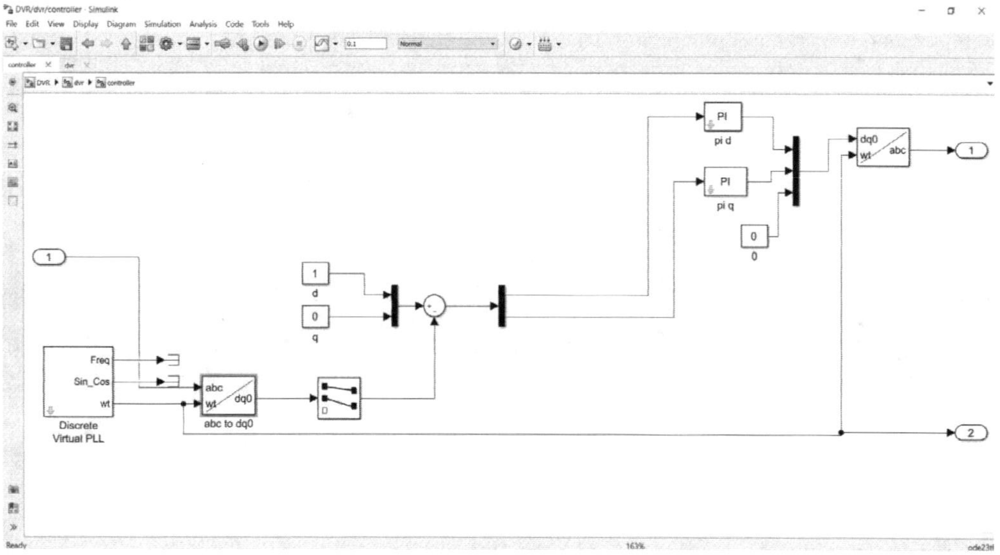

Figure 2 – PID controller

5. Simulation

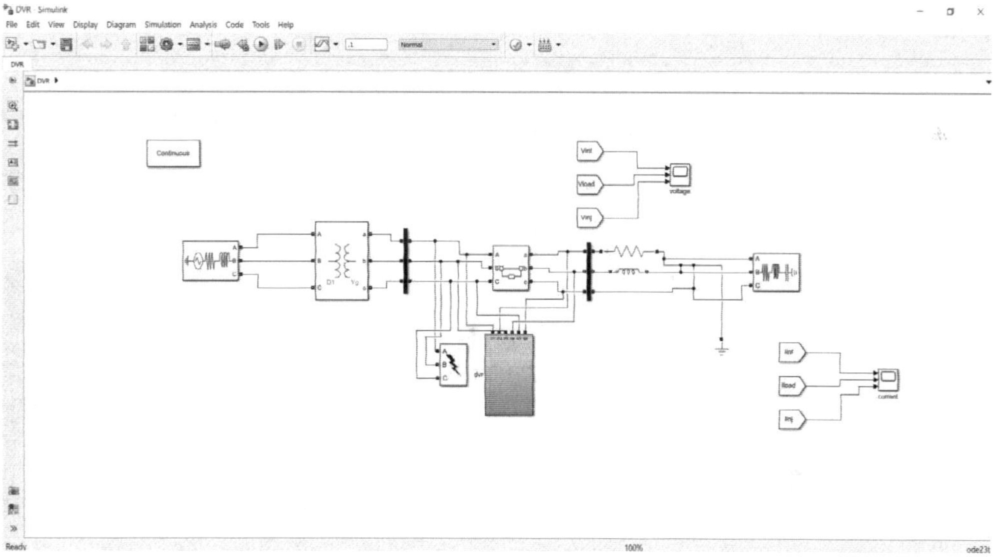

Figure 3 - Complete simulation of the distribution system with the DVR

The above figure 3 shows the complete simulation of the distribution system with the DVR.
In project take a fixed area of a small village in which the main supply is 11kv. In simulation convert, 11kv supply in 440v for small consumers and DVR are connected in between source and load. In this distribution system take a phase to ground fault and one RL load and one phase take the pure resistive load and one phase take the pure inductive load, fault time 0.016 to 0.083 sec. When supply is on then many power quality issues create each phase. when supply ON without DVR then more distortion in

supply and more loss of energy. With the help of DVR, we control the voltage magnitude and frequency and minimize the distortion supply of the desired voltage magnitude and frequency signal.

6. Result

6.1. With DVR current result

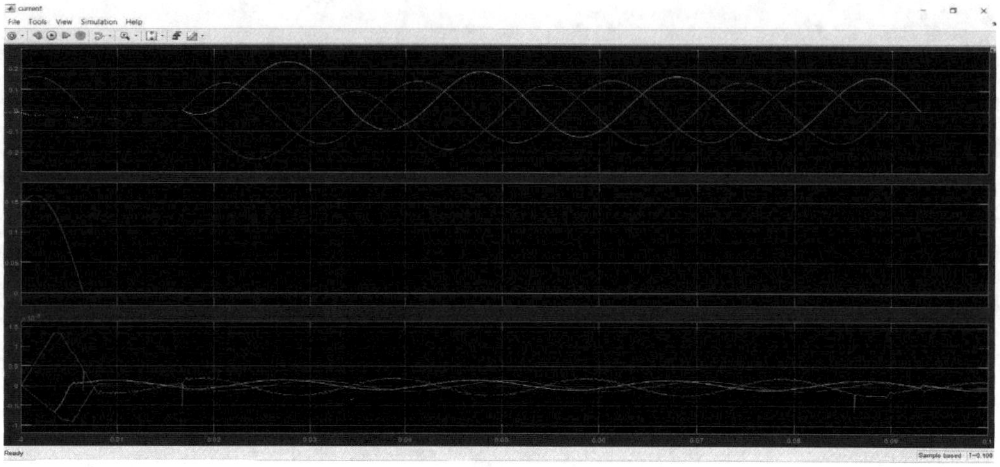

Figure 4. With DVR current result

6.2. With DVR voltage result

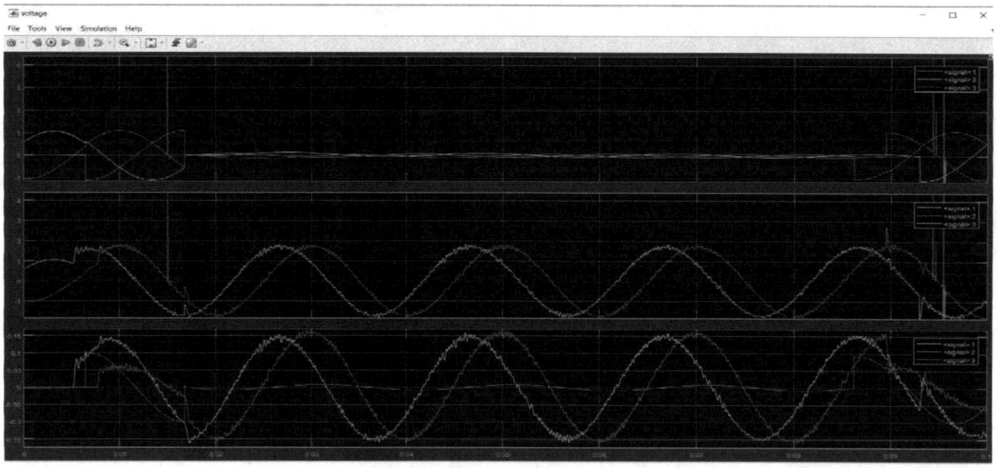

Figure 5. With DVR voltage result

6.3. Without DVR voltage result

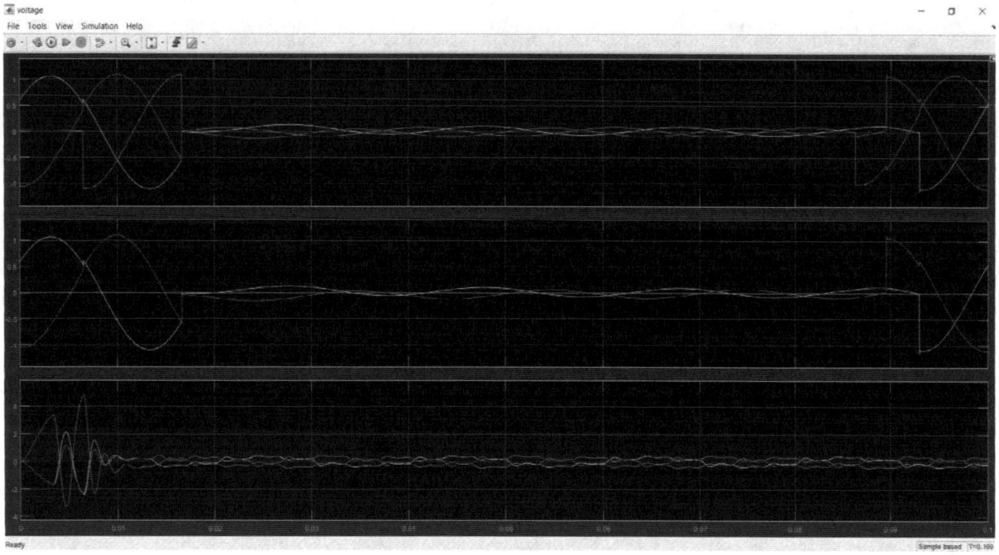

Figure 6. Without DVR voltage result

6.4. Without DVR current result

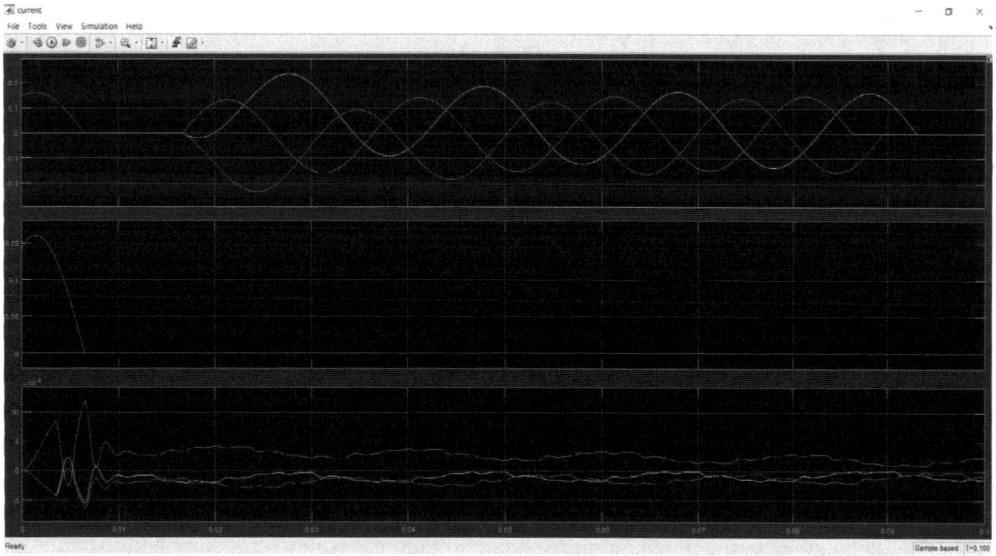

Figure 7. Without DVR current result

6.5. Result discussion

In this simulation (Figure 4,5,6,7) we see clearly what effect on distribution system due to presents of DVR. In the simulation, we consider two types of results one is with DVR voltage and current both (figure 4,5), and another is without DVR voltage and current both (figure 6,7). We easily understand when comparing both types of results that is what affects power quality when we connect the DVR in the system. When running the simulation without DVR then more distortion in voltage and current waveform means power quality issue is more creative but when DVR connect in series with PID controller in distribution system then its reduce the fluctuation and create a more smooth waveform of voltage and current and we can control the frequency according to our need by PID controller.

7. Advantages

- Use of this project a normal condition zero voltage injection.
- When occur the voltage sag and swell then DVR automatically injects the voltage in a single three-phase for a short duration.
- When a fault occurs in the distribution system then it passes away unwanted energy without any loos we can say in other words it works as a protection device.
- DVR makes the balance in the system and reduces the unbalance voltage or energy.

8. Application

- Improve the power quality in the distribution system.
- Control the voltage and frequency.
- Make a voltage balancing in the system.
- Use can like protection device also.

9. Conclusion

The power quality issue now a day create more problem in all sector urban, industrial area, a hospital in all sector where use power electronic device or non-linear load because non-linear load produces harmonics and voltage sag and swell so we work on power quality. Power quality issue means some distortion and unbalancing in the system. The above simulation uses DVR to reduce the power quality issue. DVR injects the voltage in series in the distribution system and makes voltage balancing and smooth waveform of voltage current and controller controls the voltage and frequency.
http://orcid.org/0000-0001-7360-708X

10. Future scope

Now a day all sectors like urban areas, industrial areas, medical, education sectors all sectors use power electronic devices non-linear load of these power electronic devices day by day increases fast because all sectors use more machine learning and automation. So due to more use of power electronic devices create more power quality issues because this power electronic device produces more harmonic, voltage sag, voltage swell, and distortion in the voltage-current waveform. This is all issue we reduce the help of DVR. DVR injects the voltage in series in each phase and provides the required voltage or energy.

11. Reference

[1]. Benachaiba, C., and B. Ferdi. "Power quality improvement using DVR." *American Journal of applied Sciences* 6, no. 3 (2009): 396.

[2]. Zhong, Qing-Chang, Phi-Long Nguyen, Zhenyu Ma, and Wanxing Sheng. "Self-synchronized synchronverters: Inverters without a dedicated synchronization unit." *IEEE Transactions on power electronics* 29, no. 2 (2013): 617-630.

[3]. Elnady, A., & Salama, M. M. (2005). Unified approach for mitigating voltage sag and voltage flicker using the DSTATCOM. *IEEE transactions on power delivery*, *20*(2), 992-1000.

[4]. Singh, Abinash, and Balwinder Singh Surjan. "Comparative Study of FACTS Devices for Mitigation of Voltage Sag-Swell in Distribution System." *Indian Journal of Science and Technology* 9, no. 34 (2016).

[5]. Pal, Rakeshwri, and Sushma Gupta. "State of the art: Dynamic voltage restorer for power quality improvement." *Electrical & Computer Engineering: An International Journal (ECIJ)* 4, no. 2 (2015): 79-98

[6].Yuan, Zhihui, Sjoerd WH de Haan, Jan Braham Ferreira, and Dalibor Cvoric. "A FACTS device: Distributed power-flow controller (DPFC)." *IEEE transactions on power electronics* 25, no. 10 (2010): 2564-2572.

[7]. Karthikeyan, A., DG Abhilash Krishna, A. Tejaswini, Aeshwrya Sweta Panda, and R. Laxmi Prasanna. "A comparative study of PI and PDF controllers for DVR under distorted grid conditions." In 2018 20th National Power Systems Conference (NPSC), pp. 1-6. IEEE, 2018.

[8]. Patil, Sagar S., and R. A. Metri. "Power quality improvement using shunt active power filter." In 2017 International Conference on Data Management, Analytics and Innovation (ICDMAI), pp. 152-156. IEEE, 2017.

[9]. Tulsky, Vladimir N., Artem S. Vanin, Mohamed A. Tolba, Anna Y. Sharova, and Ahmed A. Zaki Diab. "Study and analysis of power quality for an electric power distribution system-Case study: Moscow region." In 2016 IEEE NW Russia Young Researchers in Electrical and Electronic Engineering Conference (EIConRusNW), pp. 710-716. IEEE, 2016.

[10]. Kaiwart, Jyotsana, and Uma P. Bala Raju. "Techniques for power quality improvement by mitigation of harmonics." In 2017 International Conference on Energy, Communication, Data Analytics and Soft Computing (ICECDS), pp. 1532-1536. IEEE, 2017.

Comparison of Power Efficient Inverter Circuit using adiabatic technique

Mohit Kumar Goel*, Supriya Bharti, Parveen Kumar, Suman Dandotiya, Abhishek Pal, Divyank Tripathi

Department of Electronic and Communication, Lovely Professional University, Punjab, India
Mohit.16907@lpu.co.in (orcid:-0000-0001-6827-9579)

Abstract. The power decadence in conventional CMOS (Complementary Metal oxide Semiconductor) circuit can be diminished by utilizing adiabatic method. Working of Adiabatic logic depends on switching activities, which diminishes the power by giving energy dissipated as heat back to supply. Adiabatic word has come from thermodynamic framework which says that no heat transfer take place from framework to climate as well as the other way around. Adiabatic logic is well known as energy recovery logic because its main objective is to reuse energy. It reuses the energy back to the power supply along the lines diminishing the power dissipation. Adiabatic procedures depend on adiabatic logic guideline. Anu Priya has propose the technique in 2014 [3], of adiabatic procedure where energy dissipation in PMOS organization can be limited and some of energy dissipated at load capacitance can be reused rather than dispersed as heat. Yet, the adiabatic method is exceptionally reliant upon parameter variation. Expecting completely adiabatic activity from a circuit is practically not possible as it is an ideal condition, also adiabatic technology is highly dependent on parameter variations so it's quite difficult to achieve and ideal adiabatic circuit. It could be just accomplished with extremely sluggish exchanging speed. In examination, three logic families, ECRL (Efficient Charge Recovery Logic), PFAL (Positive Feedback Adiabatic Logic) and CPAL (Complementary Pass Transistor Adiabatic Logic) are contrasted and conventional CMOS logic for inverter circuit. MICROWIND technology is used while investigating different adiabatic circuits; energy utilization is broke down by variety of parameter. Result shows that adiabatic strategy is better decision in case of underpowered application in explicit frequency range.

Keywords: Adiabatic, ECRL (Efficient Charge Recovery Logic) , PFAL (Positive Feedback Adiabatic Logic), CPAL(Complementary Pass Transistor Adiabatic Logic)

1. Introduction:

Digital CMOS (Complementary Metal oxide Semiconductor) circuits have been the main reason for high execution registering and other logical and designing applications of very large scale integration (VLSI). Adiabatic logic circuits are one of the most effective advances which are being used currently to limit energy utilization. The expanding conspicuousness of compact and portable devices and the requirement to restrict power utilization in exceptionally high thickness chips have prompted fast and creative improvements in low power plan during the developing world. The main reason behind these improvements are compact gadget applications whose primary requirement is low power dissemination and high production, like journal PCs and convenient specialized gadgets. In majority of these cases, the obligation for low power utilization should be met as well as demanding goals of high chip thickness and high output should be achieved. Because of these high demands of improvement in low power design, VLSI has emerged as an extremely dynamic and innovative field. The restricted lifetime of battery in portable devices imposes limited power utilization while being outdoors. Removing the problem of limited power utilization due to heat dissipation is a challenging field which we need to overcome for better portable devices.

Adiabatic Logic families for low power consumption: The expression "adiabatic" represent the thermodynamic cycles where energy is not exchanged at all with environment, and in this manner energy is not dissipated as heat in the environment. Testing and research in this space has been predominantly powered by the fact that energy dissipation extraordinarily increases when circuits gets smaller and faster ,this issue can be resolved to an extent by adiabatic logic. Till now many research has been focused on building adiabatic logic out of CMOS since CMOS is fundamental unit of VLSI. Nonetheless, CMOS logic is way decent in energy consumption in compared with comparative technologies, where energy is dissipated as heat and for the most part it occurs while switching. Energy losses during movement of electric charges between nodes of a circuit can be limited. Implementing completely adiabatic activity on a circuit is an optimal condition. Ideal adiabatic circuit can be only accomplished with extremely slow switching speed. In real life, the digital circuit which is used to reduce energy dissipation consist of adiabatic as well as non-adiabatic part. Adiabatic part of the circuit reduces the energy dissipation by slow switching and further reused this energy from the load capacitance. While reusing the energy, adiabatic circuits uses the steady current source power supply and for lessens dissipation it utilizes the sinusoidal or trapezoidal power supply voltage. The same circuit is used while demonstrating ordinary CMOS circuits in the course of charging the output load capacitance. In CMOS circuit, steady voltage source is interchanged by consistent current source so that output load capacitance can be charged as well as released .Consequently adiabatic exchanging strategy offers the low power consumption in PMOS circuit as it repurpose the dissipated energy in the result load capacitance by switching the current source. Adiabatic Logic doesn't unexpectedly change from 0 to VDD (as well as the other way around), yet a voltage slope is utilized for recovering and charging the energy from the output. Adiabatic circuits can be used where low power circuits are required as it uses "reversible logic" to moderate energy. While this is an area of dynamic examination, current strategies depend vigorously on transmission doors and four-staged trapezoidal clocks to accomplish this objective. Different logical families of Adiabatic such as, Efficient Charge Recovery Logic (ECRL), Positive Feedback Adiabatic Logic (PFAL) and Complementary Pass Transistor Adiabatic Logic (CPAL) are examined in this paper.

2. Applying various adiabatic technique on CMOS inverter:

CMOS Inverter**:** As proposed by Pragati Upadhyay in 2016 [6], (CMOS) Complementary Metal-oxide Semiconductor inverter (NOT gate) is a circuit which generates logical outputs. It is important part of every integrated circuit as not gate is the basic designing gate. A CMOS inverter is a field-effect transistor which is designed by putting a metal gate on top of insulating layer of oxygen, which further lies on top of a semiconductor. It comprises of just two semiconductors, one N-type and other one of P-type. CMOS inverters are present in most of the electronic devices and are liable for creating information inside little circuits. When low voltage (generally denoted by 0) is passed to voltage input, PMOS is turned on and NMOS is turned off. Since PMOS is connected to VDD hence electrons starts to flow through gate resulting in high voltage .Further, when high voltage (generally termed as 1) is passed to input voltage ,PMOS is turned off and NMOS is turn ON. Here NMOS is connected to ground (VSS) so no flow of charges occur which results in low voltage at output.

Journal of Physics: Conference Series **2327** (2022) 012007 doi:10.1088/1742-6596/2327/1/012007

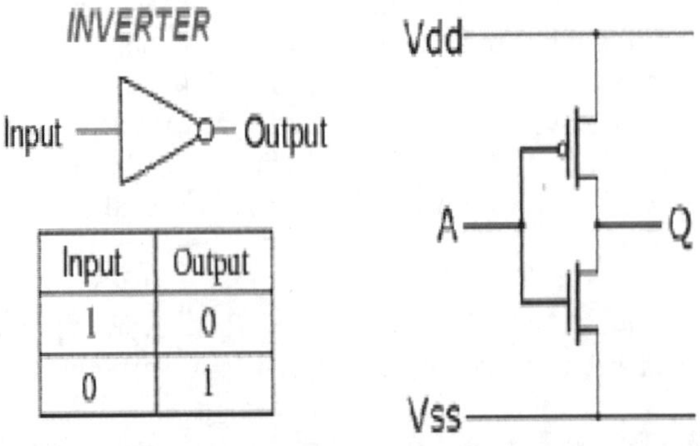

Figure 1 CMOS inverter

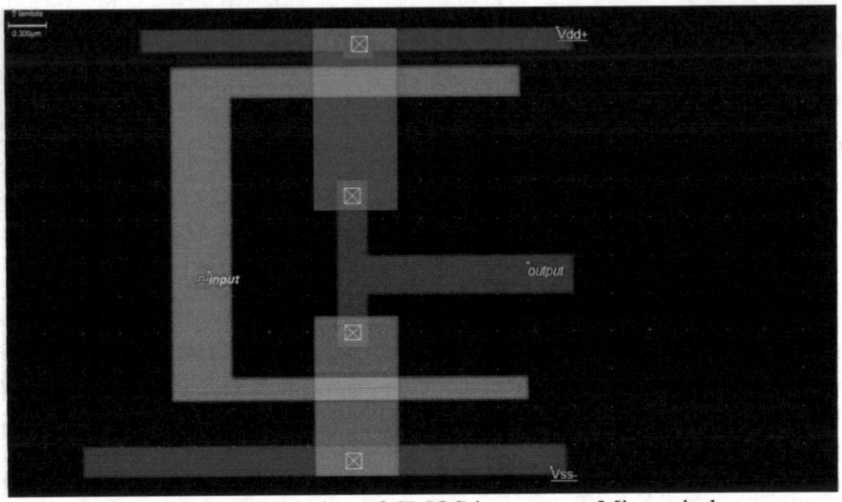

Figure 2. Layout design of CMOS inverter on Microwind

Figure 2. consists of Layout design of CMOS inverter which is drawn according to the circuit diagram of fig. 1. In layout, brown colour shows PMOS, green tone is for NMOS and the metal through which they are associated is blue in shading. Polysilicon layer is shown by red colour which is utilized to give input.

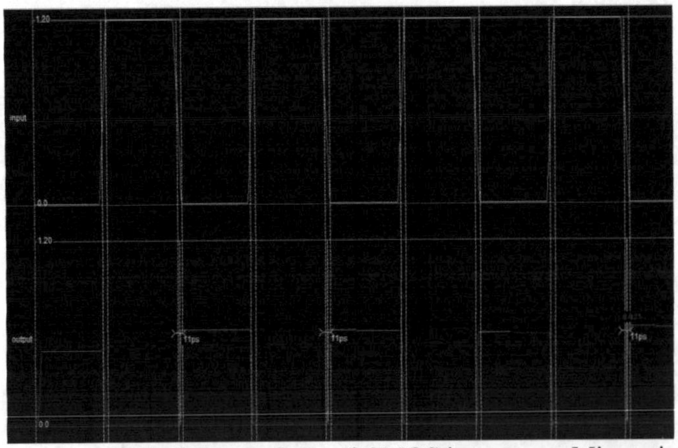

Figure 3. Simulation waveform of CMOS inverter on Microwind

From simulation waveform, it is concluded that when input is 'High' the corresponding output is 'low' and vice versa.

2.1. ECRL technique:

Idea of Efficient Charge Recovery Logic (ECRL) has been brought forward by Ashmeet Kaur Bakshi in 2013 [7] to bring improvement in adiabatic logic circuit. Energy dissipation comparison with other logic circuits is performed on an inverter chain. It embraces another strategy that performs pre-charge and assessment at the same time. It has the construction like Cascode Voltage Switch Logic (CVSL) with differential flagging. It comprises of two cross-coupled semiconductors and two NMOS semiconductors, AC power supply power are employed for ECRL doors, so that input energy can be recovered and reused, which is the fundamental objective of adiabatic technique. In ECRL circuit we have two output named out and /out, both are produced with the goal that the power clock generator can continuously propel a steady load capacitance without depending on the source signal. In both precharge and recovery stages full result swing is acquired due to the cross coupled PMOS. Also because of the edge voltage of the PMOS semiconductors, the circuits experience the ill effects of the non-adiabatic loss, both in the precharge and recuperate stages.

Now in case of ECRL inverter, two inverters are cross-coupled to one another and one inverter's feedback is other's result as well as the other way around. Working of ECRL inverter is same as that of basic ECRL.

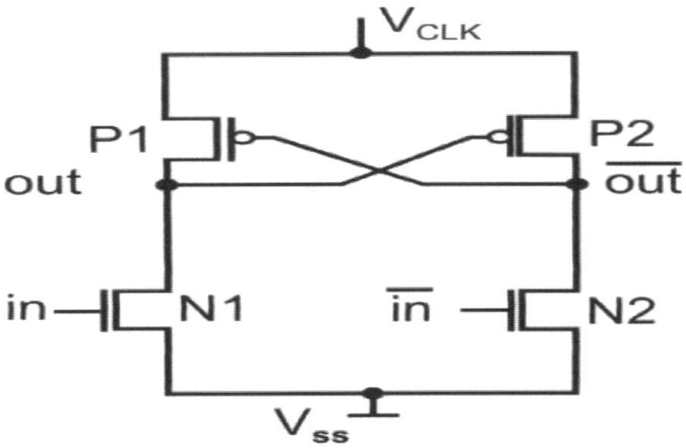

Figure 4 ECRL Inverter

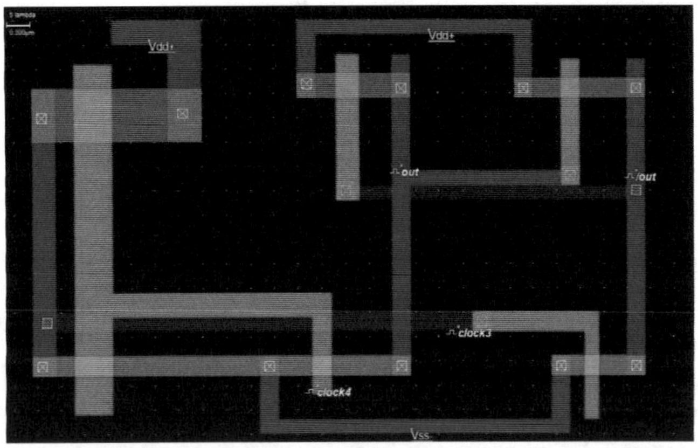

Figure 5 Layout of ECRL inverter on Micro wind

The layout design of ECRL inverter shown in Figure 5 is drawn according to its circuit diagram as shown in Figure 4.

Figure 6 Simulation graph of ECRL inverter on Microwind

When power clock (pck) rises from zero to V_{DD}, output 'out' remains at ground level. Output '/out' follows the power clock. When pck reaches at V_{DD}, outputs 'out' and '/out' hold logic value zero and VDD respectively. This output values can be used for the next stage as an inputs. Now pck falls from VDD to zero, '/out' returns its energy to pck hence delivered charge is recovered. ECRL uses four phase clocking rule to efficiently recover the charge delivered by pck.

2.2. PFAL technique

Arjun Mishra has proposed in 2012 [5] Positive Feedback Adiabatic Logic (PFAL) is being utilized here is a partially/semi adiabatic logic family. It is considered partially energy recovery logic since it consumes an exceptionally low energy when compared with that of other recognizable families. Likewise, it has a decent durability against the various parameter variations. It comprises of partial energy recovery with double rails circuit. At first, 'in' which is basically a input is kept high and input '/in' is low. At the point when power clock (pck) ascends from zero to VDD, yield 'out' remains ground level. Yield '/out' follows the pck. When pck comes to at VDD, yields 'out' and '/out' hold logical value zero and VDD separately. This result values can be utilized for the following stage as a data sources. Now pck tumbles from VDD to zero, '/out' returns its energy to pck henceforth conveyed charge is recovered. PFAL utilizes four phase clocking rule to proficiently recover the charge conveyed by pck.

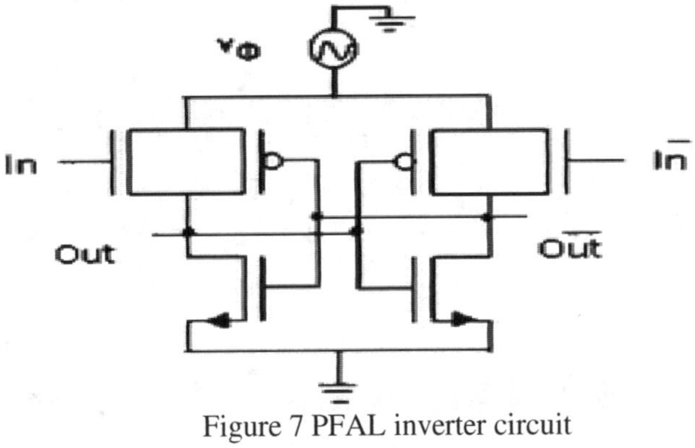

Figure 7 PFAL inverter circuit

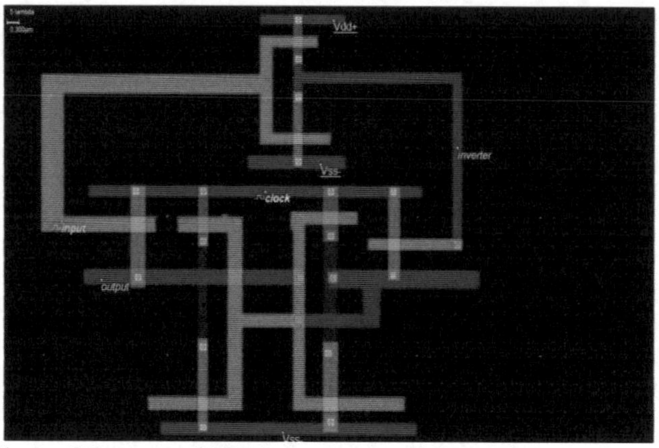

Figure 8 Layout of PFAL inverter on Microwind

The layout design of PFAL inverter is drawn according to its circuit diagram as shown in fig. 7. This simulation waveform shows that when input is 'High', output bar follows the power clock and when input is 'low', the output follows the power clock.

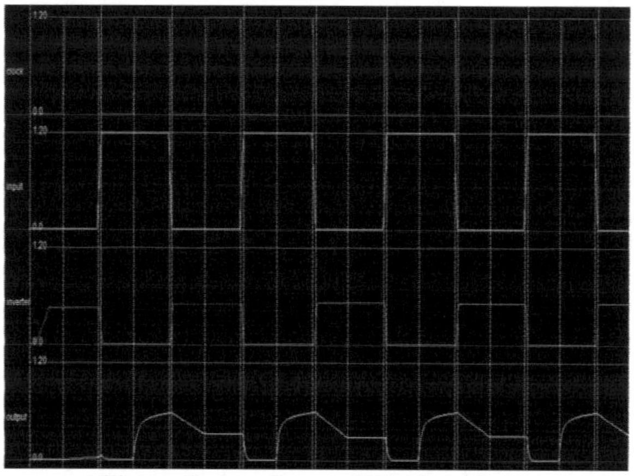

Figure 9 waveform of PFAL inverter

2.3. CPAL technique

Rakesh Kumar Yadav has proposed in 2011 [1] that CPAL circuits is more proficient in transferring and recovering energy, on the grounds that the energy loss of output load in non-adiabatic logical family has totally disposed of by utilizing complementary pass-transistor logic for assessment and transmission doors for energy-recovery. Complementary Pass Transistor Adiabatic Logic (CPAL) method utilizes either only PMOS or only NMOS semiconductors for its logical activity. The inputs are given in the reciprocal form.

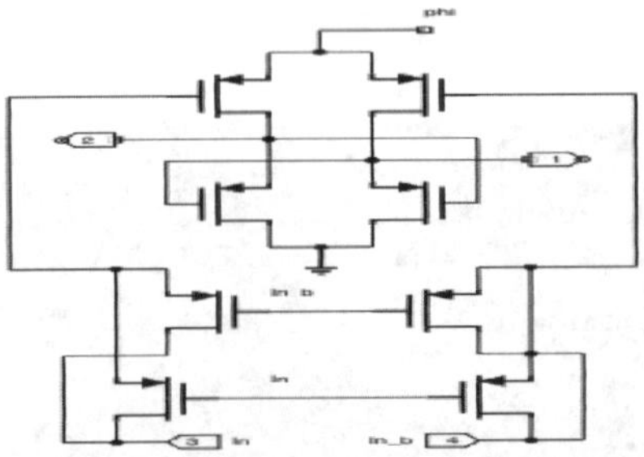

Figure 10 CPAL inverter

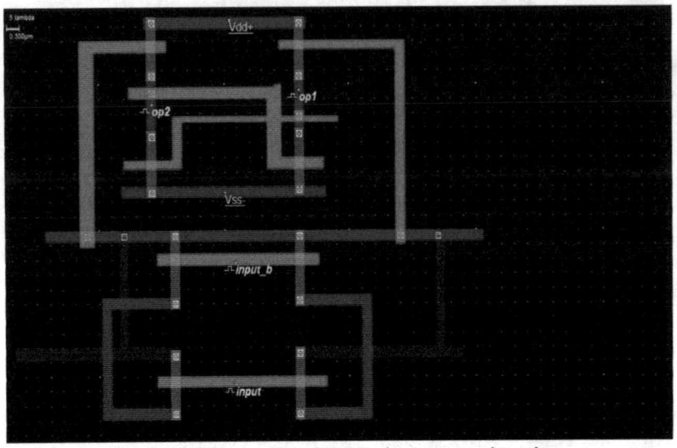

Figure 11 CPAL inverter circuit

The benefit of disposing of PMOS semiconductors from CPAL logical circuits is that the problem of parasitic capacitances related with every node of the circuit diminishes remarkably. Accordingly, the speed of the operation is higher contrasted to CMOS logical circuits. Boot-lashed strategy has been utilized to wipe out the non-adiabatic loss of the result load totally.

The layout design is drawn in Microwind of CPAL inverter is drawn according to its circuit diagram as shown in fig. 10. This simulation waveform shows that when input is 'High', output bar follows the power clock and when input is 'low', the output follows the power clock.

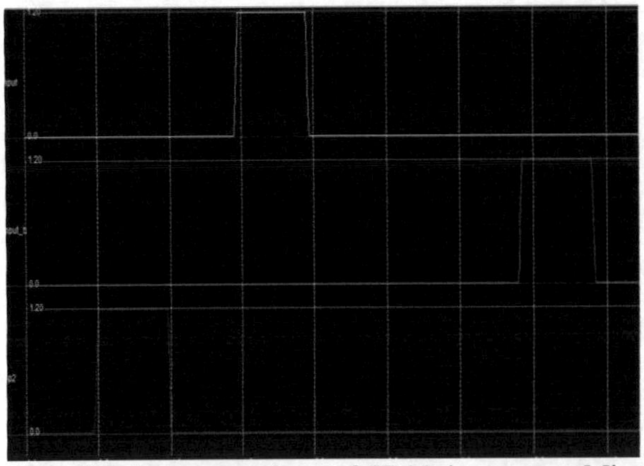

Figure 12 Simulation waveform of CPAL inverter on Microwin

3. Comparison of adiabatic circuit with CMOS inverter:

Comparison of various parameter of adiabatic technique with basic CMOS inverter to find which technique consumes less power and is more efficient.

Parameters	ECRL	PFAL	CPAL
Process(micro meter)	1.0	0.13	0.25
Supply Voltage (V)	5	1.2	2
Frequency (mHz)	10	100	100
Load Capacitor	1pF	1pF	20pF
Power dissipation in CMOS inverter	1	1	1
Power saving in adiabatic CMOS inverter	0.819	0.85	0.48

4. Conclusion:

While examining adiabatic family by varying the parameters it is found that those adiabatic logical families profoundly rely on its parameter variations. Still comparatively weary energy utilization in adiabatic families can be yet accomplished than CMOS over the broad scope of parameter variations. PFAL circuit shows preferable shavings in energy over ECRL at the high load capacitance and high frequency.ECRL is a low-energy, partially adiabatic logic. ECRL shows enormous power saving and shows the promising utilization of ECRL in a low power framework. It has been found that by making PFAL completely reversible, extensively decreased power utilization can be achieved. With the adiabatic exchanging approach, circuit energies are maintained and reused rather than dissipated as heat. This approach can now be utilized to diminish the power dispersal of digital systems as well as portable devices. The recreation results show that CPAL circuit can recover the energy dissipation in the conventional CMOS circuits. Anyway CPAL circuits can't reasonable for the application where fast activities are required. In future, more adiabatic logic is to be presented which can help in conquering the weaknesses of late adiabatic rationale circuits. We can expect minimum dissipation in realistic way since very slow switching is not quiet realistic. Study of comparison of adiabatic technology in case of other CMOS circuit such as NAND, NOR and XNOR can be done. Also further study can be done by adding MUX in CMOS circuit. Future work also includes designing and development of larger adiabatic circuits to overcome recent problem and analysis is required in higher frequencies as well.

References

[1] Rakesh Kumar Yadav, Ashwani K. Rana, Shweta Chauhan, Deepesh Ranka, Kamalesh Yadav, " Four Phase Clocking Rule for Energy Efficient Digital Circuits - An Adiabatic Concept," National Institute of Technology, Hamirpur India,2011 IEEE

[2] Vijendra Pratap Singh, Dr. S. R. P. Sinha," Design and Implementation of Adiabatic Logic for Low Power Application", Institute of Engineering and Technology, Sitapur Road, Lucknow, India,2013.

[3] Anu Priya, Amrita Rai, "Adiabatic Technique for Power Efficient Logic Circuit Design", Dept. of Electronics and Communication, RIET, Haryana, India,2014.

[4] Irfan Ahmad Pindoo, Student Member, IEEE, Tejinder Singh, Member, IEEE, Amritpal Singh, Ankit Chaudhary and P. Mohan Kumar, "Power Dissipation Reduction Using Adiabatic Logic Techniques for CMOS Inverter Circuit", Discipline of Electronics and Electrical Engineering, Lovely Professional University, 2015.

[5] Arjun Mishra , Neha Singh," Low Power Circuit Design Using Positive Feedback Adiabatic Logic (PFAL)", Amity University, Department of Electronics and Communication Engineering, ASET, Noida, Uttar Pradesh, India,2012

[6] Pragati Upadhyay, Vishal Moyal, "Implementation of Low Power Inverter using Adiabatic Logic", M.E. [VLSI Design], Dept. of ECE, SSGI SSTC (FET), Bhilai, Chhattisgarh, India,2016

[7] Ashmeet Kaur Bakshi, Manoj Sharma"Design of basic gates using ECRL and PFAL", International Conference on Advances in computing, communication and informatics, 2013

[8] Arun Kumar, Manoj Sharma, "Design and analysis of MUX using Adiabatic Techniques ECRL and PFAL", International Conference on Advances in computing, communication and informatics,2013

[9] Vojin G. Oklobd'zija, Fellow, IEEE, Dragan Maksimovíc, Member, IEEE, and Fengcheng Lin "Pass-Transistor Adiabatic Logic Using Single Power-Clock Supply" IEEE transactions on circuits and systems-II: analog and digital signal processing, vol.44, no.10, 2015

[10] C. Venkatesh, A. Mohanapriya and R. Sudha Anandhi, PERFORMANCE ANALYSIS OF ADIABATIC TECHNIQUES USING FULL ADDER FOR EFFICIENT POWER DISSIPATION, Department of Electronics and Communication Engineering, Sri Eshwar College of Engineering, India,2018

Control Strategies for Improvement of Power Quality in Grid Connected Variable Speed WECS with DFIG – An Overview

Pradeep Singh[1], Krishan Arora[2], Umesh C. Rathore[3]
[1]Ph. D. Scholar, Lovely Professional University, https://orcid.org/0000-0001-6256-4344
[2]Associate Professor, School of Electronics & Electrical Engineering, Punjab, India, https://orcid.org/0000-0003-0007-8759
[3]Professor & Head, Govt. Hydro Engineering College, Bandla, Bilaspur, HP, India https://orcid.org/0000-0003-1703-6790

Abstract—The presented paper gives a detailed observation of the PQ related problems for WE based on the EFIG or DFIG (Extra or Doubly Fed Induction Generator. Fundamental Frequency Components are the most typical reasons of disruption of power quality. On the other hand, wind loading conditions also disrupt the power quality. Numerous modes for the removing of these problems were studied to get the rid of these fundamental frequency components of the WES (Wind Energy Systems). In the presented paper, there is substantial analysis of numerous controlling techniques were presented to eliminate these PQ and different loading conditions i.e. grid conditions and wind loading conditions. Furthermore, utilizing the WECS (Wind Energy Conversion System) as an operative filter. This analysis of DFIG based WECS help research scholars to choose the adequate control procedures while utilizing the WECS as an operative filter.

Keywords—WES (Wind Energy System), DFIG (Doubly Fed Induction Generators), WECS (Wind Energy Conversion System), PQI (Power Quality Issues), OF (Operative Filter).

1. Introduction

The day by day increase in prices and the harmful effects of fossil fuel like coal, nuclear fuel etc. on our surroundings environment have encourage the whole world attentiveness to NCER (Non-Conventional Energy Resources). Variations in voltage, flicker, and harmonics are among the issues in the quality of power supply that EFIGs face when connecting wind turbines to the grid. The most prevalent wind generation technique is the extra fed induction generator (EFIG). Wind energy harvesting efficiency is improved by the ability to obtain a consistent frequency output from an EFIG powered by a variable speed turbine. To avoid unduly cautious wind power limits on the amount of wind power that can be linked to a power network, it is critical to be able to predict how wind turbines affect power quality at the common coupling point (PCC). The IEC-61400-21 standard specifies the measuring methodologies and calculation of the primary power quality characteristics parameters for wind turbines. The main problem with this alternative resource of energy is that it isn't always available when it's needed. Wind Power, a non-polluted and a sustainable source of electricity, is one of the permanent energy sources. The converter which is connected to grid side and the converter which is connected to rotor side are back-to-back coupled converters that connect the grid to the EFIG-based wind power plant (WPP) [1]. Only if the power quality parameters of the wind turbine are known is this conceivable. Harmonics are one of the most common types of power quality issues, and the IEC-61000-3 and IEEE-519 standards define the desired harmonics rules. Traditional solutions to this problem include passive filters [2]. These filters are ineffective in the frequency range where harmonic distortions occur. The authors utilized an active filter to remove harmonics and compensate for reactive power. To establish the best place for connecting active harmonic filters, simulations were run. In WESs, hybrid active harmonic filters offer benefits. This filter's passive section suppresses harmonics at the calibrated frequency, while the active section supports filtering and eliminates unwanted resonances. Some researchers have recently altered the control approaches of existing DFIG converters to eliminate harmonics and compensate for Q-power. The research looked at the GSC is used to eliminate harmonic currents caused by non-linear loads. A thorough control approach for B2B 4-leg inverter was sermonize on the load side of a WT system connected to a large or small region. In this investigation, the control method functioned well, and the THD was minimized for various operating conditions [3]. A literature overview is presented in this study, with an emphasis on EFIG, several ways for elimination of harmonic currents generated in WECS, numerous control approaches of active filters for unwanted frequency components and reorganized system for conversion of wind energy (WE) presentation as an active filter.

2. Wind Energy Conversion System

The most fundamental component of a conversion system for WE is the WT, which transforms the energy captured from the wind into useful energy i.e. mechanical energy. A gearbox is used to connect the wind turbine to the DFIG. The generator's output is fed into the electrical grid via an appropriate controller system, which helps to reduce disruptions, improve system stability, and improve system quality. Fig. 2.1.

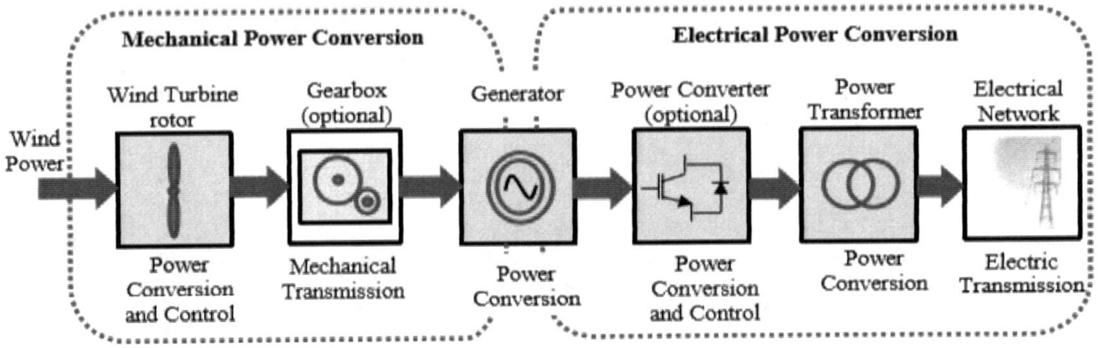

Figure 2.1 Components of Wind Energy Conversion System

2.1 System of WT with its GB (Gear Box)

The electrical power generated by a WTGs can be calculated as follows:

$$P = \frac{1}{2} A \rho V_{\omega}{}^{3} C_{p} (\beta, \lambda) \quad .... (1)$$

Where:

A = Area swept by a wind turbine (WT) in m^2,

ρ = Air Density of that area in kg/m^3,

$V\omega$ = Speed of the wind in m/s,

Cp = Co-efficient of power for the WT,

λ = The ratio of the turbine's angular speed to the maximum speed of the wind (also called as the sharp end speed)

β = The Blade's Angle of Pitch.

The relationship for the evolvement of torque is given as below:

$$T_a = \rho / \omega_t = \frac{1}{2\lambda} \rho \pi R^3 V_{\omega}{}^{3} C_p (\lambda, \beta) \quad .... (2)$$

Where:

R = Radius for the WT,

ω_t = Speed of rotation for the WT.

The gearbox connects the shaft of the WT with the rotor of the EFIG, allowing the turbine's low speed to be increased to the DFIG's greater speed. The formulae that determine this gearbox are as follows:

$\omega_t = \omega_m/G$ & $T_m = T_a/G$

The turbine speed is expressed as follows using the dynamics basic relationship:

$J\, d\omega_m/dt = T_m - T_{em} - f\omega_m$ .... (3)

J = MOI (Moment of Inertia) at the generator side,
f = VFC (Viscous Friction Coefficient) at the generator side.
T_m = Torque for the Gearbox,
T_{em} = Generator Electromagnetic Torque
ω_m = Generator mechanical speed,
λ & β is a function of rotor speed of the turbine and denotes the coefficient for the power C_p. The output power as a function of rotor speed for the WT for the numerous wind speeds (angle of the pitch β =0°).

2.2 Double Fed Induction Generator or Extra Fed Induction Generator

In recent times, EFIGs are the most prevalent type of generators utilized in WM (Wind Mills). A EFIG's fundamental configuration is presented in Fig. 2.2.

2.2.1 Model of Wind Turbines with Extra Fed Induction Generators:

The rotor (V_{dr}, V_{qr}) and stator(V_{ds}, V_{qs}) dynamics voltages in d and q frame of reference points are established via:

$V_{dr} = R_r I_{dr} + P\Phi_{dr} - n_r\Phi_{qr}$ .... (4)
$V_{qr} = R_r I_{qr} + P\Phi_{qr} + n_r\Phi_{qr}$ .... (5)
$V_{ds} = R_s I_{ds} + P\Phi_{ds} - n_s\Phi_{qs}$ .... (6)
$V_{qs} = R_s I_{qs} + P\Phi_{qs} + n_s\Phi_{ds}$ .... (7)

The rotor fluxes (Φ_{dr}, Φ_{qr}) and stator fluxes (φ_{ds}, φ_{qs}) are proportional to identical d and q currents I_{dr}, I_{qr}, I_{ds}, and I_{qs}:

$\Phi_{dr} = L_r I_{dr} + MI_{ds}$ .... (8)
$\Phi_{qr} = L_r I_{qr} + MI_{qs}$ .... (9)
$\Phi_{ds} = L_s I_{ds} + MI_{dr}$ .... (10)
$\Phi_{qs} = L_s I_{qs} + MI_{qr}$ .... (11)

The electromagnetic torque T_{em} is given by:

$T_{em} = P\,M/(L_s)\ (\Phi_{qs} I_{dr} - \Phi_{ds} I_{qr})$ ....(12)

Where:
R_r = Resistance for the rotor per phase,
L_r = Inductance for the rotor per phase,
M = Magnetizing Inductances,
n_s = Speed at the stator.
R_s = Resistance for the stator per phase,
L_s = Inductances for the stator per phase,
n_r = Rotor Speed,

2.2.2 Method of VC (Vector control) for an EFIG (Extra Fed Induction Generator):

For fixing the stator's vector of flux which is linked to the direct-axis for that resistance (R_C) for the stator has to be neglected.

$\Phi_{qs} = 0$ & $\Phi_{ds} = \Phi_s$

The electromagnetic torque becomes

$T_{em} = -P\,M/(L_s)\, I_{qr}$ .... (13)

$V_s = V_{qs} = \omega_s\Phi_s$ & $V_{ds} = 0$ .... (14)

The rotor currents are the only way to express the stator's active and reactive powers:

$P_s = V_{qs} I_{qs} = -\frac{M}{L_r} \Phi_s \omega_s I_{qr}$ .... (15)

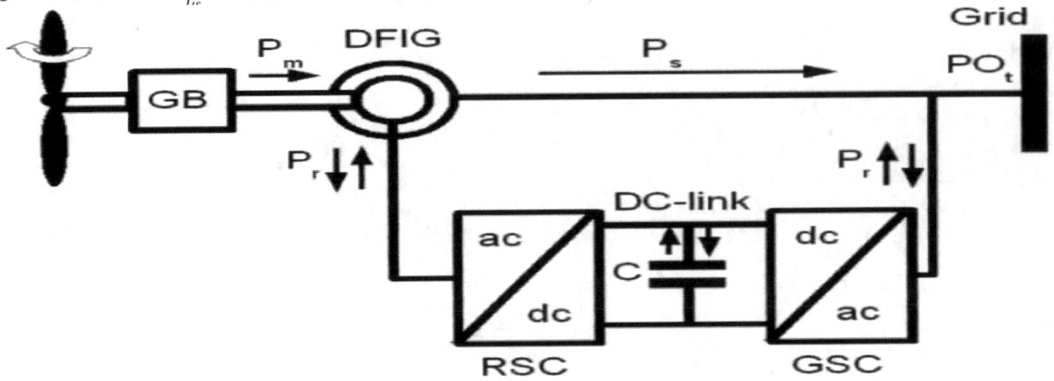

Figure 2.2 Wind Turbine with Gear Box

$Q_s = V_{qs} I_{ds} = \omega_s(\Phi\text{^}2_s)/ L_s - M/(L_s)\, \Phi_s \omega_s I_{dr}$ .... (16)

$I_{qrActive} = -\omega_s I_s / PMV_s T_{em}$ .... (17)

$I_{dr} = -(Q - (V_s\text{^}2)/ \omega_s I_s) I_s / MV_s$ .... (18)

3. Problems Related to Power Quality in WPS (Wind Power System)

Any deviation in V (voltage), I (current), P (power), or f (frequency) which causes failure in equipment's of any customer as a consequence of non-sinusoidal waveforms is mentioned as a problem of PQ [4]. The incorporation of a WES to any system of electricity causes a slew of issues, including flicker, voltage control, harmonics, instability, and other issues; however, these issues are confined to IEC and the standards of IEEE.

3.1 Variation in the Voltage

Voltages changes are created principally as a consequence of variations in the speed of the wind throughout the day and torque of the generator, which affects the changes in P and Q-power. As the WT generator turned on and off again and again, the voltages change or falls. The kind of WT also have an impact on the voltage stability. The most common type of WT is the EFIG based WT. EFIGs qualities like drooping are necessarily required for the management of V (Voltage) and f (frequency). By studying the voltage sensitivity of WECSs, the voltage at PCC may be regulated.

3.2 Voltage Sags and Swells

From half a cycle to a few milli seconds, there is a deviation in voltage happens when the quadratic mean value of the voltage dips in between 0.1 and 0.9 p. u. of pre-defined value. According to the European Standard EN 50160–IEEE 1159, voltage swell is defined as a rise in the RMS (effective) voltage from half-cycle to one minute to between 1.1 and 1.8 p. u. of nominal voltage. [5]. The operation of sensitive electronic equipment connected to the electrical grid is frequently disrupted by voltage sags and surges.vTo addresses the issues related to variation in voltage, sag and swell in voltage, observation type power devices have been proposed. A DVR (Dynamic Voltage Restorer) is one of these devices, D-STATCOM (Distribution Static Compensator, PWM-SAT (Pulse Width Modulated Type Switching Auto-Transformer and SSTS (Solid State Type Transferring Switch).

3.3 Flicker

Low-frequency voltage variations which are visible change in brightness of any lighting lamp due to sudden changes in voltage of the supply system is known as flickering in PQ [6]. The flickering produced in a variable speed load connected WT system are primarily as a consequences of fluctuations in power output, the effect of shade of tower and shear of the wind. To compensate for the flickering effect of load-connected WT, many techniques have been suggested. Some of these options are: STATCOM (Static VAR Compensator) for Q-compensation, DVR (Dynamic Voltage Restorer) for restoring the original voltage & DSTATCOM (Distribution Static Compensator) for distributed islanded lines.

3.4 Harmonics

Fundamental frequency components are the most crucial characteristic of PQ; the definition of this is as "current and voltage waveforms which are integral multiples of fundamental power frequency components" in IEEE Std.519-1992 [7]. Saturation in transformers, Arc Furnaces used in the industries and starters for the motors are the principal resources of fundamental frequency components. Harmonic distortion is now created by PE (Power Electronics) based equipment's such as Electronic components used in converters, drives with power electronic components used for varying speeds, drives used for dc machines and chargers used for charging battery charging are the most commonly employed in power systems. Overheating reduces the life of electrical components like of electrical T/F's, power generators, power factor improvement capacitors and coaxial cables used for power transfer; increased losses on the load side; resonance in the inductor of the filter; and rapid wind station outage are all difficulties caused by harmonics.

4. Harmonic Mitigation Techniques

Various fundamental frequency component reduction methodologies have been proposed in the load connected WE systems in recent years to minimize harmonics and improve WES power quality, including AHF (Active Harmonic Filters), PF (passive filters) and combinational AHF which combines the properties of active and passive fundamental frequency components filters.

4.1 Passive Harmonic Filter

Passive filters are made up of components such as R, L and C like RL, RC or LC filters. To make a passive filter, connect these parts in series or parallel. The power grid can be used to make a series or shunt connection. Shunt filters acts as a load for the generator of fundamental frequency component, absorbing harmonious current components and preventing it from flowing into the network, lowering harmonic voltage. At resonance frequency, the filter which is in series has a very high impedance for blocking, which keeps harmonics out of the power system. The low-pass filter, high pass filter, the band pass filter, notch filters (band-stop or band-reject filters) are the most often used tuned filters [8]. Low cost, easy maintenance, and reduced complexity are all advantages of employing a passive filter. The filtering characteristics of this filter are depending on the impedance of the source. It is not used for loads which are fluctuating in nature and parallel resonance effect is also include.

4.2 Active Harmonic Filter

AH filter is very essential filter used in low power, medium power and high power applications. The 1-Φ systems, 3- Φ systems and 3- Φ, 4-wires systems are the three basic forms of AH filters. Harmonics in current are often eliminated by single phase AHFs. System with three phases AHFs are a good solution for unbalanced loads because they reduce current harmonics. The AH filter power circuit typically includes a DC Voltage source (V_{dc}), an inverter circuit for converting DC into AC and a passive type filter as shown in Fig. 4.1. In SS (Steady State) mode, for elimination of LV (Low Voltage) DC oscillations in the capacitor and in transitory form, it will act as the source for storage of power. [37].

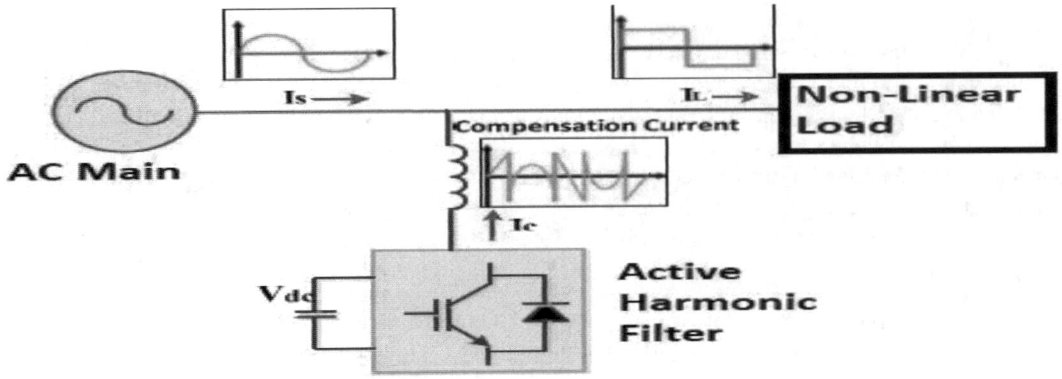

Figure 4.1 AHF with Schematic Waveforms

AHFs converters are divided into two types: current source inverters (CSI) and voltage source inverters (VSI). For the prevention of inductive effects on the system, a diode is connected with the which have the effect of voltage reversing. The AHFs converters' output inductor is utilized to eliminate the harmonic at multiple frequencies. The effects of different L and C filter compositions on switching ripple currents [38]. Harmonic cancellation and reactive power adjustment are achieved using the proposed LC passive filter [9]. When compared with the inductive filter, the study given investigated the usage of an LC type filter to provide improvements in pricing and strong performance since few L components may be employed to accomplish the required depletion of harmonics which are as a consequences of switching. Harmonic removal of both current and voltage harmonics was suggested using different types of filters like series type, shunt type and a combination of both i.e. hybrid filters. Suppression of currents which have harmonic components and Q-power correction of irregular loads have been acknowledged as effective solutions using shunt AHF.

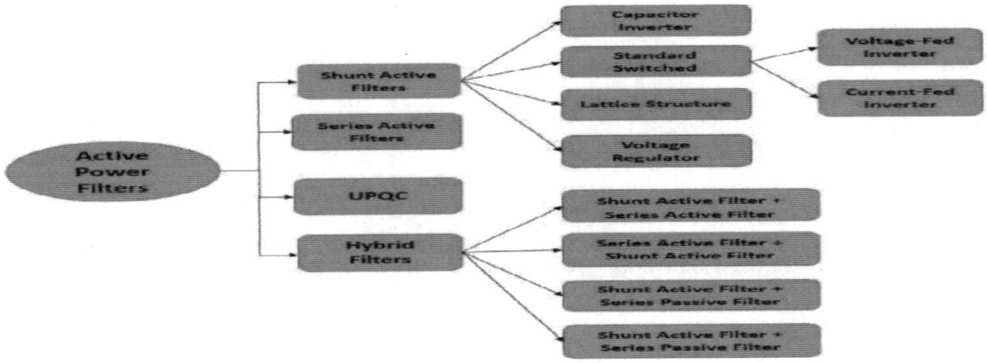

Figure 4.2 Classification of Active Harmonic Filters

The series AHF was used in the system to decrease the harmonics in the voltage and try to regulate the voltage in the range of 50/60 Hz. A hybrid AHF is the combination of series AHF and shunt LC passive filters. The LC filter adjusts for Q-power and annihilate the harmonious currents in the EFIG. The LC filter's performance is improved by using an active power filter. In comparison with the passive type filters, AHF has a faster response for control, higher tolerance, less losses in the system and reduced level of harmonics within the limits for acceptance.

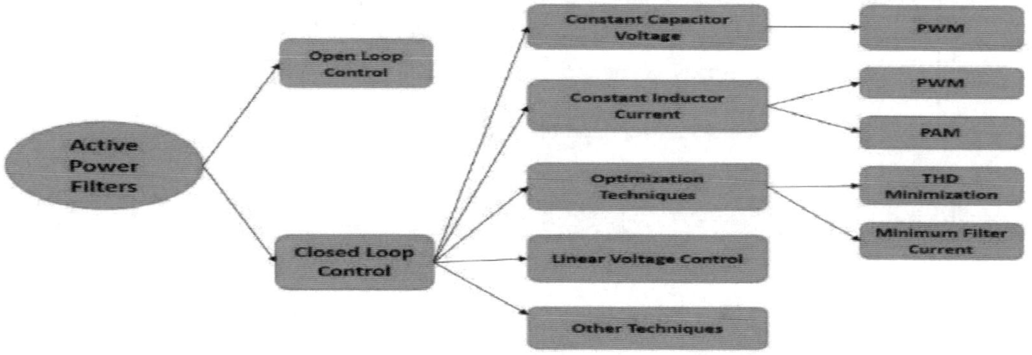

Figure 4.3 Classification of Active Filters based on Control Techniques

5. Energy Storage Systems

Grid-side energy storage directly solves one of the major challenges that renewable energy faces: renewable energy sources are not always accessible. The wind does not always blow and the sun does not always shine. Energy Storage Solutions (ESS) enable the storage and release of energy during shiny and breezy conditions. Energy storage, while it may appear to be a simple notion, may be performed in a variety of ways [10]. There are lots of energy storage devices which are used for store of excess energy in wind farms which are namely, EESS which includes SMES, SC etc. other one is MESS, TESS, ECES which may include BESS [11] Fig. 5.1.

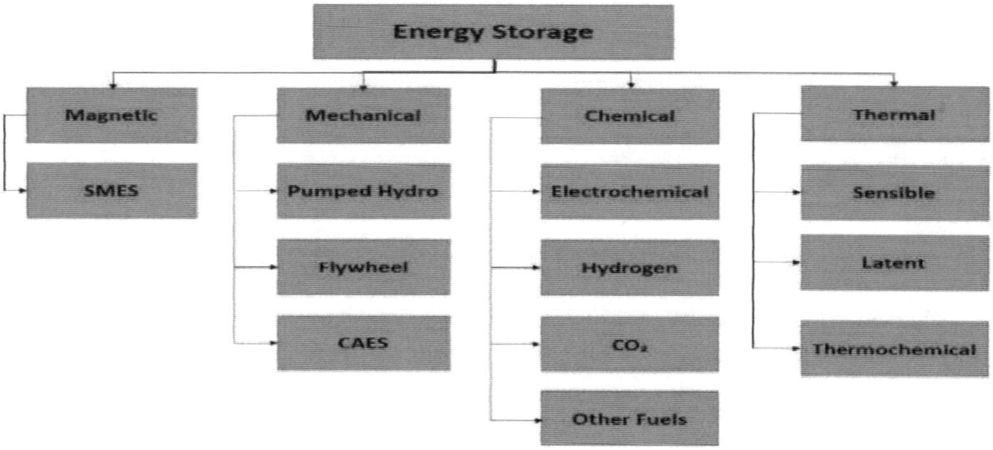

Figure 5.1 Classification of Energy Storage System

In the new energy subsector, energy storage devices play a variety of tasks. They can minimize the fluctuations in the power supply while trying to meeting the demands of the connected load; they can also compensate for the improvement of PQ of generated electricity and which will also increase the reliability at the time of operation [12]. Controllability and adjustability of new energy power generation may be increased to offer consistent power output, so dispatch control is critical while keeping the system safe and cost-effective to operate. Fig. 5.2 shows application Scenarios of ESS.

a) Smooth Power Fluctuation

Journal of Physics: Conference Series **2327** (2022) 012008 doi:10.1088/1742-6596/2327/1/012008

b) Reduce the need for peak regulation capacity in the power system.
c) Keep track of the output of a new energy power schedule.
d) It will also modulate the V and f of the electrical PS.

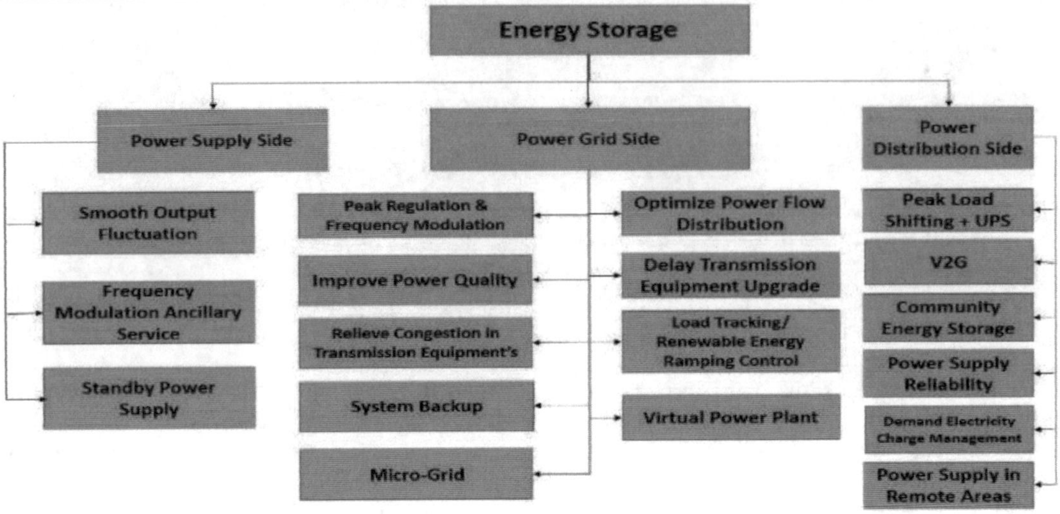

Figure 5.2 Application Scenarios of ESS

5.1 Methods for Configuring Energy Storage System

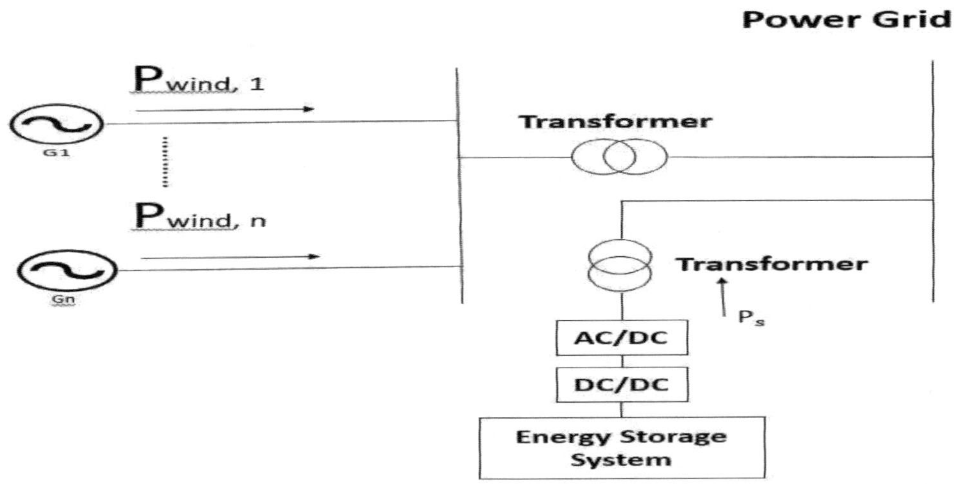

Figure 5.3 Centralized Configuration

The layout of the storage energy system in the energy load can be dissected into two types respectively in centralized and distributed Fig. 5.3 & Fig. 5.4. In wind farms, the configuration approaches are commonly used. The dispersed configuration is used on a wind turbine's excitation DC link or on each turbine's output terminal. On the wind farm's outlet, the centralized configuration is used [13].

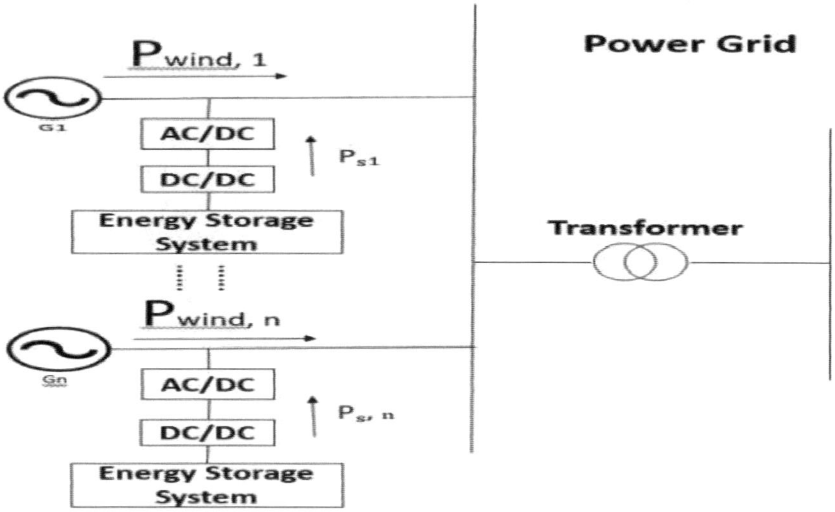

Figure 5.4 Distributed Configuration

In wind farms, the configuration approaches are commonly used. The dispersed arrangement is used on a wind turbine's excitation DC link or on each turbine's output terminal. On the wind farm's outlet, the centralized configuration is used.

6. Control Techniques

6.1 Active Harmonic Filter

Harmonic extraction topologies, program for the regulation DC capacitor voltage and program for control of current and program configuration for control of AHFs makeup. Each and every stage uses numerous control strategies are summarized above in Fig. 4.3.

6.2 Harmonic Extraction Algorithms

By isolating the fundamental frequency components from the power frequency, this restraint program feels the irregular load and develops a concurrence I & V reference signal i.e. I_{ref}/V_{ref}. Numerous fundamental frequency component removal techniques had been examined, as well as their results, Relative frequency dominion, Temporal dominion, studying and progressive control procedures are the main four approaches.

6.2.1 Relative Frequency dominion

The RFD's restriction approaches are depending upon the FT (Fourier Transform) of disturbed indications and are preferable for the 1-Φ and 3-Φ systems. The fundamental frequency component of I_l/V_l is disparate by drawing-out the preference component and then a time interval dominion reference current/voltage signal is generated using the inverse Fourier transform (IFT). Frequency domain techniques entails DFT, FFT, SMT and MFST. The lag in time in sampling system for variables controlling and calculating coefficients of Fourier Transform is a disadvantage of the FD approach.

6.2.2 Learning and advanced methods

To extract the Active Hybrid Filters reference signal, a variety of learning algorithms have been applied, including: Artificial intelligence techniques (such as artificial neural networks) are a type of ANN Artificial Neural Network), MANN (Modified Artificial Neural Network) and GA (Genetic Algorithm). The advanced procedures include making certain changes to the aforementioned method and getting the greatest results.

Changes to Low Pass Filters with Sliding Window Fourier Analysis (SWFA) are among the adjustments to the d-q approach, change of Low Pass Filters with wavelet approach. For the method of p-q, the following are the some of the changes that have been made: Integration of FLC and convert of LPF with SWFA.

6.3 Smoothness of Power Fluctuation in Energy Storage System

Random variations in wind speed, which are widely scattered over various frequency bands, causes fluctuations in wind power generations. The load side connected system will be harmed to various degrees depending upon the frequency of power alterations. As a consequence, it is necessarily to study about the frequency characteristics of the power fluctuation in order to determine which power frequency does the most of the damage and after knowing this thing we have to choose a control scheme for smoothen out these alterations [14]. WP (Wind Power) alteration can be divided into numerous sections and it is established on fluctuated power frequencies using the DFT technique, low frequencies may by 0.01 or lower, high frequency 1 or higher and intermediate frequency zone 0.01 or above. The frequency zone component can be absorbed by the rotor's inertia of WT. The intermediate frequency zone component whose power varies rapidly has a substantial influence on the load side system in a very small span of time jeopardizing the safety of our PS. When a low frequency zone component with slow fluctuation and power change rate is connected to the power system load, the AGC of the PS can respond to this [15]. The component of very low frequency area as well may be flattened due to the slow ramp rate of traditional generating units and the limited spare capacity of the PS.

The wind power smooth control method, which is based on first-order filter and SOC feedback control technique which is used for the method of smoothen control by the ESS. In the meantime, it introduces battery residual capacity SOC feedback control technique, as well as the smoothening control approach for the first order system in the output power of the Solar/Wind. Whenever the present value of feedback in the SOC increases the pre-set value, the control approach for storing will also alter the charging/discharging power of the energy storage battery in present time. This Approach always allow a battery energy storage system's goal value of power to be changed on fast note which allow the PS to function within a specified SOC range while smoothing wind power fluctuations.

6.4 Frequency Regulation of the System

The rise in supplying intermittency will harm load frequency stability for grid-connected WP due to PS insufficient structure and effective regulation of power. The switching frequency of a regular asynchronous generator based on continual speed of WT is analogous to that of a standard generator. Variable speed WT on the other hand have a monitoring system that separates angular velocity and load frequency. In case when load frequency shifts the WT are unable to generate equivalent real power, reduces the inertia of the system. As a result, when there is a power deficit, the load side frequency drastically fluctuates and drops and jeopardizing the integrity of the PS

The unit contributes in network load frequency control that implies it takes actual orders from its higher dispatching system to satisfy network power requirements and deliver sufficient energy to the AGC program based upon the storing system's fast charging and discharging rate. Inside 4 seconds, a battery energy storage unit engaging in network frequency control should react to a power demand of the better scheme [16]. A storage medium is seen to be suitable of tracing grid frequency fluctuations and assisting in frequency regulation. The energy storage system's frequency regulation is constrained due to its low storage capacity. However, related technologies can assure that the gadget can control the grid frequency [17].

7. Function of Active Power Filter in WECS

Numerous researchers had been already devised control techniques for preexisting DFIG devices to compensate for current harmonics and improve the voltage profile. A sensor-less stator space vector technique has been used to regulate the RSC and hence compensate for such power base load significant harmonics levels. To regulate the dc link voltage, the GSC of WECS had been adjusted by employing the d-q current control approach at the link. p-q could be used to detect and limit the harmonic components formed by non - sinusoidal in addition to knowing and minimizing the harmonic components caused by inductive loads [18]. Under this update, the performance of the proposed feature is being used to regulate the dc bus voltage. GSC corrections were made; the current controller was computed employing the d-q concept and the signals of switching devices was managed to use the retro current method [19]. The changes were made on GSC; the current controller was calculated using d-q concept, as well as the signals of the switching devices was adjusted by using retro current approach. The article presents theoretically and experimentally observations of THD current flow at the network in between filtration. RSC, GSC and hybridization of these both are indeed the three most popular distortion equations to describe today. PWM is used to regulate the power switches on the RSC and GSC Depending on research simulated outcomes for every control strategy, the researchers noted that almost all compensatory fundamental strategies are effective in minimizing harmonic content induced by inductive loads. However, they cause greater power losses in WECS; the main advantage of compensating harmonics through RSC-MC (Rotor Side Converter-Modulation Control) has the milder de-rating, Harmonic compensation has a number of advantages. Multiple harmonic frequencies can be eliminated via GSC controller. A most effective way to reduce all repetitive harmonics and minimize THD in I and V is to use compensating utilizing RSC and GSC control. A DFIG based WECS MATLAB/Simulink model is presented and its outputs on different wind and grid loading conditions is presented as below. The Basic model of this is presented and the average wind speed is taken in between the limits from 7 to 17 m/s. After simulate this model, the stator power and rotor power curves corresponding to different speeds is presented and the torque of this DFIG based model is also presented. The grid side conditions include voltage sagging which is also presented in the time range of 3 to 4.5 seconds & there is also the presence of crowbar circuit in the model which will protect the circuit during voltage sagging conditions. Model & all the waveforms are presented below. A 2 MW, 4-Pole, 50 Hz, 690 V (L-L), 1500rpm Wound Rotor Induction Machine, Resistance of the Stator (R_s) = 2.6 mΩ, Resistance of the Rotor (R_r) = 26.1 mΩ, Inductance of the Stator (L_s) = 2.587 mH, Inductance of the Rotor (L_r) = 2.587 mH, Mutual Inductance (L_m) = 2.5 mH, Inertia (J) = 130 Kg.m^2, Damping Friction Factor = .001 Nms, Nominal Torque (T_{em}) = 12732 Nm, Reference Frame = Rotor.

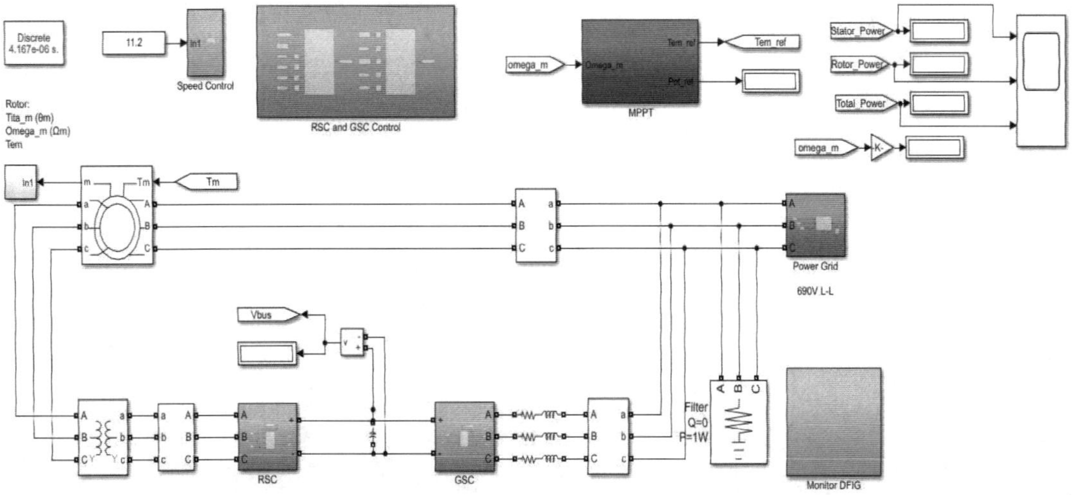

Figure 7.1 MATLAB/Simulink Model of DFIG based WECS Connected to Grid

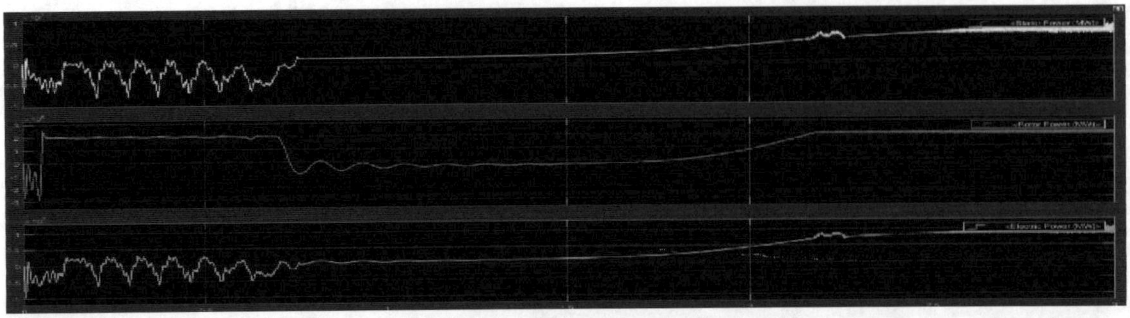

Figure 7.2 Rotor, Stator and Total Output Power

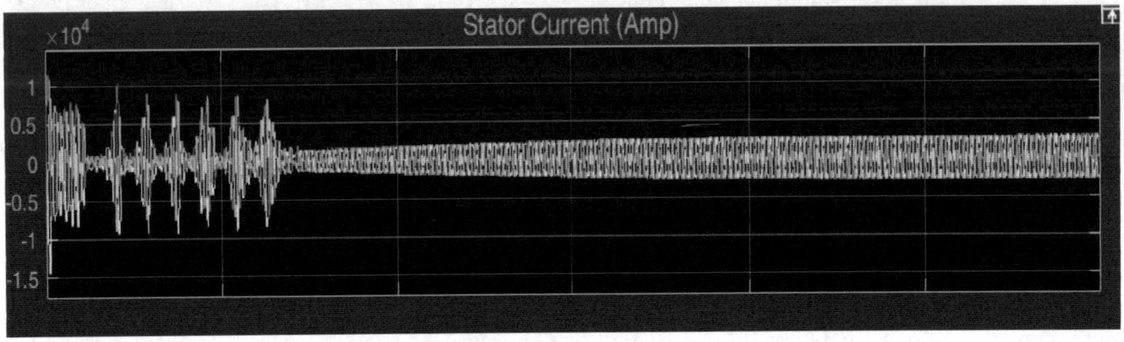

Figure 7.3 Stator Current

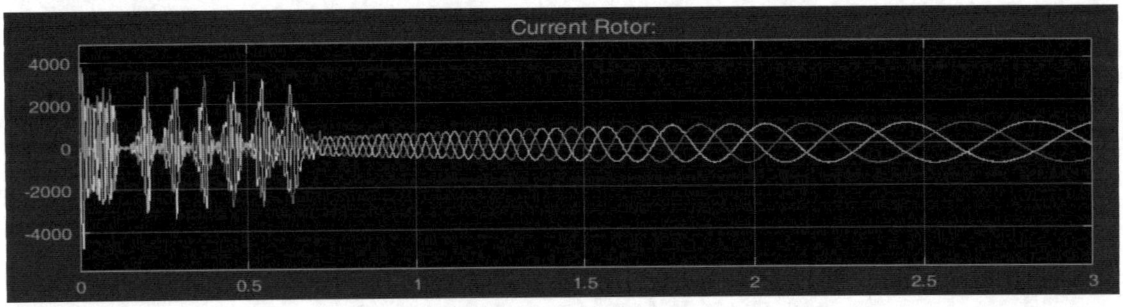

Figure 7.4 Rotor Current

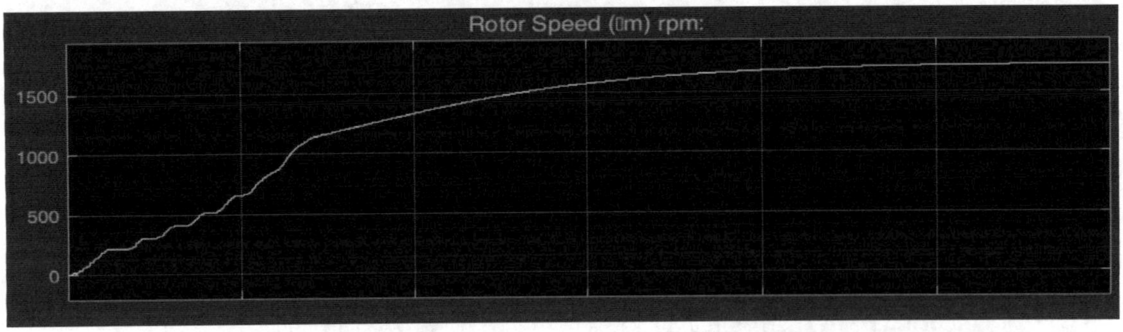

Figure 7.5 Rotor Speed with Variation in Wind Speed

Journal of Physics: Conference Series 2327 (2022) 012008 doi:10.1088/1742-6596/2327/1/012008

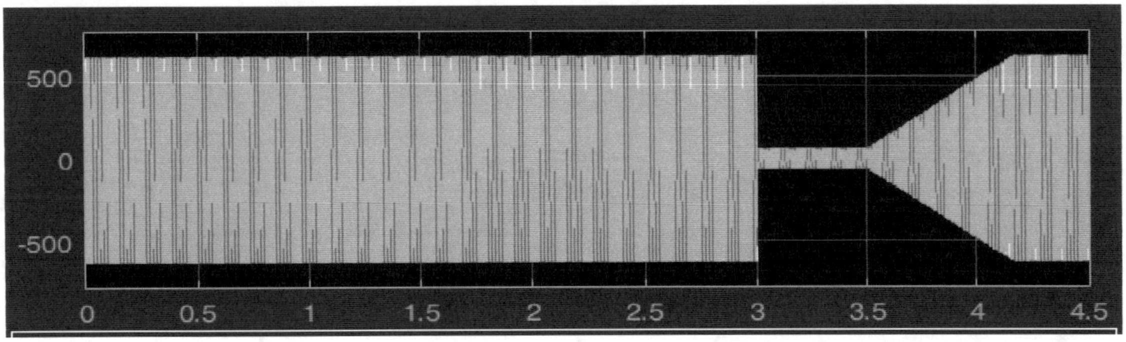

Figure 7.6 Grid Voltage due to Voltage Sag

Figure 7.7 Bus Voltage due to Voltage Sag

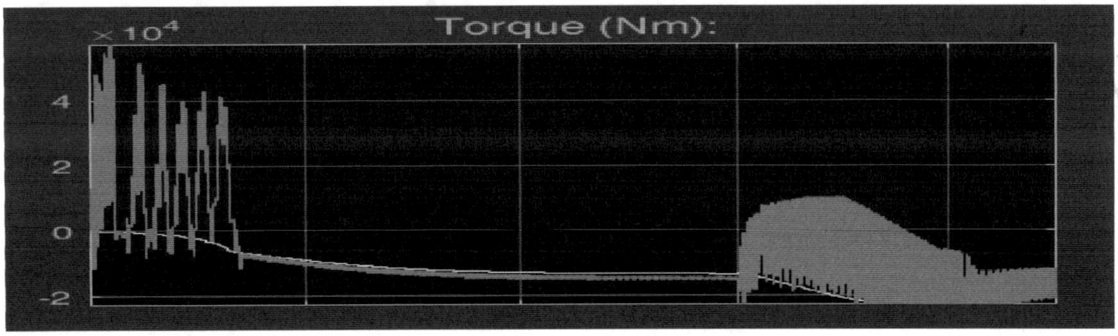

Figure 7.8 Torque Variation with the Variation in Wind Speed

8. Conclusion

A discussion of power quality challenges connected with WECSs is addressed in this work, with a focus on DFIG. The WECSs have been subjected to a variety of harmonic reduction techniques. The most effective way to solve harmonic problems is to use an active harmonic filter. This study gives a detailed description of AHF's operating principle and control algorithms, as well as the benefits and drawbacks of each control algorithm, so that scholars may grasp the foundations of AHF and several types of control algorithms for AHF are compared. Another discussion is done on Energy Storage System, different Storage System are shown in fig. 6, their application scenarios & their control strategies during different conditions like wind and grid loading conditions of voltage dips to improve the power quality. This review will assist the reader in determining the best control strategy for a WECS with active filtering and with Energy Storage Systems for long time duration during Wind gusts, low wind, Voltage dips, power quality problems and faults on the system.

References

[1] M. Latka and M. Nowak Analysis of Electrical Power Quality Parameters in Power Grid With Attached Wind Farm 2017 Progress in Applied Electrical Engineering (PAEE).

[2] A. Bubshait and A. Mortezaei Power Quality Enhancement for a Grid Connected Wind Turbine Energy System 2017 IEEE Transaction on Industrial Applications No. 3 **53** pp. 2496–2505.

[3] M. Tuka Review on Power Quality Problems on-Grid Connected Wind Power System, 2016 International Journal of Advanced Information Science & Technology (IJAIST) **5** No. 4 pp. 96–108.

[4] M. Momeni1 and A. H. Mazinan Improvement of Power Quality in Grid-Connected Inverter Through Adaptation-Based Control Strategy 2019 Energy Ecology & Environment.

[5] Eklas Hossain and Mehmet Rida Tür Analysis and Mitigation of PowerQuality Issues in Distributed Generation Systems Using Custom Power Devices 2018 IEEE Access **6** pp. 16816–16833.

[6] S. Dhakulkar et. al. Inspection of Voltage Sags and Voltage Swells Incidentin Power Quality Problems - A Review, 2017 International Research Journal of Engineering & Technology **4** No. 1 pp. 1734–1736.

[7] H. Samet and A. Asghar Bagheri Enhancement of SVC Performance in Flicker Mitigation of Wind Farm, 2017 IET Generation, Transmission & Distribution **11** No. 15 pp. 3823–3834.

[8] S. Alwin and M. Marsaline Beno Design of Hybrid Active Filter forHarmonic Reduction, 2017 Journal of Advance Research in Dynamical & Control Systems No. 7 **9** pp.106 -112.

[9] Y. Hoon et. al. Control Algorithms of Shunt Active Power Filter for Harmonic Mitigation-A Review 2017 Energies No. 12 **10** pp. 1-29.

[10] Jawad Hussain et. al. Power Quality Improvement of Grid Connected Wind Energy System Using DSTATCOM-BESS 2019 International Journal of Renewable Energy Research **9** No. 3.

[11] Subhendu Sekhar Sahooet et. al. A Coordinated Control Strategy Using Super-Capacitor Energy Storage and Series Dynamic Resistor for Enhancement of Fault-Ride Through of Doubly Fed Induction Generator 2019 International Journal of Green Energy **16** Issue. 8.

[12] Jayalakshmi N. S. et. al. Power Smoothening Method of PMSG Based Grid Integrated Wind Energy Conversion System Using BESS/STATCOM, 2019 International Journal of Power Electronics and Drive Systems (IJPEDS) **10** NO. 4.

[13] Irfan Hussain Panhwar et. al. Mitigating Power Fluctuations for Energy Storage in Wind Energy Conversion System Using Super-capacitors, 2020 Special Section on Evolving Technologies in Energy Storage Systems for Energy System Applications **8** 189747-189760.

[14] Adnan Sattar et. al. Testing the Performance of Battery Energy Storage in a Wind Energy Conversion System 2018 IEEE Industry Applications Society Annual Meeting (IAS) 23-27.

[15] Arindam Das et. al. TSBC Converter with BESS for DFIG-Based Wind Energy Conversion System, 2020 IEEE Transactions on Industry Applications **56** No. 6.

[16] Ruo Huan Yang and Jian Xun Jin Unified Power Quality Conditioner with Advanced Dual Control for Performance Improvement of DFIG-Based Wind Farm, 2021 IEEE Transactions on Sustainable Energy **12** Issue. 1.

[17] Ashwani Kumar et. al. Fuzzy Distribution Static Compensator Based Control Strategy to Enhance Low Voltage Ride Through Capability of Hybrid Renewable Energy System, 2021 Journal of Energy Sources Part A: Recovery Utilization and Environmental Effects pp. 1-18.

[18] N. Abdul Rahman et. al. Significant Insights into the Operation of DC-Link Voltage Control of a Shunt Active Power Filter Using Different Control Algorithm - A Comparative Study 2017 Turkish Journal Electrical Engineering Computer Science **25** pp. 2033–2043.

[19] Wessam A. Hafez et. al. Power Quality Issues of Grid Connected Wind Energy System Focus on DFIG and Various Control Techniques of Active Harmonic Filter: A Review 2019 International Middle East Power Systems Conference (MEPCON) pp. 1006-1014.

Journal of Physics: Conference Series 2327 (2022) 012009 doi:10.1088/1742-6596/2327/1/012009

Transients Analysis in AC Microgrid System

Umar Rashid[1], Javed Dhillon[2]

School of Electronics and Electrical Engineering, Lovely Professional University, Punjab.

umerrasheed961@gmail.com[1] javeddah@gmail.com[2]

https://orcid.org/0000-0003-4183-600X[2]

Abstract. In the modern power system, small scale versions of the power grid termed as Microgrid are being greatly used because of the practical, economic and environmental advantages over conventional structure. Microgrid as the word itself suggests (micro-small and grid-interconnection of different elements) is the advanced technology that is being currently endorsed for future power systems in electrical engineering world. Transients are produced in microgrids due to the fluctuating nature of loads and the irregular behaviour of renewable power generation, particularly whenever the transition from grid linked to islanded state happens. In this paper, for controlling the transients during the transition in islanded mode in microgrid comprising of various energy resources like Diesel Generator, Solar Photovoltaic System, Energy Storage System, Flux Coupling Type Super Conducting Fault Current Limiter (FSFCL) is used. By incorporating the FSFCL in the microgrid, transient behaviour of the microgrid is greatly improved. The model analysis is done using the MATLAB's Simulink Software and the output results in presence and absence of Flux coupling type SFCL is compared. The output results which are obtained show the supremacy of the suggested device in reducing the transients in microgrid.

Keywords. Microgrid, Super Conducting Fault Current Limiter (SFCL), Diesel Generator, Solar Photovoltaic System.

1. Introduction

With truly expansion of energy utilization in this day and age, the power system is unavoidably confronted with introducing significant expense power generating units, distribution and transmission networks, environmental issues, and environmental changes. Renewable Energy Resources (RESs) and Distributed Generation is improved and proper choices to beat this issue. These new assets are equipped for enhancing dependability for power distribution to the consumers, and decreasing the clog and loss in substations and transmission lines [1]. The advantages of renewable sources have spurred the power energy planners to remember them for the power frameworks' construction and to broadly convey Microgrids. The microgrid is limited scale networks made out from a few RESs, DGs alongside few neighbourhood loads [2]. Microgrids have become a more suitable technical notion for electricity distribution because of rising insertion of various types of distributed generating (DG) plant in the main utility grid. Microgrids can reinforce the grid and improve its performance against the outages caused by faults by making use of highly efficient distributed generation in their capacity [3-6]. In general sense microgrid primarily has two operating modes viz grid connected state and islanded state of operation.

In grid connected mode of operation microgrid is in connection with main utility grid and all loads that are within the microgrid and those loads that are connected with the local grid are supplied wholly by the local grid. In this state the voltage and frequency fluctuations are maintained by main grid. In grid connected state of operation microgrid injects power to the utility grid hence acts as current controller. Microgrid also has the capability of detaching itself from the main grid in times of faults or some sort of switching events or due to unavailability of energy resources. In islanded mode all the loads within the microgrid are supplied through the battery present within the microgrid and all the voltage and frequency variations are controlled by microgrid itself. In this mode microgrid works like voltage controller and ensures proper power sharing and balancing and voltage control.

However, when the microgrid shifts its mode of operation from grid connected to standalone mode various problems occur in microgrid. Few of them are listed below.

- In case of grid connected mode fault currents are safely negotiated because of the high ratings of protective equipment however in islanded mode of operation due to low rating of protective equipment and also fault currents having an amplitude greater than that of the distribution network can seriously damage the protection devices.
- Balancing between load demand and power generation is of the most frequently faced challenges in microgrid. Abrupt change in the load demand can incur instability in the system.
- Transition also results in huge inrush currents and as a result transients are produced in the microgrid which affects both voltage as well as frequency.
- Transients can have long lasting effect on the microgrid if left unattended. The transient effect that can occur in microgrid due to switching or any other phenomenon are
- Electronic devices present in microgrid due to transient they may perform their operation at decreased frequencies and can result in early damage to the said equipment.
- In presence of higher transient voltage, motors may run at higher temperature than normal and may result in micro-jogging.
- Motors may suffer from degradation in its insulation and windings and also produce higher noise due to transients.
- Most the distribution equipment in the microgrid are affected by transients like there is degradation in the contact surface of circuit breakers insulators switches etc.

Therefore, practical solutions that reduce high magnitude current and increase the transient performance of a microgrid are requirement of the time. Investigations and researchers have found that greater number of applications on electric utility-based equipment is going to be used in the power industry as the advancement of high-temperature based superconducting (HTS) materials continues to prosper. And one of the materials in these categories is superconducting power devices. These devices can overcome various technical challenges not only in typical power systems, such as large fossil fuel-based energy plants, but can be used in renewable based energy systems, like that in microgrids. [7-10].

Superconducting fault current limiters (SFCLs) is an advanced as well as improved technological electrical equipment that has the capability to deal with the high rising magnitude of the current particularly during any form of fault occurrence as well as during any transient conditions. They do so by making sure that high magnitude of fault level current is reduced within the acceptable limits within the first cycle of these high-level fault current. These are infact proving to be highly competitive counterforce for suppressing high magnitude transient current or we can say faulty current and protecting an electrical equipment as well as systems from short-circuit hazards. Due to the unquestionable advancements in superconducting technology, SFCLs is now becoming a significant as well as viable device for reducing the fault or high-level transient currents as well as for transitory behaviour improvement [11]. SFCl's operating principle is based on its impedance level i.e. they offer negligible or zero impedance when operating under normal condition, however in case of fault detection they tend to offer very high impedance so as to protect the equipment. Superconducting fault current limiters incorporates varying impedance that is kept in series connection mode with power system device and varies accordingly as per set operating condition. Whenever a fault occurs, the impedance values goets to that point where the fault current is lowered to a such level that the circuit breaker can manage. SFCLs has got various types like inductive, resistive type sfcl, saturated iron core sfcl [12-16]. The type of SFCL used in this project incorporates both inductive as well as resistive type sfcl.

In this research paper, inclusion of flux coupling based sfcl in mcrogrid having various types of generating as well as load units so as to enhance the transient stability of the proposed system is proposed. This paper is formulated in following order. Section 2 describes the system description of the microgrd model and the proposed flux coupling type SFCL.In Section 3 simulation analysis is discussed. In Section 4 results are discussed and conclusion in Section 5.

2. Theoretical Analysis

2.1. System Description of Microgrid Model:

Microgrids one line diagram has been shown in Figure 1. As clearly visible from the figure, microgrid comprises of Energy Storage System, Solar Array, Diesel Generator, fixed and variable loads. Fixed load is kept at 500kW while as variable load fluctuates between 200kW-500kW.The diesel generator has a rating of 1.6 MW and under normal circumstances it has a capacity generation of 500kw.Energy Storage system battery bank capacity lies around 540kWh and has a capacity generation of 100kW. The output power of a PV solar farm is expressed in the form of sinusoidal waveform with a range of 450-50 kW, which may be written as ($P_{pv} = 200\sin(Ft)+250$)kW, with $F=1/120Hz$.When microgrid is in grid connected all of the loads that come under the vicinity of microgrid are supplied by utility grid and during this time Energy Storage system gets charged. During islanding mode diesel generator and battery storage system starts delivering power to loads that are within the microgrid and battery gets discharged during this mode hence ensuring continuity of supply. Solar PV charges throughout the day and discharges during night hours.

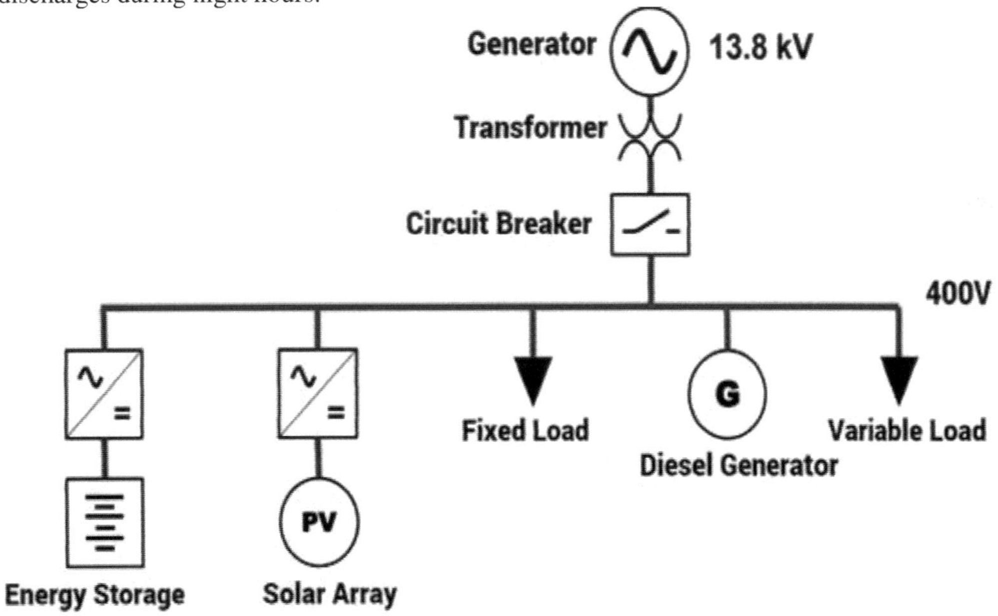

Figure 1. Microgrid Single Line Diagram

2.2. Flux Coupling Type SFCL

The proposed SFCL'S electrical structure that we have used in this research paper has been shown in the Figure 2.From the figure it is evident that the flux coupling SFCL has 3 three important elements. A coupling type transformer, a programmable switch (S_w) (shown in series connection with 10 winding of Xmer), Rs represents the resistance of a superconducting core (in series with the 20 winding of CT). For the current transformer, L1 and L2 denote the winding's self-inductance, whereas M denotes the coil's mutual inductance.The metal oxide arrestor (moa) having a resistance designated as RM is utilised in parallel combination with the CT's 10 winding to decrease these switching voltage produced in this construction. In order to acquire the important parameters of the flux coupling SFCL, equivalent circuit of the flux coupling SFCL is shown in Figure 3. This circuit is drawn taken into considerations the electric relationships that exists in these coupled inductors in the Fig 2. Under no fault or normal condition, the switch represented by SW is normally closed and the Rs (superconducting resistance) is in the superconducting state implies that during normal mode Rs has zero resistance.

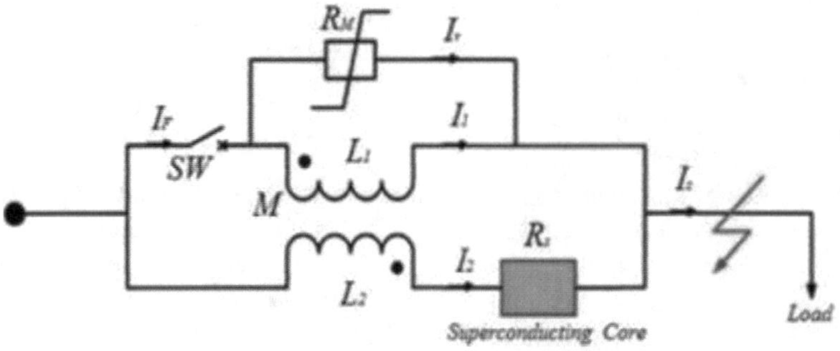

Figure 2. Electrical Structure of FSFCL

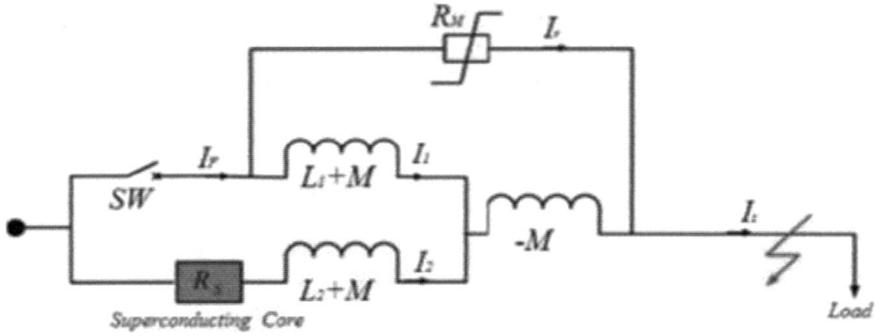

Figure 3. Equivalent Circuit of FSFCL

Therefore, before the fault happens to obtain the impedance of SFCL, we use the following equation that is shown below:

$$Z_{.FSFCL} = jw[(L_{1.} + M)\ p\ (L_{2.} + M) - M] \qquad (1)$$

Here p stands for parallel connection. Therefore, on solving equation 1 becomes

$$Z_{FSFCL} = \frac{jw(L_1 L_2 - M^2)}{L_1 + L_2 + 2M} \qquad (2)$$

Z_FSFCL plays a crucial role as long as characteristics attributes of FSCL is under consideration. Now if k represents the transformers coupling ratio and n represents the transformers transformation ration then the equation representing k and n is given below

$$n = \sqrt{\frac{L_1}{L_2}} \qquad (3)$$

$$k = \frac{M}{\sqrt{L_1 L_2}} \qquad (4)$$

Taking Equation 3 and equation 4 into consideration, Equation 2 then can be rewritten as:

$$Z_{FSFCL} = \frac{jwL_2(1-k^2)\times n^2}{n^2+2kn+1} \quad (5)$$

Having the presumption of having iron core with transformers coupling ratio k =1, then impedance under the normal operation will turn out to be equal to zero, hence FSFCL does not alter or affect the existing microgrid circuit with this presumption. It acts as if it is not present in the system. However, during case of fault or switching operation current level increases suddenly and as a result this rise in magnitude throws open the FSFCL switch and as a result fluxes associated with the winding cannot cancel each other. And as a result, the superconducting iron core then switches to a very large or high impedance state. In fault situations, the impedance of FSFCL is given as.

$$Z_{FSFCL,F} = \text{Rs} + \text{j}\omega L_2 + \text{j}\omega\text{M} + (j\omega L_1 + j\omega M + R_M)\ \text{P}\ (-\text{jwM}) \quad (6)$$

Here P in equation 6 stands for parallel connection. On solving equation 6 and incorporating *k* and *n* values equation 6 becomes

$$Z_{FSFCL,F} = \text{Rs} + \text{j}\omega L_2 + \frac{(kn\omega L_2)^2}{R_M + \text{j}n^2\omega L_2} \quad (7)$$

Making an assumption that $R_M \gg n^2\omega L_2$ equation 8 is achieved as

$$Z_{FSFCL,F} \approx \text{Rs} + \text{j}\omega L_2 \quad (8)$$

2.2.1. Flux Coupling Type SFCL Operating Principle:

When a switching or fault occurs, the current density of a superconductor exceeds the critical current density Jc. Temp (T), Magnetic Field (B), and Current Density(J) (T–B–J) properties of a superconductor material are shown in Figure 4. The zero resistance state is addressed by the most inner surface-1. A normal conducting state is shown by the surface that exists beyond surface-2, and transition state is a specified state between surface-1 and surface-2[17].The superconductor fastly approaches a high resistance state, and the high surge or faulty current is confined to a very low value, as the current density (J) increases, leading to an increase in the temp(T), magnetic flux density (B), and critical current density (Jc).

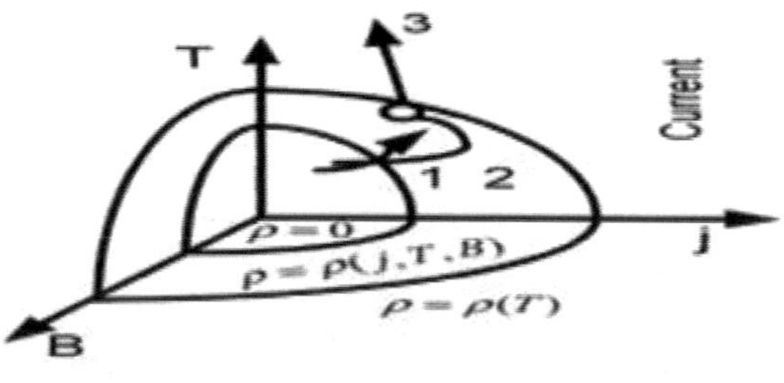

Figure 4. Properties of Semiconductor

3. Microgrid Model

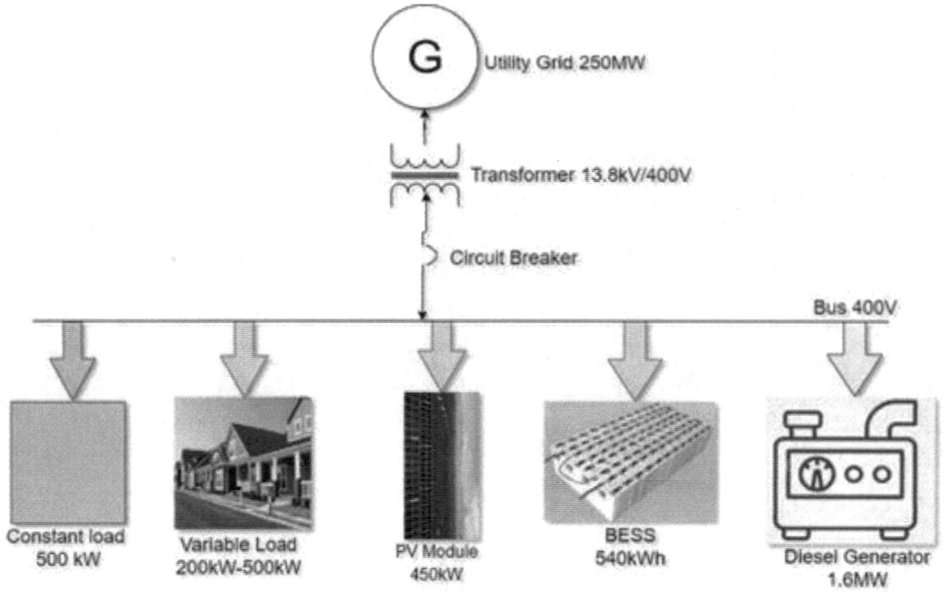

Figure 5. Microgrid Model without SFCL

The MATLAB software was used to simulate the Microgrid with and without SFCL and the results were compared. Proposed model of microgrid without FSFCL is shown in Figure 5. Exactly same model was developed in microgrid Simulink window and model was made to run for total time of 300 seconds. In this microgrid model the voltage from the utility grid is coming at level of 13.8 kV which is stepped down to 400V ac by means of the three-phase transformer for feeding single 400V ac bus. All the power system elements and all the loads are connected to this 400V ac bus. Circuit breaker is usually employed to disassociate the microgrid from the main grid by means of an appropriate step signal. Two loads are there in the microgrid and connected to 400V ac bus viz variable load whose consumption varies between 200 to 500kW and constant load whose consumption is fixed at 500kW.The energy output of a PV solar farm can be stated *(P_{pv}=200sin(2Ft)+250)* kW, with a sinusoidal waveform ranging between 450kW and 50kW. The diesel generator has rating at 1.6 MW and produces 500kW in typical operation. The ESS battery bank has a maximum capacity of 540kW-hour and is meant to supply a steady 100kW rated at 400Vac to the microgrid

In Figure 6 has been shown a microgrid model with FSFCL. FSFCL are installed and are placed on the microgrid side or load side as shown in the Figure 6.

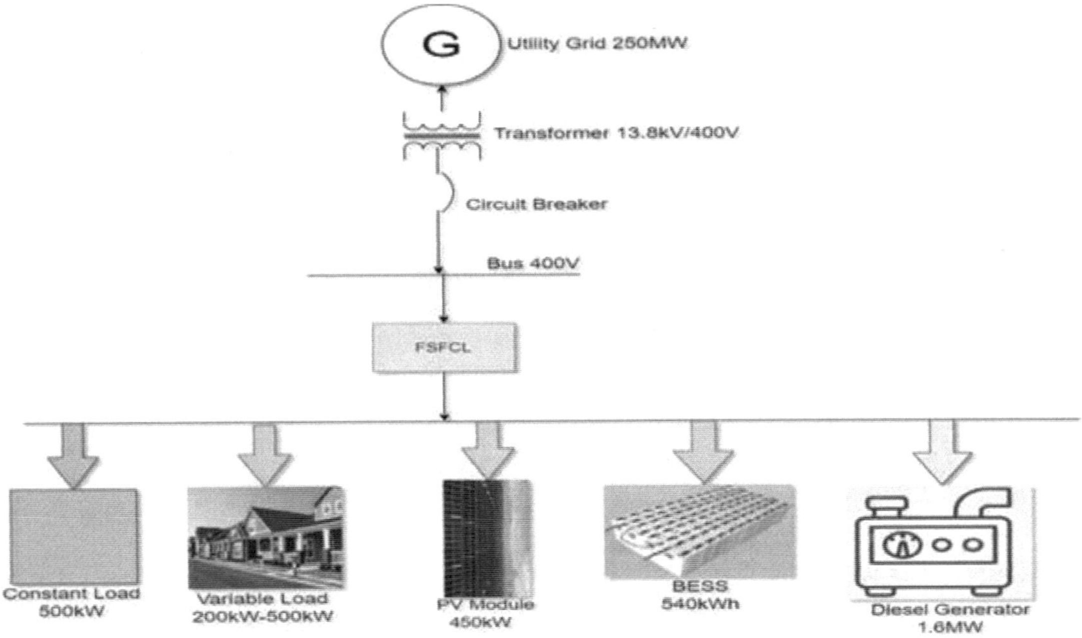

Figure 6. Microgrid Model with SFCL

In order to study the transient effect on the microgrid due to switching operation, the model was made to run for total of 300 seconds. Microgrid was intentionally islanded through circuit breaker at a step time of 150 seconds and after islanding variable load (initial load 250KW Final Load 200KW) was added at a step time of 200th seconds and the Simulink model was put into running mode .Once the compilation was complete the results were recorded for both the model i.e. with and without FSCFL. The analysis of the obtained result is done in next chapter.

4. Simulation Results

4.1. Simulated Microgrid Voltage With And Without FSFCL:

In The Figure 7 has been shown the microgrid voltage with and without SFCL. From the above graph one can observe that at 150th seconds i.e. when the microgrid is made to intentionally isolate through circuit breaker there are transients created in the microgrid voltage for a brief span of time. This is because of the fact that there is new redistribution of microgrid voltage in the microgrid. In the graph shown in Figure 7, red one represents microgrid voltage without FSFCL whereas the graph shown in blue represents microgrid voltage in presence of FSFCL. From these two graphs one can clearly visualise that in absence of FSFCL microgrid transient voltage reaches upto 2750V whereas the same transient voltage reduces to 450V in presence of FSFCL. Hence effectiveness of FSFCL during transient conditions can be clearly visualised.

Journal of Physics: Conference Series **2327** (2022) 012009 doi:10.1088/1742-6596/2327/1/012009

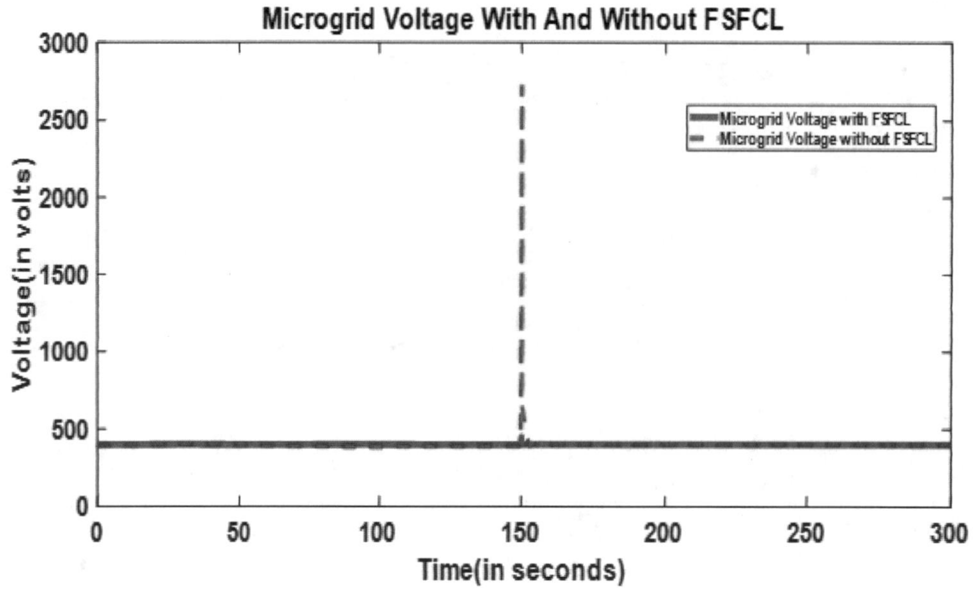

Figure 7. Simulated Microgrid Voltage With And Without FSFCL

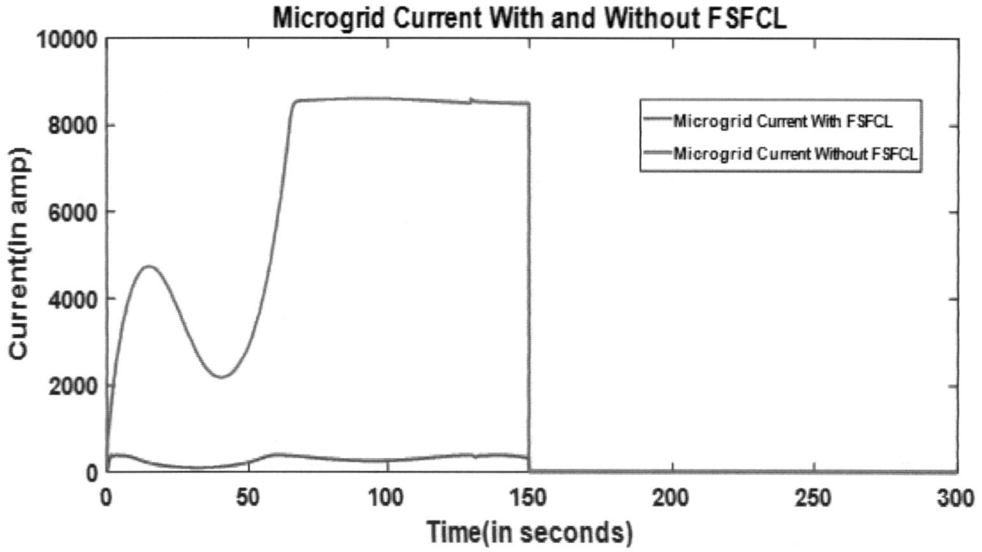

Figure 8. Simulated Microgrid Current With And Without FSFCL

In The Figure 8 has been shown the microgrid current in presence and absence of the FSFCL.From the two-graph shown in figure 8 one can observe that at 150th seconds i.e. when the microgrid is made to intentionally isolate through circuit breaker there are transients created in the microgrid currents same as in case of microgrid voltage for a brief span of time. In the graph shown in figure 8 red one represents microgrid current without FSFCL whereas the graph shown in blue represents microgrid current in presence of FSFCL. From these two graphs one can clearly visualise that in absence of FSFCL microgrid transient current reaches upto 8500A whereas the same transient current reduces to 380A in presence of FSFCL. Hence effectiveness of FSFCL during transient conditions can be clearly visualised in case of currents as well

5. Conclusion

Microgrid with DER'S during shifting from grid to standalone mode with and without FSFCL was simulated in the MATLAB. It was found that with FSCL there was the decrease in the magnitudes microgrid's current and voltage during transient condition and the reduction in magnitudes was found to be around 83 percent for voltage and 95 percent for the current. From the results it was concluded that through the inclusion of FSFCL microgrids transient stability gets greatly enhanced and hence during the transition in presence of FSFCL microgrid can make smooth transition from grid connected to standalone mode without any severe transient effects.

References

[1] F. P. Sioshansi. (2014). Distributed generation and its implications for the utility industry. Distributed generation and its implications for the utility industry.
[2] L. Chen, F. Z. (2015). Fault Ride-Through Capability Improvement of DFIG-Based Wind Turbine by Employing a Voltage-Compensation-Type Active SFCL. Canadian Journal of Electrical and Computer Engineering, pp. 132-142.
[3] M. Aurangzeb, A. X. (2020). An Evaluation of Flux-Coupling Type SFCL Placement in Hybrid Grid System Based on Power Quality Risk Index. IEEE Access, pp. 98800-98809.
[4] G. Didier, J. L. (Maay 2013). A novel approach to determine the optimal location of SFCL in electric power grid to improve power system. IEEE Transactions on Power System, pp. 978–984.
[5] L. Chen, F. Z. (2015). Application of a modified flux-coupling type superconducting fault current limiter to transient performance enhancement of micro-grid. Phys.C, Supercond. Appl., pp. 144-148.
[6] L. Chen, Y. J. (Oct 2009). ''Influence of a voltage compensation type active superconducting fault current limiter on the transient stability of power system. Phys. C, Superconductor, pp. 1760-1764.
[7] Aswathi A Nair, K. K. (July 2016). MICROGRID PROTECTION USING SUPERCONDUCTING FAULT CURRENT LIMITER. International Research Journal of Engineering and Technology (IRJET), pp. 1320-1325.
[8] Maloth Naresh, N. K. (2016). "Superconducting Fault Current Limiter for Grid Connected Power. IEEE International Conference on Industrial, pp. 576-581.
[9] N. Eghtedarpour and E. Farjah. (May 2014). "Power control and management in a hybrid AC/DC microgrid. IEEE Trans. Smart Grid, pp.1494-1505.
[10] X. Tang, W. D. (March 2014,). "Investigation of the Dynamic Stability of Microgrid. IEEE Transactions on Power Systems,, , 698-706.
[11] H. Saghafi, H. K. (2015). "Application of integrated series compensator in damping power oscillations in standalone microgrids., pp. 2-9.
[12] J. W. Bray. (January 2015). Study of a Modified Flux-Coupling-Type Superconducting Fault Current Limiter for Mitigating the Effect of DC Short Circuit in a VSC-HVDC System. Journal of Superconductivity and Novel Magnetism .
[13] al, M.Parket (2011). HTS SMES Application for the Frequency Stabilization of Grid-Connected Wind Power Generation System. Journal of Superconductivity and Novel Magnetism volume, pp. 1007-1014.
[14] Choi,Y.S.Cho, H.S. (2015). Status Evaluation Method for SMES Used in Power Grid. IEEE Transactions on Applied Superconductivity.
[15] Choi,Y.S.Cho, H.S (1749–1753,). ''Improvement on the power capacity of the transformer-type SFCL with neutral lines between secondary windings. Phys. C, Supercond., Oct. 2009.
[16] M. Abedini and M. Abedini. (2018.). Energy management and control policies of the islanded microgrids, Sustainable cities and society, pp. 714-722.
[17] M. Hamzeh, H. M. (2013). "A decentralized self-adjusting control strategy for reactive power management in an islanded multi-bus MV microgrid., pp. 18-25.

Optimal integration of renewable distributed generators and electric vehicles in a radial distributed network

V Kumar[1], S K Sudabattula[2], N Dharavat[3]

[1,2,3]School of Electronics and Electrical Engineering, Lovely Professional University

Jalandhar, Punjab, India

vivekkumar6716.vk@gmail.com, sureshsudabathula@gmail.com and

nagaraju.dnr98@gmail.com

Abstract. With the rising interest in sustainable transportation efforts, the optimal integration of Electric Vehicles (EVs) in the transportation business has come into existence very drastically. However, because of the increased power consumption, their impact on the electric power market might result in more enormous losses in the active power, a reduced voltage profile, as well as reduced voltage stability limits. In Radial Distribution System (RDS), getting the reduced effect of EVs load is crucial. It is significantly important to build charging stations for EVs and Distributed Generators (DGs) at the best locations. This article presents a combined approach to solve the optimal integration of DGs and EVs problem. First, the desired locations of DGs and EVs are found using the Voltage Stability Index (VSI) method.
Further, the resource ratings are obtained by the Firefly algorithm (FA) method. EVs charging dump and smart charging strategies are used, and comparison is done with both systems. Furthermore, the main objective of this work is to reduce the power loss and maintain a good voltage profile at each bus in the RDS. Finally, the developed approach is tested on IEEE 69 bus system.
Index Terms – Electric Vehicle (EVs), Distributed Generations (DGs), Optimal Integration, Meta-heuristic algorithm, Power loss reduction, Firefly algorithm (FA)

1. Introduction

The power system management and control have become extremely complicated due to the exhaustion of conventional energy sources (CES), non-expanding transmission and distribution networks (DN), along with increasing demand for electricity. Optimal integration of DGs and EVs in the DS has recently become a significant concern in such conditions. It is crucial to utilize DGs in distribution system (DS) and produce electricity close to the consumer's location. The DG is defined as a small local power source connected to the DN. There are several benefits of connecting DGs in the DS. The proper DG placement in the DS enhances the system's efficiency, improves the voltage profile and stability, and reduces the power loss.
Numerous researchers have given various ways for optimally assigning DGs in DN during the last few decades. The primary goal of the optimal integration of DGs is to boost the DS efficiency while maximizing the energy benefits. Despite the fact that the optimal integration of DGs in radial distribution networks (RDN) has got a huge amount of exposure in the literature, there hasn't been enough research done to take EVs into account. Because there is a risk of decreasing the efficiency of RDNs because of their optimal integration at inconvenient locations, EV optimal placement is a key aspect in providing charging infrastructure. RDN execution can be considerably enhanced by optimising DG allocation. On the other hand, several scholars have recently been motivated and drawn to the best allocation of EVs charging infrastructure. The optimal allocation of DGs is a non-linear meta-heuristic optimization problem that must be addressed simultaneously for discrete and continuous variables. Furthermore, for the allocation or placement of DGs and EVs in RDNs at the same time, in huge scale the problem might occur more complex, which requires a significant amount of processing time. Since there is no such algorithm that can handle all types of optimization problems so, new heuristic algorithms are continuously developed by the researchers and they are applied to various optimization problems.

Furthermore, while a small number of articles have merely focused on techno-environmental issues, the majority of efforts in the DG optimal allocation problem have used a multi-objective function for techno economic features. Furthermore, the charging infrastructure required to meet the growing demand of EV load which is not well targeted. Optimal integration of DGs and EVs improves the performance of DS effectively. So in this existing work with the help of meta-heuristic algorithms optimal allocation of DGs and EVs in RDN is done so that we can full fill the charging infrastructure demand and can improve the voltage stability index (VSI) improve voltage profile and we can minimize power loss by proper placement of DGs and EVs and suitable allocation of buses for the different test call we can get the proper and desired results.

Meta-heuristic algorithms are flexible; they look for a solution extensively inside a single objective or multi-objective space for an answer. As a result, they've seen a lot of application in addressing the issue of optimal electrical unit integration in distribution networks. Because meta-heuristic algorithms can create solutions that are close to perfect computationally in a responsible way, they are widely used. They've also been employed for the optimal integration of DN in electrical units to solve multi-objective optimization problems. Due to unique characteristics such as flexibility and effectiveness, meta-heuristic optimization methodologies have become increasingly common for tackling engineering issues. Meta-heuristic algorithms have the potential to solve complicated computational problems quickly, which is one of their most significant advantages. The current circumstance does not affect meta-heuristic algorithms, and their rising nature permits them to hop between various local optima.

In the literature, different authors solved this problem using various algorithms. In Table 1 Literature review for the integration of DGs to DS is shown. From the reference 1-5 DG to DS integration review is done. In table 2 from reference 6-10 EV to DS integration is done and at last in table 3 from reference 11-15 combined DG and EV to DS integration review is done. In this research, article VSI is used to identify strong and weak buses for the placement of DGs in the presence of EVs, and then the values are sent to FA to calculate the desired sizes. Finally, a traditional or dump charging approach is described, and an improved or smart charging method. The analysis considers a variety of scenarios to enhance the feasibility of improved EV charging and increasing the charging capacity.

Finally, the article's remaining sections are listed below. In section 2, the problem formulation is explained; however, in section 3, VSI and FA are described. Finally, in section 4, the results and discussion are given, followed by the conclusion of the article.

Table 1. Literature review Integration of DGs to DS

Ref.	Author	Objective	Method	Test system	Year
[1]	Suresh Kumar Sudabattula	Improved voltage profile and reduction in power loss	LSF, DFA	IEEE 69	2019
[2]	Rajesh Kumar Samala	Maximum power loss position of the RDS	PSO, ALO	IEEE 33	2020
[3]	Sirine Essallah	Active power loss minimization and voltage profile improvement	MPSO, BPSO	IEEE 33,69	2020
[4]	Nagaraju Dharavat	Aiming minimal P_{Loss}, improved VSI and voltage profile	POA	IEEE 33	2021
[5]	Mohammed Hamouda Ali	Increase the overall system Efficiency	IWHO	IEEE 33,69 and 113	2021

Table 2. Integration of EVs to DS

Ref.	Author	Objective	Method	Test system	Year
[6]	Akanksha Shukla	Multi-objective synergistic planning of EV	MOGWO	IEEE 33 and 25 node system	2019
[7]	K. Abhinaya	Fuel cost reduction and network power loss minimization	SSA	IEEE 30	2020
[8]	A. N. Archana	EVCS allocation in the distribution system	EVCS	IEEE 33	2020
[9]	Sasmita Tripathy	Total Power Loss	LVD	IEEE 13	2021
[10]	Tayenne D. De Lima	EVCS and the electrical distribution infrastructure should be planned together	AMPL and solved via CPLEX	24-node system	2022

Table 3. Integration of DGs + EVs to DS

Ref.	Author	Objective	Method	Test system	Year
[11]	Srinivasa Rao Gampa	Voltage profile Improvement	GOA	51,69 Buses	2020
[12]	Venkata K. Babu Ponnam	Optimise the electrical distribution systems, AVDI	HHOA	IEEE 33,69	2020
[13]	S. Kaveripriya	Reducing power loss	GOA &LSF	IEEE33	2020
[14]	Varaprasad Janamala	Operation and control of RDN	FSA	106,36 Buses	2021
[15]	Nagaraju Dharavat	To get improved the voltage profile as well as decrease power loss	POA	IEEE 69	2021

2. Problem Formulation

Increased EV penetration in the DS results in significantP_{Loss}. The purpose of this study is to reduce P_{Loss}.

A. Objective Function

$$min\sum_{m=1}^{24} P_{Loss}(m) \tag{1}$$

Where,

$$P_{Loss} = \sum_{m=1}^{24} I_m^2 R_m \tag{2}$$

B. Constraints

Constraints on Power Balance

$$\sum_{m=1}^{24} P_G(m) = \sum_{m=1}^{24} P_{Demand}(m) + P_{Loss}(m) + P_{EV}(m) \tag{3}$$

C. EV SOC limits

$$SoC_{min} \le SoC_m^{EV} \le SoC_{max} \tag{4}$$

Constraints on EV power charging and discharging:

$$P_{ch,m,t} \le P_{ch,m}^{max} \tag{5}$$

$$P_{disch,n,t} \le P_{disch,m}^{max} \tag{6}$$

3. Methodology

To reduce P_{Loss} A fusion of VSI, Smart Charging, and FA-based approaches is used in the network. The VSI approach is used to determine the best position in the system for allocating DGs along with EVs. FA selects the DG sizes, while smart charging finds the best time to charge the EVs.

3.1 Voltage Stability Index (VSI)

VSI method was used to determine the strong and weak buses in the test system. After using the VSI method, 61, 64 and 26 are the strong buses for the DG analysis and 2, 3 and 28 are the weak buses for the EV analysis. The system's VSI must be enhanced to solve the voltage instability problem. VSI of a system is a critical issue. Because of the continuous growth in load demand, DS operates under heavy load circumstances.VSI provides essential information regarding buses over poor voltage profiles together with levels of stability.

Equation (7) is used to calculate the VSI at node n.

$$VSI(n) = \left\{ |V_m|^4 - 4\{PX - QR\}^2 - 4\{PR + QX\} |V_m|^2 \right\} \tag{7}$$

At the receiving end load, P and Q reflect active as well as reactive power load demand. The letters R and X stand for resistance along with reactance, respectively; if VSI (k) ≥ 0 for k = 2, 3,.., n, a higher level of voltage stability is achieved.

3.2 Firefly Algorithm (FA)

FA is introduced by yang it is, a new meta-heuristic algorithm. This algorithm is based on the flashing Signal concept, which is produced by the fireflies. It is also a population-based algorithm. FA has a number of advantages over single-point search algorithms, such as the algorithm being opportunistic In terms of tuning optimal parameter values for exploration and exploitation of instabilities.

FA has a flashlight as a prominent feature. It can be used to attract potential mates as well as warn from potential threats and rivals. The firefly flashlight is mainly related to the objective function that has been solved. The light intensity I (n) varies with the distance n and is governed by the inverse square law Eq (8)

$$I(n) = \frac{I_i}{n^2} \tag{8}$$

Where, I_i is the intensity of light at source.

Also intensity of light I(n) varies with a distance n given in Eq (9) exponentially

$$I(n) = I_r e^{-\gamma n} \tag{9}$$

Where, I_r is the initial light intensity and γ is the light absorption coefficient.

The attraction of a firefly is proportional to the intensity of light as it is noticed by nearby fireflies; hence its attractiveness β is defined as

$$\beta = \beta_r e^{-\gamma n^2} \tag{10}$$

Where, β_r attractiveness at n=0

The distance between any two fireflies l and m at $d_l$and d_m is calculated as

$$n_{lm} = \|d_l - d_m\| = \sqrt{\sum_{k=1}^{x} (d_{l,k} - d_{m,k})^2} \tag{11}$$

In each generation every firefly is attracted towards the brighter firefly that is given by Eq. (12).

$$d_l = d_l + \beta_r e^{(-\gamma n^2)(d_m - d_l)} + \alpha(rand - 0.5) \tag{12}$$

Where, α is the randomization parameter, and "rand" is a random number generator with a uniform distribution in the range [0, 1].

3.2.1 Firefly algorithm Implementation for the ideal size of DGs and EVs and objective function minimization

Step 1: DS bus, as well as line data, should be established.
Step 2: Base case load flow should be initialized.
Step 3: According to VSI, the ideal allocation for DGs together with EVs is one which is suited for the input of FA.
Step 4: FA parameters are initialized.
Step 5: Each firefly corresponds to occupy the potential node position, and FA is utilised to determine the best sizes of DGs and EVs.
Step 6: The attraction is proportional to the brightness, which indicates the optimal DG sizes defined by Equation (10).
Step 7: A firefly with lower brightness, l, is attracted to a firefly with higher intelligence, m, to identify the ideal value of DGs and EVs sizes using Equation (11).
Step 8: Examine the modified solutions, i.e. the optimal values of DG and EV sizes, for the improved intensity of light.
Step 9: Determine the ideal DG and EV sizes for the firefly according to intensity of light.
Step 10: Repeat steps 4–8 until the maximum number of iterations is achieved.

4. Result and Conclusion:

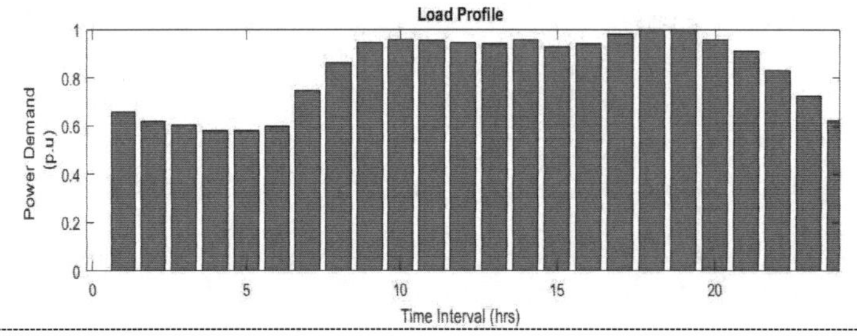

Figure 1. Power Demand of Distribution system

The outcomes of the proposed combined method are presented in this research work. The study is carried out using the 69 bus system in MATLAB (R2017b) environment. DG and EV allocation is evaluated, and the results are compared before applying the combined approach for dynamic loading situations.
The FA is utilised in this research work to solve the best allocation/placement together with the size of DGs as well as EV in a DS. With the help of this algorithm, we can reduce P_{Loss} while enhancing the VSI and voltage profile of DS. VSI and voltage profiles are suddenly enhanced by a combined approach and give the best values rather than when placed separately. The VSI and voltage profile are improved for entire simulation period when both DGs and EVs are allocated at a time. When 3 DGs and 3EVs are placed simultaneously, then we have got the most relevant results instead of other cases.
The case studies that demonstrate the effectiveness of the proposed approach methodology are listed below.

Table 4. Comparison of P_{Loss} , V_{min} and VSI_{min} values with other algorithms

Different Algorithms	Bus Number	DG Size (kW)	P_{Loss} (kW)	%dec P_{Loss}	V_{min} in (p.u)	VSI_{min} (p.u)
Base Case	N.A	N.A	225.0014	N.A	0.9678	0.8773
WOA	61 64 26	889.9726, 844.9078, 449.1367	75.5503	66.42	0.9773	0.9144
PO	61 64 26	1566.6409,191.801, 416.9132	74.1731	67.03	0.9734	0.9258
FA	**61 64 26**	**1512.1, 289.6, 406.4**	**73.2583**	**67.44**	**0.9796**	**0.9203**

Journal of Physics: Conference Series **2327** (2022) 012010 doi:10.1088/1742-6596/2327/1/012010

Table 5. DG values for the 24-hour power demand

DG1 in kW	DG2 in kW	DG3 in kW
988.2	191.3	266.3
926.7	180.2	249.9
901.3	174.8	243.0
867.8	168.4	234.0
867.8	168.4	234.0
895.2	173.6	241.3
1124.6	217.2	302.9
1300.4	250.3	349.9
1427.8	274.0	383.9
1445.0	277.1	388.5
1440.3	276.2	387.3
1427.8	273.9	383.9
1416.9	271.9	310.0
1443.8	278.1	388.5
1401.3	269.0	376.9
1416.9	271.9	381.0
1477.6	290.3	399.1
1508.9	292.4	406.4
1508.9	292.4	406.4
1444.9	277.1	388.5
1365.5	271.0	369.1
1246.3	240.0	335.5
1086.2	209.9	292.6
933.3	180.9	251.5

Base Case: The study is carried out without DGs and EVs for the 24-hour load demand indicated in Fig. 1. The total P_{Loss} For 24 hours in the base, the case is 3746.3833kW.

Case 1: The study is carried out both with and without DGs, and the outcomes are compared to those of other algorithms. Firstly, using the VSI method, we can determine the ideal locations for DG allocation. As per VSI 61, 64, 26 are the perfect locations for the DGs insertion. The FA technique is used to determine P_{Loss}, V_{min} and VSI_{min} at peak load, which are compared to other algorithms in Table 4. FA is more effective than other methods; hence it will be utilised in all contests. In addition, the demand displayed in Fig 1 is analysed using the FA methodology. Table 5 shows the DG sizes observed in 24 hours. The comparison graph of P_{Loss}, V_{min} and $VSI_{min}$with base case and DGs for 24 hours are displayed in Figs 2, 3 and 4. As per VSI 2, 3, 28 are the perfect locations for the EVs insertion and the comparison graph of P_{Loss}, V_{min} and VSI_{min} with base case together with DGs and EVs for 24 hours are displayed in Figs 5, 6 and 7. Throughout the day, the P_{Loss} It is greatly reduced, V min and VSI min is improved it can be seen through the figures.

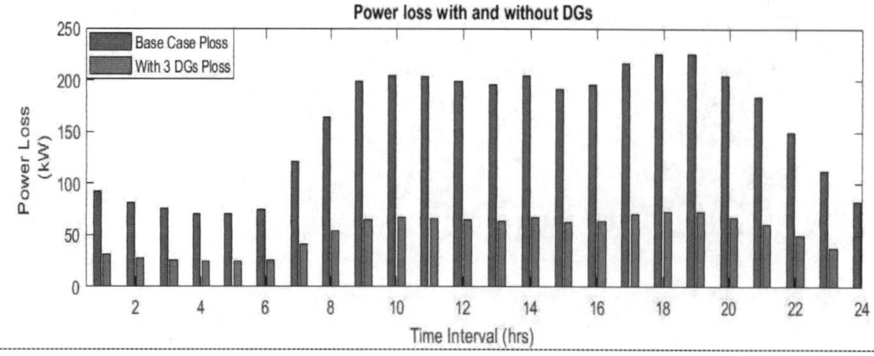

Figure 2. P_{Loss} With and without DGs

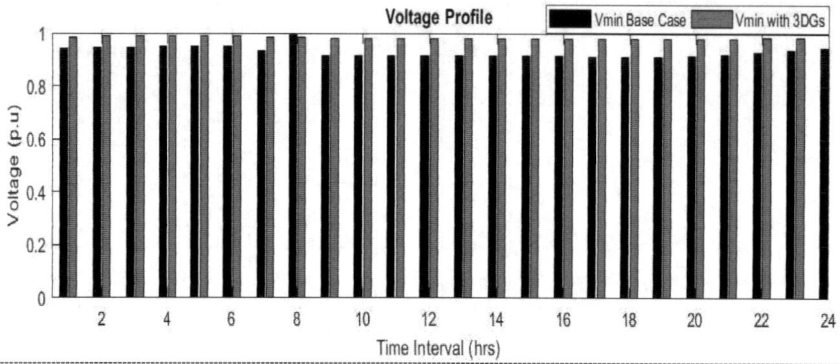

Figure 3. V_{min} With and without DGs

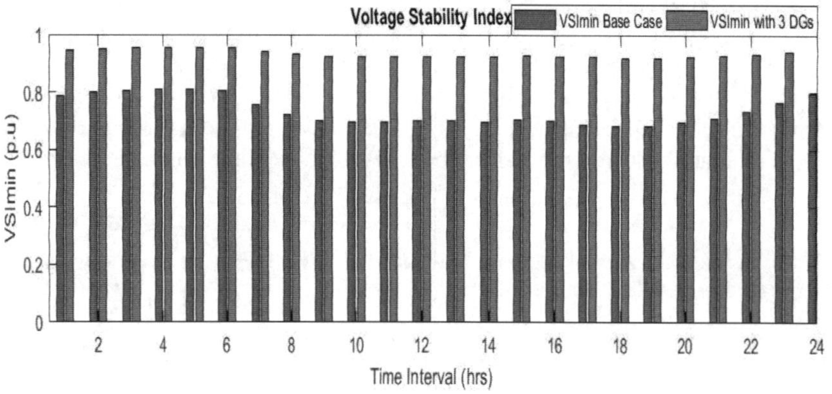

Figure 4. VSI_{min} With and without DGs

Table 6. EV Parameters

Specifications	Ratings
E.V Battery	16 Kwh
Average distance travelled by each E.V	30 KM
SOC maximum	0.9
SOC minimum	0.2
Number of E.Vs	60

Table 7. P_{Loss} Comparison with different strategies

Different Approaches	Dump Charging Strategy	Smart Charging Strategy
Base Case P_{Loss} in kW	3746.3833	3746.3833
Alone With EVs P_{Loss} in kW	3754.398	3747.592
With DGs along with EVs P_{Loss} in kW	1239.623	1236.821

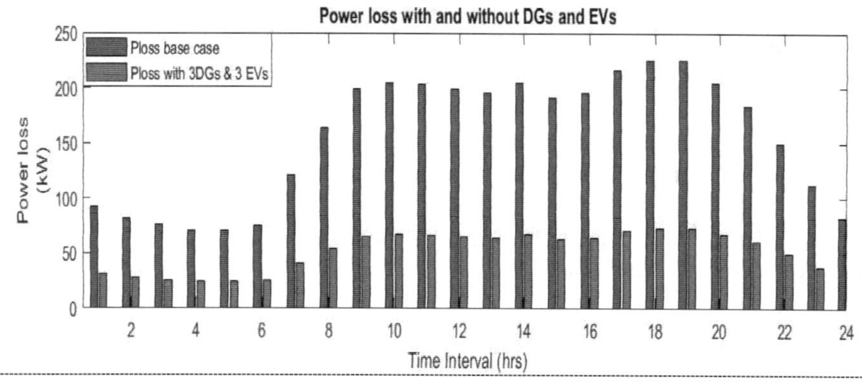

Figure 5. P_{Loss} With and without DGs and EVs

Case 2: The study incorporates both EVs together with DGs. Table 6 includes the EV parameters. The EV will leave in the morning and return in the evening. So, before their next travel, the EVs should be charged. For EV charging two methods are used that is dump charging or we can say conventional charging and other one is smart charging that is optimized EV charging method. In Dump Charging, EVs are charged without consideration for system demand immediately after their scheduled journey, enhancing the system's P_{Loss}, whereas in Smart Charging Method, The system's P_{Loss} is reduced by charging EVs during off-peak hours. Table 7 shows the P_{Loss} comparison with different strategies such as with base case, with EVs without DGs, and with EVs along with DGs. In comparison to dump charging and other algorithms, the smart charging approach combined with FA based DG allocation produces better results.

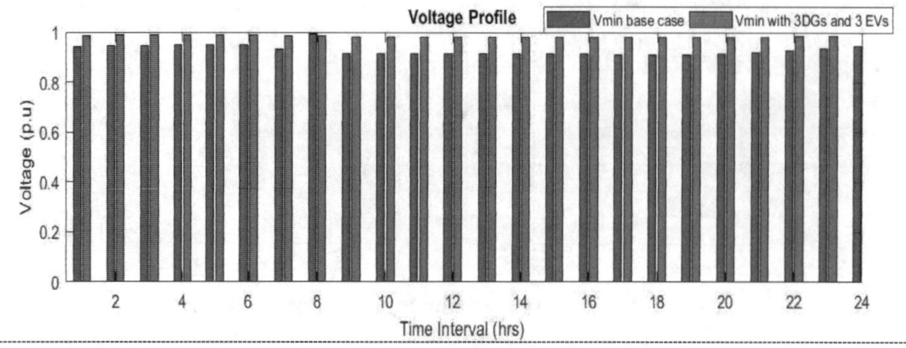

Figure 6. V_{min} With and without DGs and EVs

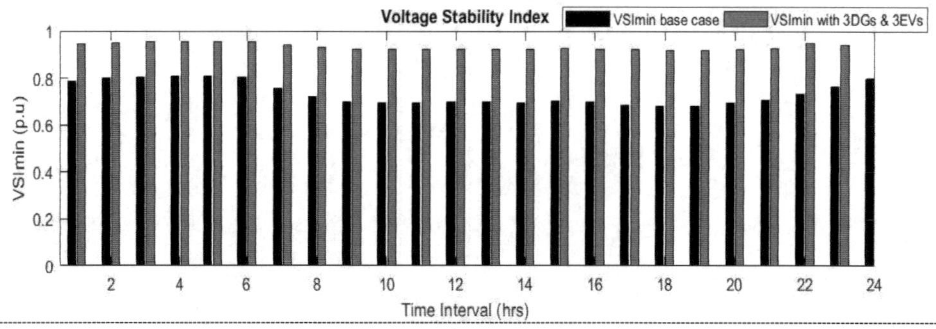

Figure 7. VSI_{min} With and without DGs and EVs

5. Conclusion

The objective achieved in this study is the allocation and sizing of DGs and EVs in the DS are solved using a combination of VSI and FA methods. These approaches can greatly reducing P_{Loss} which are 73.2583 kW, which is less than other algorithms used in the comparison, and used to enhance the VSI and voltage profile of the DS. The VSI approach selects the best sites for DGs and EVs placement and sends the values to FA to calculate the desired sizes. On an IEEE 69 bus, the analysis is carried out. Dump and smart charging strategies were used to charge EVs. In smart charging methods EVs are charged during the non-peak hours. The results suggest that using a smart charging approach and combining DGs and EVs has reduced P_{Loss} more effectively.

Orcid ID of all authors
[1]https://orcid.org/0000-0001-9119-7627
[2]https://orcid.org/0000-0002-4562-1463
[3]https://orcid.org/0000-0002-8439-1621

6. References

[1] S. Suresh Kumar, M. Kowsalya, and S. Velamuri, “Simultaneous Allocation of Distributed Generators and Shunt Capacitors in a Distribution System,” *ECTI Trans. Electr. Eng. Electron. Commun.*, vol. 17, no. 1, pp. 35–50, 2019.
[2] R. K. Samala and M. R. Kotapuri, “Distributed generation allocation in distribution system using

particle swarm optimization based ant-lion optimization," *Int. J. Control Autom.*, vol. 13, no. 1, pp. 414–426, 2020.

[3] S. Tripathy, S. N. Mohanty, S. R. Ghatak, and P. Sinha, "Modeling of EV Load and Its Impacts on the Distribution System for Static Condition," in *2021 2nd International Conference on Communication, Computing and Industry 4.0*, 2022, no. December, pp. 1–5, doi: 10.1109/c2i454156.2021.9689324.

[4] S. Essallah and A. Khedher, "Optimization of distribution system operation by network reconfiguration and DG integration using MPSO algorithm," *Renew. Energy Focus*, vol. 34, no. 00, pp. 37–46, 2020, doi: 10.1016/j.ref.2020.04.002.

[5] N. Dharavat, S. K. Sudabattula, and V. Suresh, "Optimal Allocation of Multiple Distributed Generators and Shunt Capacitors in a Distribution System Using Political Optimization Algorithm," *Int. J. Renew. Energy Res.*, vol. 11, no. 4, pp. 1479–1488, 2021, doi: 10.20508/ijrer.v11i4.12240.g8349.

[6] M. H. Ali, S. Kamel, M. H. Hassan, M. Tostado-Véliz, and H. M. Zawbaa, "An improved wild horse optimization algorithm for reliability based optimal DG planning of radial distribution networks," *Energy Reports*, vol. 8, pp. 582–604, 2022, doi: 10.1016/j.egyr.2021.12.023.

[7] A. Shukla, K. Verma, and R. Kumar, "Multi-objective synergistic planning of EV fast-charging stations in the distribution system coupled with the transportation network," *IET Gener. Transm. Distrib.*, vol. 13, no. 15, pp. 3421–3432, 2019, doi: 10.1049/iet-gtd.2019.0486.

[8] A. N. Archana and T. Rajeev, "Reliability Index based Approach for Allocating EV Charging Station in a Distribution System," *2020 IEEE Int. Conf. Power Electron. Smart Grid Renew. Energy, PESGRE 2020*, pp. 1–6, 2020, doi: 10.1109/PESGRE45664.2020.9070408.

[9] S. Tripathy, S. N. Mohanty, S. R. Ghatak, and P. Sinha, "Modeling of EV Load and Its Impacts on the Distribution System for Static Condition," *2021 2nd Int. Conf. Commun. Comput. Ind. 4.0*, no. December, pp. 1–5, 2022, doi: 10.1109/c2i454156.2021.9689324.

[10] T. D. De Lima, J. F. Franco, F. Lezama, and J. Soares, "A Specialized Long-Term Distribution System Expansion Planning Method with the Integration of Distributed Energy Resources," *IEEE Access*, vol. 10, pp. 19133–19148, 2022, doi: 10.1109/ACCESS.2022.3146799.

[11] S. R. Gampa, K. Jasthi, P. Goli, D. Das, and R. C. Bansal, "Grasshopper optimization algorithm based two stage fuzzy multiobjective approach for optimum sizing and placement of distributed generations, shunt capacitors and electric vehicle charging stations," *J. Energy Storage*, vol. 27, no. November 2019, p. 101117, 2020, doi: 10.1016/j.est.2019.101117.

[12] P. V. K. Babu and K. Swarnasri, "Multi-objective optimal allocation of electric vehicle charging stations in radial distribution system using teaching learning based optimization," *Int. J. Renew. Energy Res.*, vol. 10, no. 1, pp. 366–377, 2020.

[13] S. Kaveripriya, V. Suresh, S. Suresh Kumar, and K. Abinaya, "Optimal Allocation of DERs in Distribution System in Presence of EVs," *Lect. Notes Electr. Eng.*, vol. 688, pp. 77–88, 2021, doi: 10.1007/978-981-15-7241-8_6.

[14] V. Janamala, U. Kamal Kumar, and T. K. S. Pandraju, "Future search algorithm for optimal integration of distributed generation and electric vehicle fleets in radial distribution networks considering techno-environmental aspects," *SN Appl. Sci.*, vol. 3, no. 4, pp. 1–17, 2021, doi: 10.1007/s42452-021-04466-y.

[15] N. Dharavat, S. K. Sudabattula, and S. Velamuri, "Integration of Distributed Generation and Electric vehicles in a distribution network using Political Optimizer," *2021 4th Int. Conf. Recent Dev. Control. Autom. Power Eng. RDCAPE 2021*, vol. 3, pp. 521–525, 2021, doi: 10.1109/RDCAPE52977.2021.9633448.

High Performance Mixed Logic Decoders using MOS like GNRFET in 22nm Technology

Sneha Arora[1], Suman Lata Tripathi[2]

[1,2]VLSI Design, School of Electronics and Electrical Engineering, Lovely Professional University, Phagwara, Punjab India
snehaarora56@gmail.com, tri.suman78@gmail.com

Abstract. Graphene nano ribbon field effect transistor is an emerging field of research in VLSI technology beyond 32nm. VLSI main motive is to reduce power consumption and other parameters such as delay, PDP (power delay product) to improve the efficiency. This article discusses the requirement for high-performance applications that use little power. However, employing low-power devices for high-rank applications such as microprocessors, digital signal processors, and static random-access memory (SRAM) is very challenging. In the field of memory design and logical circuit design, it is well recognised that the Decoder plays a crucial role. As part of the planned work in 22nm technology, a technique has been developed for evaluating the parameters Power delay product, Energy delay product, Power consumption, and Delay on 14T and 15T based employing 2 to 4 decoders based on MOSFET and GNRFET. So, the final proposed circuit is obtained by application of decoders with the use of GNRFET while improving the performance parameters.

1. Introduction

As nanometre measure developments have pushed, chip thickness and working repeat have extended, making power usage in battery-worked helpful contraptions a critical concern. [1] Despite for no reduced contraptions, power usage is basic because of the extended packaging and cooling costs similarly as expected resolute quality issues. Thusly, the guideline structure objective for VLSI is to meet execution requirements inside a force spending plan. Thusly, power adequacy has anticipated extended criticalness. This errand researches how circuits subject to GNRFET (, a creating semiconductor innovation. [2]
The yearning to upgrade the structure estimations of execution, power, area, cost, and time to promote (opportunity cost) has not changed since the birthplace of the IC business. [3] In all honesty, Moore's Law is connected to propelling those boundaries. Regardless, as scaling of collecting centers progressed towards 20-nm, a segment of the channel boundaries cannot be scaled any further, especially the force flexibly voltage, the common variable in choosing interesting force. Moreover, overhauling for one factor, for instance, execution thus changed over into tremendous tradeoffs in various regions, like force. [4] Another limitation as strategies advanced toward 20-nm was how lithography was stuck at ArF lighting up source with a frequency of 193nm while the method fundamental component was pushing sub-20nm. Optical progressions, for instance, soaking lithography and twofold changing made that possible, anyway to the detriment of extended vacillation. There were moreover various improvements on the way, for instance, High-K metal gate terminal that helped – to a compelled degree – entryway spillage issues. However, the truth remained that the structure window for improving among the recently referenced arrangement factors was contracting. Arranging in GNRFET augments the structure window

eventually. Working voltage continues cutting back, essentially getting a good deal on exceptional and static force. Short channel impacts are decreased by and large, diminishing the guard banding expected to oversee alterability. Furthermore, execution continues improving stood out from planar at an unclear center point. In all honesty, at incredibly low force gracefully voltages, the execution great situation of the GNRFET stood out from its planar indistinguishable expands due to the unparalleled entryway control of the redirect in the GNRFET. [5]

It is indicated that by virtue of disregarding short out current, past strategies proposed to propel the domain of a fan-out tree may result in over the top force use. Unquestionably the dynamic mode power usage, the clock power, and the typical spillage intensity of the combinational circuits are reduced by up to 55%, 29%, and 53%, independently, while keeping up near speed and data security when diverged from the circuits executed in CMOS and GNRFET innovation. [6-8]

GNRFETs are an indispensable development in the advancement of semiconductors since mass CMOS encounters issues in scaling past 32 nm. [9-10] Usage of the graphene prompts extremely interesting arrangement openings. Rich average assortment of setup styles, made possible by free control of GNRFET gate terminal, can be used effectively to diminish supreme unique influence use LP mode circuits give an enabling trade-off among influence and zone.

Process advances of the nanoscale scale have led to a rise in the thickness and frequency of integrated circuits, increasing power usage in battery-powered portable electronics. [11] Whether it is little or big, the importance of reducing power consumption cannot be overstated. Cooling expenses grow, as well as reliability problems that may arise if no changes were ever made to the device. [12-15] Since VLSI designers are being required to meet execution requirements while maintaining under a power budget, the major design aim for VLSI is to provide a means of fulfilling those needs. Because of this, we believe that power efficacy will become more important. [16]

Because the beginning of the Electronics market, it has always been necessary to make improvements to structure measurements of execution, performance, area, expense, and time to promote (opportunity cost). [17] To be clear, Moore's Law is directly connected to the progression of specific criteria. At first, the extent to which building blocks could be scaled was tied to the technology they used to fabricate. [18] However, as the industry scaled the size of building blocks towards 20nm, the parameters used in fabricating, including power supply voltage, proved to be unable to be scaled any further. [19] Thus, improvement in any one area, for example, execution, led with tremendous compromises in other areas, which functioned in an analogous manner to control. [20]

This paper aims to implement decoder using GNRFET, an upcoming device technology that is predicted to enhance or replace CMOS mass-production technology at 22 nm and beyond, in which delay versus power may be experimented.

2. Methodology

This circuit made use of Synopsys HSPICE, which communicates with circuits and computes parameters like as power, delay, and so on via the use of SPICE-based coding. In order to do waveform analysis, Avanwaves will be used. Among the parameters to be estimated in this proposed study are delay, power, PDP, and EDP. I would like to provide a flowchart of my approach, which use to get a methodical outcome. (Figure 1). The flowchart below explains methodology in which circuit is implemented on HSPICE, Avanwaves used to verify waveforms and output is performance metrics such as power and delay.

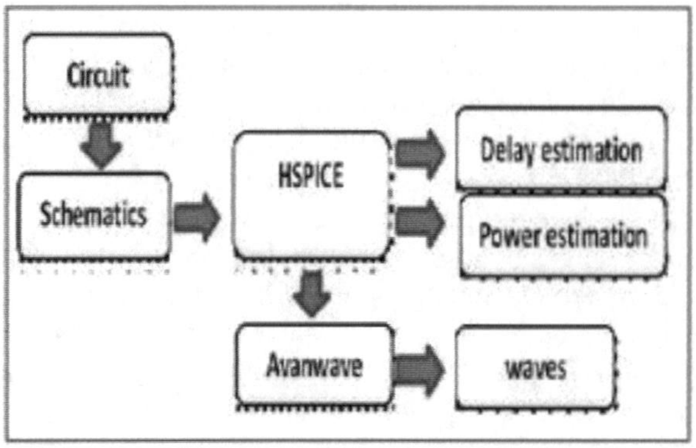

Figure 1: Flowchart of Methodology

The functioning of all decoders has been used in this paper is the same as considered it from the base circuit use different technology and evaluate their parameter to get the best result. The basic symbol of Decoder is given above (figure 2) along with the truth Table 1. So, briefly describe all 2-4(14T), 2-4(15T) decoders. The symbol represents to inputs and one enable or select and four outputs in the figure below.

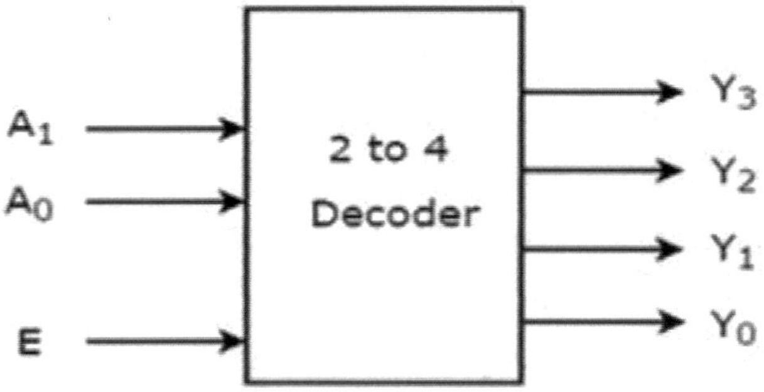

Figure 2: Symbol of 2 to 4 decoder

Table 1: Truth Table of 2 to 4 decoder

Enable	Input 1	Input 2	Out3	Out2	Out1	Out0
0	-	-	0	0	0	0
1	0	0	0	0	0	1
1	0	1	0	0	1	0
1	1	0	0	1	0	0
1	1	1	1	0	0	0

In Figure 3, 14-T based two to four decoder is represented in the GNRFET based proposed circuit. It has 2 inputs and 4 outputs with the use of MVL multi valued logic and TGL transmission gate logic. This same circuit is modified with an extra transistor as shown in Figure. The figure shows the logic implementation and output configurations.

Journal of Physics: Conference Series **2327** (2022) 012011 doi:10.1088/1742-6596/2327/1/012011

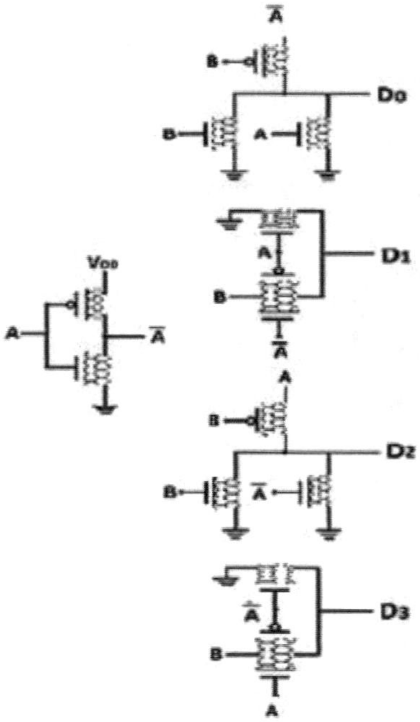

Figure 3: GNRFET 14T Decoder proposed

Figure 4 consists of 15-T and is known as high performance mode. Both 14-T and 15-T proposed circuits are made of MOS-like GNRFET hence the symbol used has Graphene Nano Ribbon based channel which reduces the short channel effects and known to give better performance.

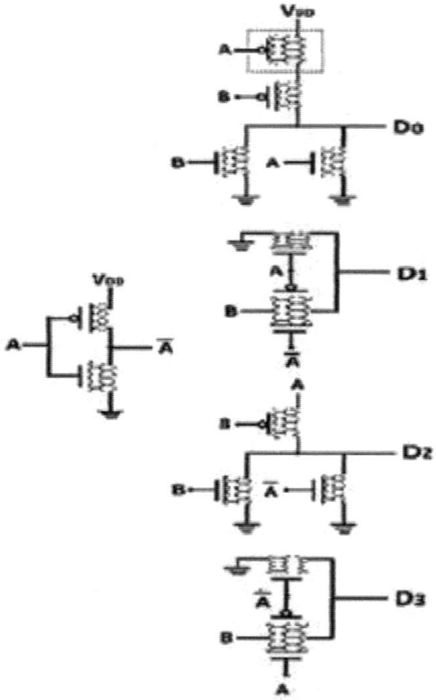

Figure 4: GNRFET 15T Decoder proposed

3. Simulation Results

This section provides the simulation results in graphical and tabular form. In Figure 5, the average power in 14-T proposed circuit using GNRFET is improved.

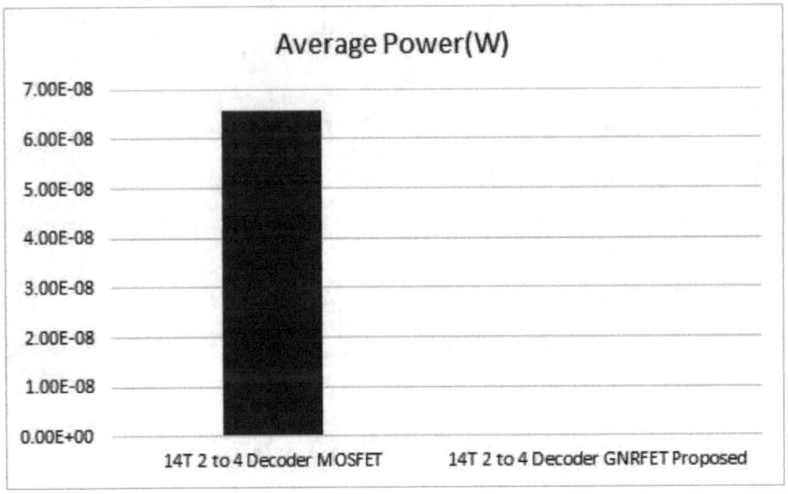

Figure 5: Average Power Output Comparison for 14-T Decoder using GNRs

In Figure 6 and Figure 7, delay and PDP for 14-T proposed circuit is better than the existing one. Delay is nearly the same in both cases.

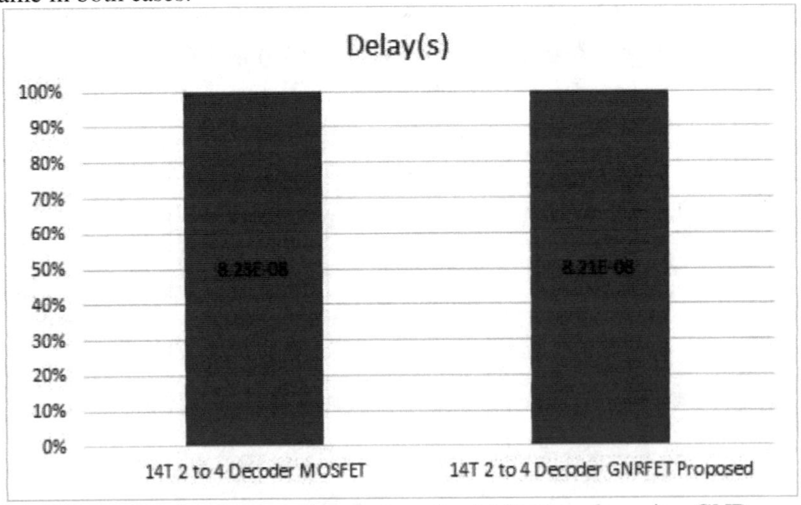

Figure 6: Delay Output Comparison for 14-T Decoder using GNRs

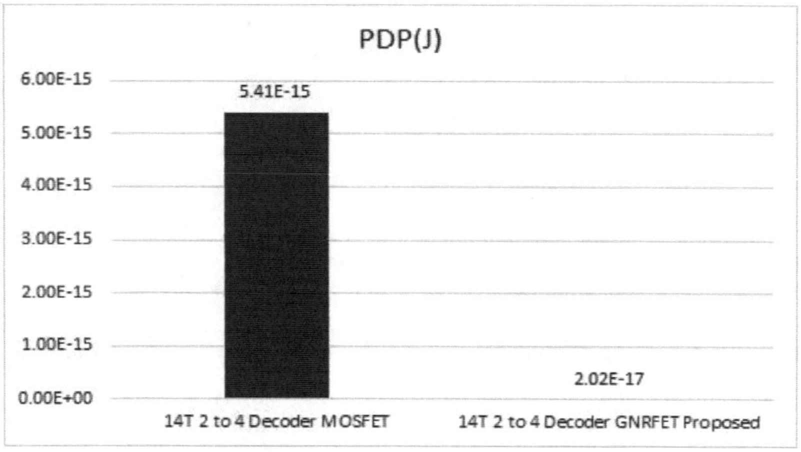

Figure 7: PDP Output Comparison for 14-T Decoder using GNRs

In Figure 8, the power dissipation for proposed decoder using GNRFET is improved with high range.

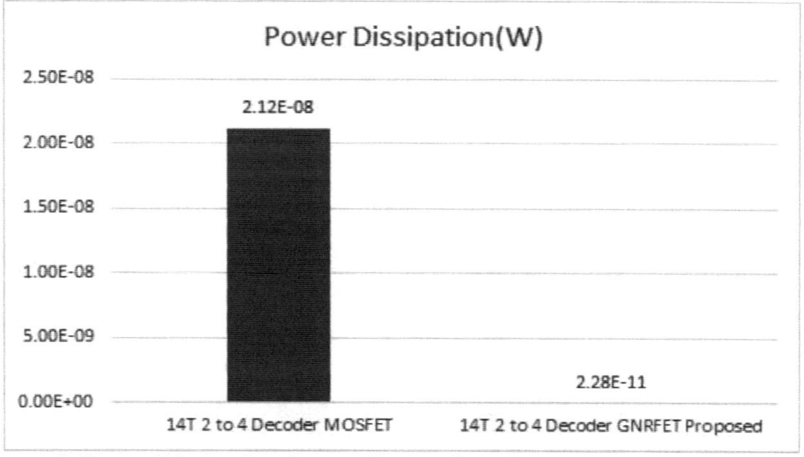

Figure 8: Power Dissipation Output Comparison for 14-T Decoder using GNRs

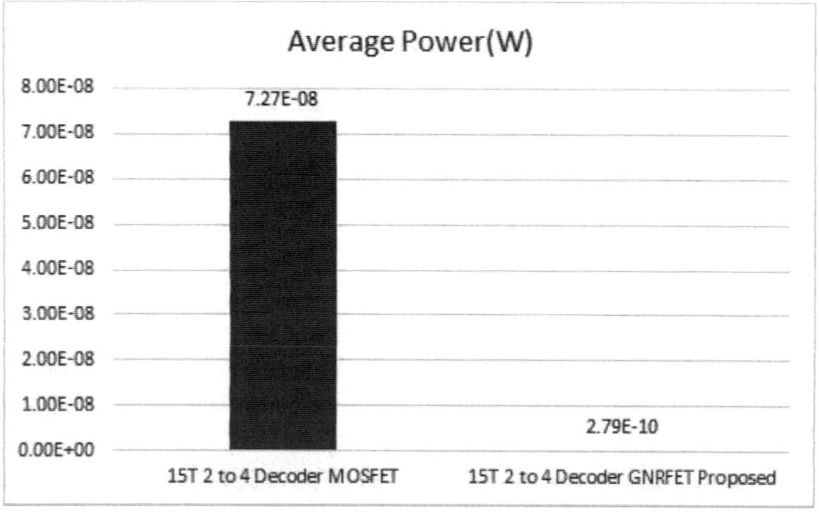

Figure 9: Average Power Output Comparison for 15-T Decoder using GNRs

Similarly, in Figure 9 even for 15-T decoder the results are improved for Average Power.

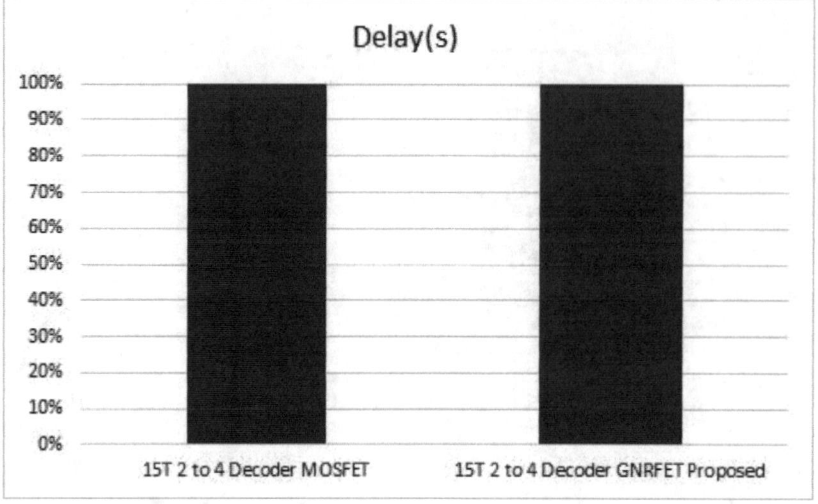

Figure 10: Delay Output Comparison for 15-T Decoder using GNRs

In Figure 10 and Figure 11, the results for 15-T delay and PDP are improved for the proposed GNRFET based 2 to 4 decoder.

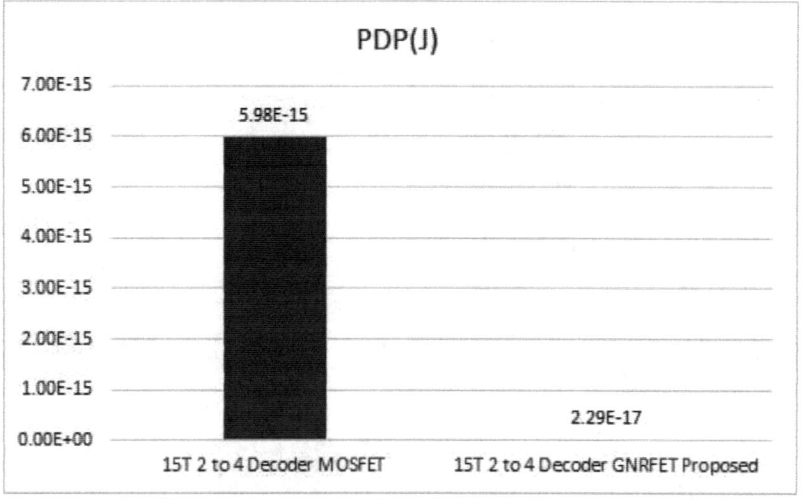

Figure 11: PDP Output Comparison for 15-T Decoder using GNRs

In Figure 12, it Is seen that the power dissipation has significantly improved with the use of GNRFETs.

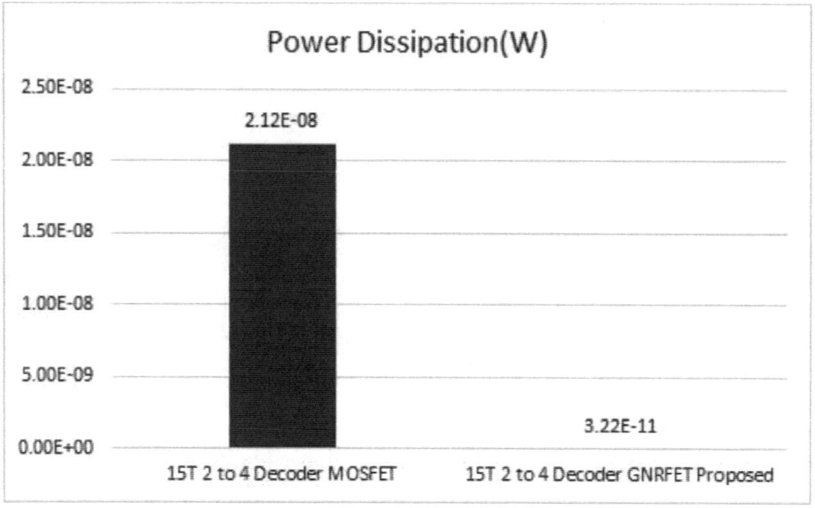

Figure 12: Power Dissipation Output Comparison for 15-T Decoder using GNRs

In Table 2, the results are shown in tabulated form. The improvements in all parameters are seen with the use of GNRFETs in the proposed work.

The proposed 15-T decoder circuit can also be explored with FinFET and compared to the existing GNR based circuit. FinFETs are vertical channel, multiple gates, low dimension, high performance FET [21-22]. FinFET has enormous potential to suppress leakage current and improve transistor drive capabilities [23-24]. FinFETs are also compatible with CMOS logic and memory design.

Table 2: Final Simulation Results

	14T 2 to 4 Decoder MOSFET	14T 2 to 4 Decoder GNRFET Proposed	15T 2 to 4 Decoder MOSFET	15T 2 to 4 Decoder GNRFET Proposed
Average Power(W)	6.58E-08	2.46E-10	7.27E-08	2.79E-10
Delay(s)	8.23E-08	8.21E-08	8.22E-08	8.21E-08
PDP(J)	5.41E-15	2.02E-17	5.98E-15	2.29E-17
Power Dissipation(W)	2.12E-08	2.28E-11	2.12E-08	3.22E-11

4. Conclusion

In this work, the objectives are fulfilled by proposing a decoder with GNRFET which has improved average power and delay. Also, by the use of GNRFETs instead of MOS the power dissipation is improved. In 22nm, the short channel effects are reduced by using MOS like GNRFETs. GNRFET ribbon structures promise better low power devices and hence a perfect substitute for the conventional MOS in lower technology length. In future work, this work can also be implemented for Multi gate field effect transistor and also the use of carbon nano tubes field effect transistor be tested. The technology nodes can be reduced below 22nm in future scope of this work. Use of Schottky barrier type GNRFET is also a possibility for improvement. The graphene structure makes it difficult in manufacturing processes.

5. References

[1] D. Balobas et. al. “Design of Low Power, High Performance 2-4 and 4-16 Mixed-Logic Line Decoders” 1549-7747 (c) pp. 1549-7747 2015 IEEE.

[2] Kazuo Taki “A Survey for Pass-Transistor Logic Technologies - Recent Researches and Developments and Future Prospects (Embedded Tutorial). ASP-DAC 1998*: 223-226*

[3] Pooja Verma et. Al. “An Optimize Approach to Design MUX Based Decoder Using Source Coupled Logic” ISSN: 1738-9968 IJHIT pp. 285-292 2016.

[4] B. Madhuri et al. “Design of Low-Power High-Performance 2-4 and 4-16 Mixed-Logic Line Decoders” IJPRES Volume 9 /Issue 5 / MAR pp-282-292 2018.

[5] D Komali et al. “Design of Low-Power High-Performance 2-4 and 4-16 Mixed Logic Line Decoders” /IJIRCCE. pp.1613-1619 2018.0602076

[6] P.Ramakrishna et al. “DESIGN OF LOW POWER HIGH PERFORMANCE 4 -16 MIXED LOGIC LINE DECODER” ISSN (PRINT): 2393-8374, (ONLINE): 2394-0697, VOLUME-5, ISSUE-4, pp. 344-349 2018

[7] SOUMEN MOHAPATRA “An Optimization of 16×16 SRAM Array for Low Power Applications” Rourkela pp.28-39 2015

[8] Y. Syamala et al. “DESIGN OF LOW POWER CMOS LOGIC CIRCUITS USING GATE DIFFUSION INPUT (GDI) TECHNIQUE” (VLSICS) Vol.4, No.5, October pp.90-95 2013

[9] G. Deepika et al. “A LOW POWER CMOS ANALOG CIRCUIT DESIGN FOR ACQUIRING MULTICHANNEL EEG SIGNALS” (VLSICS) Vol.6, No.1, February pp.25-37 2015

[10] Saikiran Sudhakar et al. “A New Approach for Low Power Decoder for Memory Array” Indian Journal of Science and Technology2 Vol 9 (29) | August pp.2-4 2016

[11] Pranay Kumar Rahi et al. “DESIGN AND SIMULATION OF 2–TO-4 DECODER USING 32nm, 45nm AND 65nm CMOS TECHNOLOGY” (IJSRET), ISSN 2278 – 0882 Volume 4, Issue 3, March pp.270-273 2015

[12] Neil H. E. Weste et al. “CMOS VLSI Design A Circuits and Systems Perspective” Addison-Wesley pp.61-95 2011

[13] M.Nirmala et al. “Design of Low Power, High Performance 2-4 and 4-16 Decoders by using GDI methodology” IJEECS ISSN 2348-117X Volume 6, Issue 12 December pp.392-397 2017

[14] Ku. Priyanka et al. “DESIGN OF LOW-POWER 2–4 MIXED-LOGIC LINE DECODERS WITH CLOCK BASED TECHNIQUE” IRJET Volume: 05 Issue: 05 | May-pp.2785-2789 2018

[15] D. Markovic et al. “A general method in synthesis of pass-transistor circuits” Microelectronics Journal 31 (2000) pp.991–998

[16] Nakka Shekhar et al. “Design of Low Power, High Performance 2-4 and 4-16 Mixed-Logic Line Decoders”ijmetmr pp.338-345 2016

[17] S. HAMEEDA NOOR et al. “LOW POWER HIGH PERFORMANCE DECODER USING SWITCH LOGIC” ijetj pp. 1-5 2018

[18] Telugu Priyanka et al. “Design of low power, high performance 2-4 and 4-16 mixed logic line decoders” IRJAET (2017) PAGES 2422 - 2426

[19] Gundaboina Shravani et al. “Design of 2-4 decoders and 4-16 decoders using GDI technique” © pp.28-33 2018, IJARIIT

[20] T. Hossain, M. S. Rahman, M. M. Rahman and A. R. A. Dibbo, "Performance Analysis of Graphene Nanoribbon Field Effect Transistor (GNRFET) based 6T and 7T SRAMs," 2020 23rd International Conference on Computer and Information Technology (ICCIT), 2020, pp. 1-5, doi: 10.1109/ICCIT51783.2020.9392713.

[21] S. L. Tripathi, R. Mishra, V. Narendra and R. A. Mishra, "High performance Bulk FinFET with bottom spacer," *2013 IEEE International Conference on Electronics, Computing and Communication Technologies*, 2013, pp. 1-5, doi: 10.1109/CONECCT.2013.6469282.

[22] S. Verma, S. L. Tripathi and M. Bassi, "Performance Analysis of FinFET device Using Qualitative Approach for Low-Power applications," *2019 Devices for Integrated Circuit (DevIC)*, 2019, pp. 84-88, doi: 10.1109/DEVIC.2019.8783754.
[23] Kumar, T.S., Tripathi, S.L., "Comprehensive Analysis of 7T SRAM Cell Architectures with 18nm FinFET for Low Power Bio-Medical Applications" Silicon (2021). https://doi.org/10.1007/s12633-021-01290-2
[24] Kumar, T.S., Tripathi, S.L. "Leakage Reduction in 18 nm FinFET based 7T SRAM Cell using Self Controllable Voltage Level Technique" Wireless Pers Commun (2020). https://doi.org/10.1007/s11277-020-07765-6.

A Brief Study on Object Detection and Tracking

Chinthakindi Kiran Kumar[1], Kirti Rawal[1,*]

[1] SEEE, Lovely Professional University, Phagwara, Punjab, 144402

*e-mail: ckkmtech11@gmail.com; kirti.20248@lpu.co.in
ORCID id: https://orcid.org/0000-0001-7758-003X, https://orcid.org/0000-0001-7724-5506

Abstract. The detection and tracking of objects in computer vision systems is a critical and challenging subject. Object identification and tracking currently has applications in a wide range of industries, including surveillance, autonomous robot navigation and vehicle navigation because to the increased accessibility of computing power and vast public datasets. When analyzing an image, the object tracking algorithm goal is to divide the region of interest, monitor its motion and location and identify any blobs. Every tracking necessitates an object detection mechanism either in each frame or whenever an object appears for the first time in the video sequence. The availability of high performance processors and low cost-high quality video cameras has enhanced the need of automated object tracking video sequences .Multiple object Identification and tracking in a dynamic environment has recently become focus of computer vision research .In this paper we discuss about the various object detection and tracking algorithms.

1. Introduction

Growing concern over safety and security is motivating an explosion of research into automated video analyzing for object Recognition and tracking has resulted from exponential raise hardware facilities such as camera processing machines and mobile phones. Object Recognition and tracking is an area of active study in computer vision that attempts to identify and monitor objects over sequences of frames and also describes the behavior of object. Today majority of the living environment including streets, parks, retail malls, metro stations, schools and homes is now monitored by technologically diverse surveillance systems .object detection and tracking is critical in the battle against terrorism, crime and public safety, as well as for effective traffic management and also used to identify the diseases accurately in medical field .Detecting object is the first stage in the video analysis process [1]. Object detection takes place in each frame or a video frame in which an object initially appears [2].In real time the images are captured under different climatic conditions may have noise that makes object detection difficult. Image processing quality has a direct impact on how well an object may be detected .A video dataset is created that takes into account many issues associated to picture quality deterioration in the imaging process, such as low image resolution, noise and blur [3].the next stage in video analysis is object classification . Object classification is a method of classifying or predicting the class of specific objects in a video frame. The objects are classified as humans, animals, birds, automobiles and moving objects [4].The final step is object tracking which refers to the association of target objects in successive video frames [5]. In other words object tracking is the task of accurately identifying the objects in a video and interpreting them as trajectories. Because of the growing technology the research took a new turn towards the automated object detection and tracking.

2. Related Works

Many authors have proposed different methods to identify and track the objects in the video frames. Few algorithms are based on image processing techniques and as the recent increase in the technology, deep learning based methods are giving promising results .In this paper we present few methods for object detection and object tracking.

2.1 Detection and tracking of Objects via Image processing techniques.

Object Detection, classification and object tracking are the three primary phases in image processing.

2.1.1 Object Detection Methods: In [6] the author has proposed background subtraction based image detection algorithm that finds the object with the dissimilarity between the current video frame and the predecessor frame in pixel terms.it can't handle a multi-model background. Each input frame has been used to iteratively update the single background model by author [7] and does not keep a background estimation buffer. Based on the temporal change of each individual pixel in the buffer that held L Video frames, the background image was created using a sliding window technique in [8] .The author [9] proposed a method for calculating the optical flow field of an image and performing a process known as clustering depending on optical flow distribution features of the image .From the above it is clear that the object detection is based on frame difference, optical flow and background subtraction.

2.1.2 Object Classification Methods: Object classification is the second phase of object tracking. In this process it identifies the objects such as humans, animals, birds, automobiles and other mobile objects in the video frame. The author [10] has proposed a motion based classification which does not require predefined patterns template but suffers in identifying the stationary objects. The motion based classification is has medium accuracy .Texture based image classification was proposed in [11] by author who calculated the gradient orientation that occurs in certain are of image, which provides better quality at the cost of longer computational time .In[12] the author has proposed shape based classification which is simple pattern matching approach which can be used with graph matching approach. X.Dong et al [13] proposed color based object detection .with this method the moving objects and its shadow can be detected in advance based on brightness and chromatic distortion in order to extract object for further use.

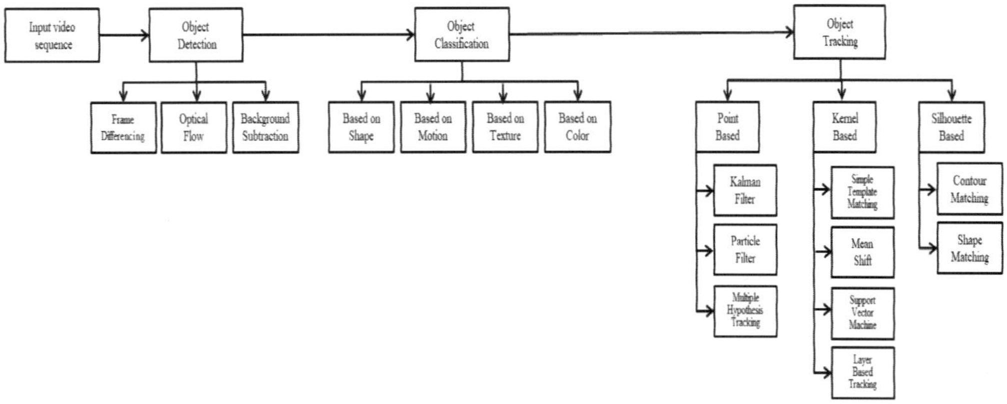

Figure 1: The general process in object detection and tracking

2.1.3 Object Tracking Methods: The final stage is tracking down the object .it is a technique for tracking moving objects in a scene .In[14] the author has given a detailed view of object tracking techniques . Point tracking, kernel tracking and silhouette tracking are the three types of object trackers. Mobile objects are represented by focal points in point tracking .Point tracking is a key issue particularly it comes to erroneous object detection and occlusion. Kernel based tracking is accomplished by calculating non stationary objects shown by elementary object region from the previous frame to next .The goal of silhouette tracking mechanism is to use an object model obtain from previous model to identify object region in each image .The detailed structure of object detection and object tracking is shown in the above figure 1.

2.2 Deep learning based Object Detection and Tracking

With the rapid development in the computer vision research, deep learning based object detection and tracking has much attentive compared to the earlier methods .Here in this we present few object detection and tracking methods.

2.2.1Deep learning based object detection: There are two different types of approaches (i) Single stage detectors (ii) two stage detectors. When compared two stage detectors are more accurate than single stage detectors, but one stage detectors are faster. Region Convolutional Neural Network (RCNN), Mask R-CNN, Fast R-CNN ,Faster R-CNN, Granulated CNN and Granulated RCNN are few among the two stage detectors .YOLO family ,SSD and M2Det[15] are the single stage detectors .various approaches of single and two stage object detection are shown in figure 2.

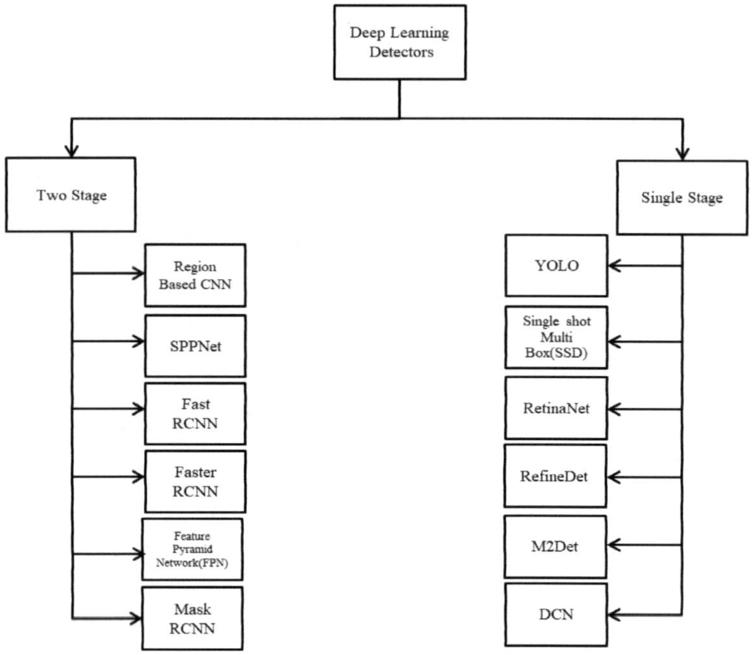

Figure 2 Deep learning single-two stage approaches for object detection

2.2.2 Deep learning based object Tracking: Deep SORT is a popular object tracking framework and one of the most widely used algorithm. Bewley et al [16] has proposed a Simple Online Real-Time Tracking (SORT). SORT tracks each detection by allocating unique id to each bounding box. When the object is lost due to occlusion, incorrect identification or other factors tracker assigns a new id and begins tracking new discovered objects. In[17] author has proposed object detection and tracking using K mean algorithm, to extract the feature points from each frame he used "good feature to track" algorithm. The Growing competitive neural network (GCNN) method developed by Lobito et al [18] was evolved from the classical competitive neural network approach and offered as a method for tracking moving objects in video surveillance. Fan et al [19] proposed method of tracking a variety of objects with focus of humans. During the offline training process convolutional neural networks are used to learn spatial and temporal features. Among the best surveys on deep learning-based object detection and object tracking are [20-21].

3. Process flow for Object Detection and tracking Using Deep Networks

In computer vision applications Object detection and object
tracking goes hand in hand .object detection is the procedure of identifying the object or locating a specific instance of interest within set of suspected frames .object tracking is the process of determining a trajectory or a path that object follows in multiple frames. The block diagram for general object detection and tracking is shown in below figure 3. To detect the objects in the frame the dataset is divided into two sections.80% of images are used for Training and 20% of images used for testing. Using deep learning techniques an image is considered to find objects within it. A bounding box is drawn around the object. The detected bounding box is sent to neural network as a reference assisting them to track. Object tracking is used to track bounding box in multiple frames.

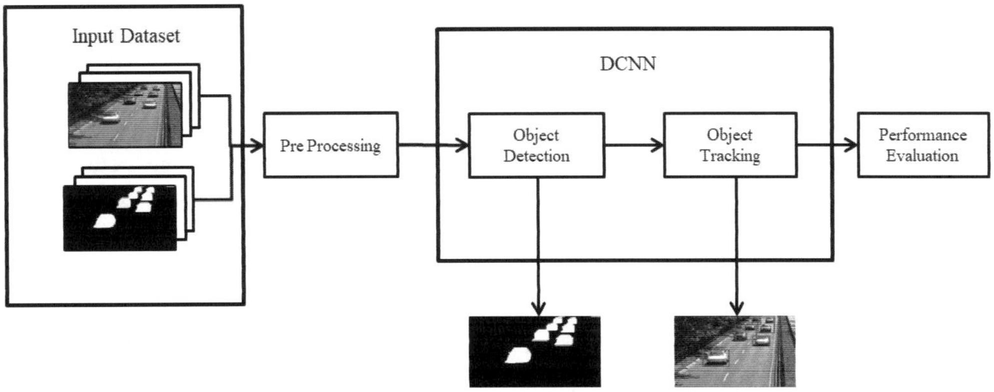

Figure 3: General block diagram of Object Detection & Tracking using Deep Networks

4. Results

In this section we summarized well know detectors and trackers and their combination which are published by different authors.Akansha Bathija et al[22] used YOLO detector and SORT tracker with 6 video sequences which gave accuracy is 85% Precision 95% and Recall 94.4% . Tianya Zhang et al [23] proposed method LiDAR detection and tracking with the overall counting accuracy is 94.74% and Precision 97.69% and Recall 70.08% in point level evaluation in range of 30-100 meters. Sankar K. Pal et al[15] conducted survey on multi object detection and tracking using MOT2015 dataset showing that in comparison to other kind of tracking assessment measures Faster R-CNN and Deep SORT combination is superior . Soumi Mitra et al[24] used KITTI 3D object detection validation set with his proposed algorithm detection and achieved sMOTA 93.28%, AMOTA 45.43%, AMOTP 77.41% for the car object and sMOTA 72.94%, AMOTA 37.95%, AMOTP 63.03.

5. Conclusion

In this paper we have included a brief analysis on image processing and deep learning based Object detection and tracking techniques. Mainly focusing on single and two stage object detectors and explained the basic object detection and tracking process. Some key observations are made considering both object detection and tracking in combination. From the results given by various authors characteristics of the objects can be more accurately extracted by using two stage detectors and Deep SORT tracking gives the promising results. Various authors used different datasets like MOT2015, KITTI 3D object detection benchmark data set, COCO dataset and INRIA Person Dataset to show the performance of the network designed and compared with state-of-the-art methods. Deep Learning based object detection and tracking is quickly thriving because of the continuous upgrading of powerful computing equipment. Object detection is prior to object tracking; as a result the accuracy of object detection over video frame is the most important factor in tracking accuracy.

Acknowledgement

I would like to express my very great appreciation to **Dr. Gaurav Sethi** (Professor, School of Electronics and Electrical Engineering, Lovely Professional University, Phagwara, Punjab, India.) for his valuable and constructive suggestions during the planning and development of this research work. His willingness to give his time so generously has been very much appreciated.

References

[1] S. R. Balaji and S. Karthikeyan, "A survey on moving object tracking using image processing," 2017 ,11th International Conference on Intelligent Systems and Control (ISCO), 2017, pp. 469-474.

[2] Ruolin Zhang, Jian Ding,"Object Tracking and Detecting Based on Adaptive Background Subtraction",Procedia Engineering,Volume 29,2012,Pages 1351-1355.

[3] L. Kong, A. Ikusan, R. Dai and J. Zhu, "Blind Image Quality Prediction for Object Detection," 2019 IEEE Conference on Multimedia Information Processing and Retrieval (MIPR), 2019, pp. 216-221

[4] Himani S. Parekh, Darshak G. Thakore,Udesang K. JaJiya, " A Survey on Object Detection and Tracking Methods",International Joulllal of Innovative Research in Computer and Communication Engineering, Vol. 2, Issue 2, February 2014.

[5] S. Ojha and S. Sakhare, "Image processing techniques for object tracking in video surveillance- A survey," 2015 International Conference on Pervasive Computing (ICPC), 2015, pp. 1-6.

[6] RupaJi S.Rakibe, Bharati D.Patil, "Background Subtraction Algorithm Based Human Motion Detection", Intelllational Journal of Scientific and Research Publications, May 2013.

[7] Rupali S.Rakibe, Bharati D.Patil, —Background Subtraction Algorithm Based Human Motion Detection International -Journal of Scientific and Research Publications, May 2013

[8] Abhishek Kumar Chauhan, Prashant Krishan, "Moving Object Tracking Using Gaussian Mixture Model And Optical Flow", International Journal of Advanced Research in Computer Science and Software Engineering, April 2013.

[9] Balaji ,S.R.;Karthikeyan,S “A Survey on moving object tracking using Image processing” 11th International Conference on Intelligent Systems and Control (ISCO),2017,469-474

[10] Hitesh A Patel, Darshak G Thakore,"Moving Object Tracking Using Kalman Filter", International Journal of Computer Science and Mobile Computing,April 2013, pg.326 – 332.

[11] Jae-Yeong Lee; Wonpil Yu, "Visual tracking by partition based histogram back projection and maximum support criteria," Robotics and Bio-mimetics (ROBIO), pp.2860,2865, Dec-2011.

[12] Salve, Suhas G.; Jondhale, Kal , “Shape matching and object recognition using shape contexts” 3rd IEEE International Conference on Computer Science and Information Technology (ICCSIT 2010) ,pp.471–474,2010.

[13] X. Dong, K. Wang and G. Jia, "Moving Object and Shadow Detection Based on RGB Color Space and Edge Ratio," 2009 2nd International Congress on Image and Signal Processing, 2009, pp. 1-5, doi: 10.1109/CISP.2009.5301770.

[14] Z. Soleimanitaleb, M. A. Keyvanrad and A. Jafari, "Object Tracking Methods:A Review," 2019 9th International Conference on Computer and Knowledge Engineering (ICCKE), 2019, pp. 282-288, doi: 10.1109/ICCKE48569.2019.8964761.

[15] Pal, S.K., Pramanik, A., Maiti, J. *et al.* Deep learning in multi-object detection and tracking: state of the art. *Appl Intell* **51,** 6400–6429 (2021). https://doi.org/10.1007/s10489-021-02293-7

[16] Kothiya, Shraddha & Mistree, Kinjal. (2015). A review on real time object tracking in video sequences. 1-4. 10.1109/EESCO.2015.7253705.

[17] A. Keivani, J. Tapamo and F. Ghayoor, "Motion-based moving object detection and tracking using automatic K-means," 2017 IEEE AFRICON, 2017, pp. 32-37,

[18] J.M.O.L. Lobato, R.M. Luque, D.L. Rodríguez, E.J. Palomo, Growing Competitive Network for Tracking Objects in Video Sequences,Springer-Verlag, Berlin, Heidelberg, 2009, pp. 109–118.

[19] J. Fan, W. Xu, Y. Wu, Y. Gong, Human tracking using convolutional neuralnetworks, IEEE Trans. Neural Netw. 21 (10) (2010) 1610–1623.

[20] Manisha Kaushal, Baljit S. Khehra, Akashdeep Sharma,"Soft Computing based object detection and tracking approaches: State-of-the-Art survey",Applied Soft Computing,Volume 70,pp.423-464,2018.

[21] S. Krebs, B. Duraisamy and F. Flohr, "A survey on leveraging deep neural networks for object tracking," 2017 IEEE 20th International Conference on Intelligent Transportation Systems (ITSC), 2017, pp. 411-418.

[22] Akansha Bathija , Prof. Grishma Sharma, 2019, Visual Object Detection and Tracking using YOLO and SORT, INTERNATIONAL JOURNAL OF ENGINEERING RESEARCH & TECHNOLOGY (IJERT) Volume 08, Issue 11 (November 2019).

[23] Zhang, Tianya and Peter J. Jin. "Roadside Lidar Vehicle Detection and Tracking Using Range And Intensity Background Subtraction." (2022).

[24] Mitra, Soumi, et al. "Towards An Optimised Vehicle Detection Algorithm For Multi-Object Tracking In Traffic Surveillance."(2020)

Sleepy Stack CSVCO with wide tuning and low power for PLL applications

Kalaka Annamma[1*], Dr. Sobhit Saxena, Dr. Govind Singh Patel[2]

[1*]School of Electronics & Electrical Engineering, Lovely Professional University, Punjab, India.

School of Electronics & Electrical Engineering, Lovely Professional University, Punjab, India.

[2] Electronics &Tele Communication Department, SITCOE, Yadrav, Kolhapur, Maharashtra.

[*]anusasmi@gmail.com[1], sobhit.23364@lpu.co.in, govindpatel1104@gmail.com[2]

ORCID ID's: 0000-0002-2101-3965, 0000-0002-6964-2058, 0000-0002-1671-8637.

Abstract. This paper describes the implementation of low power leakage technique called Sleep-Stack in a Current Starved Voltage Controlled Oscillator (VCO) a block of Phase Locked Loop (PLL). The VCO is a electronic oscillator, whose oscillation frequency is decided by the control voltage. The major issue in the VCO is the power consumption, which has an impact on the performance of the PLL. The VCO that limits the current is preferred than the other oscillator which is called as Current Starved VCO. The work i.e. the design of Current Starved VCO is implemented using low power leakage technique called Sleepy Stack. This has been implemented in 45nm CMOS Technology with a supply voltage of 1.2 V in Cadence Software. The parameters like average power, oscillation frequency, and delay are calculated for this technique. The corner analysis has been performed for all this technique to verify the stable/linearity even in the worst case scenario. It has been observed that the sleepy stack technique with gain 23.99 GHz/V, phase noise -63.8dBc/Hz, Figure of Merit value of -142.234dBc/Hz and having frequency tuning range of 77.79%.The average power obtained for the Sleepy Stack technique is 11.88nW which is less than that of the existing CSVCO.

1. Introduction

The Phase Locked Loop (PLL) is a control system used to generate the phase of output signal which is related to the phase of the input signal. The major blocks present in the PLL are Voltage controlled Oscillator and phase detector in a feedback loop. This paper depicts the one of the major blocks called VCO. The main functionality of the VCO is in the communication systems. Among the multiple types of VCOs, the ring oscillators and LC oscillators plays a major role in terms of high frequency, less noise, wide tuning range [1] and the area occupancy. The LC VCO which consists of an inductor and capacitor occupies more area. The ring oscillators consist of the delay cells which have the same functionality similar to the buffer and these cells would occupy more area. In a ring oscillator, the last delay cell output is connected to the first delay cells input [2] which is called as feedback. The delay cells can be either single ended or differential accordingly [3-4]. The increase in the number of delay cells gives rise to consumption of area [5-6]. Because of the differences in the driving capability the characteristics of the VCO become non linear which can be avoided by adding feedback circuit. By adding this feedback circuit it further leads to the increase in cost of the circuit. The current starved VCO technique was designed in order to overcome the disadvantages caused by the previous VCOs [7].This paper is organized as follows. Section 2 explains basic Current Starved VCO. Section 3 Describes Sleepy Stack leakage power reduction technique in CMOS Circuits. Section 4 Simulation results and parameters calculations. Finally Section 5 concludes this paper.

Keywords: VCO, PLL, Gain, Control Voltage, Current Starved VCO, Oscillation frequency, Tuning Range.

2. Basic Current Starved VCO

Among the different types of VCOs, ring oscillator is the efficient in terms of frequency and driving capabilities. The ring VCO has been designed with delay stage. These delay stages further when increased lead to the non linearity of the circuit. This non linearity can be overcome by increasing the number of delay cells which has increased the complexity of the circuit and also increasing the power consumption [8].

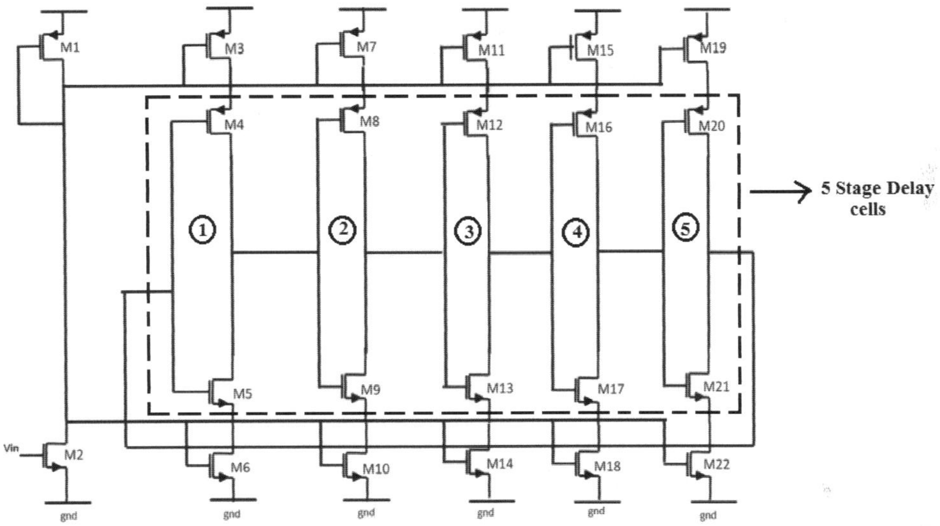

Figure 1. Current Starved VCO

In order to avoid this power consumption a VCO named current starved was designed. The output frequencies of a basic VCO are controlled by the control voltage where as the current is controlled by the current starved VCO [9-10]. The circuit consists of delay cells along with the current mirror which is used to limit the current to all the delay cells. The current starved VCO alone together has also increased the leakages and power consumption. These leakages can be further reduced by applying the low leakage techniques to the delay cells and by varying the aspect ratio of transistors which reduces the threshold voltage [11].

3. Sleepy Stack Leakage Reduction Technique implemented in VCO

The basic CSVCO as shown in figure 1 is operated with the input voltage namely control voltage (V_{ctrl}). As the V_{ctrl} decreases, the current that is to be mirrored from the transistor M1 reduces when it is passed through the different PMOS in the delay stages. Because of the active mode of transistors in the CSVCO the leakage power has been increased. In this paper, the current starved VCO has been implemented with Sleepy Stack low power leakage technique to observe the better performance of the circuit [12]. Performance analysis can be observed in terms of different parameters like frequency, delay and average power.

3.1 Current Starved Sleep stack VCO

The CSVCO circuit is implemented with Stack technique as shown in figure 2, in which both the stack and sleep techniques are implemented above and below the pull up and pull down network respectively. In normal mode, the sleep transistor would go to sleep mode, then the stack transistors reduce the leakages [13-15]. In case of the active mode the sleep transistor becomes ON and stack transistors are OFF, reducing the leakages in CSVCO circuit.

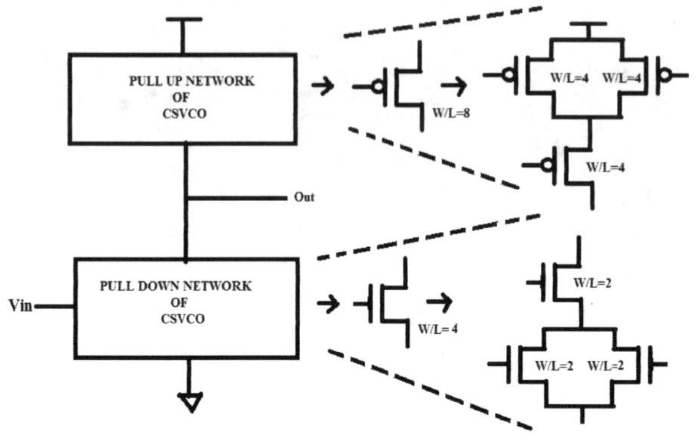

Figure2. CSVCO using Sleep Stack Technique

4. Simulation Results & Analysis

To analyze the circuit and to perform the simulation results of the CSVCO implemented with Sleepy Stack low leakage technique, the circuits has been implemented in virtuoso tool in Cadence Software in 45nm Technology with a supply voltage of 1.2 V.

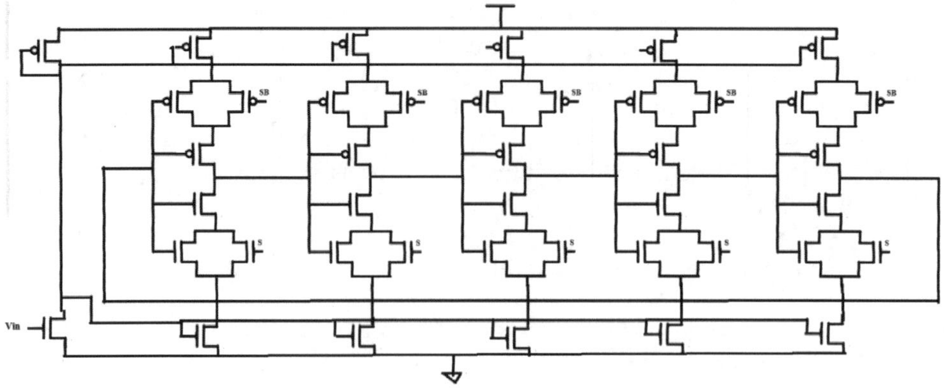

Figure3. Schematic of Sleepy Stack Technique applied to CSVCO

The simulated output of the oscillator is obtained by initializing setting the output node to either 1 or 0 to get the proper oscillations. The simulated output of a Sleepy Stack CSVCO is shown in figure 4.

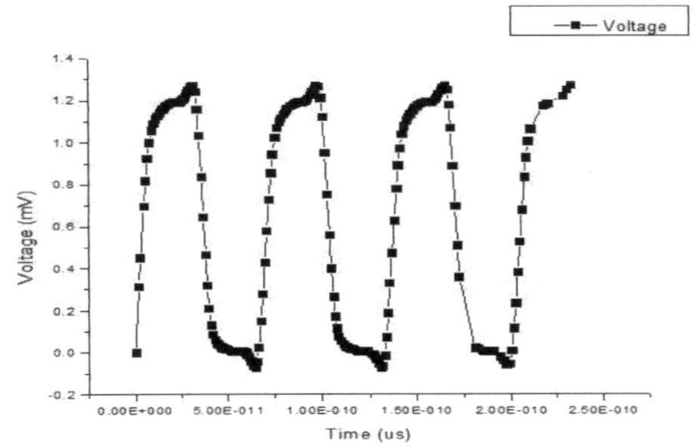

Figure4. Simulation Output of a Sleepy Stack CSVCO

The parameters like average power (P_{avg}), delay, Oscillation frequency and gain are calculated at a room temperature of 27°C and with a control voltage of 1.2 V.

4.1 Analysis of various parameters of CSVCO applied with Sleepy Stack low leakage techniques

The average power (nW), Delay (µs), Oscillation Frequency (GHz),Gain (GHz/V),Frequency Tuning Range (%),Phase noise and Figure of Merit (dBc/Hz) has been calculated for the CSVCO.

4.1.1 Gain, Average Power and Phase Noise

The ratio of the oscillation frequency (f_o) to the oscillator controlling voltage (V_c) is considered as gain and is denoted by K_{vco}[16] and is given in equation (1) by

$$K_{vco} = \frac{\Delta f_o}{\Delta V_c} \quad (1)$$

Gain obtained by the sleepy Stack VCO is 23.99 GHz/V at the supply voltage of 1.2V.The random fluctuations in the signal called noise is calculated as Phase noise in terms of dBc/Hz and at the offset frequency of 1MHz for Sleepy Stack CSVCO is -63.8dBc/Hz. The average power consumption of the Sleepy Stack CSVCO is varying with the supply voltage and control voltage and obtained value is 11.88nW.

4.1.2 Figure of Merit (FOM)

In general all the parameters performance can be calculated using the single parameter called Figure of Merit (FOM) and is given in equation (2) by

$$FOM = L(\Delta f) - 20\log\left(\frac{\Delta f}{f_o}\right) + 10\log\left(\frac{P_{avg}}{1mW}\right) \quad (2)$$

From the above given equation the figure is merit is given by the parameters like phase noise, average power, offset frequency and control voltage. The calculated figure of merit obtained by the Sleepy Stack CSVCO is -142.234dBc/Hz. The various parameter values are calculated for Sleepy Stack CSVCO and are tabulated in the Table 1.

Table 1. Parameters of CSVCO applied with Sleepy Stack Low Leakage power Technique.

S.No	Parameters	Value
1	Average power (nW)	11.88
2	Delay (μs)	0.4896
3	Oscillation Frequency (GHz)	28.79
4	Gain (GHz/V)	23.99
5	Frequency Tuning Range (%)	77.79
6	Phase noise (dBc/Hz)	-63.8
7	Figure of Merit (dBc/Hz)	-142.234

4.2 Process Corner Analysis

The process of simulating the circuit using different conditions is called corner analysis. All the proposed circuits with different low leakage techniques have their average power, oscillating frequency and tuning range calculated in process corners like nominal, slow-fast, fast-slow, fast-fast and slow-slow conditions. These conditions indicate the best and worst performance of overall circuit when both the nmos and pmos are specifically varied slowly or fast with respective to different process at different corners at 0.45 V. The parameters vary with respective to change in the input and supply voltage.

Table 2. Corner Analysis of different parameters for CSVCO applied with Sleepy Stack VCO low leakage technique.

Process corners	$P_{avg\,(nW)}$	Oscillation Frequency (GHz)	Tuning Range (%)
SS	6.204	1.178	18.03
SF	24.41	8.638	17.06
FS	33.92	2.992	4.94
FF	179.1	6.267	26.71
NN	42.71	5.889	34.91

Table 2 depicts the corner analysis and simulations of the average power, Oscillation frequency and tuning range (%) for CSVCO applied with Sleepy stack low leakage technique in 0.45 V. The performance of the current starved Sleepy Stack VCO in terms of Oscillation frequency and phase noise improves but the power consumption becomes very high with the increase in the supply voltage.

5. Conclusions

The CSVCO is applied with Sleepy stack leakage techniques in 45nm CMOS technology with a control voltage of 1.2 V. For Sleepy stack technique applied to CSVCO the 23.99 GHz/V and having frequency tuning range of 77.79%.The average power obtained for the Sleepy Stack technique is 11.88nW which is reduced by half of the existing CSVCO. It is also observed that as the temperature is increased from 27°C to 80°C then the power and delay are increased and they are decreased when the temperature is decreased to -40°C.

References

1. Prithiviraj R., Selvakumar J.(2018).A preliminary study of oscillators, phase and frequency detector, and charge pump forphase-locked loop (pll) applications. In*VLSI design circuits, systems and applications*, Springer, pp. 9–18.https//doi.org/10.1007/978-981-10-7251-2_2.
2. Prithiviraj R., Selvakumar J.(2020).Design and Analysis of Low power and High Frequency Current Starved Sleep voltage Controlled Oscillator for Phase Locked Loop Application. *In Springer Nature.*
3. Askari S., Saneei M.(2019).Design and analysis of differential ring voltage controlled oscillator for wide tuning range and low power applications. *International Journal of Circuit Theory and Applications 47(2)*:204–216.https//doi.org/10.1002/cta.2582.
4. Nitin Kumar., Manoj Kumar.(2019).Design of cmos-based low-power high-frequency differential ring VCO. *International Journal of Electronics Letters7(2)*:143-153.https://doi.org/10.1080/21681724.2018.1477181.
5. Gargouri N., Issa DB., Kachouri A., Samet M.(2015).A performance comparison of single ended and differential ring oscillator in 0.18μm cmos process. *Science Journal of Circuits Systems and Signal Processing 2(2)*:56. https://doi.org/10.11648/j.cssp.20130202.14.
6. Hwang I-C., Kim C., Kang S-M.(2004).A CMOS self-regulating VCO with low supply sensitivity. *IEEE Journal of Solid-State Circuits 39(1)*:42–48. https://doi.org/10.1109/JSSC.2003.820881.
7. Amin M., Leung B.(2015).Design techniques for linearity in time based ΣΔ Analog-to-Digital converter. *IEEE Transactions on Circuits and Systems II: Express Briefs 63(5)*:433–437. https://doi.org/10.1109/TCSII.2015.2506004.
8. Kabirpour S., Jalali M.(2019).A highly linear Current-Starved VCO based on a linearized current control mechanism. *Integration 69*:1–9.https://doi.org/10.1016/j.vlsi.2019.06.008.
9. G. K. Sharma., A. K. Johar., T. B. Kumar., D. Boolchandani.(2020). Design and analysis of wide tuning range differential ring oscillator (WTR-DRO). *Analog Integrated Circuits and Signal Processing103,* pages17–29.
10. Jovanovic G., Stojcev M., Stamenkovic Z.(2010).A CMOS voltage controlled ring oscillator with improved frequency stability. *Scientific Publications of the State University of Novi Pazar, Series A: Applied mathematics, Informatics and Mechanics 2(1)*:1–9.
11. Gargouri N., Issa DB., Kachouri A., Samet M.(2015).A performance comparison of single ended and differential ring oscillator in 0.18 μm cmos process. *International Journal of Scientific Research and Engineering Technology 3(2)*:123–128 .
12. Jang S-L., Hu C-T., Chuang Y-H.(2005).A new current source temperature compensation circuit for ring VCO. *In: IEEE international symposium on communications (ISCOM2005).*
13. M. D. Powell., S. H. Yang., B. Falsafi., K. Roy., and T. N. Vijaykumar (2000).Gated-Vdd: A circuit technique to reduce leakage in deep submicron cache memories. *In Proceedings IEEE ISLPED*, pp. 90-95.
14. Pal PK., Rathore RS., Rana AK., Saini G.(2010). New low power techniques: Leakage feedback with stack & sleep stack with the keeper. *International Conference on Computer and Communication Technology, ICCCT,* 296–301.https://doi.org.10.1109/ICCCT.2010.5640514.
15. Jun Cheol Park.(2005).Sleepy stack: a new approach to low power VLSI logic and memory. PhD. Dissertation, School of electrical and computer engineering , Georgia institute of technology .
16. Choudhury BS., Maity S.(2015).A low phase noise CMOS ring VCO for short range device application. *In: Conf. proceedings, electrical electronics signals communication and optimization, EESCO-IEEE.* https://doi.org/10.1109/EESCO.2015.7253649.

Journal of Physics: Conference Series **2327** (2022) 012014 doi:10.1088/1742-6596/2327/1/012014

Influence of fabrication parameters and optimization of supercapacitor – *A simulation analysis*

Haziqul Yaquin[1] and Dr G Raam Dheep[2*]

[1,2] Department of Electronics and Communication Engineering
School of Electronics and Electrical Engineering
Lovely Professional University
Phagwara-144001, Punjab, India

Email: hyaquin@gmial.com *raam25227@lpu.co.in

ORCID Id: https://orcid.org/0000-0003-1734-235X *https://orcid.org/0000-0001-8470-2437

Abstract. To develop comparative understanding on recent model of supercapacitor for various parameter values. The different parameters addressed here are performance variables of supercapacitor. Greenhouse gas emission mitigation, waste minimization and green technologies are need of hour. Conclusively environment protection is essential now for sustainability of life. This paper focuses on how effectively simulation software called MATLAB-Simulink be used to develop optimized model of supercapacitor before fabricating the real device. *Keywords— Supercapacitor, Modeling, Simulation, MATALB-Simulink.*

1. Introduction

Energy insecurity is major challenge to earth now as the fossil fuels like coal, petroleum and natural gas are depleting exponentially in accordance with the global demands of comfort from every sector of societies, affecting the environment and the overall health of the earth. The demands are not limited to power and transport but are now extended to comfort and luxury, even at higher costs. Metallic analog wrist watches has been converted to smart watches, telephones to smart phones, PCs to laptops and tabs, fully automatic washing machines, CFLs to LEDs, CRT TVs to LCDs and LED TVs, automatic other home appliances, conventional kick motor cycles to self-start motorcycles, remote controlled door locking systems, overhead water tank alarm system, remote controlled home appliances, smarter ACs, lightings, air filters, heaters, kitchen appliances, fire alarms, manual cars to semi and fully automatic cars, automatic railway crossing gate, underground cable fault locator systems, steam engine trains to diesel and electric engine trains, vehicles to electric vehicles, robots, medical instruments, communication systems, banking, military, etc. everywhere from common house hold to manufacturers and service providers, the demands for high power and high energy is increasing exponentially. Some of them seek high energy means long duration energy backup like laptops, mobile phones, smart watches, other electronic gadgets, and appliances, and thus are running conveniently on batteries. Some seek high power for short periods like electric fans, medical devices, some house-hold appliances, radio and telecommunication devices, protective circuits, and thus are using capacitors instead of batteries. The race towards comfort and luxury along with the basic needs have pushed the researchers and scientists to spend their brain more towards the fulfilment of these demands as much

Journal of Physics: Conference Series 2327 (2022) 012014 doi:10.1088/1742-6596/2327/1/012014

as possible in a such a way that it not only meets the demand but also become sustainable for the coming human races as our previous mankind did for us. There are a lot of such things more than what is mentioned here in terms of high energy and power demands from different sectors. The issue for our generation is not only to meet these demands but also to meet them in such a way that it becomes available and sustainable for races to come till they find their own resources and technologies to sustain life on earth and this is how the mankind has done so far. The challenge our generation is facing can be numerated as below:

1.1. Sustainable substitute of fossils fuels: energy insecurity world-wide due to depleting fossil fuels, is the biggest challenge all are trying to minimize as much as possible. The best options available for now as replacement for conventional sources (coal, petroleum, and natural gas) are renewable sources like solar, wind, hydro, geothermal, tidal and biomass. Since these fossil fuels are depleting so their replacement is a must.
1.2. Earth health and global environment protection: coal, petroleum and natural gas along-with other combustible fuels as source for energy production, vehicles and industries, have incurred a lot of environmental issues like global warning, green-house gasses emission, air pollution, water and soil pollution, ozone depletion, biosphere losses, deforestations, draught, floods, ice melting, increasing sea levels, extinction of species and birds, affected animal and plant health, new diseases, hydrosphere impacts, and as a result the overall earth health has been deteriorated which needs to be addressed seriously and for now again the best way available is the replacement of conventional energy sources by non-conventional i.e. renewable energy sources.
1.3. High energy – high power devices: majority of demands mentioned here now require not only energy for long duration like batteries but with high power also like capacitor. Electric vehicles are running on batteries but need to be of more power without compromising their energy densities. A battery cannot take the load of heavy vehicle and a capacitor cannot run an electric car for long distance as it gets discharged very fast. So, a combination of both batteries and capacitor that can minimize this gap between the two utilizing renewable sources as much to have global energy security as well as protected sustainable environment and earth health, is the recent thrust of research in the field of energy storage and power.

Some of the devices and machines and appliances have already started using both batteries and capacitors in combination for long duration high power demand meet like electric trucks but are not much efficient with respect to performance, cost, space, complexity. Scientists have developed a device called supercapacitor or ultracapacitor which have high power densities and acceptable energy densities though research are going on to optimize them in terms of high power as well as energy densities as much as possible keeping environmental impacts as low as possible by anyway. [6]

Table 1. Some of the Supercapacitors developed and tested till now.

Company	Model	Capacitance (F)	Cell Voltage (V)	Voltage (V)	ESR (m ohms)	Weight (Kg)	Reference
Econd	Pscap350	350	--	14	--	24	[02]
Econd	--	300	--	12	2	--	[03]
Nippon-Chemi-Con	ESSP08	2700	--	2.7	--	--	[04]
NessCap	ESSP18	5000	--	2.7	--	--	[04]
Maxwell	ESSP16	2600	--	2.7	--	--	[04]
EPCOS	**ESSP48**	**33**	**--**	**42**	**--**	**--**	**[04]**
Maxwell	BMOD0140	140	2.7	48.6	8	--	[05]

Table 1. It shows the list of supercapacitors developed and tested so far and are relevant here. The list contains the name of manufacturer, model name, its capacitance value in Farad (F), cell voltage in Volts, terminal rated voltage in Volts, Equivalent series resistance (ESR) in milli-ohms, its weight in Kg and reference.

Supercapacitor: A supercapacitor is a capacitor with much high capacitance and power density as compared to its conventional counterpart and as high energy density as possible to bridge the gap between a capacitor and battery. It is under the developmental stage and have achieved some remarkable milestone in two decades and have shown great potential in energy sector. [1]

Out of these devices developed so far, simulation for some of them have also been performed for various internal and external variables in leu of a better visualization of their effects on the performance of the device. Simulation is nothing but modelling of a device in controlled and desirable way to get or guess optimized output with respect to prefilled inputs that are the internal parameters responsible for the variation in the performance of the device. This is done using a number software like OrCAD, PSCAD, MATLAB-Simulink, etc. [7]

This paper is part of the ongoing project on high energy density storage device for non-conventional energy applications.[6] A review paper related to the same has already been developed and this paper includes the forecasting section of the research i.e. simulation of the device to understand the effects of various internal and external parameters on the performance of the device i.e. supercapacitor. Those parameters are Electrode material, electrolyte material, activation agent, Porosity, surface area, ESR, R_{ct}, R_{dis}, Warburg resistance Z_w, C_p, R_p, Temperature, cell voltage, terminal voltage, capacitance, charging current, discharging current and performance parameters like energy density, power density, charging time t_c, discharging time t_d, efficiency, life cycle, retention, cyclability etc. Not all but some relevant parameters have been adjusted as in previous references. MATLAB-Simulink has been used for its simulation. Initially a pre-tested supercapacitor has been simulated here for different parameters values and the results have been tabulated for reference as a relationship with the working of the device. Some of the parameters are known from the data sheet of the capacitor in test and some are obtained by screening test and rest a few are adjusted on the MATLAB-Simulink environment to optimize the model under observation. From a number of models of supercapacitor available, the best updated model till date has been modelled here in the paper alongwith the model parameters to get best results.

The parameters available with the device are: capacitance, voltage, charging and discharging current ranges.

To optimize the performance, the device must charge itself as fast as possible and discharge itself as slow as possible. Thus, maximum terminal voltage should reach soon after providing the charging current of known range and should delay in discharging when load is connected. The value of total capacitance (C') and total terminal voltage (V) of the device are known by the data sheet of the device provided by the manufacturer. The parameters obtained after testing that can be verified mathematically are shunt capacitance value (C_p), main capacitance value (C), initial voltage (u), charging current value (I) and load value (L). The parameters that can be adjusted during simulation to optimize the results, are chosen from the range of those parameters obtained from different reference papers and are shunt capacitance (C_p), Equivalent series resistance ($ESR=R_1$), self-discharge resistance ($R_{dis}=R_2$), over-voltage protection resistance (R_3), charging (t_c) and discharging time (t_d), simulation time (t_s). Also point to be noted that in this modelling paper of supercapacitor, different combination of values of variables from the ranges of parameters are tried and tested to obtain the best model till date, to obtain best optimized model of the device hypothetically to get a glimpse of effects of these variables on the performance of our device and to help in the making of the device as whole.

2. Modeling and Simulation

The supercapacitor to be modelled is ESSP48 33F 42V with 18 cells in series manufactured by EPCOS [04]

The best equivalent circuit available till now is series parallel RC circuit with nonlinear variables as shown below [04]. This model can be seen from the figure 1.

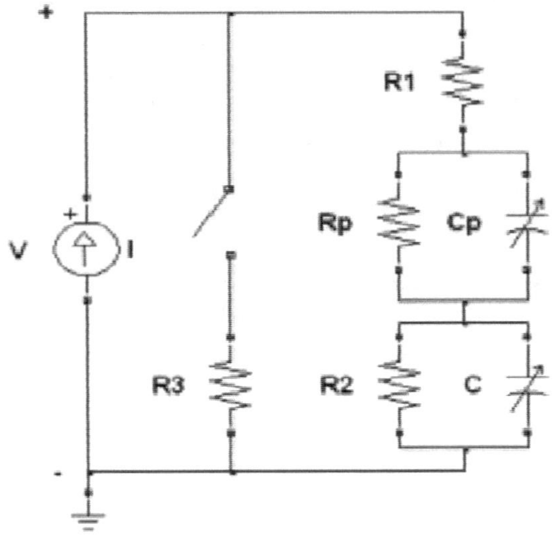

Figure 1. It shows the best possible equivalent model of a supercapacitor developed so far containing V as supply voltage, R_3 as over charging protection resistance, R_1 (=ESR) as equivalent series resistance, C as main capacitance, R_2 as self - discharge resistance and also includes the non-linear or dynamic parameters as shunt resistance R_p and shunt capacitance C_p.

Figure 1. Equivalent series-parallel RC model with nonlinear parameters

Table 2. Basic parameters available from manufacturer and screening of the device

Basic parameter	Definition	Type	Effect	Range available **[references]**
Total Capacitance (C')	As per the manufacturer	Fixed for a given capacitor	Needs to be high	**35F to 3500F**
Main capacitance (C)	Effective value after initial screening	Negligibly varies about C'	To be high	**30F to 3500F** [10] [13] [16] [17] [18]
Initial cell voltage (u)	Offset/initial voltage	Fixed	Limited	**0.01 to 3.2V** [9] [12] [14] [19] [20] [22] [23] [24]
Current (I)	For charging the capacitor	As per requirement	To be low	**5 to 20A**
Charging time (t_c)	Time for charging	As per need	To be low	**5 to 20 Min**
Discharging time (t_d)	Time for full discharging	As per need	To be high	**5 to 20 Min**

Table 2. It shows parameters that are available with device chosen with range available for experiments. Gives definition, type and effect of those parameters for device optimization.

Table 3. Modeling parameters with different available ranges from previous researches

Modeling parameter	Definition	Type	Effect	Range available **[references]**
Equivalent series resistance (ESR or R_1)	Resistance of whole internal circuitry in series (solution resistance R_s, Charge transfer resistance R_{ct}, electrolyte resistance R_e)	Variable for a given device	Reduces the performance and needs to be minimum but not zero i.e needs to be optimized	**0 to 10 ohms** [11] [15] [18] [21]
Self-discharge resistance (R_{dis} or R_2)	Resistance against self-discharge	Variable	Needs to be high but not much high i.e needs to be optimized	**100 to 18000 ohms**
Over protection resistance (R_3)	Resistance path for over voltage protection	As per need	Needs to be optimized	**0.05 to 50 ohms**
Shunt/parallel Capacitance (C_p)	Nonlinear capacitance of the whole device	Variable, depends on main capacitance C i.e (C_p=C/13)	Adds nonlinearity to the performance, needs to be optimized	**0 to 25 F**
Parallel/shunt resistance R_p	Resistance of Nonlinear capacitance circuit	Variable	Needs to be optimized	**2 to 10 m ohms**

Table 3. It shows list of parameters whose values can be implemented on the device model from the various ranges available, so are called modelling parameters. The table gives their definition, type and effect to optimize the device model as a whole with the help of experiments on Simulink model.

It contains three parts of the circuit as shown in figure 2: main circuit, non-linear circuit and protectoin circuit as given in below figures. Values of parameters used from the above available ranges are listed below for the device ESSP48 capacitor. Here also as in previous cases, a supply current of 10 A is applied untill the capacior gets charged fully and then left for 10 minutes to check the self discharge properties and then finally discharged forcefully with a negative current of -20 A working as a load, to check the compete behaviuor of the device as a whole. The parameters values taken and the Simulink model developed and also the results of charging and discharging of the device follows here. The charging and discharging of the device needs to be optimized to have the best performance output. To model the device, the following parameter values have been implemented in figure 2 and ultimately to figure 3 and figure 4:

R_1 (ESR) = 5 m ohms, R_2 (self discharge resistance) = 200 ohms, R_3 (protection res) = 10 ohms
u (initial voltage) = 0.1V, I (input supply current) = 10A, t_c (charging time) = 7 minutes
Self discharging time = Left over time for zero current supply = 10 minutes
t_d (discharging time) = 3.38 minutes, I (Discharging load = High negative input current) = -20A
C (main capacitance) = 33F, C_p (Shunt capacitance) = 33/13= 2.5F, R_p (Shunt res) = 5 m ohms
Rated Terminal voltage, V = 42V

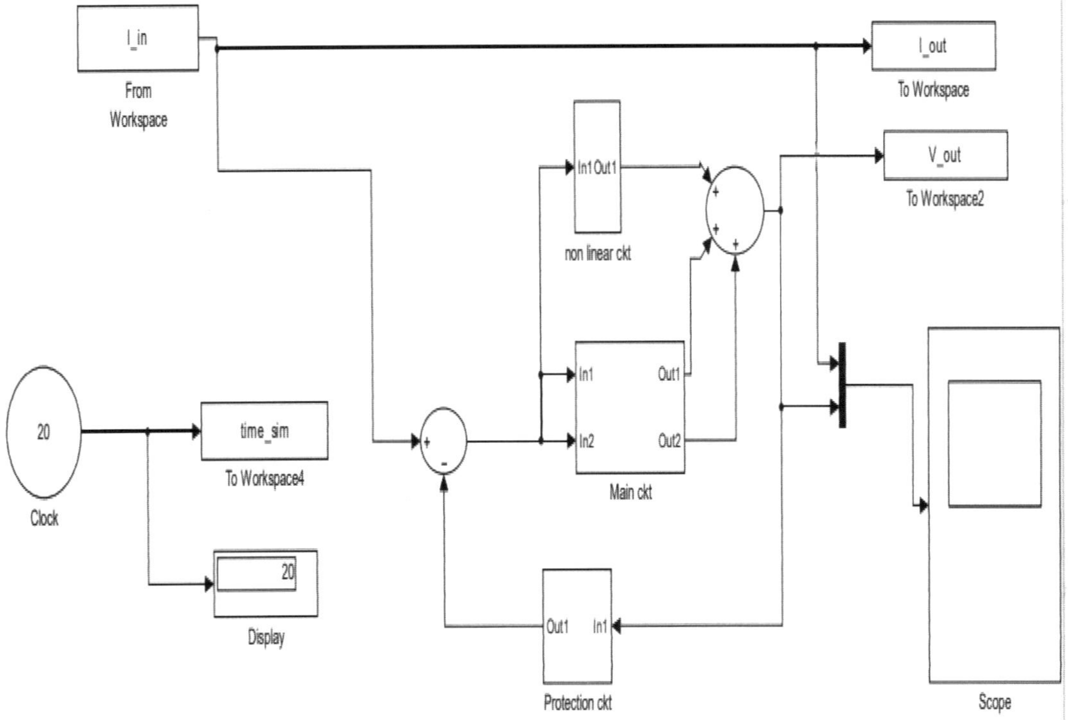

Figure 2. Matlab-Simulink full Model developed on Matlab2015

Figure 2. It shows the full model developed on Matlab-Simulink consisting of Non-linear (upper), main (middle) and protection circuit (bottom), showing input and outputs and simulation time of 20 minutes (clock-display unit left side). The upper nonlinear part consist of nonlinear parameters and lower protection part contains protection resistance, the middle main circuit contains all the main components and processes and the detailed-zoom in picture of them are given in next figures.

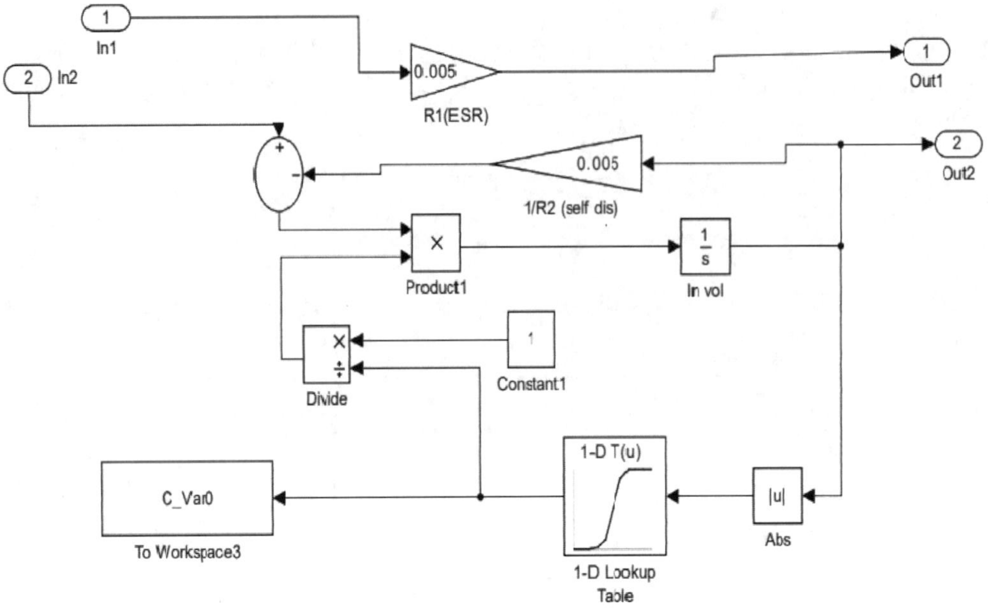

Figure 3. Main circuit of full model shown in figure 2.

Figure 3. It shows the detailed version of main circuit of full model and has 2 inputs and 2 outputs as shown in figure 2 also. Some parameters values implemented here are R_1=5 m ohms, R_2=200 ohms, C=33 F and initial voltage u=0.1 V.

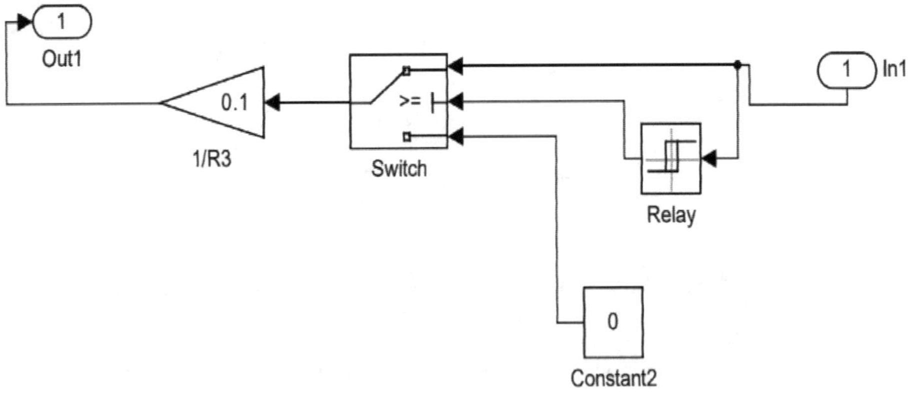

Figure 4. Protection circuit model of full model shown in figure 2.

Figure 4. It shows the detailed zoom in version of protection circuit of full model shown in figure 2, making use of relay-switch and protection resistance R_3 = 10 ohms as shown, to safeguard the device from overcharging.

Table 4. Comparison of effects of values used w.r.t previous values of simulation parameters.

Parameter	Range available	Value-used previously	This paper	Remarks
Main capacitance (C)	30 to 3500 F	33 F	**33 F**	Same capacitor
Shunt capacitance (C_p)	0 to 25 F	2.5 F	**2.5 F**	As per C/13
Shunt resistance (R_p)	2 to 10 m ohms	3 m ohm	**5 m ohm**	No accountable effect
Equivalent series resistance (ESR or R_1)	0 to 1 ohm	6 m ohm	**5 m ohm**	Negligible change but dominant parameter.
Self-discharge resistance (R_2 or R_{dis})	100 to 18000 ohms	18000 ohms	**200 ohms**	Negligible change
Initial voltage (u)	0.01 to 3.2 V	2.19 V	**0.1 V**	No effect
Protection resistance (R_3)	0.05 to 50 ohms	52 m ohm	**10 ohms**	Protection

Table 4. It gives a comparative list of all the parameters that have been modelled as per the previous latest research and the values taken in this paper for the same device and as a result come to the conclusion of effects of all those chosen parameters on the performance of the device.

All the parameters values that have been finally chosen from the available ranges and implemented in full model in figure 2 after number of simulation experiments have been tabulated here in table 4.

3. Results and Discussion

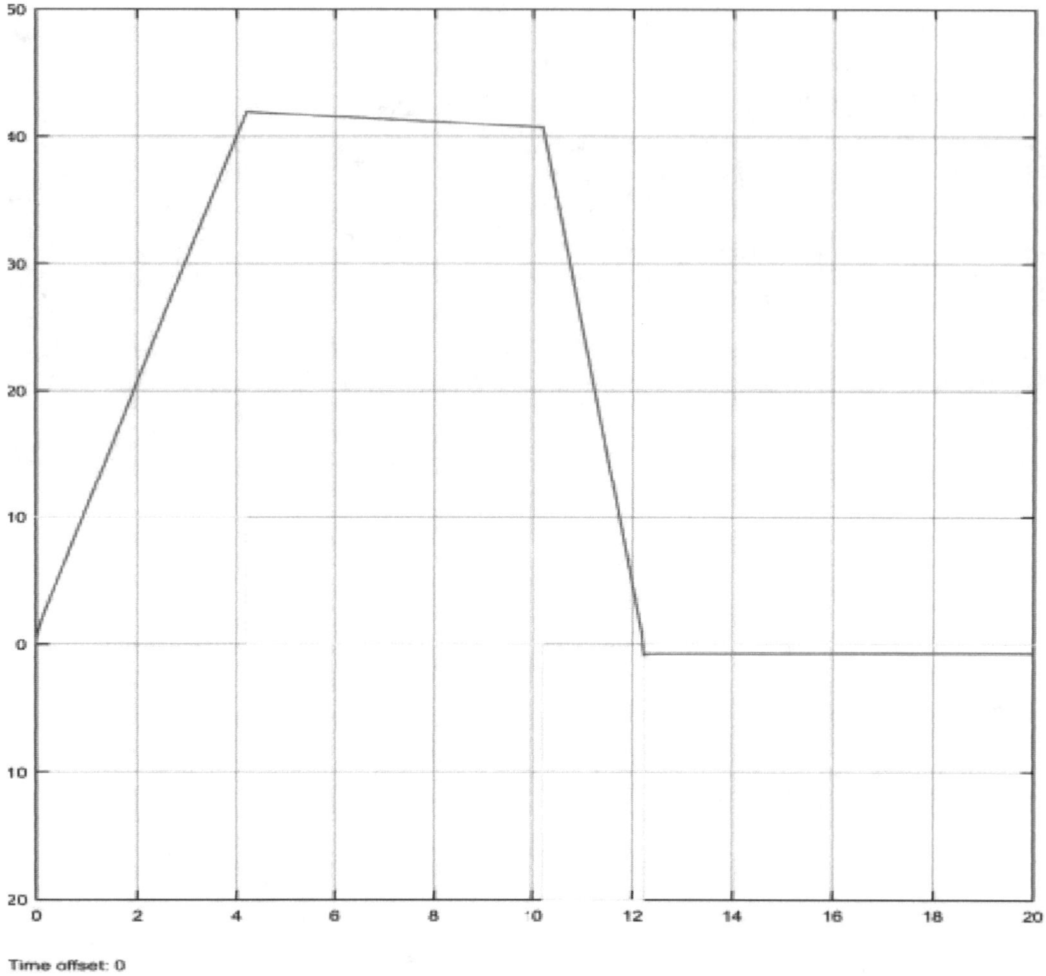

Figure 5. Current Vs Voltage curve wrt time as final output of simulation

Figure 5. It gives the output curve of charging and discharging of supercapacitor upto its full capacity of 42 volts by a charging current of 10 A for 7 minutes and discharged by the discharging current of -20 A for 3.38 minutes. In the mean while it was left with no supply for about 10 minutes to see its self-discharging behaviour which is negligible for shorter period. (X-axis is time in seconds x 100 ; Y-axis represents both current and voltage in Amperes and Voltage respectively)

As shown in the figure 5, the results of initial modeling of the above equivalent circuit is similar to what has been obtained on Orcad-Pspice environment as well as MATLAB-Simulink platform with different values of parameters which explains that the Matlab-Simulink model here is in synch with the model in reference paper [04], hence the result.

Here after so many iterations and observations it is observed that the most dominant parameter is ESR that is the equivalent series resistance. The effect of R_p, C_p and u are not much on the charging of the device. So ESR needs to be optimized. The effect of R_2 is such that the self-discharge slightly depends on R_{dis} and mostly on material of the device electrodes.

4. Conclusion

Simulation of a device can provide optimization route before its actual fabrication. Simulation also provide option to test for different independent variables that may affect the performance of the device and accordingly fabrication can be enhanced. Here also it is observed that ESR which ultimately depends on the material of electrodes and electrolyte play a vital role in the performance of a supercapacitor, thus giving an option to explore similar materials for the objective. A very low or very high ESR will be a drawback thus needs to be optimized. It is observed that self-discharge for a short period of 10 minutes is negligible but if the capacitor is left for a longer period, then it might get discharged fully. Hence it is observed that to have best results the material and methodologies of fabrication needs to be optimized. Since in previous sections of this work carbonaceous materials have been chosen, so the best carbonaceous materials need to be searched and worked upon to optimize the ESR as much as possible and the performance of the whole device.[8] Simulation opens window for unlimited number of iterations and experiments on a model to come a conclusion like here the case with ESR and saves time as well as materials, so that with minimum material samples and less time more jobs can be done in this area using the simulation methodology. Also, initial voltage doesn't increase the overall maximum voltage of the device. Best material is needed to optimize the ESR and thus the performance of the device as whole.

Here it is concluded that Equivalent series resistance is most dominant and needs to be optimized. This will help optimize the fabrication processes and avoid loss of time in laboratories searching for materials and methods to develop the device. The simulation techniques using any software gives similar results and help utilize time wisely. In the end a simulation guided laboratory fabrication will ultimately help in obtaining best possible method and material to optimize the ESR thus increasing the energy and power densities of the device resulting in the best performance possible which is ultimate goal of this research. Also, temperature parameter can be tested in simulation model before going for real time laboratory experiments to come to a theory and thus a device thus other parameters like temperature and pressure can be studied and a collective optimized analysis can be done based on all different values of explored parameters and adding those which have not been explored till now for a better anticipation of the device to be fabricated.

References

[1] Lokhande P E, Chavan U S and Pandey A 2019 Materials and fabrication methods for electrochemical supercapacitors : Overview *Electrochemical energy reviews (Shanghai university and periodicals agency of Shanghai university)*

[2] Farcas C, Petreus D, Ciocan I and Palaghita N 2009 Modelling and simulation of supercapacitors *Proc. 15th int. symp. for design and tech. of electronics packages (Gyula: IEEE)* pp 195-200

[3] Singh T P and Kumar S Y 2017 Modeling and simulation of electrochemical supercapacitor for high power delivery *Int. J. Adv. Res. 5(6)* pp 509-516

[4] Johansson P and Andersson B 2008 Comparison of Simulation Programs for Supercapacitor Modelling: Model Creation and Verification *Chalmers (Sweden: Master of Science Thesis)*

[5] Cultura A B II and Salameh Z M 2015 Modeling, Evaluation and Simulation of a Supercapacitor Module for Energy Storage Application *Proc. Int. conf. on computer info. Sys. and industrial app.* pp 876-882

[6] Yaquin H and Dheep G R 2020 Review on materials and methods for supercapacitors *Proc. 3rd int. conf. of intelligent sustainable sys. (Palladam: IEEE)* pp 1049-1055

[7] Bharti, Ashwani K, Gulzar A, Meenal G, Patrizia B, Ravikant A, Ramesh C and Yogesh K 2021 Theories and models of supercapacitors with recent advancements: impact and interpretations *Nano Express 2*

[8] Manavalan V, Ammaiyappan B S, Duggirala S R, Tata N R and Mani K 2019 Conversion of biomass waste into high performance supercapacitor electrodes for real-time supercapacitor applications *ACS Sustainable chemistry and engineering*

[9] Ezhumalai D, Arul P, Vaithiyanathan S D, Mohamed N M A and Perumal E 2020 Sustainably-derived hierarchical porous carbon from spent honeycomb for high-performance lithium-ion battery and ultracapacitors (Energy storage: Wiliey)

[10] Yajun J, Yalei D, Fei C, Zhaoqi W, Yezhan L and Zhenhao G 2020 Ultrathin Co_3O_4 nanosheets anchored on multi-heteroatom doped porous carbon derived from biowaste for high performance solid-state supercapacitor *Carbon vol 156* pp 359-369

[11] Ji Y, Jing S , Lei L , Xiao-Kai M , Jun-Hao C , Li-Zhen W and Yong Z 2020 Template-like N, S and O tri-doping activated carbon derived from helianthus pallet as high-performance material for supercapacitors *Diamond and related materials 102*

[12] Fang Y, Zihan Y, Wanru C, Qianya W, Hui W, Honglei Z and Chuang P 2019 Plane tree bark-derived mesopore-dominant hierarchical carbon for high-voltage supercapacitors *Applied surface science*

[13] Pratheepa M I and Lawrence M 2020 Eco-friendly approach in supercapacitor application: CuZnCdO nanosphere decorated in reduced graphene oxide nanosheets *SN Applied sciences 2*

[14] Subramanian N, Subramani K, Yun-Sung L, Marappan S and Vanchiappan A 2020 $Li_{0.32}Al_{0.68}MnO_2(OH)_2$ from spent Li-ion battery to build high performance supercapacitor: Waste to energy storage approach *Journal of Alloys and compounds*

[15] Yanliang W, Krzysztof K, Jiakang M, Xuecheng C, Jiang G, Ran N, Xin W, Jalal A, Ewa M and Tao T 2019 Porous carbon nanosheet with high surface area derived from waste poly(ethylene terephthalate) for supercapacitor applications *J. Appl. Polym. Sci (Wiley)*

[16] Samuel D and Thomas N 2020 Suppressed self-discharge of an aqueous supercapacitor using Earth-abundant materials *Journal of Electroanalytical Chemistry*

[17] Abdallah R, Anas M, Shaker E, Moataz S and Abou-Aly A I 2020 Polyaniline/fullerene derivative nanocomposite for highly efficient supercapacitor electrode *Int. J. of Hydrogen energy*

[18] Ajay K M and Dinesh M N 2018 Influence of various Activated Carbon based Electrode Materials in the Performance of Super Capacitor *IOP Conf. Series: Materials Science and Engineering*

[19] Natalia M, Sudhakar Y N and Selvakumar M 2013 Activated carbon derived from natural sources and electrochemical capacitance of double layer capacitor *Indian j of chem. tech. 20* pp 392-399

[20] Encarnacion R-P, Martin C and Francois B 2009 Tuning Carbon Materials for Supercapacitors by Direct Pyrolysis of Seaweeds *Adv. Funct. Mater.* pp 1032-1039

[21] Farma R, Deraman M, Awitdrus A, Talib I A, Taer E, Basri N H, Manjunatha J G, Ishak M M, Dollah B N M and Hashmi S A 2013 Preparation of highly porous binderless activated carbon electrodes from fibres of oil palm empty fruit bunches for application in supercapacitors *Bioresource technology 132* pp 254-261

[22] Seiji K, Masashi S and Daisuke T 2013 Electrical double-layer capacitance of micro- and mesoporous activated carbon prepared from rice husk and beet sugar *Electrochimica acta 114* pp 617-626

[23] Dhelipan M, Arunchander A, Sahu A K and Kalpana D 2016 Activated carbon from orange peels as supercapacitor electrode and catalyst support for oxygen reduction reaction in proton exchange membrane fuel cell *J. of Saudi Chemical Society*

[24] Liu J, Wang Z A, Wu X W, Yuan X H, Hu J P, Zhou Q M, Liu Z H and Wu Y P 2015 Porous carbon derived from disposable shaddock peel as an excellent catalyst toward VO^{2+}/VO_2^+ couple for vanadium redox battery *J. of power sources 299* pp 301-308

Power Efficient Technique for CMOS- Logic Circuits

Manoj Singh Adhikari, Shaik Sahil, Palatla Babasri Tarun Naidu, Aseervadam Thushar Roy Mandapati devendra, and Uppala Goutham

School of Electrical and Electronics Engineering, Lovely Professional University, Punjab, India

pbstarunnaidu@gmail.com

Abstract: Power outages on normal CMOS circuit is very high it can be reduced by using the adiabatic technique. Adiabatic technique is used in the pull up part of CMOS logic. Power loss of CMOS Logic is in terms of heat. Adiabatic is an efficient technique where some of the energy stored in the load and it is used for next inputs without dissipating as heat. The adiabatic technique depends largely on the parameter variation. With the help of DSCH, MICROWIND, TANNER EDA (S-edit, TWV, TSP) Software's power consumed by ECRL (Efficient Charge Recovery Logic) is compared to the power consumed by common CMOS for NAND and NOR inverter circuits. In this study it is find out that the adiabatic technique is the good choice rather than conventional technique in terms of power consumption.

Keywords: CMOS, ECRL, adiabatic

1. Introduction

The term adiabatic illustrate the thermodynamic processes in which no energy exchange with the environment, and therefore no dissipated energy loss. In VLSI, the transfer of the electric charge between nodes of circuit is considered. There are various techniques used to minimal power loss during a transient state [1]-[3]. The full adiabatic function circuit is one of the best conditions. It only reached at a very slow pace of change. In CMOS logic we get output by using the power provided to it for conventional circuit. In case of adiabatic technique the output stores the charge and the charge is used for next state this makes the adiabatic technique is power efficient than conventional circuit. The adiabatic method provides minimal power dissipation to the Pull-up network and also deploys stored power. Adiabatic Logic does not suddenly change from 0 to 1 or 1 to 0, but an electric ramp is used for charging and re-energizing the output [4]-[8].

The working phenomena of adiabatic techniques in efficient charge recovery logic (ECRL) circuit as, initially input 'in' is high or 1 and input 'in bar' is low or 0. When the power clock (pck) rises from 0 to V_{DD}, as output goes 'out' it remains at a 0. Output 'out bar' follows pck. When pck accesses V_{DD}, output 'out' and 'out bar' holding 0 and 1 respectively. These output values can be used in the next state. Now the pck falls from 1 to 0, 'out bar' returns its power to pck so remittances are received. Software used in this work is Tanner EDA (s-edit, TSP, TWV). Tanner EDA we make conventional circuits of NAND, NOR, Inverter. Adiabatic technique is remodel of conventional circuits to make power efficient [9]. Types of convention circuits are CMOS inverter, CMOS NAND, and CMOS NOR as discussed below.

2. Study of CMOS inverter, CMOS NAND, and CMOS NOR

Figure 1 shows the circuit of inverter circuit. When an input HIGH is given it returns output as LOW and for input LOW we get the output HIGH. This is same as 180^0 phase shift given to an input signal A we get output as A bar.

Figure 2. shows the CMOS Inverter Circuit and Figure 3. shows the CMOS Inverter layout. The layout of inverter is simulated in MICROWIND and it is working on 1 lamda (0.06 μm) where the diffusions of P-MOS and N-MOS are made. The length of the diffusion is 12 lamda and the breadth of the diffusion is 4 lamda is same for both of the PMOS and NMOS diffusions.

Journal of Physics: Conference Series **2327** (2022) 012015 doi:10.1088/1742-6596/2327/1/012015

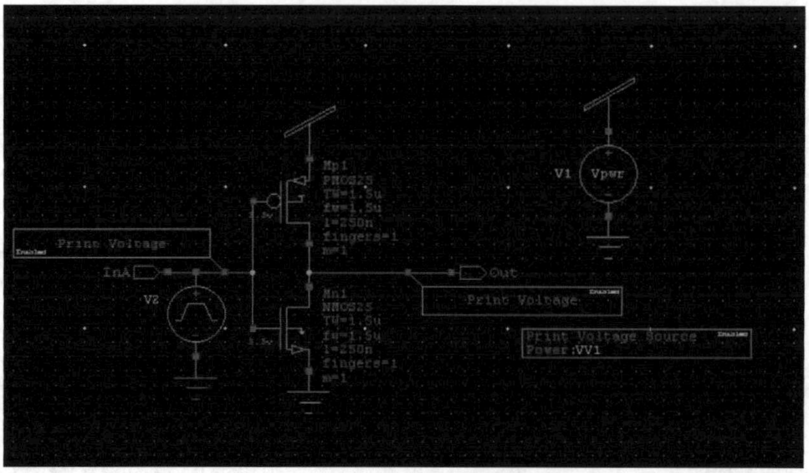

Figure 1. Inverter circuit diagram

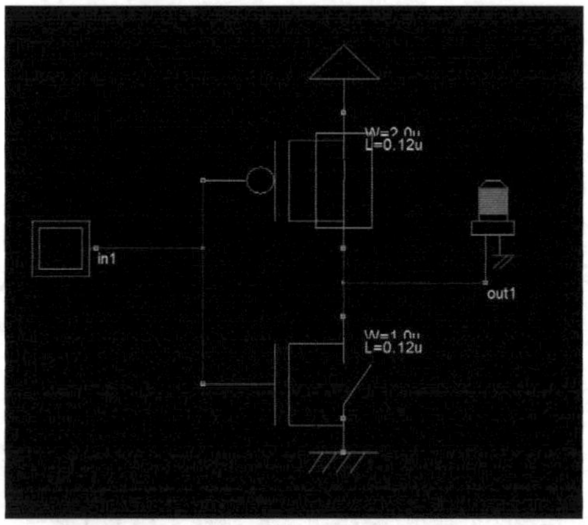

Figure 2. CMOS Inverter Circuit.

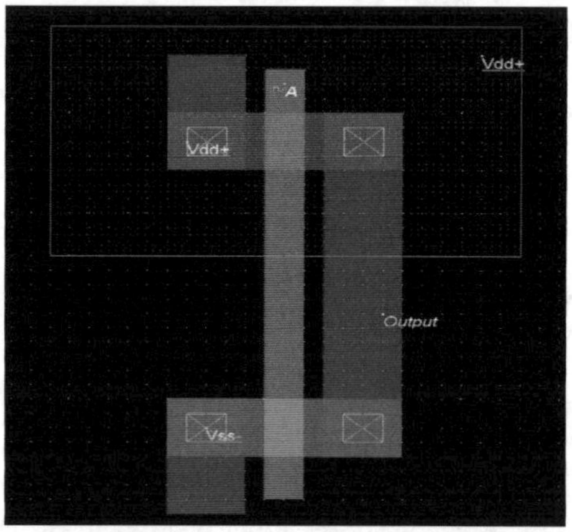

Figure 3. CMOS Inverter layout

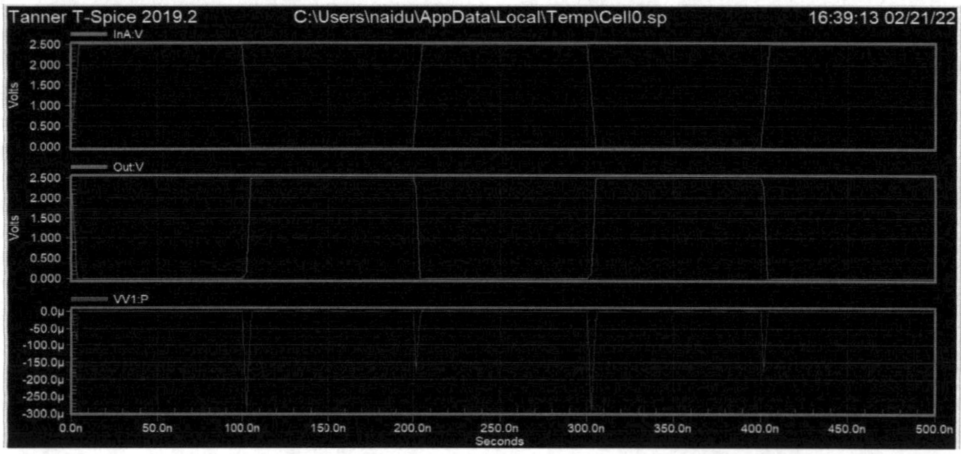

Figure 4. Waveform of inverter in TANNER EDA

Journal of Physics: Conference Series **2327** (2022) 012015 doi:10.1088/1742-6596/2327/1/012015

Figure 4. shows the waveform of CMOS inverter in TANNER EDA software. From waveforms the logic is inverter when A is 1 the output we get is 0 (vice versa). The value of power consumed in CMOS inverter is 4.8 μW. The max power used by the circuit is 2.9 x 10^{-4} W and minimum power used by inverter circuit is 5.7 x 10^{-11} W. Table 1. shows the CMOS Inverter power result.

Table 1. CMOS Inverter power result

VV1 TIME	0 to 5 x 10^{-07}	PARSING	0.06 seconds
AVERAGE POWER	4.845 X 10^{-06}	SET UP	0.01 seconds
MAX POWER	2.94 X 10^{-04}	TRANSIENT ANALYSIS	0.03 seconds
MIN POWER	5.71 X 10^{-11}	OVERHEAD	0.57 seconds
		TOTAL	0.67 seconds

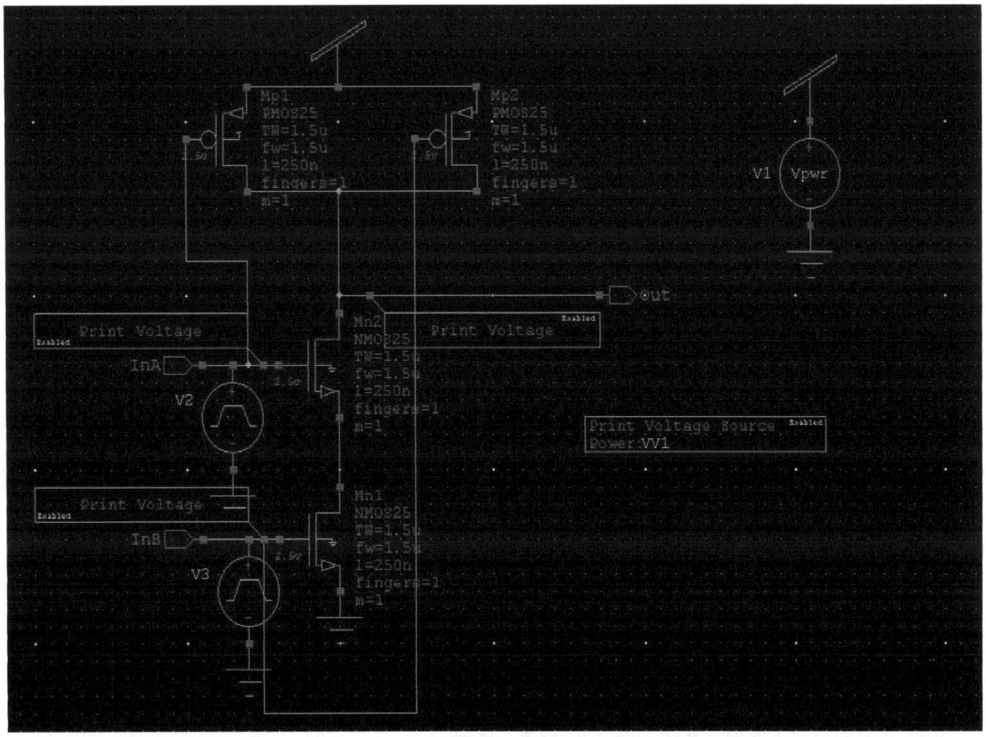

Figure 5. CMOS NAND circuit with inputs A and B

Figure 5. shows the CMOS NAND logic circuit. When input A and B are LOW it returns output as HIGH, for input A is LOW and B is HIGH (vice versa) we get output as HIGH. The only state where output is LOW is for logic HIGH.

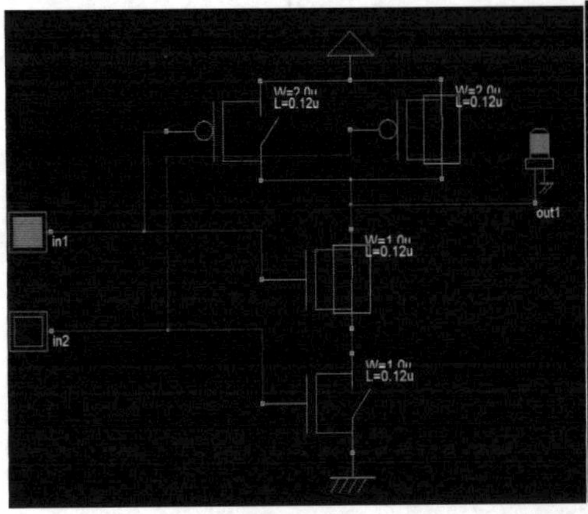

Figure 6. NAND Logic Circuit.

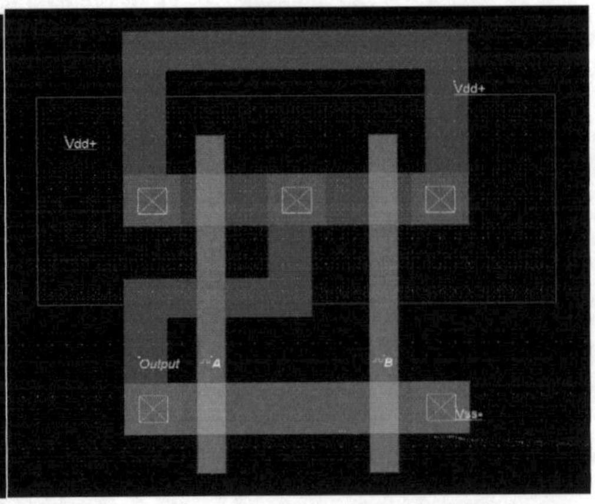

Figure 7. NAND layout on MICROWIND

Figure 6. shows the NAND Logic Circui and Figure 7. shows the layout of NAND Logic. The Blue colure layer is metal or wire used for connecting the components. The Red layer is for the inputs we give to the Circuit. The Rectangular green dotted box is N-Well on the pull-up part is to distinguish between PMOS and NMOS. This Layout is made in MICROWIND of 1 lamda (0.06 μm) technology. The PMOS and NMOS diffusion length is 24 lamda and breadth is 4 lamda. The distance between this two NMOS and PMOS diffusion is 12 lamda.

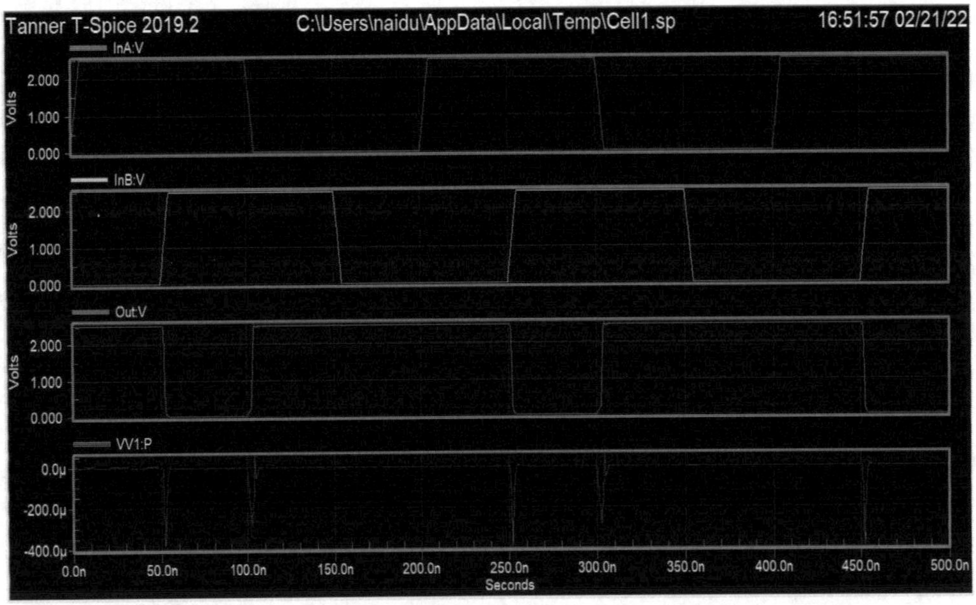

Figure 8. NAND Waveform of TANNER EDA

Figure 8. shows the NAND logic waveform. When A is LOW and B is HIGH (Vice versa) the output is HIGH. When A and B both are LOW the output is HIGH. When A and B are both HIGH the output is LOW. The value of power consumed in CMOS NAND is 6.3 μW. For power consumption graph, the power voltage source print is used to show the power in the waveform. The max power used by the circuit is 3.9×10^{-04} watts and minimum power used by NAND circuit is 1.49×10^{-11} watts.

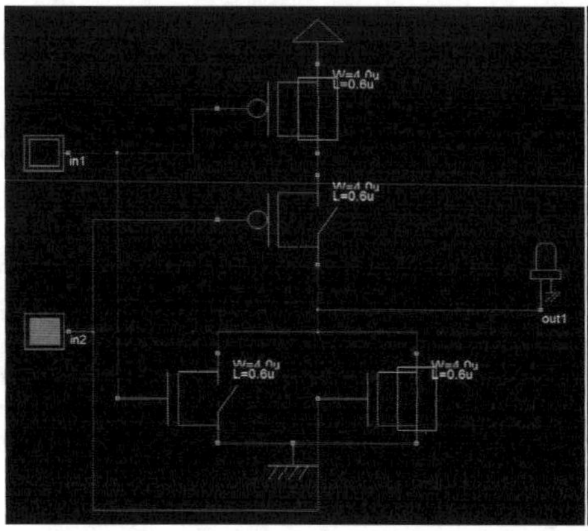

Figure 9. NOR Logic Circuit

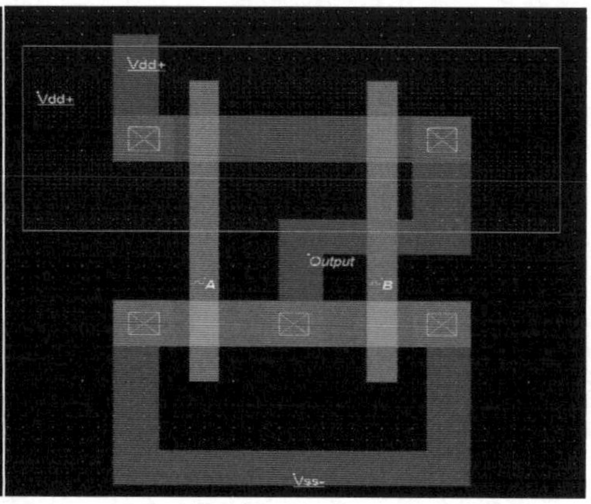

Figure 10. NOR Layout on MICROWIND

Figure 9. shows the NOR Circuit with inputs A and B. When input A and B are LOW it returns output as HIGH, for input A is LOW and B is HIGH (vice versa) we get output as LOW. If Input A and B are HIGH the result is LOW. The only state where the Output of NOR is zero is for inputs LOW.

Figure 10. shows the NOR layout.. The Blue colour layer is metal or wire used for connecting the components. The Red layer is for the inputs we give to the Circuit. The Rectangular green dotted box is N-Well on the pull-up part is to distinguish between PMOS and NMOS. This Layout is made in MICROWIND of 1 lamda (0.06 µm) technology. The PMOS and NMOS diffusion is with Length 24 lamda and Breadth 4 lamda . The distance between this two NMOS and PMOS diffusion is 12 lamda.

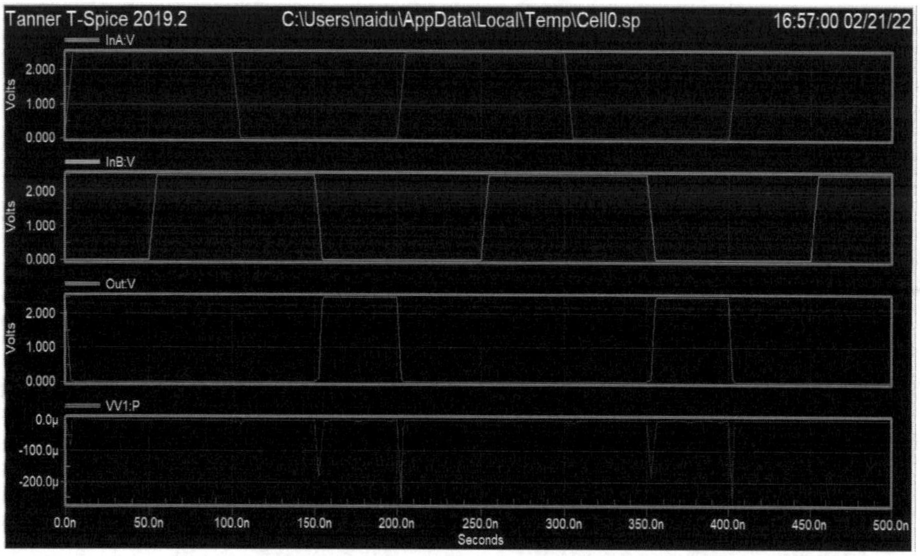

Figure 11. Waveform of NOR Logic on TANNER EDA

Figure 11. shows the NOR Logic waveform on TANNER EDA. From waveform if the value of A and B are LOW so the output is HIGH. Rest all other combinations are with the output zero or LOW. The value of power consumed in CMOS NOR is 4.66 µWatts. The max power used by the circuit is 2.68 x 10^{-04} watts and minimum power used by NOR circuit is 2 x 10^{-11} watts.

3. ADIABATIC CIRCUIT ANALYSIS

In this technique, we performed the modification in PMOS of conventional circuit. In PMOS pull-up part is introduced and pull-down part introduced in NMOS. The Power supply to this adiabatic circuit also changes. Instead of giving V_{DD} we are giving an AC power supply which is resulting in efficient power saver. This works when there is the transition of one state to the other at that instant it takes power from current state of output bar for the next state. Figure 12. shows the ECRL inverter circuit and Figure 13. shows the inverter waveform of input A on TANNER EDA.

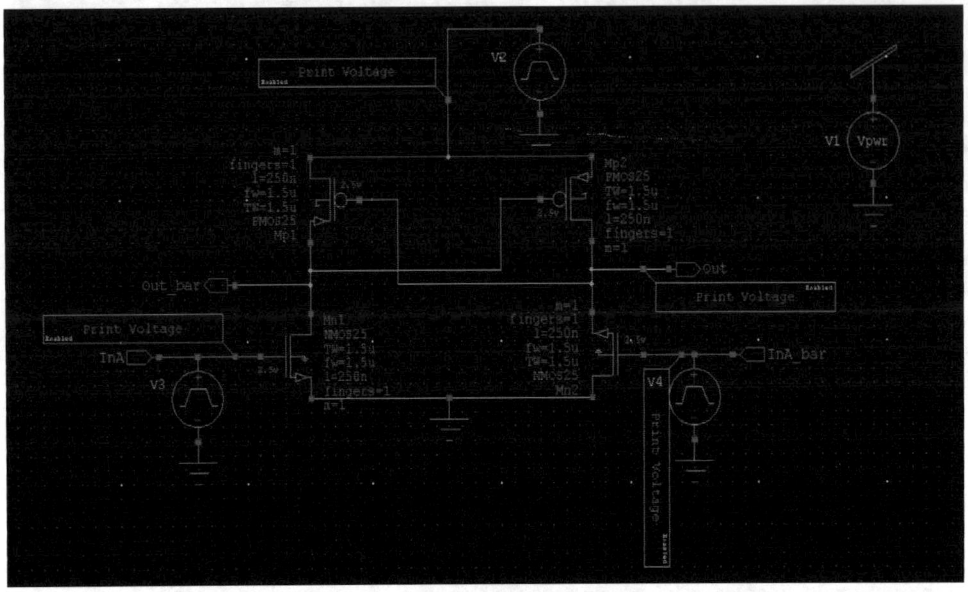

Figure 12. ECRL Inverter Circuit on TANNER EDA

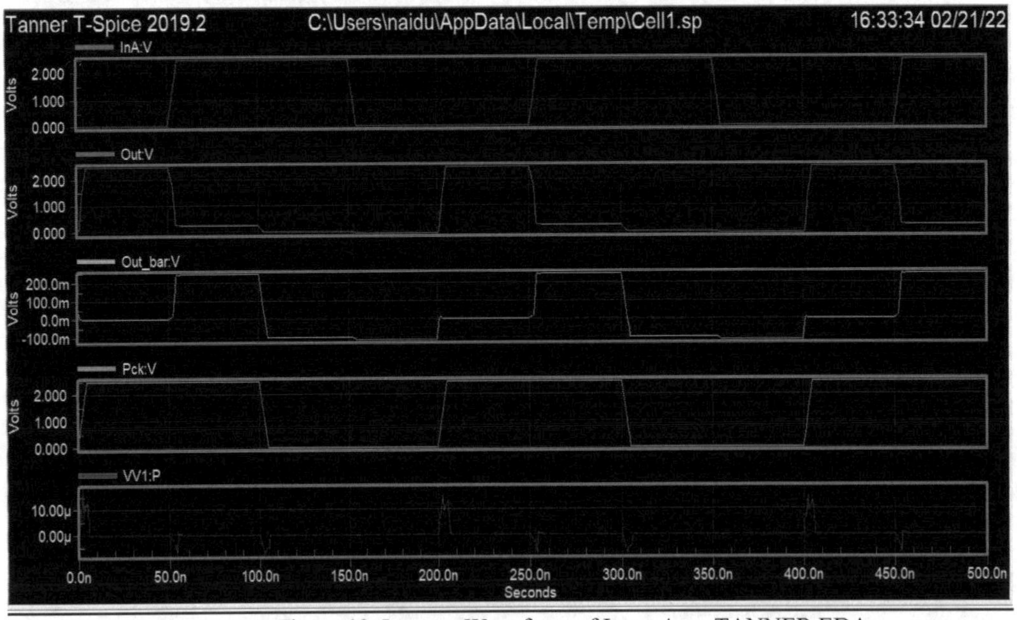

Figure 13. Inverter Waveform of Input A on TANNER EDA

Table 2. Power result of ECRL Inverter Circuit

VV1 TIME	0 to 5 x 10^{-07}	PARSING	0.19 seconds
AVERAGE POWER	1.64 X 10^{-07}	SET UP	0.20 seconds
MAX POWER	1.84X 10^{-05}	TRANSIENT ANALYSIS	0.05 seconds
MIN POWER	2.12 X 10^{-11}	OVERHEAD	0.68 seconds
		TOTAL	1.13 seconds

Table 2. describes the ECRL inverter power result. The value of power consumed in ECRL inverter is 1.64 x 10^{-07} Watts. The max power used by the circuit is 1.84 x 10^{-05} watts and minimum power used by ECRL inverter circuit is 2.12 x 10^{-11} watts.

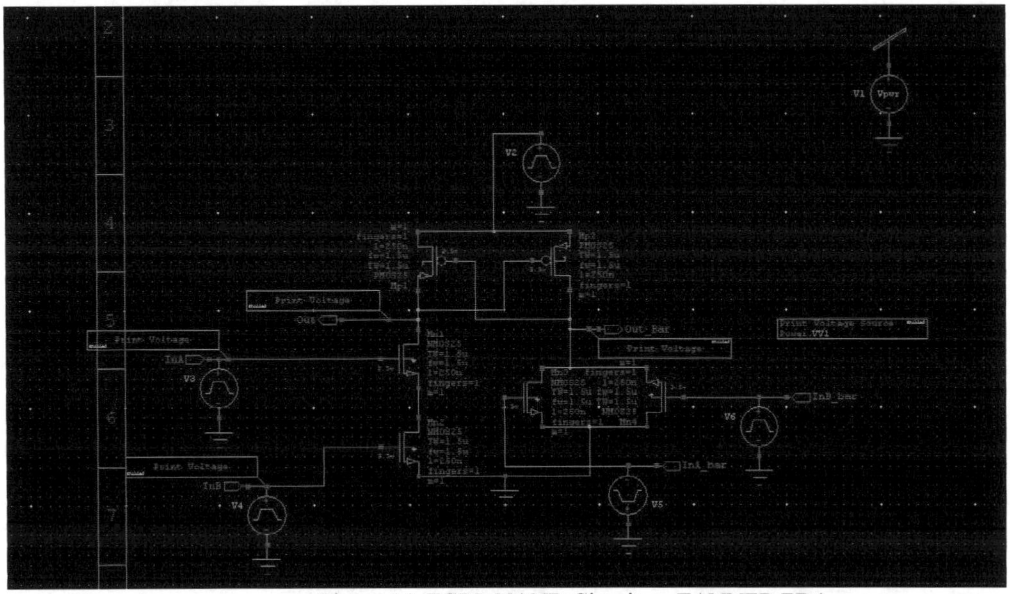

Figure 14. ECRL NAND Circuit on TANNER EDA

Table 3. Power result of ECRL NAND Circuit

VV1 TIME	0 to 5 x 10^{-07}	PARSING	0.06 seconds
AVERAGE POWER	3.69 X 10^{-07}	SET UP	0.01 seconds
MAX POWER	5.80 X 10^{-05}	TRANSIENT ANALYSIS	0.04 seconds
MIN POWER	1.57 X 10^{-11}	OVERHEAD	0.61 seconds
		TOTAL	0.72 seconds

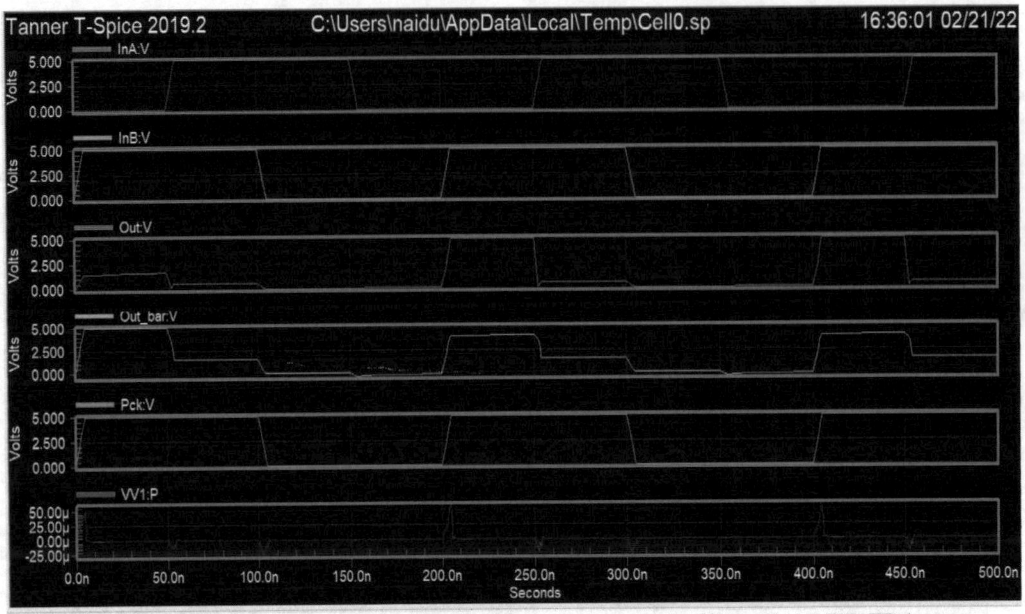

Figure 15. ECRL NAND Waveform of Input A and B on TANNER EDA

Figure 14. shows the ECRL NAND circuit and Figure 15. represents the waveform of ECRL NAND logic. The value of power consumed in ECRL NAND is 3.69 x 10^{-07} Watts. The max power used by the circuit is 5.80 x 10^{-05} watt and minimum power used by ECRL NAND circuit is 5.80 x 10^{-11} watt.

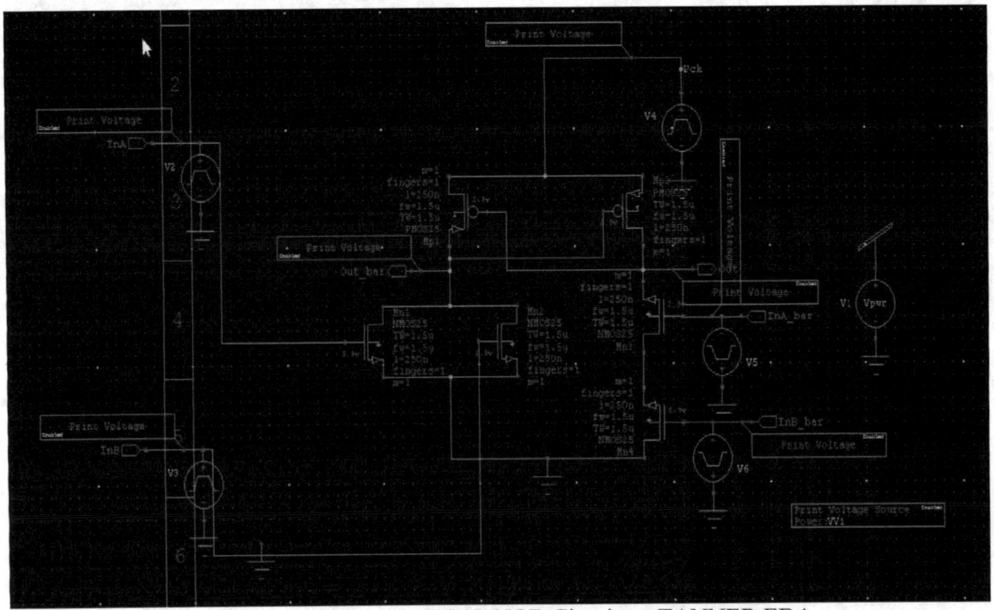

Figure 16. ECRL NOR Circuit on TANNER EDA

Figure 16. shows the ECRL NOR circuit and Figure 17. represents the waveform of ECRL NOR. The value of power consumed in ECRL NOR is 1.26 x 10^{-07} Watts. The max power used by the circuit is 1.69 x 10^{-05} watt and minimum power used by ECRL NAND circuit is 1.75 x 10^{-11} watt. Table 4. describes the power result of ECRL NOR circuit.

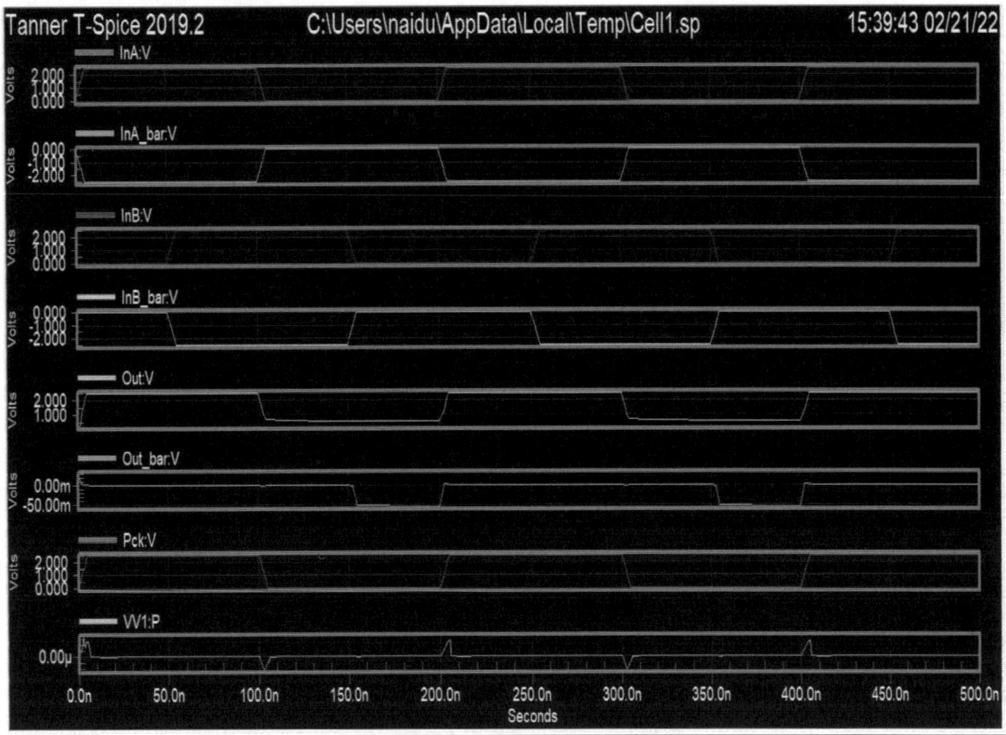

Figure 17. ECRL NOR Waveform of Input A and B on TANNER EDA

Table 4. Power result of ECRL NOR Circuit

VV1 TIME	0 to 5 x 10^{-07}	PARSING	0.06 seconds
AVERAGE POWER	1.26 X 10^{-07}	SET UP	0.01 seconds
MAX POWER	1.69X 10^{-05}	TRANSIENT ANALYSIS	0.03 seconds
MIN POWER	1.75 X 10^{-12}	OVERHEAD	0.56 seconds
		TOTAL	0.66 seconds

4. Conclusion

In this work the power consumption of conventional circuits and ECRL is observed and compared. It is find out that the adiabatic circuit is more efficient than conventional circuit. The adiabatic circuit is used as power efficient VLSI logic designing. In this research less energy consumption in adiabatic logic families achieved than CMOS logic. ECRL is a low-energy, adiabatic logic. With the help of simulation it is find out that power saving over static and other adiabatic logic families. The ECRL inverter shows near ten times power gain over a conventional inverter. ECRL gives the benefit for power saving and illustrated that the ECRL is a promising circuit in a low power system.

ORCID id
Manoj Singh Adhikari :0000-0001-7645-658X, Shaik Sahil :0000-0003-2941-2474 , Palatla Babasri Tarun Naidu :0000-0003-0372-666X, Aseervadam Thushar Roy :0000-0002-9094-8844, Mandapati devendra :0000-0001-6581-8856, and Uppala Goutham :0000-0002-2649-4411

References

[1] Kumar B D and Bharathi M 2013 Design of Energy Efficient Arithmetic Circuit using Charge Recovery Adiabatic Logic *International Journal of Engg. Trends and Technology* 4(1) 32-40

[2] Voss B and Glesner M 2001 A low power sinusoidal clock *In Proc. of the International Symposium on Circuits and Systems* 108-111

[3] Tessier R, Jasinski D, Maheshwari A, Natarajan A, Xu W and Burleson W 2005 An energy-aware active smart card *IEEE Trans. on VLSI Systems* **13 (10)** 1190-1199

[4] Shin Y, Lee C, Moon Y 2004 A Low-Power 16-bit RISC Microprocessor using ECRL Circuits *ETRI Journal* **26(6)** 513-519

[5] Arora N, Singh B P, Sharma T, Sharma K G 2010 Adiabatic and Standard CMOS Interfaces at 90nm Technology *WSEAS Transactions on Circuits and Systems* **9** 173-183

[6] Fischer J, Amirante E, Stoffi A B, and Landsiedel D S 2004 Improving the positive feedback adiabatic logic family *Advances in Radio Science* **2** 221–225

[7] Athas W C, Svensson L J, Koller J G, Tzartzanis N, and Chou E 1994 Low power digital systems based on adiabatic-switching principles *IEEE Trans. VLSI Systems* **2(4)** 398-407

[8] Chandrakasan A P, Sheng S, and Brodersen R W 1992 Low-power CMOS digital design *IEEE J. Solid-State Circuit* **27(4)** 473-484

[9] Eisele M, Berthold J, Landsiedel D S, Mahnkopf R 1997 The Impact of Intra-Die Device Parameter Variations on Path Delays and on the Design for Yield of Low Voltage Digital Circuits *IEEE Transactions on VLSI Systems* **5(4)** 360-368

Journal of Physics: Conference Series 2327 (2022) 012016 doi:10.1088/1742-6596/2327/1/012016

Detection of Security Attacks on Edge Computing of IoT Devices through NS2 Simulation

Anit Kumar[1], Dhanpratap Singh[2]

[1]Research Scholar, Lovely Professional University, Phagwara- 144411, India. Email: anit.41800692@lpu.in

[2]Associate Professor, Lovely Professional University, Phagwara- 144411, India. Email: dhanpratap.25706@lpu.co.in

Abstract. The limited resource and computation capacity of an IoT device always being a target vector for intruders to use to steal sensitive data from the device. Computational problems later solved by the engineers by introducing an edge server near to IoT device network boundary, but still it is not secure from the hands of intruders. IoT device vulnerabilities still can allow Edge server to give the control of the device to the intruders. Hence both the Edge server and the connected IoT device may fall under the control of an intruder for a misuse. In this paper we study how IoT devices in case of Edge computing falls under security threats. We have used NS2 simulator to identify a malicious node that behaves inconsistent in terms of Packet Delivery Ratio, End-to-End Delay, and Throughput.

1. Introduction

Edge Computing makes privacy and security techniques critical for the existence of IoT systems. Figure-1 is specifying how edge servers are interconnected between IoT devices used in a Smart Home, Smart Industry or Smart City. The intention of introducing an edge server with an IoT device is to provide fast response, preprocessing of the data before sending to the cloud and deal with IoT device resource constraints to enable the device to do complex computation at corresponding edge server end. The introduction of edge computing in IoT computation is decentralized and mostly placed near to the IoT devices which in turn brings many challenges not only the edge security but also the ultimate target which is an IoT device. Followings are the Edge Computing Security Risks that are most common so far.

i. Code Injection Attacks

Code Injection Attacks is used by an Attacker [46] to introduce or inject a piece of codes into a vulnerable program that results into the change of course of execution of the program. The successful result of Code Injection is always dangerous to an individual or the Organization because it can result into data loss, corruption, service denial or lack of accountability. Injection flaws generally occurs when someone sends the untrusted data to server program intentionally. These types of attacks are generally prevalent to legacy applications. It often found in XML parser, NoSQL queries, LDAP, program arguments and HTTP/SMTP headers etc.

ii. Physical Tampering & Attacks

Many semiconductor chips used in computer processor against many scientific and industrial [47] applications requires protection against the physical attacks and tampering. These types of attacks are based on establishing direction connection with electrical signals from chips. It may leads to steal sensitive information stored on the chip. However, the inbuilt security in the chip deter to protect from potential attacks to unauthorized access and get benefited from it, but due to having public measurements and configuration of the chips, intruders attempts to make security attacks on the chip.

iii. Routing Attacks

Routing Attacks is one of the serious challenge [48] in creating robust and reliable sensor networks. These attacks typically disconnect a wireless sensor network from its base station. Integration of IoT devices with the internet has shown many flaws with the underlying communication network and are exploited by the attackers to cause the routing attacks. The communication protocol are overloaded with additional challenges like unpredictable node movement and operation, dissimilar wireless access medium etc. The presence of highly dynamic topology and lack of the centralized control always facilitate the routing attacks over WSN.

iv. Distributed Denial of Service (DDoS) Attacks

These attacks are also termed as Distributed Network Attacks. This attack is based upon [49] the capacity limit to serve [49] the requests at the same time by a network resource. The DDoS attackers send multiple requests at the same time to the attacked web resource with the aim to degrade the responsiveness of the resource and halt the website to function properly. This type of attacks may occur to damage or discredit competitor's business.

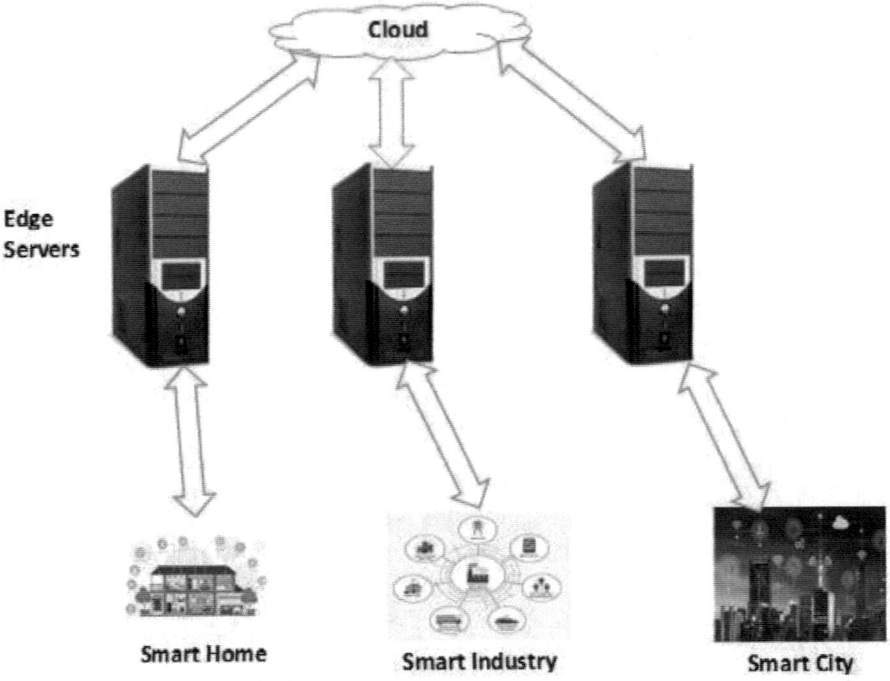

Figure-1: Edge servers for IoT netowk

Reinforcement Learning became an important method of machine learning (ML) algorithms which uses and apply the feedback architectures in the field of Computer Security and Artificial Intelligence. It covers and allows to provide sequential or dynamic responses to malware attacks that requires a limited or without advance knowledge of the surrounding environment. Internet of Things (IoT) is not a newly paradigm for anyone. In our day-to-day life we use millions of devices [1] that generate data and also capable to receive information through its sensors. These IoT data are send to the Cloud and became integral part of the day-day life of the users in many ways. Essentially, this volume of informational IoT data is processed in the cloud environment and as per needs also get stored into cloud database. This helps in improving the current knowledge, improving the process, and add valuable information for the organizations. The IoT infrastructure facilitates the integration between the physical communicating world and the communication networks. Edge Computing has emerged to minimize the costs associated with transferring, processing, and storing data from IoT environments in the Cloud. Data are pre-processed at the edge server and then only sent to the Cloud. It helps to obtain shorter response times, and making the service consistent even in the case of communication breakdowns between the IoT user and Cloud interface. It needs to make privacy and security techniques critical for the existence of IoT systems. The IoT devices with under heavy data streams are generally under security threats because of following reasons;

1 Restricted computation capacity of the underlying processor
2 Limited memory
3 Limited radio bandwidth
4 Inefficient power supply from battery
5 Lack of computational-intensive and latency-sensitive security jobs

Network Simulator– NS2 is a discrete event simulator [14] design and implemented for Network research. It provides substantial support for simulation using TCP, UDP, routing and multicast protocols. It covers both wired and wireless communication among nodes in the network. There are several parameters that are critical to measure consistent and benign behavior of nodes in a network. But we have focused on three performance metrics as below,

1 End-to-End delay
2 Throughput
3 Packet Delivery Ratio

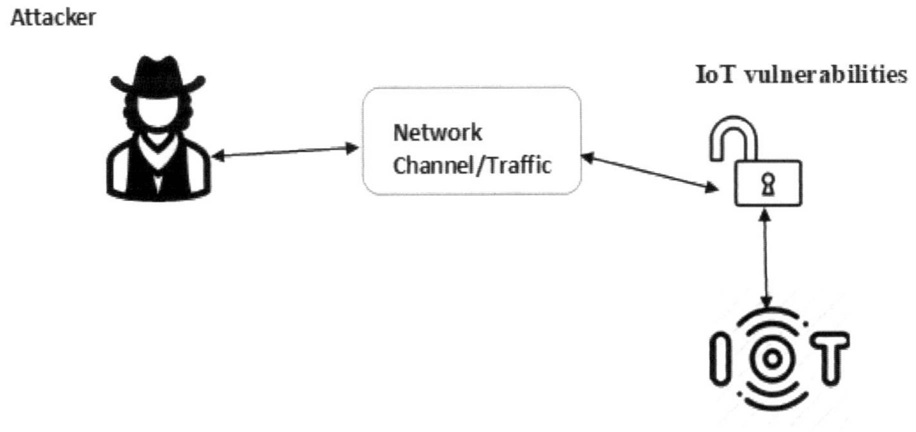

Figure-2: Attack model on IoT

Figure-2 describes that the intention of an Attacker is to find some IoT vulnerabilities from the underlying IoT network and takes benefits of it to steal sensitive data from an IoT device. There has been recent interest in utilizing Reinforcement Learning (RL) and Deep Reinforcement Learning (DRL) techniques in malware detection due to their potential to increase detection accuracy and robustness. Typically, the following types of attacks are examined for the utility of reinforcement learning and deep reinforcement learning techniques in malware detection. Attack on Edge server can affect any part of the process of working of an IoT devices and result in unreliable and inconvenience way of working. The most common process of IoT devices are as below.

1. **Sense**: Intercepts the relevant information, gather and process the data.

2. **Transport**: Transmits the data to the outside world though Internet.

3. **Store**: Stores the gathered data at some temporary or permanent storage for processing.

4. **Analyze**: Provide insights of the system from the data gathered from storage by the IoT device.

5. **Control**: IoT device uses the commands to operate.

6. **Share**: IoT data can be exchanged with third-party system.

IoT devices are gaining utmost attention in the perspective of security implementation since it became of urgent need due to unauthorized access to its sensitive data for personal benefits. Many authors presented their models to add or enhance the security layer of IoT devices. We have summarized the existing Machine Learning (ML) based security methodology for IoT devices in Table-II. IoT devices can considered as an agent, which makes the decisions whether the edge devices are required to gain computing tasks from underlying network or not. The pervasive application of IoT devices in today's communication world to provide more business opportunities in line also raise many security challenges. To overcome such security challenges needs to acknowledge the peculiar conditions under which IoT devices fails to operate as expected. These failover conditions can be as follows,

1 Low computational capacity
2 Low power capacity in battery enables IoT device
3 Unauthorized access of IoT device by Edge server
4 Distributed decision-making system in the IoT Network
5 High latency in Edge server node and IoT device
6 High response wait time

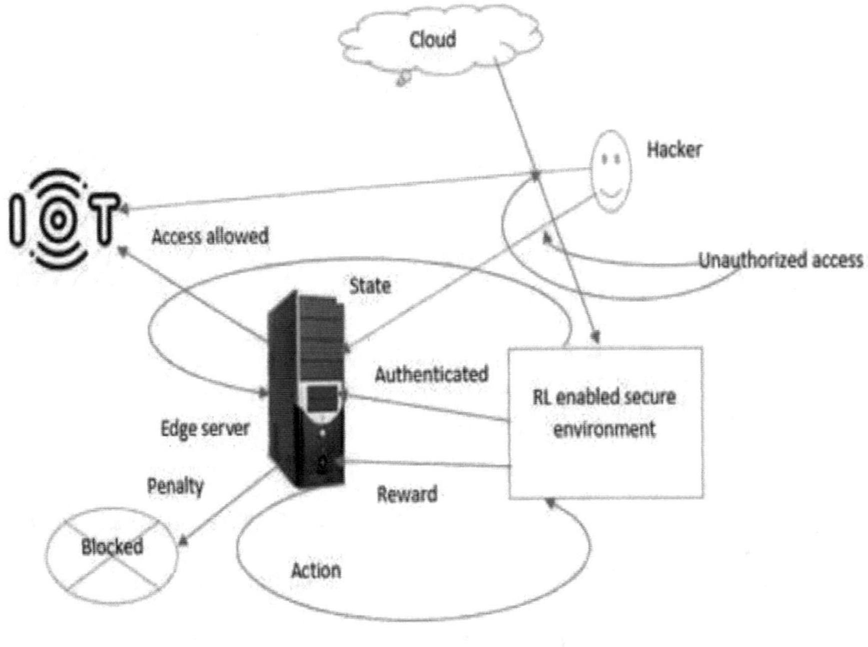

Figure-3: RL enabled secure environment to Edge server on IoT network

Any malicious edge server can take benefits of any of the above shortcomings and will try to access secure informative data from the IoT device. Hence, we will use reinforcement learning, which is a Machine Learning method to detect the behavior of any malicious node. As Reinforcement Learning deals with how software agents acts in an environment. There must be a security policy that will help to detect the behavior of the edge node while making a request of connection to access data from an IoT device. In a secure Reinforcement Learning environment, An Edge server acts as agent and will do some action with a secure policy with RL enabled environment. If by any chance any course of action on RL enabled environment is made by a malicious edge node, then its action will not bring any significant result and the state change will be "unknown" to "unauthenticated". There will be a reward like provide permission to access data from an IoT device in case of state change to "authenticate", otherwise there will be a penalty for state change to "unauthenticated" like block the requested entity for all the future times. Based upon our literature review we also propose the IoT security implementation using reinforcement learning as shown in Figure-3.

1.1. Q-Learning and Deep Q-Learning

It is known as simplest and most basic formation of RL algorithm. This algorithm calculates the Q-value [2] sometimes known as Quality values through successive iterative interactions with the environment. Based upon this calculate Q-Value, RL agent takes an appropriate action (Ai). The Q-value is the target of this learning method to determine optimal state-action values. This approach is based on some discrete and finite world and defines a computational agent that moves around [36] this world that chooses one from a finite collection of actions at every time step. At each step n, the agent makes equipped to register a step x_n € X and choose corresponding action a_n € a. The agent receives a probabilistic reward at each time step r_n with mean value $Rx_n(a_n)$ that depending upon only on the action and state. The state of the world changes probabilistically to yn according to the law:

$$\text{Prob}\ [y_n = y|x_n, a_n] = P_{XnY}[a_n]$$

The Q-Learning process generates an exact and similar matrix for the active agent. It is referred to gain the reward in maximum in the long run. Although this approach used is not perfect in itself, but this approach is only applicable into very small environments and quickly loses its stability and feasibility when the number of actions and corresponding states in the environment increases. When Q-values obtained by Q-learning have importance with respect to the values obtained in similar context, then the solution provided by Q-learning seems practical. This issue lands on to Deep Q-Learning which uses a deep neural network to more approximate the values. This approximation works efficiently as long as the relative importance is preserved.

1.2 B. Reinforcement Learning (RL)

In RL algorithm, the agent interacts [2] with the surrounding environment and make decisions. The algorithm is executed with a learning procedure in a trial manner, where it receives rewards or penalty and representation of a state. The ultimate goal to obtain an optimized solution of a problem that corresponds to achieve maximize reward for a given situation.

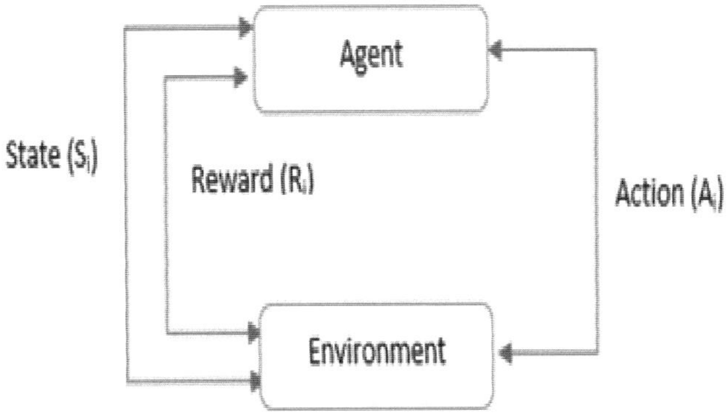

Figure-4. Agent and Environment interaction in a RL system

Figure-4 depicts that how two major elements 1) Agent, and 2) Environment in RL system interacts with each other. Environment imposes external factors or conditions on the agent that determines its state change and the reward (or, penalty).

a) **Agent**: The agent performs to learn and achieve a goal in an uncertain, potentially complex environment.
b) **Environment:** It's the surrounding to the Agent that provides a set of uncertainties and problems.

TABLE I: ABBREVIATIONS

5 G	Fifth Generation
AI	Artificial Intelligence
CRM	Cyber-resilient Mechanism
CPU	Central Processing Unit
CPS	Cyber-physical system
DDoS	Distributed Denial-of-Service
DL	Deep Learning
DRL	Deep Reinforcement Learning
DNN	Deep Neural Network
RL	Reinforcement Learning
IoT	Internet of Things
MAC	Media Access Control Protocol
MLP	Multilayer perceptron
MN	Malicious Node
TPR	True Positive Rate
UED	User Emulation Detector
SET	Security-Enhanced Traceback
ITS	Intelligent Transportation System
PPRL	Privacy-preserving Reinforcement Learning

2. Related Works

IoT devices since inception have been always a target vector for intruders. It plays a vital role for the benefits for mankind. But due to small size and rich resource incapability it possess many vulnerabilities that makes it as an attack victim. Table-II summarizes our work on the literature review done so far. We have categorized our research data in three parts namely, 1) Agent, 2) Environment, and 3) Action. Here an agent is intended to perform some action on it surrounding environment. We have compared the components used by different researchers in their research works on the ground of reinforcement learning method. Figure-5 shows how a RL based security mechanism can be applied on IoT-Edge computing.

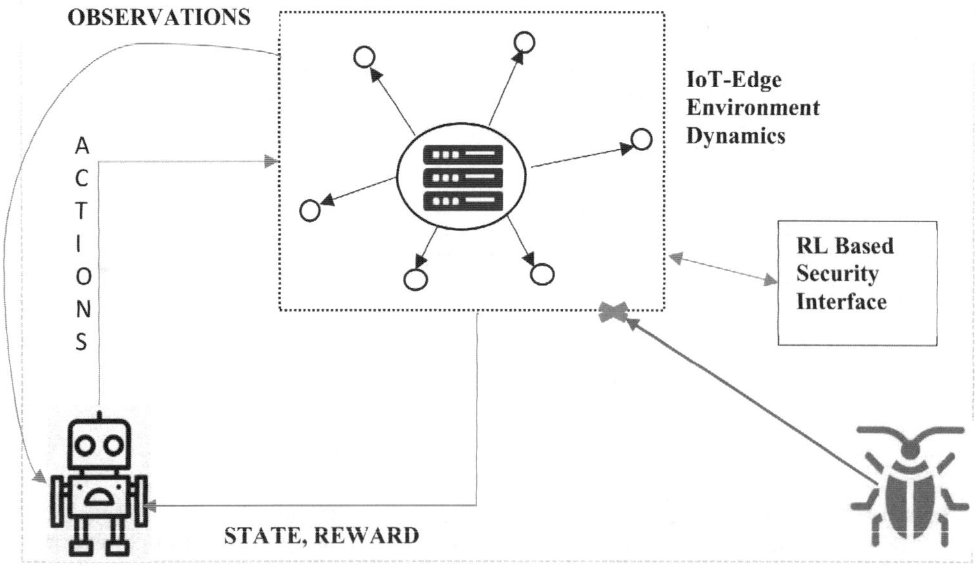

Figure-5: RL Based Security Interface in IoT-Edge Computing

Journal of Physics: Conference Series **2327** (2022) 012016 doi:10.1088/1742-6596/2327/1/012016

TABLE II: LITERATURE REVIEW ON SECURITY ATTACKS ON IoT DEVICES

Author(s)	**Agent**	**Environment**	**Action**
Qiao et al. [3]	Edge device	Intelligent Transportation Systems (ITS)	Frequently used data get stored in the Edge server and least frequently used data and not private sensitive data get stored in the Cloud.
Qu et al. [4]	Screen Image	Six Atari Games	Perturbing of images used in Game increases when more frames are counter attacked and vice-versa.
Zhang et al. [5]	Edge device	Global Industrial Model interconnected through Edge devices	Degree of load balance with the input of current state and action from the Edge device.
Yuan et al. [6]	Edge device	Multi-source Feedback Information System	Calculation of trust factors for IoT edge devices by using lightweight trust evaluating mechanism.
Huang et al. [7]	Sensor and Communication channel	Cyber System	Response and recovery mechanisms used to mitigate Cyber-attacks and strengthen the system security.
J. Park et al. [8]	End user	Cloud Computing Environment	Q-learning based RL service on Cloud platform.
Azmoodeh et al. [9]	Device OpCode	Internet of Battlefield Things (IoBT)	Selection of OpCodes sequence approach to classify malicious and benign applications
Diro et al. [10]	Social Internet of Things	Distributed Attack Detection System	Distributed IoT nodes data are sent to coordinating master node and then global update and re-propagation take place.
Hsu et al. [11]	Intrusion Detection Engine	Internet Communication Network Traffic	Result of Intrusion Detection
Shakeel et al. [12]	IoT Device Features	IoT based Health Care System	IoT devices are tested under a particular range of control before providing any access
Doshi et al. [13]	IoT-specific Network behaviors	Home Gateway Routers or other Network Middleboxes	Packet level machine learning to distinguish traffic between malware and benign data

3. Attack Detection and Problem Formulation

We have used NS2 simulation to identify a malicious node in the simulated environment. The malicious node intervenes in between a sender and a receiver nodes. Malicious node at certain time instances drops the data packets on the route between its source and destination. Simulation parameters taken for malicious node detection are as shown in Table-III as follows:

TABLE III: SIMULATION PARAMETERS

Simulation Area	800 * 541 mm^2
Number of nodes	7
MAC	Mac/802_11
Packet size	1500 bytes
Transmission rate	250 kbps
Evaluation Parameters	End-to-End delay, Throughput and Packet Delivery Ratio
Simulation time	100s

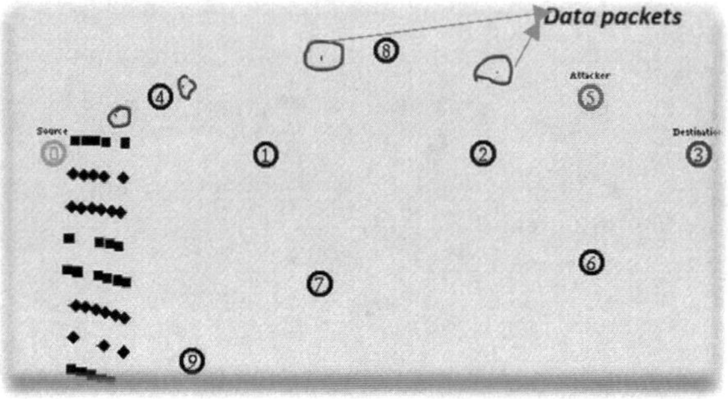

Figure-6. Network Simulation for Attack Detection

Network Simulation shown in Fig. 6 is carried for two times. First time, When all the seven nodes created are benign where node (0) is the source node and node (3) is the destination node. For Second time, Node (5) is set intentionally malicious and is starts behaving as a Attacker node that drops the data packet of communication between the Sender and the Receiver. We have summarized our experimental results in Table-IV.

TABLE IV: EXPERIMENTAL RESULTS

Parameter	**Malicious Network Simulation**	**Non-Malicious Network Simulation**
End-to-End Delay	66.2412 ms	122.62 ms
Packet Delivery Ratio	1.559079	1.398299
Average Throughput	9.745630 kbps	8.740540 kbps

4. Conclusion and Future Works

In this research paper, we have worked on the literature review to provide security on IoT devices. We have done a comparative study on different types of malware attacks thrown on IoT devices. We have worked on how an Edge server can act as an agent to learn the best policy to securely transfer data between an IoT device and the end user by using Reinforcement Learning (RL) method. Our Future work is to implement and verify how our proposed model will make an effective contribution in the era of securing IoT devices in conjunction with Edge computing by using some experimental tools.

5. References

[1] Ricardo S. Alonso, In´es Sitt´on-Candanedo, Roberto Casado-Vara, Javier Prieto, and Juan M. Corchado (2020), "Deep Reinforcement Learning for the management of Software-Defined Networks in Smart Farming" International Conference on Omni-layer Intelligent Systems (COINS)

[2] Felix O. Olowononi, Danda B. Rawat, and Chunmei Liu (2020), "Resilient Machine Learning for Networked Cyber Physical Systems: A Survey for Machine Learning Security to Securing Machine Learning for CPS", IEEE Communications Surveys & Tutorials

[3] F. Qiao, J. Wu, J. Li, A. K. Bashir, S. Mumtaz and U. Tariq (2020), "Trustworthy Edge Storage Orchestration in Intelligent Transportation Systems Using Reinforcement Learning," in IEEE Transactions on Intelligent Transportation Systems, doi: 10.1109/TITS.2020.3003211.

[4] Xinghua Qu, Zhu Sun, Yew Soon Ong, Abhishek Gupta, and Pengfei Wei (2020), "Minimalistic Attacks: How Little it Takes to Fool Deep Reinforcement Learning Policies", IEEE Transactions on Cognitive and Developmental System

[5] Rui Zhang, Hui Xia, Chao Liu, Ruo-bing Jiang, and Xiang-guo Cheng, "Anti-Attack Scheme for Edge Devices Based on Deep Reinforcement Learning", Hindawi, Wireless Communications and Mobile Computing Volume 2021, Article ID 6619715, 9 pages https://doi.org/10.1155/2021/6619715

[6] Xiaoyong Yuan, Pan He, Qile Zhu, and Xiaolin Li (2018), "Adversarial Examples: Attacks and Defenses for Deep Learning", IEEE Transactions on Neural Networks and Learning Systems

[7] Yunhan Huang, Linan Huang, and Quanyan Zhu (2021), "Reinforcement Learning for Feedback-Enabled Cyber Resilience", Elsevier Ltd.

[8] Jaehyoung Park, Dong Seong Kim, and Hyuk Lim (2020), "Privacy-Preserving Reinforcement Learning Using Homomorphic Encryption in Cloud Computing Infrastructures", IEEE Open Access Journal

[9] Amin Azmoodeh, Ali Dehghantanha, and Kim-Kwang Raymond Choo (2018), "Robust Malware Detection for Internet Of (Battlefield) Things Devices Using Deep Eigenspace Learning", IEEE Transactions on Sustainable Computing

[10] A.A. Diro, and N. Chilamkurti (2017), "Distributed Attack Detection Scheme using Deep Learning Approach for Internet of Things", Future Generation Computer Systems, Future Generation Computer Systems, http://dx.doi.org/10.1016/j.future.2017.08.043

[11] Ying-Feng Hsu and Morito Matsuoka (2020), "A Deep Reinforcement Learning Approach for Anomaly Network Intrusion Detection System", IEEE 9th International Conference on Cloud Networking (CloudNet), 2020, pp. 1-6, doi: 10.1109/CloudNet51028.2020.9335796.

[12] P. Mohamed Shakeel, S. Baskar, V. R. Sarma Dhulipala, Sukumar Mishra, and Mustafa Musa Jaber (2018), "Maintaining Security and Privacy in Health Care System Using Learning Based Deep-Q-Networks", Springer, Journal of Medical Systems (2018) 42:186, https://doi.org/10.1007/s10916-018-1045-z

[13] Rohan Doshi, Noah Apthorpe, and Nick Feamster (2018), "Machine Learning DDoS Detection for Consumer Internet of Things Devices", IEEE Symposium on Security and Privacy Workshops

[14] Accessed on 09 January 2022, “https://www.isi.edu/nsnam/ns/”

Collision Avoidance and Direction Planning for Autonomous Vehicles

Nithish Kumar Gannaram; Jitesh Kumar Nayak; Sudarsi Mahesh; Pappala Anil; Vamsi Chikkala; Asst. Prof Nitin Kumar

nithishgannaram.ng jitknayak26 sudarsimahesh anilpappala321 vamsivsvamsi nitin.14652@lpu.in

Abstract. The rapid development of information and signal processing technology has contributed significantly to the development of autonomous driving (AD) techniques, which improve driving safety while minimizing human efforts. According to analyst predictions, automatic cars will soon surpass manual cars in number as automatic driving becomes more sophisticated. Ensuring safety and reducing the road accidents is the primary concerns in automated vehicles. Recent research has shown that computer vision, deep learning, and other fields have advanced beyond the imagination. The paper covers the deep learning algorithms and techniques necessary to make a reliable and real-time collision avoidance system using concepts such as Convolutional Neural Networks (CNN), YOLOv4 (You Look Only Once), and other deep learning architectures while also reviewing the current state of the art strategies for Advanced Driver Assistance System (ADAS). By using CNN algorithms, it is possible to detect lane markings and signboards in real time as well as estimate distances between them.

1. Introduction

The main causes of accidents embrace improper road maintenance, weather condition conditions, misreading of traffic signs caused by foggy weather, light, or occluded traffic signs, etc. it makes drivers incapable of distinctive the sign boards that necessary. consistent with a survey every year one.5 million people die on the streets worldwide and within the vary of twenty to five0 million expertise the sick effects of internal wounds and harm. In addition to the planning of a driver-assistance system that detects and categorizes road signs and lane markings using a number of strategies, options the method of recognizing road signs can be subdivided into two main areas of research: classification and detection of convolutional neural networks

The traffic signs are detected by analysing the colors present on the images, especially red and blue, while the signs are classified according to their shapes and characteristics in the classification phase.
In this system in conjunction with road sign recognition we'll be specializing in Lane detection that is one important methodology within the visualization-based driver support structure and capable to be used for vehicle guiding, road congestion dodging, crash dodging, or lane departure warning in addition as different protective applications.

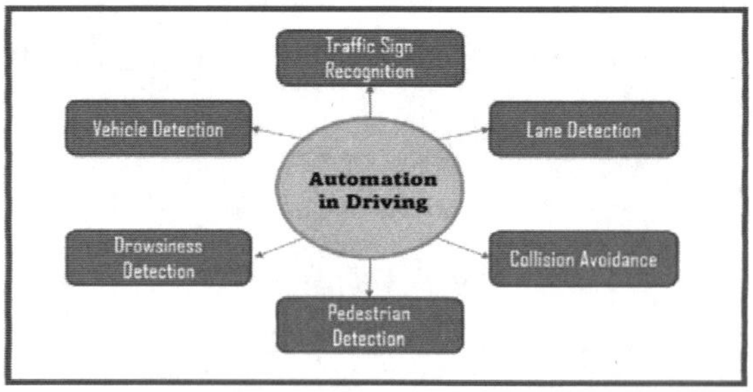

Figure 1: *Modules for automation*

Journal of Physics: Conference Series **2327** (2022) 012017 doi:10.1088/1742-6596/2327/1/012017

Autonomous or driver-guiding functions of this methodology area unit out there on all types of lanes, whether or not or not they are straight, curved, white, yellow, single, or double, additionally as on pavements or roadways. the foremost objective of the system is to develop a warning facilitate system provides the drive with fundamental quantity knowledge from lanes markers and road signs, that consist the foremost very important and troublesome tasks. second generates a warning to the drive before of any danger. This warning then permits the drive to want acceptable corrective

2. Objectives

- We need to Identify the correct sign between various traffic signs using the concept of Convolutional networks.
- And, train the multi-layer perceptron to classify the signboards efficiently based on the data provided.
- And, should detect the lanes and sign board clearly using computer vision techniques to work in any climate condition effectively.
- At the end it should be Fully Functional Automated system in a vehicle for safe and easy Access

3. Methodology

The project involves detection of lane and traffic sign detection for collision avoidance and giving the direction based on the traffic sign detected. Main Objective of the project is to prepare a robust model which detects lanes and sign boards and vehicles in front. The output should give the warning if the vehicle is out of the lane and audio output based on the sign

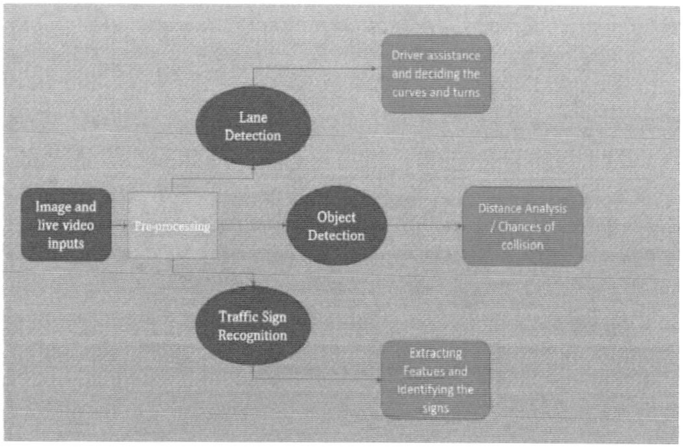

Figure 2. *Objectives for each module*

3.1 preprocessing

In the last two decades, a number of research have been made on the autonomous driving vehicles especially on the lane detection and traffic sign prediction. There were two methods traditional methods and Deep learning methods.

3.1.1 **Traditional methods** are generally by detecting the lane by using different mathematical equations to find the lanes and taking the region of interest which needed for edge detection technique in computer vison which shows the lanes. but the deep learning methods are more effective. So, researchers are preferring the deep learning methods mostly

3.1.2 Deep learning methods

with the invention and successful use of deep learning, lane detecting research has advanced to a new level. In this paper, a number of deep learning-based lane detecting algorithms are proposed. over the last few years paper contains only one of the methods to solve the task. Computer vision is a technique used to identify lane lines. A polynomial regression technique will be applied during the implementation stages to "spot" the recurved lanes and to mark them accordingly; i.e., to spot the curve lines. The result is that this technique can accurately detect lane lines under a wide variety of conditions and conditions.

a) Image Acquisition

The image or video is captured by the front facing camera mounted at the centre of a vehicle as it covers both the extremes of the road while capturing. The captured image or video is given as input to the lane and traffic sign detection.

b) Lane detection

In order to reduce the computational complexity, the image or video frames found after image acquisition are resized. Using a CNN model with SegNet architecture, the resized frame is labelled pixel-by-pixel to detect lane lines; image segmentation is used to determine the lane lines As each pixel is classified based on what it belongs to, the network will train itself to translate it into its appropriate class.

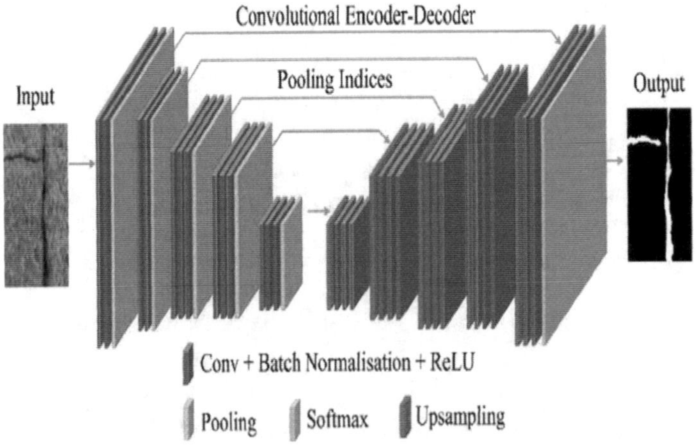

Figure 3. *Segmentation architecture*

In terms of Segnet architectures, they typically take RGB input images and assign a label. The number n denotes how many labels there are in the input image. Each pixel in a certain channel correlate to a label, and each pixel in that channel will be either 0 or 1, depending on which channel it belongs to. In lane detection, pixels with no lines are assigned a value of 0 and pixels with lanes are assigned a value of 1.

Figure 4. *Lanes*

A loss function is constructed based on the weighted cross entropy to solve the discriminative task, which can be formulated as

$$\varepsilon_{loss} = \sum_{x\in\Omega} \omega(x) \log\left(\rho_{l(x)}(x)\right)$$

c) Traffic Sign detection

Traffic Sign Boards area unit most important for safe and economical driving, therefore if we tend to area unit automating our driving expertise it should be necessary that our automobile is ready to notice signs and proceed any in keeping with it.

i) Using image processing

By using a hexadecimal threshold, the sign boards are extracted by detecting the regions in HSV color space from an image or a live video feed of the natural scene. Once the main color range used by traffic signs is determined, all the other colors are removed to create a binary image. A range of different methods, such as morphological transformation and canny detection, is used to obtain sharp and nose-free images from the feed. In this operation the continuous boundary points are used to generate contours, which are then used to crop the sign for validation.

ii) Using CNN (Convolutional Neural Network)

Colour Based Detection

A sign's color immediately catches our attention when we see it. The color of a sign immediately attracts our attention when we see one. Similarly, our signal detection algorithm is based on color. Similarly, our methods for detecting signals are based on the same principle. In the proposed algorithm, the red color is used for determining whether or not a sign is captured in a frame. A portion of a picture is checked to see if any signs are present whenever it falls into the appropriate red threshold. In order to determine the contour of the red portion of the image, the red threshold must be determined.

Shape Based Detection

Using the contours found within the previous step, the quantity of edges is calculated. this is often accomplished by mistreatment the politico Peucker algorithmic rule. After that, we'll discuss triangles and circles. The Douglas-Peucker technique finds the quantity of edges 1st, then it finds the realm of the figure. If the quantity of edges found is 3, the figure is taken into account decent, presumptuous that it meets the token condition. to boot, if there are but or adequate to six edges, and therefore the figure is giant enough, the figure is taken into account a circle of decent radius. within the 1st steps, the shapes should be determined, and once that is done, the bounding box should even be modified. Region of

Interest (ROI) and therefore the remainder of {the awrathed world the realm} are separated by a bounding box. Bounding boxes typically bit external triangles or circles of the detected figure. the surface triangle of a triangular sign bites the bounding box; the inner triangle is at intervals the external triangle and does not touch the bounding box.

Sign Validation and Recognition

A sign's bounding box must be validated once it is found. First, the threshold filter applied to the image is inverted, inverting the inner and outer triangles. The inner triangle is now black, and the outer triangle is now white. First, the threshold filter applied to the image is inverted, inverting the inner and outer triangles. The inner triangle is now black and the outer triangle is now white

Figure 5. *Traffic Sign*

And once an indication is detected, the sign should even be classified. So, we'd be victimisation TensorFlow, that is Google's open supply machine learning framework, and a convolutional neural network is enforced. Both German Traffic Sign Benchmark and Belgian Traffic Signs were used as datasets for training and testing.

Table 1. Traffic Recognition Model Performance

Algorithm	HSV-SVM	YOLOv4
Total test images	429	429
No. of traffic sign board's	410	410
Detected no of signs	221	362
Recognised no of board's	165	310
Accuracy	74.66%	85.39%

4. Future Scope

In expert systems like traffic assistance driving systems, automatic driving systems and automatic road signs identification, traffic sign detection and recognition plays an important role. The technology aids drivers and automatic driving systems in effectively identifying and recognizing traffic signs in order to reach their destinations safely. Using the advanced applications in the future, which can use the deep learning or other applications to read traffic signs, drivers will be able to have global views of the traffic signs on the road.

5. Conclusion

In order to improve the robustness of the algorithm in complex weather conditions. A deep learning way of approach is discussed here. Very large number of experiments have been done which are more successful on this area. Coming to traffic sign detection, this system is successfully capable of detecting, recognizing road signs, followed by calculating the distance of from the car to the said road sign. It has made a car as an autonomous car more intelligent at recognising and responding to real-time traffic events. The results have shown that our algorithm can detect the straight or curve lanes perfectly and the alert of the sign board is spoken perfectly.

ORCID of Authors

1. Nithish Kumar Gannaram - 0000-0002-0639-9804
2. Jitesh Kumar Nayak - 0000-0003-1117-7525
3. Sudarsi Mahesh - 0000-0002-9575-3819
4. Pappala Anil - 0000-0002-7368-0153
5. Vamsi Chikkala - 0000-0001-6177-1253
6. Asst. Prof Nitin Kumar - 0000-0002-0023-9463

References

[1] S. Guiming and S. Jidong, "Multi-Scale Harris Corner Detection Algorithm Based on Canny Edge-Detection," 2018 IEEE International Conference on Computer and Communication Engineering Technology (CCET), 2018, pp. 305-309.

[2] Y. Sun, P. Ge and D. Liu, "Traffic Sign Detection and Recognition Based on Convolutional Neural Network," 2019 Chinese Automation Congress (CAC), 2019, pp. 2851-2854

[3] K. Muhammad, A. Ullah, J. Lloret, J. D. Ser and V. H. C. de Albuquerque, "Deep Learning or Safe Autonomous Driving: Current Challenges and Future Directions," in IEEE Transactions on Intelligent Transportation Systems, vol. 22, no. 7, pp. 4316-4336, July 2021

[4] R. Kavya, K. Md Zakir Hussain, N. Nayana, S. S. Savanur, M. Arpitha and R. Srikantaswamy, "Lane Detection and Traffic Sign Recognition from Continuous Driving Scenes using Deep Neural Networks," 2021 2nd International Conference on Smart Electronics and Communication (ICOSEC), 2021, pp. 1461-1467.

[5] A. Ali, A. Hassan, A. R. Ali, H. Ullah Khan, W. Kazmi and A. Zaheer, "Real-time vehicle distance estimation using single view geometry," 2020 IEEE Winter Conference on Applications of Computer Vision (WACV), 2020, pp. 1100-1109

Dual phenomenon of surface plasmon and laser optics induced EM enhancement for rapid DNA detection with semiconductor nanostructures (SiNWs/AgNPs)

Krithikadevi Ramachandran[a*], Kais Dauodi[a,b], Soumya columbus[a], Abdelaziz Tlili[c], Mounir Gaidi[a,b]**

[a] Centre for Advanced Materials Research, Research Institute of Sciences and Engineering, University of Sharjah, P. O. Box 27272, Sharjah, United Arab Emirates

[b] Department of Applied Physics and Astronomy, University of Sharjah, P. O. Box 27272, Sharjah, United Arab Emirates

[c] Department of Applied Biology, University of Sharjah, P. O. Box 27272, Sharjah, United Arab Emirates

*Corresponding author: pt_kramachandran@sharjah.ac.ac

** Co-corresponding author: mkaidi@sharjah.ac.ae

Abstract

Recently exploring bioanalytical instrumentation principles emphasized with combination of optics and EM fields in surface enhanced Raman spectroscopy (SERS). The principles favour the rapid detection of DNA with proper fabrication of SERS sensor. 1D semiconductor nanostructures (SiNWs) functionalization with plasmonic silver nanoparticles are fabricated by PLD process and exploited for SERS detection of DNA. The DNA nucleotide bands are indexed with reference to their Raman shift as Adenine (A), Guanine (G), Cytosine (C), Thymine (T) at Raman spectral regions of 1352 cm^{-1}, 676 cm^{-1}, 1625 cm^{-1}, 1441 cm^{-1} respectively. The finest conditions for sensitive detection of DNA were investigated and their interaction mechanism are presented. The point-to-point reproducibility measurements imprinted higher degree of concordancy. The designed SERS sensor will be definitely helpful almost in all application fields with portable handheld Raman spectrophotometer in near future.

Keywords: DNA, SERS, SiNWs, silver nanoparticles

1. Introduction

Expansion of the highly sensitive detection strategies towards biomolecules like DNA, RNA, Protein etc., would create a great change in diagnosis domain for their early detection which will be much effective in application fields of medical, forensic, biological systematics and so on [1,2]. Current sensing mechanisms include nonenzymatic methods, single molecule RT-PCR screening and sequencing, mass spectrometry suffers from long time consuming, complex chemical procedures, trained professional, higher maintenance of equipment, higher cost etc., [3,4]. The prime merit of reported SERS technique is comparatively rapid, very sensitive, reliable allows direct sensing with engineered SERS sensors [5,6].

SERS devours the basic optics combined with electromagnetic fields enrichment with their sensor supports. The researchers are on underlay to provide the best design opted for higher EM fields coupling with reference to Lie and Mie scattering principle to enhance the sensitivity with reference to the chemical structure of the analyte under study [7,8]. Generally adopted superficial LSPR properties enriched plasmonic nanoparticles silver and gold are primely employed as SERS sensors [9]. Recently, semiconductor nanostructures conjunction with plasmonic nanomaterials are designed in such a way to provide extreme sensitivity in detection [10].

Herein 1D SiNWs are chosen owing to its high surface to volume ratio, to capture more molecules for detection as well to bring in suitable conditions for light trapping and confinement with plasmonic silver nanoparticles. The SiNWs are prepared by simple Metal assisted chemical etching process within the silver nanoparticles functionalization was provided by PLD technique. The **SiNWs-AgNPs (NA) sensor** characterized by analytical techniques and utilized for the direct, label free DNA detection.

2. Results and Discussion

2.1 Morphological features of NA sensor

The experimental process followed the MACE process for the fabrication of SiNWs and its functionalization is followed with PLD process (5000 NLP laser ablations) which is detailed as in our earlier reports [11]. The fabricated NA sensor morphology is depicted in Fig 1.

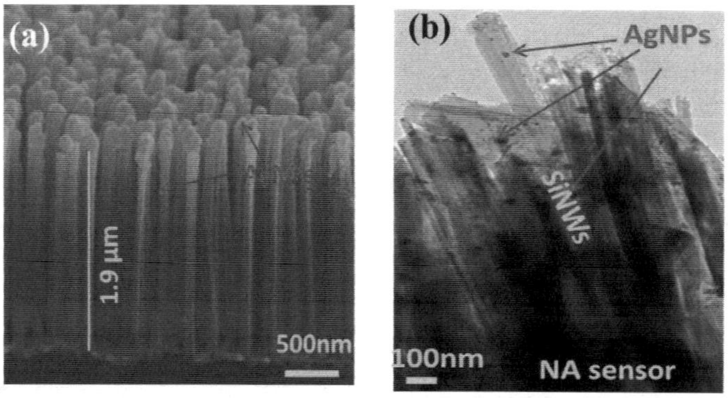

Figure 1. SEM and TEM analysis on the morphology of the NA sensor

The morphology predicted the ordered array of 1D SiNWs of length 1.9µm which is functionalized with smaller spherical AgNPs. NA sensor is exploited for systematic studies of DNA sensing with various optimization parameters of wavelength, Time and power. The commercially purchased synthetic DNA primer (two single strands) with 25 nucleotide each is chosen for SERS sensing. The aqueous DNA solution (7µL) is dropped over the NA sensor surface to cover the top face. It is allowed for drying for few minutes. Then it is kept under Renishaw Raman microscope for its detection.

2.2 Effect of wavelength in SERS DNA sensing:

Wavelength optimization studies (Fig 2 a) is preceded first to gauge its better sensitivity. The wavelength at 488nm, 633nm, 785nm have displayed the DNA spectral features with less sensitivity.

Journal of Physics: Conference Series **2327** (2022) 012018 doi:10.1088/1742-6596/2327/1/012018

It is much difficult to determine the DNA nucleotide bands from the observed spectrum. The backbone moity deoxy ribose sugar is much more prominent with the corresponding wavelengths. The wavelength with respect to NA sensor and DNA analyte were more prone to interpret the backbone molecules of DNA band. Unlike, the 514 nm wavelength excitation portrayed the specific spectral features belonging to DNA. The laser excitation provided with 514nm excitation was found to be optimum to sense the DNA.

2.3 Effect of Lasing power in SERS DNA sensing:

The further analysis is preceded with 514 nm excitation to find the optimum power and time required for sensing. At low lasing power of 0.05 μW and 0.1 μW, the sensing was very weak (Fig 2 b) to produce the resonance conditions almost likely fail to detect the DNA. But at power of 0.5 μW and 1 μW, the DNA bands spectral fingerprints are noted also at > power 1 μW, the bands tend to conjugate and appear as a big lobe in the corresponding Raman active region for DNA. The optimum power for DNA sensing with regard to NA sensor was pointed to be at 0.5 μW.

2.4 Effect of Acquisition Time in DNA SERS sensing:

In the same way, the effect acquisition time required for a complete analysis of the DNA bands are investigated (Fig 2 c). Time period was varied from 10 s to 300 s under 514 nm laser excitation (Fig 2 c) with power 0.5. At 20 s, the distinct spectral features of DNA bands are appeared whereas with the acquisition time greater than 20s only the DNA backbone is visualised with analysis.

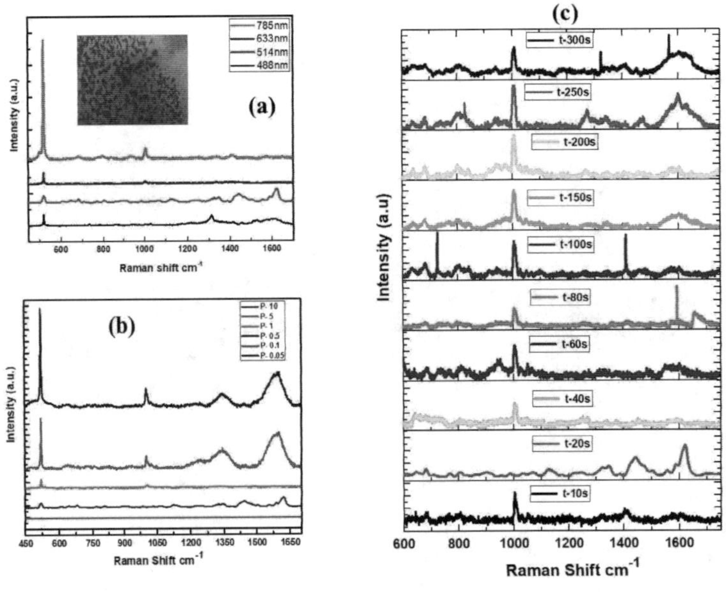

Figure 2: (a) Wavelength optimization studies for DNA detection (b) Power optimization studies (c) Time period optimization studies

2.5 Mechanism of DNA sensing in SERS

With reference to the optimization studies, the sensing conditions are optimized and presented with distinct DNA bands with their assignment (Fig 3 a). For comparison, bare SiNWs are tested with DNA sensing (Fig 3 a). The sensing lies with the context of plasmonic silver nanoparticles functionalized throughout the SiNWs. The Si peak is pronounced at 520 cm^{-1}. Adenine bands were located at Raman labels of 813 cm^{-1}, 1352 cm^{-1}, 1441cm^{-1} respectively. Guanine peaks are pointed at 676 cm^{-1}, 1342 cm^{-1} respectively. Cytosine bands are predominant at 1243 cm^{-1}, 1441 cm^{-1}, 1625 cm^{-1}. Likewise, thymine bands occupied at regions of 813 cm^{-1}, 1441 cm^{-1}. the deoxy ribose sugar and phosphate bands are present at regions of 1002 cm^{-1}, 1113 cm^{-1}, 908 cm^{-1} correspondingly. The assignments were made with reference to literatures [12, 13]. The mapping analysis portrays the excellent replica of the DNA spectrum over the sensor surface area of 25 X 25 cm^2. The 2D mapping is presented in Fig 3b as inset. The slight variations in intensity are common with respect to biomolecules sensing due to their denaturation process as well the changes on their conformation with sensor surface [14].

The mechanism of SERS assisted DNA sensing followed the Electromagnetic enhancement of fields around the region of sensing. The functionalized silver nanoparticles transform the near field to far field to bring the excitation for DNA sensing under 514nm laser wavelength. For the mentioned process the SiNWs having higher surface to volume ratio hosts higher quanta of analyte as well the silver nanoparticles for efficient light trapping and confinement with its length and area. The denser hotspots of silver nanoparticles interact with the DNA chemical moity by tying on its surface owing to adsorption process gives rise to an effective and sensitive detection [11,6]. The interaction process andplausible mechanism is presented in Fig 3b.

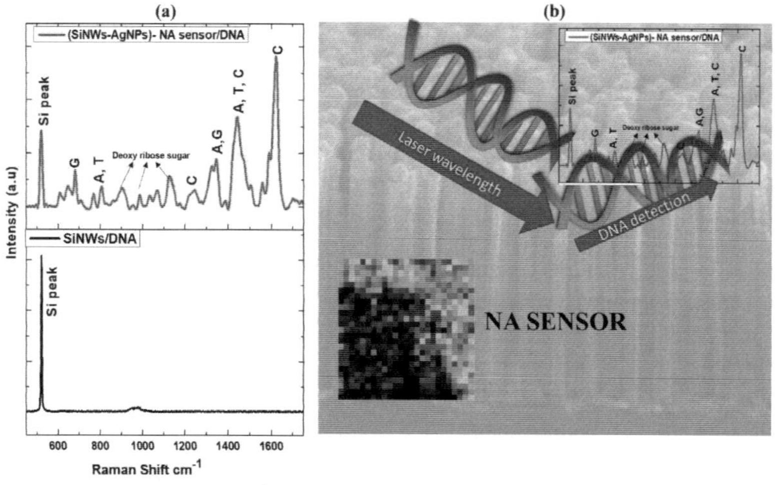

Figure 3: (a) Lower panel – DNA detection with SiNWs; Upper panel – DNA detection with NA sensor (b) Mechanism of SERS DNA sensing with NA sensor

3. Conclusion

Well-ordered SiNWs array functionalized with silver nanoparticles is exploited as an effective SERS sensor for label-free direct detection of DNA within 20 s. systematic investigations on direct DNA sensing with the designed NA sensor was analysed and optimized for finest conditions for sensing with respect to wavelength, power and time. Rapid detection and sensitive detection are the excellent features of the developed NA sensor. The reproducibility patterns imprint higher concorndancy with respect to spectral features of DNA bands. However, the study is preliminary which needs real time validation with respect to clinical samples analysis. The argued research work could be utilized well in screening of DNA without labels for various applications in forensic, health science and bio-medical diagnostics fields. The designed NA SERS sensor will grab the position of label-free biosensor which utilizes electro-optics mechanism for sensing could definitely rule the world of diagnostics in near future.

Acknowledgement

I would like to express my very great appreciation to **G. Raam Dheep** (Associate Professor, School of Electronics and Electrical Engineering, Lovely Professional University, Phagwara, Punjab, India.) for his valuable and constructive suggestions during the planning and development of this research work. His willingness to give his time so generously has been very much appreciated.

References

[1] Chip N, Ngo H T, Wang H, Fales A M and Vo-dinh T 2013 Label-Free DNA Biosensor Basedon SERS Molecular Sentinel on Nanowave Chip
[2] Bi L, Rao Y, Tao Q, Dong J, Su T, Liu F and Qian W 2013 Biosensors and Bioelectronics Fabrication of large-scale gold nanoplate films as highly active SERS substrates for label-free DNA detection *Biosens. Bioelectron.* **43** 193–9
[3] Du Y, Zhao X, Zhao B, Xu Y, Shi W, Ren F and Wu Y 2019 A novel emulsion PCR method coupled with fluorescence spectrophotometry for simultaneous qualitative, quantitative and high- throughput detection of multiple DNA targets *Sci. Rep.* 1–9
[4] Thomson D and Dietzgen R G 1995 Detection of DNA and RNA plant viruses by PCR and RT-PCR using a rapid virus release protocol without tissue homogenization **54** 85–95
[5] Han Z, Liu H, Meng J, Yang L, Liu J and Liu J 2015 Portable Kit for Identification and Detection of Drugs in Human Urine Using Surface-Enhanced Raman Spectroscopy
[6] Wang Y, Yan B and Chen L 2012 SERS Tags: Novel Optical Nanoprobes for Bioanalysis
[7] Moldovan R, Vereshchagina E, Milenko K, Iacob B, Elena A, Falamas A, Tosa N and Muntean C M 2021 Analytica Chimica Acta Review on combining surface-enhanced Raman spectroscopy and electrochemistry for analytical applications
[8] Ding S, Zhang X, Ren B and Tian Z 2013 Spectroscopy (SERS): General Introduction, Encyclopedia of Analytical Chemistry, Online © 2006–2013 John Wiley & Sons, Ltd
[9] Li C, Huang Y, Li X, Zhang Y, Chen Q, Ye Z, Alqarni Z and Bell S E J 2021 Towards practical and sustainable SERS: a review of recent developments in the construction of multifunctionalenhancing substrates 11517–52
[10] Sharma B, Frontiera R R, Henry A, Ringe E and Duyne R P Van 2012 SERS: Materials, applications, and the future Surface enhanced Raman spectroscopy (SERS) is a powerful vibrational *Mater. Today* **15** 16–25
[11] Gaidi M, Daoudi K, Tlili A, Columbus S, Leblanc-Lavoie J, Ramachandran K, Suleiman B, Alhazaa A N and El Khakani M A 2021 Fast, highly sensitive and label free detection of small geneticsequence difference of DNA using novel Surface-Enhanced Raman Spectroscopy nanostructuredsensor *Sens. Bio-Sensing Res.* **32**
[12] E. Pyrak, J. Krajczewski, A. Kowalik, A. Kudelski, and A. Jaworska, 2019 Molecules **24**, 4423
[13] Harper M M, Dougan J A, Shand N C and Faulds K 2012 Detection of SERS active labelled DNA based on surface affinity to silver 2063–8
[14] Szekeres G P and Kneipp J 2019 SERS probing of proteins in gold nanoparticle agglomerates *Front. Chem.* **7** 1–10

An Autonomous Smart Farming System for Computational Data Analytics using IoT

A I Rokade[1], A D Kadu[2], K S Belsare[2]

[1]Ph.D Research Scholar, School of Electronics and Electrical Engineering,
Lovely Professional University, Phagwara, Punjab, India.
[2]Assistant Professor,ECE, PRMIT&R,Badenera.
[*]Corresponding Email Id: rokadeashay@gmail.com

Abstract – For improved agricultural growth control, smart farming with precise greenhouses is essential, as is precision agriculture monitoring in a variety of situations. The Internet of Things (IoT) is a new era in computer communication that is gaining pace as a result of its wide variety of project development applications. Individuals may benefit from the IoT through smart and remote ways such as smart agriculture, smart environment, smart security, and smart cities. These are the latest technologies that are making life simpler in today's world. The IoT has significantly increased remote control and the variety of networked things or devices, which is a fascinating aspect. The hardware and internet connectivity to the real-time application make up the Internet of Things (IoT). The Internet of Things is made up of sensors, actuators, embedded systems, and a network connection. As a result, we'd want to develop an IoT application for smart farms. This paper demonstrated a remote parameter sensing system in smart greenhouse agriculture. The goal is to monitor greenhouse parameters like CO_2, soil moisture, temperature, humidity, and light, with adjusting actions for greenhouse windows/doors based on crops. In this experimentation, Gerbera and Broccoli is considered. The primary purpose is to adjust greenhouse conditions in line with plant needs in order to increase production and provide organic farming. As a result of the findings, it appears that the greenhouse might be operated remotely for CO_2, soil moisture, temperature, humidity and light, resulting in improved management. Overall implementation is remotely monitored via IoT using MQTT on Adafruit IO Cloud Platform and sensor data is analyzed for its normal and anomaly behavior.

Keywords – Smart Farming, Greenhouse Management, Internet of Things (IoT), Precision Farming

1. Introduction

Plant development has proved to be a novel test, given that plant field and strength are increasingly key factors for cash crops and food crops alike. The lack of knowledge about agricultural elements and information about developing breakthroughs is one of the primary challenges in modern agriculture [1]. Our forefathers eschewed employing specialized development for particular plant growth in favor of using a general marvel for all plants in the past. As a consequence of technological improvements in agriculture, plants may be produced under unusually typical natural circumstances, and specialized plants can be cultivated under specified conditions, resulting in higher yield and less compost [3].

Precision agriculture in green houses for plant development is gaining popularity as a consequence of lower-cost technology that allow farmers to re-arrive production. The greenhouse is a translucent house-like structure that can manage temperature, moisture, light infiltration, and other elements for optimal plant development. Precision agriculture is a system for detecting, monitoring, and reacting to environmental changes. It is a technique for detecting greenhouse climate, after which the data is recognised and uploaded to the cloud, and the agriculturist takes the appropriate action depending on the data. This may be observed in the current breakthrough known as the Internet of Things (IoT), which is a technology that uses web-based ways to link everything or any device to the internet. The exact agricultural framework is improving as a result of recent advancements in Wireless Sensor Networks (WSN), which is nothing more than an IoT. Precision agriculture is the most advanced cultivating technology with a vast framework that has lately emerged. In the case of a sudden shift in the greenhouse, it entails identifying, estimating, and transmitting nursery data to farmers [2] [5]. A greenhouse is a structure made of simple materials that maintains a microclimate conducive to solid plant growth, such as water stream management, temperature control, and so on. As a consequence, it protects against things like excessive light penetration, severe temperatures, illnesses, and creepy crawlies. Maintaining ideal microclimate conditions, improving irrigation and fertilization practices, controlling infection and avoiding disease outbreaks, and preventing theft and improving security are the four key benefits of a smart greenhouse system as defined in fig 1.

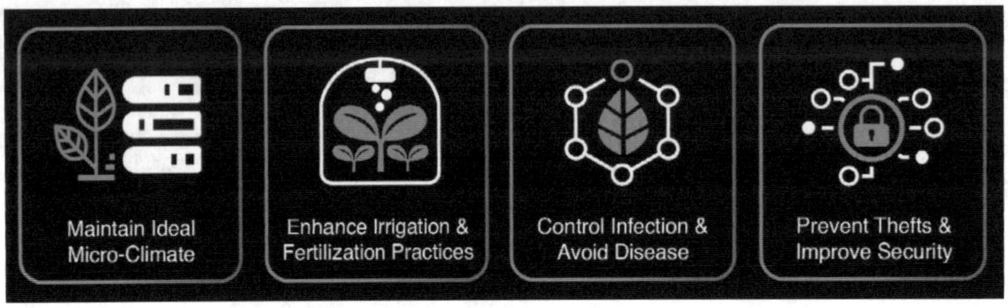

Fig 1: Advantage of Smart Greenhouse System

A farmer may grow any plant in any season by keeping the right environmental conditions. Returning to the topics of interest, there is the significance of greenhouse farming, which illustrates why they have become so popular. Because greenhouses hold moisture, they use far less water than traditional horticulture. It reduces the amount of time spent editing and expands the sorts of harvests available. Temperature and humidity are carefully controlled to meet the demands of the plants. Nurseries can also be used to develop slow-growing crops. Pests may be readily dealt with. Harvests may be made under a wide range of environmental conditions, making it incredibly versatile. The varying climatic conditions in the greenhouse will affect the plant's growth, resulting in a lesser yield at the end of the cultivation. As a result, greenhouse gases, soil moisture, temperature, and light must all be managed and monitored. This problem could be solved by implementing an Internet of Things (IoT) innovation in precision agriculture, which involves the precise application of certain greenhouse factors for optimal plant development, such as temperature management, water flow control, and light radiation, among others [4][7][10].

2. Related work

Several researchers have contributed to the development of a smart agriculture system. Initially, several sensors and actuators were utilized by the authors to achieve their goals in order to get the most efficient management of it. The following are summaries of those papers.

In complete distribution nurseries employing tolerably saline water, a flexible stage is characterized that may adjust to soilless culture conditions. The usage of IoT technology is used to create a highly adjustable perceptual framework for regulating and tracking nursery temperature [2]. The author adds [3] to continuing IoT breakthroughs in horticulture, as well as enhancements to equipment and programming frameworks. An automated FLC with IoT capabilities is built for the analysis and etymological dynamic of fertigation in a nursery [4]. A new proposal has been proposed for agricultural farmland prudence. The system employs a Raspberry Pi 3 to detect any suspicious activity or activity in the ranch area, after which the PiCam is triggered to take a photo of the situation. [5]

VegIoT Garden, energy-efficient, a low-effort and personal IoT stage for SA based on COTS devices [6], having received communications conventions, and designed to improve the planning of vegetable nurseries through acquisition, inference, and analyzation of sensor information, recognised with substantial limits of developing plants, is a limited, The suggested arrangement partially lowers [7] network inactivity when compared to conventional IoT-based horticulture and growing solutions. A monitoring agency collected data for certain harvests, then used AI to construct an alert notification which included the data and information. The architecture [9] attempts to provide a continual weed control in onions farms using a Image Processing based integrated weed control framework (WCS). Founded on the idea of LoRa and fog processing innovation, an automated agricultural platform [10] has been developed.

A fuzzy-based technique was used to study the SE and EE concerns in 5G networks and achieve an efficient symmetrical trade-off between them to improve the system's overall performance using a look-up table [11] [13]. To find the best places for these mobile nodes, a new soft computing approach called Adaptive Plant Propagation Algorithm (APPA) is used [12]. The system presents an intelligent greenhouse system for adopting monitoring and controlling in the precision agricultural domain, based on an examination of current and prior related research that examine the greenhouse system's primary aspects, benefits, and limitations [15]. Examines how artificial intelligence (AI) and telemedicine have been employed to improve continuous monitoring [16]. The purpose of this study is to examine the issues and challenges that single-carrier NOMA (SC-NOMA) technologies encounter, as well as to investigate the impact of increasing the number of SC-NOMA users on average capacity when bandwidth (BW) and power are raised [17] [18] [19].

3. Proposed work

To test the proposed smart greenhouse system, a prototype experimental model was created using an embedded system device that includes several sensors and a microprocessor, as shown in fig 2. The proposed model is designed to track several greenhouse characteristics for two crops, Gerbera and Broccoli, in different climates. A microcontroller Node MCU ESP 32 is connected to all of the needed sensors for obtaining the green house parameters. A personal computer is used to collect data serially with timestamp values for data logging of various parameters. Temperature and Humidity, Light Intensity, CO_2 and Soil Moisture are all continually monitored for 10 days on the Adafruit IO cloud Platform utilizing the MQTT protocol under day and night conditions within specific time intervals. Operational workflow of proposed system that how the system operates as defined in fig 3.

Journal of Physics: Conference Series 2327 (2022) 012019 doi:10.1088/1742-6596/2327/1/012019

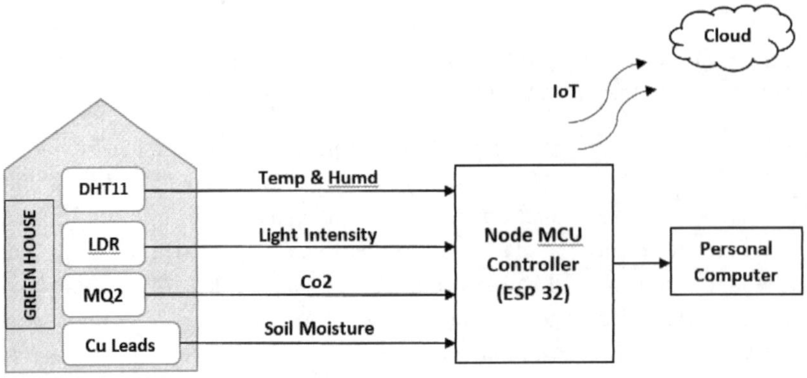

Fig 2: Proposed Experimental Model

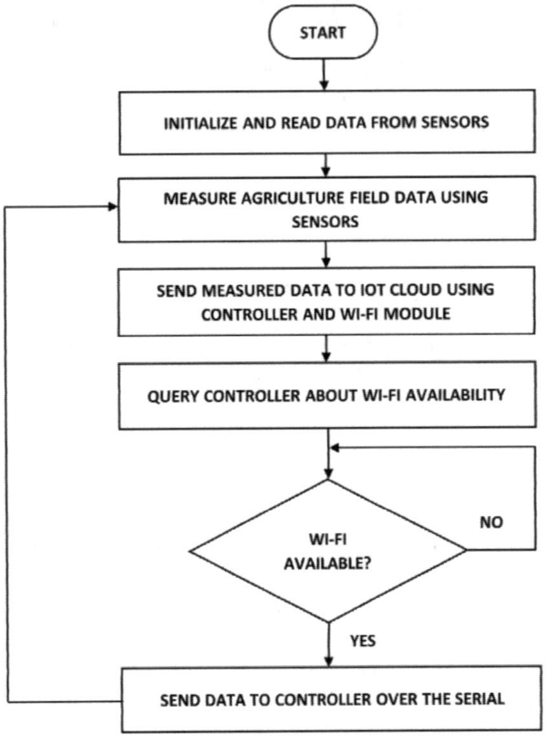

Fig 3: workflow of experimental model

Adafruit IO is a platform for visualizing, responding to, and interacting with sensor data. With the support of MQTT, the data is also kept private and safe. MQTT (Message Queue Telemetry Transport) is a TCP/IP-based lightweight publish-subscribe protocol. MQTT employs a message broker to route messages between senders who send them and receivers who are interested in receiving them. Messages can be published and subscribed to using the same client. Each message is associated with a specific subject. The message routing information is the topic, and it is essentially a string with slash separated hierarchical levels. Clients subscribe to these topics, and the broker sends them all messages with topics that match their subscriptions. Wildcards can also be used to easily subscribe to several topics. For example, as shown in fig 4, how temperature sensor data from a greenhouse system is sent.

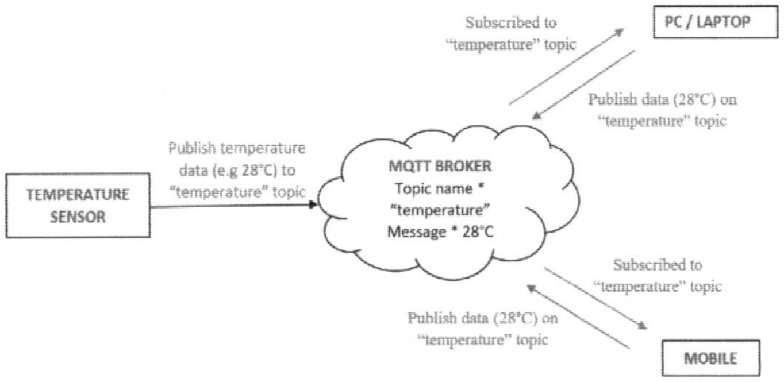

Fig 4: workflow of MQTT for sensor data

4. Experimental Results and Discussion

The proposed experimental plan is implemented on a prototype that has been tested practically as shown in fig 5 on two crops for various conditions. The two primary phases of experimentation are the creation of basic models embedded system for plant growth and feeding, the construction of a sensor net for smart greenhouse monitoring, and the automation of actuators. The proposed approach employs an embedded system to analyses greenhouse execution parameters like as CO_2, soil moisture, temperature, and plant light, yielding realistic results.

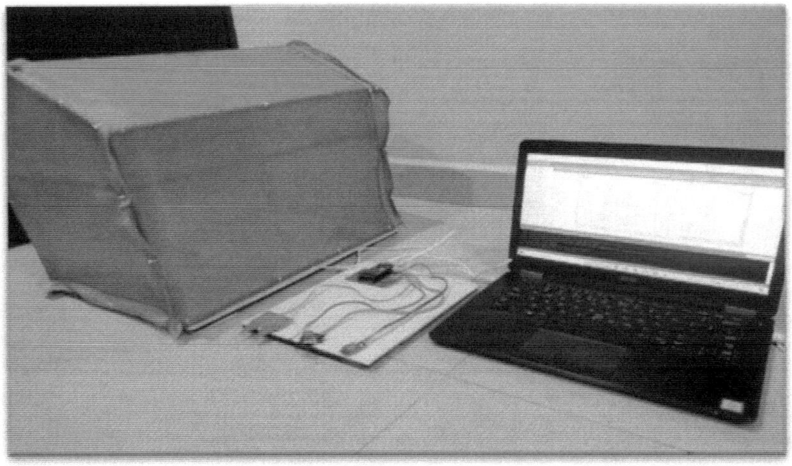

Fig 5: Experimental Test bench

All the readings are observed under several possible conditions and monitored on personal computer over serial communication. All these sensor's data is monitored over an Internet on Adafruit IO Cloud dashboard by publishing data from nodes to broker of Adafruit. Then user can subscribe this data to access it in real time.

The greenhouse doors/windows can also be rolled on/off depending on the soil moisture conditions. The plant photosynthesis process requires a high level of CO_2 concentration and water in the evenings rather than during the day; with these two sources of energy, the photosynthesis method keeps the plant cool and encourages rapid growth. Because the greenhouse absorbs CO_2 from day to night, maintain a

CO_2 level maximum at night after completing a CO_2 concentration level experiment in a green house, as illustrated in figure 6. As a result, as seen in Figure 6, the CO_2 level decreases during the day.

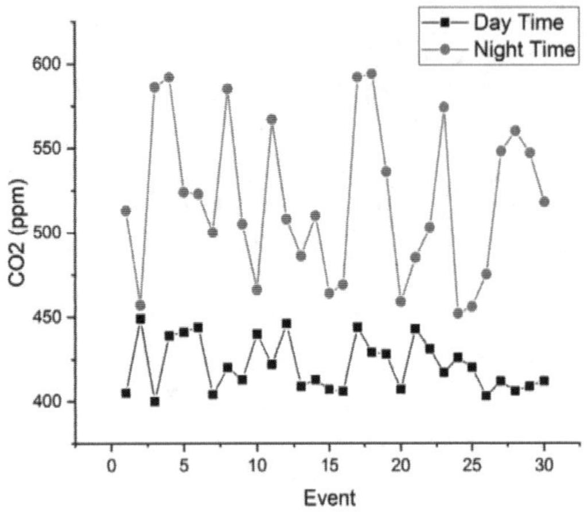

Fig 6: CO2 representation

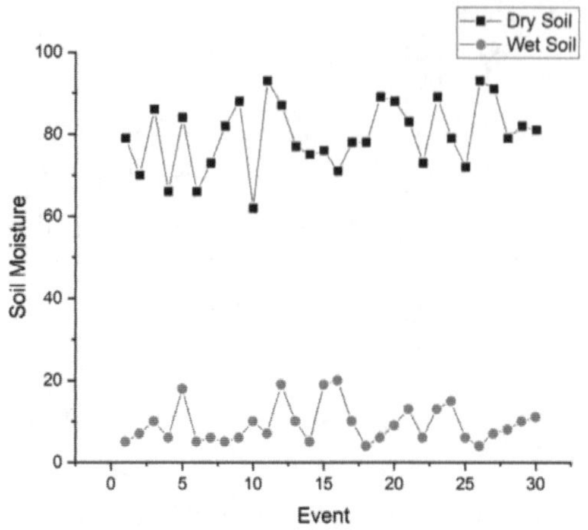

Fig 7: soil moisture representation

Because too much water can induce a fungal infection in the plant, and too little water can cause the plant to grow dry or even die, the amount of water in the soil is critical. As a result, the amount of water required by the plant is crucial. During the night, plants demand more water with CO_2 for photosynthesis. When the soil moisture sensor gives a negative value, the plants have completely covered the water, as shown in figure 7, and the greenhouse windows/doors will close automatically using a DC motor. The positive value indicates that the soil is dry and has to be re-wet, as shown in Figure 7.

Temperature is one of the most important parameters in a greenhouse; it should be kept as high as feasible. Blooming, fruiting, photosynthesis, seed germination, and other processes are all aided by warmth.

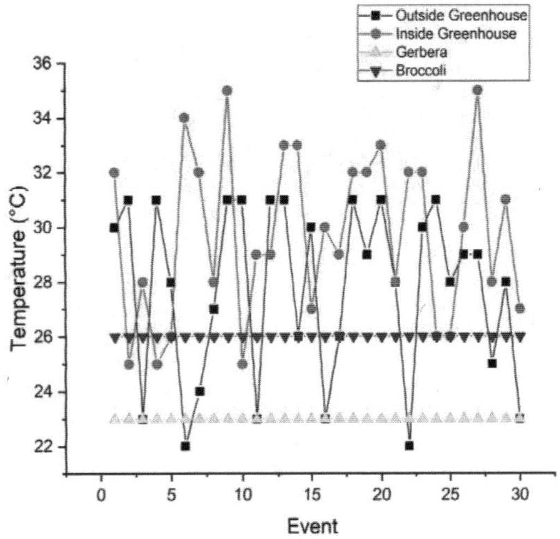

Fig 8: Temperature representation

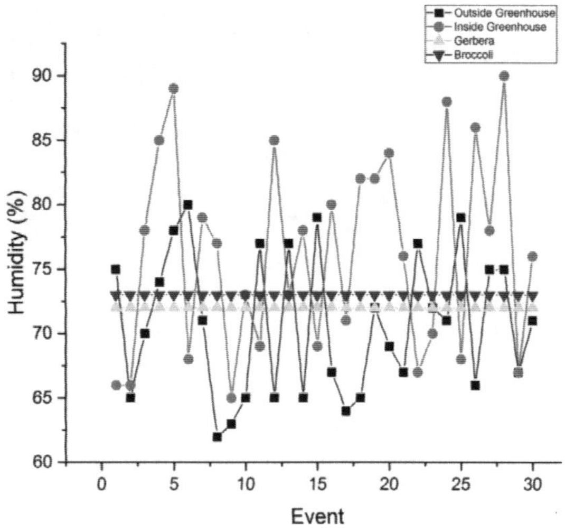

Fig 9: Humidity representation

As a result, in comparison to the temperature and humidity range in the outside greenhouse environment, the temperature and humidity range in the greenhouse was maintained to the fullest degree

Journal of Physics: Conference Series **2327** (2022) 012019 doi:10.1088/1742-6596/2327/1/012019

possible, as illustrated in Figures 8 and 9. The photosynthetic mechanism, which is found in the green component of plants and is responsible for plant development, blooming, and form, is aided by the many hues of sunlight.

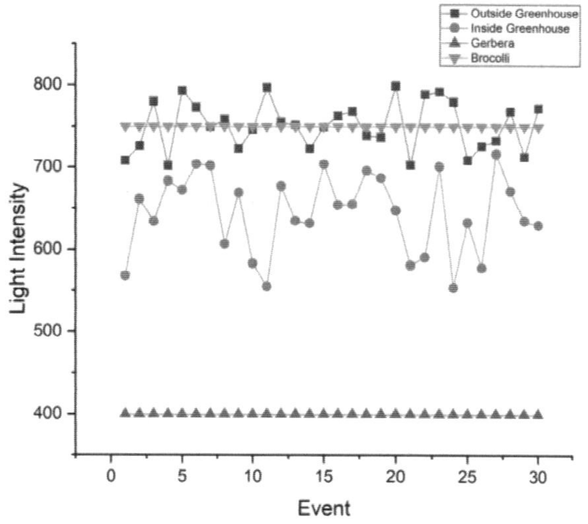

Fig 10: Light intensity representation

As a result, a sustainable amount of light penetration inside the greenhouse was maintained, as illustrated in Figure 10, as contrasted to usual light penetration outside the greenhouse.

5. Conclusion

A smart greenhouse system is a hybrid of IoT and agriculture that is driven by market demand and optimized according to a set of rules. Automation and high efficiency are required for monitoring and regulating the greenhouse environment. Because the precise information was discovered, they could be completely included in the gap analysis based on their constraints and the likelihood of a job extension. In traditional greenhouses, growers can control environmental parameters using a proportional control system that involves manual intervention, which typically leads in output loss, energy waste, and greater labor expenses. An automated greenhouse monitoring system based on the IoT comes to the rescue to handle these difficulties. The IoT is a widely used technology for linking things and collecting data. On Adafruit IO, the device is designed to remotely monitor greenhouse elements including as CO_2, soil moisture, temperature, and light. Farmers may collect this data using an Adafruit IO cloud account and an internet connection. Thus, the system will employ IoT to help farmers eliminate physical visits to the field while simultaneously enhancing yield by maintaining accurate greenhouse factors such as CO_2, soil moisture, temperature, and light. The findings for greenhouse elements such as CO_2, soil moisture, temperature, and light for broccoli and gerbera plants are investigated with the help of graphical representation based on the practical data collected by the suggested model. Furthermore, utilizing artificial intelligence, work on more particular agriculture crops with precision agriculture can be done.

Acknowledgement

I would like to express my very great appreciation to **Dr. Manwinder Singh** (Associate Professor, School of Electronics and Electrical Engineering, Lovely Professional University, Phagwara, Punjab, India.) for his valuable and constructive suggestions during the planning and development of this research work. His willingness to give his time so generously has been very much appreciated.

ORCID DETAILS OF AUTHORS

1. Ashay Rokade – 0000-0002-5695-6533
2. Ankush Kadu- 0000-0003-4738-5123
3. Karan Belsare- 0000-0003-2278-8153
4. Manwinder Singh- 0000-0002-0543-2625

6. References

[1] Zamora-Izquierdo, Miguel & Santa, José & Martinez, Juan & Martínez, Vicente & Skarmeta, Antonio. (2019). Smart farming IoT platform based on edge and cloud computing. Biosystems Engineering. 2019. 4-17. 10.1016/j.biosystemseng.2018.10.014.

[2] Subahi, Ahmed & Bouazza, Kheir Eddine. (2020). An Intelligent IoT-Based System Design for Controlling and Monitoring Greenhouse Temperature. IEEE Access. PP. 1-1. 10.1109/ACCESS.2020.3007955.

[3] Kour, Vippon & Arora, Sakshi. (2020). Recent Developments of the Internet of Things in Agriculture: A Survey. IEEE Access. 8. 129924-129957. 10.1109/ACCESS.2020.3009298.

[4] A. Carrasquilla-Batista and A. Chacón-Rodríguez, "Standalone Fuzzy Logic Controller Applied to Greenhouse Horticulture Using Internet of Things," 2019 7th International Engineering, Sciences and Technology Conference (IESTEC), Panama, Panama, 2019, pp. 574-579, doi: 10.1109/IESTEC46403.2019.00108.

[5] Shanmukhappa Angadi, Raghavendra Katagall, "Agrivigilance: A Security System For Intrusion Detection In Agriculture Using Raspberry Pi And Opencv", INTERNATIONAL JOURNAL OF SCIENTIFIC & TECHNOLOGY RESEARCH VOLUME 8, ISSUE 11, NOVEMBER 2019.

[6] Codeluppi, Gaia & Cilfone, Antonio & Davoli, Luca & Ferrari, Gianluigi. (2019). VegIoT Garden: a modular IoT Management Platform for Urban Vegetable Gardens. 10.1109/MetroAgriFor.2019.8909228.

[7] N. Ahmed, D. De and I. Hussain, "Internet of Things (IoT) for Smart Precision Agriculture and Farming in Rural Areas," in IEEE Internet of Things Journal, vol. 5, no. 6, pp. 4890-4899, Dec. 2018, doi: 10.1109/JIOT.2018.2879579.

[8] A. A. Araby et al., "Smart IoT Monitoring System for Agriculture with Predictive Analysis," 2019 8th International Conference on Modern Circuits and Systems Technologies (MOCAST), Thessaloniki, Greece, 2019, pp. 1-4, doi: 10.1109/MOCAST.2019.8741794.

[9] M. P. Arakeri, B. P. Vijaya Kumar, S. Barsaiya and H. V. Sairam, "Computer vision based robotic weed control system for precision agriculture," 2017 International Conference on Advances in Computing, Communications and Informatics (ICACCI), Udupi, 2017, pp. 1201-1205, doi: 10.1109/ICACCI.2017.8126005.

[10] M. Baghrous, A. Ezzouhairi and N. Benamar, "Towards Autonomous Farms Based on Fog Computing," 2019 2nd IEEE Middle East and North Africa COMMunications Conference (MENACOMM), Manama, Bahrain, 2019, pp. 1-4, doi: 10.1109/MENACOMM46666.2019.8988547.

[11] Khalaf, O.I., Ogudo, K.A. and Singh, M., 2021. A fuzzy-based optimization technique for the energy and spectrum efficiencies trade-off in cognitive radio-enabled 5G network. Symmetry, 13(1), p.47.
[12] Walia, G. S., Singh, P., Singh, M., Abouhawwash, M., Park, H. J. et al. (2022). Three-Dimensional Optimum Node Localization in Dynamic Wireless Sensor Networks. CMC-Computers, Materials & Continua, 70(1), 305–321.
[13] Singh, M., Kumar, M. and Malhotra, J., 2018. Energy efficient cognitive body area network (CBAN) using lookup table and energy harvesting. Journal of Intelligent & Fuzzy Systems, 35(2), pp.1253-1265.
[14] M. Hassan, M. Singh and K. Hamid, "Survey on NOMA and Spectrum Sharing Techniques in 5G," 2021 IEEE International Conference on Smart Information Systems and Technologies (SIST), 2021, pp. 1-4, doi: 10.1109/SIST50301.2021.9465962.
[15] Rokade and M. Singh, "Analysis of Precise Green House Management System using Machine Learning based Internet of Things (IoT) for Smart Farming," 2021 2nd International Conference on Smart Electronics and Communication (ICOSEC), 2021, pp. 21-28, doi: 10.1109/ICOSEC51865.2021.9591962.
[16] Kadu and M. Singh, "Comparative Analysis of e-Health Care Telemedicine System Based on Internet of Medical Things and Artificial Intelligence," 2021 2nd International Conference on Smart Electronics and Communication (ICOSEC), 2021, pp. 1768-1775, doi: 10.1109/ICOSEC51865.2021.9591941.
[17] Mohamed Hassan, Manwinder Singh & Khalid Hamid (2021) “IMPACT OF POWER AND BANDWIDTH ON THE CAPACITY RATE AND NUMBER OF USERS IN SC-NOMA”, Harbin Gongye Daxue Xuebao/Journal of Harbin Institute of Technology, 53(9), pp. 118–124. Available at: http://hebgydxxb.periodicales.com/index.php/JHIT/article/view/726 (Accessed: 15 February 2022).
[18] M. Hassan, M. Singh and K. Hamid, "Survey on NOMA and Spectrum Sharing Techniques in 5G," 2021 IEEE International Conference on Smart Information Systems and Technologies (SIST), 2021, pp. 1-4, doi: 10.1109/SIST50301.2021.9465962.
[19] Hassan, M., Singh, M., & Hamid, K. (2021). Overview of Cognitive Radio Networks. Journal of Physics: Conference Series, 1831.
[20] S. K. Roy, M. Singh, K. K. Sharma, C. Bhargava and B. P. Singh, "Mathematical Modelling of Simple Passive RC Filters Using Floating Admittance Technique," 2020 IEEE International Conference for Innovation in Technology (INOCON), 2020, pp. 1-6, doi: 10.1109/INOCON50539.2020.9298230.

Disaster Monitoring based on IoT and Long Range Assisted Framework

Swapnil Bagwari[1] , Ajay Roy[1]*, Rajesh Singh[2], Anita Gehlot[2]

[1]SEEE, Lovely Professional University, Phagwara, Punjab
[2]R&I, Uttrakhand University, Dehradun
ajoy.22652@lpu.co.in*,swapnil.15936@lpu.co.in, srajsssece@gmail.com

Abstract. We encounter so many hazardous situations which effects living species and the materials they own. Due to natural calamities and environmental disasters, there is an increase in losses of lives and materials. Landslides are caused due to heavy rains, sloping of land because of earthquakes and deforestation. The research is about the preventive landslide detection and alerting system using various sensors and Arduino. The data collected by the sensors is transferred to Arduino which is further sent to NODEMCU through Lora module. The data obtained at NODEMCU will be transferred to database server for real-time monitoring and processing. Due improvement in the field of IoT systems has made the use of low cost, low power LoRA technologies to detect the environmental disaster and prevent from dangerous situations with less damage to living beings. In view to cover larger area with reduced costs and low power consumption, we used the LoRa technology for transmitting the monitoring data and Wi-Fi communication is used for sending the alerts. The processed data will be displayed on the mobile application interface.

Keywords: Long Range, IOT, Cloud Computing, Landslide Monitoring

1. Introduction

A disaster is an incident or sequence of events that result in deaths, injuries, and disruption or destruction of property, facilities, the atmosphere, critical resources, or means of subsistence on a scale that exceeds the affected community's natural capacity to cope. A "calamitous situation which has been broken the usual course of life, eco-system and exceptional emergency efforts needed to save the environment and sustain lives" is another way to describe a disaster. The discipline in coping with and preventing both natural and man-made disasters is known as disaster management.[8] Mitigation, preparedness, response, and recovery are four phases involved in disaster management process. The mitigation process seeks to lower catastrophe costs by concentrating on long-term disaster avoidance initiatives. The creation of an action plan for an impending disaster is part of the preparedness process. When disaster occurs, the response phase involves mobilising services and relief, while the recovery phase involves restoring the affected area to its previous state [9]. Natural disasters are frequent in developing countries, and there is growing fear that they will become more common as a result of global warming. Disasters cause substantial economic harm and can result in a large number of deaths. In 2009, there were 335 natural disasters recorded around the world, killing over 10,000 people and causing over 41 billion dollars in damages[10]. Earthquakes, tsunamis, floods, bushfires, hurricanes, droughts, and heat waves have all happened at some point in history. The people have adjusted themselves to the riskier situations due to these occurrences. However, disaster impacts have recently increased significantly, partly as a result of climate change's exacerbating effects, but also due to complex growth in socio-ecological processes in a highly interconnected and globalized world. For an instance, 2017 saw a sequence of storms (Maria, Harvey, and Irma) in the Caribbean, as well as a major earthquake in Mexican City, among other incidents, resulting in the highest ever reported losses (US$ 135 billion)[12]. Around the world, geological disasters cause extensive property damage, physical injuries, and fatalities. Landslides are a dangerous geological hazard that has a significant effect on transportation development and traffic. These events are normally caused by a loss of stabilization of the solid block in a specific region. Such changes can be caused by a variety of factors, such as drought, strong wetness, snowmelt, vegetation, and so on. Rainfall, on the other hand, reported to be the most usual cause for roadway landslides.[1] Landslides are frequently caused by high rainfall in areas with a certain land slope, which is exacerbated by deforestation or earthquakes.[2] Landslide casualties are impossible to prevent because the land collapses so rapidly that residents in the area of the landslide are unable to save themselves[3].Currently, two main techniques are used for such investigation: [6]

- Geodesic approaches, which are focused on establishing punctual goals within the landslide
- Methods of remote sensing that have an extended movement field in the zone under consideration

Landslides, sinkage, and rock bursts caused by rock collapse causes massive destruction to people, industry, property and agriculture all over the world. Some landslides are devastating, and thousands of people got injured or killed as a result of landslides and the secondary disasters they cause. Landslides cause significant economic damages in the billions of dollars, as well as severe negative impacts on national economic development [4]. As a result, this study aims to build a ground-level intelligent monitoring device that can trace and track environmental activity in terms of humidity, moisture content and rainfall. The combination of sensors, microcontrollers, LoRa technology and cloud technology will have an impact on the regional communities, allowing people to access real time information at any time via a built real-time database [4]. A disastrous situation necessitates a quick and effective rescue response, particularly when an whole community and neighborhood is affected, because the danger can spread quickly, affecting many lives and properties [7].

2. Related Works

Several experiments have resulted in the development of electronic instrument designs capable of observing and detecting signs of landslides using a variety of methods. [3] Landslide monitoring has become increasingly reliable in current years, thanks to the ongoing development in the field of landslide monitoring, alerting and forecasting instruments, contents, and methods.[4] Table.5 shows the related works done by some of the researchers in the field of disaster management and monitoring. Purnomo, Fendi Aji [5] made a early land slide detection system using Arduino Module having ATMega328 Microcontroller as a central device, sliding potentiometer to detect soil movement and soil moisture sensor to detect soil moisture. The monitored data can be viewed through the computer connected or through the webpage. Tao, Zhigang [4] had made a CRD cable based landslide movement detection system. Due to the sliding force working principle of CRD cable, they detects the deep sliding force of the slope and compares it with early warning threshold. If force of slope crosses the threshold, an early warning will be detected and alerts are given using an app. Romdhane, Rihab Fekih [6] designed a landslide detection system using 4 different sensors: capacitive accelerometer, pressure sensor, magnetometer, and angular rate sensor. This system uses LoRa based communication to send data wirelessly. Fajardo, Jovilyn, et al. [9] discusses about android mobile application which serves as a medium to find out the best possible route for the volunteers and rescuers to reach the disaster prone area safely. For optimization, a genetic algorithm was used, and various parameters were modified to find the best path.

Jalooli, Ali, et al. [1] proposes a landslide technique based on the motion detection and cross movement methods. To identify the degree of landslide effect on road they have implemented the above two methods in detection phase. Roadway landslide are monitored and detected through image processing techniques and intimated to vehicles through vehicle-to-vehicle (V2V), infrastructure-to-vehicle(I2V) communications. Taqwa, Ahmad, et al. [3] designed a simple and low cost early landslide detection system using LoRa and IoT. They have used two sensors: HC-SR04 ultrasonic sensor to detect the soil movement and YL-69, a soil-moisture sensor detects the moisture content of soil. The sensor data is processed by Arduino and sent through LoRa to the nearest people for free monitoring. An android application was built to make ease of monitoring. Khadar, et al. [13] develops a Wireless Sensor Network to detect forest fires in Rieu province for the region with high risk. Deployment of sensor nodes at various location to collect the data then mathematical analysis of the data for early warnings and alerts to the representative organization for speed rescue. Herlambang, P., et al. [2] research about the four parameters: ground shift sensors, rain-gauge sensors, groundwater pressure sensors and soil-moisture sensors which are indicating parameters of landslides. The data collected from these sensors is processed by Arduino which is sent to server through NodeMCU. The obtained data is analysed and a decision-making algorithm is designed to compare the obtained values with predefined value to determine landslides. Dragulinescu, Ana-Maria, et al. [7] proposes an IoT platform which collaboratively monitors the neighbourhood for all emergency dangers then alerts the inhabitants and the rescue teams. In this paper, a testbed was designed to test the different fire scenarios. They used LoRa for the communication purpose and NodeMCU Wi-Fi module for alerting purposes. Ragnoli, Mattia, et al. [17] designed an autonomous LoRa based low-power flood detection system with a modular perspective which does not need much modifications to the proposed node. It consists of various sensors, microcontroller which are connected through LoRa module for long range data transmission. Alerts are made if flooding scenario is detected. Khan, Amina, et al. [14], this is a survey paper which provides a critical examination of existing disaster management related technologies and approaches, such as WSN, remote sensing techniques, satellite imagery, artificial intelligence, IoT, UAVs, in order to address concerns related to disaster control, identification, and management.

3. Technologies

Wireless Sensor Networks can be used for a variety of implementations, including remote environmental monitoring, remote sensing, industrial automation, and target tracking. In most cases, a WSN consists of a huge number of small nodes, which are installed in isolated and far off hostile environments or across huge geographical areas. There are huge number of small nodes of sensor which can detect changes in environment and sends report to the sensor base station or cluster head node, which then transfers data to the servers through a gate way, which should be simple to deploy and maintain [13]. Communication networks are crucial in mitigating the effect of disasters by building the required connectivity, whereas commercial network is not reliable due to a lack of redundancy; thus, these systems are at most available when they are needed the most whenever a disaster has hit [14].

Table 1 : WSN based landslide detection and monitoring systems

Ref No	Landslide detection method	Communication	Radio Module	Sensors	Merits
[6]	WSN	LoRa	iM880A/STM 32L511Cx	3 axis capacitive accelerometer, magnetometer, pressure sensor	Low Power, low cost
[22]	WSN	Zigbee (IEEE802.15.4)	ZICM2410	Soil moisture, soil temperature, rainfall, tilt angle	Low cost, low power, small size
[23]	WSN	GPRS	Texas CC2530	Air thermometer, raingauge, wind gauge, tiltmeter	Low power
[24]	WSN based (SMARTCON)	GPS trajectories	Nordic NRF24LO1	ADXL 345, Temperature and humidity sensor	Low power
[25]	WSN	WiFi/IEEE 802.11B	Micaz mote	Pore pressure transducer, moisture sensor, geophones ,rainguage, tiltmeter	Low cost sensor colume
[26]	WSN	WiFi/Zigbee/GPRS/BLE	Waspmote	Meshlium sink, precipitation sensor, accelerometer	Low energy consumption due to external database
[27]	WSN	IEEE802.15.4g	Wi-sum acceleration sensor system	Wi-sun acceleration sensor (7 sensors, slope), accelerometers	Low complexity, traffic reduction, low energy consumption
[28]	WSN and statistical model	GSM/GPRS	Micaz mote	Pressure vibration, strain	Use of fault tolerant and energy efficient routing

IoT is increasingly being used in a number of applications, such as hospital control, emergency management, and vehicle management. With the aid of various technologies, there are several artifacts, communication links, processing units, sensors, and frameworks that can be helpful for decision- making and action entreating systems in IoT. Perception, network, service, and device layer are the four layers that make up the IoT architecture. [15] Cloud servers store and process the vast majority of IoT data. Cloud computing systems are highly flexible and can be configured on demand in a "pay-per-use" mode, lowering the investment required to develop the desired analytics application. Established data analytics approaches can manage vast volumes of data stored in centralized cloud storage. Fog computing is a decentralized computing model that extends cloud services to the network's edge. Cisco introduced the fog computing idea. At the network edge, Fog is an intermediate computing layer that collects and processes time-sensitive data. Several fog nodes make up a fog network. Any computer with computing, storage, and network access can be used as a node [16]. Table3 shows the relation between the behavior of system according to specific spreading factor.

Table 2. SF Vs Bit Rate, Sensitivity, and Symbol Duration Of Lora

Spreading Factor	Bit rate	Sensitivity	Duration of Symbol
SF 12	293 (b/s)	-137 dBm	682 ms
SF 11	537 (b/s)	-134.5 dBm	365 ms
SF 10	976 (b/s)	-132 dBm	204 ms
SF 9	1757 (b/s)	-129 dBm	113 ms
SF 8	3125 (b/s)	-126 dBm	64 ms
SF 7	5468 (b/s)	-123 dBm	36 ms
SF 6	9375 (b/s)	-118 dBm	21 ms

LoRaWAN is an open-source network protocol that is governed by Long Range (LoRa) Alliance and configured to use in devices that have strict energy management. The LoRa modulation technique is used in the devices that implement this network feature. LoRa technology compromises transmission speed in order to achieve longer transmission distances, making it ideal for the implementation of WSN systems that require sending small amounts of data at regular intervals. [17] Since the system doesn't need a reference source of a clock signal which is more precise and costly, the aforementioned modulation allows using far less sophisticated transceiver devices than other 802.15.4 standard wireless communication systems used in IoT [18]. LoRa is gaining more popularity in transmitting small data packets within a short span of time over a long-range in low-power battery-packed embedded systems. Devices using LoRa allow us to configure different parameters of radio communication. As a result, a range of sensitivity levels, data rates can be selected to fulfill the requirement of our application. The transmitting band's core frequency also called carrier frequency which is under ISM (Industrial, Scientific, Medical) bands that vary by world areas, such as 868 MHz (Europe), 915 MHz (US), and 865 MHz in India. The spreading factor is the ratio between symbol rate and chip rate, which ranges from 6 to 12.

$$\text{Spreading Factor} = \text{chip rate} / \text{symbol rate}$$

SNR is increased by a higher spreading factor, enabling communication to long distances in presence of higher noise levels, time on-air increases for a low spreading factor, resulting in increased energy usage. Bandwidth is the frequency spectrum used within the transmitting band; a higher bandwidth results in higher transmission speed but it reduces sensitivity because of an increase in noise effects; a lower bandwidth, on the other hand, means more sensitivity at a lower data rate. The forward error correction rate is referred to as the coding rate (CR). Increased error safety at a longer time on-air is possible with a higher CR. The maximum output power of a transceiver over a transmission interval is known as transmission power; a long transmission range can be achieved with high output power but at the expense of increased energy consumption [17]. The following equations can be used to measure time on air in a LoRa communication network: [19]

$$T_{packet} = T_{preamble} + T_{payload}$$

Where Tpacket is the packet duration or the time on air and Tpreamble is the preamble duration and Tpayload is the payload duration. Preamble duration and Payload duration can be calculated from following equation:

$$T_{preamble} = (n_{preamble} + 4.25)\ T_s\ T_{payload} = P_{LSymb} * T_s$$

Where npreamble is preamble length, Ts indicates the duration for one symbol and PLSymb is the number of symbols contained in the payload.

$$T_s = 2^{SF} / B$$

where B denotes the bandwidth of signal

The minimum power level at which a receiver can detect and demodulate transmitted data is referred to as the receiver's sensitivity.[20] The receiver sensitivity of LoRa at room temperature is given as: [21]

$$S = SNR - 174 + NF + 10 \log 10(B)$$

where (-174) is taken because of the receiver's thermal noise at a bandwidth of 1Hz, NF represents the Noise Figure ofthe receiver, and SNR represents modulation's signal-to-noise ratio. The binary data rate represented by R_b calculated using the following

$$R_b = SF \times (B / 2^{SF})$$

Table 3. Soil Moisture Content [2]

Type of Soil	Water Content in Soil (for mm of water per m of soil)
Sand	25 - 100 (mm/m)
Loam	100 - 175 (mm/m)
Clay	175 - 250 (mm/m)

Table 4. Rate of Rainfall[2]

Rainfall	Rainfall in mm per 24hours
Very Less	less than 5
Less	5 to 20
Medium	21 to 50
Heavy	51 to 100
Very Heavy	more than 100

4. Working

Landslide detection module made with the help of NodeMCU and Arduino. The hardware consists of 2 parts – the sender and the receiver, in which the transmission of data is taken care of LORA communication. The sender is an Arduino which is interfaced with the different sensors required for the project, this consists of a Rain drop sensor (HW-028) to detect the amount of current rainfall, a Soil moisture sensor (YL-69) to check the moisture content of the soil, and an Accelerometer (GY-521) to check sudden change in position, all these values are then sent to the receiver by a LORA module (RA02). The receiver is a NodeMCU which receives the through the LORA module and then uploads it to the Thingspeak MQTT server. If the values received exceed the threshold values, the device warns the user through the serial monitor and also on the Thingspeak dashboard.

Custom app made for mobile devices with the help of MIT App Inventor. The app gathers data and services from various providers and presents an informative and easy to use platform to the consumers for easy monitoring of the environmental conditions in risky areas in real time. This app consists of a map in which the user can check his/her location and get directions to the destination too. The app also shows real time conditions of the environment in accordance to the sensor values received through the ThingSpeak server. It also consists of other functionalities that bring other useful features right at the user's disposal.

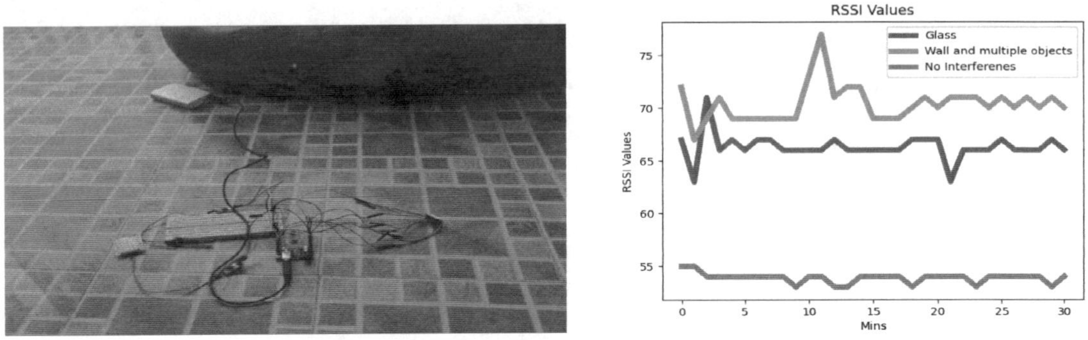

Figure 1. RSSI values in different obstacle cases

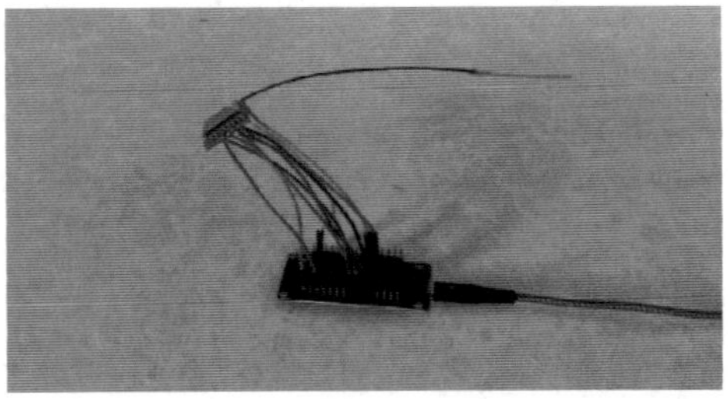

Figure 2. Receiver Fixed End Point

The testing of the components is done by placing both the devices at different places with different lengths and number of interferences between the devices to simulate different conditions that might occur when used in actual scenarios. The testing for the landslide was done by creating a dummy scenario in which we tested the different components to determine the threshold values for the soil profile we currently had. We then checked the values at all the devices to ensure that there was no loss of data. The receiver was kept at one place during all the observations.

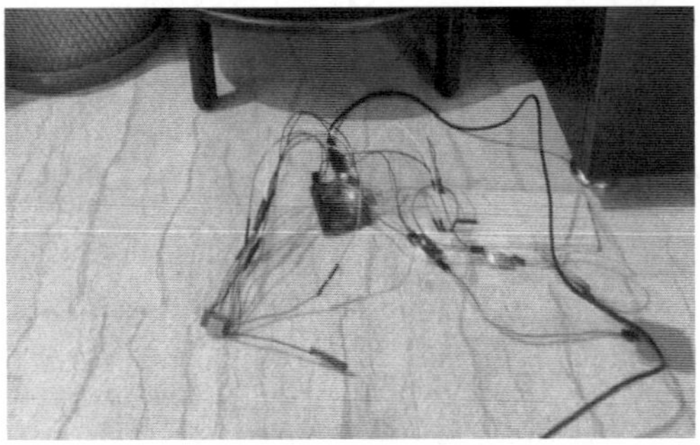

Figure 3. Wall and multiple objects

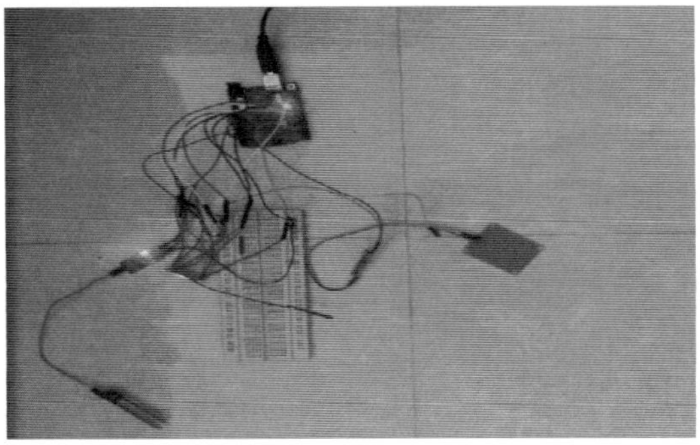

Figure 4. No Obstacles

5. Results

The app initially shows the status of connection with the ThingSpeak server on the top right corner. It shows the relevant sensor graphs below the status which update in real time. The app derives the values from the sensors through the ThingSpeak server and displays the relevant information on the screen. It automatically displays the appropriate colour coded message corresponding to the threshold set in the app for every sensor value. For example, when the Rain Sensor (HW-028) detects low values, the text "Low chances of rainfall" is displayed in the green color. Then on the basis of every sensor's value, the app calibrates whether it's safe to proceed or not and displays the corresponding message below on the screen. A "Help" button is provided in the bottom right corner which can provide clear instructions to people who face a hard time operating the app.

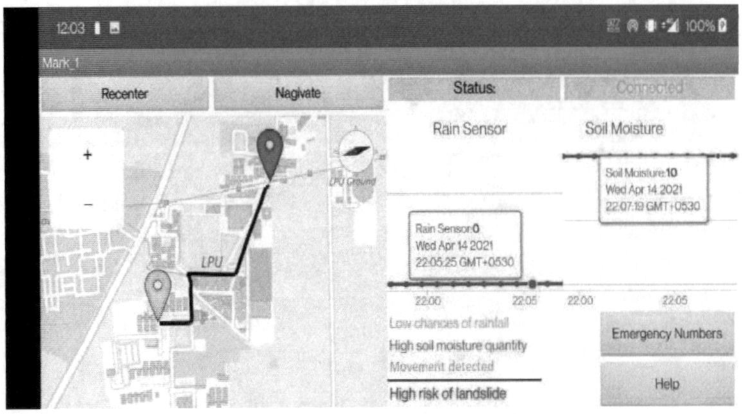

Fig 5. Mobile Application

The app uses the GPS feature on the phone to get the location of the user. When turned on, the user can click on the "Recenter" button to get his location on the map. Here, the user can set his destination and click the "Navigtion" button to get the directions with the help of "openrouteservice.org" which is an open source route planner. This helps the user use a single screen for both safety and navigation purposes so he/she does not have to keep switching between screens while driving.

6. Conclusion

A landslide monitoring system discussed in this research paper is able to detect the possible hazardous situations of landslides and sending the alerts to the neighborhood people through the mobile. The system is capable of deciding the situation severity according to the predefined values. Use of IoT, LoRa increased the scope of this system for long range communication and real time alerting. The landslide monitoring system successfully able to collect the data from sensors and send them to the Thingspeak web server. An android app is also running successfully in interpreting the available data from thingspeak server to the users and navigating them to the safest places.

References

1. Jalooli, Ali, et al. "Retracted: Public Alerts on Landslide Natural Disaster Using Vehicular Communications." International Journal of Distributed Sensor Networks 10.6 (2014): 969864.
2. Herlambang, P., A. Suharjono, and M. Mukhlisin. "Decision strategy for landslide event based on 4 parameters." Journal of Physics: Conference Series. Vol. 1517. No. 1. IOP Publishing, 2020.
3. Taqwa, Ahmad, et al. "Prototype Design of Landslide Early Detection System Using LoRa and IoT." 4th Forum in Research, Science, and Technology (FIRST-T1-T2-2020). Atlantis Press, 2021.
4. Tao, Zhigang, et al. "Design and operation of App-based intelligent landslide monitoring system: the case of Three Gorges Reservoir Region." Geomatics, Natural Hazards and Risk (2019).
5. Purnomo, Fendi Aji, Nanang Maulana Yoeseph, and Gembong Wijang Abisatya. "Landslide

early warning system based on arduino with soil movement and humidity sensors." Journal of Physics: Conference Series. Vol. 1153. No. 1. IOP Publishing, 2019.
6. Romdhane, Rihab Fekih, et al. "Wireless sensors network for landslides prevention." 2017 IEEE International Conference on Computational Intelligence and Virtual Environments for Measurement Systems and Applications (CIVEMSA). IEEE, 2017.
7. Dragulinescu, Ana-Maria, et al. "Smart Neighbourhood: LoRa- based environmental monitoring and emergency management collaborative IoT platform." 2019 22nd International Symposium on Wireless Personal Multimedia Communications (WPMC). IEEE, 2019.
8. Shukla, Shubhendu S. "Disaster Management:"Managing the Risk of Environmental Calamity"." International Journal of Scientific Engineering and Research (IJSER) 1 (2013).
9. Fajardo, Jovilyn Therese B., and Carlos M. Oppus. "A mobile disaster management system using the android technology." WSEAS Transactions on Communications 9.6 (2010): 343-353.
10. Datar, Ashlesha, et al. "The impact of natural disasters on child health and investments in rural India." Social Science & Medicine 76 (2013): 83-91.
11. Barbhuiya, Mahfuzuar Rahman, and Devlina Chatterjee. "Vulnerability and resilience of the tourism sector in India: Effects of natural disasters and internal conflict." Tourism Management Perspectives 33 (2020): 100616.
12. Rosselló, Jaume, Susanne Becken, and Maria Santana-Gallego. "The effects of natural disasters on international tourism: A global analysis." Tourism management 79 (2020): 104080.
13. Khadar, M., V. Ranjith, and K. Varalakshmi. "IOT INTEGRATED FOREST FIRE DETECTION AND PREDICTION USING NODEMCU."
14. Khan, Amina, Sumeet Gupta, and Sachin Kumar Gupta. "Multi- hazard disaster studies: Monitoring, detection, recovery, and management, based on emerging technologies and optimal techniques." International journal of disaster risk reduction (2020): 101642.
15. Zafar, Umara, et al. "Exploring IoT applications for disaster management: identifying key factors and proposing future directions." Recent trends and advances in wireless and IoT-enabled networks. Springer, Cham, 2019. 291-309.
16. Dhingra, Swati, et al. "Internet of things-based fog and cloud computing technology for smart traffic monitoring." Internet of Things (2020): 100175.
17. Ragnoli, Mattia, et al. "An autonomous low-power lora-based flood-monitoring system." Journal of Low Power Electronics and Applications 10.2 (2020): 15.
18. Institute of Electrical and Electronics Engineers. IEEE Standard for Local and metropolitan area networks—Part 15.4: Low-Rate Wireless Personal Area Networks (LR-WPANs)—Amendment 5: Physical Layer Specifications for Low Energy, CriticalInfrastructure Monitoring Networks; IEEE: Piscataway, NJ, USA, 2013; pp. 1–149.
19. Noreen, Umber, Ahcène Bounceur, and Laurent Clavier. "A study of LoRa low power and wide area network technology." 2017 International Conference on Advanced Technologies for Signal and Image Processing (ATSIP). IEEE, 2017.
20. Reda, Haftu Tasew, et al. "On the application of IoT: Meteorological information display system based on LoRa wireless communication." IETE Technical Review 35.3 (2018): 256-265.
21. Mroue, Hussein, et al. "Analytical and simulation study for LoRa modulation." 2018 25th International Conference on Telecommunications (ICT). IEEE, 2018.

22. Wang, Honghui, et al. "Research on one zicm2410-based wireless sensor network for landslide monitoring." 2011 7th International Conference on Wireless Communications, Networking and Mobile Computing. IEEE, 2011.

23. Giorgetti, Andrea, et al. "A robust wireless sensor network for landslide risk analysis: system design, deployment, and field testing." IEEE sensors journal 16.16 (2016): 6374-6386.

24. Lee, Huang-Chen, et al. "Open-source wireless sensor system for long-term monitoring of slope movement." IEEE Transactions on Instrumentation and Measurement 66.4 (2017): 767-776.

25. Kumar, Manish, and P. V. Rangan. "Wireless Sensor Network for Landslide Detection." International conference on wireless networks, Las Vegas, USA: 2009. 2009.

26. Kebaili, Mohamed Omar, et al. "Landsliding early warning prototype using MongoDB and Web of Things technologies." Procedia Computer Science 98 (2016): 578-583.

27. Liu, Zhi, et al. "Data driven cyber-physical system for landslide detection." Mobile Networks and Applications 24.3 (2019): 991- 1002.

28. Mehta, Prakshep, et al. "Distributed detection for landslide prediction using wireless sensor network." 2007 First International Global Information Infrastructure Symposium. IEEE, 2007.

Polymer Electrolyte a Novel Material for Electrochemical Devices: A Review

Palvinder Singh[1], Amit Sachdeva[2*], Cherry Bhargava[3]

[1]Department of Electronics and Communication Engineering, Chandigarh University, Gharuan, Mohali-140413, Punjab, India
[1,2]School of Electrical and Electronics Engineering, Lovely Professional University, Phagwara-144411, Punjab, India
[3]Department of Electronics and Telecommunication Engineering, Symbiosis Institute of Technology, Pune, Maharashtra, India

Abstract: Polymer Electrolyte is novel material as renewal energy sources are widely used in the industries. Electrolyte is the main parts of storing/conversion devices. In this review article polymer electrolyte novel material are discussed. In this report various polymer advantage, application, availability, categories of polymer, various approaches, and sample preparation techniques are also described. Polymer electrolyte also used in batteries, fuel cells, supercapacitors, Dye-sensitizes solar cell and energy storing/conversion devices. It is observed that polymer electrolyte is better awareness to the researchers and readers.
***Keywords**: PEMA, EC, Polymer Electrolyte, ionic conductivity.*

1. Introduction

Solid-state materials such as composite material, optical, crystals, and polyethylene electrolytes were developed in the early 1970s. Fenton et al. published the first article on polymer electrolyte in 1973, and the impact of its scientific use was identified in the early 1980s [1]. A polymer electrolyte (PE) membrane is made up of salt dissolving in a polymer matrix with a high molecular weight [2]. This solid solvent-free structure is widely used in electrochemical devices such as solid-state batteries, rechargeable batteries, and lithium-ion batteries due to its ionic conductivity property.

Table 1. Properties of Polymer electrolyte and its advantages, and application.

Polymer electrolytes		
Properties	Advantage	Applications
Transparent	Easy of fabrication	Electrochromic window
Light weight	Restricts use of corrosive bond	Solar cells, fuel cells
Solvent free	Long life span	Solid state batteries
Improved safety	Reduce flammability	Analog meter devices
High ionic conductivity	Non-toxic	Electrochromic display devices
Versatility	Low cost	Electric vehicles
Reliability	Reproducibility of the parameters	Solid –state batteries
Low self-discharge	No internal shorting	Solid –state batteries

In recent years, PEs have found new applications in advanced electrochemical and electronic devices such as superconductors, energy storage, dye-sensitize solar panels, battery banks, electrochemical devices, analogue memory units, and optoelectronic windows [2].

1.1 POLYMER HOST

For making polymer electrolyte we need on polymer material. Examples of polymer material used widely in the PEs are formed by

Table 2. Polymer host and their properties.

Polymer	Electrical conductance at normal temperature (S/cm)
Poly (acrylic acid)	2.15E4
Poly(acrylonitrile)	5.7 E4
Poly (ethyl methacrylate)	1.17 E4
Poly (ethylene oxide)	2.6 E4
Poly(ε-caprolactone)	3.8 E5
Poly (methyl methacrylate)	7.3E6
Poly (vinylidence fluoride)	4.5 E3
Poly (vinylidence fluoride-Hexafluoro propylene)	4.3E3
Poly (vinyl choride)	2.83E6
Poly (vinyl alcohol)	4.75E6

1.2 CATEGORIES OF PES

Categories of polymer electrolyte are shown in figure 2.2 after survey of literature.

1.2.1. Liquid polymer electrolyte

Liquid electrolytes are made from lithium salt blended in a natural dissolvable and a separator that keeps away from short circuits. Contingent upon the dissolvable, it isolates into two non-watery fluid non-watery fluid (NALE) and electrolyte in fluid (ALE). NALE is made by dissolving soluble salts Ethylene carbonate (EC), Propylene carbonate (PC), Diethyl carbonate (DEC), and Dimethyl carbonate in a dissolvable to deliver a strong polymer electrolyte with a high ionic conductivity (DMC). A couple of its drawing in characteristics incorporate high ionic conductivity, warm soundness in the surrounding/sub ecological, generally huge electrochemical circumstances, stable lock, and convenience[2]. Ionic conductivity is more of ALE than NALE, and it has properties like inflammability and is moderately modest in climate, alongside wellbeing, which makes it an extraordinary choice for pragmatic use, even at high temperatures.

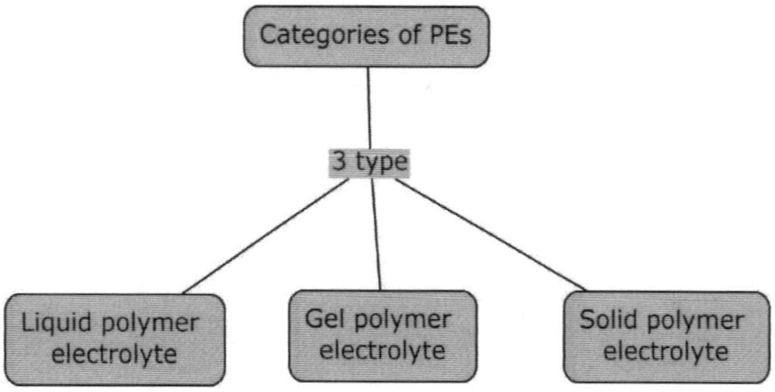

Figure 1: Categories of PEs

Ionic liquids (ILs) are commonly found in organic salts such as minerals and are present in liquid state inside the environment temperature due to inadequate ion interaction and insufficient packing atoms. The melting point of IL is lower than that of normal ambient temperature and it can be used at temperatures ranging from 300 to 400 degrees Celsius. These ionic liquids are appropriate for application due to their solvability and extensive excellent stability. Batteries, fuel tanks, Supercapacitors, and other advanced electrochemical systems are potential candidates for the development of replacement organic electrolytes[6]. Modifying the cation-to-anion ratio and using multiple cations or anions can considerably alter the properties of ionic liquid solvents. Table 3 shows the conductivity of various liquid polymer electrolytes (LPEs).

1.2.2 Gel polymer electrolytes

GPEs are third-stage materials composed mostly of an inorganic polymer matrix and a low-molecular-weight solvent for salt dissolution. Gels are classified into two types based on their manufacturing methods.:

1. Physical gels are restricted to a polymer matrix with no means of connecting the polymer to the solvent (e.g., Na/ EC / PC) (PEMA).

2. A chemical connection is formed between the polymer feature group and the cross-linker agent with the help of the chemical gels cross linker.

The GPEs exhibit increased ionic conductivity, a larger electrochemical stability window (ESW), and higher electrode reliability due to improved ionic mobility. GPEs have solid properties that are both cohesive and diffusive. In GPEs, the polymer maintains a solid-state matrix that allows for ion migration in solvents and has a conductivity of around 10^{-3}Scm-1 at room temperature. Electrolytes can operate as a plasticizer, decreasing the glass transition temperature and facilitating rapid ionic conduction. Gel polymer electrolyte is also known as hybrid polymer. Plasticizers with low evaporation solutions, such as EC, PC, DEC, and DMC, are utilised to make various organic salts. GPEs are easier to use because of features such as a variety designs, improved charging / discharging efficiency, compact size, expense, and population concentration. We also increased ionic conductivity, although this came at the expense of mechanical reliability.

Table 3. Properties of liquid polymer electrodes

Polymer Electrolyte	Conductivity	Ref.
PEO-PVdF+ LiTFSI+ Acetone/DMAc	4.9E3	[7]
PVdF-HFP+[BMIM][BF_4]+ Acetone	1.6E2	[8]
PVA+ $LiClO_4$/[EMIM][$EtSO_4$]+ DMSO	1.9E6	[9]
PVdF-HFP+ PAni-MMT/BMIMBr+ Acetone	7.7E6	[10]
PEO+ LiTFSI/BMITFSI + Solvent free	3.2E4	[11]
PEMA+ $NH_4CF_3SO_3$/BMATSFI + THF	1E4	[12]
PEO+ LiTf/EMITf + CAN	3E4	[13]
PVA+ CH_3COONH_4/(BmlMCl)+ Distilled water	7.3E3	[14]
PVA+ Li_2So_4/BMIMCI	37 E3	[15]
PEO+ $LiPF_6$/$MMPIPF_6$	1.13 E3	[16]
PEO+ EMIM-TY + Solvent free	1E4	[17]
PEO+ $LiClO_4$/[BMIM][PF_6] + Anhydrous methanol	8E6	[05]
PEO+ Al_2O_3/EMIM-TY + Distilled methanol	5.8E6	[18]
PEO+ LiTFSi [Pyr_{14}TFSI] +	1.1E4	[19]

1.2.3 Solid polymer electrolytes

For various reasons, SPEs are preferred over LPEs and GPEs, including their limited life span, specific strength, low weight, greater cell size stability, low sensitivity to the working electrode, and ability to eliminate solvent concerns. Leakage and noxious gas are reduced, and packing costs are lowered, when operations are carried out above ambient temperature. It also contains more electrochemicals and has a stronger heat stability, yet being less volatile[2]. SPEs are a type of solid polymer electrolyte with an electron donor group that degrades salt in a polymer matrix. Design freedom, miniaturisation feasibility, better safety, and gadget production ease are just a few of the benefits. In electrochemical device applications, an ionically conducting polymer membrane should have a suitable balance of electric conductivity and mechanical resistance. SPEs have the best conductivity and stability under ambient circumstances, and their underlying features could point to future uses in ionic solid state products like high-energy-density batteries, rechargeable batteries, superconductors, and fuel tanks [2].

2. APPROACHES

Various ways can be used to improve the PEs' performance and characteristics. Figure 2 depicts a variety of approaches that have been offered in the literature.[3].

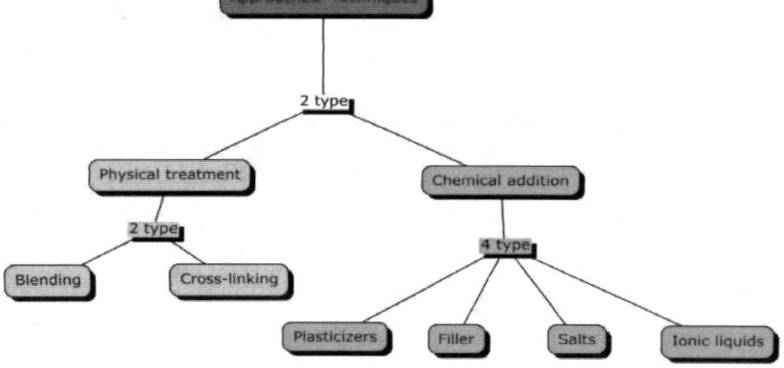

Figure 2: Techniques for improvement of polymer performance.

2.1. Blending

Among the several choices, polymer mixing is the most practical. Polymer blending is the process of combining at least two polymers, either chemically or physically. A polymer blend, as a result, is the physical blending of two or more polymers or copolymers. Due to their unique properties that outperform the component polymers, polymer blends are viable products in commercial and industrial applications. Mechanical strength, ionic conductivity, and mechanical stability are all improved with this technology. It is a common method for reducing the crystallite of a polymer matrix [2]. It is conventional technique reduce the crystallite of polymer matrix.

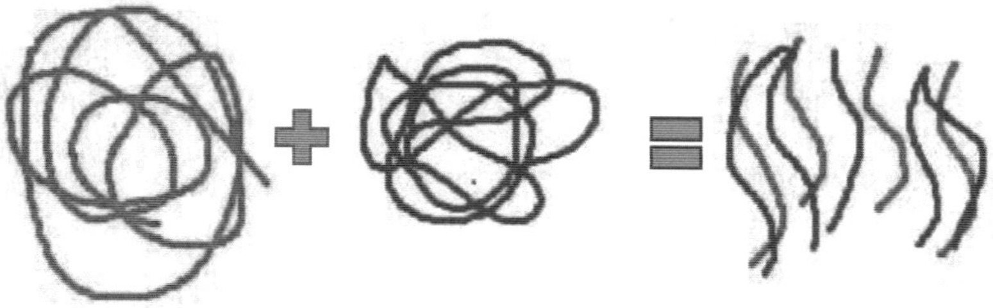

Figure 3: Polymer blends pictorial model

2.2. Cross-linking

The amorphous nature of the polymer matrix is associated with the high electrochemical performance of SPEs at room temperature. At room temperature, the cross-linked PEs have a high ionic conductivity and are completely amorphous. Cross-linked polymers, on the other hand, are frequently inflexible, rigid, and difficult to manufacture [2]. The monomer content influences the PEs' ionic conductivity and flexibility, according to the research. An example of a cross-linking polymer is shown in the following diagram.

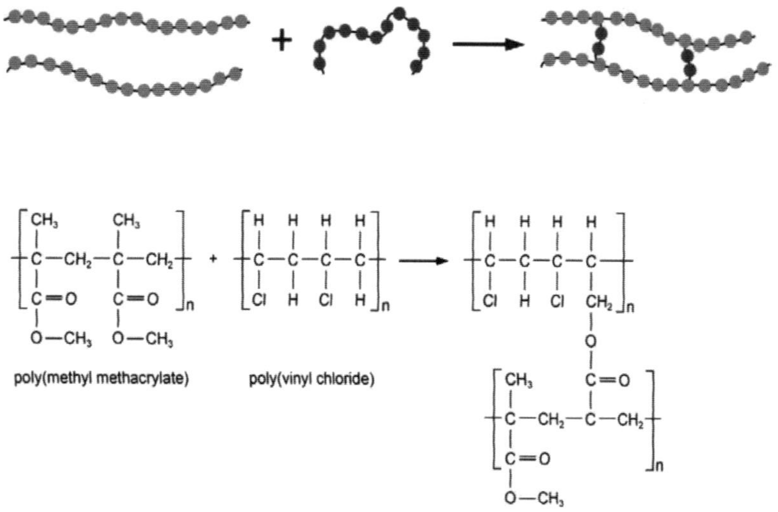

Figure 4: Polymer crosslink pictorial model[39]

2.3. Fillers:

Another way to improve SPE performance is to fill polymer matrices with tiny and small particle size nanomaterial fillers. Figure 5 depicts a visual depiction of filler doping in a polymer electrode. The electrical conductance of PEs is improved by filler doping. Surface stability, high heat stability, decreased crystallite size and glass change temperatures, amendment of the conducting polymers in amorphous zone, and enhanced interfacial stability in contact with a variety of electrode materials are just a few of the benefits. Polymers' crystallinity is reduced, which prevents them from recrystallizing, while their amorphicity is increased, which improves ionic conductivity. The fillers lower the crystalline nature of the polymers, preventing them from recrystallizing, and increase amorphicity, which improves electrical characteristics. Because smaller particles have a larger surface area to volume ratio, which minimizes crystal formation, they are more effective. Fillers have the potential to disrupt polymer matrices' lattice structure. As a result of this process, interfacial layers emerge between the filler grains. As a result, charge carriers like electrode materials become self-trapped at the polymer chains' lattice structure's grain boundary. Aluminum oxide (Al_2O_3), titanium oxide (TiO_2), zirconium oxide (ZrO_2), and hydrophobic-fumed silica are all high surface area materials. [2]

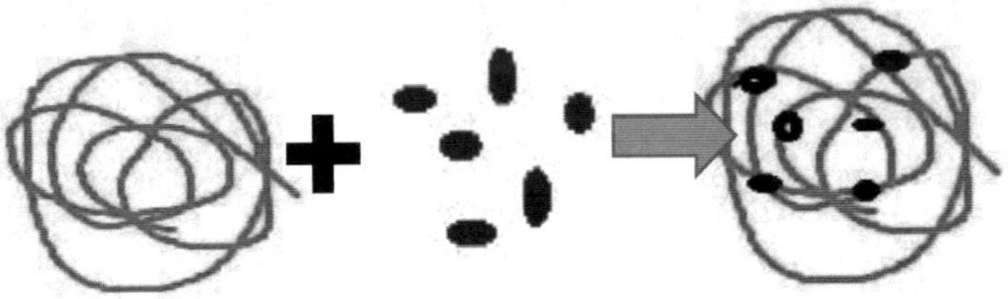

Figure 5: Example of Complex Polymer Electrolyte adding filler

2.4. Nanofillers:

Nanomaterials can be used as fillers in nanocomposite polymer electrode. The nanoparticles interaction with the polymer host boosts the PE's physical and electrical conductance[40] . Capiglia and colleagues [89, p. 2] developed a nanomaterial-based PE in 1999 by inserting silica in the form of nanomaterial (SiO_2) into PEO-$LiClO_4$ polymer electrolytes. The ionic conductivities of PEs were similarly increased by nanosized aluminiumoxide (Al_2O_3), according to another study [2].

2.5. Copolymers

A copolymer is a polymer that is made up of at least two distinct monomers. The cross-linking method can be used to make copolymers. The most commonly composite is PVdF-HFP, which is synthesized through polymerizing PVdF and HFP together, as shown in Fig. 5. Because of synergistic effects in the mixed structure, the copolymer has superior properties than the components alone. Electrolytes made of epichlorohydrin, and ethylene oxide copolymer were investigated for usage in electrochemical devices. In a lithium primary cell and a nickel/metal hybrid secondary battery, this copolymer functioned excellently [2].

3. Sample preparation technique:

The increasing use of polymer electrolytes in a variety of uses required the development of high-quality film technology. Various sample preparation techniques are listed below.

3.1 Solution casting technique

Because of its ease of fabrication, it is a traditional method for making solid polymer electrolytes bendable (Fig. 6It could also develop polymer films ranging in size from 50 to 300 micrometers.

The following are the steps in this procedure:

1. To achieve homogenous mixing, a preset amount of polymer is added to a solvent and stirred.
2. A salt solution is added to the polymer matrix, which is then agitated at atmosphere temperature until

the polymer–salt complexation develops.

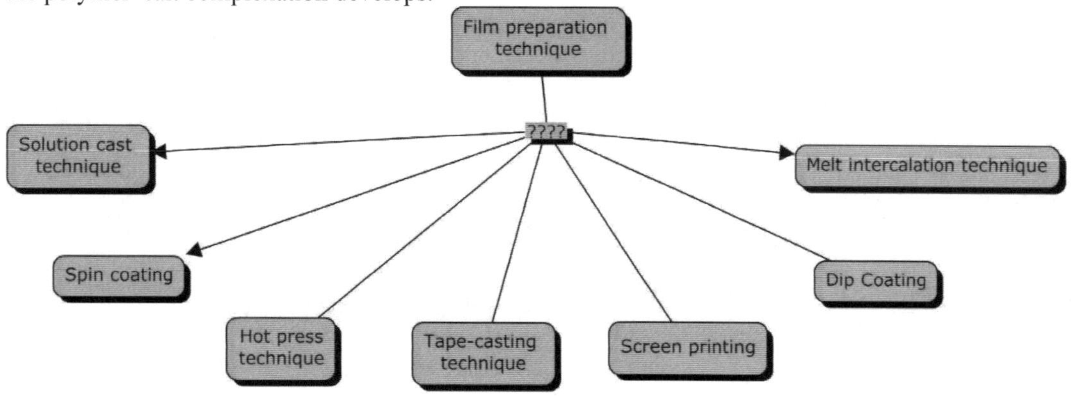

Figure 6: Sample preparation technique

3. The nanofillers/clay/plasticizer are then added, and the mixture is well mixed.
4. Pour the viscous solution into glass or Teflon Petri plates and let it sit at atmosphere temperature for 2-4 days to gently evaporate.
5. The result is a thin, uniformly thick layer.

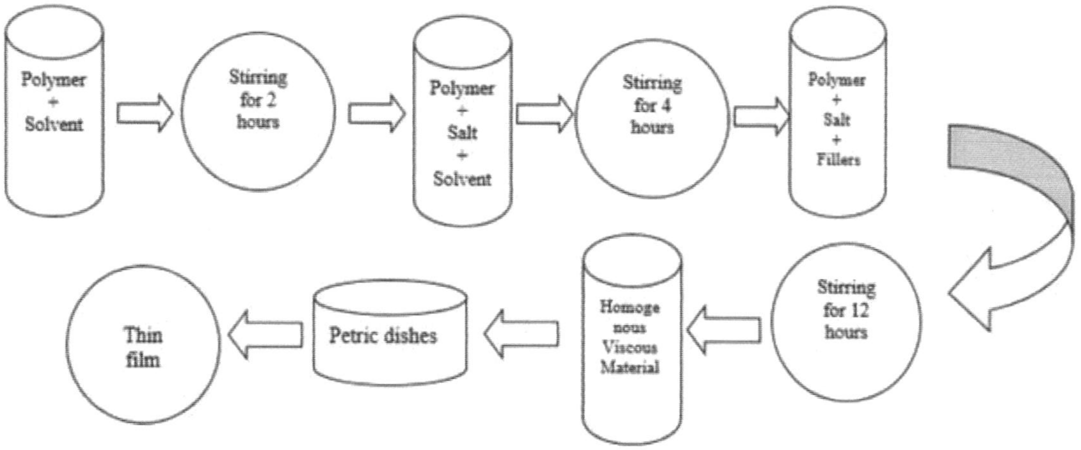

Figure 7: Solution casting Technique

3.2. Spin coating:

The spin-coating approach is like the solution casting analysis. Instead of casting the film on a surface, a little amount of the mixture is dropped on it and placed in a mechanical stirrer that can revolve at a preset speed to generate a beautiful film (Fig. 8). The substrate's rotating axis must be opposite to the surface being coated. Due to centripetal acceleration, the mixture will spread out on the substrate. The detergent is subsequently evaporated with the help of spread heating. Spin coating's simplicity and relative ease of preparation, as well as its ability to swiftly make homogenous films ranging in thickness from a few nanometers to a few microns, are all advantages.

The viscosity (concentration), adhesion strength, rotational time, and speed of rotation of the mixture can all be easily modified to adjust the film thickness. Because it has a direct impact on the force applied to the mixture, the rotating speed is critical. However, this procedure only works with highly viscosity mixtures, not very viscous ones. The spin coater's rotation is insufficient to properly distribute the gel mixture droplets and produce a thin sheet [3].

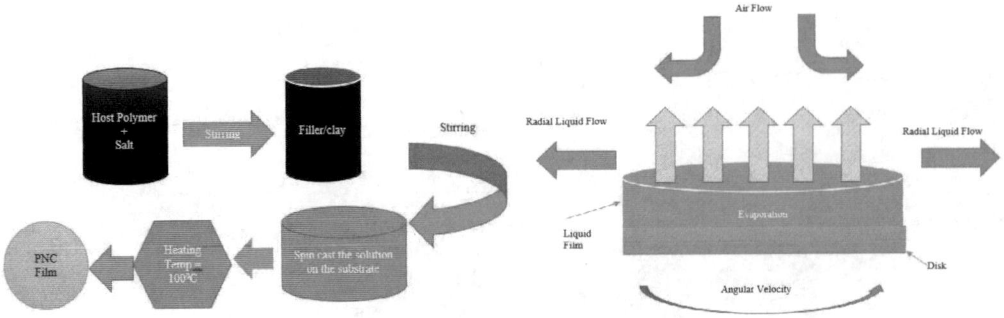

Figure 8: Spin Coating Technique and its systemic arrangement

3.3. Tape casting technique:

In a vial, stoichiometric amounts of host polymer and salt are dissolved in a suitable solvent, such as acetone or DMF, and the solution is stirred for 10–12 hours. In the polymer–salt complex, nanofiller dispersion in varying weight % relative to the host polymer is now done in the appropriate proportion and mixed for the same amount of time as before. The viscous solution is now reheated on a hot plate for a few minutes before being tape caster cast and dried at room temperature. As a result, we can make a self-supporting conductive polymer nanocomposite film with any thickness we want. The amount of slurry at the blade edge, as well as the pressure, are kept constant. The full technique is depicted in Figure 8. A gaseous metal tank is connected to a slurry-filled steel container. The slurry flow rate is controlled by the gas pressure, which keeps the amount of slurry at the scraper edge constant. A slit occurs between the edge of the scrapper and the carrier when a carrier (tape) is run through a slurry at a fixed value. After the slurry on the carrier with the size of the slit is dried, a membrane is obtained and rolled into a package. The slit can be adjusted to change the thickness of the membranes. This technology is appropriate for manufacturers because it can easily handle the size and width of the sample[3].

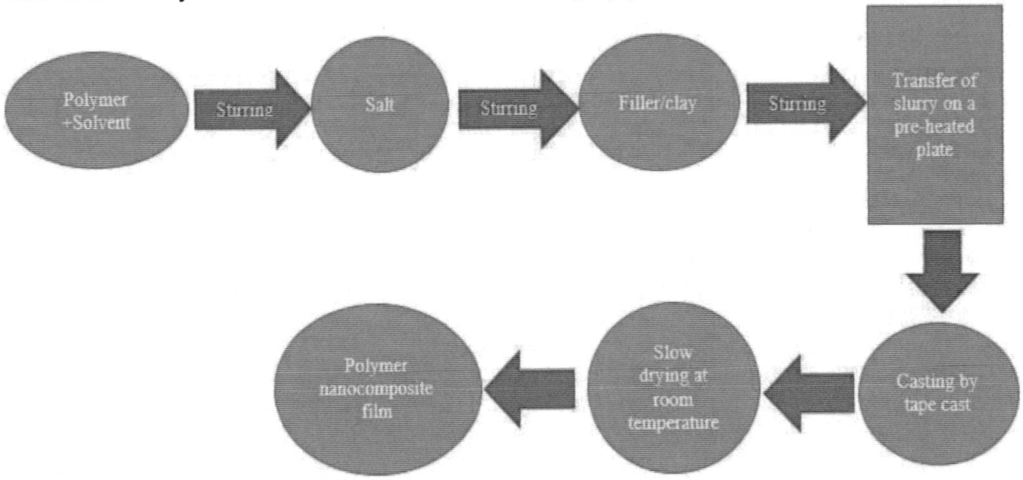

Figure 9: Tape casting technique

3.4. Hot press technique:

The hot press technique, which substitutes the standard solution cast process, is a one-of-a-kind technology that provides advantages including speed, cheap cost, solvent-free operation, and the ability to work with thick materials. This approach has the benefit of being able to complex a wide range of alkali salts and metallic particles with a wide range of polymer hosts.

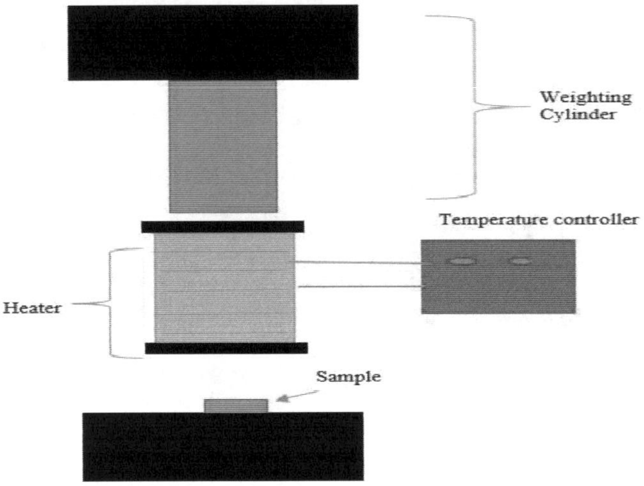

Figure 10: Hot Press Technique

A weighing cylinder, a heating chamber, and utility room, and a temperature sensor make up the equipment. In an agate pestle and mortar, a dry powder mixture containing optimum amounts of polymer, salt, and filler/plasticizer is pounded firmly for 1–2 hours at room temperature. The sample is pressed with a weighing cylinder to vary the pressure[3]. It makes a hot slurry with a mixture of polymer, salt, and filler compositions that is separated from stainless steel (SS) blocks for later usage (Fig. 10). Due to the reactive nature of lithium salts in the environment, the hot press operation is difficult for lithium ions, and special processing conditions are necessary.

4. Characterization Technique

PEs can also be categorized in several ways. These techniques are used to investigate surface, microstructural, structural, interaction in complexes, temperature properties, electrical conductivity, ionic transference number, and potential stability window. Many characterization techniques are discussed in the literature report, including X-ray diffraction (XRD), Atomic force microscopy (AFM), Transmission electron microscopy (TEM), Fourier transform infrared spectroscopy (FTIR), Differential scanning calorimetry, Impedance spectroscopy, Transference number, Linear sweep voltammetry (LSV), Thermogravimetric analysis (TGA), Dynamic mechanical analysis (DMA), Chrono

Conclusion:
According to the literature review, many polymers are available for research work, including Poly (vinyl chloride) (PVC), Poly (vinyl alcohol) (PVA), Poly (acrylic acid) (PAA), Poly (ethylene oxide) (PEO), Poly (acrylonitrile) (PAN), Poly (vinylidenedifluoride (PVdF), Poly (ethyl methacrylate) (PEMA), Poly (methyl methacrylate) (PMMA), Poly Various methodologies and sample preparation techniques are addressed in this paper; nevertheless, solution casting is a basic, commonly used technique that may be found in every research lab. Polymer Electrolyte has a wide range of applications, including energy storage/conversion devices, supercapacitors, fuel cells, dye-sensitized solar cells, and so on.

ORcID id: https://orcid.org/ **0000-0001-9670-2768**

Reference:

[1] D. E. Fenton,1997 Complexes of alkali metal ions with poly (ethylene oxide) polymer, vol. 14, p. 589, 1973.

[2] Ngai, K.S., Ramesh, S., Ramesh, K. and Juan, J.C., 2016. A review of polymer electrolytes: fundamental, approaches and applications. *Ionics*, *22*(8), pp.1259-1279.

[3] Arya, A. and Sharma, A.L., 2017. Polymer electrolytes for lithium-ion batteries: a critical study. *Ionics*, *23*(3), pp.497-540.

[4] Chaurasia, 2011 Dielectric relaxation and conductivity studies on (PEO: $LiClO_4$) polymer electrolyte with added ionic liquid [BMIM][PF6]: Evidence of ion–ion interaction - - Journal of Polymer Science Part B: Polymer Physics - Wiley Online Library." https://onlinelibrary.wiley.com/doi/abs/10.1002/polb.22182 (accessed Aug. 08, 2020).

[5] Chaurasia, S.K. and Singh, R.K. 2010. Electrical conductivity studies on composite polymer electrolyte based on ionic liquid. *Phase Transitions*, *83*(6), pp.457-466.

[6] Kim, G.T., Appetecchi, G.B., Alessandrini, F. and Passerini, S., 2007. Solvent-free, PYR1ATFSI ionic liquid-based ternary polymer electrolyte systems: I. Electrochemical characterization. *Journal of Power Sources*, *171*(2), pp.861-869.

[7] Chaurasia, S.K., Singh, R.K. and Chandra, S., 2013. Thermal stability, complexing behavior, and ionic transport of polymeric gel membranes based on polymer PVdF-HFP and ionic liquid ,[BMIM][BF4]. *The Journal of Physical Chemistry B*, *117*(3), pp.897-906.

[8] Saroj, A.L. and Singh, R.K., 2012. Thermal, dielectric and conductivity studies on PVA/Ionic liquid [EMIM][EtSO4] based polymer electrolytes. *Journal of Physics and Chemistry of Solids*, *73*(2), pp.162-168..

[9] Nath, A.K. and Kumar, A., 2014. Scaling of AC conductivity, electrochemical and thermal properties of ionic liquid-based polymer nanocomposite electrolytes. *Electrochimica Acta*, *129*, pp.177-186.

[10] Choi, J.W., Cheruvally, G., Kim, Y.H., Kim, J.K., Manuel, J., Raghavan, P., Ahn, J.H., Kim, K.W., Ahn, H.J., Choi, D.S. and Song, C.E., 2007. Poly (ethylene oxide)-based polymer electrolyte incorporating room-temperature ionic liquid for lithium batteries. *Solid State Ionics*, *178*(19-20), pp.1235-1241.

[11] Anuar, N.K., Subban, R.H.Y. and Mohamed, N.S., 2012. Properties of PEMA-NH4CF3SO3 added to BMATSFI ionic liquid. *Materials*, *5*(12), pp.2609-2620.

[12] Kumar, Y., Hashmi, S.A. and Pandey, G.P., 2011. Lithium ion transport and ion–polymer interaction in PEO based polymer electrolyte plasticized with ionic liquid. *Solid State Ionics*, *201*(1), pp.73-80.

[13] Liew, C.W., Ramesh, S. and Arof, A.K., 2014. Good prospect of ionic liquid based-poly (vinyl alcohol) polymer electrolytes for supercapacitors with excellent electrical, electrochemical and thermal properties. *International Journal of Hydrogen Energy*, *39*(6), pp.2953-2963.

[14] Zhang, X., Wang, L., Peng, J., Cao, P., Cai, X., Li, J. and Zhai, M., 2015. A Flexible Ionic Liquid Gelled PVA-Li2SO4 Polymer Electrolyte for Semi-Solid-State Supercapacitors. *Advanced Materials Interfaces*, *2*(15), p.1500267.

[15] Zhu, C., Cheng, H. and Yang, Y., 2008. Electrochemical characterization of two types of PEO-based polymer electrolytes with room-temperature ionic liquids. *Journal of the Electrochemical Society*, *155*(8), p.A569.

[16] Chaurasia, S.K., Singh, R.K. and Chandra, S., 2011. Structural and transport studies on polymeric membranes of PEO containing ionic liquid, EMIM-TY: evidence of complexation. *Solid State Ionics*, *183*(1), pp.32-39.

[17] Chaurasia, S.K., Singh, R.K. and Chandra, S., 2011. Dielectric relaxation and conductivity studies on (PEO: LiClO4) polymer electrolyte with added ionic liquid [BMIM][PF6]: evidence of ion–ion interaction. *Journal of Polymer Science Part B: Polymer Physics*, *49*(4), pp.291-300.

[18] Appetecchi, G.B., Kim, G.T., Montanino, M., Alessandrini, F. and Passerini, S., 2008. Solvent-free, PYR1ATFSI Ionic Liquids-based Ternary Polymer Electrolyte Systems. II. Battery Tests. *ECS Transactions*, *11*(29), p.119..

[19] Hellio, D. and Djabourov, M., 2006, July. Physically and chemically crosslinked gelatin gels. In *Macromolecular Symposia* (Vol. 241, No. 1, pp. 23-27). Weinheim: WILEY-VCH Verlag.

COMPARATIVE ANALYSIS OF VEDIC MULTIPLIER USING VARIOUS ADDER ARCHITECTURES

Jatin Yadav[1], Anupam Kumar[2], Shaik Shareef[3], Sandeep Bansal[4], Navjot Rathour[5]

[1,2,3]Student, School of Electronics & Electrical Engineering
Lovely Professional University, Punjab (India)
[4]Associate Professor, School of Electronics & Electrical Engineering
Lovely Professional University, Punjab (India)
[5]Associate Professor, Uttaranchal Institute of Technology, Uttaranchal University, Dehradun, Uttarakhand, India
Corresponding author: sandeep.15732@lpu.co.in[4]

ABSTRACT- ***The performance of a microprocessor depends on the efficient multiplier as it is one of the most principal component in various digital circuits. This paper reviews optimization techniques for high speed Vedic multiplier design which is based on Urdhva Tiryakbhyam Sutra of Ancient Indian Vedic Mathematics. This particular sutra is the most efficient one as it gives minimum delay for all types of complex multiplication. Adder being the most important component in a multiplier design, using the efficient adder will enhance the performance of Vedic multiplier. During the comparison, different adder topologies like Carry Look ahead Adder, Kogge Stone Adder, Carry Skip Adder are used to compare area, delay and power. The reviewed methods are implemented on 45nanometer(nm), 90nm and 180 nm CMOS technology. The results of all the prior approaches are reviewed and an efficient method out of them has been proposed.***

KEYWORDS- Kogge Stone Adder (KSA), Ripple Carry Adder (RCA), Complementary Metal Oxide Semiconductor (CMOS), Vedic Multiplier (VM), Carry Skip Adder (CSA), Urdhva Tiryakbhyam Sutra

I. INTRODUCTION

Multipliers are widely designed for the applications such as Microprocessors, communication and Digital Signal Processing (DSP). The highly efficient multiplier's demand is increasing with the gradual advancement of microprocessors. For higher order multiplications addition of partial products are extensively used by the adders [1]. Higher order arithmetic operations are crucial to extract the desired output in various real time image and DSP applications. In an efficient a digital circuit, less delay and less power usage are key requirements. Multiplication as an application is the key arithmetic operation and the rapid advancement of fast multiplier circuits has always been a matter of subject for the last couple of decades [2].Vedic mathematics relied upon DSP operations and also reduced the total processing time as compared with other counterparts. It abates roughly 40% - 60% time from inbuilt function and this very algorithm operates on the concept of Vedic multiplier [8].

II. VEDIC MULTIPLIER

The vedic multiplier is based upon the vedic sutras. The term vedic is derived from the word 'Veda' which depicts the storehouse of "knowledge". Sri Bharati Krishna Tirthaji reconstructed Vedic mathematics from vedas during the year 1911 to 1918 [3]. It is based on 16 sutras which can be used in dealing with a number of branches of mathematics. Urdhva Tiryagbhyam is the most often used sutra for the implementation and processing of various vedic multiplier designs. Being a non conventional still pretty effective, vedic mathematics is used for performing high performance multiplier. Vedic mathematics works collectively with vedic mathematics formulae and their applications for processing large arithmetical operations efficiently due to which system consumes lesser power and covers lower chip area [4].

Incredible side of vedic multiplier is it can even solve tough problems orally thus making it very fast. Applications such as ALU and MAC use many designs which are based on vedic mathematics [4]. The area and delay gets increased by using conventional methods of multiplication. The mathematical sutras are more suitable for multiplier design as it minimizes the delay and area [5]. In literature, there has been much discussion about the vedic mathematics and depicts the efficiency of sutras as it consumes less time with respect to prior methods [6].

III. VEDIC MATHEMATICS

Conventional mathematics works collectively with vedic mathematics Sutras and their applications for processing large arithmetical operations efficiently due to which system consumes less power and covers low chip area. There are basically 16 Vedic mathematics Sutras. These sutras have different applications according to the algorithm behind them and where they can be used.

One of the sixteen sutras, Urdha Tiryagbhyam which is implemented on 'vertically and crosswise' technique used for instant multiplication. It makes the computation faster and efficient.In DSP applications Multiplication of floating point numbers has been significantly applied [7].

IV. URDHVA TIRYAGBHYAM SUTRA

In Sanskrit, "Urdhva Tiryagbhyam" Sutra means crosswise and vertical. The discussed vedic multiplier uses this Sutra. In the decimal number system, the multiplication of two numbers is traditionally done by these Sutras. It is basically a algorithm which is suitable for all the cases of multiplication. In this algorithm, the partial products can be generated with the help of concurrent partial products addition [8]. Any N x N bit multiplication can be solved by this algorithm. The multiplier's processing power can be increased by providing more input and output bus widths [8]. Because of its less complex organization, it can be easily designed in an Integrated Circuit(IC). The Sutra used in this vedic multiplier design provides edge over other conventional multiplier methods as the gate delay and the area increment is slow as compared to increase in number of bits [9].

To understand this Sutra, let's take two numbers 252 x 846 which are multiplied by using the above discussed algorithm as depicted in Figure 1.

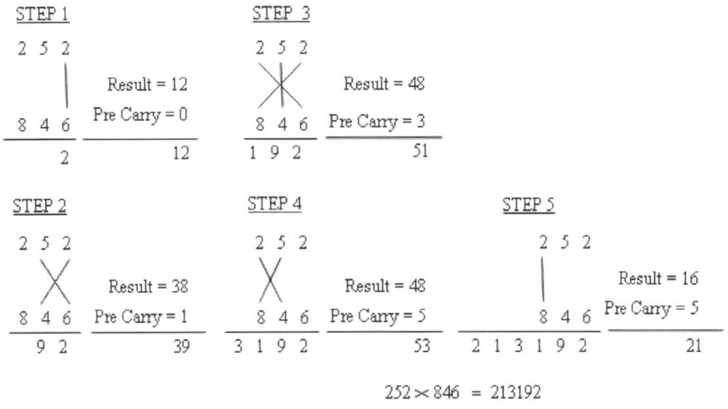

Fig 1: 2 x 2 bit Vedic mathematics calculation

V. PRIOR APPROACHES FOR MULTIPLIER DESIGN

A. 2x2 bit Vedic multiplier module

Few authors proposed the design of a 2-bit Vedic multiplier with different configurations [9]. It can be easily seen in Figure 3 that two half adders and four AND gates are used in the architecture of this particular multiplier. The Vedic multiplication of discussed design is shown in Figure 2. It was analyzed that the design of a 2 x 2 bit Vedic multiplier is identical to the design of 2x2 bit multiplier which are being used conventionally [10]. So, it is evident that a 2 bit

Journal of Physics: Conference Series **2327** (2022) 012022 doi:10.1088/1742-6596/2327/1/012022

Vedic multiplier doesn't make any major improvements in the multiplier's efficiency. So it is concluded that after calculation of final bits, total delay which is of only half adders is identical to array multiplier.

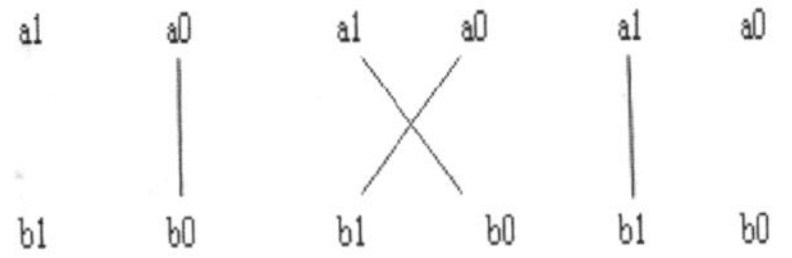

Fig 2: 2x2 vedic multiplication

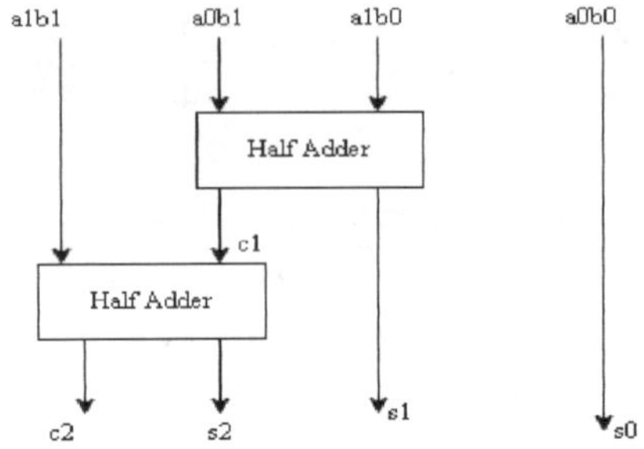

Fig 3: 2 x 2 bit vedic multiplier block diagram

B. 4x4 bit Vedic multiplier module

Most of the authors presented the design of a 4x4 bit vedic multiplier using the help of a 2x2 bit vedic multiplier. Using 2 bit multiplier blocks and 2 bits, 4x4 bit vedic multipliers can be implemented as displayed Fig.4. To get the final result, three 4 bit Ripple Carry Adders and four 2x2 bit multipliers are used which are shown in Fig.5. The delay can be reduced by following the above discussed method. The arrangements of RCA are efficient in terms of speed and delay.

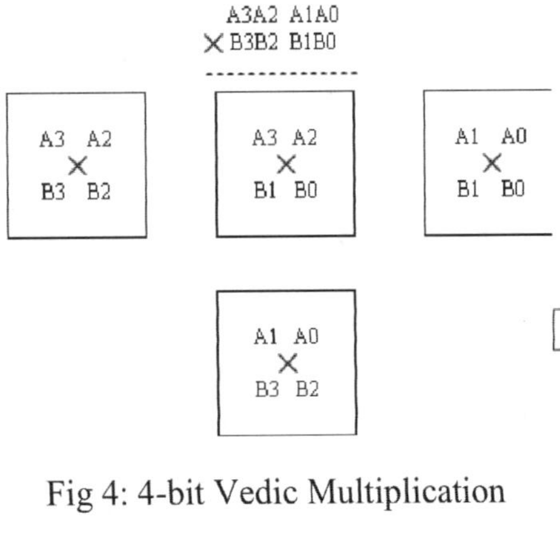

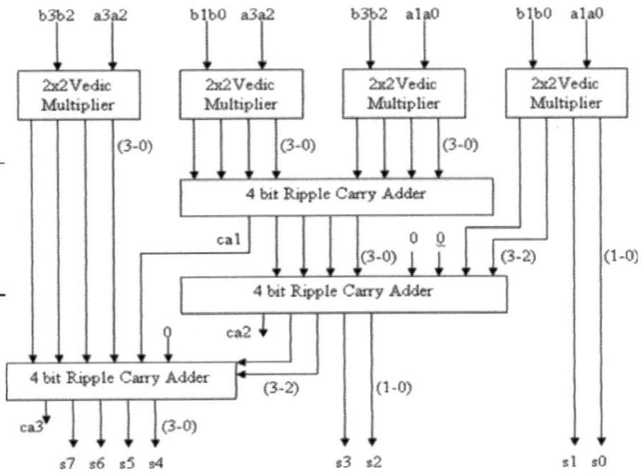

Fig 4: 4-bit Vedic Multiplication Fig 5: Block diagram for 4-bit vedic multiplier

It has to be noted that higher order vedic multipliers such as 8-bit, 16-bit etc can also be implemented using the same method but with little modifications.

VI. APPROACHES FOR ADDERS USED IN DESIGNING VEDIC MULTIPLIER

Adder is another key combinational circuit which is used to add two binary numbers. In many processors and computers, adders are used in the Arithmetic Logic Unit (ALU). To conserve energy, adders with efficient power and speed consumption play a vital role in designing of vedic multipliers [11]. Different adders which are discussed below can be replaced by the adder block in Fig 5. Some of the approaches for designing of vedic multiplier using different adder topologies are given below:

A. CARRY SKIP ADDER

Carry Skip Adder (CSA) can also be named as Carry-bypass Adder. The carry skip adder gives a solution between the RCA and Carry lookahead adder. CSA split up the bits which have to be added in the blocks. The carry and sum bit is produced within these blocks. The carry is calculated by skipping the continuous adder stages which results in less delay in Carry Skip Adder. It also consumes less power as compared to other adders because of less transistor counts and also due to length of the wire being short[12]. Figure 6 shows the Carry Skip Square and RCA. The examination of condition and skipping activity is done by Carry Skip Square [13].

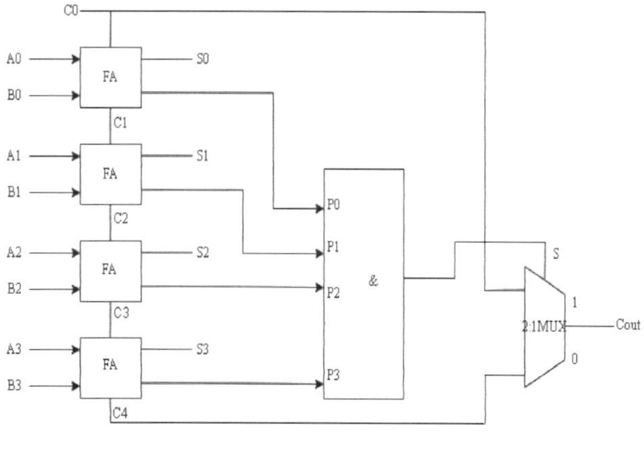

Fig 6: Block diagram of Carry Skip Adder

B. CARRY LOOK AHEAD ADDER

CLA adder [14] is also known as Fast Adder. A carry look ahead adder is speed efficient as it determines carry bits by taking less time. It requires less time to calculate higher value bits of adder by calculating carry bits before the sum. Carry look ahead adder comprises three phases. The G_i and P_i factors are produced by principal arrangement. In the second phase, it produces carry signals for the final stage. The last total yield is produced by the last stage. In Spite of

Journal of Physics: Conference Series **2327** (2022) 012022 doi:10.1088/1742-6596/2327/1/012022

taking less time, the circuit of this adder gets complicated as the number of variables increases which eventually increases the manufacturing cost as it requires more hardware than usual adder.

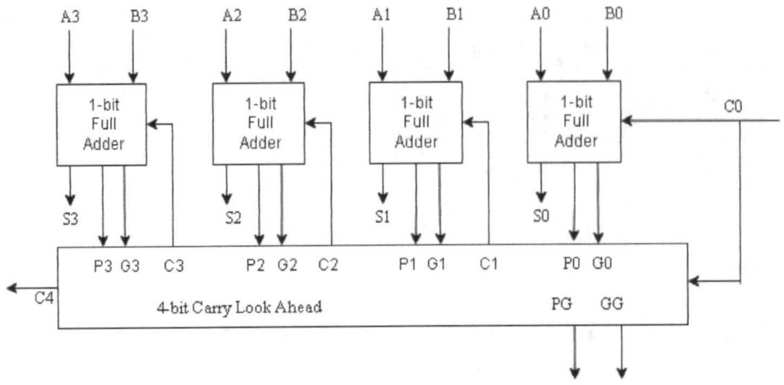

Fig 7: Block diagram of Carry Lookahead Adder

C. KOGGE STONE ADDER

KSA [15] comes under the types of parallel prefix category of carry look forward adders. It contains three computational stages. Preprocessing is the initial stage, in which propagate and generate signals of every input pair are generated. In the second stage which is the carry generation stage, carries of each and every bit is calculated for each bit. Postprocessing is the final stage in which the sum of all the bits are calculated. It is one of the high speed adders and it is mostly used for higher order arithmetic circuits. In KSA, carries are registered quickly by generating them parallelly at the cost of increased area [16]. KSA has low delay. Large area occupation is the main disadvantage of kogge stone adder.

VII. COMPARATIVE ANALYSIS

The above discussed adders are analyzed on the basis of Power Delay Product (PDP), delay, power and area by using 45nm, 90nm and 180 nm CMOS technology files. PDP can be calculated by multiplication of power (nW) and delay (nS). Though kogge stone adder occupies an increased area, it is considered fastest among other adders. Also the PDP is lowest for kogge stone adders as compared to other discussed adders. The simulation analysis of above discussed 4-bit adders is based on different CMOS technology are given in the table 1 to 3 below [17]:

Table 1: Analysis of 4-bit adders based on 45 nm CMOS technology

ADDER	POWER (nW)	DELAY (nS)	AREA (μm^2)	PDP
CSA	2152.8	953	7	2051618.4
CARRY LOOKAHEAD	2415.23	735	15	1775194.05
KSA	1936.38	606	14	1173446.28

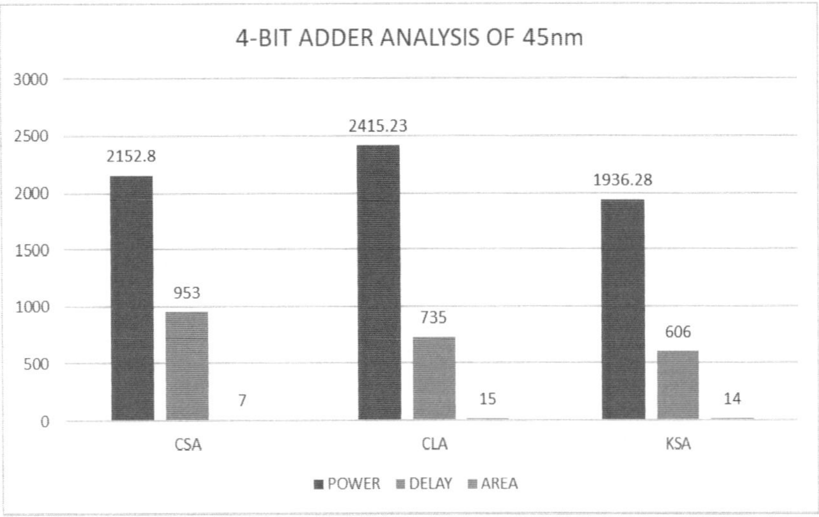

Figure 8: Graphical analysis of 4 bit adder based on 45nm CMOS technology

Table 2: Analysis of 4-bit adders based on 90 nm CMOS technology

ADDER	POWER (nW)	DELAY (nS)	AREA (μm^2)	PDP
CSA	5431.7	1132	79	6148684.4
CARRY LOOKAHEAD	4920.17	802	119	3945976.34
KSA	3710.15	705	117	2615655.75

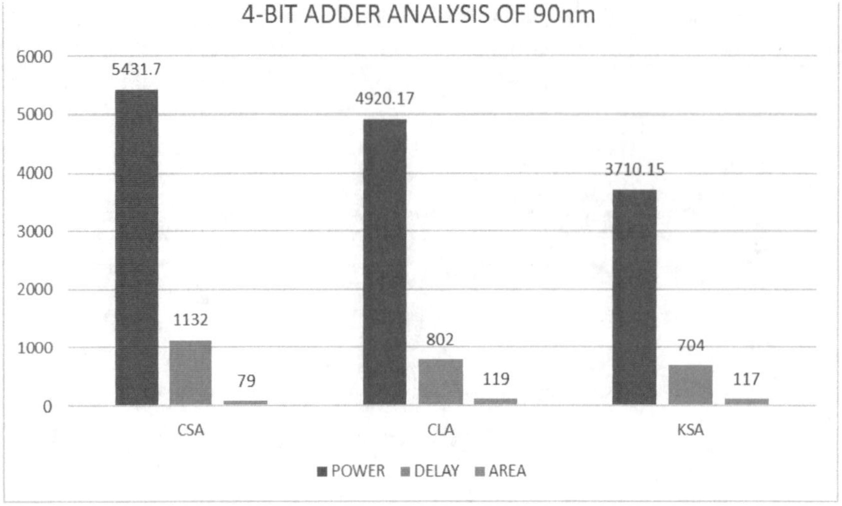

Figure 9: Graphical analysis of 4 bit adder based on 90nm CMOS technology

Table 3: Analysis of 4-bit adders based on 180 nm CMOS technology

ADDER	POWER (nW)	DELAY (nS)	AREA (μm^2)	PDP
CSA	15668.51	4417	369	69207786.6
CARRY LOOKAHEAD	16926.73	3226	389	54605637.4
KSA	33277.94	1063	807	35374449.2

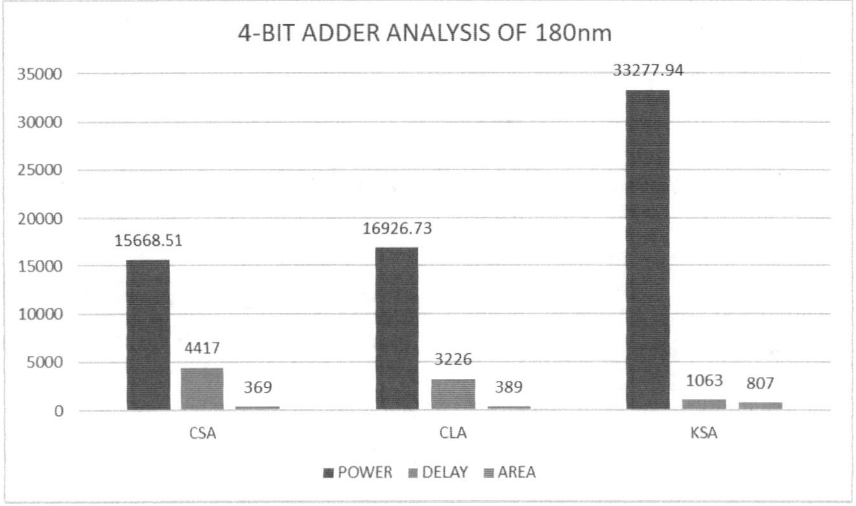

Figure 10: Graphical analysis of 4 bit adder based on 180nm CMOS technology

Journal of Physics: Conference Series **2327** (2022) 012022 doi:10.1088/1742-6596/2327/1/012022

VIII. CONCLUSION

This paper reviews the performance characteristics of Vedic multipliers which are designed using efficient techniques of Vedic mathematics "Urdhva Tiryakbhyam Sutra". This method is based on hierarchical multiplier design which provides an edge in computation compared to other conventional methods. From the comparative analysis, it is clear that Kogge Stone Adder is efficient in terms of power and delay for 45nm and 90nm CMOS technology but seems to be using high power as compared to CSA and CLA. This means with increasing minimum feature size of the transistor, KSA seems to get inefficient in terms power. CSA is more suitable for applications where area plays an major role. CLA provides the balance between all the factors. PDP is least for KSA among other adders. So, KSA is more suitable for faster operation of Vedic multiplier. Hence, Vedic Multiplier design using Kogge Stone Adder proves to be highly efficient and also provides less complexity in terms of designing the multiplier.

[ORCiD Id's- 0000-0002-3245-992X[1], 0000-0003-2366-817X[2], 0000-0002-6190-3540[3], 000-003-0975-7934[4], 0000-0002-7003-2257[5]]

REFERENCES

[1] A. K. Itawadiya, R. Mahle, V. Patel and D. Kumar, "Design a DSP operations using vedic mathematics," 2013 International Conference on Communication and Signal Processing, 2013, pp. 897-902, doi: 10.1109/iccsp.2013.6577186.

[2] Chavan, Arunkumar P., Rahul Verma, and Nishanth S. Bhat. "High speed 32-bit vedic multiplier for DSP applications." international journal of Computer applications. Vol. 135. No. 7. published by Foundation of Computer Science (FCS), 2016. 35-38.

[3] Jagadguru Swami Sri Bharati Krishna Tirthaji Maharaja, "Vedic Mathematics: Sixteen Simple Mathematical Formulae from the Veda" pp. 5- 45, Motilal Banarsidass Publishers, Delhi, 2009

[4] Wallace, C.S., "A suggestion for a fast multiplier," IEEE Trans. Elec. Comput., vol. EC-13, no. 1, pp. 14–17, Feb. 1964.

[5] Kumar, G. Ganesh, and V. Charishma. "Design of high speed vedic multiplier using vedic mathematics techniques." International Journal of Scientific and Research Publications 2.3 (2012): 1.

[6] Harsha, R., et al. "Design of vedic multiplier using Urdhva Tiryagbhyam sutra." Int. J. Adv. Res. Ideas Innovations Technol 5.3 (2019).

[7] Nirmal, N. G., and Dr DT Ingole. "Design And Hardware Implementation Of Vedic Multiplier." IJERT, Vol2 Issue4 (2013).

[8] Dagamwar, Bhagyashree V., R. N. Mandavgane, and D. M. Khatri. "Design of 16-bit Vedic Multiplier for Convolutional Encoder using VHDL." International Journal of Computer Applications 975: 8887.

[9] Verma, Pushpalata, and K. K. Mehta. "Implementation of an efficient multiplier based on vedic mathematics using the EDA tool." International Journal of Engineering and Advanced Technology (IJEAT), ISSN 2249 (2012): 8958.

[10] M. Morris Mano, "Computer System Architecture", 3 rd edition, Prentice-Hall, New Jersey, USA, 1993, pp. 346-348

[11] Jie, Lee Shing, and Siti Hawa Ruslan. "A 2x2 bit Vedic multiplier with different adders in 90nm CMOS technology." AIP Conference Proceedings. Vol. 1883. No. 1. AIP Publishing LLC, 2017.

[12] Bisdounis, L., D. Gouvetas, and O. Koufopavlou. "A comparative study of CMOS circuit design styles for low-power high-speed VLSI circuits." International Journal of Electronics 84.6 (1998): 599-613.

[13] Iynch, Thomas Walker (May 1996). "Binary Adders" (Thesis). University of Texas. Archived (PDF) from the original on 2018-04-14. Retrieved 2018-04-14.

[14] Wang, Y.; Pai, C.; Song, X., "The design of hybrid carry lookahead/carry-select adders, Circuits and Systems II: Analog andDigital Signal Processing, IEEE Transactions on Volume 49, pp.16-24, 2002

[15] P. M. Kogge and H. S. Stone,"A parallel algorithm for the efficient solution of a general class of recurrence equations,"IEEETransactions on computers, vol. C-22, no. 8, pp. 786–793, Aug. 1973

[16] Bhavani, M., M. Siva Kumar, and K. Srinivas Rao. "Delay Comparison for 16x16 Vedic Multiplier Using RCA and CLA." International Journal of Electrical & Computer Engineering (2088-8708) 6.3 (2016).

[17] R.Arun Sekar, D.Ganeshkumar, A. Rajendran, S.Sasipriya, "High Speed Multiplier Design Using Kogge Stone Adder", International Journal of Recent Technology and Engineering (IJRTE) ISSN: 2277-3878, Volume-7, Issue-6S5, April 2019

Understanding and Visualization of Different Feature Extraction Processes in Glaucoma Detection

Nanditha Krishna[1], Dr Nagamani K[2]
[1]Research Scholar, Department of Telecommunication, RVCE, Bangalore, India

[2]Professor, Department of Telecommunication, RVCE, Bangalore, India

nanditha13@gmail.com[a],nagamanik@rvce.edu.in[b]

Abstract. In the recent years the usage of mobile phone is increased and it is the major reason for cause of vision loss in several people. The continuous usage increases pressure inside optic nerve head and it leads to glaucoma disease. Also, there are lot of other reasons which leads to the cause of glaucoma. The purpose of this paper is to determine the importance of feature extraction process in glaucoma detection and implementation of different techniques for extracting convenient features for training machine learning model using pre-processed OCT (Optical Coherence Tomography) images. The two major feature extraction techniques narrated in this paper are convolutional neural network (CNN) model-based feature extraction and image processing model-based feature extraction. A performance analysis was conducted to find best feature extraction technique and both techniques performed well.
Keywords: Optical Image Coherence Tomography (OCT), Convolutional Neural Network (CNN), Glaucoma, Image Processing.

1. Introduction

According to survey conducted by WHO (World Health Organization) the second largest reason for blindness in all over the world is glaucoma [1]. It is an eye problem which leads to permanent blindness. It affects the optic nerve head and decrease blood flow through the nerves. Early detection of glaucoma and proper treatment can reduce the chance of vision loss. But the early detection of glaucoma is a difficult task and people diagnose this disease after getting severe which reduce the chances of providing treatment on time. Several surveys are conducted to find out glaucoma patients in worldwide and the reports published in 2014 says that around 64.3 glaucoma patients are there in year 2013 [2]. According to their prediction they said that the number will increase to 111.8 million by 2040. The research done in 2020 declared that there are 80 million patients in worldwide which shows that the predictions will be come true if the statistics are in same pattern [3]. By analyzing the survey reports it shows that the Asian countries like India, China, Pakistan etc have more glaucoma patients than other countries [4]. The main reason for this in Asians, anatomically the angle between sclera and iris is narrow. It leads to block the normal outflow aqueous humour from the eye. Shortage of trained doctors are one another reason for lack of early detection of glaucoma. Deployment of most modern technologies can help in diagnosis process and early detection of glaucoma. In the vision test procedure, the comprehensive glaucoma analysis in retina should be conducted. A low cost and noninvasive technique is to be implemented for the information acquisition process. In manual diagnosis process early detection of glaucoma is a very complex task. Normally glaucoma is diagnosed in clinical environment by examining the intraocular pressure of optic nerve head. A well experienced doctor can only detect the early stage of glaucoma in manual testing process. In some cases patients with glaucoma have normal eye pressure and glaucoma detection using intraocular pressure is not effective for those cases. The lack of early detection of glaucoma leads to increase of patient's day by day and millions of people are diagnosed with glaucoma. A large-scale screening process is required for this diagnosis process and it is comparatively costly. The introduction of several medical imaging techniques helps in the development of automated glaucoma diagnosis equipment's which helps doctors to detect glaucoma symptoms easily and treat patients in the beginning stage. This system reduces the man power and reduces the complexity of diagnosis. The addition of machine learning and deep learning enhances the advancement in automatic diagnosis methods [5]. The computer vision applications for diagnosis of glaucoma utilized CNN models and it was successful in the classification and segmentation process. This model has the ability to construct a model from extracted and segmented features. The model shows best results according to the availability of dataset. More dataset for training model means the model will have sufficient accuracy. The dataset may include different features of different domain [6]. In recent studies several machine learning and deep learning approaches were implemented for glaucoma diagnosis. However, the proposed models used different types of feature extraction techniques not only for glaucoma detection. Some studies point out multiple algorithms like Local binary pattern histogram (LBPH), Haar feature in feature extraction process. The combined approaches have more effectiveness than the system uses individual algorithm. The experiments conducted by the researches proves this fact [7][8]. In other words, to increase reliability and efficiency of a deep learning-based glaucoma diagnosis system, the feature extraction process has an important role. The maximum information acquired from the input image dataset leads to enhance the trained model accuracy. The image segmentation approaches mentioned in some researches using fuzzy entropy-based algorithms for segmenting information from images [9][10]. The computational speed is comparatively higher than the other approaches. Existing systems have lot of limitations in the feature extraction process and some of them are mentioned in the next section. Because of those limitations practical implementation of the work proposed by several researches don't have satisfied results. In this study the main goal is the development of a new combined feature extraction and visualization method that helps glaucoma diagnosis system to easily acquire information from the input image data. In particular the major objectives of this work are:

- To study the feature extraction techniques used in several glaucoma detection systems.
- To propose a combined technique for feature extraction.
- To visualize the extracted features from input OCT images.

This paper proposes different feature extraction methods which is used for glaucoma detection. The proposed architecture consists of two different stages. The first stage is CNN model involves in feature extraction from pre-processed OCT image. The second stage is the image processing techniques for feature extraction which concentrate on extracting colour, texture and shape features from input pre-processed images. In our experiments both models present promising result in feature extraction. Different features extracted by both techniques open ups a new path to determine glaucoma in an efficient way.

2. Recent Works

For visualizing human body's internal structure there are several imaging modalities used and some commonly known methods are OCT, Ultrasound, X-ray, CT scan, MRI Scan etc. These biomedical applications can diagnose lot of diseases and issues very easily and helps doctors to provide proper treatments on time. Among those methods OCT images of eyes helps in early detection of glaucoma. There are lot of researches and developments done across the world for detecting glaucoma. Several methods delivered better performance and some cannot perform well. In this section we are discussing the feature extraction process from OCT images which is done by different researchers. The comparative study presented by Dennies Rocky et al [7], discuss about different algorithms used for real time recognition of face. They compared Haar feature and Local binary pattern histogram (LBPH) algorithms to detect better algorithm for face recognition. According to their study they said that Haar feature based machine learning model is the effective method for object recognition. The accuracy rate of Haar feature model hit 89% and it is far better than the result of LBPH model. Rosa Andrie Asmara et al [8], developed a combined feature extraction model for vehicle detection. They mainly focus on the three models' Local binary pattern (LBP), Histogram of oriented gradient (HOG) and Haar feature in the process of detecting vehicles. They tested their experimental system with video captured by camera. Their system trains and tests model with the same dataset and it takes more execution time for HOG and LBP. Haar feature is 33.3% faster than other two algorithms they used. They suggest Haar like feature as the best technique for vehicle detection. The research conducted by Dena Nadir George et al [11], explains about the methods used for face feature crop. AdaBoost learning algorithm and Haar like features are utilized in their work for computational effectiveness. The system will crop eye, nose and mouth from the given input face image. Their framework is trained and tested with 500 colour images of faces. Their system achieved 100% accuracy in training phase and 98% accuracy for testing phase. The review proposed by Laleh Armi et al [12], discuss about different texture classification methods. They said that in object recognition and medical image analysis texture feature had an important role. Also, they show the power of computational methods of texture analysis. Texture features represents the characteristics such as regularity, smoothness, coarseness and its variations in each level. They mention advantages and disadvantages of several methods and also gathered dataset used in each method for illustration. Anne Humeau-Heurtier[13] presented a survey report based on different texture feature extraction methods. It categorizes the methods into seven classes, they are structural approach, statistical approaches, entropy-based approaches, model-based approaches, graph-based approaches, learning-based approaches, and transform-based approaches. Also, the advantages and limitations of each method are described in the report. Snehal Patel et al [14] explains various methods of texture and shape feature extraction. They discussed about the merits and demerits of those methods for finding out best algorithm. The proposed approach is to recognize fruits and its grades for agricultural field. They said that shape feature and texture feature are very important visual feature. In their work they used Fourier Transform of Boundary, Edge Detection and Boundary Tracing, fractal dimension technique etc for feature extraction. The article presented by Jianhua Liu et al [15], introduced the shape feature extraction and its importance in medical imaging applications. Common shape feature extraction methods and its characteristics are explained in detailed format in their work. The two major feature extraction techniques used in this framework is region based and boundary-based methods. Komanduri Venkata Sesha Sai Rama Krishna et al [16], developed a combined model for glaucoma detection. As per their study, they identified that feature extraction from OCT images are much complex than feature extraction from fundus images. They proposed an optimization technique using CNN algorithm for feature extraction and training. Also, they adapted hybrid firefly CNN method for decreasing processing time and it shows more effective result. Anselmo Cardoso de Paiva et al [17], developed a deep learning-based angle closure glaucoma detection system which utilizes multi-level CNN architecture. They combine the CNN model and transfer learning to extract visual feature from OCT images. They used anterior segment OCT images for training and testing. Their architecture acquires 97% accuracy while testing. The system performed well for multi-level network than normal CNN. The aim of the study conducted by Guangzhou et al [18], is to develop and efficient glaucoma diagnosis system with machine learning. They used both OCT and fundus images for prediction. Their proposed model can differentiate glaucomatous and healthy images very accurately. The information of fundus image and disc area details from OCT image helps the system to detect glaucoma easily.

Table 1 - Recent approaches techniques implemented and its issues.

Methods implemented in each work	Details of issues
Haar feature and LBPH model [7]	In their work, LBPH model is overall faster than Haar but the experiment results shows that LBPH is not robust and accurate than Haar.
LBP, HOG and Haar feature extraction [8]	Execution time for HOG feature model is 21.36s, for LBP is 19.41s and Haar is 14.56s. This proves that Haar is 46% better than HOG and 33% better than LBP.
AdaBoost and Haar like feature [11]	Their experiments shows that 98% accuracy for testing and 100% accuracy for training. Comparatively they used less amount of data for training and testing therefor they cannot say that their system is more reliable.
Texture feature methods [12][13]	Most of the methods implemented in the work are noise sensitive. Multiple methods used in the feature extraction process and some of those methods have no robustness.
Shape feature methods (Discrete Wavelet Transform, Fourier Transform of Boundary, Scale Space, Recognize boundaries of objects) [14]	Computational complexity, Loss of generality, designing complex filters are difficult. Their system cannot detect fruit shape and texture feature properly.
Shape feature extraction based on boundaries and region. [15]	Recognition rate of their proposed system is 80% and overall accuracy is 90%, they are also mentioned about the limitations of practical implementation.
Hybrid Firefly-CNN [16]	Normal CNN model takes more time for processing. External optimization technique is required.
Multilevel CNN [17] [18]	Their approach performed well, then normal CNN model. Their model cannot extract feature from iris region in some cases.

The table 1 highlights different feature extraction approaches in recent works. Major limitations and some more details for comparison are deliberated in the table.

3. Proposed Methodology and Implementation

3.1 Past Work of Pre-processing Input Data for Glaucoma Detection

In the development life cycle of automated glaucoma detection system using retinal OCT image, the pre-processing of image dataset is already successfully completed. Now the dataset is ready for feature extraction. In past work, we focused on the preparation of dataset for machine learning. The pre-processing is very important in the case of medical data because extraction of medical features is not an easy task in the case of noisy dataset. The dataset for training our system is medical OCT images. We collected different set of images from different domains for increasing dataset size. The images are taken by different machines and different environments. The size, shape, quality etc of these images are different. Before passing dataset to training field it should be prepared [19].

- Image Slicing
- Image Re-sizing
- Image Saturation Removal
- Image Flattening
- Image Filtering
- Image Augmentation

3.2 Feature Extraction for Glaucoma Detection

The dataset for extracting features is ready and the system need to extract needed information (features) from the given dataset. Two different feature extraction methodologies we implemented in this work for extracting features. Figure 1. represents block diagram of the models and methods involved in the extraction process in a diagrammatic format. Each model has its sub modules which categorize and visualize extracted features.

3.2.1 Model 1- Using CNN

- Filter Map Visualization
- Feature Map Visualization

3.2.2 Model 2- Using Image Processing

- Haar Feature
- Texture Feature
- Shape Feature

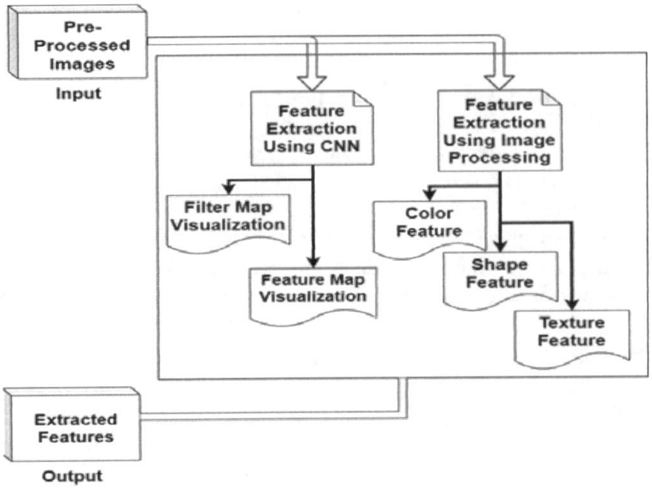

Figure. 3.1 Architecture diagram of proposed system

3.3 Feature Extraction Using CNN

Deep learning proved its ability to handle huge amount of data and it is considered as the powerful tool. Among various deep learning technique, the most established algorithm is convolutional neural network (CNN). CNN have the ability to extract higher representations automatically from image input. It was first developed in middle of 1980s and initially used for recognizing digits in handwritten format. The postal sectors utilized CNN for detecting pin codes, zip codes etc. The deep learning models requires huge amount of data for training. The lack of training dataset in early period was the major reason which affects the performance of CNN and the application of CNN were limited. But in the last of 2012 Alex Krizhevsky brought back the algorithm by providing millions of data. This opens up a new era for CNN in the field of machine learning and artificial intelligence [20]. The primary challenge of the development is applying neural networks in the feature extraction process. Also, we need to analyse how well the model performs for clinical data were collected from different machines. CNN is one of the widely used method for feature extraction. It can achieve better performance and makes the model more reliable. When compared to other models CNN have several layer architectures. This model is also used for several applications like language processing, speech recognition, audio recognition etc. The layers present in this network performs the mathematical operations for generating features map.

3.3.1 Filter Map Visualization

In neural network term we can call filter simply as weights. It has two-dimensional structure while plotting. Weights have its own relationship between each other. For visualizing filters first iterate through different layers present in the model. Then weight and bias values should be extracted from it. Next step is to normalize weight values for filter zero to one. Last process is to plot filters of each CNN layer of all channels. For gray scale images, the channel count is one and for colour image, the channel count is three (RGB).

3.3.2 Feature Map Visualization

By applying feature detectors or filters to input pre-processed images the feature map is developed. The feature map can visualize more details of internal layers of convolutional neural network. Initially a new model for visualization is created and pre-processed OCT image is given as input. The image is converted to numpy array and normalized by re-scaling. The model will extract the prominent features for classifying images. The intermediate representation of fully connected layers are plotted as feature maps.

3.4 Feature Extraction Using Image Processing

For normal image classification problem, we use local features of that image to distinguish between each image. The key features which utilized for this task are the objects edges, colour intensity, texture, shape etc. Below section describes more details about the three feature extraction methods utilized in the system for extracting information from input OCT image for glaucoma detection.

3.4.1 Haar Feature Extraction

Haar feature based model is commonly used in the object detection systems development and the model is introduced by Michael Jones and Paul Viola. For training the classifier model we need lot of positive and negative images. Then we need to extract features from the input images. In the case of Haar features it can be simply represented as rectangular regions which consist of two or more sub rectangles. Below figure 3.2 shows examples for haar features.

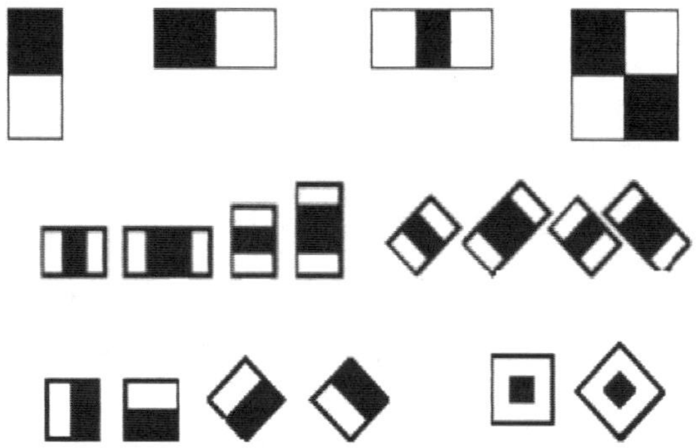

Figure 3.2 Examples of Haar- Features used in face recognition

The sum of pixels inside a rectangular region and its differences can be simply called as rectangular haar feature. It can be scaled within the same image or at any position of the image. To detect windows haar feature also consider neighbouring rectangular regions. The pixel values inside black area is added together and pixels values of white area are summed.

3.4.2 Texture Feature Extraction

In the image analysing procedure for images like medical images, remote sensing images and natural images texture feature extraction is very important. In other terms the superficial phenomenon which helps human to detect natural objects is based on texture features. All types of images have the repeated incorporating pattern in all parts of it and this can be attributed as texture. The commonly used word for texture is 'Texel'. The changes and relation of texture can be easily recognized by everyone but the defining of it is much difficult task. The texture feature can be occurred repeatedly over a region but not in the same point. By qualitative and quantitative analysis texture features can be extracted. Textures can be considered as complex visual pattern which composed of sub patterns and entities which have characteristic shape, size, brightness, colour, slope etc.

3.4.3 Shape Feature Extraction

Shape features are the visual features of an object. It explains about the shape characteristics of all objects and shapes present in the input image. For example, perimeter boundary of rectangular objects, diameter of circular objects and so on. Shape feature have two categories they are, regional characteristics based and border characteristics based. That's why the feature extraction methods for shape feature are also classified into two modes. Region based feature extraction and boundary-based feature extraction. The target in the image can be easily identified using shape feature because of its understandably and intuitiveness. The feature extraction process is not reliable in case of out of shape targets. The accuracy of shape feature extraction is depended on the segmentation and mathematical calculations.

4. Results and Discussion

Glaucoma is a blindness causing disease which is very difficult to detect in early stage. The weakness of nerve cells due to high pressure in optic nerve head decrease the capability of visual recognition. The two common imaging techniques OCT and Fundus imaging are used for glaucoma diagnosis. These techniques have the capability to identify retinal tissue structure which helps the early detection of disease. This work is focused on the reliable use of OCT image in glaucoma diagnosis. The models and techniques used for extracting features are explained in above sections and the visualization of extracted features are shown in this section. The evaluated output of each process involved in this system is represented as graphs and images. As we discussed the input for training model is pre-processed OCT image dataset. In previous work, all images contain in the dataset processed with different techniques and ready for extracting feature. The pre-processed images are called one by one from the stored folder and feature extracted. The below figure 4.1 shows the pre-processed OCT image. Before starting feature extraction process the system randomly select one image from pre-processed dataset and displayed for verification.

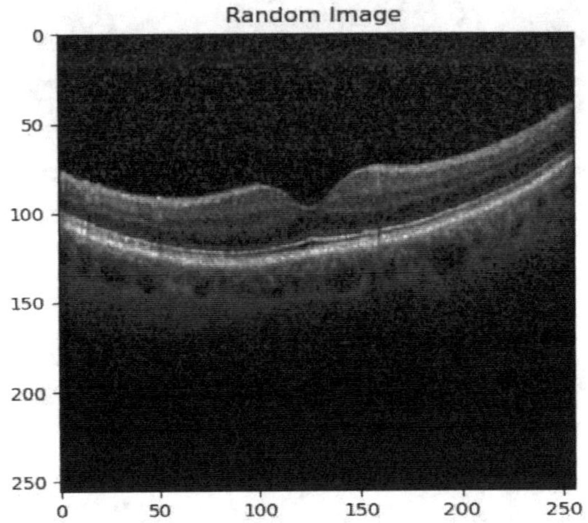

Figure. 4.1 - Randomly selected pre-processed OCT image from dataset used for feature extraction

The neural networks have complex structures and they are poor at presenting reason for predictions they made and decisions they have taken. But in the case of convolutional neural network, it is less inscrutable than other neural network techniques. The result of filters can be visualized for further learning and verification. The figure 4.2 shows some filter cases across the channels. The network terminology is based on weights, the relationship between weights is represented with two dimensional plots for different filter layers. The inhibitory or small weights are represented by dark squares and excitatory weights are represented as light squares.

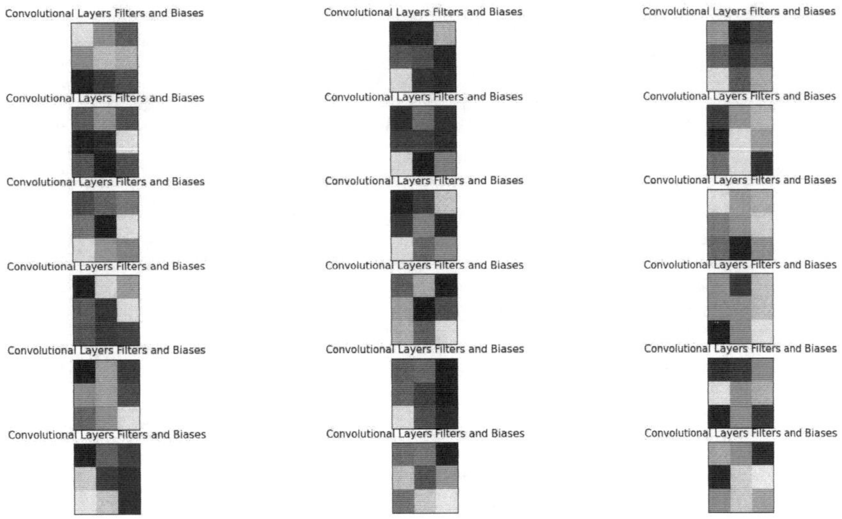

Figure. 4.2 CNN filter map visualization

The figure 4.3 shown below illustrate the feature map of convolutional neural network according to input image. This activation maps capture results by applying other filter maps. It shows more details of what features are preserved from input. All layers of feature map are shown in the figure.

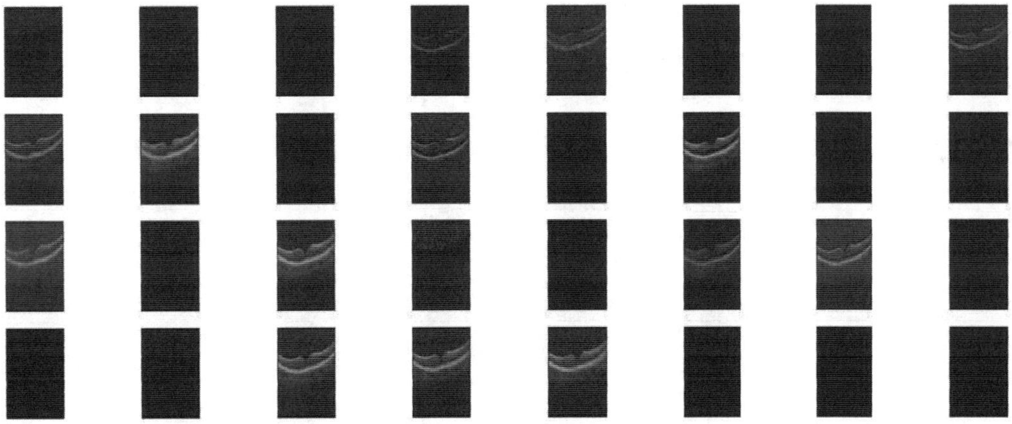

Figure. 4.3 CNN feature map visualization

The image processing techniques for feature extraction we already discussed in above sections and three major features are extracted by the system. For determining glaucoma affected scanned image these features plays an important role. The plot shows in figure 4.4 represents integral image of the extracted haar feature from input image.

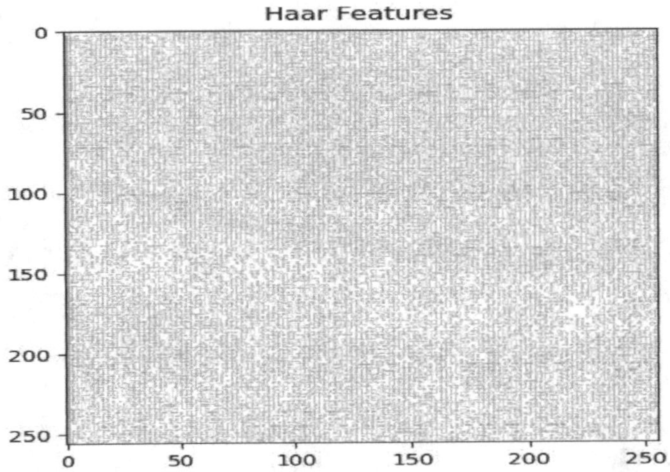

Figure. 4.4 Representation of Haar ffeature extracted from the input image

The texture feature extraction will provide more information about the input image. The object inside it, backgrounds, background context etc can be identified. The figure 4.5 shows the result after texture feature extraction from given OCT image.

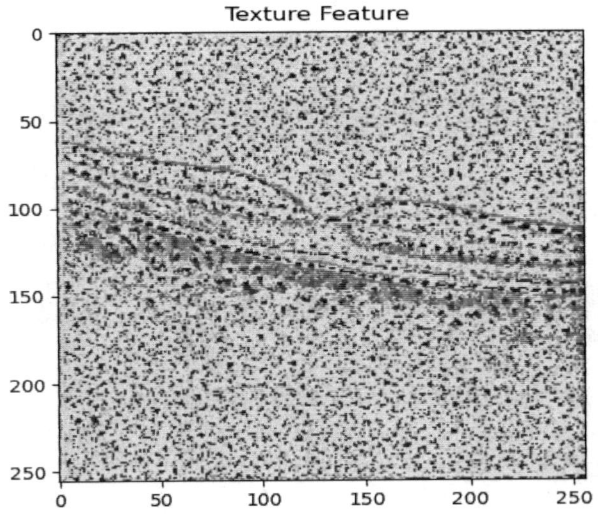

Figure. 4.5 Representation of Texture feature extracted from the input image

The below figure 4.6 represents the extracted shape features. Different shapes present in the input image is highlighted by borders in the result which shows the extracted details.

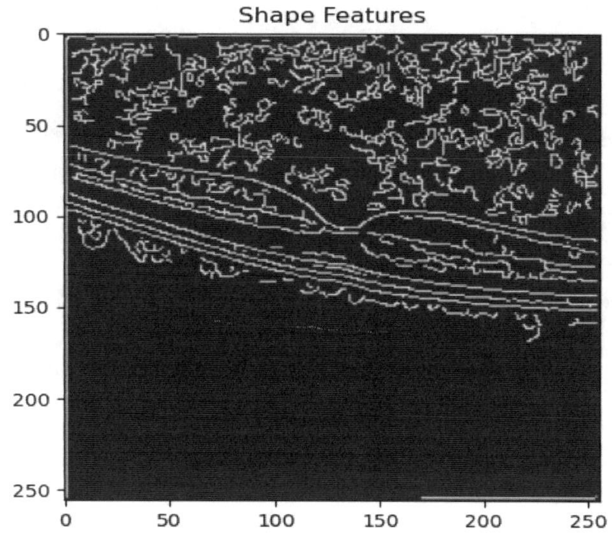

Figure. 4.6 Representation of Shape feature extracted from the input image

The above result obtained by considering random images from pre-processed dataset. The above representation for one input image gives an idea of feature extraction process of entire dataset in glaucoma diagnosis.

5. Conclusion and Future Scope

Feature extraction method is one of the most important part of machine learning and image processing. The medical information for diagnosis glaucoma is to be extracted from training dataset. Those extracted information is used for developing trained model for automatic glaucoma diagnosis. This paper presents two major different feature extraction techniques used for detecting glaucoma using OCT images. One model is based on CNN which visualizes the extracted features as filter map and feature map. Second model consist of image processing techniques which extract haar, texture and shape features. The system tested with random dataset and extracted features using two models in different perspective. The result shows both the models can extract different features which can be utilized for training better model for glaucoma diagnosis. The combined approach implemented in this work increases the reliability of the system and it also increases the percentage of accuracy in information extraction. In the proposed work the number of CNN layers used are less to avoid over-fitting due to less data for training. In future more layers can be added to CNN model for enhancing information extraction performance. But before adding layers the size of dataset is to be increased by including multiple datasets with huge number of images. Also, addition of other relevant features of image processing model can be added to retrieve exact information from images. Combining strategies of multiple models may further enhance the performance of the system.

Acknowledgement

This research is self-confessed for the works referred in the reference section for the efforts made on feature extracting techniques for the glaucoma diagnosis. We wish to acknowledge all other glaucoma diagnosis and feature extraction literature contributors for developing and maintaining this research to be in this extreme.

REFERENCES

[1]. Thylefors, B., Negrel, A.D., 'The global impact of glaucoma', Bull. World Health Organ., vol. 3, pp. 323–326, 1994.

[2]. Tham, Y.C., Li, X., Wong, T.Y., et al.'Global prevalence of glaucoma and projections of glaucoma burden through 2040: a systematic review and metaanalysis',
Am. Acad. Ophthalmol., vol. 121, pp. 2081–2090, 2014.

[3] Cook, C., Foster, P., 'Epidemiology of glaucoma: what's new?', Can. J. Ophthalmol., vol. 47, pp. 223–226, 2012.

[4]. 'Glaucoma in Asian populations',. Available [Online] at https://www.glaucoma.org/gleams/glaucoma-in-asian-populations.php, November 2017

[5]. Ali Serener, Sertan Serte, "Transfer Learning for Early and Advanced Glaucoma Detection with Convolutional Neural Networks", Medical Technologies Congress (TIPTEKNO), 2019.

[6]. Manal Al Ghamdi, Mingqi Li, Mohamed Abdel-Mottaleb, Mohamed Abou Shousha, "Semi-Supervised Transfer Learning For Convolutional Neural Networks For Glaucoma Detection", IEEE International Conference on Acoustics, Speech and Signal Processing (ICASSP), 2019.

[7]. Dennies Rocky, Aajin Roy, Anujith S, Eldho K Paul, Akas G Kamal, "Comparative Study of LBPH and Haar features in Real Time Recognition Under Varying Light Intensities", International Journal of Engineering and Advanced Technology (IJEAT), vol. 9, May 2020.

[8]. Rosa Andrie Asmara, Muh Bambang Purwanto, Cahya Rahmad, Desy Derius M. and Isa Mahfudi, "Comparison of Vehicle Detection Using Haar-like Feature, LBP and HOG Technique for Feature Extraction in Cascade Classifier", International Journal of Advanced Science and Technology, vol. 28, pp. 834-838, 2019.

[9] Shubham Mahajan , Nitin Mittal , Rohit Salgotra, Mehedi Masud , Hesham A. Alhumyani , and Amit Kant Pandit, "An Efficient Adaptive Salp Swarm Algorithm Using Type II Fuzzy Entropy for Multilevel Thresholding Image Segmentation", Hindawi Computational and Mathematical Methods in Medicine, vol. 2022, 2022.

[10] Shubham Mahajan, Nitin Mittal, Amit Kant Pandit, "Image segmentation using multilevel thresholding based on type II fuzzy entropy and marine predators algorithm", Multimedia Tools and Applications, 2021.

[11]Dena Nadir George, Naseef Husam Mohammad, Saja Talib Ahmed, "Facial Feature Crop Detection Based On Haar Feature And Adaboost Training", Journal Of Southwest Jiaotong University, vol. 55, Aug. 2020.

[12] Laleh Armi, Shervan Fekri-Ershad, "Texture Image Analysis And Texture Classification Methods - A Review", International Online Journal of Image Processing and Pattern Recognition, vol. 2, pp. 1-29, 2019.

[13] Anne Humeau-Heurtier, "Texture Feature Extraction Methods: A Survey", ACCESS, 2018.

[14] Drashti Jasani, Paras Patel, Snehal Patel, Bindiya Ahir, Khushbu Patel, Mitali Dixit, "Review of Shape and Texture Feature Extraction Techniques for Fruits", (IJCSIT) International Journal of Computer Science and Information Technologies, vol. 6, pp. 4851-4854, 2015.

[15] Jianhua Liu and Yanling Shi , "Image Feature Extraction Method Based on Shape Characteristics and Its Application in Medical Image Analysis", ICAIC, pp. 172–178, 2011.

[16] Komanduri Venkata Sesha Sai Rama Krishna, Kosaraju Chaitanya, Pedalanka Poorna Saraswathi Subhashini, Rajesh Yamparala, Satya Sandeep Kanumalli, "Classification of Glaucoma Optical Coherence Tomography (OCT) Images Based on Blood Vessel Identification Using CNN and Firefly Optimization", Journal iieta, Vol. 38, pp. 239-245, February, 2021.

[17] Marcos Melo Ferreira, Giovanna Pavani Estevel, Geraldo Braz Junior, Jo˜ao Dallyson Sousa de Almeida, Anselmo Cardoso de Paival, Rodrigo Veras, "Multilevel CNN for Angle Closure Glaucoma Detection using AS-OCT Images", Proceedings of the IWSSIP, 2020.

[18] Guangzhou An, Kazuko Omodaka, Kazuki Hashimoto, Satoru Tsuda, Yukihiro Shiga , Naoko Takada, Tsutomu Kikawa, Hideo Yokota, Masahiro Akiba , and Toru Nakazawa, "Glaucoma Diagnosis with Machine Learning Based on Optical Coherence Tomography and Color Fundus Images", Hindawi Journal of Healthcare Engineering, 2019.

[19] Nanditha Krishna, Dr Nagamani K, "Assorted Image Processing Techniques for Retinal OCT in Glaucoma Detection", Proceedings of the 2nd International Conference on IoT, Social, Mobile, Analytics & Cloud in Computational Vision & Bio-Engineering, 2020.

[20] Yu-Chieh KoI, Shih-Yu Wey, Wei-Ta Chen, Yu-Fan Chang, Mei-Ju Chen, Shih-Hwa Chiou, Catherine Jui-Ling Liu, Chen-Yi Lee, "Deep learning assisted detection of glaucomatous optic neuropathy and potential designs for a generalizable model", Plos One, 2020

SECURE PDF TEXT STEGANOGRAPHY TURNING SECRECY INTO INCONSPICUOUS ENCODING

Niharika Singh[1], Avinash Kaur[2]
[1]PG (Data science), Department of Computer Science Engineering, Lovely Professional University, Punjab (India) -144411.
[2]Associate Professor, Department of Computer Science Engineering, Lovely Professional University, Punjab (India) -144411.

Abstract- Security has become a priority in any security system. Computer applications, Internet system applications, etc. how millions of users use the Internet. This is a research project suggests an information processing algorithm Security system using PNG (Portable Network Graphics) pictures. To our knowledge, there are very few jobs available is present in the PNG archives of steganography. A Least Significant Bit (LSB) encoded stenographic system code is slightly more advanced and is built in the Matlab programming language. The proposed system is often different from other default approaches that use the LSB input method, with the next point. This: The first point about the character set of the secret message used. It automatically detects the character set for private messages. and determine whether the text is written in Latin (only 8 bits per character) or Latin/Arabic (16 characters per character). The approach presented is implemented using special Matlab code functions and encoding algorithms. Two main algorithms for deleting encrypted messages sequentially.

Keywords: Steganography, LSB insertion method, peak signal-to-noise ratio (PSNR).

1. Introduction

Information security is a hot topic these days. Steganography is the art of conveying messages securely from sender to recipient. It should ensure that no one should be aware of the existence of a confidential message. Only authorized recipients should disclose confidential information. To achieve the intended privacy, the secret message (file / object) is hidden in other media (broadcast media) and sent over the network, where intended, the recipient receives this embedded file called a stego file (stego process = data + photo Media)) where confidential information from the input area is released (unstego process).

Customization comes with the advantage of saving unnecessary cover art on both sides. Makes it possible to handle hidden Arabic text. which other researchers rarely manage. The second point that separates our approach to others is how we can choose when text is embedded between the multiple pixels of the cover image. where a given main character unit can be used only once, optional, but twice in two consecutive random selection missions. Each task has numerical results in a different order. The third point, though, deals with how the hidden text is distributed evenly in the lower areas of the image. This makes it difficult to detect the presence of hidden messages.

The LSB method starts with sending private messages and cover images to the encoder. Then inside the encoder one or more encryption protocols will be used. The key to mergers and acquisitions is required. The key can be used to reduce the likelihood that a third-party criminal will take a picture of the Stego and remove the encryption to gain access to confidential information. In this article, the coverage differs greatly from the other three areas. The first concerns the character set used. Text characterization is useful for spending less money on unnecessary parts of the overlay.

The second point is to separate the selection method where the text portion is embedded between the multiple pixels of the cover image. Where a given key sequence can be used only once, as an alternative, but twice with two consecutive random selections. Each task has numerical results in a different order. Although the third point is related to the uniform distribution of hidden text fragments in the lower areas of the image. This makes it difficult to detect the presence of hidden objects.

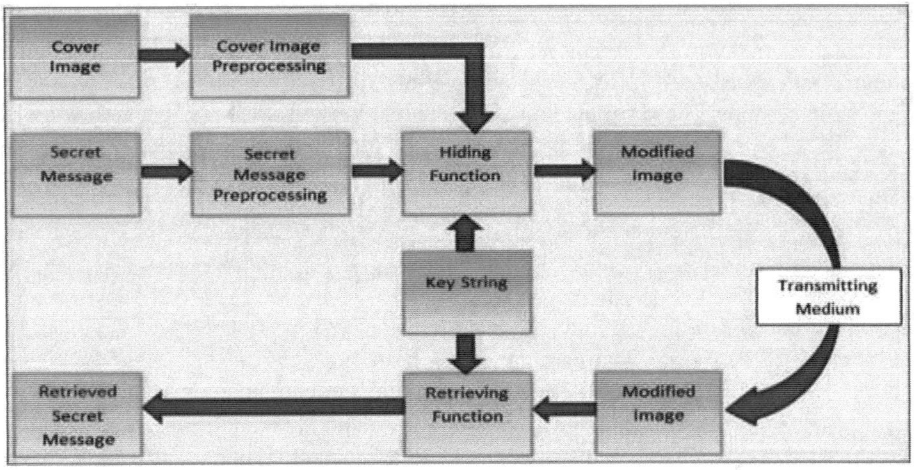

Figure 1. Steganography system

2. Literature review

Sanjive Tyagi et al. (2018), introduces a new model of PDF-based text steganography. Confidential communication is obtained by attaching confidential information to areas between text characters in PDF format. The stego text format is designed to improve embedding capacity and reduce the PDF file of stego cover cover. A method designed to develop a wide range of mounting power systems can be used in accordance with the requirements of the standard mounting capacity up to the normal mounting capacity. The editing and classification of encrypted text is achieved by creating a quadratic hashing model, a four 'creative manual that provides effective performance in complex time situations. Research results show that the established method provides useful algorithms for modern security and high level of hidden data. Demonstration and testing of the model method and other simulations provide important information to show that the test methods work better than the current methods and do not have requirements such as shadows, language, word list, voices and so on.

Security is an important part of all our lives to secure information so that no unauthorized person can steal it. For this reason, various strategies are available to protect them through security statements, secret keys, finger protection and more. However, today, in this age of digital communication on the global web, activities such as illegal authorization, exploitation and copyright infringement are on the rise. . Therefore, there is a need to protect sensitive information using cyber security guidelines that can be achieved by covering digital information and ' or translate it into an unfamiliar framework. Therefore, innovations in steganography and cryptography can be considered as important components of the use of digital information systems.

Sangita Roy et.al (2011), The word “steganography” is derived from the Greek language and means “hidden writing” from the Greek word “steganos”, meaning “to cover or protect”, and after the development of new technologies, especially technology information and computing systems in the last decade, the issue of data security has taken on special meaning. One of the key components of data security is the concept of confidential information sharing. For this purpose, various methods, including description, steganography, encoding, etc., are used. Steganography is one of the techniques in this area that has gained the most interest in recent years.

Once Steganography is complete, the main objective of Steganography is to hide information at the bottom of the media homepage so that outsiders cannot access the information contained in the media. This is the most important difference between Steganography with other methods in exchanging confidential information, for example, in computational methods People are aware of the existence of information by looking at the verified data. Although they may not be able to understand the information. No one is aware of the existence of the information in the data source. It's mostly done in movies, videos, writing, music, and audio.

Text Steganography is a more complex form of Steganography because there is no media data in the file format. While often used in image or audio files. Which can be used in Steganography Digital messages are written, stored and downloaded by computers in the same way that they are visible to the naked eye. Unlike other digital files such as images, as opposed to what is visible, stored and kept secret. Hide details about the structure. The use of information encoded in the text, from the native language to the HTML file format.

Before giving a brief overview of the text technology process. The general terminology used in text technology should be considered first on this matter. Secret messages that must be encrypted are called attachments, and secure text/audio/video used for attachments are called covers. The product after installation is called stego material. The sending and receiving of points is very important in order to align with one important marketing strategy. There is some functionality to hide information in messages. Below is a list of applications. that have been carried out and reported to date

Pritam Kumari et.al (2013), with new technological innovations to come. It is imperative to provide a way to provide confidential information online without modification and for the public to consider your audience. One way is Steganography. It comes from two Greek words steganos meaning to cover or hide and grafein‖ meaning to write. For this reason, it is common practice to hide personal information in another file called takeaway. Image files are often used as containers because of the space they take up. availability and the ability to obtain a Stego image after adding confidential information to the digital image by an algorithm. Everyone except those who wanted to get the Stego photos didn't know there was any secret information embedded in it. This method provides the highest quality while hiding private messages. This means that only the intended sender and recipient can suspect confidential information. It is also known as an instability protection scheme. Steganography and Cryptography are closely related. The word Cryptography comes from the Greek word kryptos, which means "hidden." Another way to protect information is by translating it into a readable format called numeric text. Only the person who has the secret code can scroll through it or send it in plain text.

Arvind Kumar et.al (2010), Steganography It is an example of data encryption and an attempt to hide the existence of attached data. It's a better way to protect messages than to encrypt them by simply hiding the content of the messages. not the existence of the message The first message was hidden in the warship. In order not to make any changes to the ships, in this article we will discuss how to use the technology as a message board. This document also contains information about using the tool. Other Steganography with Steganography It is a useful tool that allows data storage by means of communication. The combination of secret photos and photographs creates hidden images. Hidden images are hard to find without pulling back.

This article explores this technology in depth by introducing readers to the concepts of Steganography. A brief history of Steganography and a summary of the method Some Steganography requires Internet users to collect, send or receive personal information. The most common way to do this is to convert the data to another format. The information obtained can be understood by those who are able to restore it to its original state.

This method of protecting information is known as information. The biggest disadvantage of data is the availability of unencrypted data. The highlighted data, even if it is unreadable, is still in the log. If a person has enough time to search He or she can delete the record. One solution to this problem is camouflage. The ancient art is to hide messages from view. There are no changes or changes. The hidden message is simple but infallible for the reader. the purpose of Steganography is to conceal the existence of the message. While encryption confuses the message so it doesn't obscure it.

Shivani Kundra, Nishi Madaan (2014), Internet usage is increasing day by day. Providing security has become a major issue due to the use of the Internet. coding and Steganography It's a way to keep your information safe. Encryption is used to encrypt messages to protect them from other influences. Steganography It is a method used to hide information in frames so that no one can guess. The cover can be an image, text, audio, or video. Steganography. It comes from the Greek words 'Stegnos' which means secret and 'Graph' which means to write which usually means secret writing. purpose of Steganography is data encryption while using encryption to protect data from intruders. due to empty space Image files are therefore often used as captions.

There are two types of uninterrupted files and uninterrupted files. The original image will remain unbreakable. The compressed image is exactly the same as the original image. especially the system Steganography Because the main objective of Steganography is to hide the text in the image without affecting the quality of the GIF, BMP graphics using compression techniques. On the other hand, the pressure loss also helps to maintain the final position. It does not protect the integrity of the original image. JPEG conversion using standard compression does not imply an exact copy of the original image.

3. Proposed LSB Insertion Techniques

Normally, in LSB input mode, a random number generator is used to distribute and hide private messages with small pixel art images. To do this, usually one of the existing pixel skin images is randomly selected and used to hide the text portion of the text. This means that the random function uses the main character unit only once per pixel in the image. And because a standard color image has many pixels, which are actually a large number of pixels. Therefore, random number generators take a lot of time to complete. Moreover, generating large vectors of many specific random numbers is a difficult point function viewer. The entire area of a group of photos is logically divided or divided. are equal squares.

The correlation coefficient (NC) given in equation (1) is used to measure the similarities between the embedded and extracted secret data:

$$NC = \frac{\sum_x \sum_y osd(x,y)esd(x,y)}{\sum_x \sum_y osd^2(x,y)} \quad (1)$$

where,
osd is the actual secret data and esd is the extracted secret data.

Peak Signal to Noise Ratio (PSNR): It is used to evaluate and find the differences between original image and the cover image. Peak Signal to Noise Ratio (PSNR) indicator is given by Equation (2).

$$PSNR = 10\log\left(\frac{\max\,(ci^2(x,y)}{\frac{1}{MxN}\sum_{y=1}^{N}\sum_{x=1}^{N} ci(x,y)-si(x,y))^2}\right) \quad (2)$$

Where,
ci is the cover image and si is the stego_image.

4. System test

System core algorithm Steganography Our tests were tested on colored cover art of different sizes and content. Additionally, the test was performed only on English and English/Arabic text. The text used was of different length (any text from 1 font to 1/8 cover size) In this article, the masking method uses only one image. And the main mask algorithm can be modified to easily manipulate the tri-color channel to handle longer messages. Figures 2 and 3 show the Original and Hidden cover image in the early stages.

Journal of Physics: Conference Series **2327** (2022) 012024 doi:10.1088/1742-6596/2327/1/012024

Figures 4 and 5 show the Original and Hidden cover image in the second stages. Figures 6 and 7 show the Original and Hidden cover image in the third stages.

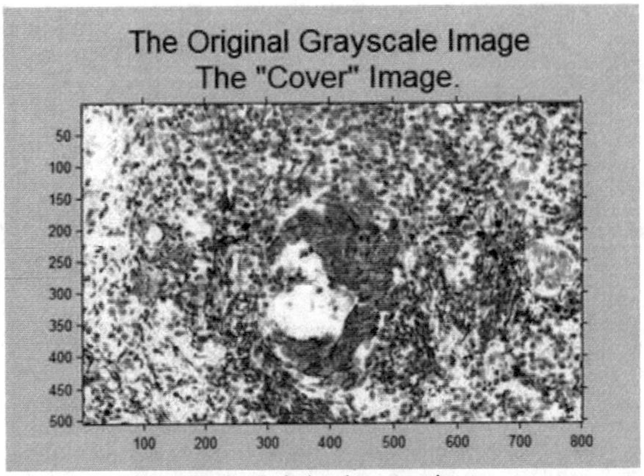

Figure2. Original cover image

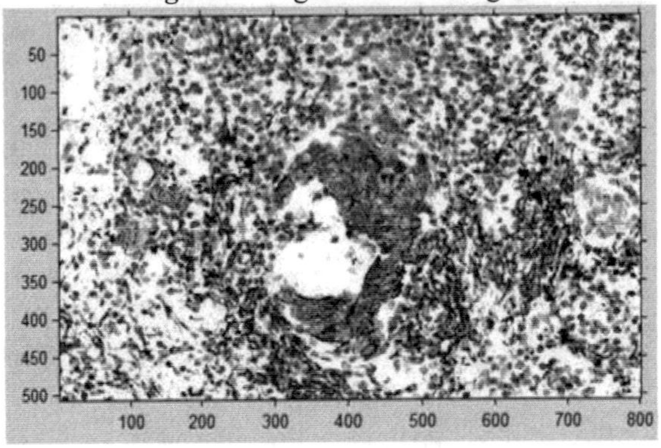

Figure 3. Hidden cover image

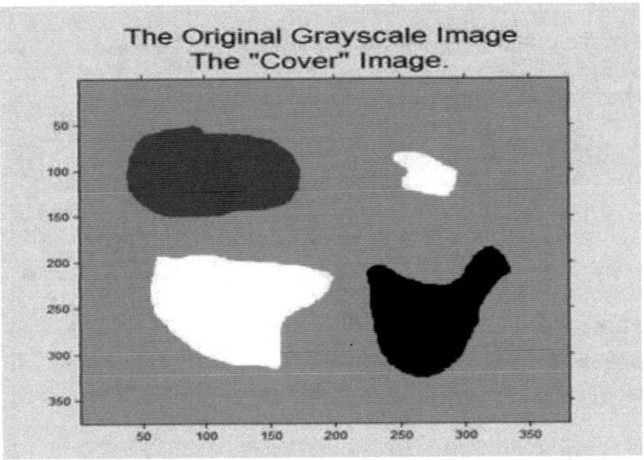

Figure 4. Original cover image
in second stage

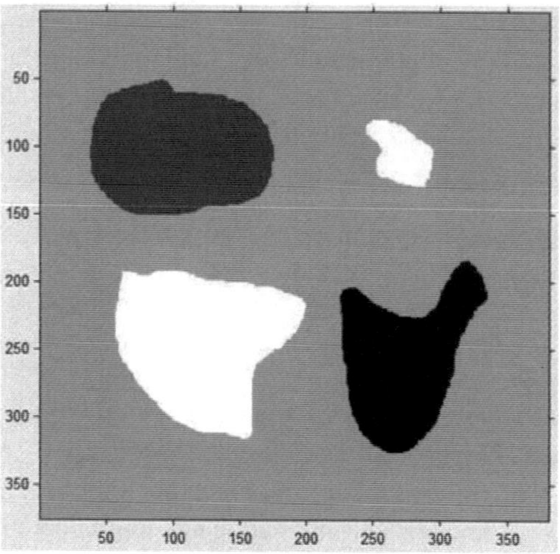

Figure 5. Hidden cover image in second stage

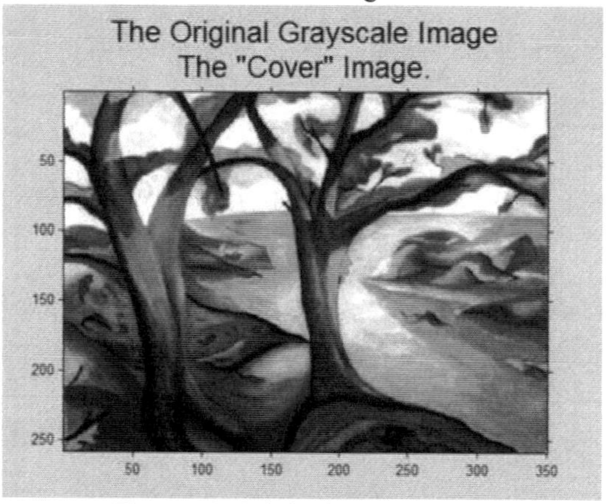

Figure 6. Original cover image in third stage

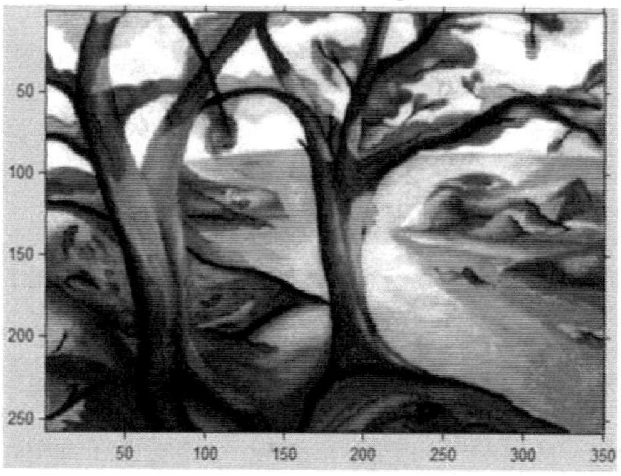

Figure 7. Hidden cover image in third stage

5. Conclusions

The LSB key input system can be recognized and designed in the Matlab programming language when testing the system. Secret texts in English and English/Arabic of different lengths will be applied in various color cover images, content and image sizes, as well as systematic measurements in the results of the stego image system. This confirms the great similarity of the stego image as compared to the original cover image. To our knowledge, the proposed steganography (MATLAB) method is the first to use PNG format cover images to combine high data volume with strong adhesive attachments. In this way, the use of the proposed method is much broader than other activities reported in this image format domain.

References

1. N. Provos and P. Honeyman, "Hide and seek: an introduction to steganography," IEEE Security & Privacy, vol. 1, no. 3, pp. 32–44, 2003.
2. S. Gael, R. Ekodeck, R. Ndoundam, "PDF steganography based on Chinese Remainder Theorem", Journal of information security and applications (2016), Elsevier, Vol. 29, 2016, pp. 1-15.
3. Chandramouli, R., Memon, N., 2001 "Analysis of LSB based image steganography techniques ", Image Processing, Proceedings. 2001 International Conference on,pub. IEEE, Vol. 3.
4. S. Samima, R. Roy, and S. Changder, "Secure key based image realization steganography," in Proceedings of the 2013 IEEE 2nd International Conference on Image Information Processing (ICIIP), pp. 377–382, Shimla, India, 2013.
5. J. Mielikainen, "LSB matching revisited," IEEE Signal Processing Letters, vol. 13, no. 5, pp. 285–287, 2006.
6. W. Luo, F. Huang, and J. Huang, "Edge adaptive image steganography based on LSB matching revisited," IEEE Transactions on Information Forensics and Security, vol. 5, no. 2, pp. 201–214, 2010.
7. B. Li, M. Wang, J. Huang, and X. Li, "A new cost function for spatial image steganography," in Proceedings of the 2014 IEEE International Conference on Image Processing (ICIP), pp. 4206–4210, Paris, France, 2014.
8. M. Boroumand, M. Chen, and J. Fridrich, "Deep residual network for steganalysis of digital images," IEEE Transactions on Information Forensics and Security, vol. 14, no. 5, pp. 1181–1193, 2018.

Generation of Single Sideband-Suppressed carrier (SSB-SC) Signal Based on Stimulated Brillouin Scattering

Rosepreet Kaur Bhogal[1], Manoj Sindhwani[2]

School of Electronics and Electrical Engineering, Lovely Professional University, Punjab.

rosepreetkaur12@gmail.com[1] , manojsindhwani324@gmail.com[2]

https://orcid.org/0000-0002-1697-4489[1] , https://orcid.org/0000-0002-6859-8622[2]

Abstract. A novel cost-effective wavelength shifter is suggested in this paper. This paper shows how to shift signals in the mm-wave area using a frequency shifter. Due to stimulated Brillouin scattering in optical fibre, the proposed approach produces a single sideband signal without the carrier signal. When compared to SSB-SC transmissions, the bandwidth efficiency of double sideband signals is lower. Because it employs an optical fibre with a performance degrading condition such as SBS as an application, the frequency shifter is entirely optical. The theory of Gain and Loss Spectrum of SBS within optical fibre is utilised to minimise the upshifted frequency component and boost the low shifter frequency component in order to generate improved carrier power and single sideband. The SBS phenomenon produces a single sideband, which is advantageous in advanced modulation systems and consumes less bandwidth. It's worth noting that the SSB-SC signal may be generated using any optical approach, and the proposed work eliminates the need for non-optical components like 90-degree hybrid couplers. SSB-SC signals are capable of combating pulse broadening difficulties and have the ability to operate at terahertz frequencies. Transmission of an SSB-SC signal is also shown to validate the performance of the produced signal. Furthermore, the Q factor of SSB-SC, SSB with carrier, DSB-SC, and DSB with carrier signals is studied at various connection lengths.

Keywords: SBS (Stimulated Brillouin Scattering), SSB (Single sideband), DSB (Double sideband) , SMF (Single mode fiber), Q-factor

1. Introduction

The future networks based on fiber optic technologies, intend significantly to boost the capacity or to pack more number of channels in the information transmission. Also data rates for each user should be high, as the demand of high speed services increasing day by day [1]. A major and prominently used method is dense wavelength division multiplexing in optical fiber communication systems to augment the capacity of the users in mobile coverage wide area [2]. With this technology of multichannel transmission in the system, large number of users can be served due to distribution of radio heads. Recent advancement in microwave photonic technology as well as optical polarization multiplexing offers and give rise to the technology called radio over fiber (RoF). Telecommunication applications like as global systems for mobile (GSM) and GPRS, which were available at the time, could accommodate lower data rates [3-4]. On the other hand, in today's world, high-capacity systems, such as WiMax and WiMAN, are required to meet user expectations, with capacity ranging from 1 to 10 Gbps and a wide coverage area [5]. However, as the number of users grows, restrictions on information flow will be imposed. One solution is to reduce the cell size to accommodate more users. This is referred to as the notion of micro-cells or pico-cells. The additional method is to employ novel bands of processes since the unlicensed ISM occurrence bands are previously crowded. Now, a lot of developers favour the millimeter-wave (mm-wave) as the novel group of operation. Optical frequency region of 40 GHz to 90 GHz, that offers higher bandwidth [6]. Some other problems come into play by utilizing these solutions. More base stations are required if cell size reduces. Increasing the frequency, motivation contributes to additional equipments, fitting and preservation costs. The components that are used in all optical systems are able to combat with the issues in fiber optic transmission. In order to generate the mm-wave, the all optical systems offers wide shift and are well suited for the transmission of high data rates. This all optical method has several advantages are used in military, radio astronomy [7]. Future communication networks need the frequency shifts in the millimetre wave range as well as the less emergence of spurious signals and more tuning capability, this is find as the tedious task by incorporating conventional mm-wave generation methods.

Optical approach or photonics can fulfil the requirements and expel the limitations by shifting operation of frequencies. The all optical approach is also offers the signals that shows better bandwidth efficiency and generate SSB signal with carrier. Generation of bandwidth efficient SSB signal in radio over fiber systems is an important method [8-10]. There are several more fields that incorporate the all optical frequency shifter such as quantum measurements, spectroscopy and the wide frequency shifters. Due to numerous advantages, all optical mm-wave signal generation in RoF systems is need of a day.

In this work, a frequency shifter is proposed using all optical method in RoF system. The shifter is realized by incorporating the SBS that expel the need of any electrical hybrid coupler or components. The main work is to enhance the low frequency component employing gain spectrum and to eliminate up shifted frequency by loss spectrum of SBS. Also modulators used are biased at minimum point to get single sideband after SBS. Achieved advance signal ie. Bandwidth efficient light signal is further transmitted over optical fiber to test its performance in terms of Q-factor and BER.

2. Principle of SBS based frequency down converter

Architecture is presented to generate the signal that has single sideband and Carrier is also eliminated. In order to accomplish the generation of aforementioned signal, a stimulated brillouin scattering is employed. To realize the SSB signal, two optical modulators are needed and an optical fiber is a root communication medium that introduces SBS. In detail discussion of SBS, it is presented that SBS has two very important phenomenons. Principle of SSB signal generation is that when the low power carrier signal enters in fiber optic, it produces the back scattered light called strokes wave and remaining signal travels to the same direction as it is intended to transmit. A new signal is also coupled to fiber optic from other end called counter direction and this signal is kept at high power. A high power signal delivers its gain to co directional signal as depicted in Figure. 1 (a) and itself experience loss of power. Gain is experienced by the low power signal that is coupled from first intensity modulator and as a result, signal is moved to lower region of frequencies as illustrated in Figure. 1 (b). Consequently, a lower frequency high power sideband is obtained and due to loss spectrum experienced by counter direction signal, is eliminated or suppressed as represented in Figure. 1 (c). The proposed method is useful in numerous applications and has wide tenability range of mm-waves. This method of SBS is attractive and getting more attention due to the advantage that mm-waves shifts can takes place early at low launched powers.

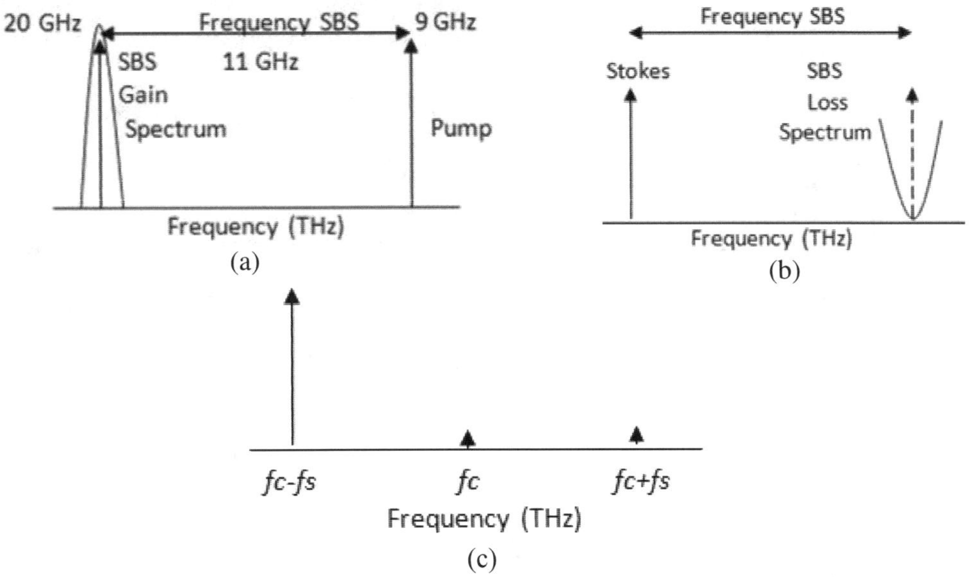

Figure 1. SBS phenomenon based (a) Gain spectrum (b) Loss spectrum of strokes and pump waves inside optical fiber and (c) Output optical spectrum

Frequency generation tunning in mm-waves is considered to be extremely important such that there is generation of less spurious signal while tunning the shift. The signal power of generated signal can be controlled or regulated by various parameters such as the loop length, launched power and modulation index of intensity modulator. SBS is also used to generate double sidebands and single sidebands. In subcarrier multiplexing, SSB with enhanced carrier signal is needed so that the carrier signal to cater the downstream users and one sideband modulated to serve upstream central unit.

3. System setup

In order to realize the bandwidth efficient single sideband suppressed carrier signal, a simulation tool Optisystem is used for analysis. In optical communication systems, laser is an important component and incorporated in the proposed architecture. Laser is operated in C band and emits the frequency of 193.1 THz. Drive of laser signal is modulated through modulator MZM and sine generator is place to generate RF signals. Modulated signal is suppressed carrier and double sideband signal due the MZM. MZM is biased at minimum point of operation. Two modulators are used and termed as upper MZM 1 and lower MZM 2. A 20 GHz sine generator frequency is provided to MZM 1 at launched power level of 0 dBm. To get a frequency shift, sine wave generator frequency in lower modulator is set to be 9 GHz. It is basically a difference of upper frequency (20 GHz) and brillouin fixed shift (11 GHz). A semiconductor optical amplifier is low cost and monolithically integrated component for amplification of weak signals. SOA is incorporated in lower modulation process to boost the signal. Lower signal is at launched power (-20 dB), however SOA amplify this to provide high power signal. Upper modulator signal is coupled to single mode fiber of link length 25 km in co-propagation direction and lower modulator signal is fed from counter direction. Co-propagation directional signal after travelling 8 km show SBS phenomenon i.e. some part of signal reflected back (stokes wave) and rest move to the same direction in which it was sent by modulator 1. Counter directional signal provide signal power to stokes wave also called gain and after distributing the gain, counter directional signal itself carry the low power. Thus these signals initiate two phenomenon of loss spectrum and gain spectrum. Thus signal at 193.11 THz due to SBS is suppressed and signal at 193.09 THz is enhance

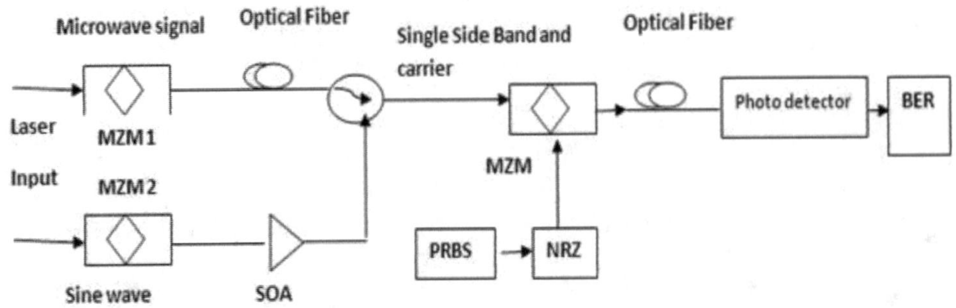

Figure 2. System setup for single sideband generation and transmission over optical fiber

For the generation of enhanced SBS phenomenon, upper modulator is biased at null point with high extinction ratio and lower modulator is fixed at less ER. Optical circulator are placed to route the signal in n+1 directional port.

After generation of SSB-SC signal, it fed to mach-zehnder modulator stage after SBS for further communication and assessment of generated signal in terms of Q factor and BER. Binary data generation in serial bits is generated from PRBS followed by NRZ pulse format. Modulated signal is transmitted over 100 km SMF. This system is unidirectional so operated in one direction by using frequency 193.09 THz due to SBS. It is received by using PIN photo-detector having 10nA dark current followed by Bessel electrical filter to remove noises. A 3R regenerator is incorporated to re sample re time and re amplify data. BER analyzer connected to regenerator to show eye diagram and also provide Q-factor and error rate.

4. Results and Discussions

For achieving the SBS in bidirectional single mode fiber (SMF), two signals are given to optical fiber in opposite directions as shown in Fig. 3. It represents that signals emerged from two different modulators are carrier less and these carrier less signals are realized with null biased MZMs (Mach-zehnder modulators). Fig. 3 (a) depicts the optical spectrum of the signal after MZM1 and Fig. 3 (b) represents the signal after lower MZM modulator. These signals are given to optical fiber from opposite direction and due to the SBS effects, it provides single side band. Fig. 4 depicts output signal from single mode fiber after the gain and loss spectrum. Single sideband signal at 193.09 THz is reported and also shows loss spectrum at upper frequency. Thus, the SSB-SC (Single sideband- suppressed carrier) signal or also referred as advanced carrier is successfully realized from proposed system architecture.

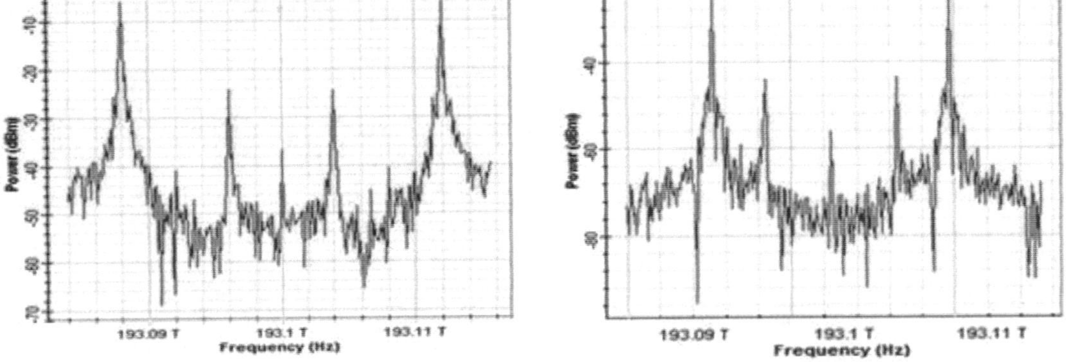

(a) (b)

Figure 3. Optical spectrums of two intensity modulators after (a) MZM 1 (b) MZM 2

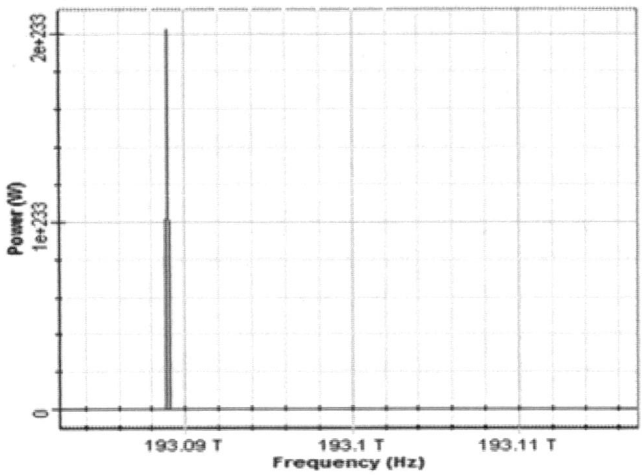

Figure 4. Single sideband signal after SBS

Further the evaluation of realized single sideband- suppressed carrier signal has been done and data from pseudo random bit sequence is modulated on the SSB-SC signal with NRZ and MZM modulator. Modulated signal is fed to transmission single mode fiber of different link lengths. SMF-28 is used for the transmission with attenuation 0.2 dB/ km and dispersion 16.75 ps/ nm/ km. Optical fiber is operated in the C-band due to less losses and scattering effects. All the nonlinear effects are considered to realize practical system. Different analysis are carried out to investigate the system such as at different distances, with and without amplifier, and comparison of SSB-SC, SSB with carrier, DSB-SC, DSB with carrier. Fig. 5(a) depicts the performance of the SSB-SC signal over optical fiber at different link lengths in terms of Q-factor. It is noteworthy that as the distance enlargement takes place, Q factor reduces due to attenuation and dispersion effects in SMF. It is also evident that the successful transmission of 90 km has been achieved over SMF at 1 Gbps with SSB-SC signal. Data rate 500 mbps introduces less dispersion and superior results as compared to 1 Gbps rate

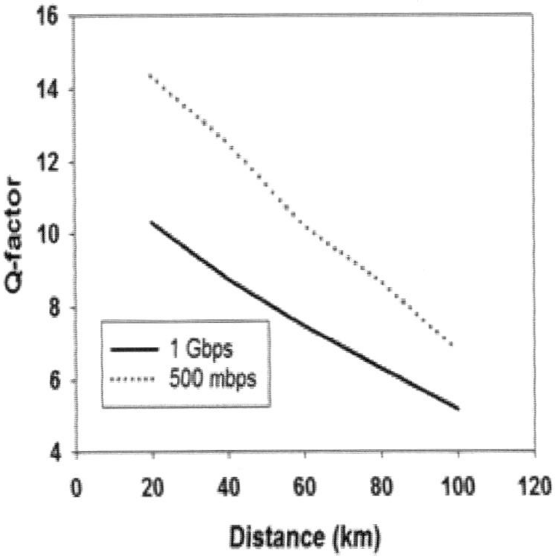

(a)

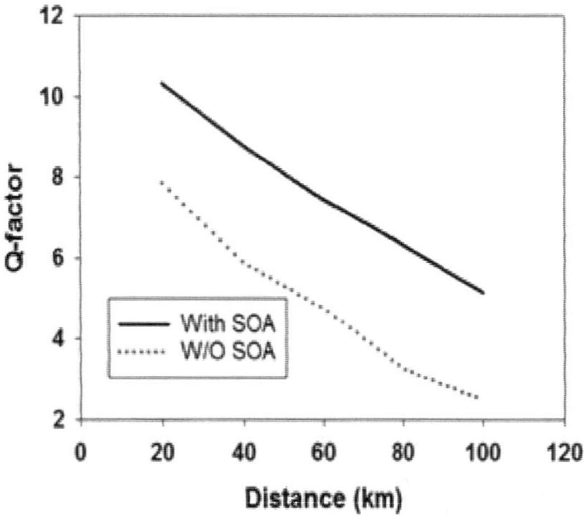

(b)

Figure 5. Distance versus Q factor (a) with SSB-SC at different rate (b) With and without SOA

Further, investigation has been done of the system by incorporating semiconductor optical amplifier and without use in the after lower MZM2 signal. Input power of lower signal is kept at minimum (-20 dB), so as to get maximum amplification from SOA. It is observed that at high launched powers, SOA introduce nonlinear effects in the system. Thus degrade the performance of the system. Fig. 5(b) depicts that when signal from lower MZM is fed to optical fiber with SOA, it exhibits better performance. However, without SOA, the generation of SSB-SC is difficult and also degrade the performance in terms of Q-factor

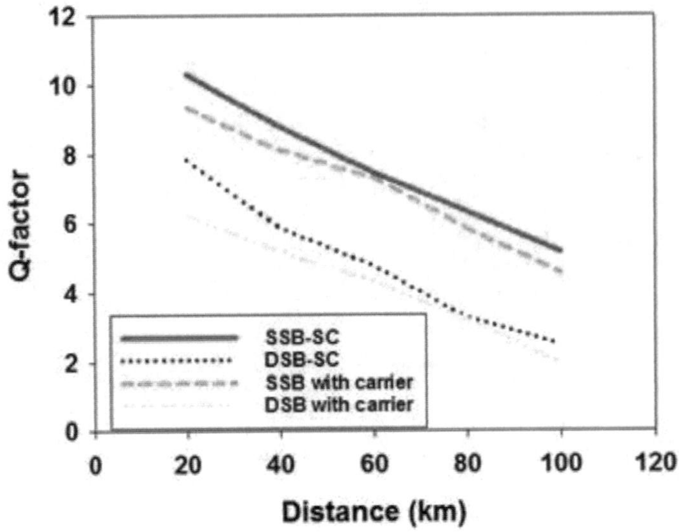

Figure 6. Comparison of different signals after SBS at varied link lengths

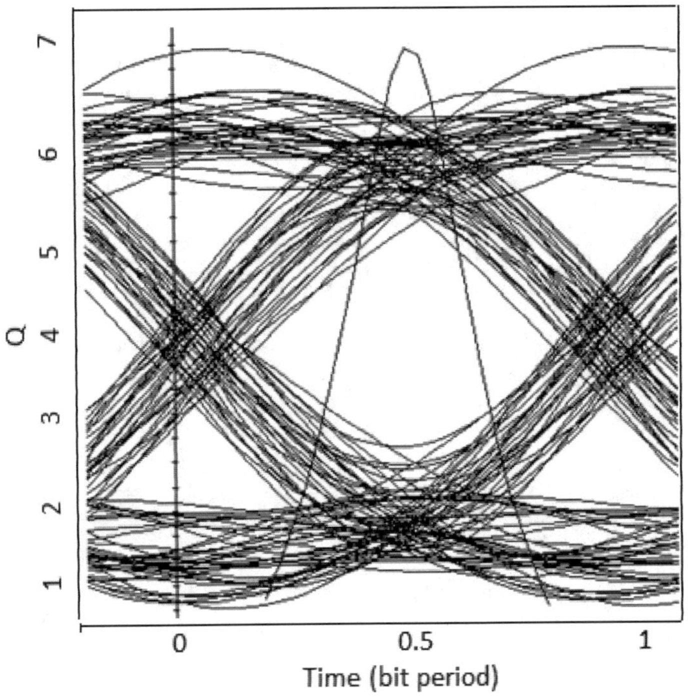

Figure 7. Eye diagram of the SSB-SC signal after 90 km at 1 Gbps

Fig. 6 depicts the performance of different signals after SBS and it is reported that single sideband-suppressed carrier signal transmission prolongs the distance and provides best performance as compared to double sideband-suppressed carrier, single sideband- suppressed carrier, double sideband with carrier signals. SSB-SC signal is comes out to be bandwidth efficient, thus provide better Q factor. However, SSB with carrier signal is also bandwidth efficient and performance is lag behind SSB-SC signal to some extent. Double sideband with carrier signal is considered bandwidth inefficient and performs worst in this transmission. Fig.7 represents the Eye diagram of the signal at 1 Gbps for 90 km distance and performance falls within the acceptable limits of BER (10-9).

5. Conclusion

In this work, a new cost effective wavelength shifter has been proposed. A millimeter range i.e 30 GHz to 300 GHz frequency shifter is demonstrated. All optical SBS generated SSB-SC signal is introduced to offer enhanced efficiency with respect to DSB with carrier signal over optical fiber. System is compared for different distances, with and without SOA amplifier and at different data rates. Results revealed that system act efficiently with SOA amplifier over different distances and at lower data rates due to less dispersion. This generation of enhanced carrier power and single sideband is reliant on the stimulated brillouin which offers gain and loss spectrum in optical fiber to expel upshifted frequency component and to enhance low frequency component. Moreover, performance of the different signal generated signals form SBS such as SSB-SC, SSB with carrier, DSB-SC and DSB with carrier signal are also investigated in terms of Q factor over single mode fiber at diverse distances. It is evident that SSB-SC is most bandwidth efficient and exhibits best performance and Q factor. A single sideband is obtained from SBS phenomenon which is useful in advance modulation formats and requires less bandwidth.

Acknowledgement

I would like to express my very great appreciation to **Ajmer Singh, Shippu Sachdeva, Javed Dhillon, Raghav Gupta** (School of Electronics and Electrical Engineering, Lovely Professional University, Phagwara, Punjab, India.) for their valuable and constructive suggestions during the planning and development of this research work. Their willingness to give their time so generously have been very much appreciated.

References

[1] Buchkremer F B J, Dumke R, Buggle C, Birkl G and Ertmer W 2000 Low-cost setup for generation of 3 GHz frequency difference phase-locked laser light *Rev. Sci. Instrum.* **71**(9) pp 3306–3308.

[2] Minasian R A, Chan E H W and Yi X 2013 Microwave photonic signal processing *Opt. Exp.* **21**(19) pp 22918–22936.

[3] Chujo W, Hanasaka T, Naganuma M and Yoneyama T 1999 A 60-GHz optical frequency shifter using coupled inverted slot lines *IEEE Trans. Microw. Theory Tech.* **47**(12) pp 2280–2286.

[4] Murata H, Anjiki M and Okamura Y 2006 Optical suppressed-carrier single side-band modulator/optical frequency shifter utilizing three-branch waveguide interferometer and polarization-reversed structure *in Proc. 36th Eur. Microw. Conf.* pp 609–611.

[5] Hisatake S, Hattori K and Nagatsuma T 2012 Continuous operation of a Bragg diffraction type electrooptic frequency shifter at 16 GHz with 65% efficiency *Adv. Opt. Technol.* **2012** pp 676785-1–676785-6.

[6] Ogiso Y, Tsuchiya Y, Shinada S, Nakajima S, Kawanishi T and Nakajima H 2010 High extinction-ratio integrated Mach–Zehnder modulator with active Y-branch for optical SSB signal generation *IEEE Photon. Technol. Lett.* **22**(12) pp 941–943.

[7] Frankel M Y and Esman R D 1998 Optical single-sideband suppressed-carrier modulator for wide-band signal processing *J. Lightw. Technol.* **16**(5) pp 859–863.

[8] Al-Dabbagh R K and Al-Raweshidy H S 2016 Photonic methods of millimeter-wave generation based on Brillouin fiber laser *Optik.* **79** pp 124-131.
[9] Khaki Z G, and Bedi H S 2012 Transient correction using EDFA: in-line optical fiber with feedback *International Conference on Computing Sciences* pp. 233-238.
[10] Bedi H S, Sharma K K and Gupta R 2020 A Review Paper on Performance Analysis of IEEE 802.11 e *1st International Conference on Computing, Communications, and Cyber-Security (IC4S 2019)* pp. 47-56

Empirical analysis of machine learning-based moisture sensing platforms for agricultural applications: A statistical perspective

1 **Neema Amish Ukani,** *2* **Saurabh S. Chakole**

1,2Lovely Professional University, Punjab,
neema.ukani@gmail.com, saurabhchakole89@gmail.com
https://orcid.org/0000-0001-6262-9005

Abstract: Modelling of accurate detection & estimation soil moisture sensors requires integration of various signal processing, filtering, segmentation, and pattern analysis methods. Sensing of moisture is generally performed via use of resistive, or capacitive materials, which change their parametric characteristics w.r.t. changes in moisture levels. These sensors are further classified depending upon capabilities of measurements, which include, volumetric sensors, soil water tensor sensors, electromagnetic sensors, time domain reflectometry (TDR) sensors, Neutron probe sensors, tensiometer-based sensors, etc. Each of these sensors are connected to a series of processing blocks, which assist in improving their measurement performance. This performance includes parameters like, accuracy of measurement, cost of deployment, measurement delay, average measurement error, etc. This wide variation in measurement performance increases ambiguity of sensor selection for a particular soil type. Due to this, researchers & soil engineers are required to test & validate performance of different moisture sensors for their application scenario, which increases time & cost needed for model deployment. To overcome this limitation, and reduce ambiguity in selection of optimum moisture sensing interfaces, this text reviews various state-of-the-art models proposed by researchers for performing this task. This review discusses various nuances, advantages, limitations & future research scopes for existing moisture sensing interfaces and evaluates them in terms of statistical parameters like accuracy of detection, sensing & measurement delay, cost of deployment, deployment complexity, scalability, & type of usage applications. This text also compares the reviewed models in terms of these parameters, which will assist researchers & soil engineers to identify most optimum models for their deployments. Based on this research, it was observed that machine learning models are highly recommended for error reduction during moisture analysis. Machine learning prediction models that utilize Neural Networks (NNs) outperform other models in terms of error performance, and must be deployed for high-accuracy & low-cost moisture sensing applications. Based on similar observations, this text also recommends fusion of different sensing interfaces for improving accuracy, while optimizing cost & complexity of deployment. These recommendations are also based on context of the application for which the sensing interface is being deployed. These recommendations must be used to further improve overall sensing performance under multiple deployment scenarios.

Keywords: Soil, Moisture, Accuracy, Delay, Cost, Machine Learning, Resistive, Capacitive, Neutron, TDR, Electromagnetic, Tensiometer

1. Introduction

Smart irrigation is a multidisciplinary field, which is aimed at improvement of farming efficiency for better yield, better grain quality, lower water consumption, optimum deployment cost, and reduced nutrient leaching in farms. A typical smart irrigation model is built using multiple sensing interfaces, a high-efficiency processing unit, and an actuation unit, which assist in feedback-based correction & control. Out of these interface, design of sensors is of primary importance, because performance of sensing interface directly controls overall efficiency of the smart farming model. For efficient smart farming, water levels in the soil must be maintained with high precision, which assists plants to grow at an optimum rate, with better yield, and crop quality. Deviation in water levels causes irreversible damages to the crop, which affects quality of the whole crop area. Thus, efficient sensing & management of soil moisture levels is of primary importance for smart farming applications. To perform this task, various moisture sensor interfaces are designed by researchers. These include, but are not limited to, volumetric sensors, soil tension measurement sensors, water depletion sensors, frequency domain reflectometry (FDR) sensors, time domain reflectometry (TDR) sensors, Neutron probes, etc. For instance, figure 1 showcases design of polymetric membrane humidity sensor, which uses a porous electric membrane along with a polymeric membrane for estimation of moisture levels.

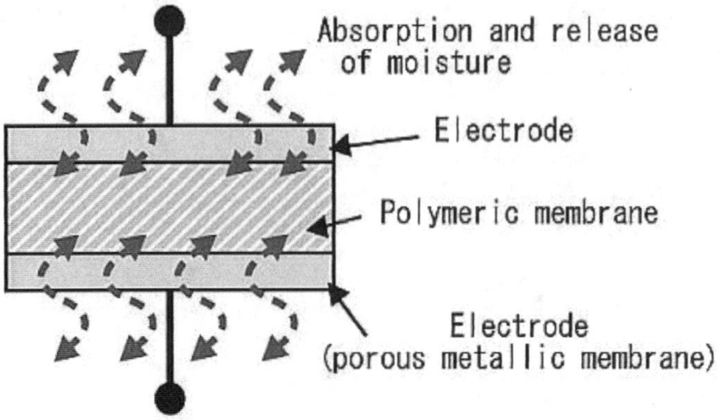

Figure 1. A typical moisture level sensor using membranes

Based on the quantity of water that passes through these membranes, they are able to estimate absorption and release of moisture levels. Similar sensing methods are used by different sensing interfaces, and are designed for different context-sensitive use cases. For instance, volumetric sensors evaluate soil water depletion (SWD) levels via equation 1, wherein Volumetric water content (VWC), permanent wilting point (PWP), and field capacity (FC) metrics are used.

$$SWD = \left[1 - \frac{VWC - PWP}{FC - PWP}\right] * 100 \ldots (1)$$

Similar measurements are made for other sensing interfaces, and assist in improving quality of moisture level estimation & analysis. Based on such measurements, researchers have proposed a wide variety of system models, and each of these models vary in terms of applicability, accuracy of sensing, delay needed for sensing & measurement, sensing error, deployment cost, & interface scalability. A discussion of these proposed models in terms of their nuances, advantages, drawbacks, & recommended research scopes can be reviewed from the next section of this text. Based on this review, researchers can identify most optimum models which are suited for their application deployments. Following this discussion, a statistical comparison of reviewed models in terms of accuracy of detection, sensing & measurement delay, cost of deployment, deployment complexity, and scalability is discussed, which will assist readers in performance estimation of these models. Upon referring this performance estimation, researchers can estimate lacunas in existing moisture sensing models, and identify fusion-based techniques that can be used to improve their performance. Finally, this text concludes with some interesting observations about the reviewed models, and recommends various methods to further improve their performance.

2. Literature Review

A wide variety of models for moisture & humidity estimation are proposed by researchers, and each of them are applicable for different application scenarios. For instance, work in [1] proposes a landslide early warning system (LEWS) that uses logistic regression (LR) with for situ soil type is implemented. The model is applied for different spatial scales, and can efficiently predict humidity values, due to which landslides are caused in certain areas. It uses a volumetric water content (VWC) value in order to estimate humidity, which is calculated via equation 2 as follows,

$$VWC = \frac{\emptyset - Min(\emptyset)}{\emptyset(99.9\%) - \mathrm{Min}(\emptyset)} \ldots (2)$$

Where, $\emptyset$ represents ratio between volume of water (V_w) & volume of soil (V_t) for the given area, and is calculated via equation 3 as follows,

$$\emptyset = \frac{V_w}{V_t} \dots (3)$$

This model has low cost, but is highly complex, and has moderate level of accuracy due to approximations in moisture estimation process. To improve this performance, work in [2] proposes use of deep learning & applies it to Fengyun-3D Medium Resolution spectral imager-II imagery, for estimation of soil moisture levels. The model uses deep belief networks (DBNs) for processing of enhanced vegetation index (EVI) & normalized difference vegetation index (NDVI) values. These values are extracted and given to temperature subnetwork & soil moisture subnetwork, which can be observed from figure 2, wherein Restricted Boltzmann Machine (RBM) is used to perform final classification. It was observed that RBM with DBN outperforms linear regression (LR), and backpropagation (BP) models in terms of accuracy of moisture detection, but has higher complexity, and slower performance when compared with these models. This performance can be improved via use of low complexity interfaces that use highly accurate sensing devices. Such an interface is discussed in [3], wherein cosmic-ray Neutron sensor (CRNS) is deployed for low-complexity measurements. The sensor was deployed using standard N_0 method, which assisted in low-complexity & highly accurate measurements. This model has better measurement accuracy when compared with electromagnetic & other sensing materials, but has higher cost of deployment, which limits its scalability performance. To improve this scalability, work in [4] proposes design of an optical sensor that uses fibre materials with Bragg Grating (FBG), which assists in lowering deployment cost, while achieving good level of accuracy for a wide variety of applications. The result of this model is evaluated via equation 4, wherein reflective index of material is combined with grating period to observe changes in reflected wavelength.

$$\lambda_{ref} = 2 * \xi_{eff} * \omega \dots (4)$$

Where, λ_{ref}, ξ_{eff} & ω represents Bragg wavelength, effective reflective index, and grating period in nano meters. The model was applied to transformers & other electric equipments, but can be extended to other applications with modifications in component compositions. Another highly efficient model for moisture estimation is discussed in [5], wherein Microwave-based moisture sensing (MWMS) is performed for granular materials via design of arbitrary shaped containers. The model is able to improve accuracy of moisture detection by 7.8% when compared with antenna array-based approach. Design of this model is depicted in figure 3, wherein different distance metrics are indicated for sensing microwave readings. The model has better efficiency than array-based or Neutron sensor-based measurements, but requires complex circuitry, which increases its deployment costs, thereby limiting real-time applicability. To improve this performance, work in [6] proposes utilization of high precision capacitive moisture (HPCM) sensors, that can be applied to polymers.

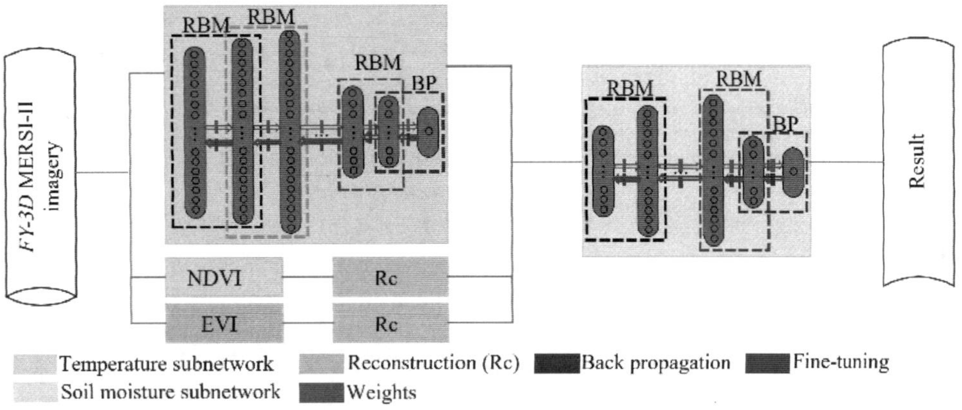

Figure 2. Deployment of RBM based DBN for moisture estimation [2]

The model evaluates temperature of the capacitor, and combines it with leakage & storage current values to obtain final moisture levels. It uses polymer characteristics like inner fringe capacitance, polymer dielectric constant, and relative variation between these values in order to estimate changes in capacitor output based on changes in moisture levels. Due to which, the model is able to achieve high-level of accuracy, with lower cost, thereby making it useful for a wide variety of applications. But to apply this model, various setup constraints are needed, which limits its scalability for real-time remote location scenarios. To improve this scalability, work in [7] proposes design of a Global Positioning System (GPS) based device, that can be used to estimate stream soil moisture levels via ground truth measurements. It uses a combination of right-hand circular polarization (RHCP) and left-hand circular polarization (LHCP) antennas for estimation of moisture levels.

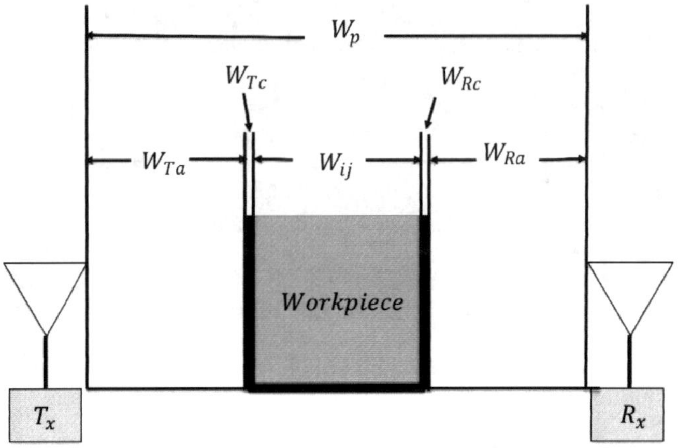

Figure 3. Microwave-based moisture detection [5]

Parameters like doppler shift, position of moving water surface, velocity of stream flow, soil moisture, soil content, sand content & clay content can be estimated via this approach. But the model is highly complex, and requires dedicated GPS antennas, which increases its deployment costs. To reduce these costs, work in [8] proposes design of passive moisture sensor that uses conductive & water-soluble yarns. To design these yarns, passive ultrahigh-frequency (UHF) & radio frequency identification (RFID) technologies are combined, and interfaced with polyvinyl alcohol (PVA) yarns. The model showcases moderate accuracy, and can be applied to real-time applications with low cost & high sensitivity performance. But it requires high power for operation, which limits its use-case to power-source enabled scenarios. To reduce this power consumption, work in [9] proposes design of High Frequency RFID passive sensor tag that is equipped with thin film sensors (HF RFID TFS) on silica substrates. The model uses a combination of different resistive & capacitive interfaces as observed from figure 4, for low power & high-speed operations. This model is capable of reducing signal complexity, and can be applied for dry soil, wet soil, and water-based scenarios, with moderate accuracy, and low complexity. It showcases better performance when compared with fringe capacitance sensor (FCS), soil impedance sensor (SIS) and TiO_2 Nano particles sensor (TNPS), thus making it useful for a wide variety of application scenarios. Performance of this method must be validated under different soil types, and can be improved via use of pulse-width modulation (PWM) [10] with capacitive sensing interfaces.

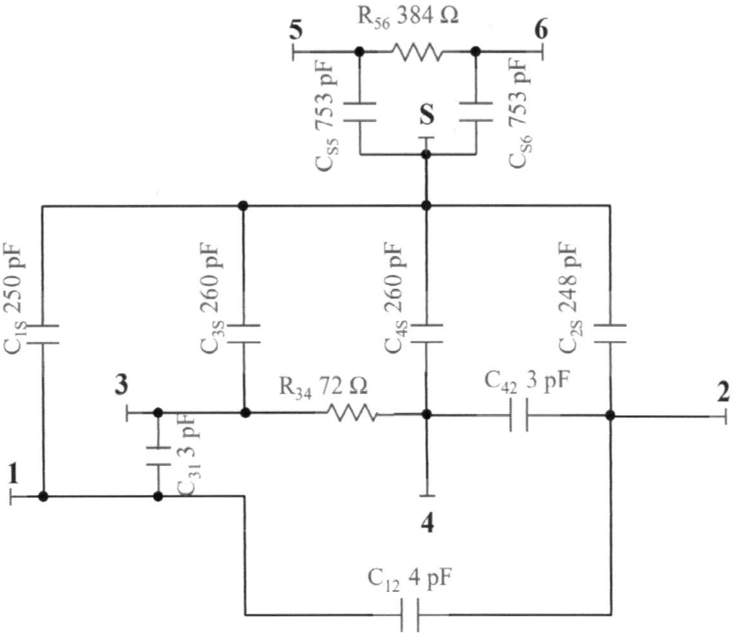

Figure 4. Design of HF RFID TFS sensor interface for moisture sensing [9]

This model uses a PWM interval as indicated in equation 5, wherein capacitive & resistive values of the circuit are used.

$$T(ON) = 1.1 * R_{load} * C_{RH} \ldots (5)$$

Where, R_{load} & C_{RH} represents load resistance & capacitance of the relative humidity sensor. From the model, relative humidity values are estimated via equation 6 as follows,

$$RH = 5.18 + 0.43 * t + 0.01 * t^2 \ldots (6)$$

Where, t represents output PWM time interval, which is evaluated for better accuracy performance. The model performance can be extended via use of effective calibration models, which assists in error reduction in estimation of continuous moisture levels. Such a model is proposed in [11], wherein error levels are used to calibrate the output moisture levels from equation 7 as follows,

$$\psi = \sqrt{\frac{\epsilon_{current}}{\epsilon_{old}\ (\emptyset * \epsilon_{current} - \epsilon_{old})}} \ldots (7)$$

Where, $\epsilon_{current}$ & ϵ_{old} represents current & old error levels for measurement of moisture via different sensors. This calibration model (CM) is able to improve accuracy of estimation by 8% when compared with various state-of-the-art methods.

Microwave sensing is one of the most effective methods for estimation of soil moisture, which is due to availability of multiple sensing bands, which can be combined to obtain accurate results. Availability of these bands is dependent upon complexity of the used microwave device. For instance, work in [12] proposes use of a P band passive microwave sensing interface, that works at a wavelength of 40 cm and frequency of 750 MHz, for sensing moisture levels of around 10-cm layer of soil. Based on this method, P-band multibeam radiometer (PPMR) & polarimetric L-band multibeam radiometer (PLMR) were designed which assist in improving efficiency of moisture detection. The P band sensor showcased better efficiency when compared with L band sensor, because of its highly dense feature extraction properties. Applications of such models are explored in [13, 14, 15], wherein battery less sensors based on RFID & UHF, multiple sensing resonators (MSR), and Poly methyl methacrylate (PMMA) sensors are discussed. These sensors assist in reducing estimation error by improving feature extraction capabilities of the sensing model. Due to which RFID & UHF is able to achieve moderate accuracy with low complexity, while MSR achieves high accuracy with moderate level of complexity, while PMMA uses a double integration model for accurate estimation of moisture dependent potential difference between electrodes. The PMMA model uses capacitive sensors to perform this task, which assist in achieving low cost & high-efficiency moisture sensing performance under different scenarios. This model is able to monitor the changes in z-component of electric field (E_z) for a ring of charge via equation 8, which assists in low complexity moisture level calculations.

$$E_z = k * \frac{Q_z}{\sqrt{z^2 + R^2}} \dots (8)$$

Where, $k, Q_z, \& R$ represents Coulomb's constant, total charge, and radius of the charged ring in Z dimensions. The model uses triangular, circular, rectangular, and point-edge based shapes for measurement of moisture levels, which are estimated via equations 9, 10, 11 as 12 follows,

$$M(triangular) = 0.00513 * E_z^2 - 0.04399 * E_z + 1.09819 \dots (9)$$

$$M(round) = 0.00127 * {E_z}^2 + 0.06885 * E_z + 1.09948 \dots (10)$$

$$M(rect) = 0.00754 * {E_z}^2 - 0.06562 * E_z + 0.81518 \dots (11)$$

$$M(point_{edge}) = 0.00719 * {E_z}^2 - 0.0295 * E_z + 1.14557 \dots (12)$$

Based on these calculations, Moisture levels from different sensors are calculated, and combined to estimate the final values. This sensor is highly efficient, but requires considerable delay for calibration & control of final moisture readings. To reduce this delay, work in [16] proposes a Fringing Field Capacitive Soil Moisture Sensor (FFCSM) which assists in level-based moisture estimations. This model is capable of low-cost & high-efficiency operation due to use of simple bridge circuits, peak detectors, and low complexity voltage measurements, which is evaluated via equation 13 as follows,

$$V_{out} = \frac{a(b * C_s - 1)}{a^2 * C_s + a * b * C_s + a + b} * V_{in} \dots (13)$$

Where, V_{in} & V_{out} represents inputs & output voltage levels, while $a, b \& C_s$ represents activation constant, bias constant & capacitance of the soil moisture sensors. These constants are evaluated via equation 14 as follows,

$$a = 2 * pi * f * R; b = \frac{1}{C} \dots (14)$$

Where, $R, C \& f$ represents standard resistance, standard capacitance, and resonating frequency of the devices. The model showcases good accuracy, with moderate level of error, which assists in deploying them for a wider variety of low complexity application scenarios. Alternatives to these models are proposed in [17, 18, 19], wherein researchers have proposed use of point heat source (PHS) method, Soil Moisture Active Passive (SMAP) method, and extension of SMAP via random forests & dense networks (SMAP RF DN) for improved moisture detection performance. It was observed that PHS model required highest calibration, and was able to produce moderately accurate results, while SMAP & SMAP with RF & DN showcased good accuracy with low calibration requirements. Performance of this model must be evaluated for different applications, and can be improved via use of Fibre Optics Sensors (FoS) [20], Multiprobe Heat Pulse Sensor (MHPS) [21], and Passive RFID-based sensors (PRS) [22], which assist in exploring different sensing interfaces for providing better estimation performance. It was observed that FoS models are low cost, and have better accuracy when compared with other material-based sensors. This performance is further improved via use of MHPS & PRS models, which estimate moisture levels with low complexity and moderate level of deployment cost.

Performance of MHPS & PRS models must be validates, and can be compared with various state-of-the-art models which are discussed in [23], wherein laboratory-based methods, bio-sensors, ion-selective membrane-based sensors, capillary electrophoresis-based methods, and spectroscopy-based methods are discussed. It was observed that bio-sensors & spectroscopy methods are highly scalable, while electrophoresis & membrane sensors have higher accuracy, but require specialized equipments for deployment & control. Due to which they are used for specialized cases, where accuracy has higher weightage than deployment capability. But these models have higher current requirements, which must be reduced for further improving their scalability. Inspired by this observation, work in [24, 25] proposes Electrical Capacitance Tomography (ECT) & Picoampere Quiescent Current Wake-up Circuit (PQCWC), which assist in low-power, and moderate accuracy moisture sensing interface designs. The ECT model uses microwave drying process for estimation of moisture distribution in polymer foams, which increases its sensitivity. While, PQCWC uses picoampere level current changes via under voltage lockout (UVLO) for low power operations. The model performance is limited in terms of accuracy, but can be used for approximating moisture levels for a wide variety of applications. To improve this performance, work in [26, 27, 28] proposes use of Internet of Trees (IoTr) with low cost & highly dispersive electromagnetic sensors (HDES), Graphene oxide for design of fibre core-offset humidity sensor (GOFCHS), and Time Domain Reflectometry (TDR) for low error, and high-accuracy moisture level estimations. These models utilize various machine learning techniques to improve quality of detection, but require high performance computational processors for effective implementation & control. Based on these models, work in [29, 30] that uses Synthetic aperture radar (SAR) for surface soil moisture level detection & moderate resolution imaging spectroradiometer (MODIS) for soil moisture index (SMI) evaluation via simplified thermal inertia-ratio vegetation index spectral feature space can be improved. These models are currently utilizing linear classification interfaces, which limits their measurement performance. This performance can be improved via use of machine learning techniques like Convolutional Neural Networks (CNNs), and Recurrent NNs (RNNs), which might increase computational complexity, but will result in better estimation of moisture levels. These SAR based methods must be further augmented via use of intelligent sensing interfaces & calibration techniques. Such techniques are discussed in [31, 32, 33], wherein Calibration of Soil Moisture & Ocean Salinity (CSMOS), hybrid symbolic aggregate approximation algorithm (HSAAA), and satellite-based soil moisture agricultural drought index (SSMDI), are proposed. These models assist in providing high accuracy moisture detection, via augmentation of sensed data, and application of machine learning to estimate & reduce sensing errors. Similarly, the work in [34, 35, 36] also proposes P Band and L Band radiometry models, fluorine-doped tin oxide (FTO) glass substrate sensor, and extended SMAP models, for estimation of moisture from different types of sites. The P Band & L Band model is highly scalable & has good accuracy, but modifies internal characteristics of soil due to use of microwaves, while the FTO model assists in estimation of moisture for gas-insulated switchgear (GIS) in an augmented non-ensemble mode of operation. The SMAP interface is highly useful for large-area applications, wherein approximate estimation of moisture level is required.

Applications of moisture sensing are discussed in [37, 38, 39], wherein researchers have deployed moisture sensors for identification of flood prone areas, use of photonic band gap (PBG) with metal organic framework, and ground level enhancement estimation via moisture sensor networks (MSNs). Here, PBG is used for high accuracy, laboratory applications, while MSNs are deployed for large-scale & accurate sensing of moisture levels, which assists in improving their applicability for a wide variety of deployment scenarios. Similar applications are discussed in [40, 41, 42], wherein researchers have deployed moisture sensors for 3D mapping of soil organic carbon content with machine learning (MSOCCML), detection of moisture for temperate forest canopies using SMAP (SMAP TFC), and Cosmic-ray neutron sensing (CRNS) for train materials. It was observed that MSOCCML is useful for general purpose applications, while SMAP & CRNS must be retuned and deployed for context-specific application scenarios. An application of these sensors is discussed in [43], wherein cosmic-ray sensors are used for kriging based robotic exploration (kCRNS) that assists in location-based mapping of soil moisture levels. This model is highly accurate, and can be used for applications that require wider sensing radius during moisture measurements. Based on this discussion it can be observed that a wide variety of sensing models are available for measurement of moisture levels, and each of them have their own applicability & performance. To further estimate performance of these models, next section evaluates them and compares their performance. This comparison will assist readers to identify best practices for moisture sensing & measurement under different application scenarios.

3. Statistical analysis

From the review it was observed that a wide variety of approaches are designed by researchers that assist in high accuracy, low complexity, and low-cost moisture level sensing & detection process. To further evaluate these models, this section compares their performance in terms of accuracy of detection (A), sensing & measurement delay (D), cost of deployment (C), complexity of deployment (CC), & scalability (S) metrics. These metrics are evaluated via investigating internal implementation details of these methods, and was quantized into low (L), medium (M), high (H), and very high (VH) ranges. Based on this estimation, statistical evaluation of these models is tabulated in table 1 as follows,

Model	**A**	**C**	**CC**	**D**	**S**
LEWS LR [1]	M	H	M	H	M
DBN RBM [2]	VH	VH	VH	M	H
LR RBM [2]	H	H	H	M	M
BP RBM [2]	M	H	H	H	M
CRNS [3]	VH	VH	M	L	L
FBG [4]	M	VH	M	L	M
MWMS [5]	H	VH	M	H	L
HPCM [6]	H	L	L	VH	L
GPS [7]	M	VH	VH	VH	M
UHF RFID [8]	M	H	H	M	M
HF RFID TFS [9]	H	L	H	M	M
PWM [10]	L	L	M	M	L
CM [11]	H	M	H	M	VH
PPMR [12]	H	M	M	H	M
PLMR [12]	M	H	M	H	M
RFID UHF [13]	M	M	H	H	M
MSR [14]	H	VH	VH	M	M
PMMA [15]	H	L	M	H	M
FFCSM [16]	H	L	H	M	M
PHS [17]	M	M	L	M	L
SMAP [18]	H	H	VH	M	H
SMAP RF DN [19]	VH	VH	VH	M	H
FoS [20]	H	M	M	L	M
MHPS [21]	L	L	M	L	L
PRS [22]	M	M	H	M	M
ECT [24]	M	L	L	M	M
PQCWC [25]	M	L	L	M	L

HDES [26]	H	VH	H	M	M
GOFCHS [27]	VH	M	M	H	H
TDR [28]	VH	M	H	M	M
SAR [29]	M	VH	VH	H	H
SMI MODIS [30]	H	VH	VH	H	H
CSMOS [31]	M	M	H	M	M
HSAAA [32]	H	L	M	H	M
SSMDI [33]	H	VH	VH	H	H
P Band & L Band [34]	VH	H	H	M	H
FTO [35]	H	H	M	L	L
eSMAP [36]	M	H	H	M	M
PBG [38]	M	H	H	M	L
MSNs [39]	H	VH	H	H	H
MSOCCML [40]	H	VH	VH	H	M
SMAP TFC [41]	M	H	H	M	L
CRNS [42]	H	M	M	H	M
kCRNS [43]	H	H	H	M	M

Table 1. Performance evaluation of different moisture sensing models

Based on this comparison, it can be observed that different models are useful for different application requirements. For instance, in terms of accuracy as observed from figure 5, DBN RBM [2], CRNS [3], SMAP RF DN [19], GOFCHS [27], TDR [28], and P Band & L Band [34] models outperform other models, thus, they can be used for highly accurate moisture detection applications. Similarly, cost of deployment & computational complexity is visualized from figure 6, wherein it is observed that HPCM [6], HF RFID TFS [9], PWM [10], PMMA [15], FFCSM [16], MHPS [21], ECT [24], PQCWC [25], and HSAAA [32] require lowest deployment cost, while HPCM [6], PHS [17], ECT [24], and PQCWC [25] have lower computational complexity when compared with other models. Due to which these models must be used for applications which require lower complexity and lower computational costs. PCA [44] is one of the methods used to reduce the dimensionality of the dataset. The use of K- means method for clustering of soil dataset. [45]

These models are highly useful, when moderate accuracy of sensing is needed, but the cost & complexity of sensing are needed to be minimized. In terms of speed of sensing, figure 7 recommends that CRNS [3], FBG [4], FoS [20], MHPS [21], and FTO [35] have the fastest response, thus must be used for low delay, and high throughput applications. These models utilize high-speed sensors, which might be costly, but provide quicker results when compared with other sensing models. Similarly, scalability of these models is also evaluated, which indicates that CM [11] has the highest scalability, which is followed by DBN RBM [2], SMAP [18], SMAP RF DN [19], GOFCHS [27], SAR [29], SMI MODIS [30], SSMDI [33], P Band & L Band [34], and MSNs [39] models.

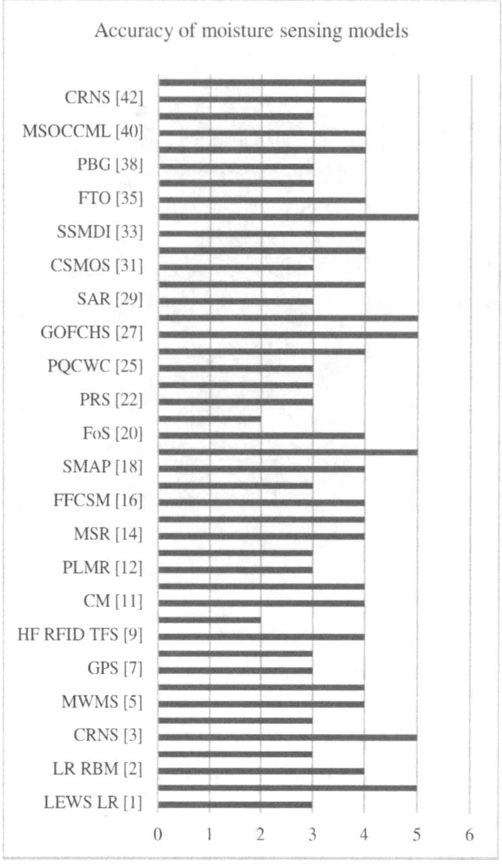

Figure 5. Comparison of accuracy for different moisture sensing models

Thus, these models must be used for applications that require general purpose moisture sensing interfaces. To fuse these performance metrics, this text proposes use of an algorithmic rank score (ARS), which is estimated via equation 15 as follows,

$$ARS = \frac{A+S}{5} + \frac{5}{C} + \frac{5}{CC} + \frac{5}{D} \ldots (15)$$

This score will allow readers to identify models that have higher accuracy, better scalability, lower complexity, lower cost & lower computational delays. Based on this ARS value, it can be observed that ECT [24], PQCWC [25], MHPS [21], FoS [20], HPCM [6], PHS [17], HF RFID TFS [9], PMMA [15], FFCSM [16], HSAAA [32], PWM [10], FTO [35], CRNS [3] and CM [11] have better overall performance, thus can be used for efficient moisture sensing applications.

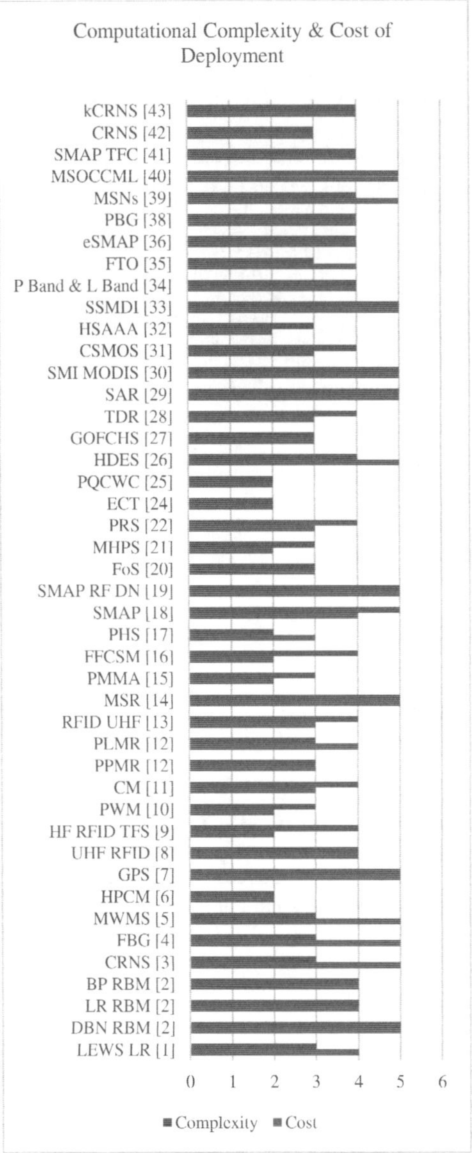

Figure 6. Computational complexity & deployment cost comparison for different moisture sensing models

Thus, based on these statistical observations, researchers can utilize identified models for their moisture sensing application scenarios.

4. Conclusion

This text compares a wide variety of moisture sensing models, and evaluates their performance in terms of accuracy, cost of deployment, computational complexity, delay needed for deployment, and overall scalability. Based on these evaluations, it was observed that machine learning models had better performance in terms of accuracy, and can be used with all kinds of sensors. These models have better scalability, and are useful in applications that require moderate level of cost & complexity. While highly accurate sensing interfaces are costly, but can be used for high-speed moisture sensing applications. Specifically, DBN RBM, CRNS, SMAP

RF DN, GOFCHS, TDR, and P Band & L Band models outperform other models; thus, they can be used for highly accurate moisture detection applications. While, HPCM, HF RFID TFS, PWM, PMMA, FFCSM, MHPS, ECT, PQCWC, and HSAAA require lowest deployment cost, while HPCM, PHS, ECT, and PQCWC have lower computational complexity when compared with other models. Similarly, CM has the highest scalability, while CRNS, FBG, FoS, MHPS, and FTO have the fastest response, thus must be used for low delay, and high throughput applications. This text also recommends use of an algorithmic rank score, which combines these metrics, and recommend that ECT, PQCWC, MHPS, FoS, HPCM, PHS, HF RFID TFS, PMMA, FFCSM, HSAAA, PWM, FTO, CRNS and CM must be used for applications that require better overall moisture sensing performance. In future, researchers must use combinations of these approaches, and integrate them with machine learning models for better applicability under different scenarios. Moreover, researchers can also utilize different machine learning models, and combine them with multiple high accuracy sensing interfaces to improve overall performance for different deployment scenarios.

Acknowledgement

I would like to express my very great appreciation to **Dr. Shelej Khera** (Professor, School of Electronics and Electrical Engineering, Lovely Professional University, Phagwara, Punjab, India.) for his valuable and constructive suggestions during the planning and development of this research work. His willingness to give his time so generously has been very much appreciated.

5. References

[1] Wicki, Adrian et al. "Assessing the Potential of Soil Moisture Measurements for Regional Landslide Early Warning". Landslides 17.8 (2020): 1881–1896. Web.
[2] Wang, Wenwen et al. "Extracting Soil Moisture from Fengyun-3D Medium Resolution Spectral Imager-II Imagery by Using a Deep Belief Network". Journal of meteorological research 34.4 (2020): 748–759. Web.
[3] Tan, Xingyan et al. "Applicability of Cosmic-Ray Neutron Sensor for Measuring Soil Moisture at the Agricultural-Pastoral Ecotone in Northwest China". Science China Earth Sciences 63.11 (2020): 1730–1744. Web.
[4] Ansari, Muhammad A., Daniel Martin, en Tapan Kumar Saha. "Advanced online moisture measurements in transformer insulation using optical sensors". IEEE transactions on dielectrics and electrical insulation: a publication of the IEEE Dielectrics and Electrical Insulation Society 27.6 (2020): 1803–1810. Web.
[5] Jiarasuwan, Seksan, Kosin Chamnongthai, en Navapadol Kittiamornkul. "A design method for a microwave-based moisture sensing system for granular materials in arbitrarily shaped containers". IEEE sensors journal 21.17 (2021): 19436–19452. Web.
[6] Moura Dos Santos, Rosane et al. "High precision capacitive moisture sensor for polymers: Modeling and experiments". IEEE sensors journal 20.6 (2020): 3032–3039. Web.
[7] Shen, Lie-Chung et al. "Stream soil moisture estimation by reflected GPS signals with ground truth measurements". IEEE transactions on instrumentation and measurement 58.3 (2009): 730–737. Web.
[8] Chen, Xiaochen, Han He, Meinan Gou, et al. "Passive moisture sensor based on conductive and water-soluble yarns". IEEE sensors journal 20.18 (2020): 10989–10995. Web.
[9] Cappelli, Irene et al. "Battery-less HF RFID sensor tag for soil moisture measurements". IEEE transactions on instrumentation and measurement 70 (2021): 1–13. Web.
[10] Kumar, Shailesh et al. "Condition monitoring of transformer breather using a capacitive moisture sensor". IEEE transactions on industrial electronics (1982) 67.11 (2020): 9779–9789. Web.
[11] Trabelsi, Samir, en Micah A. Lewis. "Calibration algorithm for rapid and nondestructive moisture sensing in in-shell nuts". IEEE sensors letters 3.12 (2019): 1–4. Web.
[12] Ye, Nan et al. "Toward P-band passive microwave sensing of soil moisture". IEEE geoscience and remote sensing letters : a publication of the IEEE Geoscience and Remote Sensing Society 18.3 (2021): 504–508. Web.
[13] Chen, Xiaochen, Han He, Zahangir Khan, et al. "Textile-based batteryless moisture sensor". IEEE antennas and wireless propagation letters 19.1 (2020): 198–202. Web.

[14] Gopalakrishnan, Sarath et al. "Battery-less wireless chipless sensor tag for subsoil moisture monitoring". IEEE sensors journal 21.5 (2021): 6071–6082. Web.

[15] Chakraborty, Moupali, Anindita Kalita, en Karabi Biswas. "PMMA-coated capacitive type soil moisture sensor: Design, fabrication, and testing". IEEE transactions on instrumentation and measurement 68.1 (2019): 189–196. Web.

[16] Protim Goswami, Manash, Babak Montazer, en Utpal Sarma. "Design and characterization of a fringing field capacitive soil moisture sensor". IEEE transactions on instrumentation and measurement 68.3 (2019): 913–922. Web.

[17] He, Fengfei et al. "Study on the mobile PHS method for soil moisture monitoring based on thermal effect". IEEE sensors journal 21.13 (2021): 15209–15217. Web.

[18] Montzka, Carsten et al. "Estimating the number of reference sites necessary for the validation of global soil moisture products". IEEE geoscience and remote sensing letters : a publication of the IEEE Geoscience and Remote Sensing Society 18.9 (2021): 1530–1534. Web.

[19] Whitcomb, Jane et al. "Evaluation of SMAP core validation site representativeness errors using dense networks of in situ sensors and random forests". IEEE journal of selected topics in applied earth observations and remote sensing 13 (2020): 6457–6472. Web.

[20] ---. "Investigation of distributed moisture and temperature measurements in transformers using fiber optics sensors". IEEE transactions on power delivery 34.4 (2019): 1776–1784. Web.

[21] de Morais Franca, Maria Bernadete et al. "A multiprobe heat pulse sensor for soil moisture measurement based on PCB technology". IEEE transactions on instrumentation and measurement 68.2 (2019): 606–613. Web.

[22] Tajin, Md Abu Saleh, William M. Mongan, en Kapil R. Dandekar. "Passive RFID-based diaper moisture sensor". IEEE sensors journal 21.2 (2021): 1665–1674. Web.

[23] Kashyap, Bhuwan, en Ratnesh Kumar. "Sensing methodologies in agriculture for soil moisture and nutrient monitoring". IEEE access: practical innovations, open solutions 9 (2021): 14095–14121. Web.

[24] Hosseini, Marzieh et al. "Electrical capacitance tomography to measure moisture distribution of polymer foam in a microwave drying process". IEEE sensors journal 21.16 (2021): 18101–18114. Web.

[25] Kargaran, Masoud, Mehdi Habibi, en Sebastian Magierowski. "Self-powered soil moisture monitoring sensor using a picoampere quiescent current wake-up circuit". IEEE transactions on instrumentation and measurement 69.9 (2020): 6613–6620. Web.

[26] Sanuade, Oluseun A. et al. "New Empirical Equation to Estimate the Soil Moisture Content Based on Thermal Properties Using Machine Learning Techniques". Arabian journal of geosciences 13.10 (2020): n. pag. Web.

[27] Wang, Zenghui et al. "Fiber core-offset humidity sensor based on graphene oxide characteristics". IEEE photonics journal 13.3 (2021): 1–8. Web.

[28] Kisekka, Isaya et al. "Spatial–Temporal Modeling of Root Zone Soil Moisture Dynamics in a Vineyard Using Machine Learning and Remote Sensing". Irrigation science (2022): n. pag. Web.

[29] Beale, John et al. "A method to assess the performance of SAR-derived surface soil moisture products". IEEE journal of selected topics in applied earth observations and remote sensing 14 (2021): 4504–4516. Web.

[30] Liu, Zhenhua et al. "Soil moisture index model for retrieving soil moisture in semiarid regions of China". IEEE journal of selected topics in applied earth observations and remote sensing 13 (2020): 5929–5937. Web.

[31] Kang, Chuen Siang, Kasturi Devi Kanniah, en Yann H. Kerr. "Calibration of SMOS soil moisture retrieval algorithm: A case of tropical site in Malaysia". IEEE transactions on geoscience and remote sensing: a publication of the IEEE Geoscience and Remote Sensing Society 57.6 (2019): 3827–3839. Web.

[32] Wang, Zhongju et al. "Soil-moisture-sensor-based automated soil water content cycle classification with a hybrid symbolic aggregate approximation algorithm". IEEE internet of things journal 8.18 (2021): 14003–14012. Web.

[33] Salvia, Mercedes et al. "The added-value of remotely-sensed soil moisture data for agricultural drought detection in Argentina". IEEE journal of selected topics in applied earth observations and remote sensing 14 (2021): 6487–6500. Web.

[34] Shen, Xiaoji et al. "Soil moisture retrieval depth of P- and L-band radiometry: Predictions and observations". IEEE transactions on geoscience and remote sensing: a publication of the IEEE Geoscience and Remote Sensing Society 59.8 (2021): 6814–6822. Web.

[35] Zargar, Zubair Hassan et al. "A linear capacitive sensor for ppm moisture measurement in SF_6 gas-insulated switchgear". IEEE transactions on instrumentation and measurement 70 (2021): 1–8. Web.
[36] Ayres, Edward et al. "Validation of SMAP soil moisture at terrestrial national ecological observatory network (NEON) sites show potential for soil moisture retrieval in forested areas". IEEE journal of selected topics in applied earth observations and remote sensing 14 (2021): 10903–10918. Web.
[37] Ahlmer, Anna-Klara et al. "Soil Moisture Remote-Sensing Applications for Identification of Flood-Prone Areas along Transport Infrastructure". Environmental earth sciences 77.14 (2018): n. pag. Web.
[38] Kim, Jun Yong, Sung-Hak Lee, en Yun Seon Do. "Optimized structure for a moisture-sensitive colorimetric sensor utilizing photonic crystals based on a metal–organic framework". IEEE access: practical innovations, open solutions 7 (2019): 85483–85491. Web.
[39] Hands, A. D. P. et al. "Detecting Ground Level Enhancements Using Soil Moisture Sensor Networks". Space weather: the international journal of research & applications 19.8 (2021): n. pag. Web.
[40] Rentschler, Tobias et al. "3D Mapping of Soil Organic Carbon Content and Soil Moisture with Multiple Geophysical Sensors and Machine Learning". Vadose zone journal: VZJ 19.1 (2020): n. pag. Web.
[41] Colliander, Andreas et al. "SMAP Detects Soil Moisture under Temperate Forest Canopies". Geophysical research letters 47.19 (2020): n. pag. Web.
[42] Schrön, M. et al. "Neutrons on Rails: Transregional Monitoring of Soil Moisture and Snow Water Equivalent". Geophysical research letters 48.24 (2021): n. pag. Web.
[43] Pulido Fentanes, Jaime et al. "Kriging-based Robotic Exploration for Soil Moisture Mapping Using a Cosmic-ray Sensor". Journal of field robotics 37.1 (2020): 122–136. Web.
[44] Abdel-Fattah, Mohamed K. et al. "Quantitative Evaluation of Soil Quality Using Principal Component Analysis: The Case Study of El-Fayoum Depression Egypt". Sustainability 13.4 (2021): 1824. Web.
[45] Hot, Elma, en Vesna Popovic-Bugarin. "Soil data clustering by using K-means and fuzzy K-means algorithm". 2015 23rd Telecommunications Forum Telfor (TELFOR). IEEE, 2015. Web.

Accelerating the Machine Learning Model Deployment using MLOps

Mandepudi Nobel Chowdary, Bussa Sankeerth, Chennupati Kumar Chowdary, ManuGupta*

Department of Electronics and Computer Engineering, Sreenidhi Institute of Science and Technology, Hyderabad, TS – 501 301, INDIA
*manugupta@sreenidhi.edu.in,manugupta5416@gmail.com
*orcid.org/0000-0003-4517-0129

Abstract. Machine Learning projects acquire a large amount of data to make predictions on the data. Deploying machine learning models is a difficult task as it is involved a lot factors such as continuous builds to improve efficiency and different libraries used for prediction. As the data grow for better predictions with proper accuracy some of the parameters are to be dynamically changed for the sake of performance tuning. To enhance the variation in the parameters it takes effort and time from the developers to make the changes in the code and deploy to the production environment. In proposed work, an end to end automation cycle to enhance the timely deployment of machine learning models with better performance is presented. This automation cycle converts the changed code with varied parameters into the container image and uploads to the container registry. Later, when the new image is uploaded it auto detects and updates the production environment with the new image data by launching a new container and shutting down the previous old one. The proposed automated model will display the updated features which are developed. The users can utilize the updated model simultaneously without any downtime to obtain the new features. The main contribution of proposed work is that it can build and deploy the machine learning models automatically, without any manual intervention. The models can also be retrained for better performance of the predictions.

Keywords: Machine Learning; Container; Deployment; Automation; Prediction.

1. Introduction

The DevOps in the IT sector helps to make the progress of deployment agile. It is very important to run the various production activities in every domain with agility. But it is very unfortunate as many numbers of the machine learning is not completed on the deadlines due to the lack of proper automation. The involvement of the DevOps activities in the line of machine learning practices will enable it to transform quickly into the production environment and can be used by the end users. Generally, machine learning is the combination of the data and algorithms to perform the operations on the data. DevOps can be combined with the machine learning resulting in the unique approach called as MLOps as shown in the figure 1. It is a complete cycle that involves the code development, model generation, orchestration and deployment. The overall life cycle is the complete combination of all the activities. MLOps helps to build the models in the most effective and efficient way with in the span of time enabling the businesses to be quick and agile [1]. Although MLOps is an intersection of the machine learning, data engineering and DevOps to enhance the better automated development equipped with the process of proper deployment. It enables the IT teams to enhance better view on the product by enabling the product deployment on the automated cycle with the concept of MLOps.

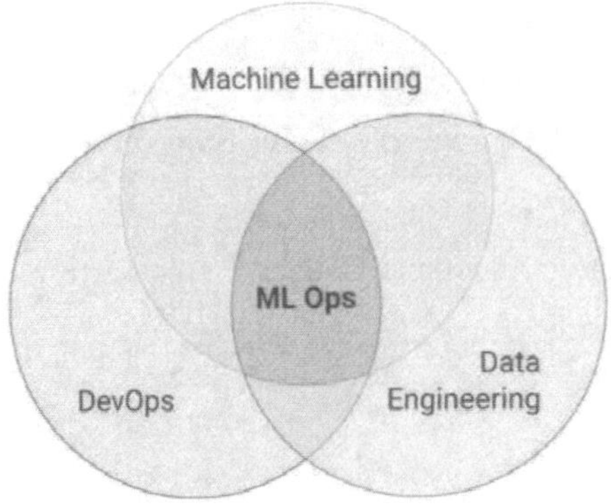

Figure 1. MLOps Intersection.

2. Literature survey

Machine Learning is divided into different categories. It starts with data collection, data processing, feature engineering, data labelling, model designing, model training and optimization by tuning the multiple parameters. Sometimes, it is quite often in the world of machine learning that data is trained multiple times recursively to enable the better accuracy of the model. This is a part of the development life cycle of machine learning products. It is very much necessary to update the features on the production environments as soon as possible by ensuring the proper working of multiple components. To enhance the faster deployment multiple DevOps tools are aligned in the line and connected with each other to enable the flow of the code from the developer to the production environment [2]. It is also necessary to deploy the code and design with the proper working and without any sort of errors.

The word MLOps defines the process of taking the experimented machine learning models to the production environment. It is the combination of work of the Data Scientists, DevOps engineers and the Machine Learning Engineer to ensure the transition from one area to the production environment. It improves the quality of the models in the production environment and enables automation to drive the business needs into better profits. It is a complete cycle that includes continuous development and continuous integration to support the multiple levels of automation.

DevOps enables the deployment on the cloud environments by transforming the code into the production builds of the product. Machine Learning is enabling the multiple software lifecycles to have the better agile stages of the life cycle and ensure proper planning with the deployment. DevOps itself contain a large range of life cycle practices that are to be optimized for better production builds. But although it can play a vital role in the development of the software product for the production builds [3]. This enables the unique concept of the MLOps to be introduced to enable the better flow of code from the developer to the deployment with the multiple levels of automation and connected by various pipelines. It is developed as the complete integration of the concept of machine learning involving a wide range of stages and combined with the stages of the deployment, thus enabling the combination as a single tool or technology for better optimization of the machine learning projects with frequent changes and tuning of the multiple parameters.

MLOps is a practice to be developed as a flow of a single entity that enables the building of better production systems. Many of the current DevOps practices are not really scalable for a large number of services as summarized in Table 1. Although the MLOps need the current practices of DevOps in the most automated way for better and faster implementation. The DevOps will enable the MLOps to be more useful and enhance better equipment of the resources. Modelling CI/CD pipelining through the usage of the various automation tools will ensure the better design of the MLOps pipeline [4]. Various kinds of environment designs are already present to overcome similar challenges in the deployment but a complete automated product or tool is necessary to speed up the ongoing machine learning projects with the overall integration of the multiple components.

Table 1. Feature Comparison

Features	Existing Model	Proposed Model
Source Code Management [5]	Manual Code Upload	Auto upload
Repository Management [6]	Webhooks	Auto trigger
Automation [2]	Script based automation	Pipelined Automation
Scalability [7]	Manual scalability	Auto scalability
Container [8]	Manual deployment	Auto deployment

3. Implementation

The implementation of the complete model with the integration of Machine Learning and DevOps require various tools of the two different domains and must be internally connected to ensure multiline automation from development to the deployment.

3.1. Machine Learning

Machine Learning is the key technology to analyse the data and get the predictions out of it. Python acts as the most used programming language for the development of various machine learning algorithms. The deeper analysis of machine learning algorithms dives into the concept of neural networks which accounts for the domain of deep learning. There are many pre-developed libraries that help in building up the model. As a part of machine learning, Keras is used for training and building the model in this project. Keras is completely integrated with TensorFlow and enhances the developer to write the code for neural networks within a few lines. As a part of diabetes prediction, Keras is used as the neural network library to design the model that can intake the data and train it to make the predictions. The model is dependent on the multiple features which are provided as the input at the time of prediction that drives to the output..

TensorFlow is also used as a part of this project development. It is an open-source platform that enables the space and tools for the developers to design complex algorithms which can be helpful in solving the challenges faced by them. In this project, TensorFlow is used to perform the numerical calculation for building the model that is accurate and predicts the output for the given set of input features. It is also convenient and capable to integrate with python and develop the model accordingly with TensorFlow. TensorFlow is another additional library used for the development of the machine learning model in this project.

3.2. Docker

The container is defined as the process on the operating system that provides a feature of another operating system running on the base operating system. The deployment of the applications will not be done directly on the base operating system in the real-time scenario due to the various reasons that might lead to different problems and failures. Hence, an extra operating system environment is created with the help of the tools like Docker and run the applications on top of it. Even during the time of failures that particular container can be shut down and start the new one which holds the same application. In this process of implementation, the base operating system remains unaffected. Apart from that container is light weighted operating system that provides abstraction across various other processes running along with it [7]. Docker acts as a tool that helps to install the container software on the system and helps to implement it in an easier manner with its architecture as shown in figure 2.

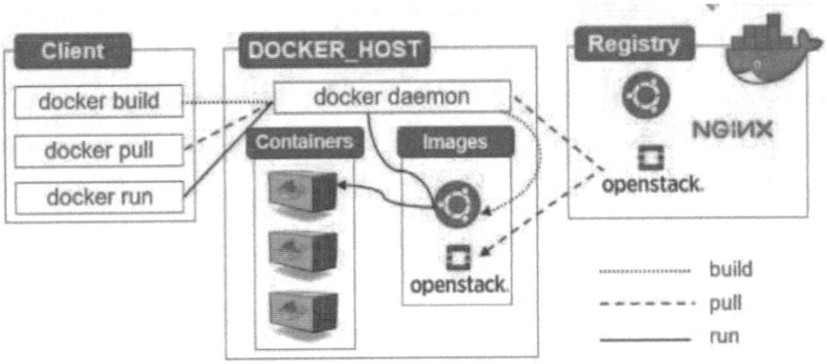

Figure 2. Docker Architecture.

Figure 2 describes the working of Docker in the client-server architecture. It enhances that the Docker host is the server and the client is the user, who can communicate to the server for various implementations. The server connects to the registry and gets the details that are necessary to fulfil the client request. In this proposed work, Docker is used in writing the Docker File that acts as an ISO image of an operating system.

3.3. Flask Framework

Flask is the framework useful in developing web applications and running them as the webserver. It is completely designed with the python programming language. Flask is used in this project to develop the GUI console for the user to enter the feature details of the machine learning that can predict the chances of diabetes for the person with those habits. The Flask application is integrated with the machine learning model, where it intakes the data and predict the output by sending the data to the model and finally displays it on the screen, where the user can receive it. Since the application is very much compatible and designed with the features of security, Flask is used for web application deployment purposes. It also enables the developer to easily integrate with the other tools, as it is developed with the well-used programming language.

3.4. Elastic Container Registry

Amazon Web Services provides an exclusive service similar to that of the Docker Hub that stores the ISO image files of the containers. This registry can be used with the various other integrated services and makes the process of deployment easier and faster. As a part of this project, ECR is used for storing the container image file which is developed and pushed by the developer that contains the various details to launch the container, run the web application and work accordingly with the mentioned ports and the protocols. It is used apart from the Docker Hub is to enhance the deployment of the complete project on the cloud environment. Hence, the inbuilt services of the cloud are used for better accuracy and availability. The image is uploaded to ECR accordingly as shown in figure 3.

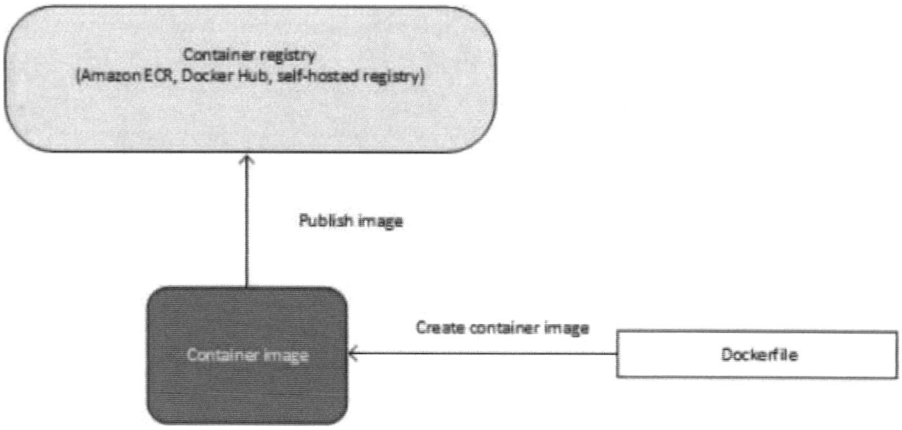

Figure 3. Image publishing using Docker to ECR.

3.5. Elastic Container Service

Elastic Container Service is provided by the AWS to deploy the containers automatically on the base operating system without managing it by the user. The cloud looks after the base operating system and its resources of it. In that case, we can easily launch the containers based on our requirements and perform the deployment of the application on the containers. As a part of this project, the ECS is useful to launch the containers that hold the web application through which the user can communicate to make the predictions with the model deployed. The ECS is completely managed by the AWS, hence there will not be any sort of management of the operating systems by us externally. The ECS distributes the launching of the containers accordingly as shown in figure 4.

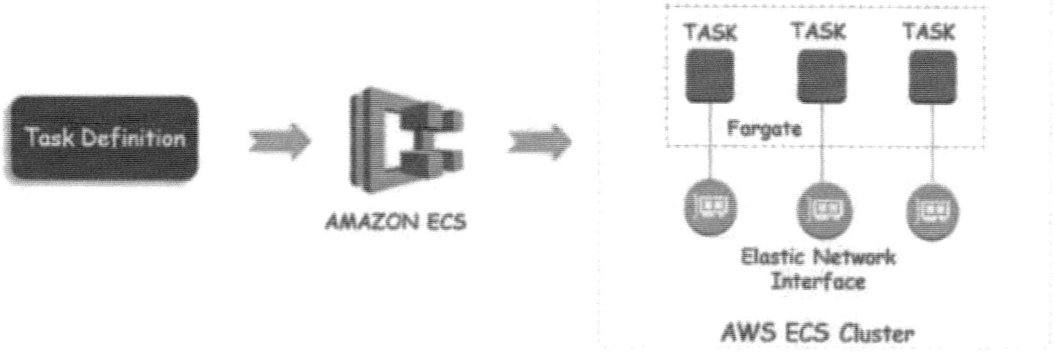

Figure 4. Task Scheduling using ECS.

3.6. Terraform

Terraform is the tool for multi-cloud deployment. It helps to deploy the resources on the cloud environment with a piece of code. It plans the infrastructure prior to the deployment and it also has the feature to destroy it completely, even the multiple services are dependent. In this work, Terraform is used for the deployment of the infrastructure on the cloud. It looks after the resources related to the ECS, ECR and deployment of the containers. It manages the deployment of the infrastructure within the span of minutes and helps in the complete automation of the deployment process on the cloud. It can be used for infrastructure deployment in various cloud environments [9].

4. Architecture

The architecture describes the flow of the process and the various integrations of both the domains through multiple pipelines. It starts from the code development and automates all the processes till deployment on the server.

4.1. Model Building

The initial step in the process of machine learning deployment is model building. The model building is performed based on the set of the lines in the Docker File. The model building requires the dataset and the pre-built libraries which are necessary for building the model. The process of model building starts with the training of the data to the algorithm and develops the model. The final model is produced as the h5 file. In this project, the model building is performed on the container environment, where it contains all the required dependencies that are mentioned through the Docker File. When the container is deployed with the Docker File, it is launched and runs the process of the model building. The flow of the code in the MLOps pipeline is defined as shown in figure 5.

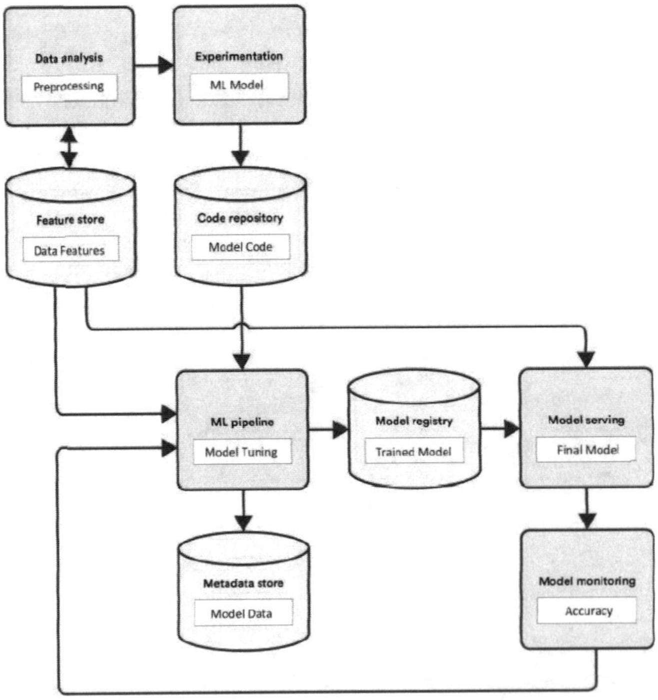

Figure 5. Block diagram of proposed model.

The dataset is read and pre-processing is done on the data. From the Keras library Sequential model, dense layer, Adam optimizer are imported. Here we used the multi-layered neural network which consists of 3 hidden layers, the input layer and hidden layers are activated by "relu" activation function while output layer is activated by "sigmoid" activation function. The "binary_crossentroy" function is used as the loss function and "adam" as the optimizer, this will give the accuracy as the metrics. The generated model has achieved accuracy of 73%, also model is repeatedly trained with the data for 100 times which helped in increasing the accuracy. At last, the model is saved with the name "cardio.h5" as shown in figure 6.

```
In [17]: model.compile(loss='binary_crossentropy', optimizer='adam', metrics=['accuracy'])

In [18]: model.fit(X,y, epochs=100)
Epoch 91/100
2188/2188 [==============================] - 2s 786us/step - loss: 0.5486 - accuracy: 0.7308
Epoch 92/100
2188/2188 [==============================] - 2s 863us/step - loss: 0.5528 - accuracy: 0.7273
Epoch 93/100
2188/2188 [==============================] - 2s 853us/step - loss: 0.5503 - accuracy: 0.7252
Epoch 94/100
2188/2188 [==============================] - 2s 776us/step - loss: 0.5477 - accuracy: 0.7298
Epoch 95/100
2188/2188 [==============================] - 2s 798us/step - loss: 0.5486 - accuracy: 0.7286
Epoch 96/100
2188/2188 [==============================] - 2s 783us/step - loss: 0.5495 - accuracy: 0.7289
Epoch 97/100
2188/2188 [==============================] - 2s 751us/step - loss: 0.5493 - accuracy: 0.7290
Epoch 98/100
2188/2188 [==============================] - 2s 810us/step - loss: 0.5500 - accuracy: 0.7277
Epoch 99/100
2188/2188 [==============================] - 2s 865us/step - loss: 0.5495 - accuracy: 0.7296
Epoch 100/100
2188/2188 [==============================] - 2s 869us/step - loss: 0.5471 - accuracy: 0.7318

In [19]: model.save("cardio.h5")
```

Figure 6. Model Training.

4.2. Creating a Docker image

The Docker Image is similar to the ISO image, which contains the operating system details and the various process to be run when the operating system is launched in the form of a container. The Docker Image can be used to contain the details of all the processes to be run in the container once it is deployed or in the state of running. Various libraries that are necessary for the container which runs the web application in this project are included in the Docker Image. The Docker image file of this project and the flask application code is shown in figure 7 and figure 8.

```
FROM docker.io/library/centos

RUN yum install python3 -y

RUN yum install python3-devel -y

RUN yum install gcc-c++ -y

RUN pip3 install --upgrade pip

RUN pip3 install Flask

RUN pip3 install pandas

RUN pip3 install h5py

RUN pip3 install keras

RUN pip3 install tensorflow

EXPOSE 80
```

Figure 7. Docker Image.

```
from keras.models  import load_model
from flask import Flask, render_template, request

app = Flask("cardio_app")
model  =  load_model("cardio.h5")

@app.route("/home")
def home():
    return render_template("form.html")

@app.route("/output" , methods=[ "GET" ] )
def dia():
        x1 = request.args.get("age")
        x2 = request.args.get("gender")
        x3 = request.args.get("height")
        x4 = request.args.get("weight")
        x5 = request.args.get("systolic")
        x6 = request.args.get("diastolic")
        x7 = request.args.get("cholesterol")
        x8 = request.args.get("glucose")
        x9 = request.args.get("smoke")
        x10 = request.args.get("alcohol")
        x11 = request.args.get("activity")

        output = model.predict([[ int(x1) , int(x2) , int

        if (round(output[0][0])) == 1:
            return render_template("positive.html")

        else:
            return render_template("negative.html")

app.run(host="0.0.0.0" ,  port=80)
```

Figure 8. Flask Application.

Journal of Physics: Conference Series **2327** (2022) 012027 doi:10.1088/1742-6596/2327/1/012027

4.3. Deploying Infrastructure using Terraform

The Docker File defines the set of the programs and the processes to be run on the container. Once the container is launched the set of commands run in the container to run the processes [8]. It is uploaded to the ECR by the developer. Once the Docker File is uploaded to the ECR registry, it triggers to run the file. It launches the containers using the ECS and runs the multiple containers as a cluster [10]. The containers hold the web application through which the users can interact. The multiple clusters of the containers are load balanced using the Elastic Load Balancer service of the AWS. The ELB runs the web application on port 80 and uses the HTTP protocol as mentioned in the Docker File. Finally, the terraform code returns the DNS name of the load balancer. Using the URL the users can reach the application.

The entire deployment process is run using the Terraform. It enhances to run the complete process and provide the output whether the implementation is successful or not. In this way, the power of infrastructure as a code is utilized with the terraform as a tool.

5. Results

The complete MLOps flow will create the production environment to deploy the machine learning model. As a part of the model, the page is displayed to take the input details from the user. The entire application is made with the Flask and make the predictions based on the data trained to the model. The homepage displayed for front-end form is shown in the figure 9. The output obtained can be positive or negative as shown in the figure 10 and figure 11 respectively. Also, the updates in the code made by the developer and pushed to the source code repository will build up the production environment within the span of 3 minutes as shown in figure 12. The proposed work's key contribution is that it can automatically construct and deploy machine learning models without any operator intervention. The models can also be retrained to improve predictions accuracy.

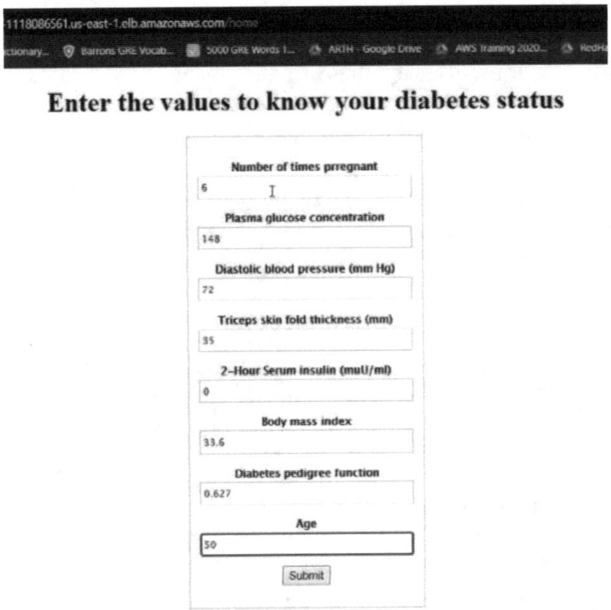

Figure 9. Front-end Form.

Journal of Physics: Conference Series **2327** (2022) 012027 doi:10.1088/1742-6596/2327/1/012027

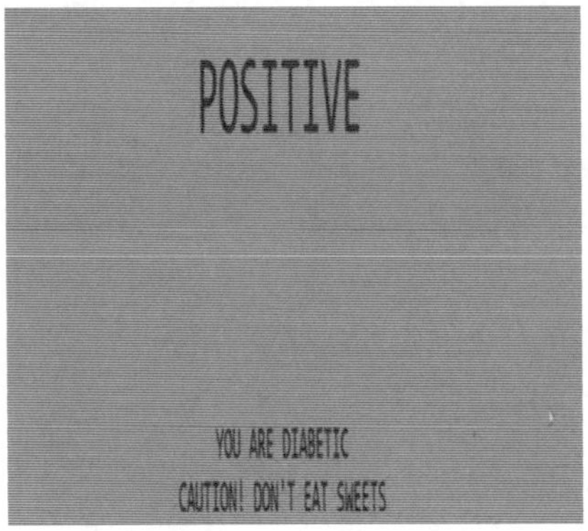

Figure 10. Positive Output.

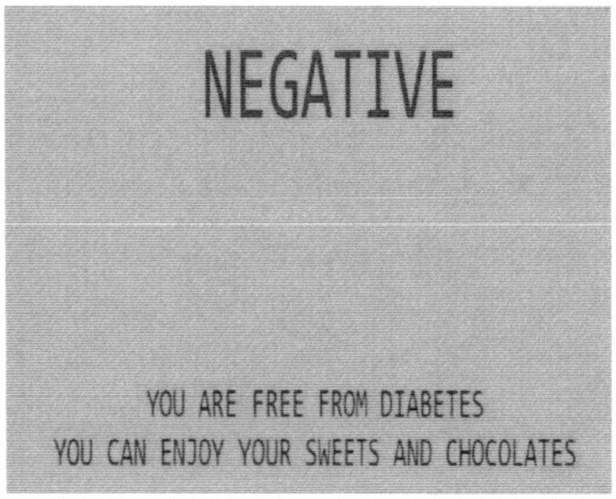

Figure 11. Negative Output.

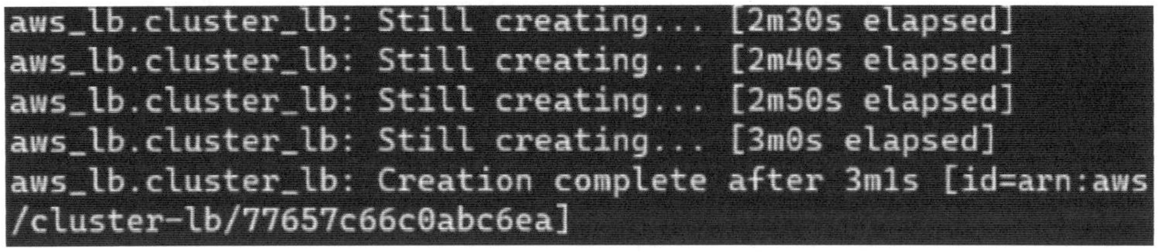

Figure 12. Timespan for Deployment

6. Conclusion and future scope

MLOps or DevOps for machine learning, empowers data science and IT groups to work together and increment the speed of model turn of events and arrangement through monitoring, validation and governance of machine learning models. MLOps is displayed on the current control of DevOps, the cutting-edge practice of effectively writing, deploying and running venture applications. DevOps got its beginning 10 years prior as a way fighting clans of software developers (the Dev) and IT tasks groups (the Ops) could team up. MLOps adds to the group the data scientists, which minister datasets and construct AI models that dissect them. It likewise incorporates ML engineers, who run those datasets through the models in focused, automated ways. When a ML model has been assembled, it should be coordinated with real-world data and the business application or front-end services. The entire application or parts thereof should be deployed without disrupting the assistance. Sending can be very difficult if the ML components aren't treated as a vital piece of the application or production pipeline [11]. The proposed model supports the model creation with automated machine learning and access incredible feature engineering, algorithm selection and hyper parameter-sweeping capabilities. This model increases group proficiency with shared datasets, notebooks, models and customizable dashboards that track all parts of the machine learning measure.

In the future, the model can be improved by performing the tuning of the parameters and analysing the accuracy of the result with the multiple parameters. In this way, it can also be integrated with tool such as Kubernetes to deploy the containers efficiently and perform the scaling of them and other features of monitoring their health and maintaining the high availability [12]. This process of MLOps will enable the data scientists to work more on the better model building and the deployment is completely automated which enhances the retraining of the model for better accuracy.

References

[1] Singla K, Bose J and Naik C 2018 Analysis of software engineering for agile machine learning projects *15th IEEE India Council International Conference (INDICON)* (Coimbatore, India) pp 1-5.

[2] Guşeilă L G, Bratu D V and Moraru S A 2019 DevOps transformation for multi-cloud IoT applications *IEEE International Conference on Sensing and Instrumentation in IoT Era (ISSI)* (Lisbon, Portugal) pp 1-6.

[3] Shafiq S, Mashkoor A, Mayr-Dorn C and Egyed A 2021 A Literature Review of Machine Learning and Software Development Life cycle Stages *IEEE Access* **9**, pp 140896-140920.

[4] Mäkinen S, Skogström H, Laaksonen E and Mikkonen T 2021 Who Needs MLOps: What Data Scientists Seek to Accomplish and How Can MLOps Help? *IEEE/ACM 1st Workshop on AI Engineering-Software Engineering for AI (WAIN)* (Madrid, Spain) pp 109-112.

[5] Gupta M, Chowdary M N, Bussa S and Chowdary C K 2021 Deploying Hadoop Architecture Using Ansible and Terraform *5th IEEE International Conference on Information Systems and Computer Networks (ISCON)* (Mathura, India) pp 1-6.

[6] Gupta M, Sanjana K, Akhilesh K and Chowdary M N 2021 Deployment of Multi-Tier Application on Cloud and Continuous Monitoring using Kubernetes *5th IEEE International Conference on Electrical, Electronics, Communication, Computer Technologies and Optimization Techniques (ICEECCOT)* (Mysuru, India) pp 602-607.

[7] Tamburri D A 2020 Sustainable MLOps: Trends and challenges *22nd IEEE International Symposium on Symbolic and Numeric Algorithms for Scientific Computing (SYNASC)* (Timisoara, Romania) pp 17-23.

[8] Bates W and Oviedo E I 2021 Software Reliability in a DevOps Continuous Integration Environment *Annual Reliability and Maintainability Symposium (RAMS)* (Orlando, FL, USA) pp 1-4.

[9] Marijan D, Liaaen M and Sen S 2018 DevOps improvements for reduced cycle times with integrated test optimizations for continuous integration *42nd IEEE Annual computer software and applications conference (COMPSAC)* vol 1 (Tokyo, Japan) pp 22-27.

[10] Liao Q 2020 Modelling CI/CD Pipeline Through Agent-Based Simulation *IEEE International Symposium on Software Reliability Engineering Workshops (ISSREW)* (Coimbra, Portugal) pp 155-156.

[11] Mysari S and Bejgam V 2020 Continuous Integration and Continuous Deployment Pipeline Automation Using Jenkins Ansible *IEEE International Conference on Emerging Trends in Information Technology and Engineering (ic-ETITE)* (Vellore, India) pp 1-4.

[12] Singh C, Gaba N S, Kaur M and Kaur B 2019 Comparison of different CI/CD tools integrated with cloud platform *9th IEEE International Conference on Cloud Computing, Data Science & Engineering (Confluence)* (Noida, India) pp 7-12.

Journal of Physics: Conference Series **2327** (2022) 012028 doi:10.1088/1742-6596/2327/1/012028

Improve Content-based Image Retrieval using Deep learning model

Suneel Kumar,
Manoj Kumar Singh, Manoj Kumar Mishra,
Department of Computer Science, Banaras Hindu University, Varanasi-221005

Abstract. The complexity of multimedia has expanded dramatically as a result of recent technology breakthroughs, and retrieval of similar multimedia material remains an ongoing research topic. Content-based image retrieval (CBIR) systems search huge databases for pictures that are related to the query image (QI). Existing CBIR algorithms extract just a subset of feature sets, limiting retrieval efficacy. The sorting of photos with a high degree of visual similarity is a necessary step in any image retrieval technique. Because a single feature is not resilient to image datasets modifications, feature combining, also known as feature fusion, is employed in CBIR to increase performance. This work describes a CBIR system in which combining DarkNet-19 and DarkNet-53 information to retrieve images. Experiments on the Wang (Corel 1K) database reveal a considerable improvement in precision over state-of-the-art classic techniques as well as Deep Convolutional Neural Network(DCNN).

1. Introduction

Image retrieval is a machine vision application that searches for relevant images in a collection of images. It is widely utilised in everyday life and it also used for a variety of additive computing machine vision fields. It has amassed thousands of methods following decades of study and development. Image retrieval based on text[1], CBIR[2], and Image retrieval based on semantics[3] are the three broad categories of image retrieval methods. Text annotations are used in text-based image retrieval. This type retrieval process is not well-executed without manual annotations, and such approaches ignore the image's visual content and hence lose sensory information. Low contrast characteristics such as color characteristics[4], edge characteristics[4], characteristics of texture[5], and GIST are commonly used in traditional content-based image retrieval. These low contrast characteristics are frequently not immediately comprehended by humans, As a result, there is a disparity between low contrast characteristics and high-level semantic notions. Image retrieval using semantics[6] tries to overcome this gap by stressing the link between visual and semantic notions by constructing a character driven hierarchy The trouble and goal of image retrieval investigation is bridging the semantic gap.. Established image retrieval approaches are still inadequate for addressing this problem entirely.

In recent years, deep learning has had an impact on and transformed nearly every discipline of computer vision, influencing and changing people's lifestyles as a result of its superior feature learning ability and hierarchical representation of features. The adoption of deep learning principles at the highest levels of research is the driving force behind this development and influence. Deep learning offers several advantages in a variety of activities requiring computer vision, including image classification[7], image recognition[8], and object detection[9]. Deep learning is rapidly revolutionising and influencing human life due to its exceptional performance in machine vision applications. Deep learning research institutes and technologies have been founded by several well-known corporations, including Google, Facebook, Microsoft, as well as some well-known start-ups. These firms have commercialized face recognition technology, vision based technology, motorist technology, video analytic technology and aided driving technologies, resulting in enhanced service and a gradual transition in people's way of life. Furthermore, deep learning has had an impact on domains other than computer vision, such as NLP, ML, etc.

Deep learning has a substantial influence on computer vision research, and this consequence has developed in tandem with deep learning's improvement, notably the performance of convolutional neural networks(CNN). CNN performance has increased in recent years as a consequence of image classification challenges on ImageNet datasets. A variety of significant CNN models have been suggested, including Alexnet, VGG, Googlenet, and Resnet. Newly suggested models update the record of image classification tasks as model depth grows. In general, pre-trained CNN using ImageNet datasets may be utilised immediately for picture feature extraction and other applications. Because the CNN has demonstrated high feature encoding capabilities, the network learning based on the ImageNet classification job has some invariance and may be used in a variety of application sectors with similar image datasets.

We describe a framework for learning visual features in CBIR using CNN, owing to the general development of deep learning and the importance of feature extraction in CBIR systems. The proposed CNN learns how to extract essential features from a given image database and then applies this knowledge to image retrieval.

2. Related Work

The deep autoencoder was trained in the first effort at deep learning image retrieval. [10]. Deep Belief Network (DBN)[11] is used to train the deep autoencoder, which is then tested on an small image dataset[11]. Deep learning, particularly the CNN[7], has since made significant advances in image classification. The CNN model has a high degree of generality, according to the researchers. The CNN model developed for image classification could be applied to other recognition tasks effectively [[12], [13], [14]]. CNN's effect on numerous aspects of computer vision has steadily grown as a result of its universality. The first image retrieval task using CNN was carried out in[15], but the results were unsatisfactory. With more spatial information, CNN's performance approaches that of BoW and VLAD encoding based on classical SIFT. Following that, "multi-scale ordereless collection" (MOP)[16] aggregates image blocks with multi scales instead of extracting CNN characteristics from the entire image, as well as enhanced retrieval performance. This technique, on the other hand, has difficulty determining the window size. Concurrently, academics have begun to investigate methods to extract CNN properties. Some pooling approaches, for example, max-pooling [17], R-MAC[18], and methods for spatial and channel-aware weighted pooling[19],were developed by reusing concepts from the classic CBIR framework. A more typical CBIR-like technique is to regard every local feature in the 3D array (along its depth dimension) as "super SIFT," then encrypt these selected features using the known VLAD or Fisher vector model and combine the encoding values on the feature map[[16], [20]]. A deep feature-based spatial search method[21] is presented as a replacement for standard CBIR.

Fine-tuning of CNN on new datasets by trimming is another technique to learn better representations[[22], [23], [24], [25]]. Although retrieval performance is good, collecting and labeling new datasets takes time. This is due to the fact that extra datasets pertaining to a given image retrieval job must be acquired in order to train the network, and rigorous data selection or cleaning techniques must be employed. The fine-tuning strategy has been used in two recent papers [22] and [23] for the retrieval of iconic building images. Both of these works rely on auxiliary datasets. [22] employs the clean version of the labeled datasets retrieved from the Web via a search engine for a landmark. This literature review presented above provide a basic comprehension of deep learning model development, image retrieval advancement, and recent advances in deep learning-based image retrieval. It aids to focusing on some critical research issues. The rest of this paper is organized as follows. We present brief overview of two model first DarkNet-19 and DarkNet-53 for feature extraction and image similarity measurement in Section 3. Section 4 presents the proposed method and then experimental result and performance discus in section 5. Finally, concluding remarks are presented in Section 6.

3. Preliminaries

Recent research has revealed that CNN-based CBIR approaches acquire better low-level feature extraction strategies make better results. Each CNN layer is made up of many neural connections with three dimensions: height, breadth, and depth.

3.1. DarkNet-19

Darknet-19 model[26] employs YOLOv2, which includes 19 conv2D layers, 5 max-pooling layers, and a soft-max layer for object classification. Table 1 depicts the Darknet-19 architecture bellow. It is a C-language and CUDA-based neural network framework. It detects objects very quickly, which is critical for real-time prediction. This network uses to classify 1000 different class images. As a consequence, the model has learned detailed feature of images to diversified set of images into classes. The network has a 256×256 image input size.

3.2. Darknet-53

In 2018, Darknet-53 was introduced as the fundamental feature extraction method in the real-time object recognition network YOLOv3(You Only Look Once). The goal of this model is to extract attributes from an input image. This network's primary concept may be expressed as a mix of YOLOv2's fundamental feature extraction and the residual module. Table 2 depicts the basic architecture of this concept. This model consist of five repeating block which has two convolution layers 1× and 3×3 in size, then one Residual layer. It has 53 convolutional layers, each of which has been followed by a batch normalisation layer and Leaky ReLU activation. The convolution layer is being used to convolve numerous filters on images, resulting in multiple feature maps. There is no pooling, and the feature maps are downsampled using a convolutional layer with stride 2. It aids in the prevention of low-level feature loss, which is frequently linked to pooling. If the goal was to conduct classification like in ImageNet, the Average pool layer, 1000 completely linked layers, and a SoftMax activation function would be added, as indicated in the image. However, in our situation, we'd like to retrieve the relevant images from collection of images.

3.3. Principal Component Analysis

Principal Component Analysis (PCA) is a dimensionality-reduction approach that is often used to decrease the dimension of vast quantities of data by reducing a big number of possibilities into a limited subset that preserves the bulk of the information in the original set. For example, data in the form of a p-dimensional variable x = $(x_1, ...x_p)$ can be expressed in lower dimension as s = $(s_1, ..., s_k)$ with $k < p$. PCA's goal is to decrease the number of possibilities in a data set while retaining as much information as feasible.

Table 1. DarkNet-19 Architecture

Type	Filters	Size/Stride	output
Conv1_1	32	3×3	224×224
Max-pool		2×2/2	112×112
Conv2_1	64	3×3	112×112
Max-pool		2×2/2	56×56
Conv3_1	128	3×3	56×56
Conv3_2	64	1×1	56×56
Conv3_3	128	3×3	56×56
Max-pool		2×2/2	28×28
Conv4_1	256	3×3	28×28
Conv4_2	128	1×1	28×28
Conv4_3	256	3×3	28×28
Max-pool		2×2/2	14×14
Conv5_1	512	3×3	14×14
Conv5_2	256	1×1	14×14
Conv5_3	512	3×3	14×14
Conv5_4	256	1×1	14×14
Conv5_5	512	3×3	14×14
Max-pool		2×2/2	7×7
Conv6_1	1024	3×3	7×7
Conv6_2	512	1×1	7×7
Conv6_3	1024	3×3	7×7
Conv6_4	512	1×1	7×7
Conv6_5	1024	3×3	7×7
Conv6_6	1000	1×1	7×7
Avg_pool		Global	1000
Softmax			

input $X \in \mathbb{R}^{n \times p}$ Here, X represent features of images. Rows of X correspond to observation and columns correspond to variables.

$$coeff = pca(X) \quad (1)$$

$$X = X \times coeff \quad (2)$$

The coefficient matrix has the dimensions p×p. Every column in coeff holds values for one major component, and the columns are sorted by element variance in descending order. PCA automatically centers the data and using the SVD technique. It requires the rectangular matrix Y (where X is n×p).

$$X = U_{n \times n} S_{n \times p} V^T_{p \times p} \quad (3)$$

where, $U^T U = I_{n \times n}$
$V^T V = I_{p \times p}$ (i.e U and V are orthogonal)
Finding the eigenvalues and eigenvectors of XX^T and $X^T X$ is the first step in calculating the SVD. The columns of V are made up of $X^T X$ eigenvectors, whereas the columns of U are made up of XX^T eigenvectors. In addition, the singular values in S are the square roots of eigenvalues from XX^T or $X^T X$.

Journal of Physics: Conference Series **2327** (2022) 012028 doi:10.1088/1742-6596/2327/1/012028

Table 2. DarkNet-53 Architecture

Type	Filters	Size/Stride	output	Repeat
Conv2D	32	3×3	256×256	
Conv2D	64	3×3/2	128×128	
Conv2D Conv2D Residual	32 64	1×1 3×3	 128×128	x1
Conv2D	128	3×3/2	64×64	
Conv2D Conv2D Residual	64 128	1×1 3×3	 64×64	x2
Conv2D	256	3×3/2	32×32	
Conv2D Conv2D Residual	128 256	1×1 3×3	 32×32	x8
Conv2D	512	3×3/2	16×16	
Conv2D Conv2D Residual	256 512	1×1 3×3	 128×128	x8
Conv2D	1024	3×3/2	8×8	
Conv2D Conv2D Residual	512 1024	1×1 3×3	 8×8	x4
Avg-pool		Global		
Connected			1000	
Softmax				

4. Proposed CBIR system

We suggested a framework for feature fusion in a CBIR system. The traditional approach first extracts the image's colour, texture, or other important qualities before manually selecting multiple elements for fusion. The downside of this strategy is that the same characteristic contributes differently to image retrieval in different datasets. It is difficult to define the features suited for the datasets and the ideal mix of features in advance in image retrieval operations. We can get better results by experimenting with different combinations over all datasets, but the time required is enormous, the efficiency is low, and the algorithm is not stable. We extract image features in this paper using both the pretrained DarkNet-19 and DarkNet-53 models, and then combine the extracted features to generate a new feature vector. In DarkNet-53, we employ activation of the last convolutional layer for feature extraction because it preserves spatial information and is connected to all input neurons. In DarkNet-19, the average pooling layer, called 'Avg 1,' is used as a feature extraction layer, yielding a feature vector.

5. Datasets and experimental results

The database consists of 10 classes in which each class contain 100 images so, total number of images in database 1000 i.e., called Corel database. All the images are in RGB space with size of 256×384 and 384×256 pixels. The classes of this database shown in table 3.

Journal of Physics: Conference Series **2327** (2022) 012028 doi:10.1088/1742-6596/2327/1/012028

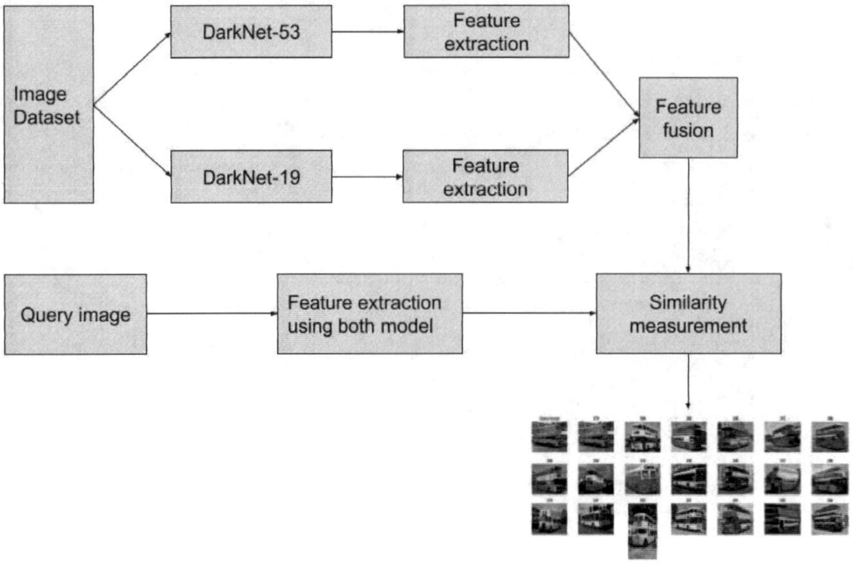

Figure 1. Proposed framework

Input: Query image denoted by Q_{Image}. Image Database denote by DB_{images}. N denotes the total number of images

Output: M Number of retrieved relevent image to Q_{Image}

(i) Total number of image N in DB_{images}.

(ii) for $i = 1$ to N

An image $I \in \mathbb{R}^{N \times M \times 3}$

Resize image into 224×224×3.

Extract high-level features using activation of both model of N images from DB_{images}.

Crate a index DB_{index} of all images using generated features.

(iii) Extract Q_{Image} features

(iv) PCA is used to reduce the image feature.

(v) for each images in DB_{index}

Using the similarity metric, compare the query image features to all images in the DB_{index}.

(vi) Sort the relevant images to Q_{Image} features.

(vii) Retrieve M images that are closed to Q_{Image}.

Algorithm 1: Image retrieval

5.1. Precision

Precision in image retrieval may be defined as the ratio of recovered image content to whole retrieved images queried.

$$Precision = A/B \tag{4}$$

Where, A is "the relevant retrieved images" and B is "the whole retrieved images.

5.2. Recall

In image retrieval, recall is defined as the ratio of recovered significant images to whole database images.

$$Recall = A/C \tag{5}$$

Where, A is "the relevant retrieved images" and C is "the whole number of relevant images in the database". Tables 3 and 4 illustrate the AP and AR of 10 classes for N=20 (number of

Table 3. Average precision(AP) of whole images using DarkNet-19, DarkNet-53 and proposed method.

Category	DarkNet-19	DarkNet-53	Proposed Method
Africans	81.75	80.10	84.15
Beaches	80.40	83.65	84.20
Buildings	95.85	94.85	96.60
Buses	100	100	100
Dinosaurs	99.70	100	100
Elephants	99.25	99.90	99.75
Roses	99.70	100	100
Foods	95.10	94.65	95.40
Horses	99.90	100	100
Mountains and glaciers	95.45	96.20	97.15
Average	94.71	94.93	95.72

Table 4. Average recall(AR) of whole images using DarkNet-19, DarkNet-53 and proposed method.

Category	DarkNet-19	DarkNet-53	Proposed Method
Africans	16.35	16.02	16.83
Beaches	16.08	16.73	16.84
Buildings	19.17	18.97	19.32
Buses	20	20	20
Dinosaurs	19.94	20	20
Elephants	19.85	19.98	19.95
Roses	19.94	20	20
Foods	19.02	18.93	19.08
Horses	19.98	20	20
Mountains and glaciers	19.09	19.24	19.43
Average	18.94	18.98	19.14

recovered images) using DarkNet-19, DarkNet-53, and the suggested technique, respectively. To calculate the AP and AR, we choose all image one by one from each classes as the query image and get the appropriate images from the dataset. In the DarkNet-19 model, the minimum

AP and AR for beaches are 80.40% and 16.08%, respectively, while the highest AP and AR for buses are 100 percent and 20%, respectively, as shown in Tables 1 and 2. In DarkNet-53, the minimum AP and AR for Africans are 80.10% and 16.02%, respectively, while the highest average precision and recall for buses, dinosaurs, flowers, and horses are 100 percent and 20%, respectively, as shown in tables 3 and 4.

In the suggested technique, we combine both methods DarkNet-19 and DarkNet-53 and then do the CBIR operation, which show the best result as compare to other methods, as shown in the table 5. In this case minimum AP and AR are 84.15% and 16.83% respectively of Africans and maximum AP and AR are 100% and 20% respectively of buses, dinosaurs, roses and horses shown in above table 3 & 4.

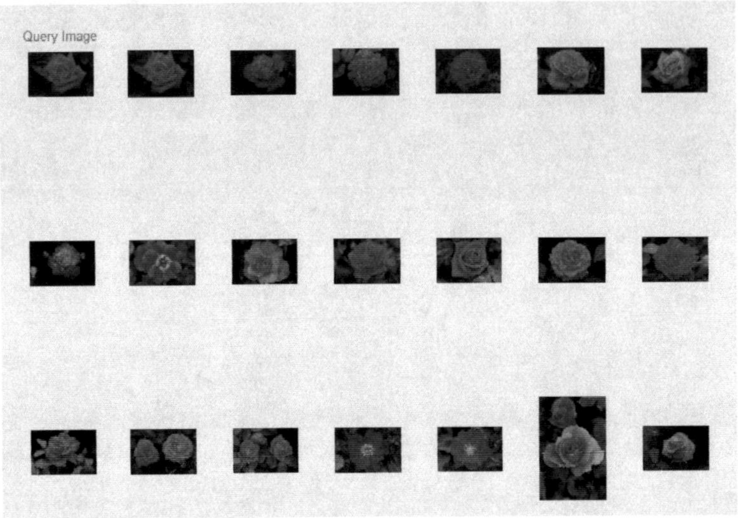

Figure 2. Roses query image, Top 20 Retrieved Images.

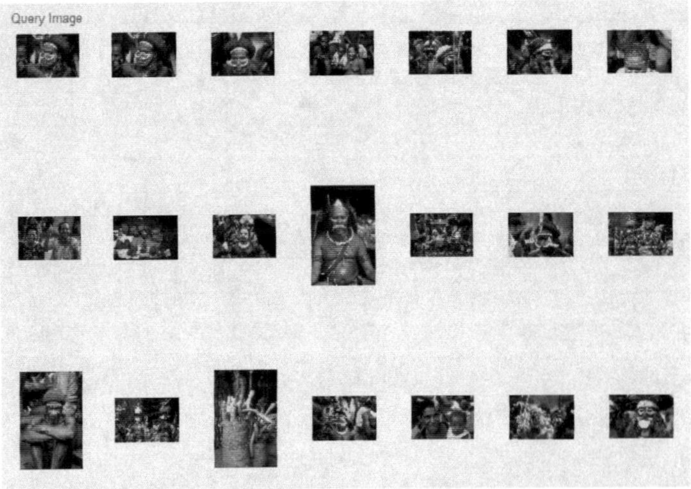

Figure 3. Africans query image, Top 20 Retrieved Images.

Table 5. Comparative analysis with other deep learning model

Performance /Method	AlexNet[14]	VGG16[7]	ResNet[27]	GoogleNet[28]	Modified AlexNet[29] +HOG+LBP	Proposed Method
Average Precision	76.22	89.08	93.27	95.61	92.37	95.72
Average Recall	41.58	58.70	71.99	79.43	70.92	19.14

6. Conclusion

The goal of this research is to create a multi-feature fusion technique for improving CNN feature representation in order to find a way to ensure that images are viewing for generally accurate images. Feature extraction and representation is a crucial problem in CBIR, as it is closely related to human perception. To retrieve related images from the image dataset, the Euclidean distance is used. In the proposed technique, we merge the feature vectors of the darknet-19 and darknet-53 models to create a new feature vector for retrieval. This experiment was carried out using the Wang dataset, which outperforms other existing state-of-the-art in terms of precision and recall.

Acknowledgement: This research is supported by seed grant under IoE, BHU [grant no. R/Dev/D/IoE/SEED GRANT/2020-21/Scheme No. 6031].

[1] Rui Y, Huang T S and Chang S F 1999 *Journal of visual communication and image representation* **10** 39–62
[2] Smeulders A W, Worring M, Santini S, Gupta A and Jain R 2000 *IEEE Transactions on pattern analysis and machine intelligence* **22** 1349–1380
[3] Bradshaw B 2000 *Proceedings of the eighth ACM international conference on Multimedia* pp 167–176
[4] Jain A K and Vailaya A 1996 *Pattern recognition* **29** 1233–1244
[5] Manjunath B S and Ma W Y 1996 *IEEE Transactions on pattern analysis and machine intelligence* **18** 837–842
[6] Mezaris V, Kompatsiaris I and Strintzis M G 2003 *Proceedings 2003 International Conference on Image Processing (Cat. No. 03CH37429)* vol 2 (IEEE) pp II–511
[7] Krizhevsky A, Sutskever I and Hinton G E 2012 *Advances in neural information processing systems* **25** 1097–1105
[8] Lee H, Grosse R, Ranganath R and Ng A Y 2009 *Proceedings of the 26th annual international conference on machine learning* pp 609–616
[9] Girshick R, Donahue J, Darrell T and Malik J 2014 *Proceedings of the IEEE conference on computer vision and pattern recognition* pp 580–587
[10] Krizhevsky A and Hinton G E 2011 *ESANN* vol 1 (Citeseer) p 2
[11] Hinton G E, Osindero S and Teh Y W 2006 *Neural computation* **18** 1527–1554
[12] Oquab M, Bottou L, Laptev I and Sivic J 2014 *Proceedings of the IEEE conference on computer vision and pattern recognition* pp 1717–1724
[13] Deng C, Ji R, Liu W, Tao D and Gao X 2013 *Proceedings of the IEEE International Conference on Computer Vision* pp 2600–2607
[14] Zeiler M D and Fergus R 2014 *European conference on computer vision* (Springer) pp 818–833
[15] Sharif Razavian A, Azizpour H, Sullivan J and Carlsson S 2014 *Proceedings of the IEEE conference on computer vision and pattern recognition workshops* pp 806–813
[16] Gong Y, Wang L, Guo R and Lazebnik S 2014 *European conference on computer vision* (Springer) pp 392–407
[17] Babenko A and Lempitsky V 2015 *Proceedings of the IEEE international conference on computer vision* pp 1269–1277
[18] Tolias G, Sicre R and Jégou H 2015 *arXiv preprint arXiv:1511.05879*
[19] Kalantidis Y, Mellina C and Osindero S 2016 *European conference on computer vision* (Springer) pp 685–701
[20] Yue-Hei Ng J, Yang F and Davis L S 2015 *Proceedings of the IEEE conference on computer vision and pattern recognition workshops* pp 53–61
[21] Razavian A S, Sullivan J, Carlsson S and Maki A 2016 *ITE Transactions on Media Technology and Applications* **4** 251–258

[22] Gordo A, Almazan J, Revaud J and Larlus D 2017 *International Journal of Computer Vision* **124** 237–254
[23] Radenović F, Tolias G and Chum O 2016 *European conference on computer vision* (Springer) pp 3–20
[24] Gordo A, Almazán J, Revaud J and Larlus D 2016 *European conference on computer vision* (Springer) pp 241–257
[25] Arandjelovic R, Gronat P, Torii A, Pajdla T and Sivic J 2016 *Proceedings of the IEEE conference on computer vision and pattern recognition* pp 5297–5307
[26] Redmon J and Farhadi A 2017 *Proceedings of the IEEE conference on computer vision and pattern recognition* pp 7263–7271
[27] Maji S and Bose S 2021 *ACM/IMS Transactions on Data Science (TDS)* **2** 1–24
[28] Szegedy C, Liu W, Jia Y, Sermanet P, Reed S, Anguelov D, Erhan D, Vanhoucke V and Rabinovich A 2015 *Proceedings of the IEEE conference on computer vision and pattern recognition* pp 1–9
[29] Shakarami A and Tarrah H 2020 *Optik* **214** 164833

A prototype model for detection and classification of landslides using satellite data

Akanksha Sharma[1], Kamal Kumar Sharma[2], Suhas Gajanan Sapate[3]

[1.] Research Scholar ,Department of Electronics and communication , Lovely Professional University, Jalandhar (P.B.), India. https://orcid.org/0000-0003-0353-2355
[2.] Professor, School of Electronics and Electrical Engineering, Lovely Professional University, Jalandhar (P.B.), India. https://orcid.org/0000-0002-6294-0829
[3.] Professor ,Department of Computer Science and Engineering, Annasaheb Dange College of Engineering and Technology, Ashta, Sangli (M.H.), India

Abstract. Landslides are natural and manmade disasters that cause threat to human life and lead to huge economic loss. Last few decade number of approaches have been developed for early detection of landslide for protecting life and saving properties. This paper proposes a prototype for an artificial intelligent model to detect and predict different types of landslides in hilly area with remote sensing techniques. All developing countries are following a steep increase in development of infrastructure like buildings, roads tunnels bridges railway tracks. Demand of connecting remote area is very high but on other side of environment it is also true that high demand of construction in morpho material area is causing many disasters like landslide. Landslide causes the loss of property and life so an early alarming will be help full for disaster management. Remotely sensed data pre-processed with artificial intelligent technologies will be helpful for landslide detection, creating landslide susceptibility map and inventory. Focus of this study is on enhancing the accuracy to detect landslide, list out the different features for extraction from satellite images and pre processing steps. This research also focuses on application of robust early prediction of type of landslide. This research will help in detection of landslide early to protect economical losses and human lives.

1. Introduction

In hilly terrains like Utrakhand , Himachal Pradesh landslides are one of the major natural disasters which take place in all the seasons , Some time because of rainy weather , some after snowfall and some time because of the fragile nature of rock forming mountains . By survey of Building Material & Technology Promotion council (BMPTC) & TARU data landslide hazard probability is divided into three categories: Low, Medium and High.[1] Landslide Hazard zonation Atlas claims that 8% of entire area of Himachal Pradesh is under high risk zone and by revised methodology Expert knowledge 3.2%area is under high risk and AHP indicate 5.65% area is under high risk zone. In mountain areas landslides are most dangerous geological hazard.[2]

Landslides are rapid movement of flow of material downward and outward. It is the movement of mass rocks, debris or earth down a slope under the influence of gravity. The size and shape of ditched mass depends on the nature of discontinuities in the rock, degree of weathering and steepness of slope. Material in landslide mass is rock, solid or both[3].Landslide can be initiated by many natural phenomena like heavy rain fall, earthquake, soil erosion, climate change, overburden of rock material, Gravity, Tsunami etc. Many Human made activities like cutting slopes for infrastructure, removal of vegetable, deforestation, road cutting etc also trigger the landslides.[4] In developing countries throughout the world landslide disasters cause loss of property and human life [5]. To reduce geological disaster risk there is a need to implement technologies for enhancing disaster protection and develop strategies on different levels. [6].

On the basis of type of movement and type of material landslide can be classified into different types. Mainly rock or soil or both is material in a landslide. In actuality, the Internal mechanism of landslide mass is displaced by the type of movement and there are five types of Landslide on this basis [7].

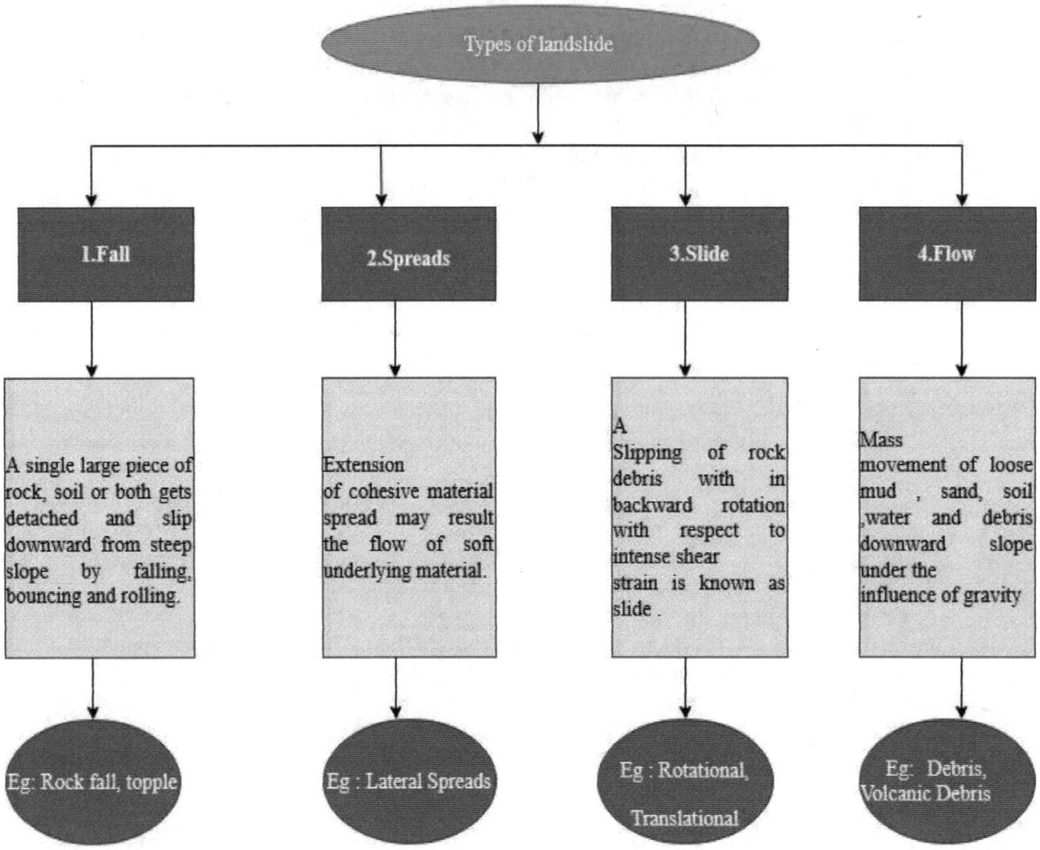

Figure 1 : Classification Of landslide based on literature review of the USGS landslide manual.[7]

Fall

A single large piece of rock, soil or both gets detached and slips downward from a steep slope by falling, bouncing and rolling. Rock fall, topple are examples of Fall type landslides . These type of landslide generally take place in

Vertical slope area, Coastal area and Rocky Banks of river . The Triggering mechanism that initiates these types of landslides are Cutting of slop either by natural or by human activities . Earthquakes

Spreads

Extension of cohesive material spread may result in the flow of soft underlying material.
Lateral Spreads is an example of spread type landslide. These types of landslides take place where flow of soil, area of seismic activity take place. The Triggering mechanisms that are helpful for spreading type landslides are Earthquake , erosion ,and snow melt .

Slide

A Slipping of rock debris with in backward rotation with respect to intense shear strain is known as slide . Displaced mass of rock debris moves relatively coherent mass along the rupture surface.
Rotational Landslide, Translational Landslide are type of slide type landslide. Area fill material associated with slope of 20 to 40 degrees will lead to such landslides. Rainfall, Rapid snow melt ,all factor that cause erosion at the base slope are few triggering mechanism that helpful for this type of landslide

Flow

Mass movement of soil, water debris ,loose mud, sand, downward slope under the influence of gravity . It is rapid in movement depending on consistency and slope angle.Debris and Volcanic Debris Flows are examples of flow landslides. Intense surface water flow, rapid snow melt , heavy rain are Triggering mechanism for this type of landslide

Landslides explain the movement of debris or earth flow or mass of rock down a slope[8] Landslides can be of different types based on material and form. Landslides can be in a number of different types of velocity and magnitude that can initiate in parts and usually take place where a landslide has occurred. [9][10].
Achieving information of landslides and cause of landslides on the basis of satellite images is very beneficial. Remote sensing techniques synthetic aperture radar, optical, and light detection and ranging measurements from spaceborne, airborne, and ground-based platforms have been used for detection and analysis of landslides and the process of landslides. Multiple and fused remote sensing techniques were used for landslide detection, inventory mapping, surface deformation monitoring, trigger factor analysis and mechanism inversion, landslide susceptibility modeling, hazard assessment, and risk evaluation. Remote sensed data like spaceborne synthetic aperture radar (SAR) and optical remote sensing, airborne light detection and ranging (LiDAR), ground-based SAR and terrestrial LiDAR, can be used for observation of environmental factors that cause landslides. [11]

2. Literature survey

A semiautomatic landslide detection and mapping model which gives the ability to separate source and run out area was designed [12]. Data set of Geo Eye 1 multi spectral images was used. The semi automated model was tested with VHR multispectral imagery and these images were acquired after the landslide incident . Object base analysis and SVM was used for classification. Result was classified into two parts, one in landslide recognition and other in internal mapping of source and runout area. Object based image classification for landslide with SVM algorithm gave commission error 26% and omission error below 24%.False negative result and false positive results value was 38& 78% for this model. This supervised approach gives 95% landslide scare detection.
A digitalized landslide inventory with satellite images was refined by using high resolution satellite imagery. Between pre and post landslide satellite images, landslide timing was constrained. Harmonic modeling of satellite images in a semiautomatic method was designed to detect landslide timing.[12] Landslides in forest leaves the scar on soil rock for some time , the recovery period for the landslides depend on landslide material and attributes. Before a landslide event, the difference of raw data and harmonic data should be positive. The proposed method was capable of determining 80% of landslide using CDNDVI and 85% of landslide after detrending CDNDVI (dCDNDVI).[13]

Landslide susceptibility and hazard mapping was catergories in three parts qualitative approach, quantitative approach and semi-quantitative approach [14]. In hilly region road slide landslide was normally a problem due to slope failure and landslide hazard map was used as an effective tool for planning and management of landslide disaster .Analytical hierarchy process was used to prepare of landslide hazard map. Main factors for landslides were slope, aspect, curvature, relative relief, fault density, drainage density, geology, topographic wetness index (TWI), distance from road and lithology. Landslide factors were further divided in to subfactors and weightage are assigned by analytical hierarchy Process [15]

Satellite SAS data processed with persistent scatterer interferometry technique used to update the landslide inventory. For landslides and classified landslides or type of landslides were explored with landslide frequency , soil , slope gradient , slope curvature and elevation etc. [16]

Satellite remote sensed data was very effective to create landslide inventory mapping during disaster risk management. Differential and multi-temporal Synthetic Aperture Radar (SAR) Interferometry can provide ground displacement radar satellite images. Landslide detection and mapping of past and present mass movements can be detected by Object-Based Image Analysis (OBIA) of optical satellite images[17].

An unmanned aerial vehicle (UAV) was used for rainfall data analysis , causative factor , post landslide behaviour and mechanism of landslide. GB-SAR(ground based synthetic aperture radar) early warning can warn 10 minutes earlier for landslides . Combination of UAV and GB-SAR will be more beneficial'[18][19]

Satellite images and convolutional neural network (CNN) can be used to identification and classification of landslide . A new method of distant domain transfer learning (DDTL) were introduced for detection and classification of landslides. In landslide task sense classification of satellite iamge were used . An attention mechanism was added to distant domain transfer learning(DDTL) (AM-DDTL) for extraction of information more accurately from satellite images. Result of AM-DDTL was comared with three existing methods CNN, pretrained model and DDTL. Experimental result shows that DDTL has besster dection themn CNN and AM DDTL has 97% classification accuracy . AM DDTL has 7% more classification accuracy then DDTL. [19]

3. Research and Gap

The Detection of debris flow is challenging as debris flow is high energy transport . Segmentation can also be a challenge and Trial-and-error approach was used to minimize over segmentation and give better results. Limited slopes give poor performance in automated mapping of landslide transport and deposition area and this can be overcome by multiple satellite acquisition geometries [12]. Detection of landslides during snow months is a challenging task so further to reduce the wrong and misclassification masking snow value from time series using normalized difference snow index can be used. [13]. Landslide causing factors play a very important role in detection of landslides .Accuracy of detection can be improved by considering more and more factors that can cause the landslide.

In DDTL algorithm feature extraction of remote sensed data is very expensive and high resolution images are very limited in open source. Google earth images, Gaofen series images are freely available.[19]

Literature survey shows that many methods and technologies were used to classify the landslide and semi automated techniques were used to identify the landslide. Still there are several challenges:

1. Achieve high accuracy for detection and classification of landslides. Different algorithms of machine learning and deep learning can be used but need professional knowledge to select more appropriate variables and dataset.
2. Interpretability is very important in machine learning and the higher the value of interpretability of a machine learning model, the easier it is for someone to comprehend why certain decisions or predictions have been made.

4. Interpretation and observation

An early alarming system for landslides will be helpful for disaster management units that can predict the event of a landslide and save lives by evacuating the predicted area. Main focus of this paper is to initiate techniques to detect and classify landslides by applying artificial intelligence for extracting features from Satellite data and predict the occurrence of landslides. There are basic four steps for Proposed automatic detection of landslide algorithms will contain four main parts : 1. Selection of area and Collection of Landslide Data from Satellite image 2. Pre-processing of images and extraction of features involved in landslides. 3. Training and testing of the algorithm 4. Calculate the result in terms of accuracy in prediction of landslide and accuracy in terms of detecting type of landslide.

The landslides are generally observed in three regions: roadside construction, river banks, hydropower projects. Area of interest in this work can be chosen from more prone area to landslide. Satellite image databases like spaceborne synthetic aperture radar (SAR) and optical remote sensing, airborne light detection and ranging (LiDAR), ground-based SAR and terrestrial LiDAR, Google earth database are considered for images of landslides . Satellite images database can be processed for more clear classification and detection. Machine learning classification algorithms such as Support vector machine, Decision tree, fuzzy based classification algorithms and fusion of algorithms will be used for best classification result.

On the basis of these four stages proposed block diagram can be shown as below:

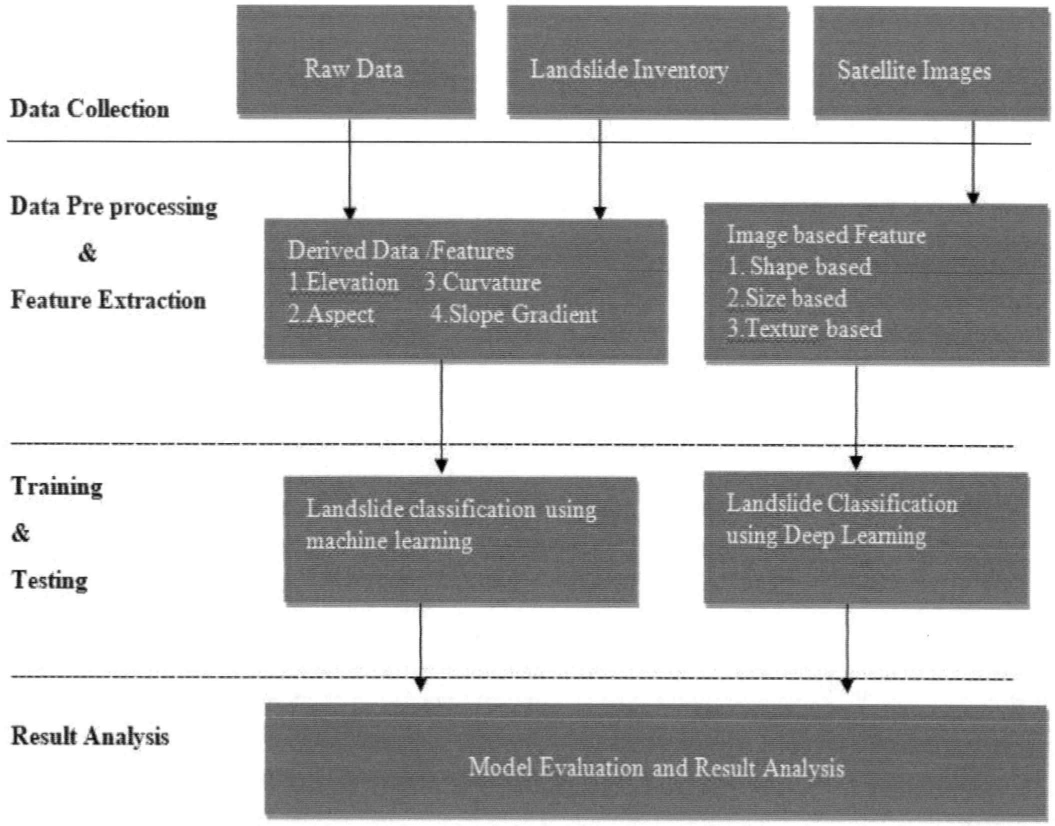

Figure 2 : Block diagram proposed work

5. Conclusion and future scope

Loss of life and infrastructure during disasters like landslides, earthquakes and other natural and manmade calamities is very huge and irreversible leading to poverty. Landslides are one of the most occurring disasters due to natural reasons and men made decisions. Loss of life and property can be reduced if early prediction of landslides can be achieved. This work will be done in maximum capacity to develop the technology which forecasts the prevalence of landslides. Many methods and algorithms will be used for pre-processing landslide data and for predicting different types of landslides. There are several challenges for designing an algorithm for automatic detection of landslides. The major challenge in satellite image processing is image classification and image segmentation. An efficient classification algorithm can be designed to achieve high accuracy in predicting the landslide. Artificial Intelligent algorithms like CNN, Support vector machine, object based classification algorithms were used but still the result for the automatic detection model is not very efficient. Machine learning and deep learning algorithms can be applied in classification and segmentation of satellite images for better performance of automatic detection models for landslide detection and prediction.

References

[1] Preparation of Hazard, Vulnerability & Risk Analysis atlas and report for the state of Himachal Pradesh. Disaster Management Cell, Department of Revenue Government of Himachal Pradesh, Shimla.(2015).1-30.

[2] Mengistu F, Suryabhagavan K V, Raghuvanshi T K and Lewi E (2019). Landslide Hazard zonation and slope instability assessment using optical and InSAR data: a case study from Gidole town and its surrounding areas, southern Ethiopia. *Remote Sensing of Land.* **3,** 1–14. DOI:10.21523/gcj1.19030101

[3] Hungr O, Leroueil S and Picarelli L (2013). The Varnes classification of landslide types, an update. *Landslides* **11**, 167–194. DOI: 10.1007/s10346-013-0436-y

[4] Lazzari M and Piccarreta M (2018) Landslide disasters triggered by extreme rainfall events: The case of Montescaglioso (Basilicata, southern Italy). *Geosciences.* **8,** 377.DOI:10.3390/geosciences8100377

[5] Alcantara-Ayala I(2002) Geomorphology, natural hazards, vulnerability and prevention of natural disasters in developing countries. *Geomorphology*, **47**, 107–124.https://doi.org/10.1016/S0169-555X(02)00083-1

[6] Gokceoglu C and Sezer E (2009). A statistical assessment on international landslide literature (1945–2008). *Landslides,* **6**, 345–351 DOI:10.1007/s10346-009-0166-3

[7] Highland L M and Bobrowsky P (2008). The Landslide Handbook—A Guide to Understanding Landslides, *Geological Survey: Reston*, Circular **1325**

[8] Hungr O, Leroueil S and Picarelli L(2014). The varnes classification of landslide types, an update. *Landslides* **11**, 167–194. https://doi.org/10.1007/s10346-013-0436-y

[9] Guzzetti F, Gariano S L, Peruccacci S, Brunetti M T , Marchesini I, Rossi M, Melillo and M (2020). Geographical landslide early warning systems. *Earth Sci. Rev.* **200**, 1029-73. DOI : 10.1016/j.earscirev.2019.102973

[10] Temme A, Guzzetti F, Samia J, Mirus and B B (2020). The future of landslides' past— a framework for assessing consecutive landsliding systems. *Landslides* [11] Zhao,C. and Lu,Z.,(2018) Remote sensing of landslides-A review.*Remote sensing 10(2)1-6.* https://doi.org/10.1007/s10346-020-01405-7

[11] Heleno S, Matias M, Pina P and Sousa JA (2015).Semiautomated object-based classification of rain-induced landslides with VHR multispectral images on Madeira Island. *Natural Hazards and Earth System Sciences* **16**,1035-48. https://doi.org/10.5194/nhess-16-1035-2016

[12] Deijns A A J, Bevington A R, Zadelhoff F V, Jong S M, Geertsema M and McDougall S (2020) Semi-automated detection of landslide timing using harmonic modelling of satellite imagery, Buckinghorse River, Canada. *International Journal of Applied Earth Observation and Geoinformation* **84.** https://doi.org/10.1016/j.jag.2019.101943

[13] Shano L, Raghuvanshi TK and Meten M.(2020) Landslide susceptibility evaluation and hazard zonation techniques – a review . *Geoenviron Disasters*, **7** ,1–19 https://doi.org/10.1186/s40677-020-00152-0

[14] Panchal S, Shrivastava A K (2022). Landslide hazard assessment using analytic hierarchy process (AHP): A case study of National Highway 5 in India. *Ain Shams Engineering Journal.* **1.** https://doi.org/10.1016/j.asej.2021.10.021

[15] Rosi A, Tofani V, Tanteri L, Stefanelli C T, Agostini A, Catani F and Casagli N (2018). The new landslide inventory of Tuscany (Italy) updated with PS-InSAR: geomorphological features and landslide distribution.*Landslides* **15,** 5–19. https://doi.org/10.1007/s10346-017-0861-4

[16] Casagli N, Cigna F, Bianchini S and Holbling D(2016).Landslide mapping and monitoring by using radar and optical remote sensing: Examples from the EC-FP7 project SAFER .*Remote Sensing Applications: Society and Environment*.**4** ,92-10. DOI:10.3390/rs13091653

[17] Liu B, He K, Han M, Hu X W, Ma G T and Wu M Y (2021) Application of UAV and GB-SAR in Mechanism Research and Monitoring of Zhonghaicun Landslide in Southwest China. *Remote Senssing.,* **13**. DOI:10.3390/rs13091653

[18] Xiang X, Chen J, Wang H, Pei L and Wu Z (2019) PS Selection Method for and Application to GB-SAR Monitoring of Dam Deformation. *Advances in civil engineering.* https://doi.org/10.1155/2019/8320351

[19] Qin S, Guo X, Sun J, Qiao S, Zhang L, Yao J, Cheng Q and ZhangY(2021). Landslide Detection from Open Satellite Imagery Using Distant Domain Transfer Learning. *Remote sensing* **13**.1-23. https://doi.org/10.3390/rs13173383

Wireless sensor networks for forest fire monitoring: Issues and Challenges

Anshika Salaria [1], **Amandeep Singh** [2] & **Kamal Kumar Sharma** [3]
[1,2,3]School of Electronics and Electrical Engineering, Lovely Professional University, Jalandhar, Punjab, India
[1] https://orcid.org/0000-0002-5324-7336 [2] https://orcid.org/0000-0001-5706-760X [3] https://orcid.org/0000-0002-6294-0829

Correspondence: amansandhu6788@gmail.com; kamal.23342@lpu.co.in

Abstract. Forest fires have recently been the most critical issue faced by the world. The huge environmental, economic, and societal damages caused by wildfires are hindrances to social development. The increasing figures of such instances since the last few years make it the need of the hour to start considering it on priority and take appropriate actions. For this, intense knowledge is required about the nature of forest fires, their causes and appropriate technology to be used. Wireless sensor networks have always been a preferable technology in such kind of disaster management and environmental monitoring scenarios. However, there are yet some prominent issues that affect the overall performance of wireless sensor networks, especially in harsh terrains like forests. In this paper, information has been provided about the causes, damages and aftereffects of forest fires. Moreover, a study has been conducted on various articles highlighting the issues and challenges being faced in wireless sensor networks. The paper further provides a simple and easily understandable analysis highlighting the types and priority of challenges. This would benefit the researchers in identifying the current research gaps in the field of wireless sensor networks, especially in applications like forest fire monitoring or environmental monitoring and disaster management.

Keywords Wireless sensor networks, forest fire monitoring, challenges and issues.

1. Introduction

The world is progressing every day technologically, economically and in many other ways. In all these enhancements what is being suffered and ignored is the environment. Somewhere we all are ignoring the fact about the consequences that one would have to face in the future, if the environmental resources are not taken care of and preserved today. No doubt several countries have initiated towards reducing the carbon content present in the atmosphere and have imposed several laws and regulations, but in developing countries like ours, India still has not been able to cope with it completely. According to a reputed newspaper Hindustan Times (Nov 13, 2021), three cities from India were listed in the top 10 most polluted cities in the world. Thus, it is visible that dealing with this issue is the need of the hour in our country. Forests as a resource, reduce the carbon content in the atmosphere, purify air pollution and provide abundant oxygen, hence these are altogether a power-packed solution. Moreover, forests are nature's way of maintaining the ecological balance. They impact the greenhouse gases, maintain the rainfalls, precipitations and prevent soil erosion [1]. On the other end, these valuable resources if encounter fire, very large amounts of carbon gases are emitted back to the atmosphere. Therefore, planting trees is not the only option to deal with the scenario in fact preventing forest fires should be treated equally important.

Damages from forest fires are well known to the world now. The world prayed together when the Australian bushfire happened, observing all the environmental and economic losses. In fact, the burning of forests starts damaging the surrounding from the first instance itself by emitting smoke and other gases into the environment. Prevailing after-effects like infertility of soil results in barren lands for ages. Ash from such fires depletes much of the nutrients from the soil, which results in less growth of flora and leads to soil erosion. Hence, it leads to disasters like landslides and floods [2]. The loss of timber and medicinal herbs due to fires results in economic losses, affecting the health and livelihood of people who depend on forests and live in the vicinity. In order to deal with this forest fires problem, an awareness of the factors responsible for such fires is required.

Forest fires can be natural or human-made. Naturally, fire can be ignited either by lightning or due to the ignitions in dry combustible materials present in forests during summers or very high temperatures while the human-made fires may be a result of carelessness, an intentional clearing up of area by burning it, leaving burning cigarettes behind, ignoring campfires, etc [1]. Therefore, a check should be kept on forest areas that are nearby human residences or tourist spots and serious laws and fines should be imposed on such human activities. The forests are not very comfortable terrains to be looked upon or monitored by human resources. Also, the major drawback of fires is that it is observed in areas like forests, only when it has spread to very large extent [1]. We cannot do much about the natural causes of forest fires but appropriate procedures should be adopted to detect such fires at the earliest in order to prevent huge damage. Though there were many such procedures adopted traditionally but somehow, they are not competent enough.

Such traditional approaches for detecting forest fires included human monitoring by watchtowers, image acquisition[3]–[5] or even satellite imaging and which is even followed today, but each method had severe drawbacks like latency, inefficiency, cost, and greater power consumption, etc [2].

Therefore, this problem should be addressed with some recent efficient technologies. Wireless sensor networks have recently been a field that is most suitable for disaster management. It has been considered a trustworthy solution for monitoring forest fires [1]. Due to its infrastructure-less nature and other useful characteristics, it becomes comparatively convenient to be used in harsh terrain areas. Therefore, a wireless sensor-based system for detecting forest fires would be a promising solution. However, the field of such networks has not been explored yet to the maximum. In order to make an efficient optimum system for detecting forest fires using wireless sensor networks, it is essential to identify the issues and challenges in wireless sensor networks at present. After that, proper solutions can be imparted with this technology to solve our purpose.

In this paper, we have surveyed articles from recent years and emphasized on

i) Identification of various challenges/issues encountered while working with wireless sensor networks.

ii) Determine the most prominent challenges as research gaps.

Since wireless sensor networks have proved to be one of the best technologies for such applications, this contribution would be helpful for researchers interested in solving forest fire, disaster management, or other environmental issues. Thus, the paper would be able to highlight various research gaps in the field for such applications and their possible solutions. The paper is structured as: in section two literature survey of state-of-the-art methods is presented. The section three presents conclusive analysis of current techniques. The section four and five presents discussion and conclusion of topic respectively.

2. Literature survey

This paper highlights various challenges being faced in wireless sensor networks by exploring relevant articles from reputed journals. A table has been prepared listing research article, challenges identified and major contributions by the authors of the articles. The challenges have been identified to highlight the research gaps faced in current methods used by researcher's worldwide to provide the early detection of forest fire. The article in [1] discusses behavior analysis of forest fires and proposes a system for the early detection. Out of two prime challenges, first is limited battery power of the nodes and the second is limited range of sensors. According to the author using wireless sensor networks in forests makes it almost impossible to replace the batteries of nodes frequently. The sensing range of sensors is an issue as smaller ranges would cover smaller areas whereas larger ranges would result in delay and inaccuracy also increasing the cost and size of node. Various machine learning software defined wireless sensor networks have been surveyed in [6]. Their current status and challenges faced are also described. A particular section is dedicated to challenges in wireless sensor networks. The challenges are broadly classified under three classes namely: Sensor node hardware, heterogeneity, and flexibility. Sensor node hardware consists of issues like limited battery of nodes especially in environments that are harsh, replacing or changing the batteries frequently is not an option. The limited size of memory and speed of computation are limitations as an enhancement in these factors would increase the size and cost of the nodes which is inappropriate f or many applications. The heterogeneity is referred to as different types of sensors, nodes and other devices in the network which cause an issue of interconnectivity both in terms of hardware interfacing and protocols. Inflexibility has been identified as a challenge related to the scalability of the network. Any increment or decrement of the number of nodes as per requirement presents many difficulties in terms of software. Making protocols and algorithms which would adapt to these modifications is a very tough task, thus identifying inflexibility as a challenge. A review on the application, challenges and future of wireless sensor networks in IoTs is presented in the article number [7] .This article has identified interoperability, coverage, energy, security and privacy as major challenges. Some possible solutions have also been proposed with respect to these challenges. Similarly, a review has been made on the state of the art models of localizations and several challenges and research gaps have been identified regarding the same [8]. Considering localization as the primary challenge several factors responsible like nonuniform deployment of nodes, noisy communication, RSSI, network scalability and non-linearity between coordinates have also been highlighted. Various algorithms, techniques and challenges of localization in wireless sensor networks have been analyzed in [9] . The challenges identified were that of localization, data transmission energy, data communication, security, fault tolerance out of which localization is considered as the major challenge. In article [10] a systematic literature survey was conducted regarding reliable, secure and energy-efficient routing in wireless sensor networks. Various challenges were identified, first and foremost being that of limited power. According to the authors where numerous works have been done regarding enhancing the lifetime of networks, there still lies the hardware market of the batteries that has not yet progressed significantly making battery life a permanent issue in sensors. Most of the surveillance applications require longer battery life. Secure routing is again being ignored as research area where as emphasis is given to performance related issues, therefore secure routing is also an issue. The network also faces fault intolerance as most of the times either the nodes are damaged or die out due to energy depletion affecting the reliable routing and overall performance of network. These factors also make the network uncertain which itself is an issue. A study has been conducted on reliability of mobile wireless networks in article [11]. The different techniques, their merits and demerits are discussed. Energy consumption, reliability and routing are considered as major issues. Article [12] presents localization as a major challenge. Routing and energy consumption have been identified as other challenges specially in applications dealing with irregular terrains. IoTs and wireless sensor networks are together analysed in article [13] in terms of methodologies, research challenges and applications. Energy, architecture, node deployment, security, quality of service, medium access control, data collection and transmission were identified as challenges. In the same way a result oriented analysis of various deployment schemes and various research issues have been discussed in article [14]. Power consumption, deployment, localisation, routing and limited storage are identified as present challenges in wireless sensor networks. A survey again has been conducted about the issues and challenges in wireless sensor networks in [15]. Therefore, the survey clearly indicates that limited energy, deployment, hardware, software, harsh environment conditions, heterogeneity, redundant data are major issues and challenges in wireless sensor networks, in recent times.

3. Analysis

The analysis shows that though tremendous amount of work is being done in the field of wireless sensor networks, some issues have not yet been resolved to desired level. The issues of limited power backup, node deployment, localization, routing, communication between different devices in the network and hardware of the nodes and other devices involved in the network are still problems. And it is important to work upon them as these challenges would be more prominent when considered in harsh terrains like forests. Therefore, while designing a system for forest fire detection it is important to make the system efficient by first dealing with these issues. Thus, this analysis would be able to highlight different challenges identified in recent years by researchers. For simplifying the analysis, a table has been prepared for identifying the type of challenges. The table consists of the reference numbers of the articles reviewed, challenges identified in respective articles and major contributions made by the authors in them.

Table 1: A summary of various challenges identified in WSNs.

S. No	Reference No	Year	Major Challenges Identified						Contributions
			Topology/ position related (localization/topology /coverage/ architecture)	Energy efficiency (network lifetime/ Limited power)	Range (sensing /communication range)	Routing	Communication (Data transmission, RSSI, Hop count,)	Hardware (Memory space/computational speed/size/ heterogeneity)	
1	[6]	2022	No	**Yes**	No	No	**Yes**	**Yes**	Reviewed various machine learning techniques and identified various challenges in WSNs
2	[7]	2022	**Yes**	**Yes**	No	No	No	**Yes**	Reviewed various papers and identified various challenges and solutions in WSNs
3	[1]	2021	No	**Yes**	**Yes**	No	No	No	Behavior analysis of forest fire.
4	[8]	2021	**Yes**	No	No	No	No	**Yes**	Identified localization as a major challenge and conducted review on various models.
5	[9]	2021	**Yes**	**Yes**	No	No	**Yes**	No	Conducted algorithm analysis and identified challenges in WSNs
6	[10]	2021	No	**Yes**	No	**Yes**	No	**Yes**	Conducted literature survey and identified challenges
7	[11]	2021	**Yes**	**Yes**	No	**Yes**	No	No	Identified the challenges of Mobile WSNs and analyzed various methods
8	[12]	2020	**Yes**	No	No	No	**Yes**	No	Identified challenges in irregular terrains and conducted Comparative analysis of various localization techniques
9	[13]	2020	**Yes**	**Yes**	No	No	No	No	Analyzed different methodologies, applications and research challenges in WSNs.
10	[14]	2020	**Yes**	**Yes**	No	**Yes**	No	**Yes**	A result-oriented analysis of wireless sensor networks deployment schemes
11	[15]	2018	**Yes**	**Yes**	No	No	No	**Yes**	Conducted a survey and identified challenges and issues in WSNs

For simplicity, similar types of issues are considered under one in the following fashion: All the node position-based issues like locating the node (localization), allocating the node (deployment), specific area covered by node (coverage) and architecture are placed under "topology". All the issues related to network lifetime or battery backups are placed under "energy efficiency". The sensing range of sensors as well as the range of transferring information or communication range are together considered under "range" issues. The routing issues are classified under "routing" itself. A separate category "communication" has been made for all issues related to communication such as data transmission, Received Signal Strength Indicator (RSSI), hop counts etc. The issues dealing with all sort of hardware, be it the size of nodes, memory size, computational speeds of processors, or heterogeneity of the devices are altogether considered under "Hardware".

The analysis clearly indicates the frequency of challenges encountered. Where energy and topological challenges were encountered very frequently the range, routing, communication and hardware issues were not identified frequently. However, range and routing can also be broadly considered under communication due to their relevancies to the latter. Also range and routing together form the basis of communication. Therefore, in order to get significant values of these similar issues, the types of challenges are again categorized into four as shown in figure 1. It is noteworthy that the most relevant challenges are broadly classified under one section.

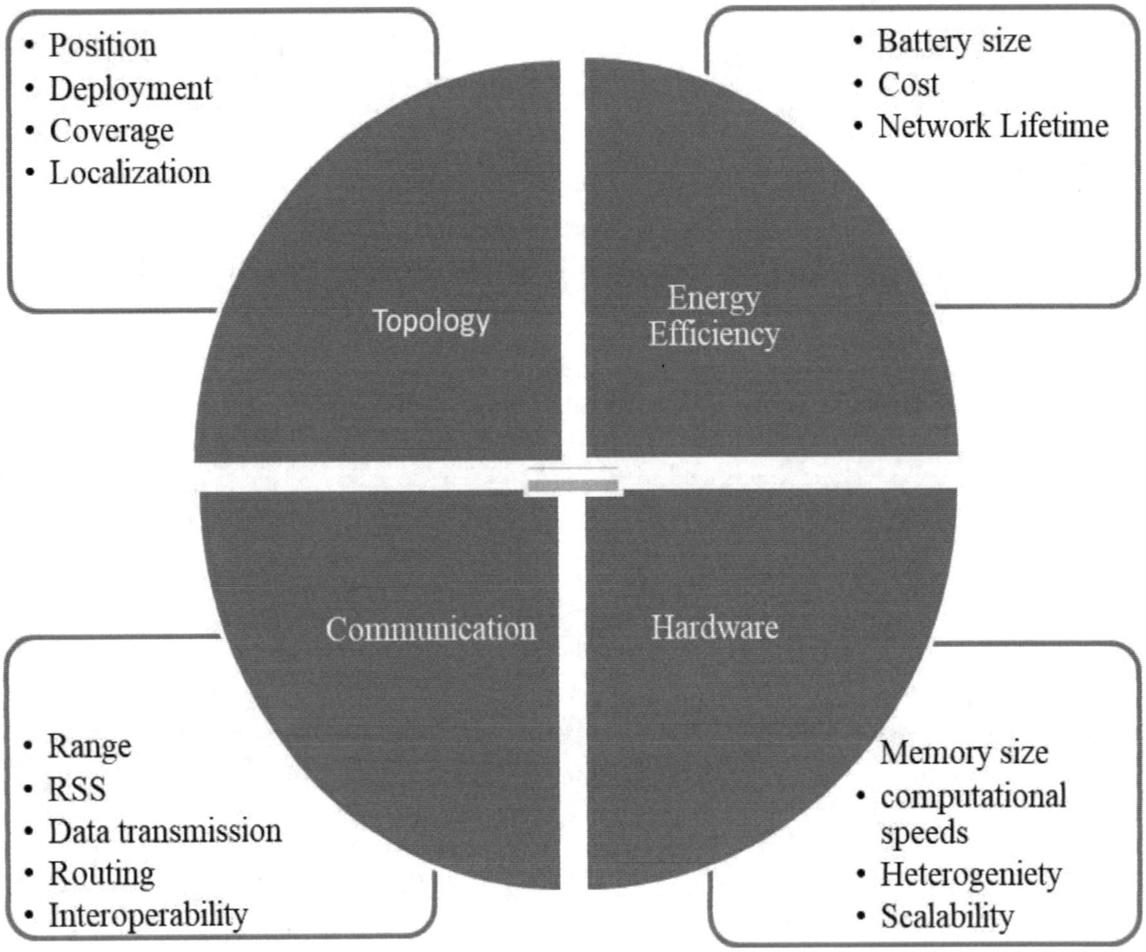

Figure 1: Similar types of relevant challenges broadly classified into four categories.

On the basis of this, a pie chart has been prepared indicating the number of times a challenge was encountered in the survey. This would be helpful in deciding the most prominent challenges and hence the research gaps in wireless sensor networks.

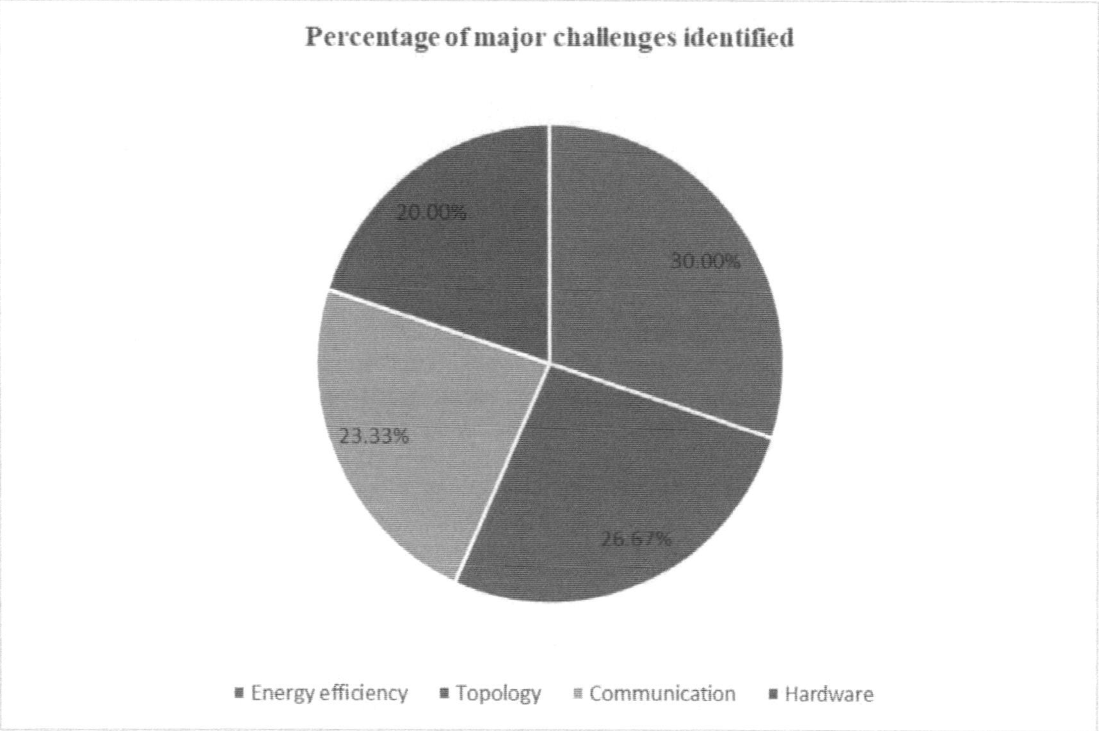

Figure 2: Percentage of challenges as per survey

4. Discussion

As per the pie chart, energy efficiency has been the most prominent issue being encountered 30% times in overall challenges. It is prominent because it is referred as the most important criteria in wireless sensor networks that would decide the actual lifetime of the network. The limited sized batteries in sensor networks no doubt present challenges. Considering the hardware part of batteries, the battery industry has not been growing significantly thereby maintaining energy efficiency as an issue [10]. Next are the topological related issues which count for 26.67 percent of the times in overall challenges. Topology deals with arrangement of nodes in the network such that they can provide efficient coverage as well as communication. Therefore, the deployment in wireless sensor networks is considered a critical issue as it affects the coverage, connectivity and many other factors in the network [14]. Another factor dealing with topology is coverage. It is also one of the prominent criteria of measuring the performance of a network. It indicates how well the deployed sensors are able to perform surveillance or monitor the given area [16].In practical geological disaster management and wildlife monitoring the terrains are not very smooth. Such irregular terrains present issues of coverage. It thus makes it a challenge to be explored about what positions should be chosen for the sensors to cover the maximum areas [17].Therefore topology still remains one of the critical issues to be worked upon. About 23.33 % of the section of the pie is covered by communication challenges. Communication related issues are more prominent in harsh environments like forests. Signal propagation in forests has been an obvious issue. Forest terrains are hostile to signal propagation. Consisting of scattered leaves, branches, tree barks, and bushes of different heights, distribution of dense as well as not so dense vegetations along with the habitat makes it a very random medium thereby resulting in absorption, scattering, and diffraction of radiations. Therefore, severe fading is encountered for radio signals in wildlife or typical forest environment making communication a major issue in such applications [18]. Remaining 20% of the challenges are about hardware. Almost all the issues encountered in the survey are related to hardware of the network. In general, the applications like disaster management, forest fire monitoring or other environmental monitoring requires sensor nodes of limited sizes. Such devices with smaller hardware would be easy to carry and install in not so easily accessible areas like forests. These limited sizes of the devices further result in limited battery, performance, computations, range, limited communication and other such issues [15].These challenges identified, clearly forms the basis of research gaps when one desires to work on forest fire monitoring or other environmental applications of wireless sensor networks. These research gaps need to be mitigated for the optimum utilization of wireless sensor networks.

5. Conclusion

Forests being the most vital resources play a very important role in our environment. The paper provides a detailed discussion on the role of forests, forest fires, their causes, impacts and damages. Since the number of forest fire incidences have increased over the years, the need of early forest fire detecting systems have been highlighted in detailed manner. Considering the irregular terrains of forests, technology of wireless sensor networks emerges as one of the most promising techniques for making the desired systems. However, various issues have been encountered in last few years in the fields of wireless sensor networks, especially for applications in irregular terrains. These challenges are required to be worked upon in order to make efficient forest fire monitoring system. The tabular survey presented in section three indicates the most prominent challenges in wireless sensor systems. Further, the percentage of challenges have been analyzed out of which energy efficiency and topology are obtained as the highest priority issues. This identification of high priority issues can be considered as research gaps in wireless sensor networks. These issues even become more important to consider while working on applications in harsh or irregular terrains. Therefore, better system performance can be achieved by involving highest priority issues in development stage of forest fire management system.

References

[1] Alkhatib A A A, Abdelal Q and Kanan T 2021.Wireless Sensor Network for Forest Fire Detection and behaviour Analysis. *International Journal of Advances in Soft Computing and its Applications* **13**, 82-104

[2] Dampage U, Bandaranayake L, Wanasinghe R, Kottahachchi K, and Jayasanka, B 2022. Forest fire detection system using wireless sensor networks and machine learning. *Scientific. Report*, **12**, 1–11.doi: 10.1038/s41598-021-03882-9.

[3] Singh A, Sethi G, and Kalra G S 2020.Spatially Adaptive Image Denoising via Enhanced Noise Detection Method for Grayscale and Color Images. *IEEE Access*, **8**, pp. 112985–113002, doi: 10.1109/ACCESS.2020.3003874.

[4] Singh A, Sethi G, and Kalra G S 2022. Amalgamation of ROAD-TGM and Progressive PCA Using Performance Booster Method For Detail Persevering Image Denoising. *Multimed. Tools Appl* **81** pp. 1719– 42. doi: 10.1007/s11042-021-11426-6.

[5] Singh A, Gaba G S and Hedabou M 2021.Robust and effective image preprocessing conglomerate method for denoising of both grayscale and color images. *J. Electron. Imaging*, **31**, pp. 1–18. doi: 10.1117/1.JEI.31.4.041203.

[6] Jurado-lasso F F, Marchegiani L, Jurado J F, Abu-mahfouz A M, and Fafoutis X, A survey on Machine Learning Software-Defined Wireless Sensor and major challenges 2022 . *IEEE Access*, 1-34 doi: 10.1109/ACCESS.2022.3153521.

[7] Jamshed M A , Ali K, Abbasi Q H, Imran M A , and M. Ur-Rehman 2022.Challenges, Applications and Future of Wireless Sensors in Internet of Things: A Review. *IEEE Sens. J.* doi: 10.1109/jsen.2022.3148128.

[8] Kagi S and Mathapati B S 2021. Localization in Wireless Sensor Networks: A Compact Review on State-of-the-Art models. *Proc. 6th Int. Conf. Inven. Comput. Technol.* (Coimbatore, India: IEEE) pp 5–12. doi: 10.1109/ICICT50816.2021.9358793.

[9] Kaur B and Prashar D, 2021. Localization in Wireless Sensor Network : Techniques , Algorithms Analysis and Challenges .*9th Int. Conf. on Reliability, Infocom Technologies and Optimization* (Noida,India: IEEE) pp 1–7. DOI**:** 10.1109/ICRITO51393.2021.9596135

[10] Biradar M, and B. Mathapathi. 2021Secure , Reliable and Energy Efficient Routing in WSN : A Systematic Literature Survey *Int. Conf. on Advances in Electrical, Computing, Communication and Sustainable Technologies (ICAECT)*(Bhilai,India:IEEE) pp1-13, doi: 10.1109/ICAECT49130.2021.9392561.

[11] S Sachan S, Sharma R and Sehgal A 2021. An Extensive Learning on the Reliability of Mobile Wireless Sensor Networks. *2021 5th Int. Conf. on Information Systems and Computer Networks (ISCON)*(Mathura, India: IEEE) pp 1–5. doi: 10.1109/ISCON52037.2021.9702503.

[12] Bhat S J and Santhosh K V 2020. Localization of Wireless Sensor Networks in Irregular Fields a Challenge. *Wireless Personal Communications*, **114**, 2017–2042. doi: 10.1007/s11277-020-07460-6.

[13] Sharma R ,Roy P and Prakash S 2020. Methodology , Applications , and Challenges of WSN-IoT. *Int. Conf. on Electrical and Electronics Engineering(ICE3)*(Gorakhpur,India : IEEE) pp 502–07. DOI**:** 10.1109/ICE348803.2020.9122891

[14] Priyadarshi R, Gupta B and Anurag A 2020.Wireless Sensor Networks Deployment: A Result Oriented Analysis. *Wirel. Pers. Commun.*, **113**, 843–866. doi: 10.1007/s11277-020-07255-9.

[15] Bala T, Bhatia V, Kumawat S and Jaglan V 2018. A survey : issues and challenges in wireless sensor network.*International Journal of Engineering & Technology* **7** 53–55.

[16] Yakıcı E and M. Karatas M. 2021. Solving a multi-objective heterogeneous sensor network location problem with genetic algorithm.*Comput. Networks*, **192**. doi: 10.1016/j.comnet.2021.108041.

[17] Lu C, Li X, Yu W, Zeng Z, Yan M and Li X. 2021. Sensor network sensing coverage optimization with improved artificial bee colony algorithm using teaching strategy. *Computing*, **103**, 1439–1460. doi: 10.1007/s00607-021-00906-0.

[18] Ferreira A E , Ortiz F M ,T. T. de Almeida T T and L. H. M. K. 2020. Costa A visitor assistance system based on lora for nature forest parks. *Electronic*. **9**, 1–24.

Charbonnier-Marchaud Based Fractional Variational Model for Motion Estimation in Multispectral Vision System

Pushpendra Kumar and Muzammil Khan

Department of Mathematics, Bioinformatics and Computer Applications
Maulana Azad National Institute of Technology Bhopal, Bhopal-462003, India

E-mail: {pkumarfma,mk.193104002}@manit.ac.in

Abstract. As we are aware that motion estimation is an active and challenging area of vision system, which leads to the applications of computer vision. In general, motion detection and tracking in the image sequence (video) is carried out based on optical flow. In the recent-past, researchers have made a significant contribution to the estimation of optical flow through integer order-based variational models, but these are limited to integer order differentiation. In this paper, a nonlinear modeling of fractional order variational model in optical flow estimation is introduced using the Charbonnier norm, which can be scaled to integer order L_1-norm. In particular, the variational functional is formulated by the inclusion of a non-quadratic penalty term, regularization term and the Marchaud's fractional derivative. This non-quadratic penalty provides effective robustness against outliers, whereas the Marchaud's fractional derivative possesses a non-local character, and therefore is capable to deal with discontinuous information about texture and edges, and yields a dense flow field. The numerical discretization of the Marchaud's fractional derivative is employed with the help of Grünwald–Letnikov fractional derivative. The resulting nonlinear system is converted into a linear system and solved by an efficient numerical technique. Several experiments are performed over a spectrum of datasets. The robustness and accuracy of the proposed model are shown under different amounts of noise and through detailed comparisons with the recently published works.

Keywords- Charbonnier norm, Image sequence, Marchaud fractional derivative, Optical flow, Variational methods.

1. Introduction

Motion estimation is known as an active and dominating area of computer vision and its recovery from a sequence of images (videos) leads to many real-time applications such as video compression and surveillance, super–resolution, robot navigation and fluid dynamics [1–4]. Motion estimation is generally performed with the help of optical flow. Optical flow is a $2D$ planar displacement vector field $\boldsymbol{u} = (u, v)^T$, which links the corresponding pixel locations in an image sequence and provides the motion information about objects in a scene [5].

Therefore, it is categorized into the class of correspondence problems in vision system. Several methods have been presented in the literature for optical flow estimation, however, differential variational methods are considered amongst the most successful approaches [5,6]. This is due to the modeling of the problem, simulation, and accuracy in the estimated results [7]. A significantly long chain of works has been formed starting from the pioneering works [8, 9]. These differential models are formulated solely with two assumptions: (1) BCA (brightness constancy assumption) and (2) smoothness assumption.

BCA results in a data term that tracks a pixel between consecutive image frames by assuming that the pixel brightness remains unchanged through one frame to another while the smoothness assumption is intended to provide a coherent motion region and preserve the motion boundary. The accurate optical flow estimation problem is a hard nut to crack due to several factors such as illumination change, occlusion, noise, and other outliers [10]. Generally, the capability of an optical flow model in preserving the flow discontinuity, large motion and accuracy is the basis of its potential assessment. Some of the models which are proposed recently such as [11] have considered convex robust data terms and some additional constraints for regularization in the inclusion of Charbonnier and Lorentzian norms. These models achieved a motivating level of performance against outliers and provide the global minima. The work presented in [12] employed total variational (TV) regularization and L_1-norm in the estimation. In addition to it, few models have also been presented based on TV regularization, L_1 and L_2-norms [13]. All these models have shown a significant amount of improvement in the optical flow estimation against outliers. However, these are all established on the basis of integer order differentiation, which is limited to its integer order. An improvement in the performance of the integer order variational models can be obtained by generalizing their differential order [14]. The objective of this work is to generalize the existing variational techniques from integer order to fractional order and provide the increased robustness against outliers, and furnish the dense and discontinuity preserving optical flow. The robustness of the model is achieved by using the nonlinear Charbonnier norm, while dense and discontinuity preserving optical flow is obtained by the fusing the variational functional of Horn et al. [8] and Nagel et al. [5]. In image processing applications, the fractional order based techniques attain a keen interest from the researchers [15]. Nevertheless, a minuscule amount of work in optical flow estimation has been made based on fractional order [16]. Despite all of these worthy contributions, no work is presented based on Charbonnier norm and Marchaud's fractional derivative in the estimation of optical flow.

Contribution

This paper proposes a novel nonlinear fractional order variational model in the estimation of optical flow. The mathematical modeling of the model is introduced with the help of Charbonnier norm and Marchaud's fractional derivative. The novelty of the proposed model is in four-fold: (i) It can be scaled to an integer order L_1-norm based model. (ii) It provides increased robustness against outliers due to the inclusion of the non-quadratic Charbonnier norm. (iii) It furnishes a dense flow field which is discontinuity preserving due to the amalgamation of the models of Horn et al. [8] and Nagel et al. [5]. (iv) It allows to generalize the existing integer order variational models. The Marchauds fractional derivative is discretized using Grünwald-Letnikov fractional derivative for numerical implementation purpose. The nonlinear system is converted into a linear system by using an outer fixed point iteration scheme. Various experiments are performed on different spectrum datasets, even if the ground truth flow is not available. The comparisons of the results with recently published works are demonstrated in terms of AAE and WE. The robustness of the model is also shown when images suffer from Gaussian noise.

2. Proposed fractional order variation model

In order to perform the estimation of optical flow through a variational model, we proposed the following model by incorporating the Charbonnier norm and Marchaud's fractional derivative as

$$E(\boldsymbol{u}) = \int_\Omega [\psi(\nabla \mathfrak{I}^T \boldsymbol{u} + \mathfrak{I}_t) + H_1(u, v) + H_2(D_+^\alpha u, D_+^\alpha v)]\, d\boldsymbol{X} \tag{1}$$

where, $\psi(z) = \delta\sqrt{1 + (\frac{z}{\delta})^2}$ is the Charbonnier norm for any variable z and δ is a small positive parameter. In the proposed model, $H_1(u, v) = \beta^2 \boldsymbol{u}^T \boldsymbol{u}$ and $H_2(D_+^\alpha u, D_+^\alpha v) = \lambda e^T (D_+^\alpha \boldsymbol{u}^T \odot D_+^\alpha \boldsymbol{u}^T) e$, where $\beta, \lambda \in \mathbb{R}^+$ and $D_+^\alpha z = (D_x^\alpha z, D_y^\alpha z)^T$. Here, $D_x^\alpha z$ and $D_y^\alpha z$ represent the Marchaud's derivative of z of order $\alpha \in (0, 1)$ with respect to x and y respectively. The notation $\odot$ represents the Hadamard product while e denotes an all-ones column vector. It can be seen that when δ tends to zero, the above model (1) reduces to L_1-norm based model, whereas for $\alpha \in \mathbb{Z}^+$, generalizes the existing integer order models.

In order to estimate the optical flow $\boldsymbol{u}$, we have to minimize the variational functional (1). Therefore, let us consider $\mathfrak{I}$, $\boldsymbol{u}$ and arbitrary functions $\eta \equiv \eta(x, y)$ and $\phi \equiv \phi(x, y)$ as the schwartz functions. Thus, the optical flow estimation can be defined with respect to the desired function $\boldsymbol{u}^* = (u^*, v^*)^T$ as

$$\boldsymbol{u} = \boldsymbol{u}^* + \epsilon\Phi \tag{2}$$

where, $\Phi = (\eta, \phi)^T$ and $\epsilon \in \mathbb{R}$. The Marchaud's derivative of the expression (2) is defined as

$$D_+^\alpha \boldsymbol{u}^T = D_+^\alpha \boldsymbol{u}^{*T} + \epsilon D_+^\alpha \Phi^T \tag{3}$$

On using the definition of ψ together with (2) and (3) into the variational functional (1), we obtain the following scalar valued function in ϵ

$$E = \int_\Omega \Bigg[\delta \sqrt{1 + \left(\frac{\nabla \mathfrak{I}^T (\boldsymbol{u}^* + \epsilon\Phi) + \mathfrak{I}_t}{\delta} \right)^2} + \beta^2 (\boldsymbol{u}^* + \epsilon\Phi)^T (\boldsymbol{u}^* + \epsilon\Phi) + \lambda e^T [(D_+^\alpha \boldsymbol{u}^{*T} + \epsilon D_+^\alpha \Phi^T) \odot (D_+^\alpha \boldsymbol{u}^{*T} + \epsilon D_+^\alpha \Phi^T)] e \Bigg] d\boldsymbol{X} \tag{4}$$

where $E \equiv E(\epsilon)$. Now, for finding the extreme value of the energy functional E, (4) is differentiated with respect to ϵ and then set $\epsilon = 0$ as per the theory given in [17], therefore

$$E'(0) = \int_\Omega \left[\left\{ \frac{(\nabla \mathfrak{I}^T \boldsymbol{u}^* + \mathfrak{I}_t)(\nabla \mathfrak{I}^T \Phi)}{\psi(\nabla \mathfrak{I}^T \boldsymbol{u}^* + \mathfrak{I}_t)} \right\} + 2\beta^2 \boldsymbol{u}^{*T} \Phi + 2\lambda \{ e^T (D_+^\alpha \boldsymbol{u}^{*T} \odot D_+^\alpha \Phi^T) e \} \right] d\boldsymbol{X} \tag{5}$$

on using Fubini's theorem [18] and the Marchaud's derivative [19] in the third term of (5), we obtain

$$\int_\Omega e^T (D_+^\alpha \boldsymbol{u}^{*T} \odot D_+^\alpha \Phi^T) e\, d\boldsymbol{X} = \int_\Omega [\{ (D_-^\alpha D_+^\alpha)^T e \} \boldsymbol{u}^* \odot \Phi]\, d\boldsymbol{X} \tag{6}$$

Thus, the Euler-Lagrange equations corresponding to (4) can be obtained using (6) into (5) and setting $E'(0) = 0$, and using the concept of calculus of variation in the resulting equation, we finally obtain

$$\frac{(\nabla\mathfrak{I}^T\boldsymbol{u}^* + \mathfrak{I}_t)(\nabla\mathfrak{I})}{\psi(\nabla\mathfrak{I}^T\boldsymbol{u}^* + \mathfrak{I}_t)} + 2\beta^2\boldsymbol{u}^* + 2\lambda\{(D_-^\alpha D_+^\alpha)^T e\}\boldsymbol{u}^* = O_{2\times 1} \tag{7}$$

where, $D_-^\alpha D_+^\alpha = (D_{-x}^\alpha D_x^\alpha, D_{-y}^\alpha D_y^\alpha)^T$ and D_{-x}^α and D_{-y}^α are the Marchaud's derivatives from the left with respect to x and y, respectively, $O_{2\times 1}$ represents a null vector of order 2×1 and $\boldsymbol{u} = (u, v)^T$. Hence, the optical flow can be obtained for each pixel position using the system of equations (7).

3. Numerical Solution

3.1. Derivative discretization scheme

Since the optical flow components have the same dimensions as those in the reference image size $m \times n$. So, u and v are discretized as follows

$$\boldsymbol{u} \equiv \boldsymbol{u}(i,j) = \boldsymbol{u}(i\Delta h, j\Delta h)$$

where, (i,j) denotes the pixel location in the given image and Δh is the mesh grid size which is equal to 1 in the operations of image processing.

Now, we discretize the Marchaud's fractional derivative of order α of the optical flow $\boldsymbol{u}$ using the definition of Grünwald-Letnikov derivative as

$$D_+^\alpha \boldsymbol{u}^T = \sum_{k=0}^{\infty} \omega_k^{(\alpha)} E_{\boldsymbol{x}}^{-k}\boldsymbol{u}^T \tag{8}$$

where, $E_{\boldsymbol{x}}^{-k} = (E_x^{-k}, E_y^{-k})^T$ and E_x^{-k}, E_y^{-k} are the shift operators in x and y directions with -k steps from the position (i,j). The notation $\omega_k^{(\alpha)} = (-1)^k\binom{\alpha}{k}$ is the coefficient in the expansion of $(1-z)^k$. On using the definition of the Left Marchaud's fractional derivative on relation (8), we have

$$D_-^\alpha D_+^\alpha \boldsymbol{u}^T = -\sum_{k=-\infty}^{0} \omega_{|k|}^{(\alpha)} \nabla\left(E_{\boldsymbol{x}}^{-k}\boldsymbol{u}^T\right) - \sum_{k=0}^{\infty} \omega_k^{(\alpha)} \nabla\left(E_{\boldsymbol{x}}^{-k}\boldsymbol{u}^T\right) \tag{9}$$

where, $\nabla\left(E_{\boldsymbol{x}}^{-k}\boldsymbol{u}^T\right) = E_{\boldsymbol{x}}^{-k}\boldsymbol{u}^T - (R\boldsymbol{u})^T$ and R is a linear operator defined by $R\boldsymbol{a} = [\boldsymbol{a}\ \boldsymbol{a} \ldots \boldsymbol{a}]$ with $\boldsymbol{a} \in \mathbb{R}^n$. Now, pre-multipying both sides in equation (9) by e^T and then approximating these expression with image processing point of view, we have

$$\{(D_-^\alpha D_+^\alpha)^T e\}\boldsymbol{u} \approx \sum_{(\bar{i},\bar{j})\in\wp(i,j)} \omega_{p_{ij}}^{(\alpha)}\{\boldsymbol{u}(i,j) - \boldsymbol{u}(\bar{i},\bar{j})\} \tag{10}$$

where, $\wp$ is the set of all the pixels in the neighborhood of pixel position (i,j) in x and y directions and $p_{ij} = \max\left[|\bar{i} - i|, |\bar{j} - j|\right]$.

The following discretized system of iteration equations is obtained for each optical flow component on using the Grünwald-Letnikov derivative discretization scheme (10) into the system (7)

$$\boldsymbol{u}^{(n+1)} = \mathcal{T}^{(n)}\left[2\lambda \sum_{(\bar{i},\bar{j})\in\wp(i,j)} \omega_{p_{ij}}^{(\alpha)}\bar{\boldsymbol{u}}^{(n)} - \frac{1}{\psi(\nabla\mathfrak{I}^T\boldsymbol{u}^{(n)}+\mathfrak{I}_t)}(\mathfrak{I}_{xy}\boldsymbol{u}_{ud}^{(n)}+\mathfrak{I}^t)\right] \tag{11}$$

where, $\boldsymbol{u}_{ud} = (v(i,j), u(i,j))^T$, $\mathcal{T}^{(n)} = (\mathcal{T}_u^{(n)}, \mathcal{T}_v^{(n)})^T$,
$\mathcal{T}_u^{(n)} = \left(\frac{1}{\psi(\nabla\mathfrak{I}^T\boldsymbol{u}^{(n)}+\mathfrak{I}_t)}\mathfrak{I}_{xx} + 2\beta^2 + 2\lambda\sum_{(\bar{i},\bar{j})\in\wp(i,j)}\omega_{p_{ij}}^{(\alpha)}\right)^{-1}$,
$\mathcal{T}_v^{(n)} = \left(\frac{1}{\psi(\nabla\mathfrak{I}^T\boldsymbol{u}^{(n)}+\mathfrak{I}_t)}\mathfrak{I}_{yy} + 2\beta^2 + 2\lambda\sum_{(\bar{i},\bar{j})\in\wp(i,j)}\omega_{p_{ij}}^{(\alpha)}\right)^{-1}$. This nonlinear system of equations is converted to a linear system by using an outer fixed point iteration technique. The resulting equations are now solved with the help of Gauss-Seidel method [27].

Figure 1. Sample reference images from the various datasets.

4. Experiments, results and discussions

4.1. Datasets

The performance of the presented model is analyzed on a variety spectrum datasets. The experiments are processed on the following datasets:

- **Synthetic dataset:** Venus, Grove, Yoesmite [20, 21].
- **Underwater dataset:** Tortoise [22].
- **Thermal dataset:** Mans [23].
- **Real dataset:** Heart [24].

The reference and target frames of the above datasets are illustrated in Fig. 1. A comprehensive discussion upon these datasets is available in [20–24].

4.2. Performance measure:-

In this paper, the performance evaluation of the algorithm is carried out by the means of average angular error (AAE) [24] for the datasets which have ground truth (GT) flow field available and by Warping error (WE) for the datasets with no GT flow available. These are defined as follows:

- $AAE = \frac{1}{MN}\sum \arccos\left(\frac{u_C u_E + v_C v_E + 1}{\sqrt{u_C^2+v_C^2+1}\sqrt{u_E^2+v_E^2+1}}\right)$

  where, M and N are the numbers of rows and columns in an image, respectively. Here, (u_C, v_C) denotes GT optical flow and (u_E, v_E) represents estimated optical flow.

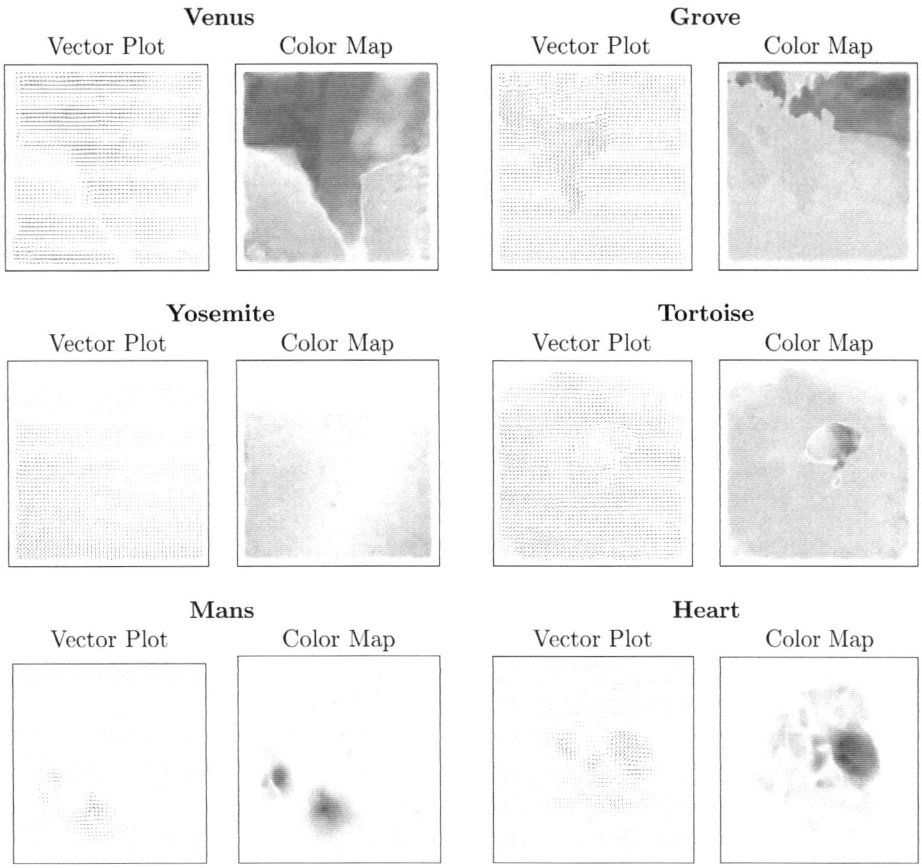

Figure 2. Estimated optical flow fields: vector representation and color maps.

- $\Lambda = \frac{\sum_{(x,y)} \chi((\mathfrak{I}_{warp} - \mathfrak{I}_{Tar}) > \partial)}{MN} \times 100$

  where, $\partial \in \mathbb{R}^+ \cup \{0\}$ and $\mathfrak{I}_{warp}$ and $\mathfrak{I}_{Tar}$ are the intensity values of warped interpolated image and target image at (x, y), respectively. Here, χ denotes the Indicator function.

4.3. Experimental discussions

In all the experiments, we consider $\beta = 0.1$, $N_{warp} = 5$, $N_{pyr} = 5$ and $N_{it} = 1.5 \times 10^3$ as per the results suitability. The output results both in qualitative and quantitative are estimated on datasets with GT, without GT and against Gaussian noise. Qualitative results are illustrated as the vector plots and color maps of the flow field. In color maps representation, different motion directions are made to correspond to different colors while homogeneous colors denote large regions moving uniformly. The estimated vector plots are also overlapped with the GT

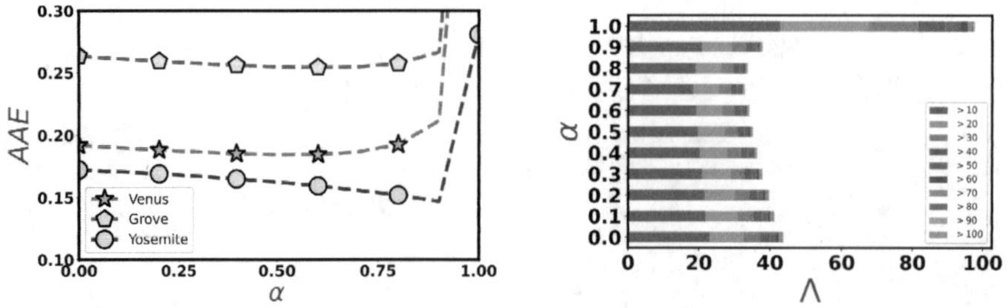

Figure 3. Effect of fractional order α on AAE (Venus, Grove, Yosemite) and WE (Tortoise). Legend shows the values of ∂.

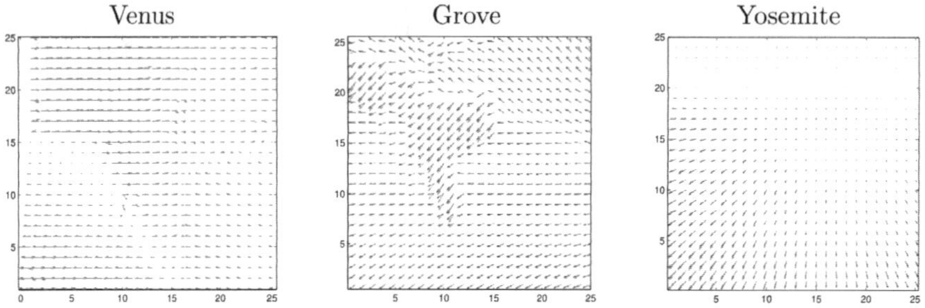

Figure 4. Estimated vector plots by the proposed model (red) through overlapping them on GT flow (blue).

vector field to illustrate the accuracy. Quantitative results are provided in terms of error measures.

In first experiment, the qualitative results are estimated for optical flow corresponding to all the datasets as shown in Fig. 1. The computed color maps and vector plots are illustrated in Fig. 2. For Venus dataset, which is a stereo image sequence, in which the foreground objects such as newspapers contain a high amount of texture and moving opposite to the background objects, we choose α order and λ as 0.7 and 7, respectively. The color maps and vector plots are dense and the flow field is smooth inside a region whereas the discontinuity at the boundaries of moving objects is non-violated. Grove image sequences contains moving trees along with a highly textured background. The moving edges in Grove introduce a triple junction near the center of the image and this triple junction is easily discernible in its color map, and for this image sequence $\alpha = 0.7$ and $\lambda = 10$. Yosemite contains large translations as well as divergent motions on the mountain region with some stationary cloudy region. For this image sequence we take $\alpha = 0.7$ and $\lambda = 10$.

In underwater scenes motion estimation is far more challenging due to various factors such as turbidity, different refractive index, low contrast, etc. Therefore, to test the performance of the proposed algorithm we have also taken underwater dataset. In Tortoise dataset, the foreground (the tortoise shell) and the background (the reefs) are difficult to be distinguished

Table 1. The effect of Gaussian noise of different amounts of STD upon AAE for Venus dataset.

STD	0.01	0.02	0.03	0.04	0.05	0.06	0.07	0.08	0.09
Horn et al. [8]	0.6502	0.6710	0.6968	0.7190	0.7425	0.7848	0.7716	0.7945	0.8022
Chen et al. [16]	0.3662	0.3934	0.4300	0.4300	0.5171	0.5488	0.5394	0.5510	0.5860
Kumar et al. [25]	0.3700	0.3968	0.4330	0.4327	0.5190	0.5515	0.5415	0.5529	0.5872
Huang et al. [26]	0.7066	0.7897	0.8721	0.8831	0.9802	1.0464	0.9828	0.9911	1.0119
Proposed model	0.3676	0.3822	0.4039	0.3991	0.4595	0.4786	0.4675	0.4810	0.4810

directly from the sequence, but in the estimated color map the tortoise can be clearly identified, here $\alpha = 0.7$ and $\lambda = 50$. Thermal dataset is also taken to validate the performance in order to keep in mind the applications of thermal imagery such as disease investigation, surveillance and sports. In thermal image sequence, multiple objects are moving and few of them are occluded due to tree. The motion can be seen clearly for $\alpha = 0.7$ and $\lambda = 2.5 \times 10^2$. The value of smoothing parameter is taken larger since this image sequence contains large amounts of noise. The Heart dataset such as heart is considered to test the accuracy on medical image sequence. The image sequence is of low contrast and pumping of heart causes a diverging motion in the scene, which is clearly represented by the vector plot and different colors of its color map.

In second experiment, we demonstrate the relationship between fractional order α and the error measures using the $2D$ plot and horizontal bar graphs as illustrated in Fig. 3, respectively. The optimal values of fractional order α for the datasets correspond to their minimum error measures. Moreover, warping error (WE) in Fig. 3 also conveys a strong indication in support of non-violation of BCA assumption. It is observed that the propounded model fairly validates the proposed formulation for all the datasets near the optimal fractional order (0.7). The precision in the estimated optical flow is also shown through vector plots in Fig. 4 by overlapping it with GT flow. This comparison exhibits that the estimated flow vectors (red) are highly congruent with GT (blue) flow vectors. Thus, the proposed model is able to produce a decent estimation of flow field under various conditions.

In the last experiment, we have compared the results of the given model with the recently published techniques. In Fig. 5, we give a comparison in terms of warping error (WE) with the models of Horn et al. [8], Chen et al. [16], Kumar et al. [25], and Huang et al. [26]. This shows the percentage of image pixels violating BCA assumption. It can be seen that the proposed model is more suitable compared to the existing model. A comparison of the proposed model in AAE with few techniques is also described in Fig. 6 for the datasets with GT flow. All these obtained results confirm the applicability and superiority of the proposed model.

5. Conclusions and future work

In this paper, a nonlinear fractional order variational model has been introduced by incorporating the Charbonneir norm and Marchaud's fractional order derivative for motion estimation through optical flow estimation. The model can be approximated to an L_1-norm based model for sufficiently small values of δ, whereas for $\alpha \in \mathbb{Z}^+$, it generalizes the existing integer order models. The flow fields are smooth on the inside of a region and are 100% dense

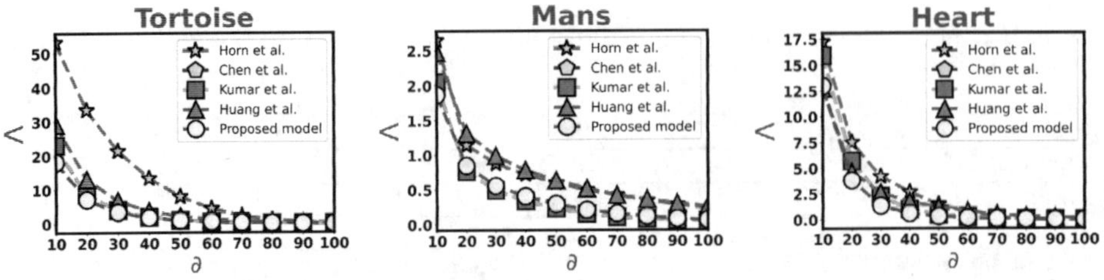

Figure 5. Comparison for the given datasets in terms of WE: proposed model vs Horn et al. [8], Chen et al. [16], Kumar et al. [25], and Huang et al. [26].

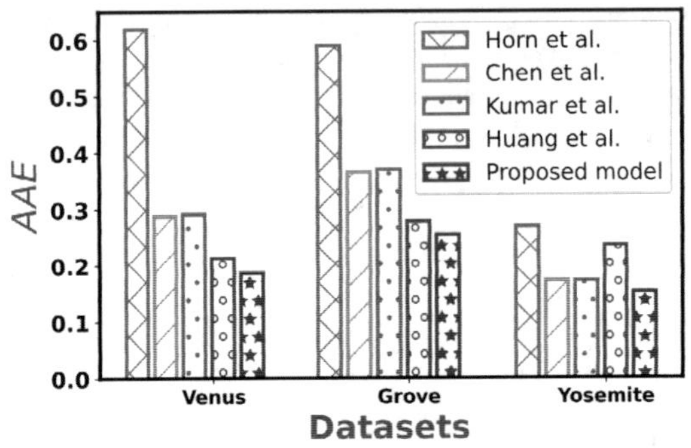

Figure 6. Results comparison in terms of AAE for the given datasets: Proposed model vs Horn et al. [8], Chen et al. [16], Kumar et al. [25], and Huang et al. [26].

whereas the discontinuities at the boundaries of moving objects are non-violated. The results are good when the fractional order corresponds to its optimal value (0.7). The optimal value of α has been chosen based on minimum error measures. The accuracy and robustness of the model have been demonstrated in the case when Gaussian noise is present in the images and by overlapping the estimated optical flow on GT flow. The comparison results of the presented model with some of the recently published techniques infer the superiority of the proposed model. In future, the presented model can be enhanced further to estimate optical flow field by incorporating some segmentation framework into the model. Also, the proposed model will be used in image denoising operations.

Acknowledgement

Pushpendra Kumar acknowledges the support of SERB, New Delhi for grant no. EEQ/2020/000154. The second author Muzammil Khan acknowledges thankfully to MHRD, New Delhi, Government of India.

References

[1] Kajo I , Malik A S and Kamel N 2016 An evaluation of optical flow algorithms for crowd analytics in surveillance system *International Conference on Intelligent and Advanced Systems* p 1–6.

[2] Ho H W, De Wagter C, Remes B D W, De Croon G C H E 2015 Optical flow for self-supervised learning of obstacle appearance *International Conference on Intelligent Robots and Systems* p 3098–3104.
[3] Makansi O, Ilg E, Brox T 2017 End-to-end learning of video super-resolution with motion compensation *German conference on pattern recognition* p 203–214.
[4] Xiong J, Idougho R, Aguirre-Pablo A A, Aljedaani A B, et al. 2017 Rainbow particle imaging velocimetry for dense 3D fluid velocity imaging *ACM Transactions on Graphics* **36**(4) 1–14.
[5] Nagel H H and Enkelmann W 1986 An investigation of smoothness constraints for the estimation of displacement vector fields from image sequences *IEEE Transactions on Pattern Analysis and Machine Intelligence*(5) 565–593.
[6] Black M J and Anandan P 1996 The robust estimation of multiple motions: Parametric and piecewise-smooth flow fields *Computer vision and image understanding* **63**(1) 75–104.
[7] Brox T, Bruhn A, Papenberg N and Weickert J 2004 High accuracy optical flow estimation based on a theory for warping *European conference on computer vision* p 25–36.
[8] Horn B K P and Schunck B G P 1981 Determining optical flow *Techniques and Applications of Image Understanding* **281** 319–331.
[9] Lucas B D and Kanade T 1981 An iterative image registration technique with an application to stereo vision *Vancouver, British Columbia.*
[10] Zimmer H, Bruhn A and Weickert J 2011 Optic flow in harmony *International Journal of Computer Vision* **93**(3) 368–388.
[11] Weickert J and Schnörr C 2001 Variational optic flow computation with a spatio-temporal smoothness constraint *Journal of Mathematical Imaging and Vision* **14**(3) 245–255.
[12] Ballester C Garrido L, Lazcano V and Caselles V 2012 A TV-L1 optical flow method with occlusion detection *Joint DAGM (German Association for Pattern Recognition) and OAGM Symposium* p 31–40.
[13] Drulea M and Nedevschi S 2011 Total variation regularization of local-global optical flow *International Conference on Intelligent Transportation Systems* p 318–323.
[14] Miller K S and Ross B 1993 An introduction to the fractional calculus and fractional differential equations *Wiley.*
[15] Hacini M, Hachouf F and Charef A 2020 A bi-directional fractional-order derivative mask for image processing applications *IET Image Processing* **14**(11) 2512–2524.
[16] Chen D, Sheng H, Chen Y and Xue D 2013 Fractional-order variational optical flow model for motion estimation *Philosophical Transactions of the Royal Society A: Mathematical, Physical and Engineering Sciences* **371** 20120148.
[17] Apostol T M 2007 Calculus, volume I *John Wiley & Sons.*
[18] Fubini G 1907 Sugli integrali multipli *Rend. Acc. Naz. Lincei* **16** 608–614.
[19] Ferrari F 2018 Weyl and Marchaud derivatives: A forgotten history *Mathematics* **6**(1) 6.
[20] Baker S, Scharstein D, Lewis J, Roth S et al. 2011 A database and evaluation methodology for optical flow *International journal of computer vision* **92**(1) 1–31.
[21] http://visual.cs.ucl.ac.uk/pubs/flowConfidence/supp/index.html.
[22] Kumar P, Kumar S and Balasubramanian R 2015 A vision based motion estimation in underwater images *International Conference on Advances in Computing, Communications and Informatics* p 1179–1184.
[23] Kumar S, Kumar S and Balasubramanian R 2013 A variational approach for optical flow estimation in infra-red or thermal images *International Conference on Image Information Processing* p 56–61.
[24] Barron J L, Fleet D J and Beauchemin S S 1994 Performance of optical flow techniques *International journal of computer vision* **12**(1) 43–77.
[25] Kumar P, Kumar S and Balasubramanian R 2016 A fractional order variational model for the robust estimation of optical flow from image sequences *Optik* **127**(20) 8710–8727.
[26] Huang Z and Pan A 2020 Non-local weighted regularization for optical flow estimation *Optik* **208** 164069.
[27] Hamming R 2012 Numerical methods for scientists and engineers *Courier Corporation.*

An Approach to Pattern Recognition for Identification of Devnagari Script Based on Fingertips and Palm

Ms. Sharvari Tantarpale [1]
Research Scholar, India, Amravati.
sharvari.tantarpale149@gmail.com

Prof. (Dr.) C. N. Deshmukh[2]
Associate Professor, PRMIT & R Badnera,
cndesh1968@gmail.com

16-digit ORCID identifier is 0000-0001-5559-7420

Abstract. In this paper, we present an finger point based signed language symbol to text identification and classification algorithm that is based upon RGB image datasets. The palm sized images based upon different sizes, backgrounds, orientation are captured to be preprocessed as per the requirements of developing a convolution neural network based algorithm. This algorithm utilizes Alexnet for the preprocessing requisites where in 47 symbols of Devanagari script are augmented based on the reference rulebook created for our requirements as highlighted in the paper. At the primary level this algorithm provides an excellent classification rate which promises upliftment for our research in the upcoming future. We have provided detailed steps and discussion on the classification parameters considered for our algorithm which is implemented on MATLAB platform with the help of machine learning solution libraries.

Keywords: convolution, Devanagri, Image processing, classification.

1. Introduction –

The true sense of expressing or communicating ones feeling, emotions, expectations or any other meaningful or non meaningful information to another in its pure form is known as a language. The barrier of such language across the world in overcome with the help of signed language/languages that utilize visual signs with some meaningful indication meant for expressing the context in mutual harmony with one and another. Signed languages not only have their own texture, representation but also have its own grammar etc. that a language requires. Signed languages though exist for the soul purpose of context exchange but still they are not universal in nature although they have tendency of many similarities by default. All the researches across world have come to acceptation that signed language along with spoken language are the two types of natural language accepted across the world which have evolved with centuries so that their true sense can-not be challenged by one with the other. Although amendments with any language at any abstract level cannot consider body language a type of nonverbal communication which is the thumb rule of any form of communication.

Fig 1: Chirogram from Chirologia, 1644, signed language, Wikipedia [1]

Across the world wherever communities of deaf and dumb or specially abled people exist then the requirement of a basic non verbal signed representation as shown in figure 1 above is adapted on the basis of many existing in the world. This need is better derived from the need of alternative communication where in not necessarily one specially abled is communication with the other specially abled but it needs to be understood by the normal as well to express and understand what the other specially abled requires or desires. The augmentative need of a sign language in regards to a number across the world is not known but each country has at least one that is utilized as its native signed language. From a Ethnologue data Indo Sign language of is ranked 151th most adapted and spoken sign language across the world [1]. One of the examples of such signed language that has achieved legal recognition is the American sign language "ASL" that is predominantly utilized across America as well as Canada [2]. ASL is a natural signed language that is utilized by the deaf communities so as to achieve communication via facial expressions, movements/motions of hand not only across North America, Anglo Canada but also West Africa and parts of Southeast Asia. In the early 19th century at the American school of deaf known as ASD situated at west Hartford, Connecticut origination of the ASL signed language took place. ASL utilized signs as the primary element of the language which involves the utilization of moment of the face, the torso and most importantly the hand.

Journal of Physics: Conference Series **2327** (2022) 012032 doi:10.1088/1742-6596/2327/1/012032

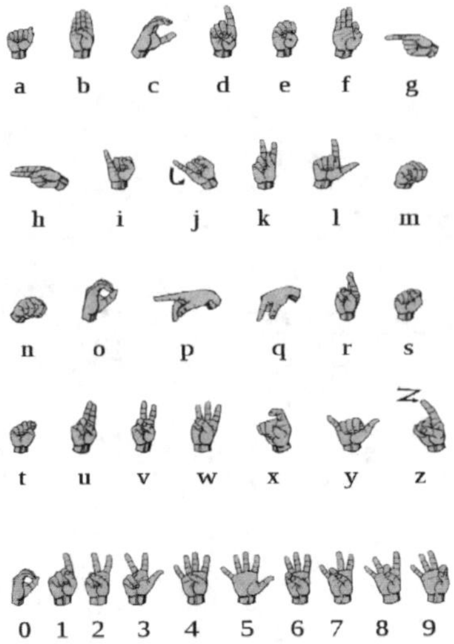

Fig 2.a **Fig 2.b** **Fig 2.c**

American Sign Language Representation and Example [2]

In this modern world of communication to remove barriers of sign language across the world, as well as for positive outcome towards society we propose an effective as well as modern machine learning based solution towards signed language. Machine learning approaches have set backs such as degree of freedom when we talk about hand movements or illumination contrast complications when we talk about skin color across the world and many more such problems towards development of machine learning based signed language. So we have not only proposed a solution to this problem but we also demonstrate a primary solution in terms of research work that shows prominent means of machine learning based sign language symbol identification and conversion of such symbols to Devanagari text. Across the world signing or utilization of signed language is primarily used by haring individuals have hearing disabilities or speaking disabilities due to certain issues in regards to mental, physical special abilities; yet no solution or literature exists that can justify that suitable sign language symbols based on the usage of figures or fingertips positions for text classification especially for Devanagari symbols. For filling this research void we propose to develop such alternative means of communication solution for the specially abled people in India.

2. Algorithm Methodology –

The main purpose of the paper is to utilize Devanagari script as it the root to various language scripts in India. So for the advancements in a country where various scripts exist one needs to start with the mother of all scripts so that our algorithm can be utilized with minimum modifications for all the scripts in near future. For faster computation in this computer based machine world where manual systems are being replaced by automated ones we also want sign detection to be made possible at finger tips of automated world. To start with we need to define the symbols that are to be utilized for creating the master dictionary. For this our work is novel in a way where Recognition of a signed symbol to its Devanagari script text representative is based on three step algorithm that involves majorly creating a dataset of images which are to be preprocessed, followed by feature extraction and finally classification of test images based on the training sets. In our algorithm on the basis of signs we have chosen finger tips of the finger where thumb is being utilized as the reference index/pointer so as to provide detection of each symbol individually. So when one uses a thumb to point a specific location on a specific finger the number of tips are restricted to 16 that is 4 points on each finger. By using a stand alone base sign where thumb does not point to any finger tip we can create a bases of doubling the number of characters which would be a unique feature in our algorithm if we talk about frame rate processing from a live video. From the point of view of creating database which is the first step one needs to first specify the points and the methods utilized for symbol detection that is the prima face rules for which the machine learning algorithm is to be developed. For the rules/ boundary development one needs to understand the overview of the devnagri symbols to be detected.

3. Devanagari Script –

Devanagari Script also known as "Nagari" is derived from the ancient Brahmi script,[3] that is vitally used in all the Indian subcontinents where each state in India has its own language main reason for utilizing the devanagari as it is derived from the Nagari script. At the preliminary level the Nagari script comprises of the 47 primary characters including 14 vowels and 33 consonants which is the first level detection of our developed algorithm as reported via this research paper. These symbols as shown in figure 3 below are to be detected with the help of sign language based moments that are the requirements of our fingertips/points rule dataset creation as shown in figure 4 below.

क ख ग घ ङ च छ ज झ ञ

अ आ इ ई उ ऊ ट ठ ड ढ ण त थ द ध न

ए ऐ ओ औ अं अः प फ ब भ म य र ल व

श ष स ह क्ष त्र ज्ञ

Fig 3. Vowels and Constants in Nagari Script [3].

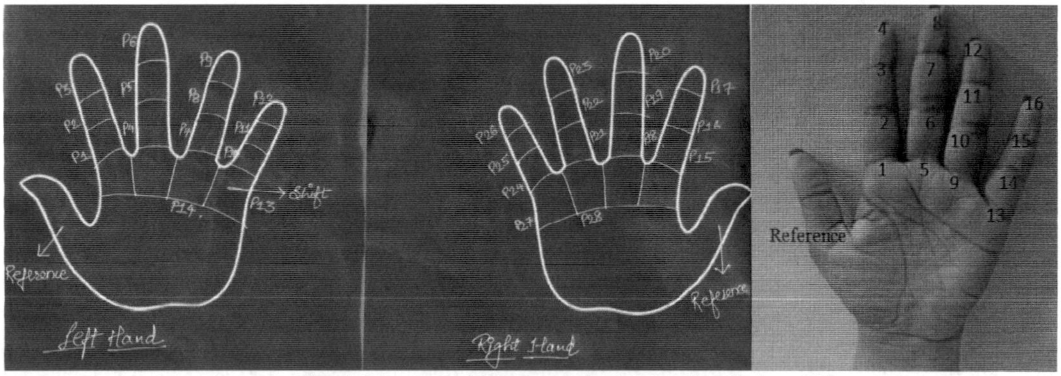

Fig 4. Finger points or symbol detection Points [3].

So considering both hands one needs to understand that in both cases thumbs are going to be used as reference points so that the thumb pointing as position of a finger becomes a point/symbol mapped towards detection. So considering both hands the number of points that can be detected are restricted to 16 points each on each hand making it total of 32 points. These points can be increased further by using certain points in coordination so that rule data base is widened [9]. The example of rule book creation is shown in the figure 5 below.

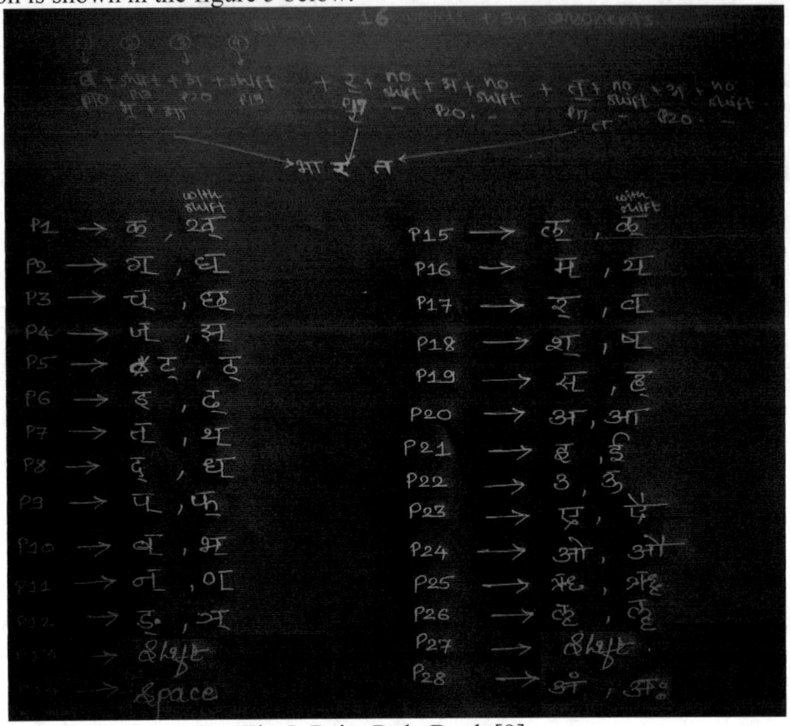

Fig 5. Point Rule Book [9].

So now one can see that a researcher needs to collect multiple images of each symbol to start creating a database based on each point to rule out multiple setbacks in terms of degree of freedom that is angles or be it set of illumination / contrast levels of background or skin etc. which one can think of before actual algorithm hindrances are seen and overcome via modification in algorithm as highlighted in the next section.

4. Database Creation –

All the images are captured with the help of cellphone camera with multiple persons or environment consideration. An Example of image based training sets creation for regards of rule book is shown in figure 6 below. The concept is that a person will use its thumb as reference and place it at the point of interest per finger to represent the symbol of identification or better known as classification in regards machine learning. Example below shows the left hand index fingers and the little fingers 4 points of importance. Similarly all the fingers of left as well as right hand are collected and identified. The data set can be collected from X various subjects, which can contain Y no. of fingertip actions for The Devanagari script, composed of 47 primary & secondary characters including 14 vowels and 33 consonants. So, there will be total $X \times Y \times 47$ image collection [11]. The data set will be captured in standard and cluttered backgrounds as well. Besides, for each gesture, the subject poses with variations in hand orientation, scale, articulation, and so forth. For capturing images standard cameras will be used of appropriate resolution.

Journal of Physics: Conference Series **2327** (2022) 012032 doi:10.1088/1742-6596/2327/1/012032

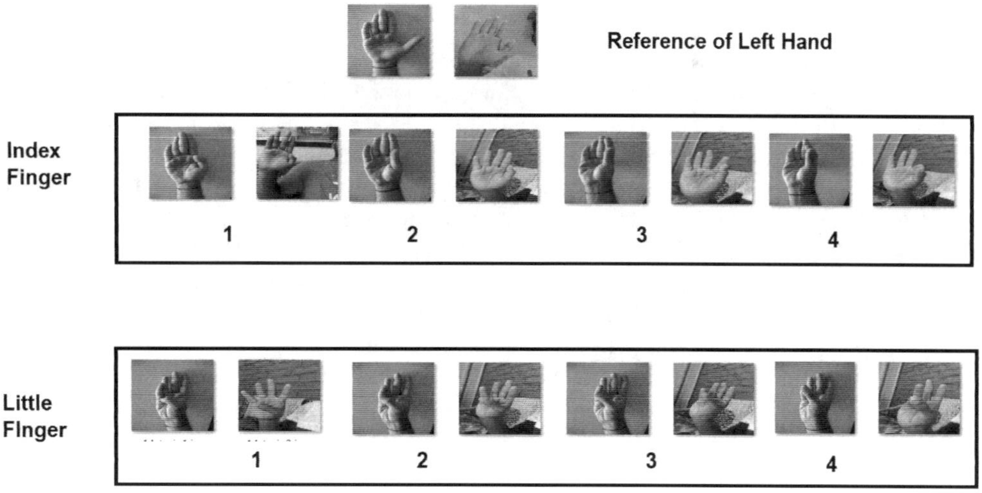

Fig 6. Database for Rule Book of images as an example.

5. Data Base Preprocessing –

At this step the compiled images for the rule book are to be processed or filtered so that all the images come at same level in regards to size, shape etc. which are required for algorithm development of the no of features of interest to be extracted for training the machine algorithm. So in regards to the requirement of data analysis; from the dataset a particular image is selected then the pre-processing is done for that selected image. Pre-processing includes cleaning, Instance selection, normalization, transformation at the first instance which can be extended as per the tuning parameters required to overcome the setbacks the algorithm might face in the near future scope. As the next step is feature extraction so the preprocessing steps are on the basis of the feature extraction sub algorithm utilized. So as per the requirements of Alexnet that we have utilized for feature extraction the images are to be resized to 277x277. For the requirements of our research we utilized MATLAB software for our algorithm, development. Example of data base preprocessing in regards to image resize is shown via a figure below:

Fig 7. Image preprocessing that is image resize in MATLAB.

6. Feature Extraction –

Various neural algorithms are utilized in machine learning, where in each algorithm has its own prerequisites when we talk about any machine learning application for that matter the application of our interest is image classification. So from the point of view of classifier selection in a neural network the first step is of feeding the database images to the feature extraction step of the algorithm development. In the algorithm all the features corresponding to images will be stored in the database that is in other words rulebook creation as shown in the above section example. Now, in regards to our algorithm development the features from the test image will be extracted and compared with those in the database using classifiers. The classifiers will then predict the class to which the test image belongs. Some of popular classification algorithms are feature matching (Euclidian distance), AdaBoost algorithm, parallel cascade structure, feed-forward back propagation neural network, kNN classifier, support vector machine (SVM) & artificial neural networks (ANN), Adapted Deep Convolutional Neural Network (ADCNN), deep convolutional neural network (CNN), neural network based on SPD, etc. Each algorithm has its merits and demerits. In the work such algorithms have been studied and the algorithms which is best suited for the proposed work in CNN. The selection of CNN is based upon the performance of the approach which is based on the following parameters:

- Classification Accuracy
- Precision and Recall
- Confusion Matrix etc.

From the point of view of CNN algorithm for feature extraction step a researcher has multiple options like that of Alexnet (architectural image shown in figure 8), Google net etc. For our research backbone of feature extraction we have chosen Alexnet because of its simplicity and numerous advantages.

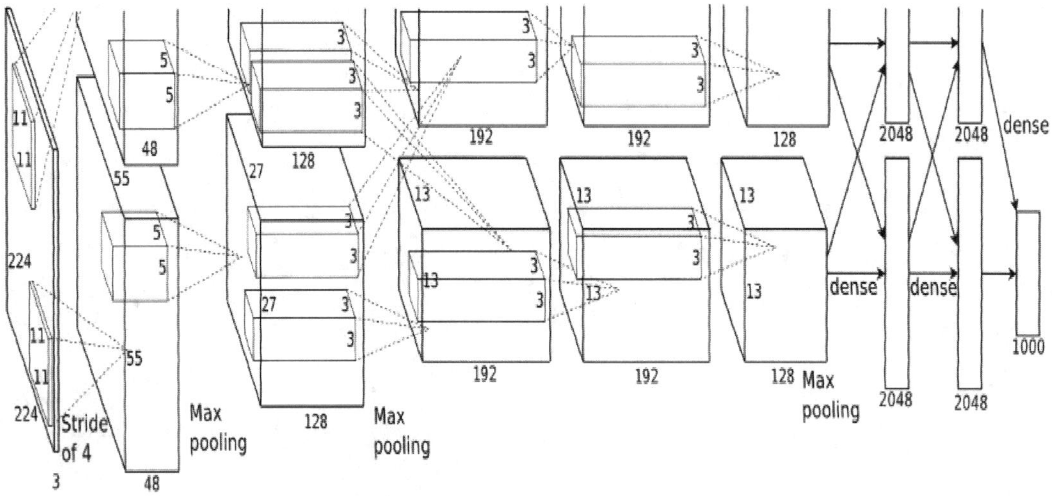

Fig 8. Alexnet Architecture

Few advantages of AlexNet comprise of that it was the first major CNN model that used GPU's for training which lead to faster training of models. Second it is a deeper architecture with 8 layers which means that is better able to extract features when compared to LeNet especially when we talk about color images. Third as Alexnet utilizes ReLu "rectified linear activation function" it promises that there is negligible loss in regards to features on extraction and finally Alexnet counters the negative output caused by the gradient feature/function hence this leads to further improvement in model training speed. Because of these at the overview we have chosen Alexnet as our feature extractor.

From the point of view of Alexnet we have defied layer number 23 as the fully connected layer and layer number 25 as the classification layer. From the point of view of feature based database creation we have kept the training feature parameters as SGDM as trainer whose learning rate is 10^{-6} having maximum epocs as 200 and batch size of 128 of for 256x256 image resized images. With the help of MATLAB this alexnet trained database is saved in mat file as shown in below figure. This net file is continuously upgradable as per the user algorithm requirements in the near future.

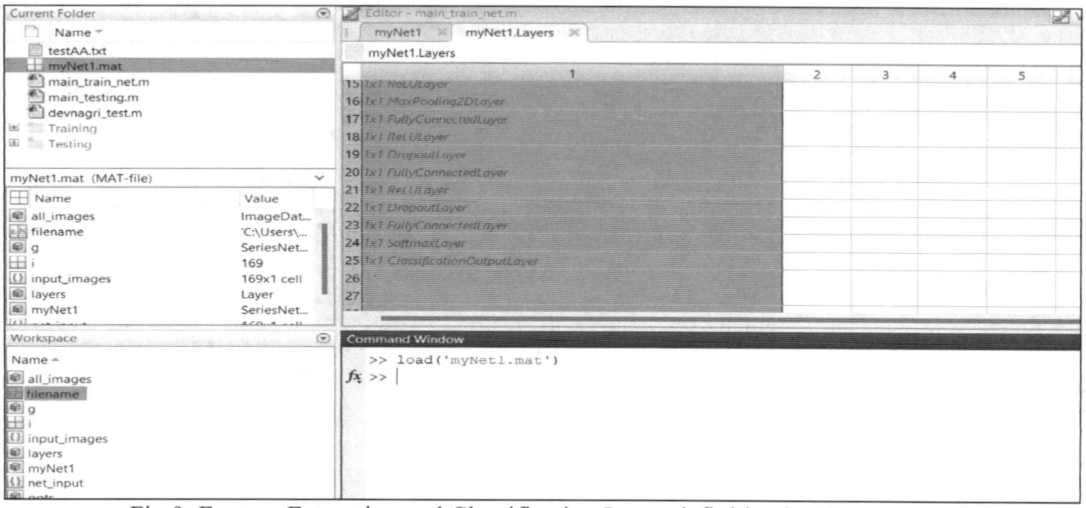

Fig 9. Feature Extraction and Classification Layer definition in ALexnet.

At this level our database creation algorithm is complete and the required features of the database are stored in the classification layer so that the algorithm can be tested with the help of images under test.

7. Classification / Algorithm Testing –

cFrom the point of view of testing we have used image capture from a video source as shown in image below that is again preprocessed in regards to size which was done for each step for database creation. Then the test images features are extracted and the extracted features are compared with the features present in the classification layer of Alexnet. The classified image features are then converted to the desired Nagari symbol as shown in the figure 11 below from the created rulebook.

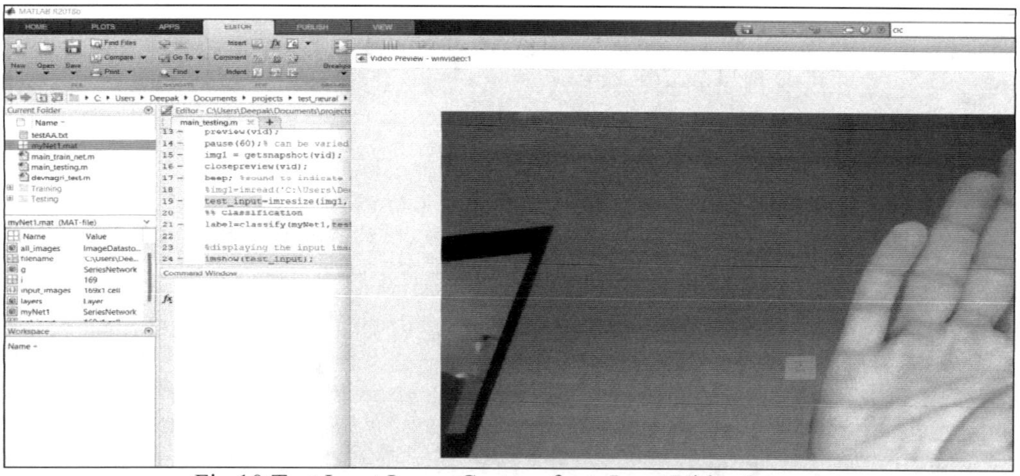

Fig 10.Test Input Image Capture from Input Video Source.

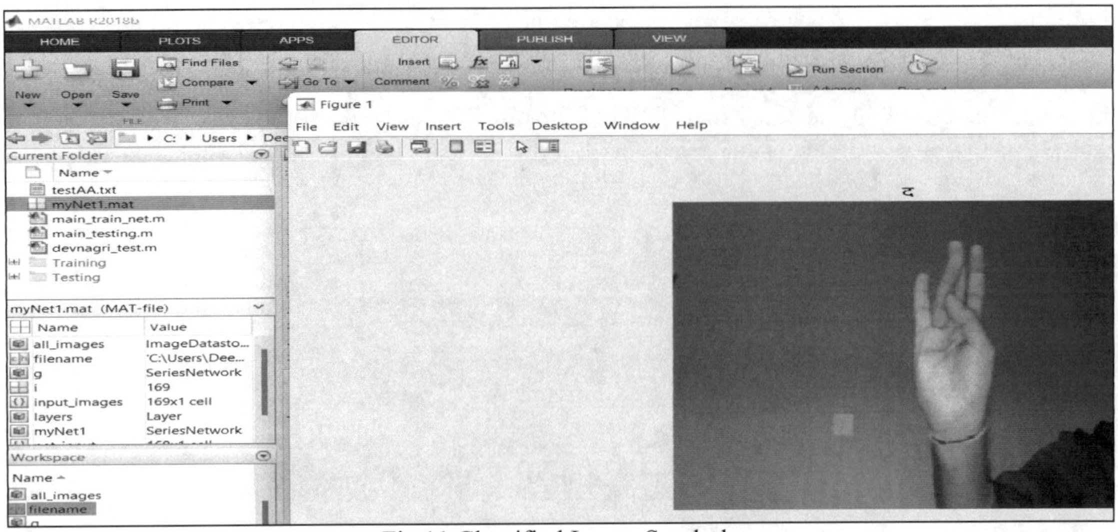

Fig 11.Classified Image Symbol.

8. Result and Discussion –

At the initial level the of identifying the 47 symbols of Devanagari have been carried out and the required database from the point of view of the algorithm has been developed successfully as shown in the image above. So with the context of utilizing our novelty individual character/sign detection has been made possible. This now needs to be utilized for multistep or rather multiple sign detection so that our novelty can make way of letter expression with multiple signs rather than keeping the algorithm restricted to binary signs which one use to do by making manual work automated. Our research work indicates successful development of CNN architecture based on Alexnet where the symbol to text classification is successfully carried out. The convolution model is to be further tuned as per requirements of the increase in number of symbol identification as per increase in number of convolution layers is desired in the near future so that a complete word level classification can be achieved subject to overcoming the amount of size and time setbacks towards completely tuned database creation is expected. The main work of proving the increase in amount of accuracy, recall and other parameters of our algorithm is the next step that we will address in the near future. This is desired to provide the bottleneck of our algorithm based on the overfitting issues during training in regards to the background variation or angle orientations that are not conquered in terms of these primary testing of our research. Our findings present a number of opportunities for the experiments herein to be continued by not only us but also the researchers striving to brighten the future of machine learning towards the betterment of human society; when we talk about signed languages to be considered as universal natural language of communication in this modern world.

References

[1] What Are The Top 200 Most Spoken Languages?, Ethnologue, Wikipedia.

[2] Armstrong, David; Karchmer, Michael (2002), "William C. Stokoe And The Study Of Signed Languages", In Armstrong, David; Karchmer, Michael; Van Cleve, John (Eds.), The Study Of Signed Languages, Gallaudet University, Pp. Xi–Xix, ISBN 978-1-56368-123-3, Retrieved November 25, 2012.

[3] R. M. K. Sinha,Veena Bansal, " ON DEVANAGARI DOCUMENT PROCESSING" , 0-7803-2559-1195$ 4.00, 1995 IEEE.

[4] Rung-Huei Liang Ming Ouhyoung, "A Sign Language Recognition System Using Hidden Markov Model And Context Sensitive Search", Proc. Of ACM VRST'96 (ACM International Symposium On Virtual Reality And Software Technology), Pp. 59-66, July, Hong Kong, 1996.

[5] Michael Boronowsky, Ansgar Marco Schmidtin, "Winspect: A Case Study For Wearable Computing-Supported Inspection Tasks", 2001 Https://Www.Researchgate.Net/Publication/221240920
[6] A Boosted Classifier Tree For Hand Shape Detection, Eng-Jon Ong And Richard Bowden Centre For Vision, Speech And Signal Processing University Of Surrey, Guildford, GU2 7XH. 2004
[7] Jian Liang, David Doermann, Huiping Li,"Camera-Based Analysis Of Text And Documents: A Survey" , IJDAR International Journal On Document Analysis And Recognition 7: 84–104, (2005)
[8] U. Pal, T. Wakabayashi , F. Kimura, "Comparative Study Of Devnagari Handwritten Character Recognition Using Different Feature And Classifiers", 2009 IEEE, 10th International Conference On Document Analysis And Recognition, 1111- 1115.
[9] Ravikiran J, Kavi Mahesh, Suhas Mahishi , "Finger Detection For Sign Language Recognition", Researchgate Proceedings Of The International Multiconference Of Engineers And Computer Scientists 2009 Vol I ,IMECS 2009, March 18 - 20, 2009,Hong Kong
[10] Pragati Garg, Naveen Aggarwal And Sanjeev Sofat , "Vision Based Hand Gesture Recognition", World Academy Of Science, Engineering And Technology 49 (2009), 972-977.
[11] Michael Van Den Bergh, Luc Van Gool, "Combining RGB And Tof Cameras For Real-Time 3D Hand Gesture Interaction," Seventh Framework Programme Of The European Commission: EU Project FP7 ICT 24314 Interactive Urban Robot (IURO) 2011.
[12] Shaowei Chu And Jiro Tanaka, "Hand Gesture For Taking Self Portrait" , Humancomputer Interaction, Part II, HCII 2011, LNCS 6762, Pp. 238–247, 2011. © Springer-Verlag Berlin Heidelberg 2011
[13] R. Jayadevan, Satish R. Kolhe, Pradeep M. Patil, And Umapada Pal, " Offline Recognition Of Devanagari Script: A Survey" , IEEE TRANSACTIONS ON SYSTEMS, MAN, AND CYBERNETICS—PART C: APPLICATIONS AND REVIEWS, VOL. 41, NO. 6, NOVEMBER 2011
[14] Chris Harrison, Hrvoje Benko, Andrew D. Wilson, " Omnitouch: Wearable Multitouch Interaction Everywhere" , UIST'11, October 16–19, 2011, Santa Barbara, CA, USA,
[15] Jagdish Lal Raheja, Karen Das, Ankit Chaudharyet. Al In "An Efficient Real Time Method Of Fingertip Detection", International Conference On "Trends In Industrial Measurements And Automation" TIMA–2011, 447-451, (2011) .
[16] Kittasil Silanon And Nikom Suvonvorn,"Hand Motion Analysis For Thai Alphabet Recognition Using HMM", International Journal Of Information And Electronics Engineering, Vol. 1 , No. 1 , July 2011, 65-71.
[17] A. Bharath And Sriganesh Madhvanath, Senior Member, IEEE, "HMM-Based Lexicon-Driven And Lexicon-Free Word Recognition For Online Handwritten Indic Scripts", IEEE TRANSACTIONS ON PATTERN ANALYSIS AND MACHINE INTELLIGENCE, VOL. 34, NO. 4, APRIL 2012
[18] Ved Prakash Agnihotri, " Offline Handwritten Devanagari Script Recognition" , I.J. Information Technology And Computer Science, 2012, 8, 37-42.
[19] Somaya Al Maadeed, Wael Ayouby,"QUWI: An Arabic And Englishhandwriting Dataset For Offline Writer Identification",Https://Www.Researchgate.Net/Publication/231438321, (2012)
[20] Sandeep B. Patil, G.R. Sinha , "Isolated Handwritten Devnagri Character Recognition Using Fourier Descriptor And HMM", Researchgate, Int. J. Pure Appl. Sci. Technol., 8(1) (2012), 69-74
[21] Jayashree R. Pansare, Shravan H. Gawande(2012) Et. Al In " Real-Time Static Hand Gesture Recognition For American Sign Language (ASL) In Complex Background", Journal Of Signal And Information Processing, 2012, 3, 364-367

Http://Dx.Doi.Org/10.4236/Jsip.2012.33047 Published Online August 2012.

[22] Chetana S.Ingulkar, A.N. Gaikwad, " Hand Data Glove: A Wearable Real Time Device For Human Computer Interaction" , International Journal Of Science And Engineering, Volume 1, Number 2 – 2013, PP-99-104 ©IJSE, ISSN:2347-2200.

[23] Hrishav Raj, Rajib Ghosh, "Devanagari Text Extraction From Natural Scene Images" , International Conference On Advances In Computing,Communications And Informatics (ICACCI), 2014.

[24] Akanksha Saran, Damien Teney And Kris M. Kitani, " Hand Parsing For Finegrained Recognition Of Human Grasps In Monocular Images" , The Robotics Institute, Carnegie Mellon University Pittsburgh, PA 15213, USA, 2015.

[25] Manisha U. Kakde , Amit M. Rawate, " Hand Gesture Recognition System For Deaf And Dumb People Using PCA", © 2016 IJESC, International Journal Of Engineering Science And Computing, July 2016

[26] Trupti A. Jundale And Ravindra S. Hegadi ,"Skew Detection And Correction Of Devanagari Script Using Interval Halving Method", @Springer Nature Singapore Pte Ltd. 2017,K.C. Santosh Et Al. (Eds.): RTIP2R 2016, CCIS 709, Pp. 28–38, 2017.DOI: 10.1007/978-981-10-4859-3 3

[27] Shilpa Chaman, Dylan D'souza, Benz D'mello , Karan Bhavsar,Jolton D'souza, "Real-Time Hand Gesture Communication System In Hindi For Speech And Hearing Impaired", Proceedings Of The Second International Conference On Intelligent Computing And Control Systems (ICICCS 2018), IEEE Xplore Compliant Part Number: CFP18K74-ART; ISBN:978-1-5386-2842-3, 2018.

[28] Nidhi Iche, Mrs. Shanthi K. Guru, "Fingertip Detection Method In Hand Gesture Recognition", International Research Journal Of Engineering And Technology (IRJET) E-ISSN: 2395-0056, Volume: 06 Issue: 01 | Jan 2019 , Www.Irjet.Net P-ISSN: 2395-0072

[29] Suprabhat Maity, Anirban Dey , Ankan Chowdhury, Abhijit Banerjee,"Handwritten Bengali Character Recognition Using Deep Convolution Neural Network", C_Springer Nature Singapore Pte Ltd. 2020, A. Bhattacharjee Et Al. (Eds.): MIND 2020, CCIS 1241, Pp. 84–92, 2020, Https://Doi.Org/10.1007/978-981-15-6318-8_8

[30] Ankit K. Sharma, Dipak M. Adhyaru And Tanish H. Zaveri "A Survey On Devanagari Character Recognition", © Springer Nature Singapore Pte Ltd. 2020, A. K. Somani Et Al. (Eds.), Smart Systems And Iot: Innovations In Computing, Smart Innovation, Systems And Technologies 141,Https://Doi.Org/10.1007/978-981-13-8406-6_4147

[31] Ayan Kumar Bhunia , Partha Pratim Roy, Aneeshan Sain & Umapada Pal, "Zone-Based Keyword Spotting In Bangla And Devanagari Documents", # Springer Science+Business Media, LLC, Part Of Springer Nature 2020.

An Approach to Fabric Defect Detection using Statistical Methods for Feature Extraction

Girish Patil[1], S. M. Deshmukh[2]

[1]Research Scholar, PRMIT & R, Badnera, Amravati, India. patilgirish213@gmail.com
[2]Professor, PRMIT & R, Badnera, Amravati, India. hod_extc@mitra.ac.in

ORCID: 0000-0003-3650-2427

Abstract. Quality control in any commodity at every stage of production not only is desired but its reproduction day in and day out is essential in this fast changing world especially when we see that shopping from a digital screen is the new trend in this era of pandemic. Quality check, control and inspection demands manual checking where in the monotonous work can make a human negligible towards work and lead to reduction in quality control. The world demands replacement of such laborious work which can only be promised by machinery custom tailored from the needs of providing accuracy, time reduction in quality checks as well as production and most importantly reduction in labour cost when commodity manufacturing is concerned. One such commodity that we cover in this paper is the textile industry where in automated fabric defect detection is harnessed as the research trend across the world with the help of image processing combined with machine learning algorithms is concerned. This area of research is challenging as the algorithm needs to cover robustness, efficiency and particularly the variety of classifiable defects in terms of surety and complexity. Out of the various techniques utilised for fabric defect detection we have utilised a threshold based image processing technique that can provide an automated stepping stone for atomization in the fabric defect detection for quality control.

Keywords: Feature Extraction, Fabric defects, Fourier transform, thresholding, etc

1. Introduction –

Defects during manufacturing in any industrial process is very costly for the industry not only in regards to time and money but moreover it is cruel in regards to reduction in customer base. In this world of industry 4.0 economic growth is vitally dependent on the business being driven by atomization in every sector where image processing and machine learning are the key providers for each and every sector. One such commodity where the finished product for a customer called fabric is a composition of material that is adjunct via process of weaving, knitting or in layman terms bonding various yarns or threads into clothing, bedsheets etc. Such fabric production is a primary need of each human being where its utilization is evidently throughout the world in day to day basis. Production of such fabric and the quality of such fabric in terms of quality check is done by a process of industrial inspection that is very time consuming and monotonous. The process of fabric defect detection in a textile industry is performed by inspection as shown in figure one below to meet the standards set by the industry which are in ordinance with the buyer specifications.

Figure 1: Manual Defect Inspection [1].

There are various defect detection techniques where in the oldest one is by doing a visual inspection of the fabric on a flat table with the help of manual eyes as shown in figure 1 or a inspection machine that is not hundred percent autonomous as shown in figure 2. A enhanced partially autonomous artificial intelligence based computerized machine has been developed in recent years yet their inadequacies in regards to accuracy when compared to natural human inspection is reported to be in the range of 45 to 60 percent [1]. Enhancement to such computer vision based machinery is in demand as detecting each type of defect known to mankind with the help of a manual human eye based fabric defect inspector is very time consuming. Such job can cause fatigue and mind numbing yet as the rate of efficiency with this tedious work is around 70 percent which is still higher compared to a computer machine.

Figure 2: Semi Autonomous Fabric Defect Inspection [1].

Computer Vision based autonomous solutions that are based upon digital image processing algorithms is under development with the aspect that the amount of laborious work and the cost associated with manual defect inspection can be reduced down tremendously. The important requirement of any fabric defect detection is to understand the type and the cause of defect in order to guaranty that such hindrances can be filtered out from industry atomization. A defect can be a mere portion of the fabric under test that does not meet the customers requirements for which quality control is to be administered in an industry [2]. There are various fabric defects that can be listed or summarized as per their types but naming few are shown in table one below. These defects are majorly caused because of machinery issues during production. The characteristic of the defect can be termed as major, minor or critical which are based upon the modifications that one has to do to the industry production unit to overcome them on the whole. The number of defects known in the area of fabric defect detection is long and variable hence providing a generalized machine based solution for all is difficult as variation exists in regards to texture, color, volume/density and variations in regards to prints and threads utilized for production. The advantage of machine based solution is early detection of fabric defect. The way of achieving this early detection is the main core of research methodology that we have utilized to provide an image algorithm based fabric detection which provides increased efficiency on comparison with manual labor fabric detection techniques as covered in next section.

1.1 Motivation

Recently, Govt of India announceAtmanirbhar Bharat Abhiyan to increase the potential for exportation of local manufacturing industries. However, the high cost of raw materials and the lack of quality control programs increase final product prices leading to loss of competitivity against companies in other countries. Identifying defects in fabric is a major concern for fabric industries, because it influences brand image and the perception of product quality. This situation leads to develop the methodology which will provide an affordable, viable, and reliable solution to the problem of expensive and unreliable manual inspection of fabrics in textile industries.

1.2 Contribution of Work

This work proposes to apply computer vision to inspect the defects that occur on the fabrics, reduce costs and increase the reliability of inspection results. Compared to the manual inspection process, computer vision allows to inspect a production process without involving human errors caused by fatigue and distractions, facilitating the quantification of quality variables. It is not to replace the role of a quality inspector, but to provide him with tools to facilitate his job.

Table 1: Fabric Defects Examples [3]

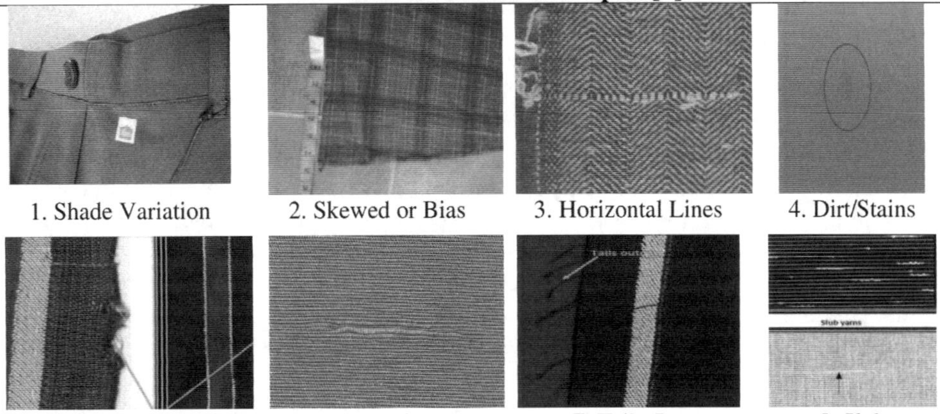

1. Shade Variation	2. Skewed or Bias	3. Horizontal Lines	4. Dirt/Stains
5. Defective selvage	6. Weft	7. Tails Out	8. Slub

2. Research Methodology

From the point of fabric inspection the fabric under test is provided to the image acquisition unit from the point of providing an image processing algorithm mechanized solution. Once the image is captured by the acquisition unit then this image is processed by the processing unit that utilizes the developed robust research algorithm comprising of various processes which are clustered into image enhancement, image restoration, image segmentation, feature extraction and recognition. From the point of view of fabric defect detection image feature extraction is the most crucial methodology that decides the power of computing that is desired to decrease the computational time of defect detection as well as increase the efficiency of algorithm [4]. The efficiency of algorithm implemented in the fabric defect system has direct relation with the feature extraction process as a efficient relation between the defect and the non defect area needs to be established. These features are the essence of classification process which is the end teem testing goal of the fabric defect system. This research paper deals with the first step of the fabric detection process that is feature extraction of an image so that a learning database can be created. The features that are extracted are periodic in nature as the fabrics follow certain symmetric structure or pattern which are to be listed in the dictionary during the process of feature extraction. Any deformity in these pattern as well needs to be enlisted so that during the classification process one can reproduce the defect from dictionary and define the type of process defect.

The process of feature extraction can be distributed across many divisions that are followed by the algorithm and any modification to these to provide better efficiency is what is sought after [5]. The divisions can be enlisted as:

- Statistical texture analysis approach
- Texture model-based approach
- Structural approach
- Signal processing-based or spectral approach.
- Hybrid or combination of one or many types.

Out of the many methodologies the statistical ones have been widely used for research as they use either the first-order or the second-order statistics to extract the features of importance. The statistical approaches have various methods to name few co- occurrence matrix, cross correlation, histogram features, edge detection, auto-correlation function and mathematical morphology methods etc. Each method have their advantages and disadvantages. From the evolution of digital image processing spatial and frequency domain analysis have been a tremendous boom for any fabric defect detection. The need of spatial is to identify the location of the defect but to find the type of defect across the surface of the fabric frequency based approaches are of prime importance [6].
For our research methodology at the prima level we provide a fabric defect feature extraction algorithm that follows hybrid approach to overcome the defect weaknesses of one method with the superior advantages of the other method. The main objective of this approach becoming popular is that researches are able to minimize the computation related complexities and improve the rate of detection; hence we follow this approach. We utilize the statistical method of auto correlation based on thresholding to calculate the mean, variance, skewness and kurtosis [7]. Based on the parameters extracted we create a image mask so that based on the masking the derived image features witness the Fourier transform method to provide surface based defect parameters. The utilization of autocorrelation function is that it is durable to noisy and illumination changes that may exist in any fabric pattern irrespective of defect. Secondly this approach has a demand of very low computational power as its complexity is very low hence provides a direct measure of similarity between the original and processed image during feature extraction. On the lower end the problems that this approach can face is that for real time applications and large size images this approach lags as the function becomes sensitive to noise interference created during framing/ masking. Hence this approach is not suitable for randomization. To remove the noise interference we utilize a thresholding based approach so that illumination based noise interference can be minimized and secondly we combine this method with Fourier transform to overcome the problem of real time computation. The Fourier transform is the frequency analysis based approach where in complete surface based defects can be detected. The advantage of Fourier transform is that the spatial features of the fabric can be compared with the frequency domain which provides us association of the common structure of fabric with the irregularities that may occur once a defect is inborn. This advantage makes sure that the pre processing done by statistical correlation can be further enhanced and noise interference is reduced to bare minimum. Spectrum analysis in regards to power provide us with preliminary results that we are looking for in terms of feature extraction and library creation to make the process of fabric defect detection completely autonomous. The hybrid based feature extraction process research methodology as covered here is implemented as covered in the next section followed by the result and future scope section.

3. Implementation

From the point of view of spatial domain the 2D Gabor and Gradient filters are utilized as they are the best suitable option for texture segmentation as well as boundary detection. Based on the filtering the angle orientation of 30 and 45 degree are utilized so that the defective textures frequency components are highlighted with the help of thresholding utilized in the algorithm. For input frequency response extraction we have kept $N = 9$ based on the various statistical parameters considered where we find that with the increase in number of samples the edge formed of the image also increases. This edge is periodic in nature which promises that the smallest convolution is detectable which becomes favorable for the defect detection. As per the test cases and input size of images considered we found that the response time is directly proportional to the size of the input image. This was concluded with the fact of various images considered 0f 512x512 size for defect detection as shown in below table 2.

Images with highly periodic and homogeneous structures are considered as in those images the textures follow a perfect pattern providing better probability of defect detection. Yet this structural limitation needs to be crossed over with the help of machine learning algorithm that can classify the type of defect as well with the increase in dataset tremendously to provide conclusive breakthrough in our research in near future.

The proposed methodology comprising of the algorithm based on the two methods is implemented on MATLAB that covers a moderate sample set so that the efficiency of the algorithm on the area coverage can be evaluated. The algorithm is formulated on the basis of following enlisted steps which can also be summarized visually with the help of figure 3.

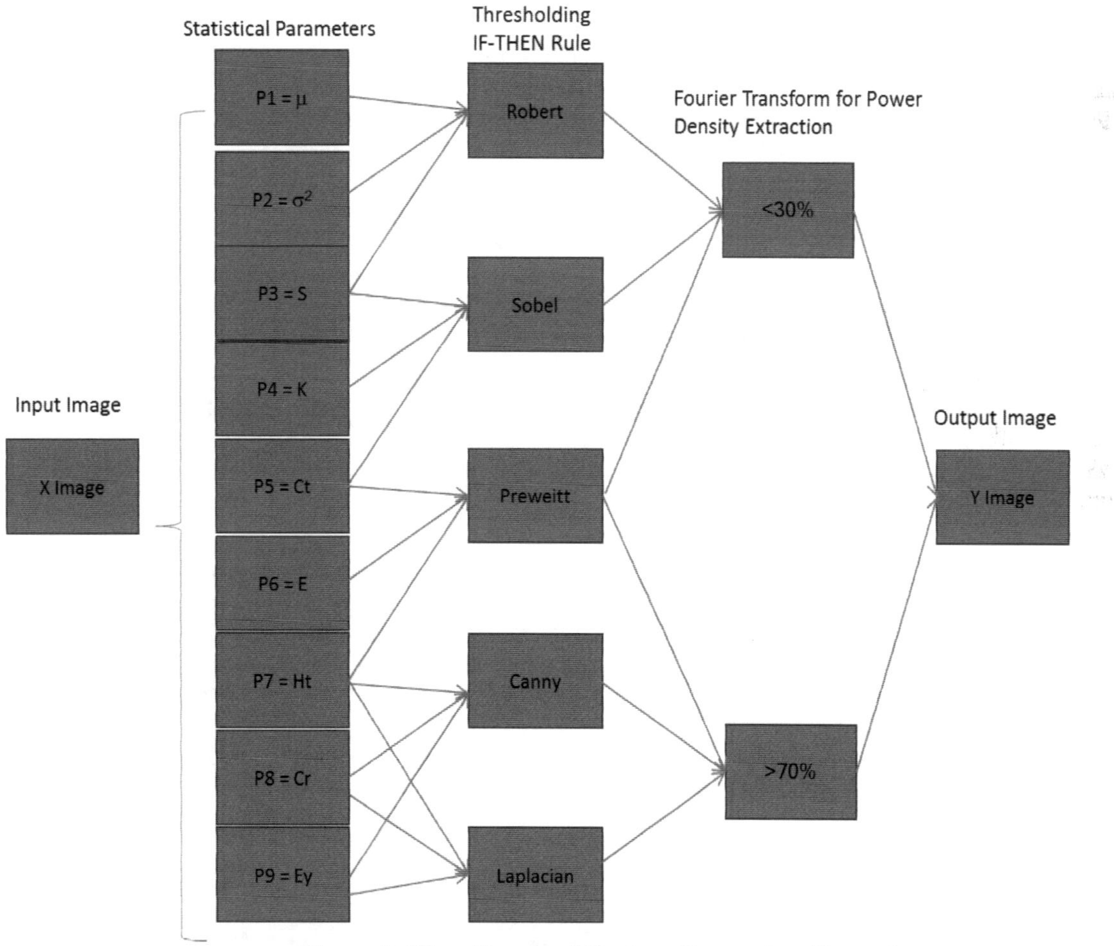

Figure 3: Flow Graph of Feature Extraction Algorithm

- Step 1: Image import for frame based segmentation based on masking (mask is calculated based on the statistical functions like example of autocorrelation) as shown in figure 4 below.

- Step 2: Image normalization for mask calculation based on thresholding (this is based on the statistical calculation of 9 different parameters that comprise of mean, variance, skewness, kurtosis etc).
- Step 3: Fourier transform based feature calculation.
- Step 4: Parameter extracted comparison based on power density of original image and the correlated image for image reconstruction..
- Step 5: Extracted feature based dictionary creation.

Based on the steps above the algorithm for image filtering requirements is based on image pattern recognition Gradient and Gabor filter banks were utilized based on the image masks for parameter extraction so that with tuning the frequency (u) and the orientation of the filters (Θ) can be extracted for masking requirements. The parameter Θ can be varied using angle orientation method with a separation angle of 30 or 45 degrees. The frequency parameter "u" is altered with the levels of 1√2, 2√2, 4√2, ..., (N / 4) √2 based on N which depends on the width of the image on which the filter is being applied. With the advent of filters respective masks as shown in example figure 4 below are applied that provide the IF else rule creation for respective parameter extraction of the image which are based on the magnitude response of filtering. This magnitude response is based on the thresholding methods as shown in figure above. These parameters are then utilized for power density extraction using Fourier Transform as shown in example figure 5 and 6 respectively.
Mathematically the Fourier Transform operation decomposes the input to provide to a sum of sinusoidal signals [8]. For the requirements of image processing application the input image or the extracted features from the image based on preprocessing are represented in the spatial domain (x,y). IF the input images width considered in N then the Fourier transform can be illustrated with the equation 1 below for the considerations of u=0 ... N-1 and v=0 ... M-1. Along with equation 1 we also require equation 2 which provides us the magnitude spectrum of the input where Fr(u) is the real part of the complex Fourier transform result and Fi(u) is the imaginary part.

$$f(u,v) = \sum_{x=0}^{N=1} \sum_{y=0}^{M=1} f(x,y) e^{-j2\pi\left(\frac{ux}{N}+\frac{vy}{M}\right)} \text{.........(1)}$$

$$M(u) = |F(u)| = \sqrt{Fr^2(u) + Fi^2(u)} \dots\dots\dots\dots\dots\dots\dots\dots\dots\dots..(2)$$

As the input image is a textured image so the magnitude spectrum not only represents the frequency components but it also provides us a way of elimination of periodic direction that is horizontal or vertical for analyzing the image. The low frequency components are eliminated based on the Otsu's method so that we get the optimal threshold value between 30% and 70 % of the power density spectrum.

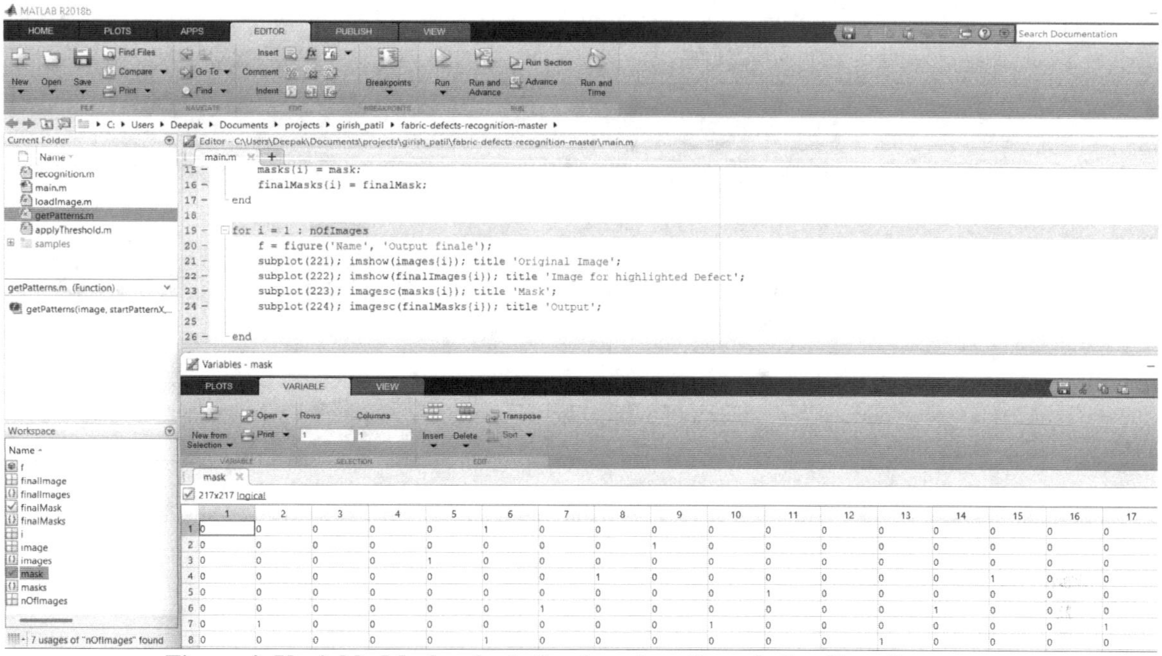

Figure 4: Variable Mask calculation based on Correlation Function.

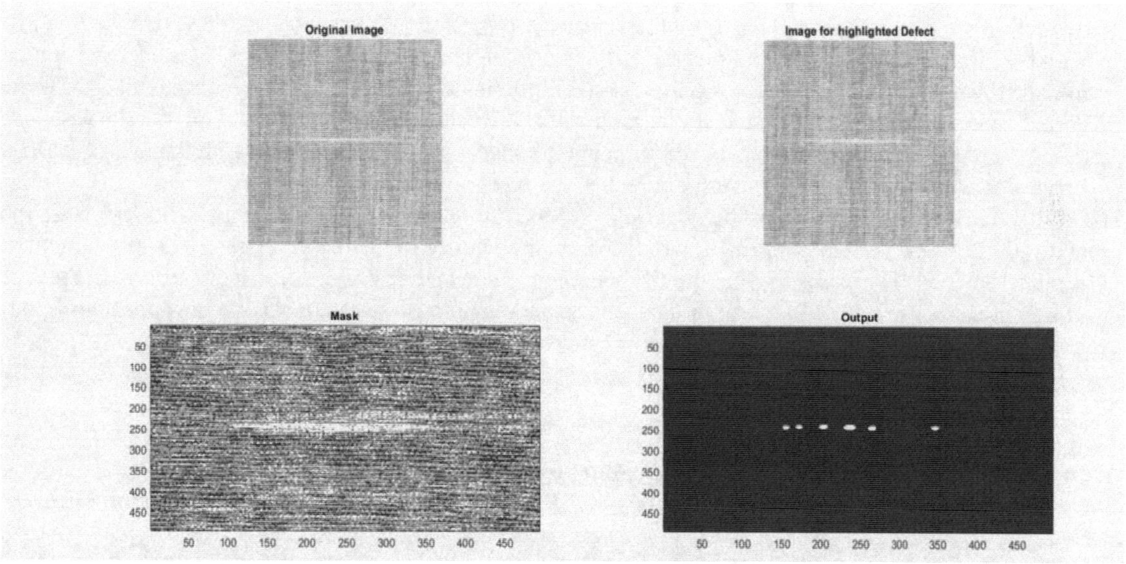

Figure 5: Power Density Based Defect highlighting sample 1.

Journal of Physics: Conference Series **2327** (2022) 012033 doi:10.1088/1742-6596/2327/1/012033

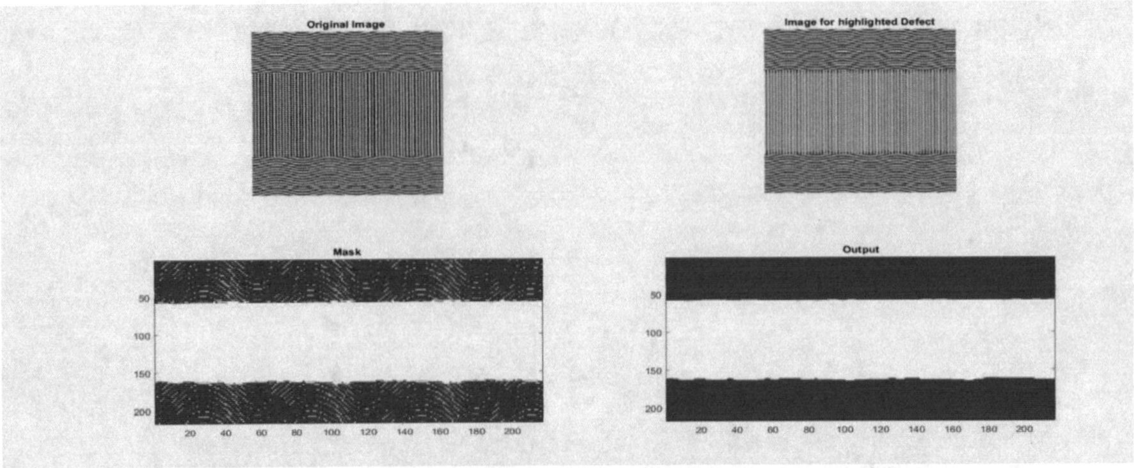

Figure 6: Power Density Based Defect highlighting sample 20.

4. Results and Conclusion

From the point of view of spatial domain the 2D Gabor and Gradient filters are utilized as they are the best suitable option for texture segmentation as well as boundary detection. Based on the filtering the angle orientation of 30 and 45 degree are utilized so that the defective textures frequency components are highlighted with the help of thresholding utilized in the algorithm. For input frequency response extraction we have kept N = 9 based on the various statistical parameters considered where we find that with the increase in number of samples the edge formed of the image also increases. This edge is periodic in nature which promises that the smallest convolution is detectable which becomes favorable for the defect detection. As per the test cases and input size of images considered we found that the response time is directly proportional to the size of the input image. This was concluded with the fact of various images considered of 512x512 size for defect detection as shown in below table 2.

Images with highly periodic and homogeneous structures are considered as in those images the textures follow a perfect pattern providing better probability of defect detection. Yet this structural limitation needs to be crossed over with the help of machine learning algorithm that can classify the type of defect as well with the increase in dataset tremendously to provide conclusive breakthrough in our research in near future.

Table 2: Fabric Defects Highlighted

Original Image	**Highlighted Image**	**Mask**	**Power Spectrum Output**
	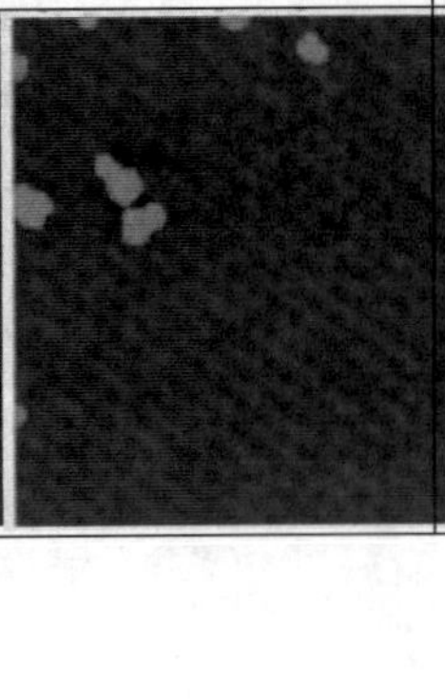	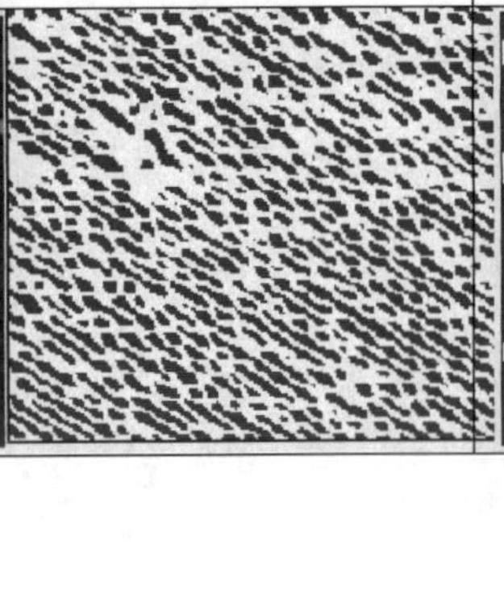	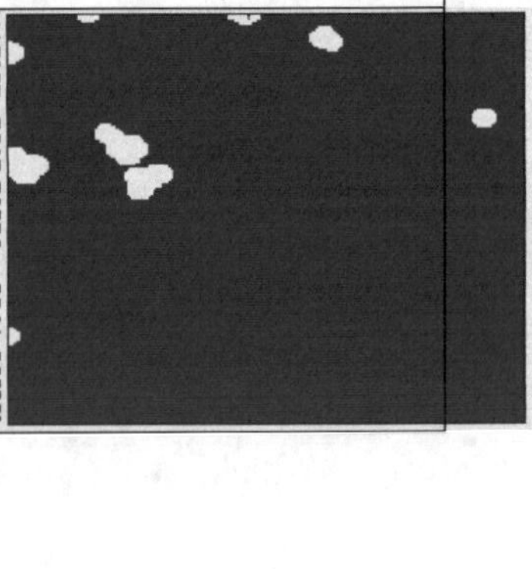

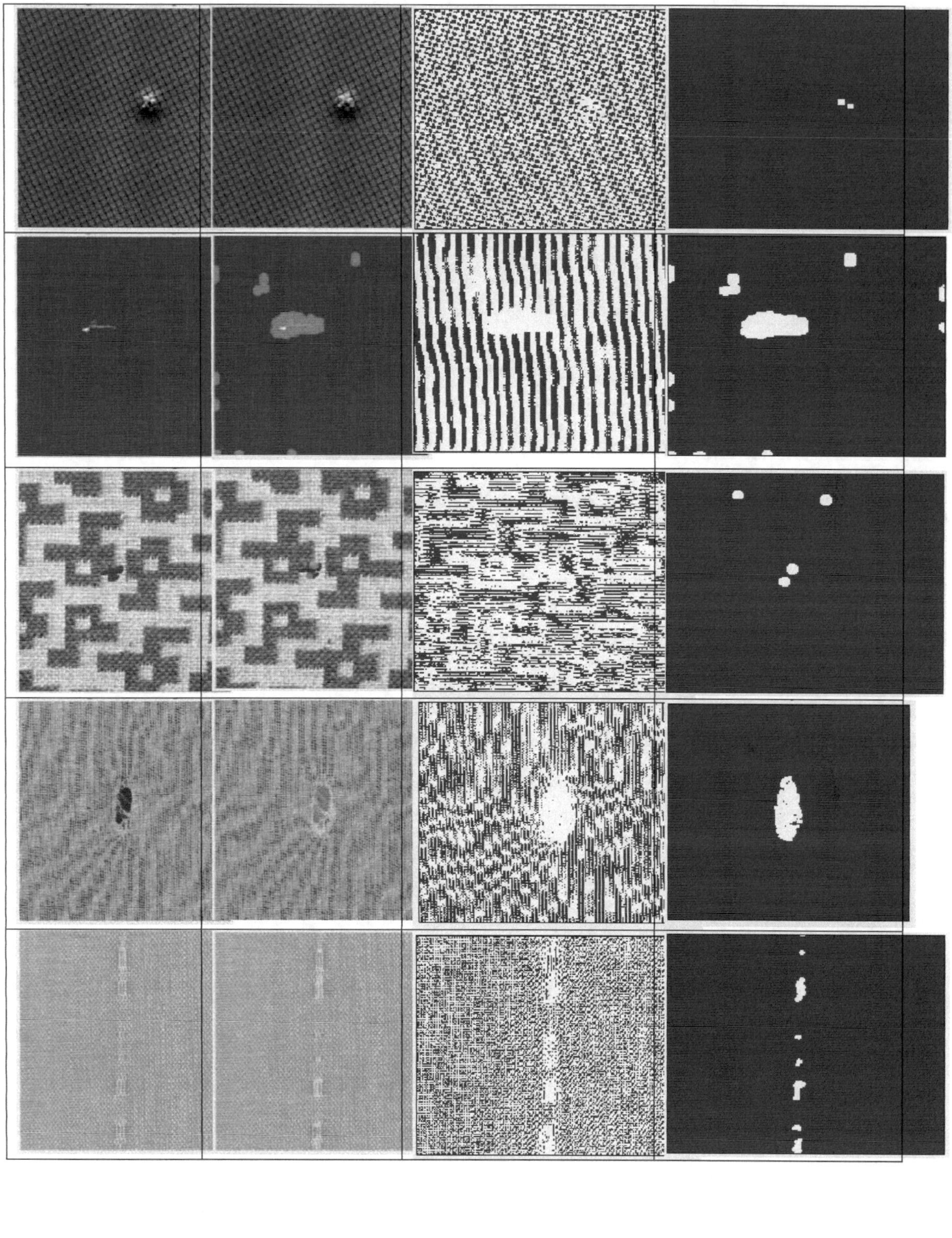

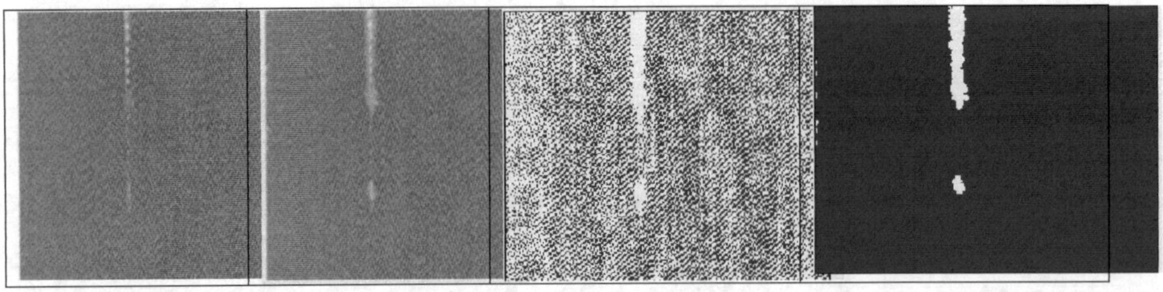

References

[1] A Abouelela, H M Abbas, H Eldeeb, A.A. Wahdan and S.M. Nassar, Automated vision system for localizing structural defects in textile fabrics, Pattern Recognit. Lett. 26 (2005) 1435–1443.

[2] Y. Yuan, Fabric defect detection using fuzzy inductive reasoning based on image histogram statistic variables, Sixth International Conference On, Fuzzy Systems and Knowledge Discovery (2009) 191–194.

[3] BLOG Website: https://www.intouch-quality.com/blog/5-common-fabric-defects-prevent

[4] H Imaoka , SInui,H Niwaya and AShibuya, Trial on Automatic Measurement of Fabric Density. J Society of Fiber Science & Technology (1988) , 44(1): 32-39,

[5] HSari-Saraf, JSGoddard and On-line optical measurement and monitoring of yarn density in woven fabrics. Automat Opt Inspect Ind, SPIE 2899: 444-452.,

[6] Hammouda K, Ed Jernigan (2006) Texture Segmentation Using Gabor Filters. University of Waterloo, Ontario, Canada, p. 1-8.

[7] C Li, J Li, Y. Li, Lingmin He, X Fu and J Chen, " Fabric defect Detection in Textile Manufacturing: A survey of the State of the Art", Hindawi Security and Communication Networks, Volume 2021, Article ID 9948808, 13 pages, https://doi.org/10.1155/2021/9948808

[8] Heideman, AM Johnson, DH Burrus and Sidney C, Gauss and the history of the fast Fourier transform. Archive for History of Exact Sciences 34(3),(1985), 265-277.

Journal of Physics: Conference Series **2327** (2022) 012034 doi:10.1088/1742-6596/2327/1/012034

Performance evaluation of various machine learning and deep learning models for gene expression

Tanima Thakur[1], Isha Batra[2] and Arun Malik[3]

[1,2,3]School of Computer Science and Engineering, Lovely Professional University, Jalandhar (Punjab) India

tanimathakur21@gmail.com

Abstract. Cancer is among those deadliest diseases that affects a mass. So, it requires the timely detection and treatment. There are number of ways through which cancer can be detected. Some of those ways are some "blood tests, CT (Computerized Tomography) scan, MRI (Magnetic Resonance Imaging), X-Ray, Ultrasound, gene expression, etc.". Genes help doctors to find whether a person is suffering or going to suffer from any disease or not. In this paper, we have gone through various ML and DL from the literature. The models are trained and tested on 5 different types of datasets. These are "Lung Adenocarcinoma (LUAD), Lung Squamous Cell Carcinoma (LUSC), Breast Invasive Carcinoma (BRCA), Kidney Renal Clear Cell Carcinoma (KIRC), and Uterine Corpus Endometrial Carcinoma (UCEC)". The models are evaluated on the different performance measures such as "MSE, Precision, Recall, F1_Score and Accuracy".

1. Introduction

Cancer is affecting large group of people. Various research is going on to timely predict the cancer so that patient's life can be saved from this deadly disease. One of that research is on gene expression. The genetic information of a human being is present in his DNA that is required for the synthesis of protein in human body [1]. "The process of transcribing a gene's DNA sequence into the RNA that serves as a template for protein production is known as gene expression" [2]. Targeted genes are identified to predict cancer. Different types of experiments are also done on genes so that specific purpose can be obtained such as predicting disease and creating hybrid species [3].

The complicated interaction between gene and environment causes cancer to develop as a disease that is dependent on several factors. As per the World Health Organization, cancer is the primary cause of mortality worldwide [30]. The detection of a cancer is a topic of concern. Lot of microarray data is available that promotes analysis of gene data and development of new techniques that have been used for diagnosis of cancer. But this type of data suffered from some problems such as high dimensions and small sample size [28]. There are various types of ML and DL models used for cancer prediction. Since our birth we are loaded with data. The sensors of the human body such as eyes, ears, nose, tongue, and nerves are generating huge amount of raw data on daily basis. This raw data has been converted by brain into important signals such as tastes, sounds, etc. We used to share these with other humans through a medium of language. This data was recorded manually in the beginning but due to lot many advancements, the process of recording data has been made automated and computerized now. For doing this task, sensors are used. We can access the stored data and can convert that into useful information. ML is a field that is responsible for creating such algorithms that are required to convert the data into useful information. ML is having its applications in most of the fields. Doctors can treat their patients with the help of ML. Engineers can build smart homes using ML and many more [4].

DL comes under ML. Lot many data and resources are required for solving various problems using DL techniques. And the present era is a source of that data and resources that act as a pavement for DL techniques [5]. ML models provide good performance in various areas. But still there are some areas such as speech and object recognition where ML is lacking behind [6]. Driverless cars are one of the outcomes of DL. DL is gaining popularity on daily basis [7].

In this paper, various ML and DL models have been reviewed and even evaluated based on different performance measures. Section II presents various methods of cancer prediction from literature. Section III presents findings and results from the literature and conclusion in Section IV.

2. Related Work

Various models of ML and DL are presented in the given section. The models are trained and tested on 5 different datasets and their performance is also evaluated. DNA microarrays is one of the cancer classification methods that is used for gene expression. It has been used for class discovery and class prediction and it is found to be as the feasible method [8]. Computational analysis and Artificial Neural Network based methods have also been used for classification of cancer in gene expression [2], [9]. One of the prediction methods based on compound covariate has been used to predict classes of tumor [10]. The features from gene expression have been selected using backward-elimination method [11]. The semi-supervised ellipsoid ARTMAP and PSO based method has been used for tumor tissue separation, but this method is having noise and dimensionality problems [12].

Tumor has been predicted using RBF-NN ("Radial Basis Function Neural Network") and RBFS ("Rough-Based Feature Selection") Method with high accuracy [13]. The core structure of multiple synthetic datasets, majorly cancerous datasets has been discovered by Disagreement and Agreement Index (DAI) [14]. The traditional method such as Differentially Expressed Genes has not extracted much knowledge from the genes, but "Reverse Engineering method-based on Bayesian-Model Average" provide that knowledge as well [15]. The segmentation of breast cancer has been done using multichannel "Markov Random Fields" [16]. The method based on Locality Sensitive Laplacian Score provides more accuracy for selecting genes than the other existing methods of gene expression [17]. The samples of tumor have been classified from gene expression using PCA (Principal Component Analysis) [18].

The breast cancer has been classified using lightweight Convolutional Neural Network (CNN), providing 98.76 accuracy [19]. "Binary Particle Swarm Optimization" along with "Decision Tree" ("BPSO-DT") and "CNN" has been used to diagnose cancer from RNA-Seq gene expression data providing accuracy of 96.90 [20]. Another CNN based method, known as GeneeXpression Network has been used to classify multi tissues of cancer [21]. The method of Partial Least Squares has been used to find genes in high dimension data [22].

There are different types of CNN such as 1D CNN and 2D CNN. 1D CNN has been used for small data but on the other side 2D CNN works for the large data. 2D CNN outperforms 1D CNN [23], [24]. ML models perform well on the small data, but for big data DL models are most preferred to get the promising results [25]. Various computational, centrality and ML methods require prior knowledge to perform different operations such as identifying important proteins, designing score functions, and selecting features respectively. But the deep learning method proposed in [26] finds the essential proteins without any prior knowledge.

The location of the branch point in gene expression has been identified using hybrid model of dilate CNN and RNN (Recurrent Neural Network) [27]. The method based on Deep Fuzzy Neural Network has been developed to diagnose various diseases by handling problems in gene data. It has been found as the reliable method than existing classification methods [28]. The clustering of genes has been done using a new method that is based on MOO (Multi-Objective Optimization) clustering along with DL methods. This method prevents the biological and statistical references of the genes during clustering [29].

3. Research Findings

This section presents the research findings from the literature. Cancer is one of the life-threatening diseases that requires utmost care and treatment. There are different types of methods that are used to detect a cancer in human body. We have compared various ML and DL models from literature. Five types of cancer have been targeted here namely "Lung Adenocarcinoma (LUAD), Lung Squamous Cell Carcinoma (LUSC), Breast Invasive Carcinoma (BRCA), Kidney Renal Clear Cell Carcinoma (KIRC), and Uterine Corpus Endometrial Carcinoma (UCEC)". Dataset has been taken from [31]. The same dataset has been used in [20] as well. We have implemented various ML models such as "Random Forest Classifier, Support Vector Classifier, Logistic Regression, Gradient Boosting Classifier, and XGradient Boosting Classifier" using Python on Jupyter Notebook in Google Colab. Various CNN and RNN models are also implemented on the same interface. CNN models implemented are VGG16, VGG19, ResNet50, and InceptionV3 along with RNN models such as GRU, LSTM, and Simple RNN. To implement our models, following steps have been taken into consideration:

1. Data has been collected first.
2. After that the dataset has been pre-processed.
3. Dataset has been broken down into training and testing.
4. Model has been trained using training data
5. Model has been tested and evaluated based on various performance measures.

There are various packages that we have used during our implementation. Some of those packages are random, re, math, numpy, sklearn, scipy, pandas, Xgboost, matplotlib.pyplot, and tensorflow. Fig 1 shows the various performance measures such as MSE, Precision, Recall, F1-Measure and Accuracy for various ML models. It has been found that Gradient Boosting is having highest MSE and Random Forest is having lowest MSE. Similarly Random Forest is having highest accuracy and Gradient Boosting Classifier is having lowest accuracy. Random Forest is performing well among all the models on all the given parameters. All other parameters are also given in the figure 1.

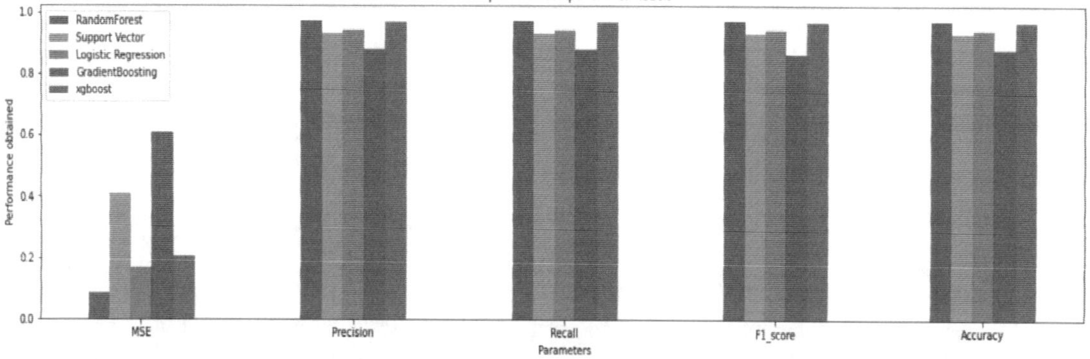

Figure 1 Performance Comparison of Machine Learning Models

TABLE I presents values of various performance measures for the different machine learning models. Random Forest is having lowest MSE of 0.08 and highest accuracy of 0.97. The highest MSE is 0.61 and it is of Gradient Boosting. As a result, Random Forest is the best ML Model out of the bunch.

TABLE I Performance Comparison of Machine Learning Models.

	RF	SV	LR	GB	XGBoost
MSE	0.08	0.40	0.16	0.61	0.20
Precision	0.97	0.93	0.94	0.88	0.96
Recall	0.97	0.93	0.94	0.88	0.96
F1 Score	0.97	0.93	0.94	0.86	0.96
Accuracy	0.97	0.93	0.94	0.88	0.96

Figure 2 shows the performance measures for various CNN Models. It has been found that VGG16 is having highest MSE and ResNet50 is having lowest MSE. Similarly, ResNet50 is having highest accuracy and VGG16 is having lowest accuracy. ResNet50 is providing best results for all the parameters.

TABLE II presents values of various performance measures for the different CNN models. ResNet50 is having lowest MSE of 1.33 and highest accuracy of 0.84. The highest MSE is 2.22 and it is of VGG16. So, it means best CNN model among all is ResNet50.

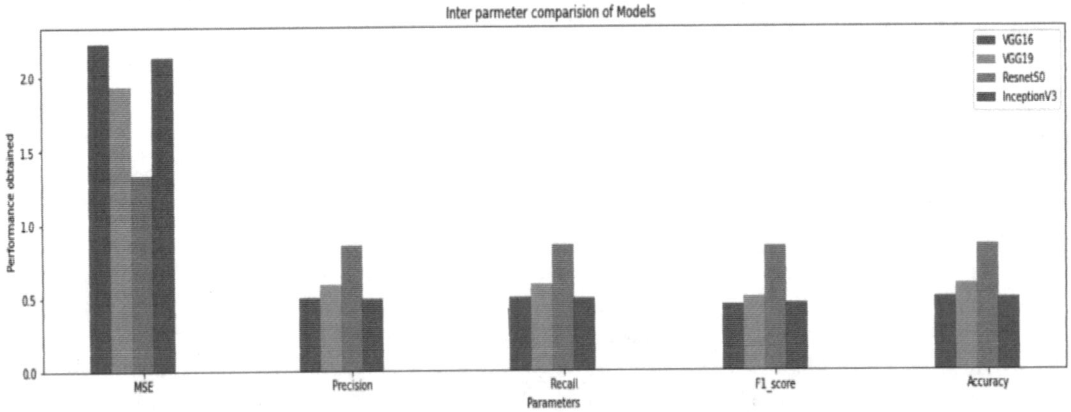

Figure 2 Performance Comparison of CNN Models

TABLE II Performance Comparison of CNN Models.

	VGG16	VGG19	Resnet50	InceptionV3
MSE	2.22	1.93	1.33	2.13
Precision	0.49	0.58	0.84	0.48
Recall	0.49	0.58	0.84	0.48
F1 Score	0.44	0.50	0.84	0.45
Accuracy	0.49	0.58	0.84	0.48

Figure 3 shows the performance measures for various RNN Models. It has been found that GRU is having the highest MSE and LSTM is having the lowest MSE. Similarly, GRU is having lowest accuracy and LSTM is having highest accuracy. So, LSTM is providing best results among all the models for the different performance measures.

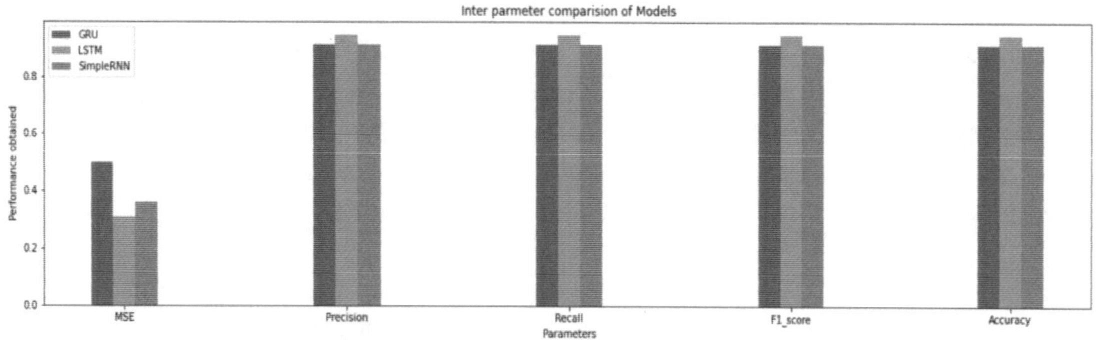

Figure 3 Performance Comparison of RNN Models

TABLE III presents the values for different performance measures for the given RNN models. LSTM is having lowest MSE of 0.30 and highest accuracy of 0.94. The highest MSE is 0.50 and is of GRU. So, it means LSTM is the best model among the given RNN models.

TABLE III Performance Comparison of RNN Models.

	GRU	LSTM	Simple RNN
MSE	0.50	0.30	0.36
Precision	0.91	0.94	0.91
Recall	0.91	0.94	0.91
F1 Score	0.91	0.94	0.91
Accuracy	0.91	0.94	0.91

4. Conclusion

Cancer is one of the life-threatening diseases that requires utmost care and treatment. It is affecting large group of people. There are different types of methods that are used to detect a cancer in human body. Various research is going on to timely predict the cancer so that patient's life can be saved from this deadly disease. One of that research is on gene expression. In this paper, we have reviewed various ML and DL models. Some of those techniques have been implemented on the dataset to classify 5 types of cancer. It has been found that Random Forest is the best model among all the given machine learning models. On the other hand, ResNet50 is the best model among the given CNN Models. It has also been found that LSTM is the best model among all the given RNN Models.

References

[1] https://byjus.com/biology/difference-between-dna-and-rna/.
[2] Slonim DK, Tamayo P, Mesirov JP, Golub TR, Lander ES. Class prediction and discovery using gene expression data. InProceedings of the fourth annual international conference on Computational molecular biology 2000 Apr 8 (pp. 263-272).
[3] https://www.khanacademy.org/test-prep/mcat/biomolecules/dna-technology/v/gene-expression-and-function
[4] Lantz B. Machine learning with R: expert techniques for predictive modeling. Packt publishing ltd; 2019 Apr 15.
[5] Salas J, de Barros Vidal F, Martínez-Trinidad F. Deep learning: Current state. IEEE Latin America Transactions. 2019 Dec;17(12):1925-45.
[6] Goodfellow I, Bengio Y, Courville A. Deep learning. MIT press; 2016 Nov 10.
[7] https://in.mathworks.com/discovery/deep-learning.html.
[8] Golub TR, Slonim DK, Tamayo P, Huard C, Gaasenbeek M, Mesirov JP, Coller H, Loh ML, Downing JR, Caligiuri MA, Bloomfield CD. Molecular classification of cancer: class discovery and class prediction by gene expression monitoring. science. 1999 Oct 15;286(5439):531-7.

[9] Khan J, Wei JS, Ringner M, Saal LH, Ladanyi M, Westermann F, Berthold F, Schwab M, Antonescu CR, Peterson C, Meltzer PS. Classification and diagnostic prediction of cancers using gene expression profiling and artificial neural networks. Nature medicine. 2001 Jun;7(6):673-9.

[10] Radmacher MD, McShane LM, Simon R. A paradigm for class prediction using gene expression profiles. Methods. 2018 Sep 1.

[11] Duan KB, Rajapakse JC, Wang H, Azuaje F. Multiple SVM-RFE for gene selection in cancer classification with expression data. IEEE transactions on nanobioscience. 2005 Aug 29;4(3):228-34.

[12] Xu R, Anagnostopoulos GC, Wunsch DC. Multiclass cancer classification using semisupervised ellipsoid ARTMAP and particle swarm optimization with gene expression data. IEEE/ACM Transactions on Computational Biology and Bioinformatics. 2007 Feb 20;4(1):65-77.

[13] Chiang JH, Ho SH. A combination of rough-based feature selection and RBF neural network for classification using gene expression data. IEEE transactions on nanobioscience. 2008 Mar 3;7(1):91-9.

[14] Yu Z, Wong HS. Class discovery from gene expression data based on perturbation and cluster ensemble. IEEE transactions on nanobioscience. 2009 Jun 2;8(2):147-60.

[15] Kim H, Gelenbe E. Reconstruction of large-scale gene regulatory networks using bayesian model averaging. IEEE Transactions on NanoBioscience. 2012 Sep 10;11(3):259-65.

[16] Ashraf AB, Gavenonis SC, Daye D, Mies C, Rosen MA, Kontos D. A multichannel Markov random field framework for tumor segmentation with an application to classification of gene expression-based breast cancer recurrence risk. IEEE transactions on medical imaging. 2012 Sep 19;32(4):637-48.

[17] Liao B, Jiang Y, Liang W, Zhu W, Cai L, Cao Z. Gene selection using locality sensitive Laplacian score. IEEE/ACM Transactions on Computational Biology and Bioinformatics. 2014 Jun 3;11(6):1146-56.

[18] Liu JX, Xu Y, Zheng CH, Kong H, Lai ZH. RPCA-based tumor classification using gene expression data. IEEE/ACM Transactions on Computational Biology and Bioinformatics. 2014 Dec 25;12(4):964-70.

[19] Elbashir MK, Ezz M, Mohammed M, Saloum SS. Lightweight convolutional neural network for breast cancer classification using RNA-seq gene expression data. IEEE Access. 2019 Dec 18;7:185338-48.

[20] Khalifa NE, Taha MH, Ali DE, Slowik A, Hassanien AE. Artificial intelligence technique for gene expression by tumor RNA-Seq data: a novel optimized deep learning approach. IEEE Access. 2020 Jan 29;8:22874-83.

[21] Khorshed T, Moustafa MN, Rafea A. Deep learning for multi-tissue cancer classification of gene expressions (GeneXNet). IEEE Access. 2020 May 6;8:90615-29.

[22] Ji G, Yang Z, You W. PLS-based gene selection and identification of tumor-specific genes. IEEE Transactions on Systems, Man, and Cybernetics, Part C (Applications and Reviews). 2010 Oct 21;41(6):830-41.

[23] Kiranyaz S, Avci O, Abdeljaber O, Ince T, Gabbouj M, Inman DJ. 1D convolutional neural networks and applications: A survey. Mechanical systems and signal processing. 2021 Apr 1;151:107398.

[24] Wu Y, Yang F, Liu Y, Zha X, Yuan S. A comparison of 1-D and 2-D deep convolutional neural networks in ECG classification. arXiv preprint arXiv:1810.07088. 2018 Oct 16.

[25] Thakur T, Batra I, Luthra M, Vimal S, Dhiman G, Malik A, Shabaz M. Gene expression-assisted cancer prediction techniques. Journal of Healthcare Engineering. 2021 Aug 19;2021.

[26] Zeng M, Li M, Fei Z, Wu FX, Li Y, Pan Y, Wang J. A deep learning framework for identifying essential proteins by integrating multiple types of biological information. IEEE/ACM transactions on computational biology and bioinformatics. 2019 Feb 5;18(1):296-305.

[27] Nazari I, Tayara H, Chong KT. Branch point selection in RNA splicing using deep learning.

IEEE Access. 2018 Dec 13;7:1800-7.

[28] Mudiyanselage TK, Xiao X, Zhang Y, Pan Y. Deep fuzzy neural networks for biomarker selection for accurate cancer detection. IEEE Transactions on Fuzzy Systems. 2019 Dec 9;28(12):3219-28.

[29] Dutta P, Saha S, Chopra S, Miglani V. Ensembling of gene clusters utilizing deep learning and protein-protein interaction information. IEEE/ACM transactions on computational biology and bioinformatics. 2019 May 23;17(6):2005-16.

[30] Pati J. Gene expression analysis for early lung cancer prediction using machine learning techniques: An eco-genomics approach. IEEE Access. 2018 Dec 20;7:4232-8.

[31] K. N. C. Ferles and Y. Papanikolaou, ``Cancer types: RNA sequencing values from tumor samples/tissues," 2018. Distributed by Mendeley. [Online]. Available: https://data.mendeley.com/datasets/sf5n64hydt/1

A Comprehensive Review of Emerging Technologies: Machine Learning and UAV in Crop Management

S.Yagnasree[1]

Anuj Jain[2]

PhD Scholar, Lovely Professional University, Punjab[1],
Professor, Lovely Professional University, Punjab[2]

Coresponding author's e-mail: a1978jain@gmail.com

Abstract. It is a known fact that India is one among the major food producers in the world. Due to the rapid urbanization and Agrochemical interventions, plants are infected by insects, pathogens, massive infestations, which leads to the deficiency of growth nutrients and ions, thereby minimizing agricultural crop yield. These infections occur in several forms like aphids, black root rot, cankers, leaf curls, rusts, anthracnose, and blights make agriculture fail to thrive and increase the impact of crop diseases. Since plant disease prevention is a continuous process that occurs regularly, agriculturists follow several strategies to keep their crops disease-free. An efficient monitoring and supporting system for continuous and long-term plant and soil health monitoring is needed to meet the needs of growing population. In this paper, existing research works in Precision agriculture, emerging technologies – GPS, GIS, Machine learning and UAVs in analyzing crop health analysis, soil health monitoring, and crop yield prediction are reviewed.
Keywords: UAV-Low altitude remote sensing, Precision Agriculture; Crop and Soil health, monitoring; Yield prediction, Machine learning.

1. Introduction

The first agricultural revolution has been initiated during the advent of increased mechanization from 1900 to 1930 as an era of industrial revolution. During this era each farmer produced enough food to meet the need of the hour, resulting growth in India's overall economic prosperity to a great extent during its early years. However, small farmers found difficulty to adopt such expensive methods, also excess and inappropriate use of fertilizers and pesticides polluted waterways and killed beneficial insects and wildlife. The immoderate irrigation practices led to eventual soil degradation. Henceforth to meet the food demand, while limiting the impact on the environment and coping with climate change, farmers and agriculture companies have turned to IoT for gathering the data by using sensors to analyze diseases, soil health and crop yield. This found to be challenging the farmers and industrialists and hence innovation in global agriculture is a central part of the solution [1]. According to forecast by FAO of the UN, food production needs to increase by 70% by 2050, placing pressure on food quality standards excluding increase in agricultural production for biofuels to meet the requirements of growing population [1]. Climate changes have certainly negative impact on crop yield. Growing water demand and rising temperatures are expected to further increase water stress in many agricultural areas by 2025 [1]. Henceforth to overcome this, emerging agricultural technologies also afford the opportunity to boost production without compromising standards [3][4].

2. Technologies used in Precision Agriculture

To optimize the overall performance in increasing production, reducing agrochemical application precision agriculture can combine multiple analysis process and technological improvements at different phases of production from sowing to harvesting like Remote sensing technologies- UAVs and satellites, data acquisition technologies- GPS systems, GIS systems, Production mapping, Soil mapping, mapping of the soil Electrical Conductivity and Variable Rate Technologies.

2.1 GPS Systems: GPS systems, generally mounted on the moving vehicle contouring the field, outputs real-time information regarding location of the areas suffering from pests and diseases, location of the soil samples captured to monitor the production status of each area identified by various sensors and to apply agrochemical doses at the affected area [9].

2.2 GIS Systems: Agricultural information provided by geo-location data recorded by GPS systems is used to create production maps [9]. These production maps carry massive amount of data which requires use of appropriate software for processing it [9].

2.3 Variable Rate Technology (VRT): Considering large land holders VRT is becoming more prevalent in collecting field information. Basically, there are two variants of VRT, map-based, which requires a vehicle that follows prescribed map and GPS for application of agrochemical products in the field whereas and sensor-based requires a vehicle outfitted with sensors that measure the properties of each region. Information collected by sensors is transmitted to computing system that calculates the needs of crop and soil [9].

2.4 Remote sensing (RS) systems: Remotely captured images/data in huge continuous narrow bands can be used for understanding the biophysical properties [6-9]. Remote sensing can detect interaction between electromagnetic radiation and soil or plant material on the Earth's surface. The amount of energy reflected from the leaves varies based on its biochemical properties.

2.5 Artificial Intelligence: AI systems work by ingesting large amounts of labeled training data, analyzing *the* data for correlations and patterns, and based on these patterns make predictions about future states. Generally, Artificial intelligence depends on three cognitive skills- Learning, reasoning and self-correction.

2.6 Machine learning: Machine learning, makes use of trained data and algorithms, a branch of artificial intelligence and computer science, and replicates the way that humans learn improving accuracy constantly [12].
Sometimes AI is interpreted to be the same as ML but, there is a distinct difference. AI uses some agents whereas ML is more or less a statistical procedure. AI uses the bigdata for learning, reasoning but also self-corrects for future actions. As an example, Chat-bot, image recognition tool. AI-powered virtual agents are efficient and always available but at the same time it is expensive, requires deep technical knowledge and limited experts available to build AI tools [14]. Main objective of ML is, to classify the data based on models developed, make predictions for future and outcomes based on these models [12]. ML is broadly classified into prediction and classification [20]. Figure 1. gives a generic Machine learning system.

- A pre-processing effort is required for converting complex raw data into a suitable state. Normally pre-processing in ML system consists of: A. Data filtering for noise cancellation, removing contradictory items, B. Data integration, when multiple-data sources exist and C. Data transformation to standardize the data, such as normalization and discretization.

Journal of Physics: Conference Series **2327** (2022) 012035 doi:10.1088/1742-6596/2327/1/012035

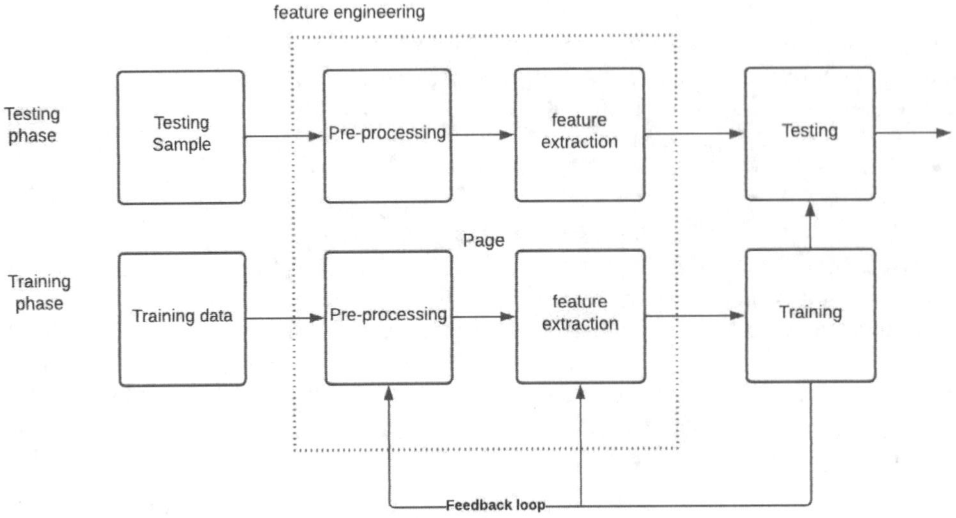

Figure 1. Machine learning system

- Feature selection focuses at identifying or/and creating the subset of features necessary for further implementation of learning model throughout the training phase.
- A feedback loop is included for any adjustments pertaining to the feature extraction unit and preprocessing unit that further improves performance of the learning model.
- Unknown data samples, called as feature vectors are given to the trained model for testing.

Machine learning can be applied where human expertise cannot be applied (mars rover to navigate), humans cannot explain their expertise, in genomics where big data-based models to be handled. Apart from these ML application can also be seen in stock market forecasting, credit card fraud detection, image recognition. Usually there are 4 types of Machine learning:

- *Supervised or inductive learning:* Training data and desired outputs with labels are given as inputs to the machine and the machine performs task of learning a function that maps an input to an output based on example input-output pairs. Two varieties of supervised learning: Regression and classification
- *Unsupervised learning:* Training data without desired outputs or unlabeled outputs are given as inputs to the machine.
- *Semi-supervised learning:* Training data with few desired outputs (labeled outputs) are given as inputs to the machine.
- *Reinforcement learning:* Rewards from sequence of actions (rewarding desired behaviors and/or punishing undesired ones)

The application of ML is continuously evolving in different sectors reducing the human induced errors. It found its way into agriculture as well, Crop Management: Involves in studies concerning crop growth prediction, disease detection, weed accumulation, crop recognition and crop quality; Water Management- associated with optimal use of water resources; Soil management- concerned with soil protection and soil management aspects; Livestock Management- Management pertaining to animal welfare and livestock production.

2.7 Low-altitude remote sensing Unmanned Aerial Vehicle:

UAV, An emerging low altitude remote sensing technology, is currently an area of thirst for researchers in precision agriculture. The earliest developed UAVs initially alarmed with military operations, but the development of emerging technologies as Inertial Measurements Unit (IMU), imaging sensors, synthetic aperture radar, Global Navigation Satellite Systems (GNSS) has initiated a path for the development of commercial UAVs, capable of assisting various support systems (e.g., geomatics, logistics).

2.7.1 UAV system architecture: UAV, an aerial remote sensing system operated as aerial-platform carry payloads such as multi-spectral, digital, and hyper spectral cameras mounted on it, capture images with high resolution. These images provide information to extract features. In general UAV system consists of a flight control system, payload system and a ground-control system. Figure 2. deploys an UAV based aerial remote sensing system.

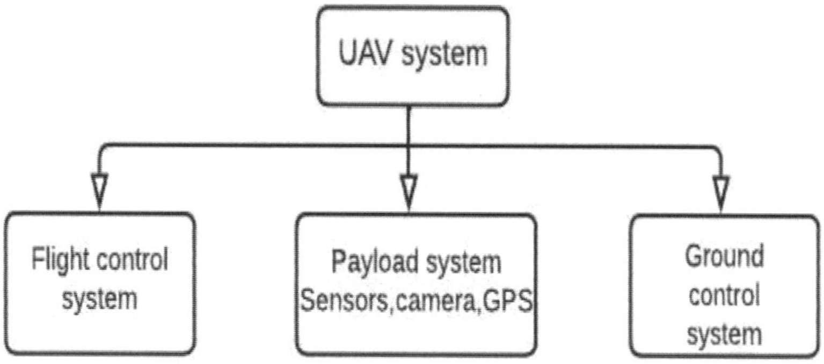

Figure 2. UAV monitoring system

2.7.2 Types of UAVs:

Different varieties of UAVs differ in size, weight, payload capability, resistant to harsh environment, power, speed, and endurance time. Agriculture crop monitoring system, generally preferred with light weight UAV class. UAVs are classified based on the type of flight mechanism as fixed wing, Multi-rotor, single rotor and fixed wing hybrid [3-10]. Table 1 projects classification of UAVs based on different parameters [31].

Table 1. Classification of UAVs based on weight, altitude and range, wings and rotors and applications [31]

Classification parameter	Types of each classification	Typical Features
Weight	Nano	Weight < 250gms
	Micro	250gms < Weight < 2kgs
	Small	2kgs < Weight < 25kgs
	Medium	25kgs < Weight < 150kgs
	Large	Weight < 150kgs
Altitude and range	Hand held	Altitude < 600m ; Range <2km
	Close	Altitude < 1500m ; Range <10km
	NATO	Altitude < 3000m ; Range <50km
	Tactical	Altitude < 5500m ; Range <160km
	MALE (Medium-Altitude Long Endurance)	Altitude < 9100m ; Range <200km
	HALE (High-Altitude Long Endurance)	Altitude > 9100m ; Indefinite range
	Hypersonic	Altitude around 15200m ; Range >200km
Wings and Rotors	Fixed wing	Fast, long endurance time, and large coverage area, no hovering and Vertical take-off
	Multi rotor	Short endurance time, small payload capacity, support VTOL and hover, and superior camera control
	Single rotor	Longer endurance times, support VTOL and hovering, good payload capability, expensive and needs to be trained effectively
	Fixed-wing hybrid VTOL	Supports VTOL and longer endurance time, imperfect forward flight or hovering
Application	Personal	Videography and entertainment

Commercial	Infrastructure surveillance, goods delivery and aerial imagery
Government and Law enforcement	Fire fighting and patrolling
Military	Surveillance and combat attacks

2.7.3 UAV applications in Agriculture:
The ability of UAV to fly at low altitude can result in high spatial resolution crop images, which can enhance the performance of the crop monitoring systems [10]. According to the literature survey, the most identified applications of Unmanned aerial vehicles for Precision agriculture are [3][10]:

- Weed detection and management
- Crop growth monitoring and yield prediction
- Vegetation health monitoring, diseases, insect pests identification
- Water management
- Spraying agrochemical doses in the field
- Natural disaster assessment

3. Literature Review

A study is conducted on research papers related to pests, weeds, and disease detection using images captured by UAVs and majority of them reviewed that subjective visual-observations of images read by UAVs are used in crop management [2]. It is also observed that, in the area of weed mapping and disease mapping, quantitative, standardized and repeatable reference data measurement techniques are lacking [2]. Hence, researchers must focus on quality and quantity of reference data to be measured and consider imaging methods that can carefully define image characteristics for further analysis [2].
In satellite based remote sensing technique the degree of overlap between X and Y during crop-growth periods is low and hence to overcome this UAV-LARS serves the best emerging technology with significant advantages as simple construction, high spatial temporal resolution [3-5][7][8]. Hence, UAV's can be deployed to deliver multispectral and hyperspectral data of the field in research for monitoring crop health [6]. UAV's can obtain the data in multi-angular directions to monitor soil health, crop health, weed detection at the early stages, pathogen infestation and pest control, nutrient estimation, irrigation management [6]. Surveys conducted on Italian vineyards with UAV, aircraft and satellite using multispectral sensors mounted on the three platforms to study the intra vineyard vegetation spatial variability. Degree of similarity been assessed using a framework and results were comparable in vineyards, featured by coarse vegetation gradients and large vegetation clusters [22].
Considering that most of the farms are located away from towns/cities, where there is a limitation in energy supply, IoT devices cannot be deployed for harvesting, spraying or yield prediction. An agriculture frame work is designed for enhancing energy harvesting rate using a sun tracker device that adjusts the angle between the solar panel and sunlight [9]. To improve the flight time by utilizing wind force a flight mode is introduced [9].
Many agronomists and researchers developed agricultural frameworks on varieties of crops. A tetracam camera placed on a multi-rotor hexa-copter was used to collect multispectral data of vineyard and tomato cultivations [10]. By using a photogrammetric pipeline to handle multispectral data, tri-band ortho-images of surveyed fields were generated [10]. Vegetation indices like Normalized Difference Vegetation Index (NDVI), Green Normalized Difference Vegetation Index (GNDVI), Soil Adjusted Vegetation Index (SAVI) extracted from ortho-images are proven to be effective sources in crop assessment [10]. Another agricultural frame work developed on vineyard cultivation using multispectral photogrammetry and UAV, demonstrates great potential to analyze the Vegetation Indices [18]. The Near-Infrared Spectroscopy (NIRS) and Digital Elevation Model (DEM) were used to collect information on the crop's vegetative condition, soil and plant moisture, and field biomass density maps. PIX4D Cloud Computing technology software is used to process the acquired data for analysis [18].

Studies were conducted on a rapeseed field to monitor and estimate performance indices (VIs) using a UAV with a consumer-grade camera mounted on it with non-modified and modified spectral ranges, as well as collecting hyperspectral data and NDVI using a Green Seeker handheld crop sensor (GS-NDVI) in conjunction with a UAV [5]. Both methods were equally accurate in predicting GS-NDVI [17]. An LSTM-based RNN reinforcement model by applying learning-based models like ML and DL, proposed, combine parametric features to calculate crop yield [11]. Finally, the manager is able to minimize the error and increase the accuracy of the gauge [11].
Considerable studies were made to analyze methods applied and features used such as temperature, rainfall, and soil type and also studied most often deployed algorithms like Artificial Neural Networks in these models [12]. From the studies, proposed suggestions for future research works. After these elaborate studies on machine learning-based papers, additionally authors also searched in electronic databases to identify deep learning-based studies, and extracted the applied deep learning algorithms out of all. Accordingly, Convolutional Neural Networks (CNN) is the most widely used deep learning algorithm in these studies, and the other widely used deep learning algorithms are Long-Short Term Memory (LSTM) and Deep Neural Networks (DNN) [12].
As discussed so far it is observed that Crop and soil images data procured from UAV to be processed to analyze and detect diseases, irresistible infections and then classify plant diseases by applying existing architectures, artificial neural networks [13]. This classification of features serves as a predictive measure. Also, to improve the efficiency of trained model, Gaussian filter is happened to be one of the predictive measures at reduced computational costs [13].
Sometimes the images of the crop were under different lighting conditions, analyzing leaf health may not be accurate. Henceforth, suitable color model can describe colors accurately and these color images are converted to gray images to evaluate plant characteristics precisely. The vertical projection method and the linear scanning method can be are combined to realize the center line of the crop rows [14]. To decrease the computational complexity, Weeds Infestation Rate (WIR) is modified, finally, Modified Weeds Infestation Rate (MWIR) is used to realize real-time decision through the minimum error ratio of Bayesian decision under normal distribution.
Though extensive research is conducted on agricultural robots, certain limitations still persist [15]. These robots must be cost effective and must be able operate at dynamic climate conditions to estimate the yield precisely [15]. Focused research must be conducted on fusing complementary sensors and sensing abilities, simple manipulators must be developed and enhanced path optimization, navigation and path guidance algorithms suitable for dynamic climate conditions must be developed [15].
A great number of fully supervised object detection methods based on CNN have been proposed for better performance [20]. But these deep learning-based methods seeking large number of samples with instance level labels is laborious and time taking process for interpreting images with many targets [20]. Therefore WSOD, Weakly supervised object detection is proposed that require only image level labels along with location information of the object [20]. WSOD model is trained by supervised information generated by a novel instance level pseudo label generation algorithm to categorize and generate labels of the proposal using localization information of the object [20]. Pseudo-labels are generated continuously applying adaptive threshold selection strategy [20]. Considering NWPU VHR-10 open datasets available conducted experiments to evaluate the performance of WSOD [20].
Features extraction, classifying the selected set of features based on crop and estimating the performance parameters to predict yield has been discussed by most of the research studies. Data Communication is equally significant in crop management to communicate the required data procured from field vehicle to Supporting system for analysis and prediction [19]. This seeks focus on the low power, long range wireless technologies in IoT [19]. Architectures, features, security protocols of different wireless communications are compared for their compatibility in echo-system applications [19]. Another important factor that effects the crop management is water management. Irrigation management is a challenging factor, considering soil type, crop variety and resources available. Few studies showing CNNs are susceptible to be texture biased, where as increasing shape bias can improve performance of CNNs [21]. To enhance the precision of center-pivot irrigation system a

simple and effective method proposed to increase shape bias in the object detection networks and experimentally validated [21].
Using digital photography, samples of six different soil types were gathered from grape soils of Chateau Kefraya terroirs in Lebanon [23]. Features such as thermal and texture features, canopy structure extracted from the data collected by UAV are used to estimate grain yield using Partial Least Squares Regression, Support Vector Regression, Random Forest Regression, input-level feature fusion based DNN and immediate-level Feature fusion based DNN [24]. Consistent performance, relatively accurate and robust estimation of yield is observed in DNN-based models over other frameworks with less spatial dependency and variations [24]. When estimating the yield of grass type, terminal plant height is certainly important characteristic for yield prediction of sorghum, advanced and early generation material and maize, optimal and late material, but has been unrealistic to measure in the field [25]. Grain production of cereal crops can be assessed as the product of average values of plant population, fertile head count, number of seeds per head, and seed mass. Analyzing these component behaviours through seasonal changes and finally their values at harvest can give better understanding of crop adaptation [26]. To resolve, a two-step machine-based image processing method is developed to detect and count the number of heads from high resolution images obtained from UAV [26].

Table 2. Summary of 8 applications of UAV based crop management.

S. No.	**Authors/ Publisher**	**Title /Year**	**Phases in crop management**				**Algorithm adopted**	**Application**
			Features extraction	**Crop Health Monitoring**	**Soil Health Monitoring**	**Crop Yield Prediction**		
1.	Chantal Saad Hajjar, Celine Hajjar, Michel Esta and Yolla Ghorra Chamoun	Machine learning methods for soil moisture prediction in vineyards using digital images -2020	Y	N	Y	N	Machine Learning methods: Nonlinear regression models- a multilayer perceptron and a support vector regression	Vineyards
2.	Maitiniyazi Maimaitijiang, Vasit Sagan, Paheding Sidike, Sean Hartling, Flavio Esposito, Felix B. Fritschi	Soybean yield prediction from UAV using multimodal data fusion and deep learning. – 2020	Y	Y	N	Y	Deep Neural Network, Partial Least Squares Regression (PLSR), Random Forest Regression (RFR), Support Vector Regression (SVR)	Soybean
3.	N. Pugh, D.W. Horne, S.C. Murray, G. Carvalho, L. Malambo, J. Jung, et al.,	Temporal estimates of crop growth in sorghum and maize breeding enabled by unmanned aerial systems. – 2018	Y	Y	N	Y	Structure From Motion, DJI Phantom 3 Professional UAS, DJI Phantom 4 Professional, 3DR Solo and fixed-wing eBee	Sorghum and Maize
4.	W. Guo, B. Zheng, A.B. Potgieter, J. Diot, K. Watanabe, K. Noshita, et al.,	Aerial Imagery Analysis-Quantifying Appearance and Number of Sorghum Heads for Applications in Breeding and Agronomy. – 2018	Y	Y	N	Y	Geolocation based plot segmentation method, two-step machine-based image processing method	Sorghum
5.	Ferhat Kurtulmuş and Ismail Kavdir,	Detecting corn tassels using computer vision and support vector machines. – 2014	Y	Y	N	N	A computer vision algorithm	corn tassels
6.	A. Kumar et al.	CIG based Stress Identification Method for Maize Crop using UAV based Remote Sensing. – 2020	Y	Y	Y	Y	Chlorophyll Index Green (CIG) vegetative index-based method	Maize
7.	Jose Ramon Saura, Ana Reyes-Menendez, Pedro Palos-Sanchez.	Mapping multispectral Digital Images using a Cloud Computing software-applications from UAV images. - 2019	Y	Y	Y	Y	multispectral photogrammetry technology and UAVs, PIX4D Cloud Computing technology software	Vineyard

8.	Zhang, J., Wang, C., Yang, C. et al. Precision Agriculture, (2020).	Evaluation of a UAV-mounted consumer grade camera with different spectral modifications and two handheld spectral sensors for rapeseed growth monitoring: performance and influencing factors. - 2020	Y	Y	Y	Y	UAV with DSLR camera (Nikon D7000) and Multispectral camera (Mica sense RedEdge 3)	Rapeseed

4. Limitations:

In spite of advantages, UAVs has certainly limitations and table-2 briefs about limitations of UAVs and future breakthroughs.

Table 3. Limitations and future breakthroughs in UAVs

Limitation	Breakthrough
Endurance capability: multiple sensors are mounted on UAV powered by lithium battery which last longs for 30 minutes, significantly affects the endurance time.	Fuel cell technology, Power management chip technology, hollow wireless charging may improve endurance capability.
Safety of UAV operation: There are no properly defined pilot management regulations till date, causing security risks and complex geographical areas, climate changes, and power grid structures resulting in UAV explosion accidents [31].	The aerial supervision department should clearly define supervision rules and regulations [31].
Effective monitoring: images captured by UAVs are irregular seeking complex data processing using professional software tools which may delay monitoring results.	Transmission bandwidth can be increased by using 5G mobile communication technology. Improved algorithms for agricultural remote sensing monitoring can develop interactive processing.

5. Conclusion:

A comprehensive review of emerging technologies, primarily UAV-based applications in the agriculture sector, was conducted with the strong conviction that agricultural production systems should be closely monitored for growth with minimal environmental impact in order to feed 9 billion people by the mid-century forecast [1]. Achieving this and compete with land users, not involved in food production such as biofuel production, urban expansion, etc. is challenging the agronomists and industrialists. Agricultural production is the result of the combined effects of natural resources like fertile land and water, biotic factors, agrochemical inputs and climate conditions. The agricultural growth rate can be improved by sufficient investments, technological development and support offered by various agricultural policies. To achieve the required growth rate and sustainable crop production, it is needed to exploit the modern tools by accumulating all available technologies into the agriculture platform. In this paper, emerging technologies- UAV and Machine learning and their applications in agriculture sector were discussed.

ORCID ID: 0000-0001-7697-0159[1]
ORCID ID: 0000-0002-5740-5127[2]

References

[1] FAO, Declaration of the world summit on Food security; FAO :2019

[2] Kaivosoja, J.; Hautsalo, J.; Heikkinen, J.; Hiltunen, L.; Ruuttunen, P.; Näsi, R.; Niemeläinen, O.; Lemsalu, M.; Honkavaara, E.; Salonen, J. "Reference Measurements in Developing UAV Systems for Detecting Pests, Weeds, and Diseases", *Remote Sensing*. 2021, **Volume 13,** 1238.

[3] ZhanTsourg, H.; Wang, L.; Tian, T.; Yin, J. "A Review of Unmanned Aerial Vehicle Low-Altitude Remote Sensing (UAV-LARS) Use in Agricultural Monitoring in China", *Remote Sensing*. 2021, **Volume 13**, 1221.

[4] Delavarpour, N.; Koparan, C.; Nowatzki, J.; Bajwa, S.; Sun, X. "A Technical Study on UAV Characteristics for Precision Agriculture Applications and Associated Practical Challenges", *Remote Sensing*, 2021, **Volume 13**, 1204.

[5] Zhang, J., Wang, C., Yang, C. et al. "Evaluation of a UAV-mounted consumer grade camera with different spectral modifications and two handheld spectral sensors for rapeseed growth monitoring: performance and influencing factors", *Precision Agriculture*, **Volume 21**, 1092–1120 (2020).

[6] Maes, W.H., Steppe, K., 2018. "Perspectives for remote sensing with unmanned aerial vehicles in precision agriculture". *Trends Plant Sci.,* **Volume 24**, pages-152_164. Feb 2019

[7] Jinrutis Radoglou-Grammatikis, Panagiotis Sarigiannidis, Thomas Lagkas, Ioannis Moscholios, "A compilation of UAV applications for precision agriculture", *Computer Networks, Elsevier,* **Volume 172,** 2020

[8] Dimosthenis C. Tsouros, Stamatia Bibi and Panagiotis G. Sarigiannidis, "A Review on UAV-Based Applications for Precision Agriculture", *Information 2019*, **Volume 10,** 349.

[9] Gao, D., Sun, Q. Hu, B. Zhang, S. "A Framework for Agricultural Pest and Disease Monitoring Based on Internet-of-Things and Unmanned Aerial Vehicles", *Sensors 2020*, **Volume 20**, 2020.

[10] Sebastian Candiago, Fabio Remondino, Michaela De Giglio, Marco Dubbini and Mario Gattelli, "Evaluating Multispectral Images and Vegetation Indices for Precision Farming Applications from UAV Images ", *Remote sensing*, **Volume 7**, 2015,.

[11] J. Patel, B. Vala and M. Saiyad, "LSTM-RNN Combined Approach for Crop Yield Prediction On Climatic Constraints," *2021 5th International Conference on Computing Methodologies and Communication (ICCMC)*, 2021.

[12] Thomas van Klompenburg, Ayalew Kassahun, Cagatay Catal, "Crop yield prediction using machine learning: A systematic literature review", *Computers and Electronics in Agriculture*, **Volume 177**, 2020.

[13] Vashisht S., Kumar P., Trivedi M.C. "Design of a Predictive Measure to Enhance Neural Network Architecture for Plant Disease Detection", *Proceedings of International Conference on Big Data, Machine Learning and their Applications*, In: Tiwari S., Suryani E., Ng A.K., Mishra K.K., Singh N. (eds), 2021.

[14] Jing-Lei Tang, Xiao-Qian Chen, Rong-Hui Miao, Dong Wang, "Weed detection using image processing under different illumination for site-specific areas spraying", *Computers and Electronics in Agriculture*, Elsevier, **Volume 122**, 2016.

[15] Avital Bechara, Clement Vigneault, "Agricultural robots for field operations: Concepts and components", *Biosystems engineering, Elsevier*, **Volume 149**, 2016.

[16] Jose Ramon Saura, Ana Reyes-Menendez, Pedro Palos-Sanchez, "Mapping multispectral Digital Images using a Cloud Computing software: applications from UAV images", *Heliyon*, **Volume 5**, 2019.

[17] Jinru Xue, Baofeng Su, "Significant Remote Sensing Vegetation Indices: A Review of Developments and Applications", *Journal of Sensors*, **Volume 2017,** 2017.

[18] Atzberger, C. "Advances in Remote Sensing of Agriculture: Context Description, Existing Operational Monitoring Systems and Major Information Needs". *Remote Sens.* **Volume 5**, 2013.

[19] W. Anani, A. Ouda and A. Hamou, "A Survey Of Wireless Communications for IoT Echo-Systems," *2019 IEEE Canadian Conference of Electrical and Computer Engineering (CCECE)", Edmonton, AB, Canada*, 2019.

[20] Wang, Hui; Li, Hao; Qian, Wanli; Diao, Wenhui; Zhao, Liangjin; Zhang, Jinghua; Zhang, Daobing. 2021. "Dynamic Pseudo-Label Generation for Weakly Supervised Object Detection in Remote Sensing Images", *Remote Sensing*. **Volume 13,** 2021.

[21] Tang, J.; Zhang, Z.; Zhao, L.; Tang, P. "Increasing Shape Bias to Improve the Precision of Center Pivot Irrigation System Detection". *Remote Sens.*, **Volume 13**, 2021.

[22] Matese, A.; Toscano, P.; Di Gennaro, S.F.; Genesio, L.; Vaccari, F.P.; Primicerio, J.; Belli, C.; Zaldei, A.; Bianconi, R.; Gioli, B. "Intercomparison of UAV, Aircraft and Satellite Remote Sensing Platforms for Precision Viticulture". *Remote Sens.*, **Volume 7**, 2015.

[23] Chantal Saad Hajjar, Celine Hajjar, Michel Esta and Yolla Ghorra Chamoun "Machine learning methods for soil moisture prediction in vineyards using digital images", *E3S Web of Conferences*, **Volume 167**, 2020.

[24] Maitiniyazi Maimaitijianga, Vasit Sagana, Paheding Sidikea,b, Sean Hartlinga, Flavio Espositoc, Felix B. Fritschid "Soybean yield prediction from UAV using multimodal data fusion and deep learning", *Elsevier*, **Volume 237,** 2020.

[25] N. Pugh, D.W. Horne, S.C. Murray, G. Carvalho, L. Malambo, J. Jung, et al., "Temporal estimates of crop growth in sorghum and maize breeding enabled by unmanned aerial systems", *The Plant Phenome Journal,* **Volume 1,** 2018.

[26] W. Guo, B. Zheng, A.B. Potgieter, J. Diot, K. Watanabe, K. Noshita, et al., "Aerial Imagery Analysis-Quantifying Appearance and Number of Sorghum Heads for Applications in Breeding and Agronomy", *Frontiers in plant science*, **Volume 9**, 2018.

[27] Ferhat Kurtulmuş and Ismail Kavdir, "Detecting corn tassels using computer vision and support vector machines", *Expert Systems with Applications*, **Volume 41**, 2014.

[28] Kumar et al., "CIG based Stress Identification Method for Maize Crop using UAV based Remote Sensing", *2020 IEEE Sensors Applications Symposium (SAS), Kuala Lumpur, Malaysia*, 2020.

[29] Ballesteros, R., Ortega, J.F., Hernandez, D. et al. "Onion biomass monitoring using UAV-based RGB imaging", *Precision Agriculture*, **Volume 19**, 2018.

[30] Corcoles, J. I., Ortega, J. F., Hernandez, D., & Moreno, M. A. (2013). "Estimation of leaf area index in onion (Allium cepa L.) using an unmanned aerial vehicle", *Biosystems Engineering*, **Volume 115**, 2013.

[31] Vinay Chamola, Pavan Kotesh, Aayush Agarwal, Naren, Navneet Gupta, Mohsen Guizani, "A Comprehensive Review of Unmanned Aerial Vehicle Attacks and Neutralization Techniques", *Ad Hoc Networks*, **Volume 111**, 2021.

Performance Investigation of Back-Compatible Integrated TWDM/GPON System Using MDM And Pulse Shapes

Ajmer Singh[1], Rosepreet Kaur Bhogal[2]

School of Electronics and Electrical Engineering, Lovely Professional University, Punjab.

ajmersinghbhinder@gmail.co[1], rosepreetkaur12@gmail.com[2]

https://orcid.org/0000-0002-5452-1181[1], https://orcid.org/0000-0002-1697-4489[2]

Abstract. Optical access networks are the prominent and promising cutting edge technology to provide high speed, and cost effective operation for meeting the requirements of ever-increasing demands of users. Multiple access techniques such as Wavelength and time division in passive optical networks are important innovations to provide high capacity. A mode division multiplexed back compatible integrated next generation PON2 (TWDM PON) and GPON is demonstrated with triple play supportability in this work. Proposed system offer pay as you grow feature and also provide enhanced performance by suppressing the inter-channel interference in TWDM-PON2. System successfully works on same optical distribution network unit for both PON standards without extra arrangements and has potential to support 62.5 Gbps in full duplex mode (Symmetrical downlink/uplink). Moreover, demonstrated back compatible system is analyzed by incorporating diverse modulations such as NRZ-DPSK, RZ-DPSK, RZ-DQPSK, NRZ-DPSK and linecodings NRZ, RZ. Received power, Optical signal to noise ratio (OSNR), BER and Q factor are noted at diverse link lengths and it is noteworthy that LP modes with NRZ in TWDM PON provide outstanding results.

Keywords. MDM, TWDM, GPON, telecommunication, optical fiber

1. Introduction

With the unpredictable bandwidth requirements due to video on demand, online games, and high-definition videos, researchers are put forward to enhance the ability of existing optical fiber access networks [1] [2]. Telecom companies have great opportunities to cater the user with value added services as the internet traffic is thriving day by day [3]. Optimal way outs to fulfil the user demands by providing economical and high-speed services, are passive optical networks. PONs has diverse standards recognized by international telecommunication union and IEEE which are in operation such as ATM-PON, BPON, GPON, XGPON, GEPON and NG-PONs. There are different data rates, distance, and users support capacity in each aforementioned PON standards. Fiber to the home networks/premises is getting attention due to their reliable and cost-effective working using Gigabit passive optical networks. GPON provide low latency and also guaranteed QOE [4] [5]. GPON basically has three diverse components such as optical line terminal, transmission medium (optical fiber), and optical network unit present at user location [6] [7] [8]. Despite the various advantages of GPON, it can provide lesser bit rates, lesser distances and also lesser users. Numerous research works for enhancing the performance of passive optical networks are reported so far such as by frequency change [9], power boosters [10], pulse shapes [11], noise removing filters [12], pulse width shortening [13] etc. However, all of the components and improvements made to improve PON performance eventually raise the system's cost and complexity.

Solution to the problem of low data rate and capacity is next generation passive optical network standard which are separated into two diverse stages such as NGPON1 and NGPON2. Supported data rate of NGPON1 is symmetrical 10 Gbps and in MGPON2 is symmetrical 40 Gbps or more. With the extensive literature review, it is perceived that NGPON2 has potential to reuse of ODN, reverse compatibility, wide bandwidth and mobility of wavelengths [14] [15]. Integration of already deployed GPON with TWDM PON increase the bandwidth per user, enhance capacity, bit rates and cater the pay as you grow services which will serve the requirements in near future.

2. Principle of MDM based Reverse compatible GPON/TWDM PON

In order to reduce the inter-channel interference among the channels of TWDM PON, mode division multiplexing is incorporated in the system. For generating the different intensity profiles, mode generators having different "azimuthal" and radial number are considered. Eq. 1 represents the mode number expression

$$Mode\ Number = A + 2B + 1 \tag{1}$$

Where, B= Radial number, A= azimuthal number

Effect of these different linearly polarized modes is such that these provide extra layering to the wavelengths which eliminate the wavelength interference effects even if mode coupling takes place.

3. System setup

A high capacity, mode division multiplexed reverse compatible 62.5 Gbps downstream and 61.25 Gbps upstream system with the incorporation of diverse modulations is depicted in Figure 1. Downstream TWDM PON has 10 Gbps data rate per channel and total channels of TWDM PON are six in both upstream and downstream. Triple play services are activated in the system and demonstration is accomplished without using optical amplifiers. Each transmitter of TWDM PON has binary data generator, pulse generator, laser and intensity modulator which are then further followed by linearly polarized mode generator. Similarly, GPON transmitter also has same components as TWDM PON except the data rate (2.5 Gbps in downstream). Wavelengths assigned to GPON for downstream is 1490 nm and in case of TWDM PON downstream, these assigned wavelengths are $\lambda 1$=1596 nm to $\lambda 6$=1601 nm with 1 nm spacing. For video signals, a sine generator pulse is overlaid at 4th channel of TWDM PON. Linarly polarized modes for TWDM PON are LP01, LP11, LP21, LP12, LP22, LP31 and in the case of gigabit passive optical network, assigned mode is LP13. Coexistence mux (CEx) is used for combining both the PON standards and then fed into few mode optical fiber. Signals are then coupled to FMF of 25 km and all the channels then passed through optical filters for separating TWDM and GPON channels. After, filtering, photo detectors, low pass filters and BER analyzers are placed to calculate Q and BER.

For upstream transmission, six channels of TWDM PON at wavelengths $\lambda 1$=1524 nm to $\lambda 6$=1529 nm with 1 nm spacing are used at data rate of 10 Gbps at each channel. GPON has upstream wavelength 1310 nm with bit rate of 1.25 Gbps. Bidirectional few mode fiber is placed in the optical distribution network and having link length of 25 km. Upstream receiver consists of optical filters for separating TWDM and GPON channels and after, filtering, photo detectors, low pass filters and BER analyzers are placed.

3.1. Diverse modulation and linecodings

RZ and NRZ linecodings are basic pulse formats which are also called on off keying and implemented over developed system for checking the performance of the system. It is perceived that return to zero modulation format or linecoding has broad spectrum as compared to non return to zero which means

RZ is not as bandwidth efficient as NRZ is. However, return to zero modulation format has prominent advantage that it is less prone to fber optical phase modulating effects (nonlinear effects). Also, NRZ and RZ both are less tolerant to dispersion effects and easily introduce errors due to interference of adjacent bits. However, in the phase shifted modulations this problem is eliminated to maximum extent. Therefore, DPSK with NRZ and RZ is implemented and analyzed over developed system. DPSK has many advantages of high dispersion tolerance, bandwidth efficiency due to phase shifting between two adjacent bits without any reference to preceding bits. Further, from the reported works, it is evident that higher the phase siftings, higher the dispersion tolerance and therefore NRZ-DQPSK and RZ-DQPSK are deployed further.

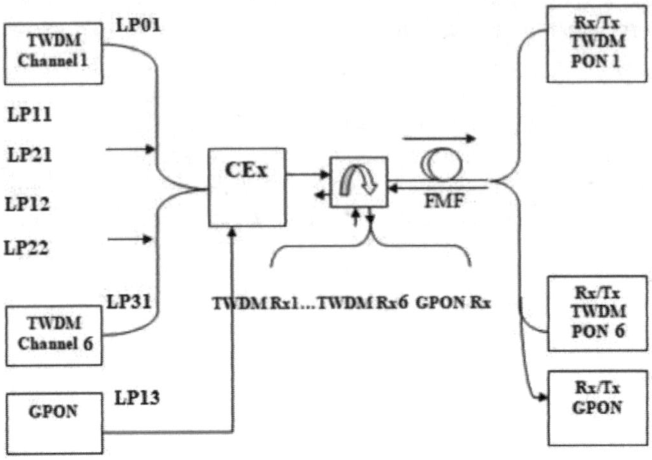

Figure 1. Illustration of proposed system

4. Results and Discussions

Investigation of the demonstrated system is accomplished for diverse link distances in terms of OSNR, BER and Q factor. An optical spectrum analyzer is employed to check the total wavelengths in the system and depicted in Figure 2. It represents the power at each carrier sognal with reference of its center frequency.

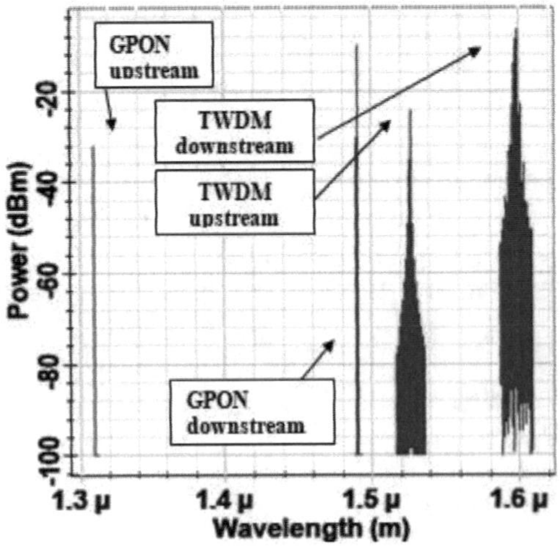

Figure 2. Illustration of proposed system

Performance of diverse intensity profiles (LP modes) are investigated at different link lengths of FMF and results observed in eh form of BER/Q factor. Figure 3 (a) depicts that deterioration increases if length of the fiber increases which is due to attenuation, dispersion and scattering effects. Results reveal that best performing mode in terms of Q is LP01 and performance is followed by LP11, LP21 respectively. Minimum Quality is seen in case of LP31 mode.

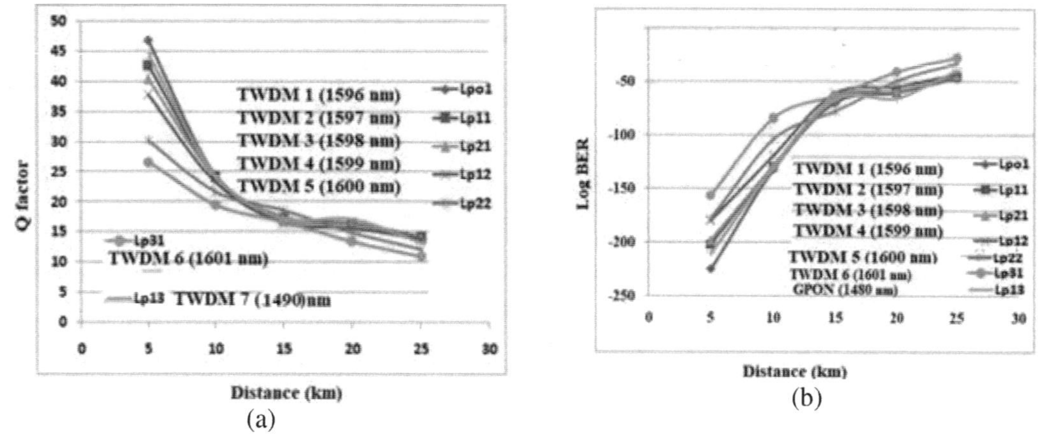

Figure 3. Justify the caption. Analysis of diverse LP modes/wavelengths for (a) Quality factor (b) Log BER at different distance of FMF

Figure 3(b) represents different LP modes and wavelength channels of TWDM and GPON BER at varied link lengths. It is perceived that minimum BER is observed in LP01 and performance is followed by LP11, LP21 respectively. Maximum BER is seen in case of LP31 mode. It is observed that basic linear modes lower order exhibits best performance. Figure 4 depicts the different LP modes for TWDM PON and GPON. All the LP modes are taken by changing the azimuthal and radial numbers.

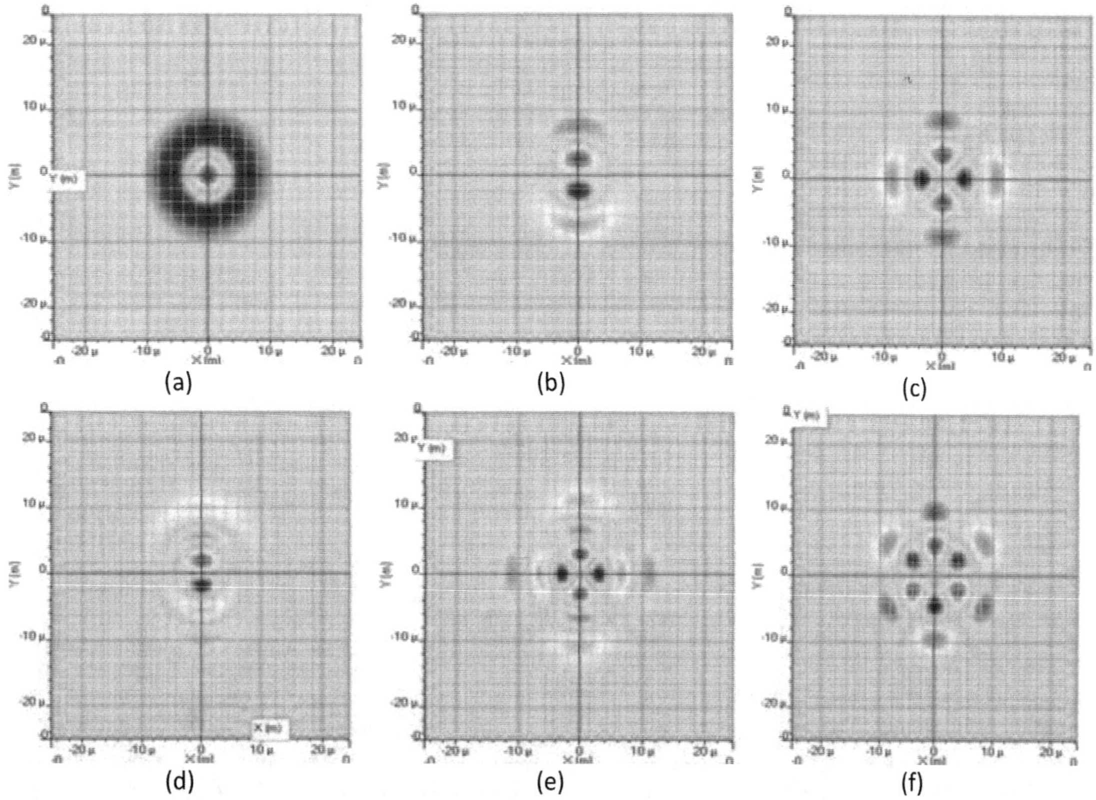

Journal of Physics: Conference Series **2327** (2022) 012036 doi:10.1088/1742-6596/2327/1/012036

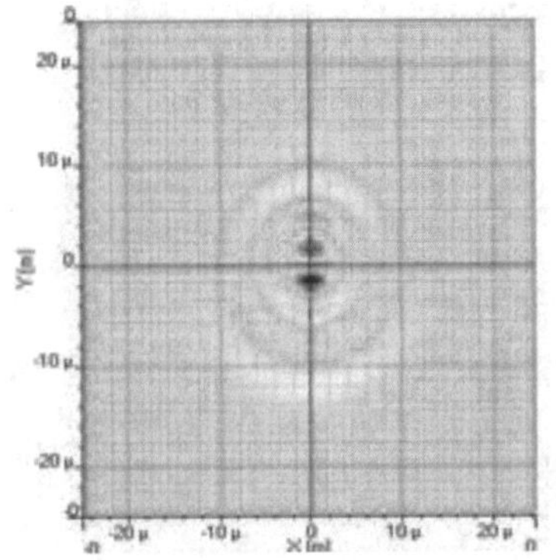

(g)

Figure 4. FMF Representation of LP profiles in TWDM PON (a) LP01 (b) LP11 (c) LP21 (d) LP12 (e) LP22 (f) LP31 and for GPON (g) LP13

Table 1 provides the values of different optical signal to noise ratio for diverse modulations and linecodings and from the results it is evident that OSNR is highest in RZ-DQPSK due to bandwidth efficient spectrum of DQPSK and non linear tolerance of RZ and second highest OSNR is of RZ-DPSK signal. In terms of received power NRZ-DQPSK is best with power 2.50 dB.

Table 1. Values of OSNR and received power for diverse modulations and linecodings

Modulations/Linecodings	OSNR	Received Power dB
NRZ	41.57	1.20
RZ	39.13	-1.89
NRZ-DPSK	28.16	2.40
RZ-DPSK	50.69	0.40
NRZ-DQPSK	34.36	2.50
RZ-DQPSK	57.21	1.40

Further, Q factor investigation of different modulations and linecodings has been done at diverse link lengths. It is evident that with the increase in the transmission distance, Q of the system decreases due to attenuations, dispersion and scattering. Modulations which are investigated are RZ-DPSK, NRZ-DPSK, RZ-DQPSK, RZ-DQPSK. It is reported that due to bandwidth efficiency and less inter symbol interference, DQPSK is optimal to use. However, it lowers the power of the system due to various insertion losses of many components which are used to generate phase shiftings. Also it is true for differentia phase shift keying which has two phase shifts.

Figure 5 depicts the Q factor performance of different modulations and linecodings at diverse link lengths. Results revealed that due to simple architecture, low losses and bandwidth efficiency, NRZ comes out to be best in terms of Q factor.

5. Conclusion

A back compatible integrated mode division multiplexed next generation PON2 (TWDM PON) and GPON is demonstrated with triple play supportability. Enhanced performance is obtained by suppressing the inter-channel interference in TWDM-PON2 and provides pay as row services. System successfully works on same optical distribution network unit for both PON standards without extra arrangements and has potential to support 62.5 Gbps in full duplex mode (Symmetrical downlink/uplink). Performance of LP01 mode is observed best due to less more coupling and TWDM PON performs better than GPON. Further different modulations and linecodings are investigated in the form of OSNR, Q and received power. Results revealed that Q factor is best in case of NRZ among all modulations and followed by RZ-DQPSK. However, maximum power is received in case of NRZ-DQPSK.

Acknowledgement

I would like to express my very great appreciation to **Manoj Sindhwani, Shippu Sachdeva, Javed Dhillon, Raghav Gupta** (School of Electronics and Electrical Engineering, Lovely Professional University, Phagwara, Punjab, India.) for their valuable and constructive suggestions during the planning and development of this research work. Their willingness to give their time so generously have been very much appreciated.

References

[1] Parol P and Pawloski M 2013 Towards networks of the future: SDN paradigm introduction to PON networking for business applications *In Proceedings of the 2013 federated conference on computer* pp 829–836.

[2] Perez G, Andrade-Lucio J A, Ibarra-Manzano O G, Alvarado-Méndez E, Trejo-Duran M and Gutiérrez-Martín H 2006 Efficient modulation formats for high bit-rate fiber transmission *Acta Universitaria* **16**(2) pp 17–26.

[3] Muhammad I A, Qureshi M S, Zafar S 2017 Demonstration of TWDM-PON Backward Compatibility with Conventional GPON *Wireless Pers. Commun.* **95**(2) pp 581-592.

[4] Sultan, D M S and Arefin, M T 2011 GPON, the ultimate pertinent of next generation triple-play bandwidth resolution *Journal of Telecommunications and Information Technology* **2** pp 53–60.

[5] Singh S, Jaiswal A K and Kumar M 2012 Transmission of voice, video and data in fiber-to-the home (FTTH) networks using OFDM *IJARCSSE* **2**(5) pp 41–45.

[6] Taheri M and Ansari N 2013 A feasible solution to provide cloud computing over optical networks *IEEE Network* **27**(6) pp 31–35.

[7] Srinath S 2014 Performance analysis of 2.5 Gbps GPON *IJAREEIE* **3**(6) pp 10148–10155.

[8] Milanovic, S., "Case study for a GPON deployment in the enterprise environment", Journal of Networks, vol. 9, no. 1, pp. 42–47, 2014.

[9] Yi L 2013 Symmetric 40-Gb/s TWDM-PON with 39-dB power budget IEEE Photonics Technology Letters **25**(7) pp 644–647.

[10] Rakkammee S, Boriboon B, Worasucheep D and Wada N 2017 Measurement of characteristic parameters of 10 Gb/s bidirectional optical amplifier for XG-PON *SPIE Third International Conference on Photonic Solutions* (Pattaya, Thailand).

[11] Winzer P J and Essiambre R J 2006 Advanced optical modulation formats *Proceedings of the IEEE* **94**(5) pp 952–985.

[12] Kaur K Randhawa R and Kaler R S 2014 Performance analysis of WDM-PON architecture using different receiver filters *Optik* **125** (17) pp 4742-4744.

[13] Eslam A El-Sahn Z and Shalaby H 2013 Coherent Long Reach OIDMA-PONs Enabled By Electronic Dispersion Compensation *Asia Communications and Photonics Conference* (Optical society of America, China) p AF2F.33.

[14] Muhammad I A, Qureshi M S and Zafar S 2017 Demonstration of TWDM-PON Backward Compatibility with Conventional GPON *Wireless Pers. Commun.* **95**(2) pp 581-592.

[15] Luo Y, Zhou X, Effenberger F, Yan X, Peng G, Qian Y and Ma Y 2013 Time-and wavelength division multiplexed passive optical network (TWDM-PON) for next-generation PON stage 2 (NGPON) *Journal of Lightwave Technology* **31**(4) pp 587–593.

Effects of Different Intensity Modulator On The Generation of MM-Waves in RoF System

Raghav Gupta[1], Ajmer Singh[2]

School of Electronics and Electrical Engineering, Lovely Professional University, Punjab.

raghav.22026@lpu.co.in[1], ajmersingh bhinder@gmail.com[2]

https://orcid.org/0000-0002-0067-5062[1], https://orcid.org/0000-0002-1697-4489[2]

Abstract. The increasing demand of wireless network has strongly take interests of new transmission systems. A phase shift of 80 GHz – 120 GHz is achieved for the millimeter wave by using SBS technique that is beneficial for the 5G networks. RoF technique requires the optical modulators in the transmitter side. Performance of different modulators such as Amplitude Modulator, Phase Modulator, Electro Absorption Modulator, and Mach Zehnder Modulator is analyzed in this demonstrated work. Performance comparison of all modulators shows that the SNR (93.32) and output power (71.56) of the system is maximum in MZM. Results shows that the system has the high power, good SNR and enhanced frequency shifts.

Keywords: RoF (Radio Over Fiber), SBS (Stimulated Briliuion Scattering), SNR (Signal-to-noise-ratio), BER, MZM

1. Introduction

Excessive progress of wireless communication increased the stipulation of capacity of users and also pace of data transmission high speed services. The use of millimeter wave frequency provides acceptable terminal mobility [1]. The radio link must be employs at high frequency carriers because at the low frequencies spectrum congestion occurs. Millimeter wave RoF technologies have arises as a contest candidate for increasing the demands of broadband multimedia services for wireless users. Many research works has been demonstrated in the generation of mm-wave signal with the radio over fiber systems [2,3]. These methods are: dual wavelength laser injection locking [3] four wave mixing [4], optical heterodyne [5]. These techniques results in the generation of mm wave but do not increase the frequency shift [6].

With the incorporation of radio over fiber communication system, more channels can be coupled at greater speed in to optical communication medium and can also be incorporated with formativeness and unsteadiness of wireless communication system. Therefore, hybrid wired and wireless optical networks are key solution to the uncontrolled rate of bandwidth demands and offer economical operations.

RoF is an appealing technique due to its low complexity at antenna site and frequency reuse. When mm wave is implicated degrading effects are induced such as dispersion effects. The dispersion shifting fiber is used to reduce or compensate the dispersion effect. The use of dispersion shifting fibers makes the system costly. Dispersion shifting fibers has high attenuation coefficient which results in the reduction of signal performance in the network. Many works has been proposed to restrain the pulse broadening

effects and a lot of techniques have been proposed for generation and transmission of radio signal. Optimal methods are essential to implement the mm-wave in RoF system to confirm that transmission these frequencies are reliable and appreciable for the generation of mm-wave. Many techniques are applied in the base station to make it fast and economical [7]. A harmonic radio frequency carrier and wideband data up conversion using single side band for radio over fiber system through Stimulated Brillouin Scattering is determined in [8].
Future networks needs data rate enhanced that can be promoted by frequency carriers shifts in optical fiber due to stimulated Brillouin Scattering. This shift is restricted to 10GHz to 40GHz [11]. The techniques introduced till now are complex or costly.
In this research work we investigate the performance of different modulators. We find that the frequency shift of 20 GHz -80 GHz is achieved and the mm wave generated by MZm possesses the highest SNR and output power. Only a single pump is used to accomplish the 20 GHz to 80 GHz frequency shifts.

2. Stimulated Brillouin Scattering

Brillouin scattering is a performance limiting factor in optical communication system. It is caused by the non linearity of the medium. Some part of the input signal scattered in the back direction. The maximum of the input signal is carried by the back scattered signal until it exceeds the threshold value. As a result, SBS decreases signal strength and allows for transmission through fibre without compromising signal quality. In optical communication systems, the deleterious impacts of SBS have been thoroughly researched.
Because of its ability to operate at low threshold power, SBS has attracted a lot of attention in the field of fibre laser and signal processing. The narrow bandwidth of the Brillouin gain spectrum was exploited in sideband selective amplifications, and the development of Brillouin stokes signal with frequency downshifted is exhibited in the expansion of multiple channel generation based on SBS in optical fibre [10].
The combination of propagating wave with input light beam creates the Brillouin effect because of the existence of microscopic imperfection. SBS becomes a powerful factor at low power level which is known as threshold of SBS.
Low threshold can be accomplished by using lasers with narrow line width. Expression for the threshold power of SBS is 1[12].

$$pth = 21\frac{K\times Aeff}{g\times Leff} \tag{1}$$

$$Leff = \frac{1-e^{-al}}{\alpha} \tag{2}$$

Where *Aeff* is effective area of optical fiber.
Leff is effective length.
K is polarization constant and *g* is gain of Brillouin.
L is the link length of SMF
α is the attenuation.
Mm-wave is generated by a co-propagating modulated signal and counter propagating pump signal in SMF.
So many techniques have been developed such as optical heterodyning, internal or external modulation for the generation of mm waves [6]. These techniques are good at mm wave generation but they are not well competent to produce a good phase shift. This work draws attention on the mm wave generation using SBS technique and maximum point biased modulator. In this work, here we are analyzing the performance of different modulators such as phase modulator, amplitude modulator, EAM modulator and MZM modulator on the basis of signal to noise ratio and output power.

3. System architecture

A continuous wave laser operating at 193.1THz frequency and electrical signal with frequency of 20GHz was modulated with external modulators such as amplitude modulator, phase modulator, MZM and EAM. All the modulators are maximum point biased.

3.1. Mach-Zehnder Modulator

It is a modulator based n the electro optic effects and fundamental principle for the working is pocker's effect. Launched signal at the input of the MZM is divided into two equal power signals and sent to diverse optical paths. These paths have different electrical voltage sources on the both paths and experience signal speed change due to voltages [9].

Electric voltage produces the variations in the refractive index according to the input voltage which produces the shifting of phase in travelling signal. Both the signals are combined at the output and the signals cause the optical wave interfere with each other.

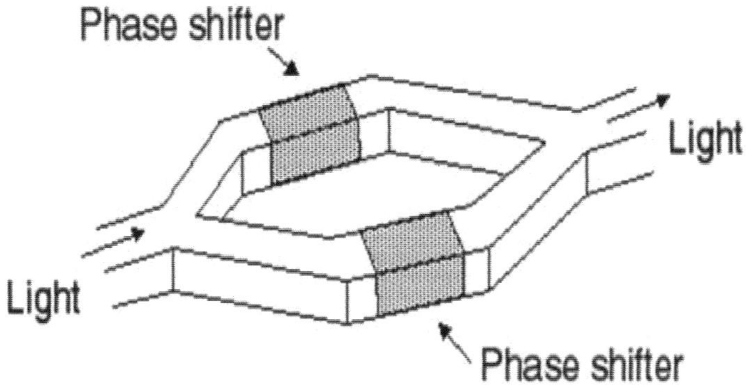

Figure 1. Depicition of MZM model

3.2. Electro Absorption Modulator

Its principle is based upon electro absorption effect. Electrical voltage in this modulator works differently such that it change the absorption region of wavelengths [9].

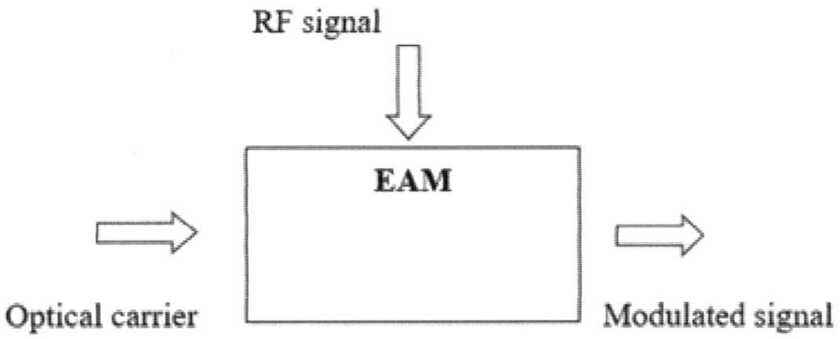

Figure 2. Block diagram of EAM

When an exterior RF signal is given to a signal travelling via a channel, the resultant electrostatic force modifies the absorbance value of the medium and raises the strength of the output signal, according to the EAM process. Franz-Keldysh effect is its basic working principle. EAM has many advantages like high data rate, low driving voltage, and high extinction ratio.

3.3. Phase modulator

It is one of the simplest sort of electro optic modulator and has pockels cells only. Phase delay of beam is introduced by an electric voltage which is operational on signal. In this, the input signal polarization is arranged with one optical axes because of this, the polarization state remains constant.

3.4. Amplitude modulator

The amplitude modulator is likewise based on the Pockels cell, which modulates the signal's polarisation state. A polarizer is used to transform the signal's amplitude and power into transmitted optical amplitude and power. Figure 3 depicts the suggested model's system architecture

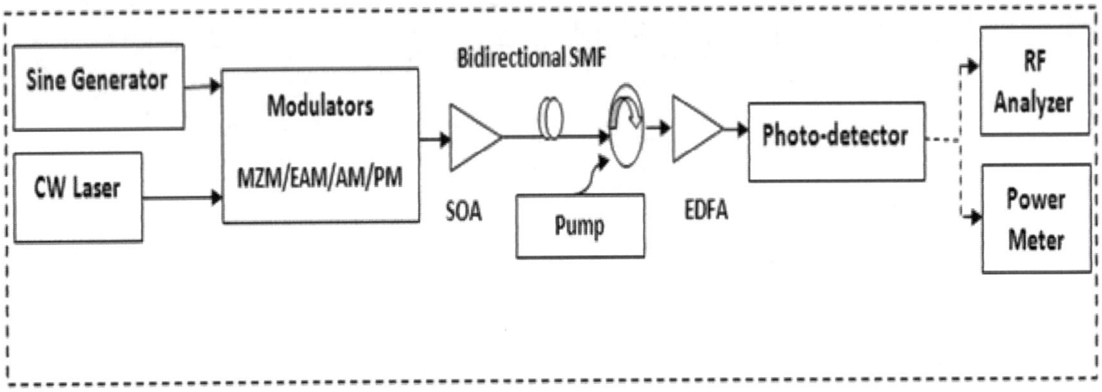

Figure 3. Block diagram of proposed architecture

Table 1. The SBS simulation model's attributes

Parameters	**Values**
Data rate	10 Gbps
Frequency used	193.1 THz
Laser linewidth	1 MHz
Optical power of laser	-30db
Attenuation of bidirectional SMF	0.2db/Km
Pulse width broadening value	16.25 ps/nm/Km
Pre amplifier	SOA
Post amplifier	EDFA
EDFA Gain	30db
Fiber length	1-50 km
SBS gain	4.6x10^{-011}m/W
Line Width	31.5MHz
Brillouin Frequency Shift	11MHz

Laser signal and sine wave signal are modulated by the modulators. The linewidth of laser signal is kept at 1 MHz and polarization of input is fixed. After the modulators, a semiconductor optical amplifier is placed to boost the signal. Modulated signals are entering to the circulator through the bidirectional single mode fiber [12-14]. The pump laser generates a shared output signal that is received at the circulator's output due to the pump laser's counter-propagating pumping wave. The SBS simulation model's attributes are provided in following table 1.
The circulator's end spectrum has both amp. Modulating signal and counter propagating signal. Optical spectrum analyzer shows the frequency shift of the input signal. It also depicts the signal power and wavelength of carrier power. Figure (3) shows the system architecture.

4. Results and Discussions

Performance of proposed system is investigated in terms of output power and output signal-to-noise ratio as well as Q factor. A CW laser generates continuous drive of light pulses centered at 193.1 THz and fed to intensity modulators. Optical spectrum of laser is depicted in figure 4 (a) and represents the carrier signal with respect to frequency (x-axis), power (y-axis).

Figure 4 (b) depicts the OSA for the mm-wave signal generated through optical fiber based on SBS. It is seen that after SBS, there is frequency shift of the newly created first order sideband and second order sideband. Maximum sift of sideband is reported in this setup of 120 GHz with the sidebands centered with the difference if 20 GHz shifts.

Figure 5 represents the performance of different intensity modulators such as MZM, EAM, AM and PM. Investigation has been done for different launched powers and performance is evaluated in terms of received power. It is seen that with the increase of input power, received power also increases. Also in proposed setup, SOA boost the signal and SBS provide a high power on the end of SMF. Comparison is carried out and it is observed that MZM exhibits best performance and followed by EAM, AM, PM respectively. Minimum output power is provided by phase modulator (PM).

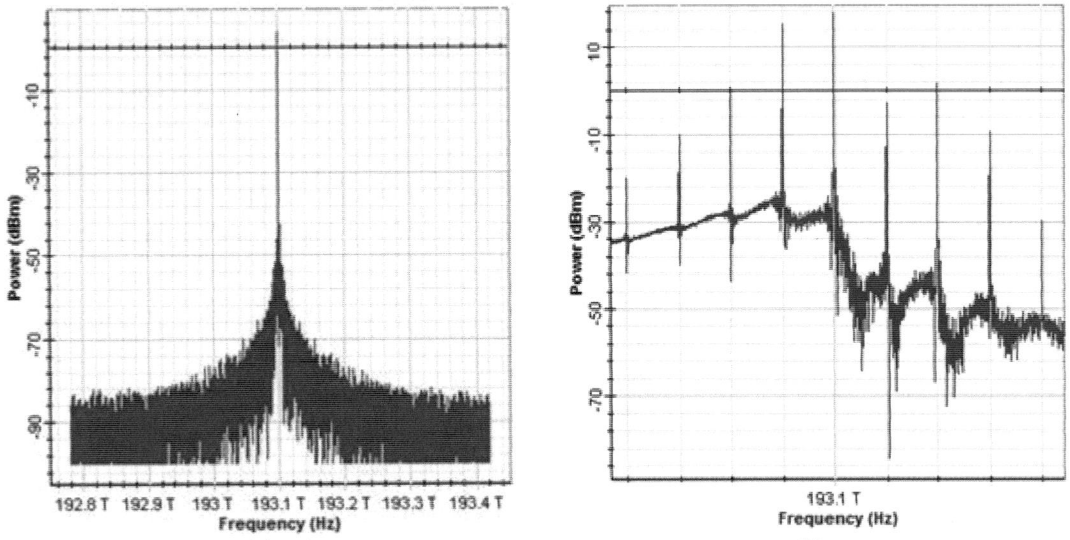

(a) (b)

Figure 4. Optical spectrum analyzer output for carrier signal (a) after laser (b) after SBS

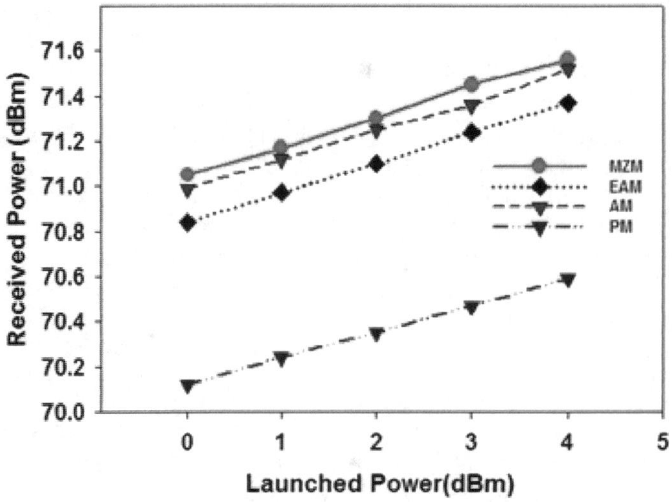

Figure 5. Graphical representation of launched power versus output received power for different intensity modulators

Figure 6 (a) depicts that signal to noise ration vary with launched power and rise with the increase of input power. SNR is the ratio of signal power to noise power and should be high for mm-waves to realize a long reach RoF system. It is reported that MZM show best performance and SNR of 93.2 is noted at 5 dBm. Amplitude modulator is found out to least suitable in case of mm-wave for proposed setup due to minimum SNR. Values of SNR at WDM analyzer with respect to power is shown in figure 7 (a), (b), (c) and (d). EDFA amplifiers gain is an important factor to be addressed and should be optimal because more gain also arise amplifier spontaneous noise. Gain is varied from 5 dB to 30 dB with the difference of 5 dB after each step till 30 dB as shown in figure 6 (b). Gain has a significant effect on the spectra after SBS and stretching the FOBS and SOBS. MZM is again comes out be best choice for future RoF systems and amplitude modulator least suitable.

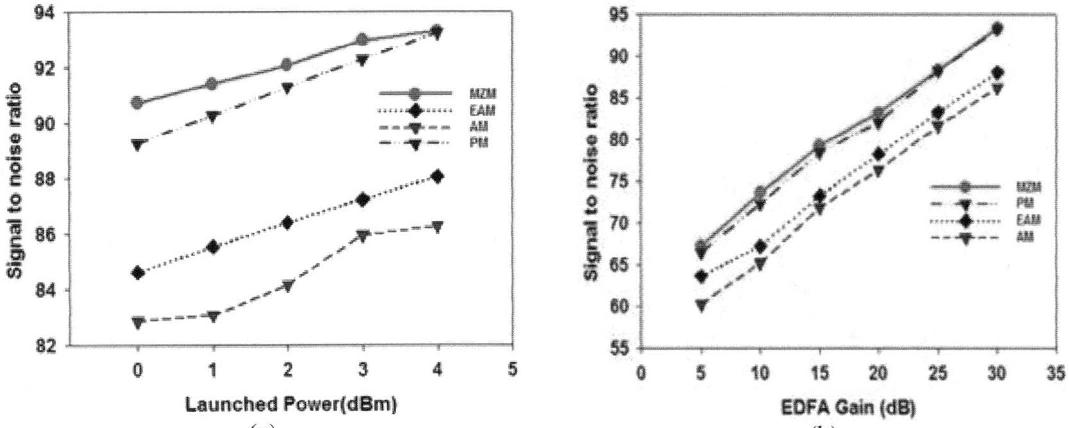

Figure 6. Depiction of proposed setup for (a) Launched power versus SNR (b) EDFA Gain versus SNR

Figure 8 represents the radio frequency spectrum for RF analyzer after photodetector to evaluate the received mm-waves. Depiction shows that mm-waves are generated at 20 GHz, 40 GHz, 60 GHz and 80 GHz with high output power.

Signal Power (dBm)	Noise Power (dBm)	SNR (dB)
32.103994	-61.218293	93.322287
32.103994	-61.218293	93.322287

(a)

Signal Power (dBm)	Noise Power (dBm)	SNR (dB)
31.02303	-57.046696	88.069726
31.02303	-57.046696	88.069726

(b)

Signal Power (dBm)	Noise Power (dBm)	SNR (dB)
30.600181	-62.652566	93.252747
30.600181	-62.652566	93.252747

(c)

Signal Power (dBm)	Noise Power (dBm)	SNR (dB)
31.318859	-54.962288	86.281147
31.318859	-54.962288	86.281147

(d)

Figure 7. Signal to noise ratio for different intensity modulators (a) MZM (b) EAM (c) PM (d) AM

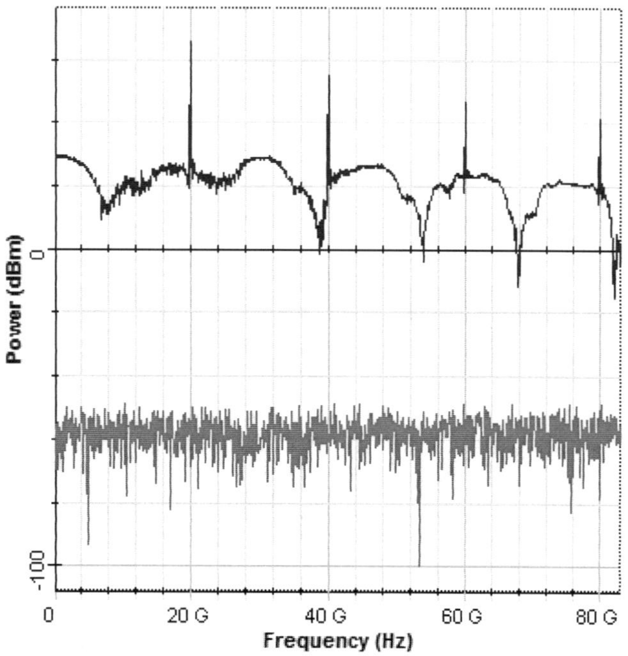

Figure 8. RF analyzer after photodetector

Maximum power is at 20 GHz and decreases as we move to further longer shifts of frequencies. Noise power level is far from signal power and there is significant difference of signal power and noise power levels. Thus, proposed system setup with MZM modulator is capable of producing enhanced mm-wave (120 GHz) with high SNR, output power.

5. Conclusion

In this work, dispersion tolerant and 20 GHz to 80 GHz phase shifts of mm-signal are achieved by using the SBS method. These enhanced frequency shifts are good for the future 5G radio over fiber system.

Performances of different modulators are investigated for the generation of the mm wave signal. All the modulators are maximum point biased. The SNR of the system is also enhanced as it is an important factor in RoF communication system to measure the sincerity of the system. The SNR of all the modulators are compared and it has been found that the SNR is maximum in the case of mach-Zehnder modulator. A low cost Rof millimeter wave generation system is proposed. We make it cost effective by using only one pump.

Acknowledgement

I would like to express my very great appreciation to **Rosepreet Kaur Bhogal, Manoj Sindhwani, Shippu Sachdeva, Javed Dhillon** (School of Electronics and Electrical Engineering, Lovely Professional University, Phagwara, Punjab, India.) for their valuable and constructive suggestions during the planning and development of this research work. Their willingness to give their time so generously have been very much appreciated.

References

[1] Andrews J G, Buzzi S, Choi W, Hanly S., Lozano A, Soong A C K and Zhang J C 2014 What will 5 G be? *IEEE J. Sel. Areas Commun.* **6** pp 1065–1082.

[2] Li X, Xiao J, and Yu J 2016 Long-distance wireless mm-wave signal delivery at W-band *J. Lightwave Technol.***34**(2) pp 661-668.

[3] Beas J., Castanon G., Aldaya I., Aragon-Zavala A., Campuzano'millimerer G. 2013 Wave Frequency Radio Over Fiber Systems: A Survey *IEEE Comm. Surv. Tutor.* **15**(4) pp 1593-1619

[4] Shee Y G, Al-Mansoori M H, Yaakob S, Man A, Zamzuri A K, Mahamd Adikan F R and Mahdi M A 2012 Millimeter Wave Carrier Generation Based On A Double-Brilliouin-Frequency spaced Fiber Laser,Opt.Express **20**(12) pp 13402-13408.

[5] Wang M, Chen C, Li Q, Huang K, Chen H, 2015 Photonic Generation Of Tunable Microwave Signal Using A Passively Q-Switched Dual Wavelength Fiber Laser, Microw.Opt. Technol. Lett. **57**(1) pp 166-168.

[6] Xu Kun, et Al.; 2011 Heterodyne Mixing And Polarization Diversity Techniques In Radio Over Fiber System With High Sensitivity And Dispersion Tolerance. Sci. China Inf.Sci. **54**(2) pp 236-243.

[7] Hirata A, Harada M and Nagatsuma T 2003 120-GHz wireless link using photonic techniques for generation, modulation, and emission of millimeter-wave signals *IEEE- Journal Of Lightwave Technology,* **21**(10) pp 2145-2153

[8] Varghese Anlony Thomas et Al.; 2015 Performance Improvement and Cost Reduction Techniques for Radio Over Fiber Communications *IEEE Communications Surveys & Tutorials*, **17**(2) pp 627-670.

[9] Prabhu K, Bose Sumanta and Kumar D Sriram 2012 Analysis of optical modulators for Radio over Free Space Optical Communication systems and Radio over Fiber systems *2012 Annual IEEE India Conference (INDICON)* pp. 1176-1179

[10] Jalal J Hamad Amen and Ismail Widad 2011 Radio Access Network Power Management Considering Radio Over Fiber Techniques For 4G Mobile System *WSEAS Transactions On Communications* **10**(3) pp 100-107.

[11] Mohamed Norliza, Sam Suriani Mohd. and Ngajikin Nor Hafizah 2016 Power Fading Effects in Millimeter-Wave Radio Over Fiber (RoF) Link *Advanced Computer and Communication Engineering Technology Springer* **362** pp 493-503.

[12] Chang, G K Et Al, 2013 Archtitecture And Applications Of A Versatile Small- Cell, Multi-Service Clous Radio Access Network Using Radio-Over-Fiber Technologies *In IEEE International Conference On Communications Workshops* pp 879-883
[13] Khaki Z G, and Bedi H S 2012 Transient correction using EDFA: in-line optical fiber with feedback *International Conference on Computing Sciences* pp. 233-238.
[14] Bedi H S, Sharma K K and Gupta R 2020 A Review Paper on Performance Analysis of IEEE 802.11 e *1st International Conference on Computing, Communications, and Cyber-Security (IC4S 2019)* pp. 47-56

Design of Broadband Passive Optical Network Incorporating Different Modulation Formats and Colorless ONUs

Shippu Sachdeva[1], Raghav Gupta[2]

School of Electronics and Electrical Engineering, Lovely Professional University, Punjab.

sachdeva.shippu@ gmail.com[1], raghav.22026@lpu.co.in[2]

https://orcid.org/0000-0002-6859-8622[1], https://orcid.org/0000-0002-0067-5062[2]

Abstract. In this work, a bidirectional broadband passive optical network (BPON) is proposed at 622.08 Mbps by incorporating colorless optical network units and different pulse shapes such as Duo-binary return to zero (DRZ), return to zero (RZ) and non-return to zero (NRZ). For varied link lengths, launching powers in terms of log BER, and Quality factor for downstream and upstream, the design of BPON is examined. The duo-binary advanced pulse shape outperformed the others in bidirectional BPON, achieving a symmetrical distance of 25 kilometres.

Keywords: BPON, DRZ, ONU, RZ, OLT, NRZ

1. Introduction

With the rapid increase in the internet services such as the online games, high definition video televisions etc are exerting the pressure on the access networks. From the last few years, the demand of the high bandwidth and data speed is augmenting which created the dramatic changes in telecom industry [1]. Telecom companies are dealing with the issues of implementing new system for high speed communication by improving their existing networks [2]. Last mile problem is a key issue in the optical fiber reliant access networks and passive optical networks (PON) are the cost efficient way to be considered nowadays [3]. In order to provide the internet access among numerous users, broadband passive optical network is an ultimate solution [4]. Therefore, PON are regarded as the premier and potential technology for fibre to the home (FTTH) access network due of its advantages of cheap expenditure on the architecture, prolonged transmission range, wide bandwidth, as well as multi-service convergence [5].

Optical modulation formats are the promising candidates to enhance the reach as well as data speed of the conventional broadband PON [6]. Numerous research works have been reported in the literature for distance and data rate enhancement in the BPON system by incorporating different linecodings such as RZ, and NRZ [7] [8]. However, aforementioned linecodings are not spectral efficient and prone to dispersion effects in the optical fiber networks. In order to combat with the pulse broadening issues and bandwidth hungry applications demands, advanced modulation formats with high tolerance to chromatic dispersion (CD), enhanced nonlinear forbearance as well as also can pass through narrowband optical filtering to provide high spectral efficiency [9]. Many code patterns are proposed so far such as using return to zero format [10], carrier suppressed RZ advanced modulation format coding, duo-decimal return to zero, RZ differential phase reversal keying [11], chip RZ. Moreover, wavelength reused colorless passive optical networks are popular to lower the cost of the system due to the elimination of intensity sources. Malhotra et al. [12] investigated the PON for downstream only. Upstream or bidirectional passive optical network is need of the day and it should be included in the BPON network design. Goyal et al. [13] also proposed a single direction PONs with less number of optical network units.

In this work, a broadband passive optical network employing three different modulation formats is proposed for bidirectional transmission. Symmetrical data speed of 622.08 Mbps is successfully achieved over 25 km over the colorless wavelength reused BPON

2. System setup

For the realization of broadband passive optical network system incorporating different modulation formats, a comprehensive and commercial simulation tool optiwave optisystem is used. Figure 1 depicts the simulation diagram of the bidirectional broadband passive optical network for eight optical network units [14-15]. Pseudo random bit sequence generator is the first and foremost component that generates the binary data bits in the form of 1's and 0's. Binary data is modulated with the different pulse shapes such as DRZ, RZ, and NRZ. A continuous wave laser of 1550 nm is used for the transmission at 0 dBm launched power. Also for the upstream colorless ONU design as shown in Figure 2, an un-modulated 1300 nm laser signal is combined with the 1550 nm modulated signal. Mach zehnder intensity modulator is employed to convert electrical data in to optical data and fed to bidirectional single mode fiber of 25 km. Optical fiber has 0.2 dB/km attenuation and also nonlinear effects are considered. Circulators are the component which is used to isolate the signals of downstream as well as upstream. After the transmission through the bidirectional single mode fiber, downstream signal is divided in two signals and one is fed to the downstream receiver, other is reserved for the upstream signal to eliminate the maintenance issues.

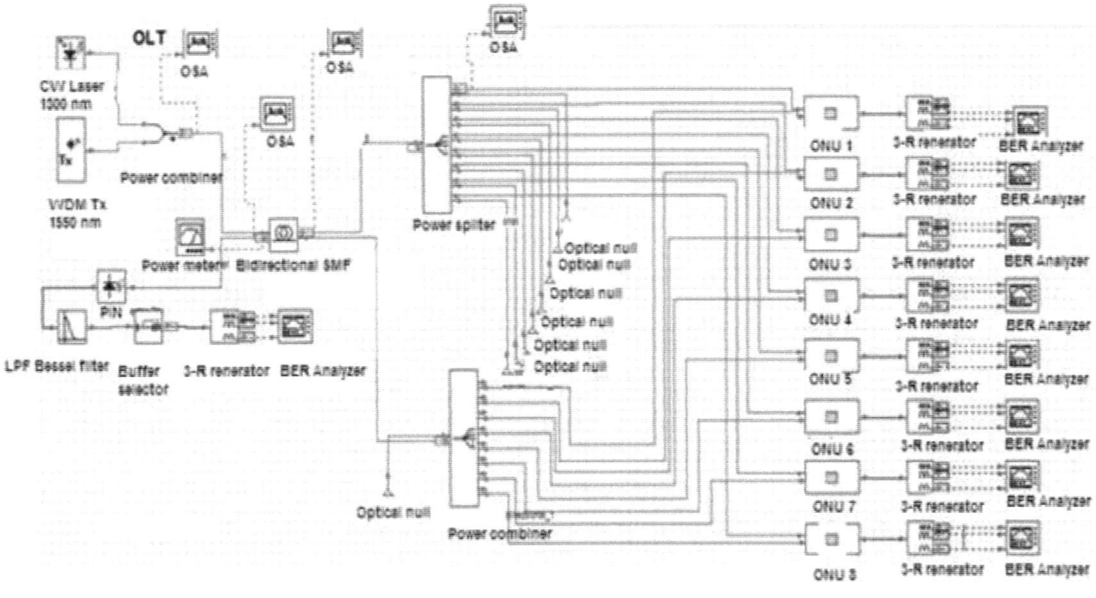

Figure 1. Simulation diagram of the proposed colorless bidirectional BPON system

Downstream receiver incorporates the photo-detector and for noise elimination, inverted rectangular filter is used followed by BER analyzer. Wavelength signal of 1300 nm is coupled to the upstream transmitter which consists of binary data generator, linecoder and intensity modulator (MZM).

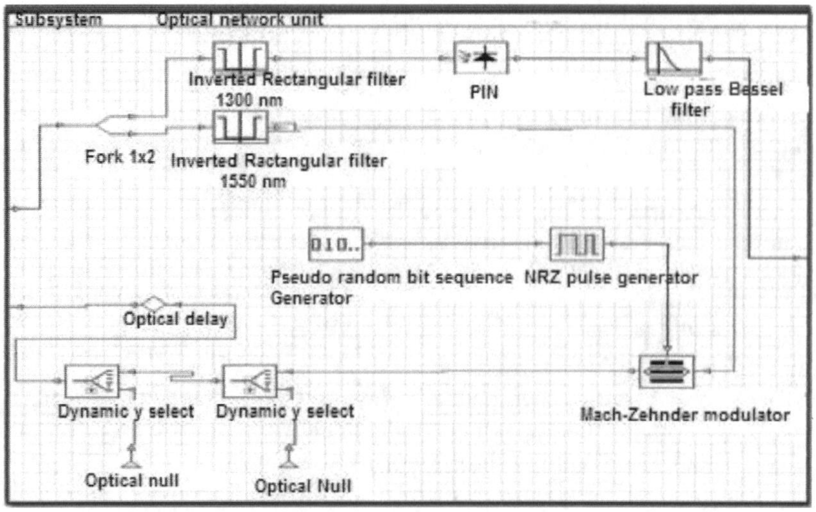

Figure 2. Illustration of the optical network unit

Time delays are inserted in upstream to reduce crosstalk inside the bidirectional single mode fiber. Upstream signal after 25 km transmission coupled to the circulator followed by photo-detector. Low pass Bessel filter employed for the noise suppression and BER for the Q factor assessment. Proposed system specifications are shown in the Table 1.

Table 1. System specification of the proposed BPON

Parameters	**Values**
Bit Rate	622.08 Mbps
Distance	25 km
Single mode fiber	SMF-28
Attenuation	0.2 dB/km
Dispersion	16 ps/nm/km
Dispersion slope	0.075 ps/nm^2/Km
Wavelengths	1550 nm, 1300 nm

3. Results and Discussions

To investigate the performance of the proposed broadband passive optical network, different distances, launched powers are considered and results are evaluated in terms of Q factor and BER. Figure 3 represents the optical spectrum analyzer that depicts the carriers of laser signals for the transmission. OSA represents the frequency/wavelength of the carriers with respect to the power.

Journal of Physics: Conference Series **2327** (2022) 012038 doi:10.1088/1742-6596/2327/1/012038

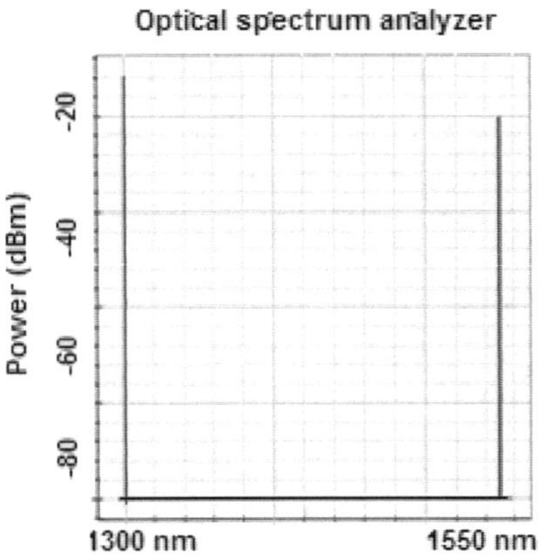

Figure 3. Representation of the optical spectrum analyzer

Figure 4 show the performance of diverse pulse shapes like NRZ, DRZ and RZ in terms of Q factor for downstream transmission. Link length is varied from 10 km to 80 km for the Q factor assessment.

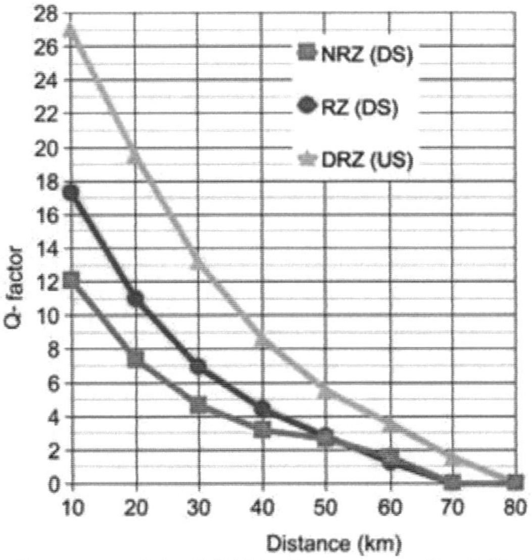

Figure 4. Performance of the BON in downstream for different distances

It is evident that as the link length prolongs, there is a significant deterioration in the Q factor. Reduction of the Q factor is because of the several factors like attenuation, dispersion and nonlinear effects along single mode fiber. Result reveals that that DRZ performs well as compared to the NRZ and RZ. Performance of the DRZ is followed by the RZ.

Figure 5 depicts the analysis of aforementioned modulation formats in the upstream transmission over different link lengths. Results revealed that Q factor decreases with the distance and DRZ exhibits the best performance in the dual directional broadband passive optical network.

Journal of Physics: Conference Series **2327** (2022) 012038 doi:10.1088/1742-6596/2327/1/012038

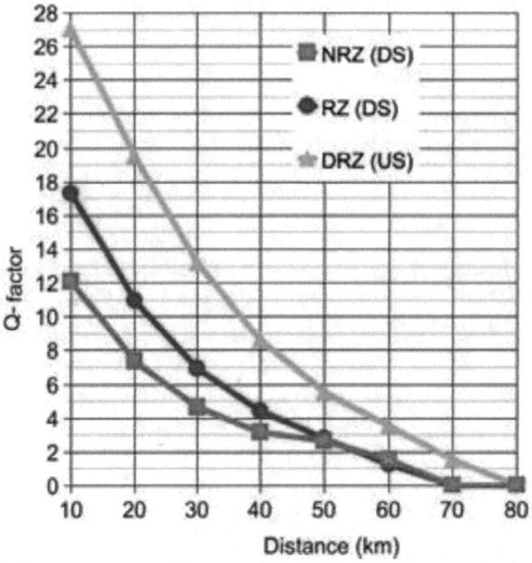

Figure 5. Q factor versus distance for upstream transmission in BPON

Figure 6 (a) (b) depicts the log (BER) with respect to the distance for downstream as well as upstream respectively for NRZ, RZ and DRZ. Bit error rate is the measure of system that computes the total false bits in the received sequence streams. Acceptable BER for the system according to international telecommunication union is 10-9.

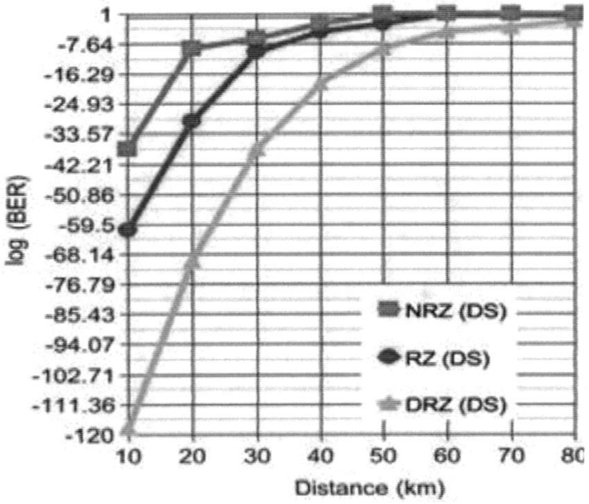

(a)

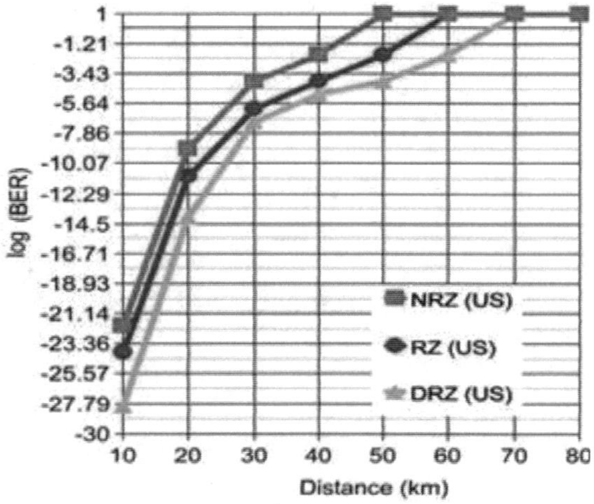

(b)

Figure 6. Graphical illustration of LoG (BER) versus link length for (a) downstream (b) upstream in proposed BPON system

It is evident that more errors in the bit streams occur when the distance prolongs. DRZ provide least errors as compared to the NRZ and RZ. Major advantage of the DRZ modulation format due to which it exhibits best performance is the dispersion tolerance and high spectral efficiency. Phase reversal is performed by DRZ format when two consecutive high bits are emerged in the sequence.

Further investigation is carried out with the variation of the input power in the system for different modulation formats. Input power of the laser is increased from -10 dBm to 10 dBm with the different of 5 dBm.

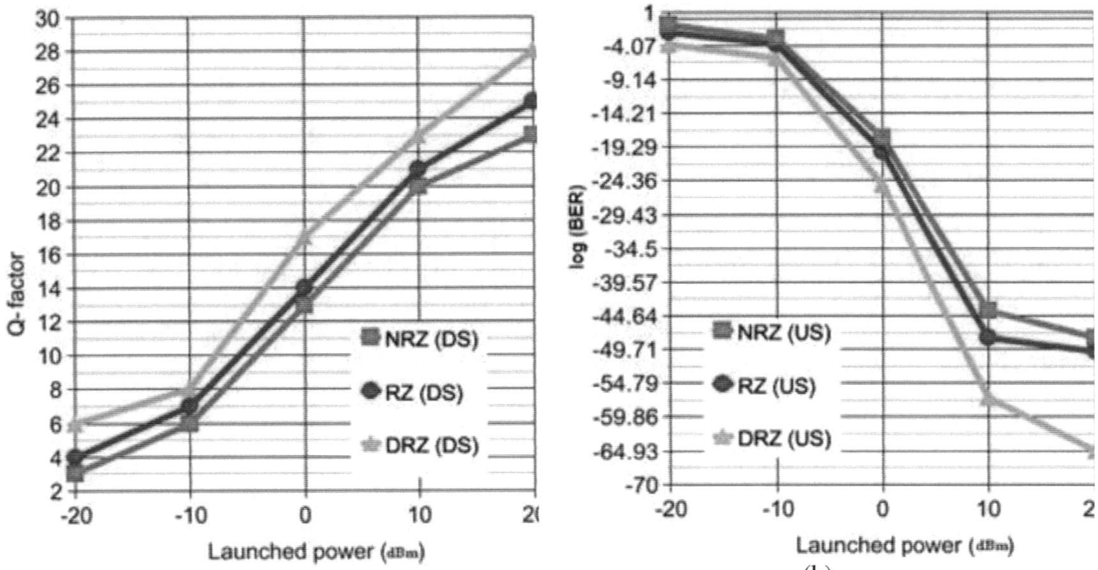

(a) (b)

Figure 7. Graphical representation of launched power versus (a) Q factor (b) LoG (BER)

Figure 8 (a) (b) illustrates the eye diagram of the bidirectional broadband passive optical network in the downstream for different modulation formats. Results reveal that the eye opening is least in the non return to zero and more in case of duo-binary return to zero. Q factor of 26.73 is observed at 25 km for DRZ, 6.43 in case of RZ and 5.23 for NRZ. It is seen that data rate of 622 Mbps is achieved for 50 km in case of DRZ, 30 km for RZ and 25 km in case of NRZ.

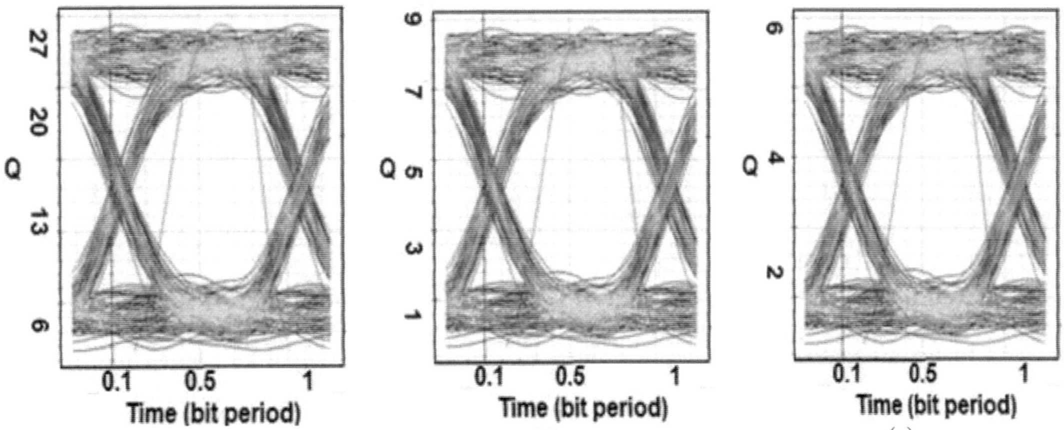

(a) (b) (c)

Figure 8. Eye diagrams of (a) DRZ (b) RZ (c) NRZ at 25 km

4. Conclusion

This research work majorly accentuated on the performance investigation of different pulse shapes such as DRZ, RZ and NRZ in BPON. Investigation is performed for different distance, launched powers in terms of BER and Quality factor. It is noteworthy that due to the extra dispersion tolerance and spectral efficiency, DRZ format performs best. RZ performs better than the non return to zero owing to the advantage of nonlinear effects tolerance. Proposed system also incorporated the colorless optical network units to make system cost efficient and for lowering the maintenance issues. Results revealed that the DRZ modulation format successfully covered the 25 km distance for bidirectional colorless PON at 622.08 Mbps.

Acknowledgement

I would like to express my very great appreciation to **Rosepreet Kaur Bhogal, Ajmer Singh, Manoj Sindhwani, Javed Dhillon** (School of Electronics and Electrical Engineering, Lovely Professional University, Phagwara, Punjab, India.) for their valuable and constructive suggestions during the planning and development of this research work. Their willingness to give their time so generously have been very much appreciated.

References

[1] Agrawal G, 2002 Fiber – Optic Communication Systems *John Wiley and Sons, USA* pp 561.
[2] Effenberger F J, et al. 2001 Advances in broadband passive optical networking technologies *IEEE Commun. Mag.* **39**(12) pp 118–122.
[3] Harstead E and Sharpe R, 2012 Future fiber-to-the-home bandwidth demands favor time division multiplexing passive optical networks *Communications Magazine, IEEE* **50**(11) pp 218-223.
[4] Zhang R and Ma J, 2017 Full-duplex hybrid PON/RoF link with 10-Gbit/s 4-QAM signal for alternative wired and 40-GHz band wireless access based on optical frequency multiplication *Optik-International Journal for Light and Electron Optics* **138** pp 55–63.
[5] ITU-T 2005 Broadband Optical Access Systems Based on Passive Optical Networks (PON) *ITU-T Recommendation* **G.983.1**.
[6] Shah S M Faizan, Gillani S, Khan M A, Shahid and M K Shahid 2015 Reach Extendibility of Passive Optical Network Technologies *Optical Switching and Networking* **18**(3) pp 211-221.
[7] Goyal R, Kaler R S and Kamal T S 2017 A Cost Effective Bidirectional Hybrid Passive Optical Network Using Common Carrier for All Optical Network Units (CCAONUs) Technique with Mitigation of Non Linearity Impact *Optik - International Journal for Light and Electron Optics* **130** pp 644–649.
[8] Li Li, Chen Yan-tao, Zhang Ji-jun, Cao Jun-feng, Li Zhi-Rui 2014 Analysis of intensity modulation performance in WDM-PON network *Optik* **125** pp 7170–7174.
[9] Latal Jan et al., 2014 Simulation of Modulation Formats for Optical Access Network Based on WDM-PON *IEEE conference*.
[10] Mizuochi T, Ishida K and Tatsuya K 2013 A comparative study of DPSK and OOK WDM transmission over transoceanic distances and their performance degradations due to nonlinear phase noise *J. Lightwave Technol.* **21**(9) pp 1933–1943.
[11] Xu C, Liu X, Mollenauer L F, 2013 Comparison of return-to-zero differential phase-shift keying and ON–OFF keying in long-haul dispersion managed transmission *IEEE Photon. Technol. Lett.* **15**(4) pp 617–619.
[12] Malhotra J S, Kumar M and Sharma A K 2013 Performance optimization of high capacity long reach 32 channel FTTH downstream link employing triple play services *Optik* **124** pp. 2424–2427.
[13] Goyal R, Kaler R S 2012 A novel architecture of hybrid (WDM/TDM) passive optical neworks with suitable modulation format *Opt. Fiber Technol.* **18** pp 518–522
[14] Khaki Z G, and Bedi H S 2012 Transient correction using EDFA: in-line optical fiber with feedback *International Conference on Computing Sciences* pp. 233-238.

[15] Bedi H S, Sharma K K and Gupta R 2020 A Review Paper on Performance Analysis of IEEE 802.11 e *1st International Conference on Computing, Communications, and Cyber-Security (IC4S 2019)* pp. 47-56

A LOW COST ALL OPTICAL PHOTONIC TECHNIQUE BASED ON BRILLOUIN SCATTERING FOR MM-WAVE GENERATION

Mohit Bansal[1], Manoj Sindhwani[2]

[1]Department of Electrical & Electronics Engineering, G.L.Bajaj Institute of Technology & Management, Greater Noida. U.P.

[2]School of Electronics and Electrical Engineering, Lovely Professional University, Punjab.

moh.bansal@gmail.com[1], manojsindhwani324@gmail.com[2]

https://orcid.org/0000-0001-8072-1462[1], https://orcid.org/0000-0002-8323-2848[2]

Abstract. Generation of millimeter wave (mm-wave) in radio over fiber (ROF) systems is suffered from nonlinearity effects in optical fiber communication. In order to mitigate the degrading effects, a low cost all optical photonic method for the generation of mm-wave is proposed in this work. Therefore, work is accentuated on the enhanced frequency shift based on stimulated Brillouin scattering (SBS) by utilizing only one pump source. Effect of length of optical fiber, launched power and photo detector quantum efficiency is investigated over proposed architecture in order to observe power fading effects. Photonically stable, mm-wave is realized productively in the cost effective way using SBS incorporating different frequencies from 20 GHz-80 GHz. Results reveals that USB (upper sidebands) and LSB (lower sidebands) generated with demonstrated system have high power and good SNR. This makes system appropriate for radio over fiber system.

Keywords: RoF (Radio-over-fiber), mm-wave (millimeter wave), SC-DSB (suppressed carrier double sideband), SBS (stimulated Brillouin scattering) , MZM (mach-zehnder modulator)

1. Introduction

Due to massive augmentation of wireless communication raised the demand of high speed and high capacity systems. Use of millimeter wave region serves architecture of integrated and cost efficient wireless millimeter wave transmissions that can provide suitable terminal mobility [1]. With the numerous communication services, there is congestion and pressure on low frequencies. Consequently, a broad spectrum of radio frequencies is desirable for the releases of load that exists on low frequencies. Millimeter-wave spectrum (30–300 GHz), introduces a prominent opportunity to augment spectrums. In radio over fiber systems, numerous studies have been demonstrated in the context of millimeter wave and their generation [2-3]. These methods can be divided as: dual wavelength laser injection locking [4], four wave mixing [5], optical heterodyne [6]. These techniques are good for mm-wave generation, but limited to enhanced frequency shift. Transmission of millimeter wave is widely investigated over optical fiber because of its large bandwidth and pico-cell sizes [7]. Radio over fiber communication has potential to provide flexible optical networks. The main feature of the integration of optical fiber and wireless network is to offer a way out for the ever increasing demands of high speed as well as plummeting the expenditure on future generation networks. RoF is an attractive candidate due its benefits such as the low complexity at antenna sites and frequency reuse. However, even there are many advantages of RoF systems, some degrading effects plays the vital role in transmission such as dispersion effects when mm-wave is implicated. In order to mitigate or compensate the dispersion effects, dispersion shifted fibers (DSF) are incorporated in the system. On the contrary, these fibers increase the total cost of the system due to their high prices. Moreover, dispersion shifted fibers has high attenuation coefficient that degrades the strength of the signal in network. To suppress the pulse broadening effects, a numerous works has been proposed to quell these effects. Also, lot of techniques has been reported for generation of radio signals as well as their communication. To confirm that transmission of mm-wave is reliable, tangible, optimal methods are requisite to implement the mm-wave in RoF [8]. RoF system having single sideband and up-conversion is demonstrated in [9]. Single sideband modulation generated through SBS is more robust and feasible to pulse broadening in the RF carrier transmission. To alleviate the pulse broadening, the combination of single sideband and single mode modulation is most efficient solution [10-11]. Next generation networks require high speed of data that can be supported through high frequency carriers shifted in optical fiber due to SBS, however, this shift is limited to 10 GHz to 40 GHz [12]. In [14], a 100 GHz shift has been observed using SBS amplification but complexity, cost and spurious signals were issues. Reported techniques that are suggested so far, are costly and complex.

Proposed research article discusses the dispersion tolerant and 20 GHz to 80 GHz sifted mm-wave signal is demonstrated based on stimulated Brillouin scattering. Proposed architecture is different from previous works because the basis of this research is all optical SBS method and null biased modulator to achieve enhanced frequency shift in mm-wave band. Local oscillator in the receiver is proposed for RoF. High frequency signals are generated both at central station and base station. In central station high carrier signal is generated through SBS and at base station by heterojunction bipolar transistor. Use of radio frequency signal as the local oscillator brings down the cost of high frequency oscillator or conventional oscillator. To achieve the 20 GHz to 80 GHz shift, only one pump is used at 1552.36 nm.

2. Principle of SBS

In optical communication, SBS is a degrading factor and limits the performance of the systems. A backward propagating wave (stokes) carries maximum part of launched energy till the Brillouin threshold. So, SBS is a barrier and limits the launched power that can fed to optical fiber without degradation of quality of signal. Degrading nature of SBS has been extensively studied and investigated in communication systems. However, SBS also have some constructive characteristics such as millimeter region generation. Numerous architectures have been reported in the context of mm-wave generation using external/direct modulation and heterodyning. Current work is accentuated on the mm-wave generation with SBS phenomenon and minimum point biased modulator. Due to the presence of microscopic defects, the interaction of propagating wave with input light beam give rise to Brillouin scattering. At the power level where SBS becomes a severe dominant factor is referred as threshold of SBS. Low threshold can be achieved by using lasers with narrow linewidth. Threshold power of SBS is expressed as [12]

$$P_{th} = 21\frac{KA_{eff}}{gL_{eff}} \qquad (1)$$

A_{eff} is referred as a affective area of optical fiber and L_{eff} is effective length. Polarization constant is K and Brillouin gain is g. [12]

$$L_{eff} = \frac{1 - e^{-\alpha l}}{\alpha}$$

Link length is L and α is the attenuation of SMF.

Mm-wave is generated through a co propagating modulated signal and a counter propagating pump signal in bidirectional optical fiber. A minimum point biased modulator derived by sinusoidal signal f_s is used for carrier suppressed signal and equation is given as (2) [13]

$$S_{strokes} = \frac{1}{2}P_0 t_{ff}[cos^2 A. j_0^2(B1). \delta(\omega + \omega_c) + sin^2 A. j_1^2(B1). \delta(\omega + (\omega_c \pm \omega s))] \quad (2)$$

where Po is the continuous wave light power into the SBS based optical frequency shifter, t_{ff} MZM insertion loss, and P_0 launched laser power, $j_m(x)$ Bessel function mth order and angular frequencies of carrier and microwave is given as ω_c and ω_s respectively.

3. System architecture

A continuous wave laser operating at low power -30 dB and 193.1 THz frequency is modulated with sine wave of 20 GHz with the external minimum biased MZM. For suppressing carrier, a MZM in minimum point operation is used by setting the bias voltages. Linewidth of laser is Laser is kept at 1 MHz and polarization is fixed. In order to boost the signal, a semiconductor optical amplifier is placed next to MZM. SOA has many advantages such as, small size, monolithically integrated, low noise figure, low power consumption, and polarization insensitive gain. A bidirectional single mode fiber with attenuation and dispersion as shown in Table 1 is considered for SBS analysis. Brillouin scattering gain is 4.6x10-011 m/W, linewidth and frequency shift is fixed for investigation. A counter propagating pump with 8 dBm power operating at 1552.36 nm is circulated through a circulator and fed to SMF. Initiation of SBS takes place and consequently, the mm-wave signal could be achieved at the port 1 of the circulator because of the pump wave of the pump laser travelling in counter direction. Due to this pump, stokes signal is shifter from 11 GHz to 20 GHz upper sideband (USB) by pump and identical shifting is observed for Lowe sideband (LSB) forward direction signal. Optical spectrum analyzer is depicter of shifted sidebands of input signal which represents the power of signal with respect to wavelengths. Figure 1 illustrates the system architecture and Figure 2 depicts the optical spectrum analyzer after SBS. Selected frequencies that are shifted after SBS operation are sent to photodetector.

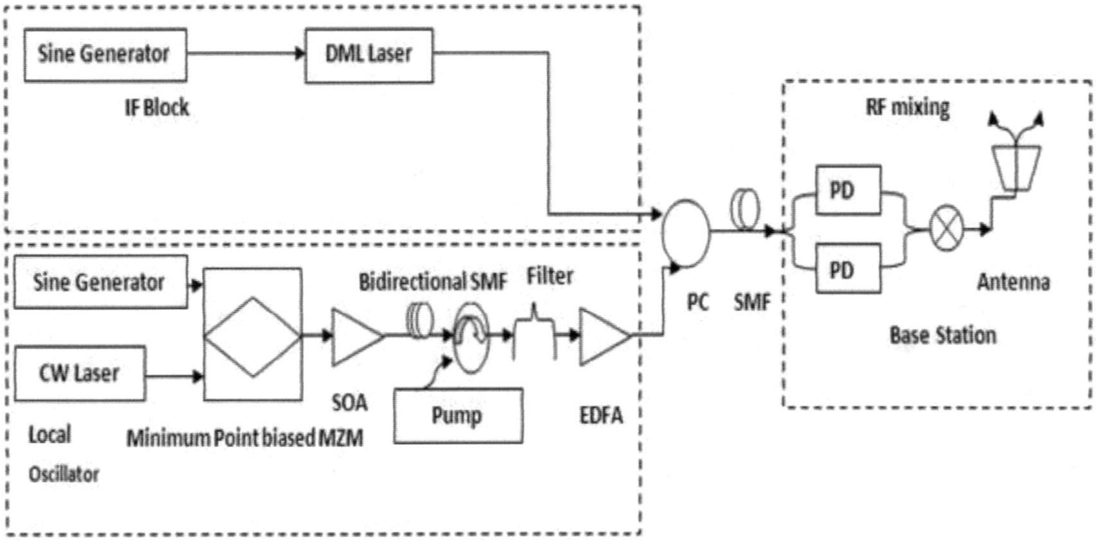

Figure 1. Proposed Radio over fiber remote optical local oscillator

4. Results and Discussions

The proposed system is assessed in terms of output power after various parameters are examined using optisystem software.

Table 1. System specification

Parameters	Values
Laser Frequency	193.1 THz
Laser linewidth	1 MHz
Laser Power	-30 dBm
Bidirectional SMF attenuation	0.2 dB/Km
Dispersion	16.25 ps/nm/Km
Amplifier prior to SBS	SOA
Amplifier after SBS	EDFA
EDFA Gain	30 dB
Noise figure	4 dB
Brillouin scattering gain	4.6x10-011 m/W
Linewidth	31.5 MHz
Brillouin frequency shift	11 GHz

4.1. Effects of SMF length on sidebands after SBS

First and foremost, effect of different distances of SMF has been studied and power of each subcarrier is accessed though power meter after filter. Parameters other than distance are fixed throughout the analysis. Length of bidirectional fiber is varied from 1Km to 50 Km in the interval of 5 Km. Erbium doped fiber amplifier plays the role of booster after SBS from the fiber. Optical power meter after filter shows the values of power levels at different distances for different frequencies (20 GHz, 40 GHz, 60 GHz and 80 GHz).

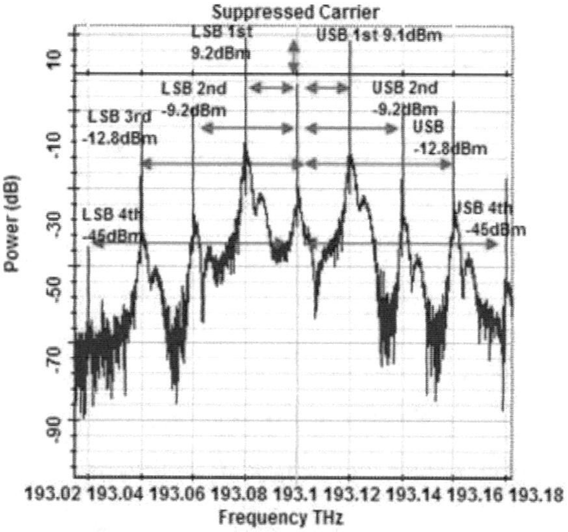

Figure 2. Optical spectrum analyzer representations after SBS

Journal of Physics: Conference Series **2327** (2022) 012039 doi:10.1088/1742-6596/2327/1/012039

Nonlinear behavior of the input signal is observed on sidebands between 10-25 Km. There are different link distance between which nonlinear effects emerges such as for 20 GHz RF signal it is at 11 Km, and at 10Km for 40 GHz, for 60 GHz it is 15 Km and 16 Km for 80 GHz. Therefore length of the optical fiber is set at 25 Km. Table 2 show the output power with variation of fiber length.

Table 2. Output power of sidebands after SBS as a function of fiber length

SMF Length (Km)	F=20 GHz	F=40 GHz	F=60 GHz	F=80 GHz
1	44.16	34.94	26.6	18.86
5	43.36	34.14	25.8	18.06
10	34.36	25.14	16.8	9.07
15	39.36	30.14	21.8	14.06
20	40.36	31.14	22.8	15.06
25	35.36	26.14	17.8	10.07
30	38.39	29.36	21.88	16.79
35	37.4	28.44	21.23	16.65
40	36.43	27.59	20.8	16.88
45	35.45	26.74	20.36	16.93
50	34.49	25.93	19.98	16.97

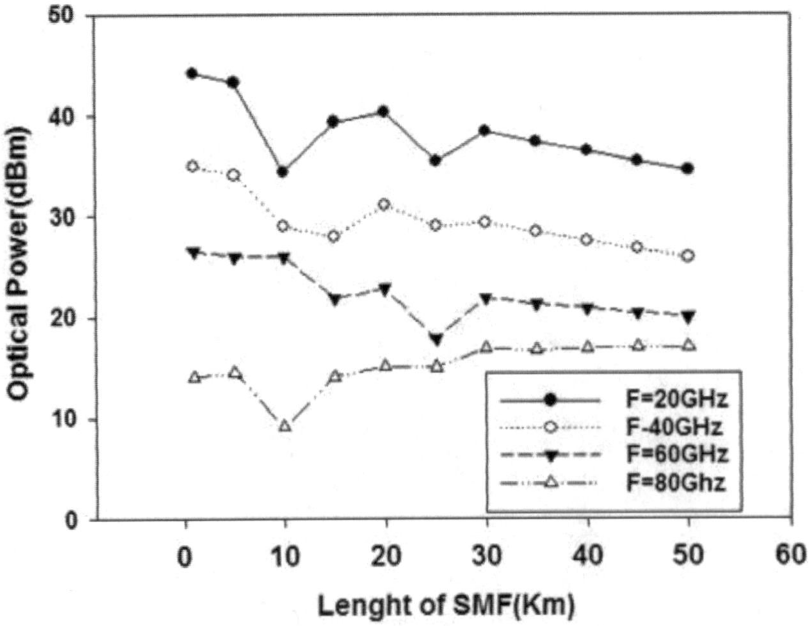

Figure 3. Graphical representations of length of SMF and optical output power of different sidebands

4.2. Effects of Launched power on sidebands

After analyzing the effect of link length, an important factor of input launched power has been investigated. Laser is biased at 193.1 THz and launched power varied from -30dB to 10 dBm in optical fiber that has fixed length 25 Km. An EDFA with 30 dB gain and 4dB noise figure is considered after SMF and signal is fed to photo detector PIN that has 10nA dark current and 80 % responsivity. Thermal, shot and ASE (amplifier spontaneous noise) noises are taken into account while analysis. Table 3 represents the values of power obtained after SBS for sidebands at different launched powers.

Table 3. Detected powers of sidebands for different launched powers

SMF Length (Km)	F=20 GHz	F=40 GHz	F=60 GHz	F=80 GHz
-30	12	-13.02	-12.74	-25.12
-25	19	-3.63	-9.88	-22.15
-20	30.59	3.5	-7.2	-19.36
-15	33.43	14.46	-4.67	-16.68
-10	46.12	17.26	-2.24	-14.72
-5	38.17	20	0.16	-11.6
0	40.8	23.9	3.54	-7.45
5	43.38	28.51	13.43	7.4

From analysis, it is concluded that maximum power is obtained for 20 GHz signal and it keeps on decreasing as increase in frequency shifts of sidebands. Minimum power is detected in case of 80 GHz sideband. With the increase of power, detected power fluctuated due to SBS nonlinear region and there is not much increase in case of 20 GHz output power from 5-10 dBm. However in case of 40 GHz, 60 GHz and 80 GHz sidebands, power increase with the increase of input launched power. Linear region of power increment where output power is vary linearly is -5 to 10dBm. For 20 GHz, 40 GHz, 60 GHz and 80 GHz detected outputs are 46.12 dBm, 29.87 dBm, 20.45dBm and 19.47 dBm respectively.

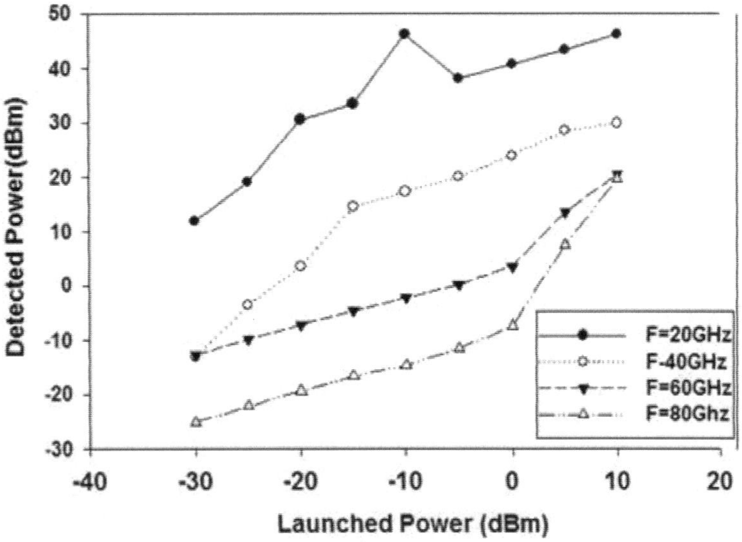

Figure 4. Variation of Detected output power (dBm) with input launched power

4.3. Effects of different Responsivities of photo detector

Photo detector is major component for the detection of signals and their efficiency depends on responsivity. Therefore, effect of different quantum efficiency is investigated for different responsivity values such as 0.1 A/W to 1 A/W. Relation between the quantum efficiency and responsivity is expressed as [15]

$$\eta = R \times \frac{hc}{\lambda q} = \frac{1.24R}{\lambda} \; A/W$$

Length of fiber as mentioned before is fixed at 25 Km and frequency at 193.1 THz of CW laser. Launched power is set to -30 dBm and EDFA gin is at 30 dB as stated before. Results reveal that, as the increase in quantum efficiency, output power also increases. There is significant change when responsivity varied from 0.1 to 0.2 A/W. Launched power of −10 dBm, from 0.1 to 0.2 A/W, the yield power is augmented enormously.

But, from 0.2 to 0.3 A/W, detected power will increase not as increased in former case, whereas 0.3-04 A/W, the increment further not as seen in first case. It is observed that as the responsivity keeps on changes from low to high, the growth rate is not as high as 0.1-0.2 A/W case. Therefore, as depicted in Figure 5, it is validated that, the responsivity of a photo detector in the suitable area that was between 0.6 and 0.9 A/W.

Table 4. Detected powers for sidebands at different responsivity of p-i-n

Responsivity of pin	F=20 GHz	F=40 GHz	F=60 GHz	F=80 GHz
0.1	20.87	2.91	-14.8	-27.18
0.2	26.89	8.93	-8.78	-21.16
0.3	30.41	12.45	-5.26	-17.64
0.4	32.91	14.95	-2.76	-15.14
0.5	34.85	16.89	-0.82	-13.2
0.6	36.43	18.47	0.76	-11.62
0.7	37.77	19.81	2.1	-10.28
0.8	38.93	20.97	3.25	-9.12
0.9	39.93	21.99	4.28	-8.1

It is projected that the result will stay stable as the responsivity increases. Table 4 represents the detected power at sidebands for varied responsivity of p-i-n. At 1 A/W detected power is maximum for the proposed system as shown in Figure 5.

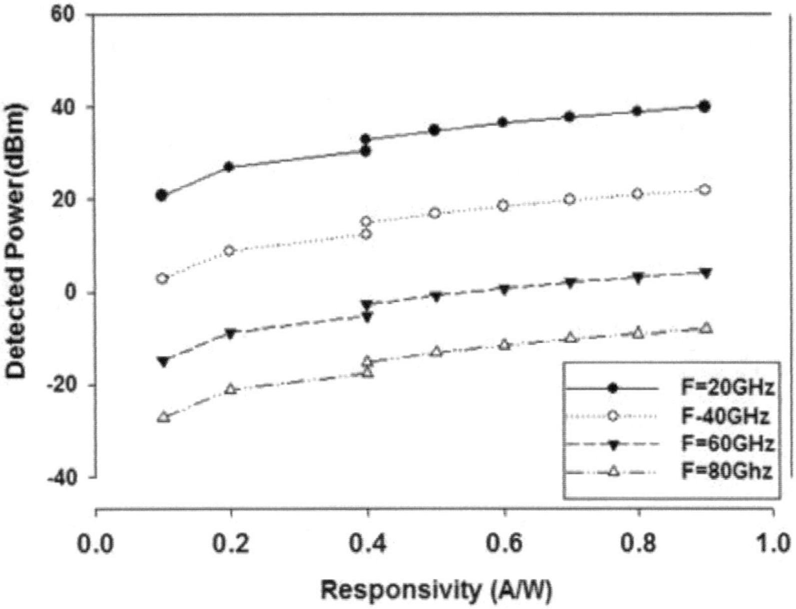

Figure 5. Graphical representation Detected power (dBm) with variation of reponsivity of PD

5. Conclusion

In this paper, a low cost radio over fiber system and mm-wave generation is presented based on SBS. A dispersion tolerant and 20 GHz to 80 GHz sifted millimeter-wave signal is demonstrated reliant on stimulated Brillouin scattering. Proposed architecture is different from previous works because the basis of this research is all optical SBS method and null biased modulator to achieve enhanced frequency shift in mm-wave band. A remote optical local oscillator system is proposed for radio over fiber system. High frequency signals are generated both at central and base station. In central station high carrier signal is generated through SBS and at base station by heterojunction bipolar transistor. Use of radio frequency signal as the local oscillator brings down the cost of high frequency oscillator or conventional oscillator. To achieve the 20 GHz to 80 GHz shift, only one pump is used at 1552.36 nm.

Acknowledgement

I would like to express my very great appreciation to **Shippu Sachdeva and Javed Dhillon** (School of Electronics and Electrical Engineering, Lovely Professional University, Phagwara, Punjab, India.) for their valuable and constructive suggestions during the planning and development of this research work. Their willingness to give their time so generously have been very much appreciated.

References

[1] Andrews J G, Buzzi S, Choi W, Hanly S, Lozano A, Soong A C K and Zhang J C, 2014 What will 5 G be? *IEEE J. Sel. Areas Commun.* **32**(6) pp 1065–1082.

[2] Chattopadhyay T, 2012 A millimeter-wave radio-over-fiber system for overcoming fiber dispersion-induced signal cancellation effect *Optoelectron. Lett.* **8**(4), pp 293–296.

[3] Lecoche, F, et al., 2009 Transmission quality measurement of two types of 60 GHz millimeter-wave generation and distribution systems *J. Lightwave Technol.* **27**(23) pp 5469–5474.

[4] Wang M, Chen C, Li Q, Huang K and Chen H, 2015 Photonic generation of tunable Microwave signal using a passively Q-switched dual-wavelength fiber laser *Microw. Opt. Technol. Lett.* **57** (1) pp 166–168.

[5] Jiang Y, Yu J, Han B, Zhang L, Wang W, Zhang L and Yang E, 2009 Millimeter-wave subcarrier generation utilizing four-wave mixing and dual-frequency Brillouin pump suppression *Opt.Eng.* **48** p 030502.

[6] Mohamad R, Idrus S M, Supaat A S, Yaakob S, Zamzuri A K and Sukito S N A, 2014 Millimeter wave carrier generation based on Brillouin fiber laser with improved tuning capability *Optik* **125**(1) pp 205–207.

[7] Kuri, T, Kitayama K, Takahashi Y, 2000 60 GHz band full-duplex radio-on-fiber system using two-port electro absorption transceiver *IEEE Photonics Technol. Lett.* **12**(4) pp. 419–421.

[8] Han, Xiuyou, et al., 2013 A new method for fiber chromatic dispersion measurement with microwave interference effect *Optical Fiber Technol.* **19**(4) pp 319–324.

[9] Gross M C, Callahan P T, Clark T R, Novak D, Waterhouse R B, Dennis M L, 2010 Tunable millimeter-wave frequency synthesis upto100GHz by dual wavelength Brillouin fiber laser, *Opt. Express* **18**(13) pp 13321–13330.

[10] Park C S, Lee, C G, Park C S, 2007 Experimental demonstration of 1.25-Gb/s radio-over-fiber downlink using SBS-based photonic upconversion *IEEE Photonics Technol. Lett.* **19**(22), pp. 1828–1830

[11] Cheng, H., et al., Single-sideband modulation based on an injection-locked DFB laser in radio-over-fiber systems *IEEE Photonics Technol. Lett.* **22**(7), pp. 462–464.

[12] Norliza Mohamed, Suriani Mohd. Sam and Nor Hafizah Ngajikin, 2016 Power Fading Effects in Millimeter-Wave Radio Over Fiber (RoF) Link *Advanced Computer and Communication Engineering Technology, springer* **362** pp. 493-503.

[13] Chan Erwin H W and Minasian Robert A, 2014 All-Optical Frequency Shifter Based on Stimulated Brillouin Scattering in an Optical Fiber *IEEE photonic Journal* **6**(2) 2014.

[14] Preußler Stefan, Wenzel Norman, Braun Ralf-Peter, Owschimikow Nina, Vogel Carlo, Deninger Anselm, Zadok Avi, Woggon Ulrike, and Schneider Thomas, 2013 Generation of ultra-narrow, stable and tunable millimeter and terahertz-waves with very low phase noise *Optical Society of America* **21**(20) pp. 23950-23962.

[15] Aoki Yasuhiro, Tajima Kazuhito, and Mito Ikuo, 1988 Input Power Limits of Single-Mode Optical Fibers due to Stimulated Brillouin Scattering in Optical Communication Systems", Journal of Lightwave Technology **6**(5) pp 710-719.

Estimation of Channel for Millimeter-Wave Hybrid Massive MIMO Systems using Orthogonal Matching Pursuit (OMP)

Shailender[1*], Shelej Khera[2], Sajjan Singh[3], Jyoti[4]
[1] Lovely Professional University, Phagwara, Punjab, India
[2] Lovely Professional University, Phagwara, Punjab, India
[3]Chandigarh group of Colleges, Jhanjeri, Mohali, India
[4]Delhi Global Institute of Technology, Jhajjar, India
*Corresponding Author Email: Shailender.41900027@lpu.in
https://orcid.org/0000-0003-4546-5066

Abstract: In a millimeter-wave (mm-Wave) large MIMO system, hybrid precoding is a critical component for lowering radio-frequency hardware costs as compared to the traditional full-digital precoding strategy. Knowledge of channels is required for hybrid precoding. Though, estimation of the channel is problematic for the mm-Wave system because of the usage of a massive array of antenna and hybrid architecture with analog precoding. Due to the extremely directed nature of wireless propagation, wireless channels have spatial sparsity. In this paper, we exploit this spatial sparsity nature of wireless channels to develop orthogonal matching pursuit technique-based estimation of channels for hybrid millimeter-wave (mm-Wave) wireless systems. Genie (Ideal) channel estimation is also performed in which we presume that actual angle of departures (AoD) and angle of arrivals (AoA) are known to us. Finally, the simulation results reveal that suggested OMP algorithm-based channel estimation has a significant advantage over conventional approaches like least-squares channel estimation.

Keywords: *Orthogonal Matching Pursuit (OMP), Energy Efficiency, Millimeter-wave, Channel estimation, Sparse Channel, Massive MIMO.*

1. Introduction: -

In wireless communication systems, spectral efficiency along with power usage are often two utmost significant performance parameters. [1,2]. The major differences from prior communication systems are the fifth generation's (5G) high-speed and low-latency features. Developing technologies, for example, physical layer technology, network densification technology, and others [3– 6] made this advancement possible. The scarcity of spectrum, on the other hand, is a major issue in today's technical development of communication networks. Millimeter-wave (mm-Wave) technology offers additional wireless communication systems' spectrum resources and meets the bandwidth demands of 5G services [7, 8]. With shorter wavelengths, mm-Wave massive MIMO technology can bundle large-scale antennas into small sizes. Thus, through beamforming with a huge antenna array, the mm-Wave system is capable of attaining required beamforming gains to compensate for high propagation losses in mm-Wave. Millimeter-wave massive MIMO technology allows for the utilization of bandwidth in the gigahertz range and huge arrays of antennas to provide more data speeds and is considered a key method in future wireless +communications [7]. When every antenna is linked to a specialized radio-frequency chain, the high hardware costs and power consumption made it unaffordable. Link reliability is improved by hybrid beamforming technology by compensating significant mm-Wave path losses.

Traditional MIMO transceiver structures can no longer achieve adequate gain for effective signal identification because of the significant signal propagation pathway-loss in mm-Wave. Consequently, antennas in number on both the transmitter and receiver end must be raised considerably to improve the received signal strength. As a result, a large MIMO transceiver design is expected to be used in future 5G wireless communication systems [9]. For fifth-generation networks, mm-Wave communication is a potential technology. In mm-Wave communication systems, hybrid precoding is used to offer a suitable trade-off approach among system performance and energy consumption [10]. Hybrid precoding strikes a balance between the best digital precoding and the most cost-effective analog beamforming [11].

Channel state information is needed in mm-Wave systems for designing precoder and combiner at transmitters and receivers. Prediction accuracy for mm-Wave MIMO channels that are currently available has a huge impact on baseband and RF precoders and combiners' performance [12], [13]. Precise knowledge of the channel is essential for attaining large gains linked with mm-Wave communication, hence effective approaches for accurate channel estimate are critical. Because of the substantially shorter wavelength of mm-Wave, the number of transmitting and receiving antennas may be substantially higher than in typical microwave massive MIMO, which makes estimation of channel for mm-Wave hybrid large MIMO system more difficult. Channel estimate in millimeter-wave frequencies differs from that at lower frequencies due to the various "virtual array" channel model typically utilized for mm-Wave large MIMO [14] and unique hardware restrictions imposed by hybrid architecture [12]. As a result, new approaches for channel estimates are needed.

However, due to the extremely directed character of mm-Wave channel propagation, spatial sparsity may be effectively used to estimate channel coefficients connected with major spatial pathways [13], [15]. As a result, sparse signal recovery algorithms [16], [17] may be used to generate reliable channel estimations for spatially sparse channels. The only angle of arrivals (AoA), angle of departs (AoD), and related path gains of dominating pathways are evaluated rather than all channel matrix elements [18]. We present an improved channel estimate technique for mm-Wave massive MIMO using the OMP algorithm in this paper. Additionally, ideal Genie-assisted channel estimation and traditional Least Square channel estimation are carried out.

The paper is arranged as follows. In Section 2, a system model is provided. Section 3 provides the mm-Wave massive MIMO channel model and solves Genie-aided and Least Square (LS) based channel estimation problems. The sparse mm-Wave channel model and proposed OMP algorithm-based channel estimation are provided in section 4. Section 5 presents simulation specifics along with performance evaluations, and finally, Section 6 brings the paper to a close.

2. System Model

To handle $N_s \leq N_{RF}$ parallel data streams, assume a single user mm-Wave hybrid massive MIMO system having N_R receiving antennas, N_T transmitting antennas, and $N_{RF} \leq \min(N_T, N_R)$ RF chain at transmitter and receiver, respectively. Figure 1 represents a hybrid MIMO architecture system [19]. An analog RF domain precoder with a series of digital MIMO baseband precoders, indicated by matrices $\mathbf{F}_{\mathrm{RF}} \in \mathbb{C}^{\mathrm{N_T} \times N_{RF}}$ and $\mathbf{F}_{\mathrm{BB}} \in \mathbb{C}^{\mathrm{N_{RF}} \times N_s}$, respectively, make up the hybrid precoder $\mathbf{F} = \mathbf{F}_{\mathrm{RF}}\mathbf{F}_{\mathrm{BB}} \in \mathbb{C}^{\mathrm{N}_T \times \mathrm{N}_s}$. Assume an mm-Wave MIMO channel that is frequency selective with delay taps of size L among transmitter and receiver. It can be expressed as $\mathbf{H}_d \in \mathbb{C}^{\mathrm{N}_R \times N_T}$, $d = 0, 1, \ldots, \mathrm{L} - 1$, where indexing of tap is indicated by d. Signal vector received $\mathrm{r}[\mathrm{n}] \in \mathbb{C}^{\mathrm{N}_R \times 1}$ at n^{th} time instant for system mentioned above is stated as

$$\mathrm{r}[n] = \sqrt{\rho} \sum_{d-0}^{L-1} \mathbf{H}_d\, \mathbf{Fs}[n-d] + \mathbf{v}[n] \quad (1)$$

where the strength of the received signal is ρ, s[n] $\in \mathbb{C}^{N_s \times 1}$ represents symbol vector to be sent and $\mathbf{v}[n] \in \mathbb{C}^{N_R \times 1} \sim \mathcal{CN}(0, \sigma^2 \mathbf{I}_{N_R})$ signifies additive white Gaussian noise vector which is circularly symmetric complex having mean of zero and matrix of covariance as $\sigma^2 \mathbf{I}_{N_R}$.

To get N_s dimensional processed signal, the receiver uses the hybrid combiner $\mathbf{W} = \mathbf{W}_{RF}\mathbf{W}_{BB} \in \mathbb{C}^{N_R \times N_s}$ as

$$y[n] = \sqrt{\rho} \sum_{d-0}^{L-1} \mathbf{W}^H \mathbf{H}_d \mathbf{F} \mathbf{s}[n-d] + \mathbf{W}^H \mathbf{v}[n]. \tag{2}$$

Matrices $\mathbf{W}_{BB} \in \mathbb{C}^{N_{RF} \times N_s}$ and $\mathbf{W}_{RF} \in \mathbb{C}^{N_R \times N_{RF}}$ denote baseband and RF combiners, respectively. It's worth noting that while $\mathbf{F}_{RF}$ and $\mathbf{W}_{RF}$ are realized by using analog phase-shifters, their components must all have the same norm.

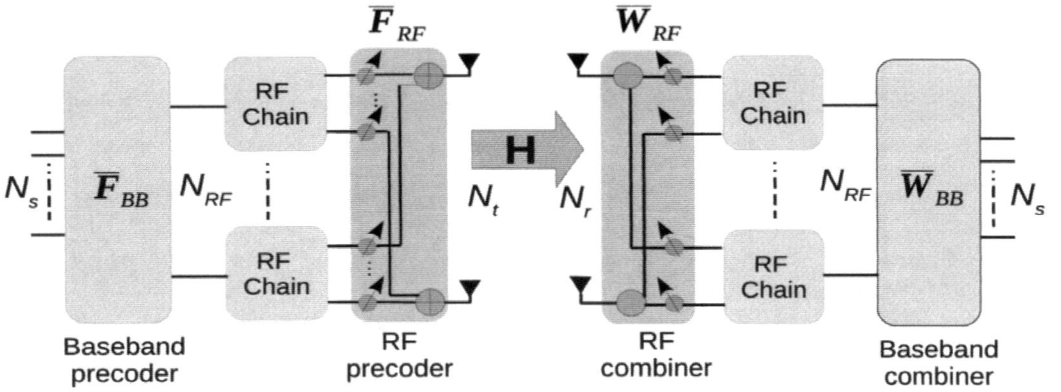

Fig 1. Hybrid mm-Wave massive MIMO signal processing [19]

3. mm-Wave massive MIMO channel model

3.1 Genie – aided Channel estimation: - mm-Wave massive MIMO channel model **H** is described as

$$\mathbf{H} = \sum_{l=1}^{L} \alpha_l \, \boldsymbol{a}_R(\theta_l^r) \, \boldsymbol{a}_T^H(\theta_l^t) \tag{3}$$

$\boldsymbol{a}_R(\theta_l^r)$ and $\boldsymbol{a}_T(\theta_l^t)$ are array response vectors corresponding to an angle of arrivals θ_l^r and angle of departs θ_l^t at receiver and transmitter respectively. L is the number of multipath or Channel sparsity level and α_l is a complex gain of lth multipath.

Further channel model may be simplified as

$$\mathbf{H} = \underbrace{[\boldsymbol{a}_R(\theta_1^r) \quad \boldsymbol{a}_R(\theta_2^r) \ldots \boldsymbol{a}_R(\theta_L^r)]}_{\overline{\mathbf{A}}_R} \times \underbrace{\begin{bmatrix} \alpha_1 & \cdots & 0 \\ \vdots & \ddots & \vdots \\ 0 & \cdots & \alpha_L \end{bmatrix}}_{\Omega} \times \underbrace{\begin{bmatrix} \boldsymbol{a}_T^H(\theta_1^t) \\ \boldsymbol{a}_T^H(\theta_2^t) \\ \vdots \\ \boldsymbol{a}_T^H(\theta_L^t) \end{bmatrix}}_{\overline{\mathbf{A}}_T^H} \tag{4}$$

$\overline{\mathbf{A}}_T$ and $\overline{\mathbf{A}}_R$ are true array response matrices containing array response vector corresponding to L different angle of departures and angle of arrivals and Ω is a diagonal matrix containing complex gain of different multipath components. In Genie aided channel estimation we presume that angle of departures and angle of arrivals are known to us and we have to estimate only gain of various multipath components $(\alpha_1, \alpha_2, \ldots, \alpha_L)$.

Genie aided channel estimation can be obtained as follow

$$\mathbf{H} = \sum_{l=1}^{L} \alpha_l \, \boldsymbol{a}_R(\theta_l^r) \, \boldsymbol{a}_T^H(\theta_l^t) = \overline{\mathbf{A}}_R \mathbf{\Omega} \overline{\mathbf{A}}_T^H \tag{5}$$

$$\overline{\mathbf{h}} = \text{vec}(\mathbf{H}) = \sum_{l=1}^{L} \mathbf{a}_T^*(\theta_l^t) \otimes \mathbf{a}_R(\theta_l^r) \alpha_l \tag{6}$$

Journal of Physics: Conference Series **2327** (2022) 012040 doi:10.1088/1742-6596/2327/1/012040

$$\overline{\mathbf{h}} = \boldsymbol{\Psi}\overline{\boldsymbol{\alpha}} \tag{7}$$

The matrix Ψ is defined as

$\boldsymbol{\Psi} = [\, \mathrm{a}_T^*(\theta_1^t) \otimes \mathbf{a}_R(\theta_1^r) \ldots\ldots\ldots\ldots \mathrm{a}_T^*(\theta_L^t) \otimes \mathbf{a}_R(\theta_L^r)]$ and $\overline{\alpha}$ is a channel gain vector written as $\overline{\boldsymbol{\alpha}} = \begin{bmatrix} \alpha_1 \\ \vdots \\ \alpha_L \end{bmatrix}$

For the pilot matrix, $\sqrt{P}\boldsymbol{I}_{N_T}Beam$ matrix **Y** of the received signal is given as

$$\mathbf{Y} = \sqrt{P}\mathbf{W}_{BB}^H\mathbf{W}_{RF}^H\mathbf{H}\mathbf{F}_{RF}\mathbf{F}_{BB}\boldsymbol{I}_{N_T}Beam + \widetilde{\mathbf{N}} \tag{8}$$

It is important to vectorize matrix **Y** of a received signal in (8) to define the channel estimation issue. **y** is used to represent vec(**Y**), therefore (8) is rewritten as

$$\mathbf{y} = \mathrm{vec}(\mathbf{Y}) = \overline{\mathbf{Q}}\overline{\boldsymbol{\alpha}} + \widetilde{\mathbf{n}} \tag{9}$$

$$\overline{\mathbf{Q}} = \left(\underbrace{\sqrt{P}\, \mathbf{F}_{BB}^T\mathbf{F}_{RF}^T \otimes \mathbf{W}_{BB}^H\mathbf{W}_{RF}^H}_{\widetilde{\mathbf{Q}}} \right)\boldsymbol{\Psi} \tag{10}$$

Therefore Genie – aided channel estimation model can be stated as

$\mathbf{y} = \overline{\mathbf{Q}}\overline{\boldsymbol{\alpha}} + \widetilde{\mathbf{n}}$ and finally estimated channel gain vector is provided by

$$\widehat{\boldsymbol{\alpha}} = \left(\overline{\mathbf{Q}}^H\overline{\mathbf{Q}}\right)^{-1}\overline{\mathbf{Q}}^H\mathbf{y} \tag{11}$$

3.2: -Channel estimation using Conventional Least Square (LS) method: -

From (8) received signal matrix **Y** is

$$\mathbf{Y} = \sqrt{P}\mathbf{W}_{BB}^H\mathbf{W}_{RF}^H\mathbf{H}\mathbf{F}_{RF}\mathbf{F}_{BB}\boldsymbol{I}_{N_T}Beam + \widetilde{\mathbf{N}}$$

After taking vectorization above equation may be written as

$$\mathbf{y} = \mathrm{vec}(\mathbf{Y}) = \left(\underbrace{\sqrt{P}\, \mathbf{F}_{BB}^T\mathbf{F}_{RF}^T \otimes \mathbf{W}_{BB}^H\mathbf{W}_{RF}^H}_{\widetilde{\mathbf{Q}}} \right)\mathrm{vec}(\mathbf{H}) + \widetilde{\mathbf{n}} \tag{12}$$

$$\overline{\mathbf{h}} = \mathrm{vec}(\mathbf{H}) \tag{13}$$

Equation (12) can be stated as $\mathbf{y} = \mathrm{vec}(\mathbf{Y}) = \sqrt{\boldsymbol{P}}\widetilde{\mathbf{Q}}\overline{\mathbf{h}} + \widetilde{\mathbf{n}}$

Now LS estimate of channel vector $\overline{\mathbf{h}}$ may be given as

$$\hat{\mathbf{h}} = \left((\sqrt{\boldsymbol{P}}\widetilde{\mathbf{Q}})^H\sqrt{\boldsymbol{P}}\widetilde{\mathbf{Q}}\right)^{-1}(\sqrt{\boldsymbol{P}}\widetilde{\mathbf{Q}})^H\mathbf{y} \tag{14}$$

4. OMP-Based Sparse Channel Estimation Model

4.1 Sparse Channel Model

Mm-Wave delivers lesser diffraction owing to a smaller Fresnel zone and resulting in a smaller number of multipath components than sub-6 GHz versions due to greater penetration losses. Massive MIMO is built into mm-Wave, resulting in increased channel sparsity. Channel model for mm-wave may be defined as follows:

$$\mathbf{H} = \sum_{l=1}^{L}\alpha_l\,\mathbf{a}_R(\theta_l^r)\,\mathbf{a}_T^H(\theta_l^t) = \mathbf{A}_\mathrm{R}\mathbf{H}_b\,\mathbf{A}_T^H$$

$$= [\mathbf{a}_R(\theta_1)\quad \mathbf{a}_R(\theta_2)\quad \ldots\quad \mathbf{a}_R(\theta_G)] \times \begin{bmatrix} h_{11} & \ldots & h_{1G} \\ \vdots & \ddots & \vdots \\ h_{G1} & \ldots & h_{GG} \end{bmatrix}\begin{bmatrix} \mathbf{a}_T^H(\theta_1) \\ \mathbf{a}_T^H(\theta_2) \\ \vdots \\ \mathbf{a}_T^H(\theta_G) \end{bmatrix} \tag{15}$$

The number of scatters or multipaths is denoted by L. At the transmitter and receiver, $\mathbf{a}_T$ and $\mathbf{a}_R$ are the vectors for array steering. l^{th} path complex gain is given by α_l. Each path has an Angle of departs θ_l^t

and an Angle of arrival θ_l^r. $\mathbf{A_R}$ and $\mathbf{A_T}$ are received and transmit array response dictionary matrices correspondingly [15].

$$\theta_i \in \Phi = \{\theta_1, \theta_2, \ldots, \theta_G\}, \text{ where } \Phi \text{ denote the angular grid.}$$

Channel representation in beamspace is provided by: -$\mathbf{H} = \mathbf{A}_\mathrm{R}\, \mathbf{H}_b \mathbf{A}_T^H$

$\mathbf{H}_b$ is a beamspace channel matrix and has a large number of zeros. Therefore, it is sparse in nature

4.2: -OMP-Based Channel Estimation

The channel estimation issue may be expressed as $min\|h_b\|_0$, which is also known as compressive sensing. This problem is a non-convex, and solving it with direct techniques is challenging. Orthogonal Matching Pursuit algorithm is a potential option to estimate and optimize sparse signals.

For the pilot matrix $\sqrt{P}\boldsymbol{I}_{N_T}Beam$ the received signal matrix $\mathbf{Y}$ from (8) is given as

$$\mathbf{Y} = \sqrt{P}\mathbf{W}_{BB}^H \mathbf{W}_{RF}^H \mathbf{H} \mathbf{F}_{RF} \mathbf{F}_{BB} \boldsymbol{I}_{N_T} Beam + \widetilde{\mathbf{N}} \quad (16)$$

Substitute $\mathbf{H} = \mathbf{A}_\mathrm{R}\, \mathbf{H}_b \mathbf{A}_T^H$ and take vectorization on both side

$$\mathrm{y} = \mathrm{vec}\,(\mathbf{Y}) = \mathbf{Q}\mathbf{h}_\mathrm{b} + \tilde{\mathbf{n}} \quad (17)$$

$$\mathbf{h}_\mathrm{b} = \mathrm{vec}(\mathbf{H}_b)\,,\, \mathbf{Q} = \sqrt{\mathrm{P}}\mathbf{F}_\mathbf{BB}^\mathrm{T} \mathbf{F}_{RF}^\mathrm{T} \mathbf{A}_\mathrm{T}^* \otimes \mathbf{W}_\mathbf{BB}^\mathrm{H} \mathbf{W}_{RF}^\mathrm{H} \mathbf{A}_\mathrm{R} \quad (18)$$

$\mathrm{h_b}$ is a sparse vector and Orthogonal Matching Pursuit technique is used for this sparse vector estimate.

OMP algorithm for channel estimation

$\mathbf{Q}_{(0)} = [\ ]$, $\mathbf{r}^{(0)} = \mathbf{y}$, $\mathbf{r}^{(-1)} = 0$, k = 1

While $\left\|\mathbf{r}^{(k-1)} - \mathbf{r}^{(k-2)}\right\| > \epsilon$

$\psi = \mathbf{Q}^H \mathbf{r}^{(k-1)}$

$i(\mathrm{k}) = = \arg\max\, [\psi\psi^H]_{l,l}$

$\mathbf{Q}_{(k)} = [\mathbf{Q}_{(k-1)} | \mathbf{Q}(:, i(k))]$

$\mathrm{h}_\mathrm{b}^{(k)} = \left((\mathbf{Q}_{(k)})^H \mathrm{Q}_{(k)}\right)^{-1} (\mathbf{Q}_{(k)})^\mathrm{H} \mathbf{y}$

$\mathbf{r}^{(k)} = \mathrm{y} - \mathbf{Q}_{(k)} \mathrm{h}_\mathrm{b}^{(k)}$

k=k+1

5. Simulation Result

Simulation performance of various channel estimate techniques for mm-Wave large MIMO system model is illustrated in this section. The model assumes that the transmitter and receiver each have 32 antennas, and BS has 8 RF chains. The angles of departure in the range 0 to 180 are calculated using 32-grid dictionary matrices. Initialization of OMP is done with 1 as a threshold value. Channel's sparsity level is assumed to be 5. Figures 2 and 3 present normalized mean square error (NMSE) comparison of Ideal (Genie)- aided channel with conventional least square (LS) and proposed Orthogonal Matching Pursuit based channel estimation respectively. Genie aided channel estimation will always be better and will provide the least NMSE because in Genie aided channel estimation, we presume that angle of departures and angle of arrivals are known to us and we have to estimate only different multipath gains only $(\alpha_1, \alpha_2, \ldots, \alpha_L)$. Suppose there are L multipaths, so we need only L pilot symbols. So, this Genie-aided channel estimation is treated as a benchmark. But in a practical scenario, it is not possible to have prior knowledge of the angle of departures and angle of arrivals, so we have to estimate AoA and AoD of dominant paths along with their gain. The proposed OMP algorithm utilizes the sparse nature of the mm-Wave channel and estimates the channel with a smaller number of pilot symbols which results in high spectral and energy efficiency of the system. The main drawback of conventional least square (LS) channel estimation is that it requires an equal or higher number of pilot symbols than transmitting antennas. ($\mathrm{N\_Beam} \geq \mathrm{N_t}$) which is a significant amount for mm-Wave large MIMO systems and causes the problem of pilot overhead and result in wastage of resources, like power and spectrum. As shown in

Journal of Physics: Conference Series **2327** (2022) 012040 doi:10.1088/1742-6596/2327/1/012040

figure 4 if we take number of pilot beams equal to number of transmitting antennas (N_Beam = N_t = 32) the performance of conventional least square (LS) channel estimation is comparable to the proposed OMP – based channel estimation. But if we reduce the number of a pilot beam (N_Beam = 24), the performance of the conventional least square (LS) channel estimation scheme dramatically degrades and channel estimation will become worse as shown in figure 5. But the performance of the proposed OMP algorithm is still good. So, it is concluded that the proposed OMP algorithm for mm-Wave massive MIMO channel estimation outperforms conventional least square (LS) channel estimation at a smaller number of pilot beams. Also, at an equal number of pilot beams and number of transmitting antennas (N_Beam = N_t = 32) the proposed OMP algorithm is slightly better than conventional least square (LS) channel estimation.

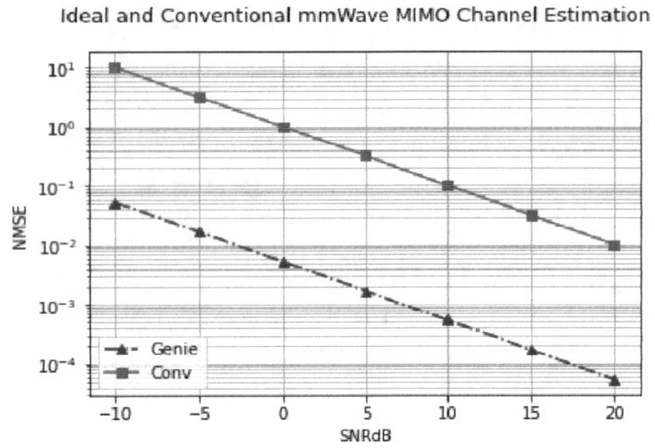

Figure 2. NMSE vs SNR with setup $N_R = N_T = 32$, $N_{RF} = 8$, L= 5, G = 32, N_Beam=32

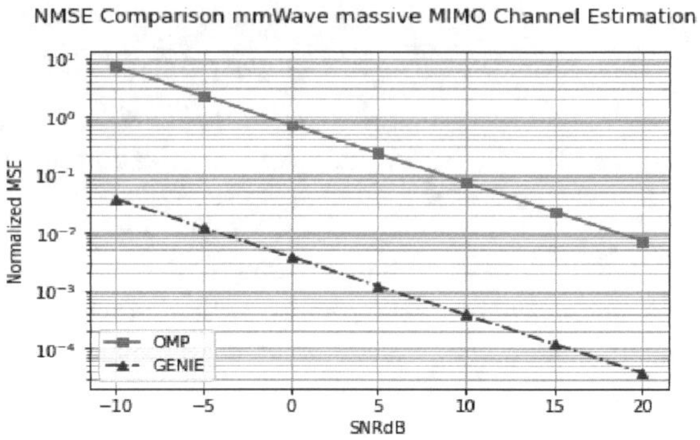

Figure 3. NMSE vs SNR with setup, $N_R = N_T = 32$, $N_{RF} = 8$, L= 5, G = 32, N_Beam=32

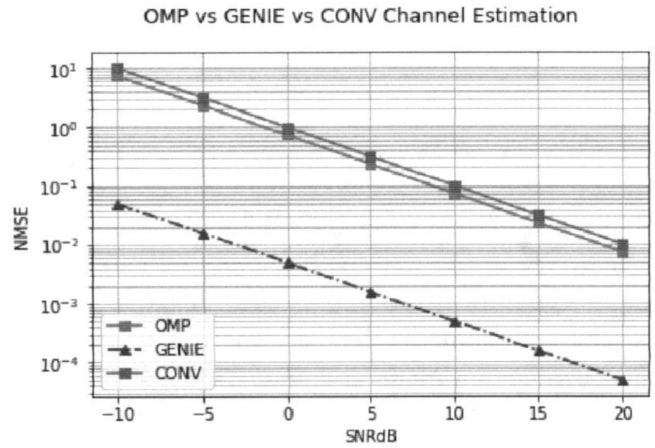

Figure 4. Error in Channel Estimation for millimeter-wave with $N_R = N_T = 32$, $N_{RF} = 8$, L= 5, G = 32, N_Beam=32

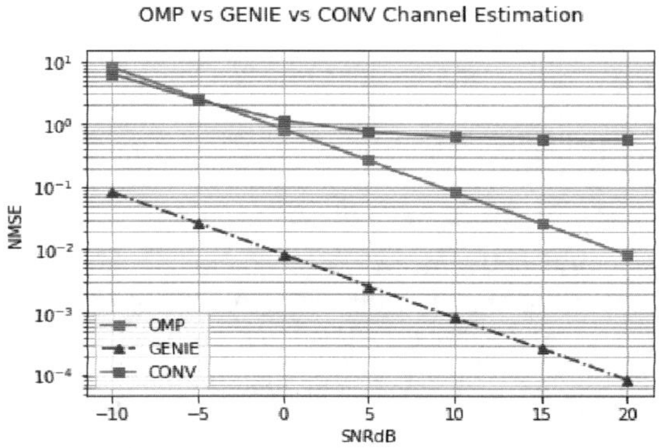

Figure 5 Error in Channel Estimation for millimeter-wave with $N_R = N_T = 32$, $N_{RF} = 8$, L= 5, G = 32, N_Beam=24

6. Conclusion

Mm-Wave massive MIMO channel has unique properties as mm-Wave delivers lesser diffraction owing to a smaller Fresnel zone and resulting in a smaller number of multipath components than sub-6 GHz versions due to greater penetration losses and lower scattering. Hybrid precoding is a potential approach for millimeter-wave (mm-Wave) systems to provide the needed array and multiplexing gain while minimizing system complexity. But hybrid precoding required accurate knowledge of channels for designing precoder and combiner. So efficient channel estimation schemes are required as conventional techniques (LS/MMSE) are not suitable for mm-Wave channel estimation as they require an equal or higher number of pilot symbols than transmitting antennas (N_Beam $\geq N_t$) which is very large for an mm-Wave large MIMO system. The proposed OMP algorithm in the paper is a promising solution for channel estimation as it exploits the sparsity nature of the mm-Wave channel and estimates channels with a smaller number of pilot symbols. However, there are some limitations of this OMP – based channel estimation as it is affected by the dictionary matrix and the halting criterion employed. This work can be expanded in the future by developing more efficient schemes for channel estimation based on sparse Bayesian learning (SBL) algorithms.

References:

[1] Li, Y., Zhu, X., Liao, C., Wang, C., & Cao, B. (2015). Energy Efficiency Maximization by Jointly Optimizing the Positions and Serving Range of Relay Stations in Cellular Networks. *IEEE Transactions On Vehicular Technology*, *64*(6), 2551-2560.

[2] Chen, S., Zhou, J., Zheng, X., & Ruan, X. (2018). Energy-Efficient Data Collection Scheme for Environmental Quality Management in Buildings. *IEEE Access*, *6*, 57324-57333.

[3] Hoydis, J., ten Brink, S., & Debbah, M. (2013). Massive MIMO in the UL/DL of Cellular Networks: How Many Antennas Do We Need?. *IEEE Journal On Selected Areas In Communications*, *31*(2), 160-171.

[4] Larsson, E., Edfors, O., Tufvesson, F., & Marzetta, T. (2014). Massive MIMO for next-generation wireless systems. *IEEE Communications Magazine*, *52*(2), 186-195.

[5] Bjornson, E., Sanguinetti, L., & Kountouris, M. (2016). Deploying Dense Networks for Maximal Energy Efficiency: Small Cells Meet Massive MIMO. *IEEE Journal On Selected Areas In Communications*, *34*(4), 832-847.

[6] Throughput and Energy Efficiency Analysis of Small Cell Networks with Multi-Antenna Base Stations. (2014), *13*(5), 2505-2517.

[7] Swindlehurst, A., Ayanoglu, E., Heydari, P., & Capolino, F. (2014). Millimeter-wave massive MIMO: the next wireless revolution?. *IEEE Communications Magazine*, *52*(9), 56-62.

[8] Roh, W., Seol, J., Park, J., Lee, B., Lee, J., & Kim, Y. et al. (2014). Millimeter-wave beamforming as an enabling technology for 5G cellular communications: theoretical feasibility and prototype results. *IEEE Communications Magazine*, *52*(2), 106-113.

[9] Wang, S., Li, L., Ruby, R., & Li, P. (2019). A general hybrid precoding scheme for millimeter-wave massive MIMO systems. *Wireless Networks*, *26*(2), 1331-1345

[10] Wang, Y., & Zou, W. (2020). Hybrid precoding design for millimeter-wave systems with the partially-connected structure. *IET Communications*, *14*(4), 561-567.

[11] Uwaechia, A., & Mahyuddin, N. (2020). A Comprehensive Survey on Millimeter Wave Communications for Fifth-Generation Wireless Networks: Feasibility and Challenges. *IEEE Access*, *8*, 62367-62414.

[12] Heath, R., Gonzalez-Prelcic, N., Rangan, S., Roh, W., & Sayeed, A. (2016). An Overview of Signal Processing Techniques for Millimeter Wave MIMO Systems. *IEEE Journal Of Selected Topics In Signal Processing*, *10*(3), 436-453.

[13] Alkhateeb, A., El Ayach, O., Leus, G., & Heath, R. (2014). Channel Estimation and Hybrid Precoding for Millimeter Wave Cellular Systems. *IEEE Journal Of Selected Topics In Signal Processing*, *8*(5), 831-846.

[14] Rossi, M., Haimovich, A., & Eldar, Y. (2014). Spatial Compressive Sensing for MIMO Radar. *IEEE Transactions On Signal Processing*, *62*(2), 419-430.

[15] Lee, J., Gil, G., & Lee, Y. (2016). Channel Estimation via Orthogonal Matching Pursuit for Hybrid MIMO Systems in Millimeter-Wave Communications. *IEEE Transactions On Communications*, *64*(6), 2370-2386.

[16] Chen, S., Donoho, D., & Saunders, M. (1998). Atomic Decomposition by Basis Pursuit. *SIAM Journal On Scientific Computing*, *20*(1), 33-61.

[17] Gorodnitsky, I., & Rao, B. (1997). Sparse signal reconstruction from limited data using FOCUS: a re-weighted minimum norm algorithm. *IEEE Transactions On Signal Processing*, *45*(3), 600-616.

[18] Ayach, O., Heath, R., Abu-Surra, S., Rajagopal, S., & Pi, Z. (2012). Low complexity precoding for large millimeter wave MIMO systems. *2012 IEEE International Conference On Communications (ICC)*.

[19] Singh, P., Srivastava, S., Jagannatham, A., & Hanzo, L. (2020). Second-Order Statistics-Based Semi-Blind Techniques for Channel Estimation in Millimeter-Wave MIMO Analog and Hybrid Beamforming. *IEEE Transactions On Communications*, *68*(11), 6886-6901.

Investigation of FSO Using Polarization Diversity Under Clear Weather, Haze and Rain

Sachin Chawla

ECE Department, Assistant Professor, Chandigarh University, Punjab, India

sachinchawla.uic@cumail.in
https://orcid.org/0000-0001-7281-4001

Abstract. Free space optical communication (FSO) is cost effective and potential technology for realization of transmission in hilly areas and for roof top communication in congested areas. In this work, a 32 channel WDM system employing polarization diversity technique is proposed in free space optical communication. Atmospheric turbulences plays vital role in system performance such as clear weather, haze and rain. Investigation has been done for different turbulences and effects are analysed. It is observed that system performs worst and better in case of rain and clear weather respectively.

Keywords: WDM, FSO, P2P, SOP, Log BER

1. Introduction

The expansion of high data needs and ever-increasing internet services has put a strain on traditional radio systems' slow speed design. FSO has properties to work other widely used methods like wireless optical communication and fiber optic [1]. FSO communication has received a lot of attention recently because of its advantages over optical cable connection [2]. Many studies have been conducted in the past to meet the wide bandwidth needs and to offer economical system [3]. Furthermore, more secure trait of FSO is due to less interception of P2P (point to point) laser transmissions. High capacity, escalated speed, license-free operation, power efficiency, and elimination of permits and trenching, and other costs associated with optical fibre transmission are all advantages of FSO communications[4][5]. FSO is a prominent possibility for geographical locations in India where optical fibre deployment is not viable, and it has received interest owing to its uses and multiple advantages. It is also perceived that FSO communication is prone to the weather changing effects such as change in density of air in summer and winter, water drops intensities in rain and fog etc [6][7]. Many studies have been conducted or proposed to reduce the consequences of atmospheric instability, as well as various ways to reduce turbulence effects [8]-[9]. In [10], FSO is analysed in the context of weather changes, and the influence of width of beam is also explored. FSO physical factors have a vital impact in total transmission as well. The sizes of the transmitter and receiver antennas determine the system's reach and reception quality. Beam divergence, along with attenuation and dispersion, is a primary performance degrading element that must be handled. Many studies have been published to date on using amplifiers to amplify weak signals, as well as beam divergence size scintillation noise correction. Although the approaches provided are effective, they either support a lower data rate or a higher level of complexity.

Therefore, in this research objective, high speed FSO system based polarization interleaved wavelength division multiplexing (PI-WDM) architecture is proposed. The most effective strategy is to meet high-speed data needs, hence the system was designed and demonstrated for 20 Gbps. Different states of polarizations are incorporated in the system to suppress inter channels and polarization crosstalk which provide enhanced results. Furthermore, prior to the FSO channel, super dense channel spacing of 50 GHz is provided between neighbouring channels to increase system bandwidth efficiency. It has been discovered that systems with polarization interleaving outperform traditional WDM FSO systems.

2. Theory of polarization diversity

The laser intensity is polarised in many, but not all, systems. This is a state of linear polarisation in which the electric field oscillates along a path orthogonal to the laser intensity propagation. Fiber lasers, on the other hand, do not produce a consistent directed electrical field. However, the laser output is not always unpolarized, since it might have the same powers in two polarizations at the same time without any amplitude correlation. The state of polarisation is not stable for a variety of reasons, including temperature changes and random flips between different orientations.

Some form of polarisation fixing mechanism is usually required to create the pure single-direction variation of electrical field vector. The polarisation extinction ratio (PER), defined as the ratio of powers in various polarisation directions and derived by recording the orientation-dependent power transmission of a polarizer, is a common way to measure the degree of specific state of polarisation. By fixing the polarisation of neighbouring channels, polarisation diversity is an efficient way to reduce the influence of inter-channel nonlinear effects. The approach is extended to developing WDM systems with orthogonally polarised bit streams in two nearby channels. The even and odd number channels are multiplexed into separate branches with orthogonal polarisation states (SOPs). To maintain channels orthogonal, polarizations are modified using linear polarisation controllers. In PI systems, the basic approach of suppression is that both cross phase modulation and four wave mixing are totally dependent on neighboring channel polarisation states, and SOPs are orthogonal. As a result, inter channel collisions produce fewer phase changes and better outcomes.

3. Simulation setup

In this work, a 32 x 10 Gbps polarization interleaved system is proposed in WDM FSO. A laser source of starting wavelength 193.1 THz is used. All the 32 channels are modulated with the MDRZ data and multiplexed through two different multiplexers. Prior to multiplexing, states of polarization are changed. Channels 1 to 31 are passed though horizontal polarization state and 2 to 32 are set at vertical polarization. This task is performed to suppress the inter channels crosstalk in the WDM FSO system. A FSO channel is considered with different attenuations such as clear weather (0.1 dB/km), haze (4 dB/km) and rain (6.27 dB/km). Figure 1 depicts the polarization interleaved free space optical system incorporating MDRZ. Atmospheric instabilities such as haze, snow, and rain are major deteriorating variables in FSO. Different attenuation levels are used while analysing the consequences of these environmental disturbances. Table 1 shows attenuations in the setting of turbulences.

Table 1. Values of amplitude degradation in different weather conditions

Atmospheric Turbulences	Attenuation (dB/Km)
Clear weather	0.11
Haze	4
Mild Rain	6.27
Medium Rain	9.64
Fog	22

After transmission through FSO link, signals are de-multiplexed and wavelengths are routed to their respective port. A receiver consists of the photo detector and low pass filter followed by 3-R regenerator and BER analyser.

4. Results and Discussions

In this work, a 32 x 10 Gbps polarization interleaved system is proposed in WDM FSO. A laser source of starting wavelength 193.1 THz is used. All the 32 channels are modulated with the MDRZ data and multiplexed through two different multiplexers. Prior to multiplexing, states of polarization are changed. Channels 1 to 31 are passed though horizontal polarization state and 2 to 32 are set at vertical

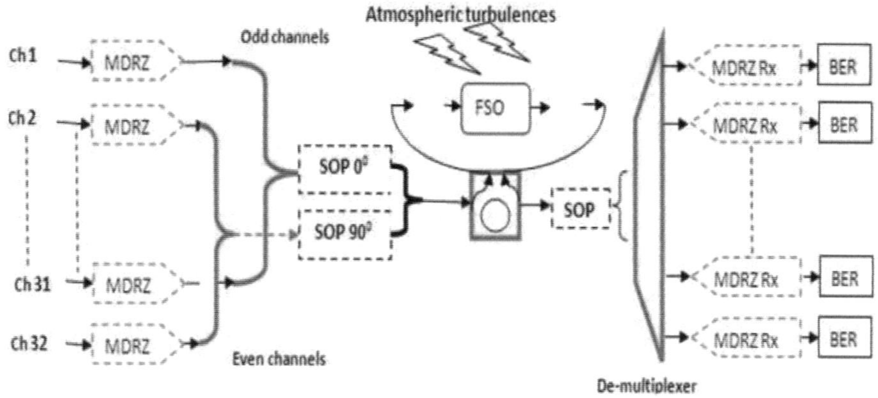

Figure 1. A 32 x 10 Gbps WDM-FSO system

polarization. This task is performed to suppress the inter channels crosstalk in the WDM FSO system. In order to investigate the system for practical scenarios, first of all, beam divergence is considered. Values of beam divergence are varied from the FSO channel such as 0.25 mrad, 0.50 mrad, 0.75 mrad and 1 mrad. Table 2 represents the values of Q factor for different SOPs evaluation at varied beam divergences. It is evident that beam divergence has very performance degrading effects on the system. Increase in divergence of beam cause decrease in Q factor.

Table 2. Values of amplitude degradation in different weather conditions

Beam divergence mrad	SOP 0	SOP 90
0.25	31.22	27.03
0.5	29.15	26.08
0.75	26.71	24.8
1	24.24	23.32

It is *vital* that you **do not add any headers, footers or page numbers to your paper**; these will be added during the production process at IOP Publishing (this is why the Header and Footer margins are set to 0 cm in table 1).

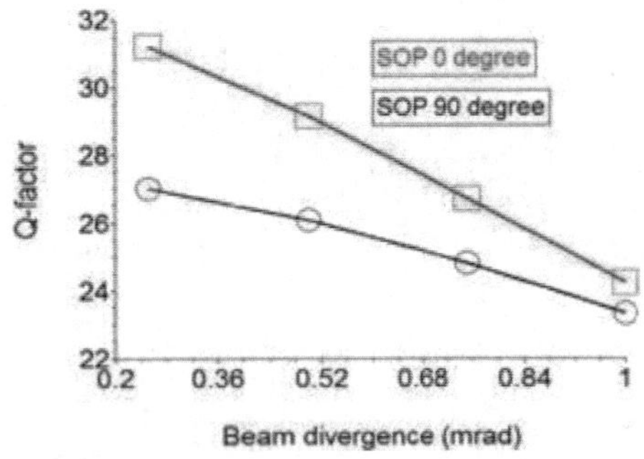

Figure 2. Effect of beam divergence on Q factor of system at two different states of polarizations

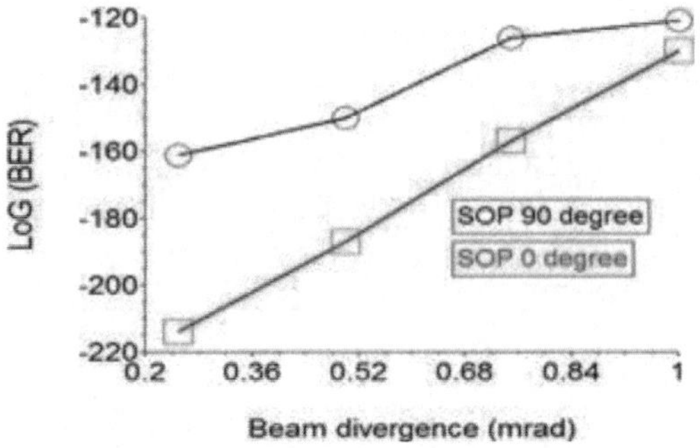

Figure 3. LoG (BER) versus beam divergence of Different states of polarizations

Two states of polarizations are investigated to find out that horizontal or vertical polarization channels perform better. Results revealed from the Figure 2 that state of polarization 0 that is horizontal polarization channels takes lead and provide better Q factor. Figure 3 represents the graph of the system for LoG (BER) versus beam divergences. It is seen that BER increases with the more broadening of light in circular diameter. BER is inversely proportional to the Q factor such that BER is maximum where Q is minimum. SOP 0 is reported better from the analysis as it exhibits the less errors that the state of polarization 90 degree.
Furthermore, the length of the FSO link is adjusted from 1 to 7 kilometres in order to assess the quality of reception under various air degradation scenarios. In clear weather circumstances, the longest distance is seen, whereas in fog, the shortest distance is measured. Figure 4 shows a graphical depiction of several turbulences. The Q-factor for clear weather varies from 19.15-14.05 for 1-7 kilometres, as can be shown. When it comes to haze and rain, the figures are 18.92-0 and 16.19-0, respectively. For varied turbulences, System covered >9 Km distance for clear weather, 4 Km for haze, 2 Km for mild rain. So, it is obtained from the results that the clear weather is optimal condition for free space optical communication and at this attenuation system performs better than haze and rain weather instabilities.

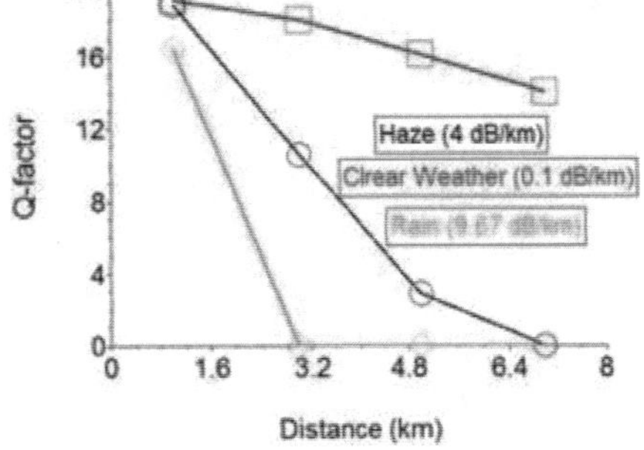

Figure 4. Effect of different atmospheric Turbulences on system at different distances

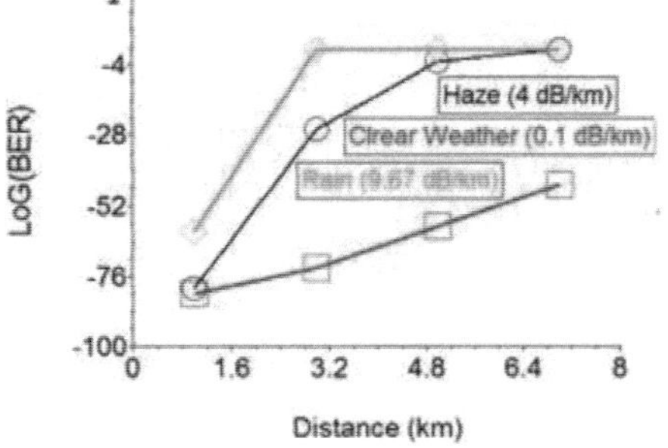

Figure 5.LoG (BER) versus distance for different weather conditions

Figure 5 represents the performance of system at different distances for atmospheric instabilities. It is seen that there is maximum BER in case of rain, less BER in case of haze and least BER in clear weather case.
Figure 6 shows the system's performance in terms of signal to noise ratio over various distances. The ratio of received signal strength to noise is known as the signal to noise ratio. For improved Q factors and lower BER, SNR should be at its maximum. As can be seen in Table 3, the signal to noise ratio decreases as distance increases. Due to the advent of noise effects, SNR is acquired greater in the polarisation interleaved system up to a distance of 9 km and less beyond that. Also graph depicts the SNR values for different atmospheric turbulences. It is observed that Clear weather give best SNR.

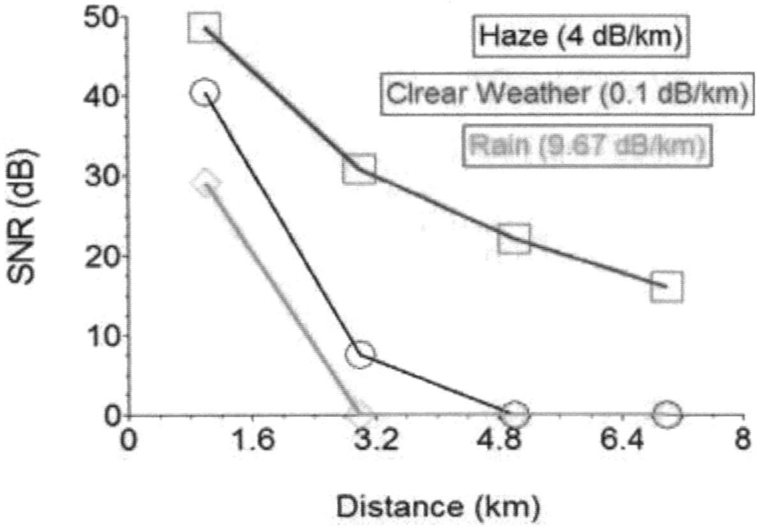

Figure 6. Distance versus Signal to noise ratio of proposed system

Table 3. Values of SNR (dB) at different distances for different weather conditions

Distance (km)	Clear weather	Haze	Rain
1	48.7 dB	40.48 dB	29.14 dB
3	30.84 dB	7.55 dB	0 dB
5	22.02 dB	0 dB	0 dB
7	15.99 dB	0 dB	0 dB

Results revealed that as the distance increases from 1 km to 7 km, signal to noise ratio decreases. This is due the noise impairments to the signal that travel through free space. More noise emerges at the longer distances.

5. Conclusion

In this work, a 32 x 10 Gbps polarization interleaved system is proposed in WDM FSO. A FSO channel is considered under different weather conditions and results revealed that for the distance of 1 km and 7 km in FSO, value of Q-factor is 19.15-14.05 respectively under clear weather. As in case of haze and rain is 18.92-0 and 16.19-0 respectively. System covered >9 Km distance for clear weather, 4 Km for haze, 2 Km for mild rain. So, it is obtained from the results that the clear weather is optimal condition for free space optical communication and at this attenuation system performs better than haze and rain weather instabilities. Moreover, effect of beam divergences analyzed and results revealed that at low divergence system performs better such as in the case of 0.25 mrad.

Acknowledgement

I would like to express my very great appreciation to **Manoj Sindhwani, Shippu Sachdeva** (School of Electronics and Electrical Engineering, Lovely Professional University, Phagwara, Punjab, India.), **Ruchika** (Chandigarh University, Punjab, India) for their valuable and constructive suggestions during the planning and development of this research work. Their willingness to give their time so generously have been very much appreciated.

References

[1] Isaac I. Kim, Bruce McArthur, and Eric Korevaar, "Comparison of laser beam propagation at 785 nm and 1550 nm in fog and haze for optical wireless communications", 2012.

[2] Muhammad Saleem Awn, Erich Leitgebi, Marzuki, Thomas Plank, "A Study of Fog Characteristics using Free-Space Optical Wireless Links", *Journal of Radio Engineering*, Vol. 19, No. 2, pp. 213-222, June 2010

[3] A. M. Zin, M. S Bongsu, S. M. Idrus, and N. Zulkifli (Member of IEEE), "And Overview of Radio-over-Fiber Network Technology" *International Conference on Photonics*, ISSN Number 978-1-4244-7187-4, 2010

[4] M. Grabner, V. Kvicera, "On the Relation between Atmospheric Visibilit drop size distribution of Fog for FSO Link Planning", ECOC 2009

[5] Ondrej Fiser, Vladimir Schejbal, "Comparison of formulae estimating fog attenuation on Free Space Optics Link", IEEE, 2009

[6] Sergejs Makovejs, Giancarlo Gavioli, Vitaly Mikhailov, Robert I. Killey and Polina Bayvel, "Experimental and numerical investigation of bit-wise phase-control OTDM transmission" *Optics Express*, Vol. 16, No. 23, Nov. 2008

[7] Pei-Lin Chen, Shenq-Tsong Chang, Shuen-Te Ji, Shu-Chuan Lin, Han-Hsuan Lin, Ho-Lin Tsay, Po-Hsuan Huang, Wei-Chieh Chiang, Wei-Cheng Lin, San-Liang Lee, Hen-Wai Tsao, Jin-Pu Wu1 and Jingshown Wu, "Demonstration of 16 Channels 10 Gb/s WDM Free Space Transmission Over 2.16 km ", IEEE, pp.235-236, 2008

[8] HU Guo-yong, CHEN Chang-ying and CHEN Zhen-qiang "Free-Space Optical communication using visible light", *Journal of Zhejiang University Science* A; ISSN 1009-3095(Print);ISSN1862-1775(Online),pp.186-191,2007

[9] Klaus-Dieter Langer and Jelena Grubor Fraunhofer, "Recent Developments in Optical Wireless Communications using Infrared and Visible Light", IEEE, pp. 146-150, 2007

[10] Zednek Kolka, Otakar Wilfert, Dalibor Biolek, Viera Biolkova, " Availability model of Frre Space Optical Dta Link", *International Journal of Microwave and optical technology*, Vol.1, No.2, august 2006

Investigation of Inter-Satellite OWC Incorporating Dense WDM and Polarization Interleaving (PI)

Sachin Chawla

ECE Department, Assistant Professor, Chandigarh University, Punjab, India

sachinchawla.uic@cumail.in
https://orcid.org/0000-0001-7281-4001

Abstract. Inter-satellite communication based on lasers (LIC) is a possible choice for future satellite networks because to ultra-high capacity and speed systems that may be achieved using compact, light-weight, and low-power optical equipment. The performance of a 64 × 40 GB/s optical wireless system with polarisation diversity for dense WDM is investigated in this research paper. Using a DWDM system with a channel spacing of 50 GHz, a high-speed system is achieved to meet the broad bandwidth needs of data services in satellites. A polarisation diversity approach is implemented in the system to decrease polarisation crosstalk and nonlinearities among nearby WDM channels. Furthermore, a comparison of the polarisation interleaved and traditional DWDM-OWC systems is performed. In addition, it wasdiscovered that a system with varying states of polarisation across neighbouring channels performs better in terms of Q-factor and BER than a traditional system.

Keywords- Optical wireless channel (OWC), Polarization interleaving (PI), State ofpolarization (SOP), Bit error rate (BER), Differential quadrature phase shift keying (DQPSK),

1 Introduction

Optical wireless communication is a fascinating technique in the world of communication that allows us to transmit data faster than microwave and radio frequency technologies [1]. Optical wireless communication is a type of optical communication that employs a wireless channel to send data to its destination. It is based on the same principle as laser communication technology [2]. Laser communication sends the data or signal at several data rates over long distances using laser. As the numbers of users are increasing, the demand for bandwidth and high speed is also increasing and there is a wide research in this forte [3]. Therefore, Optical wireless communication is the only way to meet these demands because of the high bandwidth, high transmission capacity [4]. Infrared light region is widely operational for transmission of data over long distances. Use of OWC in space is termed as Intersatellite optical wireless communication (IsOWC) [5]. IsOWC offers high bit rate, high quality and more security [6]. Although the Optical communication that works in vacuum is considered to be lose less but the transmitted signal suffers different path losses that increases with the communication distance [7]. Integration of Dense Wavelength Division Multiplexing (DWDM) with IsOWC provides high capacity and pacy network. IsOWC reaches up to several Tbps with incorporation of DWDM, which is the necessity of inter-satellite communication. Receiving power is shown as [8].

$$P_R = P_T \eta_T \eta_R \left(\frac{\lambda}{4\pi z}\right)^2 G_T G_R L_T L_R \quad (1)$$

Where GT, GR is gain of transmitter and receiver antenna and LT, LR are pointing errors respectively. Z, λ are distance and wavelength respectively along with ηT, ηR optical efficiencies of TX RX. Beam divergence should be less and only then link can be of longer distances. Loss of pointing is expressed as

$$L_T = \exp(-G_T \theta_T^2) \quad (2)$$

$$L_R = \exp(-G_R \theta_R^2) \quad (3)$$

azimuth pointing error of transmitter and receiver are θT, and θR respectively.

Numerous researches has been reported on intersatellite optical communication systems employing different modulations [9], wavelengths [10], polarization states [11] etc. Comparison of phase shift keying reported in [12], and results revealed that Quadrature phase shift keying is superior to differential phase shift keying in terms of long haul transmission. Analysis for linecodings such as Nonreturn to zero (NRZ) and Return to zero (RZ) in high speed system is presented in [13]. Major limitation of reported systems are less speed data carrying ability, low number of WDM channels, more channel spacing (100GHz- 200 GHz), high cost, Bandwidth inefficient modulation formats. However, to suppress crosstalk among WDM channels, techniques are proposed several times in the past such as low power system, unequal channel spacing [14], transceiver diversity [15] , hybrid modulation [16], but are not very much efficient to quell effects of interference and demonstrate ultra- high speed IsOWC system.

In this work, The performance of a 64×40 Gb/s optical wireless system with polarisation diversity for dense WDM is investigated. Differential quadrature phase modulation is considered in proposed architecture due to its advantages such as reliable data transfer, robust against transmitter/receiver frequency discrepancies and non-coherent detection. Moreover, comparison is established between polarization interleaved and conventional DWDM OWC system.

1.1 Differential Quadrature Phase Shift Keying

Differential quadrature phase shift keying is a good choice for improving system endurance to nonlinearities and achieving a longer transmission reach. DQPSK, also known as real multilevel modulation, has piqued the interest of academics in recent years due to its multiple benefits. In this higher modulation, it generate four phase shifts (0, +π/2,-π/2, π) at symbolrate of half the total data rate. A DQPSK is most appropriately and suitably implemented with convenience by two parallel Mach-zehnder modulators functional as phase modulators.

As shown in Figure 1, a transmitter consisting of a continuous Laser source, a power divider with two paths of identical energy, two mach-zehnder modulators, 900 phase shifter in one arm or path and power combiners to accumulate signals. This depiction represents the schematic of parallel MZM generation of the multilevel modulation. However, serial MZMs are also included in place of parallel configuration for DQPSK generation. In diagram, signal transition and constellations of upper/lower arms or paths is also represented. Aforementioned arrangement exhibit the π shifts by modulators irrespective of the drive signal overshoot. Subsequently, this architecture needs binary bits only and is easy to generate at high speeds rather than waveforms. A pulse carver is also incorporated in the System to generate return to zero DQPSK. It has a compressed spectrum, which is useful for achievinghigh spectral efficiencies in wavelength division multiplexing systems, improved tolerance to pulse

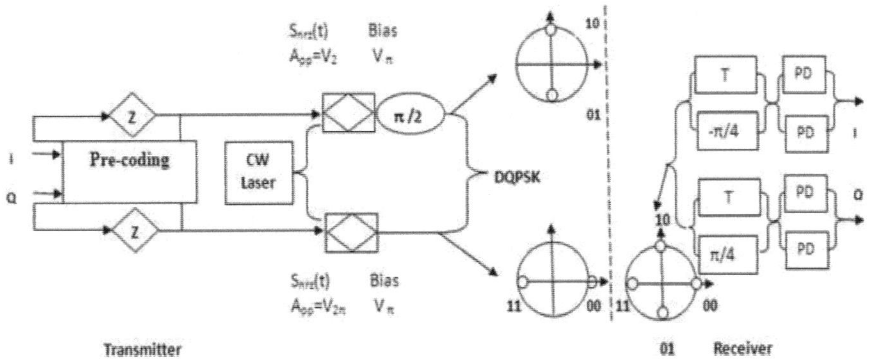

Figure 1. Depiction of Differential Quadrature Phase Shift Keying Transmitter and Receiver

broadening, and a longer symbol period than binary modulation formats, which makes DQPSK more resistant to polarisation mode dispersion (PMD). Reception of the signals is carried out by splitting input signal and fed to balanced receivers. Different biasing in delay interferometers is to demodulate binary data and delay is two time the duration of bit. Figure 2(a) illustrates the optical spectrum of modulation and represents the carrier frequency with respect to power of the signal. Optical spectrum of DQPSK signal is narrow, thus suitable for DWDM systems.

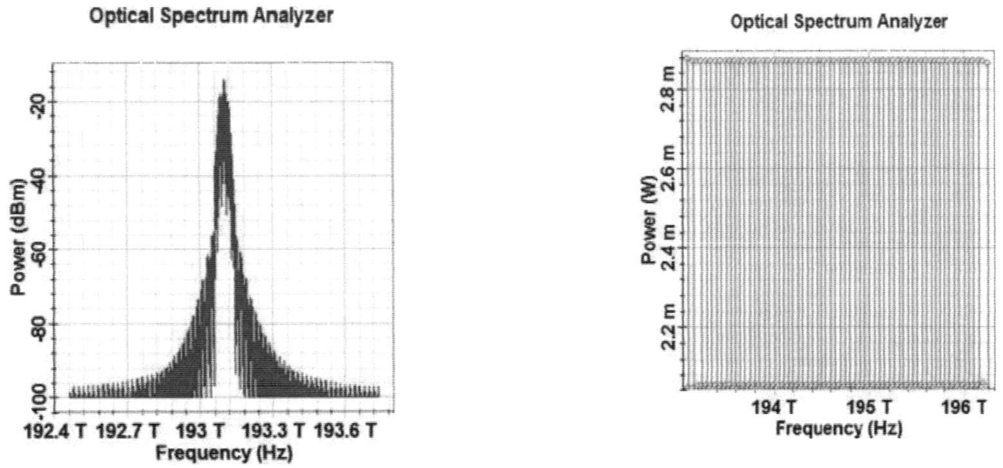

Figure 2. Representation of (a) single DQPSK transmitter (b) 64 DQPSK WDM transmitters

2. Proposed Architecture

For the accomplishment of proposed architecture a leading and compendious simulation tool that facilitates users to implement, investigate, and simulate communication system in optical domain, Optiwave's OptiSystemTM is used. In this research article, a 64x40 Gb/s WDM system is demonstrated at ultra dense spacing (50 GHz) among adjacent channels to realize a bandwidth efficient system. A continuous wave laser array signals with starting frequency 193.1 THz and power 30 dBm are fed to a DQPSK modulation as depicted in Figure 3. DQPSK modulation consisting of pseudo random bit sequence generator and MZMs as elaborated in section (i) for phase shifts. In order to investigate the impact of nonlinear degradations on the system, the polarisation of even and odd channels is modified. The linear polarizer transmits only the linear polarisation component that is parallel to the polarizer's transmission axis and rejects the orthogonal component. The johns matrix describe the transfer function of polarization as

$$\begin{bmatrix} cos^2(\theta) \\ cos(\theta).sin(\theta) \end{bmatrix} \begin{bmatrix} cos(\theta).sin(\theta) \\ cos^2(\theta) \end{bmatrix} \quad (4)$$

Same SOP 0 degree is carried out by λ1- λ63 odd channels in linear polarizer. Similarly, Same SOP 90 degree is carried out by λ2- λ64 even channels in linear polarizer. The investigated range for the proposed link is 1250 Km with two erbium doped fiber amplifiers (EDFA) amplifiers; one is in preand other in post configuration. An EDFA gain is set to 25 dB with 4 dB noise figure in order torealize a practical system as shown in Table 1. An optical wireless channel with 250 Km link reach is considered for each loop and total loops are set to 5 to get 1250 Km distance. Power amplitude degradation for OWC is fixed to 0.14 dB/Km [10] and transmitter, receiver aperture diameter are 20cm and 25 cm respectively. A 1:64 DE multiplexer route the specific frequency to their respective output port with the help of reference frequency and filtering of Gaussian filter. Receiver section consisting of optical couplers with coupling ration of 50:50 and divide signal for two pairs of photodetectors that receives drive with time delay and phase shift to input signal as shown in Figure 1. A p-i-n photo detector with 100% responsivity is placed in the receiver by considering shot, thermal and ASE (Amplifier spontaneous noise) distortions. Electrical bias is provided to electrically subtracted output of balanced photodetectors followed by 3-R regenerator followed by eye diagram analyser.

3 Result and discussions

Figure 3 depicts the proposed architecture of 64 channel DWDM system over optical wireless channel with non-coherent detection and polarization diversity. Investigation has been carried out at 40 Gb/s data rate for two cases (1) without polarization diversity (2) with polarization diversity at 50 GHz dense spacing in WDM system. Mainly, the regions that affect or influence the transmitting beams are troposphere and ionosphere. Reach of troposphere is 10 Km-15 Km with respect to ground and ionosphere spreads up to 60 Km-1000 Km. The influence of atmospheric instabilities is less till the height of 2 Km on free space communication channel. All the work is considered for clear weather and attenuation corresponding to it is fixed as 0.14 dB/Km.

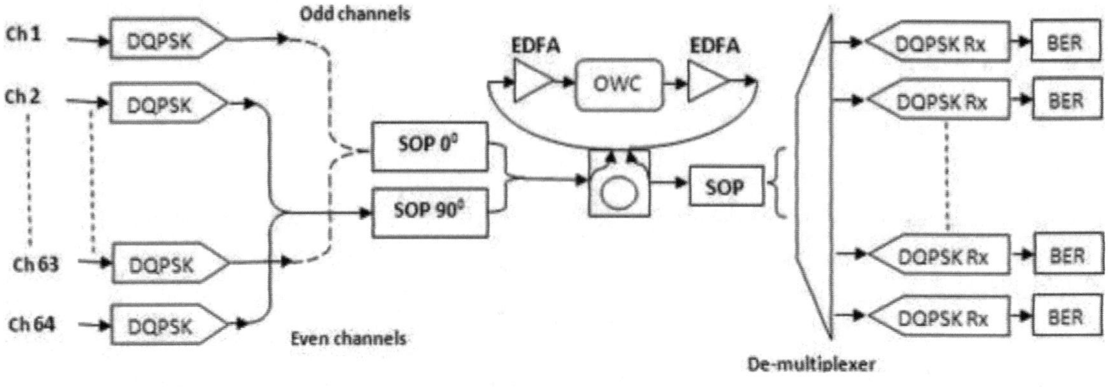

Figure 3. Proposed system of 64 channel DWDM with polarization diversity

DWDM system over optical wireless channel is analyzed for with and without polarization diversity in terms of Q-factor and bit error rate (BER). First and foremost, a loop length that consists of EDFA-OWC-EDFA is varied and Q factor is observed at 250 Km- 1250 Km.

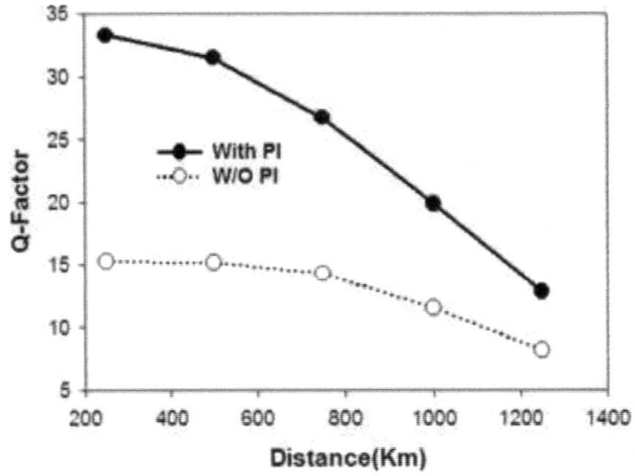

Figure 4. Graphical representations of Q factor and distance for with and without polarization interleaving

Figure 4 represents the graphical depiction of Q-factor with respect to distance of two setups (1) with polarization diversity (2) without polarization diversity and it is observed that with the increase of link length, Q-factor decreases due to attenuation and adjacent channel crosstalk. Minimum crosstalk and maximum Q factor is reported for polarization interleaved channels having different polarization for odd and even channels. Link length increased in the span of 250 Km and maximum distance is 1250 Km, values of Q-factor for these ranges are 33.28-12.86 respectively in case of the incorporation of polarization diversity as shown in Table 2.

Table 2. Q-factor values with Distance (Km)

Distance (Km)	With Polarization Diversity	Without polarization Diversity
250	33.28	15.26
500	31.49	15.19
750	26.72	14.29
1000	19.9	11.53
1250	12.86	8.16

Table 3. BER values with Distance (Km)

Distance (Km)	With Polarization Diversity	Without polarization Diversity
250	$3.1x10^{-243}$	$6.09x10^{-53}$
500	$5.4x10^{-218}$	$1.7x10^{-52}$
750	$1.2x10^{-157}$	$1.2x10^{-46}$
1000	$1.7x10^{-88}$	$4.5x10^{-31}$
1250	$3.6x10^{-38}$	$1.5x10^{-16}$

Figure 5 shows the performance of the system in terms of BER at different distances and it is found that there is significant increase in the BER with increase in link distance. Comparison of with and without polarization interleaved system is established and minimum BER is obtained at 250 Km and for polarization interleaved system. This is due to reason that by altering the polarization, there is suppression of crosstalk and adjacent channel impairments in DWDM system because of un-identical state of polarizations. As represented in Table 3, BER obtained at 250 Km and 1250 Km for with

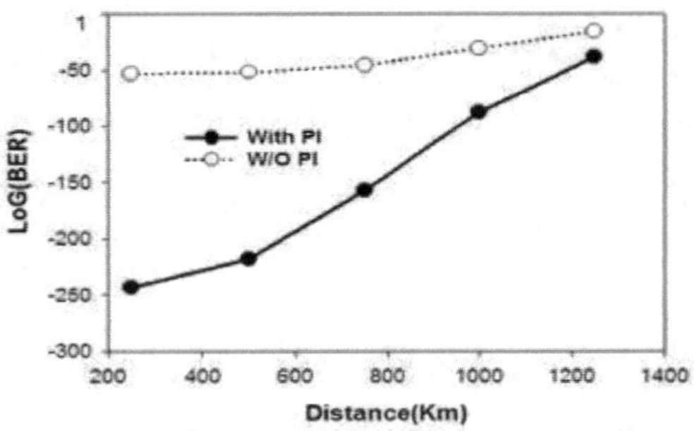

Figure 5: Values of BER with respect to distance of DWDM-OWC with and without polarization diversity

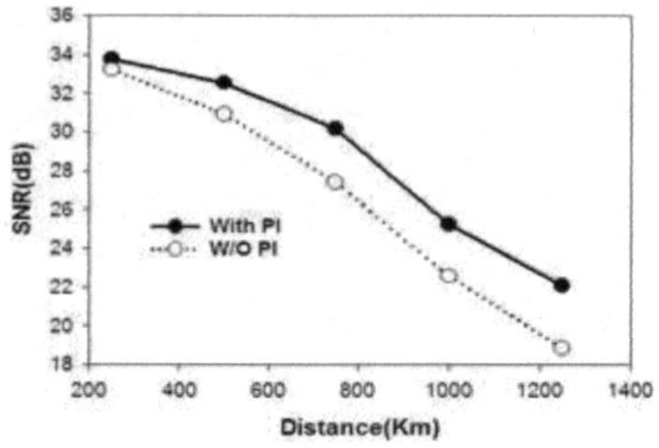

Figure 6: Signal to noise ration versus distance illustration

polarization interleaving and without are 10-243, 10-38 and 10-53, 10-16 respectively. It is clear from the error rate that polarization diversity is superior to conventional DWDM OWC systems.
Figure 5. Values of BER with respect to distance of SNR of 33.75 for PI system is reported and value of without polarization diversity system obtained 33.22 at 250 Km. It is seen that PI system exhibits better performance than without polarization diversity due to less adjacent channel noises. Increase in link length introduces more noises and as a result SNR degrades with the link reach. Figure 7 shows the comparative analysis of Q factor versus bit period that is time of any transmitter bit for two proposed cases. This comparison is carried out for conventional system and polarization diversity system. It is prominently observed that maximum Q factor for PI system is 33.28, 31.49, 26.72, 19.9 and 12.86 at distances of 250 Km with double distance each time. For without PI system the values of Q are 15.26, 15.19, 14.29, 11.53 and 8.16. DQPSK with PI exhibits enhanced performance and curveis far better than without PI conventional DQPSK DWDM system. Signal to noise ration, a frequent situation that some beam of intensity impinges a detector, which generates a photocurrent when opticalpower incident on it along with some electronic noises such as quantum and ASE noises. Performance of system or SNR limits due to shot and thermal noises or by noise generated by the detector electronics. Figure 6 illustrates the signal to noise ration at different distances for with and without PI system.

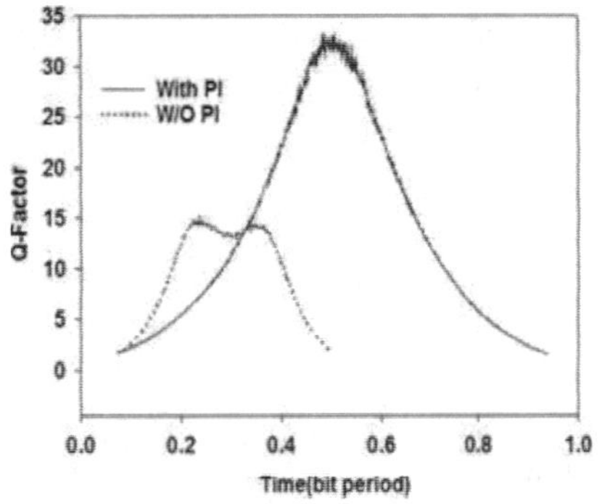

Figure 7. Maximum Q factor versus Time (bit period) for with and without polarization diversity

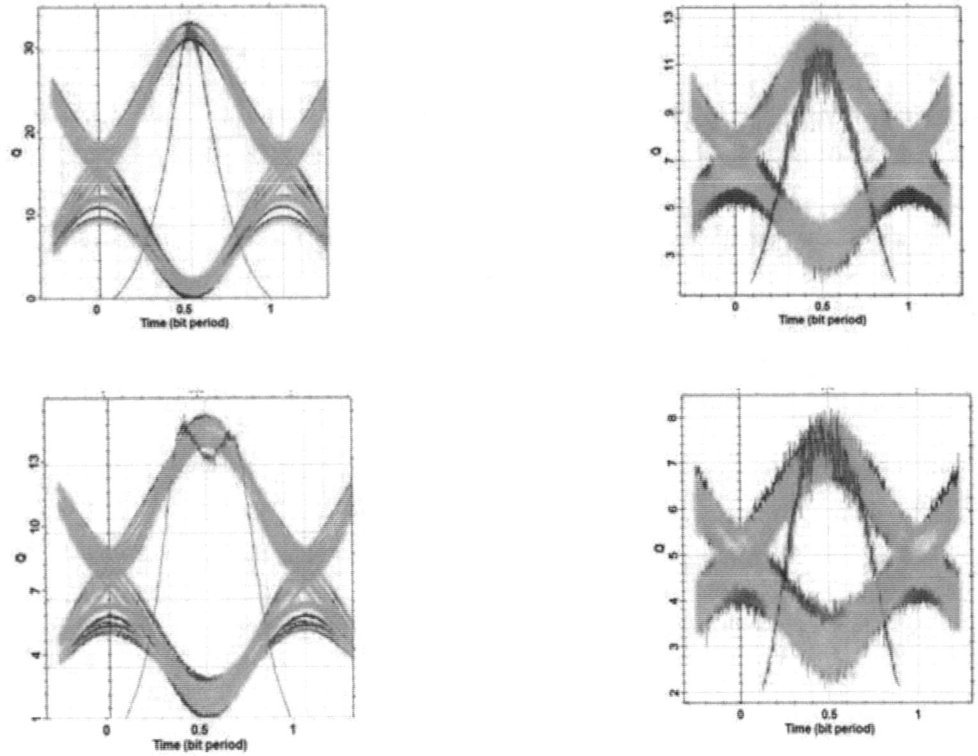

Figure 8. Eye diagram at 250 Km and 1250 Km respectively for (a)(b) With PI (c)(d) Without PI

Figure 8 depicts the Eye diagram of polarization diversity incorporated system and conventional DWDM. Figure 8(a) (b) represents Eye diagrams of DQPSK signal at 250 Km and 1250 Km which allow user to calculate Q factor, BER, eye opening and eye closer etc. Greater eye opening is observed at shorter distance and in case of polarized interleaved system. Figure 8(c)(d) illustrate Eye diagram of the system at aforementioned distances for conventional system and it is seen that more eye closer exists in conventional systems due to more adjacent channel interference. It is noteworthy that to design a dense and high capacity DWDM optical wireless system, Polarization diversity is right candidate to exhibit enhanced performance in terms of Q factor and BER at 40 Gb/s.

4 Conclusion

In this research article, a high speed DWDM enables 64×40 Gbps Is OWC system is demonstrated. The high speed system is accomplished to cater the bandwidth hngry demands of information services in inter-satellites at 50 GHz frequency spacings. A polarisation diversity approach is used in the system to reduce polarisation crosstalk and nonlinearity among nearby WDM channels. In addition, a comparison is made between the polarisation interleaved and regular DWDM optical wireless systems.At 40 Gb/s, it is noticed that a system with varying states of polarisation across neighboring channels performs better than a traditional system.

Acknowledgement

I would like to express my very great appreciation to **Manoj Sindhwani, Shippu Sachdeva** (School of Electronics and Electrical Engineering, Lovely Professional University, Phagwara, Punjab, India.), **Ruchika** (Chandigarh University, Punjab, India) for their valuable and constructive suggestions during the planning and development of this research work. Their willingness to give their time so generously have been very much appreciated.

References

[1] A. Alipour, A. Mir, A. Sheikhi, “Ultra high capacity inter-satellite optical wireless communication system using different optimized modulation formats”, Optik, vol. 127, no. 19, pp. 8135-8143, 2016.

[2] Xiaorui Wang, Lei Guo, Yejun Liu and Lincong Zhang, “Analysis of atmosphere channel for space-to-ground optical communications”, Optics communication, vol. 306, pp. 42-48, 2013.

[3] Bijayananda Patnaik, P.K. Sahu, “Ultra high capacity 1.28 Tbps DWDM system design and simulation using optimized modulation format”, Optik, vol. 124, no. 13, pp. 1567-1573, 2013.

[4] Z. Zhu, Y. Li, S. Zhao, X. Li, K. Qu, and J. Ma, “Broadband linearized analog intersatellite microwave photonic link using a polarization modulator in a Sagnac loop,” Applied optics, vol. 55, no. 5, pp. 1022–1028, 2016.

[5] Y. Liu, S. Zhao, Z. Gong, J. Zhao, C. Dong, and X. Li, “Displacement damage in bit error ratio performance of on–off keying, pulse position modulation, differential phase shift keying, and homodyne binary phase-shift keying-based optical intersatellite communication system,” Applied optics, vol. 55, no. 11, pp. 3069–3076, 2016.

[6] N. Kumar, “Enhanced performance analysis of inter-satellite optical-wireless communication (IsOWC) system,” Optik-International Journal for Light and Electron Optics, vol. 125, no. 8, pp. 1945–1949, 2014.

[7] Jin Zhu, Dayan Wang, and Wanqing Xie, “Packet error rate analysis of digital pulse interval modulation in intersatellite optical communication systems with diversified wavefront deformation,” Applied optics, vol. 54, no. 6, pp. 14711-1477, 2015.

[8] S. Arnon, “Performance of a laser satellite network with an optical preamplifier”, J. Optical Society of America, vol. 22, no. 4, pp 708-715, April 2005.

[9] J. Singh and N. Kumar, “Performance analysis of different modulation format on free space optical communication system,” Optik, vol. 124, no. 20, pp. 4651–4654, 2013.

[10] A. Z. M, H.A.Fayed, A.A.El Aziz and M.H.Aly, “The Influence of Varying the Optical Wavelength on ISL Performance Recognizing High Bit Rates”, IOSR Journal of Electronics and Communication Engineering, vol. 9, no. 1, pp. 64-70, 2014.

[11] Somanath Pradhan , P. K. Sahu, Rajat Kumar Giri, Bijayananda Patnaik, “Inter-satellite optical wireless communication system design using diversity techniques”, proc. ICMOCE IEEE, 2015.

[12] A Penchala Bindushree, Nataraju A. B., Vijesh T V, Laxmiprasad, “Design And Simulation Of QPSK Modulator For Optic Inter Satellite Communication”, int. journal of sci. and tech., vol.3, no. 8, pp. 402-408, 2014.

[13] Supinder Kaur, Simarpreet Kaur, “Comparative Analysis between NRZ and RZ Coding of WDM System,” int. jour. adv. res. com. comm. engg., vol. 1, no. 6, 214-218, 2016.

[14] Wing C. Kwong, and Guu-chang yang,”An algebraic Approach to the Unequal-Spaced Channel-Allocation Problem in WDM Light wave systems,” IEEE Transactions on Communications, vol. 45, no. 3, pp. 352-359, March 1997.

[15] I. Aggarwal, P. Chawla and R. Gupta, “Performance Evaluation of Intersatellite Free Space Optical Communication System with Varied Parameters and Transceiver Diversity” Advances in Electronic ad Electric Engineering, vol. 3,ISSN 2231-1297, pp. 847-852, 2013.

[16] S. Robinson, S. Jasmine, R. Pavitra, “ Investigation on Hybrid WDM (DWDM + CWDM) free space optical communication system, vol. 6, no. 4, pp. 1187-1192, 2015.

Journal of Physics: Conference Series **2327** (2022) 012043 doi:10.1088/1742-6596/2327/1/012043

Bidirectional Passive Optical Network Employing Different Line codings/Modulation Formats in Wavelength Reused RSOA Based Symmetrical 10 Gbps

Ruchika

CSE Department, Assistant Professor, Chandigarh University, Punjab, India

ruchika.e10474@cumail.in

https://orcid.org/0000-0002-4251-3993

Abstract. Passive optical networks (PONs) with the integration of multiple channels are the high capacity systems to cater the large number of channels in order to fulfill demands of internet applications. To realize a crosstalk suppressed and cost effective system, an integrated multi channel PON for mitigation of the crosstalk between upstream and downstream channels using different pulse shapes such as Return to zero (RZ), non RZ (NRZ), compressed spectrum RZ (CSRZ), Differential phase shift keying (DPSK) is proposed. Use of reflective semiconductor optical amplifier (RSOA) has been proposed for upstream modulation to make system cost efficient. 2-channel downstream and 2-channel upstream wavelengths are used to exhibits total symmetrical data rate of 10 Gbps. Results revealed that proposed system can support 10 Gbps data rate up to 20 km

Keywords. PON, BER, WDM, ONU, TDM.

1. Introduction

Nowadays, because of the high data rate demands, researchers are giving more attention on the development of the current passive optical networks or access networks for next generation [1]. Architecture of access networks based on the PONs are taken as a ultimate wayout to offer a low cost systems by serving multiple end users [2], high suppleness in addition to scalable network [3]. There are two major categories that classify PONs on the basis of multiplexing such as time division multiplexing (TDM) [4] and WDM [5]. The employment of the time delay or time division multiplexing has been reported for passive optical networks to cater the wide bandwidth requirements of internet services [6]. But, the time division causes the issue of bandwidth sharing and consequently led to wastage of time as well as cost [7]. On contrary, WDM is a prominent and impending technology to overcome these limitations. For FTTH services, WDM-PON is considered as the ultimate solution to offer wide bandwidth and fast communication [8]. Bidirectional or full duplex passive optical networks are needed to serve the ever increasing demands of current generation internet services. However, bidirectional WDM PON networks suffers from a serious issue of crosstalk between downstream and upstream channels. In order to suppress this crosstalk, numerous investigations are proposed so far using similar and dissimilar linecodings/modulations for upstream and downstream. At present, use of the NRZ is popular due to simple generation. But, in near future, advanced modulations such as CSRZ, DPSK etc. are needed due to their high spectral efficiency and has power to combat with dispersions effects. Moreover, wavelength reuse is an important factor to consider in passive optical networks to design a low cost WDM bidirectional system. Many research articles are reported so far to incorporate wavelength reuse. However, they used expensive intensity modulators, which is itself an issue. Consequently, we have designed a WDM passive optical network by considering these important factors.

2. Theory of Modulation

Modulation is the technique of superimposing the information contents of a modulating signal on a high-frequency carrier signal by modifying the carrier signal's characteristic according to the modulating signal. The many types of digital modulation systems are categorized as follows:

2.1. Non-return-to-zero On-Off Keying (NRZ-OOK)

In NRZ modulation, zeros and ones are represented by binary low and high states respectively for 100% duty cycle. In communication systems, it is a most important modulation and spectral efficient modulation. This format offers a number of advantages, including the fact that it is less susceptible to laser phase noise and allows for more effective spectrum transmission over greater distances. Due to its simplicity and cost effectiveness, NRZ linecoding has also been used in the past. Figure 1 shows a representation of NRZ with their waveforms.

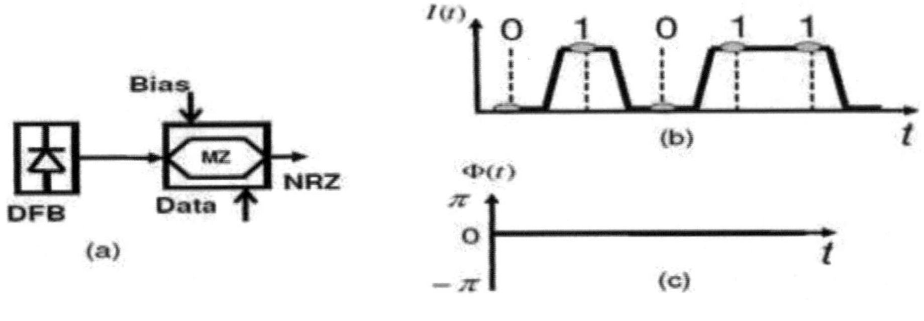

Figure 1. NRZ modulation scheme: (a) NRZ transmitter (b) waveform for Intensity I(t) and (c) phase Φ(t)

2.2. Non-return-to-zero On-Off Keying (NRZ-OOK)

In RZ modulation, zeros and ones are represented by binary low and high states respectively for 50% duty cycle and are conveyed by voltage pulses with specific properties in RZ digital data transmission. In fibre optic communication, the RZ modulation format has the advantage of being more resistant to nonlinear influences. Although it is created without the need of a pulse carver, the RZ pulse format takes up more space inside the fibre. As a result, it is inefficient in terms of bandwidth.

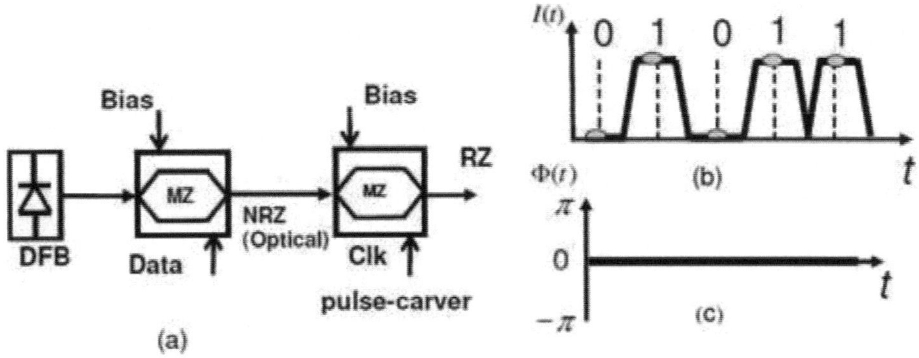

Figure 2. RZ modulation scheme: (a) NRZ transmitter (b) waveform for Intensity I(t) and (c) phase Φ(t)

2.3. Carrier suppressed returned to zero (CSRZ)

It's a type of pseudo-multilevel modulation. The optical field's sign is reversed at each bit change, which distinguishes it. Every bit transition causes sign reversals, which are fully independent of the signal's information-carrying portion. Figure 3 depicts a common CSRZ transmitter configuration. CSRZ may be readily created by sinusoidally operating a MZM pulse carver at half the data rate between its transmission maxima.

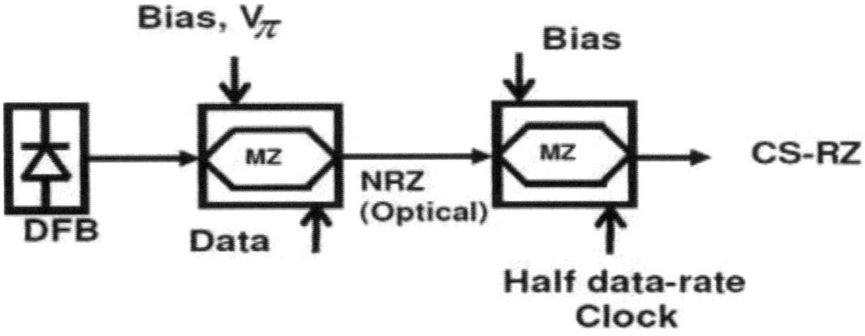

Figure 3. Compressed spectrum return to zero modulation generation

2.4. Differential phase shift keying (DPSK)

It's a sort of phase modulation that changes the phase of the carrier wave to transmit data. A digital "1" is represented in the optical carrier by a pi phase shift between successive data bits, whereas a digital "0" is represented by no phase shift between successive data bits.

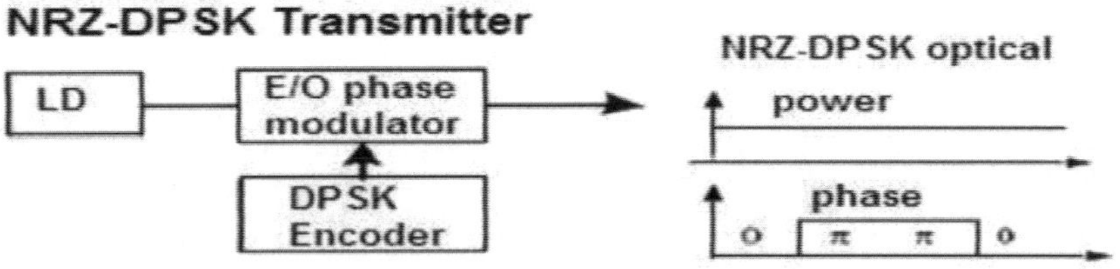

Figure 4. NRZ-DPSK generation

3. System Setup

Figure 5 illustrates the WDM-PON at 10 Gbps employing different modulation formats in downstream and NRZ in upstream. Novelty of the system is use of CSRZ/RZ/NRZ (downstream) and NRZ in upstream combination to combat with crosstalk issue at 10 Gbps. System proposed is symmetrical and also centralized lightwave that supports 2 WDM channels in this work. A laser light wave source at 193.1 THz frequency and 0 dBm input power is used. Major region to used C-band frequency is the low losses and scattering effects of this band over fiber optic cables. Two WDM channels are modulated in central office with aforementioned modulations one by one. From this point, downstream communication in bidirectional passive optical network initiates. The modulated WDM signals are accumulated with the multiplexer of 2 x 1, as well as the multiplexed signal is transmitted over 40-km SMF-28 (single-mode fiber). After transmission of 40 km, signal is de-multiplexed by 1 x 2 wavelengths with frequency spacing of 100 GHz. Receiver section consisting photo-detectors, filters. A p-i-n photo-detector with 100% responsivity and 10 nA dark current is placed in the receiver by considering shot, thermal and ASE (Amplifier spontaneous noise) distortions. Photo-detectors followed by 3-R regenerator.

The eye diagram visualizer is final component that calculates the received signal's final quality factor, signal to noise ratio, and BER, among other things. One half of downstream DQPSK signal is demodulated and other half is provided to the re-modulation. Reflective amplifier is modulated the data of NRZ signal over four frequency channels and multiplexed to travel for upstream. After 40 km transmission, signal is de-multiplexed and detected through PIN photo-diode followed by low pass filter and Eye diagram analyzer. Here, we have considered four different cases such as case 1 for NRZ in downstream, case 2 is RZ, case 3 is CSRZ and case 4 is DPSK. All cases are compared and results are evaluated to suggest the optical configuration to provide better crosstalk suppression.

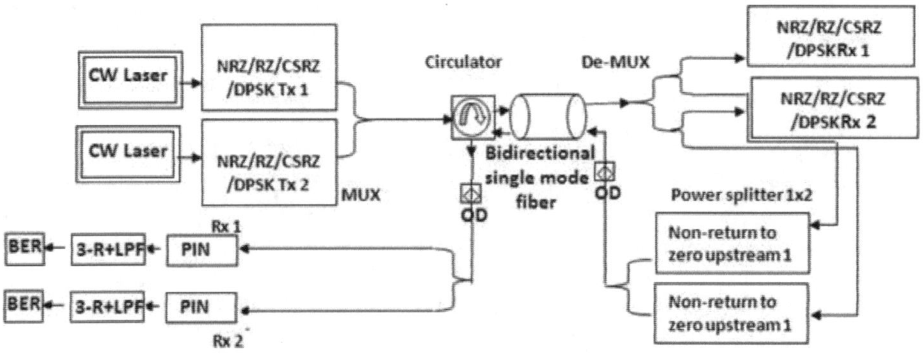

Figure 5. Block diagram of proposed work

4. Results and Discussions

WDM passive optical network model using different linecodings/modulation formats has been successfully simulated using optisystem 7 software. The results are investigated and analyzed accurately. The model has been constructed by following the general parameters in Table 1.

Table 1. System specification

Parameters	Values
Data rate	10 Gbps
Pulse shapes	NRZ/RZ/CSRZ/DPSK

No. of WDM channels	2
Launched Power	0 dBm
Link Length	20 km
Software	Optisystem 7
Photodetector	PIN
Fiber length	20 km

As shown in table 1, for proposed system setup, data signal with 10 Gbps bit rate is modulated by different modulation formats one by one. The reference wavelength used by the channels is 1550 nm. The sequence length is considered 64. Comparison has been proposed for the different link lengths in WDM passive optical network that incorporates diverse modulations for the system analysis. Power is varied from -5 dBm to 4 dBm with the difference of 1 dBm. Figure 6 illustrates the performance of diverse linecoding in the form of Q factor. It is evident that as we increase the input power emergence of attenuation is combated by signal strength. Consequently the Q factor increases as we go to low power to high power. Results also revealed that Q factor of CSRZ increases to maximum extent as compared to other modulation formats with the increase in power. It is due to reason that in bidirectional CSRZ-NRZ system, crosstalk is induced less as compared to other formats based WDM passive optical network. DPSK exhibits performance less than CSRZ but, better than NRZ and RZ.

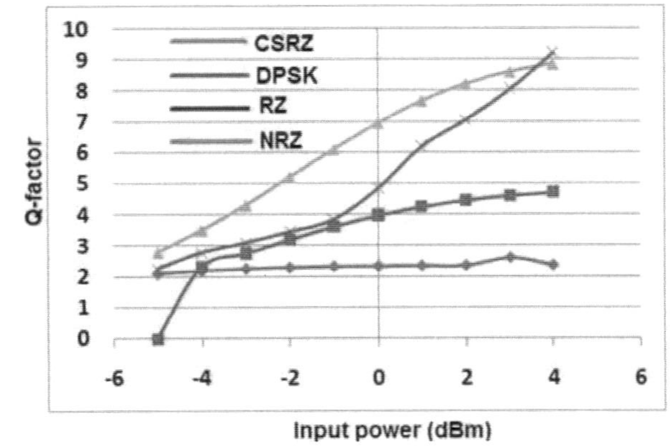

Figure 6. SBS Representation of input power versus Q-factor for downstream

At power of 10 dBm, CSRZ provides Q factor of 8.85. It is observed that CSRZ-NRZ system works for 20 km only with good Q factor at 10 dBm input power.

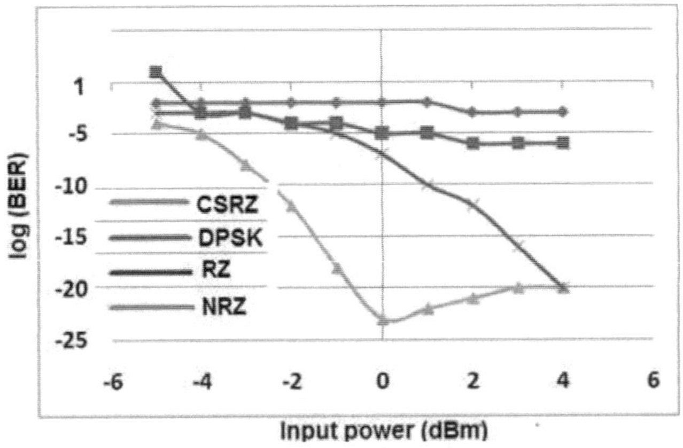

Figure 7. Representation of input power versus log (BER) for downstream

Bit error rate is most important measure for any communication system. It provides the information that whether communication system is valid or not. It is basically a ratio of the errors in bits after transmission at receiver to the total bits transmitted from central office. Acceptable range of BER is 10-9 for optical serial modulation. Figure 7 depicts the variation of LoG (BER) with power. It is seen that Q factor and BER varies inversely. BER increase, that is, more errors are observed at low input power levels rather than high launched power. BER is more in the NRZ in downstream system and least in CSRZ system. Reason, again are the different linear and nonlinear factors that emerges in fiber optic communication along with effects of crosstalk. Eye diagram is the decision making of signals and provide results of Q factor, bit error rate etc. Performance of CSRZ based passive optical network is studied further for different distances in case of downstream as well as upstream in term of Q factor as shown in Figure 8.

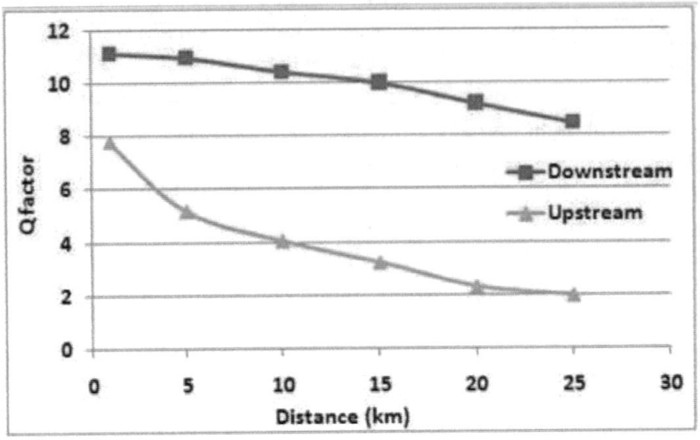

Figure 8. Performance of CSRZ-NRZ based PON system at different distances in terms of Q factor for DS and US

It is observed that as distance prolongs, more and more noises and interferences indulge in the system. As a result of this, Q factor decreases in both the cases such as upstream and downstream. However, performance of the downstream signals transmission is better than upstream. Figure 9 depicts the log BER versus distance for aforementioned cases also and it is clear from the figure that error increases with the enhancement of distance.

Figure 10 represents the effect of pulse broadening is performance deteriorating of the system. With the distance increase, pulse broadens more and this will lead to the extra errors in the system. Performance of downstream transmission is superior to the upstream.

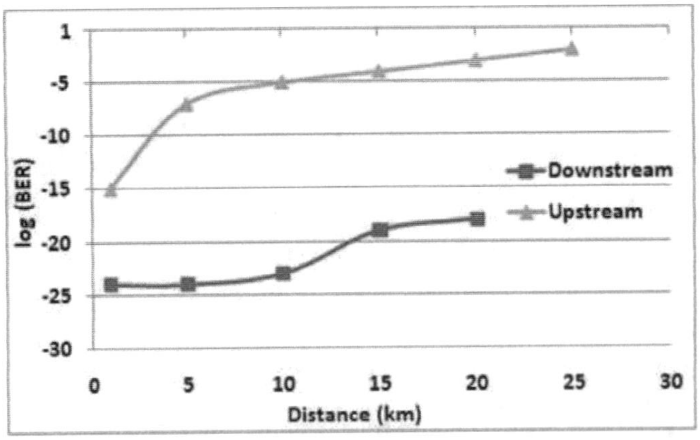

Figure 9. Performance of CSRZ-NRZ based PON system at different distances in terms of log BER for DS and US

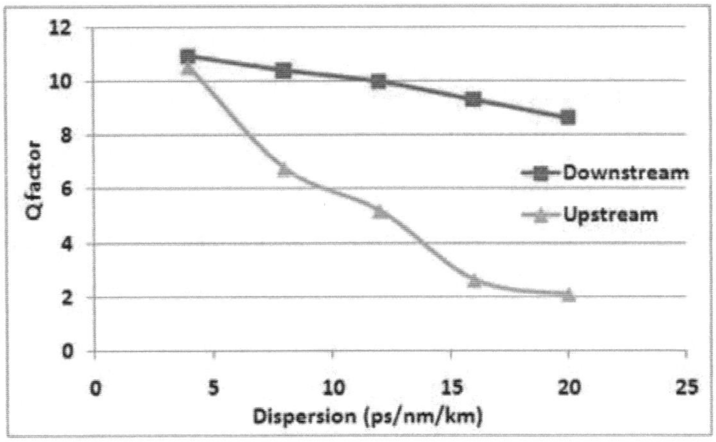

Figure 10. Graphical representation of dispersion versus Q factor for DS and US in case of CSRZ-NRZ PON

5. Conclusion

In this research article, we accentuated on the design of WDM PON system incorporating different linecodings/modulations for upstream and downstream in order to suppress crosstalk. A 20-km-long colorless symmetrical WDM-PON with diverse linecodings/modulations in downstream and NRZ in upstream is investigated to reduce crosstalk. It is observed that the compressed spectrum RZ in downstream and NRZ in upstream performs best followed by the DPSK-NRZ. RSOA is employed for re-modulation and also makes system cost efficient. It is observed that same modulation format in downstream/upstream (NRZ-NRZ) is prone to maximum crosstalk than other formats.

Acknowledgement

I would like to express my very great appreciation to **Manoj Sindhwani, Shippu Sachdeva** (School of Electronics and Electrical Engineering, Lovely Professional University, Phagwara, Punjab, India.), **Sachin Chawla** (Chandigarh University, Punjab, India) for their valuable and constructive suggestions during the planning and development of this research work. Their willingness to give their time so generously have been very much appreciated.

References

[1] Monika Amit, Wason Amit and Kaler R S, 2013 Investigation of four wave mixing effect with different number of input channels at various channel spacing *Opt. Int. J. Light Electron.* **124** (20) pp 4227–4230.

[2] Kaur Manpreet and Bansal Dipti, 2017 A Long Haul Carrier Generated Ultra Dense Passive Optical Network Incorporating Low Cost VCSEL *International Journal of Engineering Development and Research (IJEDR)* **5**(2) pp 1554-1559.

[3] Zhao Zhou, Hai-tao NIE and Yao-jun Wang, 2016 Long reach DWDM-PON with 12.5 GHz channel spacing based on comb source seeding *Optoelectronics letters* **12**(4) pp. 304-307.

[4] Tiwari Umesh and Saini Chhavi, Saini Than Singh and Bhardwaj Manish, 2016 Performance Analysis of UDWDM Passive Optical Network with Enhanced Span Length using FBG based Dispersion Compensation *13th International Conference on Fiber Optics and Photonics, Kanpur India.*

[5] Wang C H, Shih F Y, Yeh C H, Chow C and Chi W S, 2009 10 Gb/s TDM passive optical networks using four wavelengths multiplexed channels *Opt. Commun.* **282** pp. 2476–2479.

[6] Lee Ju Han, Lee Kwanil, Lee Sang Bae and Kim Chul Han 2009 Extended-reach WDM-PON based on CW supercontinuum light source for colorless FP-LD based OLT and RSOA-based ONUs *Opt. Fiber Technol.* **15** pp 310–319.
[7] Goyal Rakesh and Kaler R S, 2012 A novel architecture of hybrid (WDM/TDM) passive optical networks with suitable modulation format *Opt. Fiber Technol.* **18** pp 518–522.
[8] Neto B, Klingler A, Reis C, Dionísio R P, Nogueira R N, Teixeira A L J and André P.S., 2011 Enhanced optical gain clamping for upstream packet based traffic on hybrid WDM/TDM-PON using fiber Bragg grating *Opt. Commun.* **284** pp 1354–1356.

EFFECTS OF ORTHOGONAL MODULATIONS ON FWM IN MULTICHANEL SYSTEMS

Ruchika

CSE Department, Assistant Professor, Chandigarh University, Punjab, India

ruchika.e10474@cumail.in

https://orcid.org/0000-0002-4251-3993

Abstract. In multichannel systems, four wave mixing (FWM) is a conspicuous and significant performance degrading element. Optimized modulation formats are necessary to decrease FWM effects and boost the data rate supportability of multichannel systems. By combining 3-D and 2-D orthogonal modulation formats, multichannel system with eight channels and each operating at 10 Gbps is suggested in this paper. Polarization shift keying is an important modulation that has been examined bay taking Q-factor into consideration at various distances. Furthermore, FWM power is examined and a comparison for 3-D, 2-D, and 1-D modulations at different frequency spacings such as 100 GHz, 50 GHz, and 25 GHz has been established. Because 2-D orthogonal modulation has the least amount of FWM, it is suggested for usage in WDM systems with ultra-dense spacings.

Keywords. PolSK, DPSK, DQPSK, FWM, WDM, Single mode fiber (SMF)

1. Introduction

Because of the increasing fiber optic links, the need for high bandwidth demands has increased due to an increase in internet-based services, and cloud computing [1]. Multichannel system is a promising technique for meeting the needs of numerous users by cramming more and more channels into a smaller space. The combination of WDM and orthogonal modulation formats [2] is a promising and capable option for meeting user needs. Orthogonal modulation schemes use less bandwidth and are less susceptible to noise caused by phase shifts. As a result, it enables the WDM system to accommodate additional channels and accumulation at dense spacings [3]. Orthogonal modulation formats provide a number of benefits, including increased security, data carrying capacity, and flexibility. Nonlinear effects such as FWM is major Kerr's effect dependant nonlinearities that deteriorate the Kerr's effect [4]. The power dependency of the optical fiber's refractive index is the primary source of nonlinear effects. These issues must be solved in order to achieve long-distance transmission. FWM is a key performance-degrading nonlinearity in multi-channel systems that originates from the power buildup of multichannel system inside a SMF [5]. When more than one frequency travels through fibre optics at the same time, the FWM effect causes the refractive index to fluctuate, resulting in additional frequency signals that restrict the overall performance of multichannel system. Non uniform frequency spacings [6], hybrid modulators [8], non-zero dispersion fibre (NZ DCF) [7,] and management of dispersion [9] approaches have all been claimed to mitigate the FWM effects in mmultichannel systems. Author in [10] proposes a 2-channel system that suppresses four wave mixing by employing orthogonal formats at low launching power levels. However, the ever-increasing need for high data rates has prompted us to expand the system's capacity and examine the formation of FWM in multi-channel WDM systems.

In this paper, we look at the Q factor and FWM power of an 8 x 10 Gbps WDM system combining orthogonal modulation formats at varied channel spacings (100 GHz, 50 GHz, and 25 GHz) and distances.

2. Four Wave Mixing in WDM Systems

FWM is a type of optical parametric oscillation based on the Kerr's effect concept. In wavelength division multiplex systems, this is a key source of performance degradation. Crosstalk and performance deterioration are introduced by four wave mixing, which may be understood by noticing the side frequency peak that appears owing to phase mismatch of travelling optical pulses, as illustrated in (1). This occurs when the three frequencies propagate through the optical cable at the same time, as seen in (2).

$$\omega_{ijk} = \omega_i + \omega_j - \omega_k \quad (1)$$

$$\omega_i, \omega_j, \omega_k \quad (2)$$

When considering WDM with equal channel spacing, coherent in-band crosstalk arises. The fact that the newly created frequency corresponds with previously existing frequencies is a major cause of modulation of phase M are the products of FWM, N are the multi channels, and given as (3)

$$M = \frac{1}{2}(N^3 - N^2) \quad (3)$$

As a result, as the number of channels grows, FWM will play an increasingly important role. As a result, it is critical to consider four wave mixing when developing large capacity and extremely dense WDM systems.

3. System Setup

In order to investigate the proposed wavelength division multiplexed system employing orthogonal modulation format, a simulation tool Optiwave's optisystem is used to realization and to comprehend the system. We accentuate on the major degrading factor FWM,that is analyzed at different channels spacings and for 3-D, 2-D and 1-D modulations. A 8 x 10 Gbps WDM system is demonstrated over 40 Km single mode fiber (SMF) and FWM is observed for varied link lengths. Figure 1 (a), (b), (c), (d) depicts the system architecture of proposed WDM incorporating 3-D, 2-D and 1-D modulation formats respectively.

A PRBS (pseudo random bit sequence generator) is a binary bits tributaries generator in the form of ones's and zeros's and transmitting bits at the rate of 10 Gbps per channel. A continuous wave laser operating in C-band (1530 nm- 1570 nm) and at power level of 0 dBm. Laser linewidth is kept at 10 MHz to make system more practical. Further the NRZ linecoding is used to provide energy to binary data stream. In case of 3-D orthogonal modulation, aforementioned signal is further encoded with DQPSK signal and fed to the linear polarizer to shift the polarization state to 45 degrees. Polarization shifted signal is encoded with data such as to realize the polarization shift keying modulation. For 2-D orthogonal modulation, DQPSK encoding stage as shown in fig 1 (b) is eliminated and signal direct fed to polsk encoding. A DPSK modulation is used for 1-D modulation without any polSK as depicted in fig 1 (d). Decoding is also illustrated in the fig 1 (b), 1 (c) and 1 (d) for 3-D, 2-D and 1-D modulations.

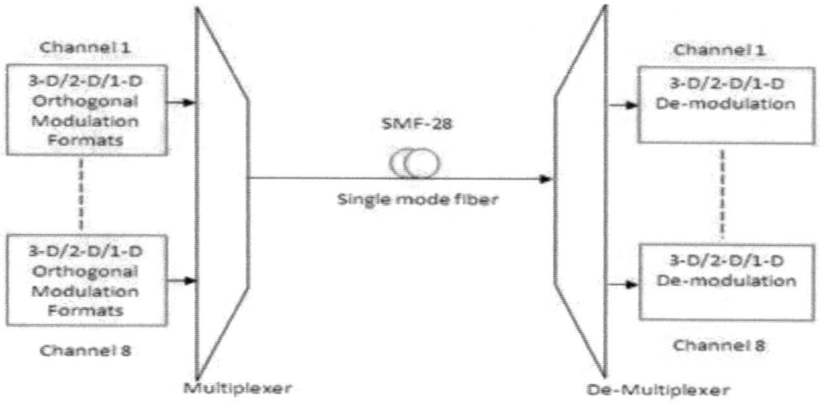

Figure 1(a). Representation of 8 x 10 Gbps wavelength division multiplexed system incorporating 3-D/2-D/1-D modulation formats

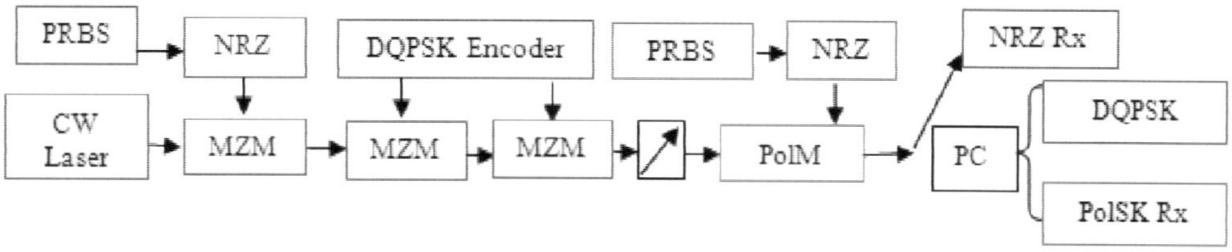

Figure 1(b). Internal structure of 3-D orthogonal modulation

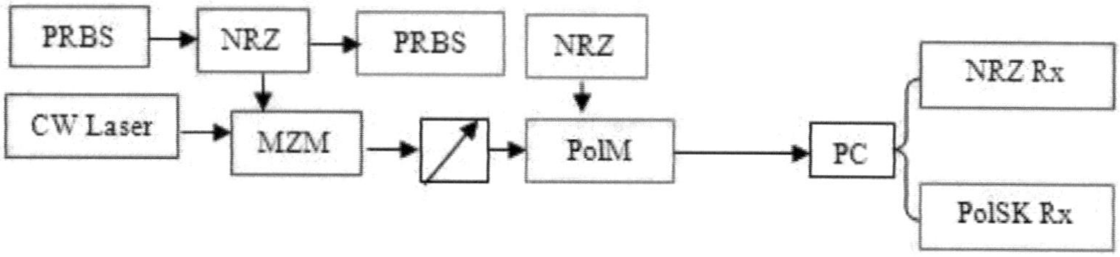

Figure 1(c). Internal structure of 2-D orthogonal modulation

Figure 1(d). Differential phase shift keying 1-D modulation

4. Results and Discussions

To investigate the Q factor in WDM system incorporating DQPSK, NRZ and PSK, diverse system parameters taken are different distances, different frequency spacings, and launched power levels.

Figure 2 (a) (b) (c) illustrate the Q factor for different WDM channels and their effects on FWM. Figure 2 (a) represents the comparison of DQPSK, NRZ and POLSK under FWM at 100 GHz channel spacings. Similarly, Figure 2 (b) and 2 (c) depicts the performance and comparison at 50 GHz and 25 GHz. Because of pulse width increase and amplitude degradation, as well as nonlinear effects, the Q factor drops as the length of the optical fibre increases. Modulations are compared and it is evident that at 100 GHz and 50 GHz channel spacings, PSK is found out to be best modulation followed by NRZ. It means that in order to obtain high Q factor, polarisation shift keying is recommended under the effects of FWM. Further, there is slight change observed as the channel spacing decreases fro 50 GHz to 25 GHz, DQPSK performs best. So, at ultra dense channel spacings, it is better to use DQPSK instead of PSK modulation.

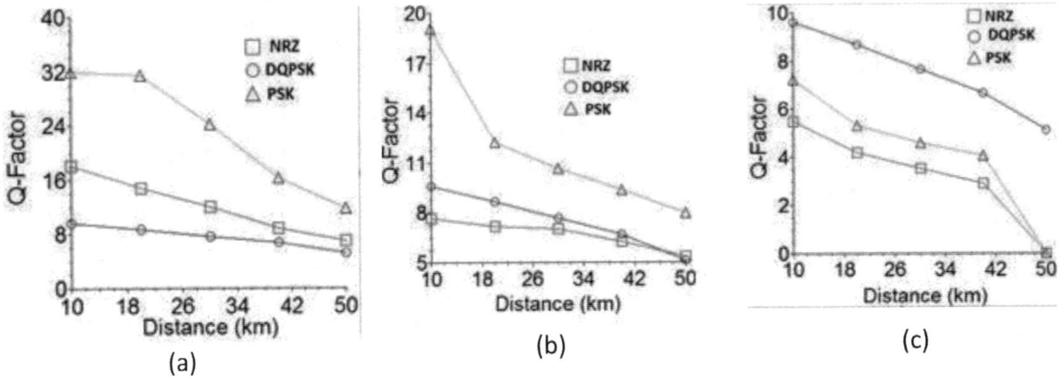

Figure 2. Represents the Q-factor versus Distance at (a) 100 GHz (b) 50 GHz (c) 25 GHz for 3-D orthogonal modulation

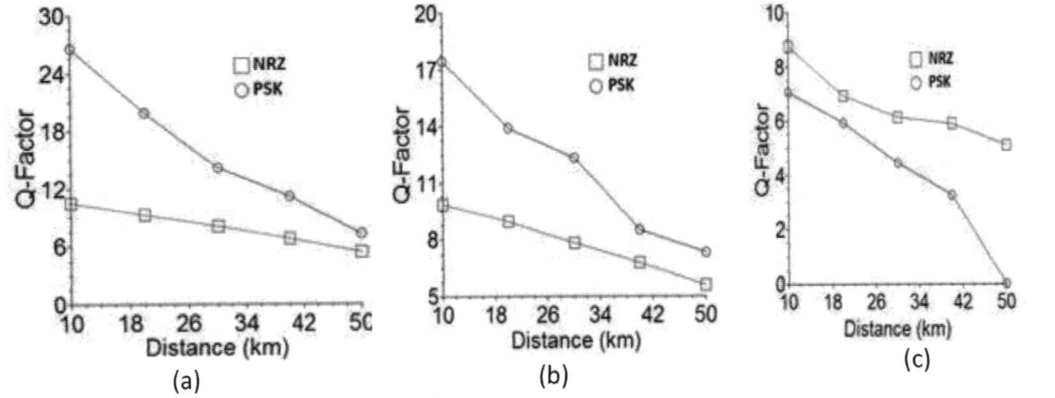

Figure 3. Represents the Q-factor versus Distance at (a) 100 GHz (b) 50 GHz (c) 25 GHz for 2-D orthogonal modulation

Figure 3 represents the performance of 2-D orthogonal modulation for WDM system at different link lengths and frequency spacings. Results revealed that at 100 GHz and 50 GHz channel spacings, PolSK performs best and at 25 GHz, NRZ crosses the PolSK performance and thus well suited for ultra dense WDM. Figure 4 depicts the 1-D DPSK modulation performance at different frequency spacings and varied link lengths. Distance increase causes the degradation in Q factor and at lower spacings. Further, Four wave mixing is investigated for all modulation proposed in this work at different frequency spacings and distances as illustrated in fig 5 (a), (b) and (c) for 3-D, 2-D and 1-D respectively. It is seen that maximum FWM emerges at 25 GHz and minimum for 100 GHz for all modulations. So, with the decrease in frequency spacings, FWM increases. Figure 6 illustrates the comparison of three modulations at different distances in terms of FWM generated. 3-D orthogonal modulation is generating the maximum power FWM followed by 1-D modulation. For WDM systems at ultra dense frequency spacings, 2-D orthogonal modulation is suggested as it produces minimum power FWM components.

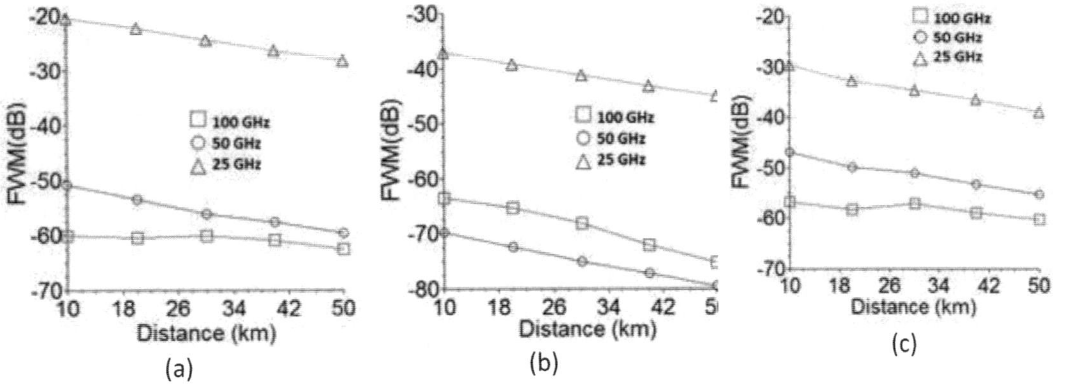

Figure 4. Represents the Q-factor versus Distance at (a) 100 GHz (b) 50 GHz (c) 25 GHz for 1-D modulation

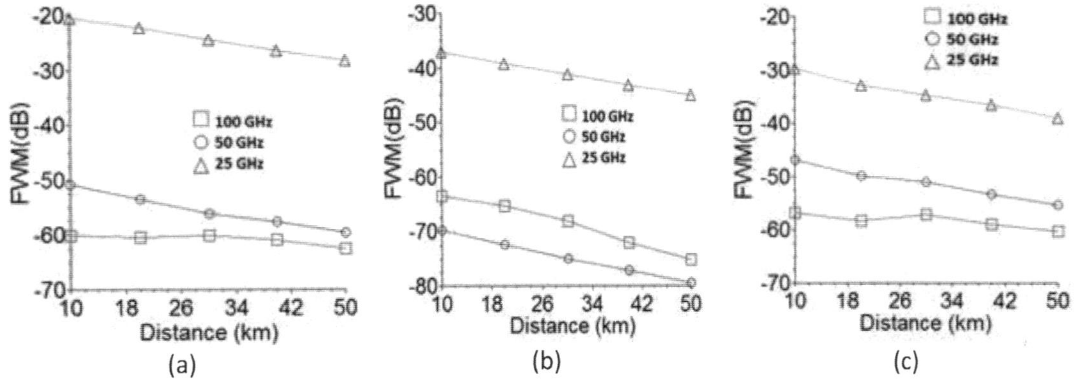

Figure 5. Four wave mixing versus distance at different frequency spacings for (a) 1-D modulation (b) 2-D (c) 3-D orthogonal modulations

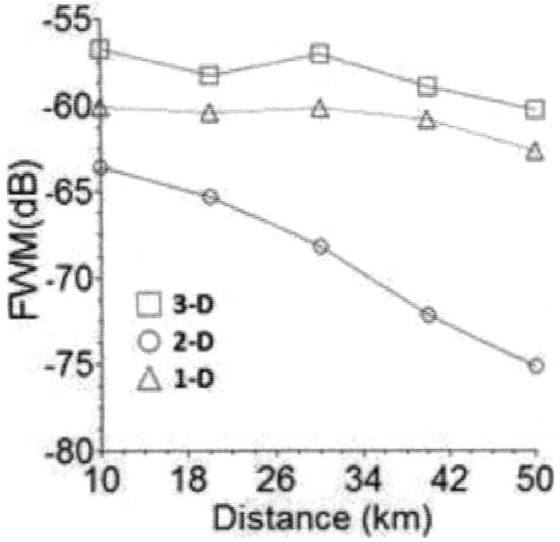

Figure 6. Comparison of all proposed modulations

5. Conclusion

In this research paper, a capacity enhanced 8 x 10 Gbps WDM system is proposed and investigation of four wave mixing has been done at different frequency spacings. Orthogonal modulation formats 3-D (NRZ-DQPSK-PolSK) and 2-D (NRZ-PolSK) are analyzed along with 1-D modulation at 100 GHz, 50 GHz and 25 GHz to evaluate the FWM. Results revealed that for all modulations, with the increase of distance, value of Q factor decreases and also there is significant degradation as channel spacing reduces. 3-D orthogonal performs best in terms of Q factor and 1-D performs least. Also comparison has been done amid different proposed modulations and it is observed that FWM emerges minimum in case of 2-D orthogonal modulation and maximum for 3-D orthogonal modulation. However, for 1-D modulation, four wave mixing components exceed the power than 2-D but lower to 3-D orthogonal modulation formats. So, 2-D orthogonal modulation is suggested for ultra dense WDM systems under the effects of four wave mixing.

Acknowledgement

I would like to express my very great appreciation to **Manoj Sindhwani, Shippu Sachdeva** (School of Electronics and Electrical Engineering, Lovely Professional University, Phagwara, Punjab, India.), **Sachin Chawla** (Chandigarh University, Punjab, India) for their valuable and constructive suggestions during the planning and development of this research work. Their willingness to give their time so generously have been very much appreciated.

References

[1] Shao Y and Chi N, 2012 High spectral-efficiency 100 Gb/s transmission using DRZ, DQPSK and PolSK three dimension orthogonal modulation *Opt. Commun.* **285**(6) pp 1049–1052.

[2] Boffi Pierpaolo, et al., 2004 20 Gb/s differential quardrature phase shift keying transmission over 2000 Km in 64-channel WDM system *Opt. Commun.,* **237** pp 319–323.

[3] Sheetal A, Sharma A K and Kaler R S, 2010 Simulation of high capacity 40 Gbps long haul WDM system using different modulation formats and dispersion compensated schemes in presence of Kerr's effect *Optik* **121** (8) pp 739–749.

[4] Shao Y et al., 2008 Novel optical orthogonally modulation schemes for superimposing DPSK Signal on dark RZ signal *Opt. Commun.* **281**(4), pp 3658–3667.

[5] Inoue K and Toba H, 1995 Fiber four wave mixing in multi- amplifier systems with non uniform chromatic dispersion *J. Lightwave Technol.* **13** pp 89–93.

[6] Forghieri F, Tkach R W and Chraplyvy A R, 1995 WDM systems with unequally spaced channels *J. Lightwave Technol.* **13**(5) pp. 889–897.

[7] Nikodem Michal P, Zurawski Wojciech and Abramski Krzysztof M, 2008 Four-Wave Mixing in Non-Zero Dispersion Shifted Fibers *Transparent Optical Networks, 2008. ICTON 2008. 10th Anniversary International Conference, Athens, Greece* pp 342-345

[8] Sugumaran S, Sharma N, Chitranshi S, Thakur N and Arulmozhivarman P, 2013 Effect of Four-wave Mixing on WDM System and its Suppression Using Optimum Algorithms *International Journal of Engineering and Technology* **5**(2), pp. 1432-1444.

[9] Matsuda T., Murakami M. and Imai T., 1998 Dispersion management scheme for suppressing FWM in DWDM transmission *Lasers and Electro-Optics Society Annual Meeting* (*LEOS '98. IEEE*). pp. 372-373

[10] Singh Surinder and Singh Sukhbir, 2016 Investigation on four wave mixing effect in various optical Fibers for different spectral efficient orthogonal modulation Formats *Optics & Laser Technology*, **76**, pp. 64-69.

Journal of Physics: Conference Series **2327** (2022) 012045 doi:10.1088/1742-6596/2327/1/012045

AN ECONOMICAL AND ULTRA DENSE WDM-PON SYSTEM WITH POLARIZATION INTERLEAVING AND PWR

Shippu Sachdeva[1], Mohit Bansal[2]

[1]Department of Electrical & Electronics Engineering, G.L.Bajaj Institute of Technology & Management, Greater Noida. U.P.

sachdeva.shippu@ gmail.com[1], moh.bansal@gmail.com[2]

https://orcid.org/0000-0002-6859-8622[1], https://orcid.org/0000-0001-8072-1462[2]

Abstract. In this paper, a cost effective and easy maintenance based dispersion compensation technique is proposed in ONU of the wavelength reused wavelength division multiplexing reliant passive optical network. Moreover, in order to suppress intra-channel crosstalk DQPSK is employed for downstream and for inter-channel crosstalk suppression, polarization interleaving is used in the system. Furthermore, for the investigation of proposed system in terms of dispersion reduction, three different scenarios are considered such that system with only DCF, linearly chirped FBG and joint module of DCF+FBG. System has total 4 channels and each has bit rate of 20 Gb/s with 25 GHz channel spacings. It is observed that joint module of DCF+FBG has maximum ability to compensate dispersion and linearly chirped FBG has least performance. Proposed joint technique for pulse width reduction in ONU is cost effective, highly efficient to combat with pulse width reduction issues and also needs very less maintenance.

Keywords: Pulse width reduction (PWR), Pulse width Broadening (PWB), DCF, Linearly chirped FBG, DCF+FBG

1. Introduction

With the explosive increase in the bandwidth hungry internet applications, pressure is continuously increasing on the optical networks [1]. Passive optical networks (PON) are key technologies to cater the demand of high speed services with low cost and by supporting multiple users [2]. There are different variants of passive optical networks standardize by international telecommunication union [3]. Time division multiplexing is an attractive variant of PONs that has been employed to fulfill the demands of broadband services. However, time division multiplexing suffers from the issue of bandwidth sharing [4] [5]. In order to utilize the bandwidth of the optical fibers, researchers worked on the wavelength division multiplexing based PON. WDM based PON systems are well competent and has potential to support large distance and users. It emerges as a key technology for the next generation fiber to the home (FTTH) technology [6]. Wavelength reuse in the WDM PONs is widely employed because it saves the cost of the system by eliminating the requirement of laser source in the optical network unit [7]. Signals from the central office towards ONU is termed as downstream and are re-modulated in uplink i.e. transmitted back to OLT (optical line terminal) [8] [9]. Advantage of wavelength reuse is that downlink as well as uplink uses same wavelength for the operation and enhance the efficiency of wavelength utilization [10][11][12]. Modulation of the signal also plays a vital role to decide the performance of the optical networks [13]. Differential Quadrature phase shift keying provide high bandwidth or spectral efficiency and less prone to pulse broadening effects. Therefore competent enough to prolong the transmission distance between OLT and ONU. Phase shifting modulation offers high extinction ratio, less prone to dispersion and nonlinear effect [14][15]. Transmission distance enhancement at high data is limited to short distance due to effects of dispersion. Numerous works has been reported to compensate pulse broadening inside optical fiber. Dispersion compensation fiber [16], fiber bragg gratings [17] and optical phase conjugations [18] are widely employed in the optical networks. However, DCF has high efficiency to compensate pulse width broadening but comes at the high cost and increase the overall cost of the system. PWR efficiency of FBGs is less but they are cost efficient [19]. Similarly, optical phase conjugations (OPCs) make system complex and expensive. High speed optical networks are severely vulnerable to PWB (pulse width broadening) issues and limit the total reach of the system. In this work, a cost effective and easy maintenance based dispersion compensation technique is proposed in ONU of the wavelength reused ultra dense WDM-PON. Also, inter-channel and intra-channel crosstalk are suppresses by using polarization interleaving and DQPSK respectively. A cost effective and high efficiency PWR joint module is proposed for the ONUs.

2. Simulation setup

Proposed ultra dense wavelength reused WDM PON employing polarization interleaving is depicted in Figure 1. To realize the proposed system, an important and compendious tool Optiwave's OptiSystemTM is used. Data rate of the proposed work is fixed to 20 Gbps and differential quadrature phase shift keying modulation as shown in Figure 2 is employed for downstream data transmission. To reduce the interference hetero modulations are given to downstream channels and upstream channels. For uplink, NRZ is incorporated and also to reduce inter-channel interference, polarization interleaving done among the different channels of WDM. Total 4 WDM channels in C-band are considered and frequency starts from 193.1 THz. Input power of the system is 0 dBm and it comes from the CW laser. Diverse polarizations are given to even and odd channels by linear polarizer that coincides with the transmission axis of the polarizer and eliminates the orthogonal component. State of polarization of odd channels $\lambda 1$- $\lambda 3$ is horizontal and even channels $\lambda 2$- $\lambda 4$ is vertical. These four PI channels are modulated with DQPSK and multiplexed with 4x1 mux. Multiplexed signal is fed to 40 km SMF-28 which has 0.2 dB/km attenuation and 17 ps/nm/km dispersion. After signal transmission inside optical fiber, it is demultiplexed at the specific frequencies according to the transmitter.

Each signal then passed through the PWR unit. Three different cases are used in PWR module (1) only DCF (2) Linearly chirped FBG and (3) joint module of DCF+linearly chirped FBG. System specifications of proposed model and PWR modules are shown in Table 1 and Table 2 respectively. After dispersion compensation, signals are decoded using DQPSK receiver which consists of photo-detectors, phase delay, MZI switch, 3-R regenerator and BER analyzer. One half of downstream DQPSK signal is demodulated and other half is provided upstream block for re-modulation. For uplink, 4 NRZ signals are modulated by wavelength reused signal and RSOA. Upstream signal made to pass from same bidirectional optical fiber and received by NRZ receiver.

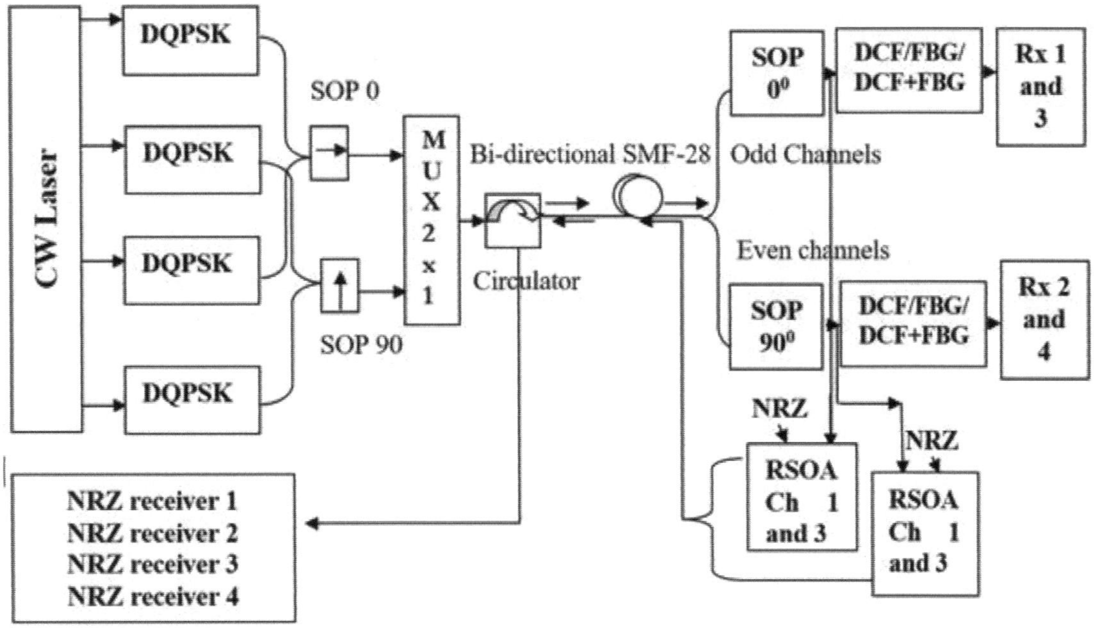

Figure 1. Proposed ultra dense and polarization interleaved DQPSK-NRZ WDM PON system

Table 1. System specification

Parameters	Values
Data rate	20 Gbps
WDM Channels	4
Freq. spacing	25 GHz
Modulation DS	DQPSK
Modulation US	NRZ
Wavelength Re-use component	RSOA
SOP 1	0 degree
SOP 2	90 degree
Photo-detectors	PIN

Table 2. PWR Module specification

Parameters	Values
DCF attenuation	0.5 dB/km
DCF Dispersion	-85 ps/nm/km
FBG length	2 mm
FBG effective index	1.45
FBG chirping	Linear
Linear parameter	0.0001 um

Journal of Physics: Conference Series **2327** (2022) 012045 doi:10.1088/1742-6596/2327/1/012045

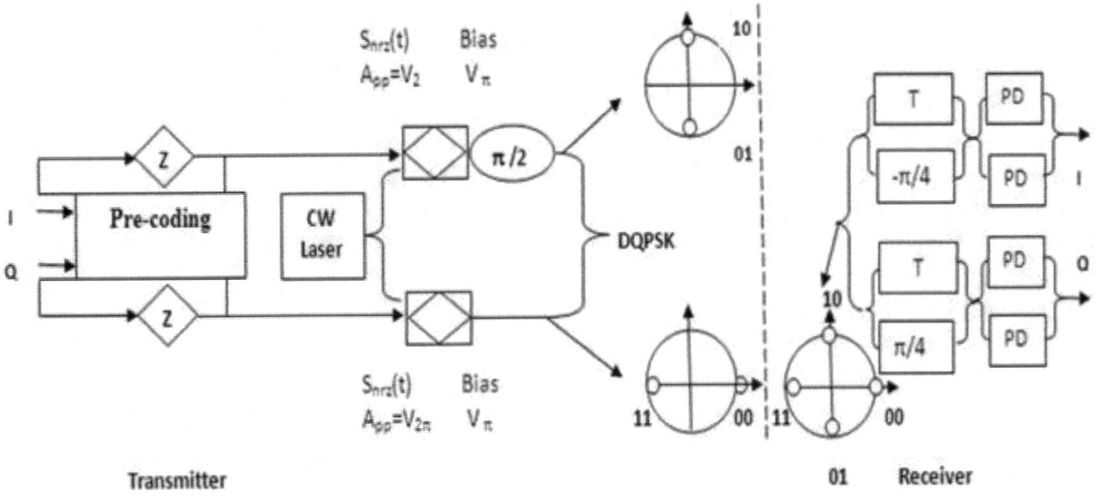

Figure 2. Differential Quadrature Phase Shift Keying (DQPSK) Transmitter and Receiver

3. Results and Discussions

First and foremost, proposed system is investigated to find the pulse width reduction efficiency of the different techniques such as dispersion compensation fiber, linearly chirped FBG and joint module of DCF+FBG over the distance of 75 km of Bi-directional SMF-28. Data rate of the each channel is 20 Gbps and time of one bit is calculated as the 0.5 ns or 50 ps. Optical time domain analyzer is a component which used to represent the bits according to the time and power. An OTDV is placed after the channel 1 and output is as shown in Figure 3 (a).

In order to check the effects of pulse width broadening, OTDV is placed after the optical fiber and prior to the receiver of each channel i.e in optical network unit for downstream transmission. A pulse of time 100 ps is seen after 75 km of link distance as depicted in Figure 3 (b). For the calculation of the pulse width reduction efficiency of three different modules in ONU, they are employed in the ONU. First of all we consider the case of linearly chirped FBG. Filtered 193.1 THz signal is made to pass through FBG and it reflects the desired signal and compensates the effects of PWB. Time of the bit is observed 60 ps and efficiency of FBG comes out to be 40 % as shown in Figure 3 (c). Further dispersion compensation fiber is investigated and it provides the time of pulse 55 ps and thus it is 45 % efficient. A DCF and linearly chirped combined module is placed in this case and it is noteworthy that maximum pulse width reduction is observed for the length of 6 km and 2 mm linearly chirped FBG. Length of DCF is iterated for different values from 5-20 km but optimal results are observed for 6 km. This joint technique provides the pulse of time 52 ns and thus it is 48 %. All the efficiencies are calculated from the formula as expressed in (1)

$$PWR\% = \frac{\text{Time of Broadened pulse}-\text{Time of original pulse}}{Time\ of\ brodened\ pulse} \quad (1)$$

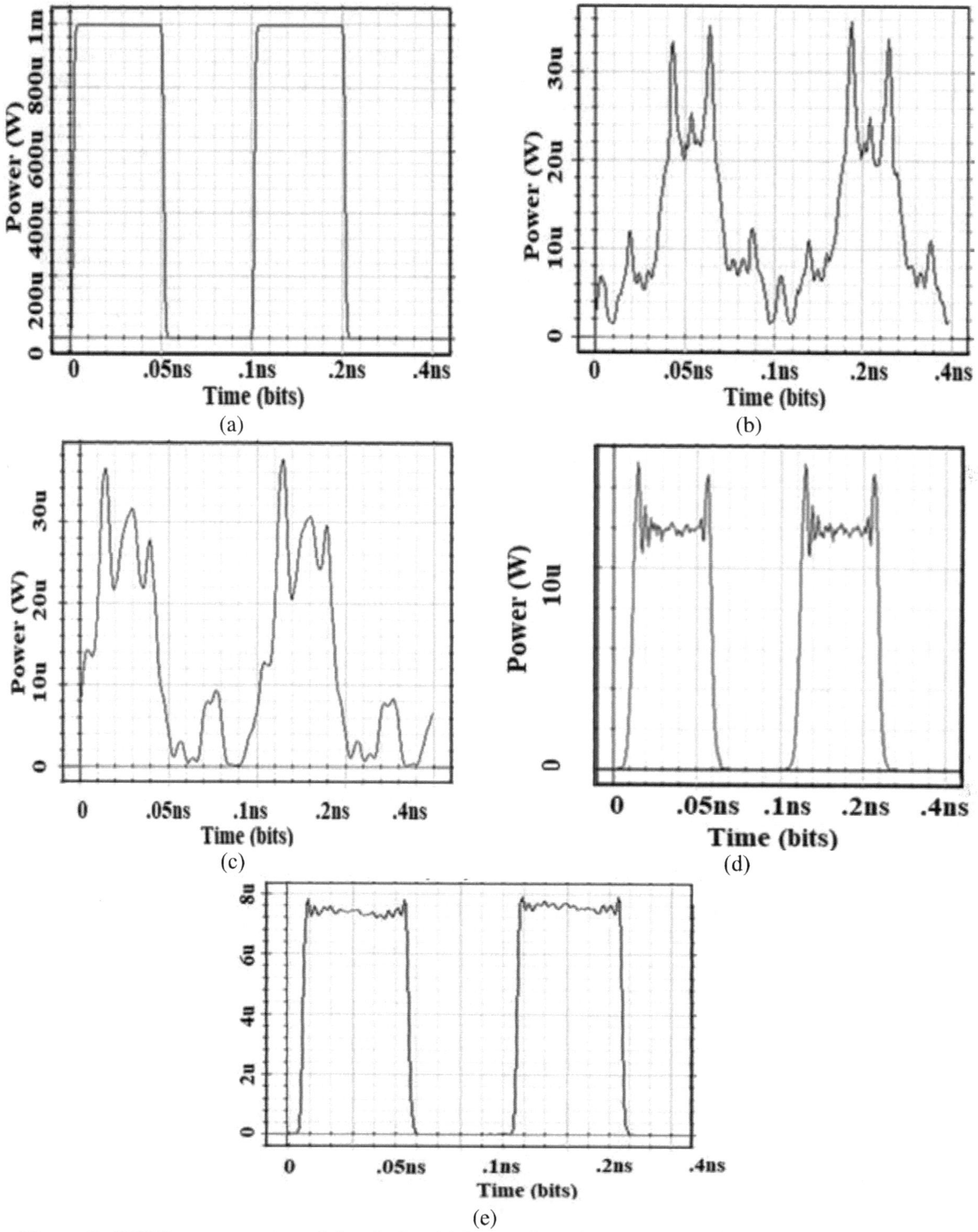

Figure 3. OTDV representation of signal after (a) Transmitter (b) 75 km (c) FBG (d) DCF (e) DCF+FBG

Figure 4 represents the performance of the polarization interleaved ultra dense WDM PON in terms of Q factor at different distances for three different PWR cases. Distance is varied from 15 km to 75 km and DCF, FBG and DCF+FBG modules are employed in the ONU. Results reveals that Q factor of the downstream decreases with the increase in the link length. As mentioned in the former analysis, due to the high efficiency of pulse width reduction, joint module of DCF+FBG provide maximum Q factor and followed by the performance of dispersion compensation fiber. Least Q is obtained in the case of linearly chirped FBG due to least PWR efficiency. However, use of these dispersion compensation modules decreases the maintenance issues that are needed when they are placed in the transmission line Similarly, performance of the proposed system is evaluated in terms of Log BER at different distances for three different PWR cases. It is evident that Log BER of the downstream increases as the distance prolongs. Due to the high efficiency of pulse width reduction, joint module of DCF+FBG provides least Log BER. Performance of DCF is somewhat below the performance of joint module. Maximum Log BER is obtained in the case of linearly chirped FBD due to least PWR efficiency as depicted in Figure 5. Therefore, joint module is suggested to use for the dispersion compensation and for the easy maintenance of the WDM PON systems.

Figure 6 represents the graphical representation of Q factor versus distance for with and without polarization interleaving in WDM-PON. It is evident from the investigation that because of the different states of polarization in even and odd channels of the proposed system, Q factor is obtained better in case of polarization than without polarization interleaving

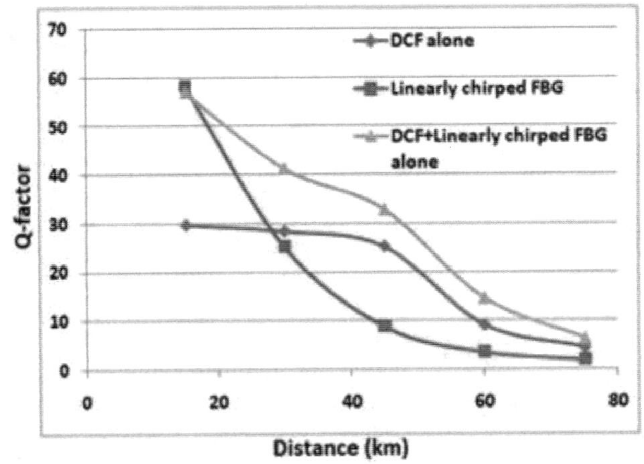

Figure 4. Representation of Q factor versus distance for different cases of PWR

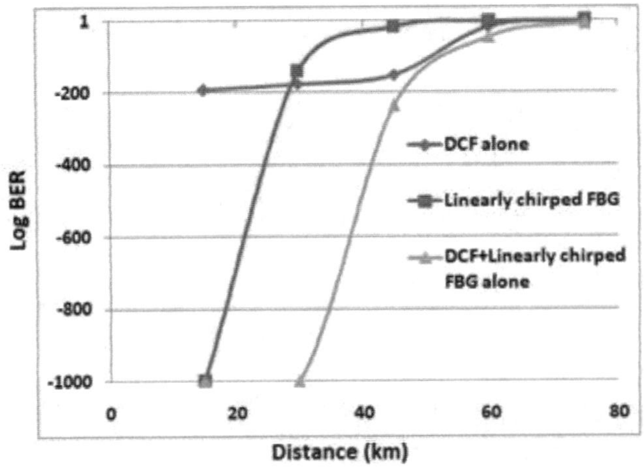

Figure 5. Representation of Log BER versus distance for different cases of PWR

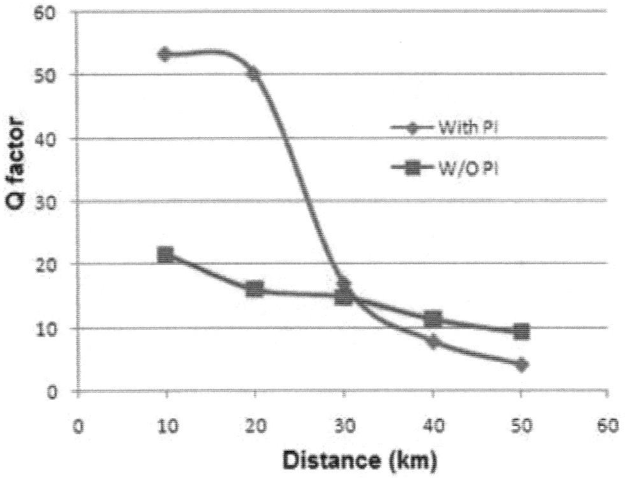

Figure 6. Comparison of the proposed with and without polarization interleaving

Figure 7 depicts the Eye diagrams of the proposed system. It is decision component which calculates the Q factor and BER of the system. BER represents, the average number of ones and zeros, in the bit stream, and also average amplitudes. It has property that more Q factor be seen from the wide opening of the diagram. Figure 7 (a) (b) is diagram of the signal after 1 km and 75 km respectively in the case of linear chirped FBG. It is perceived that eye closure is more at 75 km thus shows very distorted eye. However, eye opening is wide for 1 km analysis using FBG alone. Eye opening is little bit more at 75 km for dispersion compensation fiber as shown in Figure 7 (d). Q factor for the joint technique of DCF+FBG comes under acceptable limit and therefore this module provides best results as shown in Figure 7 (f).

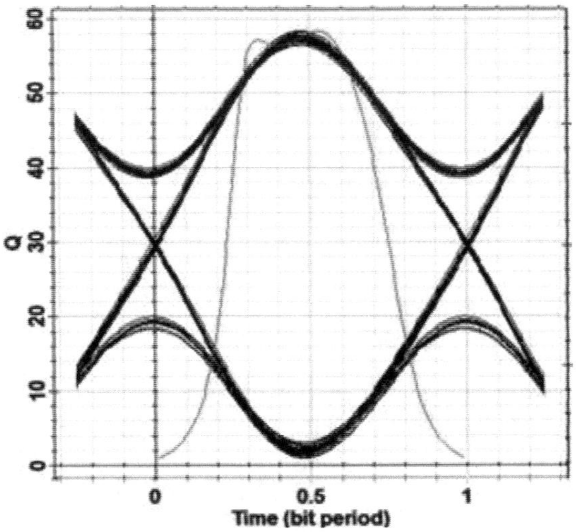

(a)

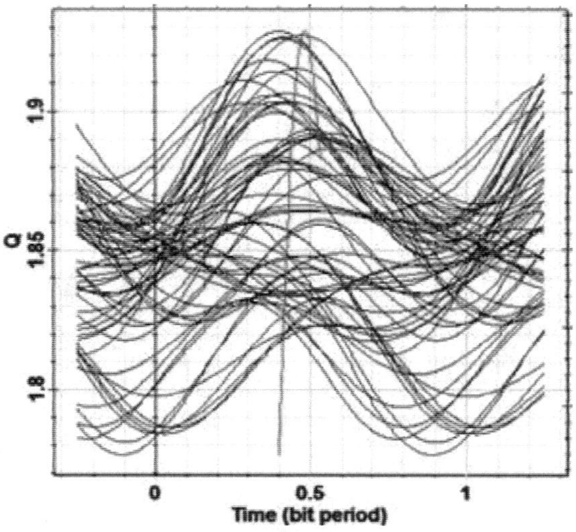

(b)

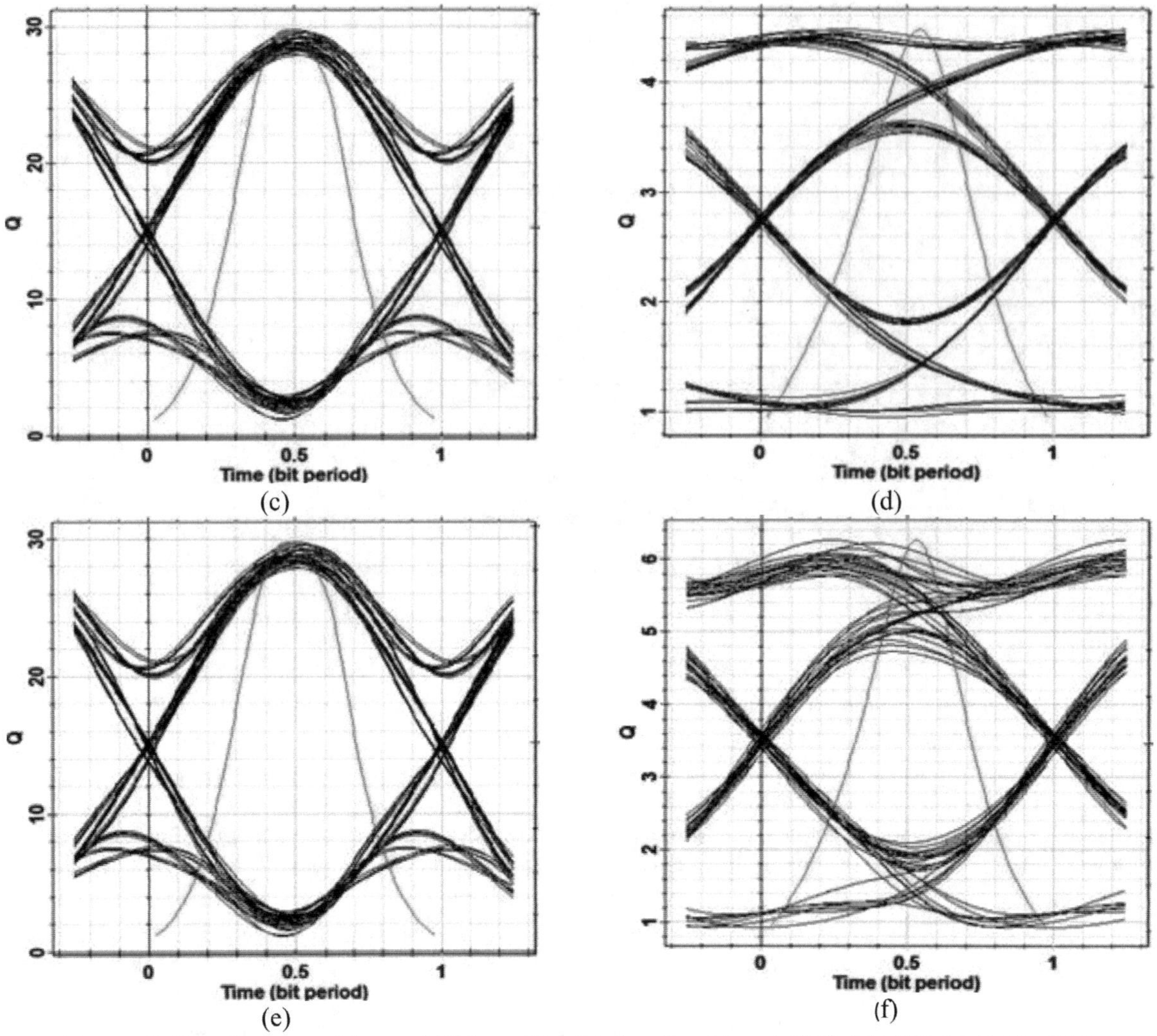

Figure 7. Eye diagram at1 km and 75 km of (a) (b) lineally chirped FBG (b) DCF (c) DCF+FBG

4. Conclusion

This work is accomplished though simulation design of a cost effective and easy maintenance based wavelength reused WDM passive optical network with dispersion compensation in ONU. Inter-channel and intra-channel interferences are suppressed by using polarization interleaving and DQPSK respectively. Pulse width reduction efficiency and performance of three different modules is analyzed such as DCF alone, linearly chirped FBG alone and DCF-FBG jointly. It is observed that due to maximum (48%) PWR% of joint module of DCF+FBG, it is preferred over DCF (45% PWR%) and FBG (40% PWR%). Moreover, proposed combined module is cost effective due to use of lesser length of DCF fiber in the joint module then required in DCF alone system. For the optical fiber of distance 40 km, DCF is needed 8 km for maximum dispersion compensation in systems using DCF alone, however, for optimal PWR in joint module, only 6 km DCF is required. Thus 25% of cost of DCF is saved in the ONU of the proposed system. Q factor and Log BER are also observed for all the three cases of dispersion compensation and results revealed that DCF+FBG perform best. Comparison of system for polarization interleaving and without PI is also investigated and it is perceived that WDM PON system with PI performs superior to normal passive optical networks. Joint module of DCF+FBG attains significant PWR in ONU of the WDM-PON with less maintenance and moderate cost.

Journal of Physics: Conference Series **2327** (2022) 012045 doi:10.1088/1742-6596/2327/1/012045

Acknowledgement

I would like to express my very great appreciation to **Manoj Sindhwani and Javed Dhillon** (School of Electronics and Electrical Engineering, Lovely Professional University, Phagwara, Punjab, India.) for their valuable and constructive suggestions during the planning and development of this research work. Their willingness to give their time so generously have been very much appreciated.

References

[1] Martínez E F A, Leguizamón G A P and Fajard C A S, 2016 Towards a New Generation of Passive Optical Networks **21**(1) pp 49-62.

[2] Goyal R, Kaler R S and Kamal T S, 2017 A cost effective bidirectional hybrid passive optical network using common carrier for all optical network units (CCAONUs) technique with mitigation of non linearity impact *Optik* **130** pp. 644-649.

[3] Ansari N and Zhang J, 2013 Media Access Control and Resource Allocation: For Next Generation Passive Optical Networks *Springer Briefs, in Applied Sciences and Technology*, **chapter 2** pp 11-22.

[4] Elmagzoub M A, Mohammad A B, Shaddad R Q and Al-Gailani S A 2014 Physical layer performance analysis of hybrid and stacked TDM–WDM 40G-PON for next generation PON *Optik - International Journal for Light and Electron Optics 125*(20) pp 6194-6197.

[5] Kaur A, Kaur B and Singh K, 2017 Design and performance analysis of bidirectional TWDM-PON employing QAM-OFDM for downstream and re-modulation for upstream *Optik* **134** pp 287-294.

[6] Xue X, Ji W, Huang K, Li X and Zhang S, 2018 Tunable Multi-wavelength Optical Comb Enabled WDM-OFDM-PON With Source-Free ONUs *IEEE Photonics Journal* **10**(3).

[7] Chen X and Yao J, 2016 Wavelength Reuse in a Symmetrical Radio Over WDM-PON Based on Polarization Multiplexing and Coherent Detection *Journal of Lightwave Technology* **34**(4) pp 1150-1157.

[8] Xu M et al., 2015 Wavelength Sharing and Reuse in Dual-Band WDM-PON Systems Employing WRC-FPLDs *IEEE Photonics Technology Letters* **27**(17) pp 1821-1824.

[9] Guo Q and Tran A V, 2013 Demonstration of a 40 Gb/s wavelength-reused WDM-PON using coding and equalization *IEEE/OSA Journal of Optical Communications and Networking* **5**(10) pp A119-A126.

[10] Kim B, 2008 RSOA-Based Wavelength-Reuse Gigabit WDM-PON *Journal of the Optical Society of Korea* **12**(4) pp 337-345.

[11] Cui W, Shao T and Yao J, 2014 Wavelength Reuse in a UWB Over WDM-PON Based on Injection Locking of a Fabry–Pérot Laser Diode and Polarization Multiplexing *Journal of Lightwave Technology* **32**(2) pp 220-227.

[12] Choudhury P K and Khan T Z, 2016 Symmetric 10 Gb/s wavelength reused bidirectional RSOA based WDM-PON with DPSK modulated downstream and OFDM modulated upstream signals *Optik* **372** pp. 180-184.

[13] He J, Yang D and Chen L, 2011 A full-duplex radio-over-fiber system with differential phase-shift keying signals *Journal of Physics Conference Series* **276**(1) p 012062.

[14] Wei C, 2017 High-Capacity Carrierless Amplitude and Phase Modulation for WDM Long-Reach PON Featuring High Loss Budget *Journal of Lightwave Technology* **35**(4) pp 1075-1082.

[15] Shao T et al., 2012 Convergence of 60 GHz Radio Over Fiber and WDM-PON Using Parallel Phase Modulation With a Single Mach–Zehnder Modulator *Journal of Lightwave Technology* **30**(17) pp 2824-2831.

[16] Düendorfs V, Spolitis S and Bobrovs V, 2017 Comparison of dispersion compensation methods for 40Gbit/s WDM-PON transmission systems *Progress In Electromagnetics Research Symposium - Spring (PIERS), Russia* pp. 2431-2436.
[17] Düendorfs V, Spolitis S and Lvanovs G, 2012 Comparison of chromatic dispersion compensation techniques for WDM-PON solution *2nd Baltic Congress on Future Internet Communications, Lithuania* pp. 64-67.
[18] Aleksejeva M and Spolitis S, 2017 Performance investigation of dispersion compensation methods for WDM-PON transmission systems *Progress In Electromagnetics Research Symposium - Spring (PIERS), Russia* pp. 2437-2442.
[19] Mohammed N A, Solaiman M and Aly M H, 2014 Design and performance evaluation of a dispersion compensation unit using several chirping functions in a tanh apodized FBG and comparison with dispersion compensation fiber *Applied optics* **53**(29) pp 239-247.

Bandwidth Enhancement Technique for Microstrip Antenna for Higher X-Band and Ku-Band for Next Generation Wi-Fi and Wi-max Applications

[1]Praveen Tiwari, [2]Praveen Kumar Malik and [3]Amandeep Kaur

[1]Research Scholar, SEEE, Lovely Professional University, Punjab, India
[2]Professor, SEEE, Lovely Professional University, Punjab, India
[3]Assistant Professor, SEEE, Lovely Professional University, Punjab, India

[1]pravti@gmail.com, [2]pkmalikmeerut@gmail.com, [3]aman.dhaliwal18@gmail.com
Corresponding author's e-mail : pravti@gmail.com

Abstract. This paper presents the novel design and fabrication of a square shaped array antenna for transmission in the higher X band and Ku-band. The influence of the constituent square's size factor over the signal quality with the squares positioned in various array configurations is recorded and reported using a discretized design approach. The design was fabricated and tested to meet the needed standards after being optimized using the antenna simulator software. A novel DC shorting approach has been devised and reported to improve antenna performance. Major antenna performance parameters are being observed and found in accord with the literature.

Keywords: Bandwidth Enhancement, Microstrip Antenna, Wi-Fi, and Wi-max Applications

1. Introduction

In the last few decades, with exponential growth in wireless communication technology and Internet services, the demand for high data rate services has increased. Owing to this, the number of users rose tremendously and it can be widely seen that in the future, huge traffic congestion will be experienced in communication networks. Good infrastructure is the most difficult challenge for manufacturers and service providers to provide more capacity in networks in order to achieve efficient communication services. compactness of devices is another big issue. There are some serious challenges faced in wireless communication services, like multipath fading, co-channel interference, and delay spread, which degrade the signal quality. With the explosive growth of wireless communication applications, new design specifications for integrated devices pave the way for new design specifications for integrated devices that demand greater compactness, low profile, and low cost.The antenna is important in wireless signal transmission because it converts electrical signals into electromagnetic waves and acts as a transducer on both the transmitter and receiver sides.To reduce overall circuit dimensions for RF components, antenna miniaturisation is becoming essential to obtain an optimised design for handheld wireless communication gadgets. To accomplish this, PCB technology-based microstrip patch antennas are becoming a buzz word, gaining research attention for their more compact size with good gain and bandwidth characteristics. Moreover, wireless devices like mobile phones operate on different technologies like the ISM band for Wi-Fi, Bluetooth BLE and Wi-MAX, GSM, CDMA, etc. Conventional antennas mainly operate on a single band of frequency, but now there is a need to design multiple band antennas which can resonate on different frequency bands and omit the need for multiple antennas in one device. To achieve high data transmission rates, antennas must meet the minimum band requirements set by the FCC for every wireless standard. Due to the inherent advantages of large channel capacity and high data rate communication, demand for the future technology for fifth-generation (5G) wireless communication has been rapidly expanding [1]. An antenna with a wide bandwidth characteristic is required to support these capabilities. There are numerous wideband or ultra-wideband antennas [2-6], but their construction is time-consuming. Because of its small size, light weight, low cost, and ease of fabrication and integration, the microstrip patch antenna is one of the most fascinating structures that is always chosen for use [7, 8]. The main disadvantage is the limited impedance bandwidth, which is often less than 5% [9]. As a result, many papers have been published on how to increase the bandwidth of a microstrip patch antenna by using different feeding strategies or material composites. The feeding mechanism is only addressed by a technique known as proximity coupled-feed while the patch structure is still being fixed. The bandwidth can be on the order of 8%, but the antenna structure must be thicker than a normal one [10].

2. Experiment and method of design

Antenna design is a phenomenal and iterative process that entails multiple steps, including antenna software design or simulations, fabrication, and testing. This section discusses the methods and tools used to design an antenna, analyse it, fabricate it, and test it. While the process of designing a simple antenna, such as a dipole, is straightforward and the current distribution in the antenna structure can be calculated and measured with reasonable accuracy, this makes it simple for designers to calculate antenna performance parameters such as gain, bandwidth, VSWR, and radiation patterns, among others. However, for complex antenna structures, the current densities are difficult to calculate, necessitating the use of simulation tools to analyse antenna performance. Numerous commercial and open source tools for antenna design and analysis are available. FDTD, Methods of Moments (MoM), and the finite element method are the most frequently used techniques (FEM). The design of the proposed 4x4 Array X band & Ku band planar microstrip patch antenna is shown in figure 1. The proposed antenna is resonating at 12.1GHz to 13.05GHz with total bandwidth of operation is 950MHz. It consists of patch dimension of 7.11 x 7.11 mm, patch to patch dimension (C to C) is 16.2mm or 0.68 λ, Substrate PCB dimension (LxWxh) is 65x65x0.8mm, Transmission feed line is 0.44mm, Impedance transformer is 0.6 X 4 mm and Height of dielectric (h) is 0.8mm as shown in figure 1. The corporate feeding network is used new in this design. We are using PTFE Teflon woven with d_K=2.2 and loss tangent =0.009 because this substrate PCB material has low dissipation loss hence gives good performance and Gain.

To make it practically, we have created a geometric model by selecting the modeller type. Following that, an appropriate dielectric material such as FR4 Epoxy or Rogers RT Duroid 5880 with the desired thickness and dielectric constant is assigned to the designed antenna structure. The following steps provide assignment of sources (port line lumped and wave port) and boundary conditions such as perfect E to patch, ground, and Radiations to Radiation box. After defining the boundary conditions for all sheets and solids in the HFSS simulator, a port, either wave or lump, is required to excite the antenna structure. Following structure modelling and validation, the structure solution is configured. Following that, the solution frequency is assigned and a frequency sweep is added to generate the solution frequency over the specified frequency range. Far field setup is included to calculate parameters such as antenna gain and radiation patterns in the far field. After analysing the design, antenna performance parameters such as S11, VSWR, gain, 2D-3D radiation patterns, and directivity can be calculated graphically, via a table, or via a smith chart.

2.1 Antenna Design steps:

1. Establish the ground plane
2. After calculating with mathematical expressions, create the substrate and assign dimensions.
3. Assign an insulating material
4. Create the patch and assign dimensions to it in order to capture the results of mathematical calculations.
5. Construct a feed line with the appropriate dimensions.
6. Connect the structure together, i.e. patch with feed-line
7. Assign patch, ground plane perfect E boundary conditions
8. Assign an antenna structure a port (Lumped or Wave) to couple electromagnetic energy
9. Create a radiation box and associate it with a radiation boundary.

3. Literature review

Young-Bae Jung et al. [11] describes triband antenna for Ka-K band. Author uses a Cassegrain reflector with integrated feeder for the k band to design modified hybrid antenna (HA). A microstrip antenna is used that consist of four linear sub-arrays and placed on the sub-reflector of the HA for Ku band services to provide high gain and to achieve rapid 2-D scanning ability at cheap cost for satellite applications. So, antenna is designed using a novel beam- steering through sub-reflector i.e. rotational and flat. A hexagonal structure is designed that uses feeder consist of 20 dual band horn elements those provide dual polarization also to maximise the efficiency. To reduce size, transmitter and receivers are connected with feeder horn elements. Antenna was tested for both indoor and outdoor environment and provides an EIRP of 55.2dBW and GT of 16.8dB for k band and 7.6db for ku band. As microstrip patch antenna has narrow bandwidth that is serious drawback of microstrip patch antenna. Author Yikai Chen et al. [12] uses concept of distributed L and C circuit to enhance bandwidth of E-shaped conventional patch antenna. In conventional E shape antenna, low inductance occurs due to probe fed, so LC circuit is used to reduce the thickness of air acting as substrate. A new resonant frequency introduced by LC circuit, near to that of E-shape increases the bandwidth of antenna without compensating the gain. Designed antenna operates on AMPS band from 824 to 894GHz frequency band. Simulation results shows that designed antenna has an impedance over 9% for Voltage Standing Wave Ratio (VSWR) over the frequency band is less than 2 but shows satisfactory performance in radiation pattern within the operated bandwidth. Hussein Attiam et al. [13] discusses that microstrip antenna gain can be enhanced under some resonance conditions when antenna is covered with superstrate with proper distance in free space. Resonance conditions can be deduced by using proper transmission method and the cavity model to achieve the highest gain. Resonance effect can be changed by adjusting the spacing between antenna's superstrate and substrate and by varying thickness of the substrate. Depending upon these resonant lengths the permittivity, permeability and superstrate are determined and characteristics impedance can also be determined of the multilayer structure. Antenna performance is verified using simulation results and analytical methods also. From simulation results author concludes that proposed method enhances the antenna performance by 50% when compared to previous methods. In research article [14] Ramona Cosmina Hadarig et al. proposes use of electromagnetic band-gap (EBG) structures and artificial magnetic conductors (AMC) due to their advantages use to improve antenna performance like better efficiency, high gain, and low back lobe and less side lobe levels. He designed antenna using combination of EBG with patch antenna in single layer and combination of AMC with patch antenna in two different layers. Due to more compactness and robustness antenna is compatible with plane antenna fabrication technology because it not needs via holes. Proposed antenna works in RFID 2.48GHz frequency band, also it shows better radiation properties without increasing antenna size and thickness. Proposed antenna provided bandwidth of 34MHz with EGB and of 46MHz with AMC over 23MHz for simple antenna, as well as gain improved from 4.6bB to 5.576 dB over the operating frequency band. Feeding methods play important role as antenna efficiency also depends upon how power is transferred to the radiating element directly or indirectly. In research article [15] Soumyojit Sinha et al. describe advantages of microstrip antenna as low cost, low profile and conformal antenna. Different methods are used for antenna excitation like by coaxial probe and by microstrip line directly. Various indirect methods for excitation are also used like aperture coupling and electromagnetic coupling as there is not direct metallic contact between the patch and the feed line. Input impedance of antenna totally depends on type of feeding method used. So, author designed a rectangular patch antenna that works at 2.21 GHz frequency band using 1E3D simulator. Antenna feeding is done using probe. From simulation results antenna performs well in terms of radiation for wireless applications like mobile handheld radios, global positioning systems (GPS), radar for missiles. To improve antenna bandwidth Y. Sung in [16] proposes printed wide slot line fed compact microstrip patch antenna with parasitic centre patch to further increase the bandwidth of conventional patch antenna. A 50ohm microstrip feed line is used to excite slot and rotated square slot resonator is considered as basic patch design. Resonator shows two frequencies i.e.f1 low resonant and f2 high resonant frequency. From simulation results author investigates that low resonance frequency deceases and high resonant frequency increases by placing parasitic patch in centre of rotated square slot. Simulation results show improvement in bandwidth more than 1GHz. Also proposed structure provides impedance bandwidth of 80% for 2.23-5.35 GHz bandwidth. To improve microstrip antenna bandwidth a pair of parasitic patches is used by S. T. Fan et al. in [17]. Antenna dimensions taken are 37mm by 37mm, and fabricated using FR4 material with thickness of 1.6mm, relative permittivity of 4.4. To improve band, microstrip feed with 16 mm length and width 2.5 mm is used with a pair of semi-circular parasitic patches on both sides of it to provide additional resonance in the circuit. To provide strong coupling with feed line, gap of 0.5mm is maintained between patch and feed line. Antenna design is simple and easy in fabrication. Simulation results shows that bandwidth improvement is 136% for VSWR less than 2 for 2.1 to 11.1 GHz bandwidth. From simulation results the proposed antennas can be used for ultra-wideband modern wireless Communication applications. P. A. Ambresh et al. [18] proposed a new micro-strip patch antenna with reduces size using FR4 dielectric substrate with slots for wireless applications. Proposed antenna works at two resonant frequencies 3.55 GHz and 4.99 GHz. From simulation results antenna shows 18.2% compactness for 270MHZ impedance bandwidth. Antenna shows very good performance in terms of return loss (RL) upto -36.14dB with the VSWR below 2. Author concludes from results that antenna show better performance with linear polarization and also broadside radiation pattern is achieved with less

cross polarization. The suggested antenna has a simple shape and is compact in size, measuring 45 35 0.79 mm3. [19].
A novel microstrip patch antenna array with a double meander dipole construction with good gain and wide bandwidth is proposed. The twin meander dipole array construction was able to produce two in-phase resonant frequencies in the Ku-band (12–18 GHz), resulting in increased impedance bandwidth without the need for additional design sections. Furthermore, by using a double-layer substrate approach, the array's gain was increased by 2 dBi over the full working frequency range. The E33 type LPKF prototyping PCB machine was used to make the suggested antenna. [20]. According to test results, the antenna can cover 1.49 to 1.60GHz, 1.87 to 2.51GHz, and 4.63 to 5.34GHz, and generate three primary frequencies: 1.57, 2.15, and 5.06GHz. [21].
The term "multiple-input multiple-output" (MIMO) refers to a system that uses more than one antenna to broadcast and receive data packets. It increases channel capacity without adding more power. Due of the limited space available in modern compact devices, MIMO antenna elements must be arranged near together. An unfavourable coupling occurs between the closely spaced antennas, causing the antenna properties to deteriorate. An ultra wide-band (UWB) MIMO antenna system with increased isolation is given in this research. The system operates over a frequency range of 2–13.7 GHz. The antenna elements are close together, with a 3 mm gap between them. The mutual coupling between the antennas is decreased by using slotted stub in addition to the system's UWB feature. The isolation has been enhanced and is now less than 20 dB across the whole operating range. The antenna's primary performance metrics are unaffected by the addition of the decoupling network. The system is built on a FR-4 substrate, which is affordable and widely available. The corresponding circuit model is also offered to help understand how the suggested system works. Different radiating modes and inter-mode interaction are studied and calculated to effectively depict the proposed system. [22].
This article shows how to make a highly compact ultra-wideband multiple-input multiple-output (UWB MIMO) bi-port antenna with good isolation using FR4. The antenna is merely 18 28 1.6 mm3 in length. The MIMO antenna is made up of two identical antenna elements that are symmetrically arranged in opposite directions on the same dielectric substrate. High isolation and good impedance matching are accomplished by plugging three crossed stubs between two disconnected ground planes. This UWB MIMO antenna's working frequency band is 1.9–14 GHz, and the isolation is preserved at 20.2 dB throughout the analysis frequency band [23].

4. Antenna Design

The proposed antenna is resonating at 12.1GHz to 13.05GHz with total bandwidth of operation is 950MHz. It consists of patch dimension of 7.11 x 7.11 mm, patch to patch dimension (C to C) is 16.2mm or 0.68 λ, Substrate PCB dimension (LxWxh) is 65x65x0.8mm, Transmission feed line is 0.44mm, Impedance transformer is 0.6 X 4 mm and Height of dielectric (h) is 0.8mm as shown in figure 1.

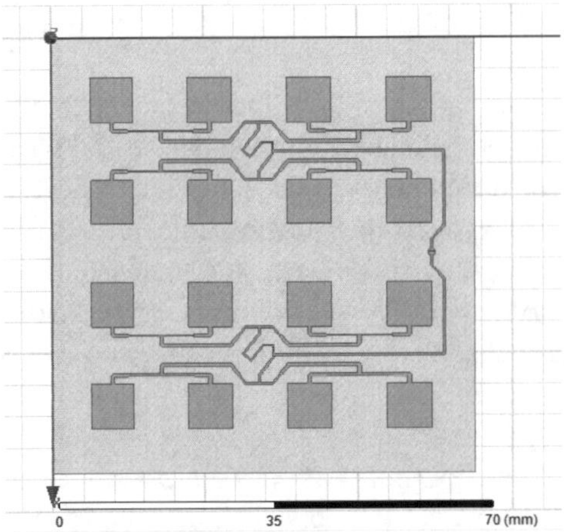

Figure 1: Dimensions of the proposed 4x4 antenna array (65mm X 65mm)

Proposed design can be easily interfaced to the Wi-Fi & Wi-Max devices by using suitable SMA or MMCX interconnections depends on the board input type as shown in figure 2.

Figure 2: Application of the proposed antenna array

5. Results and discussions

The proposed broadband antenna idea has been constructed and evaluated using virtual network analyzer (VNA) and high frequency simulator structure (HFSS) software to achieve the desired performance. For the simulation result, the moment-based simulator HFSS is employed. The details of the planned antenna characteristic are explained in the next section. Proposed antenna is fabricated using the PCB manufacturing technique and tested using the Vector network analyzer and anechoic chamber as well. Figure 3 shows the fabricated antenna using the PCB technology.

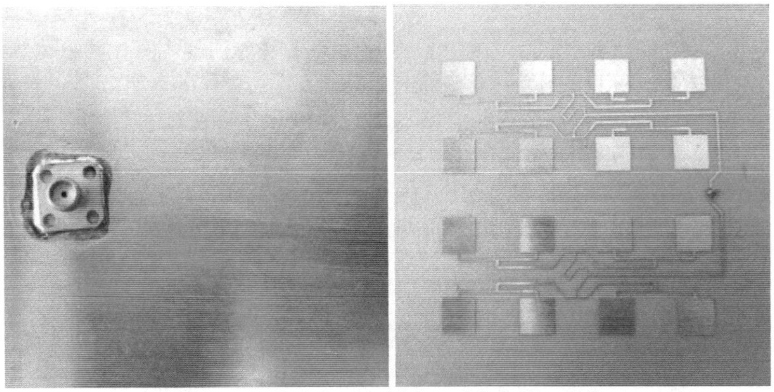

Figure 3: Front and back side of fabricated antenna

Figure 4a and 4b shows the testing of the proposed antenna for return loss.

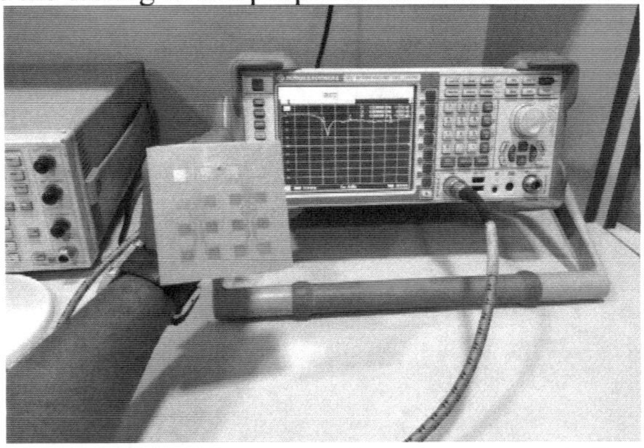

Figure 4a: Testing of the antenna for measurement of return loss

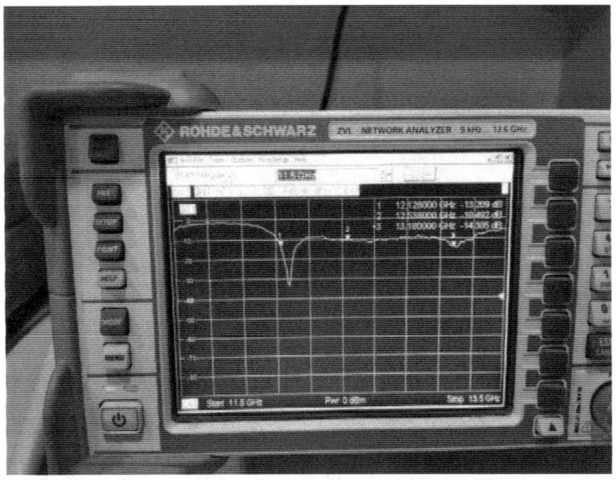

Figure 4b: Testing of the antenna and result for return loss

Typical values of reflection coefficient are in between $0 < \Gamma < 1$. Ideally Z_{in} should be equal to Z_0 such that there is no impedance mismatch and maximum power can be transmitted from the antenna i.e. $Z_0 = Z_{in}$. Magnitude of S_{11} is known as return loss. It can be defined as

$R_{Li/p} = 20\log_{10}|S_{11}|dB$. The return loss of an antenna is used to determine whether it is single band or multi band. It can also tell you what the antenna's bandwidth is. In most cases, we compute the bandwidth below -10dB. It is clearly evident form the figure 4b and figure 5 that value of the return loss is less than -10dB from the range of 12.01GHz to 13.05 GHz in the simulated and practically measure as well. This clearly indicates that over this range of the frequency antenna will radiate efficiently and minimum signal will be return back to the antenna.

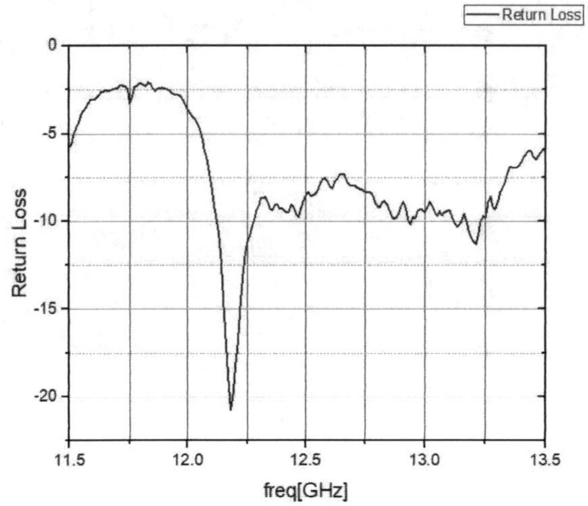

Figure 5: Measurement of simulated return loss value using software

VSWR is also directly correlated with the coefficient of reflection (S_{11}) and it is defined as $VSWR = \frac{1+|\Gamma|}{1-|\Gamma|}$. Owing to the back and forth traversal of the signal/current within the wiring, a static wave is formed during the transmission of the signal from the antenna. The ratio between the maximum e of a standing wave measured along the transmission line can be used to compute VSWR. i..e $VSWR = \frac{V_{max}}{V_{min}}$. The typical value of the standing wave ratio for better signal efficiency and transmission should be $1 < VSWR < \alpha$.

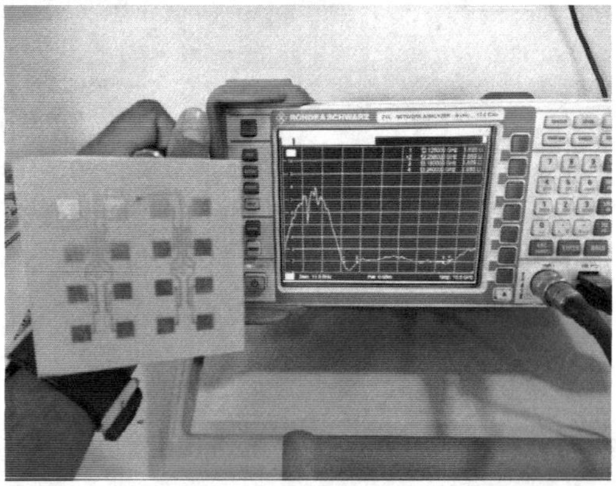

Figure 6: Testing of the antenna for measurement of voltage standing wave ratio

Journal of Physics: Conference Series **2327** (2022) 012046 doi:10.1088/1742-6596/2327/1/012046

From the figure 6 (Testing of the antenna) and figure 7 (Simulated value of the antenna) it is clear depicts that the there is an accord of the value of the standing wave ratio between the measured and simulated one. Practically and during the simulated it has been observed that the value of the VSWR is in the range of 2 only within the frequency range of 12.01GHz to 13.05 GHz.

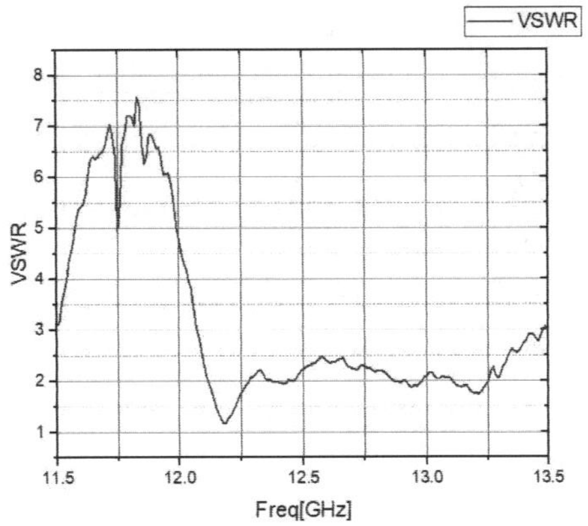

Figure 7: Measurement of simulated standing wave ratio using software

The detectors are not 100% efficient because radiation can travel long distances between contacts in the detector material before detection is feasible. The effectiveness of a detector is one of the most essential features of a detector in radiation measurement. Gamma spectrometry is one of the most extensively used detector systems in this industry, and its performance is directly related to the detection efficiency knowledge. The detection efficiency is the percentage of radiation detected by a given detector out of the total yield emitted by the source. Table 1 shows the designed antenna electrical parameters.

Table 1: Designed antenna electrical parameters

Sl. No	Parameters	Obtained Values	Required value
1	Frequency	12.1-13.05GHz	More than 10GHz
2	Bandwidth	950MHz	500MHz
3	Gain	18dBi	10dBi
4	VSWR	< 2	< 2
5	HPBW	18 Deg. (A & E plan)	----

Multiple additional factors can influence the signal, including but not limited to the dimensions of the detector , the gap between the source and detector, their locations, and their quality of cross-sectional material absorption .

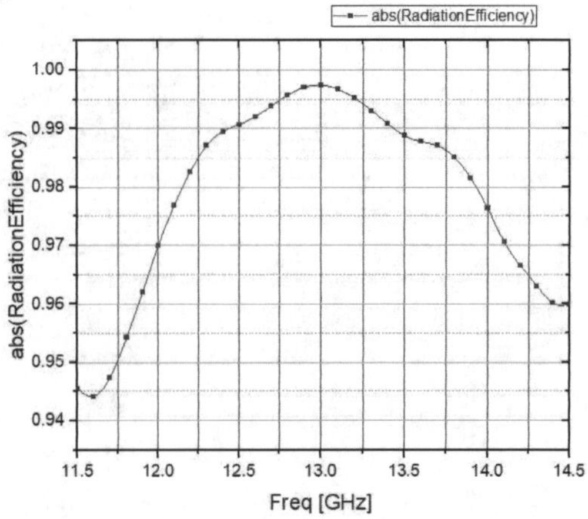

Figure 8: Measurement of absolute radiation efficiency

Figure 8 shows that measurement of the absolute radiation efficiency, and it is also found that it is in accord with the concept of antenna theory within the frequency range of 12.01GHz to 13.05 GHz.

It might be difficult to accurately measure antenna designs and gain. It necessitates not just costly test equipment, but also highly experienced and qualified personnel. A third issue stems from the test equipment itself: it is impossible to eliminate interactions between test equipment, test fixtures, and/or equipment holders. Most microwave component testing is significantly different from antenna pattern and gain measurements. This is due to the fact that it must be performed in an open space setting with minimal reflections. Far Field Range, Near Field Range, and Compact Range are the three types of frequently used antenna pattern and gain measurement techniques. Regardless of frequency band, these devices are inherently unwieldy and costly. The cost becomes expensive when a Vector Network Analyzer (VNA) or Synthesizer and Spectrum Analyzer are included.

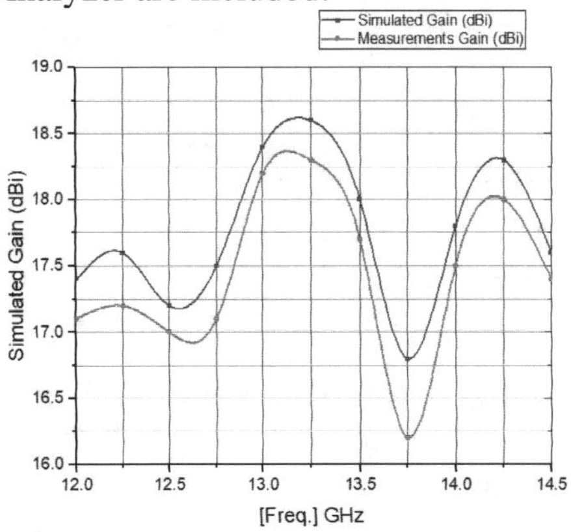

Figure 9: Measurement of the simulated and measured gain

Figure 9 shows the plot of the comparison of the simulated and measured gain. From the graph it is evident that due to the calculated geometry of the proposed antenna gain of the antenna is very good in the desired range of the frequency. In our design intuitively we are targeting the frequency range of 12.01GHz to 13.05 GHz, and it has been observed that gain is more than almost 15dBi between the specified ranges of the frequency.

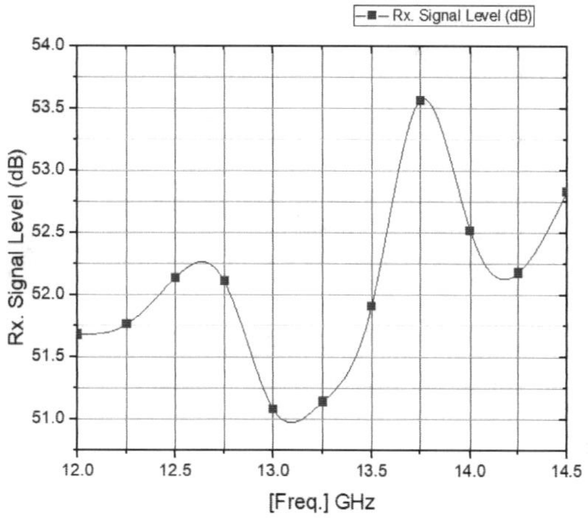

Figure 10: Measurement of received signal level

the magnitude of intensity of the signal that arrives at the reciever's en post traversal through the medium is called recieved signal strength (RSS). It is a dependent variable influenced by The magnitude of the transmitted wave, the size of the traversed medium, and the radio environment. The received signal intensity is a difficult arbitrary measurement that is heavily connected with the transmitter's position. RSS can be used to distinguish the attacker and the victim if they are separated by a reasonable distance to identify spoofing attempts. Several access points (APs) in fixed locations will serve as RSS sensors, measuring RSS readings of wireless traffic emitted from the wireless nodes, in the RSS-based spoofing attack detection approach. Figure 10 depicts the value of received signal with respect to the frequency and it can be observed that the received signal is also in the agreement of the conceptual value. Proposed antenna is also compared with a reference antenna [28], and table 2 shows the obtained value of the antenna performance parameters as compared to the reference literature.

Table 2: Designed antenna electrical parameters vs reference literature.

Sl. No	Parameters	Designed antenna	Reference literature
1	Frequency	12.1-13.05GHz	14GHz
2	Bandwidth	950MHz	536MHz
3	Gain	18dBi	17dBi
4	VSWR	< 2	< 2
5	HPBW	18 Deg. (A & E plan)	18 Deg. (A & E plan)

Satellite whereabouts are discerned using variables called the the azimuth and elevation . Azimuth is the angular direction pointing to the satellite location from observer. It is an angular measurement in a spherical coordinate system ,specifically the angle between the projected and a reference vector on the reference plane measured in degrees from 0 to 360. elevation is the size of gap between observer and the satellite. On a scale of one to ten, both are rated. It all starts with North at 0 degrees. As we move to the right (clockwise), you'll face East (90°), then South (180°), then West (270°), and finally North (360° and also 0°). So, if your satellite's Azimuth is 45 degrees, it means it's to the northeast of you. Elevation is sometimes expressed as a number of degrees. At 0° Elevation, a satellite just rising over your horizon, and at 90° Elevation, a satellite straight overhead.. The vector from a point of interest to an observer (origin) is projected perpendicularly onto a reference plane. When used as a celestial coordinate, the azimuth is the horizontal direction of a star or other astronomical object in the sky. The star is the object of interest, and the reference plane is the area around an observer on Earth's surface, using true north as the reference vector. On the horizontal plane, the azimuth is the angle formed between the north vector and the star's vector. Azimuth is usually measured in degrees (°). Figure 11 and 12 shows the graphical representation of azimuth and elevation plane radiation of the proposed antenna, and it can be evident from the figure that it is more directional. Figure also depicts that there is a good accord between the measured and tested values of the pattern. Yellow color graph shows the simulated or measure value and grey color graphs show the tested value of the radiation pattern at different interval of the angle.

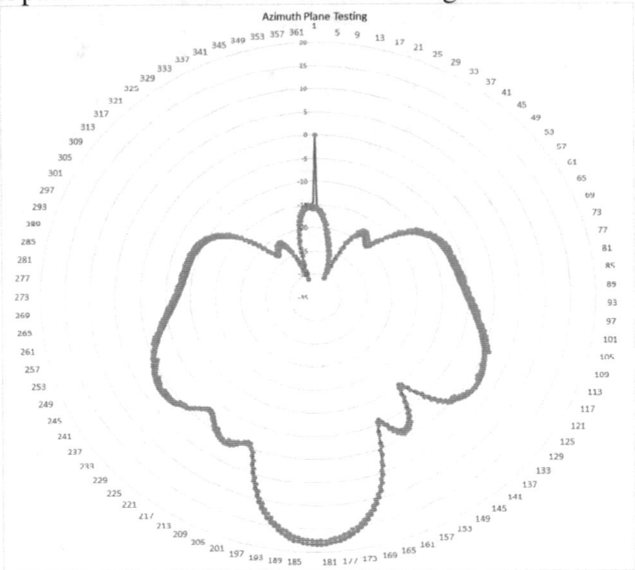

Figure 11: Illustration of azimuthal plane radiation (Measured and tested)

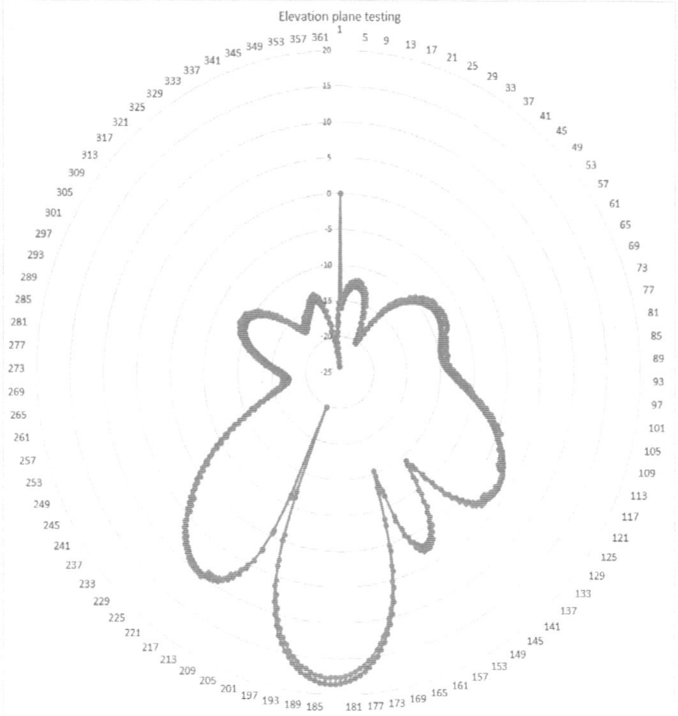

Figure 12: Illustration of elevation plane radiation (Measured and tested)

Novelty of the design can be described in terms of its frequency, bandwidth and gain. Proposed design is having operating frequency range of 12.1 to 13 GHz with 950 MHz bandwidth. All other antenna performance parameters are also in accord with the theory and practical comparison. Gain of the proposed design is 18dBi with VSWR less than 2. the comparative analysis between the proposed design and prior artwork has been presented in table 3.

Table 3: Comparison of the proposed work with recent work

Ref.	Substrate Used	Operating Frequency Band (GHz)	Gain (dB)	VSWR	Return Loss	Impedance Bandwidth
19	Rogers RT 5880	11.85–15.94	10.2dBi	1 to 2	-40.32dB	23.019GHz
20	RT Duroid 5880 LZ	12–18	7.9 dBi	NA	below −10 dB	5.2 GHz
22	FR-4	2–13.7	1.1–4.3 dB	NA	>20 dB	Large bandwidth
23	FR-4	1.9–14	0.4-4.8 dBi.	NA	>20.2 dB	NA
Present work	RT/Duroid 5880	12.5- 13.05	18 dBi	VSWR<2	S11<-10 dB	950 MHz

6. Conclusion and future work:

Article presented a 4x4 arrays, X-band and Ku-band planar microstrip antenna design with matching almost all the antenna performance parameters. 4x4 array architecture is designed to enhance the gain of the antenna keeping other performance parameters in accord. Designed antenna is operating efficiently in the higher X-band and Ku-band with desired gain, absolute efficiency, and received signal as well. Total operating bandwidth of the antenna is around 1GHZ, and proposed antenna can be used for the applications of Wi-Fi and/or Wi-max of next generation. With reference to the simulation results, it can be concluded that the suggested design is suitable for Wi-Fi and/or Wi-max, particularly in the frequency range of 12GHz to 13.05GHz, and has appropriate bandwidth. Between these frequencies, the antenna's performance parameters are all within the intended range. As a result, it can be stated that the proposed antenna would have good radiation properties if employed in a practical communication application. Designing of the antenna above the frequency range 2.4 GHz is more toil and tedious as far as design, fabrication and testing is concern. In near future we will try to design antenna with mMIMO or more complex array design to acquire more fruitful results in terms of antenna parameters.

Declarations:

- **Availability of data and material:** All the data concerned to this research is available with the author. Data will be required as and when required.
- **Competing interests:** There is no any competing interest between the authors. First author is sole responsible for the publication of the article.
- **Funding:** The research work is not funded by any organization.
- **Authors' contributions:** The first author writes the article after completing his research work and the second author does formatting of the article.
- **Acknowledgement**: The author expresses his gratitude to his supervisor for his technical and moral assistance. Also, a big thank you to Lovely Professional University for helping me write this paper. The author would also want to express gratitude to his friends for encouraging him to write the review article.

References:

[1] Zheng L. et al., 2003, Diversity and Multiplexing: a Fundamental Tradeoff in Multiple-antenna Channels, *IEEE Transactions on Information Theory*, vol. 49, no. 5, pp. 1073-1096.
[2] Wen S. et al., 2020 A Low-profile Wideband Antenna with Monopolelike Radiation Characteristics for 4G/5G Indoor Micro Base Station Application, *IEEE Antennas and Wireless Propagation Letters*, vol. 19, no. 12, pp. 2305-2309.
[3] Kuroptev P. D. et al.,2017, Modified 0.6-50 GHz Ultra-wideband Double-ridged Horn Antenna Design for Parameters Improvement, in *47th European Microwave Conference (EuMC)*, pp. 1-4.
[4] Limpiti T. et al., 2016, Design of a Printed Log-periodic Dipole Antenna (LPDA) for 0.8-2.5 GHz Band Applications, in *13th International Conference on Electrical*

Engineering/electronics, Computer, Telecommunications and Information Technology (ECTI-CON), pp. 1-4.
[5] Tang X. et al., 2016, Design of a Wideband Circularly Polarized Strip-helical Antenna with a Parasitic Patch, *IEEE Access,* vol. 4, pp. 7728-7735.
[6] Yousefi T. et al., 2016, A Wideband Multimode Permeable Conformal Antenna Thinner than l/75 Using Advanced Ferromagnetic Laminate Composite Materials, *IEEE Antennas and Wireless Propagation Letters*, vol. 15, pp. 1931-1934.
[7] B. J. Niu, et al., "Bandwidth-enhanced Four-antenna MIMO System Based on SIW Cavity," Electronic Letters, vol. 56, no. 13, pp. 643-645, 2020.
[8] Huang H., et al., 2020, A Low-profile Single-ended and Dual-polarized Patch Antenna for 5G Application, *IEEE Transactions on Antennas and Propagation,* vol. 68, no. 5, pp. 4048-4053.
[9] Lee K. F., et al., 2011, Microstrip Patch Antennas. London: *Imperial College Press.*
[10] Rathi V., et al., 1996, Improved Coupling for Aperture Coupled Microstrip Antennas, *IEEE Transactions on Antennas and Propagation*, vol. 44, no. 8, pp. 1196-1198.
[11] Jung Y B, Eom S Y, and Jeon S I, 2010, Novel Antenna System Design for Sat.llite Mobile Multimedia Service, *IEEE transactions on Vehicular technology*, Vol-59, pp(s): 4237-4247, Issue-9.
[12] Chen Y, Yang S, and Nie Z, 2010, Bandwidth Enhancement Method for Low Profile E-Shaped Microstrip Patch Antennas, *IEEE Transactions on Antennas and Propagation*, Vol.58, Issue- 7, pp(s):2442- 2247.
[13] Attia H, Yousefi L, Ramahi O,2011, High-Gain Patch Antennas Loaded with High Characteristic Impedance Superstrates, *IEEE Letters on Antenna and Wireless propagation*, Vol.10, pp(s): 858-861.
[14] Hadarig R C, et al., 2011, On the Bandwidth Enhancement of Patch Antenna Using EBG/AMC Structures, *IEEE 6th European Conference on Antennas and Propagation (EUCAP)*, pp(s): 2853-2857.
[15] Sinha S, Begam A, 2012, Design of Probe Feed Micro-strip Patch Antenna in S-Band, *International Journal of Electronics and Communication Engineering*, Vol.5, Issue-4, pp(s): 417-423.
[16] Sung Y., 2012, Bandwidth Enhancement of a Microstrip Line-Fed Printed Wide-Slot Antenna With a Parasitic Center Patch, *IEEE Transactions on Antennas and Propagation*, Vol.60, Issue-4, pp(s): 1712-1716.
[17] Fan S. T., Yin Y. Z., Lee B., Hu W., Yang X., 2012, Bandwidth Enhancement of a Printed Slot Antenna With a Pair of Parasitic Patches, *IEEE letters on Antennas and Wireless Propagation,* Vol.11, pp(s): 1230-1233.
[18] Ambresh P. A., Pushpanjali G. M., Sujata A. A., Hadalgi P. M., Hunagund P. V., 2013, S-Band compact microstrip antenna with slots, *National conference on Challenges in research & Technology in the coming Decades(CRT 2013),* pp(s): 1-3.
[19] Paul L C, Islam M M, 2021, A Super Wideband Directional Compact Vivaldi Antenna for Lower 5G and Satellite Applications, *International Journal of Antennas and Propagation*, vol. 2021, Article ID 8933103, 12 pages, 2021. https://doi.org/10.1155/2021/8933103
[20] Awl H N., et.al., 2021, Double Meander Dipole Antenna Array with Enhanced Bandwidth and Gain, *International Journal of Antennas and Propagation,* vol. 2021, Article ID 9936781, 8 pages, 2021. https://doi.org/10.1155/2021/9936781

[21] Xie T, et.al., 2021, A Novel Pattern Branch Antenna for 3G\ 4G\ WLAN\ Bluetooth\ Navigation Applications, *International Journal of Antennas and Propagation*, vol. 2021, Article ID 1559519, 7 pages, 2021. https://doi.org/10.1155/2021/1559519
[22] Ahsan A , et.al., 2020, Isolation Improvement in UWB-MIMO Antenna System Using Slotted Stub, *Electronics* 9, no. 10: 1582. https://doi.org/10.3390/electronics9101582
[23] Wang M, et.al., 2021, High-Isolation UWB MIMO Antenna with Multiple X-Shaped Stubs Loaded between Ground Planes, *International Journal of Antennas and Propagation*, vol. 2021, Article ID 1155471, 13 pages, 2021. https://doi.org/10.1155/2021/1155471

Evaluation of the effect of 5G Hardware Impairments on achieving high Energy Efficiency

Bogabathina Vaishnavi[1] and Kamal Kumar Sharma[2]
[1] Student, Lovely Professional University, Phagwara, Punjab, INDIA
[2] Professor, Lovely Professional University, Phagwara, Punjab, INDIA
*Corresponding author's email: kamal.23342@lpu.co.in
Orcid ID: [1]https://orcid.org/0000-0001-7102-1974, [2]https://orcid.org/0000-0002-6294-0829

Abstract. The most demanding constraints for next-generation wireless communication systems (5G) are meeting ever-increasing requirement for high data rates while providing a seamless quality of service across the whole network. The deployment of huge Massive Input Massive Output array of antenna and Small cell spatial densification are significant facilitators of high information output, broader scope, as well as increased energy efficiency. However, there are certain critical drawbacks with this technology that must be solved, including performance decrease owing to hardware limitations. To that end, this essay investigates Hardware flaws and their consequences for massive multiple-input multiple-output. In addition, The EE expanding point of operation may be negligible in terms of spectral efficiency, and by increasing the number of base stations, energy may be considerably enhanced.

Keywords. Ultra-Dense Network, Massive MIMO, Hardware Impairments, Energy Efficiency

1. Introduction

The best assuring technologies in 5G to handle the network system capacity issue is an ultra-dense network (UDN). Malleable and extreme formation of minor Base Station can enhance spatial reuse. Increasing the traffic density in areas like Airports and large shopping malls requires both outdoor and indoor deployment to provide seamless coverage and capacity. Small cells were used in Heterogeneous Networks (Het-Nets) to introduce the macro layer, paving the way for UDNs. Nodes located far away to macro-BS might couple with small BS, resulting in a higher received signal quality.

Although there is a lot of research being done on massive MIMO, the impact of transceiver hardware failures on these systems has gotten little attention so far, despite the fact that it is important. Large arrays may appeal only to network deployment, if each antenna element is made up of low-cost components. Compensation algorithms [14], which may be implemented through analog and digital signal processing, are often used to offset the impact of hardware limitations. These solutions cannot totally eliminate the impairments. However, residual impairments exist because time-varying hardware features cannot be fully quantified and approximated, due to the presence of various noises. In the high-power scenario, transceiver variations are known to fundamentally restrict capacity [6], [24], as there is only very few literature that evaluate the performance in the large number of antenna regime. A low peak-to-average power ratio in a transmitter can lessen the influence of amplifier non-linearity's.

2. UDN Energy Efficiency Techniques

In recent years EE had become a burning topic internationally [6]. Various techniques for power efficient community layout for wireless mobile network operations have been presented in the literature. In [3] grouped EE approaches into five categories, the majority of which coincide with those defined in [4] and [6]. These techniques are categorized into the five groups stated below.

- Network Design and Implementation
- Radio Transmission Process Optimization
- Sleeping Strategy for Base Stations
- Hardware Remedy
- Energy Capture

The various strategies presented in the literature have some advantages and some disadvantages. They investigated some characteristics individually; additional mixed approaches had presented in which the characteristics of two or more operations are combined to increase the system EE. It is postulated in this study that the base station describes SBS and MBS.

2.1. *Network Design and Implementation*

Energy-efficient cellular network planning and deployment might be characterized by categorization of BS planning along with famous HetNets approaches. The techniques used were described more below under the following headings. BS Density Optimization: Schematic design for concern areas covering network is a significant component considered during the development phase of mobile cellular networks [3]. Once the coverage area has been determined, The BSs required for providing a specific degree of QoS may be approximated based upon the area of coverage [3]. In comparison to the planning and estimate procedure in homogeneous networks, SBS implementation is not planned in compact HetNet [4].

2.2. *Radio Transmission Process Optimization*

EE Network scheme has become expensive in enhancing radio transmission variables to increase the area of coverage, reduce the Intervention and allotment of natural resources. An efficient way for suppressing intervention is Beamforming [11]. Because intervention in WN degrades the EE of a network [4] beamforming approach is a promising way for enhancing the EE. The beamforming process is heavily influenced by the type of RF precoding units used. Beamforming and precoding are intuitively identical. However, in the context of beamforming, the spotlight is on techniques used for processing of signals.

2.3. *Sleeping Strategy for Base Stations*

Several methods have been experimented to save energy when the usage is minimum. Instead of totally shutting down a BS, Figure 2 depicts a model network sketch in which flipping off BS strategy has been used. When all tiny cells are inactive, the MBS might lower its signal strength and finally switch off. However, Despite of MBS's greater transmission power, if the procedure has been chaotic user equipment may choose to associate with MBS. Further efforts in this approach have studied issues regarding arbitrary selection upon BSs to UEs [4], creating awareness towards traffic [14], output delay [4], and allotment of resources to improve on existing methods.

2.4. *Hardware Remedy*

By lowering the hardware complexity of BSs and UEs, low circuit power consumption may be obtained. Nevertheless, wireless mobile networks, energy reductions at the BSs are more important. Notable research in this field focuses on administration of network based on cloud and delineate hardware of RF. The component of Antenna has numerous interconnection topologies with the chains of RF is being offered on the RF hardware development, as mentioned within massive MIMO deployment. The theme of energy amplifier efficacy is another significant benefaction.

2.5. Energy Capture and Transfer
Ultra-high data rates have been identified as a key aim goal for future production of wireless mobile networks. Successively to meet with increasing information transfer tariff, mobile device power consumption increases [5]. SWIPT is being offered as a constant source of power for cellular workstations, therefore boosting the standards of customer's comfort, as a result of the restricted battery lifespan of cellular gadgets. [8] Provides further knowledge about SWIPT. PPP is commonly used in traditional HetNet to create the formulas. However, to construct an energy-efficient SWIPT HetNet, statement showing the quantity and charge of power gathered must be included.er energy from radio waves to charge their internal batteries.

3. Hardware Impairments
Because massive MIMO systems include a huge quantity of antennas near base station, to afford this system one should employ inexpensive chains of RF. As a result, the system's performance suffers as a result of hardware flaws. Some of these hardware impairments will be discussed more below.

3.1. Carrier Frequency Offset (CFO)
Using low-cost local oscillators makes it more likely of oscillator instability and creates carrier frequency offset, which is caused by a discrepancy between the transmitter's and receiver's oscillators. Because we have several users in huge MIMO systems, one need numerous Carrier Frequency Offsets for approximation and to adjust. Because the BS antennas utilize a common oscillator in integrated huge MIMO com-plex, the quantity of CFOs that must be calculated is as good as the quantity of users. These findings also suggest that this technique out-performs [16] in terms of performance and complexities.

3.2. *Distortion of Power Amplification*
Distortion of Power amplification occurs due to nonexistence of intake transmitted signal within the range of power amplifier. [15]. Power amplifiers with a wide linear range are costly, therefore employing cheap chains of RF exacerbates the Distortion of Power Amplification problem within enormous MIMO complexes. To address this issue, the authors of [16] presented another solution that use an approximation message forwarding mechanism. The results demonstrate that this approach has a lower PAPR than [16] but a higher PAPR than [15]. The developers of [21] exploited the system's degrees of freedom to lower the PAPR.

3.3. *I/Q Imbalance*
In-phase/quadrature imbalance is distinctive equipment flaws in digital transmission systems. In an ideal system the magnitude of achievement of the blenders for factual and hypothetical components are same. The CSI knowledge is insufficient for huge MIMO complexes with In-phase/quadrature imbalance, and this I/Q imbalance must be approximated and adjusted. [16] Suggested a novel precoder that analyses the real and imaginary parts of the signal independently and accounts for multithreading intervention and imbalance of I/Q. The results indicate that the effect of I/Q imbalance persists in large MIMO systems, in spite of the quantity of antennas in BS approaches ubiquity, regularized ZF (RZF) precoders are used.

4. Simulation Results
The Simulator used here is MATLAB**.** Fig 1 depicts EE lower bound as a function of number of Base Station Antennas M and active UE'S K with optimized value of pilot reuse factor β. When the value of M is 95 and K is 11, the Global EE is achieved at Maximum. With a large number of Base Station Antennas (M=95) and huge β value inter-cell interference in dense deployments is effectively controlled.

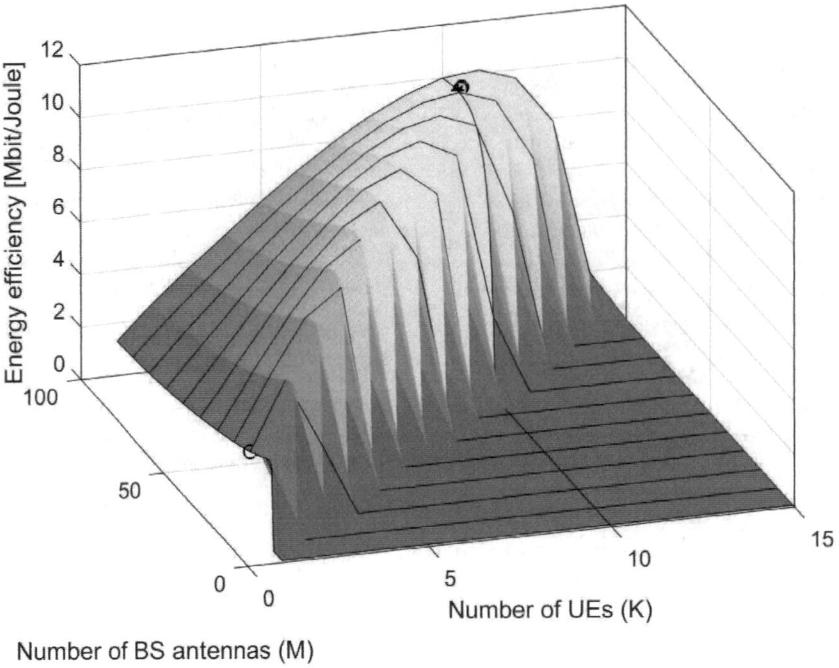

Figure 1. Optimal Design of EE for Number of BS antennas and Number of UEs

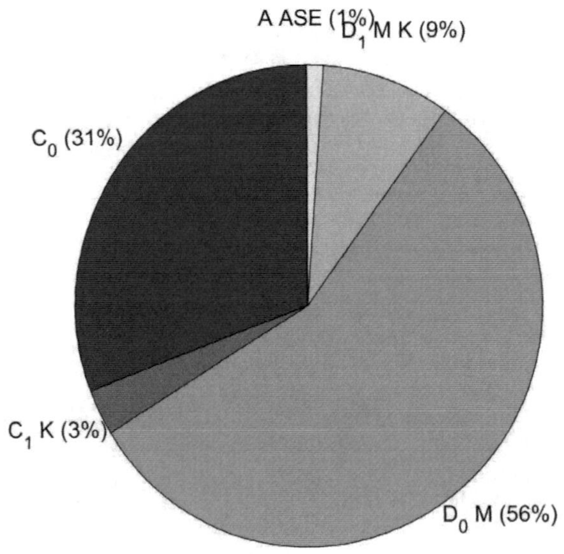

Figure 2. PIE Chart representation of each component in APC

To learn much more about the global optimum, check out the pie chart in which the proportional magnitude of each component in the APC is illustrated in Figure 2. The static power utilization C_0 and the utilization D_0M of the BS transceiver chains are predominating components. These appear to be the most important aspects to address in order to achieve more Energy Efficient Hardware.

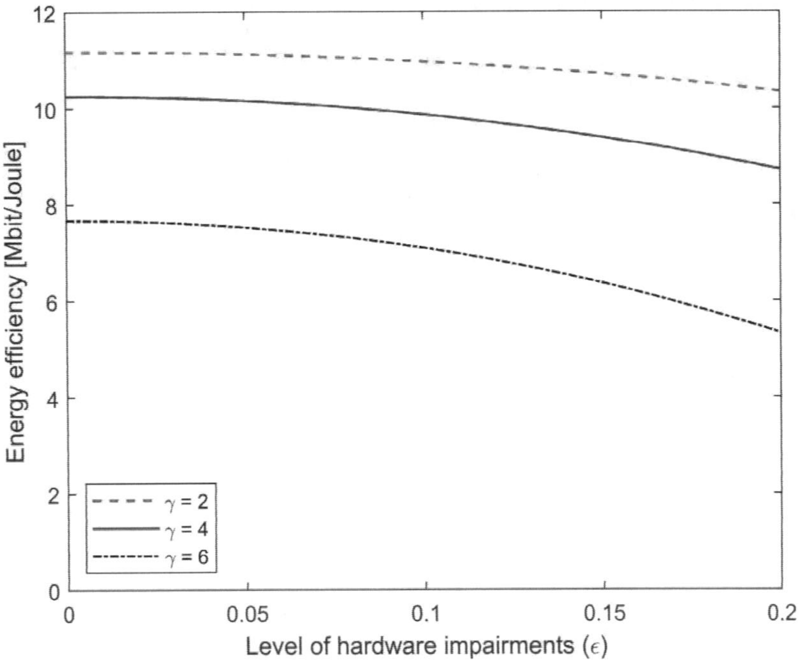

Figure 3. Effect of Transceiver Hardware Impairments on EE

The EE is plotted as a function of the amount of Hardware Impairments in Figure 3, where ϵ refers to the level of Hardware impairments for various SINR Values γ. The Values of SINR considered in this simulation are $\gamma = [2, 4, 6]$ respectively. It is observed that EE is reduced with the level of impairments of Hardware. For $\gamma = 2$, the impact is negligible, but with increase in γ when $\gamma = 6$ the loss is considerably high. Therefore the findings imply that the Hardware impairments have a significant influence on Channel Capacity when SINR Value increases. $[\gamma = 6]$ Whereas when the same SINR Value is low for $\gamma = 2$ their impact is minimal.

5. Conclusion

This article examined current advances in EE approaches in 5G UDN. The adoption of ultra-dense HetNets poses new obstacles, such as high energy needs for indoor networks and network administration complications. Numerous efforts have been made to overcome these concerns. This study provides a summary of the hardware impairments and its effect on channel capacity in massive MIMO, highlighting the key research in this subject. The simulation covered in this study demonstrated that moderate level of hardware impairments have no influence on networks intended for high EE. As the values of SINR increases, it is clearly evident that EE declines.

References

[1] Fu, Yu, Mohammad Dehghani Soltani, Hamada Alshaer, Cheng-Xiang Wang, Majid Safari, Stephen McLaughlin, and Harald Haas. "End-to-end energy efficiency evaluation for B5G ultra dense networks." In 2020 IEEE 91st Vehicular Technology Conference (VTC2020-Spring), pp. 1-6. IEEE, 2020.

[2] Chen, Xin, Xu Liu, Ying Chen, Libo Jiao, and Geyong Min. "Deep Q-Network based resource allocation for UAV-assisted Ultra-Dense Networks." Computer Networks 196 (2021): 108249.

[3] Marabissi, Dania, Lorenzo Mucchi, and Simone Morosi. "User-cell association for security and energy efficiency in ultra-dense heterogeneous networks." Sensors 21, no. 2 (2021): 508.

[4] Venkateswararao, Kuna, and Pravati Swain. "Binary-PSO-based energy-efficient small cell deployment in 5G ultra-dense network." The Journal of Supercomputing (2021): 1-22.

[5] Perdomo, José, Mårten Ericsson, Mats Nordberg, and Karl Andersson. "User Performance in a 5G Multi-connectivity Ultra-Dense Network City Scenario." In 2020 IEEE 45th Conference on Local Computer Networks (LCN), pp. 195-203. IEEE, 2020.

[6] Alamu, Olumide, Abiodun Gbenga-Ilori, Michael Adelabu, Agbotiname Imoize, and Oluwabusayo Ladipo. "Energy efficiency techniques in ultra-dense wireless heterogeneous networks: An overview and outlook." Engineering Science and Technology, an International Journal 23, no. 6 (2020): 1308-1326.

[7] Venkateswararao, Kuna, and Pravati Swain. "Traffic aware sleeping strategies for small-cell base station in the ultra-dense 5G small cell networks." In 2020 IEEE REGION 10 CONFERENCE (TENCON), pp. 102-107. IEEE, 2020.

[8] Zhang, Qi, Xiaodong Xu, Jingxuan Zhang, Xiaofeng Tao, and Cong Liu. "Dynamic load adjustments for small cells in heterogeneous ultra-dense networks." In 2020 IEEE Wireless Communications and Networking Conference (WCNC), pp. 1-6. IEEE, 2020.

[9] Bulashenko, Andrew, Stepan Piltyay, and Ivan Demchenko. "Energy efficiency of the D2D direct connection system in 5G networks." In 2020 IEEE International Conference on Problems of Infocommunications. Science and Technology (PIC S&T), pp. 537-542. IEEE, 2020.

[10] Xiong, Dong, Ling Dai, and Xiaoping Zeng. "Research on Area Spectrum Efficiency and Energy Efficiency in 3-tier Ultra-Dense Heterogeneous Cellular Networks." In 2020 IEEE 6th International Conference on Computer and Communications (ICCC), pp. 256-261. IEEE, 2020.

[11] Zhang, Yao, Meng Zhou, Haitao Zhao, Longxiang Yang, and Hongbo Zhu. "Spectral efficiency of superimposed pilots in cell-free massive MIMO systems with hardware impairments." China Communications 18, no. 6 (2021): 146-161.

[12] Mokhtari, Zahra, Maryam Sabbaghian, and Rui Dinis. "A survey on massive MIMO systems in presence of channel and hardware impairments." Sensors 19, no. 1 (2019): 164.

[13] Papazafeiropoulos, Anastasios, Emil Bjornson, Pandelis Kourtessis, Symeon Chatzinotas, and John M. Senior. "Scalable Cell-Free Massive MIMO Systems: Impact of Hardware Impairments." IEEE Transactions on Vehicular Technology (2021).

[14] Dey, Sauradeep, Ekant Sharma, and Rohit Budhiraja. "Hardware-Impaired Rician-Faded Massive MIMO FD Relay: Analysis and OptimizationSINGLECOL." IEEE Transactions on Communications (2021).

[15] Zheng, Jiakang, Jiayi Zhang, Luming Zhang, Xiaodan Zhang, and Bo Ai. "Efficient receiver design for uplink cell-free massive MIMO with hardware impairments." IEEE Transactions on Vehicular Technology 69, no. 4 (2020): 4537-4541.

[16] Elhoushy, Salah, and Walaa Hamouda. "Performance of distributed massive MIMO and small-cell systems under hardware and channel impairments." IEEE Transactions on Vehicular Technology 69, no. 8 (2020): 8627-8642.

[17] Kazemi, Mohammad, Abbas Mohammadi, and Tolga M. Duman. "Analysis of DF Relay Selection in Massive MIMO Systems with Hardware Impairments." IEEE Transactions on Vehicular Technology 69, no. 6 (2020): 6141-6152.

[18] Belaoura, Widad, Khalida Ghanem, Mohummed Zeeshan Shakir, and Khalid Qaraqe. "Impact of hardware impairments on the performance of millimeter-wave massive MU-MIMO systems with distributed antennas." In 2020 IEEE Eighth International Conference on Communications and Networking (ComNet), pp. 1-4. IEEE, 2020.

[19] Demir, Özlem Tugfe, and Emil Björnson. "Admm-based one-bit quantized signal detection for massive MIMO systems with hardware impairments." In ICASSP 2020-2020 IEEE International Conference on Acoustics, Speech and Signal Processing (ICASSP), pp. 9120-9124. IEEE, 2020.

[20] Lim, Byungju, Won Joon Yun, Joongheon Kim, and Young-Chai Ko. "Joint Pilot Design and Channel Estimation using Deep Residual Learning for Multi-Cell Massive MIMO under Hardware Impairments." arXiv preprint arXiv:2108.04485 (2021).

[21] López-Pérez, David, Antonio De Domenico, Nicola Piovesan, Harvey Baohongqiang, Geng Xinli, Song Qitao, and Mérouane Debbah. "A survey on 5G energy efficiency: massive MIMO, lean carrier design, sleep modes, and machine learning." arXiv preprint arXiv: 2101.11246 (2021).

[22] Mathonsi, Topside E., and Tshimangadzo M. Tshilongamulenzhe. "Intelligent Energy Efficiency Algorithm for the 5G Dense Heterogeneous Cellular Networks." In 2020 International Conference on Computational Science and Computational Intelligence (CSCI), pp. 144-149. IEEE, 2020.

[23] Shuvo, Md Shahin Alom, Md Azad Rahaman Munna, Sujan Sarker, Tamal Adhikary, Md Abdur Razzaque, Mohammad Mehedi Hassan, Gianluca Aloi, and Giancarlo Fortino. "Energy-efficient scheduling of small cells in 5G: A meta-heuristic approach." Journal of Network and Computer Applications 178 (2021): 102986.

[24] Inamdar, Mohammad Azharuddin, and H. V. Kumaraswamy. "Energy efficient 5G networks: Techniques and challenges." In 2020 International Conference on Smart Electronics and Communication (ICOSEC), pp. 1317-1322. IEEE, 2020.

[25] Ahmed, Ali Hameed, Ahmed Thair Al-Heety, Belal Al-Khateeb, and Alaa Hamid Mohammed. "Energy Efficiency in 5G Massive MIMO for Mobile Wireless Network." In 2020 International Congress on Human-Computer Interaction, Optimization and Robotic Applications (HORA), pp. 1-6. IEEE, 2020.

[26] Lorincz, Josip, Zvonimir Klarin, and Dinko Begusic. "Modeling and Analysis of Data and Coverage Energy Efficiency for Different Demographic Areas in 5G Networks." IEEE Systems Journal (2021).

[27] Abdullah, Sawsan N., Farhad E. Mahmood, and YE Mohammed Ali. "Comparative Study the Effect of Antenna Selection on Capacity and Energy Efficiency of 5G Systems." In IOP Conference Series: Materials Science and Engineering, vol. 1152, no. 1, p. 012013. IOP Publishing, 2021.

[28] Gui, Jinsong, Xiangwen Dai, and Naixue Xiong. "Design and Analysis of Network Behaviors for Optimizing Network Energy Efficiency in 5G mm Wave Systems." IEEE Transactions on Network Science and Engineering (2021).

[29] Sachan, Smriti, Rohit Sharma, and Amit Sehgal. "SINR Based Energy Optimization Schemes for 5G Vehicular Sensor Networks." Wireless Personal Communications (2021): 1-21.

[30] Shurdi, Olimpjon, Luan Ruci, Aleksander Biberaj, and Genci Mesi. "5G Energy Efficiency Overview." European Scientific Journal, ESJ 17, no. 1 (2021): 1.

[31] Thirunavukkarasu, Ramya, and Ramachandran Balasubramanian. "An Efficient Code Domain NOMA Scheme with Enhanced Spectral and Energy Efficiency for Networks Beyond 5G." Wireless Personal Communications (2021): 1-25.

Journal of Physics: Conference Series **2327** (2022) 012048 doi:10.1088/1742-6596/2327/1/012048

Improvement in Depth of Reflection Co-efficient below -20 dB for Millimeter–Wave Antenna

Simerpreet Singh[1]

Gaurav Sethi[1]

Jaspal Singh Khinda[2]

Department of Electronics and Electrical Engineering, Lovely Professional University, Phagwara, Punjab, India[1]
Department of Electrical Engineering, Bhai Gurdas Institute of Engineering and Technology, Sangrur, Punjab, India[2]

E-mail: s.simerpreet@gmail.com[1], gaurav.11106@lpu.co.in[1], jskhinda@gmail.com[2]

Abstract. A low cost microstrip antenna is designed for *mm*–wave applications on FR4 substrate. It consist of patch, etched with an open polygon slot (OPS) at lower central edge of it and partial ground plane. This etching of OPS is used to improve depth of reflection co-efficient below -20 dB which is equivalent to mismatch loss (ML_{dB}) $\leq$ 0.04 dB. This proposed *mm*–wave antenna also provides 3 dBi– & 5 dBi–gain-bandwidth of 6 GHz (34.72—40.72GHz) & 3.91 GHz (34.72—38.63GHz) respectively with a peak gain of 12.4 dBi at 35.1 GHz and acceptable radiation efficiency of 58.05 %. Simple design, low manufacturing cost and ease of fabrication makes proposed antenna suitable for high end practical *mm*–wave applications.

1. Introduction

The *mm*-wave communication will be realistic in the nearby future after resolving the key issues related to it [1–3]. The *mm*–wave antennas are needed for high data rate communication as discussed in [4]. Moreover, the improvement in depth of reflection co-efficient (S_{11}) [5, 6] also be a considerable parameter. This improves the power delivery from source to the patch, that will ensures the maximum radiations from the antenna. Also wide band antennas [7–9] are also playing significant role modern communications systems. In past, many *mm*–antennas [11–16] were developed to achieve *mm*–wave frequency range.

However, these antennas were state of art, though none of them improves the depth-of-reflection-coefficient below -20 dB for a wide bandwidth in *mm*-wave frequency range to ensure 99 % power delivery from source to patch for wide band applications. With this motivation, a wide band *mm*–wave microstrip antenna is designed to address above mentioned short comings. In this paper, the designed antenna provides a wide *mm*–wave range bandwidth of 2.12 GHz (36.13–38.25GHz) with the mismatch loss (ML_{dB}) $\leq$ 0.04 dB which is equivalent to $S_{11} \leq$ -20 dB. Also, -10 dB–impedance-bandwidth and 3 dBi–gain-bandwidth of 6 GHz (34.72–40.72GHz) and average radiation efficiency (RE %) of 58.05 % is achieved by the proposed antenna design.

Journal of Physics: Conference Series **2327** (2022) 012048 doi:10.1088/1742-6596/2327/1/012048

2. Design analysis

The purposed antenna is evolved on FR4 Substrate having dimensions of $L_s \times W_s \times h$ (i.e. 10 $mm \times 12\ mm \times 1.6\ mm$). The initial dimensions of design are approximated according to [10] at frequency of 7.5 GHz. The patch of dimensions $L_1 \times W_1$ (i.e. 6 mm × 12 mm) is placed on the substrate, partial ground place of dimensions $L_g \times W_g$ (i.e. 3.8 mm × 12 mm) is place on back side of the substrate and is fed by a 50 Ω inset microstrip feed-line width (W_f) of 3 mm & length (L_f) of 4 mm as illustrated in Figure 1. In order to obtain impedance bandwidth less than or equal to -10 dB & -20 dB, an open polygonal slot (OPS) is etched at center lower edge of the patch. Also, the length of feeding line L_f increases to 6.16 mm from its previous value of 4 mm. These arrangements makes it inset fed *mm*–wave microstrip antenna as illustrated in Figure 2. These adjustments improves the impedance matching patch to a level that 99 % power will be delivered to it. Henceforth, the mismatch loss decreases below value of 0.04 dB for a wide bandwidth of 2 GHz. The dimensions and placement of etched slots in the ground place are adjusted by parametric study with pattern search technique in HFSS. The final dimensions of the proposed structure are depicted in Table 1.

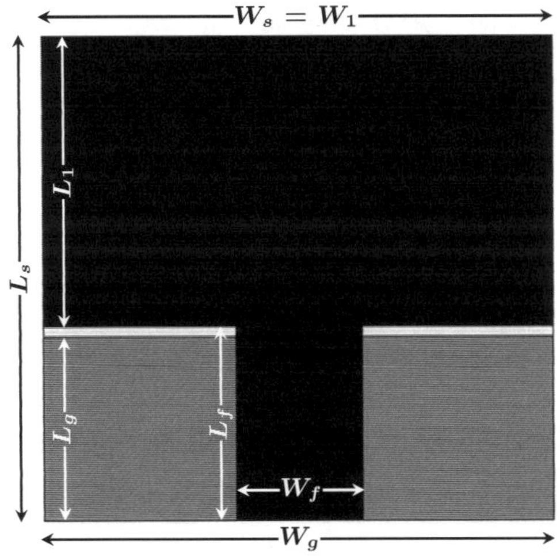

Figure 1: The Initial *mm*–wave Antenna Design.

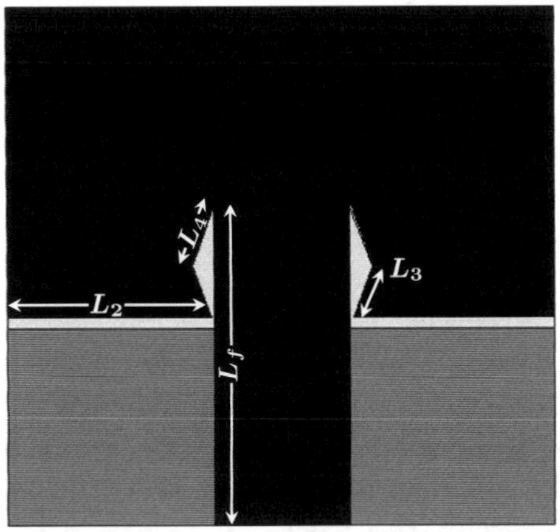

Figure 2: The Final *mm*–wave Antenna Design.

Table 1: Dimensions of proposed *mm*–wave antenna design.

Label	Value	Label	Value	Label	Value
L_s	10	W_1	12	L_f	6.159
W_s	12	L_2	4.434	W_f	3
h	1.6	L_3	1.082	L_g	3.8
L_1	6	L_4	1.254	W_g	12

Figure 3 compares the reflection co–efficient (S_{11}) of initial and final proposed design of *mm*–wave antenna. It is depicted that there is enormous enhancement in the S_{11} due to improvement in matching of impedance. The -10 dB impedance bandwidth of initial design is 4.9 GHz (34.72–39.62GHz) and it is enhanced to 6 GHz (34.72–40.72GHz). This demonstrates the enhancement of bandwidth is 1.1 GHz over initial design. Moreover, -15 dB & -20 dB impedance bandwidth is 3.19 GHz (35.75–38.94GHz) & 2.12 GHz (36.13—38.25GHz) respectively. Therefore it can be concluded that mismatch loss (ML) are significantly improved for proposed *mm*–wave antenna design. The maximum improvement in depth of S_{11} is -36.0289 dB & -32.8156 dB at 36.95 GHz & 37.92 GHz respectively.

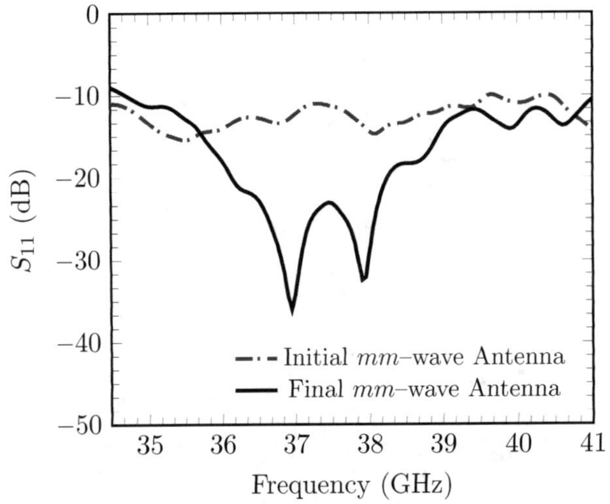

Figure 3: The reflection co–efficients comparison of Initial and Final *mm*–wave antenna design.

Figure 4 compared the gain–plots of initial and final *mm*–wave antenna design. It is noticed that there is tremendous enhancement in gain and gain–bandwidth. The 3 dBi & 5 dBi gain–bandwidths are 6 GHz (34.72–40.72GHz) & 3.91 GHz (34.72–38.63GHz) respectively. There are five peaks of gain in the frequency of interest i.e. 12.4 dBi, 8.39 dBi, 10.61 dBi, 6.7 dBi & 10.24 dBi at 35.1 GHz, 36.25 GHz, 37.35 GHz, 38.25 GHz & 39.85 GHz respectively. As detailed in the Section 2, this improvement in the gain for final design is due to inset feeding of current to the central parts of the patch. It is uniformly spread all around the patch, hence, the overall patch is participated in the radiations.

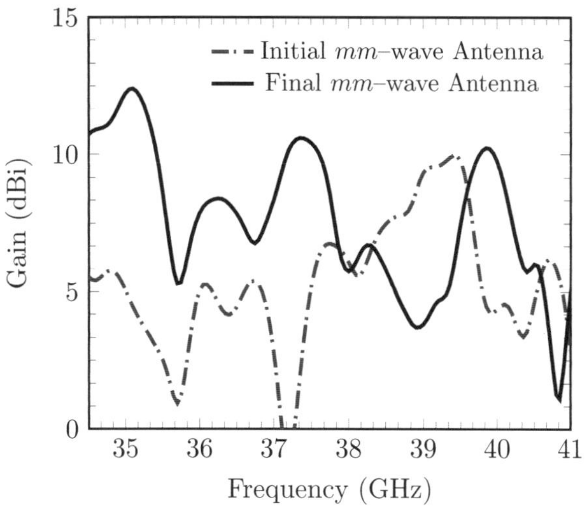

Figure 4: The gain–plot comparison of Initial and Final *mm*–wave antenna design.

2.1. Parametric Analysis of mm–wave antenna structure

In order to optimizes the *mm*–wave antenna parameters, the parametric study is performed by varying the width, position and size of the feed-line width, position of etched polygon and length of ground plane.

2.1.1. Effect of W_f: Figure 5 illustrates the structures of final *mm*–wave antenna designs of feed-line width W_f at $W_f = 1$ mm, $W_f = 2$ mm and $W_f = 3$ mm and its effect on S_{11} and gain is demonstrated in Figure 6. The significant effect of the W_f is seen on both S_{11} anf gain–plots w.r.t. frequency. The S_{11} is deteriorates with decrease in the value of W_f which results mismatching of impedance between feeding line and patch. Similarly, The gain-plots are also deteriorates as exhibited in Figure 6 (b).

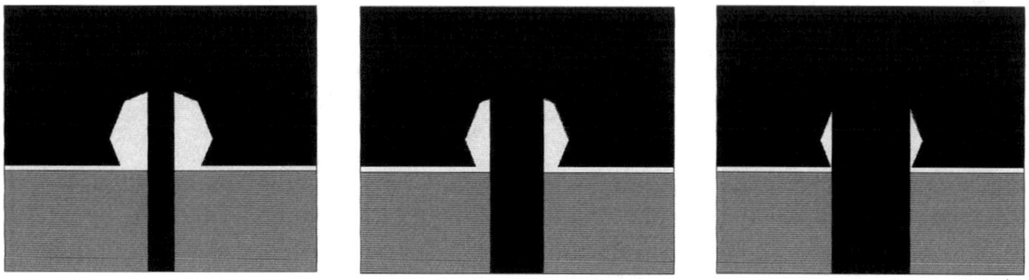

(a) $W_f = 1$ mm (b) $W_f = 2$ mm (c) $W_f = 3$ mm

Figure 5: Final Antenna structures at (a) $W_f = 1$ mm (b) $W_f = 2$ mm (c) $W_f = 3$ mm.

2.1.2. Effect of Polygon positions: Figure 7 illustrates the *mm*–wave antenna structures for various position of etched polygon with 8-sides using HFSS and there effects on S_{11}–plot and gain–plot are demonstrated in Figure 8. It is seen that both reflection co-efficient and gain-plots are deteriorated with increase value of position from 1.4 mm to 1.6 mm.

Journal of Physics: Conference Series **2327** (2022) 012048 doi:10.1088/1742-6596/2327/1/012048

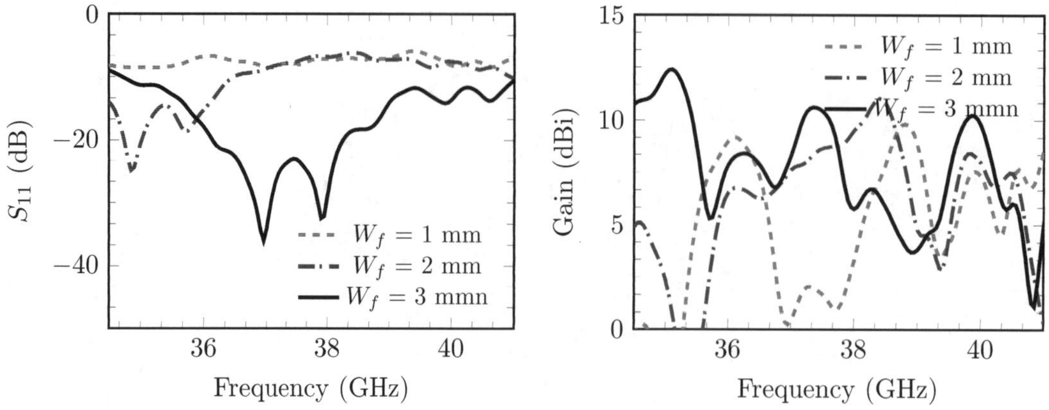

(a) (b)

Figure 6: (a) The effect of W_f at W_f = 1 mm, W_f = 2 mm & W_f = 3 mm for Final *mm*–wave antenna design on (a) reflection co–efficients (b) gain–plots.

(a) Position = 1.4 mm (b) Position = 1.5 mm (c) Position = 1.6 mm

Figure 7: Final Antenna structures at (a) Pos = 1.4 mm (b) Pos = 1.5 mm (c) Pos = 1.6 mm.

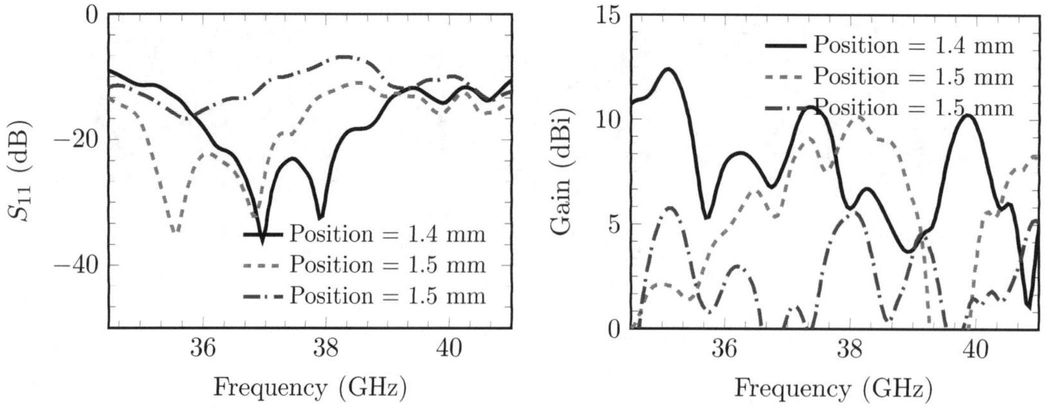

(a) (b)

Figure 8: (a) The effect of Position = 1.4 mm, Position = 1.5 mm, Position = 1.5 mm for Final *mm*–wave antenna design on (a) reflection co–efficients (b) gain–plots.

Journal of Physics: Conference Series **2327** (2022) 012048 doi:10.1088/1742-6596/2327/1/012048

3. Results and Discussions

Figure 9 demonstrates the mismatch loss (ML) for Initial and Final mm–wave antennas. The values of ML_{dB} are extracted from the VSWR as per Equation (1)

$$ML_{dB} = -10log_{10}\left(1 - \left(\frac{VSWR-1}{VSWR+1}\right)^2\right) \quad (1)$$

It is observed that the ML for final mm–wave antenna design are decreased below value of 0.04 dB for bandwidth of 2.12 GHz (36.13–38.25GHz) which is corresponding to S_{11} = -20 dB. This insures the 99 % power is delivered to the radiating patch and only 1 % power is reflected back. However, if S_{11} = -10 dB, the power delivered is 90 %. It is also noticed that initial mm–wave antenna attains higher ML_{dB} values for said frequency of interest.

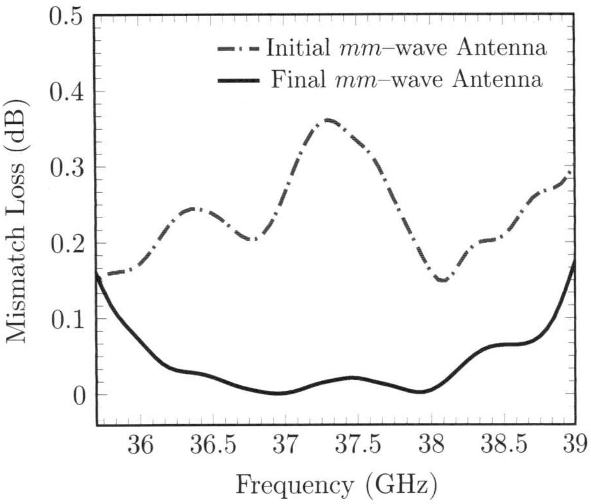

Figure 9: The mismatch loss (ML) comparison of Initial and Final mm–wave antenna design.

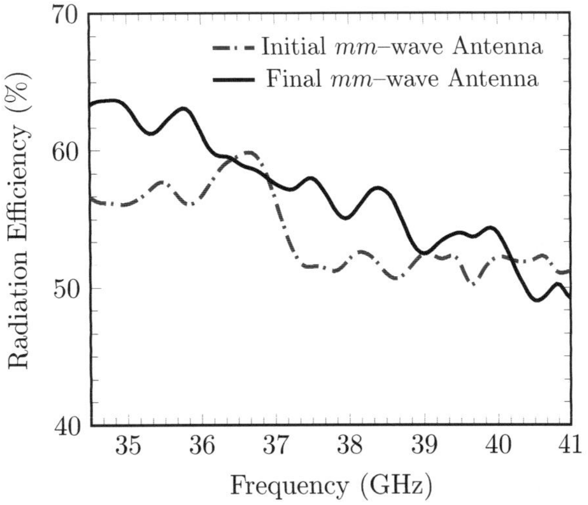

Figure 10: The radiation efficiency comparison of Initial and Final mm–wave antenna design.

Figure 10 shows radiation efficiency (RE (%)) of initial and final mm–wave antennas. It is depicted that radiation efficiency is noteworthy enhanced for final design. The average RE (%)

for initial and final designs are 54.02 % & 58.05 % respectively and it is 4.02 % more for final design than that of initial one.

Table 2 compares the fractional bandwidth (FB (%)) of mismatch loss (ML), gain–bandwidth and radiation efficiency (RE (%)). The FB for $ML_{dB} \leq 0.04$ dB is enhanced to 5.73 % from its nil value for initial design. Also, the comparison of gain for ≥ 3 dBi & ≥ 5 dBi is also shown for both antenna designs. The average RE (%) is also increased to 58.05 % from 54.02 %.

Table 2: The comparison of various parameters for initial and final *mm*–wave antenna design.

Antenna	$ML \leq 0.45$		$ML \leq 0.04$		Gain Bandwidth		RE (%)
	Bandwidth	FB (%)	Bandwidth	FB (%)	≥ 3 dBi	≥ 5 dBi	
Initial	34.7–39.62	13.24	0	0	9.18	5.95	54.02
Final	34.72–40.72	15.9	36.12–38.25	5.73	15.9	10.66	58.05

Table 3 exhibits the comparison of various parameters i.e size, -10 dB & -20 dB bandwidth with the proposed structure. It can be easy depicted that antennas [11] to [15] are designed using a costly materials and have very low tangent loss (tan δ). Also, overall size of antennas are larger than proposed structure except [13] & [15], However, these antenna offers a narrow band -10 dB–impedance-bandwidth. It also noticed that [11] to [16] offers narrow -20 dB bandwidth.

Table 3: The parameters comparison of various state-of-art *mm*–wave antenna with proposed final design.

#	Substrate	tan (δ)	Size	BW $\leq$ -10 dB	%age	BW $\leq$ -20 dB
[11]	RO3003	0.004	15 × 15	35.2–38.4	8.7	narrow band
[12]	RO5880	0.0009	20.5 (Radius)	23.1–24.9	7.5	narrow band
[13]	RO5880	0.0009	6.8 × 6.8	28/38	—	narrow band
[14]	RO4003	0.0027	13 × 13	26.1–40	42.05	narrow band
[15]	RO3880	0.0009	8.6 × 9.2	39.4	—	narrow band
[16]	FR4	0.02	10 × 10	30.77–46	39.68	narrow band
This	FR4	0.02	10 × 12	34.72-40.72	15.9	36.12–38.25

4. Conclusion

The *mm*–wave antenna proposed in this paper offers a wide -10 dB–impedance-bandwidth and 3 dBi–gain-bandwidth of 6 GHz (34.72—40.72GHz). The designed antenna is small in size and having low manufacturing cost makes this antenna suitable for high end *mm*–wave applications. The mismatch loss are also achieved here for value of ≤ 0.04 dB that represents the 99 % power is delivered to patch from its source and only 1 % is reflected back. Thus, a simple design is modified for improvement in depth of reflection coefficient below -20 dB to obtain a wide bandwidth of 2.12 GHz (36.13—38.25GHz). The final antenna design have multiple peaks of gain 12.4 dBi, 8.39 dBi, 10.61 dBi, 6.7 dBi & 10.24 dBi at 35.1 GHz, 36.25 GHz, 37.35 GHz, 38.25 GHz & 39.85 GHz respectively, 5 dBi–gain-bandwidth of 3.91 GHz (34.72—38.63GHz) and moderate average radiation efficiency of 58.05 %. High peak-gain and wide gain-bandwidth represents that this antenna can be used for a number applications in the *mm*–wave frequency range.

References

[1] Singh S, Sethi G, Khinda JS. A Historical Development and Futuristic trends of Microstrip Antennas. International Journal of Computing and Digital Systems (IJCDS) 2022, http://dx.doi.org/10.12785/ijcds/050301

[2] Malik PK, Wadhwa DS, & Khinda JS. A survey of device to device and cooperative communication for the future cellular networks. International Journal of Wireless Information Networks 2020, 1–22 .

[3] Wadhwa DS, Malik PK, and Khinda JS. High gain antenna for n260-& n261-bands and augmentation in bandwidth for mm-wave range by patch current diversions. World Journal of Engineering 2021.

[4] Singh S, Sethi G, Khinda JS. A Review Analysis on Microstrip Patch Antenna. AIJR Proceedings 2021, 297-303.

[5] Khinda JS, Tripathy MR, Gambhir D. Multi-edged wide–band rectangular microstrip fractal antenna array for C–and X-band wireless applications.Journal of Circuits, Systems and Computers 2017, 26(04).

[6] Khinda JS, Tripathy MR, Gambhir D. Improvement in depth of return loss of microstrip antenna for S-band applications. Journal of Circuits, Systems and Computers,2018, vol. 27(04).

[7] Jassal J, Sappal AS, and Khinda JS. Triangular Microstrip Antenna for C–band and X–band applications. IEEE Conference SPIN–2019, Amity University 2019.

[8] Khinda JS, Tripathy MR, Gambhir D. Low Cost Ultra–Wide Band Microstrip Fractal Antenna for C–Band Applications. Proceedings of IEEE International Conference on Innovations in Control, Communication and Information Systems (ICICCI–2017) 2017.

[9] Khinda JS and Tripathy MR. Wideband Triangular Antenna design with enhanced Gain Bandwidth for Wireless Communication Applications. Proceedings of 2015 RAECS UIET Panjab University Chandigarh 2015.

[10] Balanis CA. Chapter 11: Frequency independent antennas, antenna miniaturization and fractal antennas.Antenna Theory: Analysis and Design, John Wiley and Sons 2016, 641–650.

[11] Kun W, Kornprobst J, Thomas FE. Microstrip fed broadband mm-wave patch antenna for mobile applications. IEEE International Symposium on Antennas and Propagation (APSURSI), IEEE 2016.

[12] Hong K et al. A High-Performance Radome for Millimeter Wave Antenna Applications. Cross Strait Quad-Regional Radio Science and Wireless Technology Conference (CSQRWC). IEEE 2019.

[13] Aliakbari H et al. A single feed dual-band circularly polarized millimeter-wave antenna for 5G communication.10th IEEE European conference on antennas and propagation (EuCAP) 2016.

[14] Jain U et al. Wide-Band Microstrip Patch Antenna for Millimeter Wave-band Applications. IEEE-HYDCON 2020.

[15] Saeed A et al. A low Profile Multiband Microstrip Patch Antenna For 5G Mm-Wave Wireless Applications. IEEE International Conference of Technology, Science and Administration (ICTSA) 2021.

[16] Wadhwa DS, Malik PK, and Khinda JS. mm-wave patch antenna for high data rate communication applications in 2021 International Conference on Computing, Communication, and Intelligent Systems (ICCCIS). IEEE, 2021, pp. 778–781.

Journal of Physics: Conference Series **2327** (2022) 012049 doi:10.1088/1742-6596/2327/1/012049

Efficient Cluster Head Selection and Trust Based Routing in MANET

Meena Bharti[1],*, Shaveta Rani[2], and Paramjeet Singh[3]

[1] Research Scholar, I.K. Gujral Punjab Technical University, Kapurthala, Punjab, 144603, INDIA

[2,3] Professor, Maharaja Ranjit Singh Punjab Technical University, Bathinda, Punjab, 151001, INDIA

E-mail: meenabharti89@gmail.com

Abstract. Mobile Adhoc Network (MANET) faces different security challenges due to its collaborative and open communication behavior and restricted availability of resources. Routing is hard and vulnerabilities issues are regularly exposed to highly dynamic nature and numerous communication breakdowns induced by mobile nodes. Thus, security measures that can minimize the effects of various attacks must be included in the MANET's routing protocols. Secure communication with optimal energy is a big challenge in such networks. This paper presents a secure optimal routing scheme. The objective of the scheme is to find the most optimal secure path with low energy consumption using the proposed function. This scheme is divided into two algorithms; the first algorithm finds cluster head effectively and the second computes the trust value of each path for secure communication. The trust value is computed using the trust-based scheme using packet forwarding information. On the basis of the trust value, this algorithm detects the malicious nodes and removes them from the network and then the other proposed routing scheme computes the optimal path. Then path with less hop count and minimum energy usage is selected optimally.

1. Introduction

MANET is a network in which mobile devices are connected by wireless channels and form an unpredictable layout. The routers' movement is offered randomly and they organize themselves haphazardly. MANET could be used for a variety of things, including meetings, search-and-rescue missions, and battlefield. MANET is a very dynamic network that allows nodes to easily migrate and connect, with no permanent infrastructure, making it vulnerable to malevolent users [1]. These threats are virtually hard to detect, making it difficult to develop defence against them. MANET is used in various application areas due to its ability of swiftly building networks. The networks will function properly if all of the devices involved believe and cooperate with each other. Routing is hard and vulnerabilities issues are regularly exposed highly dynamic nature and numerous communication breakdowns induced by mobile nodes. Thus, security measures that can minimise the effects of various attacks must be included in the MANET's routing protocols [2]. Due to the MANET's vulnerabilities and characteristics, a security solution that provides security services such as accessibility, reliability, secrecy, and authentication is required. The security of MANET is divided into two parts: routing functional security and data transmission protection. Security-based routing systems use various encryption algorithms to enable secure routing as well as provide account credentials that prevent against alteration and replay of routing data transmission. Basic routing methods, detecting and isolating-based secure routing mechanism, incentive-based routing mechanisms, and trust-based routing are the multiple kinds of security routing techniques [3,4]. A routing algorithm describes how to connect routers with one another and share data that allows them to choose paths among any pair of nodes network. The routing system's principal goal is to choose the best route for data delivery among devices. There are 3 kinds of routing algorithms in MANET: proactive, multicast and reactive. Each technique has its own use but there are so many technical issues these routing techniques such as unrestricted mobility, flexibility, efficiency, stability, constructing optimal path, providing fundamental support for secure transmission, and reducing network load with less energy consumption. This paper presents such routing solutions that find the most optimal secure path with low energy consumption. Contributions of this paper are as follows:

(i) Presentation of efficient cluster head selection scheme to do load balancing on cluster heads.

(ii) Calculation of trust value to detect whether the node is malicious or legitimate.

(iii) An algorithm is presented for a selection of paths based on trust value, residual energy and minimum hop count.

The organization of the paper is as Section 1 presented Introduction to the MANETs, Section 2 includes Literature review related to the field of security in MANETs, Section 3 is for proposed work while Section 4 presents experimental results and evaluation and at last Section 5 presents the conclusion of the paper.

2. Related work

The remote system's adaptability and mobility allowed it to be used in a wide range of applications. Because of its ability to self-organize and maintain itself, the MANET has emerged as a key topic among today's distant systems. In any event, MANET is encountering multiple security vulnerabilities due to its open communication and changeable layout. So, various researchers have developed secure routing schemes and count measures from various attacks [5–9]. To mitigate the risks posed by nodes and to improve network security, Kumar and Narsimha proposed a technique in which calculation of trust values are based on direct and indirect values. The secret keys are distributed by the certificate authorities to all of the nodes. The offending nodes are removed [1]. To improve the efficiency of node reliability assessment a hierarchical system is proposed. The reliability is managed by the trust-based node, and it incorporates the quality of messages as well as the number of messages. The reliability of transmitted data is increased via node-to-node mutual authentication without CA. In this scenario DSN monitoring is done to recognise anomaly nodes [2]. To improve security as well as the framework's execution another technique has been proposed in which Bayesian technique, and Dempster Shafer Theory, are bound together a trust administration scheme that estimates the trust of the watched hub. This works conspire coordinate technique in which the aggressor hub is identified as the hub with the lowest trust esteem. This work plot improves the administration framework's execution of brought together confide [10]. In another research, a goal framework for secure routing is created utilising a blended optimising method called M-LionWhale. M-LionWhale is an optimisation technique that combines the lion algorithm (LA) with the whale optimising algorithm (WOA) to achieve the best route decision in MANET. Many quality of service parameters are taken into account by the number of co- optimisation model, including bandwidth, link lifespan, latency, energy, and trust. A fitness model is designed for the optimum route choice using the predicted number of co parameters. At maximal simulated duration, the suggested M-LionWhale method can achieve the best results with 24.1313 percent residual energy, 0.3051 PDR and 0.2966 kbps throughput [11]. Several fuzzy-based MANET techniques have been developed recently, spawning a new set of issues [6].

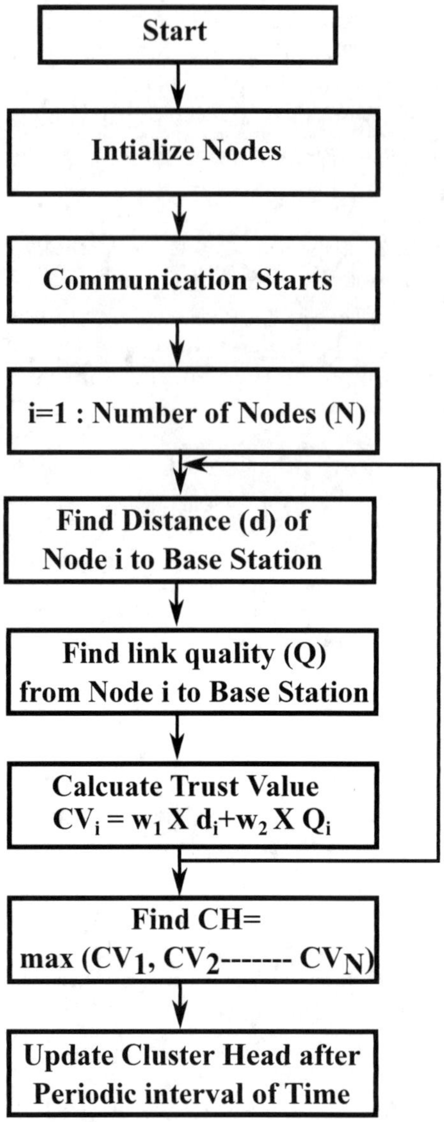

Figure 1. Algorithm of cluster head selection

3. Proposed Work

The proposed routing scheme is divided in two phases:

(i) Initially cluster heads are selected
(ii) After finding secure routing paths, an optimal route is chosen. The Algorithm 1 for choosing cluster head is shown in Figure 1.

Cluster head selection is done on the basis of link quality and distance of node from base station. Cluster head has more work load as compared to other nodes. So, its energy consumption is high. To do load balancing the cluster heads are updated after regular interval of time. The Algorithm 2 for calculation of trust is shown in Figure 2.

The calculation of trust is based ratio of number of packets forwarded by node to total packets

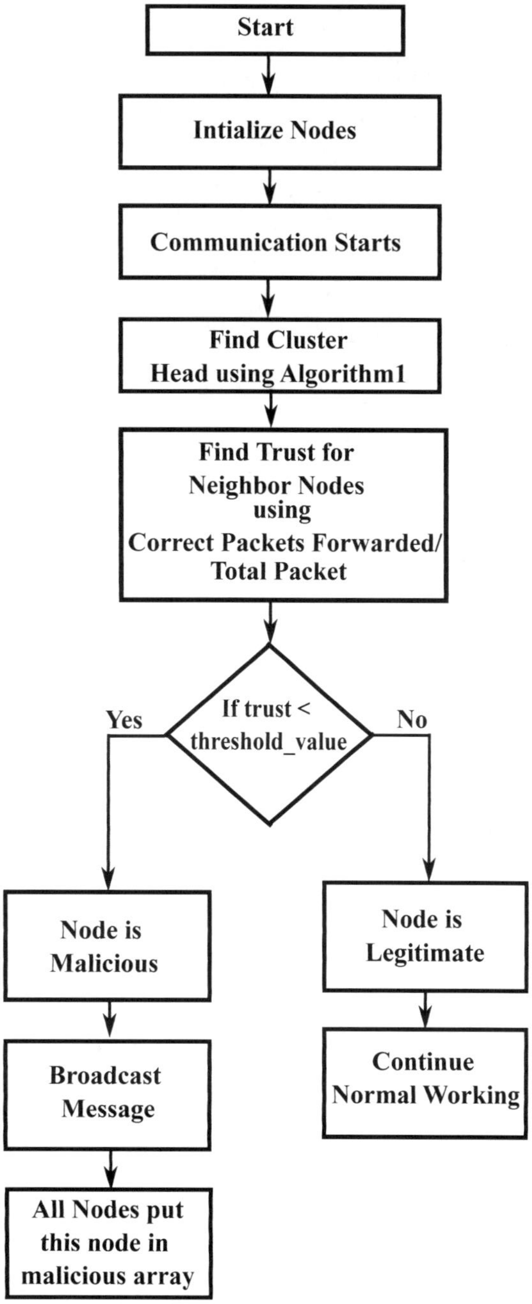

Figure 2. : Algorithm for calculation of trust

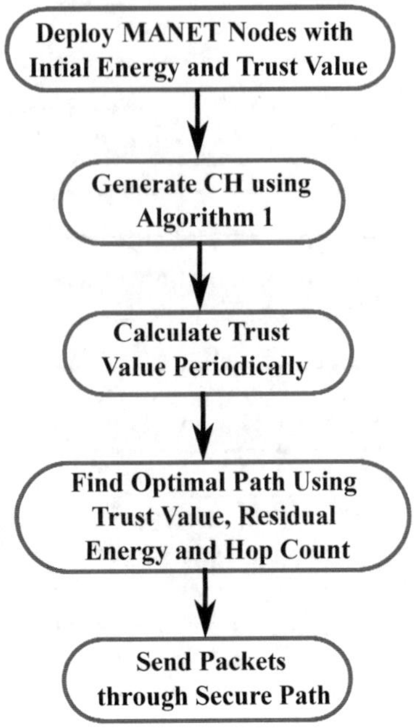

Figure 3. :Algorithm to find secure path

send to it. Node which has trust value higher than threshold value is considered as legitimate node while node having trust value lower than threshold value is considered as malicious node. As soon as malicious node is found the message is broadcasted and all other nodes put malicious node in malicious node array. The algorithm 3 for finding secure path is shown in Figure 3. To find secure path trust value, hop count and energy is taken in consideration. The nodes in path are checked for these parameters and the path having higher value is chosen for sending packets.

4. Performance Evaluation

The proposed RL routing scheme has been implemented on the MANET network with the nodes from 100 to 400 and then Energy consumption has been computed with and without the proposed scheme as shown in Figure 4. Thus, there is reduction in energy consumption with the proposed routing scheme. Then throughput is also computed with and without proposed scheme and high throughput has been achieved with the proposed scheme as shown in Figure 5.

5. Conclusion

Today is the world of high-speed data transmission. There is tremendous growth in internet and wireless devices. Everyone carries smart devices like mobile phones, tablets etc. These sensor nodes sometimes carry some sensitive information. There is a high need to provide security to the data passing through these sensor nodes. In this study, a scheme is presented to evaluate an individual network node's trust value and, as a result, a route optimization algorithm is introduced. As a result, the scheme provided a secure and reliable path throughout the system. When building a safe and stable, optimal route that is less vulnerable to path failures during the data relay period, the notion of nodes' faith is applied. The proposed algorithm computes the optimal path with secure transmission of data. The work presented in this paper is compared with the network in which no secure routing algorithm is applied in terms of energy consumption

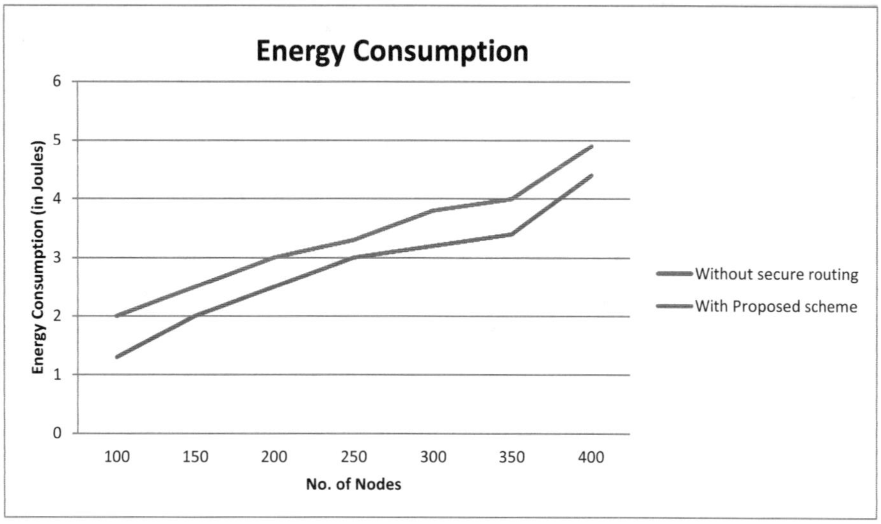

Figure 4. : Energy Consumption with and without RL routing scheme

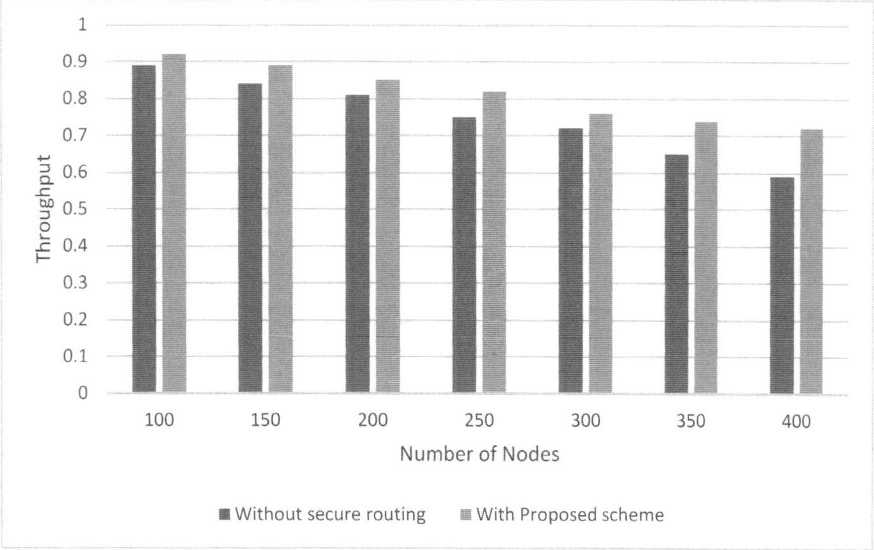

Figure 5. : Throughput with and without RL routing scheme

and throughput and the results show that secure routing provides less energy consumption and high throughput. The proposed scheme can be simply incorporated with a variety of additional network applications in the future. To improve secure and speedy path planning, we'd like to enhance this scheme.

6. Acknowledgments

The authors wish to thank many anonymous referees for their suggestions to improve the paper. Bharti, M. would like to thank I.K. Gujral Punjab Technical University for offering the Ph.D. course in Computer Science & Engineering and providing support to access the resources for research.

7. References

[1] Rajkumar B, Narsimha G. Trust based certificate revocation for secure routing in MANET. Procedia Computer Science. 2016;92:431-41. ISBN: 1877-0509 Publisher: Elsevier.

[2] Yang H. A study on improving secure routing performance using trust model in MANET. Mobile Information Systems. 2020;2020. ISBN: 1574-017X Publisher: Hindawi.

[3] AlRubaiei M, sh Jassim H, Sharef BT, Safdar S, Sharef ZT, Malallah FL. Current vulnerabilities, challenges and attacks on routing protocols for mobile ad hoc network: a review. Swarm Intelligence for Resource Management in Internet of Things. 2020:109-29. Publisher: Elsevier.

[4] Wadhwani GK, Khatri SK, Muttoo SK. Critical evaluation of secure routing protocols for MANET. In: 2018 International Conference on Advances in Computing, Communication Control and Networking (ICACCCN). IEEE; 2018. p. 202-6.

[5] Desai AM, Jhaveri RH. Secure routing in mobile Ad hoc networks: A predictive approach. International Journal of Information Technology. 2019;11(2):345-56. Number: 2 ISBN: 2511-2112 Publisher: Springer.

[6] Garg MK, Singh N, Verma P. Fuzzy rule-based approach for design and analysis of a Trust-based Secure Routing Protocol for MANETs. Procedia computer science. 2018;132:653-8. ISBN: 1877-0509 Publisher: Elsevier.

[7] Karlsson J, Dooley LS, Pulkkis G. Secure routing for MANET connected Internet of Things systems. In: 2018 IEEE 6th international conference on future internet of things and cloud (FiCloud). IEEE; 2018. p. 114-9.

[8] Ponguwala M, Rao S. E 2-SR: a novel energy-efficient secure routing scheme to protect MANET-IoT. IET Communications. 2019;13(19):3207-16. Number: 19 ISBN: 1751-8628 Publisher: IET.

[9] Rajendran N, Jawahar PK, Priyadarshini R. Cross centric intrusion detection system for secure routing over black hole attacks in MANETs. Computer Communications. 2019;148:129-35. ISBN: 0140-3664 Publisher: Elsevier.

[10] Sarbhukan VV, Ragha L. Establishing secure routing path using trust to enhance security in MANET. Wireless Personal Communications. 2020;110(1):245-55. Number: 1 ISBN: 1572-834X Publisher: Springer.

[11] Chintalapalli RM, Ananthula VR. M-LionWhale: multi-objective optimisation model for secure routing in mobile ad-hoc network. IET Communications. 2018;12(12):1406-15. Number: 12 ISBN: 1751-8628 Publisher: IET.

Ideal congestion control model to enhance the quality of service of VANET network

Rajkamal Gogoi[1] and Sandeep Kumar Arora[2]

[1] Student, Lovely Professional University, Phagwara, India
[2]Assistant Professor, Lovely Professional University, Phagwara, India
*Corresponding author's email: sandeep.16930@lpu.co.in
Orcid ID: [1]https://orcid.org/0000-0002-6037-7737, [2]https://orcid.org/0000-0002-7818-5827

Abstract: Vehicular Adhoc Technology (VANET) is a mobile ad-hoc network where each vehicle combines with a communication link. Well fascinated and proper connectivity between the vehicle nodes are very essential to make the communication efficient. But high mobility and rapid change in locations of the vehicles often hampers its performance. Also often disconnection and overloading issues inside the network create an extra burden to vanet. So it is very important to make the network free from these factors so that smooth communication can happen between the network. In this paper, we have proposed our algorithm to control the network from congestion occurrence. Which ultimately enhanced the performance and QoS of the network
Keywords: Adhoc, delay time, QoS, throughput, vanet.

1. Introduction

VANET is a branch of a mobile Adhoc network where each vehicle is a node. Although there are few differences from MANET like, vehicles travel through a previously identified road, and there are various factors which determine the velocity of vehicles like the density of vehicles, constructions, conditions and additionally it has to follow traffic rules and safeties.

Rapid changes in position, wireless links, number of vehicles create a big challenge in front of the vanet communication. It varies from city to rural areas, traffic environment with a great number of vehicles to highway environment with few numbers of vehicles where often connection failure and low network can fusible in presence of less number of vehicles and low-density areas. Therefore, a well-developed infrastructure, routing protocol, network model are required to make VANET communication more prominent in our day-to-day life. However, researchers coming up with new developments every day can be assured that within a few years VANET network will be everywhere.

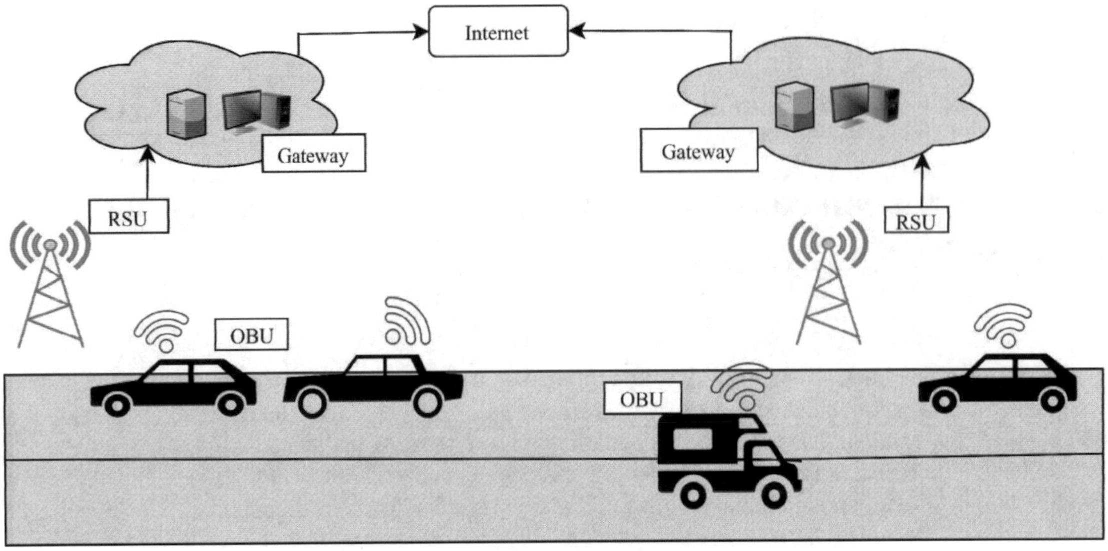

Figure 1. Vanet architecture

Quality of service defines how efficiently a network can be operated. Various factors like throughput, delay, packet loss, time of arrival are all mixed along to produce associate best network for communication. Whenever it will be able to provide a good amount of output that will be the day when vanet becomes a useful tool for public communication. Among its advantages, there are many challenges in the VANET environment which creates a big question mark on the reliability of this network. These challenges are required to be solved in a fast-paced manner to provide stable and reliable routing in VANET. Hence a lot of research and developments are required to organize to make this network more suitable.

Research and developments are in progress in the field of VANET. Every day researchers come with newly developed, highly advanced solutions to make this network reliable. Enormous efforts and intelligence have been spent on the advancement of road conditions, related systems tasks such as traffic system, message sharing, distortions, and secure accident reporting. In VANET, each vehicle is attached with onboard sensors which intelligently exchange messages with other connected devices either by V2V or V2I transmission. In the coming time, 5G technology promises to be a very strong technology that will allow vehicular connections a secured, fast and reliable communication.

2. Literature Review

Vehicular Adhoc Network (VANET) will be an intelligent transportation mode that combines both vehicle technology with internet technology. To introduce VANET service, in a realistic public environment, there should be some modifications done to the network to make it a more reliable and secured mode of communication for the people [2][3]. Along with the signal side, the physical side of the system is also required to develop as it also plays a significant role in performance. With the betterment of the road construction and roadside units, communications get fluent and time accurate [6].

Congestion in a communication network is very often. It creates a huge challenge in front of vanet communication [4]. The vehicle moves at very high speed and connections get often disconnected due to this high mobility. To reconnect broken link network should be available. And this availability of the network is possible only if it is out of congestion [5]. VANET applications mostly have delay problems or quality of service requirements. These days' vehicles are having high speeds so communication in such a rapid topology change environment is quite challenging [8][9]. Moreover, the variation of the vehicle density from zero or few vehicles up to terribly dense with over a hundred vehicles per kilometer in town areas arises a deep challenge on the dependability of the communication network [7].

It is enormous that the routing protocol considers parameters like link quality, density, and a few different effective metrics throughout packet forwarding closer to destination spot. Security is likewise one of the primary problems in VANET [11]. Cooperation amongst inter-vehicular networks and sensor networks located inside the automobiles or alongside the street has to be investigated and analyzed. As the range of automobiles increases uninterrupted communication exchange among the nodes ought to additionally be maintained for easy transmission [13]. In this paper, a thorough study and comparison of the performance of routing protocols in heavily loaded networks are provided. For comparison, we have designed a system model and simulated it for different time slots so that we could reach a fine solution. The focus of this paper is to analyze the performance of these protocols in a heavily loaded condition and to determine an effective protocol that provides a better result.

3. Research Methodology

In this section, we have discussed the research methodology that we have identified to proceed with our work. Going through several research works and simulations, we have decided to design our algorithm to maintain the congestion inside the network. With three kinds of analysis strategies, a groundwork is often organized. quantitative, qualitative, and mixed technique whereas in the mixed methodology approach, each qualitative and quantitative approach are combined. To implement our program, we have modified the existing programs inside the simulation software so that we can get our desired output. To verify the results of our proposed model, we have designed various environment scenarios with a varying number of vehicle nodes which will help us to distinguish the performance of the proposed model in changing environmental conditions.

Therefore, we tend to be determined to go with the mixed methodology approach. It will provide us with both theoretical knowledge and practical knowledge. Thus, the collected data and information will help us to represent our final results.

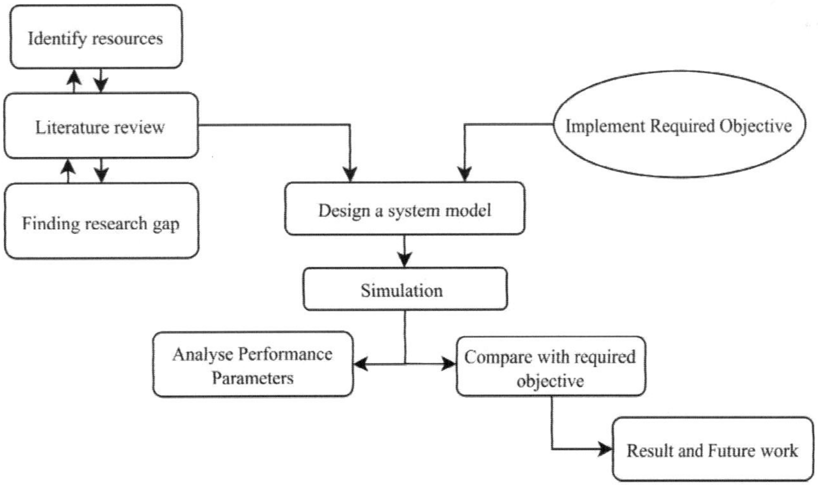

Figure 2. Flowchart of research methodology

4. Proposed model

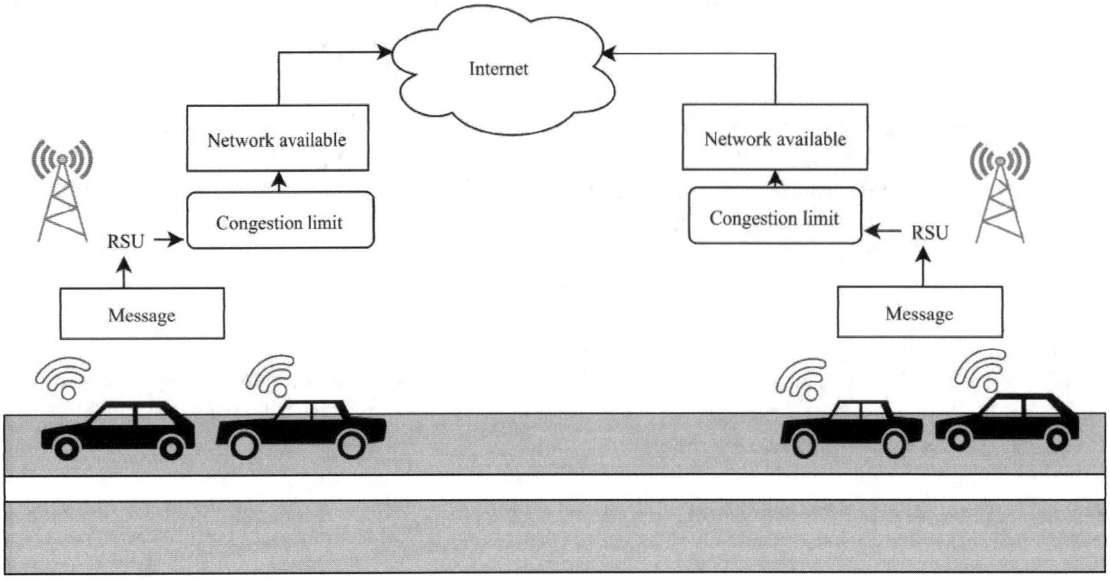

Figure 3 Proposed model architecture

Roadside units will store all the information and updates about the network. Whenever it receives a message from the source node, it verifies whether the network is out from congestion or not. If it finds that the network already reached its congestion limit, it delays the forwarding and searches for the next node where the network is available. Otherwise, it directly forwards the message to its next node if the network is out of congestion.

The main focus of our proposed model is to reduce the congestion inside the network so that smooth communication can happen. In a normal congestion control algorithm, the slow start approach is mainly used. In this approach what happens is that the sender sends the first segment to its receiver node wait for its acknowledgment. After receiving the acknowledgment, the congestion window increases exponentially i.e. 1 to 2. Then two gets double and it becomes 4 likewise.

The increase in the windows segment value also increases the extra burden to the network. Because the source node sends n amount of message segments then it has to wait for the acknowledgment of n messages. So it increases the latency. Also after expiring the timeout value, the sender again sends the same messages towards the receiver which results in network overload. So in our proposed model, we have proposed an algorithm to control the congestion limit so that smooth communication can happen between mobile nodes. In this algorithm, RSUs (Roadside units) are the ultimate storage of all the required information about the network and keep updated the other nodes about the network from time to time.

5. Algorithm

Roadside units are the ultimate storage of the network information and keep updated regularly with other nodes. So whenever RSU receives any messages from source nodes, it is RSUs responsibility to figure out the best routing path.

RSU updates its information to other nodes through table sharing which includes source information, destination, distance, sequence number, etc. In this algorithm, when RSU receives the message from the sender, it looks for network congestion. Here we introduced a constant threshold value which is 20. So according to the algorithm, while RSU finds that the network is going out of congestion limit i.e., >20,

RSU sets the message into wait mode until a network is figured out where the congestion limit has not reached yet. And when there are available network where congestion has yet not occurred, RSU informs all other nodes about the network and forward the packet to its next node towards the destination.

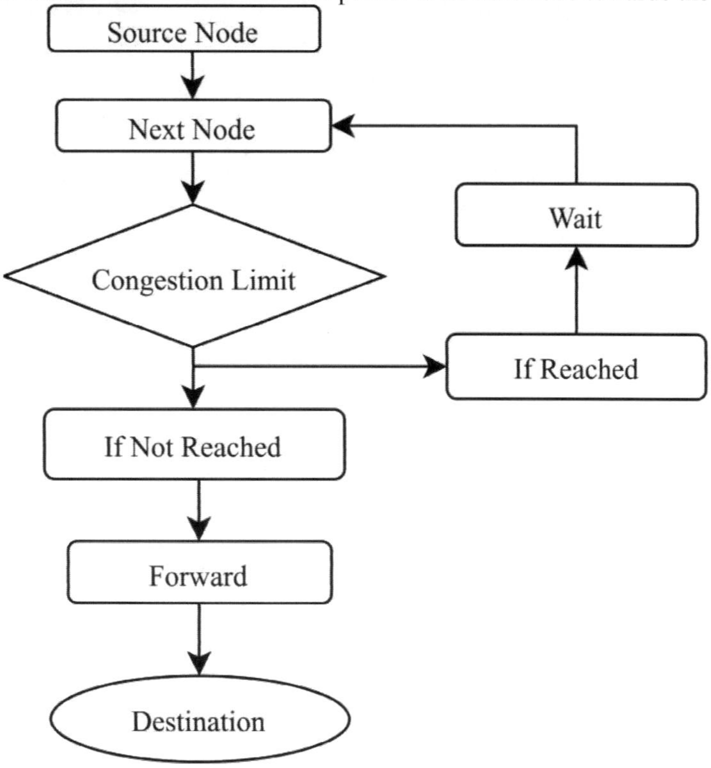

Figure 4 Algorithm of the proposed model.

6. Estimation of metrics

6.1 RSUs determine the congestion in the network through RTT (Round trip time) of messages and acknowledgment. More RTT value means more congestion in the network. These RTT values are used to set the timeout value for packets.

6.2 When connection begins with the sender to receiver, the source node starts sending segments in slow start mode and takes congestion window (cwnd_) to 1 Maximum Segment Size, which means that the sender only can send 1 segment at a time.

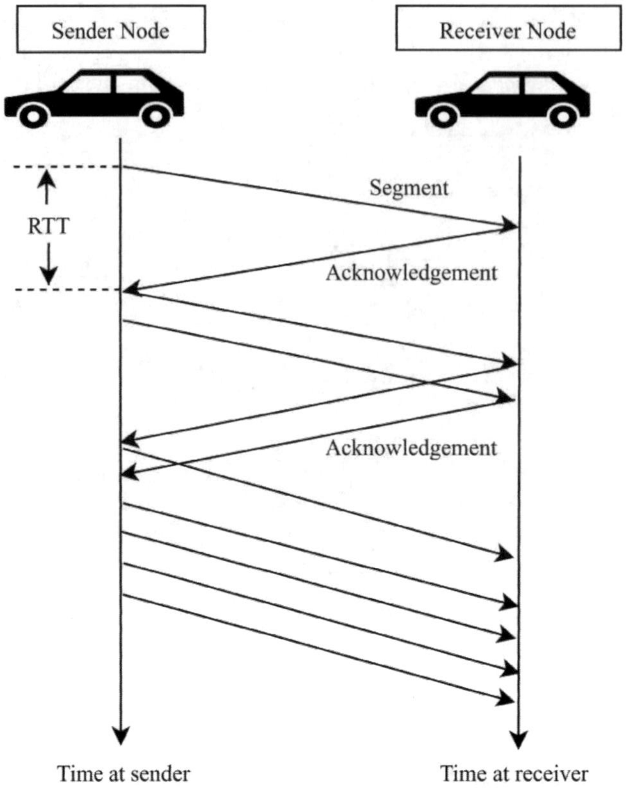

Figure 5 working of the congestion window

6.3 After receiving acknowledgment from the receiver node, MSS increases exponentially i.e. 2. Then it goes increasing like 4, 8.. etc. After reaching a particular threshold or window size, TCP starts congestion avoidance which results in increased congestion window additively instead of exponentially. At a certain point when losses occur, congestion window segments reduce to one, and also threshold value reduces to half of its previous value known as multiplicative decrease.

6.4 In our proposed model we have limited this threshold value at a certain constant point at the source file and made it updated. By doing so we will get a constant output in the congestion window threshold. The First step is to add new commands in the source file i.e TCP.cc inside ns2.35 and add a new command in TCPAgent class to fix a constant value for the congestion window (cwnd_).

6.5 The modifications that are made should be updated by running the command -$make. To instruct the TCP agent to use our new TCP algorithm, introduce an unused value of cwnd-_option. We can check the list in opencwnd().

6.6 Now to bind both C++ with the variable name in the TCL file, we are required to put the following statement in TcpAgent::TcpAgent(), which is the constructor of the class TcpAgent).

6.7 In the ns TCL file, by default congestion window code is not available so we need to add it on our own. In the termination procedure, we need to add a global variable ns and set the congestion window. This command will combine both the Tcl variable with the TCP source file. Final readings will be saved in the file introduced in the program declaration.

```
proc plotWindow {tcpSource outfile} {
  global ns
  set now [$ns now]
  set cwnd [$tcpSource set cwnd_]
# data is recorded in a file called congestion.xg
  puts $outfile "$now $cwnd"
  $ns at [expr $now+0.1] "plotWindow $tcpSource $outfile"
}
set outfile [open "congestion.xg" w]
$ns at 0.0 "plotWindow $tcp0 $outfile"
```

Figure 6 Program for TCP congestion window

```
*set tcp0 [new Agent/TCP/Vegas]

proc plotWindow {tcpSource outfile} {
  global ns
  set now [$ns now]
  set cwnd [$tcpSource set cwnd_]
# data is recorded in a file called congestion.xg
  puts $outfile "$now $cwnd"
  $ns at [expr $now+0.1] "plotWindow $tcpSource $outfile"
}
set outfile [open "congestion.xg" w]
$ns at 0.0 "plotWindow $tcp0 $outfile"
```

Figure 7 Program for Vegas congestion window

7. Result and analysis

7.1. Simulation Parameters

For the simulation, we have used NS-2 software, which is most preferable for urban environment creation. With the help of NSG2.1, within NS2, we have created three different scenario models. All three scenarios were designed with different vehicle nodes i.e. 10 vehicle nodes, 51 nodes, and 71 nodes. The traffic source type is FTP and the data packet size is 512 bytes. The rate of data sending packets is 1500 packets/ second. The total area is considered as 500m x 500m of square area and during the simulation, each node is moving from a random initial point to a random destination point. In our simulation, all the nodes are moving with 25 m/s randomly in 500m x 500m of the square area. This process helped us to create mobile nodes to clearly understand the performance of each protocol in a continuously varying environment. For the comparison, we have selected three parameters which are throughput, packet delivery ratio, and end-to-end delay. Total we have created three scenarios and in each scenario, we have simulated three times for each parameter, one for 10 seconds, 25 seconds, and 50 seconds. So a large number of simulation results we got, helped us to distinguish it clearly.

Table 1. Simulation parameters

Parameters	Values
Simulator	NS-2, NSG-2.1
Mobility model	Random waypoint
Area of map	500m x 500m
Vehicle speed	25 m/s
PHY/MAC protocol	AODV / DSDV / DSR
Network traffic	FTP
Simulation time	10 sec, 25 sec, 50 sec

In a normal congestion control algorithm, after reaching the congestion threshold value TCP follows an additive increment procedure where window size increases linearly instead of exponentially. And also the threshold value was reduced to its half the previous value. To transmit the next segment, the source node must have received acknowledgment from the receiver node. After an expired period or after getting more than three acknowledgments from the receiver, the source node retransmits the segment which ultimately creates an extra burden to the network.

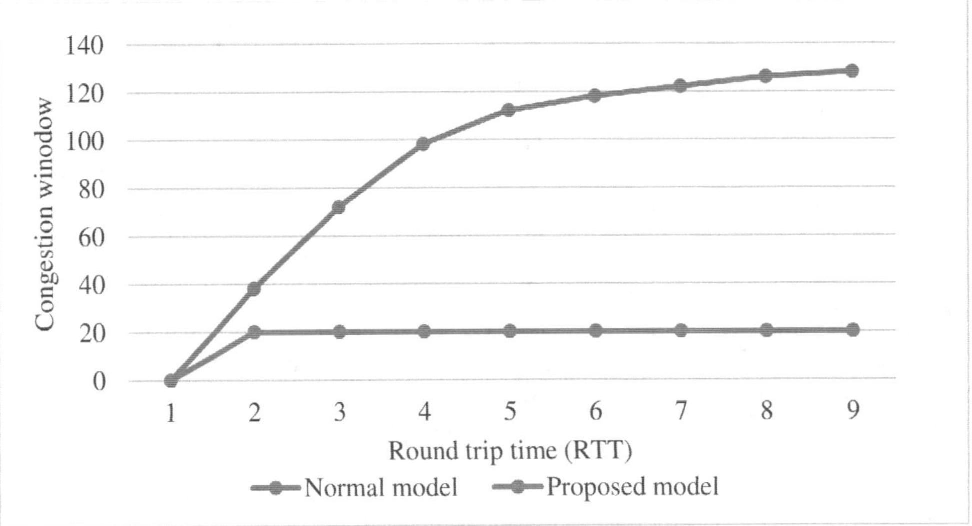

Figure 8 Comparison of normal and proposed TCP congestion window

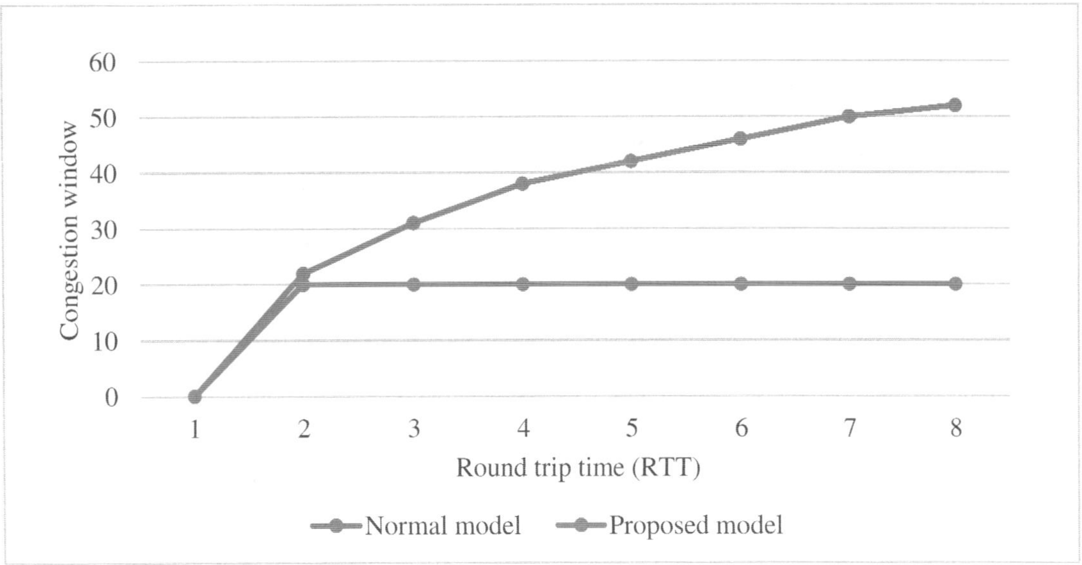

Figure 9 Comparison of normal and proposed Vegas congestion window

From the above graph, we can see that in a normal congestion control algorithm, the congestion window increases exponentially. And it reaches a very high value which ultimately shows the possibility of more packet drops. Because more the number of the congestion window, lead to higher RTT (Round Trip Time). Higher RTT value keeps source node waiting to communicate next segment.

In comparison to the normal congestion algorithm, our proposed algorithm shows better satisfactory results. The graph shows that after reaching the threshold value, the congestion window was not further incremented and it stayed at the same point which ultimately the network formed more congestion. The advantage of this limited congestion value is that the window size will not be further incremented. Lower the segment value, lower the congestion in the network. Which results in lesser RTT value and more packet delivery throughout the network.

7.2. Packet delivery ratio

It is the ratio between the number of packets received at the receiver side to the total number of message packets sends from the transmitter side. To figure out the PDR value through normal TCP congestion protocol and our proposed congestion protocol, we have used AODV, DSDV, and DSR routing protocol for the communication model.

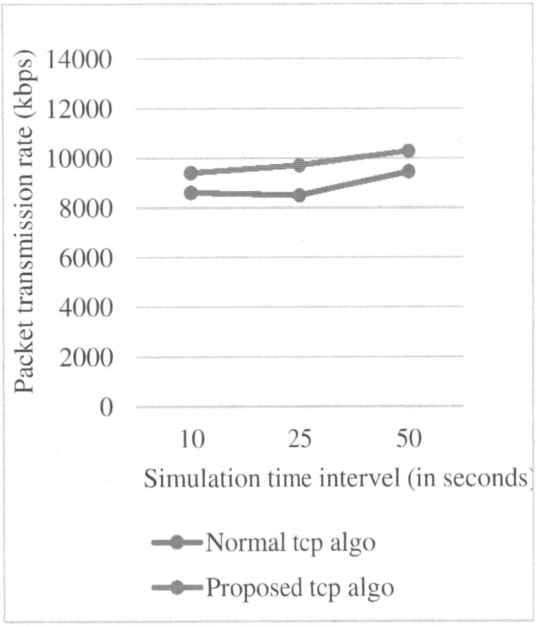

Figure 10 Pdr of AODV routing protocol

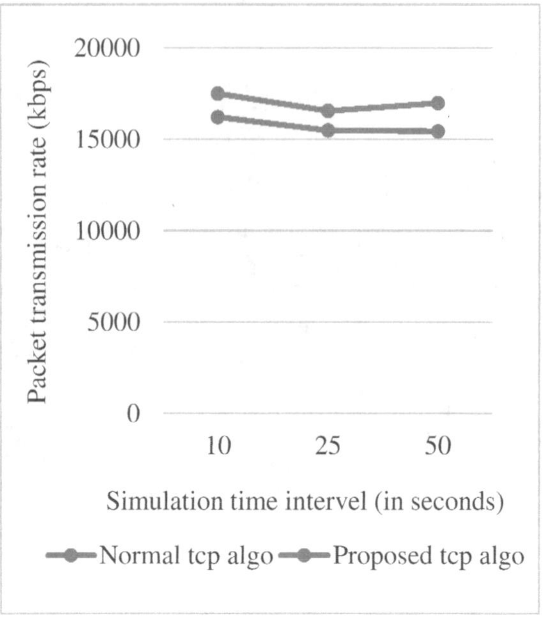

Figure 11 Pdr of DSDV routing protocol

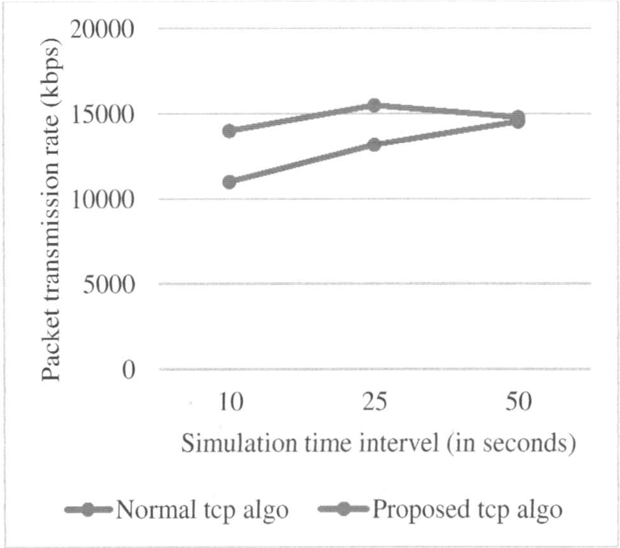

Figure 12 Pdr of DSR routing protocol

From the above-collected data of simulation, we got the graphs as follows. On the x-axis, it is representing different simulation times and, on the y-axis, it is representing packet transmission rate in kbps. It is clear from the graphs that our proposed TCP algorithm provides a better packet delivery ratio in comparison to normal TCP congestion.

7.3. Average throughput

The throughput of a signal is defined as the actual amount of data a receiver received at a particular time. For calculating the average throughput, simulation was done for varying simulation times, for 10 seconds, 25 seconds, and 50 seconds. This step repeated to all three protocols.

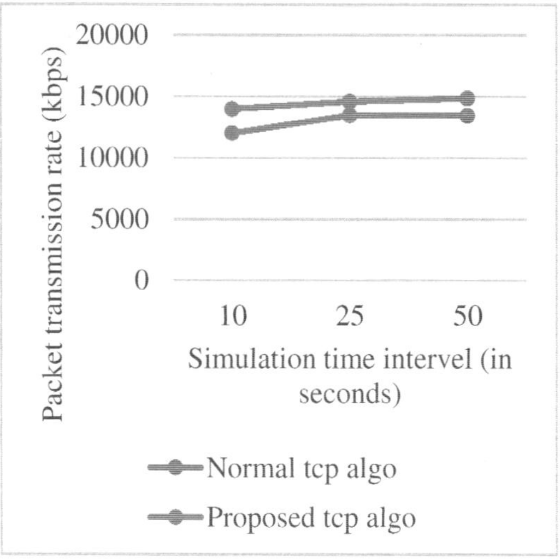

Figure 13 Avg. throughput of AODV

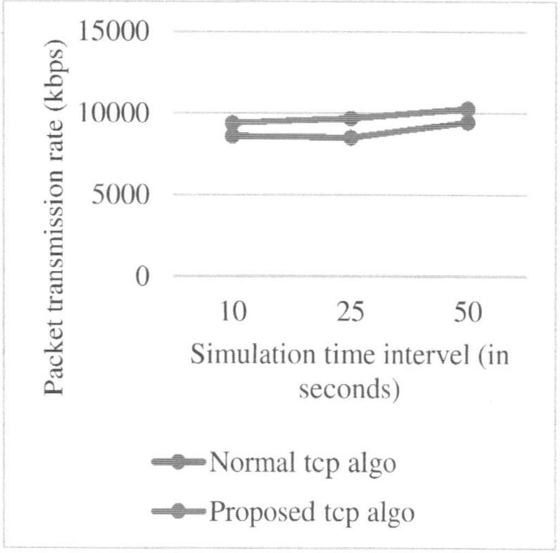

Figure 14 Avg. throughput of DSDV

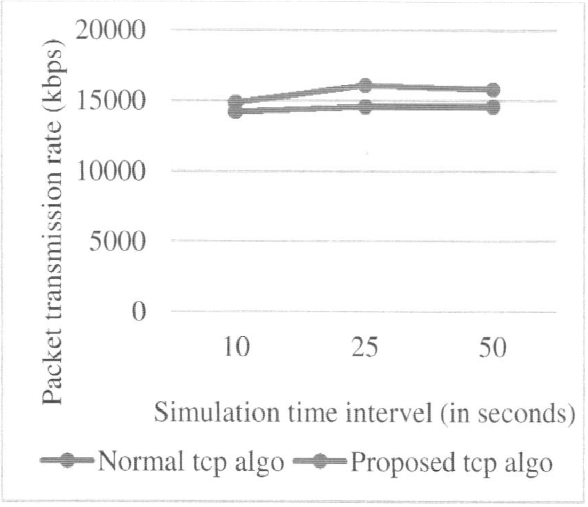

Figure 15 Avg. throughput of DSR

From the above graph, we can see new proposed TCP algorithm is providing better average throughput than compared to default congestion algorithm.

7.4 End-to-end delay

End-to-end delay is defined as the time taken to reached the packet from source to destination across the network. To calculate the end-to-end delay we applied the same procedure to all three routing protocols and measured the simulation outcome.

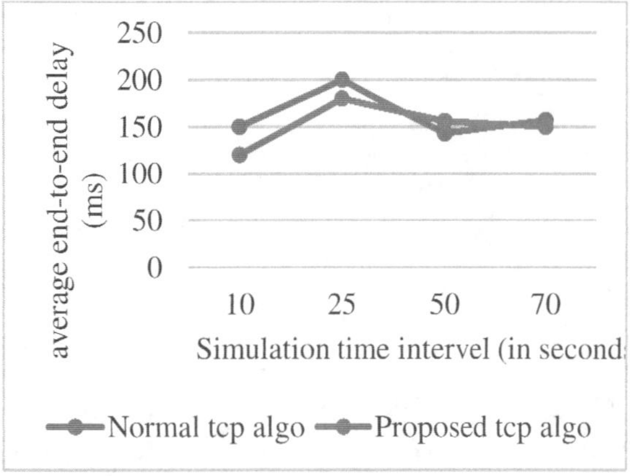

Figure 16 End-to-end delay of AODV

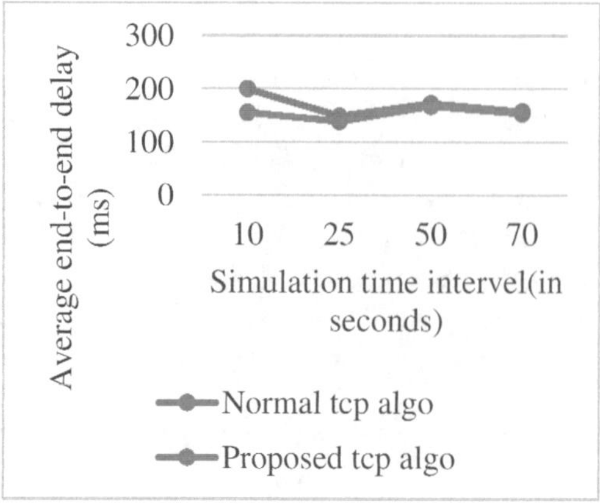

Figure 17 End-to-end delay of DSDV

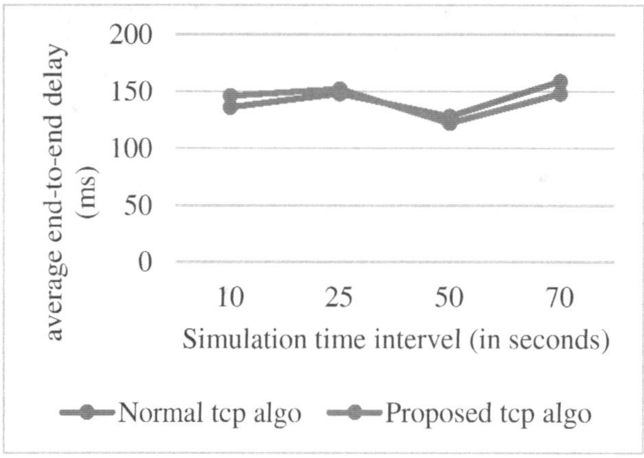

Figure 18 End-to-end delay of DSR

From the above graph, we can see that our proposed TCP algorithm shows better results than that of normal TCP congestion. From all these simulation results we can now assure that by fixing a congestion window value, pdr and throughput value can be increased along with the less end-to-end delay of packet transmission.

8. Conclusion

In this paper, we have proposed an ideal congestion control mechanism that can maintain the congestion at a very low and constant level. As a result of this algorithm, the network experienced a congestion-free transmission and it leads to a huge difference in its performances. A constant window value reduces overload inside the network and offers free paths between the nodes to communicate and hence it reduces the time taken by the packets to move from source to destination. While the end-to-end delay between nodes decreases, its packet delivery ratio, and throughput automatically get increase. Simulations are carried out to verify the performance of the proposed technique and this shows that the proposed technique provides a high rate of throughput and efficient pdr value. In future, the work on scalability and latency can be performed when the number of vehicles goes increases. It is believed that by applying a stabilized congestion control technique the vehicles can communicate with their neighbor easily and it also helps to build an efficient communication system.

References

[1] Mingwei Li and Yuanwei Jing, "Feedback congestion control protocol for wireless sensor networks", IEEE 2012, 24th Chinese Control and Decision Conference (CCDC) - Taiyuan, China, 2012.05.23-2012.05.25, 4217–4220, doi:10.1109/CCDC.2012.6244675

[2] Kang Qiaoyan, Wang Jianfeng, and Meng Xiangru, "Vegas-like Expert-Controlled Multicast Congestion Control Algorithm for Wireless Networks", IEEE 2009, International Conference on Artificial Intelligence and Computational Intelligence -Shanghai, China,2009.11.7 - 2009.11.8, doi:10.1109/aici.2009.182

[3] Li Tao, Wu Lingyun, Liu Feng, Wang Yanwu and Wu Zhang, "Stability and control of network congestion control system", IEEE 2009, Chinese Control and Decision Conference (CCDC) - Guilin, China, 2009.06.17-2009.06.19, 381–384, doi:10.1109/ccdc.2009.5195085

[4] Yang Xinhao and Li Ze, "Congestion Control based on node and link in wireless sensor network", IEEE 2016, 35th Chinese Control Conference (CCC)-Chengdu, China, 2016.7.27-2016.7.29, 8383–8386, doi:10.1109/ChiCC.2016.7554692.

[5] Naumov V and Gross T, "Connectivity-Aware Routing (CAR) in Vehicular Ad-hoc Networks", IEEE INFOCOM 2007, 26th IEEE International Conference on Computer Communications, doi:10.1109/infcom.2007.223.

[6] Genping Liu, Boon Seet, Kai Juan, and Wong K, "A Routing Strategy for Metropolis Vehicular Communications," International Conference on Information Networking, vol. 2, pp. 533-542, February 2004/ 10.1007/978-3-540-25978-7_14

[7] Karp B and Kung H, "GPSR: greedy perimeter stateless routing for wireless networks", 6th Annual international conference on Mobile Computing and Networking, Massachusetts, United States, August 06 - 11, 2000, pp. 243-254.

[8] Zhang Yueping and Loguinov Dmitri, "***Local and Global Stability of Delayed Congestion Control Systems***", IEEE-2008, 53(10), 2356–2360, VOL. **53**, NO. 10, doi:10.1109/TAC.2008.2007135

[9] Sepulcre M, Gozalvez J and Härri J, "***Application-Based Congestion Control Policy for the Communication Channel in VANETs***", IEEE-2010, 14(10), 0–953. VOL. **14**, NO. 10, doi:10.1109/lcomm.2010.091010.100345

[10] "A Comparative Performance Analysis of AODV, DSR and DSDV Protocol for Mobile Ad-hoc Networks", IJARCCE, ISO-3297, 2007 Certified Vol. **6**, Issue 6, June 2017, DOI10.17148/IJARCCE.2017.6636

[11] Kashif Naseer, Farooq Islam, "Improved Road Segment-Based Geographical Routing Protocol for Vehicular Ad-hoc Networks", Received: 2 May 2020 / Revised: 25 July 2020 / Accepted: 30 July 2020 / Published: 4 August 2020/ doi.org/10.3390/electronics9081248

[12] José Alfonso, Aguilar Calderón and Luis Urquiza, "Survey on Routing Protocols for Vehicular Ad Hoc Networks Based on Multimetrics", Received: 3 August 2019; Accepted: 9 October 2019; Published: 17 October 2019

[13] "***Performance Comparison of AODV, DSDV and I-DSDV Routing Protocols in Mobile Ad Hoc Networks***", European Journal of Scientific Research, 2009, ISSN 1450-216X Vol.**31** No.4, pp.566-576.

[14] Laith Al, Mohammed Alreshoodi and Abdelhabib Bourouis, "QoE-Aware Geographic Routing Protocol for Video Streaming over Vehicular Ad-hoc Networks" Received: 1 April 2020; Accepted: 26 May 2020; Published: 31 May 2020, doi:10.3390/computers9020045

Journal of Physics: Conference Series **2327** (2022) 012051 doi:10.1088/1742-6596/2327/1/012051

Performance Analysis of Massive MIMO under pilot contamination.

Ragit Dutta

Student, Lovely Professional University, Phagwara, India

*Corresponding author's e-mail: ragragitit1996@gmail.com
Orchid ID: https://orcid.org/000-0002-8276-6741

Abstract. 5g is fast becoming a reality in the modern world. Telecom companies all around the world are already rolling out 5G in various parts of the world. As of now there are 3 key technologies that will enable us to implement 5G in our surroundings. Massive MIMO is one of them. Massive MIMO focuses on the aspect of improving the spectral efficiency. Tremendous data rates are achievable with spectral efficiency. However there are several bottlenecks in the implementation of Massive MIMO. Pilot Contamination is one of the most prominent issues in massive MIMO. This paper is about addressing the issue of Pilot Contamination in Massive MIMO. This paper also focuses on the various mitigation schemes proposed by researchers in the past.

1. Introduction

Enhanced data speed is something everyone desires. Widespread 5G is becoming a reality in the near future. Many telecom companies have already tested and implemented 5G in different parts of the world. Many mobile companies are already rolling out 5G phones. However, we will have to completely change the technical and software setups to implement it. Previous mobile network generations have subsequently improved themselves by exploiting the two degrees of freedom; frequency and time. Three prominent techniques are being discussed that will enable 5G; Ultra densification of cells and millimeter wave both have major problems that are hard to sort out[9].The focus of this paper is on massive MIMO. In simple terms, this technique refers to adding antennas at both receiver and transmitter sides. This is done so as to exploit the gains of spatial directions. However, pilot contamination is a major issue that acts like a bottleneck in Massive MIMO. This issue arises while estimating the channel state information of the network.

Data Usage has exponentially grown in the last decade. The amount of data handled by wireless networks in 2010 was just under 3 Exabyte. The same number had grown to 190 Exabyte in 2018. After the corona virus pandemic, it has become 600 Exabyte [9]. Video file transmission is being seen as the driving force behind this exponential increase. Moreover, in recent times, smart appliances and IOT integrated environments are becoming extremely common all over. All these require that the present network capacities be at least increased by a thousand times. Massive Multiple Input Multiple Output is a strong solution to this requirement. "1000 X" network goal can be achieved by this technique. As a

Journal of Physics: Conference Series **2327** (2022) 012051 doi:10.1088/1742-6596/2327/1/012051

result, it has attracted attention from industries and academia. Spectral efficiency (SE) improvement is done by it. And hence all important parameters like data rate, cost, energy efficiency etc get positively changed.

Channel state information is absolutely necessary to reap the benefits of Massive MIMO. CSI (Channel state information) is not known before hand. CSI is acquired with the help of training sequences. Training sequences are signals of known quantity. The BS transmits them during the uplink phase. Pilot sequences is the term used to refer such signals. Theoretically, the uplink sequences being used have to be orthogonal in nature. But, owing to the high mobility of the User terminals, the coherence time is very low. Low coherence time puts restriction on the no of orthogonal pilot sequences available to the BS. The channel hardening filtering scheme demands that only orthogonal signals are utilized in the channel estimation[8]. When this condition is not met, then it is rendered inefficient. Interferences arise in the channel estimation when non orthogonal sequences are used. This interference arises from the adjacent cell and are difficult to eliminate. This phenomenon is called Pilot contamination, also known as inter cell interference. Considered a secondary with systems having limited antenna arrays, Inter cell interference is regarded as a grave issue when the number of antenna arrays[11][13]. It is even more prominent in a massive Multiple Input Multiple Output network ecosystem, because it saturates the spectral efficiency.

The main focus of this report is assessing the impact of plot contamination in Massive MIMO. Prior works on pilot contamination mitigation have been thoroughly studied and analyzed. Overall, we are interested in mitigating this contamination so that the potential of massive MIMO can be totally released.

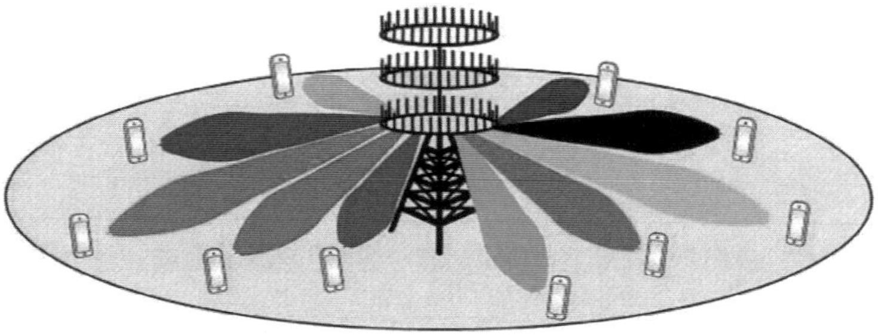

Fig 1: Massive MIMO Network

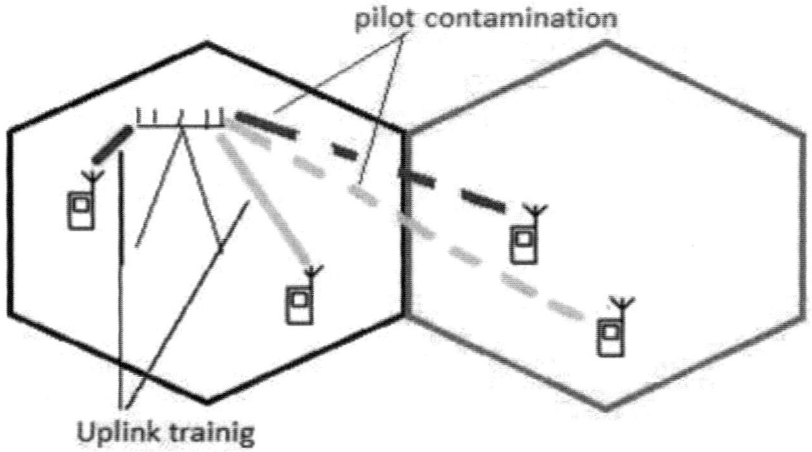

Fig 2: Pilot Contamination

Journal of Physics: Conference Series **2327** (2022) 012051 doi:10.1088/1742-6596/2327/1/012051

2. System Overview

Through this entire report, the system is a multi user Multiple Input Multiple Output (MIMO) system. In it, each BS has M antenna, serving K single user terminals. M is much larger than K. The environment has flat frequency fading. The following linear model, with additive Gaussian noise is taken into consideration. Fig 3 is the pictorial representation of the linear model [14]

$Y = Hx + (1/\sqrt{\rho})n$

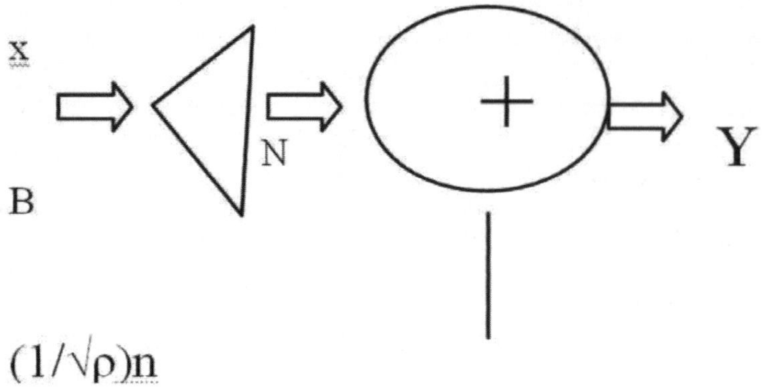

Fig 3: Linear MIMO channel mode with additive noise

n is the noise vector. n € C^N (C^N is the set of complex numbers). n is assumed to be i.i.d (independently identically distributed). Also the Signal to Noise ratio (SNR) is taken to be constant.

H € $C^{N \times B}$ is the channel matrix. Its each and every elements denote the gain of the channel between the BS antenna and each user. $H = [h_{1.}, h_2, \ldots, h_n]$ represents the magnitude, where h_i € C^N denote the individual user channels between BS and UTs. They are taken as mutually independent and constant entitities within a coherence time of T channel accesses. x € C^B is the input signal vector. Its entries are also taken as i.i.d (Independently identical Distributed) having unitary variance. Hence, the received signal y € C^N means that it is a complex Gaussian distributed having zero-mean and covariance matrix [8]

$C_y = E[yy^H] = \quad + (1/\sqrt{\rho})\, I_n$

The channel, noise and transmitted signal are mutually independent to each other; they are uncorrelated

2.1. Uplink

In uplink phase, the transmission from user to antenna takes place. K transmitters transmit pilot sequences towards the M antennas. B is the no of user terminals and N is the number of antennas at the base station M. H_{ul} is the channel matrix of size M X K, and x_{ul} is the transmitted size K X 1. n_{ul} and y_{ul} are the additive noise and received signal vector respectively. Both of size M.

2.2. Downlink

The roles get reversed in the downlink phase. In this phase, the transmission takes place from antennae to the user terminals. Therefore, we define the received signal as

$Y_{dl} = H_{dl} x_{dl} + (1/\sqrt{\rho_{dl}})\, n_{dl}$

with H_{dl} € $C^{K \times M}$, x_{dl} € C^M, n_{dl} € C^K and y_{dl} € C^K.

2.3 Channel reciprocity

Time division duplexing (TDD) and Frequency division duplexing are currently considered in 2 way communication of digital systems. The key difference between TDD and FDD is channel reciprocity. By reciprocity, it means that when a channel knows about the uplink data, then it will know about the downlink data simultenously. Reciprocity is observed in TDD, but not in FDD. In FDD, the channel needs feedback in downlinking, to estimate the channel state.

2.4. Law of Large Numbers

This law is an important law in statistics and probability. In order to understand massive MIMO, one must know this law. This section gives a brief explanation to the Law. The law states that "as a sample size grows, the mean gets closer to the average of the population."

3. Massive MIMO

Massive MIMO is the scalable option of mu MIMO. The number of antennas in the arrays of BS is minimum 64 or 128. Massive MIMO can be mistaken for MIMO with very high numbers of antennas. It is not so. There are differences between MIMO and massive MIMO in terms of parameters like duplexing mode, channel acquisition, pre coding, performance etc. The key difference is the relation between the number of antennas and user terminals. M>>>K in Massive MIMO and it is not so in MIMO. Also, large numbers of antennas and user terminals are seen in massive MIMO. Massive MIMO improves Spectral Efficiency, which is why it can form as the backbone of a massive MIMO[10].
The pros of Massive MIMO can be summed up in the following bullet points:-

- Multiplexing gain: The network capacity can be increased by 10 times in Massive MIMO. The reason being the aggressive spatial multiplexing used in massive MIMO.
- Spectral efficiency: Spectral efficiency is observed in massive MIMO systems due to the enormous numbers of antenna Arrays used in the massive MIMO systems
- Energy efficiency – Massive MIMO systems, when they meet the orthogonality condition in channel state estimation are proven to be highly energy efficient. At the presence of accurate CSI, the uplink and downlink transmit power can be efficiently utilized in the network. Each BS antenna array will use the minimum power
- Simple linear processing: Because BS station antenna is much larger than the UT antenna (M>> K), simplest linear pre-coders and detectors are optimal
- Increased robustness and Reliability – Due to the high diversity covered by the propagation channel, link performance and better data rates are observed in Massive MIMO systems.

4. Pilot Contamination

As mentioned before, accurate and reliable estimation of Channel State Information is needed in a massive MIMO network. The estimation is done by pilot sequences. Pilot sequences are known signals that are sent across the channel during uplinking. In an ideal scenario, each antenna of each BS should be given orthogonal training sequence. Practically, this is very hard to achieve. The main reason behind this limitation is the low coherence interval. Coherence interval is low because the user terminals are very mobile, and are constantly moving from cell to another. As a result, training sequences have to reused during the channel acquisition phase. This act is known as pilot contamination. The term "Contamination" is used as it is seen as an undesirable signal. This problem leads to the saturation in the network gain of systems having very high numbers of antennas. [10]

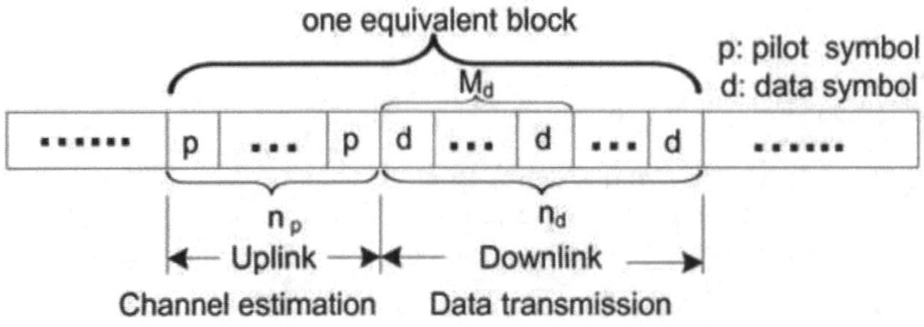

Figure 4: Time division duplexing sequence

4.1. Source of Pilot Contamination

This section throws light on the all known causes of pilot contamination in massive MIMO. All of these reasons are thoroughly summarized in a sequential manner.

1) Usage of pilot sequences that are not orthogonal – The user terminals in a cell are very mobile. Hence the coherence time is very low in a cell. This low coherence time acts as an obstacle in allocating a large number of non orthogonal pilot symbols. Theoretically, as orthogonal pilots would need length of least K * L symbols (K= total number of UTs in a cell and L = total number of cell in the system) owing to frequency reuse factor of one. This is not possible, owing to the short coherence time of user terminals. Hence, non orthogonal sequences have to be used in estimating the Channel State Information. These causes a phenomenon known as pilot contamination (inter cell interference). The issue is not desired, as because of it, the gain of the network gets saturated; it stagnates and does not increase with ascending numbers of antenna. The disturbance is brought about by the interference from adjacent cells. The Signal to Interference ratio (SIR) is made finite.

2) Hardware Impairment: Hardware bugs can also cause it. Various impairments like Quadrature imbalance (I/Q), quantization noise, Amplifier non linearity etc can be seen in the hardware of radio frequency chain. [11]

4.2. Techniques to mitigate Pilot Contamination

- Protocol Based Technique:- It is the simplest method of mitigating this issue. It simply means decreasing the no of users that use orthogonal sequences. Technically, it decreases the frequency reuse factors. Slight betterment is seen by this method. But generally, since fewer user are served simultaneously, frequency reuse may make little difference.
- Precoding Technique:- This technique seeks to reduce the minimum mean square error (MMSE) of the user and the interference. A particular base station will have precoding matrix in it. Precoding matrix contains algorithms that will reduce the MMSE within the process Multicell cooperation is the basis of this technique. It has limitations due to the ascending information overhead exchange.
- ‘Blind Technique:- Subspace partitioning is the basis of this technique. Systems with power controlled handoff strategy use it. Blind technique will randomly allocate operational duration to the sequences [12]

5. Literature Review

While preparing for this paper, several papers have been studied. Many notable researchers have done research on pilot contamination. The authors Jeffrey G. Andrews and Stefano Buzzi explored the needs and requirements of 5G in the near future. 5g won’t just be a generational upgradation of mobile broadband services. Many technologies like virtual reality, Internet of things, smart appliances will be made possible by 5G. Unlike 4G, 5G will be very integrative. They pointed out three prominent

technologies that will enable 5G; ultra densification, millimeter wave and massive MIMO. Also, there are issues of implementation and standardization.
Erik G. Larsson, Ove Edfors and Fredrik Tufvesson pointed out the various advantages over conventional point-to-point MIMO; like use of inexpensive single-antenna terminals, non requirement of rich scattering environment, and simplified resource allocation. Pilot contamination is one issue that they identified in their paper. Their article presents an overview of the massive MIMO concept and contemporary research on the topic [10].

Elijah, in his paper emphasized on the pros and cons of massive MIMO. According to him, precise Channel State Information (CSI) amongst the Base stations is vital for the success of massive MIMO. He specifically classified the pilot contamination mitigation schemes into pilot-based approach and subspace-based approach. Finally, he summarized the open issues such as training overhead, deployment scenario, computational complexity, use of channel reciprocity etc. [11]

Moving on, in another paper, a certain Jakob Hoydis proposed a multicell MIMO uplink channel. In it, each base station (BS) has a huge number of antennas N. A general assumption in the paper is that the Base stations perform channel estimation with the training sequences of the User Terminals. According to him, as N tends to infinity, (i) The matched filter (MF), becomes optimal. Matched filtering is the simplest mode of signal detection (ii) The magnitude of transmit power per user can be made negligible, (iii) Pilot contamination is the major obstacle of system performance. His paper primarily revolved on knowing the maximum number of antennas upto which the first two major conclusions can be true [1].

Deli Qiao explored the various additional gains that can be achieved by modifying the distribution of antennas at multiple levels. Area spectrum efficiency (ASE) and Area Energy Tradeoffs were recognized as two factors that influence the performance of these scenarios. Distribution of antennas in cells can be a factor in coordinating the operation of cells, equipped with substantial antennas [2].

According to Lu Lu, MIMO network means equipping base stations (BS)with a very large number of antennas, and it is seen that exponential improvement in spectral and energy efficiency can be achieved using simple processing. The main focus of the paper are The speculatated advantages of massive multiple input multiple output (MIMO) are the main focus of the paper. Also the various implementation problems related to communication channel estimation, pre-coding method and detection are addressed in the paper [3].

Marios Kountouris in his paper worked on integrating massive MIMO base stations with TDD based HetNet(Heterogenous Network). His focus was on enhancing the area spectal efficiency and coverage probability. He worked on finding the correct duplexing mode of Single carrier. [4][5]

All these extensive literature review has helped me a lot in gaining good knowledge about the problems faced in implementation of 5G. A lot is yet to be researched and studied about pilot contamination. Unless not solved, it will be a difficult task to achieve the goal of superfast integrative internet

Journal of Physics: Conference Series **2327** (2022) 012051 doi:10.1088/1742-6596/2327/1/012051

6. Results and Discussion

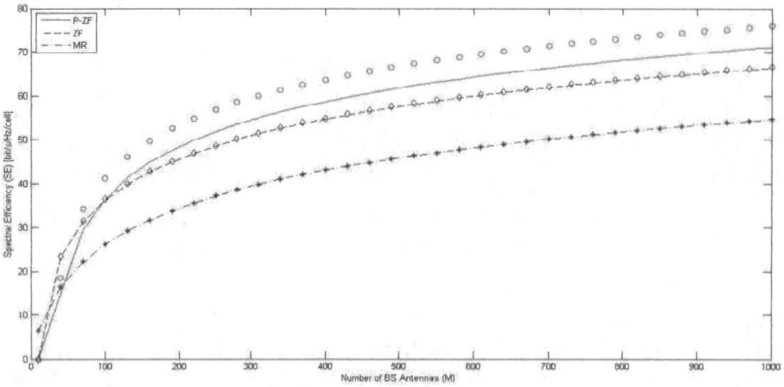

Fig 5: Graphical representation of Spectral efficiency vs No of Base station antennas in pilot contaminated system

Fig 5 of the simulation denotes the Graphical representation of Spectral efficiency vs No of Base station antennas in pilot contaminated system. It is observed that there is saturation in the spectral efficiency, even though there is increase in number of antennas in Base stations. This stimulation was done to see the comparison of spectral efficiency across the various filtering schemes used like Pole Zero forcing(P-ZF), Zero forcing(ZF) and Maximum Ratio(MR). It is desired that linear growth of spectral efficiency is obtained. However, it is observed that there is saturation of Spectral Efficiency in all the filtering schemes. This is not desired.

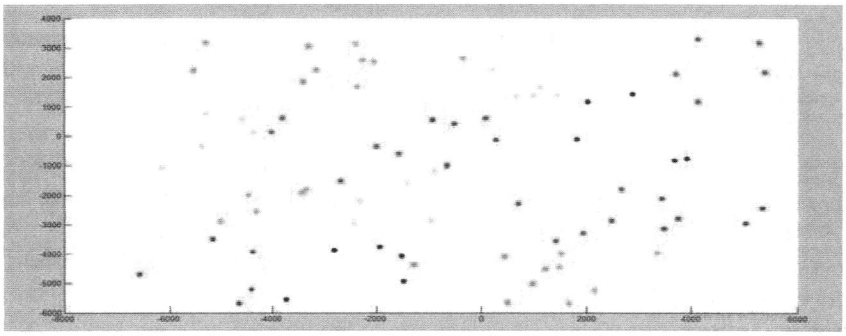

Fig 6: Graphical representation of users in the network

Figure 6 denotes the distribution of users in the network. It is the simulation part of the userplacement module which I worked on. In it, the module randomly places the users in cells of 1.6 km. The module takes in the Base station co ordinates and defines the hexagonal cells

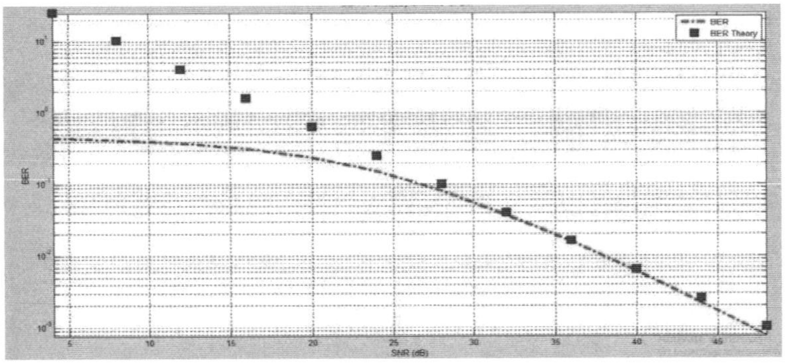

Fig 7: Graphical representation of BER v/s SNR graph

Figure 7 denotes the representation of BER v/s SNR graph. As observed, in both theory and practical terms, the BER continuously decreases with time.

7. Conclusion

We observe that 5g is a reality in the new future. Massive MIMO is one technology that will enable to implement 5G. Massive MIMO suffers from the issue of Pilot contamination. Various studies have been done to mitigate pilot contamination. Yet a lot is left to be desired in solving this issue. This paper is a summary of the various past research work done on this field. It also analyses the various mitigation schemes.

Acknowledgement

I would like to express my very great appreciation to **Dr. Ajay Roy** (Associate Professor, School of Electronics and Electrical Engineering, Lovely Professional University, Phagwara, Punjab, India.) for his valuable and constructive suggestions during the planning and development of this research work. His willingness to give his time so generously has been very much appreciated.

References

[1] Jakob Hoydis et. al. “Massive MIMO: How many antennas do we need”, IEEE-2011.

[2] Deli Qiao, “Massive MIMO architecture for 5G networks: Co-located, or distributed”, IEEE-2014.

[3] Lu Lu, Student Member, IEEE, Geoffrey Ye Li, “An Overview of Massive MIMO Benefits and Challenges”, IEEE journal of selected topics in signal processing, vol. 8, no. 5, october 2014.

[4] Marios Kountouris, Nikolaos Pappas, “HetNets and Massive MIMO: Modeling, Potential Gains, and Performance Analysis”, 19 sep 2013.

[5] Tadilo Endeshaw Bogale, “Massive MIMO and Millimeter Wave for 5G Wireless HetNet: Potentials and Challenges”, IEEE vehicular technology magazine, 21 oct 2015.

[6] T. L. Marzetta, “Multi-cellular wireless with base stations employing unlimited numbers of antennas,” in Proc. UCSD Inf. Theory Applicat.cWorkshop, Feb. 2010.

[7] H. Q.Ngo,E.G.Larsson, and T. L.Marzetta, “Energy and spectral efficiency of very large multiuser MIMO systems,” IEEE Trans. Commun.,cvol. 61, no. 4, pp. 1436–1449, Apr. 2013.

[8]https://www.avnet.com/wps/portal/abacus/solutions/markets/communications/5g-solutions/understanding-massive-mimo-technology/

[9] J. G. Andrews et al., "What Will 5G Be?," in IEEE Journal on Selected Areas in Communications, vol. 32, no. 6, pp. 1065-1082, June 2014, doi: 10.1109/JSAC.2014.2328098.

[10] E. G. Larsson, O. Edfors, F. Tufvesson and T. L. Marzetta, "Massive MIMO for next generation wireless systems," in IEEE Communications Magazine, vol. 52, no. 2, pp. 186-195, February 2014, doi: 10.1109/MCOM.2014.6736761.

[11] O. Elijah, C. Y. Leow, T. A. Rahman, S. Nunoo and S. Z. Iliya, "A Comprehensive Survey of Pilot Contamination in Massive MIMO—5G System," in IEEE Communications Surveys & Tutorials, vol. 18, no. 2, pp. 905-923, Secondquarter 2016, doi: 10.1109/COMST.2015.2504379

[12] Mishra, Prof. (2016). REVIEW OF THE PILOT CONTAMINATION PROBLEM FOR MASSIVE MIMO AND POSSIBLE SOLUTION. 10.5281/zenodo.57041.
[13] https://youtu.be/L985zYsIuwc
[14] Y. Wu, T. Liu, M. Cao, L. Li, and W. Xu, "Pilot contamination reduction in massive MIMO systems based on pilot scheduling", EURASIP Journal on Wireless Communications and Networking, vol. 2018, no. 1, Jan. 201

Intelligent Reflecting Surface Aided MIMO - A Review

Praveen Srivastava[1], Nishant Tripathi[2],Vikas Srivastava[2], Utkarsh Pnadey[2]

[1, 2] School of Electronics and Electrical Engineering
[1, 2] Lovely Professional University, Punjab, India
[1, 2]Department of Electronics & Communication Engineering
[1, 2] Pranveer Singh Institute of Technology, Kanpur, India
praveensri4u@gmail.com[1], nishant.tripathi@psit.ac.in[2], utkarsh.pandey08@gmail.com[2]
ORCID Id 0000-0002-7962-8815[1]

Abstract. Intelligent Reflecting Surface (IRS) is a transformative and innovative technique for enhancing the energy efficiency of modern day communication for 5G and beyond networks. Multiple input multiple output (MIMO) is widely used technology as it enhances the energy and spectral efficiency of network. IRS in combination with MIMO is widely used to reduce the non-Line of sight communication losses and also to increase the efficiency. IRS is a tool which changes the phase of the incoming signal by transforming it to a discrete phase and retransmits it towards user. This paper present a deep insight to IRS aided MIMO communication network technology and latest work done in this area. A comparative analysis of works of different authors is represented in a tabular form so as to have a better insight into this area.
Keywords: Intelligent reflecting surface, MIMO, Energy efficiency, beamforming, LOS, Optimization

1. Introduction

In modern time, communication have become an integral part in common man's life, it help people to transfer real time information to remote areas. Wireless communication is emerging to new heights with increasing number of services offered by telecommunication companies like speed, flexibility and network efficiency. It has become an important tool for technology driven new generation of communication by facilitating easy information sharing and boosting productivity. One can roam in any part of world freely and stay connected with people without worrying about internet connectivity. One of the major problems faced by user especially in metropolitan cities is the blockage of signal due to the high rise obstacle like tall buildings or lying in the shadow area of any obstacle. Intelligent reflecting surface (IRS) plays a major role in eliminating these problems by directing the signal coming from base station towards the user, IRS is commonly deployed on the outer surface of tall buildings thus it reflects the signals towards the user. This eliminates the multipath distortion, shadowing effects and blockage of LOS signals. IRS not only reflects the signals but it also changes the phase of the incoming signals by shifting it to discrete phase levels, thus improving the quality of signal which is reaching towards users. MIMO is latest technology which is used to increase spectral and energy efficiency of the network, it also increases the channel capacity thus helping in meeting demands of large number of users and their variety of needs. With the growing number of users and formation of small cells in a network MIMO is a leading technology for 5G and beyond generations of communications. IRS in combination with MIMO would become a game changer technology for future energy hungry networks.

There are several generation of wireless technology like 1G,2G,3G,4G and beyond. Presently we are using technology up to fourth generation but research is going on for 5G and beyond technology. A new kind of services with better internet connectivity, speed and low latency would be offered by 5G and coming generation. Mobile wireless communication is one of the major game changer in industrial development and economic growth in modern societies. Its ability to provide cutting edge technologies to end users has made it to be essential part in human life. Thus IRS together with MIMO would lead the incoming 5G and beyond generations. The present work is a backbone for setting up infrastructure for future generation network in 5G/6G system.

1.1 Motivation

During the last few decades researchers were focused on mitigating the channel related impairments but in last few years with the growth of meta materials and meta surfaces which ultimately led to the development of IRS, the propagating environment has been reconfigured for wireless communication. The advantages of MIMO like improved spectral and energy efficiency has led to the development of IRS assisted MIMO network. Hence nowadays researchers are analyzing and focusing on development of IRS and MIMO communications.

1.2 Beneficence of the Article

Major beneficence of this work is listed as follows:

- Literature review of latest research works done in this area is presented in an organized manner. Background of MIMO communication and IRS system is highlighted here.
- This paper present an extensive review of different methodologies used in latest works like optimization algorithms, simultaneous optimization of base station beamforming and passive beamforming at IRS etc.
- Representation of latest works is presented in a tabular form with methodologies and outcomes/limitations.

1.3 Related Works and Existing Surveys

The updated research works done in the area of IRS-MIMO is summarized in table4. These works are latest to the best of author's knowledge and shows a comparison of existing surveys in IRS-MIMO.

1.4 Article scheduling

The remaining part of paper is classified as follows. Fundamental concepts of IRS are presented in section 2. Section 3 shows the combination of IRS with MIMO technologies thus utilizing the best part of both the world. Section 4 presents the latest work done in the field of intelligent reflecting surface aided with MIMO communication. Various techniques, methods and optimization algorithms have been discussed with the existing technologies. Section 5 summarizes the latest works in a tabular form and showed various strategy/ methodologies with the outcomes and limitations. The conclusion is presented in section 6.

2. Intelligent Reflecting Surface

Large intelligent reflecting surface consist of a set of active or passive reflecting elements which changes the phase of the incoming signal so that it combines constructively at the user. Deployment of IRS changes the overall propagation environment of the communication system and by using MIMO it increases the throughput as compared to without using IRS. The propagation paths between base station and IRS is configured which changes the efficiency. The IRS is advantageous in enhancing the SINR of present communication system without requiring any additional power or changes in infrastructures.

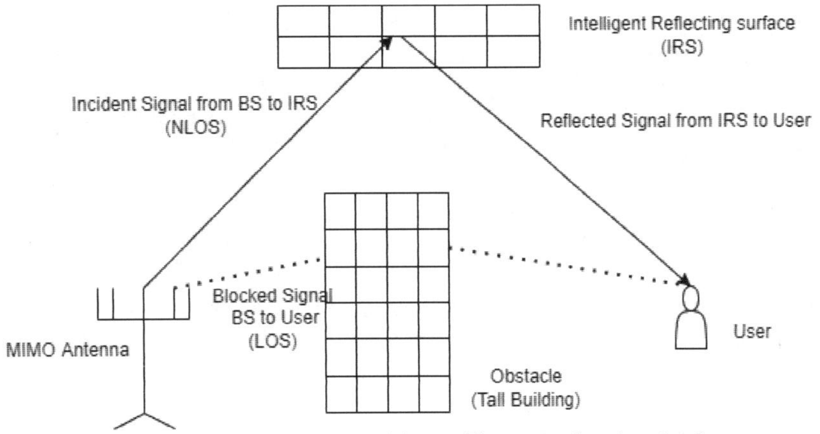

Figure 1: Communication with Intelligent Reflecting Surface

The basic role of IRS elements is to modify the features of incoming signal like phase, frequency, amplitude and polarization, which results in highly focused beam-forming towards the user. IRS aided future generation networks. The critical issue faced by incoming generation networks is the blockage of Line of sight signal.IRS plays a pivot role here due to its ability to correct propagation environment. Hence IRS modifies characteristics of environment into a useful resource which enhances the network performance. The 3dimension deployment of IRS in network removes the dead-spot. Thus IRS configures the wireless channel, to assist in sensing user. As the number of users grows in any network, the network should become more spectrum efficient and energy efficient. Intelligent reflecting surface innovate and optimizes the existing network for future 5G and beyond networks [1]. MIMO has the advantage that simultaneously many signals could be transmitted and received thereby increasing the capacity of the system. MIMO network can handle several users while maintaining good quality of services (QoS).

3. Combined IRS –MIMO Communication

The development of MIMO technology has helped a lot to improve the energy and spectral efficiency of network. The scattering environment affects the performance of communication system. Signal blockage, diverse frequency effect and complex system increases requirements of energy. This demands the need of low power consumption and low complexity network. To resolve these issues IRS becomes a beacon of hope, by emerging as front runner to meet the requirements of coming generation communication networks. Jointly using IRS and Multiple input multiple output system will increase the spectral efficiency with little additional complexity which can serve multiple purposes like IRS aided MIMO communication, point to point communication, SWIPT MIMO , Multicell communication and Cognitive Radio system.

4. Related Works

Authors in [1] presented exhaustive review on intelligent Reflecting Surface aided multiple input multiple output (MIMO) system. They showed that use of IRS enhances performance of network and its application with MIMO is considered to be lead runner in 5G/6G and beyond networks. They have discussed the basic concepts and key issues which needed to be addressed like Channel estimation protocols, IRS positioning strategies, applications, algorithms and design considerations of IRS aided MIMO networks were presented. Authors in [2] investigated the wireless transmission of signals in harsh environment. They optimized the spectral efficiency and power control in a MIMO network. They have considered uplink and downlink transmission in a MIMO network. Spectral efficiency and power control are optimized using ensemble method and performance compared with gradient descent method, Newton method and genetic algorithm. Artificial Intelligence (AI) based ensemble method, used on the neighborhood field optimization method. They have compared results with benchmark schemes and also shown that proposed algorithm significantly improved the spectral efficiency in a MIMO network as compared to conventional methods. Authors in [3] investigated the resource allocation algorithm design for IRS aided simultaneous wireless information and power transfer (SWIPT) system. They adopted physics based IRS structure which measures the angle of incidence and reflection on IRS. They partitioned IRS into set of different tiles and jointly designed the phase shift of each element into different transmission modes and minimized the transmitted power by using penalty based method, successive approximation and semi definite relaxation algorithm to solve the joint optimization problem of transmitted power and reflection coefficients of IRS elements.

As in [4] authors examined the deployment of intelligent reflecting surface (IRS) in covert wireless environment. They formulated the jointly design of transmitted power at base station and reflection coefficients of reflecting surface for covert communication considering global channel state information (CSI) without wardens instantaneous CSI. They developed a penalty successive convex approximation (PSCA) algorithm and derived the analytical expression for transmit power and reflection coefficients of IRS. They also showed that significant growth can be gained by using IRS in covert communication. In [5] authors discussed the channel modeling for SISO wireless system considering channel to be linear. Authors optimized the channel parameters for IRS assisted wireless networks and showed that wireless communications using IRS has given a paragon shift in the design, as compared to the commonly used one in wireless communication ,where only active components were considered while hybrid architecture containing of both active and passive components will work together in a smart manner. Although IRS aided communication is in a nascent stage, this work showed the fundamentals of IRS assisted networks. Authors in [6] investigated the passive beamforming and IRS placement design for IRS used full duplex system. They aimed to reduce the transmitted power of the base station and UL user by jointly optimizing their transmitted power and reflection coefficients of the IRS elements. They further investigated the minimum power, required in IRS supported wireless system for different IRS deployment design along with its comparison with half duplex system. They also solved the power minimization problem by converting it into measurable form and proposed an efficient algorithm to solve it which is based on BCD method.

In [7] authors showed that their exist an optimal IRS and Base station density ratio which can enhanced the system throughput. They deployed IRS based on stochastic geometry and found a cost effective deployment in terms of throughput per unit cost. They model the random BS/IRS locations by independent homogeneous Poisson point Processes (HPPP) and proposed a practical UE to IRS association rule. They unveiled the existence of optimal IRS and BS density ratio for hybrid system under a given cost of deployment. Authors showed that the IRS/BS density ratio could be used for providing cost effective approach to achieve sustainable capacity growth for future wireless networks. In [8] authors improved the beamforming gain for MIMO system by using Semi Definite Relaxation (SDR) technique. Authors proposed a quantization algorithm for discrete phase shifts at IRS. They converted the non-convexity problem of beamforming design into converging form by using SDR technique. In [9] authors investigated the different estimation technique of channel and designing of shifting phase of IRS element used MIMO for one user. They proposed Single Path approximated Channel (SPAC) with emphasis on rank one matrix (SEROM) for channel estimation; they also developed solution for designing of phase shift element by converting the non-convex problems into set of linear equations which could be solved easily. They developed combined framework for estimation of channel and designing of IRS element.

In [10] authors investigated estimation of channel and low cost beamforming by developing cooperative beam training method to help channel estimation by using IRS. They designed two hierarchical codebooks for the proposed training procedures, which is balancing between robustness against noise and searching complexity. They further proposed two hybrid beamforming (HB) designs for single and multiuser cases. Their simulation results proved that the projected joint beam training and hybrid beamforming is achieving results comparable to fully digital beamforming implemented under perfect channel state information. In [11] authors investigated the complexity problem of IRS and used compressive sensing and deep learning tools for solving it. They proposed IRS architecture based on sparse channel sensors where few elements of the passive IRS are made active(less than 1%). They then developed two design solutions of IRS matrices with negligible overheads. This increases the spectral efficiency and energy efficiency. They designed the codebook which is based on deep MIMO dataset. In [12] the IRS aided simultaneous wireless and power transmission techniques for MIMO system is developed by changing the phase shifts of elements of IRS, reflected signal by the IRS element can be added constructively at the ERs and IRs.Weighted Sum rate maximization problem of the IR's and energy harvesting requirements of ERs are considered under non-convex unit-modulus constraints. They considered BCD algorithm for optimization of precoding matrices at the BS and phase shift matrix of the reflectors. For sub problems they developed low complexity iterative algorithms which increases the performance of the SWIPT system, their proposed algorithms converges rapidly hence is suitable for practical implementations.

In [13] authors studied the possibility of IRS deployment for increasing the beamforming gain of the uplink in MIMO. They showed that maximal ratio combiner (MRC) beamforming strategy could not be used in IRS aided system because orthogonality of two signals is lost .They proposed a new strategy of zero forcing in beamforming design. They have also shown that IRS used MIMO system achieves higher throughput as compared to communication without IRS.In [14] authors increased the cell-edge user's performance in multicell communication by deploying an IRS at the cell boundary. They cancelled the intercell interference by jointly optimizing precoding matrices at the BS and passive shifts of phase at IRS. Authors proposed a pair of algorithms for solving the optimization problem of phase shifts known as Majorization-Minimization algorithm (MM) and Complex Circle Manifold (CCM) method.

In [15] authors joined IRS with physical layer security (PLS), for increasing secrecy rate performance of MIMO channels. They proposed optimization algorithms which could work for both known CSI and unknown CSI. They proposed majorization-minimization algorithm combined with One-by-One (OBO) method for solving power minimization problem. Authors presented optimization algorithm for multiple input single output system (MISO) which could be extended to MIMO system.

5. Tabular Review

Outline of related work with IRS-MIMO system

Reference	Publication Year	Strategy/ Methodology	Outcomes/ Limitations
[1]	2021	Review of various key issues of IRS assisted MIMO technology	Presented extensive review about IRS assisted MIMO network.
[2]	2021	Artificial learning based ensemble method is used to enhance spectral efficiency and power control of MIMO in harsh environments.	Spectral efficiency and power control are optimized using ensemble method and performance compared with gradient descent method, Newton method and genetic algorithms.

[3]	2021	Partitioned IRS into different tiles. For different transmission mode phase shift elements of each tiles is jointly designed .	Resource allocation for IRS aided SWIPT system. Used physics based IRS. Penalty method, successive convex approximation and SDR algorithms used for solving optimization problem.
[4]	2021	Jointly designed transmit power and IRS reflection coefficients by showing covertness for IRS assisted delay constrained covert communication.	Improved Performance gain of IRS used covert communication. Proposed penalty successive convex approximation (PSCA) algorithm for solving design problem.
[5]	2021	IRS reflection modeling and optimization	Channel modeling for SISO system done
[6]	2021	Developed BCD based algorithms for passive beam forming design.	IRS deployment cases for FD wireless system considered. WSR of transmitted power of the Access point and Uplink user is minimized.
[7]	2021	Showed the existence of IRS and Base station density ratio which can enhance the system throughput.	Showed the spatial throughput of IRS deployment based on stochastic geometry and found a cost effective solution.
[8]	2021	Converted the non-convexity problem of beam forming design into converging form by adopting SDR technique.	Beam forming Gain improved for MIMO system by using SDR technique. For discrete phase shifts at reflecting surface, proposed quantization design for IRS elements.
[9]	2021	Proposed Single path approximated channel (SPAC) method and Selective emphasis on rank one matrix (SEROM) technique for channel estimation. Developed a solution for designing of IRS phase shift by converting it into set of linear equations.	Channel estimation methods and designing of phase shift for IRS aided one user MIMO system.
[10]	2021	Used ternary tree hierarchical beam search at Base station and users as compared to conventional binary search method	Channel estimation with low cost beam forming design dealt with. Designed two hierarchal codebooks for proposed training procedure.
[11]	2021	Used machine learning/deep learning tools for problem solving. Accurate ray tracing based Deep MIMO dataset is used for evaluation.	Sparse Channel Sensors architecture proposed for IRS. Deep learning based solutions is developed. Only few (less than 1%) elements of IRS are active rest all passive.

[12]	2020	IRS assisted SWIPT system	Performance of ER's & IR's enhanced.
[13]	2020	Illustrated limitations of Maximal Ratio Combiner (MRC) scheme and proposed Zero Forcing (ZF) design at BS and reflection coefficients design of IRS.	Feasibility of IRS deployment for MIMO system considered. Intruder interference reduced by using ZF scheme.
[14]	2020	WSR maximization solved by optimizing precoding matrices at Base station and phase shifts at the reflecting surface, considering the constraint.	Developed Majorization-Minimization algorithm and Complex Circle Manifold method for optimization.
[15]	2020	Combined IRS with Physical Layer Security (PLS) to increase secrecy performance of MIMO channels.	Enhanced the secrecy rate in the channel with full CSI and no CSI. Proposed two algorithms for optimization of transmit covariance matrix at transmitter and phase shift coefficient at IRS.

6. Conclusion & Discussion

This paper includes the latest research work done on the IRS-MIMO. Fundamentals of IRS as well as the technologies like mutual beamforming at base station and reflecting surface, optimization algorithms, semi definite relaxation method, BCD algorithms, Successive convex approximation algorithms, gradient descent method, Newton method, genetic algorithms, channel estimation, majorization-minimization algorithms, IRS deployment strategies and complex circle manifold method have been dealt with. An extensive survey with comparative summary is presented in a tabular form which gives a better insight into the area. All the latest work done in IRS-MIMO is dealt with emphasise on technology used and limitations /outcomes. As indicated through latest research work that system performance using IRS with MIMO is far better than without use of IRS, hence IRS-MIMO will be the future backbone of 5/6G and beyond networks, although still in a nascent stage a lot of research works is required in this area.

Acknowledgement

I would like to express my very great appreciation to **Dr. Shelej Khera** (Professor, School of Electronics and Electrical Engineering, Lovely Professional University, Phagwara, Punjab, India.) for his valuable and constructive suggestions during the planning and development of this research work. His willingness to give his time so generously has been very much appreciated.

7. References

[1] S. N. Sur and R. Bera, "Intelligent reflecting surface assisted MIMO communication system: A review," *Phys. Commun.*, vol. **47**, p. 101386, Aug. 2021, doi: 10.1016/J.PHYCOM.2021.101386.

[2] X. Zhang, H. Qi, X. Zhang, and L. Han, "Spectral Efficiency Improvement and Power Control Optimization of Massive MIMO Networks," *IEEE Access*, vol. **9**, pp. 11523–32, 2021, doi: 10.1109/ACCESS.2021.3051170.

[3] Y. Hu, M. Liu, and Y. Feng, "Resource Allocation for SWIPT Systems with Nonlinear Energy Harvesting Model," *Wirel. Commun. Mob. Comput.*, vol. **2021**, 2021, doi: 10.1155/2021/5576356.

[4] X. Zhou, S. Yan, Q. Wu, F. Shu, and D. W. K. Ng, "Intelligent Reflecting Surface (IRS)-Aided Covert Wireless Communications with Delay Constraint," *IEEE Trans. Wirel. Commun.*, pp. 1–14, 2021, doi: 10.1109/TWC.2021.3098099.

[5] Q. Wu, S. Zhang, B. Zheng, C. You, and R. Zhang, "Intelligent Reflecting Surface-Aided Wireless Communications: A Tutorial," *IEEE Trans. Commun.*, vol. **69**, no. 5, pp. 3313–51, 2021, doi: 10.1109/TCOMM.2021.3051897.

[6] Y. Cai, M. M. Zhao, K. Xu, and R. Zhang, "Intelligent Reflecting Surface Aided Full-Duplex Communication: Passive Beamforming and Deployment Design," *IEEE Trans. Wirel. Commun.*, pp. 1–30, 2021, doi: 10.1109/TWC.2021.3095939.

[7] J. Lyu and R. Zhang, "Hybrid Active/Passive Wireless Network Aided by Intelligent Reflecting Surface: System Modeling and Performance Analysis," *IEEE Trans. Wirel. Commun.*, pp. 1–17, 2021, doi: 10.1109/TWC.2021.3081447.

[8] C. Feng, W. Shen, X. Gao, and J. An, "Passive beamforming design for intelligent reflecting surface assisted MIMO systems," *China Commun.*, vol. **18**, no. 3, pp. 18–28, 2021, doi: 10.23919/JCC.2021.03.002.

[9] S. Kim, H. Lee, J. Cha, S.-J. Kim, J. Park, and J. Choi, "Practical Channel Estimation and Phase Shift Design for Intelligent Reflecting Surface Empowered MIMO Systems," pp. 1–14, 2021, [Online]. Available: http://arxiv.org/abs/2104.14161.

[10] B. Ning, Z. Chen, W. Chen, Y. Du, and J. Fang, "Terahertz Multi-User Massive MIMO with Intelligent Reflecting Surface: Beam Training and Hybrid Beamforming," *IEEE Trans. Veh. Technol.*, vol. **70**, no. 2, pp. 1376–93, 2021, doi: 10.1109/TVT.2021.3052074.

[11] A. Taha, M. Alrabeiah, and A. Alkhateeb, "Enabling Large Intelligent Surfaces with Compressive Sensing and Deep Learning," *IEEE Access*, vol. **9**, pp. 44304–21, 2021, doi: 10.1109/ACCESS.2021.3064073.

[12] C. Pan *et al.*, "Intelligent Reflecting Surface Aided MIMO Broadcasting for Simultaneous Wireless Information and Power Transfer," *IEEE J. Sel. Areas Commun.*, vol. **38**, no. 8, pp. 1719–34, 2020, doi: 10.1109/JSAC.2020.3000802.

[13] Z. Wang, L. Liu, and S. Cui, "Intelligent Reflecting Surface Assisted Massive MIMO Communications," *IEEE Work. Signal Process. Adv. Wirel. Commun. SPAWC*, vol. 2020-May, no. **2018**, **2020**, doi: 10.1109/SPAWC48557.2020.9154291.

[14] C. Pan *et al.*, "Multicell MIMO Communications Relying on Intelligent Reflecting Surfaces," *IEEE Trans. Wirel. Commun.*, vol. **19**, no. 8, pp. 5218–33, 2020, doi: 10.1109/TWC.2020.2990766.

[15] L. Dong and H. M. Wang, "Enhancing Secure MIMO Transmission via Intelligent Reflecting Surface," *IEEE Trans. Wirel. Commun.*, vol. **19**, no. 11, pp. 7543–56, 2020, doi: 10.1109/TWC.2020.3012721.

LOAD BALANCING USING FIXED GEOMETRIC ARRANGEMENTS OF FIXED AND MOBILE SMALL CELLS IN MILI-METER NETWORK

Shekhar Verma[1] , Vaibhav Pandey[2] , Preeti Khurana[3]

School of Electronics and Electrical Engineering, Lovely Professional University, Punjab.

shekharverma05@gmail.com[1], vpandey9999@gmail.com[2] , preeti.Khurana@lpu.co.in[3]

Abstract. The upcoming generation of cellular networks is going to make extensive use of mmWave for communication. Hence, there will be a need for small cells to counter the loss incurred due to the more energy dissipation of mmWave. Small cells contain transmitters and receivers and hence there will be the need to balance the load efficiently so that no one is overwhelmed with functions to perform while others are relatively idle. In this paper, the performance of the user association algorithm is analysed when it is subjected to different scenarios like micro-urban, macro urban, suburban and rural. these scenarios are subjected to different frequency bands of mmWave and are compared for the values of the load balancing index. The load balancing algorithm is subjected to different small cell deployment techniques and the comparison is made for best deployment strategy among Quadrature based Approach (QBA) and Random deployment. Simulation results show that the sub-urban scenario has the maximum load-balancing index. On comparing QBA and random deployment approach QBA has a higher load balancing index in the suburban and rural scenarios and random deployment has a higher load balancing index in Indoor environments.

Keywords. mmWave communication, Load Balancing, macro base station, picocells, user association, energy, efficiency.

1. Introduction

Throughout his path, wireless contact has come a long way. The evolution of wireless communication began in the late 1970s when the earlier versions of 1 G were launched on the market. Since then, a large number of users in different eras have benefited many generations. Users can access the internet together with phone calls and the data rate is faster in each generation to come. This has become possible because of the technological changes that are being used in various generations. The need to move from the existing 4G network has become a requirement now with the introduction of a greater number of users globally. The upcoming 5G generation has the vast potential to solve the shortcomings which are in LTE networks like the increase in data rate and accommodating a large number of users. Technologies like mmWave, massive MIMO, NOMA (Non-orthogonal Multiple Access), small cell, beamforming and MEC (Mobile Edge Computing) form the backbone of 5G. As we know that the spectrum range between 1 GHz-6GHz is overcrowded in today's scenario and is being used by every other technology like GPRS, HSPA, LTE, satellite communication, C-band communication, and many others. The unused and unlicensed band of 30 GHz-300 GHz is proposed for the working of 5G known as mmWave [1]. Also, we know that the higher the frequency of signal smaller will be the size of the antenna since the relationship is inversely proportional therefore in 5G we will witness the substantial growth of transmitters and receivers in tiny areas. This technique is known as Massive MIMO. Earlier the generations from 1G to 3G used time, frequency and code as the multiple access technique. 4G LTE uses an orthogonal frequency division multiple access (OFDMA) scheme. 5G introduces the concept called NOMA (Non-Orthogonal Multiple Access) which makes the use of user equipment power level to access the base station. This allows different signals to use the same channel but with a higher sum rate. When it comes to accessing the information through clouds there were a few problems like latency due to the distance of the server physically located far from the user. Another problem was to first download the application and the accessing it which can be inefficient for the devices with low memory and processing power. In MEC we have cloud servers located nearer to the users so that the application running can be split to run one half in the device and the other half in the cloud server itself. This makes applications like video conferencing, streaming run seamlessly with any ping issues. Because original voice-oriented wireless services are converted into data-oriented services and more indoor user equipment, more connection budget and expansion of coverage are required to meet more user requirements [2]. The design of a cellular network infrastructure tends strongly towards heterogeneous elements and deviates significantly from traditional BS of high power to operate this cost-effectively. These elements include primarily micro, pico, femto and relay BSs, which differ primarily in transmitting power, physical size, backhaul, reliability, ease of use and propagation characteristics [3]-[5]. The appropriate deployment of low-power BSs would help to eliminate holes in the traditional cellular network and increase hot spot capacity [6]. Nonetheless, since the signal strength consumer obtained from different BSs depends heavily on the transmitting power of base stations, the coverage area of small BSs will be lower than higher power BSs if we only follow the maximum signal strength association strategies. To make the best use of new low-power infrastructures, we should find a proper user association algorithm that aggressively pushes mobile users onto the small, lightly loaded BS that provides higher rates over time by delivering more physical resources to mobile users.

1.1 Prior work

The authors in [7] analysed how each BS extends over a definite geographical region which is referred to as a cell. The radius of the cell can be altered depending on urban and rural scenarios. Channel assignment to cells and mobile stations is essential to the functioning of a network, and to ensure that the capacity of call carrying and load balancing is increased, efficient channel distribution techniques are needed. The different channel assignment techniques provide the efficient usage of the spectrum which in turn helps to reduce the call blocking probability. It often happens in the cells that there is a traffic imbalance leading to one cell handling more functions and others are ideal. Moreover, as the generation upgrades, the need for data rate also goes high and such scenarios lead to the system being inefficient. If the mobile stations are not abundantly available across the geographical region this will also lead to the load imbalance and several of the users may not get the desired services. It is necessary to transfer overloaded traffic from hot cells to neighbouring cooler cells to balance the load between different cells. Ad hoc relay stations (ARS) are employed for the iCAR technique in the mobile network to efficiently manage traffic loads and to monitor the sharing of channels between cells through the primary and secondary relay [8]. The unlicensed band of ISM is used to run those ARS which consequently help to eradicate cellular band interference. The authors proposed an iCAR-based algorithm for a k-tier system that is built regardless of the number and location of hot cells. Effective load balancing is shown and the chances of call blocking of the busy cells can be significantly reduced. The authors in [9] suggested a scheme using smart BS antennas which dynamically adjusts the shape and size of cellular coverage depending on the distribution of regional traffic. A tactical negotiation approach for real-time smart cellular coverage control is also discussed in this study. The authors in [10] proposed a self-organizing load balancing that offers policies for load balancing which is self-optimizing to enhance the adaptation and robustness of cellular networks based on Fixed Relay Station (FRS). A game-theoretical analysis is proposed that significantly increases capacity utilization in the network even when the cells do not behave cooperatively [11]. The quantity of load to be accepted or unloaded shall be deemed independently to be decided by each cell. The authors in [12] proposed a scheme for services that are data-oriented and delay resistant with the help of mobile relay nodes (vehicle relays), which is an alternative for energy-saving balancing of the load. It is shown that significant energy savings are observed by delaying the transmission of messages while simultaneously reducing the imbalance of cell load. A new scheme for load balancing is introduced to relocate overloaded traffic from busy cells to nearby cooler cells where an online algorithm is proposed to control dynamically RSS associations with BS and MU associations with RSS and BSs [13][38-39]. This enhances the performance of boundary users and the fairness of users without hampering the overall system performance. The authors in [14] proposed a traffic load balancing strategy based on cell clusters is proposed to solve the cell congestion problem in cooperative cellular networks. Firstly, a cell cluster is built dynamically using a hot cell and neighbouring chosen cold cell so that the cell clusters traffic load is averaged and can be lowered than the predetermined chosen value and the average probability of blocking is calculated to reduce the traffic shifted from the hot cell to each non-congested cell cluster. Eventually, the spectral efficiency of all potential transmissions in the cell cluster is calculated to ensure maximum spectral efficiency when an MS from one of the busy cells is assigned to a corresponding cold cell.

Apart from different load balancing techniques, the small cell deployment methods are also been explored in previous research. The authors in [15] proposed a quadrant-based approach for cell setup. The new mathematical approach proposed helps to determine the optimum number of small cell access points and the optimal location of these small cell access points for the specific deployment. Apart from this, the authors in [16] examine the topologies by analysing the efficiency of the network in terms of the spatial outage and the effect of a completely random topology compared to that of a perfectly regular topology.

1.2 Contributions

Inspired by the maximum probability criteria we have explored the use of serl association algorithm to work for four different geographical demographics:

1. Micro urban scenario
2. Macro urban scenario
3. Sub-urban scenario and
4. Rural scenario

The users are distributed arbitrarily and the radius of the macro base station changes according to the four scenarios. The idea is to find the optimal load balancing index for different values for the macro base station coverage region. The number of small cells also varies as the radius for the macro base station changes.

1.3 Organization

The remainder of the paper is structured as follows. The association algorithm, as well as the system model for load balancing, is explained in section II. The results derived from simulation and its deciding parameters are elaborated in section III. Section IV comprises the conclusion

2. SYSTEM MODEL AND ASSOCIATION ALGORITHM

The authors in [17] formulated their scheme as a weighted problem of maximization mathematically, and it was considered the assumption that all sub-bands of each BS are allocated with equal power. Transforming the problem into a convex problem of optimization, they relaxed the association indicator variables and then applied a gradient descent method to locate the optimized solution on the boundary of the optimization function's feasible area. Users then picked the BS that indicates peak association.

For user k, we obtain rom BS n the achievable rate [in Kbps] according to equation (1)

$$r_{nk} = W \ log\,2\,(1 + \ SINR_{nk}) \qquad (1)$$

Where the bandwidth is denoted by W of sub-band (180KHz) and the SINR received for user k ϵ K from BS n ϵ N is denoted by $SINR_{nk}$ on one sub-band. According to each user's practical rate requirement, the number of resources utilized for each BS is given according to equation (2):

$$S_{nk} = d_k / r_{nk} \tag{2}$$

where d_k is the practical rate of user k.

$$max_{x,y} \sum_{n \epsilon N} \sum_{k \epsilon K} x_{nk} s_{nk} \log r_{nk} - \sum_{n \epsilon N} y_n \log y_n$$

$$s.t. \textstyle\sum_{n \epsilon N} x_{nk} = 1, \forall k \epsilon K \tag{3}$$

$$\textstyle\sum_{k \epsilon K} x_{nk}\, s_{nk} = y_n, \forall n \epsilon N$$

$$0 < y_n \leq M, \forall n \epsilon N$$

$$0 \leq x_{nk} \leq 1, \forall n \epsilon N, \forall k \epsilon K$$

where $X = \{x_{nk,} \ \forall n \epsilon N \ \forall k \epsilon K\}$; $\sum_{k \epsilon K} x_{nk} s_{nk}$ is the total load of BS n.

$$\sum_{k \epsilon K} x_{nk} s_{nk} \leq M$$

if the constraint for a total resource of eps.

Because the hierarchical architecture that solves the problem of convex optimization leads to high computational complexity and low reliability, a distributed algorithm with low overhead and complexity is desirable [17]. We use the Lagrange dual decomposition method [18] in this section to build a distributed algorithm that is compatible with the Ye's [19]. Instead of solving the original convex optimization problem directly, users and BSs can solve two sub-problems separately that i te the dual problem.

Algorithm: Association Algorithm

Estimating $r_{nk}, \forall n \in N, \forall k \in K$ by initializing the step size ξ.
if t=0 **then**
Initialize association probability x_{nk}^0 and Lagrange multipliers including μ_n^0, v_{nk}^0, λ_k^0.
else
Apply KKT conditions and calculate the factors x_{nk}^{t+1} and y_n^{t+1}.
if iteration reaches the convergence precision of x_{nk} or the maximal iteration number **then**
Set the maximal value of optimal association probabilities $(x_{n:}^{t+1})$ of each BS(n) to 1, and transmit it to the corresponding BS(n).
else
Update Lagrange multipliers via information x_{nk} and y_n.
$t \leftarrow t + 1$
end if
end if

3. SIMULATION RESULTS

Table I shows different parameters for simulation purposes. To observe the changes in the load balancing index for different scenarios the radius of the macro base station has to be varied from 250m to 1732m.

the number of small cells will have a radius of 75m. The SINR will be variable depending on each base station. The number of users distributed randomly over an area can also be varied.

TABLE I. SIMULATION PARAMETERS

Parameters	**Values**
The radius of picocell	75m
The radius of Macro Base Station	250m,500m,1299m,1732m
Number of UE	20,40,60,80,100,120
Number of picocells	variable
SINR depending on the number of base stations	Variable for every base station
Macrocell transmit power	46 dB
Pico cell transmit power	30 dB
System Bandwidth	20*10^6 MHz

The results obtained are for four different scenarios:

3.1 Load balancing in micro-urban scenario

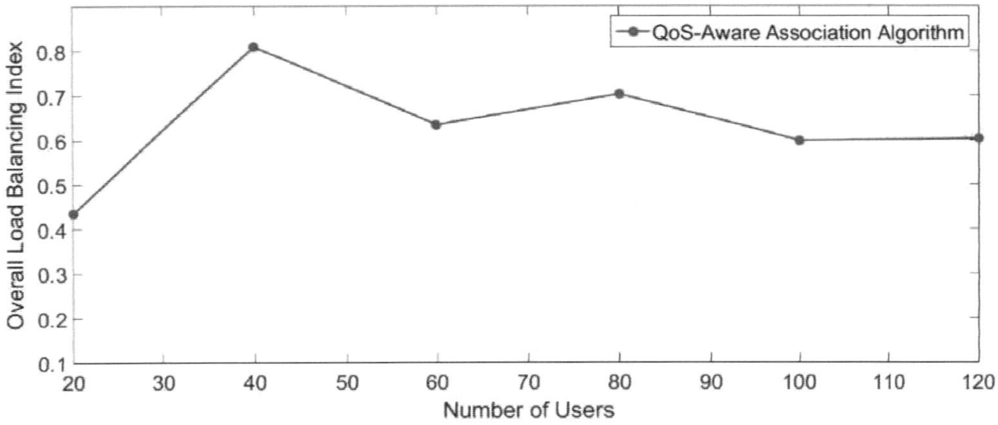

Fig. 1. Overall Load balancing index in HCNs (250m)

Fig.1 shows the load balancing index variation concerning the number of users. The radius of the macro base station is taken as 250m. It is observed that as the number of users increases the value of the load balancing index also increases.

3.2 Load Balancing in the macro urban scenario

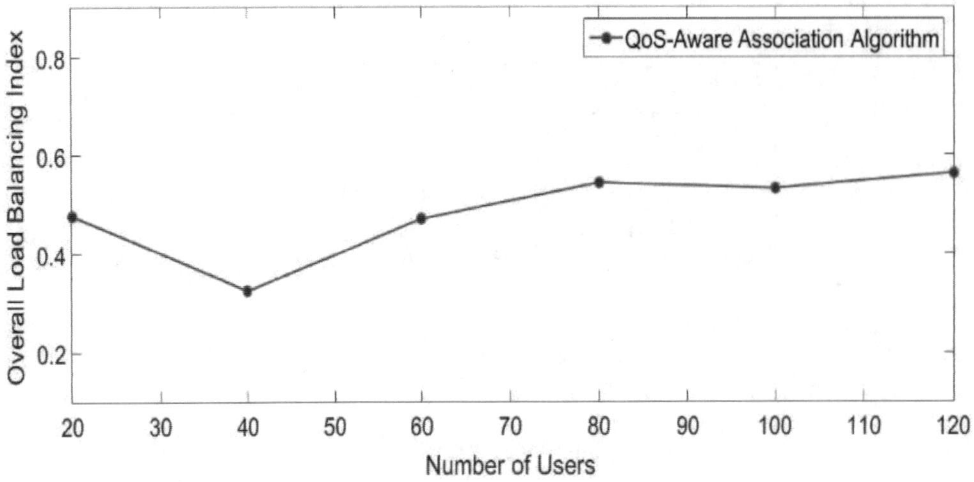

Fig. 2. Overall Load balancing index in HCNs (500m)

Fig.2 shows the variation of the load balancing index for the macro urban scenario when the radius of MBS is increased to 500m. There is an improvement in the average value of load balancing as the users located within the region will not be deprived of the resources compared to the micro-urban scenario.

3.3 Load balancing in a suburban scenario

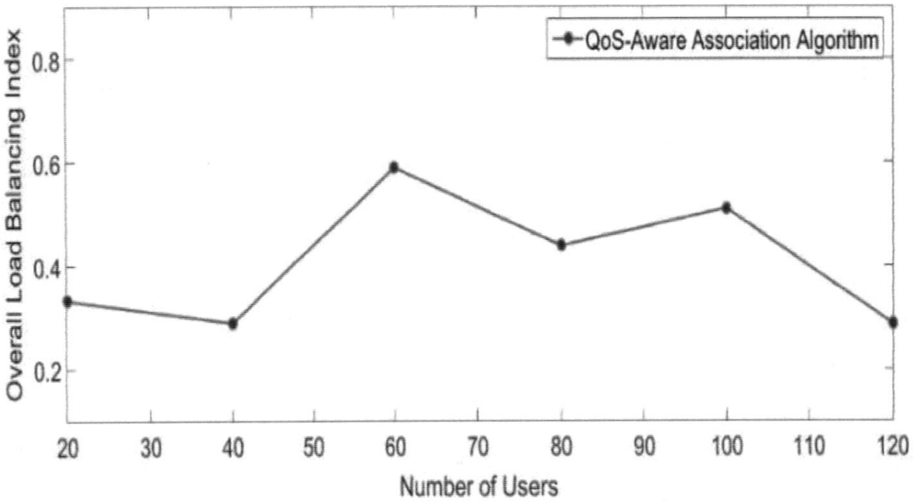

Fig. 3. Overall Load balancing index in HCNs (1299m)

The coverage area of MBS is further increased to 1299m and the results obtained in Fig.3 clearly show the constant variation in the load balancing index. The variation is due to the non-uniformity in the distribution of users.

3.4 Load balancing in the rural scenario

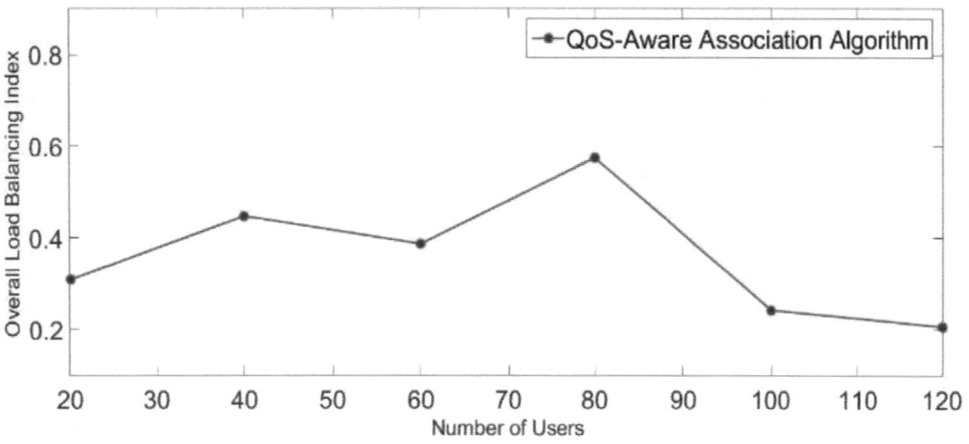

Fig. 4. Overall Load balancing index in HCNs (1732m)

Fig.4 shows the dip in the load balancing index as the users will be sparsely located from the MBS and therefore will be difficult for the base station to provide the resources to the users. For the cellular networks to run seamlessly and efficiently in a rural deployment scenario there has to be an optimal load balancing association probability algorithm.

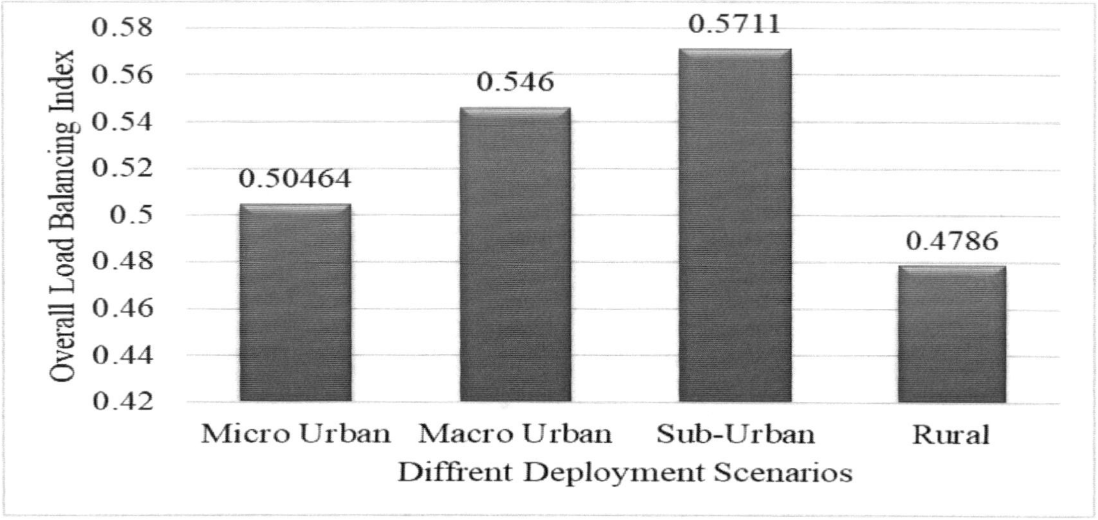

Fig. 5. Comparison between different scenarios

From the above Fig.5, it is clear that the rural areas will have the lowest load balancing index as the distribution of resources will not be efficient due to the sparsely and distantly located users. On the other hand, the micro-urban scenario will have a large number of users located very near to each other and there might be a possibility of the requirement of resources at the same time. The macro base station may not be able to provide equal resources to all the users at the same time and hence some of the base stations might be deprived of the facilities offered by the macro base station. The optimal load balancing scenario happens to be in the suburban area where the base stations get an equal share in bandwidth in comparison with the other three scenarios.

3.5 Comparison between different small cell deployment algorithms

3.5.1 Sub-Urban Scenario

Fig.6 shows the comparison in a sub-urban scenario when the radius of the macro base station is equal to 1299m. The Quadrant based deployment approach has a higher overall Load Balancing Index than the random deployment. The average value achieved for the QBA is 0.416 of the Load Balancing Index and 0.356 in random deployment. The QBA algorithm is 16.8% more efficient than the random deployment scheme.

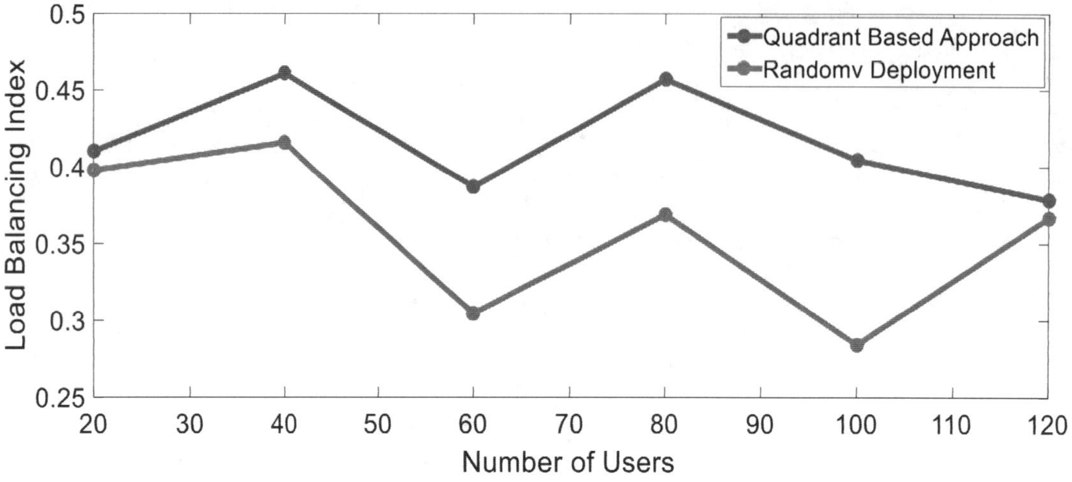

Fig.6. Comparison of different algorithms in Sub-Urban Scenario

3.5.2 Rural Scenario

The comparison between different algorithms in the rural scenario is shown in Fig.7. The radius of consideration in this scenario for the macro base station is 1732m. In this case, also the QBA approach has the highest overall Load Balancing Index followed by a random approach. Simulation results show that the average Load Balancing Index for the QBA algorithm is 0.4256. The value for random deployment setup is 0.3264 respectively. The QBA algorithm is 30.3% more efficient than the random deployment setup.

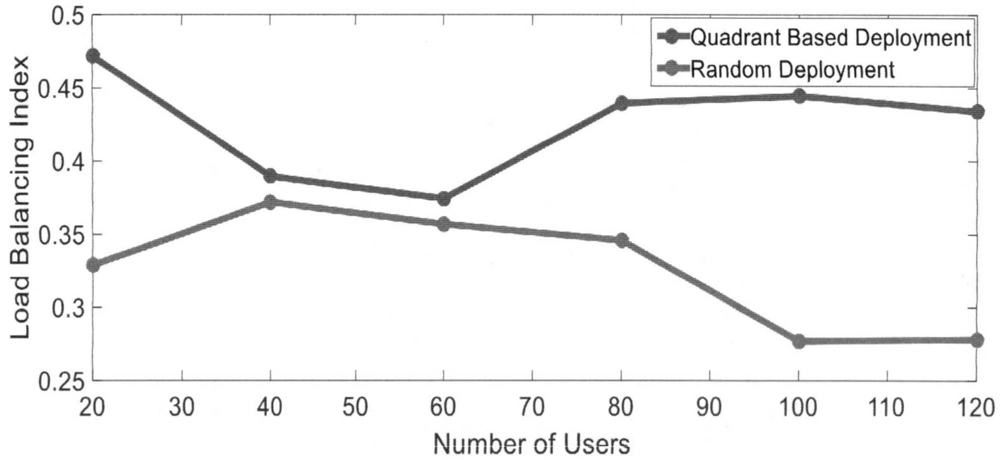

Fig.7. Comparison of different algorithms in Rural Scenario

3.5.3 Indoor Environment

An indoor complex environment is one of the most viable areas for mmWave communication. The frequency used in mmWave communication is supported heavily by the geographic conditions of the indoor complex as the attenuations are supposed to be very less and hence the transmission occurs without much loss as compared to the outside environment. The traditional workplace inside an indoor complex setting involves enclosed rooms, walled buildings, empty spaces, halls etc. The comparison of algorithms for an indoor environment is shown in Fig.8. Unlike the other two cases, the overall Load Balancing index for the random deployment comes out to be on top of the QBA approach. The average value for the random deployment is 0.508 and 0.3432 for the quadrant deployment scheme. The random scheme came out to be 48% more efficient than QBA.

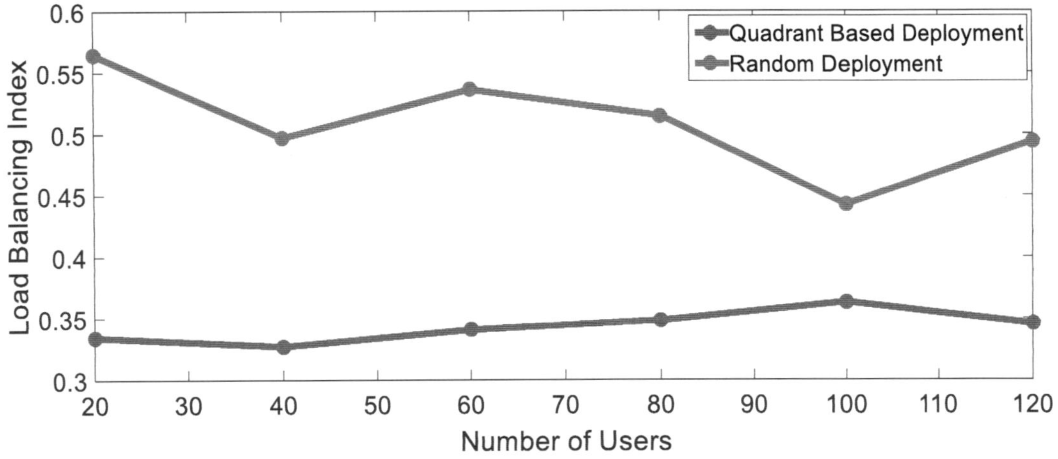

Fig.8. Comparison of different algorithms in Indoor Environment

4. CONCLUSION AND FUTURE SCOPE

In this paper, we have made a comparison in the load balancing index for different values of the radius of the macro base station. The paper considers a practical user deployment scenario that results in higher user density in some regions and lowers user density in others. Furthermore, it is observed that the small cells for each scenario also keep on varying. Comparing all the scenarios it can be clearly stated that the suburban area will have the optimal distribution of load among the base stations. Apart from these, different small cell deployment algorithms are also compared in which the QBA approach is more favourable for suburban and rural areas whereas random deployment is favourable for indoor environments. The work can be extended by proposing an algorithm related to loading balancing that can be extended by making the radius of picocells vary according to the density of the users in that geographical area. Subsequently, the distance between the two picocells can also be varied. This work can also be modified and explored in different areas of wireless communication by following the mentioned research papers [10-27].

Acknowledgement
I would like to express my very great appreciation to **Dr. Akhil Gupta, Harpreet Singh Bedi, Manmohan Sharma, Raghav Gupta (School of Electronics and Electrical Engineering, Lovely Professional University, Phagwara, Punjab, India.)** for their valuable and constructive suggestions during the planning and development of this research work. Their willingness to give their time so generously have been very much appreciated.

References

[1] B. A. Bjerke, "LTE-advanced and the evolution of LTE deployments," *in IEEE Wireless Communications*, vol. 18, no. 5, pp. 4-5, October 2011.

[2] E. Yanmaz, O. K. Tonguz, S. Mishra, Hongyi Wu, and Chunming Qiao, "Efficient dynamic load balancing algorithms using iCAR systems: a generalized framework," *Proceedings IEEE 56th Vehicular Technology Conference, Vancouver, BC, Canad*a, 2002, pp. 586-590 vol.1..

[3] Z. Yang and Z. Niu, "*A New Relay Based Dynamic Load Balancing Scheme in Cellular Networks,"* 2010 IEEE 72nd Vehicular Technology Conference - Fall, Ottawa, ON, 2010, pp. 1-5.

[4] X. Wang, H. Tian, F. Jiang, X. Li, X. Hong and T. Li, "Cell-Cluster Based Traffic Load Balancing in Cooperative Cellular Networks," 2010 *7th IEEE Consumer Communications and Networking Conference*, Las Vegas, NV, 2010, pp. 1-5.

[5] Gupta, Akhil, and Rakesh Kumar Jha. "Power optimization using optimal small cell arrangements in different deployment scenarios." *International Journal of Communication Systems 30.13 (2017): e3279.*

[6] P. Jain and A. Gupta, "Energy-Efficient Adaptive Sectorization for 5G Green Wireless Communication Systems," in *IEEE Systems Journal*, vol. 14, no. 2, pp. 2382-2391, June 2020.

[7] Chopra S.R., Gupta A., Monga H. (2019) Performance Analysis of Space-Time Trellis Codes in Rayleigh Fading Channel. In: Yadav N., Yadav A., Bansal J., Deep K., Kim J. (eds) Harmony Search and Nature Inspired Optimization Algorithms. *Advances in Intelligent Systems and Computing, vol 741. Springer, Singapore.*

[8] Gupta, Akhil, Shiwani Dogra, and Ishfaq B. Sofi. "Performance evaluation of spatial multiplexing using different modulation techniques in MIMO system for small and large scale fading channel." *International Journal of Sensors Wireless Communications and Control* 9.2 (2019): 188-202.

[9] Gupta, Akhil, Rakesh K. Jha, and Reeta Devi. "Security architecture of 5g wireless communication network." *International Journal of Sensors Wireless Communications and Control* 8.2 (2018): 92-99.

[10] Sofi, Ishfaq Bashir, Akhil Gupta, and Rakesh Kumar Jha. "Power and energy optimization with reduced complexity in different deployment scenarios of massive MIMO network." *International Journal of Communication Systems 32.6 (2019): e3907.*

[11] A. Gupta and S. R. Chopra, "Designing and Testing of Quadrifilar Helix Antenna for NOAA Weather Satellite," 2019 *6th International Conference on Signal Processing and Integrated Networks (SPIN)*, 2019, pp. 785-790.

[12] S. R. Chopra, A. Gupta and H. Monga, "An LTE Approach with MIMO by Using Suboptimal Selection of Antenna," *2018 International Conference on Intelligent Circuits and Systems (ICICS)*, 2018, pp. 167-172.

[13] Goyal, Krittika, and Akhil Gupta. "A literature survey on different types of pulse-based sensor for acquisition of pulse." International Journal of Control Theory and Applications 9.10 (2016): 361-365.

[14] R. Kausar, A. Gupta, I. B. Sofi and K. Arora, "Bit Error Rate Based Performance Evaluation of LTE OFDMA System," *2018 International Conference on Intelligent Circuits and Systems (ICICS), 2018,* pp. 161-166, DOI: 10.1109/ICICS.2018.00042.

[15] Anand, Pankaj, Akhil Gupta, and Sonam Bhagat. "Bit Error Rate Assessment of Digital Modulation Schemes on Additive White Gaussian Noise, Line of Sight and Non-Line of Sight Fading Channels." *International Journal of Engineering Science Invention 3.8 (2014): 6-10.*

[16] Gupta, Akhil, and Krittika Goyal. "An analytic review on antenna modelling." *International Journal of Control Theory and Applications* 9.10 (2016): 319-336.

[17] Gupta, A., Jha, R.K. Power optimization with low complexity using scaled beamforming approach for a massive MIMO and small cell scenario. *Wireless Netw* 26, 1165–1176 (2020).

[18] I. B. Sofi, A. Gupta and R. Kausar, "Performance Evaluation of Different Channel Estimation Techniques in MIMO System for Hata Channel Model," 2018 *International Conference on Intelligent Circuits and Systems (ICICS)*, 2018, pp. 155-160, DOI: 10.1109/ICICS.2018.00041.

[19] R. T. Kamurthi, S. R. Chopra and A. Gupta, "Higher-Order QAM Schemes in 5G UFMC system," *2020 International Conference on Emerging Smart Computing and Informatics (ESCI), 2020*, pp. 198-202, DOI: 10.1109/ESCI48226.2020.9167619.

[20] Raina, Ambica, Akhil Gupta, and Rakesh Kumar Jha. "Analysis of AODV and DSR Routing Protocols for a Static Ad Hoc Network Using Qualnet Network Simulator." *International Journal* 4.1 (2014).

[21] Sunday, A.P., Jha, R.K. & Gupta, A. Energy and spectral efficiency optimization using probabilistic based spectrum slicing (PBSS) in different zones of 5G *wireless communication network. Telecommun Syst* 73, 59–73 (2020).

[22] M. Singh, K. Arora and A. Gupta, "Equalization in WIMAX System," 2018 *International Conference on Intelligent Circuits and Systems (ICICS*), 2018, pp. 330-333, DOI: 10.1109/ICICS.2018.00074.

[23] Jain, Preksha, and Akhil Gupta. "Adaptive NOMA towards 5G green wireless network." *Transactions on Emerging Telecommunications Technologies* 31.3 (2020): e3887.

[24] H.S. Bedi, K.K. Sharma, and R. Gupta, "A Review Paper on Performance Analysis of IEEE 802.11 e", *1st International Conference on Computing, Communications, and Cyber-Security (IC4S 2019*), 2020, pp. 47-56.

[25] Z.G. Khaki, and H.S. Bedi, "Transient correction using EDFA: in-line optical fibre with feedback", *International Conference on Computing Sciences, 2012,* pp. 233-238 [http://dx.doi.org/10.1109/ICCS.2012.76].

[26] Bedi, Harpreet, Kamal Kumar, and Gupta Raghav. "A review paper on improving the network efficiency of IEEE 802.11 e networks." *Intelligent Circuits and Systems* (2021): 206-211.

[27] Harpreet Bedi, Kamal Kumar Sharma; Analyses of CSMA/CA Protocol Without Using Virtual Channel Sensing in DCF Mode, *Global Emerging Innovation Summit (GEIS-2021) (2021)* 1: 484. https://doi.org/10.2174/9781681089010121010058.

Performance Improvement of Next-Generation Network with Cognitive Radio Network and Coordinated Multipoint Joint Transmission

Disashi Valentin[1], Akhil Gupta[2], Manmohan sharma[3]

School of Electronics and Electrical Engineering, Lovely Professional University, Punjab.

valentindisashi@gmail.com[1] , akhilgupta112001@gmail.com[2], manmohan.21137@lpu.co.in[3]

Abstract. Cognitive Radio Network (CRN) is a favourable technology for the future of wireless networks. It aims to seamlessly exploit spectrum resources by opportunistically using the unused or underused radio spectrum that is licensed. Cognitive Radio Network is a technology that is planned to be used in the 5th Generation of wireless communication. This technology is belied to be the answer to the unused and underused radio spectrum in the Radio environment. Whereas Coordinated Multipoint Joint Transmission (CoMP JT) is a technology that has already been used, first used in 3GPP LTE-Advanced. Coordinated Multipoint Joint Transmission aims to improve the performance of the cellular network. In this report, Cognitive Radio Network is being implemented along with Coordinated Multipoint Joint Transmission techniques. An inclusive and layered architecture of Cognitive Radio Network along with Coordinated multipoint Joint Transmission is being designed in this report, showing different technologies and different steps involved in the concept of this combination. Then the impact of CoMP JT on CRN is being studied. By implementing CoMP JT on CRN, it has been able to show how the performance of CRN is changing when implemented alone and when implemented with CoMP JT and some of the significant results are being depicted to better understand the changes in CRN with CoMP JT. Lastly, a study is performed on the combination of Cognitive Radio Network and Coordinated Multipoint Joint Transmission. As far as the investigation is being performed, two factors have been considered, which is: Signal to Interference plus Noise Ratio (SINR) and Total Average Throughput (TAT). The detailed study is being performed in this report and is being implemented in three different environments. Hence the results are being compared and a conclusion is being drawn based on these results.

Keywords. Cognitive Radio Network, Coordinated Multipoint Joint Transmission, Signal to Interference plus Noise Ratio, Spectrum management, Throughput, Energy, Efficiency.

1. Introduction

This part of the paper is going to give a presentation about different technologies used in this work, which are Cognitive Radio Network and Coordinated Multipoint Joint Transmission. The motivations for combining them with the expected outcomes are explained in this introduction part which is section one as well. Followed with the objectives that are going to be fulfilled in this work, there are 3 of them and they are explained in the following lines. The research methodology is being explained along with the software/tools used. And finally, the thesis organization is being presented to show different sections.

Cognitive Radio Network: With the advancement of technology and the availability of communication devices, the number of users is multiplied continuously, causing the demand for frequencies for communication and wireless channels to increase at a very high rate. Frequency reuse and spectral efficiency improvement techniques were implemented to solve this problem, hence increasing the communication system bandwidth and spectral efficiency [1]. However, these techniques have a drawback, which is the underutilization of the frequency band being assigned to the users. This drawback comes into the picture while the user is not using or communicating in the allocated channel; that means when the user is idle as far as the particular channel is concerned.

In the traditional radio system, the approach to spectrum management is not flexible enough in the way that any other user cannot access the allocated channel of a user that is not communicating, it is being wasted and the loss is considered when all the unutilized channels are combined, hence giving a considerable band of frequencies wasted. The problem can be overcome by finding a way to access those unused channels (Spectrum holes) dynamically when its licensed user is not communicating. Thus the introduction of Cognitive Radio Network (CRN) is a Radio system that has the capability of sensing the unused spectrum from the licensed users and accessing it without causing interference or by limiting the interference level at a tolerable level. In CRN, the licensed users of the spectrum are called Primary Users (PU), and the unlicensed users trying to access the spectrum hole of the primary users are called Cognitive users or Secondary Users (SU) [2]. This technology is a promising solution for the complete management of the radio spectrum in the communication domain. It is being studied by several researchers and proposed to be implemented in 5G (fifth generation of cellular mobile communication) [3]. The secondary user analyses observe and learn; hence the CRN adapts itself to the conditions of the environment and takes future decisions based on the analyses performed.

The ability to sense a spectrum hole and temporarily use it by keeping the interference at a tolerable level are two critical components needed for the success of CRN. Besides these two key components, the CRN has two main characteristics that are:

- Cognitive capability: Cognitive Capability is the capability to capture or to sense the information from the radio environment in radio technology. Joseph Mitola first explained it: the cognitive capability in terms of the cognitive cycle, "a cognitive radio continually observes the environment, orients itself, creates plans, decides, and then acts."
- Re-configurability: Cognitive capability offers spectrum awareness; however, Re-configurability is the radio capability to change the functions, enabling the cognitive radio to be programmed dynamically following radio environment (frequency, transmission power, modulation scheme, communication protocol) [4].

Improvement of spectrum efficiency is the main aim of CRN, and it is being achieved by several techniques that are discussed in this report. Techniques out of which the four basic are: Spectrum sensing: that allows the SU to detect whether the spectrum is vacant or occupied by PU; by far, the most important step in the CRN. Spectrum decision: which is implemented on the SU, to decide what the best possible spectrum that can be used is; without causing interference to PUs.

This technique is mainly based on algorithms that allow decisions to be made, out of which Machine Learning and Artificial Intelligence play a crucial role. Spectrum sharing: Once the SU has gained access to some spectrum resources, it is going to share that resource with other SUs that require communication. This technique allows the maximization in the use of available spectrum resources. Spectrum mobility: It is the last step in the spectrum management of CRN. Spectrum management aims the improvement of spectrum efficiency. The spectrum mobility is mainly based on algorithms and techniques that allow SUs to shift from one radio environment to another using handoff techniques that prevent the interruption of communication. It also uses some algorithms that predict the behaviour of Pus in other radio systems [5]. Added to spectrum management, there are security aspects and other optimization techniques to ensure the proper functioning of CRN.

In the current scenario, the wireless network is operating on the policy of static spectrum allocation, in which governmental institutions assign a statistically wireless spectrum to licensed clients in a vast geographical region for a certain or uncertain duration of time. However, with the increase in demand for the wireless spectrum, this technique causes some shortages in the spectrum for certain bands. The reason is: some portion of this allocated spectrum is not continuously used hence leading to an underutilization of the spectrum allocated and creating spectrum holes.

CRN is proposed to solve this problem and is envisioned to give high spectrum efficiency to its users as well as dynamic spectrum access techniques. But due to high fluctuation in the available spectrum in

Journal of Physics: Conference Series **2327** (2022) 012054 doi:10.1088/1742-6596/2327/1/012054

terms of frequency and time, and multiple Quality of Service (QoS) required, accessing those spectrum holes is very challenging for cognitive users or Secondary users [5]; so they must be able to:

- Observe which part of the spectrum is unused at a particular time and frequency. (Spectrum Sensing)
- Decide which one is the best available spectrum hole and access it. (Spectrum Decision)
- Allow other SU to access the same spectrum hole. (Spectrum Sharing)
- Evacuate the spectrum hole without causing interference when the PU is about to use it. (Spectrum mobility)

These steps are the foundation of spectrum management in CRN and are shown in Fig. 1. [5]

Fig. 1 Steps of spectrum management in CR

Coordinated Multipoint (CoMP) is a technique in which the base station is being coordinated with each other in a radio system. CoMP had been implemented into the 3GPP LTE-Advanced to mitigate the inter-cell interference (ICI) in the cellular wireless radio network [6]. The aim is to bundle all the neighbouring base stations (BSs) and hence create a coordinated multipoint with the cooperating set.
In CoMP, two main approaches are coordinated beamforming (CB) and joint transmission (JT), The difference is that the data stream for each Mobile Station (MS) is only available at and transmitted from a single BS in CB; whereas, for the JT, the data for each mobile station is distributed among multiple cooperating base stations and can hence be simultaneously transmitted [6].

In terms of overall system throughput enhancement, the JT approach is the best suited since it simply forms a kind of virtual large antenna array for each mobile station, but causes a high signalling overhead through the backhaul. This high signalling overhead is incurred by disseminating each user's data to multiple BSs in the cooperating set. In a low-mobility environment, however, the amount of data to be carried out in the backhaul is significantly greater than the channel state information (CSI). To alleviate the backhaul requirement, it is desirable to disseminate the user data only to a subset of the cooperating BSs. This technique is called BS clustering [6].
Cognitive radio (CR) is considered to be a promising technology for future wireless networks to make opportunistic utilization of the unused or underused licensed spectrum. Though it has not been used in any technology yet, it is being proposed as the solution for spectrum management in the next generation of wireless communication. Meanwhile, coordinated multipoint joint transmission (CoMP JT) is another technology already implemented. It was first performed in 3GPP LTE-Advanced and proved to be efficient by improving the network throughput. Hence, it is a promising technique to enhance the performance of cellular networks.

Both Cognitive Radio Network and Coordinated Multipoint Joint Transmission are two highly promising concepts for future cellular networks. Multiple types of research are being conducted on CRN based on CoMP JT transmission. A combination of these two technologies promises to be resulting in a better performing Cognitive Radio Network, the thing that is vital for the consistency of the technology itself and the services of the Next Generation of wireless communication.

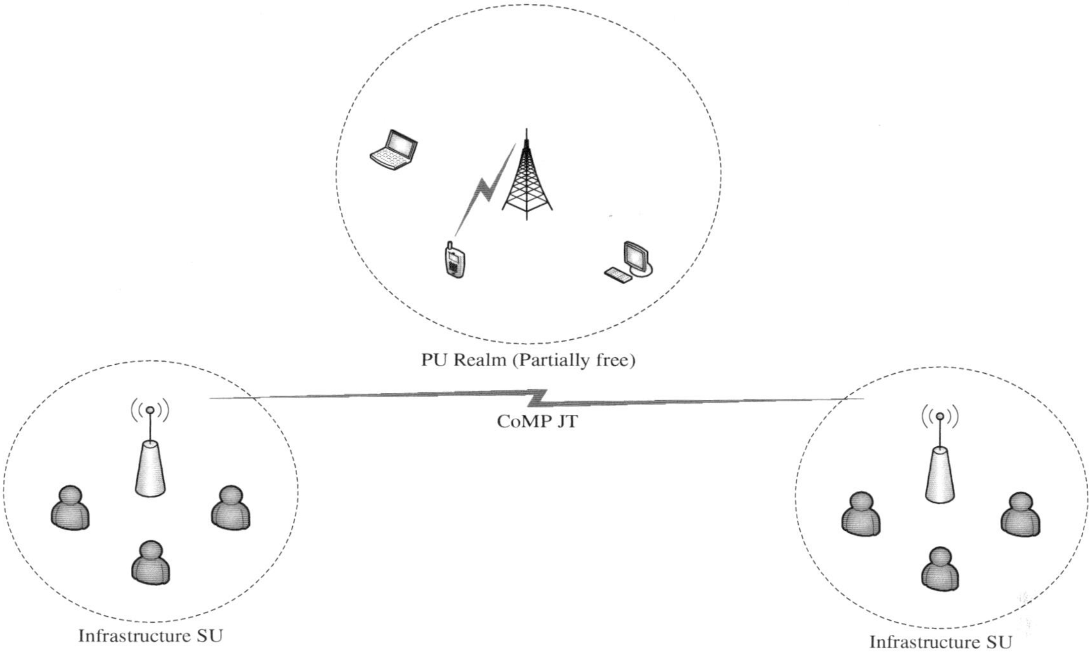

Fig. 2 CNR scenario with CoMP JT

Fig. 2 depicts a Cognitive Radio system having 1 PU Transmitter and 2 SU Transmitters. These 2 SU TX are communicating with each other using Coordinated Multipoint Joint Transmission. Each PU TX and SU TX are having the respective Receivers that receive the services from them.
The main problem in the JT technique is the synchronization of all the distant sites, multi-cell channel estimation, feedback of the channel state information (CSI) and the synchronous exchange of data. However, using the CoMP JT in CRN has a significant impact which is the increase of the Signal to Interference plus Noise Ratio (SINR) [7].

1.1 Background research

The challenges of defining and keeping spectrum efficiency are depicted, along with all the parameters to be considered to achieve the benchmarks of efficient spectrum management. Then an overall survey is performed in Cognitive Radio Network. Depicting the significant aspects of the network that include spectrum sensing, spectrum decision, spectrum sharing, and spectrum mobility. Along with the spectrum access and assignment algorithms [1]-[2].

A study of Cognitive Radio Network is being studied in the fifth generation of the cellular network. In this 5G, the spectrum sharing is being analyzed and considered to be implemented in 5G [3] and then a Matlab simulation is performed on Cognitive Radio Network. Simulations are covering all the significant aspects of Cognitive Radio Network and show results based on the work performed. Hence, giving a better understanding of the concept of Cognitive Radio Network [3] – [4].

Another part focuses on the spectrum management in Cognitive Radio Network. All the spectrum management techniques, including spectrum sensing, spectrum decision, spectrum sharing, and spectrum mobility. Along with the spectrum access and assignment algorithms [5].

The design of Coordinated Multipoint Joint Transmission with Multi-antenna Base station and Single Antenna User for a limited backhaul data transfer. Aiming to minimize the overhead on the backhaul

system. Inspired by continuous sensing, two algorithms are proposed to overcome the limited backhaul problem. Then Coordinated Multipoint Joint Transmission is referred to as multiple cooperative Input Multiple Output and it has been shown that it is more efficient than MIMO-OFDM. Then some technical challenges are being identified and addressed. Some of them are feedback design, synchronization and backhaul traffic. Radio over Fiber technology is used to implement a Coordinated Multipoint transmission in the small cell having millimetre waves. The aim is to reduce Inter-Cell Interference, increase the reception quality and infrastructure utilization as well as improve the chance of Line of Sight links. The focus is pointed towards Joint Transmission in the Coordinated multipoint Transmissions. Having a Base station from different networks, the heterogeneity of the system is being considered, and this problem is overcome. Cognitive Radio Network is implemented alone and analyzed, then it is implemented along with Coordinated Multipoint Joint Transmission and the changes are being analyzed. It has been found that with Coordinated Multipoint Joint transmission, the Signal to interference plus Noise Ratio (SINR) has been increased at the Primary User. [6 - 8], [9 - 19].
Learning problems in Cognitive Radio Network are being highlighted and the matter of using Artificial intelligence is pointed out. The two main points of focus are feature classification and decision making. Different Machine Learning algorithms are being proposed for Cognitive Radio networks to reach a high throughput with minimal interference and optimal use of the spectrum resource. Highlight the flaws in hardcoded Cognitive Radio systems. They are hence proposing algorithms with soft coded engines that enable the radio system to remember learning results to apply them continuously in the radio environment [8], [20 – 21].

The spectrum management aspect of the cognitive radio network. Designing intelligent spectrum management algorithms and intelligent spectrum mobility management to improve, in the long term, the performances of the system such as delays in throughput. The spectrum sensing in Cognitive Radio Network. Working on the cooperative spectrum sensing, Unsupervised Machine Learning techniques are applied to it to classify the spectrum into available ones and non-available one. Hence the training time is being optimized to decrease the overall sensing latency. [21-25][45-46]. The comparison of literature with the existing work is mentioned in Table I.

1.2 Contribution

The two technologies that are Cognitive Radio Network and Coordinated Multipoint Joint Transmission have been combined both from an Architectural point of view, aiming to design an inclusive layered architecture including all the techniques and technologies used in this combination. And also, from a simulation point of view, a simulation has been implemented in the Matlab tool. The architecture proposed in this paper includes the four steps of spectrum management in CNR and shows all the transitions happening from one to the other. It also shows the use of CoMP JT and other technologies.

TABLE I COMPARISON OF LITERATURE WITH THE EXISTING WORK

Ref.	Objective	Layered Architecture	CRN + CoMP JT	SINR improvement with JT	Throughput improvement with JT
[7]	Throughput improvement in CRN	No	Yes	No	Yes
[6]	Limited Backhaul Data Transfer	No	No	Yes	No
[19]	CoMP in Heterogeneous Network	No	No	No	No
This Paper	Next-Generation Network improvement	Yes	Yes	Yes	Yes

1.3 Organization and notation

The remainder of the paper is structured as follows. The algorithm, as well as the system model, is explained in section II. The performance optimization in different deployment models is shown in section III. The results derived from simulation and its deciding parameters are elaborated in section IV. Section V comprises the conclusion

2. SYSTEM MODEL

The below image Depicts a Cognitive Radio system having 1 PU Transmitter and 2 SU Transmitters. These 2 SU TX are communicating with each other using Coordinated Multipoint Joint Transmission. Each PU TX and SU TX are having the respective Receivers that receive the services from them.
The main problem in the JT technique is the synchronization of all the distant sites, multi-cell channel estimation, feedback of the channel state information (CSI) and the synchronous exchange of data. However, using the CoMP JT in CRN has a significant impact that is the increase of the Signal to Interference plus Noise Ratio (SINR)

2.1 Algorithm

```
1.   Initialize values
ns                  // Number of SU
np                  // Number of PU
COM = 0             // Number of PU communicating
SU1[ns]
SU2[ns]
BS_SU1
BS_SU2
PU1[np]
BS_PU1
2.   for 1:np
3.   if PU1[np] = = 1
4.   then COM ++
5.   if length of PU1[np] == length of COM
6.   then (All COM busy)
7.   return to line 2.
8.   else
9.   then (SU use non busy COM)
10.  SU1, SU2 can use PU1 channel
```

Where ns: number of SU, np: number of PU, COM: number of communicating PU (busy PU), SU1 and SU2: number of SU in the first base station and second base station vicinity respectively, BS_SU1 and BS_SU2: first and second SU base stations, PU1: number of PU in the PU base station vicinity, BS_PU1: PU base station.
This algorithm depicts the general working of the cognitive radio network considering the primary technique that is spectrum sensing. At first, the variable is initialized. A count is initiated, where the number of PU communication is being incremented for each PU that for busy or communicating, shown inline 2 to 4. After that, the number of PU communicating is being compared to the initial number of PU, and if it is equal, it means there is no free channel to be used by SU because all the PU are busy, and in the other case; the SU would be able to use the unused PU channels for their communication.

3. METHODS

This section is going to talk about the improvement in the performance of Cognitive Radio Network as far as throughput and Signal to Interference plus Noise Ratio are concerned. Three deployment models or environments are to be considered in this, which are: Urban environment, Suburban environment, and rural environment.

3.1 Total Average Throughput

The Total Average Throughput (TAT) is being calculated in a Cognitive Radio Network that is implemented alongside Coordinated Multipoint Joint Transmission. The TAT is calculated by first finding the Throughput of different networks. Considering that the number of available PU nodes is equal to the number of networks to which SU nodes are going to be connected. Since SU nodes are going to borrow spectrum from PU nodes.
Hence, for all these networks, the throughput is being calculated, and an average of all of them is found out, which will be the final TAT. The operation is repeated three times for the three different environments that are the urban environment, the Suburban environment and the Rural Environment. The difference in these environments is the transmission range or the radius of each base station. They vary according to the environment and their values are 0.5 Km for Urban cells, 1.299 Km for Suburban cells, and 1.732 Km for Rural cells [26].
Throughput = (Number of bits transmitted) / (given Time Frame)
The TAT is being improved with the inclusion of the Coordinated Multipoint Joint Transmission technique to Cognitive Radio Network, and this is going to be explained with the simulation results in the coming section [7].

Signal to Interference plus Noise Ratio

(a) INR for Cognitive Radio Network:

$$\text{SINR} = \frac{\text{SU tx power } \left(\frac{\text{Distance between PU and SU}}{\text{d0}}\right)^{-n}}{\text{Noise}}$$

Where n is the path loss exponent, and d0 is the far-field cross over distance.

The above formula shows how to find SINR in a Cognitive Radio Network. In this, the data or information is not shared between different SU nodes, and multiple based stations are not transmitting them in a coordinated manner like in Joint Transmission. Hence we have only one node accessing the radio channel of the PU network at a time. This technique avoids interference between different SU nodes. This technique is being implemented for the three different environments and the results are being discussed in the next section [7].

(b) SINR for Coordinated Multipoint

$$\text{SINR} = \frac{\text{SU tx power } \left(\frac{\text{Distance between BS and User}}{\text{do}}\right)^{-n}}{\text{Noise}}$$

The above formula shows how to find SINR in a Coordinated Multipoint scenario, where Cognitive Radio Network is not being implemented. In this, the data or information is not shared between different SU nodes in a Cognitive Radio Network though multiple nodes or base stations are transmitting them in a coordinated manner since this is a Joint Transmission technique. Hence we have not only one node but different nodes accessing the radio channel of another network at a time. With a strong network backbone like Cognitive Radio Network, it is expected that the SINR in this technique will be poor compared to if it is combined with Cognitive Radio Network. This technique avoids interference between different SU nodes. And is being implemented for the three different environments and the results are being discussed in the next section [7].

(c) SINR for CRN + CoMP JT

$$SINR = \frac{\text{SU tx power}\left(\left(\frac{\text{Distance between BS and User}}{\text{do}}\right)^{-n} + \left(\frac{\text{Distance between BS and User}}{\text{do}}\right)^{-n}\right)}{\text{Noise}}$$

Joining these two technologies, according to their functionality explained in section III, is going to improve the values of SINR because it is being implemented on a network having a strong backbone and architecture that is Cognitive Radio Network, alongside a technology that is proven to be efficient in terms of overall information sharing and throughput.
The combination of Cognitive Radio Network and Coordinated Multipoint Joint Transmission is promising to give better results in terms of both throughput and SINR, and the effect of the three different environments on this is to be discussed in the next section [7].

4. Results

After implementing all the discussed technologies, formulas, and concepts explained in the above lines, some results have been generated from the simulation performed on MATLAB R2015a and they are discussed below.

4.1 Total Average Throughput

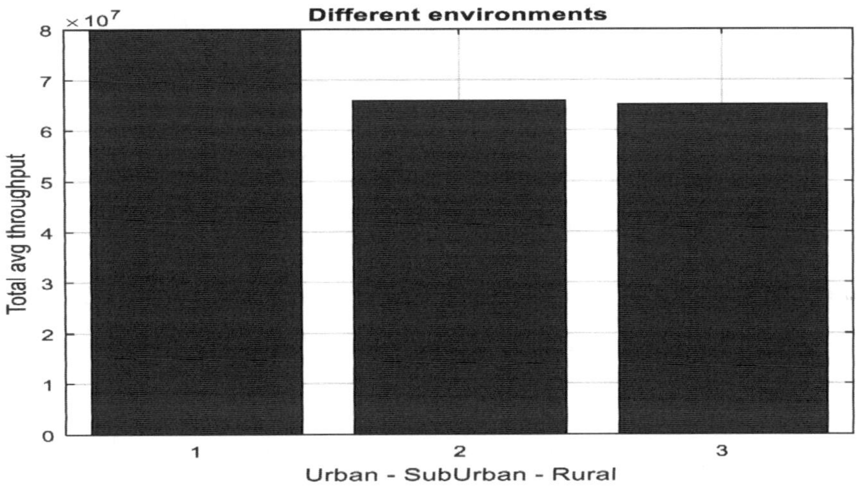

Fig. 3 TAT in different environments

This TAT graph is being calculated from the combination of CRN and CoMP JT, and it is being implemented in the three different environments as shown in Fig.3. The results show that in the Urban Environment, the throughput is higher, and it is due to the reason that multiple cell towers are deployed because of the high concentration in the building. The thing has done to compensate for the missing open long-range for data communication when compared to the suburban environment and rural environment.

4.2 Signal to Interference plus Noise Ratio

It is shown in Fig. 4. How SINR is varying in different environments, and all these have been implemented in the two technologies combined. They are Cognitive Radio Network and Coordinated Multipoint Joint Transmission. The reason for the lower SINR in the urban environment is the excess in interference, fading and other shadowing effects due to building. These buildings are entirely missing in rural. However, the distances are long and the terrains are unsure, hence causing more attenuation. In

Suburban, however, it is the perfect environment where there are less interference and short distances, thus the higher SINR.

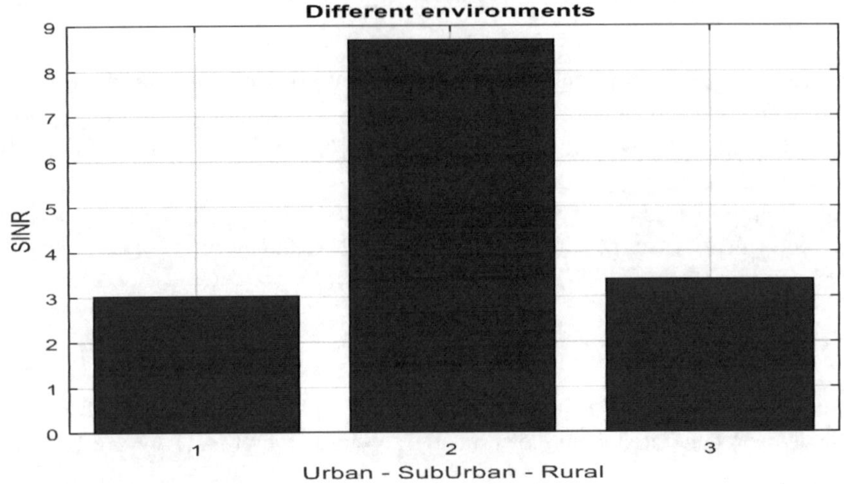

Fig. 4 SINR in different environments

4.2.1 Urban environment

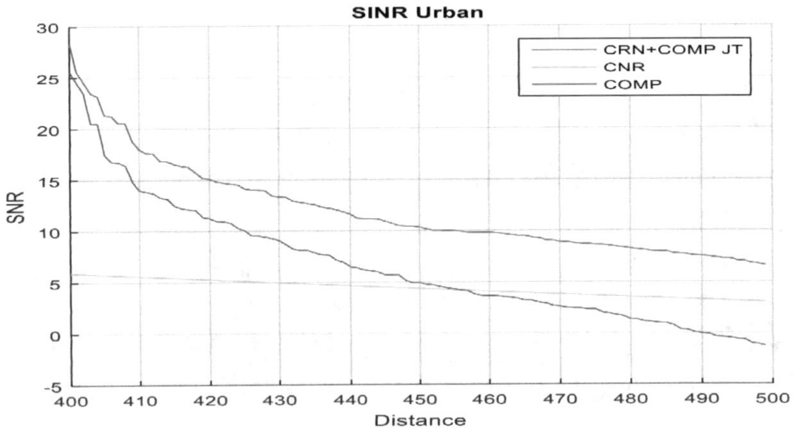

Fig. 5. SINR in the urban environment

4.2.2 Suburban environment

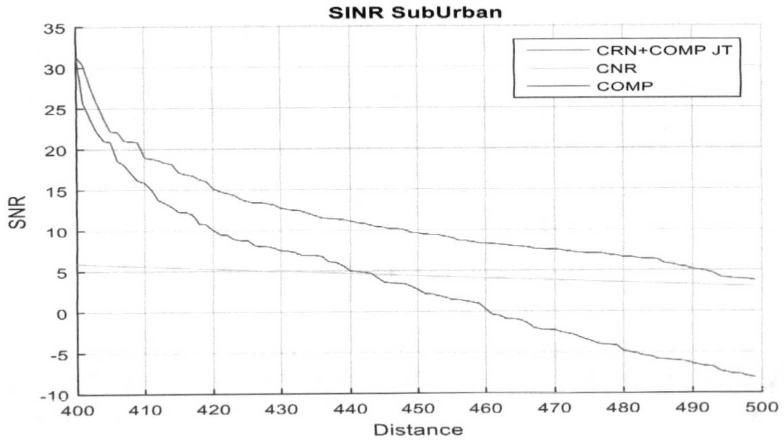

Fig 6. SINR in a suburban environment

4.2.3 Rural environment

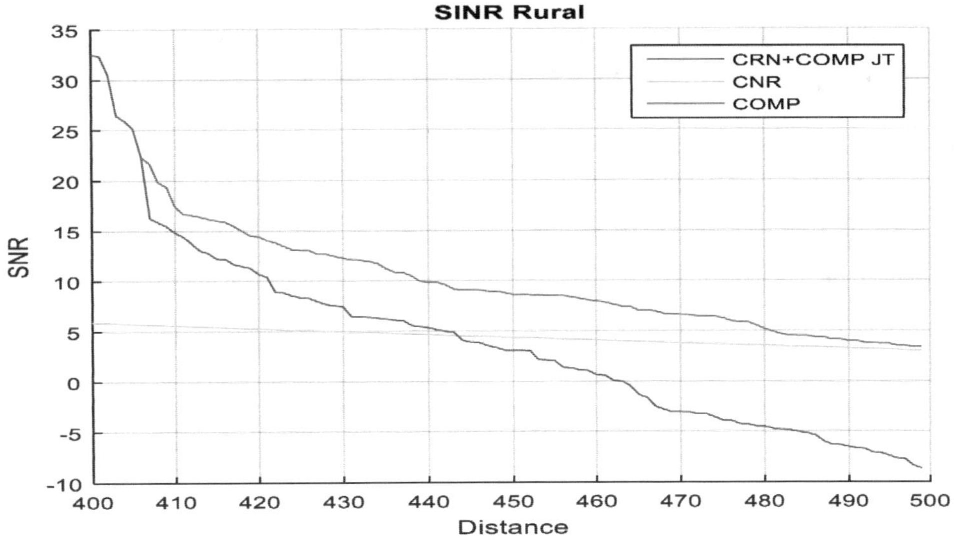

Fig. 7. SINR is n rural environment

These last three graphs are represented in Fig. 5, Fig. 6 and Fig.7. It can be seen how SINR is changing in different environments, for the same reasons explained in Fig. 4. However, added to this, it is shown the variation of SINR for different techniques/technologies used. CRN, CoMP and CNR + CoMP JT. These three technologies exhibit a similar behaviour overall, which means the SINR value is decreasing with the increase in the distance of SU nodes from the PU.

As far as CNR + CoMP JT is concerned, the reason for the lower SINR in the urban environment is the excess in interference, fading and other shadowing effects due to building. These buildings are entirely missing in rural. However, the distances are long and the terrains are unsure, hence causing more attenuation. In Suburban, however, it is the perfect environment where there are less interference and short distances, thus the higher SINR.

Considering now the case for CRN alone, the SINR value is having minor variation with the increase in the distance. The downside to this technology (CRN) being applied solely without CoMP JT is the lower values of SINR despite the stable like variations. This reason is due to the missing communication between different nodes in the SU network, hence less information to be shared amongst themselves.

The poorest of SINR values comes from the part where CoMP techniques have been applied alone without combining it with CRN. This causes the whole system to work with an uncertain and weak backbone network. They are hence reducing the vales of SINR even to less than zero at a specific range.

5. CONCLUSION

Cognitive Radio Network is a promising technology that aims to improve spectral efficiency by efficiently using the licensed user spectrum by unlicensed users, without causing interference to the licensed one, or by keeping the interference at a tolerable level. Though this technology is a promising one, it can still be improved by an already existing technology that is Coordinated Multipoint Joint Transmission.

In this report, these two technologies have been combined both from an Architectural point of view, aiming to design an inclusive layered architecture, including all the techniques and technologies used in this combination. And also, from a simulation point of view, a simulation has been implemented in the Matlab tool.

The architecture proposed in this report includes the four steps of spectrum management in CNR and shows all the transitions happening from one to the other. It also shows the use of CoMP JT and other technologies.

The other work performed in this thesis is the implementation of SINR behaviours in different environments and the study of Throughput behaviour implemented in CRN along with CoMP JT in all three different environments. Results are compared, and it is shown that the combination of Cognitive Radio Network and Coordinated Multipoint Joint Transmission offers the best performance to the system. As far as throughput and SINR are concerned in CNR + CoMP JT, the urban scenario offers the highest Throughput, and the Suburban scenario offers the highest SINR.

This work can also be modified and explored in different areas of wireless communication by following the mentioned research papers [27-46].

Acknowledgement

I would like to express my very great appreciation to **Harpreet Singh Bedi, Shekhar Verma, Preeti Khurana, Raghav Gupta (School of Electronics and Electrical Engineering, Lovely Professional University, Phagwara, Punjab, India.)** for their valuable and constructive suggestions during the planning and development of this research work. Their willingness to give their time so generously have been very much appreciated

ORCID Id of author

Akhil Gupta: - https://orcid.org/0000-0002-0054-3801.

REFERENCES

[1] P. Rysavy, "Challenges and Considerations in Defining Spectrum Efficiency," in *Proceedings of the IEEE, vol. 102*, no. 3, pp. 386-392, March 2014.

[2] D. M. Alias and Ragesh G. K, "Cognitive Radio networks: A survey," 2016 I*nternational Conference on Wireless* Communications, Signal Processing and Networking (WiSPNET), Chennai, 2016, pp. 1981-1986.

[3] L. Zhang, M. Xiao, G. Wu, M. Alam, Y. Liang and S. Li, "A Survey of Advanced Techniques for Spectrum Sharing in 5G Networks," in *IEEE Wireless Communications, vol. 24, no. 5, pp. 44-51, October 2017.*

[4] Ghosh G, Das P, Chatterjee S, "Simulation and analysis of cognitive radio system using Matlab," *International Journal of Next-Generation Networks,* Vol.6, Issue. 2, pp. 31-45, June 2014.

[5] I. F. Akyildiz, W. Lee, M. C. Vuran and S. Mohanty, "A survey on spectrum management in cognitive radio networks," in *IEEE Communications Magazine*, vol. 46, no. 4, pp. 40-48, April 2008.

[6] J. Zhao, T. Q. S. Quek, and Z. Lei, "Coordinated Multipoint Transmission with Limited Backhaul Data Transfer," *in IEEE Transactions on Wireless Communications*, vol. 12, no. 6, pp. 2762-2775, June 2013.

[7] Reasat T, Saha A, Uddin MF. Cognitive radio network with the coordinated multipoint joint transmission. *Int J Commun Syst.* 2017;30:e3310.

[8] M. Bkassiny, Y. Li and S. K. Jayaweera, "A Survey on Machine-Learning Techniques in Cognitive Radios," in *IEEE Communications Surveys & Tutorials, vol. 15, no. 3, pp. 1136-1159, Third Quarter, 2013.*

[9] R. Irmer et al., "Coordinated multipoint: Concepts, performance, and field trial results," in *IEEE Communications Magazine,* vol. 49, no. 2, pp. 102-111, February 2011.

[10] R. K. Sharma and D. B. Rawat, "Advances on Security Threats and Countermeasures for Cognitive Radio Networks: A Survey," in *IEEE Communications Surveys & Tutorials, vol. 17, no. 2, pp. 1023-1043*, Second quarter of 2015.

[11] J. N. Soliman, T. A. Mageed and H. M. El-Hennawy, "Taxonomy of security attacks and threats in cognitive radio networks," 2017 Japan-Africa *Conference on Electronics, Communications, and Computers* (JAC-ECC), Alexandria, 2017

[12] Z. Wei and B. Hu, "A Fair Multi-Channel Assignment Algorithm With Practical Implementation in Distributed Cognitive Radio Networks," in *IEEE Access, vol. 6, pp. 14255-14267, 2018.*

[13] E. Z. Tragos, S. Zeadally, A. G. Fragkiadakis and V. A. Siris, "Spectrum Assignment in Cognitive Radio Networks: A Comprehensive Survey," in *IEEE Communications Surveys & Tutorials, vol. 15, no. 3, pp. 1108-1135, Third Quarter, 2013.*

[14] J. Unnikrishnan and V. V. Veeravalli, "Algorithms for Dynamic Spectrum Access With Learning for Cognitive Radio," in *IEEE Transactions on Signal Processing*, vol. 58, no. 2, pp. 750-760, Feb. 2010.

[15] T. M. C. Chu, H. Phan and H. Zepernick, "Dynamic Spectrum Access for Cognitive Radio Networks With Prioritized Traffics," in *IEEE Communications Letters, vol. 18, no. 7, pp. 1218-1221, July 2014.*

[16] G. Hattab and M. Ibnkahla, "Multiband Spectrum Access: Great Promises for Future Cognitive Radio Networks," in *Proceedings of the IEEE*, vol. 102, no. 3, pp. 282-306, March 2014.

[17] E. W. M. Wong and C. H. Foh, "Analysis of cognitive radio spectrum access with finite user population," in *IEEE Communications Letters, vol.* 13, no. 5, pp. 294-296, May 2009.

[18] L. Cheng et al., "Coordinated Multipoint Transmissions in Millimeter-Wave Radio-Over-Fiber Systems," in *Journal of Lightwave Technology*, vol. 34, no. 2, pp. 653-660, 15 Jan.15, 2016.

[19] G. Nigam, P. Minero and M. Haenggi, "Coordinated Multipoint Joint Transmission in Heterogeneous Networks," in *IEEE Transactions on Communications*, vol. 62, no. 11, pp. 4134-4146, Nov. 2014.

[20] L. R. Martins Pinto and L. H. Andrade Correia, "Analysis of Machine Learning Algorithms for Spectrum Decision in Cognitive Radios," 2018 *15th International Symposium on Wireless Communication Systems (ISWCS),* Lisbon, 2018

[21] C. Clancy, J. Hecker, E. Stuntebeck and T. O'Shea, "Applications of Machine Learning to Cognitive Radio Networks," in *IEEE Wireless Communications*, vol. 14, no. 4, pp. 47-52, August 2007.

[22] K. A.M, F. Hu and S. Kumar, "Intelligent Spectrum Management Based on Transfer Actor-Critic Learning for Rateless Transmissions in Cognitive Radio Networks," in *IEEE Transactions on Mobile Computing*, vol. 17, no. 5, pp. 1204-1215, 1 May 2018.

[23] K. M. Thilina, K. W. Choi, N. Saquib and E. Hossain, "Machine Learning Techniques for Cooperative Spectrum Sensing in Cognitive Radio Networks," in *IEEE Journal on Selected Areas in Communications,* vol. 31, no. 11, pp. 2209-2221, November 2013.

[24] T. Yucek and H. Arslan, "A survey of spectrum sensing algorithms for cognitive radio applications," in *IEEE Communications Surveys & Tutorials*, vol. 11, no. 1, pp. 116-130, First Quarter 2009.

[25] J. Li, Z. Feng, Z. Feng and P. Zhang, "A survey of security issues in Cognitive Radio Networks," in *China Communications*, vol. 12, no. 3, pp. 132-150, Mar. 2015.

[26] Gupta, Akhil, and Rakesh Kumar Jha. "Power optimization using massive MIMO and small cells approach in different deployment scenarios." *Wireless Networks* 23.3 (2017): 959-973.

[27] P. Jain and A. Gupta, "Energy-Efficient Adaptive Sectorization for 5G Green Wireless Communication Systems," in *IEEE Systems Journal*, vol. 14, no. 2, pp. 2382-2391, June 2020.

[28] Chopra S.R., Gupta A., Monga H. (2019) Performance Analysis of Space-Time Trellis Codes in Rayleigh Fading Channel. In: Yadav N., Yadav A., Bansal J., Deep K., Kim J. (eds) Harmony Search and Nature Inspired Optimization Algorithms. *Advances in Intelligent Systems and Computing, vol 741. Springer, Singapore*.

[29] Gupta, Akhil, Shiwani Dogra, and Ishfaq B. Sofi. "Performance evaluation of spatial multiplexing using different modulation techniques in MIMO system for small and large scale fading channel." *International Journal of Sensors Wireless Communications and Control 9.2 (2019)*: 188-202.

[30] Gupta, Akhil, Rakesh K. Jha, and Reeta Devi. "Security architecture of 5g wireless communication network." *International Journal of Sensors Wireless Communications and Control 8*.2 (2018): 92-99.

[31] Sofi, Ishfaq Bashir, Akhil Gupta, and Rakesh Kumar Jha. "Power and energy optimization with reduced complexity in different deployment scenarios of massive MIMO network." *International Journal of Communication Systems 32.6 (2019): e3907.*

[32] A. Gupta and S. R. Chopra, "Designing and Testing of Quadrifilar Helix Antenna for NOAA Weather Satellite," 2019 *6th International Conference on Signal Processing and Integrated Networks (SPIN)*, 2019, pp. 785-790.

[33] S. R. Chopra, A. Gupta and H. Monga, "An LTE Approach with MIMO by Using Suboptimal Selection of Antenna," *2018 International Conference on Intelligent Circuits and Systems (ICICS), 2018, pp. 167-172.*

[34] Goyal, Krittika, and Akhil Gupta. "A literature survey on different types of pulse-based sensor for acquisition of pulse." *International Journal of Control Theory and Applications 9.10 (2016): 361-365.*

[35] R. Kausar, A. Gupta, I. B. Sofi and K. Arora, "Bit Error Rate Based Performance Evaluation of LTE OFDMA System," *2018 International Conference on Intelligent Circuits and Systems (ICICS), 2018, pp. 161-166*, DOI: 10.1109/ICICS.2018.00042.

[36] Anand, Pankaj, Akhil Gupta, and Sonam Bhagat. "Bit Error Rate Assessment of Digital Modulation Schemes on Additive White Gaussian Noise, Line of Sight and Non-Line of Sight Fading Channels." *International Journal of Engineering Science Invention* 3.8 (2014): 6-10.

[37] Gupta, Akhil, and Krittika Goyal. "An analytic review on antenna modelling." International *Journal of Control Theory and Applications* 9.10 (2016): 319-336.

[38] Gupta, A., Jha, R.K. Power optimization with low complexity using scaled beamforming approach for a massive MIMO and small cell scenario. *Wireless Netw* 26, 1165–1176 (2020).

[39] I. B. Sofi, A. Gupta and R. Kausar, "Performance Evaluation of Different Channel Estimation Techniques in MIMO System for Hata Channel Model," 2018 *International Conference on Intelligent Circuits and Systems (ICICS),* 2018, pp. 155-160, DOI: 10.1109/ICICS.2018.00041.

[40] R. T. Kamurthi, S. R. Chopra and A. Gupta, "Higher-Order QAM Schemes in 5G UFMC system," *2020 International Conference on Emerging Smart Computing and Informatics (ESCI)*, 2020, pp. 198-202, DOI: 10.1109/ESCI48226.2020.9167619.

[41] Raina, Ambica, Akhil Gupta, and Rakesh Kumar Jha. "Analysis of AODV and DSR Routing Protocols for a Static Ad Hoc Network Using Qualnet Network Simulator." *International Journal 4.1 (2014).*

[42] Sundan, A.P., Jha, R.K. & Gupta, A. Energy and spectral efficiency optimization using probabilistic based spectrum slicing (PBSS) in different zones of 5G *wireless communication network. Telecommun Syst* 73, 59–73 (2020).

[43] M. Singh, K. Arora and A. Gupta, "Equalization in WIMAX System," 2018 International Conference on *Intelligent Circuits and Systems (ICICS), 2018, pp. 330-333, DOI: 10.1109/ICICS.2018.00074.*

[44] Jain, Preksha, and Akhil Gupta. "Adaptive NOMA towards 5G green wireless network." *Transactions on Emerging Telecommunications Technologies* 31.3 (2020): e3887.

[45] H.S. Bedi, K.K. Sharma, and R. Gupta, "A Review Paper on Performance Analysis of IEEE 802.11 e", *1st International Conference on Computing, Communications, and Cyber-Security (IC4S 2019), 2020, pp. 47-56.*

[46] Z.G. Khaki, and H.S. Bedi, "Transient correction using EDFA: in-line optical fibre with feedback", *International Conference on Computing Sciences,* 2012, pp. 233-238 [http://dx.doi.org/10.1109/ICCS.2012.76].

[47] Bedi, Harpreet, Kamal Kumar, and Gupta Raghav. "A review paper on improving the network efficiency of IEEE 802.11 e networks." Intelligent Circuits and Systems (2021): 206-211.

[48] Harpreet Bedi, Kamal Kumar Sharma; Analyses of CSMA/CA Protocol Without Using Virtual Channel Sensing in DCF Mode, Global Emerging Innovation Summit (GEIS-2021) (2021) 1: 484. https://doi.org/10.2174/9781681089010121010058.

Bandwidth Optimization for Vertical Handoff using Hybrid Cuckoo Search and Genetic Algorithm Scheme

Preeti Khurana[1], Keshav Jha[2], Akhil Gupta[3]

School of Electronics and Electrical Engineering, Lovely Professional University, Punjab.

preeti.Khurana@lpu.co.in[1], keshavjha775@gmail.com[2], akhilgupta112001@gmail.com[3],

Abstract. Future remote systems should have the option to handle the gain in a different system state. One of the most of the time tried issues for coordination is the vertical handoff, which is an option to handoff portable hubs between different systems. While customary handoff codes are based on quality checks, vertical handoffs must assess additional factors, for example, funding costs, proposed governance, systematic location, and customer learning. In this paper, some improvements are proposed for the implementation of vertical handoff option calculations aimed at increasing the nature of the administration experienced by each customer. Firstly, talk about the idea of alignment-based handoff is done. At that point, cost efficiency can be achieved by sending decisions on the target systems depending on the classification of the customer and the measurements provided by the system. Finally, a demonstration investigation suggests that the proposed reforms can achieve significant gains like governance and the productive use of assets. In the present work, two other half-schemes are proposed using the cuckoo search calculation and the genealogical calculation. In the two mixing plots, the calculation consists of two steps: the primary stage, which examines the location of the cuckoo search (CS), or the genetic algorithm (GA). In the subsequent stage, to improve worldwide pursuit and dispose of catching into a few neighbourhood optima is done. Reproducibility results and testing with old-time CS and GA calculations confirm the feasibility of the proposed computation in designing various benchmark correction capabilities.

Keywords. Handover, Bandwidth, Location management, Energy Efficiency, Hybrid model.

1. Introduction

Massive testing for remote systems of the fourth era (4G) and fifth era (5G) involves the establishment of a wide range of systems for remote correspondence [1]. For example, information management such as high transmission capability 3G cell and GPRS is currently being enhanced by the wireless local area network (WLAN) organization, IEEE 802.11, and hyper LAN 3rd generation partnership project (3GPP) is now acting. Universal Mobile Telecommunication System (UMTS) collaborates with Hyperlane, counting the open issue of when, where, and how to dispatch handoffs between the 3G Cell Framework and WLAN.

Remote innovations are developing toward broadband data access over numerous systems administration stages, to give pervasive accessibility of interactive media applications. Late patterns show that wide-zone cell systems dependent on the 5G principles and remote Local Area Networks (WLANs) will coincide to offer mixed media administrations to end clients. These two remote access innovations have attributes that impeccably supplement one another. By deliberately joining these advancements, a merged framework can give both all-inclusive inclusion and broadband access. Thusly,

the combination of heterogeneous systems is required to turn into a primary concentration in the advancement toward the cutting edge remote systems.
Handoff is a way to handle the dynamic associations of a versatile customer while moving within a remote system [2]. The Vertical Handoff, or Framework Handoff, uses the handoff between different systems [3], [4]. Generally, a handoff look is made on the evaluation of the received signal strength (RSS) in a versatile hub. However, traditional RSS correlations are not sufficient to settle on the vertical handoff option because they do not consider different connection options for multifunctional users. Other related works on vertical research have been introduced in late research writing. In [5], problems related to multi-order structures have been investigated and neural-based vertical handoff calculations have been developed to meet client data transfer capability requirements. In [4], a robust handoff option calculation is proposed with a cost efficiency that considers some of the previously mentioned factors. In this paper, while extending the work on vertical handoff calculus and cost efficiency, we propose some improvements to the handoff selection process and commit together: (1) internalizing handoff cost work progress records that increase dynamic characteristics. Vertical handoff, (2) fuse of the system settlement process in vertical handoff metric, increase handoff figure to reduce estimated reflexes and reduce handling power, and (3) increase proposed handoff cost functions for various types of the hub. Customer administration is spread across different systems. Section II depicts the alignment-based handoff idea. At that time, in Section III and IV, cost efficiency is based on the decision of the target systems, the customer, and the system based on the measurement. Finally, in Section V, the presentation test shows that the proposed breakthrough can achieve significant benefits like governance and the efficient use of assets. Section VII completes the paper.
There are two primary situations in VHO: moving out of the favoured system (MO) and moving into the favoured system (MI). In the merged model, it is exceptionally alluring to connect the MT with the favoured system, as long as the favoured system fulfils the client application. This can improve the asset use of both access systems, just as improving the client saw QoS. Moreover, this handoff ought to be consistent with the least client intercession, while progressively adjusting to the remote channel state, arranging layer attributes, and application prerequisites.

Growth is an area of applied science that maintains extreme expectations of efficiency in the space of definition, subject to various constraints on variable properties [1]; such problems can be organized into two classes: unrelated and compelling problems. The interface between the enthusiasm for software engineering and the activities that solve global development problems has led to incredible growth [2, 3]. There are two classifications of correction procedures: exact and heuristic. The accurate methodology ensures that the ideal system is completed and works wonderfully for some problems. However, for countless selected complex problems or factors, exact methods may require high computational costs [2]. Many complex problems fall into the category of such complex problems, so an alternative method is needed to measure them intelligently over time [2]. For these problems, meta-heuristic calculations can be considered a productive tool for obtaining ideal arrangements [4]. Two important features of meta-heuristics are strengthening and expanding. Reinforcement means using data from current best practices. This process looks around the area of the current best fit and selects the best ones. Broadly speaking, this addition is called probing, which ensures that the computational bait location can be more efficiently evaluated by simple randomization. The original advance jumps the morphology from any nearby optima and ensures that the new arrangement can be seen differently [5]. These strategies have received critical scrutiny because they are supportive, rigid, mild, and often unknown, regardless of some parameter tuning [4]. If so, the application of such single techniques is highly forbidden here and there, especially for high dimensional and non-linear problems. On this basis, these methods usually require extensive measurements of computational possibilities and are often implicated in neighbouring optima. Of late, various strategies for integrating meta-heuristics with neighbouring search methods are a useful solution to defeat the loss of moderate combustion rate and unilateral evolution of meta-heuristics [6].
Of these half and half, the nearest research methods are embellished with meta-heuristics to manage them, especially in the neighbourhood neighbouring minima, and in particular to further their union. Lost in the previous stage, it has recently become clear that consolidation on a single meta-heuristic is

very much forbidden. Combining meta-heuristic with other correctional systems, the perceived half-breed meta-heuristic gives increased productivity and greater adaptability when managing genuine world and enormous scale issues. On this basis, crossover meta-heuristics use different calculation properties [6], [7], [8–25]. All half and half meta-heuristic methods can be performed in mass alloys or even in integrated mixtures [8]. Of late, Yang [17] has produced another cuckoo discovery census (CS) through the breeding parasite of some cuckoo species, without putting eggs in the homes of other host-winged organisms (regularly other species). In this paper, two crossover schemes are excluded from the standard CS and Genetic Algorithm (GA). The proposed calculations, CS-GA and GA-CS, deal with the uncontrolled problems of progress worldwide. Both variants are tried on many uncontrolled problems. The search results show the accuracy and rapid implementation of the proposed calculations that correlate with the other best in square measurement.

2. CUCKOO SEARCH ALGORITHM

Cuckoo Search (CS) enumeration is a meta-heuristic probe enumeration that has been proposed lately by Yang and Deb [17], which relies on a set of accredited parameters: (1) each egg of a cuckoo is placed and it is taken unilaterally. (2) The best houses with high egg capacity (calibration) will be maintained for the following ages. (3) The amount of host homes that can be accessed is fixed and the host can find the outer egg with probability. For this situation, the host-winged animal may abandon the egg or leave the home to create a completely new home in another area. Levy is presented with the plane, creating new answers for it Cuckoo.

$$x^{(t+1)}{}_{i} = x^{(t)}{}_{i} + \alpha\ \theta\ Levy(\lambda) \qquad (1)$$

Where λ is the progression size, which ought to be identified with the size of the issue of intrigue. The item implies the passage insightful increases. We consider a Levy trip in which the progression lengths are circulated by the accompanying likelihood appropriation:

$$Levy\ u = t^{-\lambda},\ 1 < \lambda <= 3 \qquad (2)$$

This has a boundless fluctuation. Here the sequential hops/steps of a cuckoo structure an irregular walk process that complies with force law step length dispersion with an overwhelming tail.

3. GENETIC ALGORITHM

GA, proposed in the mid-1970s [4] is a worldwide hunting technique that mimics the metaphor of general organic assessment. GA starts with the population (age) of randomly born people. Each individual (chromosome) responds unchanged to a viable problem. Eligible individuals are selected by referring them to a previously prepared welfare function rated for each individual in this population. People with lower wellness honours have a better response to this problem. Following this underlying process, hybrid and transformation functions are used, where the population of the current population produces youngsters. These children were excluded from the wellness score. After the election, hybrids and transitions are applied to the underlying population, narrowing down another population and expanding the generation counterpart one by one. This process of hybridization and transformation continues until the final state is reached [4].

4. Methods

The available bandwidth capacity to the MT relies upon the extent of time that the Mobile Terminal (MT) remains in the WLAN and the 5G organize, just as the WLAN state when the MT is associated with the WLAN. To the MT, the WLAN is in one of two states: WLAN Up and WLAN Down. The WLAN Upstate speaks to the occasion that the WLAN signal got at the MT is over the affectability level

α. WLAN Down is the converse case. Let p[k] be the likelihood that the WLAN is in the Upstate at time k[32]. Unmistakably,

$$p[k] = \Pr\{RSS[k] > \alpha\} \quad (3)$$

$$=Q((\alpha-\mu[k])/\beta). \quad (4)$$

In the proposed handoff calculation, the MO separation changes; thusly the caught WLAN Up spans do as well. For the remainder of the investigation, we are keen on assessing the framework execution during the progress locale, which is characterized as the scope of separation between the moment that the RSS begins to sway around the interface affectability and the WLAN edge. The progress locale assurance is comparable to a long-standing complex level intersection issue that is diagnostically manageable just for a couple of straightforward cases and is typically illuminated numerically for complex cases.

4.1 System Analysis

Step 1: For example, let's take uninhabitable versatile centres and they are highly impacted by the deposition of different systems, which consider only six individual proxies, including reproductive processes. For example. Code division multiple access (CDMA), Global system for mobile communication (GSM), WLAN, Wireless Fidelity (WiFi), and so on.

Step 2: Number of cuckoo birds: This discovery is related to the number of cuckoo winged creatures in the area.

Step 3: During the time spent on the CSA (Cuckoo Search Algorithm), the size of the step follows the separation of the cuckoo's winged organism for a certain number of cycles or cycles. It tries to take the average foreground shape to achieve a compelling fit. If the volume of progress is too large or too small,

Step 4: This refers to another computation called hereditary or genealogical calculation and searches for a population source that originates from CS (cuckoo search) and finds the ideal setting. The versatile hub is the time to expand another function.

Step 5: The initial development plan at that time takes the social relevance of the fitting framework from the open Framework, in which the critical number of the framework is estimated by taking into account the parameters that are not applicable, and the case is sent for genetic computation, a different framework.

Step 6: Similarly, if we mimic the same situation with advanced parameters, faster cuckoo search calculation, users can find the best meeting framework at the right time and it will get the best framework after genealogy calculation.

Step 7: Using the cuckoo-searching half-breed approach with hereditary calculations, we reduce the time it takes to find better organisms, thus improving transmission efficiency.

4.2 .Flowchart

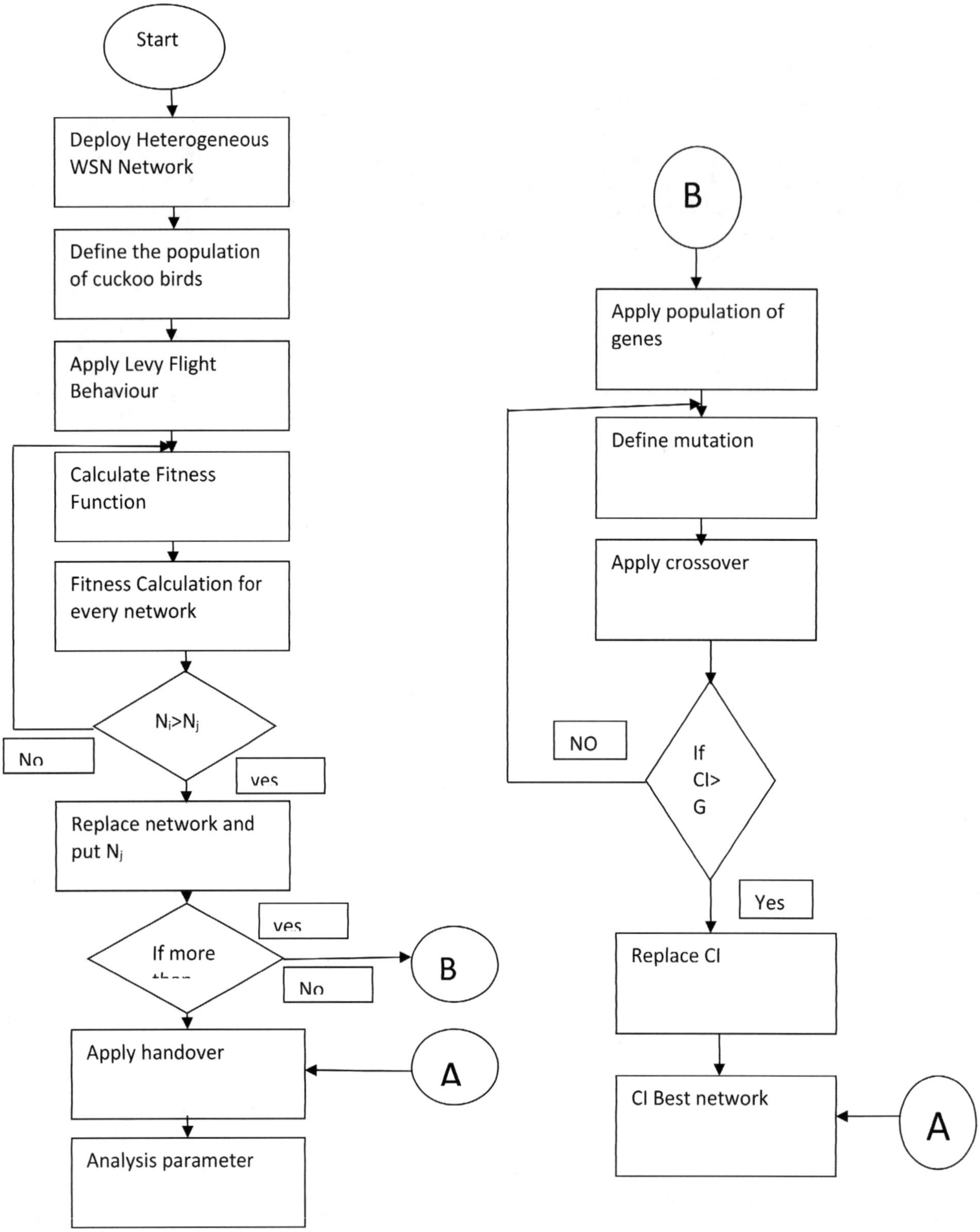

The proposed CS-GA calculations are a shared mixture of CS and GA strategies. Half of this species, at an early stage, checks the location of the bait to isolate the most sot cracked spots of the CS (or GA) haunting. The second is the addition of GA (or CS) to the search location (initiated by the CS) to improve the bait globally and avoid getting caught in the nearby optima. And invented new improved systems. Thus, the CS begins with the population, which is closer to the ideal system. Furthermore, CS becomes the best model parameter vector.

Proposed Algorithm: Hybrid algorithm of Cuckoo Search and Genetic Algorithm

---

Step 1: Network Initialization

Procedure Next_one (location, destination)
Check velocity and position for MN (Mobile Node)
a ← false
Query_MIIS_DBC (location, destination, MIIS_Pkt)
If (min_cover_time (loc, DIS, MIIS.Pkt)>0) then
If (DiS_ReC_Probability(loc, DIS, MIIS.Pkt)>0)
Then a ←true
Initialize the population of the host nest
Return (a)
While iteration<Max_network do
Fori=1……N do
For (updating fitness functionality by Levy flight)
Update velocity and position for Mobile node
updateNext_one (loc, dest)
Evaluate Threshold (TH)
Update fitness function

Step 2: Apply Genetic fitness function

Procedure Next_one (time, Prediction window, sense prediction)
Evaluate TH
Filter network by Genetic Fitness value
Update velocity and position of Mobile Node

Step 3:The hybrid Cuckoo search-Genetic Algorithm phase

Procedure check_next(time, Prediction window, sense prediction)
Fori=i……N do
Calculate the probability of crossover and
Evaluate TH
if converge then go to step 4
else change next position and apply
Step 2 till converge
After convergence find
Step 4:Evaluate parameters
TH value ←Cuckoo_search_GA

Calculate Cost function (TH value)

Procedure Cost Function
Repeat
Update application request list (MIIS_weights)
Update list of Network Preference
Update QOS ←QOS_req(MIIS_weights)
TH ←TH (app_req, user, QoS)
Return

5. RESULTS

In this section, numerical reproducibility is performed to identify the proposed reproducibility, relying on some important improvement problems [9, 13, and 17]. In these problems, the basic parameter of CS is the number of households n = 100, the outer egg/calibration exposure is Pa = 0. 25. The mix count result consists of 20 autonomous runs per digit. Each test was performed on the Windows 7 Ultimate 64-Piece Working Framework, Processor. The simulation parameters used for the research work in this paper are mentioned in Table I. The simulation parameters are identified with the help of one base paper [32]. Notwithstanding the above examination, we have recreated the VHO calculations utilizing MATLAB. Table 1 shows the reenactment boundary esteems. The WLAN boundaries are utilized as in [32], which are reasonable to demonstrate open-air rural (for example with tree and low structures along the street side which is like the qualities of the business WLAN administrations) and indoor areas with wide zones, (for example, lodging anterooms and grounds). These boundaries bring about a WLAN inclusion of 100 meters around. The information rates that appeared in the table are utilized for execution assessment just and have no impact on the handoff choice. Furthermore, a basic versatility model has expected in which an MT moves ceaselessly from the WLAN passage in an orderly fashion at a consistent speed V(km/hr). Furthermore, the proposed calculation will work with any versatility design since the calculation powerfully adjusts to the MT speed, and the calculation time goal is adequate to follow portability design variety, particularly for low-speed MTs.

TABLE I. SIMULATION PARAMETERS

Parameter	Value
Transmit power	(6,10,12,4)Mw
Bandwidth factor	(2,2.3,3,4)Mbps
The bandwidth of weight factor	(3,3.4,4,4.5)
A signal strength weighting factor	(0.4,0.23,0.5,0.45)
Number of mobile nodes	(20,40,32,60)
Communication rate	1000kb/s

This unit presents the results of the proposed structure. The structure proposed in this paper is analyzed against four methods:

1. Random method
2. RSS method
3. Cuckoo Search method

4. Genetic method

The results demonstrate that the novel proposed technique has the best advantage. Thus, it helps achieve the objectives of the research study i.e. Bandwidth.

Research in Fig. 1 shows that the residuals of the data transfer capability are at different speeds of the portable hub. This probe uses a variety of remote channels with different functions of the hub. The versatile focal point (portable hub) speeds up the hub's exposure and enhances the system-based positioning effect. The handoff position varies if different speed events occur, but it is difficult to determine which type of sensor bunch data transmission is versatile. This option uses static, dynamic, and advanced basis calculations. This includes improving the data transfer capability by selecting a successful sensor channel or by selecting vertical transmission capability through cross-hand strain and single enhancement draw methods. In the proposed approach, the use of half and half correction can help improve static base calculation, for example, RSS. The use of RSS helps to determine the static threshold and can be enhanced by static edge advancement calculation. In most cases, the enhancement mechanism is not found, and it provides an arbitrary answer that minimizes the performance of the data transfer capability. In the proposed method, cuckoo search and genealogical calculation are additionally used. This system improves the performance of cuckoo innovation. The diagram shows a significant improvement in data transmission using the proposed half and half method when contrasted with the current method and other single streamlined strategies.

In met remote frameworks, proficient vertical handoff of the executives between heterogeneous systems is basic to the general framework execution. We have introduced an application-explicit sign quality tuning component to a cross-layer versatile VHO approach, which considers the remote channel variety, arrange layer inertness, and application QoS requests. We have proposed an expository structure to assess the exhibition of VHO dependent on transfer speed. The versatile VHO approach has been appeared to improve the framework asset usage by expanding the dependence of the MT on the WLAN, just as rationing the assets of the 5G organised for clients situated outside the WLAN. All the more significantly, the proposed half breed model gives a way to adaptable framework structure [26-32].

The figure shows a great match between examination and reenactment. Figure 1 shows the enhancement for the accessible bandwidth by utilizing half and half cuckoo search and Genetic calculation. Obviously, from an unadulterated transfer speed perspective, it is ideal for the Mobile Terminal to perform handoff just a single time at the WLAN edge, even though the RSS can briefly go beneath the Mobile Terminal affectability level in the progress area.

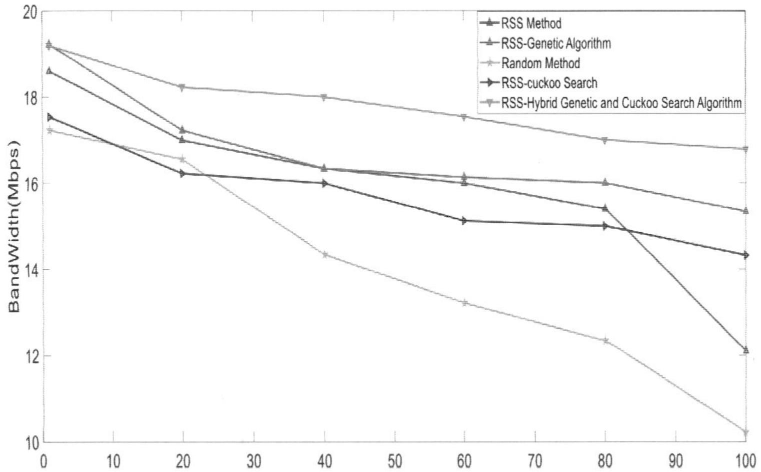

Fig.1 Bandwidth optimization using proposed technique

6. LIMITATIONS OF HYBRIDISATION METHOD

In the Cuckoo search algorithm we have used random step size and random input which reduce the quality of the solution so how can we give an efficient input to a Genetic algorithm for initializing by an efficient solution using without bothering about cuckoo search output, how can this problem be resolved because cuckoo search output is input for Genetic Algorithm. So, this is the biggest challenge between the hybridization of the cuckoo search and the Genetic Algorithm. In terms of the number of instances, the complexity of cuckoo search O (n) and GA complexity is O (n log). So total complexity O (n + n log n), where n is the number of populations.

7. CONCLUSIONS

Increasing governance through the use and coordination of individual systems is testing the creation of a more skilful, versatile, and canny vertical handoff meeting. In this paper, new advances have been made for vertical handoff option computation to maximize the benefits of handoff for both client and system. Integrates progress system end highlight to enhance data transfer capability. The hybridized investigation has demonstrated considerable utility like the administration and the progressively effective use of the property from the proposed improvements.

This work can also be modified and explored in different areas of wireless communication by following the mentioned research papers [33-54].

Acknowledgement

I would like to express my very great appreciation to **Harpreet Singh Bedi, Manmohan Sharma, Shekhar Verma, Raghav Gupta** (School of Electronics and Electrical Engineering, Lovely Professional University, Phagwara, Punjab, India.) for their valuable and constructive suggestions during the planning and development of this research work. Their willingness to give their time so generously have been very much appreciated.

ORCID Id of author

Akhil Gupta: - https://orcid.org/0000-0002-0054-3801.

REFERENCES

[1] Rao, Singiresu S. Engineering optimization: theory and practice. John Wiley & Sons, 2019.

[2] Liberti, Leo, and Nelson Maculan, eds. Global optimization: from theory to implementation. Vol. 84. Springer Science & Business Media,2006.

[3] Raouf, Osama Abdel, and Ibrahim M. Hezam. "Solving fractional programming problems based on swarm intelligence." Journal of Industrial Engineering International 10.2 (2014): 56.

[4] Yang, Xin-She. Engineering optimization: an introduction with metaheuristic applications. John Wiley & Sons, 2010.

[5] Parpinelli, Rafael S., and Heitor S. Lopes. "New inspirations in swarm intelligence: a survey." International Journal of Bio-Inspired Computation 3.1 (2011): 1-16.

[6] Wang, Gaige, and Lihong Guo. "A novel hybrid bat algorithm with harmony search for global

numerical optimization." Journal of Applied Mathematics 2013 (2013).

[7] Roeva, Olympia. "Genetic algorithm and firefly algorithm hybrid schemes for cultivation processes modelling." Transactions on computational collective intelligence XVII. Springer, Berlin, Heidelberg, 2014. 196-211.

[8] Atanassova, Vassia, et al. "Generalized nets, ACO-algorithms and genetic algorithm." Monte Carlo Methods and Applications (2012): 39-46.

[9] Abdel-Baset, Mohamed, and Ibrahim M. Hezam. "An effective hybrid flower pollination and genetic algorithm for constrained optimization problems." Advanced Engineering Technology and Application An International Journal 4 (2015): 27-27.

[10] Abdel-Baset, Mohamed, and M. Hezam Ibrahim. "An improved flower pollination algorithm for ratios optimization problems." Applied Mathematics & Information Sciences Letters An International Journal 3.2 (2015): 83-91.

[11] Kanagaraj, Ganesan, S. G. Ponnambalam, and Wei Chen Esmonde Lim. "Application of a hybridized cuckoo search-genetic algorithm to path optimization for PCB holes drilling process." 2014 IEEE International Conference on Automation Science and Engineering (CASE). IEEE, 2014.

[12] Lim, W. C. E., Ganesan Kanagaraj, and S. G. Ponnambalam. "A hybrid cuckoo search-genetic algorithm for hole-making sequence optimization." Journal of Intelligent Manufacturing 27.2 (2016): 417-429.

[13] ABDEL-BASET, M. O. H. A. M. E. D., and IBRAHIM M. HEZAM. "An improved flower pollination algorithm based on simulated annealing for solving engineering optimization problems." Asian Journal of Mathematics and Computer Research (2015): 149-170.

[14] Pradeep, S. Akila, and R. Manavalan. "Analysis of cuckoo search with genetic algorithm for image compression." International Journal of Engineering Research 2.6 (2013): 386-392.

[15] Rashad, M. Z., et al. "Genetic Cuckoo Optimization Algorithm (GCOA)." International Journal of Computer Applications 90.3 (2014).

[16] Zheng, Hongqing, Yongquan Zhou, and P. Guo. "Hybrid genetic-cuckoo search algorithm for solving runway dependent aircraft landing problem." Research Journal of Applied Sciences, Engineering, and Technology 6.12 (2013): 2136-2140.

[17] Gandomi, Amir Hossein, Xin-She Yang, and Amir Hossein Alavi. "Cuckoo search algorithm: a metaheuristic approach to solve structural optimization problems." Engineering with computers 29.1 (2013): 17-35.

[18] Akyildiz, Ian F., et al. "Mobility management in next-generation wireless systems." Proceedings of the IEEE 87.8 (1999): 1347-1384.

[19] Pahlavan, Kaveh, et al. "Handoff in hybrid mobile data networks." IEEE Personal Communications 7.2 (2000): 34-47.

[20] Stemm, Mark, and Randy H. Katz. "Vertical handoffs in wireless overlay networks." Mobile Networks and Applications 3.4 (1998): 335-350.

[21] Wang, Helen J., Randy H. Katz, and Jochen Giese. "Policy-enabled handoffs across heterogeneous wireless networks." Proceedings WMCSA'99. Second IEEE Workshop on Mobile Computing Systems and Applications. IEEE, 1999.

[22] Pahlavan, Kaveh, and Prashant Krishnamurthy. Principles of wireless networks: A unified approach. Prentice-Hall PTR, 2011.

[23] Zheng, Jun, and Abbas Jamalipour. Wireless sensor networks: a networking perspective. John Wiley & Sons, 2009.

[24] Yavatkar, Raj, Dimitrios Pendarakis, and Roch Guerin. "A framework for policy-based admission control." (2000).

[25] Rappaport, Theodore S. Wireless communications: principles and practice. Vol. 2. New Jersey: prentice hall PTR, 1996.

[26] Rizkallah, Jacky, and Nadine Akkari. "SDN-based vertical handover decision scheme for 5G networks." 2018 IEEE Middle East and North Africa Communications Conference (MENACOMM). IEEE, 2018.

[27] Dhivya, J. Josephine, and M. Ramaswami. "Ingenious method for conducive handoff appliance in cognitive radio networks." International Journal of Electrical and Computer Engineering 8.6 (2018): 5195.

[28] Goudarzi, Shidrokh, et al. "Hybridisation of genetic algorithm with simulated annealing for vertical-handover in heterogeneous wireless networks." International Journal of Ad Hoc and Ubiquitous Computing 24.1-2 (2017): 4-21.

[29] Yew, Hoe Tung, et al. "New Vertical Handover Method to Optimize Utilization of Wireless Local Area Network in High-Speed Environment." Plus one 11.11 (2016): e0165888.

[30] Seo, SungHoon, SeungChan Lee, and JooSeok Song. "Policy-based intelligent vertical handover algorithm in heterogeneous wireless networks." 2007 International Conference on Convergence Information Technology (ICCIT 2007). IEEE, 2007.

[31] Gohil, Zarna M., D. N. Khandhar, and K. M. Pattani. "Implement & analysis of RSS threshold-based vertical handoff decision algorithm for LTE & WLAN." International Journal of Innovative Research in Computer and Communication Engineering 5.1 (2017): 311-317.

[32] Zahran, Ahmed H., Ben Liang, and Aladdin Saleh. "Signal threshold adaptation for vertical handoff in heterogeneous wireless networks." mobile Networks and Applications 11.4 (2006): 625-640.

[33] P. Jain and A. Gupta, "Energy-Efficient Adaptive Sectorization for 5G Green Wireless Communication Systems," in IEEE Systems Journal, vol. 14, no. 2, pp. 2382-2391, June 2020.

[34] Chopra S.R., Gupta A., Monga H. (2019) Performance Analysis of Space-Time Trellis Codes in Rayleigh Fading Channel. In: Yadav N., Yadav A., Bansal J., Deep K., Kim J. (eds) Harmony Search and Nature Inspired Optimization Algorithms. Advances in Intelligent Systems and Computing, vol 741. Springer, Singapore.

[35] Gupta, Akhil, Shiwani Dogra, and Ishfaq B. Sofi. "Performance evaluation of spatial multiplexing using different modulation techniques in MIMO system for small and large scale fading channel." International Journal of Sensors Wireless Communications and Control 9.2 (2019): 188-202.

[36] Gupta, Akhil, Rakesh K. Jha, and Reeta Devi. "Security architecture of 5g wireless communication network." International Journal of Sensors Wireless Communications and Control 8.2 (2018): 92-99.

[37] Sofi, Ishfaq Bashir, Akhil Gupta, and Rakesh Kumar Jha. "Power and energy optimization with reduced complexity in different deployment scenarios of massive MIMO network." International Journal of Communication Systems 32.6 (2019): e3907.

[38] A. Gupta and S. R. Chopra, "Designing and Testing of Quadrifilar Helix Antenna for NOAA Weather Satellite," 2019 6th International Conference on Signal Processing and Integrated Networks (SPIN), 2019, pp. 785-790.

[39] S. R. Chopra, A. Gupta and H. Monga, "An LTE Approach with MIMO by Using Suboptimal Selection of Antenna," 2018 International Conference on Intelligent Circuits and Systems (ICICS), 2018, pp. 167-172.

[40] Goyal, Krittika, and Akhil Gupta. "A literature survey on different types of pulse-based sensor for acquisition of pulse." International Journal of Control Theory and Applications 9.10 (2016): 361-365.

[41] R. Kausar, A. Gupta, I. B. Sofi and K. Arora, "Bit Error Rate Based Performance Evaluation

of LTE OFDMA System," 2018 International Conference on Intelligent Circuits and Systems (ICICS), 2018, pp. 161-166, DOI: 10.1109/ICICS.2018.00042.

[42] Anand, Pankaj, Akhil Gupta, and Sonam Bhagat. "Bit Error Rate Assessment of Digital Modulation Schemes on Additive White Gaussian Noise, Line of Sight and Non-Line of Sight Fading Channels." International Journal of Engineering Science Invention 3.8 (2014): 6-10.

[43] Gupta, Akhil, and Krittika Goyal. "An analytic review on antenna modelling." International Journal of Control Theory and Applications 9.10 (2016): 319-336.

[44] Gupta, A., Jha, R.K. Power optimization with low complexity using scaled beamforming approach for a massive MIMO and small cell scenario. Wireless Netw 26, 1165–1176 (2020).

[45] I. B. Sofi, A. Gupta and R. Kausar, "Performance Evaluation of Different Channel Estimation Techniques in MIMO System for Hata Channel Model," 2018 International Conference on Intelligent Circuits and Systems (ICICS), 2018, pp. 155-160, DOI: 10.1109/ICICS.2018.00041.

[46] R. T. Kamurthi, S. R. Chopra and A. Gupta, "Higher-Order QAM Schemes in 5G UFMC system," 2020 International Conference on Emerging Smart Computing and Informatics (ESCI), 2020, pp. 198-202, DOI: 10.1109/ESCI48226.2020.9167619.

[47] Raina, Ambica, Akhil Gupta, and Rakesh Kumar Jha. "Analysis of AODV and DSR Routing Protocols for a Static Ad Hoc Network Using Qualnet Network Simulator." International Journal 4.1 (2014).

[48] Sundan, A.P., Jha, R.K. & Gupta, A. Energy and spectral efficiency optimization using probabilistic based spectrum slicing (PBSS) in different zones of 5G wireless communication network. Telecommun Syst 73, 59–73 (2020).

[49] M. Singh, K. Arora and A. Gupta, "Equalization in WIMAX System," 2018 International Conference on Intelligent Circuits and Systems (ICICS), 2018, pp. 330-333, DOI: 10.1109/ICICS.2018.00074.

[50] Jain, Preksha, and Akhil Gupta. "Adaptive NOMA towards 5G green wireless network." Transactions on Emerging Telecommunications Technologies 31.3 (2020): e3887.

[51] H.S. Bedi, K.K. Sharma, and R. Gupta, "A Review Paper on Performance Analysis of IEEE 802.11 e", 1st International Conference on Computing, Communications, and Cyber-Security (IC4S 2019), 2020, pp. 47-56.

[52] Z.G. Khaki, and H.S. Bedi, "Transient correction using EDFA: in-line optical fibre with feedback", International Conference on Computing Sciences, 2012, pp. 233-238 [http://dx.doi.org/10.1109/ICCS.2012.76].

[53] Bedi, Harpreet, Kamal Kumar, and Gupta Raghav. "A review paper on improving the network efficiency of IEEE 802.11 e networks." Intelligent Circuits and Systems (2021): 206-211.

[54] Harpreet Bedi, Kamal Kumar Sharma; Analyses of CSMA/CA Protocol Without Using Virtual Channel Sensing in DCF Mode, Global Emerging Innovation Summit (GEIS-2021) (2021) 1: 484. https://doi.org/10.2174/9781681089010121010058.

Optimization of Cell-Free Massive MIMO System

Manmohan Sharma[1], **Sunny Verma**[2], **Shekhar Verma**[3]

School of Electronics and Electrical Engineering, Lovely Professional University, Punjab.

manmohan.21137@lpu.co.in[1] vermasunny273@gmail.com[2], shekharverma05@gmail.com[3],

Abstract. As an innovative implementation, Cell-Free Massive Multiple Input Multiple Output (MIMO) has appeared in typical Cellular Massive MIMO Networks. This protocol doesn't recognize cells, as shown by its name, even though a significant number of APs operate on the same frequency/time resources. Connection from multiple distributed access points through joint signal processing is called Cell-Free Massive MIMO. The Cell-Free Massive MIMO System, a contrast between Cell-Free Massive MIMO Systems and Distributed Massive MIMO, the prime focus in this thesis is on Cell-free Massive MIMO and, along with this discussion, on Cell-free Massive MIMO signal processing, Channel Estimation, Uplink Signal Detection, Cumulative Distribution, Spectral Efficiency & Ubiquitous Cell-Free Massive MIMO Model. Ubiquitous Cell-free Massive MIMO contributes to a Massive MIMO system, a distributed system that implements consistent user-centre distribution to solve that constraint of mobile phone interferences as well as to introduce macro-diversity. We investigated the Cell Radius at different locations in CDF with Spectral Efficiency [bits/s/hertz]. Cell-Free Massive MIMO is an evidence-based preventive of massive MIMOs with distributed high percentage APs that serve even lower margins. The cell-free model is not segregated into cells and any individual is concurrently represented by every Access point.

Keywords. Cell-free massive MIMO, Distributed Massive MIMO, Energy, Efficiency, Signal Processing in Cell-free Massive MIMO System.

1. Introduction

There 5G seems to be the latest generation telecom infrastructure. This is the latest craze for 1G, 2G, 3G and 4G telecommunications infrastructure. 5G needs to get a new wireless network to link mostly all machines, objects and tools. The wireless 5G technology offers quick data speeds, ultra-low delay, greater performance, broadly available bandwidth, wider availability and added user interaction. Enhanced usability, better quality as well as a new most efficient application layer enable revolutionary touch screens to organizations to encourage markets [2]. It has been built to assist modern reaction rates and provide digital programs for consumer interfaces of the next decade. 5G would expand the range of smartphones to new places, rates, high reliability and low latencies [3]. 5G would effects all the industries, helping to strengthen carry, wireless laboratories, new crops, computer-controlled services and therefore more. 5G is currently here, or in late 2019 new 5G networks will be implemented by multinational carriers. For certain countries, the national availability of 5G broadband technology is expected through 2020. Today, both of Microsoft's top suppliers encourage 5G phones. 5G would eventually be available to far more users. Although it is difficult to determine that everybody has links to 5G, we expect solid 5G technologies to move within a year and hope that by and above 2020, the government also introduced might begin its 5G networks.

The process Multiple Input Multiple Output (MIMO) is a multiple antenna approach that is used in communication. Multiple inputs and outputs are used to periodically communicate some knowledge. MIMO process benefits from a broadcast interaction multi trajectory, at which data transmitted bounces off buildings, ceilings and many other materials, disrupting the receiving antenna numerous times in previous locations and at different points in time. Inside the latter process, several telescopes are referred to as the sender and are sent by several antennas to the receiver [5] to increase the transmission rate. The 'methodology of plurality' including 'effective data rate' was called multiple-antenna methods. In this way, multiple antennas can be used to provide diverse benefits and enhance [6] the performance of cellular connections while both transmitting and reciprocating diversity. During the spread spectrum numerous different flows can be perfected, thus increasing available bandwidth. There is indeed a range of different uses of directional sensors for wireless technology.

The cell-free Massive MIMO, in which many connection sources in a huge area typically serve many users in much the same spectrum channel, is amongst the most ambitious developments for such a new standard for wireless devices in the next period. While several factors have already been assiduously investigated in terms of allocation of resources, model layout, including massive cell-free MIMO efficiency measurements, mac layer safety in cell-free massive MIMO was not taken advantage of. The study aims to change to Network MIMO, but now the massive MIMO is indeed a very consumer product under its full title Cell-Free Massive MIMO. Each main new feature is the robust sample of Insufficient CSI of neutrino mass spectral efficiency (SE), but the MIMO platform is technically unusual. Using the attributes between both MIMO and the MIMO service, the cell-free massive MIMO system includes a good spectrum, power consumption and exposure. Moreover, cos of the additional production of multichannel systems, the Structural model Multicasting is an engaging case for the Large MIMO Framework. Improvements to games, newspaper articles, even digital signage such as group email. Main MIMO efficiency without cellular. Significant statistical effectiveness, improved quantitative versatility or low efficiency of the trace can be accomplished by cell-free massive MIMO. In terms of several improvements, the massive cell-less MIMO damages from either a substantial percentage of dispersed access points (APs) or aircraft emissions caused by uncontrolled broadband electricity consumption through multiple access. After this, we have implemented an effective, consumer optimization that requires routing and aggregating the various APs to improve the power quality contributing significantly to a specific MU. This mixture probability variational phenomenon catches up with the complexities about one FL phase among time- preparation and execution and the calculation of education changes.

2. CELL-FREE MASSIVE MIMO

Cell-Free Massive MIMO is a system, with a large number of access points spread over a wide area that reliably supports a large number of consumer terminals. This concentrates on cellular frequencies. It is possible to use millimetre-wave bands as a combination with cellular frequency bands, however. No concepts of cells or cell boundaries occur here [3].

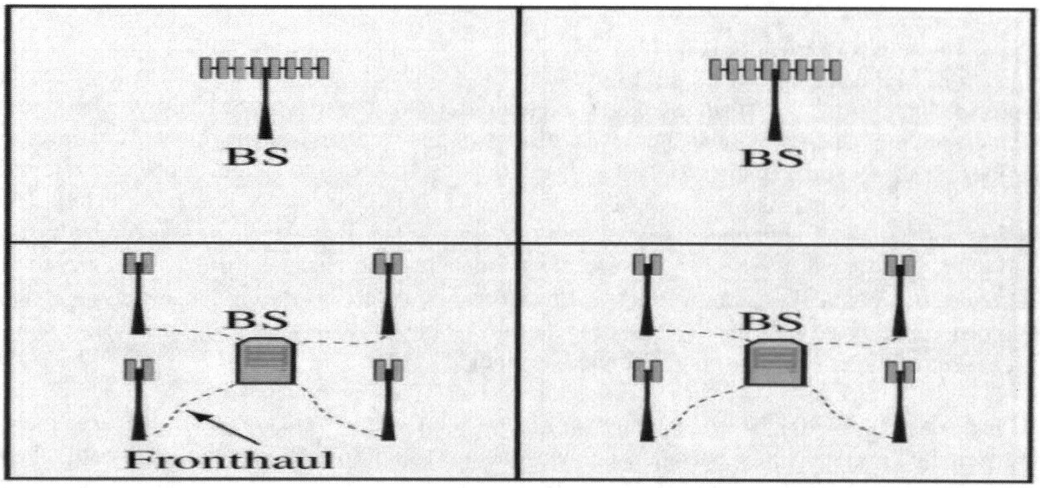

Fig.1. Massive MIMO BS cellular networks of either co-located arrays (top) or distributed arrays (bottom) [3].

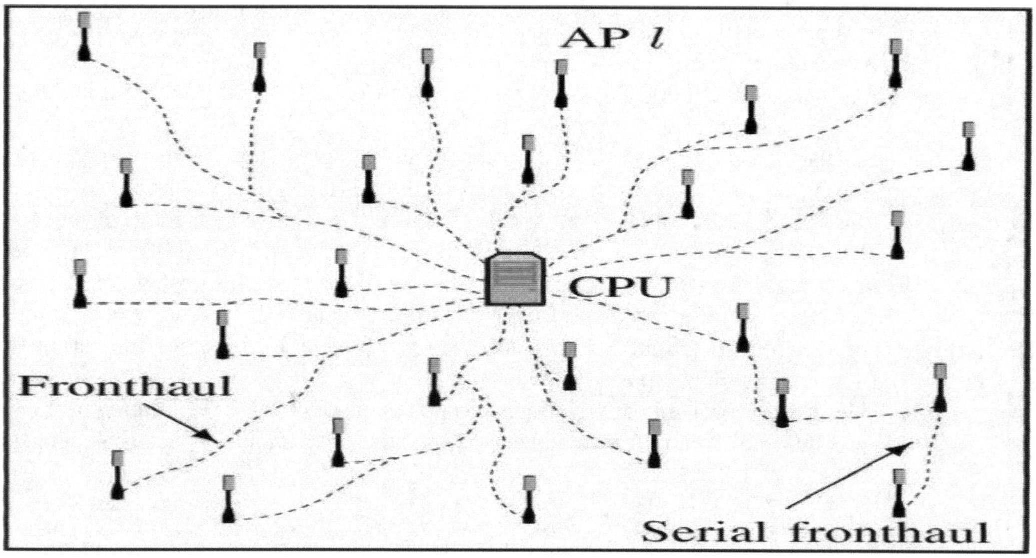

Fig.2. Cell-free Massive MIMO network [3].

Cell-free Massive MIMO is comparable with traditional cellular Massive MIMO and its primary purpose is to identify the most competitive cell-free implementation. Both topologies of the network are shown in Fig. 1. The major variations make the analysis non-trivial and give insightful results.

Inputs into the architecture of networks beyond-5G. Channel hardening and spatial interference reduction help cellular Massive MIMO, but cell-edge UEs may have bad channel conditions. Cell-free Massive MIMO, on the other hand, gains from broad macro-diversity, but its potential for intrusion suppression relies strongly on how it is run.

3. DISTRIBUTED MASSIVE MIMO

Free Massive MIMO is different from distributed Massive MIMO. Both systems have so many service antennas to support many consumer terminals in a distributed manner, and they're not completely the same. The base station antennas are distributed within each cell with distributed Massive MIMO [4]. and these antennas support only user terminals within that cell. For comparison, there are no cells in the Cell-Free Massive MIMO. Both the service antennas support all the user terminals coherently.

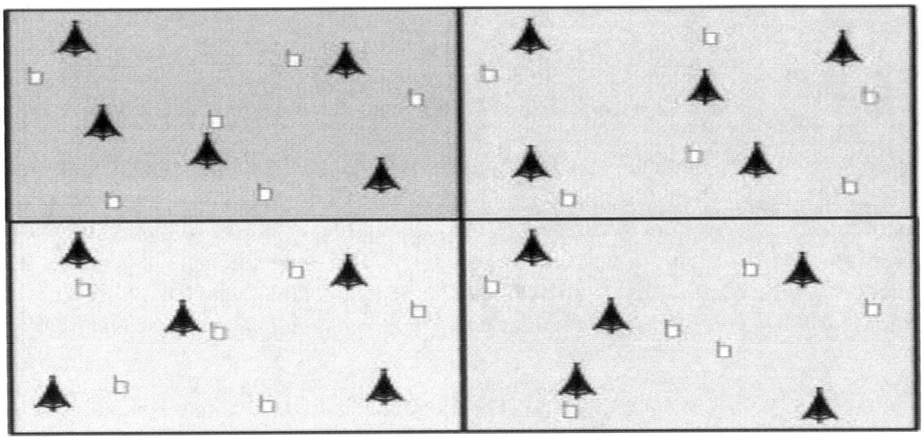

Fig.3. Distributed Cell-free Massive MIMO

4. SIGNAL PROCESSING IN CELL-FREE MASSIVE MIMO SYSTEM

The signal processing techniques are very important in CF massive MIMO systems, which vary from those of consolidated Massive MIMO systems in terms of their distributed antenna architectures, particular precoding, and signal detection to reduce inter-user interference, channel fading models, and relatively discrete fronthaul network constraints [7].

4.1 Channel Estimation

To support effective precoding and signal detection to reduce inter-user interference, accurate channel estimation is important. In the Uplink, Based on the classic MMSE criterion, for example, in a decentralized manner [7].

Therefore, for instance, only the average channel envelope at the UE must be understood to accurately detect the received signals without downlink pilots. In comparison, the channel does not always provide valuable hardening in CF large MIMO structures. Therefore, by merely using E, the output degrades, because the instantaneous channel gain fluctuates. For CF large MIMO systems, a downlink training method was suggested and the realistic output was dramatically increased by the advocated Fairness-based power control of max-min. Just in the case of mutually orthogonal downlink pilots, however, is included in this analysis [7].

4.2 Identify Uplink Signal Detection

During the UL signal detection process, the UEs transfer their uplink data signals to the APS. Because one should derive the closed-form SE expressions, taking into account the effects of all incomplete CSI, spatially correlated fading, and/or hardware impairments, the use of matched filtering locally at each AP was suggested by the seminal papers. However, it has been shown that UL processing at the CPU, partially or fully consolidated, would increase the SEE Considerably. Four different collaboration principles for the UL signal significantly. Four separate standards of collaboration for UL signal identification have been implemented in this sense. With channel statistics, by refining the UL receiver weighting coefficient vector, the CPU can optimize the SE. This technique was referred to in.

Cellular Massive MIMO systems as large-scale fading decoding (LSFD), and it was shown that the SE of the LSFD receiver in CF massive MIMO networks is much greater than that of the receiver of the maximum-ratio (MR). On the other hand, the improvement of the coefficients of the receiver filter at the CPU was considered, which significantly improved the UL consumption rate. However, the matter is non-convex and highly complex. the receiver philtre at the CPU was considered, which increased the UL usage rate dramatically. The question, however, is non-convex and has a high complexity of computation, which is impractical in large MIMO networks in CF. In conclusion, at the time of counting, the investigation of poor compliance, but optimal distributed uplink signal detection systems is still an open issue [7].

4.3 Modeling

(a) Large- Scale Modeling

In the scenario of irreligious radio transmission in the positions of the AP, the shadowing has shown that these locations tend to be Poisson-distributed to every user. This strategy sharpens as the shadow strengthens and is particularly accurate for interest values [15]. Using this result we randomly position the APs and consumers around the system so certain their positions are in line with respective (mutually independent) binomial point processes. It relates to 1000 channel snapshots for each CDF in this article in such a draped around path loss with 200 APs, guarantee wide-scale solute 95% reliability interval of 0.3% [15].

Signals are subject to pathloss with exponent η, giving a wide scale channel gain $G_{n,k} = d_{n,k}^{-\eta}$between the nth AP and the consumer, limited by $d_{n,k}$.

$$\mathrm{SNR}_{n,k} = Except \tag{1}$$

Across both locations, you could connect large-scale SNRs

$$\mathrm{SNR}^{\mathrm{r}}_{n,k} = \frac{\mathrm{SNR}_{n,k}}{\rho} \tag{2}$$

(b) Designing on small scale

With the exception of $G_{n,k}$, there is a small-scale, fading flag $\boldsymbol{h}_{n,k}$. to the opposite network of kth consumer to the nth AP which requires $\boldsymbol{h}_{n,k} \sim \mathcal{N}_{\mathbb{C}}(\boldsymbol{0}, \boldsymbol{I}_M)$, Forward and connection among nth AP and kth users is $\boldsymbol{h}_{n,k}$. due to various mutual recognition.

4.4 Quantization Scheme

(a) Optimum system for Quantizing

According the principle of Buss-gang [17], a non-linear modulation performance could be described as a normal distribution:

$$Q(z) = h(z) = az + n_d, \forall k \tag{3}$$

For which a function is fixed and n_d belongs to the noise of distortion external to quantizer data, z. The phrase an is used

$$a = \frac{\mathbb{E}\{zh(z)\}}{\mathbb{E}\{z^2\}} = \frac{1}{p_z}\int_Z zh(z)f_z(z)dz \tag{4}$$

where $p_z = \mathbb{E}\{|z|^2\} = \mathbb{E}\{z^2\}$is the power of a z and the significance level we decrease is that z is a positive number and f z (z). Pre-processor of,

$$b = \frac{\mathbb{E}\{h^2(z)\}}{\mathbb{E}\{z^2\}} = \frac{1}{p_z}\int_Z h^2(z)f_z(z)dz \tag{5}$$

The Signal to Noise Ratio (SDNR) is then established,

$$\mathrm{SDNR} = \frac{\mathbb{E}\{(az)^2\}}{\mathbb{E}\{n_d^2\}} = \frac{p_z a^2}{p_z(b-a^2)} = \frac{a^2}{b-a^2} \tag{6}$$

(b) Optimal Quantization Scheme

We consider fronthaul connections with Rm = R bits, ∀m ∈ {1,...,M}, that are equal to the L = 2R stage of quantization to simplify our studies. We then use A scalar L-level quantizer Q as an AP application,

$$Q(x) = \sum_{l=0}^{L-1} q_l T_l(x) \quad (7)$$

where $T_l(x)$is equal 1 for$x_l < x \leq x_{l+1}$ and 0 otherwise. We regard Q as a standardized, constant phase quantizer $\Delta = x_{l+1} - x_l$ and rehabilitation value

$$q_l = \left(l - \frac{L-1}{2}\right)\Delta \quad (8)$$

We represent the exact and image component independently, complex-valued signal x as a function of C.

$$x_q = Q(x) = Q(\mathbb{R}\mathrm{e}\{x\}) + iQ(\mathbb{I}\mathrm{m}\{x\}) = q_l^R + iq_{l'}^I \quad (9)$$

Where q_l^R and $q_{l'}^I$ are also the restoration together Actual and innovative part values with the couple

$$(q_l^R, q_{l'}^I) \in \{q_l^R, \ldots, q_{l-1}^R\} \times \{q_{l'}^I 0, \ldots, q_{l-1'}^I\}. \quad (10)$$

4.5 MAX-MIN ALGORITHM

As a technology approach, we utilize max-min optimization to refine the likely 95 % confidence reliability criterion, which ensures fair results for all users on the network but does not offer the optimal 95 % confidence performance possible. The problem of locating the AP locations at the grid points is just as shown in: provided a goal per user t SNR: b = $t\{1, \ldots, 1\}^T$

$$A\boldsymbol{x} \geq \boldsymbol{b}, \boldsymbol{x} \in \{0,1\}^N \quad (11)$$

Where ‖ $xthe\ end$u is the-null element number is in x. The general sparsity, which attempts to rebuild a sparse signal x, resembles with assumptions of K<N and N>>M.

To maximize the amount, we use the following inequality following from the inequality of the Jensen

$$\frac{1}{K}\sum_{i=1}^{K} \log_2(1 + \mathrm{SNR}_i) \leq \log_2\left(1 + \frac{1}{K}\sum_{i=1}^{K} \mathrm{SNR}_i\right) \quad (12)$$

which provides an upper bound for the actual problem.

5. UBIQUITOUS CELL-FREE MASSIVE MIMO

Ubiquitous Cell-free Massive MIMO is really a hierarchical Massive MIMO requirement the proliferation of old one sites to resolve cells intervention limitations and to increase the eco density of mobile networks. These characteristics distinguish the overall cell-free massive MIMO to previously integrated remote distribution networks, together with device optimizations for Massive MIMO architectures. Utilizing Massive MIMO, i.e. large bandwidth performance, manageability and virtually perfect vertical compression Universal Cell-Free Massive MIMO, you can upgrade that standard COMP-JT. And have a target group of a cell-free MIMO new paradigm. The incorrect graph shows that, because of strong uplink activity, SEs become poor on the circuit level of a data connection. The correct graph shows that cell-free routing can avoid disruptions by coprocessing across a cell-free channel has a more reliable customer production. The SE is related both to errors in communication networks.

Figure 5. Illustrates user efficiency at multiple positions in a region of nine APs: the left figure indicates that due to heavy inter-cell interaction, the SEs in a cellular network is weak at the cell edges, while the right figure shows that a cell-free network can prevent interference by co-processing over the APs and provide users with more uniform performance. Only signal transmission losses are limited to the SE [9].

SE is obtained by UEs in different areas in a region protected by nine APs implemented on a regular grid. Ignore also that total SE, comparison to unsecured 257- QAM, as stated as 8 bit/s/Hz.

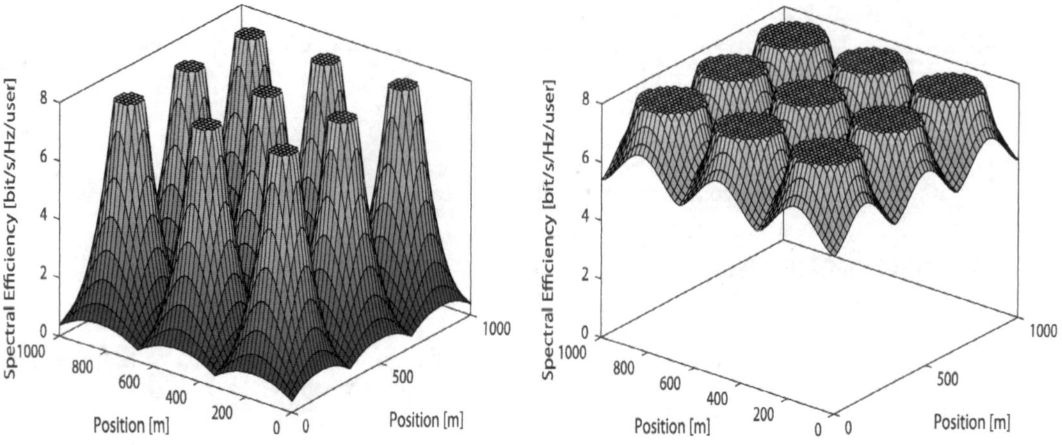

Fig.4. Cellular Network & Cell-free Massive MIMO Network.

6. THE SCALABILITY PROBLEM

The TDD operation makes it possible to make the overhead measurement independent of the number of APs relative to previous distributed wireless networks, Besides, Conjugate Beamforming is used for (e.g. maximal transmission ratio, MRT), which has been particularly advocated in the literature, which allows for easy and fully distributed processing [2].

There are three primary issues of scalability:

(a) Data Processing-

Put-up up data for each UE on the network. And it will need to be sent to any AP from the Processor. This data would be impractical at each AP, moving the computing complexity [2];

(b) Network Topology-

The involution of the CPU interconnection. And as a CPU, it does not size. Only one link to any AP on the network is needed;

(c) Power Control-

Estimate of the coefficients of power control. Even putting computational problems aside, it does not scale;

6.1 User-centric vs Cell-centric Clustering

Installing AP clusters and constraining the co-processing of the signal within the cluster is the usual approach to attempting to correct the scalability problem. There are two proposed methods in the literature:

(i) Cell-Centric Clustering

Consisting of the deployment of fixed disjoint clusters of APs where only UEs residing in their shared coverage area act as Apps in a cluster;

(ii) User-Centered Clustering

Consisting of the deployment of complex (possibly partly overlapping) AP clusters depending on the specifications of each UE served;

The canonical method for the collection of cooperating APs is cell-centric clustering. Each cluster is served by one CPU, and a cluster would be created by the APS linked to a given CPU. Instead, these clusters should coordinate with each other or they will have to communicate by reliable transmission, taking up the problem of scalability. There is no clear cooperation on data or communication on power control, between the CPUs in such a device. It suffers from low efficiency.

The MIMO, COMP-JT, and multi-cell MIMO cooperative networks are conventionally implemented in a cell-centric way. In contrast to that, the concept behind user-centric clustering is that only a limited number of APs should be involved in the operation of the UE with each UE. Properly, a close-by AP cluster represents each UE, see [2]. User-centric transmission can be seen as a special transmission. In the case of the typical large MIMO cell-free configuration[63-64], Where, except those connected to the nearest APs, all coefficients of power control of a given UE k are limited to zero. The user-centric solution includes the absence of inter-cell interference. so it does better than the cell-centric approach.

7. SIMULATION PARAMETERS AND RESULTS

7.1 Numerical Parameters

In this section of Cell-Free Massive MIMO, different parameters have different separate values at different locations. Evaluate the Cumulative Distribution and Spectral Efficiency for cell-free Massive MIMO system Model.

Description	Parameters	Values
Cell Radius	Indoor Hotspot	60 m
	Ultra-Dense Urban	200 m
	Dense Urban	250 m
	Urban	500 m
	Sub Urban	1299 m
	Rural	1732 m

Table.1. Cell Radius for Different Parameters

7.2 Simulation Result-

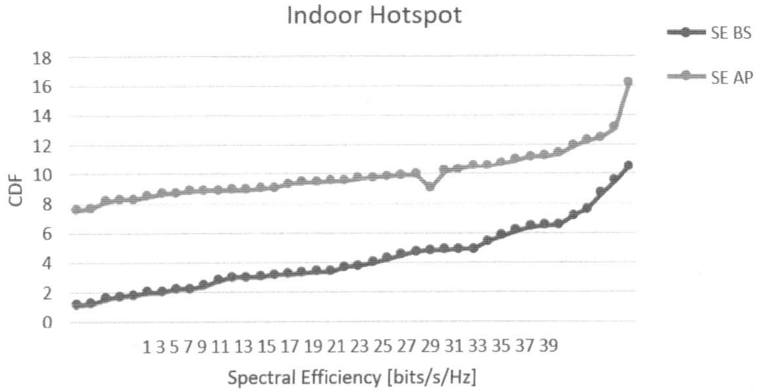

Fig.5. Cumulative Distribution with Spectral Efficiency in Indoor Hotspot Deployment Scenarios.

Figure 5 shows the variation in Cumulative Distribution concerning spectral efficiency [bits/s/Hz] using SE BS and SE AP shape approach for different Cell radii. For Indoor Hotspot Deployment Scenarios, the Cell radius is 60 meters. It is the minimal Cell Radius.

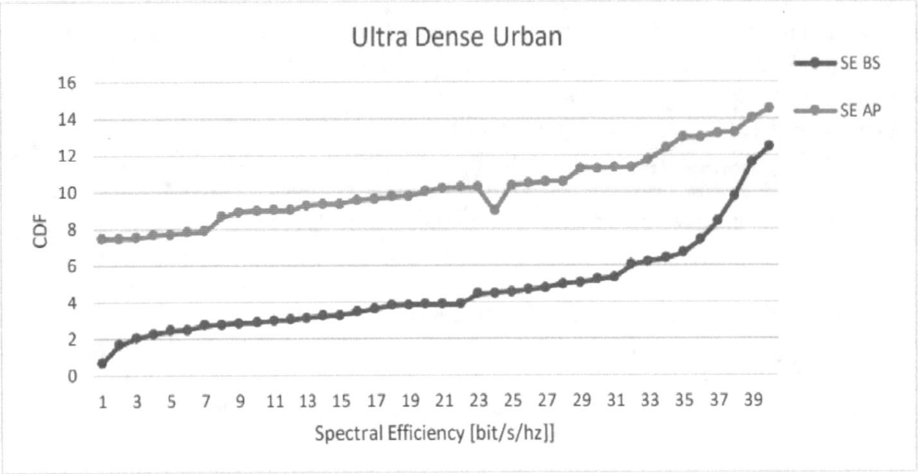

Fig.6. Cumulative Distribution with Spectral Efficiency in Ultra-Dense Urban Deployment Scenarios.

Figure 6 shows the variation in Cumulative Distribution concerning spectral efficiency [bits/s/Hz] using SE BS and SE AP shape approach for different Cell radii. For Ultra-Dense Urban Deployment Scenarios, the Cell radius is 200 meters. Its Cell Radius is more than Indoor Hotspot.

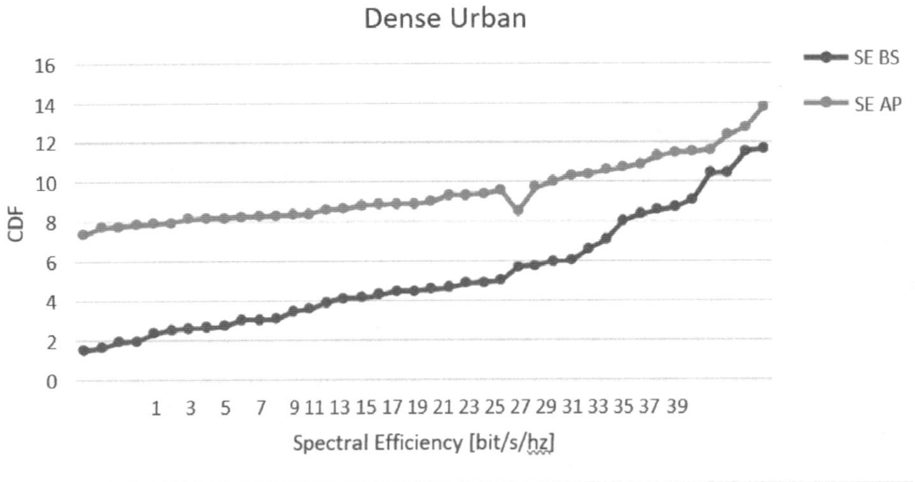

Fig.7. Cumulative Distribution with Spectral Efficiency in Dense Urban Deployment Scenarios.

Figure 7 shows the variation in Cumulative Distribution concerning spectral efficiency [bits/s/Hz] using SE BS and SE AP shape approach for different Cell radii. For Dense Urban Deployment Scenarios, the Cell radius is 250 meters. Its Cell Radius is more than Ultra-Dense Urban.

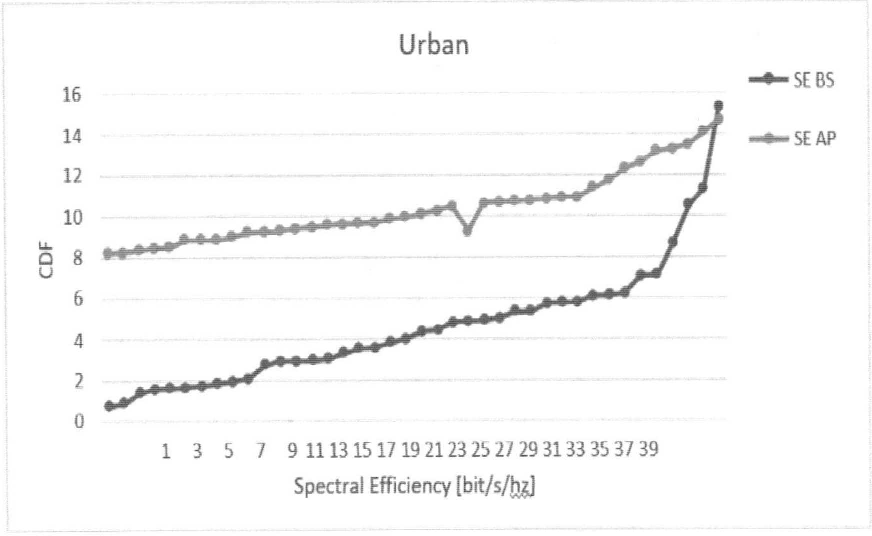

Fig.8. Cumulative Distribution with Spectral Efficiency in Urban Deployment Scenarios.

Figure 8 shows the variation in Cumulative Distribution concerning spectral efficiency [bits/s/Hz] using SE BS and SE AP shape approach for different Cell radii. For Urban Deployment Scenarios, the Cell radius is 500 meters. Its Cell Radius is more than Dense Urban.

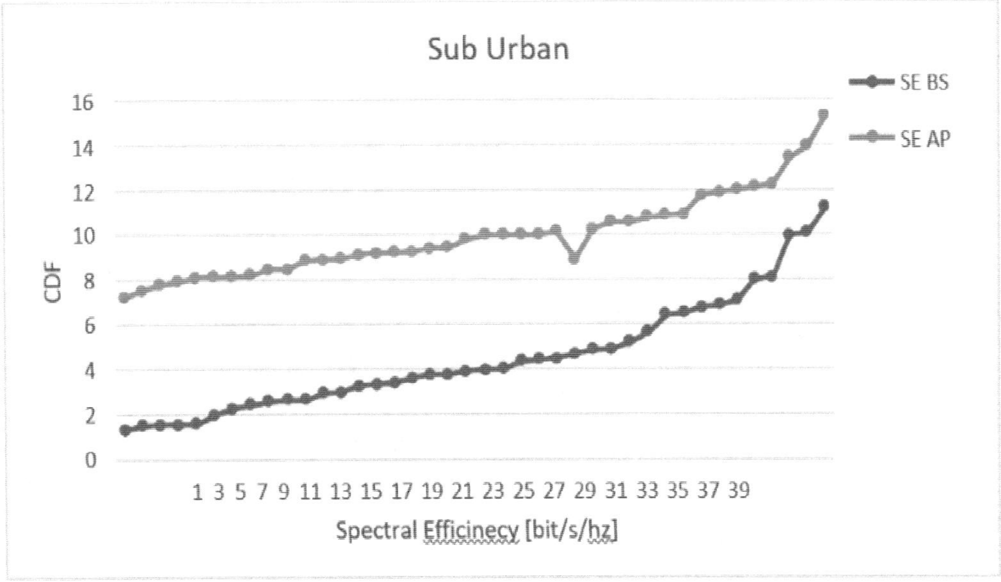

Fig.9. Cumulative Distribution with Spectral Efficiency in Sub Urban Deployment Scenarios.

Figure 9 shows the variation in Cumulative Distribution concerning spectral efficiency [bits/s/Hz] using SE BS and SE AP shape approach for different Cell radii. For Sub Urban Deployment Scenarios, the Cell radius is 1299 meters. Its Cell Radius is more than Urban.

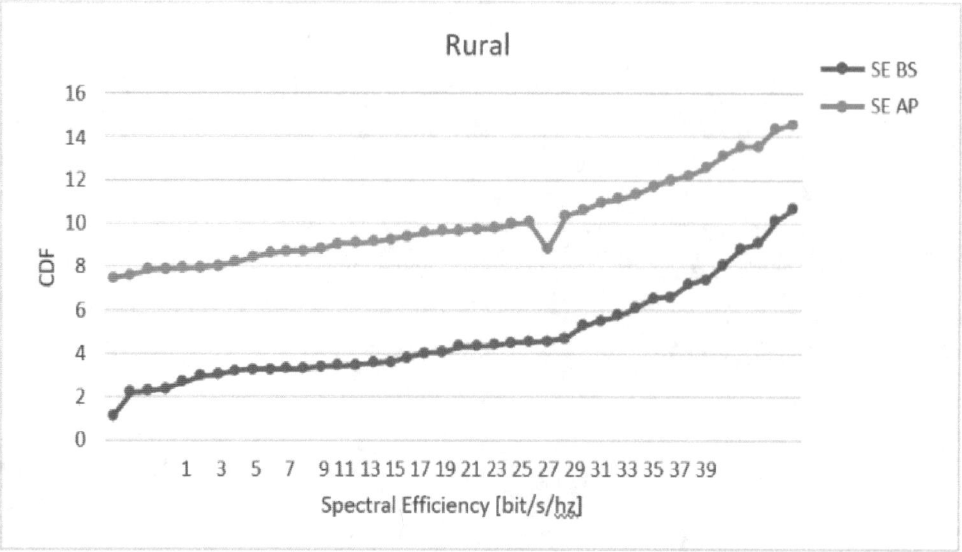

Fig.10. Cumulative Distribution with Spectral Efficiency in Rural Deployment Scenarios.

Figure 10 shows the variation in Cumulative Distribution concerning spectralefficiency [bits/s/Hz] using SE BS and SE AP shape approach for different Cell radii. For Rural Deployment Scenarios, the Cell radius is 1732 meters. Its Cell Radius is more than Sub Urban.

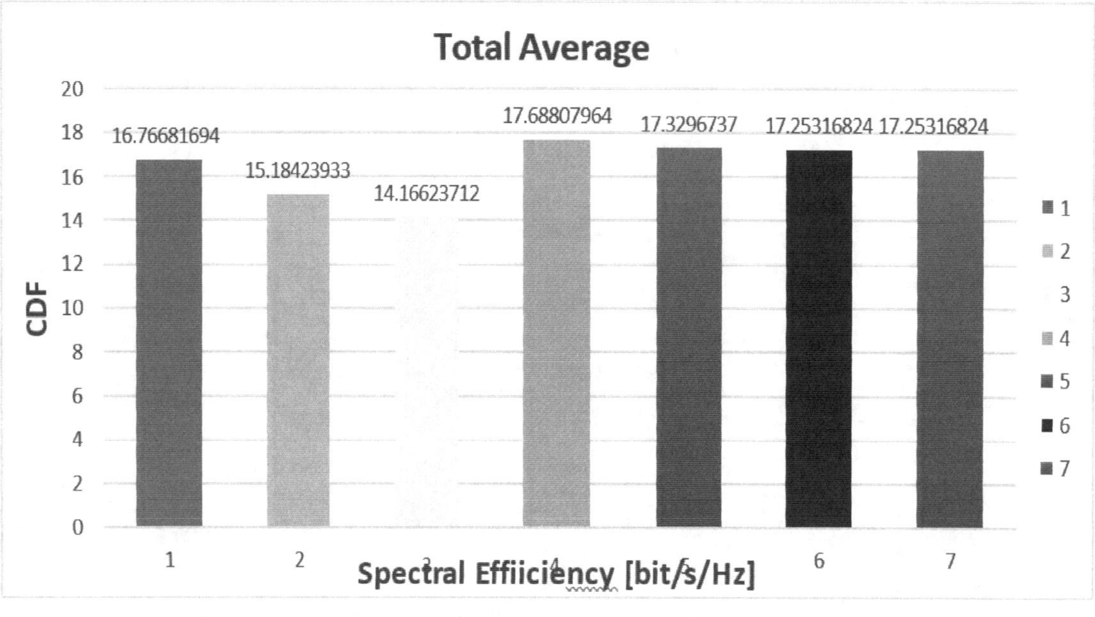

Fig.11. Total Average Value between CDF and Spectral Efficiency.

Figure 11 shows as Total Average Value between CDF (Cumulative Distribution) with SE (Spectral Efficiency). This is the Total Average Value of Cell Radius at Different Location for Cell-Free Massive MIMO. And This Graph is showing the maximum average value of Urban Deployment Scenarios, and the minimum average value of Dense Urban Deployment Scenarios.

8. BENEFITS OF CELL-FREE MASSIVE MIMO

CF's massive MIMO systems are some certain attributes powerful technologies for the repression of macro-diversity and intra interference. As a large number of serving APs are highly likely to be around each EU, both EUs are well-positioned [53] [6]. More simply, Table 1 explains the benefits of massive MIMO approaches by CF and details some of them.

8.1 Large Energy Efficiency

In future, since energy efficiency is an integral output matrix, the wireless network should be constructed. It is also usual for the centralized MIMO to achieve high energy efficiency as far as M bits/Joule are concerned (EE). And At such a particular step the rate fluctuations within each transmitter could be greatly reduced and overall energy consumption. Regardless of the high number of advantages from unified massive MIMO solutions. This transforms into an improved EE [53]. And another advantage of MIMO cell-free is the high-rate, low-delay link to front-end, the coherent communication between both the distributed APs [6] will reduce the intercell interaction [52]. The typical EE of MIMO network solutions however is constrained by the energy use of front ends.

Architecture	Centralized	Network	Cell-free
Number of Antennas	Large	Moderate	Large
Deployment Cost	High	High	Low
Macro Diversity	Small	Moderate	Large
Channel Hardening	Strong	Weak	Moderate
Favourable Propagation	Strong	Weak	Moderate
Uniform Coverage	Bad	Moderate	Good
Energy Efficiency	Large	Small	Very Large
Channel Estimation	Global	Global	Local
Fronthaul Resource	Small	Large	Moderate

Table.2. Comparison Between Centralized Massive MIMO, Network MIMO & Cell-free Massive MIMO.

8.2 Flexible and Cost-Efficient Deployment

The Downlink massive MIMO principle supports package transmission based on the established computationally efficient pre-processing communication, including the maximum TPC ratio to almost optimal performance. And reverse, by invoking a mixture of receivers, the signal received from each EU can be decoded in the UL. The signal required can be concentrated on each EU by invoking DL-based beamforming transmission precoding. And it is an advantage of spatial variety, to ensure that their antennas fade separately by more than half a wavelength ($\mu/2$) [52] [54]. hence a massive MIMO scheme based on a 100-element UPA spaced $\mu/2$ uniform planar array (UPA) which is used by traditional BS is 0.5 to 0.5m, but cannot be used in space-restricted environments. However, only a small range of antennas are available to the cell-free AP. They are also ideal for areas that are spatially confined such as working places or are mounted on the lamps and light systems [6].

8.3 The Channel Hardening and The Favourable Propagation Conditions

Concerning Massive MIMO multi-cell, the core concept of Cell-free Massive MIMO is sometimes Millions or billions of antennas. The key characteristics of massive MIMO can effectively be used through getting fewer antennas in each region unit than users for scaling applications [52]. The distributed CF massive MIMO antennas will, however, still establish some space channel interface between the signals that reach each other at a particular UE receiver.

8.4 Appealingly Uniform Quality of Services

A lot of other distributed antennas in CF MMIMO will minimize the nearest antennas between average distance and a random UE. An intuitive result is that cell-free solids are much more efficient in contrast with a limited network of cells whereby the consumer Just with is provided with the nearest one antenna [52].

The authors compared the above-mentioned cell net rate with CF massive MIMO systems, which helps the reliability of channel measuring loss, the pilot pollution and practical power management to be taken into account.

9. CONCLUSION

For this section, we are investigating the improvement of cell-free massive MIMO schemes, most often in cell-free massive MIMO, with both the support among the most appropriate Process Control Methods. Every AP is linked through a computer Processor in a cell-free massive MIMO. Fronthaul's sequential format. The processor organizes each group and functions autonomously. The AP is split through cell-centred cluster formation. Multiple Processors contribute to the less challenging implementation, each one addressing a cell-centric set. The clusters, all Cell-focused and Node Users, are therefore in favour of the EU. The approach they access broadband connections will (or should) centralize Mobile network internet devices. Another of the major challenges is just the provision of ubiquitous wi-fi services by providing Gbit/s rates for just any customer but which also has a reasonable skill set for malicious smartphone owners.

10. Future Scope

A distributed deployment of the Wavelength channels depends on the CF massive MIMO structures. New experimental experiments have improved our philosophical understanding but have dramatically uncovered new topics that need more study.

This work can also be modified and explored in different areas of wireless communication by following the mentioned research papers [46-65].

Acknowledgement

I would like to express my very great appreciation to **Dr. Akhil Gupta, Harpreet Singh Bedi, Preeti Khurana, Raghav Gupta** (School of Electronics and Electrical Engineering, Lovely Professional University, Phagwara, Punjab, India.) for their valuable and constructive suggestions during the planning and development of this research work. Their willingness to give their time so generously have been very much appreciated.

REFERENCES

[1] Zhang, Y., Cao, H., Zhou, M. and Yang, L., 2019. Cell-free massive MIMO: Zero forcing and conjugate beamforming receivers. Journal of Communications and Networks, 21(6), pp.529-538.

[2] D'Andrea, C., Garcia-Rodriguez, A., Geraci, G., Giordano, L.G. and Buzzi, S., 2020. Analysis of UAV communications in cell-free massive MIMO systems. IEEE Open Journal of the Communications Society, 1,pp.133-147.

[3] Ngo, H.Q., Ashikhmin, A., Yang, H., Larsson, E.G. and Marzetta, T.L., 2017. Cell-free massive MIMO versus small cells. IEEE Transactions on Wireless Communications, 16(3), pp.1834-1850.

[4] Zhang, Y., Cao, H., Zhou, M. and Yang, L., 2019. Cell-free massive MIMO: Zero forcing and conjugate beamforming receivers. Journal of Communications and Networks, 21(6), pp.529-538.

[5] Van Chien, T., Björnson, E. and Larsson, E.G., 2020. Joint power allocation and load balancing optimization for energy-efficient cell-free massive MIMO networks. IEEE Transactions on Wireless Communications, 19(10), pp.6798-6812.

[6] Buzzi, S. and D'Andrea, C., 2017. Cell-free massive MIMO: User-centric approach. IEEE Wireless Communications Letters, 6(6), pp.706-709.

[7] Doan, T.X., Ngo, H.Q., Duong, T.Q. and Tourki, K., 2017. On the performance of multigroup multicast cell-free massive MIMO. IEEE Communications Letters, 21(12), pp.2642-2645.

[8] Interdonato, G., Björnson, E., Ngo, H.Q., Frenger, P. and Larsson, E.G., 2019. Ubiquitous cell-free massive MIMO communications. EURASIP Journal on Wireless Communications and Networking, 2019(1), pp.1-13.

[9] Bashar, M., Cumanan, K., Burr, A.G., Ngo, H.Q. and Poor, H.V., 2018. Mixed quality of service in cell-free massive MIMO. IEEE Communications Letters, 22(7), pp.1494- 1497.

[10] Hoang, T.M., Ngo, H.Q., Duong, T.Q., Tuan, H.D. and Marshall, A., 2018. Cell-free massive MIMO networks: Optimal power control against active eavesdropping. IEEE Transactions on Communications, 66(10), pp.4724-4737.

[11] Li, Y. and Baduge, G.A.A., 2018. NOMA-aided cell-free massive MIMO system. IEEE Wireless Communications Letters, 7(6), pp.950-953.

[12] Nawaz, S.J., Sharma, S.K., Mansoor, B., Patwary, M.N. and Khan, N.M., 2020. Non-coherent and backscatter communications: Enabling ultra-massive connectivity in the era beyond 5G. arXiv preprint arXiv:2005.10937.

[13] Mai, T.C., Ngo, H.Q., Egan, M. and Duong, T.Q., 2018. Pilot power control for cell-free massive MIMO. IEEE Transactions on Vehicular Technology, 67(11), pp.11264- 11268.

[14] Björnson, E., Hoydis, J. and Sanguinetti, L., 2017. Massive MIMO Networks: Spectral, energy, and hardware efficiency. Foundations and Trends in Signal Processing, 11(3- 4), pp.154-655.

[15] Attarifar, M., Abbasfar, A. and Lozano, A., 2019. Modified conjugate beamforming for cell-free massive MIMO. IEEE Wireless Communications Letters, 8(2), pp.616-619.

[16] Bashar, M., Cumanan, K., Burr, A.G., Debbah, M. and Ngo, H.Q., 2019. On the uplink max-min SINR of cell-free massive MIMO systems. IEEE Transactions on Wireless Communications, 18(4), pp.2021-2036.

[17] Alonzo, M., Buzzi, S., Zappone, A. and D'Elia, C., 2019. Energy-efficient power control in cell-free and user-centric massive MIMO at millimetre wave. IEEE Transactions on Green Communications and Networking, 3(3), pp.651-663.

[18] Femenias, G. and Riera-Palou, F., 2019. Cell-free millimeter-wave massive MIMO systems with limited fronthaul capacity. IEEE Access, 7, pp.44596-44612.

[19] Zhang, X., Guo, D., An, K., Ding, Z. and Zhang, B., 2019. Secrecy analysis and active pilot spoofing attack detection for multigroup multicasting cell-free massive MIMO systems. IEEE Access, 7, pp.57332-57340.

[20] Zhang, J., Chen, S., Lin, Y., Zheng, J., Ai, B. and Hanzo, L., 2019. Cell-free massive MIMO: A new next-generation paradigm. IEEE Access, 7, pp.99878-99888.

[21] Interdonato, G., Ngo, H.Q., Frenger, P. and Larsson, E.G., 2019. Downlink training in cell-free massive MIMO: A blessing in disguise. IEEE Transactions on Wireless Communications, 18(11), pp.5153-5169.

[22] Özdogan, Ö., Björnson, E. and Zhang, J., 2019. Performance of cell-free massive MIMO with Rician fading and phase shifts. IEEE Transactions on Wireless Communications, 18(11), pp.5299-5315.

[23] Hu, X., Zhong, C., Chen, X., Xu, W., Lin, H. and Zhang, Z., 2019. Cell-free massive MIMO systems with low-resolution ADCs. IEEE Transactions on Communications, 67(10), pp.6844-6857.

[24] Alrabeiah, M. and Alkhateeb, A., 2019, November. Deep learning for TDD and FDD massive MIMO: Mapping channels in space and frequency. In 2019 53rd Asilomar Conference on Signals, Systems, and Computers (pp. 1465-1470). IEEE.

[25] Liu, H., Zhang, J., Zhang, X., Kurniawan, A., Juhana, T. and Ai, B., 2019. Tabu-search-based pilot assignment for cell-free massive MIMO systems. IEEE Transactions on Vehicular Technology, 69(2), pp.2286-2290.

[26] Zhang, Y., Zhou, M., Cheng, Y., Yang, L. and Zhu, H., 2020. RF impairments and low-resolution ADCs for nonideal uplink cell-free massive MIMO systems. IEEE Systems Journal.

[27] Zhang, Y., Cao, H., Zhou, M. and Yang, L., 2019, May. Spectral efficiency maximization for uplink cell-free massive MIMO-NOMA networks. In 2019 IEEE International Conference on Communications Workshops (ICC Workshops) (pp. 1-6). IEEE.

[28] Interdonato, G., Björnson, E., Ngo, H.Q., Frenger, P. and Larsson, E.G., 2019. Ubiquitous cell-free massive MIMO communications. EURASIP Journal on Wireless Communications and Networking, 2019(1), pp.1-13.

[29] Björnson, E. and Sanguinetti, L., 2019. Making cell-free massive MIMO competitive with MMSE processing and centralized implementation. IEEE Transactions on Wireless Communications, 19(1), pp.77-90.

[30] Ngo, H.Q., Tran, L.N., Duong, T.Q., Matthaiou, M. and Larsson, E.G., 2017. On the total energy efficiency of cell-free massive MIMO. IEEE Transactions on Green Communications and Networking, 2(1), pp.25-39.

[31] Zhang, Y., Zhou, M., Cao, H., Yang, L. and Zhu, H., 2019. On the performance of cell-free massive MIMO with mixed-ADC under Rician fading channels. IEEE Communications Letters, 24(1), pp.43-47.

[32] Bashar, M., Cumanan, K., Burr, A.G., Ngo, H.Q., Debbah, M. and Xiao, P., 2019. Max– min rate of cell-free massive MIMO uplink with optimal uniform quantization. IEEE Transactions on Communications, 67(10), pp.6796-6815.

[33] Alonzo, M., Buzzi, S. and Zappone, A., 2018, July. Energy-efficient downlink power control in mmWave cell-free and user-centric massive MIMO. In 2018 IEEE 5G World Forum (5GWF) (pp. 493-496). IEEE.

[34] Ngo, H.Q., Ashikhmin, A., Yang, H., Larsson, E.G. and Marzetta, T.L., 2015, June. Cell-free massive MIMO: Uniformly great service for everyone. In 2015 IEEE 16th international workshop on signal processing advances in wireless communications (SPAWC) (pp. 201-205). IEEE.

[35] Zhu, W.P., Ajib, W., Amarasuriya, G. and Kusaladharma, S., 2019, May. Achievable rate analysis of NOMA in cell-free massive MIMO: A stochastic geometry approach. In ICC 2019-2019 IEEE International Conference on Communications (ICC) (pp. 1- 6). IEEE.

[36] Mukherjee, S. and Lee, J., 2018, December. Offloading in edge computing-enabled cell-free massive MIMO systems. In 2018 IEEE Globecom Workshops (GC Wkshps) (pp. 1- 6). IEEE.

[37] Ngo, H.Q., Ashikhmin, A., Yang, H., Larsson, E.G. and Marzetta, T.L., 2017. Cell-free massive MIMO versus small cells. IEEE Transactions on Wireless Communications, 16(3), pp.1834-1850.

[38] Kim, S. and Shim, B., 2018, June. FDD-based cell-free massive MIMO systems. In 2018 IEEE 19th International Workshop on Signal Processing Advances in Wireless Communications (SPAWC) (pp. 1-5). IEEE.
[39] Abdallah, A. and Mansour, M.M., 2019, July. Angle-based multipath estimation and beamforming for FDD cell-free massive MIMO. In 2019 IEEE 20th International Workshop on Signal Processing Advances in Wireless Communications (SPAWC) (pp. 1-5). IEEE.
[40] Björnson, E. and Sanguinetti, L., 2019. Making cell-free massive MIMO competitive with MMSE processing and centralized implementation. IEEE Transactions on Wireless Communications, 19(1), pp.77-90.
[41] Hajri, S.E., Denis, J. and Assaad, M., 2018, June. Enhancing favourable propagation in cell-free massive MIMO through spatial user grouping. In 2018 IEEE 19th International Workshop on Signal Processing Advances in Wireless Communications (SPAWC) (pp. 1-5). IEEE.
[42] Huang, Q. and Burr, A., 2017, May. Compute-and-forward in cell-free massive MIMO: Great performance with low backhaul load. In 2017 IEEE International Conference on Communications Workshops (ICC Workshops) (pp. 601-606). IEEE.
[43] Karlsson, M., Björnson, E. and Larsson, E.G., 2018. Techniques for system information broadcast in cell-free massive MIMO. IEEE Transactions on Communications, 67(1), pp.244-257.
[44] Zhang, Y., Cao, H., Zhou, M. and Yang, L., 2019. Cell-free massive MIMO: Zero forcing and conjugate beamforming receivers. Journal of Communications and Networks, 21(6), pp.529-538.
[45] Femenias, G. and Riera-Palou, F., 2019. Cell-free millimeter-wave massive MIMO systems with limited fronthaul capacity. IEEE Access, 7, pp.44596-44612.
[46] P. Jain and A. Gupta, "Energy-Efficient Adaptive Sectorization for 5G Green Wireless Communication Systems," in *IEEE Systems Journal*, vol. 14, no. 2, pp. 2382-2391, June 2020.
[47] Chopra S.R., Gupta A., Monga H. (2019) Performance Analysis of Space-Time Trellis Codes in Rayleigh Fading Channel. In: Yadav N., Yadav A., Bansal J., Deep K., Kim J. (eds) Harmony Search and Nature Inspired Optimization Algorithms. Advances in Intelligent Systems and Computing, vol 741. Springer, Singapore.
[48] Gupta, Akhil, Shiwani Dogra, and Ishfaq B. Sofi. "Performance evaluation of spatial multiplexing using different modulation techniques in MIMO system for small and large scale fading channel." International Journal of Sensors Wireless Communications and Control 9.2 (2019): 188-202.
[49] Gupta, Akhil, Rakesh K. Jha, and Reeta Devi. "Security architecture of 5g wireless communication network." International Journal of Sensors Wireless Communications and Control 8.2 (2018): 92-99.
[50] Sofi, Ishfaq Bashir, Akhil Gupta, and Rakesh Kumar Jha. "Power and energy optimization with reduced complexity in different deployment scenarios of massive MIMO network." International Journal of Communication Systems 32.6 (2019): e3907.
[51] A. Gupta and S. R. Chopra, "Designing and Testing of Quadrifilar Helix Antenna for NOAA Weather Satellite," 2019 6th International Conference on Signal Processing and Integrated Networks (SPIN), 2019, pp. 785-790.
[52] S. R. Chopra, A. Gupta and H. Monga, "An LTE Approach with MIMO by Using Suboptimal Selection of Antenna," 2018 International Conference on Intelligent Circuits and Systems (ICICS), 2018, pp. 167-172.
[53] Goyal, Krittika, and Akhil Gupta. "A literature survey on different types of pulse-based sensor for acquisition of pulse." International Journal of Control Theory and Applications 9.10 (2016): 361-365.

[54] R. Kausar, A. Gupta, I. B. Sofi and K. Arora, "Bit Error Rate Based Performance Evaluation of LTE OFDMA System," 2018 International Conference on Intelligent Circuits and Systems (ICICS), 2018, pp. 161-166, DOI: 10.1109/ICICS.2018.00042.

[55] Anand, Pankaj, Akhil Gupta, and Sonam Bhagat. "Bit Error Rate Assessment of Digital Modulation Schemes on Additive White Gaussian Noise, Line of Sight and Non-Line of Sight Fading Channels." International Journal of Engineering Science Invention 3.8 (2014): 6-10.

[56] Gupta, Akhil, and Krittika Goyal. "An analytic review on antenna modelling." International Journal of Control Theory and Applications 9.10 (2016): 319-336.

[57] Gupta, A., Jha, R.K. Power optimization with low complexity using scaled beamforming approach for a massive MIMO and small cell scenario. Wireless Netw 26, 1165–1176 (2020).

[58] I am. B. Sofi, A. Gupta and R. Kausar, "Performance Evaluation of Different Channel Estimation Techniques in MIMO System for Hata Channel Model," 2018 International Conference on Intelligent Circuits and Systems (ICICS), 2018, pp. 155-160, DOI: 10.1109/ICICS.2018.00041.

[59] R. T. Kamurthi, S. R. Chopra and A. Gupta, "Higher-Order QAM Schemes in 5G UFMC system," 2020 International Conference on Emerging Smart Computing and Informatics (ESCI), 2020, pp. 198-202, DOI: 10.1109/ESCI48226.2020.9167619.

[60] Raina, Ambica, Akhil Gupta, and Rakesh Kumar Jha. "Analysis of AODV and DSR Routing Protocols for a Static Ad Hoc Network Using Qualnet Network Simulator." International Journal 4.1 (2014).

[61] Sundan, A.P., Jha, R.K. & Gupta, A. Energy and spectral efficiency optimization using probabilistic based spectrum slicing (PBSS) in different zones of 5G wireless communication network. Telecommun Syst 73, 59–73 (2020).

[62] M. Singh, K. Arora and A. Gupta, "Equalization in WIMAX System," 2018 International Conference on Intelligent Circuits and Systems (ICICS), 2018, pp. 330-333, DOI: 10.1109/ICICS.2018.00074.

[63] Jain, Preksha, and Akhil Gupta. "Adaptive NOMA towards 5G green wireless network." Transactions on Emerging Telecommunications Technologies 31.3 (2020): e3887.

[64] H.S. Bedi, K.K. Sharma, and R. Gupta, "A Review Paper on Performance Analysis of IEEE 802.11 e", *1st International Conference on Computing, Communications, and Cyber-Security (IC4S 2019), 2020, pp. 47-56.*

[65] Z.G. Khaki, and H.S. Bedi, "Transient correction using EDFA: in-line optical fibre with feedback", *International Conference on Computing Sciences,* 2012, pp. 233-238 [http://dx.doi.org/10.1109/ICCS.2012.76].

[66] Bedi, Harpreet, Kamal Kumar, and Gupta Raghav. "A review paper on improving the network efficiency of IEEE 802.11 e networks." *Intelligent Circuits and Systems (2021): 206-211.*

[67] Harpreet Bedi, Kamal Kumar Sharma; Analyses of CSMA/CA Protocol Without Using Virtual Channel Sensing in DCF Mode, *Global Emerging Innovation Summit (GEIS-2021)* (2021) 1: 484. https://doi.org/10.2174/9781681089010121010058.

Enhance the performance of AODV routing protocol in mobile ad-hoc networks

Ali Abdulmalek Abdullhaj Saif[1], Krishan Kumar[2]
1 M.Tech. Student, SEEE, LPU, Punjab, India
2 Professor, SEEE, LPU, Punjab, India
E-mail: alisaif200q@gmail.com

Abstract. Enhancing the quality of service is still facing various challenges due to nature and configuration of mobile ad-hoc network (MANET) such as free movement of nodes, unpredictable changes in network topology as no infrastructure exist in the network. The large usage of mobile ad-hoc network in various applications raises the demand to enhance the performance as well as reduce loss of packets through various factors. The routing protocols play the most important aspect in MANET. There are three kinds of routing protocols that are: reactive, proactive and hybrid that are utilized to optimize the network's functionality. Ad-hoc on-demand Distance victor (AODV) routing protocol is the popular used to achieve best output result. The most important objectives of this paper are to enhance the performance of mobile ad-hoc network by using AODV protocol. This paper gives some changes in AODV routing protocols that's assist the performance enhancement of AODV in mobile ad-hoc network. The simulation is carried out on the network simulator NS2 and the output result show the improved AODV has excellent performance in term of packet delivery ratio, secondly throughput, third is the routing overhead and finally is end to end delay.

Keyword: Aodv, MANET, RREQ and RREP

I. INTRODUCTION

The following is a definition of a mobile ad-hoc network that can be describe as group of devices which are connected together to form a network without any structure designed before. The nodes in MANET move from one position to another position randomly and all that's node are connected by wireless links therefore it is easy to arrange the position of the nodes. A MANET is a self- configure network in which the topologies are change randomly and that's why the nodes are free to move whiten the network. The infrastructure property of MANET makes it more flexible and can work at any place.

In the last years ago, many studies have been done to maximize the performance of MANET by using different protocols. In a mobile ad-hoc network, there are three main types of routing protocols: first one is proactive, second one is reactive (on demand) and third one is a hybrid routing protocols. Routing protocols that are proactive in which every hope (node) store and keep (maintain) the data packet to other node within the network. But the reactive (on- demand) routing protocol, the path of transmission of data packets is established when the source node want to send that packet to the target instead of accumulating to data routing tables. The hybrid routing protocols gather the features of both two others protocols mentioned. Throughput decreases and data packet ratio become unbalanced, that's lead to decrease the performance of t mobile ad-hoc network. therefore, the significance to change throughput and packet delivery ratio and Oher's parameter to get high performance become more compulsory. This paper gives anew modifications in AODV that support to enhance the output result in network simulator NS2.

2. ROUTING PROTOCOLS

Mobile ad hoc network has a variety of routing protocols to decorate the overall performance of the network, these protocols are classified as shown in Fig. 1

a) Proactive routing protocols

The proactive routing protocols have another name is named table driven routing protocol. In this type of protocols, the information's will send to the neighbors by the mobile node and every node have to sustain a routing table which contains the reachable and number of nodes [2]. Due to that in case of a big networks when the number of nodes increases then the overhead will also increase and these the much more major limitation of proactive routing protocol [3].

b) Reactive routing protocols

The reactive routing protocol, which is also known as on-demand routing protocol in which the rout is established only when it is required. Due to that the bandwidth will increase on demand routing protocol. These kind of protocol works with two main process which are Rout discovery process rout maintenance process [4].

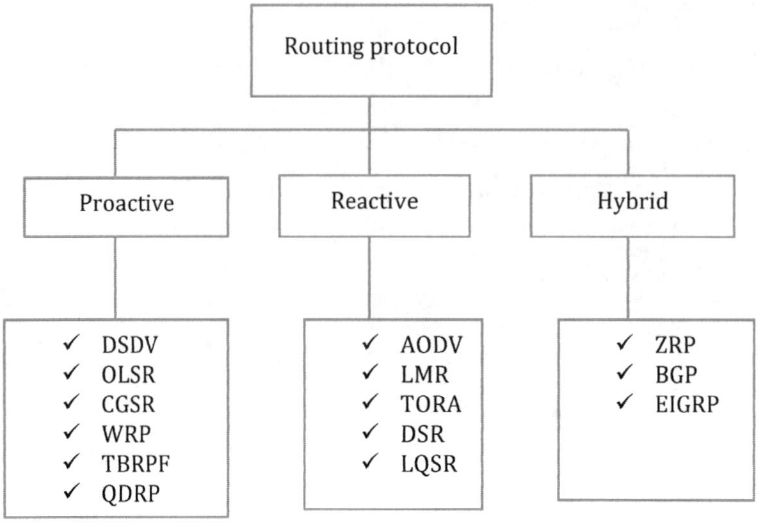

Figure.1 Routing Protocol

The AODV is the most common routing protocol used as reactive routing protocol that use two kinds of message
RREQ and RREP message [5]

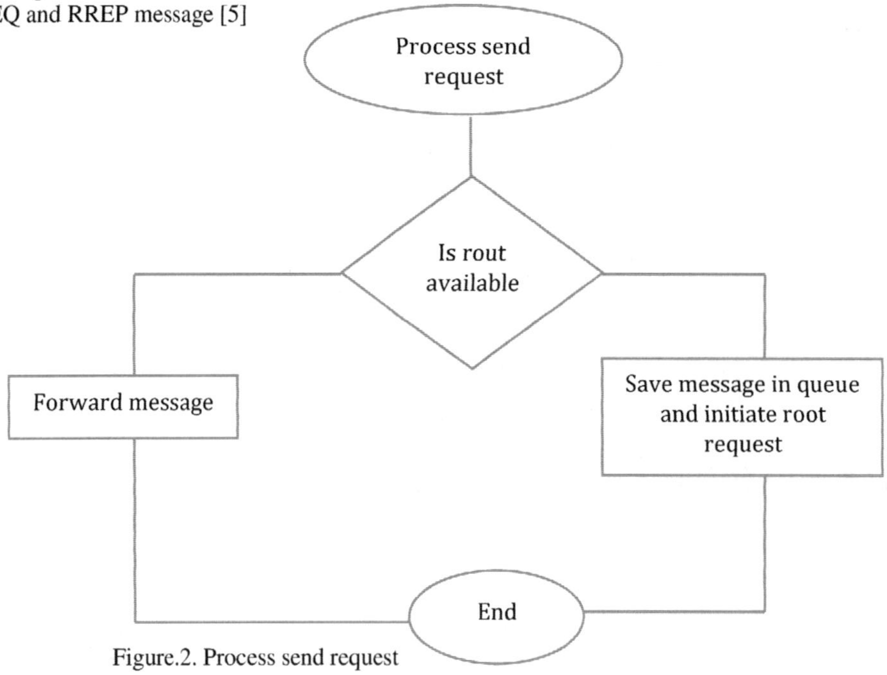

Figure.2. Process send request

The routing table consist the following things

1. The address of the destination
2. Address of the second node
3. The sequence number of target node
4. Hop count

3. AODV Routing Protocol

AODV support unicast and multi-cast routing and it deal with two main process which are path discovery and path maintenance [6].

- AODV Route Discovery:

In path discovery process, when a source nod requires to connect with another node, it broadcasting a message or request to its neighbors and this request called rout request (RREQ), Then every neighbor node has two options [7]. Either the rout satisfied the request (RREQ) by sending rout reply (RREP) to the source node, otherwise that node forward a request to the neighbor's node till the request get the destination node or the required node [8].

- AODV Route maintenance:

A path discovered in in rout discovery process between the two node which are the source node and target

(destination) node is kept up while it is needed by the source node. The break's upstream node sends a message rout error (RERR) to its neighbors. [9]. Similarly, this node also sends the rout error to the neighbor's node. So that process doesn't stop till the source node is reached. At the time the rout error message is arrived to the source hope (node), then the source node has two options, either its re-initiates the rout discovery process by sending the rout request message (RREQ) if sending is still needed otherwise the source node stop sending the data [10]

4. PERFORMANCE ANALYSIS

Looking for ways to improve the routing protocol's performance based on various factor called simulation metrices such as routing overhead, The packet delivery ratio, end-to-end node latency, packet loss rate, and throughput are all factors to consider. The parameter that can be varied such as number of nodes, number of packets, dimension, total simulation duration maximum and minimum speed mobility and pause time.

4-1 Packet delivery ratio is the ratio between the sum of packet sent to the sum of successful received packet. that's mean the ratio of the total number of data packet which are successfully received by the destination node to those data packet which are sent by the source node.

(Total number of packets received / total number of packets sent) is the packet delivery ratio formula.

4-2 End to End Delay

The end-to-end latency is known as the period it takes for a data packet to move from the initial node (source node) to the target node (destination node). This factor includes the whole delay in time while the packets transmitted from the start node (source node) to the end node (destination node) consisting the all kind of delay such as manipulating delay, queuing delay, transmitting delay and terminating system processing delay [11].

4-3 Throughput

One of the most important factors that affects network performance is throughput, it is defined as the rate of successfully packet which arrived to the required node or the destination node from the source node. It is the rate of sending the packets of data to thee target node per unit of time [12].

4-4 Routing overhead

Links between the nodes of mobile ad-hoc network is broken frequently or repeatedly due to the random and free node movement so the network topology change.

The routing overhead defend as the total number of routing packets that sent from the source node for two main reasons either for routing discovery to select the suitable path or for the routing maintenance [13].

5. EXPERIMENTAL ANALYSIS

Table 1. Parameters used in simulation

PARAMETERS	VALUE
Operation system	Ubuntu 20.04
NS2 version	V1.15
Channel type	Wireless channel
Number of nodes	50
Data type	UDB/CBR
Simulation time	100
MAC protocol	MAC/802.11
Area of simulation	500*500
Data packet size	512
Radio propagation model	TwoRayGrou nd
Routing protocols	AODV

A) Changes of packet delivery ratio with different nodes

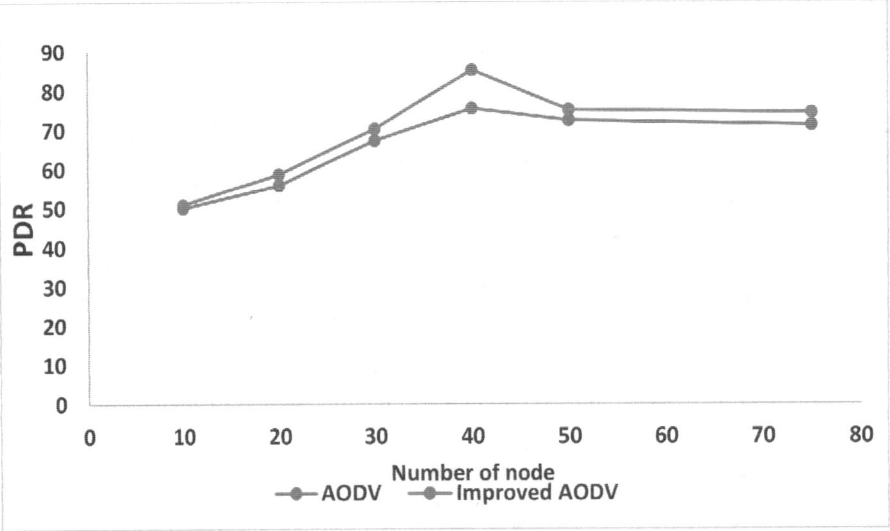

Figure 3. Packet delivery ratio

From Fig.3 it describes the packet delivery ratio of the new improved AODV in the beginning is same as the AODV that's because many reasons such as no establishment of routing topology and delay. when there is a smaller number of nodes then it increases gradually. The overall performance of PDR for improved AODV is better than AODV.

2) Throughput with time

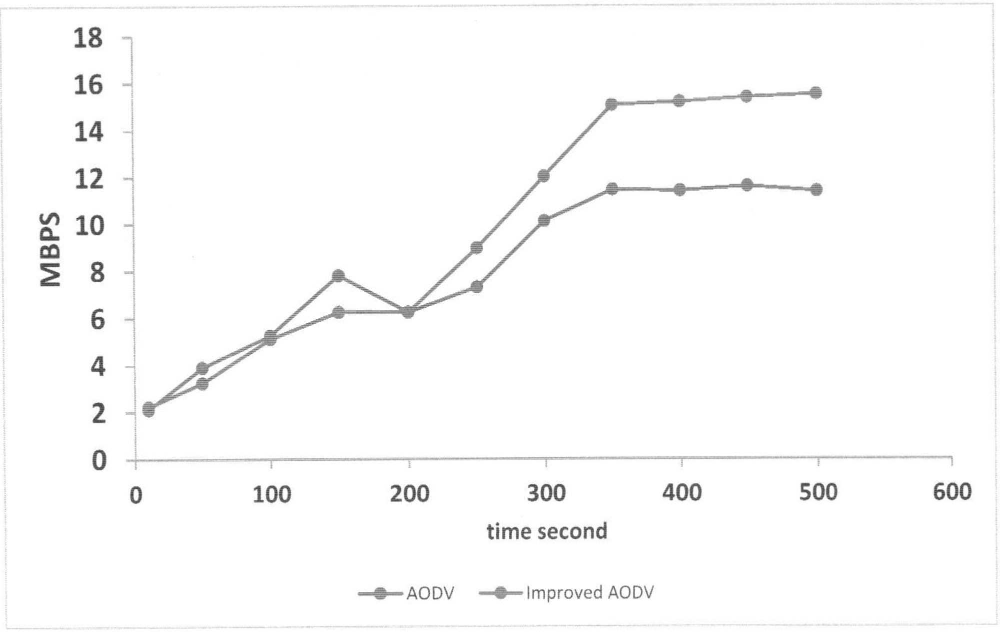

Figure.4. Throughput

The throughput is the number of successfully packet received per second. From fig (6) the throughput in the beginning nearly at 140 second it is very low due to the data delay is a mostly high because the routing topology is still not established.

After some time, the throughput will increase due to relative stability of delay. So, the improved AODV works well comparing with AODV and the number of successfully packet.

3) Changing the End-to-end latency (delay) with different nodes

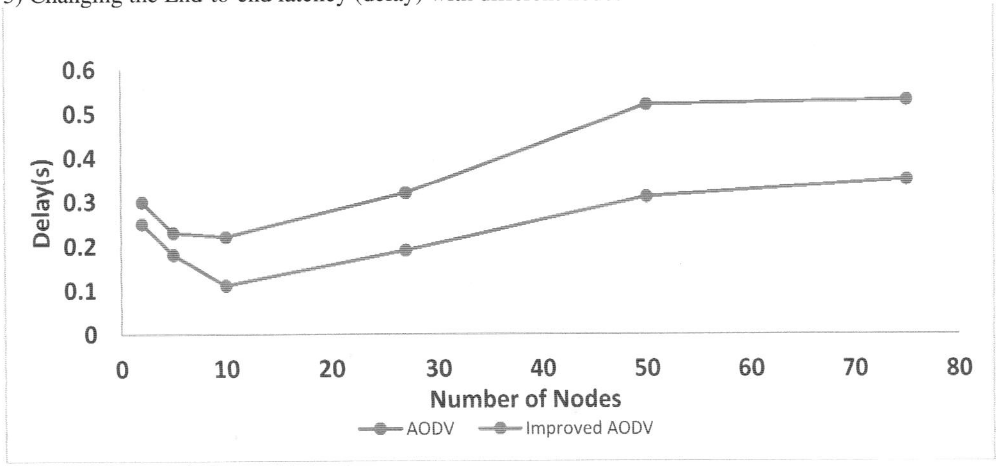

Figure.5. end to end delay vs number of node with total number of nodes are 80 nodes.

The end-to-end latency (delay) is the whole duration of time it takes for a data packet to flow from the starting node (source node) to the target node (end node). In the beginning as showed in fig (5)

the delay for both protocols are near to each other's in value and the number of data packet sent per second is low. By increasing and changing the to number of nodes further then the end-to-end delay will reduce. So, the improved AODV is working well comparing with AODV protocol.

4)Routing overhead

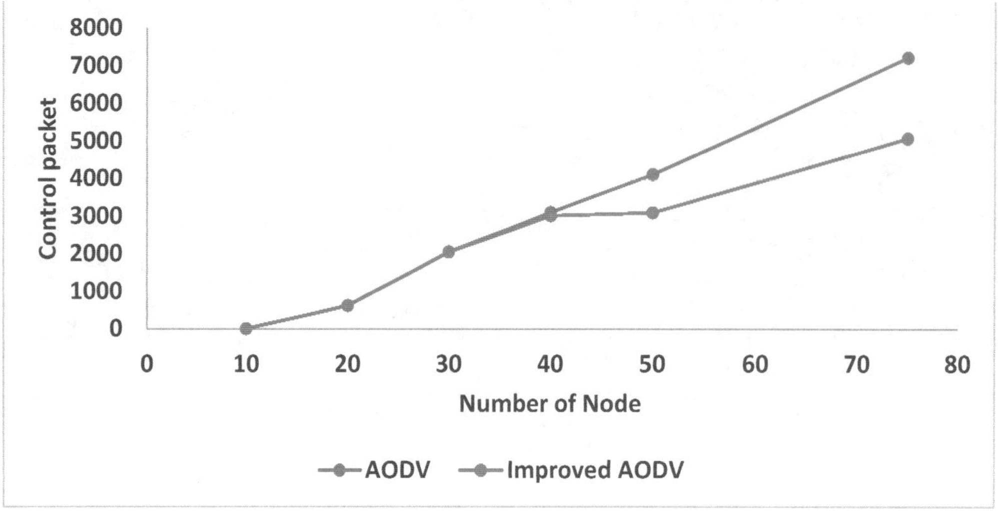

Figure.6. Routing overhead

In the beginning when the number of nodes is less thee control packets that are sent from the source node for two main reasons either for routing discovery to select the suitable path or for the routing maintenance are almost having same value. With increasing number of nodes, the routing overload for the improved AODV is well comparing with the AODV.

6. CONCLUSION

Different performance matrices are used in this study to optimize the effectiveness of the AODV routing protocol in a mobile Ad - hoc networks, specially by maximize the packet delivery ratio and the second is the throughput as well, also reducing the packet overhead and end to end delay while transmission the data packet. The ad-hoc network has many challenges in routing due to change in the structure as the free movement of nodes. Based on that this work introduced an improved AODV protocol to enhance the performance of routing According to varying some parameters while transmission such as number of nodes, number of packets, dimension, total simulation duration and pause time. This ensures the improved protocol works with quality of services (QOS). The output of simulation shows in NS2 a better feature of improved AODV protocol in term of overall performance. We are still doing extra works on improving this field and observe the performance and test the result that will assist to go further and expand the direction.

REFERENCES

[1] Chavan, A. A., D. S. Kurule, and P. U. Dere. "Performance analysis of AODV and DSDV routing protocol in MANET and modifications in AODV against black hole attack." *Procedia Computer Science* 79 (2016): 835-844.

[2] Sridhar, S., R. Baskaran, and P. Chandrasekar. "Energy supported AODV (EN-AODV) for QoS routing in MANET." Procedia-Social and Behavioral Sciences 73 (2013): 294-301.

[3] Sharma, Sanjay, and Pushpinder Singh Patheja. "Improving AODV routing protocol with priority and power efficiency in mobile ad hoc WiMAX network." international journal of computer technology and electronics engineering (IJCTEE) 2.1 (2012): 87-93.

[4] Malek, AL-Gabri, et al. "Improved the energy of Ad Hoc on-demand distance vector routing protocol." IERI Procedia 2 (2012): 355-361.

[5] Malek, AL-Gabri, et al. "Improved the energy of Ad Hoc on-demand distance vector routing protocol." IERI Procedia 2 (2012): 355-361.

[6] Mungare, S. A., and S. P. Rekha. "Review of Enhanced AODV Routing Protocol for Mobile Ad hoc Wireless Networks." International Journal of Advanced Research in Electronics and Communication Engineering 1.5 (2012): 1-6.

[7] Abu-Ein, Ashraf, and Jihad Nader. "An enhanced AODV routing protocol for MANETs." International Journal of Computer Science Issues (IJCSI) 11.1 (2014): 54.

[8] Chavan, A. A., D. S. Kurule, and P. U. Dere. "Performance analysis of AODV and DSDV routing protocol in MANET and modifications in AODV against black hole attack." Procedia Computer Science 79 (2016): 835-844.

[9] Peng-cheng, G. U. O. "Research of AODV Routing Protocol for Ad Hoc Networks." Science Technology and Engineering 18 (2013): 023.

[10] Qiankun, Zhu, et al. "A mobile ad hoc networks algorithm improved AODV protocol." Procedia Engineering 23 (2011): 229-234.

[11] Venkatasubramanian, S., R. Mohankumar, and C. Hemalatha. "Improved AODV routing protocol for manet to enhance the network performance." International Journal of Advanced Research in Computer Science & Technology (IJARCST 2014)\({\copyright}\) (2014).

[12] Jacob, Jaya, and V. Seethalakshmi. "Performance analysis and enhancement of routing protocol in MANET." International Journal of Modern Engineering Research 2.2 (2012): 323-328.

[13] Preetha, K. G., and A. Unnikrishnan. "Improving the routing performance of mobile ad hoc networks using domination set." Procedia Computer Science 46 (2015): 1209-1215.

High Gain and Bandwidth Array Antenna for Satellite Application

Mohd. Wasim, Shelej Khera and K Malaisamy

[1]School of Electronics and Electrical Engineering Lovely Professional University, Punjab, India

[2]Department of Electronics and Communication Engineering, Saranathan College of Engineering, Trichy, Tamil Nadu, India

Abstract. A MIMO array antenna with a resonance frequency of 12.29 GHz is proposed for Ku-band applications. An 8-element antenna array is proposed and parameters like as return loss, VSWR, gain, directivity, and radiated power are utilized to validate model and test findings. The proposed antenna array has a bandwidth of 136.4MHz and the maximum gain in the Ku Band at a resonance frequency of 12.29GHz. It also has a gain of 8.89 dBi and a radiation efficiency of 100%.

Keywords: Microstrip feed, antenna array, Microstrip patch, CST.

1 Introduction

Antennas are the most essential component of any wireless communication system. Signal distances are greater because geostationary satellites are located at higher altitudes. As a result, route loss increases and signal power levels decrease.[1] An antenna's directivity and gain should fulfil the criteria of a satellite. In addition, the antenna’s size needs be reduced in order for it to be mounted on the satellite [2]. The problem of limited bandwidth is solved by using a "cutting a notch" approach, such as an E-structure patch Microstrip antennas have several benefits, such as small weight, compactness, resilience, and so on. [3] However, they have major drawbacks, such as restricted bandwidth, low strength (almost 6 dB), and large losses, making them unreliable for satellite communication [4]. One of the greatest applications for supporting a high mobility satellite communication device in the Ku band may be Very Tiny Aperture Terminal systems, which employ a small microstrip antenna.

The use of cluster radio wires in current and future remote communication frameworks has expanded significantly, permitting the arrangement of isotropic, Omni, twofold, and circularly polarized radiation designs. These antennas can be designed for the different configurations of the array to achieve high gain. microstrip radio wires are utilized to construct clusters since they can broadcast an adjusted flag over a wide recurrence range. Microstrip fix arrays may too be utilized to overcome issues caused by clashing signals, holding association, and quality, which is accommodating for adherent communication. The components of a radio wire are made by optimizing the antenna's properties into single-band, multi-frequency, and broadband, which may be fulfilled by adding a variety of major radiation components [5, 6].

Satellite communication is becoming increasingly popular in today's modern world. This means that in order to effectively cover the required operational bands, the future satellite communications terminal antenna must fulfil multi-band or wideband standards. The difficulty of array antenna design, on the other hand, doubles exponentially as the number of operating frequency bands increases. Furthermore, the array antenna must be compact enough to be effectively positioned within the system [7]. In today's modern world, greater bandwidth is necessary to send and receive more information, hence bandwidth augmentation and decreased array antenna size is a hot study subject. Because of its low-profile construction and smaller size, microstrip patch array antennas are becoming more desirable in satellite applications [8]. As a consequence, they are fully compatible with embedded array antennas in portable devices like telephone, satellite TV, mobile satellite technology, satellite broadband, and so on. On missiles, tracking and communication antennas must be narrow and conformal, and they are widely utilized. Patch antennas constructed of microstrip. Their main advantages are small weight and minimal volume, as well as a low profile planar arrangement array antenna [9-16].

This array antenna supports both linear and circular polarization and may be readily made conformal to the host surface and produced in huge quantities. Because of its tiny construction, this antenna can be readily integrated with MICs. Because of its simple construction, single resonant frequency, low profile, and high data rate, the suggested Microstrip patch array antenna has been widely employed in recent years. This array antenna's return loss is reduced at the resonance frequency level, and each element of that array antenna is small in size. The antenna had a better radiation pattern, a greater gain, and was more efficient. An array antenna consists of eight components

This work describes a revolutionary tiny size microstrip array antenna for Ku band satellite applications, which is built on a FR4 substrate. A microstrip array antenna has a high gain and operates on a balanced feed. The microstrip array antenna was studied for array topologies with a 14-array configuration in order to improve overall gain. Because of its high gain, light weight, and controlled operation, it is suitable for satellite communication. Various approaches are used to construct various types of antennas for satellite applications. In wireless applications, a series fed dipole antenna built for base band applications with a gain of roughly 16 dBi was observed spanning a frequency range of 1.7 GHz to 2.5 GHz. [17].

Antenna parameters such as directivity, gain, and bandwidth are exceedingly poor, according to the prior literature research. Many steps have been taken to construct an antenna for satellite applications, with array elements added either in the parallel or serial arrangement. As a consequence, the microstrip array antenna was chosen as an array element and examined in a 2x4 array configuration. Various array topologies have been recognized for their great directivity and increased signal-to-noise ratio.

The study is prepared as follows; Section II discusses the geometry and technique for developing the microstrip patch antenna. Section III describes in detail the approach to be followed for creating the equivalent circuit design. Section IV presents the patch's outcome and debate. Finally, Section V brings the paper to a close.

2. ARRAY ANTENNA DESIGN

The patch structure is made up of an array of 8 element antennas. Because of its cheap material cost and simplicity of manufacturing, as well as its potential to provide optimal performance for a wide range of applications [19], FR4 is employed as a substrate in this design. The design values are given in Table 1: dielectric constant, substrate height, loss tangent, patch length, patch width, and thickness.

Table 1 Design values of the single element antenna

SI.No.	Parameters	Values
1	Dielectric constant	4.4
2	Substrate Height	1.6 mm
3	Loss tangent	0.007 mm
4	Patch length	72 mm
5	Patch width	50 mm
6	Thickness	1.6 mm

In this part, the suggested 8-element array antenna is referred to as a "direct feed array antenna" since it is developed using coaxial feed and quarter wave transform, as illustrated in Fig.2. Ls, L1, L2, L3, L4, L5, L6, L7, L8, W1, W2, W3, W4, W5, W6, and are the array antenna's dimensional parameters, as indicated in Table 2

Table 2. Dimension of proposed antenna

S. No	**Parameter**	**Value (in mm)**
1.	L_S	6.8
2.	L_1	0.7
3.	L_2	0.5
4.	L_3	2.0
5.	L_4, L_6, L_8	5.0
6.	L_5	1.4
7.	L_7	1.584
8.	W_s	7.0
9.	W_1	0.4
10.	W_2	20.4
11.	W_3	4.9
12.	W_4	40.8
13.	W_5	2.8
14.	W_6	4.9

Figure 1 depicts the schematic construction of a single element patch antenna. The dielectric constant of the design substrate is 4.4 and it is 0.635mm thick. Because there are four elements on the left and four elements on the right, the feed point radiation is located in the common section of the entire 8-element structure, and the input power operates as a quarter wave transform structure. The array antenna is

designed to retain gain while also operating in the broadside direction [4, 5] and is matched to the system impedance Z0=50. It has a frequency range of 11.0GHz to 15.0GHz.

The practical width and length can be calculated by basic antenna equations,

as shown in below

$$W = \frac{c}{2fr\sqrt{\mu_0 \in_0}}\sqrt{\frac{2}{\in_r + 1}} = 7.0mm$$

$$L = \frac{c}{2fr\sqrt{\in_{eff}}\sqrt{\mu_0 \in_0}} - 2\Delta L = 6.8mm$$

where, length extension of the patch antenna [11 -13],

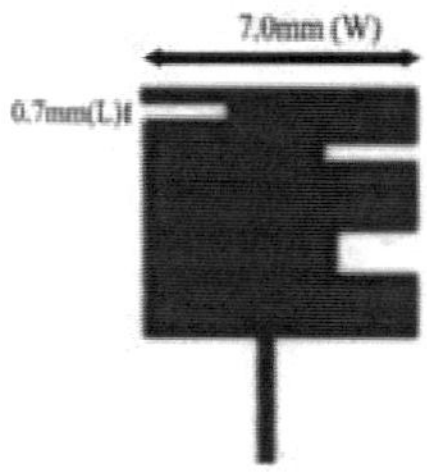

Fig 1. Schematic structure single patch element

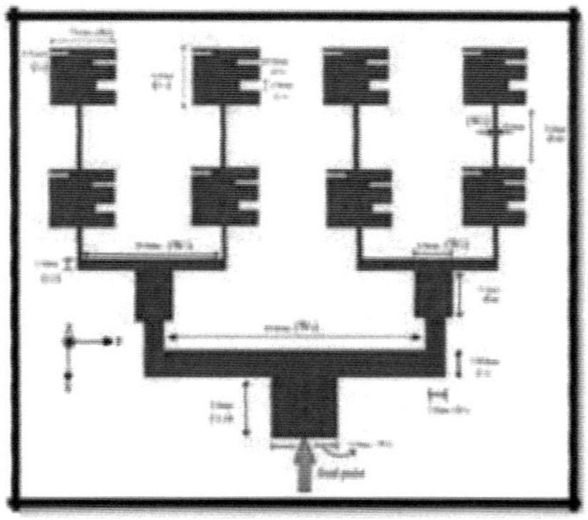

Fig 2. Schematic structure of eight element antenna array

The CST tool is used to create an 8-element Microstrip patch array antenna. Because it employs conducting strips to connect that patch and may be thought of as an extension of the patch, this array antenna is one of the easiest to design and construct [6]. By choosing z1, the input impedance Zin may be changed until it equals Z0, at which time the array antenna impedance is matched. Antenna parameter z1 may be expressed by varying the width of the quarter wave length of the

Microstrip; it can get wider and have lower characteristic impedance (Z0). The goal is to match the transmission line's input impedance (Zin) to its antenna (Z0). If the array antenna's impedance is ZA, and eventually it displays as an input impedance of the array antenna's commencing quarter wave length [7, 8].

3. RESULT AND DISCUSSION.

CST momentum was used to construct the Microstrip patch array antenna, and the array antenna may be manufactured. Following design, the planned antenna was simulated, and the directivity, gain, and returns loss were acquired, as well as a 3D picture of the developed patch array antenna's far field emission. Figure 3 depicts the antenna construction. A narrowband Microstrip patch antenna at 12.29GHz for satellite applications was built using the CST tool (software).

Practically, the antenna was tested with a VNA model no: E5063A ENA, with a frequency range of 12GHz-18GHz. The antenna has a frequency band of 11.0 GHz to 15.0 GHz and resonates at 12.29GHz. As a result, the generated bandwidth is 136.4MHz, which is a broad band covered by an array antenna and ideal for satellite applications. Figure 6 depicts S (1, 1) [deg] (Phase S (1, 1) Vs operating frequency).This graph depicts the fitted (phase) value of -165.656, and the overall performance of the array antenna is reflected by its higher gain and directivity.

For antennas, the VSWR is a real and positive value. The minimal VSWR is 1.0, and no power is reflected from the antenna in this scenario, which is optimal. The VSWR for the centre frequency of 12.29 GHz is 1.097, which is about optimum. Thus the computer simulation and testing results are obtained. Table 3 discusses about the comparison result statement of simulated and measured / tested parameter values.

From tested, the observed some variations in results are discussed below:

- In the KU band, both the simulated and manufactured results resonated at 12.29GHz. The array antenna performance in terms of return loss is practically same, while the bandwidth of the fabricated results is greater.
- The return loss plots demonstrate that the resonant frequencies for all designs have varied in magnitude from the designed frequency.

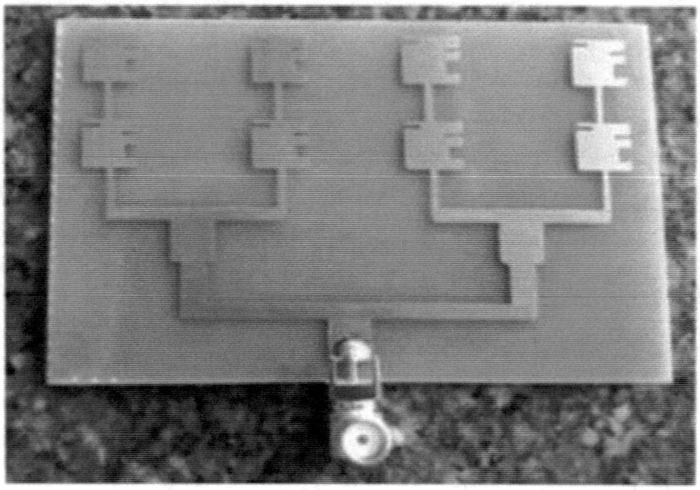

Fig.3. Fabricated structure of proposed antenna

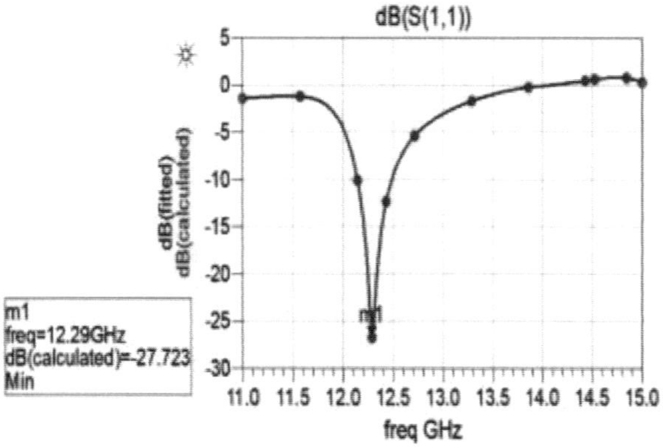

Fig.4. Return Loss

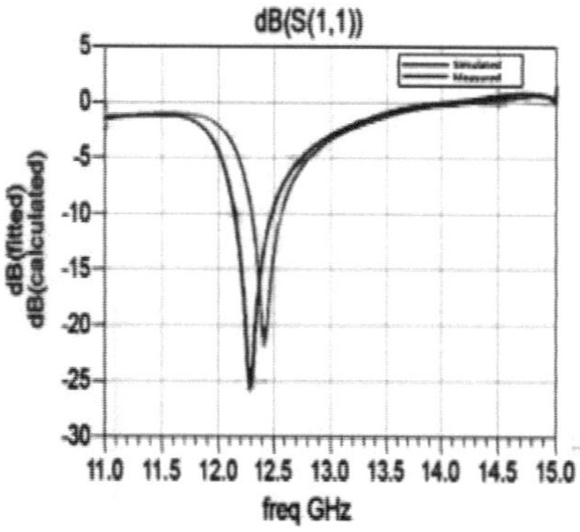

Fig. 5. Return Loss
(Simulated and Measurement)

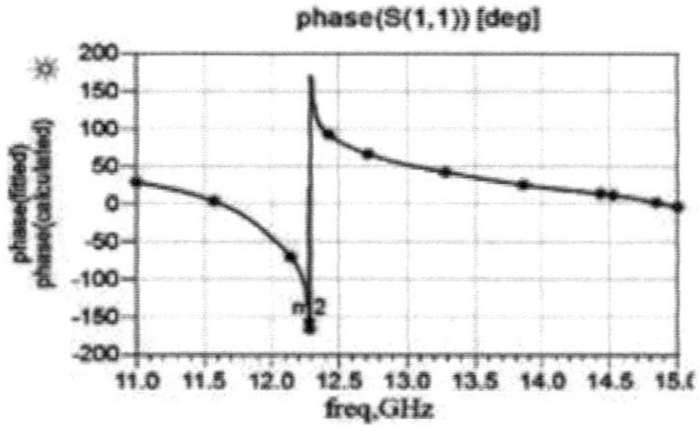

Fig.6. Proposed antenna phase output

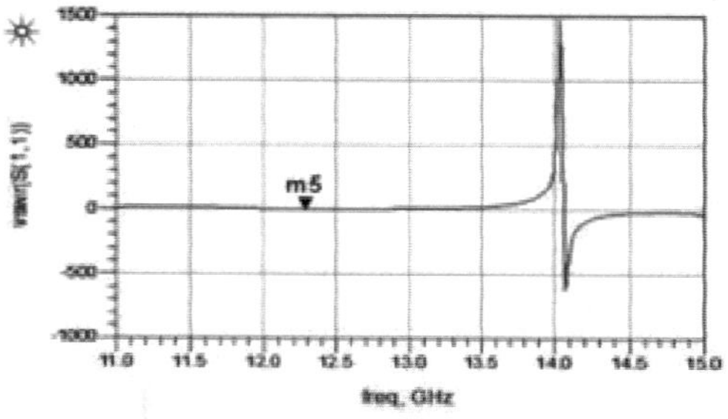

Fig.7. VSWR results of proposed antenna

In the actual world, a material with varying length, width, and height will cause the resonance frequency to change; nonetheless, it is considered to remain constant during simulation. Other parameters impacting etching accuracy, such as the chemical used, surface cleanliness, and metallization thickness, may cause the resonance frequency to change. The modelling application also has a constraint for modifying the return loss, resonance frequency, and bandwidth, implying that the conductor is difficult to draw beneath the substrate. Because the Probe feed antenna requires the feed to be on the same layer as the ground plane, an extra substrate is specified so that the feed may be drawn on it. In simulation, the design is ideal, with no air gap between the patch and the ground plane. As a result, no electromagnetic
connection to the environment is portrayed.

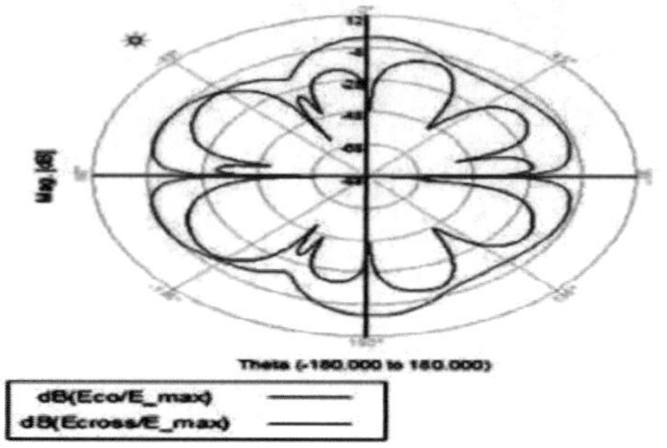

Fig. 8. Linear Polarization

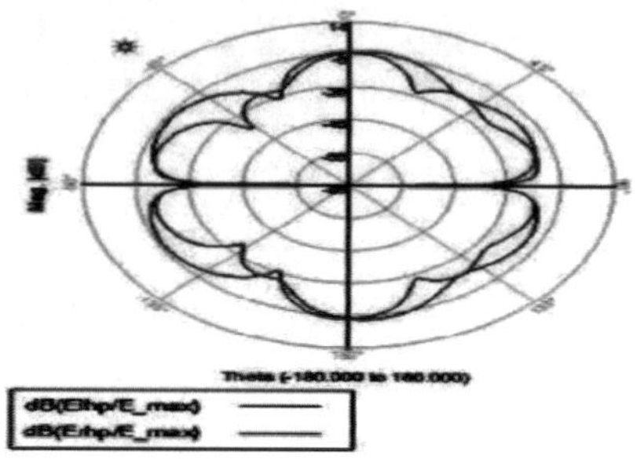

Fig. 9. Circular Polarization

4. CONCLUSION

A rectangular Microstrip Patch single band Array Antenna with a bandwidth of 175MHz was constructed. It resonates at 12.29GHz. Using computer modeling and testing, an array antenna with a relatively tiny area of just (50mm * 72mm) and eight elements was created and examined. As a consequence, this array antenna might be used in satellite applications (Ku band). For each array antenna layout, antenna properties such as VSWR, bandwidth, return loss, radiation pattern, gain, and efficiency are computed. The resonant frequency of the single band array antenna is 12.29 GHz. Furthermore, the manufactured antenna has a wider bandwidth than the simulated antenna. The frequency range covered by simulated results, which were used for satellite applications, and constructed results in an array antenna span a wide range of frequencies and have certain uses, such as television and radio, as well as Earth observation and weather. The constructed antenna's bandwidth is also boosted when compared to the simulated array antenna.

REFERENCES

[1] A. Singh, S. Vijay and R.N. Baral, "Performance Analysis of High Gain Beam Forming Conformal Array For Avionic Application", International Journal of Systems Control and Communications, Vol. 9, No. 3, pp.266-278, 2018.

[2] S. Hasegawa, T.Yasuzumi, O.Hashimoto and Y.Kazama, "A Phased Array Antenna using Cross Dipole Antenna for Mobile Satellite Communications", International conference on Electromagnetics in advanced Applications, pp.465-468, 2010.

[3] J. Angel, D.D. Ozada and Su Donglin, "Microstrip Antenna for Satellite Communications", 8 International symposium on Antennas, Propagation and EM Theory, pp. 320-322, 2008.

[4] Settapong Malisuwan, Jesada Sivaraks, Navneet Madan and NattakitSuriyakrai, "Design of Microstrip Patch Antenna for Ku-Band Satellite Communication Applications", International Journal of Computer and Communication Engineering, Vol. 3, No. 6, pp. 413-416, 2014.

[5] Shohei Hasegawa, Takenori Yasuzumi, Yasuhiro Kazama and Osamu Hashimoto, "Phased Array Antenna with Crossed Dipole Antenna for Mobile Satellite", Microwave and Optical Technology Letters, Vol.53, No. 12, pp.2805-2809, 2011.

[6] M. Dogan, K. Özsoy and I.Tekin, "Printed Dipole Array Fed with Parallel Stripline for Ku-band Applications", CiteSeer, 2009.

[7] Khalid Hati, et al, (2017), A Novel multiband patch antenna array for satellite applications, Procedia Engineering, 181, pp 496 – 502.

[8] C. Divya and V. Koushick, "Design and simulation of dual band Microstrip feed dual Yagi Uda antenna for C / X Band applications," 2017 1st International Conference on Next Generation Computing Applications (NextComp), 2017, pp. 109-114, doi: 10.1109/NEXTCOMP.2017.8016184.

[9] Koushick, C. Divya, (2017), Design and Simulation of Common Feed with Double Microstrip Patch Antenna for Dual Band Satellite Applications, International Journal of Engineering Research and Modern Education, Vol.2, Iss.1, pp 155 – 161.

[10] V. Koushick, C. Divya, (2017), Design and Analysis of Array of Slot Antenna for S-Band Application, International Journal of Advanced Trends in Engineering and Technology, Vol. 2, Iss. 1, pp 115 – 121.

[11] K. Malaisamy, B. K. Balaji, "Design of Microstrip Array Antenna for KuBand Applications",. International Journal of Communication and Networking System Volume: 05, Issue: 01, pp 29-32,2016.

[12] Hau Wah Lai, et al., (2016), Broadband Circularly Polarized Patch Antenna Arrays with Multiple Layers Structure, IEEE Antennas and wireless propagation letters, pp 1 – 4.

[13] Rachana Yadav et.al, (2015), Reconfigurable Single and Dual Band Microstrip Patch Antenna for Satellite communications, IEEE, pp 1565 – 1570.

[14] Huayan Jin, et al., (2014), Differential-Fed patch antenna arrays with low cross polarization and wide bandwidths, IEEE Antennas and wireless propagation letters, Vol. 13, pp 1069-1072.

[15] Lida Kouhalvandi, Selcuk Paker , H. Bulent Yagci, (2015), Ku - band slotted rectangular patch array antenna design, IEEE.

[16] M. M. Bilgic, K.Yegin, (2014), Wideband Offset Slot Coupled Patch Antenna Array for X/Ku-Band Multimode Radars, IEEE Antennas and wireless propagation letters, Vol. 13, pp 157-160.

[17] Wasim, M., & Khera, S. (2021). Cross-orthogonal series-fed two-dipole array antenna with parasite directors for the 5G radome applications. Materials Today: Proceedings.

[18] Wasim, M., & Khera, S. (2021, October). 1X4 Cross-Orthogonal Series Fed Two Dipole Array Antenna with Parasite Directors for the 5G Radome Applications. In 2021 6th International Conference on Signal Processing, Computing and Control (ISPCC) (pp. 62-65). IEEE.

[19] Wasim, M., & Khera, S. (2021, September). Antenna Array System with Enhanced Gain Using Cross Dipole for the LTE. In 2021 9th International Conference on Reliability, Infocom Technologies and Optimization (Trends and Future Directions)(ICRITO) (pp. 1-4). IEEE

Energy Harvesting Technique for Massive MIMO Wireless Communication Networks

Sandhya Bolla[1] and Manwinder Singh[2]

[1]Research Scholar, SEEE, Lovely Professional University, Punjab ORCID Id:0000-0001-9178-9681
2Associate Professor, SEEE, Lovely Professional University, Punjab ORCIDId: 0000-0002-0543-262
E-Mail: sansriindu@gmail.com, manwinder.25231@lpu.co.in

Abstract. Massive For wireless data and energy transfer, MIMO, a key component of future 5G systems, is favourable. A big part of the reason for this is that it has the ability to narrow down the distribution of energy. The scope of something enormous Multiple-Input in 5G, harvesting may be used to boost system capacity and efficiency via multiple-output multiple access. This was followed by the introduction of the Simultaneous Wireless Information and Power Transfer (SWIPT) method, commonly known as energy harvesting optimization. Specific to each user's needs, we connected a splitter on the receiver, and we measured their output using power beacons. In order to achieve the lowest possible communication rate, the user's total gathered power is boosted. Simulations show that the proposed technique is more successful than the other options out there. There are two sources of energy for D2D transmitters: specialised beacon signals, and cellular users (CUs) radio frequency (RF) interference. In D2D transmitters, inverse power control techniques are used for power regulation. We employ Poisson Point Processes(PPP) to model the proposed hybrid network's CU, D2D user (DU), and PB locations. Each slot in the time period is assigned a different beam forming weight. Using the alternating minimization technique, we provide a solution approach for a non-convex quadratic ally limited linear problem. Long-distance wireless power transmission requires the use of energy beam generation technologies in a large-scale MIMO system in order to make sure that the transmission is safe. In other words, this letter tries to maximise the energy efficiency of information transmission (bits per joule) while keeping the required quality of service (QoS), which is a delay constraint, in mind. This is done by maximising the transfer length and transmits power, among other things. In this case, we try to maximise the total amount of energy at the relays, use block diagonalization (BD) at the source, and try to minimise interference between the relay-destination channels. As a last step, we also run simulations of the power consumption of a generic circuit. As a supplement to this work, there are some basic online rules for all possible situations. Numerical results show that a near-optimal online technique can do the same job as its offline counterpart when the maximum power consumption of the circuit is taken into account.
Keywords: Energy harvesting, massive MIMO;RF interference, hybrid networks, device-to device, power beacons.

1 Introduction

Recent improvements in the design of highly efficient RF energy harvesting hardware offer many advantages, such as complete control over energy transfer, a wide range of charging range, and an extremely small form factor [1, 2]. Wireless power transfer makes it possible to recharge batteries in places like the battlefield, under water, and in body area networks, for example. This makes wireless networks last longer in places like these. Because of this, wireless power transfer has been getting a lot of attention from both academia and industry.

When transmitting electricity wirelessly, electromagnetic propagation is the most common method. There is propagation loss in both wireless power transfer and wireless data transmission, which includes route loss, shadowing, and quick fade. As a result, wireless power transfer efficiency is a very important and difficult problem. In an effort to improve the efficiency of the transfer, multi-antenna wireless power transmission techniques are being used. These techniques use beams of energy to move power. In MIMO broadcast systems, the best ways to design transmit beams for power transmission were found. In addition, [8] came up with a way to shape energy beams based on channel state information that isn't as accurate as the information at the power source. When the transmission distance is not very short, a conventional multi antenna system cannot meet practical energy efficiency requirements because there are only a few antennas close to the power source [9].

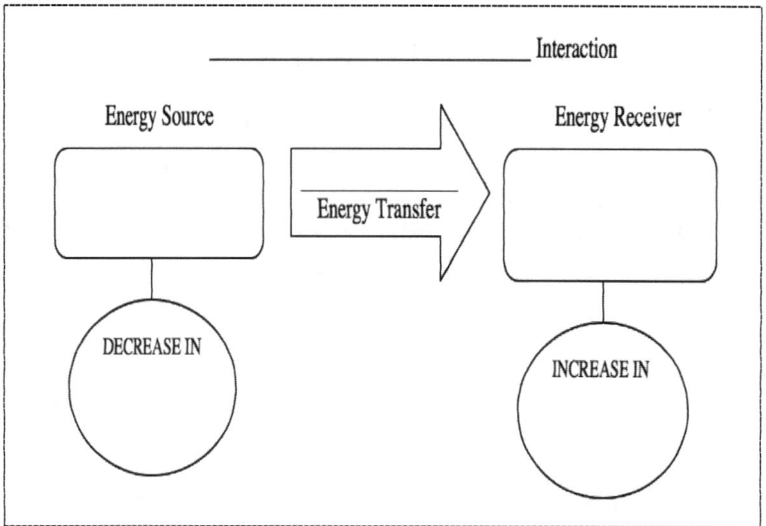

Figure 1. Energy Transfer from Transmitter to Receiver

A lot of antennas at the transmitter could make a big difference in how well the transmission works in large-scale MIMO [10] and [11]. Because tiny antennas can be used in the 60GHz WiFi 802.11/ac standard, large-scale antenna beam shaping is used to change the direction of the beam.

Journal of Physics: Conference Series **2327** (2022) 012059 doi:10.1088/1742-6596/2327/1/012059

Wireless power transmission performance may be considerably improved by using large-scale MIMO. Wireless energy transmission's ultimate purpose is to meet the receiver's need for labour, which makes intuitive sense. Medical implants may be charged wirelessly and transmit medical data to a remote receiver using the acquired energy, for example. In contrast, the vast majority of previous initiatives have evaluated and designed wireless power transmission without taking into account the intended application of the gathered energy. In this letter, we look at large-scale MIMO systems that use energy beam forming to transmit information while also collecting energy, a process known as wireless powered communication. In order to avoid not being able to harvest and send at the same time, time slots should be divided into parts for harvesting and parts for transmitting. In order to improve performance, it is important to find the best time switch point, which means allocating time resources at the right time. The transfer of energy at the power source, which is still an important resource, has an impact on how well the system works in the end. [12] A rise in the popularity of green communication, which is communication that doesn't use a lot of energy, has been caused by energy shortages and climate change.

These letters are used to improve energy efficiency, which is how many bits of information are given for each joule of energy that is collected. Three things are accomplished in this letter. To begin, it tackles the tough problem of transmitting data and electricity across large distances wirelessly. Second, it has a significant impact on energy efficiency. The third advantage is that it ensures a high degree of Quality of Service for all customers (QoS). The rest of this letter is shown below. When we look at wireless information and power transmission in large-scale MIMO systems for the first time in Section II, we will see how they work. Finally, in Section III, we come up with an energy-efficient way to allocate resources. Recent attention has been paid to the potential of energy harvesting in wireless communication networks to prolong the network's lifespan while simultaneously offering environmentally friendly communications. As far back as time immemorial, people have relied on the natural world for their energy. SWIPT, the simultaneous wireless transport of data and power, is now possible because to recent developments in energy harvesting employing RF electromagnetic waves [1]. As a consequence of this, wireless communication methods may not be ideal for simultaneously transmitting both data and electricity at the same time. When it comes to transmitting information and electricity, wireless communication systems must be adapted. Using a single-input single-output (SISO) system, SWIPT's first study [2] looked at the fundamental trade off between the speeds at which energy and reliable information may be sent. When discussing access and multi-hop channels in this article, we shall use the words “energy” and “power” interchangeably. In the previous work, the rate-energy area was defined and the transmit signal's covariance matrix was designed.

SWIPT downlink systems with multiuser multiple-input, single-output (MISO) architecture have been studied in [7]. For MISO multicasting systems, SWIPT was investigated in [8]. A SWIPT system that uses MIMO spatial multiplexing hasn't had any work done on the practical design of the source precoder and receivers yet, according to the author's understanding. There was a big trade-off between the EH receiver's harvested energy and the information rate at the destination node when there was real-time channel state information (CSI).

We also found that there was a trade-off between the EH receiver's harvested energy and its outage risk, which we call the outage-energy trade-off. This is what we call the outage-energy trade-off. When designing a two-hop SISO DF relay system, the PS and TS approaches were employed to determine the practical throughput and ergodic capacity of the system. The purpose of this study is to discuss SWIPT in the context of multiuser MIMO BCs. The capabilities of BCs have been investigated in [7] and [8]. The use of dirty paper coding (DPC) in MIMO BCs has been shown to approach capacity in [9] experiments. Despite this, DPC is very computationally intensive, making it difficult to use in practise. Zero forcing (ZF) and block diagonalization (BD) are two examples of simple yet inefficient suboptimal strategies that may be implemented by anybody. There are problems with offline scheduling in both perfect and non-perfect circuit power consumption situations when total power is defined as being of the water-filling type with a ceiling level. This makes it difficult to maximise the weighted sum throughput without a network connection. We can derive some theoretical inferences about energy transfer and transmission power distribution over time based on the KKT data. Many heuristic strategies for online scheduling have been devised and tested in simulations under both conditions. When circuit power consumption is more than zero, the performance of the online methodology is equivalent to that of the offline technique.

An investigation of the same kind looked into the energy efficiency of large MIMO systems that were charged wirelessly. It looked at how much energy they used. [15] looked into the possibility of making a wireless data and power transmission system for a single person. The length of the power transfer and the power that was used to send it were both carefully thought out in order to come up with an energy-efficient resource allocation method that was limited by a delay. It has been looked into in other research [5] how to improve the throughput of large MIMO wireless information and power systems. In the large-antenna regime, [5] came up with a way to use resources that maximises throughput while still keeping things safe. In this paper, we look into the energy efficiency of a large MIMO wireless information and power transmission system. We use a scalable power consumption model to do so. People who have done past research on wirelessly powered large MIMO systems have either not paid attention to or treated the power consumption of node circuits as a fixed thing, which is not true. In systems like [12], base station on/off methods that use less power from the grid may be shown to work. A lot of micro grids are likely to power the BSs. This is based on the assumptions made in [13]. The authors give algorithms for how to use resources efficiently so that electricity costs can be cut as much as possible. In [14], the idea of using renewable energy sources to power cell phones is looked at, as well as the possibility of doing so. A model for deploying energy harvesting sources that takes into account how energy harvesting sources are spread out across the country is being worked on. In [15], you can find the best settings for small-cell access points to use when they're not connected and when they are connected. Wired communication systems have recently received significant scientific interest for a method of concurrently transferring data and collecting energy using radio frequency (RF) waves. Both flat fading and frequency selective channels were pioneered by references [1] and [2] for SWIPT in SISO systems.

2. System Model and Assumptions

Cognitive radio systems may simultaneously improve energy and spectrum efficiency by including energy harvesters into their structure. Large antenna arrays at base stations (BSs) help both the amount of energy that is radiated and how efficiently the radio waves use up the radio waves.

This is because large antenna arrays help me spatially combine the different data streams and (ii) focus the radiated energy on a large number of user nodes that are spread across a lot of different places in the world (MIMO). For example, the goal of this letter is to talk about the possible sum rate for cognitive large MIMO systems that use wireless energy harvesting to get their power (WEH). Study: In [3,] the slotted-mode energy harvesting cognitive radio system is looked at. Then, by optimising energy harvesting, spectrum sensing, and data transmission, the secondary system's throughput can be as high as possible while the primary system's safety is kept in check. When there are a lot of MIMO relays that communicate in both directions, it looks at wireless energy harvesting options. When it looks at two receiver designs, it talks about two of them: the power split receiver and the time-switch receiver.

MIMO systems can be made to work better so that SWIPT can be done more quickly. This essay tries to do this. Recently, scientists have been able to figure out why. There is a big trade-off when you use a single input/single output (SISO) connection. Many receiver (RX) topologies were defined in [11] in order to show why the SWIPT benefits that were theoretical aren't possible to achieve in real life. Some parts of the received signal are used for EH and some for Information Decoding in every design [14] that we've seen so far. One part of the signal used for EH is different from another used for Information Decoding (ID). MIMO SWIPT system transmitter (TX) pre coding techniques were described in [12]. There has been considerable discussion about spatial switching (SS) [16], which divides MIMO channels into its spatial Eigen channels before allocating some to energy transmission and others to information transfer.

Assume that one access point has m antennas, whereas all of the other nodes in the network have r each. An energy source that can supply radio frequency energy to users powers the access point. In order for the users to be entirely self-sufficient, the radio frequency radiated by the access point is utilised to generate electricity for them. When we do our first analysis, we make the assumption that user n's energy demand stays the same and is equal to En for n 2 [1; N]. This makes it easier to see the benefits of energy harvesting scheduling. It is our goal to reduce the harvesting time as much as possible in order to satisfy these energy needs. When the access point is activated in time slot S, this is what is output, where the energy beam-forming vector is marked by vs 2 cm, and the energy carrying signal by ys is designated by the symbol y.

M antennas are placed in a hybrid beam-forming system with L radio chains and K phase shifters for each RF chain. For base station antennas in current cellular networks, this layout works well. There is, however, no single enabling technology that can serve all 5G application demands throughout the standardisation process for 5G networks. Since 5G-based solutions offer a safety net, researchers have already started creating Beyond 5G (B5G) technologies. The Internet of Things (IoT) and Low Earth Satellite (LES) systems, both of which use existing 5G technology, are now possible because to the integration of existing and new communication technologies. As a consequence of the development of 5G/B5G technology, wireless network traffic has increased and energy consumption has increased dramatically, resulting in significant environmental and economic impacts.

Cellular networks use 60 billion kWh of electricity annually, with base stations accounting for 60 to 80 percent of that (BSs). Every year, this produces roughly 100 million tonnes of carbon dioxide—almost 13 percent of the total global transportation emissions—and is predicted to keep increasing at a rate of 6 percent each year [3].[4] To achieve a more sustainable state of development for the global communication network, green communication is viewed as an economic and social imperative in tackling global climate change.

3. Performance analysis

As required by FCC base station requirements, the base station has sixteen antennas, each with a PA of 40 dBm maximum power. We show that massive MIMO is possible by harvesting electromagnetic energy from the environment. First, we use a base station implementation of simultaneous wireless information and power transfer (SWIPT) to send information and power at the same time (BS). To make 5G mm Wave systems with many MIMO antennas work better, the UDN says that EE optimization is one of the problems. There are some ideas for how to do this because of the non concave shape of the goal function. The Dinkelbach method is used to come up with these ideas. CoMP with maximum energy efficiency (Max EE) has been suggested as a way to get the best possible energy efficiency (EE).

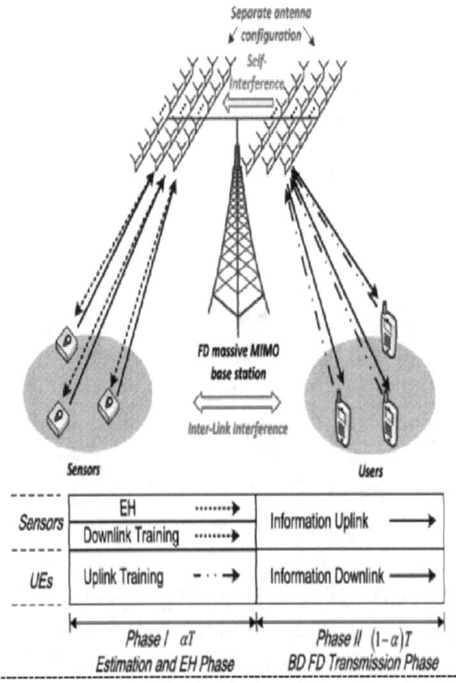

Figure 2: Illustration of full duplex Massive MIMO SWIPT system and data frame (Referred from EURASIP Journal on Wireless Communications and Networking)

The simulation results show that the proposed method has a much faster rate of convergence and is a lot more successful than current methods, as shown by the simulation results. The construction of specialised power stations by D2D transmitters, on the other hand, may allow them to accumulate adequate energy for future transmissions (QoS). People who looked at the literature came up with a new model for a hybrid network that is easy to work with. There are a lot of different parts to the hybrid network idea, like massive MIMO, D2D, and wireless energy harvesting. D2D transmitters get power from the radio frequency interference caused by cell phone users' radio frequency interference in the air around their phones. We came up with a way to dynamically allocate power for downlink multi-user MIMO-NOMA. It is important to perform different power allocation operations for communication between clusters and between clusters and between clusters. When a lot of people are in a cluster, the power that each cluster gets is based on how many people are in the cluster (see Figure 2).

4. Energy Harvesting Maximizing

The addition of a power splitter for each user at the receiver to assist with SWIPT deployment would increase energy collection while maintaining a moderate communication rate, according to the researchers. In the literature, this approach is referred to as SWIPT with power split [26], and it may be seen in Figure 4. Each user receives two separate signals at the same time.

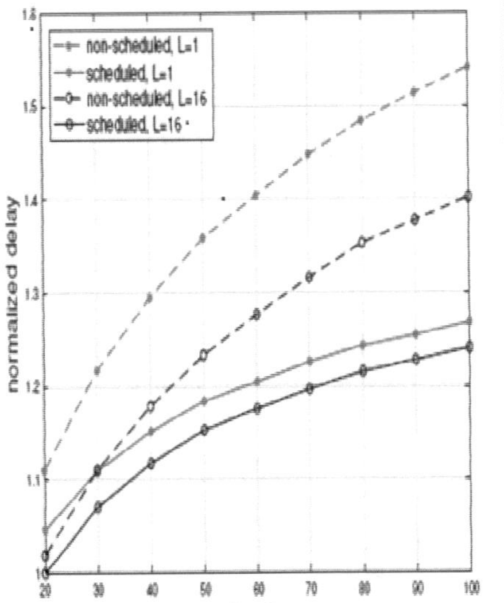

Figure 3. Time required to gather energy for a certain number of users

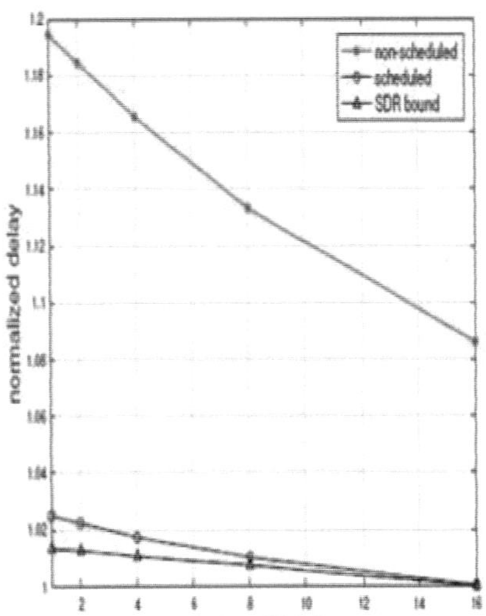

Figure 4. Time required to collect energy in a 60-node network with a variable number of RF links

When the information decoder produces an output, Energy Harvesting gets a portion of it, while the remainder is relayed back to the information decoder for processing (EH). If the nth beam is used by the nth user, the received signal to EH may be created at that beam. Each person has a total of sixteen antennas that they can use. As the wavelength of the carrier frequency used for the transmission increases, so does the distance between the antennas that are used to send and receive it. PL(d) = PL(d0), where PL(d) is the path loss at the distance of distance d in decibels, PL(d0) is the path loss at the reference distance of 1 m, is the path loss exponent, and PL(d0) is the path loss at the reference distance of d0. This is how the path loss at distance d is calculated. With line of sight, transmission in the 28 GHz range has PL(d0) = 61:39 decibels [13], and PL = 1:2 decibels, and z = 1:7 decibels. People think that the energy collecting rate is 0:2. Each person who uses n 2 [1; N] needs en = 106J of power. As shown in Fig. 1, the amount of time it takes to get the energy from different users changes. Study: When there are more than 30 people who use the radio, low-RF-chain plans (L = 1) outperform high-RF-chain plans (L = 16) Value of scheduling energy harvesting grows as more people use it. Most of this may be attributed to better use of the additional latitude provided by larger networks in terms of scheduling. A 60-node network's normalized energy harvesting time for varied RF chain counts is shown in Figure 2. Energy collection may be completed in a fraction of the time it takes without the use of scheduling 8. Due to the scheduling greater compensation for a limited number of RF links, scheduling advantages get more apparent as the number of RF chains reduces.

5. Conclusion

It has been shown that optimizing the time of RF energy harvesting can improve the performance of a wireless communication network that has a base station with a lot of antennas and a lot of users with a lot of antennas. We tried out a new way to divide power. For figuring out how much power each person uses on average, the new method is better than the old method. Finally, for the massive MIMO system, we add SWIPT to it. Assuming that each person can talk at a slow pace, we came up with a way to figure out the best way to split the power. If we want to cut down on the number of RF chains we need in a hybrid beam-forming structure, we can use energy harvesting scheduling to do that. Improved energy harvesting can be done by cutting down on the size of the network and the number of RF chains. Because energy harvesting research has become a new trend in the development of wireless networks, this study shows how to send both information and power at the same time in a massive MIMO system that is both energy efficient and effective way. In this case, we looked at the bare minimum amount of power transfer, data transfer, and the bare minimum amount of energy used for the task. As a result, the non-convex problem that was previously talked about has been turned into a more manageable convex problem. With the help of a simulation, it was found that the suggested approach is more efficient in terms of power use and time use, as well as service quality, while still achieving the maximum energy efficiency. A D2D(Device to Device) underlay big MIMO hybrid network is being looked at, and it makes use of the energy harvesting method. When sending data, the D2D transmitters take in energy from both the PBs and the RF interference that comes from the receivers at the same time. D2D communication is only possible when the transmitters have enough strength to reverse channel inversion in favor of receivers. Using stochastic geometry and different PPPs to indicate the locations of network components, it is feasible to construct a high enough chance that a D2D transmitter harvests enough power.

References

[1] L. Xie, Y. Shi, Y. T. Hou, and A. Lou, "Wireless power transfer and applications to sensor networks," IEEE Wireless Communications, vol. 20, pp. 140–145, August 2013.

[2] J. Xu, L. Liu, and R. Zhang, "Multiuser miso beamforming for simultaneous wireless information and power transfer," IEEE Transactions on Signal Processing, vol. 62, pp. 4798–4810, September 2014.

[3] J. Tang, D. K. C. So, A. Shojaeifard, K. K. Wong, and J. Wen, "Joint antenna selection and spatial switching for energy efficient mimoswipt system," IEEE Transactions on Wireless Communications, vol. 16, pp. 4754–4769, July 2017.

[4] R. Morsi, D. S. Michalopoulos, and R. Schober, "Multiuser scheduling schemes for simultaneous wireless information and power transfer over fading channels," IEEE Transactions on Wireless Communications, vol. 14, pp. 1967–1982, April 2015.

[5] H. Ju and R. Zhang, "Throughput maximization in wireless powered communication networks," IEEE Transactions on Wireless Communications, vol. 13, pp. 418–428, January 2014. [6] L. Liu, R. Zhang, and K. C. Chua, "Multi-antenna wireless powered communication with energy beamforming," IEEE Transactions on Communications, vol. 62, pp. 4349–4361, December 2014.

[6] O. T. Demir and T. E. Tuncer, "Antenna selection and hybrid beamforming for simultaneous wireless information and power transfer in multi-group multicasting systems," IEEE Transactions on Wireless Communications, vol. 15, pp. 6948–6962, October 2016.

[7] Y. Sadi and S. C. Ergen, "Minimum length scheduling with packet trafc demands in wireless networks," IEEE Transactions on Wireless Communications, vol. 13, pp. 3738–3751, July 2014.

[8] S. Han, I. Chih-Lin, Z. Xu, and C. Rowell, "Large-scale antenna systems with hybrid analog and digital beamforming for millimeter wave 5g," IEEE Communications Magazine, vol. 53, pp. 186–194, January 2015.

[9] Z. q. Luo, W. k. Ma, A. M. c. So, Y. Ye, and S. Zhang, "Semi definite relaxation of quadratic optimization problems," IEEE Signal Processing Magazine, vol. 27, pp. 20–34, May 2010.

[10] M. Grant and S. Boyd, "CVX: Matlab software for disciplined convex programming, version 2.1," Mar. 2014.

[11] O. T. Demir and T. E. Tuncer, "Antenna selection and hybrid beam forming for simultaneous wireless information and power transfer in multi-group multicasting systems," IEEE Transactions on Wireless Communications, vol. 15, pp. 6948–6962, Oct 2016.

[12] S. Deng, M. K. Samimi, and T. S. Rappaport, "28 ghz and 73 ghzmillimeter-wave indoor propagation measurements and path loss models," in 2015 IEEE International Conference on Communication Workshop (ICCW), pp. 1244–1250, June 2015.

[13] S. Yang and L. Hanzo, "Fifty Years of MIMO Detection: The Road to Large-Scale MIMOs," IEEE Communications Surveys & Tutorials, 2015, 17(4), pp. 1941-1988.

[14] Khalaf, O.I., Ogudo, K.A. and Singh, M., 2021. A fuzzy-based optimization technique for the energy and spectrum efficiencies trade-off in cognitive radio-enabled 5G network. *Symmetry*, *13*(1), p.47.

[15] Singh, M., Kumar, M. and Malhotra, J., 2018. Energy efficient cognitive body area network (CBAN) using lookup table and energy harvesting. *Journal of Intelligent & Fuzzy Systems*, *35*(2), pp.1253-1265.

[16] M. Hassan, M. Singh and K. Hamid, "Survey on NOMA and Spectrum Sharing Techniques in 5G," 2021 IEEE International Conference on Smart Information Systems and Technologies (SIST), 2021, pp. 1-4, doi: 10.1109/SIST50301.2021.9465962.

[17] Mohamed Hassan, Manwinder Singh & Khalid Hamid (2021) "IMPACT OF POWER AND BANDWIDTH ON THE CAPACITY RATE AND NUMBER OF USERS IN SC-NOMA", *Harbin GongyeDaxueXuebao/Journal of Harbin Institute of Technology*, 53(9), pp. 118–124. Available at: http://hebgydxxb.periodicales.com/index.php/JHIT/article/view/726 (Accessed: 15 February 2022).

[18] M. Hassan, M. Singh and K. Hamid, "Survey on NOMA and Spectrum Sharing Techniques in 5G," 2021 IEEE International Conference on Smart Information Systems and Technologies (SIST), 2021, pp. 1-4, doi: 10.1109/SIST50301.2021.9465962.

[19] Hassan, M., Singh, M., & Hamid, K. (2021). Overview of Cognitive Radio Networks. *Journal of Physics: Conference Series, 1831.*

[20] S. K. Roy, M. Singh, K. K. Sharma, C. Bhargava and B. P. Singh, "Mathematical Modelling of Simple Passive RC Filters Using Floating Admittance Technique," 2020 IEEE International Conference for Innovation in Technology (INOCON), 2020, pp. 1-6, doi: 10.1109/INOCON50539.2020.9298230.

[21] Cisco, "Cisco visual networking index: Global mobile data traffic forecast update, 2014-2019 white paper," Feb 2015.

[22] J. G. Andrews, S. Buzzi, W. Choi, S. V. Hanly, A. Lozano, A. C. K. Soong, and J. C. Zhang, "What Will 5G Be?," IEEE Journal on Selected Areas in Communications, 2014, 32(6), pp. 1065-1082.

[23] X. Jia, P. Deng, L. Yang, and H. Zhu, "Spectrum and Energy Efficiencies for Multiuser Pairs Massive MIMO Systems WithFullDuplex Amplify-and-Forward Relay," IEEE Access, 2015, 3, pp. 1907-1918.

[24] Y. D. Lin and Y. C. Hsu, "Multihop cellular: A new architecture for wireless communications," in Proc. IEEE INFOCOM, 2000, vol. 3, pp. 1273–1282.

[25] X. Zhou, R. Zhang, and C. K. Ho, "Wireless Information and Power Transfer: Architecture Design and Rate-Energy Tradeoff," IEEE Transactions on Communications, 2013, 61(11), pp. 4754-4767.

[26] L. Yuanwei, W. Lifeng, S. A. R. Zaidi, M. Elkashlan, and T. Q. Duong, "Secure D2D Communication in Large-Scale Cognitive Cellular Networks: A Wireless Power Transfer Model," IEEE Transactions on Communications, 2016, 64(1), pp. 329-342.

[27] K. Huang, V. K. N. Lau. "Enabling wireless power transfer in cellular networks: architecture, modeling and deployment," IEEE Transactions on Wireless Communications, 2014, 13(2), pp, 902-912

[28] C. Ma, J. Liu, X. Tian, H. Yu, Y. Cui, and X. Wang, "Interference Exploitation in D2D-Enabled Cellular Networks: A Secrecy Perspective," IEEE Transactions on Communications, 2015, 63(1), pp. 229-242.

[29] M. Haenggi and R. K. Ganti, Interference in Large Wireless Networks. Norwell, MA, USA: Now Publishers, 2009.

[30] S. Shalmashi, E. Bjornson, M. Kountouris, S. Ki Won, and M. Debbah, "Energy efficiency and sum rate when massive MIMO meets device-to-device communication," in Communication Workshop (ICCW), 2015 IEEE International

Conference on, 2015, pp. 627-632.

[31] H. Anqi, W. Lifeng, M. Elkashlan, C. Yue, and W. Kai-Kit, "Spectrum and Energy Efficiency in Massive MIMO Enabled HetNets: A Stochastic Geometry Approach," IEEE Communications Letters, 2015, 19(12), pp. 2294-2297.

[32] Dai, L.; Wang, B.; Ding, Z.; Wang, Z.; Chen, S.; Hanzo, L. A Survey of Non-Orthogonal Multiple Access for 5G. IEEE Commun. Surv. Tutor. 2018, 20, 2294–2323.

[33] Frenzel, L. "Millimeter Waves Will Expand the Wireless Future. [online] Electronic Design", 2017.

[34] S. Deng, M. K. Samimi, and T. S. Rappaport, "28 GHz and 73 GHz millimeter-wave indoor propagation measurements and path loss models," in IEEE International Conference on Communications Workshops (ICCW), pp. 1244–1250, Jun. 2015.

[35] C. U. Bas, et al., "28 GHz microcell measurement campaign for residential environment" IEEE GLOBECOM, Dec. 2017.

[36] S. Sun et al., "Investigation of prediction accuracy, sensitivity, and parameter stability of large-scale propagation path loss models for 5G wireless communications," IEEE Trans. Veh. Technol., vol. 65, no. 5, pp. 2843–2860, May 2016.

[37] M. K. Samimi and T. S. Rappaport, "3-D millimeter-wave statistical channel model for 5G wireless system design," IEEE Trans. Microw. Theory Techn., vol. 64, no. 7, pp. 2207–2225, Jul. 2016.

[38] Hong S, Pan C, Ren H, Wang K, and Nallanathan A 2020 Artificial-noise-aided secure MIMOwireless communications via intelligent reflecting surface **68(12)**

Utilization of a Wireless Network Performing CSMA/CA with Random Backoff Algorithm

Harpreet Singh Bedi
SEEE, Lovely Professional University, Punjab.
harpreet.17377@lpu.co.in
https://orcid.org/0000-0003-1389-4333

Abstract. Optimizing WLAN performance in IEEE 802.11 network with extremely different working environments is a difficult task. We must use a network that can provide the most benefits to its end-users to improve network strength and data transmission. To increase performance in this congested situation, we construct an adaptive backoff algorithm for the IEEE 802.11 DCF that enhances in congested network conditions, system throughput is improved, the collision probability is reduced, and good fairness is maintained. This paper deals with the utilization of networks for different backoff strategies for different network conditions. The results obtained were comparatively much better than the existing methods.

Keywords. CSMA/CA, DCF, EDCF, LAN, RTS, ENERGY, EFFICIENCY.

1. Introduction

WLANs are effectively becoming the most popular and widely utilized networks. The low operational cost and the powerful and effective medium access control (MAC) protocol with resource sharing capabilities are the reasons for its success. WLANs based on IEEE 802.11 has become the most widely used wireless networking standard. Collisions can happen in wireless LANs, just like they can in wired networks when multiple stations communicate over the same transmission medium at the same time. PCF is a highly centralized MAC protocol in which an access point (AP) transmits polling messages to different stations to supply collision-free services. The open system interconnections (OSI) standard model for data transmission contains a sublayer called medium access control in the data link layer (MAC) [2]. It is in full control of transmission medium control design and multiplexing.

1.1 Mac Layer

It is the most important data link layer, and it is in charge of data transmission. It is responsible for the medium and grants it control over data access and transmission from the sender to the receiver. In wireless transmission, this system employs an access control mechanism and provides a full-duplex arrangement. This channel or layer can communicate with each other as well as other network devices in a single or dual-direction manner

1.2. Distributed coordination function(DCF)

It is the most widely used method in IEEE 802.11, which is based on the CSMA protocol. It follows a transmission schedule and uses the first-in-first-out mechanism. If the size is doubled, the backoff algorithm will be examined. All of Mac's features will be tested and performed, depending on the bit

Journal of Physics: Conference Series **2327** (2022) 012060 doi:10.1088/1742-6596/2327/1/012060

frame and CW size.

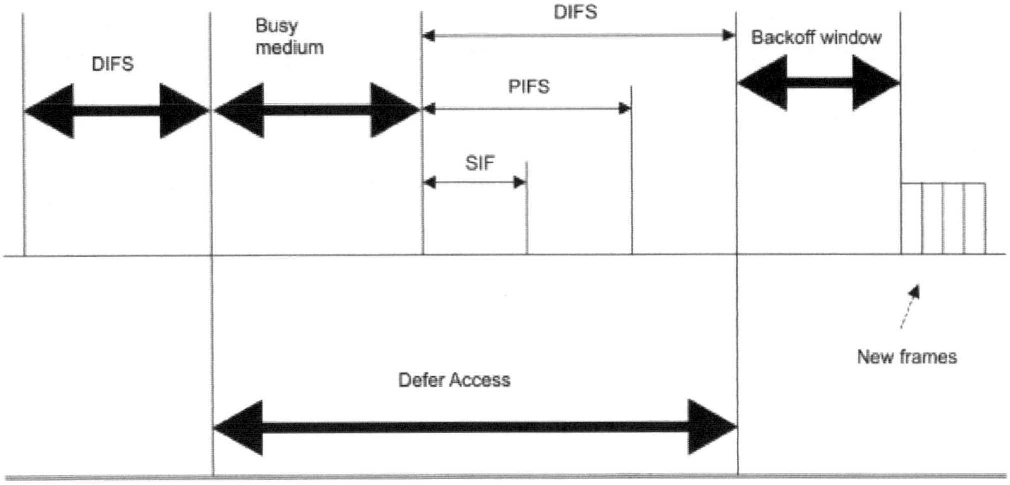

Figure 1. Timing relationship for DCF.

1.3 Enhanced distributed coordination function(EDCF)

With the help of rising DCF, EDCF is utilised to monitor and organise QoS. It aids in separation and data transmission for QSTA, which requires eight stations to make UP adjustments. The system will employ the FIFO method to obtain access to the access category, which will aid in the transmission of data to the stations.

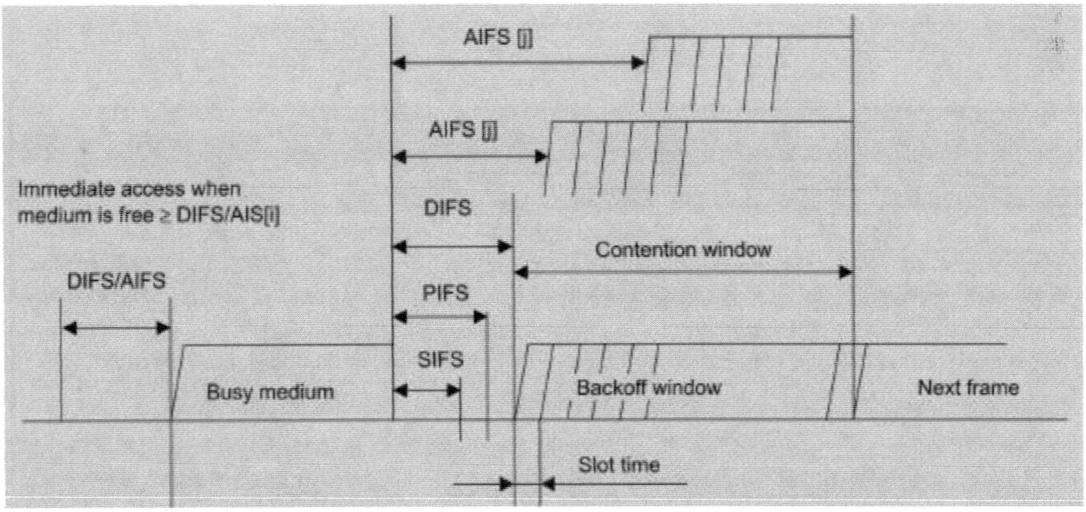

Figure 2. Timing relationship for EDCF.

The "listen before speaking" technique is used in CSMA, and each station must feel the medium for potential transmission [4]. If the medium has been idle for a particular period, the station broadcasts the packet. Otherwise, the station would wait until the medium is free before transmitting. The backoff algorithm is then used [5]. For quite some time, the CSMA/CA protocol has been documented in the literature. Many articles have been written that analyze the protocol's performance issues and suggest ways to improve it [6]. Since it specifies how long stations must pause before transmitting when the medium is considered busy, the backoff algorithm has a significant impact on CSMA/CA efficiency [8]. As a result, if there are considerable delays that do not provide a direct advantage, the channel will be underutilized, wasting bandwidth. If the interval between retransmission trials is short, on the other hand, there will be an increasing amount of collisions, resulting in low channel utilization and bandwidth waste [9]. The research presents a two-stage approach to increase the efficiency of the CSMA/CA protocol's backoff mechanism, and hence the performance of the IEEE802.11 WLAN. This method works similarly to the standard [10].

1.4. Carrier Sense Multiple Access with Enhanced Collision Avoidance(CSMA)

An IEEE 802.11 station senses the channel for a distributed inter-frame space when it joins the CSMA contention for the channel time (DIFS). If the station detects that the channel is idle, it transmits. The transmission will be delayed if the channel is identified as being busy. The initial value for a backoff counter is:

$B \sim U[0, CW - 1],$

where U is the uniformly distributed random distribution and CW denotes the contention window value, which is set to CWmin for the initial transmission attempt. While the channel is sensed busy, the backoff counter countdown is frozen. [3] In the following section, we analysed a model in which three types of slots are differentiated: empty, success, and collision.

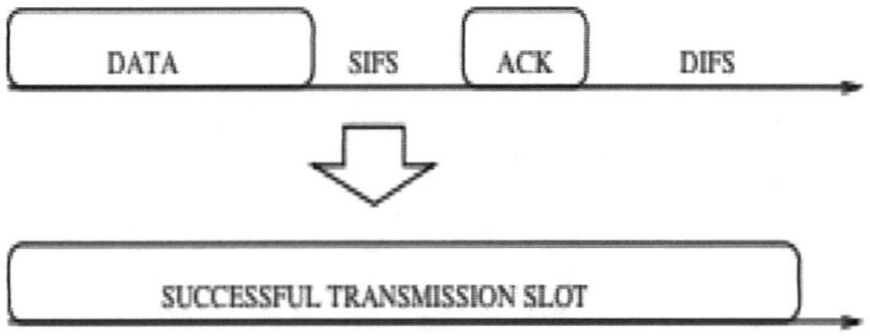

Figure 3. slot level.

We'll suppose that sending two or more packets at the same time always results in a collision and that all of the packets involved are lost. .

Fig. 4 depicts a scenario in which two stations join the contention (indicated by a little arrow) and a third station transmits. The first backoff is chosen at random by the two additional stations.

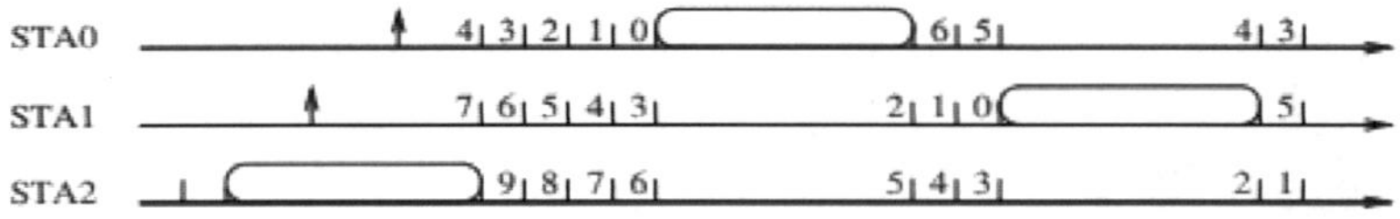

Figure 4. Example of CSMA/CA.

Fig. 5 depicts two stations colliding, then selecting random backoff values and successfully transmitting.

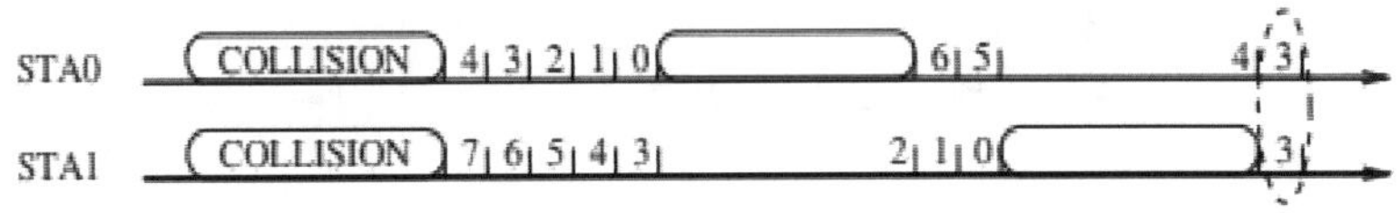

Figure 5. Example of CSMA/CA. A random backoff is used after successful and unsuccessful transmissions

Table 1 Time durations of important parameters used in various IEEE 802.11 protocols

Description	Value			
	IEEE 802.11b	IEEE 802.11a	IEEE 802.11g only	IEEE 802.11g +legacy
Slot Time	20 µs	9 µs	9 µs	20 µs
SIFS	10 µs	16 µs	10 µs	10 µs
DIFS	50 µs	34 µs	28 µs	50 µs
Size of RTS frame	20 Bytes	20 Bytes	20 Bytes	20 Bytes
Size of RTS frame	14 Bytes	14 Bytes	14 Bytes	14 Bytes
Size of RTS frame	14 Bytes	14 Bytes	14 Bytes	14 Bytes
CW_{min} (Slot Times)	31	15	15	15
CW_{max} (Slot Times)	1023	1023	1023	1023

Results and Discussions

The remainder of this article will be dedicated to deciding the proper values for V and CWmin. To evaluate the system's performance, we'll utilise two metrics: efficiency and fairness, whereas efficiency is the percentage of time spent on complete transmissions, which is influenced by the duration. We'll use IEEE 802.11b's default parameter values and a frame length

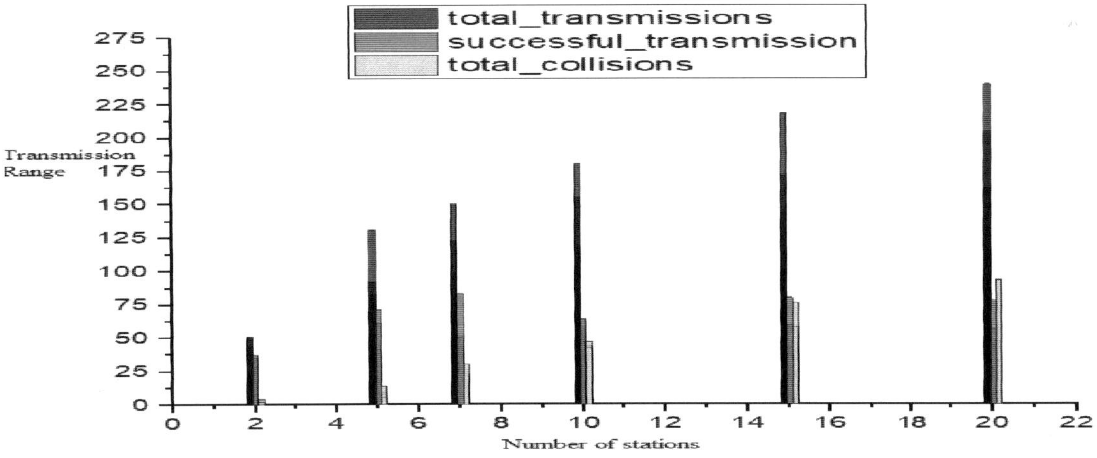

Figure 6. Describes the relationship of stations

For a particular station, the transmission range is varied. If we increase the number of stations, the total transmissions will be increased with the probability of maximum successful transmissions but with the least collisions and vice versa.

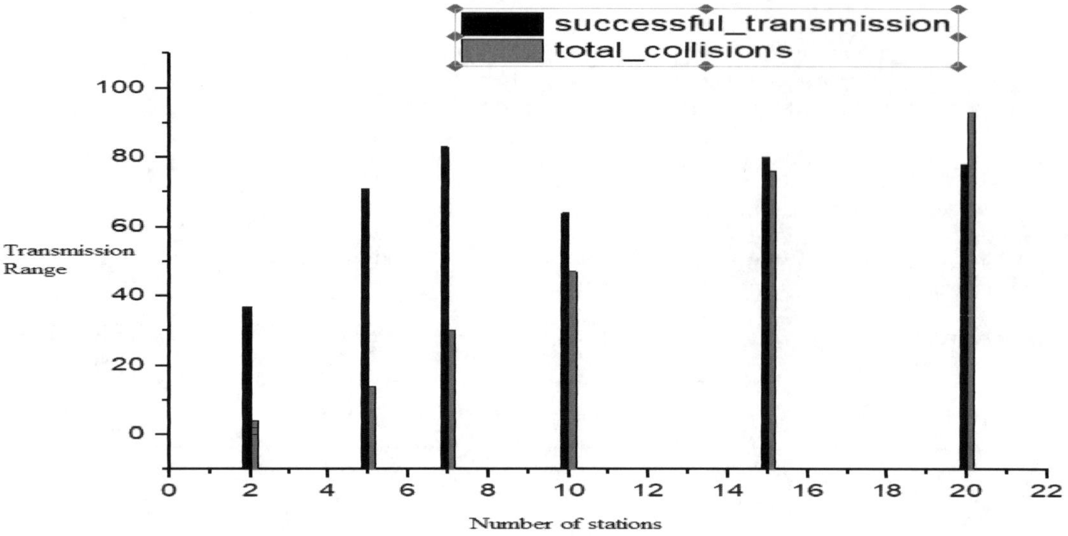

Figure 7. Number of stations with Transmission range for maximum 20 stations

If we have a less number of stations, the probability of successful transmission is more with less number of collisions. As we increase the nodes/stations, several collisions will be increased and will collide with other nodes or stations as per the availability of the channel or medium.

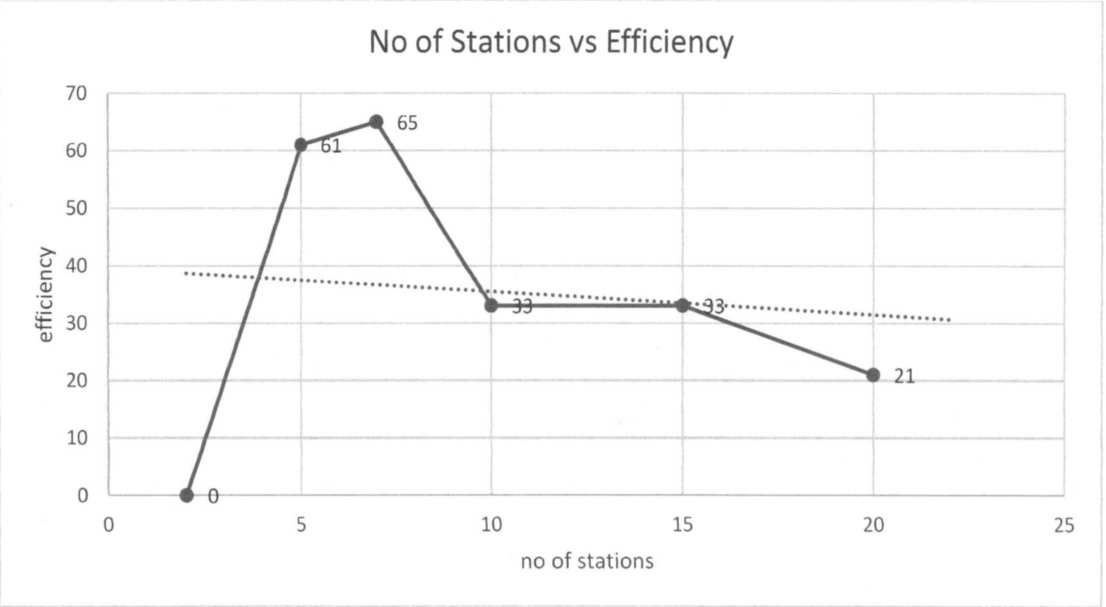

Figure8. No.of stations Vs Efficiency.

Fig 8 tells us that as we increase the number of stations the efficiency of the station is decreased or lowered. The efficiency of the node can be increased by some parameters.

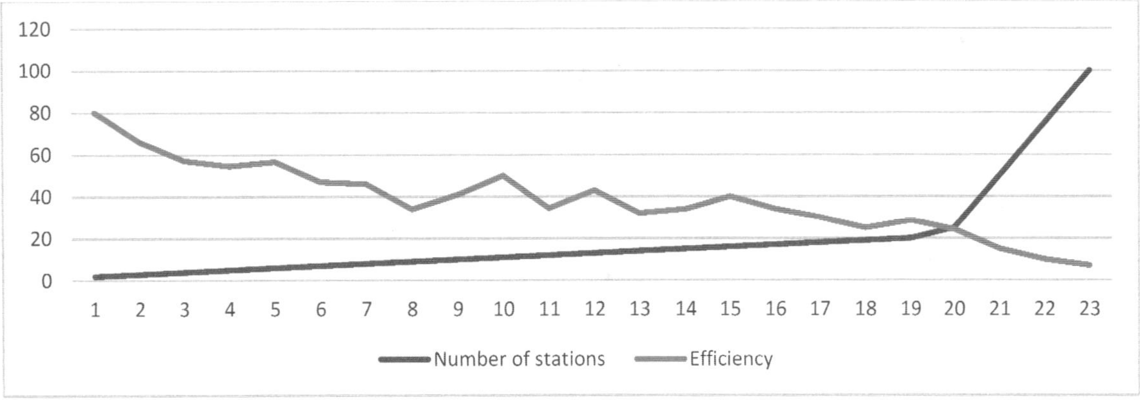

Figure 9 No/ of stations with the utilization of the network

Utilization of the network or node can be increased by prioritizing activities based on the type of service they provide Another common cause of slow Internet connection in businesses is a lack of bandwidth. (1) reducing the average network transmission distance and (2) reducing the time.

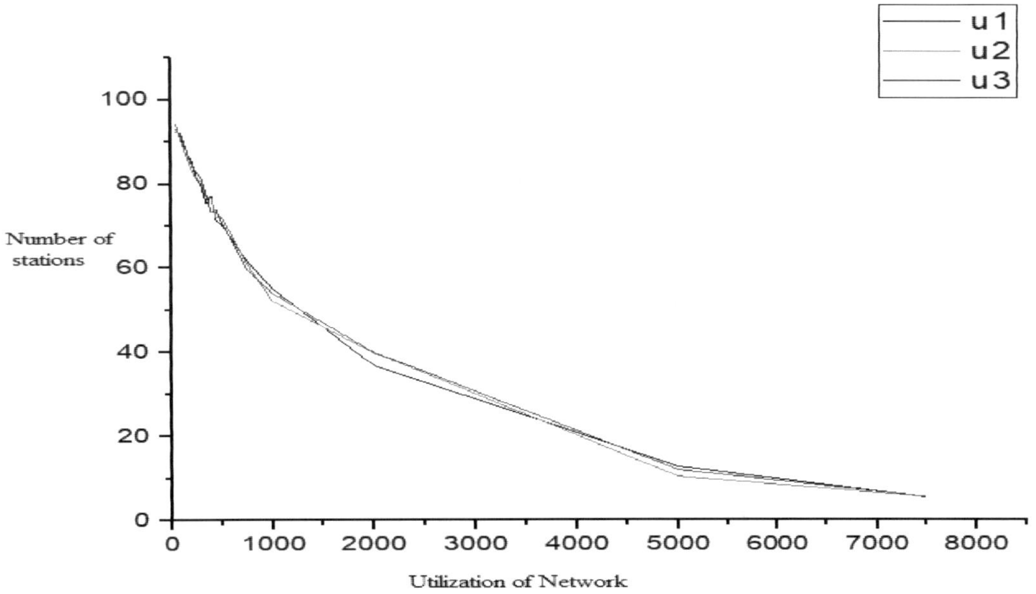

Figure 10. Number of station vs Utilization of the network for different conditions

Fig. 10 shows the relationship between the number of stations with the utilization of the nodes. In this, we have taken different network scenarios with the utilization of the network.

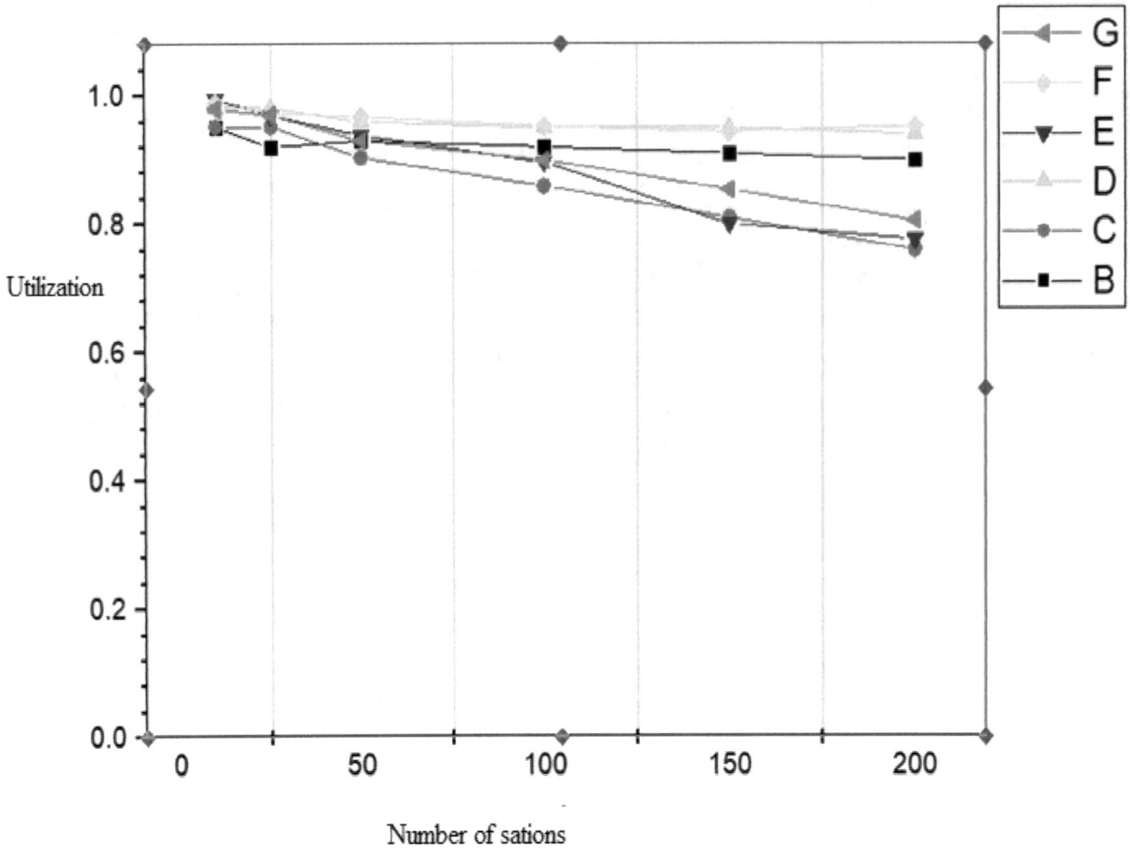

Figure 11. improved utilization of the network with random backoff algorithm

In this Fig.11, we have Computed the utility for the following backoff strategies for 802.11 networks with Additive increase multiplicative decrease along with Multiplicative increase multiplicative decrease with the contention windows size along with the variations in the size of the network.

Conclusion

IEEE 802.11b, 802.11a 802.11g are the most popular standard protocols used in WLANs. Since most of the existing backoff algorithms have been analyzed for IEEE 802.11b protocol to compare with, the performance of the proposed algorithm is also analyzed considering the parameters of the IEEE 802.11b protocol. The performance of a wireless network is highly dependent on several parameters such as the wireless channel model, data rate, packet length, number of nodes, etc. The CFB has been proved to improve the performance of the system at the cost of a delay increase for specific forms of traffic. Based on the network traffic and supported applications, one should enable. short simulations to depict the protocol's behaviour in a highly dynamic network. We used this information to develop an adaptation process that ensures CWmin is within the acceptable range for high performance. We used simulation to verify that the suggested protocol works, and we supplied an example of how the protocol operates.

Acknowledgement

I would like to express my very great appreciation to **Dr. Kamal Kumar Sharma** (Professor, School of Electronics and Electrical Engineering, Lovely Professional University, Phagwara, Punjab, India.) for his valuable and constructive suggestions during the planning and development of this research work. His willingness to give his time so generously has been very much appreciated.

References

[1] W. Pattara-Atikom, P. Krishnamurthy, and S. Banerjee, "Distributed mechanisms for quality of service in wireless LANs", *IEEE Wireless Communications* vol. 10, no. 3, pp. 26-34, 2003.

[2] D. Chen, and P.K. Varshney, "QoS Support in Wireless Sensor Networks: A Survey*", International conference on wireless networks*, vol. vol. 233, 2004, pp. 1-7

[3] K. Langendoen, and G. Halkes, "Energy-efficient medium access control", Embedded systems handbook, vol. 6000, 2005.

[4] T. Wark, P. Corke, P. Sikka, L. Klingbeil, Y. Guo, and C. Crossman, "Transforming agriculture through pervasive wireless sensor networks", IEEE Pervasive Computing vol. 6, no. 2, 2007.

[5] O. Tsigkas, and F-N. Pavlidou, "Providing QoS support at the distributed wireless MAC layer: a comprehensive study", *IEEE Wireless Communications* vol. 15, no. 1, pp. 22-31, 2008.

[6] J.S. Vardakas, and M.D. Logothetis, "End-to-end delay analysis of the IEEE 802.11 e with MMPP input-traffic", *International Symposium on Autonomous Decentralized Systems*, 2009, pp. 1-6

[7] X. Liu, and W. Zeng, "Throughput and delay analysis of the IEEE 802.15. 3 CSMA/CA mechanism considering the suspending events in unsaturated traffic conditions", 6th *International Conference on Mobile Adhoc and Sensor Systems*, 2009, pp. 947-952

[8] N.S. Chauhan, and L. Kaur, "Implementation of QoS of different multimedia applications in WLAN", *International Journal of Computer Applications (0975 – 8887)*, vol. 62, no. 8, 2013.

[9] S. Singh and A. Kumar, "Performance analysis of adaptive clipping technique for reduction of PAPR in alamouti coded MIMO-OFDM systems in *6th International Conference on Advances In Computing & Communications, ICACC 2016,* 6-8 September 2016, Cochin, India., vol. 93, pp. 609-616, 2016.

[10] I. Syed, and B-h. Roh, "Delay analysis of IEEE 802.11 e EDCA with enhanced QoS for delay-sensitive applications", *35th International Performance Computing and Communications Conference* (IPCCC), 2016, pp. 1-4.

[11] H.S. Bedi, K.K. Sharma, and R. Gupta, "A Review Paper on Performance Analysis of IEEE 802.11 e", 1st International Conference on Computing, Communications, and Cyber-Security (IC4S 2019), 2020, pp. 47-56.

[12] Z.G. Khaki, and H.S. Bedi, "Transient correction using EDFA: in-line optical fibre with feedback", International Conference on Computing Sciences, 2012, pp. 233-238 [http://dx.doi.org/10.1109/ICCS.2012.76].

[13] Bedi, Harpreet, Kamal Kumar, and Gupta Raghav. "A review paper on improving the network efficiency of IEEE 802.11 e networks." *Intelligent Circuits and Systems* (2021): 206-211.

[14] Harpreet Bedi, Kamal Kumar Sharma; Analyses of CSMA/CA Protocol Without Using Virtual Channel Sensing in DCF Mode, *Global Emerging Innovation Summit (GEIS-2021) (2021)* 1: 484. https://doi.org/10.2174/9781681089010121010058.

Advancement of Full-Duplex Cognitive Radio Network: A Review

Vikas Srivastava[1, 2], Parulpreet Singh[1], Praveen Srivastava[1, 2]

[1]Lovely Professional University, Phagwara
[2]Pranveer Singh Institute of Technology, Kanpur
orcid id- 0000-0003-2165-5543[1]

Abstract- Communication systems, with various pervasive applications, have become a crucial piece of day to day lives. Wireless systems request an ever-increasing number of phantom assets to help with the regularly expanding quantities of clients. As indicated, the present crunch can be fulfilled by use of cognitive radio networks (CRNs). At a given time, secondary users (SUs) in half-duplex (HD) CRNs can either transmit or detect the signal range. This activity confines the secondary users' throughput because the secondary user cannot transmit during range detection. With advancement in self-interference separation (SIS) technique, full-duplex (FD) CRNconsider synchronous range detection and transmission at a given time on any channel. This FD activity builds throughput and diminishes crashes as compared to HD-CRNs. This paper presents a complete review of FD-CRN correspondences. This paper spread the supporting system models and transmitted and received radio wire structures. This paper describes various SIS approaches in FD-CRNs. Here, authors discuss range detection methods and security requirements for FD-CRNs and review significant advancement in full-duplex medium access control (FD-MAC) conventions as open issues, difficulties, and future research bearings to help FD activity in CRNs.

Keywords: Half duplex, Full Duplex, Cognitive radio Network, Machine Learning, Self-Interference separation

1. Introduction

Two activities exist in cognitive radio: authorized (Primary) band activity and unlicensed (Secondary) band activity. An unlicensed user uses the assigned gap for a licensed user when incidentally unused range. As a result, unlicensed users are aware of the spectrum gap between licensed users and unlicensed users. The unlicensed user will intermittently screen RF (spectrum sensing) and select the best accessible option (spectrum decision). The gap detecting process is utilized with range forecasting. A range-detecting strategy exists over numerous measurements, for example, time, space, recurrence, code, and kind of sign. An intellectual client has 2 significant attributes: (1) cognitive capacity to make sense of a bit of new range. (2) Reconfigurability-A cognitive client gradually modifies their transmission parameters in response to detected condition variety.

Cognitive radio capacities are classified as the accompanying: range detecting (discovering range blank areas), range choice (choosing the best working parameters), range sharing (helping out different clients), and range versatility. Adaptation, detection, thinking, and examination are part of the intellectual cycle[1][2].

Section 2 describe about spectrum sensing. Section 3 gives information about spectrum analysis. Section 4 describe about spectrum sharing. Section 5 gives information about full-duplex cognitive radio network. Section 6 gives multiband spectrum sensing and sliding window approaches for full-duplex cognitive radio networks. Section 7 describes machine learning based CRN for cooperative spectrum sensing. Section 8 describes the conclusion.

Journal of Physics: Conference Series **2327** (2022) 012061 doi:10.1088/1742-6596/2327/1/012061

2. Spectrum Sensing

CU senses the spectrum status and PU activities periodically to present an ideal channel. Spectrum sensing has five types

2.1 Centralized

All Cognitive Users in system sense spectrum exclusively and send their detected data to combination centre.[3]

2.2 Distributed

All the CU, sense data and trade inside themselves to make a choice.

2.3 Non-Cooperative

Non-cooperative spectrum sensing do not need additional resources to detect the spectrum and take a spectrum access decision[3]

2.4 Cooperative

Detection reliability is improved in cooperative spectrum sensing. However, it Increases energy and computational costs, extra transmission costs, end-to-end delay, sensing time as it uses an extra spectrum band to transmit the sensing results between various nodes towards the fusion centre in centralized networks and distributed networks.[4]

2.5 Interference based spectrum sensing

For spectrum sensing, the performance metrics are the probability of detection (P_d) and the probability of false alarm (P_{fa}). P_d should be high for better spectrum sensing, and P_{fa} should be low. The frame structure is a combination of data transmission and sensing time. Our main aim is to improve the throughput of a secondary user by reducing sensing time.

$P(t) = h\,m(t) + n(t)$ H1 (when primary user is present)

$n(t)$ H0 (when primary user is absent)

where n(t) is additive white gaussian noise, m(t) is message signal of PU, and h is channel gain.

There are 3 types of states-

(1) Wait state- If PU is active during sensing phase, the CU has to wait for next frame to sense other channels.
(2) Switch state- During data transmission when PU is detected as active the SU has to shift its transmission to other inactive channels.
(3) Stop state: When the primary user is active during the data transmission period. If all other channels are active Cognitive users has to stop communication.

If sensing state is active and the channel state is idle, it is a false alarm and is denoted as probability of false alarm (P_{fa}). If the sensing state is idle and the channel state is active, it is a missed detection and is denoted as the probability of missed detection ($1\text{-}P_d$). [5][6]

There are different techniques used for spectrum sensing-

Table 1. Different Spectrum sensing Technique.

Type	Test statistics and feature	Advantage	Disadvantage
Energy detection	In a given bandwidth and time interval, the energy of SU is measured and compared with threshold. It determines the	Easy to implement: does not require prior knowledge of the primary signal Low computational complexity	High P_{fa} Very unreliable in low SNR due to shadow effect and multipath fading Cannot differentiate between primary and secondary users

	presence/absence of PU's.[7]		
Feature detection	The received signal's spectrum density function, as well as the mean's periodicity and autocorrelation function[5]	Better detection in low SNR systems Differentiated among various types of primary and transmission systems	The primary signal is associated with a cyclostationary feature. There is no OFDM-based communication. More sensing time is required. Power consumption rises as computation becomes more complex[5].
Matched filter[6]	Received sampled signal necessitates prior knowledge of PU signal characteristics such as operating frequency, available bandwidth, order, modulation type, pulse shaping, PN scheme and frame structure.	Best detection in low SNR Require fewer signal samples to achieve good detection Improved spectrum efficiency	High complex Synchronize error Accessible spectrum structure must be known to SU. For demodulation of each signal type, a dedicated receiver is needed.
Multiband based sensing scheme [8]	By minimizing spectrum handover frequencies the challenges of channel maintenance and network throughput for SU are met.	Reduction of detection time enhances the system performance.	Sub-bands and PU's signal may be correlated
Waveform based sensing scheme[9]	Coherent sensing consists of hopping sequences, patterns of transmitted pilot signal, spreading sequences, preambles, and midambles.	Receiver performs synchronization.	Waveform patterns can completely know PU's presence.

Cyclostationary means signal features such as autocorrelation, mean, periodicity. By analyzing the cyclostationary features of detected SU signal, transmission of PU can be known [5]. PU transmission (H0) stops when cyclic autocorrelation becomes zero. Otherwise, it shows that PUs occupy the channel (H1)[5]. There exist probability of interference in interference aware based spectrum sensing hence new metrics exist for spectrum sensing. The radio identification-based sensing scheme fully knows PU's spectrum characteristics transmission technology. In wideband compressive spectrum sensing, the signal bandwidth is more significant than the coherence bandwidth. Its limitations are high sensing time, computational complexity, and high sampling rate. Another technique for spectrum sensing is statistical covariance-based sensing (here, use eigenvalues), filter-based sensing, fast sensing (reduce delay), learning-based sensing, and measurement-based sensing. Receiver sensitivity is not dependent on sensing performance. A conventional sensing scheme deals with a single frequency band. Multiple frequencies should be used to decrease sensing time.[3]-[6]

3 Spectrum Decision

Here, cognitive users get information about the spectrum hole and then decide by optimizing the system performance metrics (time, frequency, and power). [4][5]

4 Spectrum Sharing

Avoid collisions between licensed and unlicensed users. Spectrum opportunities for PU's activities are obtained as early as possible by reducing sensing interval and sensing time by MAC layer [6]. MAC protocol is used to access the spectrum. Extra feedback channel is needed in link adaptation for receiver information about received signal quality. Accordingly transmitter modifies its parameters for obtaining good quality report from receiver [6].In CR-based distributed networks, the MAC layer chooses efficiently access the spectrum after the spectrum gaps are found more significant than 1 Cognitive user.[6].

The CR system performs system mobility. System mobility is a combination of spectrum monitoring and spectrum prediction. In Spectrum prediction, a channel's future state (idle or active) can be forecast based on historical information about the channel and, at a particular time, the emergence of a PU can be predicted. In spectrum monitoring, cognitive users detect the primary user's appearance during the cognitive user's data transmission by exploring the received signal characteristics at the cognitive user. [4][5]

In spectrum sharing, licensed users allocate spectrum to cognitive users, and cognitive users optimize performance such as power and frequency according to the spectrum owner. It is generally divided into four models based on time duration and cognitive user efficiency.

4.1. Long-term model

A licensed user assigns spectrum to a cognitive user for a set time.

4.2. Dynamic model

A small batch of the spectrum is given to a cognitive user for a short time.

4.3. Spectrum Common Model

All cognitive users have the same right to access the radio spectrum.

4.4. Shared user model

The primary and cognitive users share the spectrum on an opportunistic basis. Here, spectrum exchange is used to avoid a collision. It is classified as:

4.4.1 Cooperative spectrum sharing- Interference of different cognitive users is identified, interference information is transferred with different cognitive users, and spectrum allocation is taken.

4.4.2 Non-cooperative spectrum sharing- The interference on a specific cognitive user is found, and a message is broadcast to all cognitive users. Spectrum sharing inside the cognitive radio network is called "internetwork" and spectrum sharing among two or more networks is called "internetwork".

4.5 Spectrum sharing is classified as

4.5.1 Interweave spectrum access: Cognitive users have to exploit unused frequency band in a particular geographic area. It is defined in the temporal domain and the spatial domain. Their parameters are P_D, P_{FA}, interference efficiency, throughput, data loss and energy efficiency. Limitation of Interweave Spectrum Access: The cognitive user has to wait until the primary user uses the spectrum. The CU will end communication if the primary user needs the spectrum [3].

4.5.2 Underlay Spectrum Access: Both PU and CUuse the spectrum simultaneously at the same location and frequency. Its primary focus is to control power of the cognitive user to mitigate interference. Underlay spectrum access is associated with throughput, cognitive user's transmission power level, and interference temperature at the primary user. If the interference temperature is below threshold, communication is started in the presence of a primary user. Its advantage is that those cognitive users can communicate without waiting for spectrum holes. Its limitation is a low data rate (optimal power allocation) with a power limitation. The SUs will send in the spectrum band while PUs is not idle in the spectrum underlay, but the PU's interference temperature limit governs each SU's transmitter. [10]

4.5.3 Overlay spectrum access: The cognitive and primary users transmit on the same frequency band, and advanced coding strategies the secondary transmission shares the spectrum. It is complex due to high cognitive demand. The overlay parameters are the complexity approach, the maximum transmitted power by a primary and cognitive user, and throughput.[4]-[6]

4.5.4 Hybrid Spectrum Access: Hybrid spectrum combines interweave and underlay, removing limitations of interweave and underlay. It is less complex than the overlay. In the absence of a primary user, cognitive users can access the channel with maximum power and power gets reduced in the presence of PU.

Table 2. Comparison between Interweave, Underlay and overlay technique

Parameter	Interweave	Underlay	Overlay
Coexistence with PU	No	Yes	Yes
Power adaption	Not required	Required	Not required
Required knowledge	Spectrum opportunity	Interference level	Complete understanding of the PU message and codebook

5. Spectrum Mobility

When a licenced user is spotted, a cognitive device vacates its channel and reconstructs a transmission link on a separate channel to ensure smooth transmission. Spectrum mobility is the method that allows this movement from one channel to another with minimal performance deterioration.

6. Full-Duplex Cognitive Radio Network

In half-duplex, the secondary users perform spectrum sensing, or it transmits the data at a given time, limiting the throughput of the secondary user. Challenge of full-duplex wireless communication is self-interference. So self-interference should be eliminated and double the spectral efficiency. There are different types of method by which self-interference is cancelled- passive separation, analog cancellation and digital cancellation. Sometimes for improving spectral efficiency, we can use MIMO and OFDM. There is high isolation between transmit and receiver antenna to suppress self-interference. So isolation is high, then path loss exists, and path loss affects the performance. Generally, phase noise limits analog and digital cancellation. There are different types of challenges in a half-duplex: Presence of hidden terminal, Loss of throughput due to congestion, and Large end to end delay [1][2].
Self-interference may degrade the integrity of the full-duplex mode. These problems can mitigate by the FD MAC layer protocol. For suppression of SI, there are 2 types of protocol used, i.e. amplify and forward (AF), Decode and forward (DF). Parameter for this is outage probability and BER. Self-interference means there is a narrow gap between the transmitter and receiver antenna. So we can use a single antenna to suppress self-interference.

Table 3. Comparison between Half Duplex and Full duplex

Parameters	Half Duplex	Full Duplex
Throughput gain	low	High
Collision avoidance	suffered	mitigate
hidden node problem	suffered	mitigate
congestion with the aid of MAC scheduling (throughput less due to congestion)	high	Low due to FD MAC scheduling
end to end delay	High	Low
Primary user's detection quality	low	high
Performance overcome by SI	avoided	Not avoided
Link reliability	high	low
Buffer size requirement	small	low
Spectrum utilization	low	high
Overall noise capacity	low	high
Energy consumption	low	high
Hardware complexity	low	high

We can replace HD radio with FD because 1- HD-CRN devices will not sense and access the spectrum simultaneously. 2- HD-CR devices use two different orthogonal channels for transmission and reception. [1]

There are different types of methods for self-interference suppression cancellation-

(1) Passive cancellation- It is 3 types- (a) Antenna separation
(b) Antenna cancellation
(c) Directional Passive cancellation

(2) Active cancellation- It is 3 types-

(a) Analog cancellation- it is 2 type-
(i) SI inversion based
(ii) Dynamic adaption against changing environment

(b) Digital cancellation- It is 2 types-
(i) Linear SI cancellation
(ii) Nonlinear SI cancellation

(c) MIMO aided aggregate analog and digital cancellation- It is 3 types
(i) Time-domain cancellation
(ii) Spatial suppression
(iii) Natural isolation

(3)Impairment Imposed on SI cancellation- The main goals are primary user detection and higher SU throughput, proposing an optimal policy for shifting between full-duplex and half-duplex. PU outage probability can be reduced by transmitting and scan mode. So a different type of self-interference technique canbe used. Different parameters are used for this, like secondary user collision probability, primary user outage probability, secondary user throughput, and power control problems.[11][12]

Sometimes, sensing duration is increased, then sensing efficiency improves. There should be a delay is to make the decision. Compared with HD, FD decrease conflicts and increase throughput[13]. To improve the capacity of the next-generation wireless system, we study the combination of Non-orthogonal multiple access and full duplex[14].

7. Multiband Spectrum Sensing and Sliding Window approach for full-duplex Cognitive Radio

Energy-detection and waveform based spectrum sensing are two major research approaches in FD-CRNs. However, these approaches have some problems like-Low robustness against SNR, High complexity, Low detection performance of PU's. To overcome these limitations, we can use multi-band based spectrum sensing and a sliding window sensing-based approach. In a multiband based scheme, our secondary user network throughput increases.

While considering the time, frequency and space dimensions, different approaches for sensing in FD-CRNs have been proposed. So, in this case, many research papers use code and angle dimensions. In frequency, time, and space, secondary users sense new frequency bands, spectrum holes at specific times, and PU's locations. Now author takes code and an angle that senses PU's spreading code sequence and PU's direction. So our main aim is to cancel self-interference as much as possible in FD CR.[15][16]

8. Machine Learning based CRN for improved cooperative spectrum sensing

By machine-learning algorithm, detecting a busy primary user is possible. Cognitive radio users grouped before the cooperative sensing process. Grouping algorithms divide normal and abnormal users into 2 groups. In practice, detection performance is often reduced by multipath fading, shadowing, and hidden primary users. Cooperative spectrum sensing utilizes spectral diversity and integrates cognitive radio from multiple geographic locations. There are different types of technology being used in a cognitive radio network. Many research papers can use fuzzy logic technology; for interference and power management, we can use neural networks; for resource allocation, we can use game theory. There are different challenges, such as interference issues due to CR's coexistence with primary users and seamless communication. Many research paper have to optimize different parameters by particle sworn optimization (PSO). So many research papers can use multiband cooperative sensing. In a fuzzy logic-based detection method, scan time will be minimized in low SNR. Throughput will be small if sensing time of CR is longer which results in less time for transmission. But if transmission time is longer which will cause interference to the primary user [17]. In FD communications, transmit and sense on same frequency band that develop awareness capability of spectrum. So Yirun Zhang et al suggest a machine learning-based cooperative spectrum sensing framework in FD-CRN, robust with accuracy against interference[18]. "This paper proposes an artificial neural network-based hybrid spectrum sensing technique, which considers sensing a binary classification problem to detect whether the primary user is idle or busy. The proposed scheme utilizes energy detection and likelihood ratio test statistic as features to train the neural network"[19].

9. Conclusion

Synchronous detection and transmission of the authorized groups with a full-duplex cognitive radio network (FD-CRNs) can upgrade throughput. Impedance to primary users (PUs) brought about by the restricted detecting time in regular half-duplex (HD) CRNs will be limited by persistent detecting in FD-CRNs. Self-impedance (SI) during full-duplex (FD) activity in CRNs has been limited through dynamic, aloof, or crossbreed self-interference suppression (SIS) approaches. This paper has given a forward-thinking study on research in FD-CRNs. This article has grouped FD-CRN models as indicated by the essential CR activity into underlay, overlay, and interweave FD-CRNs. This paper secured the FD-CRN radio perspectives, including the various reception apparatus plans, receiving wire pairings in transmission modes. Range detecting in FD-CRNs has additionally been reviewed into vitality identification based, waveform-based and cyclostationary approaches. The author investigated overall efforts to modify existing CRN MAC conventions to help the FD activity. This paper has additionally given insights regarding work on security and protection in FD-CRNs. This paper had completed review with frameworks for open issues, difficulties and future research directions.

References

[1] L. Song, R. Wichman, Y. Li, and Z. Han, "Full-Duplex Communications and Networks," *Full-Duplex Commun. Networks*, 2017, doi: 10.1017/9781316662106.

[2] "Amjad Muhammad, Fayaz Akhtar, Mubashir Husain Rehmani, Martin Reisslein, Full-duplex communication in cognitive radio networks: A survey, IEEE Commun. Surv. Tutor. **19** (4) (2017) 2158–2191

[3] B. Wang and K. J. R. Liu, "Advances in cognitive radio networks: A survey," *IEEE J. Sel. Top. Signal Process.*, vol. **5**, no. 1, pp. 5–23, Feb. 2011, doi: 10.1109/JSTSP.2010.2093210.

[4] D.M. Alias, Cognitive radio networks: A survey, in: Proceedings of Wireless Communications, Signal Processing and Networking (WISPNET), International Conference, IEEE, Chennai, India, 2016, pp. 1981–1986

[5] Y.C. Liang, K.C. Chen, G.Y. Li, P. Mahonen, Cognitive radio networking and communications: An overview, IEEE Trans. Veh. Technol. **60** (7) (2011) 386–407 -

[6] T. Jiang, Z. Wang, and Y. Cao, "Cognitive radio networks: Efficient resource allocation in cooperative sensing, cellular communications, high-speed vehicles, and smart grid," *Cogn. Radio Networks Effic. Resour. Alloc. Coop. Sensing, Cell. Commun. High-Speed Veh. Smart Grid*, pp. 1–127, Jan. 2015, doi: 10.1201/B18302/COGNITIVE-RADIO-NETWORKS-TAO-JIANG-ZHIQIANG-WANG-YANG-CAO.

[7] "Atapattu Saman, Chintha Tellambura, Hai Jiang, Energy Detection for Spectrum Sensing in Cognitive Radio, Springer, New York, NY, USA, 2014

[8] Z. Quan, S. Cui, A. H. Sayed, H. Vincent Poor, Z. Quan, and A. H. Sayed, "Optimal Multiband Joint Detection for Spectrum Sensing in Cognitive Radio Networks," *IEEE Trans. SIGNAL Process.*, vol. **57**, no. 3, 2009, doi: 10.1109/TSP.2008.2008540.

[9] Amjad Muhammad, Mubashir Husain Rehmani, Shiwen Mao, Wireless multimedia cognitive radio networks: A comprehensive survey, IEEE Commun. Surv. Tutor. **20** (2) (2018) 1056–1103

[10] Zhang Ning, Hao Liang, Nan Cheng, Yujie Tang, Jon W. Mark, Xuemin Sherman Shen, Dynamic spectrum access in multi-channel cognitive radio networks, IEEE J. Sel. Areas Commun. **32** (11) (2014) 2053–2064

[11] Y. Liao, L. Song, Z. Han, Listen and Talk: Full-Duplex Cognitive Radio Networks, Springer, 2016

[12] T. Wang, Y. Liao, B. Zhang, L. Song, Joint spectrum access and power allocation in full-duplex cognitive cellular networks, in: Proceedings of IEEE International Conference on Communications (ICC), London, UK, 2015

[13] F. Sun *et al.*, "Research on Key Technology of Full Duplex Cognitive Radio Network," *J. Phys. Conf. Ser.*, vol. **1920**, no. 1, p. 012035, May 2021, doi: 10.1088/1742-6596/1920/1/012035.

[14] M. Mohammadi *et al.*, "Full-Duplex Non-Orthogonal Multiple Access for Next Generation Wireless Systems," *IEEE Commun. Mag.*, vol. **57**, no. 5, pp. 110–116, May 2019, doi: 10.1109/MCOM.2019.1800578.

[15] Kim Dongkyu, Haesoon Lee, Daesik Hong, A survey of in-band full-duplex transmission: From the perspective of PHY and MAC layers, IEEE Commun. Surv. Tutor. **17 (4)** (2015) 2017–2046

[16] A. Youness, N. Kaabouch, A comprehensive survey on spectrum sensing in cognitive radio networks: Recent advances, new challenges, and future research directions, Sensors **19 (1)** (2019) 126

[17] Khalfi, B.; Zaid, A.; Hamdaoui, B. When Machine Learning Meets Compressive Sampling for Wideband Spectrum Sensing. In Proceedings of the International Wireless Communications and Mobile Computing Conference, Valencia, Spain, **26–30** June 2017; pp. 1120–1125

[18] Y. Zhang, Q. Wu, and M. R. Shikh-Bahaei, "On Ensemble Learning-Based Secure Fusion Strategy for Robust Cooperative Sensing in Full-Duplex Cognitive Radio Networks," *IEEE Trans. Commun.*, vol. **68**, no. 10, pp. 6086–6100, Oct. 2020.

[19] D. K. Patel, M. López-Benítez, B. Soni, and Á. F. García-Fernández, "Artificial neural network design for improved spectrum sensing in cognitive radio," *Wirel. Networks 2020 268*, vol. **26**, no. 8, pp. 6155–6174, Jul. 2020, doi: 10.1007/S11276-020-02423-Y.

Journal of Physics: Conference Series **2327** (2022) 012062 doi:10.1088/1742-6596/2327/1/012062

Indoor Navigation System for Public Evacuation in Emergency Situation

Manoj Wakchaure[1], Mubin Tamboli[2],Shrinivas Sonkar[3]

[1−3]Amrutvahini College of Engineering, Sanagmner, Maharashtra, India

E-mail: [1]manoj13apr@gmail.com

Abstract. Today Technology has achieved wide scope of range. Nowadays we can see the different type of emergency application but most of them for outdoor navigation like GPS etc. AR in indoor navigation has become a good technology to show the right direction efficiently. In this paper, we are going to develop one application that focuses on indoor navigation using Augmented Reality technology. Using this application user will scan a QR code, then the system will recommend the shortest and efficient pathway. Generally, our system is useful in multi-floor buildings like malls, institutes, etc. Our pathfinder application will use a sensor inside the smartphone for navigation which will direct users to the emergency exit path.

1. Introduction

In 2009 according to the USA (emergency management agency), The fire that started inside a corporate building cause almost 1500 casualties and many life losses due to that emergency. The first anniversary of the Japanese Fukushima Daiichi nuclear power plant accident also caused injury to many people [1].

If the fire spread, all over the building it releases harmful gases like the burning of plastic etc. Inhalation of these gases may cause death or another side effect will show up. Civilian may get injured due to wall collapsing, the fire may spread all over the floor so they should escape before the fire burn them, or even due to a crowd of people they crushed each other to escape themselves first. Here is an important role is about time because in this kind of situation evacuation must be done as early as possible to save the life. We focus on developing an android app for the user so that the user can save his life during any unwanted situation that happens inside the building such as a fire, an earthquake [2].

This application requires a navigation sensor called an accelerometer. This sensor is inbuilt into the smartphone by using it can calculate user position [3]. This android pathfinder app is helpful to the user who is unknown about the emergency exit of building or even user can use the same application for his destination inside building such as a shop in the mall etc. Although there are already some current localization structures, which include Wi-Fi or Radio Frequency Identification (RFID), that also can assist to discover and music customers. Those structures are pretty expensive. Such structures require putting in a luxurious infrastructure of loads or lots of Wi-Fi or RFID devices. They additionally do now no longer offer Augmented Reality (AR) assisted steering to customers, as ours does. The proposed system makes use of already current smartphones, in conjunction with cloud servers, to localize customer's internal homes to offer the essential data and the most desirable evacuation course to observe throughout emergency situations.

Further, paper is organized as follows: Section 2 describes Related Work, Section 3 focused on Proposed System, section 4 shows Evacuation Path Recommendation followed by mathematical model in section 5. Then section 6 highlights on Application and then results in section 7. Finally paper is concluded with Conclusion.

2. Related Work

RFID labels are investigated [4] as a method for non Global Positioning System (GPS) limitations. RFID labels, stationary positioned in the structure in advance with exact area information will be utilized as navigation routing points. The prerequisite to introducing alike, a foundation inside surroundings is a significant restriction of this methodology.

Now in smartphones, one of the favored techniques is the Received Signal Strength of Wi-Fi network signals [5-7]. Strategies utilized are multilateration (MLAT) and fingerprinting. MLAT needs at any rate three passageways with a piece of exact information on their areas to locate a client's area through Wi-Fi receiver signal strength (RSS) estimations. The fingerprint method requires the client to plan the Wi-Fi network signal engendering attributes of an environment heretofore, making a selective warmth map that might be counselled to look at Wi-Fi RSS readings got by the client. A proposed methodology requires whole earlier information on climate so its application is less summarized.

Pedometer Dead Reckoning (PDR) methods give broader arrangement by not needing any earlier change or information on the navigational environment. Given a known beginning area, (for example, that given by GPS at the section highlight a structure), clients can probably decide their navigation route through stride recognition and heading assessment. Such methods are now restricted to the cell phone sensor arrangement and the cell phone sensor quality. The most precise outcomes need outside cell phone sensor situation from the client's point of view [5] [8-9] for example giving 0.7 m - 0.95 m exactness's in 9863 m2 of three-storied building space [9]. Evacuation through Microsoft HoloLens or head-mounted display (HMD) is a required device that is costly and every user can't find in an emergency situation [10-12]. Utilizing just the versatile advanced cell phone gadget offers a more summed up arrangement, however, experiences restricted sensor getting to, helpless micro-electro-mechanical system (MEMs) cell phone sensor quality, and looser bonding between the client's development and detecting ability. CompAcc [13] can likewise limit a versatile client by utilizing a guide coordinating with strategy alongside GPS, an accelerometer sensor with the compass, however just in open air zones. This current creator's strategy can't be utilized to confine a client inside a structure at first since it requires a GPS association with at first restrict the client. Regardless of whether the client was at first restricted inside the structure, this current creator's strategy couldn't be utilized, on the grounds that it needs the intermittent utilization of GPS to address for limiting blunder errors. Step determination is done statically previously, adding to a lot of mistakes, contingent upon the changeability of a client's walk. Dynamic step determination has been investigated; however, cell phone sensors are regularly placed on the leg to catch leg swing. Work to make up for this mistake utilize the Kalman filter [14] [4] Weinberg articulation [15] and zero velocity updates (ZUPT) [8]. Hence, these three procedures limit inertial drift and they can anticipate activities dependent on earlier occasion information. Specifically, the Kalman filter has given wide use, and the ZUPT has given amazing corrective accuracy.

3. Proposed System

Proposed system as shown in figure 1. The system consist of five modules described as below.

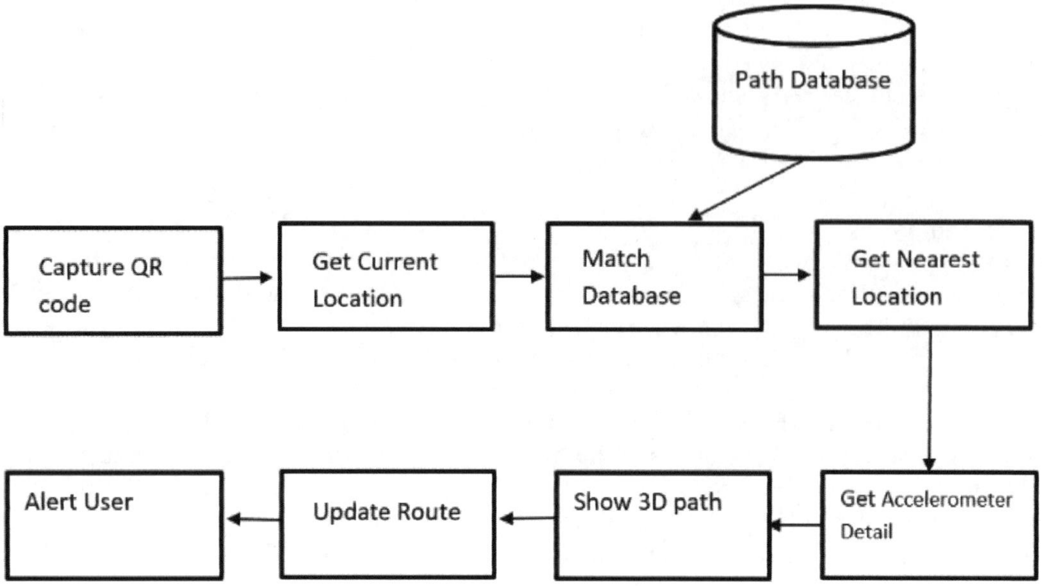

Figure 1: System Architecture

3.1. Modules

3.1.1. Map Building Firstly, the system will be designed as per Mall or as per any other location. We are providing a predefined path to our application.

3.2. Get Current Location

The user accesses the application and scans the quick response (QR) code. Get the current location of the user based on QR code and show the next location.

3.2.1. Route Finding Get Accelerometer Values then user's smartphone camera will open and it will identify the path and using the Dijkstra algorithm the route is identified.

3.2.2. 3D Model Update location details as per the input and show the 3D model view. Generally, we are going to use Arrow as a 3D model which shows the path.

3.2.3. Navigate User The application shows the alert to the user using navigation arrows as per movement in the path. If any turn is there in the path it will show accordingly. If the user takes the wrong path it will alert the user.

4. Evacuation Path Recommendation

Our Pathfinder application will recommend the best emergency or destination path for the user. It will search for a short path [16] from the user's current location to his final destination. Consider a scenario in which any unwanted situation happens at a building like a mall, etc. If the user new to that location he just has to scan the QR code nearby wall [19]. After scanning the QR code user's current location is identified then it sends information to the server after that the server calculate possible exit time for the location. The mathematical equation for the

Journal of Physics: Conference Series **2327** (2022) 012062 doi:10.1088/1742-6596/2327/1/012062

'exit time' estimation is shown in equation 1, then Application will provide options for the user whether he wants to take an emergency exit or even another destination in a mall-like shop will be shown up. The user has to decide its path then accordingly system will show a path with an Arrow sign which helps to reach the destination. One important thing is we have to give the shortest path to the user.

$$ExitTime = Walking_{D}istance/Walking_{S}peed \quad (1)$$

The AR Pathfinder application assists users to stay away from a congested area in a multi-floor building. Consider whenever there is a greater number of individuals around an exit emergency door, trying to go away as soon as possible, most of them could not able to do make themselves safe due to time. Even Pathfinder application initially recommended one exit path but unfortunately, the crowd is also following the same exit so there maybe they will require more time to evacuate so if, they have more than one emergency door our pathfinder will redirect to a new path. When it finds a crowded area at the path which was selected before. The pathfinder will again find the new shortest path to that person, after that, it will impart this data to other users through the application server. Hence, we have improved the earlier equation to oblige the present condition. The application will measure the delay time of peoples trying to go away. Now, this currently measured new exit time uses the initial old exit time from the starting point to the Group Point (the position at which the user's speed becomes slow) in addition to the delay time, as the equation 2.

$$ExitTime = Walking_{D}istance(Start, Group_{P}oint)/WalkingSpeed + DelayTime \quad (2)$$

The application calculates a new delay time for the user i.e. measured by the distance covered from Group point to the suggested emergency exit door of that building and detection in the drop in speed. In particular, to determine delay time we have to divide the "exiting" distance from the Group Point (the user's location) by the slowed speed as in equation 3.

$$DelayTime = Walking_{D}istance(Group_{P}oint, Exit)/Reduced_{W}alking_{S}peed \quad (3)$$

4.1. Augmented Reality
When AR and virtual reality (VR) combined are known as Mixed Reality [17]. We have use AR on an application that helps for the path-based tags to the destination [18]. The 3D path is shown to the user while he is going through his recommended path. We have to use the AR core library for AR technology. The OpenGL library is one of the important parts for rendering the 3D model.

5. Mathematical Model
System Description:

$S = (I, O, F)$
Where,
$S : System.$

$I = SI, LM$ are set of Inputs
Where,
1. SI: Sensor Inputs
2. LM: Location Map

$F = F1, F2, F3, F4$ are set of Function
Where,
$F1$: Map Building
$F2$: Get Current Location
$F3$: Get Accelerometer Details
$F4$: Route Finding
$O = O1$ are set of Output
Where,
$O1$: Navigate User
Success Conditions:
Proper inputs will provide proper output.
Failure Conditions:
No database, internet connection.

6. Application
Indoor navigation using AR technology applications for smart phone devices are being more important and needed for people who want to find inside building destinations or emergency evacuation such as Shopping Malls, Theatres, Hospitals, car parking and even we can use this in Corporate buildings or Educational Institute, Museums. All such Multi floor building are real examples where we can use Indoor navigation application.

7. Results
The developed application pathfinder was tested in the building of Amrutvahini college of engineering. 35 nodes and 39 links were designed for the demo test. We did this experiment on 3 users out of which 2 users had our application and 1 user is not having an application. All 3 users are unknown about the place. A fire was organized on the building for the experimental study, since the real fire could not be started in the building it's just assumed that between node 24th to 25th the fire is initiated. In the organized scenario, the fire first starts in between 24th and 35th nodes) and it spreads towards the 17th nodes as shown in figure 2. The application was tested on three user types. The personal features of these people is provided in below Table 1.

Table 1: The personal features of individuals.

User Detail	User-A	User-B	User-C
Age	19	34	40
Gender	Male	Female	Male
Familiarity with Building	No	No	No
Pathfinder application installed	Yes	Yes	No
Scan QR code	Yes	Yes	No
Fire Protection Outfits	No	No	No

The emergency evacuation path of User-A is shown in figure 2a when the fire initiated, User-A is near the 30th node then he scans the nearby QR code and our application shows the shortest escape Path for user-1 and length is 45 m similarly in 2b User-B is at the 25th node, and the fire was also initiated in-between node 24th to 25th. So the user had our application she scans the QR code and application shows an escape path which was 38m long. Now, in figure 2c User-C is

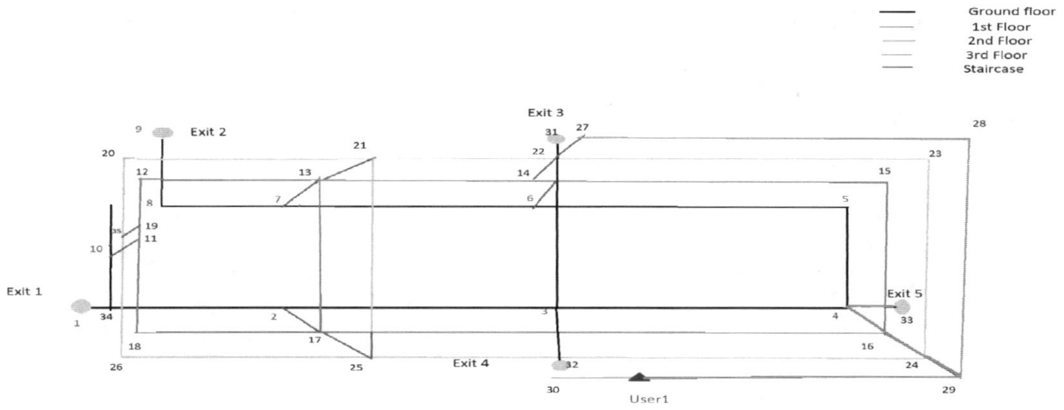

(a) Path of User-A

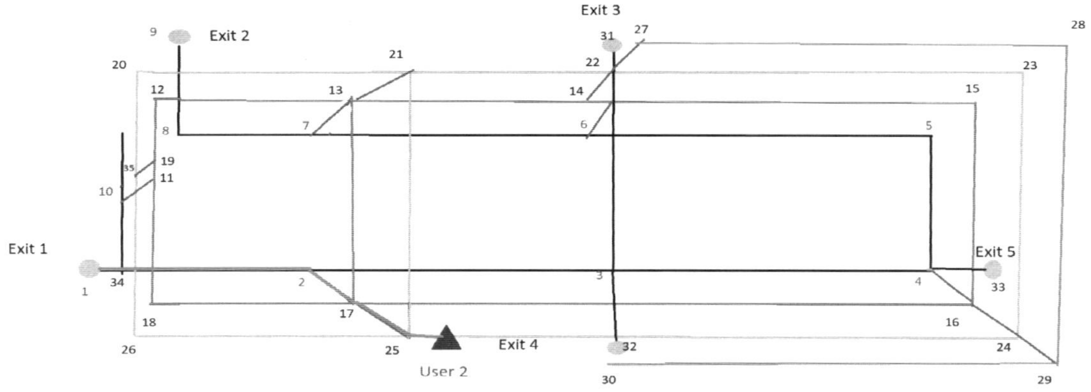

(b) Path of User-B

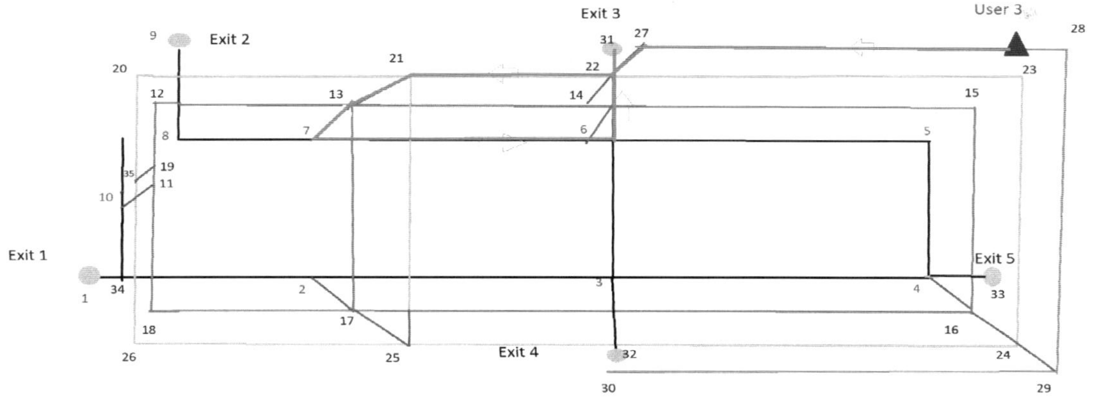

(c) Path of User-C

Figure 2: The Separate Evacuation path for each user

Journal of Physics: Conference Series **2327** (2022) 012062 doi:10.1088/1742-6596/2327/1/012062

(a) Image 1

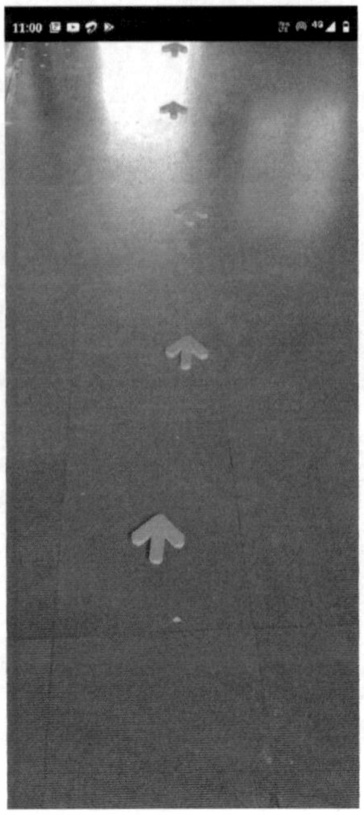

(b) Image 2

Figure 3: Application Images

at nearby node 28th who had not our application he just run randomly through corridors User-C took the 82m longest path and also the time required to escape was more which may cause into a risk.

Some instance screenshots that were taken while the user was escaping through the mobile application during the evacuation process are shown in figure 3.

The following table 2. Shows the Evacuation detail for all three users. As we can see from

Table 2: An Escape path detail of users

User Detail	User-A	User-B	User-C
Time required to escape	2:00 Mins	1:30 Mins	5:00 Mins
Path length for escape	45 meters	38 meters	82 meters
Use Pathfinder Application	Yes	Yes	No

table 2. Those users who use this application have to evacuated earlier than those who have not used our application. User-A and User-B use our application so within 2 min our application evacuated them to exit. User-C gets panic while finding the path and he could not find the accurate path he took almost 5 min to evacuate out of the building and the max time is almost risky.

8. Conclusion

This paper has presented, a unique application that makes use of pedometry, indoor navigation, and augmented reality, to propose an excellent exit pathway with the shortest path and minimum time to its users wanting to leave a multi-storied building in an emergency situation. Hence, with an assessment of this cell phone application, we have proven how this application leverages the sensors available on a cell phone, in conjunction with customized day by day walking step length assessment and in emergency information, to help a well-timed evacuation operation. These are practical pedometry set of rules with customized step estimation that provides excessive positioning with great accuracy.

References

[1] Tsai M. K. and Yau N. J. 2013 Enhancing usability of augmented-reality-based mobile escape guidelines for radioactive accidents *Journal of environmental radioactivity* **118** 15-20

[2] Ortakci Y., Atila U., Demiral E., Ozacar K. and Karas I.R. 2017 Comparison of a Smart Mobile Application-Supported Evacuation System and Traditional Evacuation Systems *IAG Workshop on BIM and GIS Integration*

[3] Ahn J. and Han R. 2011 Rescueme: An indoor mobile augmented-reality evacuation system by personalized pedometry *IEEE Asia-Pacific Services Computing Conference* 70-77

[4] Guerrieri J. R., Francis M. H., Wilson P.F., Kos T., Miller L.E., Bryner N.P., Stroup D.W. and Klein-Berndt, L. 2006 RFID-assisted indoor localization and communication for first responders *First European conference on antennas and propagation* 1-6

[5] Fuchs C, Aschenbruck N, Martini P and Wieneke M 2011 Indoor tracking for mission critical scenarios: A survey *Pervasive and Mobile Computing* **7**(1) 1-5

[6] Miluzzo E., Lane N.D., Fodor, K., Peterson R., Lu H., Musolesi M., Eisenman S.B., Zheng X. and Campbell A.T. 2008, Sensing meets mobile social networks: the design, implementation and evaluation of the cenceme application *6th ACM conference on Embedded network sensor systems* 337-350

[7] Miluzzo E., Cornelius C.T., Ramaswamy A., Choudhury T., Liu Z. and Campbell 2010 Darwin phones: the evolution of sensing and inference on mobile phones *Proceedings of the 8th international conference on Mobile systems, applications, and services* 5-20

[8] Beauregard S. 2007 Omnidirectional pedestrian navigation for first responders *4th Workshop on Positioning, Navigation and Communication* 33-36

[9] Woodman O. and Harle R. 2008 Pedestrian localisation for indoor environments *10th international conference on Ubiquitous computing* 114-123

[10] Stigall J., Bodempudi S.T., Sharma S., Scribner D., Grynovicki J. and Grazaitis P. 2018 Building evacuation using microsoft HoloLens *27th International Conference on Software Engineering and Data Engineering* 8-10

[11] Mitsuhara H., Iguchi K. and Shishibori M. 2017 Using Digital Game, Augmented Reality, and Head Mounted Displays for Immediate-Action Commander Training *International Journal of Emerging Technologies in Learning* **12** (2)

[12] Mitsuhara H., Shishibori M., Kawai J. and Iguchi K. 2016 Game-based evacuation drills using simple augmented reality *IEEE 16th International Conference on Advanced Learning Technologies (ICALT)* 133-137

[13] Constandache I, Choudhury RR and Rhee I. 2010 Compacc: Using mobile phone compasses and accelerometers for localization *IEEE INFOCOM* 1-9

[14] Seco F., Jimenez A.R., Prieto C., Roa J. and Koutsou K. 2009 A survey of mathematical methods for indoor localization *IEEE International Symposium on Intelligent Signal Processing* 9-14

[15] Ladstaetter S., Luley P., Almer A. and Paletta L. 2010 Multisensor data fusion for high accuracy positioning on mobile phones *12th international conference on Human computer interaction with mobile devices and services* 395-396

[16] Dijkstra E. W 1959 A note on two problems in connexion with graphs *Numerische mathematik* **1**(1) 269-71.

[17] Li X., Yi W., Chi H.L., Wang X. and Chan A.P. 2018 A critical review of virtual and augmented reality (VR/AR) applications in construction safety *Automation in Construction* **86** 150-162

[18] Lovreglio R. and Kinateder M. 2020 Augmented reality for pedestrian evacuation research: promises and limitations *Safety science* **128** 104750

[19] Chaudhari S. 2022 QR code: Scan QR code with android phones *beaconstac*

To evaluate the performance of machine learning algorithms in predicting student dropout on MOOC platforms

Gaurav Kumar[1], Amar Singh[2] and Ashok Sharma[3]

[1]Research Scholar, Lovely Professional University, Phagwara, India
[2]Professor, Lovely Professional University, Phagwara, India

[3]Assistant Professor, Department of Computer Science and IT, Bhaderwah Campus, University of Jammu, Jammu, India

Corresponding Author's Email : gauravk6in@yahoo.com

Abstract. Online learning using Massive Open Online courses(MOOCs) has gained a lot of hype in recent years due to its great potential in having the widest reach in delivering the state-of-the-art resources to the unlimited number of online learners without limiting itself to any geographical boundary. Along with gaining popularity, MOOCs have been facing challenges like high attrition or dropout rate since its birth. The main motivating factor behind the study is to fill the gap which has been there because of very limited literature available there to find the real cause behind these challenges. The current study tries to find the solution of the said challenges by finding the significant contributing factors which highly affect the target variable in the study which is number of certified students in this case. The dataset used in this paper is publicly available in dataverse repository of Harvard university. The dataset is a compilation of student clickstream log data consisting of 641138 instances of enrolled students in various MOOC courses of Harvard and MIT. The study evaluates machine learning models like logistic regression, decision tree, random forest, K-Nearest Neighbor to determine their efficiency in predicting the student dropout. The results of this study can be used to create a framework for recommending necessary actions to the at-risk students to reduce the dropout rate.

Keywords: MOOCs, dropout, attrition rate, learning analytics

1. Introduction

Online learning has gained a lot of popularity and acceptability because of its vast scalability and accessibility of state-of-the-art resources to the learners which wouldn't have been possible without the advent and the evolution of modern day information and communication technologies. The Massive open online courses(MOOCs) share largest share of percentage in the success of online education. Along with having lots of potential, there are many problems and challenges which have been faced by MOOCs since their origin. The major problem which is being faced in MOOCs is its high attrition rate. There is still not enough literature available to explore the possibilities of finding the solution to know the exact reasons behind high dropout rate of MOOCs. The current study is just an initial step towards finding the factors responsible for high dropout rate. In the past lots of researchers have tried to find the factors contributing in students' dropout from MOOCs. This paper extends few more steps in the direction of exploring the reasons behind student dropout along with the timely prediction of at-risk students having highest probability of getting dropped so that the instructors could use that information well in time to reduce the dropout rate up to some extent by making some changes in their pedagogy.

2. Literature survey

Xing W et al. reported that massive open online courses have got great potential as well as high dropout rates in online education. Their study established a mechanism to identify the students at-risk early and accurately. An approach based on temporal modelling was designed to identify the students who were more likely to dropout. This approach acted as helping aid for the instructors who could give timely support to the at-risk students. A principle component analysis based approach was also proposed which showed better results than temporal modeling approach[1]. Fei M et al. proposed recurrent neural network model by using long short-term memory cells to predict the at-risk students. The study showed some significant results compared to some other studies which used simple machine learning based methods[2]. Simpson O compared the graduating data of the distance education students of the UK open university with the student data of some other conventional institutions and found a deficit in the distance education courses. The reason behind deficit was found to be more emphasis on teaching rather than learning[3].Xing W et al. proposed a deep learning based temporal prediction mechanism to further optimize the performance of prediction model in predicting individual student dropout probability. The results showed better performance compared to some baseline methods of machine learning[4]. Tan M et al. proposed students' personal characteristics and their academic performance as input for predicting student dropout. Various models like Artificial Neural Network, Bayesian Networks and Decision tree were used. The results showed better performance in case of decision tree in predicting the student dropout in comparison to other methods[5]. Liang J et al. used several supervised classification models like SVM, logistic regression, random forest and gradient boosting decision tree for predicting the student dropout. Gradient boosting decision tree showed better accuracy in prediction[6]. Cambruzzi W et al. proposed learning analytics system which incorporate data visualization, text analysis, dropout prediction and suggestive pedagogical actions to reduce the dropout. The prediction method was experimented in a Brazilian university with 87% precision and showed the reduction of 11% in dropout rate by recommending changes in pedagogies used in teaching[7].Normandi et al. worked on the problems faced by the MOOCs such as their acceptance and learning outcomes[8]. Coussement et al. benchmarked logit leaf model with eight other algorithms. The logit leaf model outperforms all other algorithms in predicting the student dropout[9]. Srilekshmi et al. proposed a prediction model based on association analysis by using previous student data. The model proposes a real-time feedback to the students at-risk of getting dropped out[10].

3. Research gap

The previous studies conducted in the field of MOOCs and online learning are not providing enough information regarding the contributing factors responsible for motivating a student to dropout. The current study tries to fill that void in finding out the significant factors responsible for dropping out of a student from a MOOC course by first finding the correlation among all the variables and then finding out the significant variables which contribute maximum towards target variable. As the number of enrollments are generally very high in a MOOC course but there is a significant reduction in the number of registered as well as certified students. The target variable is certified and has binary values only.

4. Research questions

- RQ1: To find the correlation among all the dependent and independent variables of student log data.
- RQ2: To find the significant features contributing maximum towards certified target variable.
- RQ3: To evaluate the performance of various prediction algorithms using error rates.

5. Methodology

The dataset for the proposed methodology is obtained from Harvard's Dataverse repository. The dataset contains the clickstream student log data of MOOC courses conducted on Harvard and MIT platform. The dataset contains exactly 641138 instances of various enrolled students. The dataset is collected by getting prior permission for using it from the Harvard repository. After collecting the dataset, the dataset pre-processing is performed to filter the data by data cleaning or data wrangling to get rid of some missing or invalid values. The dataset structure is explained in Table 1. The dataset used in the current study is accessible from the link https://dataverse.harvard.edu/dataverse/mxhx

Table 1. Dataset explanation.

Attribute	Explanation
course_id	Identifies institution name and semester e.g. HarvardX/CB22x/2013_Spring"
userid_DI	First portion Identifies dataset and second portion represents any random ID number. Example userid_DI: MHxPC130442623".
Explored	0/1: anyone who accessed at least half of the chapters in the courseware.
Certified	0/1: anyone who earned a certificate.
final_cc_cname_DI	During de-identification some countries names were replaced with the corresponding continent/region name. Examples: "other south asia"
LoE	Highest level of education
YoB	Year of birth
gender	Possible values male and female or others
nevents	Number of interactions with the course. Example: "502"
ndays_act	No. of unique days student interfaced with the course. Example: "16"

nplay_video	Number of play video events within the course. Example: "52"
nchapters	Number of chapters with which student interacted. Example: "12"
nforum_posts	Number of posts to the discussion forum. Example: "8"

The dataset usage policy can also be found at the link https://dataverse.org/best-practices/sample-dua.The correlation among all the dependent and independent variables is first computed. After computing the correlation, the feature measurement is performed to find out the most significant factors responsible for contributing maximum towards target variable which is named certified. Various machine learning algorithms like logistic regression, decision tree, random forest, K-Nearest Neighbor were trained and tested for prediction purpose. The classification efficiency of each model through precision, recall, F1-score and support values is computed. The evaluation of all the models is done in terms of error rates and by using ROC curve method. The proposed methodology of the study is depicted in Figure 1.

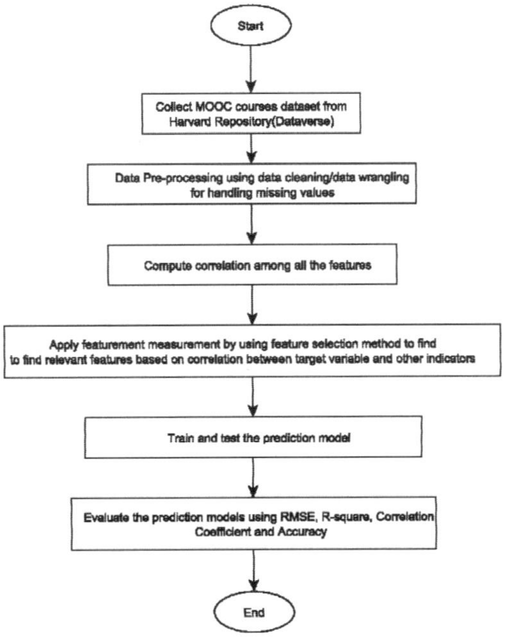

Figure 1. Proposed Methodology

The correlation matrix is computed and is shown in Table 2. After applying feature selection method, the selected features are shown in Table 3. The classification efficiency of logistic regression is shown in Table 4. The classification efficiency of decision tree is shown Table 5. The classification efficiency of random forest is shown Table 6. The classification efficiency of K-Nearest Neighbor is shown in Table 7. The Table 8 displays the comparison of all the four machine learning models in terms of error rates. The Fig 2. shows the performance evaluation of all the four machine learning models using ROC curve method.

Table 2. Correlation matrix among all variables.

	course_id	userid_DI	explored	certified	final_cc_cname_DI	LoE_DI	YoB	gender	nevents	ndays_act	nplay_video	nchapters	nforum_posts
course_id	1.000000	-0.003828	-0.117864	-0.117864	-0.093450	0.027643	0.077293	0.139462	0.273069	-0.004330	-0.071580	0.003653	0.023304
userid_DI	-0.003828	1.000000	0.004814	-0.001555	0.003481	0.002316	0.004056	-0.006964	0.000110	0.002283	-0.001947	0.002304	0.003884
explored	-0.117864	0.004814	1.000000	0.719572	-0.011290	0.006106	-0.018157	-0.045326	0.714618	0.748083	0.459078	0.862880	0.109359
certified	-0.093450	-0.001555	0.719572	1.000000	0.000406	0.014123	0.000521	-0.048066	0.693849	0.716512	0.414949	0.711822	0.092742
final_cc_cname_DI	0.027643	0.003481	-0.011290	0.000406	1.000000	-0.015281	-0.286984	-0.087489	0.032876	0.028414	0.010474	-0.014422	0.007492
LoE_DI	0.077293	0.002316	0.006106	0.014123	-0.015281	1.000000	0.144916	0.033274	0.011028	0.014957	0.001435	0.024247	0.016280
YoB	0.139462	0.004056	-0.018157	0.000521	-0.286984	0.144916	1.000000	0.052871	-0.026457	-0.028291	-0.034884	0.003312	0.045346
gender	0.273069	-0.006964	-0.045326	-0.048066	-0.087489	0.033274	0.052871	1.000000	-0.028929	0.002821	-0.026472	0.004556	0.011287
nevents	-0.028114	0.000110	0.714618	0.693849	0.032876	0.011028	-0.026457	-0.028929	1.000000	0.874185	0.726988	0.793291	0.125695
ndays_act	-0.004330	0.002283	0.748083	0.716512	0.028414	0.014957	-0.028291	0.002821	0.874185	1.000000	0.592528	0.834771	0.141927
nplay_video	-0.071580	-0.001947	0.459078	0.414949	0.010474	0.001435	-0.034884	-0.026472	0.726988	0.592528	1.000000	0.503482	0.077212
nchapters	0.003653	0.002304	0.862880	0.711822	-0.014422	0.024247	0.003312	0.004556	0.793291	0.834771	0.503482	1.000000	0.132425
nforum_posts	0.023304	0.003884	0.109359	0.092742	0.007492	0.016280	0.045346	0.011287	0.125695	0.141927	0.077212	0.132425	1.000000

Table 3. Selected features after feature selection.

Features	Correlation value against target variable
explored	0.719572
certified	1.000000
nevents	0.693849
ndays_act	0.716512
nplay_video	0.414949
nchapters	0.711822

The Table 3 shows selected features after applying the feature selection method on the results of the correlation as shown in the Table 2.

Table 4. Classification efficiency of logistic regression.

	precision	recall	F1-score	support
0	0.98	0.98	0.98	36424
1	0.77	0.75	0.76	3033
Accuracy			0.96	39457
Macro avg	0.87	0.86	0.87	39457
Weighted average	0.96	0.96	0.96	39457

The Table 4 shows the classification efficiency of logistic regression classifier in terms of precision, recall, F1-score and support.

Table 5. Classification efficiency of decision tree.

	precision	recall	F1-score	support
0	0.98	0.98	0.98	36424
1	0.71	0.71	0.71	3033
Accuracy			0.96	39457
Macro avg	0.84	0.84	0.84	39457
Weighted average	0.96	0.96	0.96	39457

The Table 5 shows the classification efficiency of decision tree classifier in terms of precision, recall, F1-score and support.

Table 6. Classification efficiency of random forest.

	precision	recall	F1-score	support
0	0.98	0.98	0.98	36424
1	0.75	0.77	0.76	3033
Accuracy			0.96	39457
Macro avg	0.86	0.87	0.87	39457
Weighted average	0.96	0.96	0.96	39457

The Table 6 shows the classification efficiency of random forest classifier in terms of precision, recall, F1-score and support.

Table 7. Classification efficiency of K-Nearest Neighbor Method.

	precision	recall	F1-score	support
0	0.97	0.98	0.98	36424
1	0.72	0.69	0.70	3033
Accuracy			0.96	39457
Macro avg	0.85	0.83	0.84	39457
Weighted average	0.95	0.96	0.95	39457

The Table 7 shows the classification efficiency of K-Nearest Neighbor classifier in terms of precision, recall, F1-score and support.

Table 8. Error rates.

Model	RMSE	Corr	R^2	Accuracy
Logistic Regression	0.1922280370219179	0.73667383	0.47925928122521355	0.9630483817827001
Decision Tree	0.21065949663934677	0.68640908	0.37461111208871667	0.9556225764756571
Random Forest	0.19419562193776596	0.73686495	0.46854445162079406	0.9622880604202043
KNN	0.21161977282618802	0.67948363	0.368896536299693	0.9552170717489926

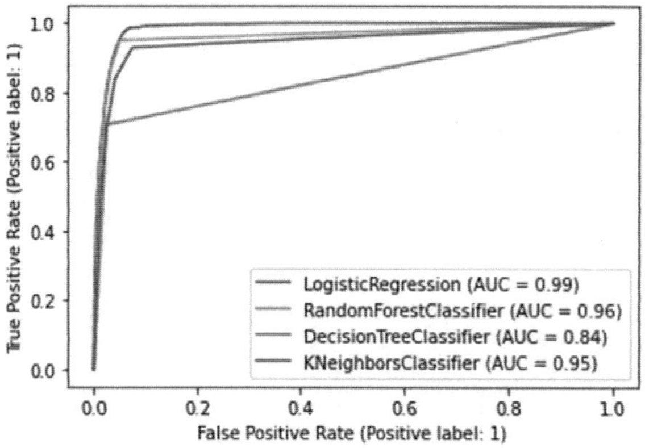

Figure 2. Performance evaluation using ROC curve.

The Figure 2 shows the performance evaluation of all the classifiers applied on the dataset in terms of Area under the curve where logistic regression classifier shows maximum area under the curve.

6. Conclusion

As shown in Table 8., The error rates in terms of Root means square error, correlation coefficient, R-square, and accuracy for all the four machine learning models used for prediction are displayed. The algorithm with lowest RMSE, highest correlation coefficient, highest R-square value and highest accuracy is considered to be the best in prediction. Logistic regression is found to be the best among these four models in prediction of student dropouts. The Fig 2 also confirms the same as it shows maximum AUC value for logistic regression in comparison to all other machine learning models used for prediction. In the previous studies such a high accuracy on such a large voluminous data has not yet seen.

7. Future scope

The results of the current study can be used for creating a framework to recommend the necessary timely actions to the students at-risk well in time from the end of instructors to reduce the dropout rate of the online learners.

References

[1] W. Xing, Temporal predication of dropouts in MOOCs: Reaching the low hanging fruit through stacking generalization, Comput. Human Behav. 58 (2016) 119–129. https://doi.org/10.1016/j.chb.2015.12.007.

[2] M. Fei, Temporal Models for Predicting Student Dropout in Massive Open Online Courses, Proc. - 15th IEEE Int. Conf. Data Min. Work. ICDMW 2015. (2016) 256–263. https://doi.org/10.1109/ICDMW.2015.174.

[3] O. Simpson, Student retention in distance education: are we failing our students?, Open Learn. J. Open, Distance e …. (2013). https://doi.org/10.1080/02680513.2013.847363.

[4] W. Xing, D. Du, Dropout Prediction in MOOCs: Using Deep Learning for Personalized Intervention, J. Educ. Comput. Res. 57 (2019) 547–570. https://doi.org/10.1177/0735633118757015.

[5] M. Tan, P. Shao, Prediction of student dropout in E-learning program through the use of machine learning method, Int. J. Emerg. Technol. Learn. 10 (2015) 11–17. https://doi.org/10.3991/ijet.v10i1.4189.

[6] J. Liang, J. Yang, Y. Wu, C. Li, ..., Big data application in education: dropout prediction in edx MOOCs, … Multimed. Big Data …. (2016). https://ieeexplore.ieee.org/abstract/document/7545065/.

[7] W. Cambruzzi, S.J. Rigo, J.L. V Barbosa, Dropout prediction and reduction in distance education courses with the learning analytics multitrail approach, J. Univers. Comput. Sci. 21 (2015) 23–47. https://www.scopus.com/inward/record.uri?eid=2-s2.0-84933041536&partnerID=40&md5=4db95bcdc38b59e6855095cccc31c969.

[8] L. Normandi Atiaja Atiaja, R. Segundo Guerrero Proenza, MOOCs: Origin, characterization, principal problems and challenges in higher education, J. E-Learning Knowl. Soc. 12 (2016) 65–76. https://www.scopus.com/inward/record.uri?eid=2-s2.0-84956664221&partnerID=40&md5=9e0f5d31d7fca87419cebfdb4c02e01c.

[9] K. Coussement, M. Phan, A. De Caigny, D.F. Benoit, ..., Predicting student dropout in subscription-based online learning environments: The beneficial impact of the logit leaf model, Decis. Support …. (2020). https://www.sciencedirect.com/science/article/pii/S0167923620300804.

[10] M. Srilekshmi, S. Sindhumol, ..., Learning analytics to identify students at-risk in MOOCs, … Educ. (2016). https://ieeexplore.ieee.org/abstract/document/7814823/.

Journal of Physics: Conference Series **2327** (2022) 012064 doi:10.1088/1742-6596/2327/1/012064

Classification of Sleep using Polysomnography

Hemu Farooq[1], Anuj Jain[2] and M.K. Shukla[2]

[1] Electronics and Communication Engineering Department, Bhagwant University, Rajasthan, India,
[2] Electronics and Electrical Engineering Department, Lovely Professional University, Punjab, India

Abstract. Sleep is often recognized as a necessary component of a person's well-being and is an extremely vital component of a healthy person's well-being. Sleep is a state in which a person is both physiologically and psychically at ease. The sleep conception is appraised exceedingly unusual, and it has piqued the interest of researchers all around the world. The stages of sleep are examined in order give benefits for studying sleep utilized for the purpose of research. The ability to diagnose sleep disorders has been demonstrated by carefully examining the sleep score system and its various stages. As can be seen, manual sleep stage classification is a time-consuming method that requires adequate measure for sleep professionals to undertake statistical and quantitative analysis. Furthermore, errors and abnormalities in the categoization of the similar facts can occur frequently. As a result, the adoption of an autonomous scoring system to enable trustworthy classification is becoming increasingly popular.The scheduled task teaches you how to apply an automatic system based on EEG (Electroencephalogram), EMG (Electromyogram) and EOG (Electrooculogram), which is known as a polysomnography test or PSG test. For a total of 30 seconds, the recording was measured in length segments. The standard collection of parameters, which gives the distinct stages of sleep, was obtainedfrom several hospitals in sleep laboratories. Sleep waking, Non-Rapid Eye Movement (Stage 1, Stage2,Stage3), and Rapid Eye Movement are the stages. In clinics, the procedure can be highly effective, especially for neurologist detecting sleep problems.

1. Introduction

1.1 Sleep

It is defined as a successfully repeatedly process which involves physical, mental, and emotional state of the body marked by a shift in consciousness. All voluntary muscles are inhibited during sleep, resulting in minimal engagement with the outside world [1]. Sleep has been shown to be a extremely economized characteristic throughout biological and organic change in animals. It is the state where an individual transitions from an aware to a subconscious condition. Sleep is viewed as a necessary phase in many physiological functions, including experience processing and memory consolidation. People with sleep problems such as insufficient sleep, sleep narcolepsy, and breathing problems while sleeping, among others, do not receive enough sleep and must be treated accordingly. The importance of sleep is thus highlighted by the symptoms that persons who suffer from sleep problems encounter [2].

The mechanism of scoring procedure is done in front of a sleep expert by evaluating and examining the PSG test (polysomnography). In the field of sleep medicine and research, a sleep scoring mechanism is regarded as a necessary aspect. The polysomnoraphy is a process that records the transference of several bio signals such as occulogram signal, encephalogram signal and myogram signal. All of the signals that were captured are gathered into an epoch, which is a short period of time that is carefully examined to identify numerous types of sleep [3].

Typically, scoring system of sleep consists of different stages such as stage one as waking, stage second as rapid eye movement (REM), and stage third as non-rapid eye movement (NREM). The first stage is the stage in which a person prepares for sleep is known as waking. Also, the person is relaxed and in a soothing situation, and his or her EMG is most likely upraised or standardized system. The third stage non-rapid eye movement is also parted into four catagories: first category happens after a short period of waking and EMG declines, second cateogory occurs when a person is in the stage of relax and myography value continues to decline, third category occurs when myography is further declined due to muscle relaxation, and fourth cateogory occurs when myography is moderate and the subject is referred to as asleep. EMG displays muscle tones known as twitches in the REM state, which is defined by rapid or random eye movements [4].

Sleep is a vital feature of a person's overall health and is seen as a critical component of a person's mental and physical abilities. Sleep is a state of virtual isolation from one's environment. When an individual's mental and physical states are in a state of rest, they are said to be sleeping. The concept of sleep is meant to be incredibly intriguing, and it has piqued the interest of scientists all over the world [5]. Sleep deprivation or disturbances cause a variety of diseases of the kidney, heart, and other organs, as well as increasing the risk of diabetes, high blood pressure, and strokes. Passive diseases are what they're called. As a result, disorganization be the cause of an insufficient sleep which must be addressed [6].

The sleep analysis is very important in the examination and studying of sleep research and medicine and has become well-known in the field of diagnosis. The visual examination of polysomnograms was used to develop an automatic grading system (PSG). The Polysomnography test involves the transmission of numerous bio signals such as EOG (Electrooccularogram), myogram, and encephalogram that, when studied for a brief period of time known as an epoch and processed, might reveal distinct sleep stages [7,8].

1. 2 Polysomnography

Electroencephalogram (EEG): An electroencephalogram (EEG) test can be used for evaluating electrical impulses in brain. The connection between brain cells via electric impulses is known as stimulated exercise of the brain. EEG is used to trace and list the waves of the brain pattern. The electrodes that are used to analyze electric stimulus are attached in front of head under the aegis of cables. These electrodes are little flat discs that are utilized to record brain signals and communicate them to a computer. In EEG recordings, the impulses will display wavy lines, dips, and peaks. Doctors can determine if the patterns are normal or pathological using these waveforms [9].

Electromyogram (EMG): It's a methodical process for assessing and managing the well-being of muscularity and sensory cells generated from sensory neurons. The EMG converts the input data into featured data or graphs, assisting clinicians with their diagnosis. If a person presents to the doctor with syndrome of disorganization of a muscles and nerve cells, such as itching, insensibility, or unrecognized deficiency of the parts of body, the doctor will recommend a myography test because the outcome will help the doctor diagnose disorders of the contracting and relaxing of muscles, nerve cells, or disorders affecting the connection between the muscular strength and neurons [10].

Electrooculogram (EOG): This approach is utilized to determine the corneo-retinal potential, which exists between the rear and front of the human eye. The electrooculogram is the name given to the resulting signal. EOG is commonly used to diagnose ophthalmological problems as well as to record eye movements. Electrodes are placed below or above the eye, or to the left or right of the eye, to measure eye movement [11]. When the eye moves from the centre to particular side of the electrodes, it sees the positive side of the retina while seeing the negative side of the retina on the other electrode, resulting in a potential difference between the electrodes. The eye location will be measured by the potential that is recorded.

2. Material and Method

For the development of the research, real-time implementation was used.

2.1 Equipment's and devices

There was a connection between the polysomnography and the computers.

- Data for the study of sleep stages was acquired using a device known as a polysomnography device. For each patient, data was collected for a seven-hour period. These disparate data were initially gathered together by sleep experts and a plysomnography software gadget.
- A biomedical device that is designed to compute physiological signals, particularly during sleep stages.

Journal of Physics: Conference Series **2327** (2022) 012064 doi:10.1088/1742-6596/2327/1/012064

2.2 Materials

Electroencephalogram (EEG)

Electromyogram (EMG)

Electrooculogram (EOG)

- Sleep Electroencephalogram Recognition NN (SRNN): allows for the detection of a number of important characteristics in EEG waveforms.
- Sleep Stage Diagnosis NN (SSNN): allows for the diagnosis of various sleep states.
- Contextual Diagnosis NN (CDNN): to increase the correct judgments rate by taking into account contextual factors.
- The relationship between the performances of characteristic waves.

3. Results

Sleep has an important role in maintaining a person's physical and mental health throughout their lives, resulting in a higher quality of life and greater health. Many people suffer from sleep disorders that have a negative impact on their everyday lives. Sleep apnea, insomnia, and narcolepsy are examples of sleep disorders. The analysis of sleep stages is thus beneficial for recognizing and treating such illnesses, and it must be implemented.

Different standard data bases of physiological signals were collected in front of sleep specialists from various hospitals, particularly from sleep laboratories. Electroencephalograph data were recorded in real time utilizing conventional electrodes on the skin's surface. Electromyography signals were obtained using electrodes placed beneath the individual's chin, two of which were used. Electrodes were inserted beneath or top of the eye, or to the far or wide of the eye, to obtain electrooculography signals. Different stages of sleep were comprehended through viewing physiological data. The distinct procedure of the methods are as following:

3.1 Signal Processsing:

It is necessary to alter the original signal obtained from EMG in order to use the ANN. The signals are processed to remove faulty signals, retrieve important data, and forecast utility. The signals generated from physique are frequently distorted by waves that are unrelated to them. These waves are caused by biological or technical difficulties. The physical body generates its own electrical activity, which can be detected using the equipment listed below.

- ECG artifacts, such as spikes occurring during the QRS complex stage.
- Blinking and eye movements cause EOG artifacts.
- High-frequency artifacts in EMG

.

Filtration with filters like high pass, low pass and band-pass filters, depending on the artifacts present in the signal, is a common approach for removing these artifacts. The signal reprocessing is done to normalise the EMG signal and improve it. The EMG signal is divided into thirty second epochs, with each epoch representing a different stage of sleep. Butterworth filtering is used to filter the signals in this case. Filter with a low pass (LPF).

.

3.2 Results of proposed model

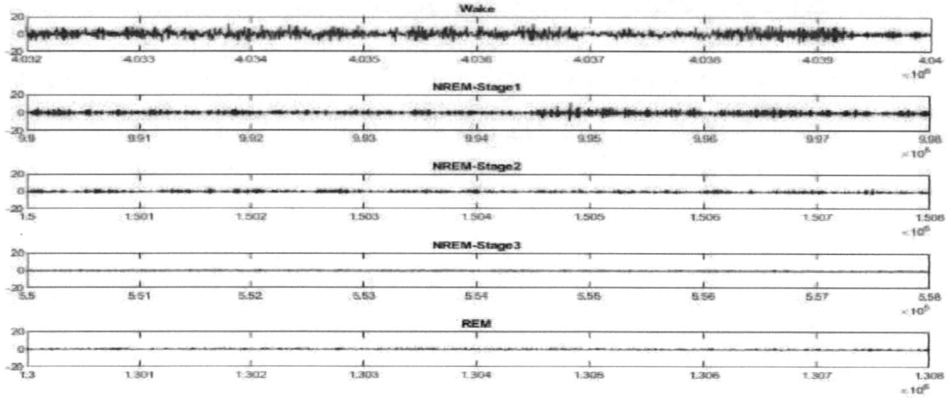

Figure 1: EMG signal acquisition

The figure 1 indicates the distinct sleep stages with the activity of the brain.

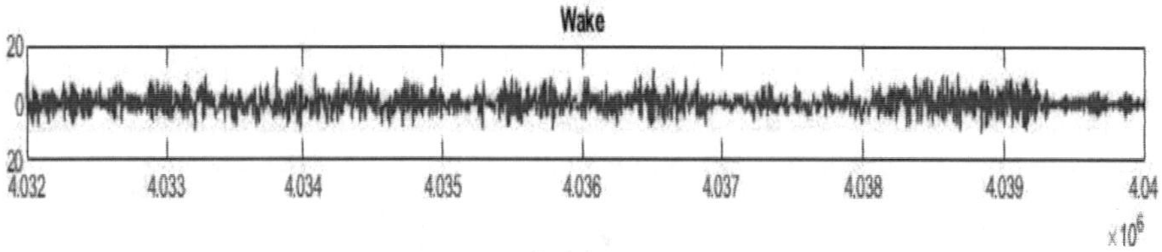

Figure 2: Activity during wakefulness

Wakefulness (Stage 0): It is defined as the condition of being immediately before falling asleep, with the eyes closed and the body relaxed. During this transient period, the EEG typically shows a beta wave with an amplitude of 10–34 V and a bandwidth of 16–25 Hz, as well as an alpha wave with a larger amplitude but a bandwidth of 8–12 Hz. The 4–8 Hz theta wave dominates as the patient falls asleep; (b) the EMG can be strong or moderate determining intensity of muscular tension; and (c) the REMs can be rigorous or rare determining the velocity of perceptible inspection.

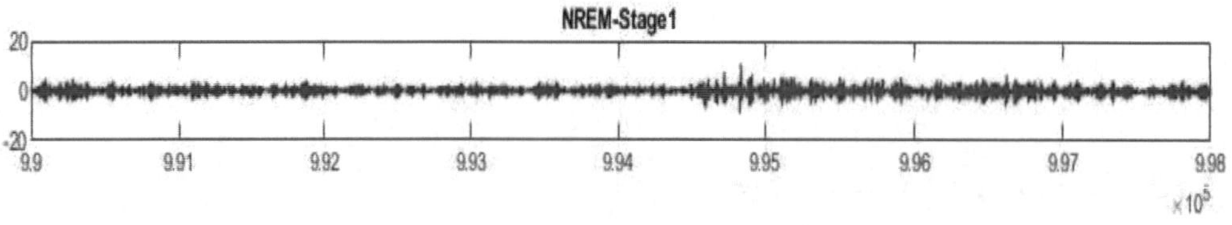

Figure 3: Activity during drowsy

Stage first: The period of this stage lasts instantly, and (a) the encephalogram shows repressed rhythm of alpha waves (50 percent of the time) but with small magnitude short term harmonics with a spectrum ranging from 3–7 Hz [16]; (b) myogram signal, on the other hand, is fitting smaller than stage 0; and (c) the REMs are generally disappearing except for slow rolling eye movements.

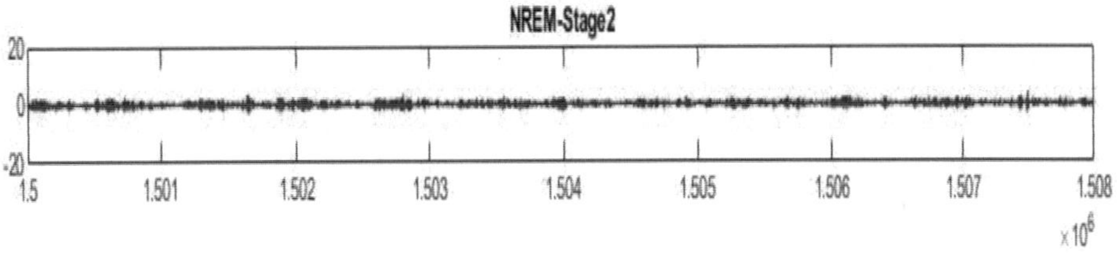

Figure 4: Activity during light sleep

Stage 2: this stage consists of, (a) the encephalogram, in distinction to a continual surroundings of relative low voltage, consists of breach of multiple frequencies in the range of 2–16 Hz, and this phase includes sleep spindles and high voltage biphasic waves known as K complexes; (b) the myograms comes up with a reduced muscle tonus; and (c) the rapid eye movement fade away.

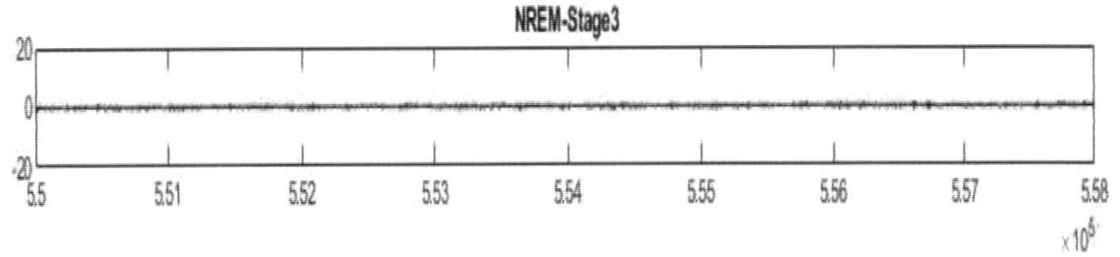

Figure 5: Activity during Deep sleep

Stage 3: (a) The encephalogram reveals high amplitude (>75 V), slow (0.5–4 Hz) waves called delta waves; (b) the myogram signal shows a further reduction in muscle tonus; and (c) the rapid eye movements are becoming increasingly rare, except as contaminated artifacts owing to other activities like EEG.

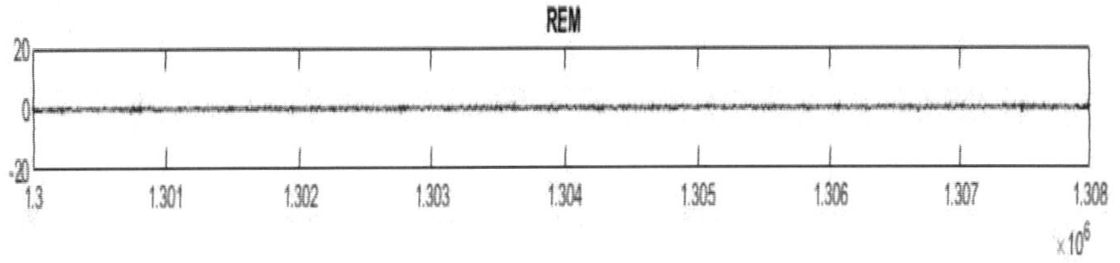

Figure 6: Activity during REM

REM sleep: this phase of sleep has, (a) the encephalogram retrogress to a low voltage with a spectral content similar to that of stage 1 and wakefulness; (b) the myograph shows muscle tones known as "Twitches," as well as the occasional short tones of distal muscles and transient activities that are most likely caused by heart frequency or blood pressure; and (c) the EOG shows bursts as prominent rapid eye movements.

4. DISCUSSION

The association between sleep stages and physiological signal has been proven to be an effective proposal using a built classifier called ANN classifier. The classifier uses a mathematical tool to help identify the different stages of sleep. When compared to Hanaoka et al. [14], Tagluk et al. [12] obtained a 74.7 percent accuracy on scoring capabilities, which was judged acceptable enough. The goal of this research is to determine sleep stages utilizing physiological signals (EEG, EMG, and EOG) obtained at the same time.

Synchronizing EEG, EMG, and EOG signals has proven to be an excellent alternate method for determining sleep stages. The use of myography and occulurogram to remove artifacts from synchronous encephalography signals worked well, and it could be useful in the clarifying of all encephalography signal researches. Several methods are being implemented, but using an automated system always has an advantage over standardized traditional methods and terminology for estimating distinctive stages of sleep. Most of the terminology would possess physical tensions on an individual which is under observation, as well as the utilization of pricey polygraphs and specialists with advanced levels of knowledge. R- K (Rechtschaffen and Kales) estimated sleep stages at Eighty Two Point Six percent for NREM stages, Thirty Eight Point Three percent for REM stages, and Seventy Point Five percent for wakeful stages [11]. In this situation, the REM stage appears to be under-estimated. An attempt [12] to approximately define the relationship linking different sleep phases and heart rate is in fact not contemplated valid sleep stage categorization. Methods [13] for calculating sleep stages based on the link of different sleep stages and body motions were thought to be unrealistic. Equivalent to non interfering solutions [14] that rely on a static charge sensitive bed do not account for all sorts of sleep stages. The methods based on body motions detected by infrared sensors and large classifier was successful in categorizing sleep stages. Akin et al. [15] and Kurt et al. [16] attempted to calculate the assertive level and the drowsy level, respectively. However, the goal of the proposed study is to characterize sleep stages utilizing Polysomnography data collected simultaneously.

5. CONCLUSION

The goal of the study was not only to look at the results of comparative methodologies, but also to offer a generalship that uses encephalograph, occulugram and myogram signals to train a high-precision classifier. The current study suggests different stages of sleep based on method known as polysomnography. In the presence of a sleep analyzer, the data base was gathered from several hospitals. By using a normalizing method, the acquired signals were converted to appropriate signals.

The goal of the study was not only to compare the results of different methodologies, but also to offer a strategy that included EEG, EMG, and EOG signals. The major goal of this study is to estimate sleep stages utilizing simultaneous recordings of physiological signals. Scientists may find the method valuable in evaluating the wake-sleep relationship and identifying sleep disorders

References

[1] HJ. Park, JS. Oh, DU. Jeong, KS. Park "Automated Sleep Stage Scoring Using Hybrid Rule- and Case-Based Reasoning.", Computers and Biomedical Research , 33, pp. 330-349, 2000.

[2] David T. Krausman, Bel Air, Richard P. Allen, "Sleep Scoring apparatus and method. Google Patents," 12 April 2005.

[3] Penzel, T., Conradt, R., Computer based sleep recording and analysis. Sleep Med. Rev. 4:131-138, 2000.

[4] Tagluk, M. E., Sezgin, N., and Akin, M., "Estimation of Sleep Stages by an Artificial Neural Network employing EEG, EMG, EOG", Journal of Medical System, 2010, 34(4), pp.717-725.

[5] ZT. Yeh, RP. Chiang, SC. Kang, CH. Chiang "Development of the Insomnia Screening Scale based on ICSD-II.", Int J Psychiatry Clin Pract , 16(4), pp. 259-267, Oct. 2012.

[6] David T. Krausman, Bel Air, Richard P. Allen, "Sleep Scoring apparatus and method. Google Patents," 12 April 2005.

[7] Vni Rao, M.D., Alyssa Bergey, B.A., Hugh Hill, M.D., David Efron, M.D., and Una McCann, M.D. Clinical Research Reports, Sleep Disturbance After Mild Traumatic Brain Injury: Indicator of Injury? he Journal of Neuropsychiatry and Clinical Neurosciences. 23: 2; 201-205, 2011.

[8] Penzel, T., Conradt, R., Computer based sleep recording and analysis. Sleep Med. Rev. 4:131-138, 2000.

[9] Tagluk, M. E., Sezgin, N., and Akin, M., "Estimation of Sleep Stages by an Artificial Neural Network employing EEG, EMG, EOG", Journal of Medical System, 2010, 34(4), pp.717-725.

[10] Mousmita Sarma, Kandarpa Kumar Sarma, "Fundamentals consideration of ANN", Phoneme-Based speech segmentation using hybrid soft computing frame work, April 2014; 47-75.

[11] Kulkarni, P., Ade R., "Incremental learning from unbalanced data with concept drift and missing features: a review", International journal of data Min knowledge management process, 2014; 15-29.

[12] Shimada, T., Tamura, K., Fukami, T., & Saito, Y. (2010, July). The effect of using Elman-type feedback SOM for sleep stage diagnosis. In *Complex Medical Engineering (CME), 2010 IEEE/ICME International Conference on* (pp. 165-170). IEEE

[13] Tagluk, M. E., Sezgin, N., & Akin, M. (2010). Estimation of sleep stages by an artificial neural network employing EEG, EMG and EOG. *Journal of medical systems*, *34*(4), 717-725.

[14] Watanabe, T., and Watanabe, K., Noncontact method for sleep stage estimation. IEEE Trans. Biomed. Eng. 51 (10)1735–1748, 2004. doi:10.1109/TBME.2004.828037

[15] Hanaoka, M., Kobayashi, M., and Yamazaki, H., Automated sleep stage scoring by decision tree learning, 23rd Annual EMBS International Conference, 1751–1754, 2001

[16] Akin, M., Kurt, M. B., Sezgin, N., and Bayram, M., Estimating vigilance level by using EEG and EMG signals. Neural Comput. Appl. 17 (3)227-236, 2008. doi:10.1007/s00521-007-0117-7.

[17]. Kurt, M. B., Sezgin, N., Akin, M., Kirbas, G., and Bayram, M., The ANN- based computing of drowsy level. Expert Syst. Appl. 36 (2)2534-2542, 2009. doi:10.1016/j.eswa.2008.01.085.

Journal of Physics: Conference Series **2327** (2022) 012065 doi:10.1088/1742-6596/2327/1/012065

Autonomous Assistance System for Visually Impaired using Tesseract OCR & gTTS

Tushar Khete[1], Aditya Bakshi[2]

[1,2]School of Computer Engineering, Poornima University, Jaipur, Rajasthan

Abstract. OCR makes machines to recognize text automatically. It not only saves retrieval times but also helps in digitalization of important documents. In this paper different techniques are used for converting textual matter from an paper document or an image into machine readable form (OCR text output) and then into audio output using gTTS (Google Text-to Speech, a Python library and CLI tool to interface with Google Translates text-to-speech API). In proposed model, the content writes spoken mp3 data to a file, a file-like object (byte string) for further audio manipulation with flexible pre-processing and tokenizing. The approach results into a graspable and comprehensible text output & then fed into audio output for inference with 99 % accuracy. This paper outlines the stages of development, the major challenges with some interesting result comparison with different state of approaches.

1. Introduction

OCR is defined as the process of electronic or mechanical conversion of documents or text embedded in images to computer understandable ASCII characters. Actually, text embedded in images or documents is converted into machine code non variation. It Also minimizes human effort and increases data reliability.

With OCR, you'll be able to convert any physical paper document, or an image into an accessible electronic version with text. For example, if you scan any image or document with a scanner, the output will most likely be a digital file (Which isn't machine-encoded) with ".jpg" or ".pdf" format, but on loading the identical with an OCR tool (Tesseract engine, for example). The tool recognizes the text and saves the output as an editable & computer-understandable text file.

OCR involves two important steps, viz- text/entity detection, in which the text embedded in an images or document is localized. Followed by text recognition, text extraction from image. OCR can be used for both handwriting recognition and printed text recognition. Handwriting character recognition is a complex task because of variety of writing styles and letter strokes of different users. This SLR is purely based upon printed character recognition system. Any discussion related handwriting OCR is outside the scope of this research paper.

There are various types of OCR engines used in existing research studies and text extraction, such as: Desktop OCR, Server OCR, Web OCR, OmniPage Ultimate, Readiris, etc. But only a few of them are free and open source. Their accuracy varies from 70% to 98%, depending upon document images & noise in them. In this research study, we've utilized Tesseract OCR engine.

Journal of Physics: Conference Series **2327** (2022) 012065 doi:10.1088/1742-6596/2327/1/012065

Tesseract is an open-source text recognition (OCR) Engine, available under the Apache 2.0 license. It can be used directly, or (for programmers) using an API to extract printed text from images[9-13]. The advantage of using this engine is that it is flexible in terms of supporting multiple languages, handwritten text and could be compiled to run on different platforms. In recent years, OCR has been a hot topic for research, and especially in Character Recognition, extracting text from different pictures & converting them into audio.

There are about 285 million visually impaired persons in the world, out of which 30% are blind. The approach is to make reading documents, papers, webpages, etc for such people accessible. Recent advancements in Computer science and image processing operations have made this achievable. So, this research paper deals with assisting visually impaired individuals with latest technologies [14-16]. Thus, this OCR based assistance system can perceive the text embedded in any image, in whatever language it may be, and convert it into audio output.

2. Background

The Tesseract OCR (Optical Character Recognition) is an open-source software, devised by Hewlett-Packard (HP) in 1984-94.

It started as a PhD thesis project at HP Labs, in Bristol, and succeedingly developed into a software/hardware system for HP's scanners. The commercial OCR engine tools of the day were way too inefficient and, failed miserably. This motivated the development of a superior counterpart, the Tesseract OCR. Tesseract engine had a significant lead after a collaborative project between HP's scanner division, Colorado and HP Labs Bristol. But it has never been a product.

It's advancement at successive stage was a compressor tool, in HP Labs Bristol, from where more focus was laid on enhancing its rejection efficiency as compared to base-level accuracy. The progress ended in 1994, after the end of the research work [16-18]. The project was sent to the UNLV for Annual OCR Test in 1995, for showcasing its importance wrt OCR scanners of that time. Tesseract OCR was released an open source project, by HP at the end of 2005.

3. Proposed Model: Autonomous Assistance System for Visually Impaired using OCR

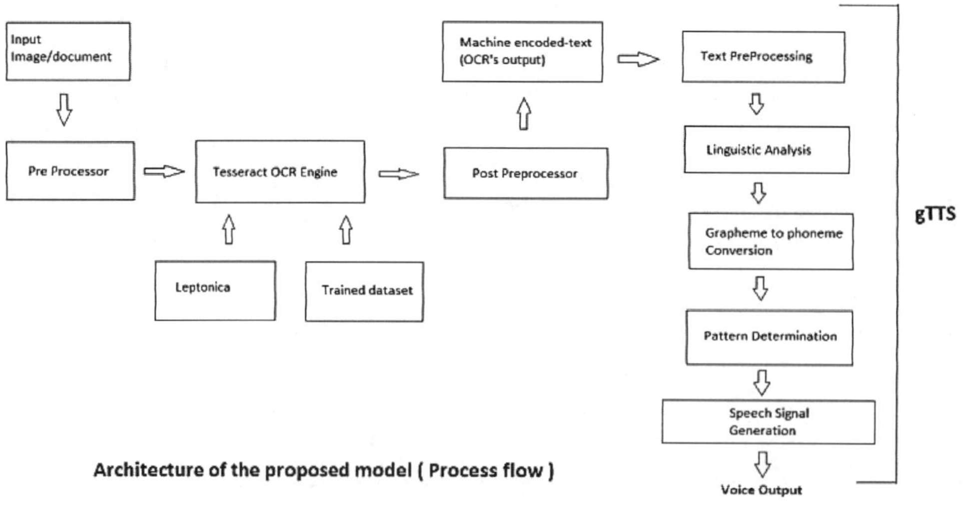

Fig.1. Proposed model.

Journal of Physics: Conference Series **2327** (2022) 012065 doi:10.1088/1742-6596/2327/1/012065

Optical Character Recognition (OCR) technology has been greatly upgraded over the past few decades, thanks to the research work, more computing power and advanced machine learning methods. The proposed model is shown in Fig.1.

3.1. Loading the Image (input):
Load any image (any image with text) from the computer or download it and load the same. It could also be live captured from webcam. For output to be accurate, good quality images must be ensured. Make sure that the input image is not blurry, noisy or discoloured [19]. If the input source contains any of these, our output will most probably comprise errors. Hence, superior the image quality, higher will be the OCR's accuracy.

3.2. Image Processing:
Image processing deals with enhancing the quality of input image. Various OCR engines provide built-in processing steps, but not every step matches to our input image case [20]. Let's go through some pre-processing filters to boost our OCR's accuracy:

3.2.1. Rescaling:
Images/documents rescaled to at least 300 DPI (Dots per inch) are recommended. Increasing DPI above it only increases output file size without any enhancement in file quality, whereas DPI below it results in noisy & undesirable result.

3.2.2. Contrast and sharpness increasing:
Contrast & sharpness are important factors in determining the accuracy of OCR. It is recommended to increase the contrast between input image and its background. It gives more accurate result. If sharpness of input is more, it gives more clarity in the OCR'ed text.

3.2.3. Image Binarizing:
Binarizing involves concerting a multicolored image (RGB) to a black and white image. Use OpenCV library features like simple thresholding and more sophisticated zonal analysis to convert image to white and black. Tesseract OCR does binarization internally, using OSTU algorithm. But results in an inferior result if input image is of uneven darkness. In Fig.2 and Fig.3 binarization of an image has been shown.

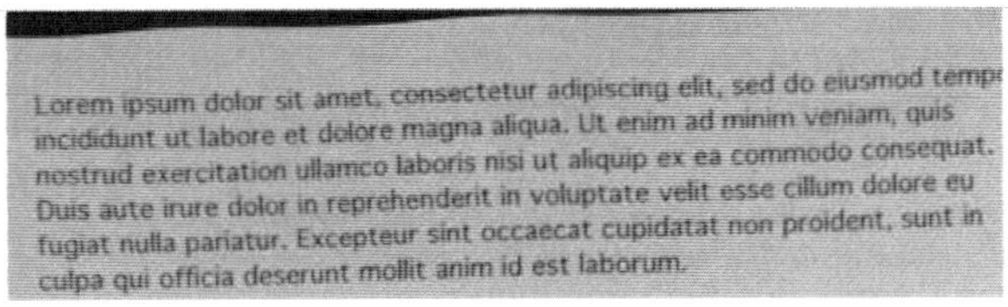

Fig. 2. Binarizing input

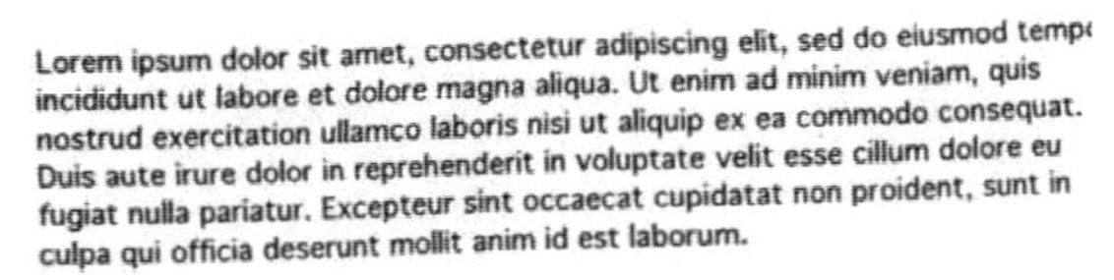

Fig. 3. Binarizing output

Journal of Physics: Conference Series **2327** (2022) 012065 doi:10.1088/1742-6596/2327/1/012065

3.2.4. Noise Removal:
Noise can be random changes in image density seen as grainy on film and element-level discrepancy in digital images.. It's an equally vital issue as sharpness. Tesseract OCR isn't cent per cent economical in noise removal, this causes quality degradation. The Image Noise Removal is shown in Fig.4.

Fig. 4. Image Noise removal

3.2.5. Zonal Analysis:
"Zone" here refers to area of interest, & zonal analysis means recognition of areas of interest for words to be recognized correctly. Layout analysis is an important process in extracting information from documents or images. The main requirement is to always produce the same document image format provided regardless of which OCR is used to process an input image/document.

The information in documents/images depends on field positions in input, regardless of the OCR used. In simple words, the extractor knowledge is not necessary to train our extractor, with OCR processed samples. The proposed model can process the documents with complex structure.

3.2.6. Rotation also referred as, deskew ensures text to be perfect horizontal, & not tilted at any angle. If an image is tilted, it can be deskew by rotating it clockwise or anti clockwise direction. A skewed image severly degrades OCR's outcome. The Deskewing or rotation with output is shown in Fig.5.

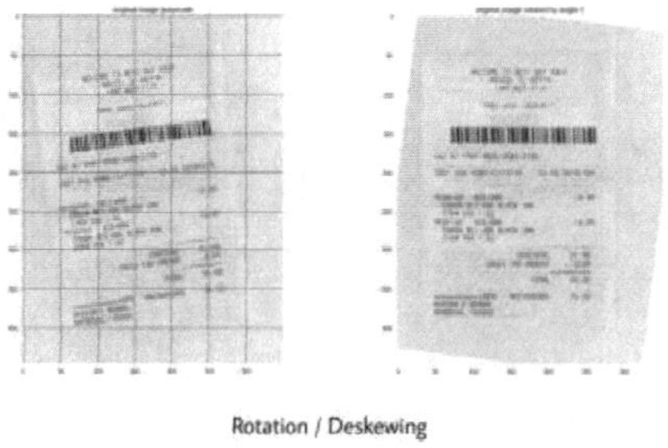

Fig 5. Rotation/Deskewing

For getting result, machine-encoded text output from OCR to gTTS for speech conversion. The audio output obtained is exceedingly clear & can also be translated to various other languages, depending upon user.

4. Literature Review
In literature review, text recognizing and other OCR based assistance systems and its approaches are presented. There are various aid approaches like voice stick, mobile applications, brick pi reading and pen reading. These systems work on creating datasets for speech. Assistance via mobile application isn't

convenient for a visually impaired. To sort out these cons, a more aggregable approach has been developed for voice-assistance, that works of various English fonts & produces an extremely accurate audio output. In Table 1, different approaches on OCR with its advantages and disadvantages are shown. Any blind person can assist in reading any document or website or e-Book.

Table 1. Different Approaches on OCR

Ref. No.	Author Name	Approach	Advantages	Disadvantage	Accuracy
[1]	Abhishek Mathur, Akshada Pathare, Prerna Sharma and Sujata Oak	Suggested an OCR based smartphone application for visual assistance.	Recognizes text in English ,and also any other numeral.	Not friendly methodology for a blind individual.	N/A
[2]	Chucai Yi and Yingli Tian	Scene text recognition from detected text regions, using an approach to combine scene text detection and scene text detection agorithms.	Successfully detects text regions from natural scene image/video and recognizes text information from the detected text regions.	Lags in accuracy.	62.8%
[3]	Pooja Chandran, Aravind S, Jisha Gopinath and Saranya S S	Image to speech conversion using optical character recognition, in MATLAB	The application developed is user friendly, cost effective and applicable in the real time.	Limited testing was performed.	N/A
[4]	Julinda Gllavata, Ralph Ewerth and Bemd Freisleben	Detects, localizes, and extracts colors and text from input images. This approach was much focused on being robust comparatively to other text appearances.	The software is written entirely in JAVA for easy parallel execution on heterogeneous network computing platforms.	Low quality input image/video files complicate the text detection process, so the detection efficiency of the algorithm should be improved.	Recall of 88.7% and a precision of 83.9%.

[5]	Sonia Bhaskar, Nicholas Lavassar and Scott Green	Algorithm for accurate text recognition on a business card, given an Android mobile phone camera image of that card in varying environmental conditions.	It was possible to secure a majority of the accuracy gains by correcting for only illumination and rotation in the constrained environment of an Android user. The result is a highly functional and fun to use mobile application.	Some of preprocessing steps designed in MATLAB, were not implemented on the mobile device.	N/A
[6]	D.Sasirekha and E.Chandra	Proposed a "Text to speech synthesiser", a software that recognizes and reads text embedded in an image aloud automatically.	Text to speech system (TTS) was a simplified approach and in step by step way.	No practical demonstration of methodology and accuracy evaluation.	N/A
[7]	Aparna.A and Prof.I.Muthumani	Developed a system for recognizing handwritten English characters in cursive using the Hidden Markov Model (HMM) algorithm.	Applies median filter to avoid errors due to noise in the scanned image.	Limited to just cursive font.	N/A
[8]	Yamini D.Patil and Nandakishor C.Patil	Developed a system through which a visually impaired person can grasp the knowledge of words they can't see, using Optical character recognition (OCR) & speech synthesis.	The Microsoft Win 32 SAPI library allows selecting the device for audio/video input for use. The meythodology was cost effective , real time applicable and user friendly.	OCRed output voice conversion is tcomplicated and tedious as compared to gTTS.	N/A

5. Results and Discussion

This portion comprisesof model evaluation on different samples of images. It has been observed that model's accuracy greatly depends on input-image quality. On grounds of character-level accuracy, the model is about 99% accurate (1 out of 100 characters recognized as "uncertain"). The better the image, more precise will be the output. In Table 2, the character level accuracy with different test cases is shown.

Table 2. Test cases: input vs output (Images)

Test Cases	Input image	Output
1.	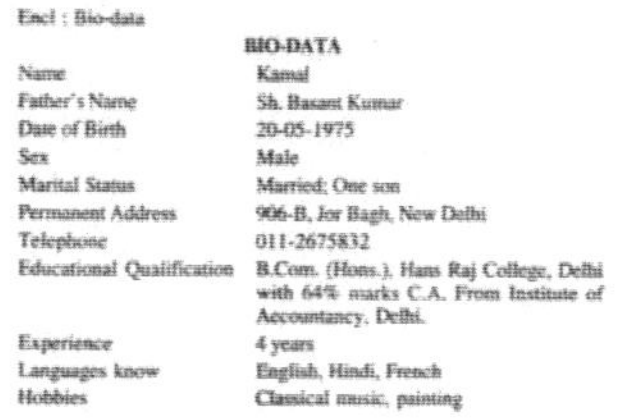 Encl : Bio-data BIO-DATA Name Kamal Father's Name Sh. Basant Kumar Date of Birth 20-05-1975 Sex Male Marital Status Married; One son Permanent Address 906-B, Jor Bagh, New Delhi Telephone 011-2675832 Educational Qualification B.Com. (Hons.), Hans Raj College, Delhi with 64% marks C.A. From Institute of Accountancy, Delhi. Experience 4 years Languages know English, Hindi, French Hobbies Classical music, painting	Encl : Bio-data BIO-DATA Name Kamal Father's Name Sh. Basant Kumar Date of Birth 20-05-1975 Sex Male Marital Status Married; One son Permanent Address 906-B, Jor Bagh, New Delhi Telephone 011-2675832 Educational Qualification B.Com. (Hons.), Hans Raj College, Delhi with 64% marks C.A. From Institute of Accountancy, Delhi. Experience 4 years Languages know English, Hindi, French Hobbies Classical music, painting
2.	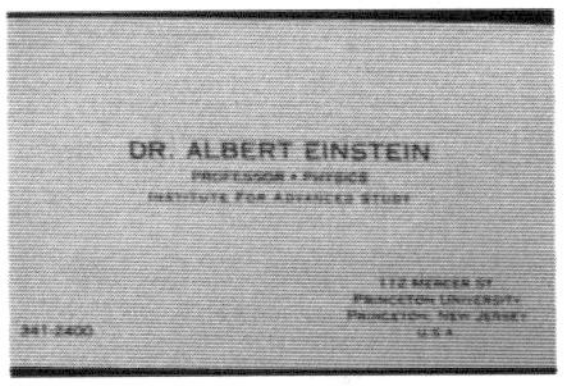 DR. ALBERT EINSTEIN 341-2400	DR. ALBERT EINSTEIN PROFESSOR " PHYSICS INSTITUTE FOR ADVANCED STUD' 112 MERCER ST. PRINCETON UNIVERSITY PRINCETON, New JERSEY 341-2400 USA
3.	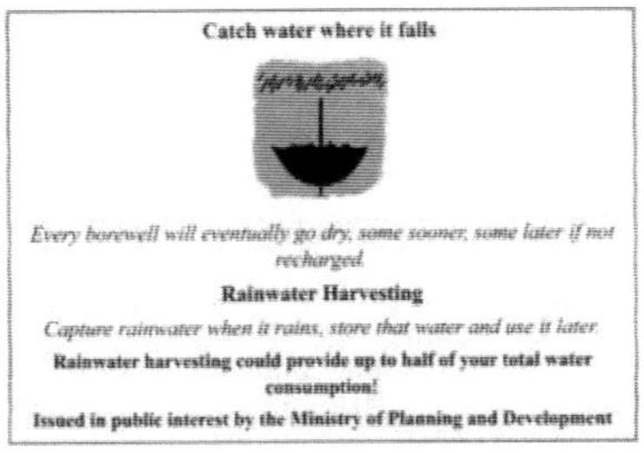 Catch water where it falls *Every borewell will eventually go dry, some sooner, some later if not recharged.* **Rainwater Harvesting** *Capture rainwater when it rains, store that water and use it later.* **Rainwater harvesting could provide up to half of your total water consumption!** **Issued in public interest by the Ministry of Planning and Development**	Catch water where it falls Every borewell will eventually go dry, some sooner, some later if not recharged. Rainwater Harvesting Capture rainwater when it rains, store that water and use it later. Rainwater harvesting could provide up to half of your total water consumption! Issued in public interest by the Ministry of Planning and Development

4.	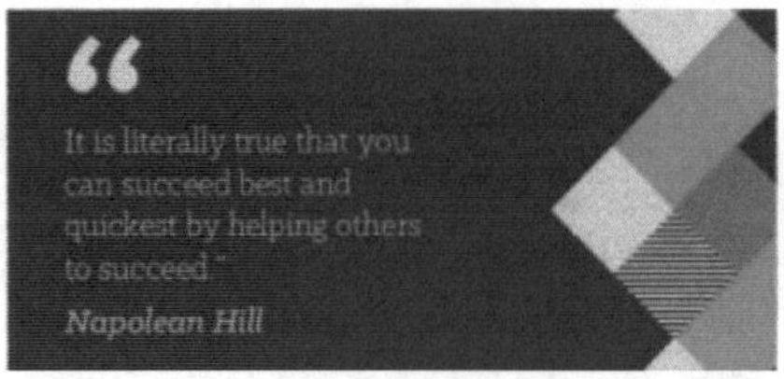	It is literally true that you can succeed best and quickest by helping others to succeed." Napolean Hill

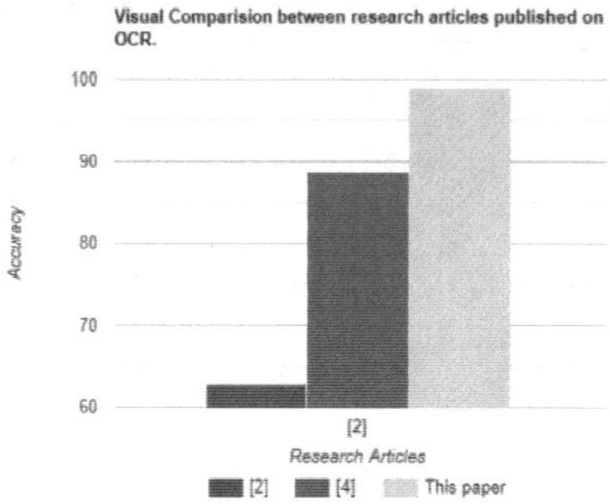

Fig. 6. Visual Comparison with efficiency of OCR-Test Case

In Fig.6, visual comparison between [2], [4] and proposed work depict the respective OCR-test case efficiencies; [2], show by blue region, successfully detects textual content areas from any input image/video , and transcribes that text into machine understandable text for further processing with an efficiency of 62.8%. For [4], shown by red region, recall is calculated to be 88.7% and precision of 83.9%. Comparatively, approach used in proposed work is advantageous and more efficient, found to be ~99% accuate in test cases.

6. Conclusion

In this paper, a model has been created with gTTS(Google text to speech) - based OCR application using Tesseract OCR engine. The application successfully recognizes text of various fonts from different input images and transforms the OCR output into an audio output, which is extremely accurate. We have implemented Tesseract Engine because it is the most powerful and open source software , and also it does not require any license/investment. It will assist visually impaired people so that they can read the document using this aiding methodology. Experiments have been performed by evaluating visual comparison of OCR-Test Cases and good results have been achieved by calculating the precision of the model. The model has been a far better methodology in OCRing any input document or image and transcribing it into audio output. The proposed system eases the digital experience of people having learning disabilities, reduced vision or those with literary difficulties.

Orcid-id: 0000-0003-1637-4237
Orcid-id: 0000-0001-5550-081X

7. References.

[1] Abhishek, Akshada, Prerna and Sujata 2019 AI based Reading System for Blind using OCR , Third International Conference on Electronics Communication and Aerospace Technology [ICECA 2019] IEEE Conference Record # 45616

[2] Chucai Yi & Yingli Tian, 2014 Scene Text Recognition in Mobile Applications by Character Descriptor and Structure Configuration, IEEE TRANSACTIONS ON IMAGE PROCESSING, VOL. 23, NO. 7, JULY 2014

[3] Pooja Chandran, Aravind S, Jisha Gopinath and Saranya S S, 2014 Design and Implementation of Speech Generation System using MATLAB , International Journal of Engineering and Innovative Technology (IJEIT) Volume 4, Issue 6, December 2014.

[4] Julinda Gllavata', Ralph Ewerth' and Bemd Freisleben' 2003 , A Robust Algorithm for Text Detection in Images, Proceedings of the 3rd International Symposium on Image and Signal Processing and Analysis (2003).

[5] Sonia Bhaskar, Nicholas Lavassar and Scott Green, Implementing Optical Character Recognition on the Android Operating System for Business Cards, EE 368 Digital Image Processing.

[6] D.Sasirekha and E.Chandra, 2012 Text to Speech: A Simple Tutorial , International Journal of Soft Computing and Engineering (IJSCE) ISSN: 2231-2307, Volume-2 Issue-1, March 2012.

[7] Aparna.A and Prof.I.Muthumani, Optical Character Recognition for Handwritten Cursive English characters , International Journal of Computer Science and Information Technologies, Vol. 5 (1) , 2014, 847-848.

[8] Yamini D.Patil & Nandakishor C.Patil, August-2015 Text to Speech Synthesis for Blind Person, Gjra - Global Journal for Research Analysis Volume-4, Issue-8 ,ISSN No 2277 - 8160.

[9] Avanika Vaghasiya, Dr. Amit Thakkar and Sanket Suthar, 2020 Real Time Optical Character Recognition (OCR) for Regional Languages, International Journal of Advanced Science and Technology Vol. 29, No. 03, , pp. 7670 - 7677.

[10] Anand Shinde, Parvinder Singh, Jay Patil, Jaideep Singh and Trupti Baraskar, 2021 Text Extraction from Images using Tesseract, International Research Journal of Engineering and Technology (IRJET) Volume: 08 Issue: 07 | July 2021.

[11] Himank Dave, Aryaman Gobse , Aryika Goel and Swati Bairagi, 2020 , OCR Text Detector and Audio Convertor, International Journal for Research in Applied Science & Engineering Technology (IJRASET) Volume 8 Issue V May 2020.

[12] Swati Tomar and Amit Kishore, 2018 A Review: Optical Character Recognition, International Journal of Engineering Sciences & Research Technology.

[13] Jamshed menon , Maira Sami , Rizwan Ahmed Khan and Mueen Uddin, 2020 Handwritten Optical Character Recognition (OCR): A Comprehensive Systematic Literature Review (SLR) , IEEE August 14, 2020.

[14] Aditya Bakshi, Sunanda Gupta et al., "3T-FASDM: Linear Discriminant Analysis based 3-Tier Face Anti-Spoofing Detection Model using Support Vector", International Journal of Communication Systems, Wiley, 2020, vol 33, issue 12.

[15] Aditya Bakshi, Sunanda Gupta "An Efficient Face Anti-Spoofing and Detection Model Using Image Quality Assessment Parameters" in Multimedia Tools and Applications, 2020.

[16] Shakti, Aditya Bakshi "An Optimal Energy Efficient Spatial-Temporal Correlation Method for Data Aggregation in Wireless Sensor Networks" published in International Journal of Control Theory and Applications, ISSN : 0974-5572,Number 45(2016).

[17] Lalramhluri Khawlhring, Aditya Bakshi "Detection of Black Hole using Time Difference and Neighborhood nodes" published in International Journal of Applied Engineering Research, ISSN: 0973-4562, Number 402(2015).

[18] Rausheen Bal, Aditya Bakshi and Sunanda Gupta" Performance Evaluation of Optimization Techniques with Vector Quantization used for Image Compression" published in International Conference on Harmony search, Soft Computing and Applications,2018, ASIC 2018.

[19] Aditya Bakshi and Sunanda Gupta" A Comparative Analysis of Different Intrusion Detection Techniques in Cloud Computing" published in 2nd International Conference on Advanced Informatics for Computing Research ,2018, CCIS 956, pp. 358–378.

[20] Aditya Bakshi, Sunanda Gupta "A Taxonomy on Biometric Security and its Applications" International Conference on Innovations in Information and Communication Technologies, ISSN. 2522-8714, pp.

Cause Effect Analysis of Ecological Pollutants on Internal Physique of Human Subjects Using Radial Recurrent Neural Network Approach

Y Pimpale[1], **S GUPTA**[2*], **R Kanday**[3]
[1,2,3]Lovely Professional University, Phagwara, Punjab, 144411, India
yogitaspimpale@gmail.com[1]
sachin.23305@lpu.co.in[2*]
rajeev.13544@lpu.co.in[3]

Abstract: Globally, ecosystems are changing at an unprecedented rate. Ecosystem management include natural resources and the biophysical environment, but it also requires consideration of all anthropogenic aspects, including social, economic, and cultural factors .Environmental factors are thought to be responsible for almost half of the worldwide burden of disease. Ecosystem changes are increasingly recognized as having an impact on human health and playing an important part in the onset and re-emergence of an expanding variety of diseases. Ecological and environmental Imbalances negatively affect on human health, food security and global economic geopolitical stability. In this study, a cohort-based data set of Ecological pollutants and Physiological signals such as ECG and anthropogenic data of human subjects were extracted from Maharashtra. A hazard ratio based on neural networks was developed and found to be deplorable in both the unhealthy and healthy categories of human individuals. This research is crucial in shedding insight on the influence of interactions between natural and anthropogenic variables on human health. Suchinitiatives might contribute to a better knowledge of the human health consequencesof accelerated environmental change, as well as better decision-making in the fieldsof environmental conservation, public health policy, and new management frameworkdesigns.

Keywords: Anthropogenic factors, Recurrent Neural network, Human health, ECG signal

1. Introduction

The ecology of the environment is made up of a number of interrelated elements that are linked to demographic and geographical behavior. Ecosystems are changing more quickly and widely as a result of degradation of natural systems' services, which can have a negative impact on human health, well-being, and economy [1]. Environmental concerns such as an increase in anthropogenic and industry-related unethical practices of fuel burning and waste management are affecting naturalbalances, causing the onset of some types of difficulties in human beings' health. This ecological imbalance is a key contributor to the development and re-emergence of diseases, as well as poor health and life quality. Cancer, Arrhythmia, Hypertension, Asthma, Atherosclerosis, Stroke, and Diabetes are among the diseases that are fast rising in the population on several continents. China and

India are two of the most important locations where the number of cases has been steadily growing over the last several decades. Polluted ecological vulnerability is now one of the most essential jobs for academics to grasp the modeling scenario in order to prepare solution-based protocols and frameworks for humanity's development [2,3]. Climate change, agricultural and animal husbandry, building materials, energy and water availability are the key driving elements behind ecosystem changes. In developing countries, these changes are occurring at a faster and more intense pace.Figure 1 illustrates how an ecological approach to human health is the result of a process of adaptation between an individual and his or her environment, which includes social, cultural, and economic environments as well as access to health services, and how anthropogenic pressure on ecosystems can harm human health. Table 1 emphasizes the loss of fundamental ecological services and the consequences for human health. The effects of physical, biological, chemical, and socialvulnerabilities in the environment caused by changed ecosystem processes and services have a direct impact on health. Health effects are mediated through more complicated and indirect ways as a result of ecological processes. According to the survey, contaminants in everyday necessities such as pure food, water, ambient environment, and healthy housing have a negative impact on human functionality. Overall, research that collectively modal the causes of all of these ecological elements are needed to scale the consequences on human health organs. Due to a lack of research in this field, the project has been directed at the increasing issue through a bilateral cohort study in India's rural areas to better understand the consequences with many specific and temporal elements. [4,5]. The physiological signals of human individuals were evaluated in this cohort study, and neural network- based modeling was used to interpret the elevated concern using independent ecological, demographic, and socioeconomic aspects. To back up the findings, a multilayer Artificial Neural Network-based framework has been created. The study design, methodology, and facts are detailed in the next part, with clear conclusions. The accuracy of modals is also evaluated in terms of their abilityto overcome confounders and variables. The ecosystem approach to human health relies on participatory and interdisciplinary research that is attentive to the needs and aspirations of many socialgroups, including gender differences. As we will see, research is critical in providing light on the influence of natural and manmade factors on human health. We will demonstrate that research is critical in any endeavor to enhance human health through increasing ecological health.

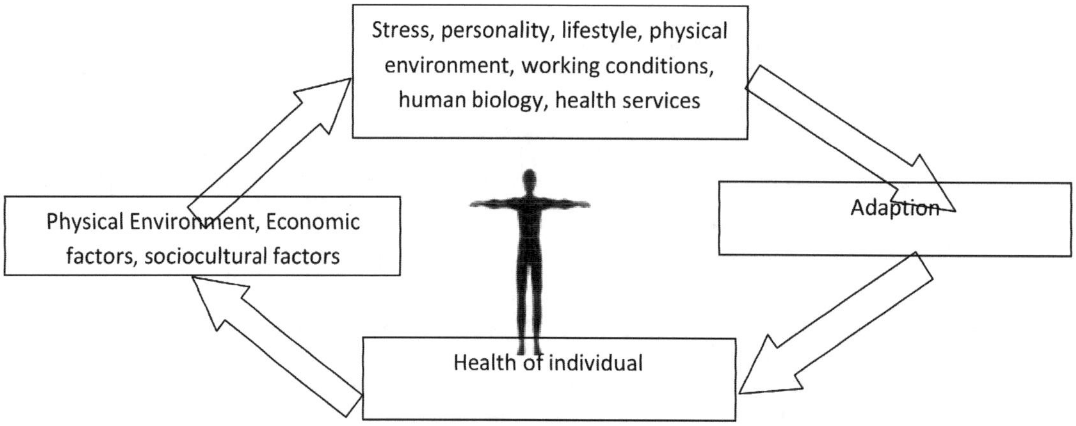

Figure 1. Ecological approach to human health

Table 1: Overview of ecosystem changes affect on health of Human subjects estimated in reported

S No.	Determinant type	Descriptors for the Developing World	Mechanism of Human body Organs and affected part	Diseases and Disorders
1.	Physical factors	Climate variations: Temperature Precipitations Extreme events Housing ,Noise Exposure to ultraviolet radiation	Heart (Indirectly harming the body by causing systemic inflammation and oxidative stress in the circulatorysystem), lung, liver, kidney, cardiactissue, skin, eyes, skin and hair,reproductive system.	Myocardial infarction, Cardiac arrhythmias, ischemic stroke, Vascular dysfunction, hypertension , atherosclerosis Systemic inflammation, oxidative stress, disturbance autonomic nervous system, blood pressure.
2.	Biological factors	Disease vectors: Mosquitoes Fleas Flies Rodents Viruses and microorganisms Biomedical wastes	Muscle, bone ,joint pain, stomach, bleeding from nose or gums, headache, pain behind eyes, swollen glands, rash, Nausea, vomiting,	Malaria, dengue, yellow fever, chikungunya, plague, leishmaniasis, filariasis, plague, tyhus, lyme diseases, relapsing fever.
3.	Chemical factors	Agrochemical products: Pesticides , Fertilizers, Heavy metals(mercury, pollution) Atmospheric pollutants , Dust	Nervous, respiratory, cardiovascular, hematopoietic, immune, endocrine, hepatic, renal system , human cell lines, gastrointestinal tract, skin.(Mercury exposure affects cardiac parasympathetic activity and heart rate variability, and pollution can cause epigenetic alterations and genetic abnormalities)	Atherosclerosis, hypertension, arrhythmia and diabetes, cancer coronary heart disease, myocardial infarction, carotid artery obstruction, mitochondrial damage, inflammation lung function, pro-inflammatory responses, endothelial dysfunction.
4.	Social factors	Employment working conditions ,Culture ,religion ,Income, Education Racial and gender discrimination	Heart, brain, lungs.(The consequences of severe early life stresses that generate recurrent biological stress responses, social variables interact with environmental factors through bio-behavioral pathways that influence human health)	Increases adulthood risk for cardiovascular disease, stroke, diabetes, autoimmune disease, and certain cancers, depression

2. Materials and Methods

2.1. Data Collection

For data related to these aspects, Local pollution control organizations and the Indian Meteorological Department (https://mausam.imd.gov.in) provided information on these areas. Physiological data from human individuals, such as ECG, PPG, and other socio-economic survey data, were obtained in Maharashtra from Pravara Medical Trust, Loni, Maharashtra (https://www.pravara.com). Patients were treated for free by physicians and practitioners at these NGO's under sanctioned camps. Human individuals of various ages were subjected to regular signal measures, with healthy and non-healthy people undergoing physiological signal assessments [5,6,7]. The doctor monitored and documented physiological signals such as ECG, PPG, and other anthropogenic information with the individuals' agreement for research purposes. This will aid research into improving human health through enhancing the ecosystem's health. These organizations assessed anthropogenic parameters such as body weight, height, mass, index, socioeconomic status, history, and symptoms of illnesses through routine check-ups and questionnaires. The ECG signal was recorded using a professional six-channel system that was available at all clinics that followed a conventional medical practice. The PPG signal was measured using a conventional infrared-based meter for heart-related parameter monitoring. This aid in comprehending the effects of natural and anthropogenic factors has on human health. All physiological measures were performed non-invasively, without the use of any drugs or synthesizers.

2.2. Modeling based on Artificial Neural Networks (ANNs)

The neural network can take into account unpredictable and dynamic covariates in respect to independent factors that can influence the dependent variable. Another challenge in the modeling framework was the modification of variables and estimation of confounder weight age, are the significant issues to straighten the hypothesis of results. In comparison to existing statistical models, ANN is one of the evolutionary models for multi-parameter adjustment with over 90% testing accuracy for prediction. Practicability, Simplicity, and accuracy are the main advantages of theproposed ANN models [5,6,7,8]. The processing of physiological data under the confluence of ecological factors has been divided into five steps, as indicated in Figure 2.

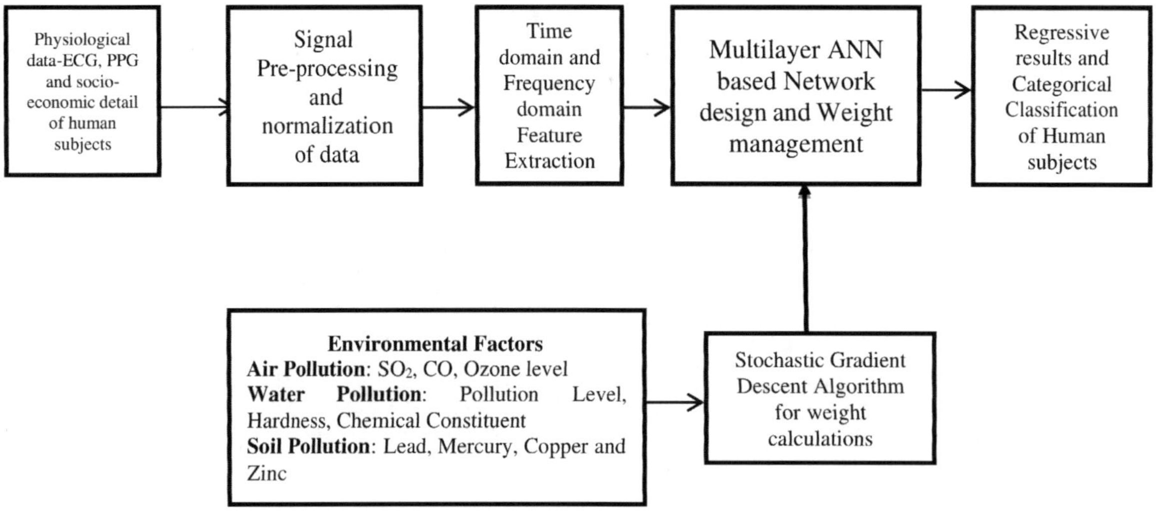

Figure 2. Development and Process Modelling for MLANN for Hazard ratio calculations and Estimation models

The time domain and frequency domain characteristics of physiological data from ECG signal were retrieved. There are numerous features in each category, however during initial correlation procedures, two features were chosen based on their instance probability index.

2.3. Time Domain Features

2.3.1. Spectral Entropy: It is the computation of the spectral power distribution as well as the time-series signal's forecasting capacity. This parameter is used to compute the spread of energy in a signal with a high probability index and is used to calculate the distribution of spectral power over time for module based peak variations. Shannon and information entropy in the information data are used to calculate this entropy. The signal's spectral entropy is given by,

$$X(K) = \sum_{n=0}^{N-1} x(n) W_N^{kn}, 0 \leq K \leq N-1 \quad (1)$$

Where, $W_N = e^{\frac{-j2\pi}{N}}$

2.3.2. Skewness: Skewness is the degree of departure from symmetry which depends on the comparative positions of mean, median and the mode. In the computation of variation under the mean of a specified part, this time domain parameter is utilized to calculate the averaged cubed deviation in respect to the signal's standard deviation. It is determined mathematically as indicated in equation

$$Skewness = \frac{\sum_{i=1}^{N}(Y_i - \overline{Y})^3}{(N-1)s^3} \quad (2)$$

Where $\overline{Y}$ -is the mean
S -is the standard deviation
N -is the number of data points

2.3.3. Kurtosis: kurtosis is measure of the peakednes of a distribution. The higher the kurtosis, the more peaked the distribution will be. The kurtosis is determined and provided as a relative or absolute value. Kurtosis is calculated as. Kurtosis is measured by β_2.

$$\beta_2 = \frac{\sum_{i=1}^{n}[\frac{x_i - \mu}{\sigma}]^4}{N} \quad (3)$$

2.4. Frequency Domain Features: Power spectral density parameters

- Ultra low frequency (ULF): 0.0001 Hz ≥ ULF < 0.003 Hz
- Very low frequency (VLF): 0.003 Hz ≥ VLF < 0.04 Hz
- Low frequency (LF): 0.04 Hz ≥ LF < 0.15 Hz
- High frequency (HF): 0.15 Hz ≥ HF < 0.4 Hz

The spectral power of the RR tachogram is separated into four bands. These spectral characteristics demonstrate strong discriminating between persons with abnormalities and those who are healthy.

3. Results and Discussion

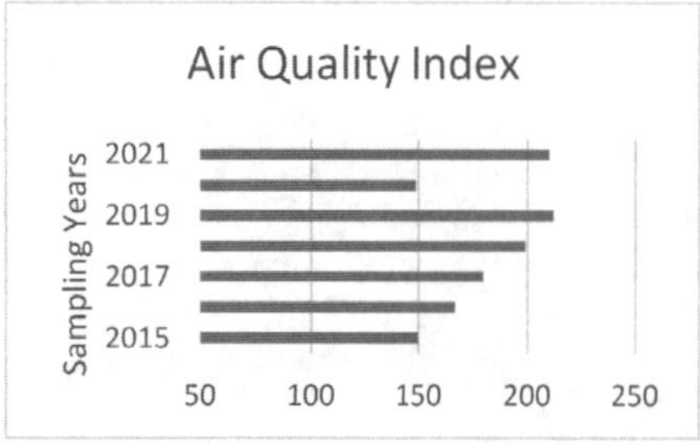

Figure 3. Air Quality Index at sampling site

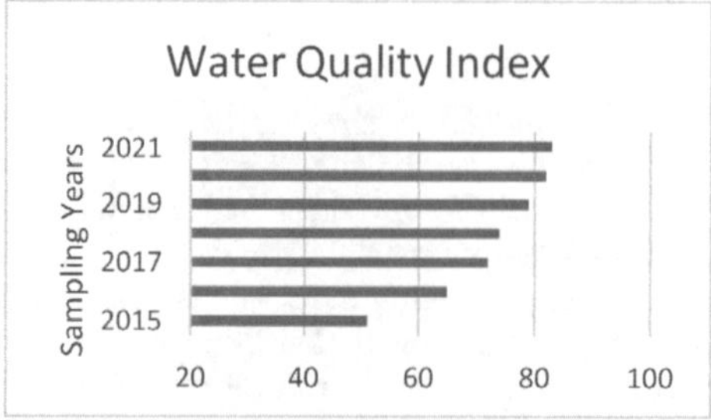

Figure 4. Water Quality Index at sampling site

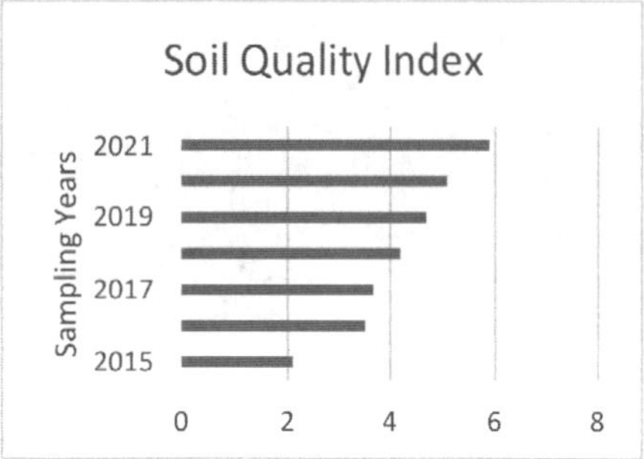

Figure 5. Soil Quality Index at sampling site

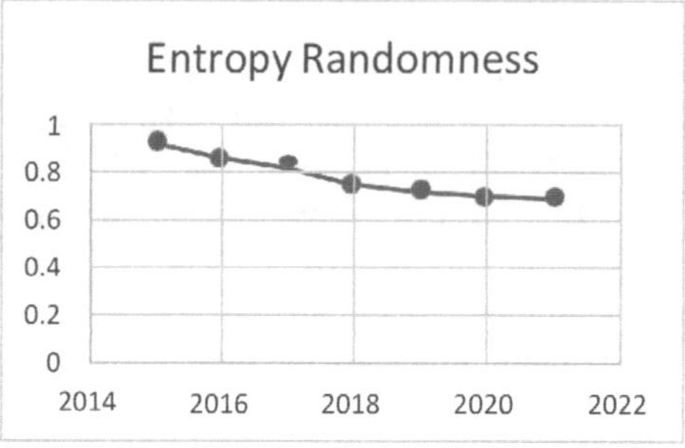

Figure 6: Entropy Randomness' feature of ECG

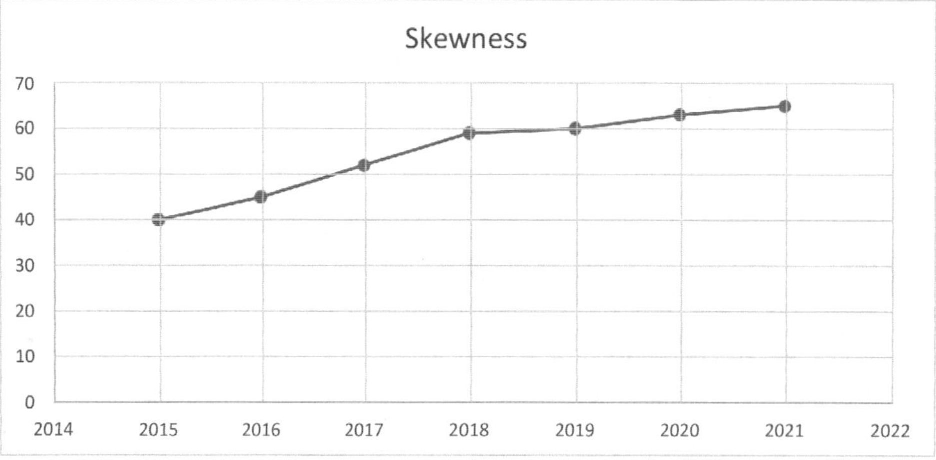

Figure 7. Skewness based deviation of ECG

Figure 3 to Figure 5 shows the variation of levels of pollution in selected sites for the sampling year 2019 to 2020. At site 1, the pollution level varies from 21µg to 29 µg of cubic sample of soil with ±1.3 µg of standard deviation. At site 2 the level of Pollution varies from 21 to 31µg with standard deviation of ±1.7 µg. At site 3 the maximum level of pollution was increased to 34µg with ± 2.3 µg of deviation which is 2.6% more than other sites. At site 4, the variation of Pollution is from 25 µg to nearly 34µg. At site 5, the level of Pollution is varied from 26µg to 36 µg in the standard of soil sample.

During sampling years, the level of Pollution in the soil of sampled sites varied from 23% to 76% higher than standard permitted level. Such variation clearly shows the vulnerability of inhabitants due to their practice of burning Dunk cake cakes and residue of plants and crops. As per WHO reports, such level of Pollution hypothetically responsible of increases in health issues in various organs of human and animals in the respected areas [3,4,5]. As per discussion with doctors, clinicians and Head of villages, it has been observed that there are various aspects that are responsible for releasing of such toxic in the soil of this area such as use of anthropogenic sources in an unorganized manner and requirement of fuel in the area. Other chemical factors such as use of Pesticides, growth of hybrid crops such as paddy are also responsible for increased level of Pollution in the soil of these areas. The recycling water in these areas are also one of the main reasons for increased level of Pollution in the area. At present, in this region excess use of pollution pro founded products is on higher demand due to their cheap cost and to maintain soil turbulence. At sites, thelevel of pollution is much higher than the permitted levels which cause the likelihood of health-relatedissues in the inhabitants of sites. The overall variation at each site increased by 23.45% from previous years which indicates a constant growth in level of pollution in the soil samples of each site. There is no such comparative difference in the level of Pollution due to more or less the same practice of pollution generation in concerned areas. Due to paddy as main crop of these areas, the use of waterand chemically oriented pesticides, urea-based fertilizer sand ash based domestic fertilizers gradually contribute a main role in contamination of soil. The average rain in these areas also contributes insuch practices for travel of Pollution from one area to other through various ways of interconnected sewerages. So, there was a need of this study to analyses ground-based aspects and they're in relation with disorders of human subjects. To investigate the worst effects of Pollution on the human subjects, the collected data from clinicians and other authentic sources is bifurcated in various categories for clear analysis. The machine learning based algorithm was designed for high probability index and incorporation of non-linear and confounding factors for high mean square effort value. In further results, the MSE, RMSE, MAPE and MABE based results are produced for exhaust analysis. Fig. 7 is showing the physiological and dermis-based effects on both male and female gender in the selected sites.4. In first quarter of 2019, the pollution effect was observed on 75 to 110 persons over 1000 cases reported in that duration. The cases of dermic laps were more in women's than men by 6.87%. Similarly, the cases were extended by 12% in the second quarter of year 2019 [5,6,7]. The cases of skin etching and faint of skins were more in both categories. The skin related issues were observed to be more in female candidates. In second quarter the effects were increased from previous quarter by 5.34%. So, overall, the increment was 8.43% from the base period of study. In the first half of year 2020, the cases of dermic related were going to increase from 100 to 250 persons out of 1000 reportedcases in the areas. In this duration the cases were still increasing with the rate of 7.48% from the previous and base years due to exposure of pollution generation activities in the areas. The report of cases was more in female candidates than men by 14.21%. Similarly in the end of year 2020, the casesof skin related diseases were increased from 100 to 350 persons per reported 1000 cases. As per collected data and questionnaire-based survey, it has been observed from the data that the females were more prone to pollution exposure due to their more play period in such activities like cooking, making domestic fertilization, washing of utensils with ash of dunk cakes than men, the men who were reported pollution contaminated were maximally working in the brick construction factories and farming where the sources of pollution generation are also more. The results clearly showed the trend of increment of cases in the areas in relation to more cases in female candidates than men due to occupation related exposure periods.

In the first half of selected sampling years, the cases of high BMI persons varied from 120 to 150 but for other categories it varied from 70 to 90. Similarly in next half year of selected sampling years, the cased were increased by 8.87% in High and Low BMI persons. Similarly in the next consecutive years, the rate of cases was increased by 5.63% from the previous year rate. It has been observed fromthe BMI results that the BMI of human subjects have also a great importance to judge the health statusin relation to affect modifiers. There are various toxic metals which are coming in contact with humansubjects from various activities and directly or indirectly affecting the health of living inhabitants by their redox and oxidation processes. As per report, pollution is one of the most common and toxic metal which is generated from various human activities during day-to-day tasks [10,11,12]. There are various activities from where pollution is originated but the use of dunk cakes in household for fire and cooking is one of the significant sources which came in to picture. There is no consideration of BMI and socio-economic factors in the analysis part of such kind of exposure related study. In this study the use of ANN for association of Covariates and confounders produces strong likelihood of results due to adjustment of nonlinearities in data with minimum uncertainty [13,14,15,16]. This study reveals the practices which were used by weak economic persons for their survival. In future, such kinds of studies are required where ANN has to incorporate instead of conventional statistical methods for strongest of hypothesis. The results in the study may enhance the efforts of NGO's and local government to do the needful for these kinds of inhabitants.

4. Conclusion

The undue ecological burden of raised level in AQI, WQI and SQI is a global issue for pollution monitoring and control agencies to profound some valuable actions. The unethical practices in the ecology exponential raised the burden and with time, this burden is increasing due to low mixing rates and resistance of natural sources to dissolve then. Consequences, the health parameter of human subjects are depreciating. The hypothesis of these burden rendering parameters is observed by measuring the time domain-based features from ECG signal at sampling sites. The results are showing a steep decrement in skewness and entropy randomness parameters. The trend of ECG signal variation quartiles is a clear indication of health effects on physiological structure of human subjects. In future, there is need to solicitate such kind of environmental engineering-based studies to understand the depth of issue. Our results will raise the concern of pollution control agencies to accommodate such methods to reduce the ecological burden on human subjects.

Acknowledgement

Authors are thankful to Pravara Medical Trust, Loni, and Maharashtra for providing the data for generation of perspectives.

Conflict of Interest

Authors have no conflict of Interest.

https://orcid.org/0000-0003-4457-7410
https://orcid.org/0000-0003-3515-8601
https://orcid.org/0000-0003-2569-1891

References

[1] Abdul K S M, Jayasinghe S S, Chandana, E P, Jayasumana C and De Silva P M C 2015 Arsenic and human health effects: A review *Environmental toxicology and pharmacology* **40(3)** 828-846

[2] Alvarez H.A O, Kubzansky L D, Campen M J and Slavich G M 2018 Early life stress, air pollution, inflammation, and disease: an integrative review and immunologic model of social-environmental adversity and lifespan health *Neuroscience & Biobehavioral Reviews* **92** 226- 242

[3] An Z , Jin Y, Li J, Li W and Wu W 2018 Impact of particulate air pollution on cardiovascular health *Current Allergy and Asthma Reports* **18(3)** 1-7

[4] Chen J, Su Y, Lin F, Iqbal M, Mehmood K, Zhang H and Shi D 2021 Effect of paraquat on cytotoxicity involved in oxidative stress and inflammatory reaction: A review of mechanisms and ecological implications *Ecotoxicology and Environmental Safety* **224** 112711

[5] Cosselman K E, Navas-Acien A and Kaufman J D 2015 Environmental factors in cardiovascular disease *Nature Reviews Cardiology* **12(11)** 627-642

[6] Cunningham G B, Wicker P and McCullough B P Pollution 2020 health and the moderating role of physical activity opportunities *international journal of environmental research and public health* **17(17)** 6272

[7] Du Y, Xu X, Chu M, Guo Y and Wang J 2016 Air particulate matter and cardiovascular disease: the epidemiological, biomedical and clinical evidence *Journal of thoracic disease* **8(1)** E8

[8] Ferdous Z and Nemmar A 2020 Health impact of silver nanoparticles: a review of the biodistribution and toxicity following various routes of exposure *International journal of molecular sciences* **21(7)** 2375

[9] Fiordelisi A, Piscitelli P, Trimarco B, Coscioni E, Iaccarino G and Sorriento D 2017 The mechanisms of air pollution and particulate matter in cardiovascular diseases *Heart failure reviews* **22(3)** 337-347

[10] Gautam D, and Bolia N B 2020 Air pollution: impact and interventions *Air Quality, Atmosphere & Health* **13(2)** 209-223

[11] Genchi G, Sinicropi M S, Carocci A, Lauria G and Catalano A 2017 Mercury exposure and heart diseases *International journal of environmental research and public health* **14(1)** 74

[12] Hasan M K, Shahriar A and Jim K U 2019 Water pollution in Bangladesh and its impact on public health *Heliyon* **5(8)** e02145

[13] Izah S C , Chakrabarty N and Srivastav A L 2016 A review on heavy metal concentration in potable water sources in Nigeria: Human health effects and mitigating measures *Exposure and Health* **8(2)** 285-304

[14] Lehner R , Weder C, Petri-Fink A and Rothen-Rutishauser B 2019 Emergence of nanoplastic in the environment and possible impact on human health *Environmental science & technology* **53(4)** 1748-1765

[15] Munawer M E 2018 Human health and environmental impacts of coal combustion and post-combustion wastes *Journal of Sustainable Mining* **17(2)** 87-96

[16] Shahrbaf M A, Akbarzadeh M A, Tabary M and Khaheshi I 2021 Air pollution and cardiac arrhythmias: A comprehensive review *Current Problems in Cardiology* **46(3)** 100649

Journal of Physics: Conference Series 2327 (2022) 012067 doi:10.1088/1742-6596/2327/1/012067

Co-Resident Attack and its impact on Virtual Environment

Sudibyajyoti Jena, Likhet Kashori Sahu, Debahuti Mishra, Madhuri Rao, Kulamala Vinod Kumar

Department of Computer Science & Engineering, Siksha 'O' Anusandhan Deemed to be University, Bhubaneswar, Odisha, India.
madhurirao@soa.ac.in

Abstract. Cloud computing (CC) is one of the emerging computing models that potentially transform the IT industry. The cloud computing is defined as a computing paradigm that enables suitable, universal, on-demand network access of various computing resources like data storage, OS (Operating System), computer applications and software as well. A Cloud model essentially is characterized with features like on-demand self-service, better network access, resource distribution, elasticity and measured services. Cloud computing is also known as Internet Computing and is indeed reshaping the way industries and organisations function. It facilitates the service taker or the client to use various IT resources like OS (Operating System), middleware and various applications without the need of downloading it. In this paper a brief view of this computing paradigm and the underlying cloud infrastructure is presented. We essentially present an analysis of various security challenges that are inherent in the Infrastructure as a Service(IaaS) cloud model and explain terms like virtualization, VM (virtual machine), SLA (Service level agreement), etc. Here, CloudSim simulator is explored to study the impact of variation of cache size of the Virtual Machine on the availability of data centres ,that occurs due to co-resident attack. Further, a brief view of the co-resident attack is further presented and some of the feasible countermeasures are discussed further.

Keywords: Hypervisor, Virtual Machine, Co-resident attack, Infrastructure as a Service (IaaS)

1. Introduction of the Cloud Infrastructure

Cloud computing is vastly used in both large and small industries and organizations. According to [1], a cloud model consists of 5 essential characteristics like on-demand self-service, better network access, resource distribution, elasticity and measured services. So it is important to understand its infrastructure as well as service and deployment model. There are several components in cloud architecture like client infrastructure, runtime cloud, cloud storage, internet and data management. In 1961 John MacCharty in his speech said that computing can be sold to the client or user as any other utility. Though at that time might it seemed impossible but the evolution of technology made this possible. The computer architecture defines as an infrastructure that provides on-demand services over the internet like a data storage facility, applications, testing environment, etc for performing various tasks [2]. Based on the deployment model cloud can be of four types as follows.

Private Cloud: It is also named as internal cloud and it is that cloud infrastructure which is accessible by the users within an organization. This deployment model is appropriate for those organization for whom data security and confidentiality is the priority. This model provides better performance, storage as well as high data security.

Public Cloud: This deployment model is also known as pay and use model. In this cloud infrastructure, the service provider supervises the resources available over the cloud to the consumer. Some of the examples of this model are IBM Smart cloud Enterprise and Windows Azure Service Platform.

Community cloud: This cloud infrastructure provides services to one or more organisations. This is mostly used in organisations like the health care community, universities, etc. One of the major advantages of this model is that it provides better data services over public cloud.

Hybrid Cloud: This is the mixture of different cloud infrastructures like public cloud, private cloud and community cloud. The major advantage of this model is that it reduces the risk while providing better accessibility to the user. The two major disadvantage of this model is that data security completely depends on the service provider and managing different types of cloud infrastructure is difficult. Some of the examples of this model are Gmail, Google Drive, Microsoft Office, etc.

Based on service provided by cloud there are three types of service models:

SaaS (Software as a Service): In this service model the cloud services provide software service or application that runs on cloud infrastructure and is accessible to the client through a web browser. In this service model data security and system update completely relies on the service provider [2].

PaaS (Platform as a Service): This service model avails the client or user with tools (like a virtual platform) to build, test and deploy higher-level services [3]. This model provides services like OS, application software, development and testing environment to the developer for building services with the help of programming language. And the client access to those services produced by the developer with the help of an API (like Google App Engine).

IaaS (Infrastructure as a Service): In this service model the service provider delivers the computer infrastructure (like storage, processing, and other computing resources) based on use and pay scheme. The user of this service model can manage and control the operating system, data storage and application but has no control over the cloud infrastructure [1]. The two main features of IaaS are (1) physical components like hardware, storage, and (2) cloud software like API which provides the user with the accessibility to the cloud service. One of the major advantages of using this paradigm is on-demand scalability. Some of the examples of IaaS provider are AWS (Amazon Web Service), RC (Reliance Communication), etc.

2. IaaS Model Security Issues

This section explains the important technologies (i.e., Virtualisation, Virtual Machine, and Virtual Machine Monitor) used in this service model and various security vulnerabilities (like co-location attack) that are possible in this paradigm. Virtual machine, also called virtual servers is a software program which act as emulation of actual computer system. Virtualization is defined as a technology that enables partitioning of a mainframe computer to many logical instances (called virtual machine) that runs and managed by a single computer mainframe [4]. The advantage of this technology is its scalability and its ability of partitioning that allows several applications to run at the same time minimizing the overhead and maximizing the efficiency [5]. The CVI (Cloud Virtual Infrastructure) comprises of three major components [6]: hypervisor (or, Virtual Machine Monitor), virtual switch and Virtual Machine (VM). The virtual machines are the most vulnerable to attacks. These hosted virtual machines are prone to attack as the cloud service provider has no control over it [7]. The major security threats on this service model includes data breach, system based attacks, insecure interface, malicious insider, etc, that affects data confidentiality, data integrity and data availability. Some of the basic features expected in terms of security in cloud domain are-

Confidentiality- This requires to ensure that data when in transit and in storage is accessible only to authorized entities. An un-authorized party should not be able to even read a message that a cloud consumer issues as a cloud service.

Integrity - It is very important to ensure that the data transmitted by a cloud service is received exactly as a cloud service on the receiver end. Data integrity requires to look into aspects of how data is stored, retrieved and importantly how it is processed.

Authenticity- It is the very important to ensure that any data accessed or viewed is made available from a source that is authorized. This further ensures that none of the parties deny an interaction that may have happened between them at a given instant.

Availability- A cloud service needs to be available seamlessly so that it could be accessed and be used.

2.1 Motivation and our contribution-

A cloud based interaction must facilitate that all above aspects of security are attained in its each and every interaction that takes place between a consumer and a provider. The motivation of this research is to explain how a cloud interaction may not appear unreliable at all but at the same time data breach may happen. An hacker can cause a co-resident attack, which needs to be detected and avoided. In this paper so-resident attack is explained. A simulation study explaining a scenario of how un-noticed this attack could be, is explained. Further some possible countermeasures are briefed.

2.2 Platform Virtualisation:

For understanding the co-location attack and its consequences it is necessary to know various security threats in platform virtualisation and how they affect the cloud infrastructure. Virtualisation environment is prone to all types of common attacks and when more systems run in a virtual environment the risk of attacks increases to a higher extent as adversary or attacker get more entry point and interconnection for launching attacks [8]. The attack on platform virtualisation classifies into three type's hypervisor based attacks, VM based attacks and VM image based attacks.

2.2.1 Attacks on Hypervisor:

A hypervisor (virtual machine monitor) is a portion of a software that produces as well as operates one or more virtual machines and act as an interface between the host machine and hardware. VMM can be be categorized as type-1 hypervisor (Native hypervisor) which run directly on the hardware and as type-2 hypervisor. The type-2 hypervisor run on the user operating system itself. One of the major attacks on the hypervisor is hyper jacking. In this attack, the adversary tries to install adversarial VMM that operates secretly and get access to critical data. [9] Introduces a new resource freeing attack that has three major players (i.e., beneficiary, helper and victim). In this, the helper tries to maximize the usage of other resources until it reaches the bottleneck. In this way the victim cannot get further access to the target resource.

2.2.2 Attacks on Virtual Machine:

In cloud computing paradigm VM is an emulation of the computer system that provides functionality similar to a computer system in the absence of hardware infrastructure. One of the major characteristics of virtualisation is isolation; in case isolation is compromised the then the VM in the same environment becomes vulnerable to various attacks. For example, feature like shared clipboard is used to transfer data among several VM but this can create a gateway for the attacker to access sensitive data [10]. In some other attacks (co-resident attack) the malicious user construct side channel between VMs on same virtual environment and get access to sensitive data.
VM Escape: Another type of attack on VM is VM Escape attack. This attack usually happens when the isolation feature between host and VMs is compromised. The attacker takes advantage of any fault in isolation feature and tries to get access to the host machine.
VM Cloning: In this type of attack the attacker copies the virtual machine program files from the host operating system and then makes a clone or copy of the user virtual machine to access a sensitive data.
VM Memory Dump (VMM): In VMMD attack the attacker tries to gain information from the user by dumping the memory assigned to the target virtual machine.

2.2.3 Attack on Virtual Machine Image:

A VM Image is a clone of the virtual machine. Generally, the cloud server provides the customers with the ability to create VM images and avail them to users. Some of the attacks on VM are the following.
Information leakage: The VM image consists of the current state virtual machine when it is copied. So the user of the virtual images can have the risk of losing their sensitive information like password and sensitive data.
Malicious Code Injection Attack: In this attack, the attacker or the adversary tries to inject fake code in the virtual machine and make it public. These virtual machines can successfully move into network unnoticed by any detection mechanism.

3. Co-resident Attack:

In this attack, the malicious user tries to collocate the VMs on the same server and built various side channels to get sensitive data from the victim. This attack is also known as co-location attack or co-residency attack. In cloud computing VMs run on the same physical environment, to get maximum utilization of hardware platforms and are isolated from one another. But the adversarial user overcomes the logical isolation to gain data from the actual user [11]. In real-world cloud environment launching such attack is arduous, but as most of the paper on this attack showed how this attack can be potentially harmful so the risk is high in any cloud platform in near future. Before going for the steps on how this attack is launched we will explain what side-channel attack is. A side-channel attack focuses on gathering information about faults in the application of the computer system and on its basis launch the attack. The various side-channel attacks that work along with co-locating VMs are power-monitoring attack, timing attack, cache attack, etc. In power monitoring attack the adversary tries to gain information on the degree of power consumption during computation [12]. In the cache attack, the attacker launches the attack based on cache access. While timing attack is based on the variation of computation time of different activities. For example, the attacker can deduce encryption key by observing accesses made by the victim. In co-location attack to verify that two or more VMs are co-locate, the attacker performs TCP traceroute operation to get Dom0 IP address and then compare the Dom0 IP address of those VMs (same Dom0 IP address mean they are co-resident) to know if they are co-resident. To achieve co-residence the attacker uses brute force strategy. Fig-1 describes how this attack is done with the help of flow chat.

The steps for launching this attack are:

1. The attacker set target VMs
2. The attacker starts some adversarial virtual machines.
3. Check that new VM is co-residence with the target VM.
4. Repeat steps 2 and 3 until the attacker co-locate the target VM
5. Once the VM is co-located the attackers apply various side channels to gain sensitive information.

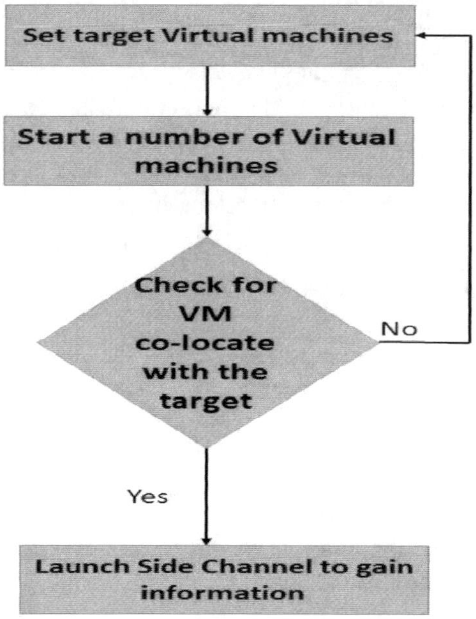

Fig-1: Flowchart for the description of Attack

4. Simulation Study & Analysis

In this a simulation study was carried out in cloudsim simulator. The considered scenario represents a situation that arises due to co-resident attack. Let us assume that there are two data centres in an organization that are obviously in two different geographical locations. Essentially by being in different locations, data centres ensure that the software systems being accessed and used is rather more reliable and fault tolerant. However, if Virtual Machine running on one data centre is under colocation attack, it would facilitate the attackers' tasks and jobs to run on the virtual server as if there was never an attack. It is very difficult to trace the attack by the behaviour and availability of the VM and datacentres. During a colocation attack an attacker can create job or task that could conveniently be assigned for execution in any of available VM of the data centres. The attacker could create jobs of different time requirements. If the VMs available on of a given data centre is having different MIPS specifications, it makes the detection process very challenging. A simulation study was undertaken to explain how the behaviour of the cloud system is not visibly affected by a co-location attack.

Cloudsim simulator is a JAVA based simulator where Data Centres are mapped to Virtual Machines and Cloudlets or jobs are executed on Virtual Machine with the help of a broker. Fig 2 depicts the output obtained when one Data centre is considered with two Virtual Machines A and B of different configurations are considered to be available. A broker further assigns the cloudlets to these VMs. It can be noted from figure 2 that the execution of cloudlet 1 running on VM1 takes 50% less time than Cloudlet 0 that runs on VM0. Though both the VMs are running on the same data centre and also are assigned with cloudlets of same length, the overall throughput of the system is affected as VMs are not of same configurations. Hence, how cloudlets are assigned has a huge impact on the behaviour of the cloud system. Cloudlets can be doubled by side attack and therefore for a system with varying observations, prediction of possible attack is challenging. The only way to deal with it is to use some countermeasures for avoiding the attack to take place.

```
Datacenter_0 is starting...
Broker is starting...
Entities started.
0.0: Broker: Cloud Resource List received with 1 resource(s)
0.0: Broker: Trying to Create VM #0 in Datacenter_0
0.0: Broker: Trying to Create VM #1 in Datacenter_0
0.1: Broker: VM #0 has been created in Datacenter #2, Host #0
0.1: Broker: VM #1 has been created in Datacenter #2, Host #1
0.1: Broker: Sending cloudlet 0 to VM #0
0.1: Broker: Sending cloudlet 1 to VM #1
8.1: Broker: Cloudlet 1 received
16.1: Broker: Cloudlet 0 received
16.1: Broker: All Cloudlets executed. Finishing...
16.1: Broker: Destroying VM #0
16.1: Broker: Destroying VM #1
Broker is shutting down...
Simulation: No more future events
CloudInformationService: Notify all CloudSim entities for shutting down.
Datacenter_0 is shutting down...
Broker is shutting down...
Simulation completed.
Simulation completed.

========== OUTPUT ==========
Cloudlet ID    STATUS    Data center ID    VM ID    Time    Start Time    Finish Time
    1         SUCCESS         2              1        8        0.1          8.1
    0         SUCCESS         2              0        16       0.1          16.1
CloudSimExample3 finished!
```

Fig 2: Simulation output for one datacentre with 2 Virtual machine with different specifications.

Cross Tenant Information leakage is one often occurring outcome due to attacks and security breaches in cloud platform. It is likely that a Cloud Service Provider is hosting two different customers on the same data centre. The Virtual Machines running on the data centre are essentially housed by the help of hypervisor mechanism. Here often there are chances of information leakage or information being tampered between the customers. It is also possible that one of the company claims that his information is leaked, though without any technical observations but only based on how his competitor is performing. It is not wise for a Cloud Provider to ignore the impact of sharing and multiple accessibility that cloud promises to provide.

Another common scenario of colocation attack is when a set of data centres accommodating one of VMs each and require executing services of more than one consumer for a given network topology on them. It is likely that these VMs may be sharing common cache and power resources, making it vulnerable to such attacks. It is likely that one of the VM is malicious and intends to affect the cloud consumers businesses adversely.

5. Some Countermeasures Against Colocation Attack:

The countermeasures for co-location attack focuses on different techniques to prevent this attack as shown in Fig 3. The countermeasures for this attack can be classified into two types where the first one focuses on preventing the attacker from achieving co-residence and the second one focuses on preventing side-channel attacks and information leakage.

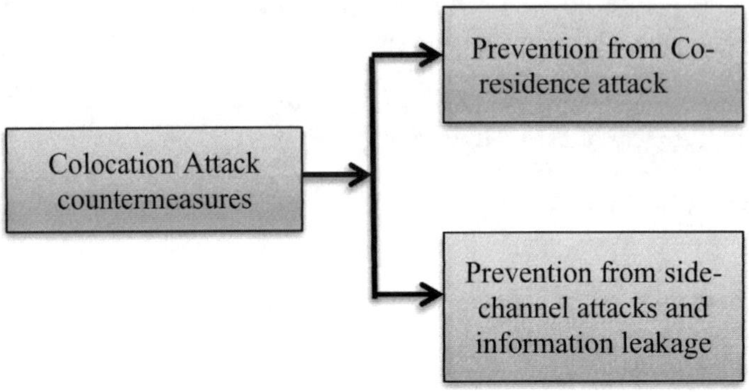

Fig 3. Classification of countermeasures for colocation attack.

5.1 Preventing Side Channel Attacks (SCA):

Side-channels are used to transfer sensitive data between two or more VMs in the same virtual environment. Preventing side-channel attacks is important for data confidentiality of the authentic user. This countermeasure is hardware-based and costly. Most of the SCA in the cloud is based on high-resolution clocks. [13] Proposed a method of eliminating internal clock reference to protect sensitive information.

5.2 Migrating VMs Periodically:

This method helps to counter the co-resident attack. VCG (Vickrey-Clarke-Groves) mechanism is one that provides a periodical migration of virtual machines. The drawback of this method is that frequent migration of VMs can cause performance degradation, power consumption, etc.

5.3 Preventing Verification of Co-residence:

We already discussed that the attacker can verify co-residence of two or more virtual machines by checking their Dom0 IP address while other techniques that the attacker can use to detect co-residence are very difficult to achieve. So, the cloud security provider confines the Dom0 IP address from the malicious user.

5.4 Detecting Co-location Attacks:

The security provider can detect colocation attacks by detecting various abnormalities. In [14] Yu et al observes that during the co-resident attack there are irregularities in CPU and RAM utilization and cache miss behaviour.

5.5 VMs Allocation Policy:

This countermeasure focuses on increasing the probability of non-co-resident VMs. In CLR (Co-location Resistance Algorithm) a fixed number of servers open and randomly allocate a new VM to one of these servers. This algorithm basically tries to ensure that a sense of VM isolation is maintained.

6. Discussions

Currently as per [15] it is was anticipated that IaaS would see a growth by 13.4% in the beginning of 2021 itself. Cloud services have a huge demand now and many businesses are trying to upgrade to cloud platform. However, cloud does have some very serious threats and security challenges, which the entire technical community is trying to address. IaaS elevates and leverages business to adopt to the changing scenarios and reduce cost of operations tremendously but at the same time it secure and reliable delivery is dependent not just on the cloud delivery partner but indeed is a shared responsibility. The cloud architecture and deployment methodology also have a huge role to play when it comes to how cloud is ensuring scalability, reliability, elasticity and on demand usage of resources for the cloud customer. Cloud facilitates pay as you go, but no customer wants his data to be breached or compromised and even eavesdropped. The potential harm co-resident attack could do is significant and cannot be ignored. Many new cloud customers are not well aware of the potential of a Service Level Agreement and often don't see how their trust on provider is just a myth. Many applications that adhere to utility computing are also prone to be attacked by hackers who intend to use services without paying. Sometimes the novice cloud customer exposes his authentication keys without knowing the potential harm it can cause to his business. Hence, security and authentication have a huge role in how successfully a cloud consumer explores the potential of cloud technology.

7. Conclusion

This paper demonstrates the security and privacy issues in cloud computing paradigm and presents detail about cloud architecture, virtualization, VMM and VM. CO-location attack and side-channel attacks are explained here. These attacks manipulate the logical isolation feature and affect data confidentiality like encryption key and password. This paper further explains how co-location attack occurs in a step by step procedure. A simulation study is carried out to explain how challenging this attack is and how it can go unnoticed. Understanding how the attack happens is essential in order to develop a feasible solution to prevent it. Some of the countermeasures currently available are also explained in this paper. The countermeasures are basically of two types and are based on either preventing a co-residence attack from occurring or by preventing a side channel attack. Our study shows how the system performance cannot be used as an indicator to identify such a threat. Tracing IP packets and observing the behavior of the traffic are essentially the only ways to avoid it at some point of time. Detecting the colocation attack based on cache miss could be a possible wat to detect the attack, which we intend to explore in our future work.

ORCID ID

Sudibyajyoti Jena: 0000-0003-0432-5884, Likhet Kashori Sahu: 0000-0002-7404-9937, Debahuti Mishra: 0000-0002-6827-6121, Madhuri Rao: 0000-0003-1976-0627, Kulamala Vinod Kumar: 0000-0002-3718-5606

References

[1] Puthal, D., Sahoo, B. P., Mishra, S., & Swain, S. (2015, January). Cloud computing features, issues, and challenges: a big picture. In *2015 International Conference on Computational Intelligence and Networks* (pp. 116-123). IEEE.

[2] Krishna, B. H., Kiran, S., Murali, G., & Reddy, R. P. K. (2016). Security issues in service model of cloud computing environment. *Procedia Computer Science*, *87*, 246-251.

[3] Ahmed, A., & Zakariae, T. (2018). IaaS cloud model security issues on behalf cloud provider and user security behaviors. *Procedia computer science*, *134*, 328-333.

[4] Reuben, J. S. (2007). A survey on virtual machine security. *Helsinki University of Technology*, *2*(36).

[5] Ibrahim, A. S., Hamlyn-Harris, J., & Grundy, J. (2016). Emerging security challenges of cloud virtual infrastructure. *arXiv preprint arXiv:1612.09059.*

[6] Ibrahim, A. S., Hamlyn-Harris, J., Grundy, J., & Almorsy, M. (2011, September). Cloudsec: a security monitoring appliance for virtual machines in the iaas cloud model. In *2011 5th International Conference on Network and System Security* (pp. 113-120). IEEE.

[7] Bedi, H. S., & Shiva, S. (2012, August). Securing cloud infrastructure against co-resident DoS attacks using game theoretic defense mechanisms. In *Proceedings of the international conference on advances in computing, communications and informatics* (pp. 463-469).

[8] Mishra, S. K., Sahoo, B., & Parida, P. P. (2020). Load balancing in cloud computing: a big picture. *Journal of King Saud University-Computer and Information Sciences*, *32*(2), 149-158.

[9] Varadarajan, V., Kooburat, T., Farley, B., Ristenpart, T., & Swift, M. M. (2012, October). Resource-freeing attacks: improve your cloud performance (at your neighbor's expense). In *Proceedings of the 2012 ACM conference on Computer and communications security* (pp. 281-292).

[10] Kirch, J. (2007). Virtual machine security guidelines. *The Center for Internet Security*.

[11] Han, Y., Chan, J., Alpcan, T., & Leckie, C. (2014, June). Virtual machine allocation policies against co-resident attacks in cloud computing. In *2014 IEEE International Conference on Communications (ICC)* (pp. 786-792). IEEE.

[12] Sahoo, S., Mishra, S. K., Sahoo, B., & Turuk, A. K. (2019). Co-resident Attack in Cloud Computing: An Overview.

[13] Aviram, A., Hu, S., Ford, B., & Gummadi, R. (2010, October). Determinating timing channels in compute clouds. In *Proceedings of the 2010 ACM workshop on Cloud computing security workshop* (pp. 103-108).

[14] Yu, S., Gui, X., & Lin, J. (2013, January). An approach with two-stage mode to detect cache-based side channel attacks. In *The International Conference on Information Networking 2013 (ICOIN)* (pp. 186-191). IEEE.

[15] https://dzone.com/articles/most-common-iaas-security-issues-and-ways-to-mitig

Journal of Physics: Conference Series **2327** (2022) 012068 doi:10.1088/1742-6596/2327/1/012068

Machine learning based design of reinforced concrete shear walls subjected to earthquakes

Praveena Rao[1*], Hemaraju Pollayi[1] and Madhuri Rao[2*]

[*] Corresponding Authors

[1] Department of Civil Engineering, GITAM University, Hyderabad, TS, India

[2] Department of Computer Science Engineering, Siksha 'O' Anusandhan Deemed to be University, Bhubaneshwar, Odisha, India

E-mail: raopraveena6@gmail.com, dpollayi@gitam.edu, madhurirao@soa.ac.in

Abstract. Civil engineering structural components are classified according to their projected structural performance in the present building code regulations and design standards. These building design codes are largely based upon previous experimental results of thousands of samples tested to failure and validated with analytical solutions. Machine Learning techniques (ML) is a subset of Artificial Intelligence (AI) that facilitates classification and prediction of structural performances for a broad spectrum of complex structures with greater accuracy. Machine learning models have the potential to make reliable predictions with the help of algorithms. Thereby, saving a tremendous amount of time and resources invested in experimental investigations of large structural components such as shear walls and columns. The ML algorithms can learn from the available data, deduce underlying inter-relationships, make inferences and detect patterns based on previous experience. In the present work, various ML algorithms were implemented to identify the influence of geometrical as well as mechanical characteristics. Database of 393 specimens of reinforced concrete shear walls with rectangular (R), flanged (F) and barbell (B) cross-sections are adopted for the analysis. Shear walls are fundamentally classified into four failure categories which include flexure or due to bending, shear, intermediate flexure-shear and sliding due to shear. The objective of this paper is to classify and predict the shear strength, flexural strength as per the Indian standard code provisions and failure modes of shear walls with the help of ML techniques. Algorithms such as KNearest Neighbors, Naive Bayes, Decision Tree, Random Forest, AdaBoost, LightGBM, XGBoost and Cat-Boost is implemented using Python. Highest accuracy of 85% is achieved on the test set by Random Forest, 83% by CatBoost and 81% by LightGBM boosting algorithms. It is observed that input variables such as aspect ratio (lw/tw), characteristic strength of concrete in compression (f_{ck}), characteristic yield strength of steel (f_y), percentage of steel (ρ), web vertical reinforcement, horizontal reinforcement, boundary element reinforcements play a vital role in governing the shear strength (V_u) and flexural strength (M_u) of shear walls.

1. Introduction

The present work is aimed at predicting the ultimate strength of shear walls in flexure due to bending and shear, as well as the modes of failure based on mechanics driven equations conforming to Indian standard building codes. General provisions for earthquake resistant buildings from IS 1893 (Part I): 2002 and reinforcement ductile detailing in concrete structures for seismic safety from IS 13920: 1993 have been considered in this model. Shear walls are vertical components found in multi-storey high rise buildings, typically designed for the purpose

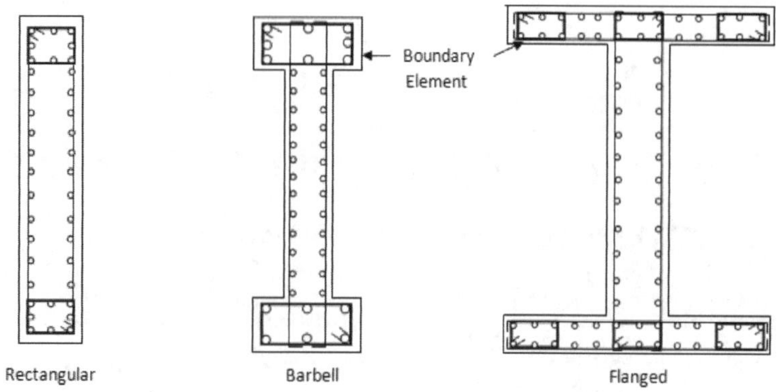

Figure 1. Types of shear wall sections.

of resisting forces acting laterally [1]. They are mainly provided for seismic safety against strong ground motions due to earthquakes or wind action owing to its exceptional lateral stiffness and strength [2]. The entire compression load of buildings acting downwards is resisted by the columns and the load-bearing structural walls, thereby balancing the forces to keep the system in equilibrium [3]. Whereas, the lateral forces subjected due to high velocity winds, random dynamic earthquake ground motions, even lateral earth pressure or hydrostatic forces are sustained by shear walls [4]. These are efficient and economic structural systems that are highly recommended in countries experiencing massive earthquakes frequently in order to avert severe damage to structures and loss of lives. Ideally, a set of shear walls are placed in symmetrical positions around the central axis of buildings or as core walls encasing the building elevator system with thickness varying from 150mm to 400mm [5]. The geometrical and mechanical properties of materials are the major factors that influence the behaviour of shear walls. Thus, a machine learning framework is implemented to demonstrate the failure or damage modes of shear walls based on the available databases of experimental investigation results collected from different literatures. The input parameters corresponding to the shear wall design factors such as aspect ratio, web reinforcements, boundary element reinforcements, characteristic strength of concrete in compressive, steel tensile yield strength , flexural and shear strength are some of the critical factors that are of paramount importance in governing the modes of failure [6, 7]. In this work, an open source data base has been taken from the literature and modified with new input parameters by incorporating the empirical equations and regulations given by the "Indian Standard code of practice for concrete structures". The data set consists of 393 experimental results of shear wall specimens of rectangular (R), flanged (F) and barbell (B) cross-sections tested for determining the relationship between the input variables, strength of walls and their failure pattern or mode [8]. With the advent of machine learning techniques, it has been successfully applied in several fields for predicting the future trends or patterns based on some inter-relationships drawn between the different input parameters from large data sets of previous studies. Thus, the ML algorithm learns from the previous data provided and predicts the future response in the form of output. To understand the underlying characteristics an exploratory analysis is performed on the features and four types of failure modes are identified and labelled as flexural failure (F), shear failure (SF), shear-flexure failure (SFF) and sliding shear (S) failure modes. The ML model does not depend on any assumptions regarding any input variables rather deduces the relative importance of every input feature on the output labels [9]. Feng et al. have predicted the shear strength of squat shear walls with the help of boosting algorithms [10]. The authors have developed a reasonable XGBoost model for shear strength which gave a ratio of 1.0 for predicted over test data. To rectify the discrepancies in the results obtained by analytical models versus the experimental results for predicting the behaviour of shear walls, genetic programming (GP) based algorithms were deployed. The authors have developed and trained GP models collaborating with some vital mechanics driven equations from design standards (ACI) for achieving higher accuracy of prediction [11]. An artificial neural network (ANN) hybrid ML model integrated with algorithms for best optimization, named adaptive harmony search (AHS) has been applied by the authors. It has been proposed that the accuracy achieved by this hybrid model has proven superior performance over existing empirical derivations for predicting the ultimate shear strength of shear walls [12]. Zhang et al., have proposed a probabilistic model for the prediction and classification of a 4-storey moment-resisting frame to map the response and failure patterns for a building subjected to earthquake. Random Forest and regression trees deployed have high prediction accuracies of 91% and 88% to evaluate the structural safety of a building damaged due to seismic forces [13]. Analytical models with the help of an artificial neural network (ANN) is implemented for predicting the shear strength of masonry shear walls strengthened with fibre reinforced mortar. The ANN model has shown a substantial balance between accuracy and precision while identifying the correlation between mechanical and geometrical properties controlling the strength of the walls [14].

2. Methodology

Test size of 0.3 is adopted such that the data set is conventionally split into 70% training and 30% testing sets. ML model thus learns by deducing some underlying patterns or relationships among the input variables upon the failure modes and strength parameters from the training data, subsequently making predictions on the test data. The database consists of a collection of previous experimental data including geometrical features of shear wall specimens, mechanical properties of construction materials used and the results of experimental laboratory tests. The input parameters selected include wall properties that influence the response of shear walls are as given in Table 1. There are totally 393 specimens of shear walls experimentally tested collected from different sources are assembled in the database. Out of which 152 specimens were observed to fail in flexural failure, 96 in flexure-shear intermediate failure, 122 in shear failure and 23 in sliding shear failure modes at the time of experimental testing. The flexural strength and shear strength capacities of the specimens are computed as per the Indian standard code regulations and taken as one of the inputs for predictions.

3. Indian Standard Code Provisions

As per the Indian Standard code provisions, shear walls may be provided as rectangular cross-sections or monolithic walls in between two columns, known as barbell sections may also be provided. These two end columns serve the purpose of boundary elements and help in increasing the flexural as well as shear strength capacities considerably [17]. It is of paramount importance to remember during the design phase that the behaviour of shear walls subjected to seismic forces greatly depends on the height-to-width ratio also known as aspect ratio [18]. Walls with smaller aspect ratios develop large shear stresses resulting in diagonal cracking in the direction of loading. Such walls are termed as squat shear walls and failure is mainly dominated by the ultimate shear carrying capacity of the wall due to inadequate shear reinforcement. Provisions for "design and ductile detailing of reinforced concrete shear walls" conforming to IS 13920: 1993, are as follows:

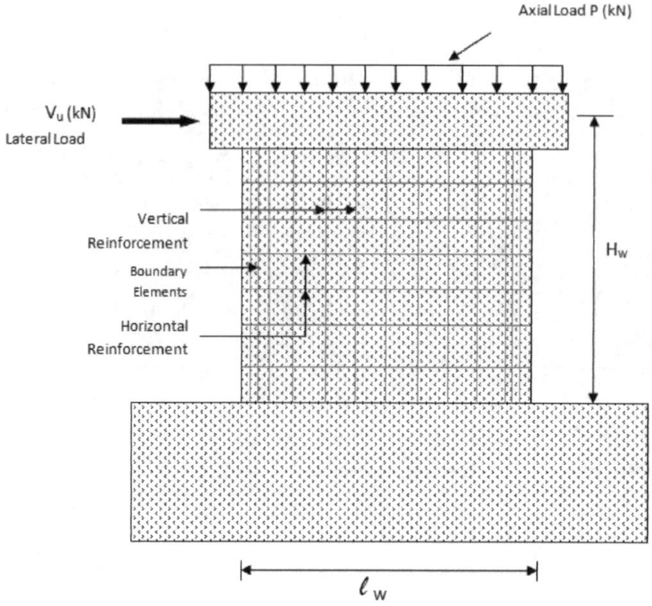

Figure 2. Ductile detailing of reinforced concrete shear walls.

Table 1. Shear wall design parameters.

Nomenclature	Parameters
l_w	Wall length
t_w	Wall thickness
f_{ck}	Characteristic compressive strength of the concrete
M_u	Ultimate flexural strength
V_u	Shear strength capacity
ρ_{vw}	Web vertical steel reinforcement ratio
ρ_{hw}	Web horizontal steel reinforcement ratio
ρ_{vc}	Boundary element ratio of vertical steel reinforcement
ρ_{hc}	Boundary element ratio of horizontal steel reinforcement
$f_{y,vw}$	Web vertical steel reinforcement yield strength
$f_{y,hw}$	Web horizontal steel reinforcement yield strength
$f_{y,hc}$	Boundary element horizontal steel reinforcement yield strength
$f_{y,vc}$	Boundary element vertical steel reinforcement yield strength
A_b	Area of boundary element
A_g	Wall gross cross-sectional area

3.1. Shear Strength Criteria

The nominal shear stress τ_v in the wall section may be computed as,

$$\tau_v = \frac{V_u}{t_w d_w} \tag{1}$$

The value of τ_v shall not exceed τ_{cmax} is given in table 19 IS 456:2000 [19].

3.2. Flexural Strength Criteria

For slender rectangular wall sections the moment of resistance M_{uv} shall be computed as given in IS 456:2000 for axially loaded compression members or columns under the influence of combined axial and bending loads. For uniformly distributed vertical reinforcements, the moment of resistance or flexural strength of the shear wall section may be estimated as given below:

(A) For $\frac{x_u}{l_w} \leq \frac{x_u^*}{l_w}$:

$$\frac{M_{uv}}{f_{ck}t_w l^2{}_w} = \phi\left[\left(1+\frac{\lambda}{\phi}\right)\left(\frac{1}{2}-0.416\frac{x_u}{l_w}\right)-\left(\frac{x_u}{l_w}\right)^2\left(0.168+\frac{\beta^2}{3}\right)\right] \tag{2}$$

The values of $\frac{x_u}{l_w}$, $\frac{x_u^*}{l_w}$, ϕ and λ shall be calculated by the following equations:

$$\frac{x_u}{l_w} = \left(\frac{\phi+\lambda}{2\phi+0.36}\right) \tag{3}$$

$$\frac{x_u^*}{l_w} = \left(\frac{0.0035}{0.0035+0.87\frac{f_y}{E_s}}\right) \tag{4}$$

$$\phi = \left(\frac{0.87 f_y \rho}{f_{ck}}\right), \qquad \lambda = \left(\frac{P_u}{f_{ck}t_w l_w}\right) \tag{5}$$

Where, ρ is ratio of vertical reinforcement provided, given by the following equation:

$$\rho = \frac{A_{st}}{t_w l_w}, \qquad \beta = \frac{0.87 f_y}{0.0035 E_s} \tag{6}$$

A_{st} is the total area of vertical reinforcement, E_s is the modulus of elasticity of the steel bars and P_u axial compression load acting on the wall.

(B) For $\frac{x_u^*}{l_w} < \frac{x_u}{l_w} < 1.0$:

$$\frac{M_{uv}}{f_{ck}t_w l^2{}_w} = \alpha_1\left(\frac{x_u}{l_w}\right) - \alpha_2\left(\frac{x_u}{l_w}\right)^2 - \alpha_3 - \frac{\lambda}{2} \tag{7}$$

where,

$$\alpha_1 = \left[0.36+\phi\left(1-\frac{\beta}{2}-\frac{1}{2\beta}\right)\right] \tag{8}$$

$$\alpha_2 = \left[0.15+\frac{\phi}{2}\left(1-\beta-\frac{\beta^2}{2}-\frac{1}{3\beta}\right)\right] \tag{9}$$

and

$$\alpha_3 = \frac{\phi}{6\beta}\left(\frac{1}{\frac{x_u}{l_w}}-3\right) \tag{10}$$

The x_u/l_w values shall be determined by solving the following quadratic equation:

$$\alpha_1\left(\frac{x_u}{l_w}\right)^2 + \alpha_4\left(\frac{x_u}{l_w}\right) - \alpha_5 = 0 \tag{11}$$

where,

$$\alpha_4 = \left(\frac{\phi}{\beta}-\lambda\right); \alpha_5 = \left(\frac{\phi}{2\beta}\right) \tag{12}$$

The roots for the quadratic equations shall be calculated as:

$$\left(\frac{x_u}{l_w}\right)_{Root1,\ Root2} = \frac{-\alpha_4 \pm \sqrt{\alpha_4^2 - 4\alpha_1(-\alpha_5)}}{2\alpha_1} \tag{13}$$

4. Machine Learning Models

Major part of machine learning is the classification of observations for deriving valuable insights. A brief description of ML algorithms implemented in this work to establish a best-fitting model for the classification of failure modes in shear walls is given in this section.

4.1. Naives Bayes

This classifier is based on the Bayes theorem with Naives assumptions that each every feature in the data set is independent and bears no relation with any of the other features. It also states that all features have equal weights i.e., they contribute equally towards the outcome. It is given by the following equation:

$$P_r(x) = P_r(Y = f|X = x) = \frac{\pi_f f_f(x)}{\sum_{i=1}^{K} \pi_i f_i(x)} \tag{14}$$

4.2. Decision Trees

The model of decision trees resembles a tree structure flowchart with branches representing the outcome of the test on attributes and terminal nodes representing class labels. The algorithm asks questions and learns from data features based on simple decision rules. This process repeats until the subsets reach the same values as target variables and the predictions are no longer improving. The first step of tree building is to split the training set into subsets by branching with non-overlapping regions.

4.3. Random Forest

It comprises a collection of decisions trees that operate as an ensemble forrest of uncorrelated individual trees with the help of bagging and feature randomness techniques. Thus, the prediction accuracy of the forrest model is higher compared to individual model predictions due to the low correlation between the individual tree models in the ensemble.

4.4. K-Nearest Neighbors

This algorithm has been successfully implemented in large scale regression and classification problems. It works on the assumption that similar samples or data points exist in close proximity or space, thus classification of a data point is executed by looking at its nearest data point. The shortest distance between items in the training set and the object to be classified is computed. Thus each element keeps two data values pertaining to the distance from the object to be classified and distance from its neighbour class.

4.5. Boosting Methods

The working principle of boosting algorithms emphasises sequential model training for improving or boosting the prediction power by training a series of weak models. Where each model plays the role of compensating for the weakness of its predecessors by identifying the data points miss-classified by the preceding model and increasing their weights. Some of the boosting algorithms commonly used are XGBoost, AdaBoost, CatBoost and Light-GBM. Initially, all points are assigned equal weights and subsequently, points that are classified incorrectly shall be given more weights in order to increase the accuracy of the ensemble model.

5. Model Implementation and Prediction

In this work, eight ML techniques have been used to establish an optimally suitable model for strength prediction and failure modes identification of shear walls. Experimental data for different shear walls are taken from literature and corresponding shear strength, as well as flexural strength values are computed. The influence of different input parameters is assessed with exploratory data analysis by plotting the correlation matrix which quantifies the inter-relationship among the different input parameters. The correlation among the individual input

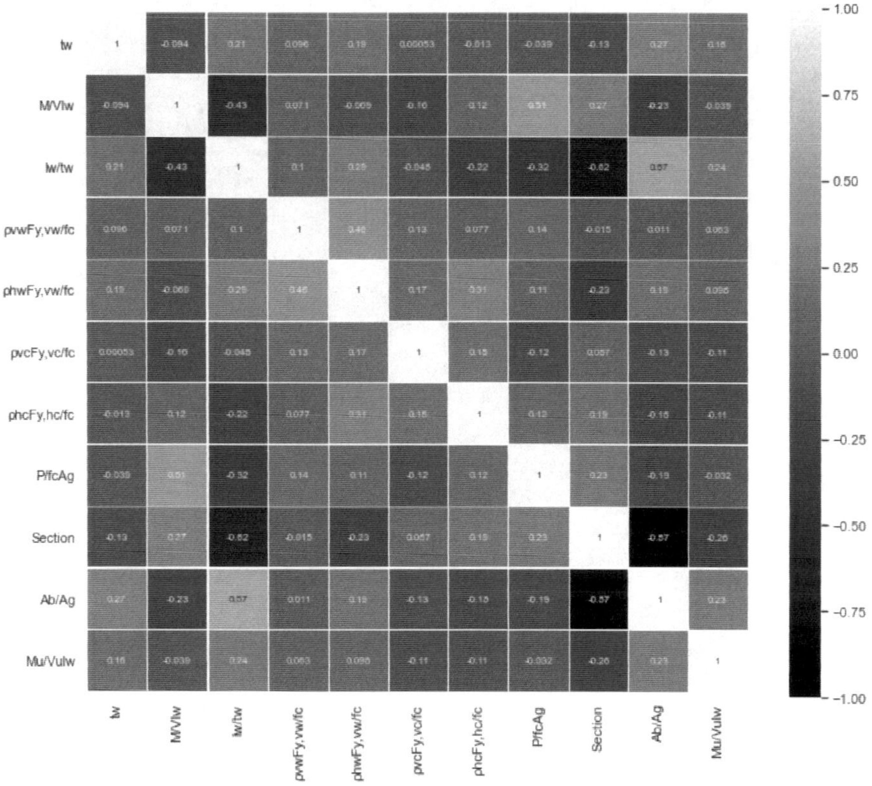

Figure 3. Correlation matrix for input design parameters.

parameters can be demonstrated with the help of the correlation matrix. The coefficient of correlation expresses the degree of inter-relationship within two parameters' relative movements. It may be observed from Figure 3 that a few parameters bear a higher degree of correlation to each other while some have a negligible correlation. As noted from the figure, the coefficient of correlation for cross-section shape and Ab/Ag is seen to be −0.57, denoting a strong negative relationship. Similarly, the coefficient of correlation for l_w/t_w and section shape (or Ab/Ag) is −0.62 (or 0.57); coefficient of correlation for M/Vl_w and $P/f_{ck}Ag$ is found to be 0.51; also coefficient of correlation for M_u/V_ul_w and l_w/t_w is found to be 0.24. Thus, it may be inferred that correlations between M/Vl_w and section shape (or Ab/Ag)are extremely weak or in other words, has the least influence on each other. It may be stated, that nonlinear classification algorithms are required as the current study is performed on a highly limited set of input parameters. Thus, the importance of the correlation matrix is to delineate the factors that bear maximum impact upon the strength and dominate failure modes of shear walls. Figure 4 shows the quantitative features including M/Vl_w, l_w/t_w, $\rho_{vw}F_y$, ρ_{vw}/f_ck, $\rho_{hw}F_y$, ρ_{vw}/f_ck and M_u/V_ul_w for all failure modes, where the values of M_u/V_ul_w are computed as per the IS 13920:1993 code provisions. Thus the results of the experimental database are compared with the empirically derived solutions. Pair plot is an effective data exploration tool with the help of which the distribution of any one variable can be shown along with the relationships between

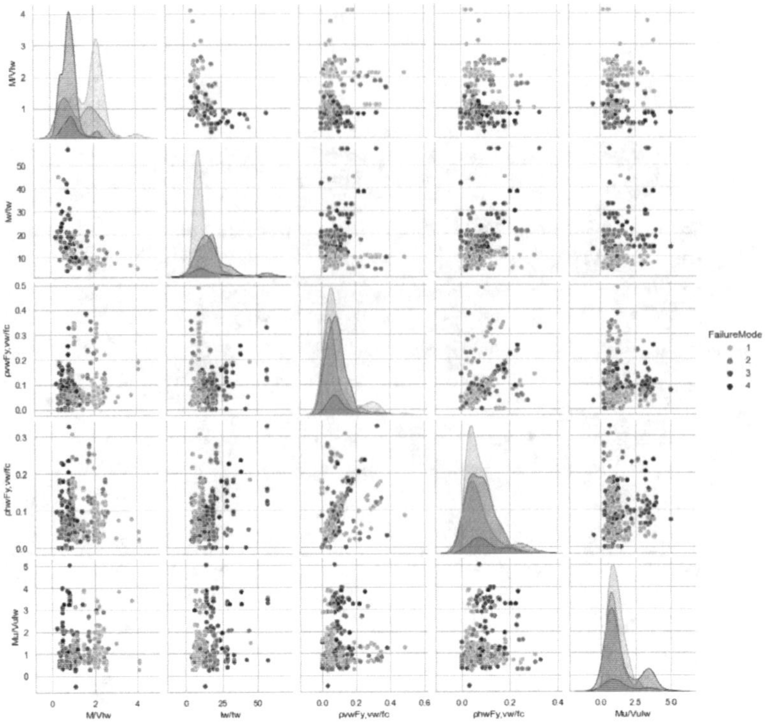

Figure 4. Pair plot of quantitative features of failure modes.

any two variables. The pair plot of the quantitative features shows the distribution of all input parameters along the diagonal while the remaining scatter plots demonstrate the relationships among the variables for the four categories of failure modes. The database is split into 70% training set and 30% testing set for evaluating the model performance. The accuracy of all the ML models are evaluated with the help of a confusion matrix, which plots the observed values versus the predicted values of each failure mode. It consists of a table with i number of rows and j number of columns equal to the failure modes (4*4) and the number of elements along the diagonal depicts the number of failure modes correctly predicted while the elements represented off-diagonal are the ones classified incorrectly by the ML models. The confusion matrix forms the base for all forms of evaluation metrics such as accuracy, precision, recall and F1 score. The accuracy denotes the percentage of failure mode predictions made correctly and is calculated by the average of the diagonal values lying across the confusion matrix. It is computed by the following equation:

$$Accuracy = \frac{TruePositive + TrueNegative}{TotalSamples} \tag{15}$$

Where true positive and negative depict the number of actual output that are predicted correctly. The accuracy of eight ML techniques are evaluated with the help of the confusion matrix and the accuracy achieved by the individual models are summarized in Table 2. From the table above it may be noted that the best-optimized model for classification of shear wall failure modes is achieved by the random forest technique, where the accuracy of the training set is as high as 100% and accuracy achieved on the test set is 85%. Whereas, the K-Nearest neighbours technique shows 81% accuracy on the training set but 86% accuracy on test set.

Table 2. Performance evaluation of eight ML techniques using the training and test set.

Sl.No.	ML Technique	Accuracy on Training Set	Accuracy on Test set
1	Naives Bayes Technique	71%	78%
2	Decision Tree	100%	81%
3	K-Nearest neighbors	81%	86%
4	XGBOOST	100%	79%
5	Random Forest	100%	85%
6	ADABOOST	78%	69%
7	Light GBM	100%	81%
8	Cat BOOST	100%	83%

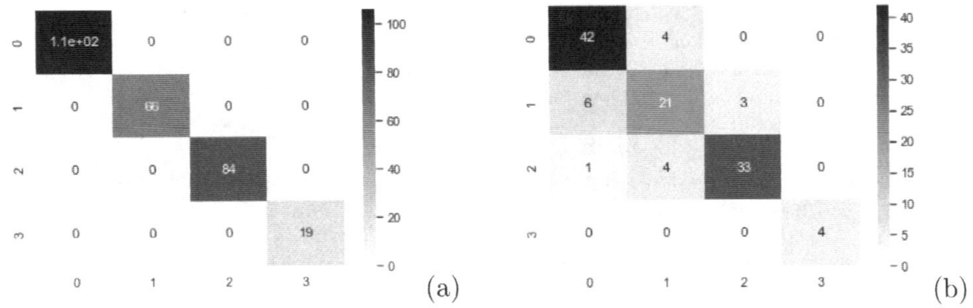

Figure 5. Random forest confusion matrix of (a) training set and (b) test set.

The XGBOOST technique has achieved 100% accuracy on training set yet only 79% accuracy is achieved on test set which demonstrates insufficient training of the model. Figure 5 represents the confusion matrix for the optimum ML technique achieved by random forest technique on training and test sets.

6. Conclusions

This paper describes a methodology for the dynamic analysis of reinforced-concrete shear-wall buildings subjected to earthquakes within a machine learning framework. This work establishes an ML model for the classification of failure mode and strength prediction of reinforced concrete shear walls. The performance of shear walls under loading is substantially impacted by factors such as the ratios of steel, the shear-span ratio of the wall, as well as the design load which bear maximum importance in the design and detailing of shear walls. The experimental database consists of 393 shear walls of rectangular, flanged and barbell sections where 152 specimens fail in flexural, 122 solely fail in shear, 96 undergo flexure-intermediate-shear failure and 23 fail in sliding due to shear. Eleven input parameters have been adopted that integrate the geometrical parameters, material characteristics, ductile and shear reinforcement detailing of shear walls into the model for best-optimized results. These inputs are validated with shear strength and flexural strength criteria conforming to IS 13920:1993 code. The prediction model is built on the training set and the model's performance is assessed on the test set. In the present work, eight ML techniques including K-Nearest Neighbors, Naive Bayes Classifier, Decision Tree, Random Forest, XGBoost, AdaBoost, CatBoost, and Light-GBM are implemented. Global accuracy methods are used to access the model performance which signifies the accuracy with which the model predicts for unknown data points. Based on the performance and efficiency of the different models it may be inferred that the highest accuracy of 85% is achieved on the test set by Random Forest, 83% by CatBoost and 81% by Light-GBM. This work demonstrates the rapid failure mode identification and strength prediction of shear walls. The present machine learning framework can be extended by using International Building Code, IBC and establish models for structural health monitoring (SHM) of the reinforced-concrete shear-wall buildings for future study.The impact of computation complexities of the above ML algorithms can also be explored in future work.

References

[1] Jain S K and Medhekar M S 1993 Seismic behaviour, design and detailing of RC shear walls, Part I: behaviour and strength *Indian Concrete Journal* July pp 311-318
[2] Siam A, Ezzeldin M and El-Dakhakhni W 2019 Machine learning algorithms for structural performance classifications and predictions: application to reinforced masonry shear walls *Structures* **22** pp 252-65
[3] Lefas I D, Kotsovos M D and Ambraseys N N 1990 Behavior of reinforced concrete structural walls: Strength, deformation characteristics, and failure mechanism *ACI Structural Journal* **87(1)** pp 23-31
[4] Huang S, Huang Y, He A, Tang X, Chen Q, Liu X and Cai J 2018 Experimental study on seismic behaviour of an innovative composite shear wall *J. Constr. Steel. Res.* **148** pp 165-179
[5] Zhang H, Cheng X, Li Y and Du X 2022 Prediction of failure modes, strength, and deformation capacity of RC shear walls through machine learning *J. Buil. Engg.* **50** 104145
[6] Medhekar M S and Jain S K 1993 Seismic behaviour, design and detailing of RC shear walls, Part II: design and detailing *Indian Concrete Journal* September pp 451-457
[7] Oesterle R G, Aristizabal J D, Fiorato A E, Russell H G and Corley W G 1979 Earthquake Resistant Structural Walls - Tests of Isolated Walls - Phase II *Report to the National Science Foundation* Construction Technology Laboratories, Portland Cement Association, Skokie, Illinois
[8] Mangalathu S, Jangb H, Hwangc S and Jeonc J 2020 Data-driven machine-learning-based seismic failure mode identification of reinforced concrete shear walls *Engg. Struc.* **208** 110331
[9] Mangalathu S and Jeon J 2019 Stripe-based fragility analysis of multi-span concrete bridge classes using machine learning techniques *Eart. Engg. and Struc. Dyn.* **48(11)** pp 1238-55
[10] Feng D C, Wang W J, Mangalathu S and Taciroglu E 2021 Interpretable XGBoost-SHAP machine-learning model for shear strength prediction of squat RC walls *J. Struc. Engg.* **147** 04021173
[11] Gondia A, Ezzeldin M and El-Dakhakhni 2020 Mechanics-guided genetic programming expression for shear-strength prediction of squat reinforced concrete walls with boundary elements *J. Struc. Engg.* **146** 04020223
[12] Keshtegar B, Nehdi M L, Kolahchi R, Trung N T and Bagheri M 2021 Novel hybrid machine leaning model for predicting shear strength of reinforced concrete shear walls *Engg. Comp. Preprint*
[13] Zhang Y, Burton H V, Sun H, Shokrabadi M 2018 A machine learning framework for assessing post-earthquake structural safety *Struc. Safety* **72** pp 1-16
[14] Cascardi A, Micelli F and Aiello M A 2016 Analytical model based on artificial neural network for masonry shear walls strengthened with FRM systems *Composites B* **95** 252e263
[15] Deger Z T and Kaya G T 2021 Glass-box model representation of seismic failure mode prediction for conventional RC shear walls *Geophy. Research Letters Preprint*
[16] Lagaros N D and Papadrakakis M 2012 Neural network based prediction schemes of the non-linear seismic response of 3D buildings *Advances in Engineering Software* **44** pp 92–115
[17] Bureau of Indian Standards, IS 1893 (Part 1):2002 *Indian standard criteria for earthquake resistant design of structures* part 1 general provisions and buildings(Fifth Revision) BIS 2002
[18] Bureau of Indian Standards, IS 13920 : 1993 (Reaffirmed 2003) *Indian standard ductile detailing of reinforced concrete structures subjected to seismic forces - code of practice* (Third Reprint November 1996) BIS 1993
[19] Bureau of Indian Standards, IS 456 : 2000 (Reaffirmed 2005) *Indian standard plain and reinforced concrete code of practice* (Fourth Revision April 2007) BIS 2000

Privacy Preservation of Digital Health Records in Cloud

Kundan Munjal[1], **and Rekha Bhatia**[2]

[1] Research Scholar, University College of Engineering, Punjabi University Patiala, 147001 and Assistant Professor, AIT-CSE, Chandigarh University, Gharuan
[2] Associate Professor, Department of Computer Science, Punjabi University Patiala, 147001

E-mail: Coressponding Author:kundanmunjal@gmail.com

Abstract. The variety and amount of patient healthcare digital data is rapidly expanding. The field of artificial intelligence has fast proven revolutionary for healthcare, allowing for unprecedented speed and precision in data analysis. Many hospitals have already transitioned to electronic health records (EHRs), a digital version of paper medical information. Information Technology has already aided in simplifying operations in this area, making the process far more efficient and patient-centered than in the past. A patient's trust may disintegrate in the face of recurrent incidents if there is no precise control and safety in place. As a result, it limits the potential for digital health to promote an age of more accessible, connected, and individualized treatment. When it comes to picking a cloud solution, healthcare providers are most concerned about security and privacy. In this research, we seek to demonstrate various ways that aid patient data privacy while simultaneously allowing cloud services to be used.

1. Introduction

Electronic health records (EHRs) are (1) digital standard health information records. These records can be generated by any healthcare provider. Digitization of medical records not only increase the quality of care but also it is cost effective. It will also increase accessibility as health records cab be accessible whenever and wherever possible. Reducing the communication gap between health-care providers, with lower costs is one of the main objective of Electronic Health Records. The three (2) (3) very basic and main goals of security is (CIA) i.e. confidentiality, Integrity and availability.

(i) Confidentiality: Only authorized person should have an access to the information.
(ii) Integrity: Information should be correct and it should not be altered by an unauthorized person.
(iii) Availability: Information should be accessible, available and usable at any time but only by an authorized entity.

A shared (1) EHR with complex layout of sensitive data includes Medical test reports, patient's demographic information, patient health history and other medical records. There is a need for a fully secured model that complies with the terms and conditions set forth by legal and regulatory agencies. As a result, only organisations with a justified need-to-know privilege assigned by the patient have access to sensitive information. Unless and until a specific treatment alternative is given, a patient may choose to openly hide his health information connected to an HIV/AIDS treatment. As a result, a simple, dependable, efficient, and reliable system is required in which patients can simply permit a number of health associates to access their sensitive information in its entirety or in part. *Pseudonymisation* and *Anonymisation* techniques can be used to preserve the patient health record. Anonymisation is the method of converting data in a manner that it can no longer be linked to the individual to whom it relates. Anonymized data can never be used to identify the person to whom it belonged. Any advance processing of the data will not be able to identify any person from the data. Pseudonymisation, on the other hand, substitutes a pseudonym for any recognizable data attributes. The essential difference between pseudonymisation and anonymisation is with pseudonymisation, it is able to identify an individual by reviewing the associated data. Because of the type or view of the data, this is not always important to apply anonymisation techniques. Pseudonymized data lend themselves well to processing and analysis. There is other approach that is helpful in preserving privacy of data and allows processing of data in encrypted form. This type of encryption approach is known as homomorphic encryption. In section 2 related work in the area of privacy of healthcare data is presented followed by section 3 that focuses on deep learning in healthcare with various challenges and finally in section 4 role of homomorrhic encryption in providing data privacy and security is presented.

2. Background

2.1. Privacy in EHR

Identity-based encryption (IBE) was developed by (4) the authors, that enabled some polynomial number of independent entities to observe parameters and disseminate private keys. An encryptor can pick a value "dk" and a set of parametersfor each authority in this scheme. As a result, encrypt a message so that a client can only decipher it if he possesses at least "dk" of the specified parameters out of each authority "k". The advantage of this scheme is that it can tolerate multiple number of corrupt authorities. In (5) authors presented and implemented a patient telemonitoring system (PTS) that features location privacy at network layer. In this paper, an architecture was proposed that improves PTS by means of location privacy and data encryption. A point-to-point protocol called Mist was also used.The method was implemented and tested through various operating parameters to determine the data transmission burden, and results showed that the overall operational costs was minimal even when the data update percentage is high. In (6) the authors have provided a method for automating privacy standards for the exchange of patient data throughout Europe on a health sector, as well as ensuring their implementation both internally and externally. This approach has also formulated a series of regulations that will be implemented at the grid's process level. It ensures that privacy is respected before data access is provided. The main drawback of the work was the requirement of patient consent. Therefore, specification of privacy obligations as enforceable policies need to be extended by applying anonymization, encryption and pseudonymization in the proposed work. In (7) authors have evaluated common pseudonymization approaches. After analyzing it with respect to legal and technical criteria it has been concluded that personal information is highly sensitive that demands protection policies. in the paper Pseudonymization was chosen as perhaps the promising feasible option from a list of possibilities that included role-based access control, anonymization, depersonalization, and encryption in this research. Several approaches have been discussed such as PIPE *(Pseudonymization of Information for Privacy in e-Health)*, eGK *(Electronic health card)* and HIPAA *(Health Insurance Portability and Accountability Act)* and have been concluded that PIPE and eGK have been proved to provide similar type of security. In (2) authors have discussed DICOM security that relies on the encryption of the communication channels. In this paper, a data model for systems using DICOM based on the PIPE pseudonymization approach in conjunction with HIPAA have been presented. The proposed word have been compared with existing approaches and various benefits of the architecture have been discussed such as access management and security. In (8) authors offer a research strategy that combines conceptual-analytical, artifact-building, and artifact-evaluation techniques. A new pseudonymization technique for health data has been proposed and implemented, which preserves health data decoupled from the respective patient-identifying details, allowing privacy-preserving supplementary use of the patient records in clinical tests without additional anonymization steps. The deployed solution provides a one-of-a-kind solution that ensures data privacy (e.g., in accordance with HIPAA) while also allowing primary and secondary data use at the same time. In (9) the studies have proposed a nationwide healthcare framework idea with multi-level data protection in order to increase the data security of medical information when it is stored and exchanged/shared. A trustworthy external server has maintained personally identifiable data and related pseudonyms in this article, whereas a medical information server has controlled medical data assigned to pseudonyms. In (10) the authors looked at published privacy techniques such as a fast de-pseudonymization procedure and a patient-based pseudonym recently as they applied to genetic data research and warehousing. An asymmetric encryption system that separates the tasks of pseudonymization and de-pseudonymization has been presented. A working software prototype has been constructed to demonstrate the feasibility of the proposed transnational, data privacy-preserving method. In (11) authors have proposed an unlinkable pseudonym system that allows the usage of domain specific pseudonyms, where the cross domain relationship between both the pseudonyms is known only by a centralized authority. This proposal has circumvented the centralized authority's necessity to be a completely trusted entity, allows information to be maintained and transmitted in a decentralised fashion. This work provides strong assurances in terms of uniformity and stability, even if the converter is corrupt. Authors in (12) have solved the issue of minimizing the requirement for manual record verification induced by data pseudonymization. The major goal of this research was to see how machine learning approaches may be used to connect records in epidemiology cancer registries. The solution has been found by encoding pseudonymizeddata so that it can be interpreted by a classifier. In (13) a pseudonym scheme have been proposed that addressed domain-specific pseudonyms in 2015 and discussed unlinkable pseudonyms that can be converted by a central authority. In previous paper transparency have been surrendered which is the crucial part because Users may no more enquire about and examine the transit of their private information with the converter. In this paper,seemingly conflicting requirements have been addressed and offered a method in which the ignorant converter generates client-centric audit reports while keeping all privacy features. How to withstand the converter's complete malicious behavior, for example, by disseminating the logging method or using current developments in the blockchain sector, is future work that has yet to be completed.

2.2. Health Informatics and Security

In (14) authors uses a single data structure for predicting clinical outcome $(i.e., death)$, standard measure of amount of care $(i.e., re - admissions)$, a measure of resource consumption (i.e., length of stay) and measure of understanding patients problem $(i.e., diagnoses)$. All electronic health records were de-identified and EHRs were taken from two hospitals containing demographics, diagnoses, procedures etc. and have used Fast Healthcare Interoperability Resources (FHIR) for representing events in patient timeline. Same modelling algorithms were used in both the hospitals but each hospital was treated as a separate dataset. As patients data vary in density of data points as well as length of data so they considered three deep learning neural network architectures. *First* is recurrent neural network, *Second* attention based time aware and *third* boosted time based decision stumps and final model is a collection of predictions from all three model architectures. Three types of score were used based upon certain attributes related to patients health aEWS score, mHOSPITAL score and mLIU score. In (15) authors have introduced the concept of cryptonets i.e neural networks applied to encrypted data and used secured cloud computing for training as well as prediction purpose. It showed capability of cloud service for applying neural networks on encrypted and also generating predictions in encrypted form and achieved 99 % accuracy with predictions of 59000 per hour on a single pc. It uses Homomorphic encryption with neural network and by combining both cryptography and machine learning they were able to create a setup in which both security as well as accuracy was achieved. A review (16) due to the high dimensionality and variability of healthcare data, using deep learning for healthcare poses a number of obstacles. It also discussed latest and imminent applications of deep learning in medicine and highlighted important aspects that effects healthcare. Various applications of deep learning discussed were Clinical Imaging, EHR(Electronic Health Records), Genomics and Mobile with challenges and opportunities like data volume data quality , temporality , domain complexity and interoperability. Lastly it concluded that deep learning can be used as a guiding principle to organize both exploratory investigation and hypothesis driven research in clinical domain. In order to achieve this medical tasks and statistical tasks must be integrated at all the levels which includes study design to data interpretations. The first identified (17) problem of traditional encryption process and the problem that was identified was nformation system that is working with the encoded information or encrypted data was only responsible for storage and retrieval of data for user. But if system needs to perform certain operation on the data it first has to decrypt the data and at the end security of the data is lost. But there are various applications in which not only data is stored for storage purpose but also for operation purpose. Solution provided for this problem was privacy homomorphism i.e the encryption function used will allow the information system to operate on the encrypted i.e no need of decryption for applying the operations. Simple data i.e unencoded and operations to be done on it were assumed to be drawn from some algebraic system and at the end ensured privacy of data which can be operated. Based upon all the papers read it has been found that various papers has been published in security and privacy of health data. As per literature survey health informatics is playing a vital role in clinical science and very less amount of papers has been published in security concern of health informatics. Table 1 categorizes three types of selected papers from some reputed journals with good number of citations category one related to security in healthcare category two related to only health informatics and category three related to security in health informatics.

3. Deep Learning in Healthcare

Health care data is high dimensional and heterogeneous and generally unstructured that includes EHR i.e electronic health records like image files, sensor data, text data and other complex data. Deep Learning in healthcare consists of three modules *first* Fig 1 is related to *clinical science* that includes the effectiveness of healthcare through Genomics, Radiology, and Oncology. *Second* module is *Data Science* which includes applying various Machine Learning, Data Mining algorithms and finally, *Third* module is *healthcare* which will be used to diagnose, analyze for treatment and survival. Conventional mining and statistical approaches require to carry out feature engineering approaches (18) to extract important as well as robust features and then build predictive inferences on top of them. Predictive analysis on neural networks or machine learning has not been widely applied in healthcare field as data in healthcare is complex. However, deep learning approaches have not been used in healthcare as that could benefit from its functional capabilities. Various aspects of deep learning that are extensively required in medical field like potential of handling complex and multi-dimensional data, superior performance and so on. While applying deep learning there are various challenges related to healthcare like heterogeneity, time dependency and noisy data Traditional ANNs are different from deep learning. The major difference between deep learning and Artificial neural (18) networks is ANNs are usually limited to three layers and are optimized for performing a specialized task and cannot be generalized.

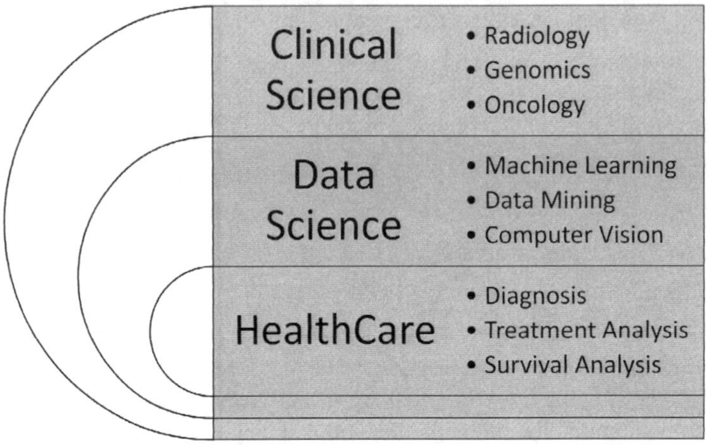

Figure 1: Deep Learning in healthcare

3.1. Various Challenges of Deep Learning in Healthcare

3.1.1. Data Sufficient in Volume? Deep learning is basically an intensive model.Only with the provision of a large volumes of information can the best results be reached. in such intensive model but volume of data is challenging in healthcare and won't be able to train a deep learning model with as many patients as needed. The amount of patient data or health (18) care data that is needed to train an efficient and effective deep learning model would be high in volume.

3.1.2. Quality of Data Healthcare data is highly heterogeneous, noisy and sometimes incomplete. Training deep learning model with these problems is a challenging task.

3.1.3. Data Updations (Temporality) Diseases take new path and keep on changing overtime. Non determinism is there. In order to achieve this goal we have to design a deep learning model by keeping time factor in mind so that temporal healthcare data can easily be handled.

3.1.4. Security of Data As far as healthcare industry is concerned security of data is the prime issue. Various regulations have been passed regarding privacy of data mainly in the field of healthcare. HIPPA (Health Insurance Portability and Accountability Act) of USA and GDPR(General Data Protection Regulation) in Europe. In 2013 Ministry of Health & Family Welfare have notified standards for electronic health records (EHR) for India the main need for EHR was to record each information of a child health as his/her life progresses. It was found (19)that countries considered four areas where they have made extensive efforts: standards & interoperability, research and education, Policy and guidelines as well ICT infrastructure. Various essential activities were (20) recognized in respective domains. A path is proposed based on this, which involves the establishment of secure health infrastructures. The present state of India's healthcare industry is marked by low government expenditure (1 percent of GDP), large out-of-pocket costs (71 percent), excessive anaemia among young females (56 percent), and infant mortality rates (47 per 1000 live births) and maternal deaths (212 per 100,000 live births). In healthcare, the nation lags behind the other emerging nations such as Bangladesh and Sri Lanka. The prime objective of this finding is to develop a standard for maintaining electronic health records of India and as per IT ACT, 2000 by Ministry of Law, Justice and Company Affairs has secured access to any electronic record and discloses such information to third party without the consent of concerned person would be convicted with a sentence of imprisonment of up to two years and a fine of one lakh rupees. In order to perform machine learning for medical predictions we must be sharing data to third party. Third party will then perform analysis and provide predictions. Main Problem in healthcare with deep learning is that how a user can employ a predictive model without compromising personal or private information as hospitals might use cloud services for predictions and due to regulations patients medical data cannot be revealed.

4. Homomorphic Encryption: A Solution to Patient Data Privacy

Privacy and security of patient digital health record is a big challenge. The challenge is further complicated by privacy restrictions and worries about the hazards of releasing personal and sensitive health and genomic data. From the last ten years, advances in cryptography have delivered an approach known as fully homomorphic encryption, that may be used to encrypt information in such a way that it can be stored in an untrustworthy cloud while also being calculated in a significant fashion in an encrypted format without access to decryption keys. Thus, *Homomorphic encryption is an advanced cryptographic process that allows users to compute encrypted data and receive encrypted results. Homomorphic encryption allows for the processing of encrypted data while also maintaining privacy.* Various Steps involved in homomorphic encryption in healthcare are described in Fig 2. In (21) authors cloud service provider contribute huge investments in data centres to provide services to small companies which helps these companies to cut their costs. But one of the hurdle to take these services from big companies is concerned over privacy and confidentiality of data being taken care of by cloud so there is a need of cloud service which offers confidential handling of machine learning assignments for various applications. In (22) authors proposed Homomorphic encryption scheme allows certain computation to be performed on encrypted data by manipulating encrypted data without the need to decrypt them. As compare to this a fully homomorphic scheme (21) allows various arbitrary operation over encrypted data as a result any function can be computed. However, fully homomorphic encryption schemes are inefficient for practical use. Comparison of various papers in the field of privacy of patient data is done in Table 1.

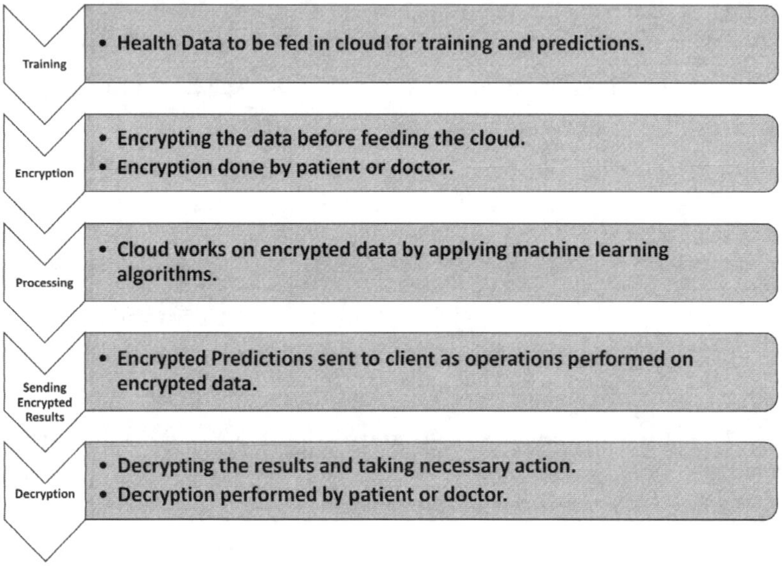

Figure 2: Steps of Homomorphic Encryption in Healthcare

Table 1: Comparison of various papers based upon Security in Health Care, Health Informatics and security with Health Informatics

Paper	Summary	Privacy Achieved	Health Informatics	Privacy With Health Informatics
(22)	(i) Used the concept of cryptonets i.e crytpography with neural netwoks in field of health care (ii) Used Homomorphic encryption for encrypting learning data	Yes	Yes	Yes
(15)	(i) Main contribution was the accuracy of neural networks with homomorphic encryption (ii) High level of throughput was achieved instead it uses cryptography, machine learning and engineering. (iii) Works requires good encoding schemes that allows for small attributes and faster homomorphic computation	Yes	Yes	Yes
(7) (23)	(i) E-health pseudonymization of data was used for privacy, and a framework for centralised deidentification of critical data was provided. (ii) The approach provides for unassociated preservation of health information and patient personally identifiable information while also allowing unauthorised users to reestablish the patient and health data association.	Yes	No	No
(1)	(i) Approach supported selective sharing of virtual composite electronic health records. (ii) Various key difficulties, such as dispersed data integration and privacy concerns, were taken into account. (iii) In the healthcare context, an e-consent method was used to facilitate improved patient data exchange.	Yes	No	No
(24)	(i) Developed a private convolutional deep brief network. Model can be applied to energy based deep neural networks and also able to take large scale datasets	Yes	Yes	Yes
(25)	(i) Comprehensive review for deep learning in health informatics and provided a critical analysis of appropriate merits and drawbacks of techniques.	No	Yes	No
(16)	(i) Identified the problem of traditional encryption schemes related to storage and retrieval encryption scheme and proposed a homomorphic privacy scheme for operation based encryption which can be used in cloud , machine learning and neural networks i.e where user has to share their data.	Yes	No	No
(21)	(i) It assigns the implementation of a machine machine learning approach to a third-party computation service provider while maintaining the security of both training and testing data using a levelled homomorphic encryption method.	Yes	No	No
(26)	(i) Proposed a fully homomorphic encryption mechanism. The scheme avoids modulus switching and keeps the size of encrypted data to a single ring element	Yes	No	No
(27)	(i) Designed an architecture of deep learning containing auto encoders and softmax output layer to diagnose Alzhimer's disease at an early stage and showed performance gain in categorization of domain groups.	No	Yes	No
(28)	(i) A unique approach which learns spatial image characteristics required in supervised classification for training purpose wholly from unlabeled data and automatically learns various features of images with slight efforts. Method also uses deep learning for featured learning and random forest for supervised classification.	No	Yes	No.
(29)	(i) With three two-dimensional CNNs, the method leveraged learning with convolutional neural networks for voxel categorization. These have a one-to-one relationship with the three-dimensional CNNs' xy, yz, and zx planes. They test their system with 114 unseen knee MRIs, each with roughly 2 million voxels to sort through.	No	Yes	No
(30)	(i) Developed a generic model named as Doctor AI uses recurrent neural networks and uses medical conditions and medications used. RNN applied to longitudinal time stamped health records of over 8 years containing 260000 patients.	No	Yes	No
(31)	(i) Convolutional neural networks were used with the free software tool basset to understand the design and operational of DNA sequences linked to genomics.	No	Yes	No

5. Conclusion

Maintaining electronic health records is an important step in medical progress since it will make the diagnosing process easier. Artificial Intelligence in electronic health records is critical for predictive modeling in a variety of sectors relating to healthcare, however electronic health records include sensitive data that must be protected. The data needed for training must be encrypted for security reasons, and the gap that will be filled in this study is that the machine learning algorithm will function on encrypted data and provide encrypted predictions. The user will subsequently be able to decipher the predictions and take the appropriate actions depending on the information provided. User will then be able to decrypt the predictions given and take necessary steps based upon prediction received.

References

[1] Jing "Jin, Gail-Joon Ahn, Hongxin Hu, Michael J. Covington, and Xinwen" Zhang. "patient-centric authorization framework for electronic healthcare services". *"Comput. Secur."*, "30"("2–3"):"116–127", "mar" "2011". ISSN "0167-4048". doi: "10.1016/j.cose.2010.09.001". URL "https://doi.org/10.1016/j.cose.2010.09.001".

[2] Daniel Abouakil, Johannes Heurix, and Thomas Neubauer. Data models for the pseudonymization of dicom data. In *2011 44th Hawaii International Conference on System Sciences*, pages 1–11, 2011. doi: 10.1109/HICSS.2011.136.

[3] Sebastian Haas, Sven Wohlgemuth, I. Echizen, Noboru Sonehara, and Günter Müller. Aspects of privacy for electronic health records. *International journal of medical informatics*, 80:e26–31, 10 2010. doi: 10.1016/j.ijmedinf.2010.10.001.

[4] Melissa Chase. Multi-authority attribute based encryption. In Salil P. Vadhan, editor, *Theory of Cryptography*, pages 515–534, Berlin, Heidelberg, 2007. Springer Berlin Heidelberg. ISBN 978-3-540-70936-7.

[5] Ilias Maglogiannis, Leonidas Kazatzopoulos, Konstantinos Delakouridis, and Stathes Hadjiefthymiades. Enabling location privacy and medical data encryption in patient telemonitoring systems. *IEEE Transactions on Information Technology in Biomedicine*, 13(6):946–954, 2009. doi: 10.1109/TITB.2008.2011155.

[6] Hanene Boussi Rahmouni, Tony Solomonides, Marco Casassa Mont, and Simon Shiu. Modelling and enforcing privacy for medical data disclosure across europe. *Stud. Health Technol. Inform.*, 150:695–699, 2009.

[7] Thomas Neubauer and Johannes Heurix. A methodology for the pseudonymization of medical data. *International Journal of Medical Informatics*, 80(3):190–204, 2011. ISSN 1386-5056. doi: https://doi.org/10.1016/j.ijmedinf.2010.10.016. URL https://www.sciencedirect.com/science/article/pii/S1386505610002042.

[8] Johannes Heurix and Thomas Neubauer. Privacy-preserving storage and access of medical data through pseudonymization and encryption. pages 186–197, 08 2011. doi: 10.1007/978-3-642-22890-2_16.

[9] Stefan Benzschawel and Marcos Silveira. Protecting patient privacy when sharing medical data. 02 2011.

[10] Harald Aamot, Christian Kohl, Daniela Richter, and Petra Knaup. Pseudonymization of patient identifiers for translational research. *BMC medical informatics and decision making*, 13:75, 07 2013. doi: 10.1186/1472-6947-13-75.

[11] Jan Camenisch and Anja Lehmann. (un)linkable pseudonyms for governmental databases. In *Proceedings of the 22nd ACM SIGSAC Conference on Computer and Communications Security*, CCS '15, page 1467–1479, New York, NY, USA, 2015. Association for Computing Machinery. ISBN 9781450338325. doi: 10.1145/2810103.2813658. URL https://doi.org/10.1145/2810103.2813658.

[12] Yannik Siegert, Xiaoyi Jiang, Volker Krieg, and Sebastian Bartholomäus. Classification-based record linkage with pseudonymized data for epidemiological cancer registries. *IEEE Transactions on Multimedia*, 18(10):1929–1941, 2016. doi: 10.1109/TMM.2016.2598482.

[13] Jan Camenisch and Anja Lehmann. Privacy-preserving user-auditable pseudonym systems. In *2017 IEEE European Symposium on Security and Privacy (EuroS P)*, pages 269–284, 2017. doi: 10.1109/EuroSP.2017.36.

[14] Alvin Rajkomar, Eyal Oren, Kai Chen, Andrew Dai, Nissan Hajaj, Peter Liu, Xiaobing Liu, Mimi Sun, Patrik Sundberg, Hector Yee, Kun Zhang, Gavin Duggan, Gerardo Flores, Michaela Hardt, Jamie Irvine, Quoc Le, Kurt Litsch, Jake Marcus, Alexander Mossin, and Jeff Dean. Scalable and accurate deep learning for electronic health records. *npj Digital Medicine*, 1, 01 2018. doi: 10.1038/s41746-018-0029-1.

[15] Ran Gilad-Bachrach, Nathan Dowlin, Kim Laine, Kristin Lauter, Michael Naehrig, and John Wernsing. Cryptonets: Applying neural networks to encrypted data with high throughput and accuracy. In Maria Florina Balcan and Kilian Q. Weinberger, editors, *Proceedings of The 33rd International Conference on Machine Learning*, volume 48 of *Proceedings of Machine Learning Research*, pages 201–210, New York, New York, USA, 20–22 Jun 2016. PMLR. URL https://proceedings.mlr.press/v48/gilad-bachrach16.html.

[16] Roberto Somolinos, Adolfo Muñoz, M. Elena Hernando, Mario Pascual, Jesús Cáceres, Ricardo Sánchez-de Madariaga, Juan A. Fragua, Pablo Serrano, and Carlos H. Salvador. Service for the pseudonymization of electronic healthcare records based on iso/en 13606 for the secondary use of information. *IEEE Journal of Biomedical and Health Informatics*, 19(6):1937–1944, 2015. doi: 10.1109/JBHI.2014.2360546.

[17] Ronald L. Rivest and Michael L. Dertouzos. On data banks and privacy homomorphisms. 1978.

[18] Riccardo Miotto, Fei Wang, Shuang Wang, Xiaoqian Jiang, and Joel T Dudley. Deep learning for healthcare: review, opportunities and challenges. *Briefings in Bioinformatics*, 19(6):1236–1246, 05 2017. ISSN 1477-4054. doi: 10.1093/bib/bbx044. URL https://doi.org/10.1093/bib/bbx044.

[19] Sunil Kumar Srivastava. Adoption of electronic health records: A roadmap for india. *Healthc. Inform. Res.*, 22(4):261–269, October 2016.

[20] Pradeep Deshmukh. Design of cloud security in the ehr for indian healthcare services. *Journal of King Saud University - Computer and Information Sciences*, 29(3):281–287, 2017. ISSN 1319-1578. doi: https://doi.org/10.1016/j.jksuci.2016.01.002. URL https://www.sciencedirect.com/science/article/pii/S1319157816300118.

[21] Thore Graepel, Kristin Lauter, and Michael Naehrig. Ml confidential: Machine learning on encrypted data. In Taekyoung Kwon, Mun-Kyu Lee, and Daesung Kwon, editors, *Information Security and Cryptology – ICISC 2012*, pages 1–21, Berlin, Heidelberg, 2013. Springer Berlin Heidelberg. ISBN 978-3-642-37682-5.

[22] Pengtao Xie, Misha Bilenko, Tom Finley, Ran Gilad-Bachrach, Kristin Lauter, and Michael Naehrig. Crypto-nets: Neural networks over encrypted data, 2014.

[23] Thomas Neubauer and Mathias Kolb. Technologies for the pseudonymization of medical data: A legal evaluation. In *2009 Fourth International Conference on Systems*, pages 7–12, 2009. doi: 10.1109/ICONS.2009.48.

[24] NhatHai Phan, Xintao Wu, and Dejing Dou. Preserving differential privacy in convolutional deep belief networks, 2018.

[25] Daniele Ravì, Charence Wong, Fani Deligianni, Melissa Berthelot, Javier Andreu-Perez, Benny Lo, and Guang-Zhong Yang. Deep learning for health informatics. *IEEE Journal of Biomedical and Health Informatics*, 21(1):4–21, 2017. doi: 10.1109/JBHI.2016.2636665.

[26] Joppe W. Bos, Kristin Lauter, Jake Loftus, and Michael Naehrig. Improved security for a ring-based fully homomorphic encryption scheme. In Martijn Stam, editor, *Cryptography and Coding*, pages 45–64, Berlin, Heidelberg, 2013. Springer Berlin Heidelberg. ISBN 978-3-642-45239-0.

[27] Siqi Liu, Sidong Liu, Weidong Cai, Sonia Pujol, Ron Kikinis, and Dagan Feng. Early diagnosis of alzheimer's disease with deep learning. In *2014 IEEE 11th International Symposium on Biomedical Imaging (ISBI)*, pages 1015–1018, 2014. doi: 10.1109/ISBI.2014.6868045.

[28] Guorong Wu, Daoqiang Zhang, Luping Zhou, Youngjin Yoo, Tom Brosch, Anthony Traboulsee, David Li, and Roger Tam. Deep learning of image features from unlabeled data for multiple sclerosis lesion segmentation. 09 2014. ISBN 978-3-319-10580-2. doi: 10.1007/978-3-319-10581-9_15.

[29] Adhish Prasoon, Kersten Petersen, Christian Igel, Francois Lauze, Erik Dam, and Mads Nielsen. Deep feature learning for knee cartilage segmentation using

a triplanar convolutional neural network. volume 16, pages 246–53, 01 2013. ISBN 978-3-642-38708-1. doi: 10.1007/978-3-642-40763-5_31.

[30] Edward Choi, Mohammad Taha Bahadori, Andy Schuetz, Walter F. Stewart, and Jimeng Sun. Doctor ai: Predicting clinical events via recurrent neural networks. In Finale Doshi-Velez, Jim Fackler, David Kale, Byron Wallace, and Jenna Wiens, editors, *Proceedings of the 1st Machine Learning for Healthcare Conference*, volume 56 of *Proceedings of Machine Learning Research*, pages 301–318, Northeastern University, Boston, MA, USA, 18–19 Aug 2016. PMLR. URL `https://proceedings.mlr.press/v56/Choi16.html`.

[31] David Kelley, Jasper Snoek, and John Rinn. Basset: Learning the regulatory code of the accessible genome with deep convolutional neural networks. page 028399, 10 2015. doi: 10.1101/028399.

Journal of Physics: Conference Series **2327** (2022) 012070 doi:10.1088/1742-6596/2327/1/012070

Medical Image Encryption using Latin Image Cipher Algorithm

Siju John[1], S.N Kumar[2]

[1] Research Scholar, Lincoln University College, Jalan Lembah Sireh, 15050 Kota Bharu, Kelantan, Malaysia
[2] Department of Electrical and Electronics. Amal Jyothi College of Engineering, Kottayam, Kerala, India - 686518
https://orcid.org/0000-0002-0803-7639, https://orcid.org/0000-0002-2530-1454

Abstract. Image processing has significant applications in the health care sector, medical data processing, analysis, storage, and transfer. The Latin Image Cipher algorithm was proposed in this work for the encryption of medical images. The encryption algorithm proposed in this research work comprises Latin square whitening, substitution, and permutation. The efficiency of the algorithm was also validated by inducing noise in the input images. The performance validation of the proposed algorithm was validated by the histogram analysis and correlation analysis. The information entropy measure also validates the proficiency of the encryption/decryption algorithm. The algorithm was developed on MATLAB 20021 and tested on COVID 19 CT images.

1. Introduction

Medical image encryption plays a vital role in teleradiology applications for data transfer. A desirable image encryption strategy method is implemented to obtain excessive gadget performance. Arnold cat map was employed in [1] for the encryption of images, mixing matrix generated were encrypted by the independent component analysis. Chaotic systems are appropriate for image encryption owing to their numerous intrinsic characteristics. The 2D logistic tent map was utilized in [2] for the encryption of color images, the cross-plane permutation and non-sequential diffusion overcome the limitation of classical chaotic algorithms. The logistic chaotic map and data hiding principle was put forward in [3] for ROI-based encryption, computational complexity was observed to be low and fit for real-time applications.

Hybrid chaotic algorithms along the number theory approach were found to be efficient for the encryption of DICOM medical images, thereby improving the security aspect [4]. Medical ethics states the importance of security for medical images and in today's era, telehealth gains prominence and stresses the need for a proficient encryption model. The dynamic chaotic map along with the Brownian motion principle was incorporated in [5] for the image encryption.

Retinal image encryption was carried out in [6], the pseudo-random numbers were generated from the chaotic maps and transmitted. The hybrid chaotic algorithm deployed in this work yields efficient results when compared to the classical chaotic models. The hybrid chaotic algorithm along with the artificial neural network was used in [7] for image encryption. The performance of the cryptosystem was enhanced by the DNA sequence mapping along with the hybrid chaos for the encryption scheme [8], robust results were generated, when compared with the classical models. The discrete wavelet transform decomposition along with the edge detection scheme was deployed in [9] for the encryption of images, the system was found to be efficient in immunity against attacks. A novel 1D chaotic map along with the double stage classical scrambling and a third stage highly efficient scrambling was highlighted in

Journal of Physics: Conference Series **2327** (2022) 012070 doi:10.1088/1742-6596/2327/1/012070

[10] for image encryption. Section 2 describes the Latin image cipher encryption algorithm with results and discussion highlighted in section 3, finally, the conclusion was drawn in section 4.

2. Proposed Methodology

2.1. Latin Square

A Latin Square of order M is a matrix of size MXM, comprised of M unique elements. Each symbol appears once in a row or column. Motivation from the mathematician Leonhard Euler, the Latin Square was formulated.

The Latin Square of order M is a function gL of (m, n, i) expressed as follows:

$$gL(m,n,i) = \begin{cases} 1, L\ (m,n) = Si \\ 0, otherwise \end{cases} \tag{1}$$

Where m represents the row element in L with m ∈ M, n the represents the column element in L with n ∈ M

For a Latin Square of order M,

$$\textstyle\sum_{m=0}^{M-1} gL\ (m,n,i) = 1 \tag{2}$$

and

$$\textstyle\sum_{n=1}^{M-1} gL\ (m,n,i) = 1 \tag{3}$$

The typical examples of Latin Squares are represented in figure 1. The sudoku puzzle is also a typical case of Latin Squares. There are different techniques for the generation of Latin Square. The typical algorithm for the generation of Latin Square is as follows:

```
Algorithm 1: Generation of Latin Square generator L= LSG(q1,q2).
Inputs: q1 and q2 are M length squares.
Output: L is a Latin Square of order M
Q seed = Sort map(q1)
Q shift = Sort map(q2)
for m = 0:1:M-1 do
        L (m, :) = rowshift (q1seed, q1shift) (m)
end for.
```

The typical example of generating a 4X4 Latin Square is represented in figure 1.

3	2	4	1
1	4	2	3
4	3	1	2
2	1	3	4

Figure 1. Typical example of 4×4 Latin Square

The q1 and q2 are M length squares generated from a pseudo-random number generator. The sort map function estimates the index map between q1 and the sorted version of q1*. The row shift function, row shift(w,v) shifts the element q1 with v elements to left.

2.2. Probabilistic Encryption with Key Translation

The plain text refers to the input in a cryptography system and ciphertext is the output. Substitution and permutation play vital roles. A typical N round architecture of substitution permutation network is depicted in figure 2. The S Box represents the substitution, maps one input byte to another in a deterministic manner. The p box represents the permutation box that shuffles bit positions of the input in a deterministic manner. The decryption process is the inverse process of encryption. The property of substitution and permutation is similar to confusion and diffusion. The proposed encryption or decryption algorithm utilizes the block size of 256 × 256 with eight rounds. Probabilistic encryption creates random values in cipher. The objective behind randomness is to ensure security. The random feature is done by including noise in the LSB plane of the input medical image. The XOR operation is performed between the LSB plane of the input image and the randomly generated 256 × 256-bit plane. Figure 2 depicts the encryption scheme using the Latin Image Cipher algorithm.

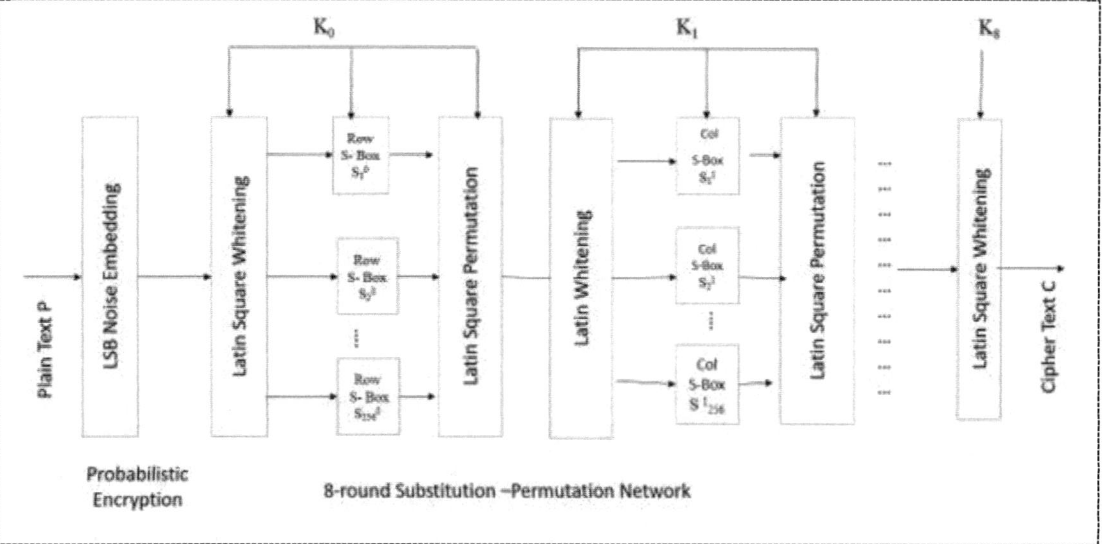

Figure 2. Encryption scheme using Latin Image Cipher algorithm

In the classical cipher system, the keys are used without translating Pseudo-random, in this encryption scheme, the key translation is done in such a manner up to a value of 8.

The steps in the key translation are as follows
Split the encryption key 'K' into 8 32 bits subkeys such that
K= [K0, K1, …., K7].
Pseudo random number generator is used to generate random sequences $(q1^0, q2^0)$, $(q1^1, q2^1)$,…….,$(q1^7$, utilizing the subkeys.
Latin Squares are formulated from the generated random sequences.
LM= LSG$(q1^M, q2^M)$
The mechanism of the classical whitening stage involves the mixing of plain text with a round key using an XOR operation.

Let x= I(m,n) represents the pixel value of the plain text in an image and d= L(m,n) represents the respective elements in the Latin Square.
The pixel-level equation is as follows

$$\begin{cases} C(m,n) = [SR(l(m,n),[D]3) + L(m,n)]2^8 \\ I(m,n) = [SR\ ([C(m,n) + L(m,n)]2^8,[D]3 \end{cases} \quad (4)$$

Where y= L(m,n) represents the cipher text, D is the rotation parameter and SR represents the spatial rotation function with respect to different values of direction 'd'

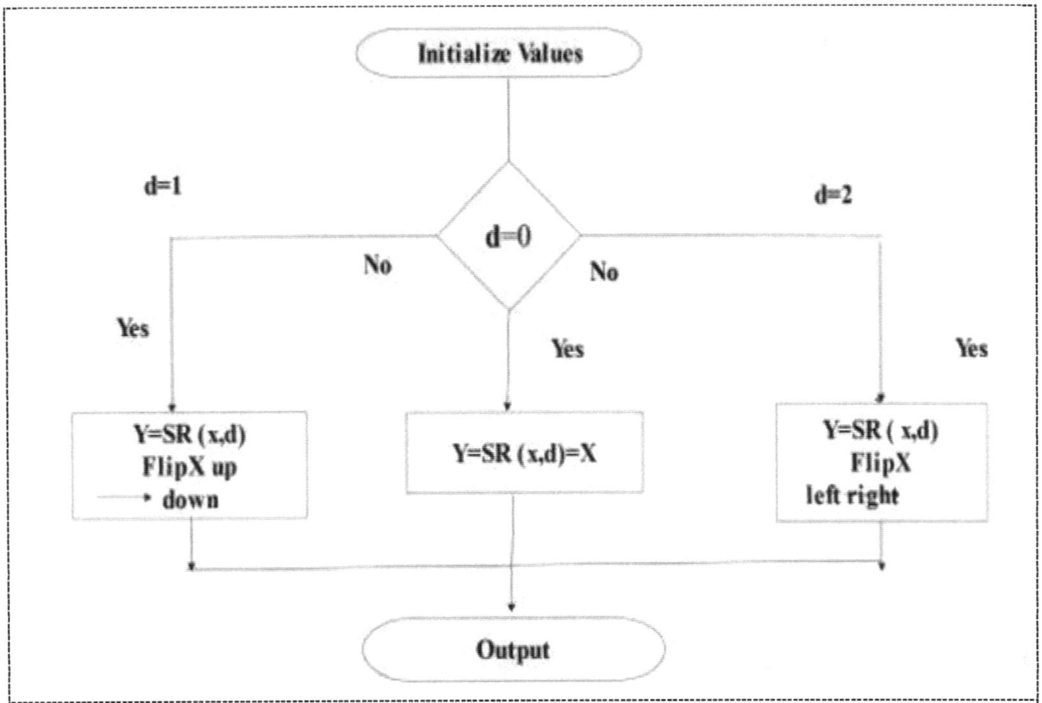

Figure 3. Probabilistic encryption procedure

The Latin Square whitening (LSW) in the image level is expressed as follows

$$LSW = \begin{cases} C = E(L,I,D) \\ I = D(L,C,D) \end{cases} \quad (5)$$

The plain text image 'I' was recovered from the ciphertext image 'C' using the above equations. The rows and columns of the plain text image 'I' become unfamiliar after Latin Square Row substitution and Latin Square Column Substitution. The permutation box performs byte shuffling and scrambling.

3. Results and Discussion

This chapter proposes medical image encryption using the Latin square image cipher algorithm. The algorithms are developed in MATLAB 2020a, processes specification as follows; Processor: Intel(R) Core (TM) i7-10510U CPU @ 1.80GHz 2.30 GHz, RAM: 16.0 GB, System Type: 64-bit operating system, x64-based processor. The 6 medical data sets are utilized in this work, D1-D6 [11]. The validation of results is vital to judging the performance of encryption techniques. The following analysis is performed in this work on the encrypted image; histogram, correlation, and entropy. The decrypted image was validated by the performance matrix peak to signal noise ratio and mean square error. Figure 4 depicts the encryption/decryption results corresponding to the medical data set (D1). In figure 4, (a) depicts the input image, (b) depicts the encrypted image, (c) depicts the decrypted image, (d) depicts the histogram of the input image, (e) depicts the histogram of the encrypted image and (f) depicts the histogram of the decrypted image.

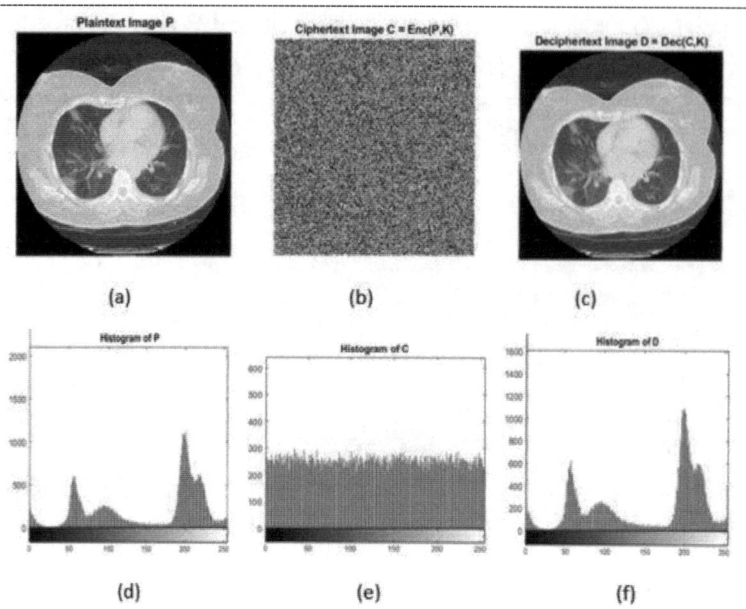

Figure 4. Encryption/ Decryption of medical image (D1) [12] corresponding to Latin image cipher algorithm

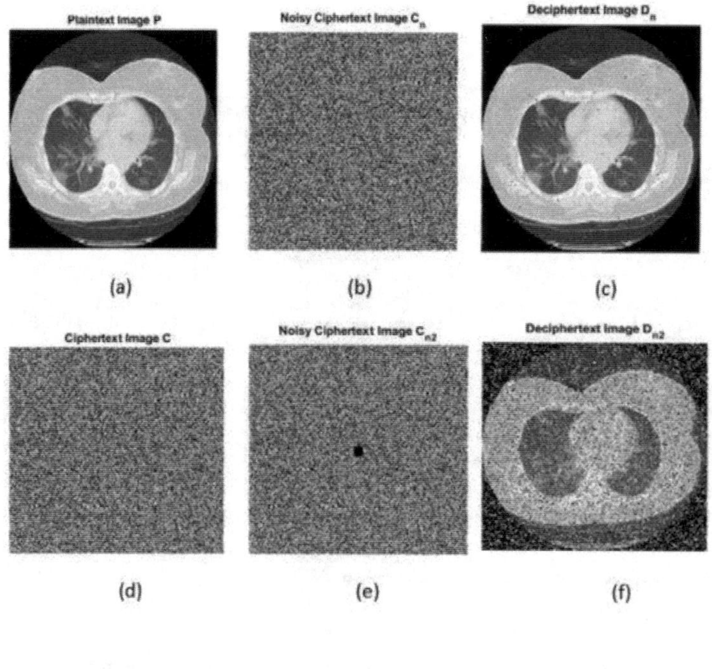

Figure 5. Effect of noise in the encryption/decryption algorithm on the medical image (D1) corresponding to Latin image cipher algorithm

For an encrypted image, the ideal plot of the histogram as the flat shape with no isolated peaks. The flag shape of the histogram indicates randomness. The histogram of the input image has several peaks, while the histogram of the encrypted image is flat. The flat shape of the nature of the histogram ensures immunity against the statistical attack. The figure 4 results reveal that, the encrypted image histogram is flat, thereby it ensures the efficiency of the algorithm. The effect of noise in the input image was analyzed from the results depicted in figure 5. In figure 5, (b) depicts the noisy cipher image corresponding to 0.025% of noise and (e) depicts the noisy cipher image corresponding to 0.22% of noise, (c) and (f) depicts the decrypted image corresponding to the cipher images in (b) and (e). Figure 6 depicts the encryption/decryption results corresponding to the medical image (D2).

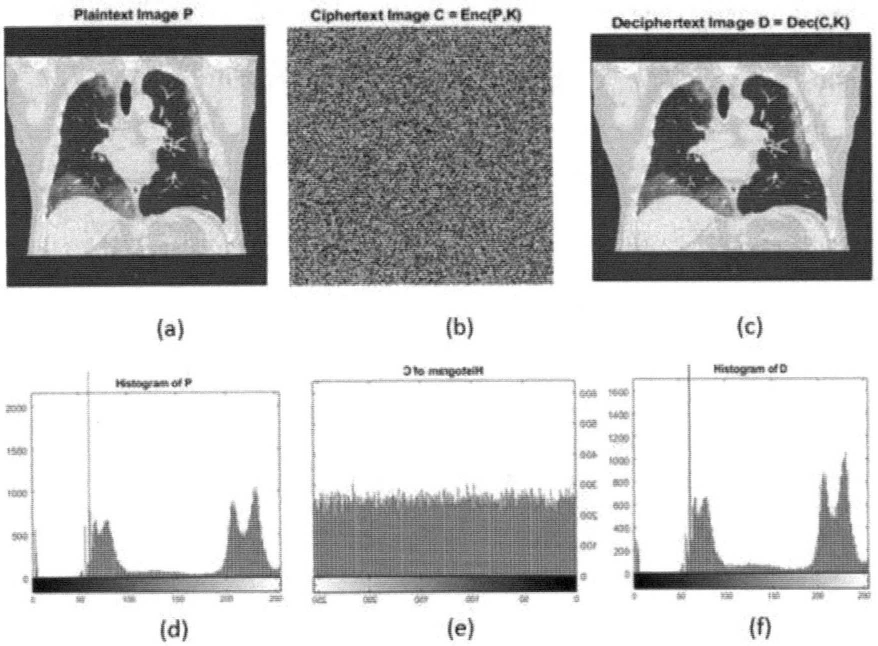

Figure 6. Encryption/ Decryption of medical image (D2) [13] corresponding to Latin image cipher algorithm

The correlation measures are estimated between the input image and the encrypted image. The encryption algorithm efficiency was determined from the correlation value. Lowering the correlation coefficient value higher the efficiency of the algorithm ensures the immunity against the statistical attack. Apart from the estimation of correlation between the pixels in the input image and encrypted image, the horizontal, vertical, diagonal correlation coefficient values of the encrypted image are also estimated.

$$E(p) = \frac{1}{N}\sum_{i=1}^{N} p_i \tag{6}$$

$$D(p) = \frac{1}{N}\sum_{i=1}^{N}[p_i - E(p)^2 \tag{7}$$

$$Cov(p,q) = \frac{1}{N}\sum_{i=1}^{N}[[p_i - E(p_i)][q_i - E(q_i)]] \tag{8}$$

$$\gamma(p,q) = \frac{Cov(p,q)}{\sqrt{D(p)}\sqrt{D(q)}} \tag{9}$$

The information entropy is a proficient measure for evaluation the cryptosystem. Lower the value of entropy indicates the security thread to the encryption system. For an 8-bit grayscale image, the ideal value of information entropy should be 8. Higher the value of entropy indicates the resistance to information outflow.

$$H\left(I = \sum_{i=0}^{2^N-1} p(I_i)\, log_2 \frac{1}{p(I_i)}\right) \tag{10}$$

Figure 7 depicts the encryption/decryption results corresponding to the medical image (D3).

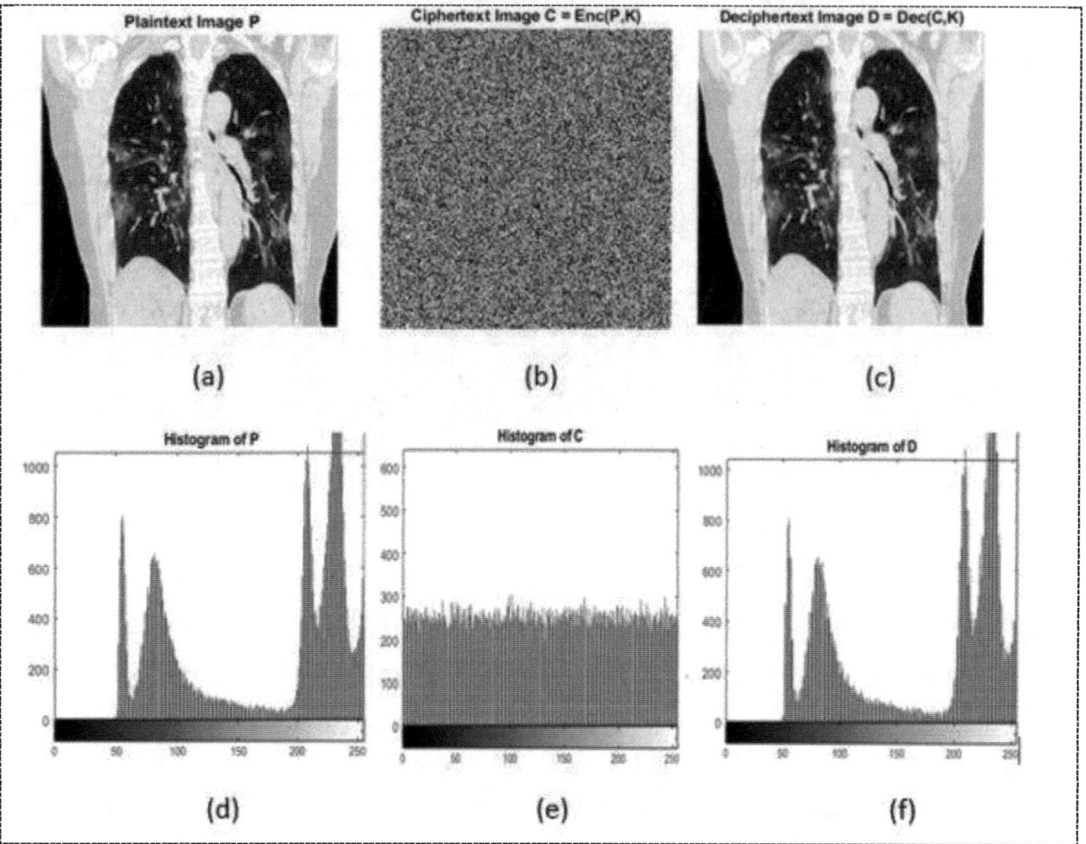

Figure 7. Encryption/ Decryption of medical image (D3) [14] corresponding to Latin image cipher algorithm

Peak to signal noise ratio (PSNR) is estimated between the original input image and the decrypted image. Higher the value of PSNR and lower the mean square error (MSE) indicates the robustness of the encryption algorithm.

$$MSE = \frac{1}{M \times N} \sum_{i=1}^{M} \sum_{j=1}^{N} (I_{ij} - I_{0ij})^2 \tag{11}$$

$$PSNR = 10 \times log_{10} \left(\frac{255^2}{MSE}\right) \tag{12}$$

Figure 8 depicts the encryption/decryption results corresponding to the medical image (D4). Figure 9 depicts the encryption/decryption results corresponding to the medical image (D5).

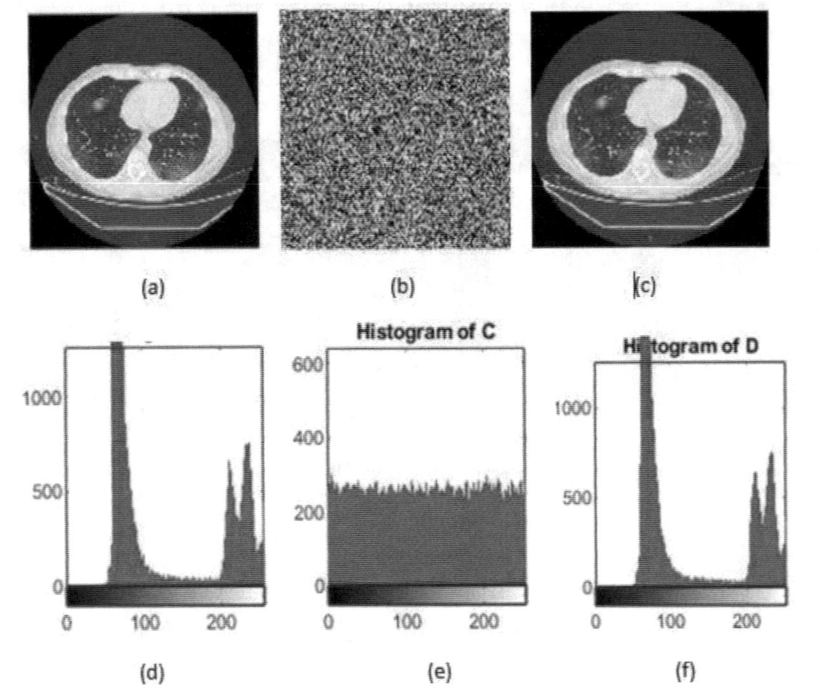

Figure 8. Encryption/ Decryption of medical image (D4) [15] corresponding to Latin image cipher algorithm

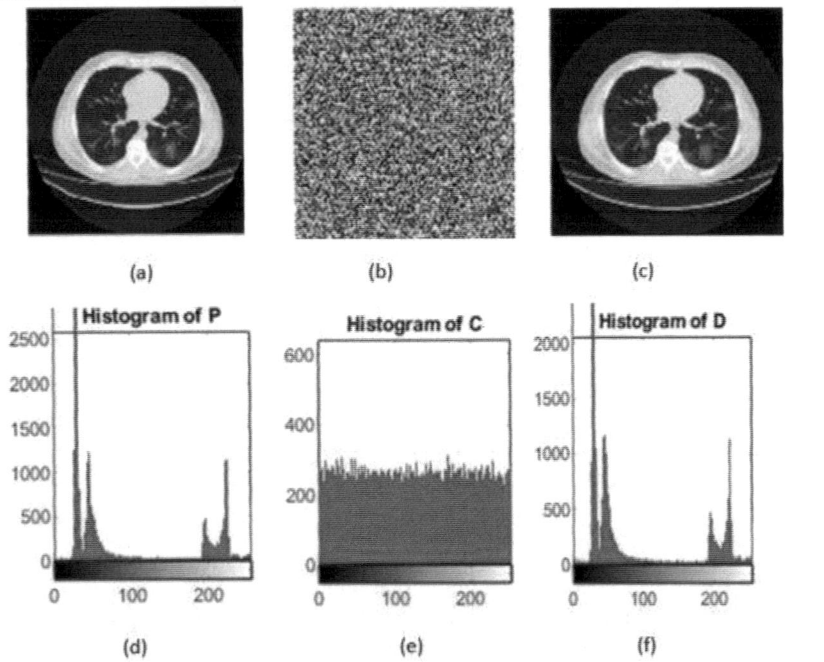

Figure 9. Encryption/ Decryption of medical image (D5) [16] corresponding to Latin image cipher algorithm

Journal of Physics: Conference Series **2327** (2022) 012070 doi:10.1088/1742-6596/2327/1/012070

Figure 10 depicts the encryption/decryption results corresponding to the medical image (D6). The correlation analysis for the validation of the encryption algorithm is represented in table 1. The lower value of correlation favors the efficiency of the encryption algorithm.

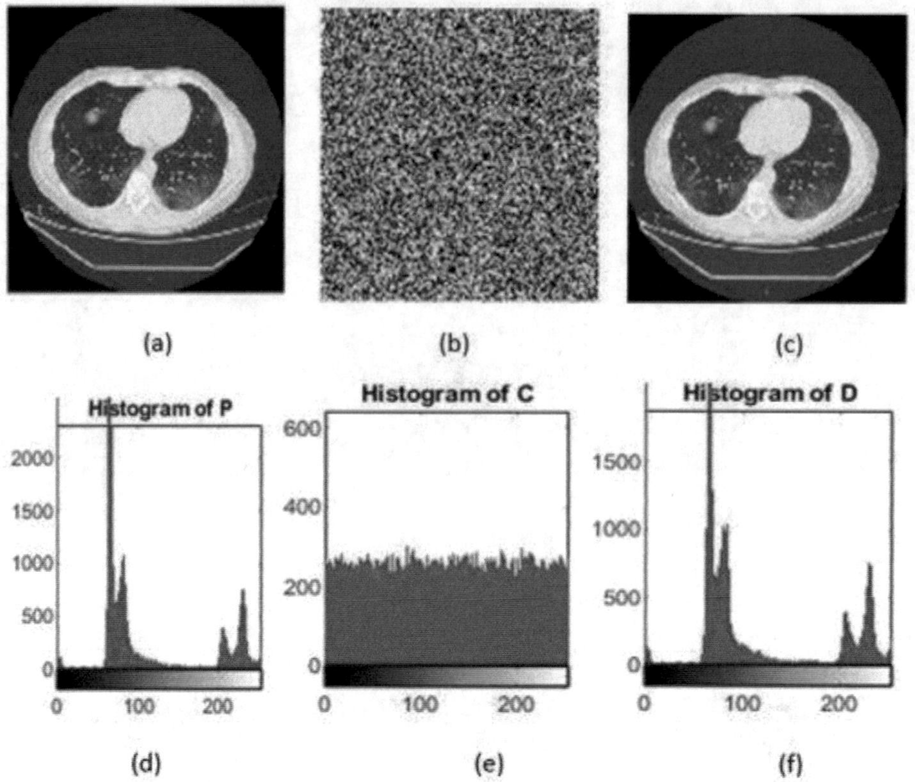

Figure 10. Encryption/ Decryption of medical image (D6) [17] corresponding to Latin image cipher algorithm

Table 1. Correlation analysis for the validation of encryption algorithm.

Image Details	Correlation analysis	Horizontal correlation	Vertical correlation	Diagonal correlation
D1	0.02	2.146328e-03	2.903699e-03	2.510431e-03
D2	0.03	1.644944e-03	1.020309e-04	4.086496e-03
D3	0.02	-4.041685e-03	-3.746232e-03	-1.452418e-03
D4	-0.01	2.631857e-03	2.351230e-04	-1.075435e-03
D5	0.01	-8.574951e-03	1.816777e-03	-1.645526e-03
D6	0.00	2.123373e-03	5.145050e-03	2.139022e-03

The information entropy of the input and encrypted images are represented in table 2. A higher value of entropy of the encrypted image favors the robustness of the encryption technique.

Table 2. Entropy values for the validation of the encryption algorithms.

Image details	The entropy of the input image	The entropy of the encrypted image
D1	7.9973	8
D2	7.9969	8
D3	7.9974	8
D4	7.9968	8
D5	7.9973	8
D6	7.9972	8

The PSNR and MSE values are estimated between the input image and the decrypted image. A higher value of PSNR values and lower values of MSE favors the efficiency of the encryption/decryption algorithm.

Table 3. PSNR and MSE values of the encryption/decryption technique.

Image Details	PSNR	MSE
D1	51.17	0.203949
D2	51.13	0.203751
D3	51.15	0.203873
D4	51.11	0.256454
D5	51.12	0.254654
D6	51.14	0.204254

The performance metrics validation reveals the proficiency of the medical image encryption using Latin image cipher algorithm, in contrast with the classical approaches. The PSNR was found to be substantially good with low value of PSNR. The correlation and entropy analysis also proves the efficiency of the proposed encryption approach in ensuring security for teleradiology applications.

4. Conclusion

In this research work, the Latin image cipher algorithm that relies on a probabilistic encryption scheme was proposed for the medical images. The security aspect of medical data gains importance while transferring it, especially through the cloud network for teleradiology applications. There is an urge for the development of a robust encryption technique there by ensuring security. The Latin image cipher algorithm was validated on COVID 19 CT images and the performance was judged by the performance metrics. The noise was induced in the input image to determine the proficiency of the algorithm against noise attack. The performance metrics results validation was found to be satisfactory and is a best choice for the encryption of medical images for the telemedicine application. The algorithm can easily be implemented in software and hardware; the future work will be the hardware implementation of the encryption algorithm in an embedded processor. The pre-processing algorithm based on the medical imaging modality along with the encryption model implementation in the embedded processor will be the future work.

References

1. Abbas NA. 2016 *Egypt. informatics J.* **17**: 139-146.
2. Hua Z, Zhu Z, Yi S, Zhang Z, Huang H, Sun J, Liao X, Chen X and Guo S. 2021 *Inf. Sci. (Ny).* **27**: 1063-1083.
3. Sun J, Liao X, Chen X and Guo S. 2017 *Int. J. Bifurc. Chaos* **27**: 1750073.
4. Chandrasekaran J and Thiruvengadam SJ. 2017 *Secur. Commun. Networks* **2017**.
5. Khan M, Masood F, Alghafis A, Amin M and Batool Naqvi SI. 2019 *PLoS One* **14**: e0225031.
6. Ebrahim SMA. 2020 *Int. J. Netw. Secur. Its Appl. Vol* **12**.
7. Pan S, Wei J and Hu S. 2020 *Multimed. Tools Appl.* **79**: 9163-9176.

8. Guesmi R and Farah MA. 2021 *Multimed. Tools Appl.* **80**: 1925-1944.
9. Jeevitha S and Amutha Prabha N. 2021 *J. Ambient Intell. Humaniz. Comput.* **12**: 3373-3388.
10. Pankaj S and Dua M. 2021 *Netw. Model. Anal. Heal. Informatics Bioinforma.* **10**: 1-19.
11. Hollett M and Bell D. 2020 In Radiopaedia.Org. Radiopaedia.org, 2020.
12. Sheikh Y and Rasuli B. 2020 In Radiopaedia.Org. Radiopaedia.org, 2020.
13. Bell D and Nicoletti D. 2020 In Radiopaedia.Org. Radiopaedia.org, 2020.
14. Weerakkody Y and Macori F. 2020 In Radiopaedia.Org. Radiopaedia.org, 2020.
15. Feger J and Cetinoglu K. 2020 In Radiopaedia.Org. Radiopaedia.org, 2020.
16. Shah V and Kovratko M. 2020 In Radiopaedia.Org. Radiopaedia.org, 2020.
17. El-Feky M and Nicoletti D. 2020 In Radiopaedia.Org. Radiopaedia.org, 2020.

Acknowledgments
The authors would like to acknowledge the support from the Schmitt Centre for Biomedical Instrumentation, Department of EEE, Amal Jyothi College of Engineering, Kanjirappally.

Journal of Physics: Conference Series **2327** (2022) 012071 doi:10.1088/1742-6596/2327/1/012071

Prediction model of heart diseases based on hybrid model

Rimpy Morya[1], Someet Singh[2]

[1,2]Department of Electronics and Electrical Engineering, Lovely Professional University, Phagwara, India, 144411
Email: rimpymorya1947@gmail.com[1], someet.17380@lpu.co.in[2]

Abstract. Heart diseases are crucial cause of global death. Premature detection of heart diseases and uninterrupted analysis decreases the death rate. The detection of heart diseases using various machine learning algorithm is implemented for analysis of the heart attack in minimum time. Machine learning plays a very important role in the prevention of diseases and huge comprehension a health record. Recent trend in machine learning methodology being wield in contemporary evolution in different sector of internet of things. There is different algorithm in machine learning, which is analysis of heart diseases along with different accuracy. The purpose of this research to develop machine learning model to detect the heart diseases. In this a hybrid algorithm is proposed using combination of KNN and logistic regression algorithm. Furthermore, a protype is developed which consist of set of sensors to monitor the timely for prediction or analysis of heart attack.

1. Introduction

The heart is the main performer of our body. The function of heart pumping an oxygenated blood to the other part of body. Thus, we should save heart from diseases. There are many factors for heart illness such as blood pressure, cholesterol and bad habits like smoking. Many scientists have reported in their research that the dominant peril influence of heart diseases is excessive consumption of alcohol, unhealthy food and eating tobacco. The prognosis and cure of heart diseases exceedingly arduous when connoisseur and contemporary technology are not accessible. The effectual prognosis and apttherapy can save numerous person life. Thus, the obtainable of data and data mining methodology predominantly machine learning and premature monitoring of heart disease thatcan aid patients. The significant figure of data that indispensable for therapy of medical maintenance territory. Medical tutelage is requisite, which provide convalescent craving for his future new life.

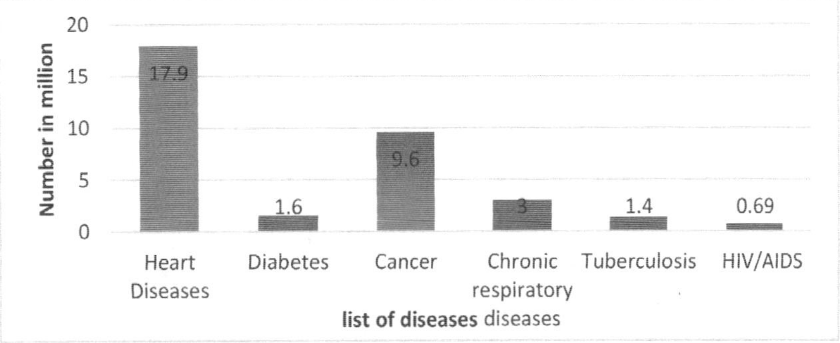

Figure 1. Death rate of different type diseases in million

In 2017 of all projected worldwide deaths, 23 million are anticipate to be because of heart disease[1]. As a matter of fact, that cardiovascular diseases exclusive source of death. The ailment of heart is determined through the sinus rhythm which is provided by ECG. The different type of heart condition detecting through ECG. The low-cost screening test of heart diseases done through the help of ECG. The contemporary progression in Machine Learning algorithm (MLA)the technology in medical science is also significantly improve over decade. Python integration with machine learning have been supported various type of libraries like sci-kit learn, matplotlib and numpy which help in heart diseases detection and analysis. Machine learning and internet of thing has become fascinating topic in the field of research. Machine learning have played a very major role in department for prevention of diseases and huge comprehension a health record. Recent trend in machine learning methodology being wield in current evolution in different sector of internet of things. IOT connected the device each other via internet. There are lot of application of IOT but in field of health care monitoring IOT play a vital role. The purpose of this research to develop a real time system for analysis the risk of heart attack. Our proposed system is cost effective and with less electronic component. This approachbased on the parameter which we have consider from machine learning. There is different algorithm in machine learning, which is analysis of heart diseases along with different accuracy.

2. Literature Review

Few past years lot of researchers are described various models for prediction of heart disease. As this topic related to the health of the human and heart disease is crucial disease now a day so this problem needs to resolved so in this section, we explain some existing solution of this problem.
B. S. Rao et al.[2] proposed system on data set by their weight and trained the neural network the outcome of data prescribes the fitness of each person, according to result some predefined operations are performed, then trained neural network detect the heart diseases. A. Shetty et al. [3] proposed the system using data integration and data cleaning from dataset to predict the heart disease. Record of patient medical record from hospital are considered while suggesting the framework. A comparative study is made [4] on different machine learning algorithm, author experiments gives the high accuracy rate of preventing and controlling of chronic disease made a comparative study. The naïve bayes has the maximum efficiency. Sharma et al. [5] forecast the identification of cardiovascular diseases, with the help of neutral networkto scrutinize and enhance the serval machine learning algorithm also using weight initializing technique. The result of research approach indicate that AdaGrad algorithm has 85% accuracy and provide fastest result as compared to another. Another intersecting approach of MLA of hard votingensemble model by R. Atallah et al. [6] combine stochastic gradient descent, k nearest neighbors, random forest, logistic regression algorithm using ensemble hard voting classifier. The accuracy of model is 90%. Mohan et al. [7] analysis the disease with hybrid model by combining random forest with linear model to detect disease with accuracy of 88.7%. A Gupta et al. [8] developed the algorithm based on machine learning for prediction of cardiac infarction and built the porotype that integrated serval type of sensor, whichmonitoring the human health that generate by heart diseases and correlate k nearest neighbors with data mining, random forest, decision tree and naïve bays, recognized the k nearest neighbors algorithm are more accurate and efficient than other. Ali et al. [9] designed the algorithm with the help of X^2 statistical model for refinement of characteristic and used deep neural network for categorization. They are differentiating the X^2 statistical model with artificial neural network anddeep neural network and precision of their result is 93.33%. The author in [10] their model determined the cardiovascular diseases employing various algorithm method and identifying whether the person suffering from heart diseases or not, find the accuracy and efficiency of different algorithm. Reddy et al. [11] designed the algorithm for detection of heart disease by fuzzy logic with adaptive genetic algorithm. The dataset taken from UCI heart disease dataset model.

3. Proposed Method

This section provides solution for detection of diseases. The designed model based on machine learning. The explanation of all the part discussed as below:

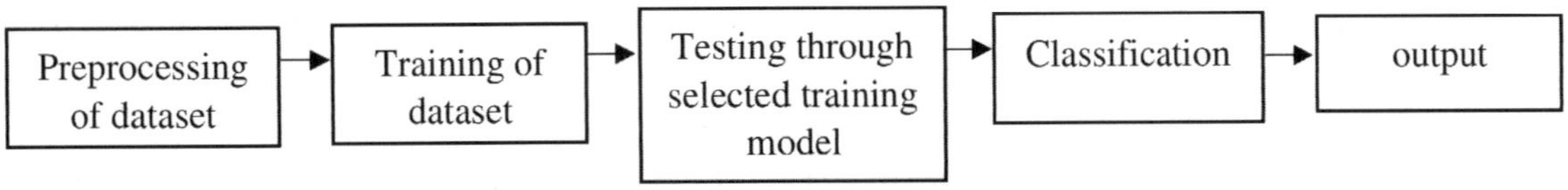

Figure 2. Block diagram of prediction model

3.1. Collection of Data

The collection of data taken from Kaggle. The collected data have an attribute: the person age (age), resting blood pressure (trestbps), chest pain (cp), fasting blood sugar (fbs), cholesterol (chol), maximum heart rate (thalach), electrocardiographic result (restecg), ST depression including (oldpeak), excersie included angina (exang), major vessel (ca), Peak exercise (slope), thalassemia (thal). The percentage of normal and abnormal patients in data set are represented by below pie plot.

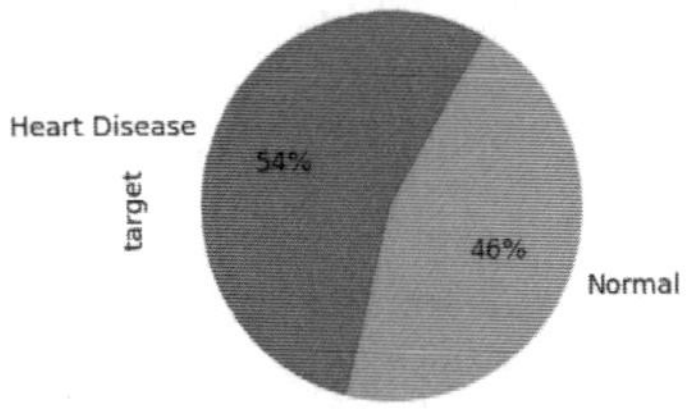

Figure 3. Percentage of patient in dataset

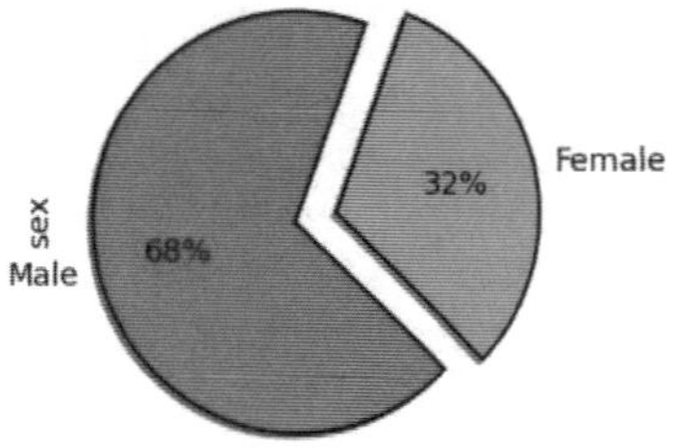

Figure 4. Distribution of gender in dataset

The gender distribution of dataset is shown in pie chart. Chart represents the dataset have 68% male and 32% female. Furthermore, analyse the data we have make the distribution graph of age, thalach, chest pain, and trestbps.

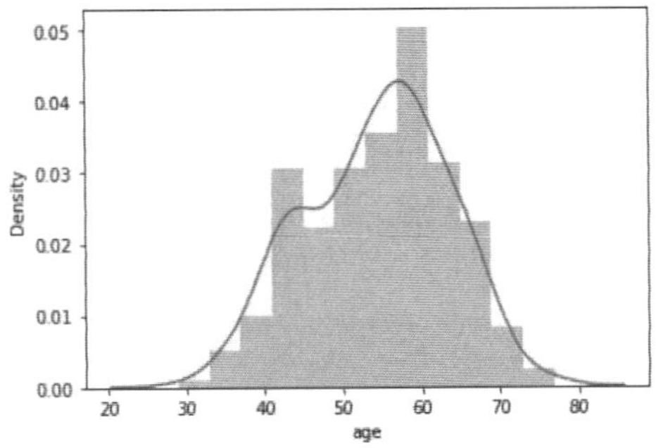

Figure 5. Age distribution

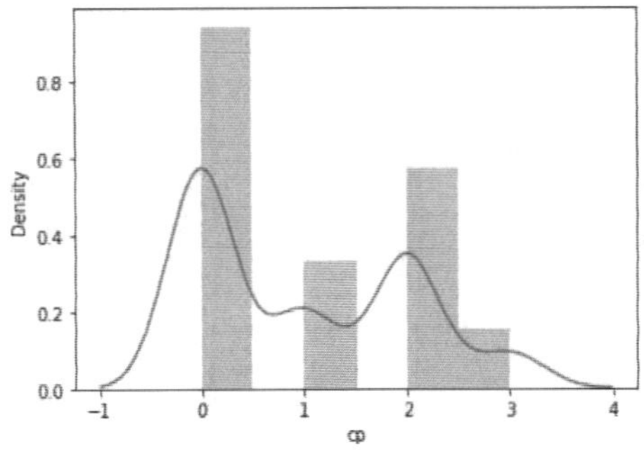

Figure 6. Chest pain distribution

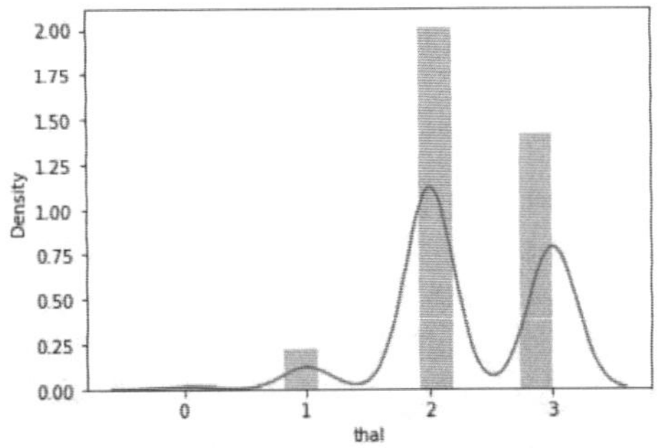

Figure 7. Thalach distribution

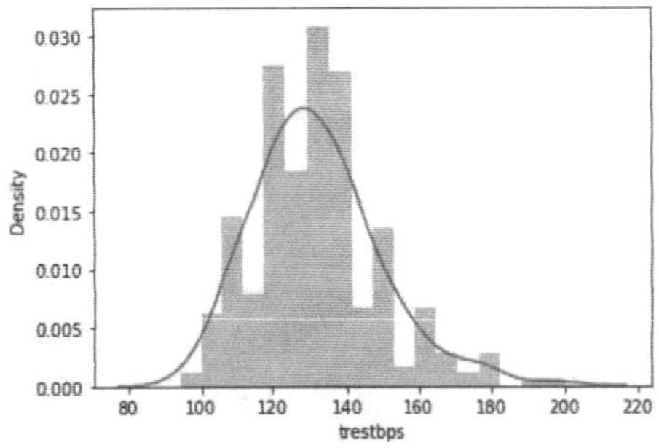

Figure 8. Resting blood pressure distribution

3.2. Addressing the missing values

In our dataset we have find the missing value by using df null().sum but we have not find any missing value.

3.3. Feature Selection

For data analyzation we have calculated correlation value between each attribute and target diagnosis. From correlation value it is cleared that Cp, Thalach, Exang and oldpeak are highly correlated with target attribute. For more clarification of correlation between each feature and target showing in heatmap.

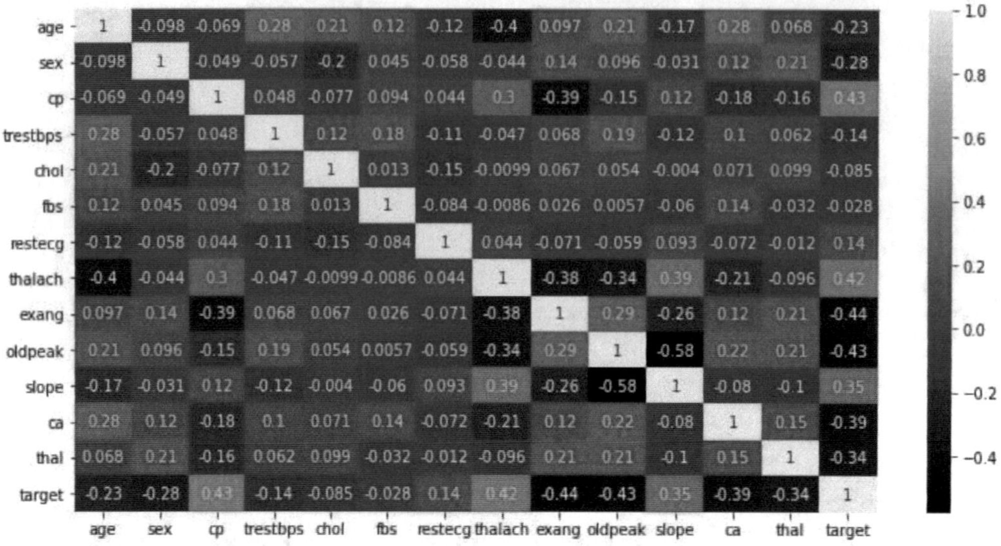

Figure 9. Heat map of correlation value

3.4. Model Training

This phase includes splitting of dataset in order to test and train. 80% dataset for training used as a input to machine learning algorithm and rest of 20% used for testing. We exploited the k-nearest neighbors (KNN) and logistic regression and hybrid of KNN and logistic regression using stacking technique. Explanation of algorithm given below

3.4.1. K-Nearest Neighbors

Supervised learning model KNN is used in the classification problem. The distance between the new instance finding by this classifier. Then nearest k data point selected from the predefined K number for the new instance. Finally, the majority class of k data point responsible for classification. Euclidean distance is used to calculated difference between the continuous attribute. For example, first point (x_1, x_2, x_3 ... x_n) and second point (y_1, y_2, y_3, ... y_n) formula to calculate the distance is shown below.

$$\sqrt{(x_1-y_1)^2+(x_2-y_2)^2+\cdots\cdots\cdots\cdots(x_n-y_n)^2}$$

Euclidean distance formula is responsible for problem that is large value frequency swamp small value. For Example, dataset in heart measure a cholesterol in range between 100 and 190 and age measure 40 and 80. The value of cholesterol will be greater as compared to age. To solve this issue distance measures between same instances is achieved by normalizing the continuous attributes. KNN work with both continuous and discrete attribute. In case of discrete attribute if attribute value of x_2 and y_2 are different then different between them is one else zero.

First model is build using K-Nearest Neighbor classifier with default parameter. The accuracy of model is 71.4%. The result obtained in the form of confusion matrix is show below

Journal of Physics: Conference Series **2327** (2022) 012071 doi:10.1088/1742-6596/2327/1/012071

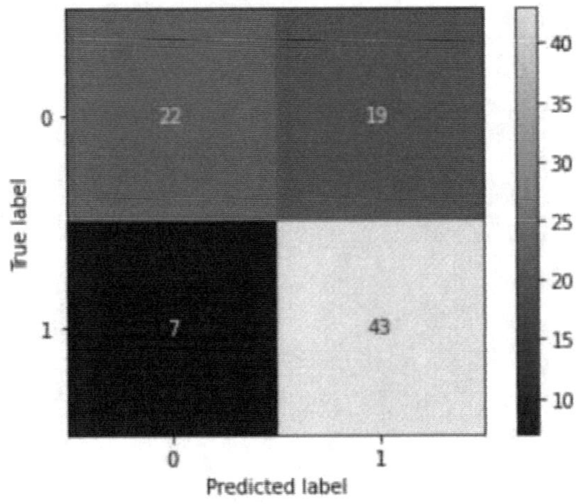

Figure 10. Confusion matrix of KNN

3.4.2. Logistic Regression
Binary variable is used as target variable in this analysis technique. Logistic regression represents the relationship between one or more independent variable and one dependent variable. The probability of target value predicted by logistic regression.

The second model build with the logistic regression using default parameters. The accuracy of this model 87.9%. The confusion matrix of logistic regression shows below

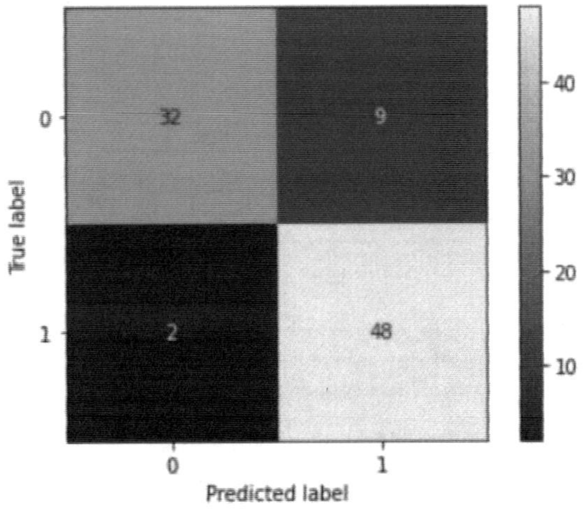

Figure 11. Confusion matrix of Logistic regression

3.4.3. Hybrid Model
Hybrid model using stacking is designed by combining k nearest neighbors and logistic regression model. SGD classifier is used as meta classifier to build a hybrid model. As performance of model gets improved using stacking. The accuracy of stacking hybrid model 89%. The confusion matrix of hybrid model shows in figure 12.

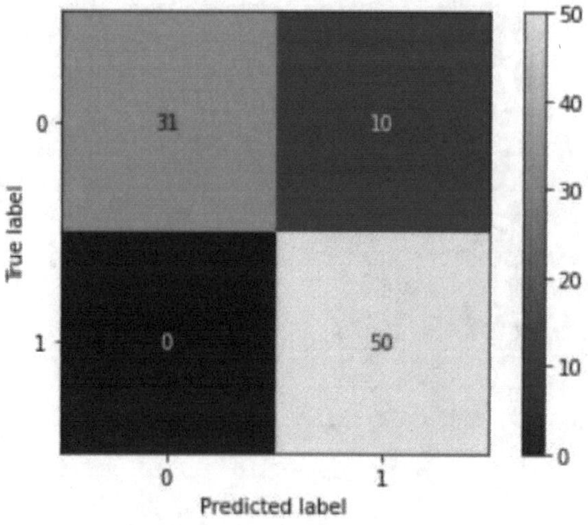

Figure 12. Confusion matrix of hybrid model

4. Result and discussion

In Python 3.9, experiment is implementation KNN and Logistic regression algorithm used in this experiment for heart disease detection. The result of shows disease detection using KNN and Logistic regression algorithm is effective. KNN attained 71.4% accuracy, Logistic regression attained 87.9% accuracy and Hybrid algorithm attained 89%.

Table 1. Results

Learning Model	Prediction percentage
KNN	71.4%
Logistic Regression	87.9%
Hybrid	89%

5. Conclusion

The very important and challenging task in medical field is prediction of heart disease because death rate is significantly controlled in case of early detection of disease. The processing of heart disease data will help in saving the human lives. Machine learning algorithms were used in this research to process the data to get the information. The proposed hybrid approach is used which is made by combining the KNN and logistic regression. Hybrid model show better result as compared to individual single algorithm. The future work of this research include development of protype model with integration of deep learning algorithms. As deep learning algorithm plays an important role in health care and provide better outcome in disease detection.

6. References

[1] Organization W H 2017 Cardiovascular diseases

[2] Rao B S, Rao K N and Setty S P 2014 An approach for heart disease detection by enhancing training phase of neural network using hybrid algorithm *Souvenir 2014 IEEE Int. Adv. Comput. Conf. IACC 2014* 1211–20

[3] Shetty A and Naik C 2016 Different data mining approaches for predicting heart disease *Int J Innov Res Sci Eng Technol* **5** 277–81

[4] Deepika K and Seema S 2016 Predictive analytics to prevent and control chronic diseases *2016 2nd International Conference on Applied and Theoretical Computing and Communication*

Technology (iCATccT) (IEEE) pp 381–6

[5] Sharma V, Rasool A and Hajela G 2020 Prediction of Heart disease using DNN *2020 Second International Conference on Inventive Research in Computing Applications (ICIRCA)* (IEEE) pp 554–62

[6] Atallah R and Al-Mousa A 2019 Heart disease detection using machine learning majority voting ensemble method *2019 2nd international conference on new trends in computing sciences (ictcs)* (IEEE) pp 1–6

[7] Mohan S, Thirumalai C and Srivastava G 2019 Effective heart disease prediction using hybrid machine learning techniques *IEEE access* **7** 81542–54

[8] Gupta A, Yadav S, Shahid S and Venkanna U 2019 HeartCare: IoT based heart disease prediction system *2019 International Conference on Information Technology (ICIT)* (IEEE) pp 88–93

[9] Ali L, Rahman A, Khan A, Zhou M, Javeed A and Khan J A 2019 An Automated Diagnostic System for Heart Disease Prediction Based on $\chi2$ Statistical Model and Optimally Configured Deep Neural Network *IEEE Access* **7** 34938–45

[10] Nashif S, Raihan M R, Islam M R and Imam M H 2018 Heart Disease Detection by Using Machine Learning Algorithms and a Real-Time Cardiovascular Health Monitoring System *World J. Eng. Technol.* **6** 854–73

[11] Reddy G T, Reddy M, Lakshmanna K, Rajput D S, Kaluri R and Srivastava G 2020 Hybrid genetic algorithm and a fuzzy logic classifier for heart disease diagnosis *Evol. Intell.* **13** 185–96

Fatigue Detection Post Physical Activity Using Machine Learning Algorithms

Ashish Sharma[1], Gaurav Sethi[2]
[1]Research Scholar, Lovely Professional University, Jalandhar, India
ashish.41800512@lpu.in
[2]Associate Professor, Lovely Professional University, Jalandhar, India
gaurav.11106@lpu.co.in

Abstract. Construction work is purely effortful and the prevention of injuries at construction job sites is essential for encouraging worker's well being and health which is generally overlooked at the construction sites. World's construction industry is one amongst those having unsatisfactory work health issues. A large number of laborers and construction workers have to undergo fatigue risk at their job place. This paper describes the current state of the art of the research carried out in case of fatigue assessment after performing some physical activity providing an insight into fatigue, its detection and an overview of the causes of risk fatigue and its countermeasures. A number of subjective and objective fatigue assessment approaches have been used that have further stimulated the inclusion of latest and advanced approaches for fatigue detection. Although individual's knowledge regarding the fatigue detection approaches has been enhanced, there is as yet minimal research in the field of fatigue detection post physical activity. In this study the stress would be on objective fatigue detection techniques where the acquisition of physiological data of construction workers is required. The study provides a ground for detection of physiological fatigue based on motion capture data in the form of videos.

Keywords: Fatigue, construction workers, physical activity.

1 Introduction

The construction and labor sites record a considerable number of on-site injuries in India and abroad. Also, another major cause of nonlethal injuries that is caused mainly due to having a break from construction work, as well as a huge counterbalance costs and medical costs with frequent work dysfunctions [1].

Fatigue is one of the major causes for accidents happening in the construction sites. The construction industry often exposes workers to massive workloads, extended work timings, and iterative tasks making body fatigue unavoidable [2,3]. Physical fatigue can also birth to continual health problems as well as task related body and joint disorders, inveterate fatigue problems, and immune system dysfunctions [4,5]. Thus fatigue detection and assessing technology that permit for mediation in advance of fateful effects to workers' protection, fitness, and efficiency are worth studying.

Fatigue in general terms is defined as the reduction in the transient performance of the muscles to employ optimal force after undergoing a physical activity. Fatigue is a symptom, not a condition. Fatigue develops moderately in steps over a time period for works that do not demand continued employment of maximum possible force. As per a study on construction workers, a worker undergoes five to six hours of average walking in individual timeshift and around 40-44% of the individuals carrying out these jobs are reported to be exhausted due to body fatigue [6].

Traditional approaches to assess the physical exposure experienced by the workers at construction sites are generally based on ocular observation carried out by a skilled observer or it may rely on any other subjective approaches such as questionnaires. However, as per the nature of fatigue detection, certain fatigue indicators may vary among different workers with distinct social and ethnic backgrounds [7,8]. This is the reason that subjective techniques of fatigue detection are generally implemented to the task specific environment and the target individuals.

Physical and body related measurements, encompassing pulse rate, oxygen intake and outer body temperature are employed to conquer the limitations of subjective detection methods [9,10]. As above said method may be clumsy and difficult to implement, sometimes secondary measures have been implemented to assess physical exhaustion, enhanced jerk, and joints related issues [11]. These detection methods are based on motion data collected using optoelectronic sensors, which is the pioneering approach for non-invasive physical locomotion capture in different studies [12,13]. Practically, the above technique is limited to employ as it requires great skillfulness, a huge processing cost and it is also a difficult task to deploy these methods on most job sites[12].

2 Fatigue, Fatigue Characteristics and Corresponding Fatigue Detection Measures

2.1 Fatigue and Its Characteristics

Defining fatigue in terms of composite interplay of the biological phenomena, physiological processes, and behavioral demonstration is a tedious task.

Authors in [14] define fatigue as functional organ failure. Excessive energy consumption may be the leading cause for such a functional organ failure. Fatigue in other words may be called as a decay in efficiency and productivity due to partial healing from last physical activities [15]. As per [16], fatigue is defined as a fall in any tissue 's ability to apply force as a result of carrying out a physical activity.

Author in [17] proposed to differentiate acute and chronic fatigue. Acute fatigue is generally related to an individual cause, and exists in healthy individuals, is sensed generally as normal, has a rapid start and short but severe course, is generally relieved by taking rest, diet, exercise, and stress management, and has a minute effect on the daily performance. But, Chronic fatigue, is specified as having an unspecified function, and it mainly affects persons with poor clinical health, having random causes. Chronic fatigue may worsen with any further physical or mental activity and there is little improvement even after taking a rest.

Ream et. al [18] defined fatigue as: "A subjective, unpleasant symptom which incorporates total body feelings ranging from tiredness to exhaustion creating an unrelenting overall condition which interferes with individuals' ability to function to their normal capacity".

After overall discussion, we can now view physical fatigue as a knowledge about lessened capacity for performing physical activity due to lack in the resources required to do so.

2.2 Fatigue Detection Methods

The detection of fatigue has always been a tedious task for researchers.. Often the assessment of fatigue is related to conditions where fatigue is observed, thereafter curbing the generalization of the results and observations. Detecting fatigue has been a tedious task not merely because of its pervasive nature but also because fatigue is a symptom and, as such, its subjectivity presents additional detection difficulties. As per the statement provided the author in [19] " The absence of an overall definition of fatigue preempts any scientific basis for measuring the condition, because logically, that which cannot be defined cannot be measured, and is not understood. The absence of a relevant means to detect and measure fatigue have seriously limited efforts to synthesize common knowledge about fatigue.

Numerous efforts to detect and measure fatigue have evolved in the present state-of-the-art despite all above concerns. Early work in fatigue detection was aimed at fatigue in the workplace and was undergone by industrial psychologists and military scientists. Generally healthy population was targeted for these studies at a specific time for detection.

Although several techniques have now been evolved for fatigue detection and measurement, the findings can be variable in nature and are dependent upon a number of intrinsic and extrinsic factors. Intrinsic factors account mainly for attributes like body health and experience. Extrinsic factors may include cumulative workload and task complexity.

There are several methods of fatigue detection used for laboratory studies and field studies. Under this study, we will focus on the measures that can be deployed in field studies.

2.3 Subjective Methods

In the subjective approach of fatigue detection, the system relies on the surveys and questionnaires based on self perceived exertion scales, conducted on selected and suitable subjects. A consent for the same is taken from the subjects. Although these have the advantage of being inexpensive, they lack reliability [20].

Benjamin et al. performed a study on fatigue detection in an office environment. The study design was guided by a specific theory of change. Authors expected that training would increase office ergonomic knowledge and that would help workers to restructure their workspace, which in turn would further influence working postures and behaviors. Here, the authors collected the dataset through online mode after a training method named 'theory of change" as shown in Fig. 1.

Data collection occurred two months and one month before the change employed and two, six, and twelve months after the change theory practice. A short daily symptom survey was conducted at the start, middle and end of each day in a 5-day workweek. The level of pain or discomfort was rated by the subjects on a scale of 0 for no pain to 10 for extremely severe pain for different body parts.

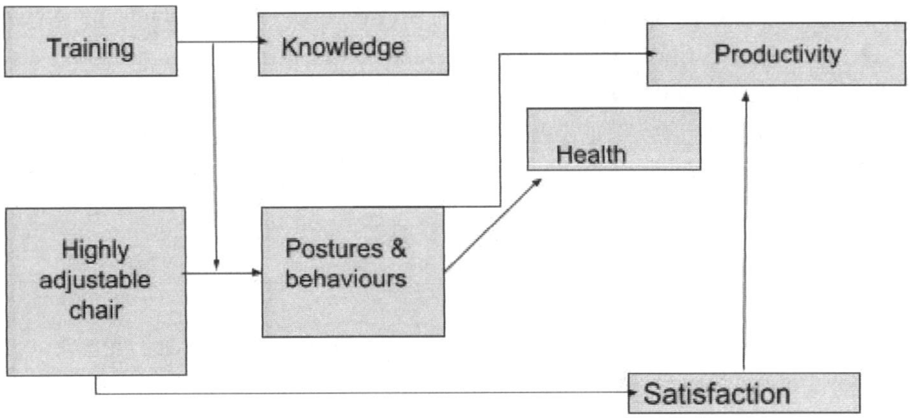

Fig. 1. Theory of Change [21]

The sum of all the ratings by an individual subject ranging from 0 (no pain anywhere in the body) to 90 (extremely severe pain in full body) was considered as the primary variable. In the questionnaire, thirty different covariates and potential co-founders were observed.
After performing primary and secondary hypotheses, the fatigue was observed as per Fig. 2.

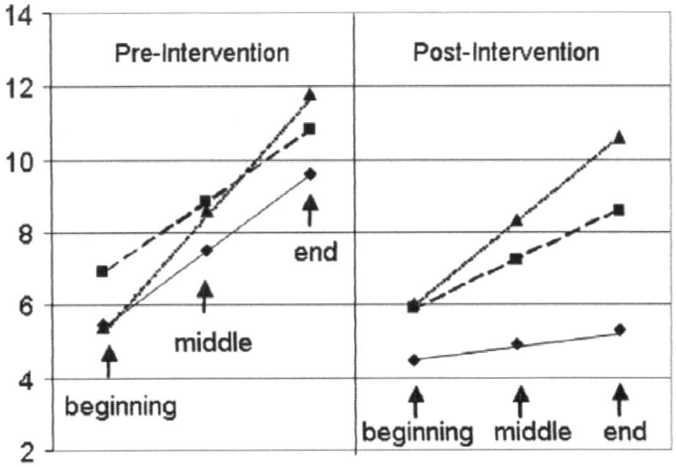

Fig. 2. Bodily pain at the beginning, middle, and end of the day before and after training [21]

Park et. al conducted another study on each five working days of a selected week. Authors surveyed working hours of the subjects for complete weeks and observed body health etc. during the last month. Secondly, subjective fatigue before leaving for work in the morning at residence and during afternoon performing work at office was surveyed. And lastly the stress by self reported questionnaire was surveyed. Authors deployed fatigue questionnaires and the stress questionnaire. The subjects were inspected for pre-study health conditions. The subjective fatigue was observed two times at home before leaving to work in the morning and during duty time in the afternoon.

Journal of Physics: Conference Series 2327 (2022) 012072 doi:10.1088/1742-6596/2327/1/012072

Authors used One way analysis of covariance (ANCOVA) to adjust for differences in different age groups. It was concluded after performing this research that the JAIH subjective fatigue questionnaire was able to distinguish between fatigued and non-fatigues subjects.
Several other subjective fatigue detection studies have been conducted so far and it can be concluded that these methods are easy and inexpensive but lack reliability.

2.4 Objective Methods

Objective approaches of fatigue detection are based on physiological processes (movement of different body parts) and are more reliable. The on-site implementation of these systems is a cumbersome and expensive task. In the current state-of-the-art fatigue detection and physical exertion assessment has been implemented mostly by using Electromyography (EMG) and with the inception of cheap and reliable wearable motion sensors it has been performed using inertial measurement units (IMUs).
In [22], the authors have performed the task using EMG. The study was performed on ten female participants. None of the subjects were reported for any kind of motor health issues. The study was performed on the production-line workers. The task was to pick a product followed by visual control and then placing it back as illustrated in Fig. 3.

(a) (b) (c)

Fig. 3. Task performed in [22]. (a) picking the product; (b) visual control; (c) placing the product.

The speed of the sprayer was used as the measure for work pace. The task was performed by the subjects for 1 week with 9 hours a day working period. The subjects were allowed a 10 minute short break every hour. There was a lunch break for 30 minutes. Measurements were carried out on Monday and Friday.
Detection of fatigue was carried out using EMG. It was observed by authors that the amplitude of EMG increased during the day time. Local perceived discomfort increased in the upper body during duty time. Fig. 4 shows the EMG test contraction method used in [22].

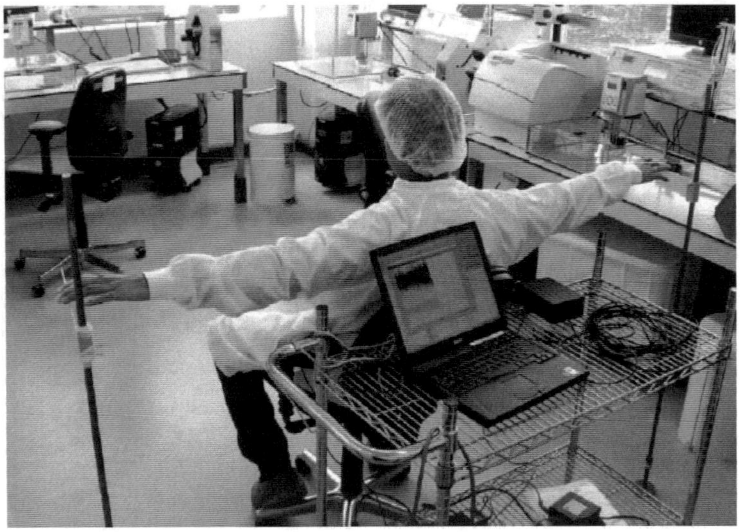

Fig. 4. Electromyography test contraction method [22]

Karg et. al, 2014, performed another study by employing optical markers to record the data set. The data acquisition took place at the Tokyo University of Agriculture and Technology, Tokyo. A set of 35 passive optical markers were used to capture a dataset with 10 motion cameras. The subjects performed squat exercise. Seven subjects provided with the description of the experiments performed repeated sets of five squats. After a set of five squat exercises each time, the extent of fatigue of each subject was recorded on a 6 points Likert-scale. Table 1 shows an illustration of every level employed in [24].

TABLE 1
Different Fatigue Levels and Their Description

Level	Fatigue	Fitness
1	Not tired at all	Not fit at all
2	Not really tired: can do many more	Barely fit
3	A little tired: can do a few more	Practice sport sometime
4	Tired: can do a little more	Practice sport often
5	Very tired: can hardly endure more	Practice sport regularly

6	Extremely tired: cannot endure more	Athlete

It was observed by the authors that fatigue during squat exercises influenced different body parts and joints. A Hidden Markov Model technique was applied to incorporate the findings that the fatigue keeps on increasing continuously.
Another study incorporated in [20] was based on the use of sensors on different body parts. Inertial Measurement Units (IMUs) were used to capture whole body kinematics of the subjects. The study was performed at the Canadian Masonry Design Centre, where 32 male workers with distinct experience levels in bricklaying job were observed. Different IMU units were placed as shown in Fig. 5.
The data collected was processed using MATLAB. The fatigue here was observed in terms of jerk (rate of change of acceleration) values. The resultant acceleration of the eleven body segments were calculated out of their cartesian components. It was observed that the jerk values kept on increasing with fatigue.

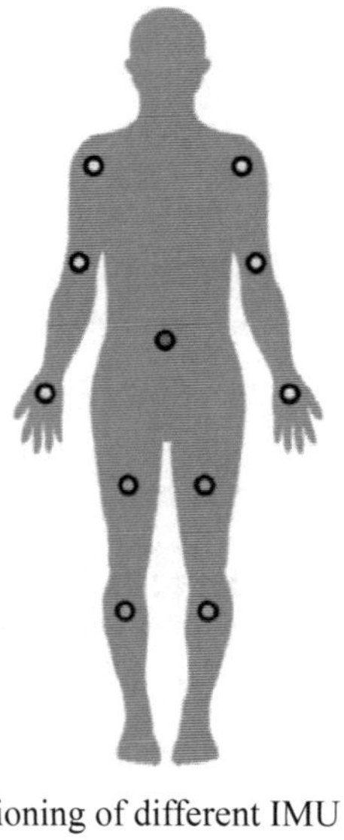

Fig. 5. Positioning of different IMU units in [20]

3 Proposed Method for Fatigue Assessment

Subject based assessment of fatigue for different actions of workers like loading or unloading building materials, mixing building materials, pitching action etc. is carried out. Fig 6 below presents an overview of different steps in the proposed methodology for fatigue detection (fig 6):

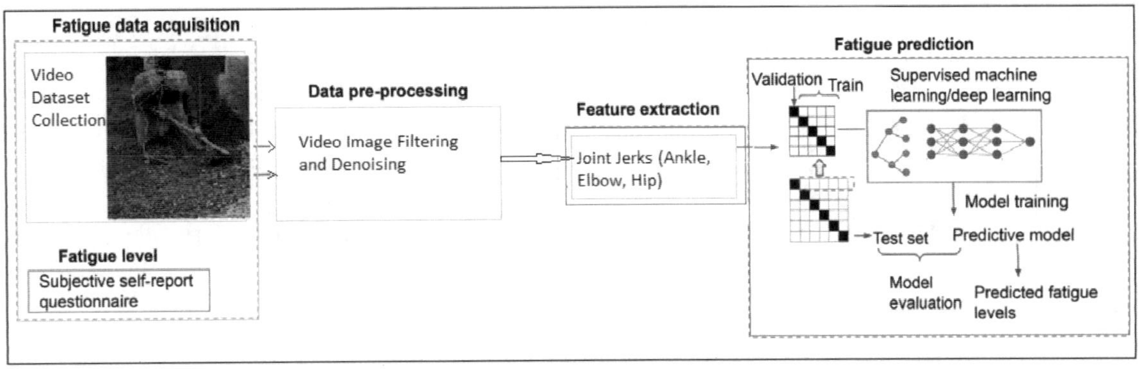

Fig. 6. An overview of the proposed method

The fatigue dataset is collected at a construction site using a high resolution motion camera from multiple angles and at a predefined distance so as to have uniformity in the dataset. The number of subjects used for the study so as to have maximum accuracy. A subjective questionnaire is required to be filled by the subject before being part of the study and only those subjects are chosen who are having no chronic illness that may lead to inaccurate results.

The dataset video images are preprocessed using suitable filtering and denoising techniques. Filtering usually refers to the method of removing some redundant information as per the requirements of image processing, and preserving the useful information. By filtering the images are made more suitable for ocular as well as objective assessment systems as per the application. Images are left blurry when these are attacked by some noisy components.Sometimes the noise is introduced at such a level that it may destroy the useful information and affect further feature extraction from these images.

The resultant joint and muscle jerks are extracted from the video images using some suitable model for the calculation of pixel jerk in the consecutive images. The jerk j from image pixel positions can be calculated as given below

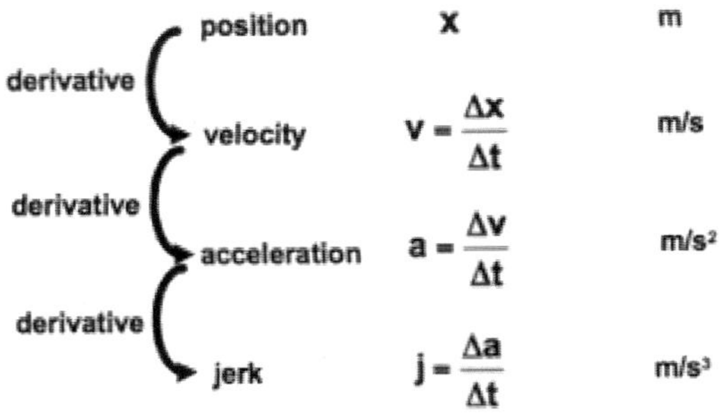

Fatigue is predicted by training the regression models. Any suitable training model like Random Forest (RF) or Convolutional Neural Network(CNN) is used. The test dataset is classified between fatigued and non-fatigued subjects.

4 Conclusion

Fatigue can be physiologically, intellectually, or psychologically damaging. Detecting fatigue is more important than curing it because once it is recognized, treatment becomes easier. After having a review of research carried out in fatigue detection it can be concluded that in using subjective as well as objective approaches, there is always a requirement of large data sets. There is always a variation in the collected parameters in terms of demographic and other individual differences in the development. Although the objective approach is much more reliable and accurate, it also adds to the cost. This study also concludes that work exposes individuals to fatigue that they would rarely experience without work. A significant population across the world is exposed to fatigue risk due to their busy work schedules and other non-work related causes. So, there is always a need to assess fatigue effectively for fatigue risk management.

ORCID of ASHISH SHARMA : 0000-0002-2750-5960

ORCID of Dr. Gaurav Sethi: Not Available

References

1. K. Yang, C. R. Ahn, and H. Kim. "Validating ambulatory gait assessment technique for hazard sensing in construction environments." Automation in Construction, vol. 98, pp. 302-309, Feb. 2019.
2. A. Aryal, A. Ghahramani, and B. Becerik-Gerber, ”Monitoring fatigue in construction workers using physiological measurements,” Automation in Construction, vol 82, pp.154-165, oct. 2017.
3. L. Zhang, "Jerk as a Method of Identifying Physical Fatigue and Skill Level in Construction Work." Master's thesis, University of Waterloo, 2019.
4. E. Valero, A. Sivanathan, F. Bosché, and M. Abdel-Wahab, “Analysis of construction trade worker body motions using a wearable and wireless motion sensor network,” Automation in Construction, vol 83, pp .48-55, Nov. 2017.
5. Z. Maman, S. Yazdi, L. A. Cavuoto, and F. M. Megahed, “A data-driven approach to modeling physical fatigue in the workplace using wearable sensors,” Applied ergonomics,vol. 65, pp. 515-529, Nov. 2017
6. A. J. Dittner, S. C. Wessely, and R. G. Brown, “The assessment of fatigue: a practical guide for clinicians and researchers,” Journal of psychosomatic research, vol. 56(2), pp. 157-170, Feb. 2004.
7. L. S. Aaronson, C. S. Teel, V. Cassmeyer, G. B. Neuberger, L. Pallikkathayil, J. Pierce, ... and A. Wingate, “Defining and measuring fatigue,” Image: the journal of nursing scholarship, vol. 31(1), pp. 45-50, Mar. 1999.
8. Y. Zhu, R. R. Jankay, L. C. Pieratt, and R. K. Mehta, “Wearable sensors and their metrics for measuring comprehensive occupational fatigue: a scoping review,” In Proceedings of the Human Factors and Ergonomics Society Annual Meeting, Vol. 61,pp. 1041-1045, Sept. 2017
9. S. Hwang, J. Seo, J. Ryu, and S. Lee, “Challenges and opportunities of understanding construction workers’ physical demands through field energy expenditure measurements using a wearable activity tracker,” Construction Research Congress, pp. 2730-2739, 2016.
10. U. C. Gatti, S. Schneider, and G. C. Migliaccio, “Physiological condition monitoring of construction workers,” Automation in Construction, vol. 44, pp. 227-233, Aug. 2014.

11. S. A. A. N. Bolink, H. Naisas, R .Senden, Essers, H. Heyligers, I. C. Meijer, and B. Grimm, "Validity of an inertial measurement unit to assess pelvic orientation angles during gait, sit–stand transfers and step-up transfers: Comparison with an optoelectronic motion capture system," Medical engineering & physics, vol. 38(3), pp. 225-231, Mar. 2016.
12. C. M. Bauer, F. M. Rast, M. J. Ernst, J. Kool, S. Oetiker, S. M. Rissanen, ... and M. Kankaanpää, "Concurrent validity and reliability of a novel wireless inertial measurement system to assess trunk movement," Journal of Electromyography and Kinesiology, vol. 25(5), pp. 782-790, Oct. 2015.
13. P. J. Berger, L. McCutcheon, M. Soust, A. M. Walker, and W. M. Wilkinson, "Electromyographic changes in the isolated rat diaphragm during the development of fatigue" European journal of applied physiology and occupational physiology, vol. 62(5), pp. 310-316, Sept. 1991.
14. P. Gander, L. Hartley, D. Powell, P. Cabon, E. Hitchcock, A. Mills, and S. Popkin, "Fatigue risk management: Organizational factors at the regulatory and industry/company level," Accident Analysis & Prevention, vol. 43(2), pp. 573-590, Mar. 2011.
15. A. J. Dittner, S. C. Wessely, and R. G. Brown, "The assessment of fatigue: a practical guide for clinicians and researchers," Journal of psychosomatic research, vol. 56(2), pp. 157-170, Feb. 2004.
16. B. F. Piper, "Fatigue: current bases for practice. In Management of Pain," Fatigue and Nausea, pp. 187-198. Macmillan Education UK, 1989.
17. E. Ream, and A. Richardson, "Fatigue in patients with cancer and chronic obstructive airways disease: a phenomenological enquiry," International journal of nursing studies, vol. 34(1), pp. 44-53, Feb. 1997.
18. D. Eidelman, "Fatigue on Rest" and associated symptoms (headache, vertigo, blurred vision, nausea, tension and irritability) due to locally asymptomatic, unerupted, impacted teeth. Medical hypotheses, vol 5(3), pp. 339-346, Mar. 1979.
19. L. Zhang, M. M. Diraneyya, J. Ryu, C. T. Haas, and E. M. Abdel-Rahman, "Jerk as an indicator of physical exertion and fatigue," Automation in Construction, vol. 104, pp. 120-128, Aug. 2019.
20. B. C. Amick III, M. M. Robertson, K. DeRango, L. Bazzani, A. Moore, T. Rooney, and R. Harrist, "Effect of office ergonomics intervention on reducing musculoskeletal symptoms," Spine, vol. 28(24), pp. 2706-2711 Dec. 2003.
21. T. Bosch, M. P. De Looze, and J. H. Van Dieen, "Development of fatigue and discomfort in the upper trapezius muscle during light manual work," Ergonomics, vol. 50(2), pp. 161-177, Feb. 2007.
22. J. Park, Y. Kim, H. K. Chung, and N. Hisanaga, "Long working hours and subjective fatigue symptoms, "Industrial health, vol. 39(3), pp. 250-254, 2001.
23. M. Karg, G. Venture, J. Hoey, and D. Kulić, "Human movement analysis as a measure for fatigue: A hidden Markov-based approach," IEEE transactions on neural systems and rehabilitation engineering, vol. 22(3), pp. 470-481, Jan. 2014.
24. F. A. Barbieri, Dos P. C. R. Santos, E. Lirani-Silva, R. Vitório, L. T. B. Gobbi, and J. H. Van Diëen, "Systematic review of the effects of fatigue on spatiotemporal gait parameters," Journal of back and musculoskeletal rehabilitation, vol. 26(2), pp. 125-131, Jan. 2013.

Journal of Physics: Conference Series **2327** (2022) 012073 doi:10.1088/1742-6596/2327/1/012073

A Review on the Detection of the Post COVID-19 Symptoms for Long Term Diseased Patients using Machine Learning Algorithms

Anitha Patibandla
Research Scholar, Dept.of ECE, Lovely Professional University, Punjab, India
anithakamal.ghanta@gmail.com, (0000-0002-1657-6966)

Abstract. Long term diseases require continuous monitoring, sometimes periodic monitoringto verify if any serious concern requires an attention. In recent years, it is noticed that the COVID-19 pandemic has triggered serious concern towards the long-term diseased individuals. As the mortality rate of the COVID-19 clearly indicates that the highest percentage of deaths reflect in the individuals suffering from long term diseases such as diabetes, pneumonia, cardiovascular and acute renal failure. Though they are tested for COVID negative through conventional apparatus, it doesn't confer that they are completely out of post consequences. Hence a periodic, if necessary continuous monitoring needs to be aided, which in current scenario is a challenging task. Hence, our current article reviews the use of machine learning algorithms to detect and diagnose pre and post COVID-19 effects onlong term diseased patients.

1. Introduction

Machine learning algorithms received significant attention in the scientific community due to their intriguing nature of predicting the physical phenomena through historical data. The learning basically adopts in two phases, supervised [33] and unsupervised learning [39]. Supervised learning requires regression and classifiers to process the data, whereas unsupervised learning requires clustering of data to predict the phenomena. In the current context, we would like to emphasize on the COVID-19prediction and diagnosis through these machine learning algorithms. However, there is an ample amount of research cited in the literature to realize the feasibility of using the machine learning algorithms. Currently, supervised learning algorithms are in application, but most of them are still in laboratories and haven't received clinical compliance yet. The complexity of the clinical compliance is due to the mutation of variants displaying different symptoms and hence, it is still observed to be a lacuna for now [3].Since machine learning algorithms have feasibility of processing large amount of data that makes the prediction model more accurate, we stick our current interest towards prediction and diagnosis of post COVID-19 effects on the long-term diseased patients. Long term diseases patients suffering from diabetes mellitus [35], cardiovascular [37], pneumonia and acute renal failure [31] display a complex behavior over a period of time which is difficult to predict that root cause is due to COVID-19. Hence a continuous monitoring of the patient and data isrequired. Works pertaining to detection of pre COVID- 19 symptoms have been well cited in the literature using demographic [4], standard symptoms [2] and physical characteristics [8] of the individuals. The post symptoms of covid-19 on a healthy individual is not reported elsewhere and for a diseased individual, it is very scarcely cited and recent inclination in this area enabled us to write the current review. The article emphasizes on the various long-term diseases and their symptoms, COVID-19 effect on these diseases, datasets available and machine learning algorithms.

2. Review on long term diseases

Long term diseases are those which may not be infectious but slowly decline the inherent body characteristics over time resulting in death. These individuals have very low immune system, though not clinically evidenced but over a long time, they are prone to get effected easily by infectious diseases such as COVID-19 in the current case. Majority of the long-term diseased patients have foreseen diabetes mellitus which is most commonly seen disease in the current era, followed by pneumonia, acute renal kidney failure and typical cardiovascular diseases. We shall now discourse each of them individually so that the nature of the disease allows us to classify the features to be used for a machine learning model.

2.1 Diabetes mellitus
Diabetes is observed to be the most common chronic disease with a characteristic of blood glucose being abnormal which is caused due to defective insulin secretion. Diabetes is divided into two categories referred to as type I and type II [28]. Long term diabetes leads to chronic failures such askidneys, tissues, blood vessels and heart. Type I diabetes basically poses symptoms such as increased thirst, frequent urination, high blood and glucose levels Few machine learning algorithms such as support vector machine (SVM) [9], decision tree (DT) [24], logistic regression (LR) [21] and linear discriminant analysis (LDA) [5] have been reported in literature. However most widely used one is SVM and recently neural networks have shown a huge interest by the researchers because of its technical feasibility and performance. The data set basically requires for the prediction of diabetes mellitus includes physical examination indexes: age, pulse, breathe, left and right systolic pressure (RSP, LSP), left and right diastolic pressure (LDP, RDP), fasting glucose, height and weight. Apart from this other include physique index, waistline, low- and high-density lipoprotein (LDLP, HDLP)and retinopathy etc. In literature most of the data with respect to the Indian demographics basically observed from Indian Diabetic Retinopathy Image Dataset). Few works in this context in relation with the dataset can be inferred in [34, 19, 23]. Our interest in this context would be derived to use this dataset for validation to see the severity of disease growth through retinopathy and it serves as a standard for validation.

2.2 Pneumonia
Pneumonia is the most common disease observed in individuals which can have cause of symptomsin various forms. Literature suggests distinctively more than 30 symptoms and it is very typical disease that causes lung infection in no time result in death. The air sacs formed due to infection may cause lungs filled with fluid or puss resulting in fever, fatigue and unease in breathing [25]. The types of pneumonia include bacterial pneumonia, viral pneumonia, mycoplasma pneumonia and various other variants. The most commonly seen bacteria is streptococcus, the influence of this bacteria completely depends upon the age and immune system of the individual. Otherwise, these bacteria may move down creating air sacs in lungs resulting in inflammation and filling pus thus pneumonia. The symptoms of this pneumonia may include high fever, low appetite, sweating a lot, fast breathing, chilling, sharp chest pain during cough and breathe [38]. Similarly, viral pneumonia also displays symptoms such as fever, dry cough, chest pain during breathing and cough, uneasiness in breathing [26]. There are several machine learning algorithms reported in the literature to predict or diagnose the pneumonia. Most of the machine learning algorithms use chest X-ray as the common dataset to predict the pneumonia. Apart from that, one can use spirometry, oximetry and blood test datasets also to determine the pneumonia. Relevant works with respect to the aforementioned datasets can be found in NIH clinical center, America's Research Hospital. Few works relevant to the current context can be cited in weakly supervised classification [10], deep convolutional neural networks [16], Convolutional neural networks [11], direct neural network [12], decision tree [13] and random forest [6] models. In the current context, for the detection pneumonia, there is complexity to contrast with COVID-19 data, since, few symptoms of the COVID-19 match with the pneumonia such as cough, unease in breathing, fever, fatigue and chilling. As a result, there is an intricacy to determine the preCOVID-19 symptoms, but post COVID-19 symptoms are similar in pneumonia patients and hence ithas become easy in some cases.

2.3 Cardiovascular
Cardiovascular diseases include various disorders in heart and blood vessels that include coronary artery disease, high blood pressure, Arrhythmia, cardiac arrest, stroke, congestive heart failure, congenital heart diseases. As of now such diagnostic tools or predictive machine learning algorithms is available in the literature. However, our current motivation is to study the post COVID-19 symptoms on the patients with cardiovascular medical history. The abnormalities for the post COVID-19 on the cardiovascular may have huge impact. Various machine learning algorithms have been proposed to identify various cardiovascular diseases and out of them, it is consistently reported in the literature that naïve Bayes algorithm was found to be most suitable algorithm for the prediction of the heart diseases. However other algorithms such as SVM [29] and ANN [30] when used in combination with boosting algorithms have displayed better results. The classifications include age,gender, chest pain type, resting blood pressure, serum cholesterol, fasting blood sugar, resting electrocardiograph and maximum heart rate achieved. Works pertaining to this database include machine learning model for classification of Arrhythmia

detection [17], ECG is used as a biomarker for detecting the heart failures and brief of this context is discussed in [14] with respect to machine learning. Other interesting works on detection of acute coronary syndrome using 12- lead ECG through machine learning algorithm is predicted in [15].Our current interest is to identify the risk inthe patients with cardiovascular problems infected with COVID-19 on their long-term sustainability. As of now, there are very limited number of datasets is available in the literature, hence a lot of focus on choosing appropriate machine learning is difficult. A thorough amount of research is required by performing the data prediction using various machine learning algorithms. Though there might be ample research is ongoing, but still at laboratory level. The dataset for aforesaid classifiersis retrieved from Kaggle database.

2.4 Acute renal kidney failure

Acute renal kidney failure is dysfunction of single or both kidneys and basically seen in long term diseases such as diabetes mellitus. This happens in healthy persons within a days or weeks without a prior notice to an individual. This is probably due to lack of awareness in the public interest, but precautionary symptoms indicate diarrhea, severe dehydration, vomiting, swelling in legs and fluctuations in blood pressure may be an early indicator [36]. The side effects of the dialysis include severe malnutrition, weight fatigue, protein loss (serum creatinine), pancreatitis, bacterial infections,anemia and severe diarrhea [32].There are machine learning algorithms recently developed in the scientific community in predicting the early symptoms of the acute kidney renal failure. It was observed from the literature that effect of COVID-19 on the acute renal failure has not been reported yet. One interesting paper that cites acute kidney injury using recurrent neural network (RNN) was predicted [18]. The dataset basically retrieved from the MIMIC III database. Using the same database following works have been instigated and presented in the literature. For instance, [1] has predicted acute kidney disease using machine learning algorithms i.e., recurrent neural network and decision trees for detection of sepsis acute kidney disease. Currently no such investigations or databases are available for the detection and prediction of the acute kidney failure diseased individuals who are infected with COVID-19 and hence it requires lot of attention in the current scientific community. This requires an implicit work to get the latest dataset of these periodic treatments report and hence there is still a scope for research in this area.

3. Methodology

We would like to predict the types of diseases based on the type of classifiers and features produced by the amount of data availability. It is intuitive to understand that, through the same data, it is realizable to tailor various characteristics via various machine learning algorithms. For instance, consider there is an availability of the chest X ray, if we were to identify the effect of cardiovascular disease, we might go with features relevant to cardiac muscles and heartbeat, whereas in case of identifying the effect of pneumonia on the heartbeat, we need specifically heart beat and same chest X ray but a correlative data from the lungs CT scan. So, it is imperative to identify a method that can realize the features in a way distinct for each algorithm respectively.

Briefly, machine learning algorithms are those algorithms which are trained for learning through an historical data and predict/forecast the behavior meeting the clinical diagnostic procedure. Each of these models have their own algorithms such as: in supervised learning, the data is categorised asclassification and regression, wherein classification refers to a predictive model for a given input to produce an output using an algorithm Whereas the regression is to predict the output for the continuous values, i.e., for continuous input. Most widely used classifiers are Naive Bayes and support vector machines and most of the problems which discussed earlier in the literature review uses these classifiers as they produce more reliable data prediction than other algorithms do.. In our current proposed research, it is equally important to identify an optimised classifier and regression models with booster gradients to reduce the computational time and power to predict the model data. The processing methodology adopted in the current context would flow in the following way as shown in the figure.1 below. However, during the research, a more genuine dataset with high precision will be identified.

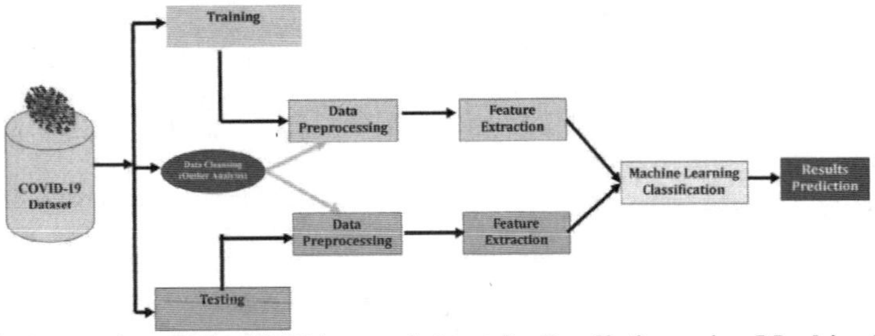

Figure 1: Processing of Covid-19 image dataset for Prediction using Machine learning

Further the data is processed for training, depending on the type of the data type, in the current context, we have CT scan images and chest X ray images, a numerical data for pulse, heartbeat and blood sample data. The processing of the images requires an image processing technique followed by choosing suitable machine learning algorithms depending on the desired characteristic feature. Similarly, for the numerical dataset, a statistical based algorithm is preferred so that computational time can be reduced and high data prediction accuracy can be achieved. The feature extraction completely depends on the number of columns and hence the number of classes. Since ample data and variables for particular diseases are available in various forms, optimization of the feature extraction is desired. The optimization is performed to reduce the computational memory and power. We intend to explain a sample of how an artificial neural network with Navier Bayes, artificial neural network with convolutional neural network, and statistical ensemble with a suitable metaheuristic algorithm as an optimization technique can give better results. However, during the research, a more distinct model will be chosen depending on the performance and prediction parameters.

4. Datasets

Various datasets pertaining to the aforementioned long-term disease based COVID-19 infected patients can be retrieved from [7] for Indian Diabetic Retinopathy Image Dataset, [27] for acute renal failure at MIMIC III, [22] for pneumonia at NIH clinical center, America's Research Hospital, [20] for cardiovascular at Automated Cardiac Diagnosis Challenge (ACDC). This will be tailed for hospital specific data upon the performance in confusion matrix.

5. Conclusion

It is implicative that there is limited data available after the COVID-19 to study the impact on the long-term diseased individuals. But it is very crucial to monitor these individuals to ensure that they completely recover without any complications. But, however with the limited data availability, it is desired that developing algorithms with an optimization technique is of primary importance and hence the current article would enable researchers to appraise this new area of studying post COVID- 19 symptoms.

Acknowledgement

I would like to express my very great appreciation to **Dr. Kirti Rawal** (Associate Professor, School of Electronics and Electrical Engineering, Lovely Professional University, Phagwara, Punjab, India.) for her valuable and constructive suggestions during the planning and development of this research work. Her willingness to give her time so generously has been very much appreciated.

6. References

1. He, J., Lin, J. and Duan, M., 2021. Application of Machine Learning to Predict Acute Kidney Disease in Patients With Sepsis Associated Acute Kidney Injury. *Frontiers in Medicine*, *8*.
2. Zoabi, Y., Deri-Rozov, S. and Shomron, N., 2021. Machine learning-based prediction ofCOVID-19 diagnosis based on symptoms. *npj digital medicine*, *4*(1), pp.1-5.
3. Roy, B., Dhillon, J.K., Habib, N. and Pugazhandhi, B., 2021. Global variants of COVID- 19: Current understanding. *Journal of Biomedical Sciences*, *8*(1), pp.8-11.Zhang, J., Jun, T., Frank, J., Nirenberg, S., Kovatch, P. and Huang, K.L., 2021. Prediction of individual COVID-19 diagnosis using baseline

demographics and lab data. *Scientific Reports*, *11*(1), pp.1-8.

4. Alharan, A.F., Algelal, Z.M., Ali, N.S. and Al-Garaawi, N., 2021, September. Improving Classification Performance for Diabetes with Linear Discriminant Analysis and Genetic Algorithm. In *2021 Palestinian International Conference on Information andCommunication Technology (PICICT)* (pp. 38-44). IEEE.
5. Akgundogdu, A., 2021. Detection of pneumonia in chest Xray images by using 2D discretewavelet feature extraction with random forest. *International Journal of ImagingSystems and Technology*, *31*(1), pp.82-93.
6. Porwal, P., Pachade, S., Kokare, M., Deshmukh, G., Son, J., Bae, W., Liu, L., Wang, J., Liu, X., Gao, L. and Wu, T., 2020. Idrid: Diabetic retinopathy–segmentation and grading challenge. *Medical image analysis*, *59*, p.101561.
7. Assaf, D., Gutman, Y.A., Neuman, Y., Segal, G., Amit, S., Gefen-Halevi, S., Shilo, N.,Epstein, A., Mor-Cohen, R., Biber, A. and Rahav, G., 2020. Utilization of machine-learning models to accurately predict the risk for critical COVID-19. *Internal andemergency medicine*, *15*(8), pp.1435-1443.
8. Kopitar, L., Kocbek, P., Cilar, L., Sheikh, A. and Stiglic, G., 2020. Early detection oftype 2 diabetes mellitus using machine learning-based prediction models. *Scientific reports*, *10*(1), pp.1-12.
9. Hu, S., Gao, Y., Niu, Z., Jiang, Y., Li, L., Xiao, X., Wang, M., Fang, E.F., Menpes- Smith,W., Xia, J. and Ye, H., 2020. Weakly supervised deep learning for covid-19 infection detection and classification from ct images. *IEEE Access*, *8*, pp.118869- 118883.
10. Jain, R., Nagrath, P., Kataria, G., Kaushik, V.S. and Hemanth, D.J., 2020. Pneumonia detection in chest X-ray images using convolutional neural networks and transfer learning. *Measurement*, *165*, p.108046.
11. Sirish Kaushik, V., Nayyar, A., Kataria, G. and Jain, R., 2020. Pneumonia detection using convolutional neural networks (cnns). In *Proceedings of First International Conference on Computing, Communications, and Cyber-Security (IC4S 2019)* (pp. 471-483). Springer, Singapore.
12. Yee, S.L.K. and Raymond, W.J.K., 2020, September. Pneumonia diagnosis using chest x- ray images and machine learning. In *Proceedings of the 2020 10th International Conference on Biomedical Engineering and Technology* (pp. 101-105).
13. Hammad, M., Iliyasu, A.M., Subasi, A., Ho, E.S. and Abd El-Latif, A.A., 2020. A multitier deep learning model for arrhythmia detection. *IEEE Transactions on Instrumentation and Measurement*, *70*, pp.1-9.
14. Al-Zaiti, S., Besomi, L., Bouzid, Z., Faramand, Z., Frisch, S., Martin-Gill, C., Gregg, R., Saba, S., Callaway, C. and Sejdić, E., 2020. Machine learning-based prediction of acutecoronary syndrome using only the pre-hospital 12-lead electrocardiogram. *Nature communications*, *11*(1), pp.1-10.
15. Sirazitdinov, I., Kholiavchenko, M., Mustafaev, T., Yixuan, Y., Kuleev, R. and Ibragimov,B., 2019. Deep neural network ensemble for pneumonia localization from a large-scale chest x-ray database. *Computers & electrical engineering*, *78*, pp.388-399.
16. Alfaras, M., Soriano, M.C. and Ortín, S., 2019. A fast machine learning model for ECG-based heartbeat classification and arrhythmia detection. *Frontiers in Physics*, *7*, p.103.
17. Koyner, J.L., Carey, K.A., Edelson, D.P. and Churpek, M.M., 2018. The development of amachine learning inpatient acute kidney injury prediction model. *Critical care medicine*, *46*(7), pp.1070-1077.
18. Porwal, P., Pachade, S., Kamble, R., Kokare, M., Deshmukh, G., Sahasrabuddhe, V. andMeriaudeau, F., 2018. Indian diabetic retinopathy image dataset (IDRiD): a database for diabetic retinopathy screening research. *Data*, *3*(3), p.25.
19. Bernard, O., Lalande, A., Zotti, C., Cervenansky, F., Yang, X., Heng, P.A., Cetin, I., Lekadir, K., Camara, O., Ballester, M.A.G. and Sanroma, G., 2018. Deep learning techniques for automatic MRI cardiac multi-structures segmentation and diagnosis: is the problem solved?. IEEE transactions on medical imaging, 37(11), pp.2514-2525
20. Wu, H., Yang, S., Huang, Z., He, J. and Wang, X., 2018. Type 2 diabetes mellitus prediction model based on data mining. *Informatics in Medicine Unlocked*, *10*, pp.100- 107.
21. Wang, X., Peng, Y., Lu, L., Lu, Z., Bagheri, M. and Summers, R.M., 2017. Chestx-ray8:Hospital-scale chest x-ray database and benchmarks on weakly-supervised classification and localization of common thorax diseases. In *Proceedings of the IEEE conference oncomputer vision and pattern*

recognition (pp. 2097-2106).
22. Bhatia, K., Arora, S. and Tomar, R., 2016, October. Diagnosis of diabetic retinopathy using machine learning classification algorithm. In *2016 2nd international conference on next generation computing technologies (NGCT)* (pp. 347-351). IEEE.
23. Habibi, S., Ahmadi, M. and Alizadeh, S., 2015. Type 2 diabetes mellitus screening and risk factors using decision tree: results of data mining. *Global journal of health science*,*7*(5), p.304.
24. Gereige, R.S. and Laufer, P.M., 2013. Pneumonia. *Pediatrics in Review*, *34*(10), pp.438-456.
25. Ruuskanen, O., Lahti, E., Jennings, L.C. and Murdoch, D.R., 2011. Viral pneumonia. *The Lancet*, *377*(9773), pp.1264-1275.
26. Johnson, A., Pollard, T., and Mark, R. (2016) 'MIMIC-III Clinical Database' (version 1.4), *PhysioNet*.
27. Ozougwu, J.C., Obimba, K.C., Belonwu, C.D. and Unakalamba, C.B., 2013. Thepathogenesis and pathophysiology of type 1 and type 2 diabetes mellitus. *J Physiol Pathophysiol*, *4*(4), pp.46-57.
28. Mythili, T., Mukherji, D., Padalia, N. and Naidu, A., 2013. A heart disease prediction model using SVM-decision trees-logistic regression (SDL). *International Journal of Computer Applications*, *68*(16).
29. Murthy, H.N. and Meenakshi, M., 2013. ANN model to predict coronary heart disease basedon risk factors. *Bonfring International Journal of Man Machine Interface*, *3*(2), pp.13-18.
30. Bellomo, R., Kellum, J.A. and Ronco, C., 2012. Acute kidney injury. *The Lancet*, *380*(9843), pp.756-766.
31. Qian, Q., Nath, K.A., Wu, Y., Daoud, T.M. and Sethi, S., 2010. Hemolysis and acute kidneyfailure. *American Journal of Kidney Diseases*, *56*(4), pp.780-784.
32. Cunningham, P., Cord, M. and Delany, S.J., 2008. Supervised learning. In *Machine learning techniques for multimedia* (pp. 21-49). Springer, Berlin, Heidelberg.
33. Rema, M. and Pradeepa, R., 2007. Diabetic retinopathy: an Indian perspective. *Indian journal of medical research*, *125*, pp.297-310.
34. Mellitus, D., 2005. Diagnosis and classification of diabetes mellitus. *Diabetes care*, *28*(S37), pp.S5-S10.
35. Van Biesen, W., Vanholder, R. and Lameire, N., 2006. Defining acute renal failure: RIFLE and beyond. *Clinical journal of the American Society of Nephrology*, *1*(6), pp.1314-1319.
36. Nabel, E.G., 2003. Cardiovascular disease. *New England Journal of Medicine*, *349*(1), pp.60-72.
37. Mosier, D.A., 1997. Bacterial pneumonia. *Veterinary Clinics of North America: Food Animal Practice*, *13*(3), pp.483-493.
38. Barlow, H.B., 1989. Unsupervised learning. *Neural computation*, *1*(3), pp.295-311.

An LoRa based WSN for monitoring targeted Audience with Heart Diseases

Narenthira Prasath D[1], Dushyant Kumar Singh[2*]
Lovely Professional University, Phagwara, Punjab, 144411

Orcid ID author 1: https://orcid.org/0000-0002-9762-1545
Orcid ID author 2: https://orcid.org/0000-0003-1859-9122
Email*: dushyant.kumar@lpu.co.in

Abstract
Health 5.0, health digitization is growing rapidly with the help of Internet of Things (IoT) and Internet of Medical Things (IoMT). The growth in this sector motivated us in learning the loopholes in the advancements. The various communication technologies and strategies are discussed based on the power consumption, speed and range. The LoRa is seen to work on low power, high range with less data transmission. Investigating the fields of Wireless Sensor Network, LoRa is seen as a reliable protocol that can best fit for Wireless sensor Network. The applications based on the capability of the system is reviewed considering the wide scale properties of each component. The best fit processing unit based on cost with the transmission protocols is discussed to give a clear vision on the technologies. Various transmission protocols are reviewed and LoRa is taken ahead in exploring and enhancing the features considering the knowledge gap. The speed of LoRa is simulated using Pothos, and the range is tested on the real ground. The data distribution based on heavy data transmission is discussed. The role of IoT and IoMT in the growth of the sector is visualized as the major contribution is given by the technologies. The Future scope of research in healthcare is concluded motivating the researchers giving a wide range opportunity to work on.

Keywords: Health 5.0; Internet of Things (IoT); Internet of Medical Things (IoMT); Wireless Sensor Network (WSN); Sustainable Development; LoRa; Pothos.

1. Introduction

If Wealth is lost, nothing is lost; If health is lost, everything is lost. This shows the significance of health sector in the daily life. the advancements in this sector are seen to be comparatively high with other sectors. The growth of the sector is due to the demanding need of the advancements to serve better. The healthcare infrastructure stays in the competitive phase between countries as the sector has the potential researchers in every corner of the world. A continuous data monitoring is done digitally to collect a wide range of dataset to automate the working mechanism. The Machine Learning (ML) and Artificial Intelligence (AI) brought a hike in the sector but demands a proper dataset from user. The Wireless Sensor Network is a solution technology provided to collect data in an efficient way. The different communication systems are available. The Bluetooth, BLE, Wi-Fi, Zigbee, LoRa has their own standards. The capabilities and shortages in the self, made them application specific. The brief about IoMT, WSN, Comparison of various networking technologies, Literature survey based on LoRa technology is detailed in the article.

The research study is done to identify the practical range of the transmission LoRa device for an application for Wireless Sensor Network implementation of health band for a selected group of people. The study also covers the data transmission techniques for heavy data transmission using Lora for long distance transmission. The study is proposed for places where the lack of cellular network or the need of special network for limited access is seen.

1.1. Motivation of this paper

Humans can never afford to ignore their health. A better lesson has been taught by this pandemic regarding the healthcare sector. It showed both the potential of the healthcare sector as well as the demanding improvements in the field. This gained the attention of many researchers. The need of advancement in the healthcare sector motivated the authors in learning the existing technologies to propose a solution for the modern problem statement. The authors found LoRa to be the efficient way to send data for long ranges without the physical interference. But the deficiency in the data transmission rate is seen and the research to enhance the speed is proposed. The problem in the forests and rural region where the lack of network is seen for effective health data communication is hard. One of the major problems in developing countries in the villages is the lack of a comprehensive system capable of both health data transmission and heavy data transmission for a long range. Authors developed the system to deal with a major problem encountered by the country.

1.2. IoMT in Healthcare

Internet of Medical Things came up bringing the solutions in medical field using Internet of Things. Solutions on smart health monitoring, pandemic detection, smart elderly people guidance, social unity based on healthcare are proposed and advancements are going on. The IoMT mainly focuses on digitization of data which includes collection of data from sensor data, data segmentation, data storage and data analyzing. The smartness in IoMT requires smart communication technologies which enables higher range and less power. The low power consumption and high-range covering helps in enabling the feature of smartness in the system. The Figure 1 shows the comparison of data rate and distance coverage, and power consumption. The figure shows the NFC, WLAN, Bluetooth which has increasing data rate and less distance coverage. Zigbee and 6LoWPAN has a bit less data rate with more distance coverage. Cellular communication has the highest data rate with highest coverage. On the other hand, LoRa and Sigfox has lowest data rate with highest distance coverage. The power consumption increases with the data range is clearly seen. The LoRa is selected as the power consumption is less and data coverage is more despite less data range. In case low data rate sensors LoRa will be the advanced solution. Further, based on applications, the network technology can be decided.

Every Network Technology requires a specific topology for serving better the purpose. The frequency is another major role-playing character to decide a networking technology [1]–[4]. The networking technology with their corresponding frequency and Topology is stated in Table 1. The data collected shows the LoRa as a satisfying technology which has a wide range of operating frequency and could be built on star or mesh topology. The diverse opportunities in the advanced technology opens a path to the researchers to select the technology based on the application requirement. A vital signal-based Health monitoring system using IoT is developed using three different networking technologies and the study concluded saying the LoRa to be the advanced one [5]. The results are compared between GSM, Wi-Fi and BLE. The Lora is proved to have better reliability.

Table 1. Topology and frequency of the communication technologies

Network Technologies	Topology	Radio Frequency (MHz)
LoRa	Star/Mesh	433, 860-1020 MHz
Wi-Fi	Star	2.4 GHz
BLE	Adhoc	2.4 GHz
ZigBee	Mesh	868.3 MHz, 902-928 MHz, 2.4GHz
SigFox	Star	862-928 MHz

1.3. Communication Technologies for Remote Data

Various researches are reviewed critically to highlight the key notes. The literature work is studied on various networking technologies. The LoRa is selected to the further research as it had a good scope in application point of view. Considering the Wireless Sensor Network, the relay nodes are to be placed for efficient data collection[6]. The relay positioning is done through Optical Relay Placement Algorithm (OPRA) is proposed by the author. Authentication stands next as the privacy is concerned. The health data could govern the country if it is mishandled. The three-way authentication system is proposed to ensure the privacy of data[7]. The WSN on scattered nodes is detailed to collect data precisely. The efficient use of WSN on cattle health monitoring system is focused and the benefits are proposed by the author[8].

ZigBee based solutions on health monitoring is widely seen, the Transmission range is a drawback which is again highlighted. The ECG and heart rate monitoring of 17 males of 25 years old is taken into consideration for transmitting data through WSN[9]. The data transfer speed to the server is critically reviewed by the author the prove the advantages of ZigBee technology. More than the technology on transmission, the technology on the outfit and data collection is reviewed. The wearables-based sensor is widely suggested by researchers as it is user friendly and can be a plug and play system[10], [11]. The human posture detection and the ECG, EMG, temperature is detected across WSN.

The LoRa is widely used than Bluetooth is reviewed by the author and is stated in his work[12]. LoRa in the field of Air Quality monitoring from different nodes is also widely seen as the advantage[13]. Implementation of similar technology replacing the sensors based on the counterfeit action would help the society better. The motivation and suggestion of other researchers in moving on to the advanced technology forces us to move on with LoRa. The LoRa is used nowadays by various sectors. The topologies and the various sensors that are capable to work on is the knowledge gap identified[14]. The need to know the combination of technological to bring it in action with better efficiency is stated as a result of critical research. LoRa is best fit with Star and Mesh topology and has a wide range of operating frequency. The energy efficiency is stated to be very less when compared with ZigBee and BLE[15]. The research concludes saying the demerits of the BLE and ZigBee providing a solution with the advancements.

The Fog computing based WSN on Health monitoring is reviewed to check the computational speed and the efficiency of the system[16]. The fog computing enables the components of edge computing to make it work efficiently. Elderly monitoring in a smart city is reviewed[17], [18]. The LoRa in the sector of health enables the smart city which makes it a good go to the technology. The Elderly monitoring for the specific problem statement taken can be addressed only by LoRa as it has the lower power and high range specifications. LoRa is user friendly and beneficiary module to serve the future.

The workers' health monitoring in a construction site by monitoring both the atmosphere and the body health using LoRa based network is reviewed[19]. The System proves the efficiency of LoRa based system showing the capability of the system. The connected safety and interconnection are proposed by the Wireless Sensor Network. The communication between the sensor nodes is to taken care by the LoRa for better efficiency is viewed reviewing multiple research papers. Table 2 shows the application areas of different reviewed Networking technologies.

Table 2: Applications based on Networking Technologies is explained

Networking Technology	Application	Paper
BLE, Wi-Fi, GSM	Temperature and Blood Pressure monitoring	[5]
ZigBee	ECG and Heart Rate	[9]
LoRa	System energy efficiency detection based on LoRa	[15]
LoRa	Bolted Joints Health monitoring	[18]
LoRa	Cattle health monitoring	[8]
LoRa	Air Quality Monitoring	[13]
LoRa	Connected safety and healthcare monitoring	[19]
LoRa	Wearable IoT sensor-based health monitoring	[11]
LoRa	Walk stick-based elders monitoring	[17]

As the result of various researches reviewed critically, the LoRa module is the advanced technology module with low power consumption and high distance coverage. The networking technology is a major part of any IoT enabled system. The decision has to be wise considering the applications usage. The healthcare is a growing sector which needs attention when the IoT is considered. The modern solutions are proposed on IoT, but the applications that comes to market is less due to the lack of awareness of the communication protocol. For example, if a prototype is designed working with Bluetooth, bringing into the application major of BLE must be used. The protocols used till date in the healthcare sector has smaller communication space. The wireless transmission protocols used in the ages are limited to communication range and data rate. The security of data transmission is another major issue seen in the technologies. Troubleshooting all the issues with a compatible protocol to serve the health sector is seen in LoRa. The application-oriented prototyping would help in marketing the technology. The research lag is found in the technology areas where the data rate of transmission is less. The parallel synchronization of bigger data (EMG, CT scan, MRI) through different LoRa modules to sync could help better. No research is found stating the synchronization. The data rate is found to be a drawback in LoRa as less data can be transmitted at a moment. The alter of combing two networking technologies to improvise system can also be carried on research. The advancements in the LoRa are open in health sector which has a bright market value.

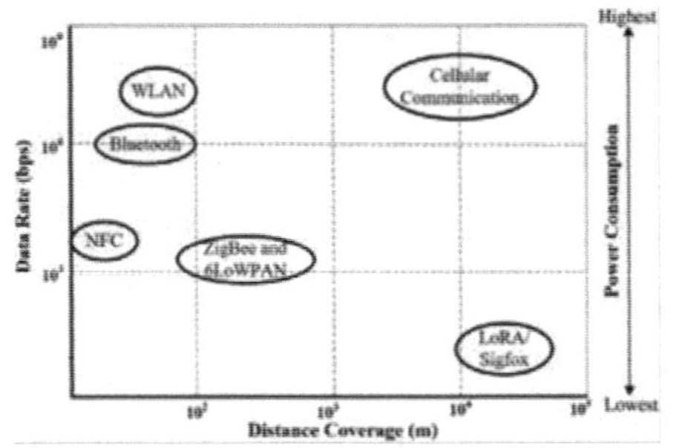

Fig. 1. The communication technologies classified based on data rate, distance coverage and power consumption

2. Lora in Healthcare

Chirp is a dump in frequency. The chirp is of two types. High to Low or Low to High. The high to low chirp states the frequency dump in which the frequency dumps from higher frequency to lower frequency. The low to high chirp states the increase in frequency from low frequency to high. Spreading Factor (SF) helps in deciding the performance of the LoRa system. The SF of system always lies between 7-12. Larger SF increases the communication range with more power communication with less data and vice versa. The pothos application is taken in simulating the LoRa system to check for the characteristics of data transmission. The chirp is given as the input data to the simulated system and the power spectrum is seen on the graph on the graph. The output data in the continuous natural number is seen as the transmitted signal through the LoRa module. The loss in the signal can be found in the output box. The demodulated Frequency and dechirped demodulated frequency are seen on two different graphs as output. The Fig.2 shows the Pothos application window with the above prescribed functionalities.

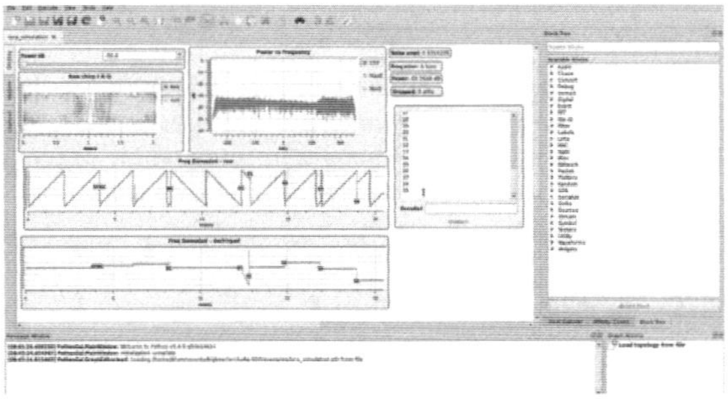

Fig. 2. The Pothos application window with the functionalities.

2.1. Pothos simulation for LoRa test

On the platform, several tests are performed to assess the LoRa modem's ability to transmit high data rates in a shorter amount of time. The authors began by lowering the Spread Factor for any given time instance, the chirps are doubled half the number of samples; achieving a greater data transfer in less time. When the SF is reduced, the data transmission rate is increased by means of faster chirps. The results of the lowering the SF is seen in Fig 3a. The second test is done on increasing the Spread factor for the same instance of time. The samples are doubled for the half the number of chirps which significances the lower data transfer in more time. The increase in SF reduces the data transmission rate with the high active time in air which leads to better performance in case of transmission range. The results of higher SF are seen in the Fig 3b. In terms of data transmission rate and range, the Spread factor has a major impact. If Higher data rate is needed, SF has to be reduced; which apparently the range is reduced. If higher range is needed, SF have to be increased; which significantly reduces the data rate.

Journal of Physics: Conference Series **2327** (2022) 012074 doi:10.1088/1742-6596/2327/1/012074

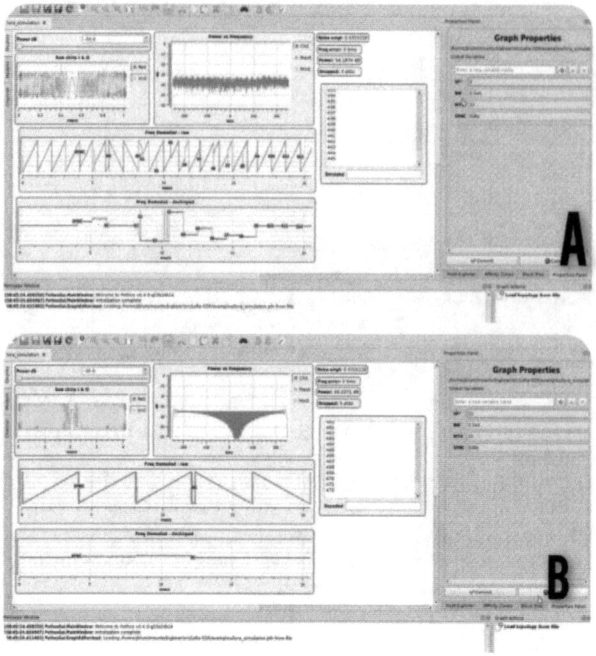

Fig. 3. (a) Simulation in Pothos with reduced Spread Factor. (b) Simulation in Pothos with increased Spread Factor.

Noise, an environmental disturbance that affects the transmission of signal is simulated to check the system efficiency. Simulating the idle system performance does not provide a full understanding. So, the application where the noise is fed in the environment to detect the change in the data rate and range is checked. The system is simulated with a considerable amount of minimum noise. The waveform is smeared still the data packets are received. The flat edges of the frequency wave are seen in Fig 4. Further increasing the noise, decreases the power of the signal which can be seen over frequency power graph. The loss of signal is also seen. The Fig 5a shows the significances of increase in noise simulated condition. When the SF is increased in the heavy noise environment, the data packet losses are avoided. The role of spread factor during the noisy environment can be seen in the simulation in Fig 5b.

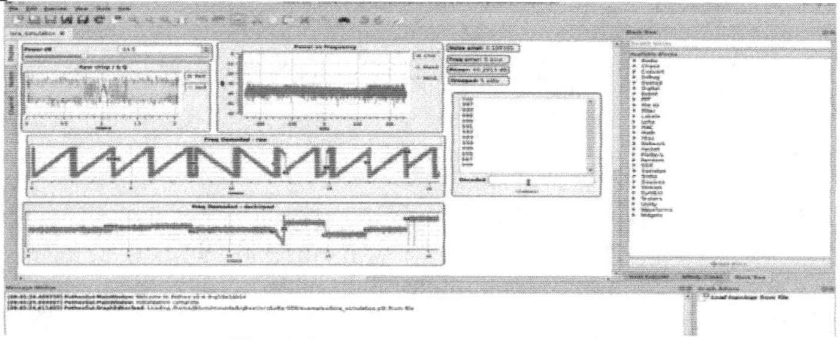

Fig. 4. System incorporated with Noise

Journal of Physics: Conference Series **2327** (2022) 012074 doi:10.1088/1742-6596/2327/1/012074

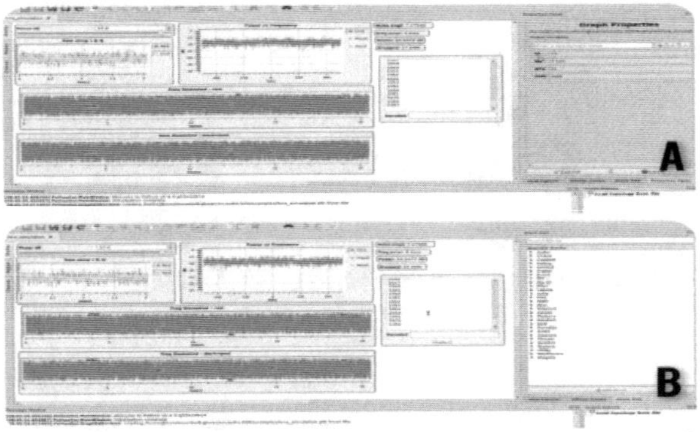

Fig. 5. (a) Effects of heavy noise in the Frequency vs Power graph and data losses in the system. (b) The data loss is reduced in increasing the SF.

2.2 Range Test

The Lora module with titled version "E32 433T30D (TTL 1W)" with the antenna is used in the complete research work. Fig 6 shows the LoRa transmitter and receiver module. The module is tested for range and the data transfer rate. Most of the times, the basic (temperature, BP, sugar, oxygen level) health data is really small and the advanced (MRI, CT images) health data is really huge. Two applications are developed considering the need of the system. The similarities in the solutions are totally different but the range and data rate plays a major role. Complete research on considering the different applications based on different mandatory functionalities is discussed.

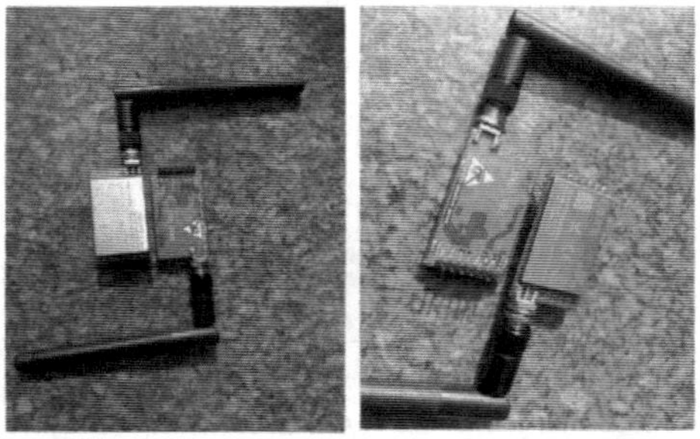

Fig. 6. LoRa module "E32 433T30D (TTL 1W)" is displayed

The climatic conditions play a major role in wireless communication system as the radio waves are often affected with the environmental conditions, to make the study transparent the climatic conditions at which the tests are performed is shown in Fig 7. The Lora setup is seen in Fig 8. Fig 8a shows the Transmitter system, 8c shows the receiver system and 8b shows the data received.

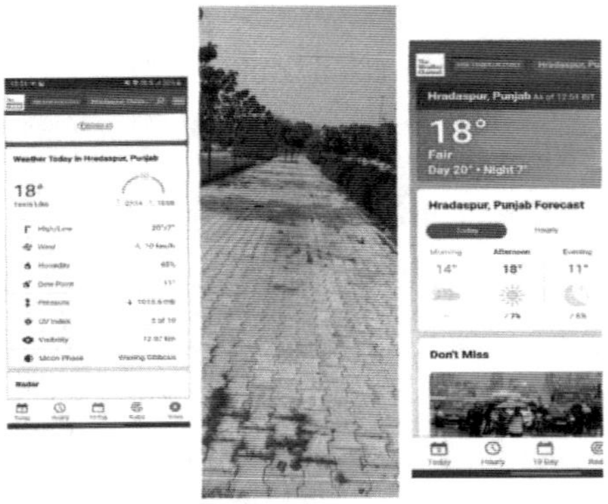

Fig. 7. The climatic conditions including temperature, humidity, wind, pressure, UV index, dew point, visibility.

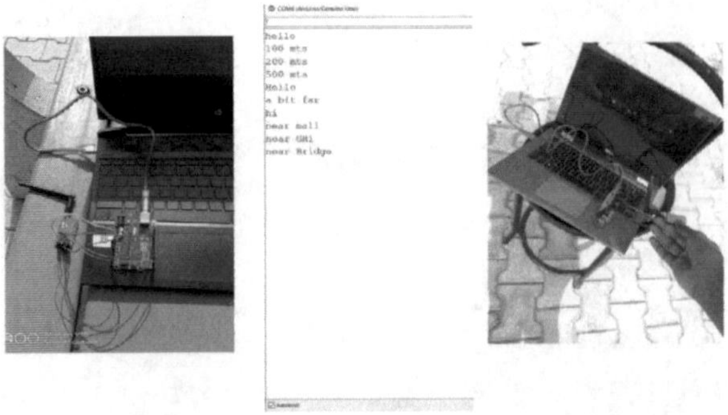

Fig. 8. (a) Transmitter System (b) Received Signal (c) Receiver System

The authors went on with the first test in the open field, The ground in Lovely Professional University is selected for the test. The ground is mostly clear with few trees in the sides. The Fig 9a shows the range of around 1000 mts from "block 26" to block "30 parking". Fig 9c shows the range of approx. 1500 mts from "Law gate" till "block 14". Fig 9b shows the receiver setup. The images are taken from the WhatsApp location calculating the distance between two centers. The results were really satisfactory that the system gave perfect results in the locations with same altitude. The authors went on with the further test on higher altitude. The Transmitter system is set on the 9th floor of the building (approx. 85ft) and the receiver system is set to move far in the lower altitude. The range is seen for around 3000 mts considering the buildings on the way. The location indication is seen near the "Uni hospital" and the "Block 56". The Fig 10a shows the google map image of the receiver and transmitter whereas the 10c indicates the WhatsApp location of the system. The area with the buildings is shown in Fig 10b in which the test was held.

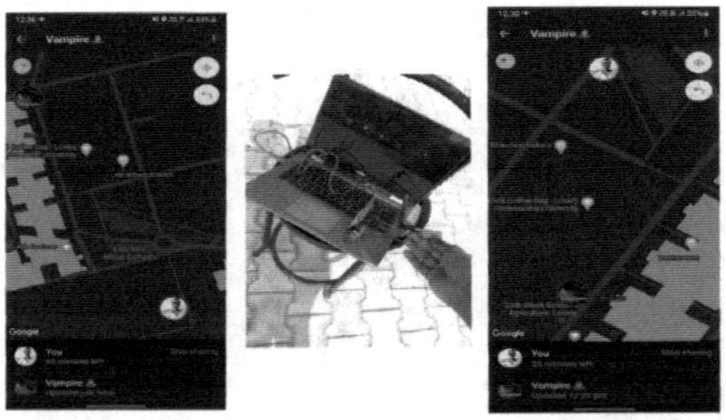

Fig. 9. (a) Covering range of over 1000 mts (b) receiver system in the ground (c) Covering range of approx. 1500 mts.

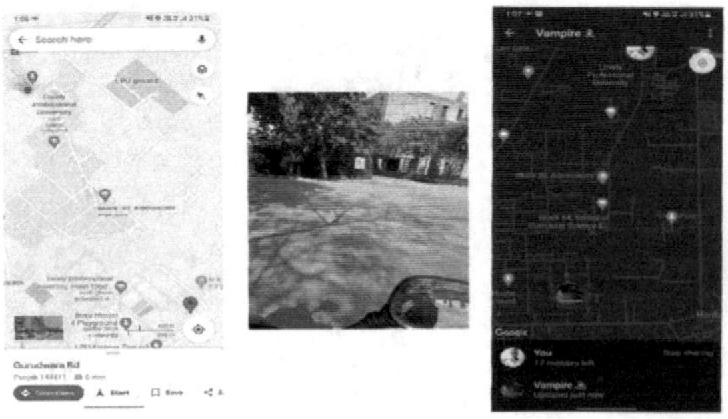

Fig. 10. (a) Google map imaging of System (b) The Scenery at which the test was conducted (c) WhatsApp imaging of Transmitter and Receiver

2.3 Data Rate Test

The LoRa module has 4 modes in which the communication between the LoRa and controller is established. The modes are enabled used 2 dedicated pins Mo and M1. The Mode 0 states the Transmission Mode. The UART communication is enabled all time to send and receive data. When user sends a data, module sends using UART protocol and in idle state wireless receiving is enabled to receive data. During configuration mode, Data can be received normally, using the same receiving function as in mode 0. In WOR mode, an automatic preamble is added before transmitting when a party is defined as a transmitting party. Data can be received normally; the receiving function is the same as mode 0. In Deep Sleep Mode, wireless receiving and transmitting is disabled. The module will configure when the mode is switched from Deep sleep mode to any other mode. The functionality of LoRa module is detailed in Table 3.

Table 3: The different modes of the LoRa module

S. No	M0	M1	Status
1	0	0	Transmission Mode
2	0	1	Configuration Mode
3	1	0	WOR mode
4	1	1	Deep Sleep Mode

The official tool by LoRa named "LoRa Modem Calculator Tool" helps in determining the bitrate based on the predefined background formulas. Various tests are performed to see the bitrate difference. First authors kept SF as 12 and CR as 1 is seen in Fig 11. The bitrate was found to be 732.42 bits per second with preamble duration of 83 ms. The time on air was 215 ms. The transmit mode timing is 299 ms and receiver mode timing is 215 ms with sleep time of 3156 ms. Transmit Charge Consumption is seen really high 26910 micro-Columb. Fig 12 shows the results for SF 10 with CR 2. The bitrate is found to be 3255 bits per second with the time on air as 61.95 ms. The permeable duration is 20.99 ms and symbol time is 2 ms. The transmit mode time is 70.7ms and receiver mode timing is 62 ms with sleep time of 3412 ms. The transmit charge consumption is seen to be less around 6359 micro-Columb as compared to the previous scenario. Comparing both the situation gives an idea of the CR and SF on the system. The increase in CR significantly reduces the data rate and the decrease in SF increases the data rate with the significant power loss and the time on air.

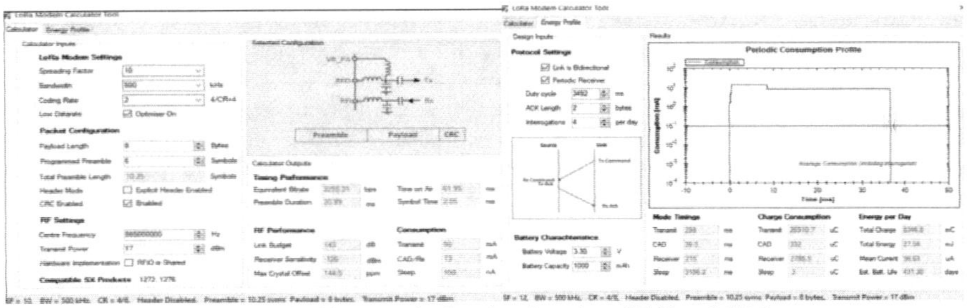

Fig. 11. The bitrate and other data based on SF 12 and CR 1.

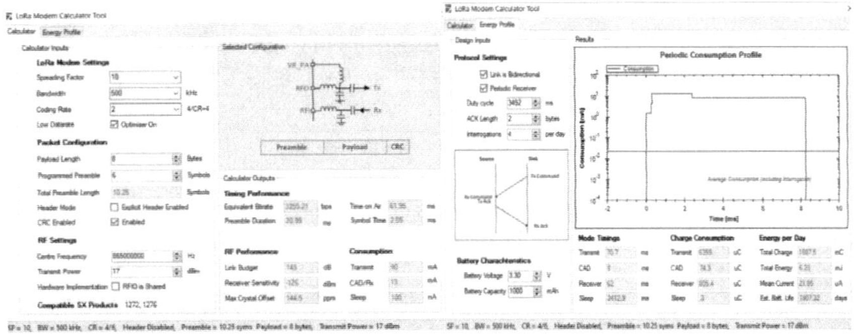

Fig. 12. The bitrate and other data based on SF 10 and CR 2.

3. Wireless Sensor Network in an open network with targeted audience of ECG monitoring

The ECG band is enabled with the controller system holding the LoRa module as the transmission node. Heart attack is a kind of disease that cannot be diagnosed in prior notice. The diagnosis can be done few hours before the affect. The system would help in regular monitoring of patient remotely in the areas where the network is limited or the need of private network is needed. The rural area people with the poor health condition can have the device to remotely and timely monitor of the heart. The fig 13 shows the transmitter and receiver setup for the enabling of targeted audience monitoring for the specific data using IoT Technologies. The system is capable to livestream the data to the distant doctor. People with heart diseases needs a regular monitoring of heart using Electro Cardiogram. This system would help in fetching the data from sensor and send through Lora transmitter. The doctor on the other end can receive the signal using LoRa and decode it in the controller and display the ECG data on the monitor. Series of data is sent through LoRa, which is procured from the sensor setup. Moreover, the basic health data such as body temperature, Blood Pressure, Heart rate are also transferred to the distant doctor using the system.

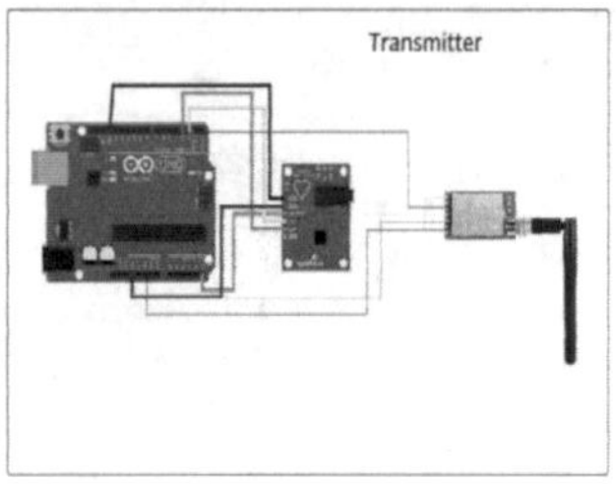

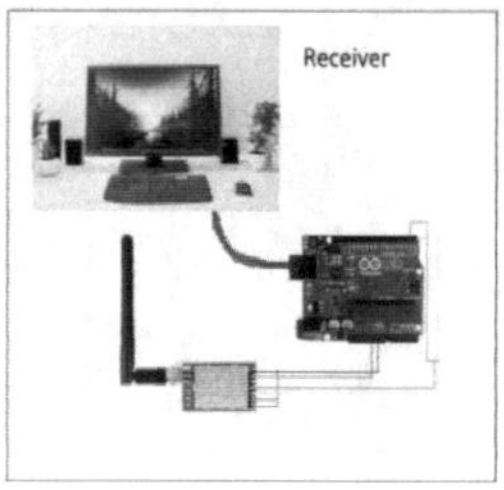

Fig. 13. The receiver and transmitter system with the sensor setup.

4. Conclusion

The various communication technologies existing on various health domain literature is reviewed and the LoRa is found to be a better device for the targeted group monitoring in the remote areas. The range and data rate can be managed to develop an application for healthcare sector. The simulation of LoRa is performed to check the noise deviation and data rate. Another simulation is done to determine the bitrate, power losses, transmit mode timing and various other parameters. The lora is developed for the transmission of ECG signal using LoRa for distant doctor. The results of using LoRa in the remote monitoring and for the specific purpose is appreciable. The advancements in the technology with Machine Learning and Artificial Intelligent techniques would help us in building the intelligent system with capabilities to serve the health domain. The early prediction of heart attack and other deadly diseases with the dataset developed by testing on thousands of people would help in achieving the further goal. The system to ensure the health of heart from a remote distance and to analyze the data with enhanced techniques to ensure the good health of the human is a demanding need and can be achieved using the system.

Acknowledgements

I would sincerely thank my mentor for providing an extended support on my work for the successful completion. His interest and motivation inspired me in working on the idea. This research was supported partially by Lovely Professional University, Phagwara, Punjab, India.

References

[1] M. Shahidul Islam, M. T. Islam, A. F. Almutairi, G. K. Beng, N. Misran, and N. Amin, "Monitoring of the Human Body Signal through the Internet of Things (IoT) Based LoRa Wireless Network System," *Appl. Sci.*, vol. 9, no. 9, p. 1884, May 2019.

[2] B. Vejlgaard, M. Lauridsen, H. Nguyen, I. Z. Kovacs, P. Mogensen, and M. Sorensen, "Coverage and Capacity Analysis of Sigfox, LoRa, GPRS, and NB-IoT," in *IEEE Vehicular Technology Conference*, 2017, vol. 2017-June.

[3] E. Morin, M. Maman, R. Guizzetti, A. Duda, and É. Morin, "Comparison of the Device Life-time

in Wireless Networks for the Internet of Things," *IEEE Access*, no. 5, pp. 7097–7114.

[4] D. Ismail, M. Rahman, and A. Saifullah, "Low-Power Wide-Area Networks: Opportunities, challenges, and directions," in *ACM International Conference Proceeding Series*, 2018, pp. 1–6.

[5] K. N. Swaroop, K. Chandu, R. Gorrepotu, and S. Deb, "A health monitoring system for vital signs using IoT," *Internet of Things*, vol. 5, pp. 116–129, 2019.

[6] Y. Li, C. S. Chen, K. Chi, and J. Zhang, "Two-tiered relay node placement for WSN-based home health monitoring system," *Peer-to-Peer Netw. Appl.*, vol. 12, no. 3, pp. 589–603, 2019.

[7] P. Soni, A. K. Pal, and S. H. Islam, "An improved three-factor authentication scheme for patient monitoring using WSN in remote health-care system," *Comput. Methods Programs Biomed.*, vol. 182, p. 105054, 2019.

[8] B. Sharma and D. Koundal, "Cattle health monitoring system using wireless sensor network: A survey from innovation perspective," *IET Wirel. Sens. Syst.*, vol. 8, no. 4, pp. 143–151, 2018.

[9] T. S. Sollu, Alamsyah, M. Bachtiar, and A. G. Sooai, "Patients' Heart Monitoring System Based on Wireless Sensor Network," *IOP Conf. Ser. Mater. Sci. Eng.*, vol. 336, no. 1, 2018.

[10] L. Gao, G. Zhang, B. Yu, Z. Qiao, and J. Wang, "Wearable human motion posture capture and medical health monitoring based on wireless sensor networks," *Meas. J. Int. Meas. Confed.*, vol. 166, 2020.

[11] J. Wan *et al.*, "Wearable IoT enabled real-time health monitoring system," *Eurasip J. Wirel. Commun. Netw.*, vol. 2018, no. 1, 2018.

[12] A. Rahaman, M. M. Islam, M. R. Islam, M. S. Sadi, and S. Nooruddin, "Developing iot based smart health monitoring systems: A review," *Rev. d'Intelligence Artif.*, vol. 33, no. 6, pp. 435–440, 2019.

[13] S. Liu, C. Xia, and Z. Zhao, "A low-power real-time air quality monitoring system using LPWAN based on LoRa," *2016 13th IEEE Int. Conf. Solid-State Integr. Circuit Technol. ICSICT 2016 - Proc.*, pp. 379–381, 2016.

[14] M. S. Islam, M. T. Islam, A. F. Almutairi, G. K. Beng, N. Misran, and N. Amin, "Monitoring of the human body signal through the Internet of Things (IoT) based LoRa wireless network system," *Appl. Sci.*, vol. 9, no. 9, 2019.

[15] C. Bouras, A. Gkamas, and S. A. K. Salgado, "Energy efficient mechanism for LoRa networks," *Internet of Things*, vol. 13, p. 100360, 2021.

[16] J. Kharel, H. T. Reda, and S. Y. Shin, "Fog Computing-Based Smart Health Monitoring System Deploying LoRa Wireless Communication," *IETE Tech. Rev. (Institution Electron. Telecommun. Eng. India)*, vol. 36, no. 1, pp. 69–82, 2019.

[17] A. Lachtar, T. Val, and A. Kachouri, "Elderly monitoring system in a smart city environment using LoRa and MQTT," *IET Wirel. Sens. Syst.*, vol. 10, no. 2, pp. 70–77, 2020.

[18] M. Sidorov, P. V. Nhut, Y. Matsumoto, and R. Ohmura, "LoRa-Based Precision Wireless Structural Health Monitoring System for Bolted Joints in a Smart City Environment," *IEEE Access*, vol. 7, pp. 179235–179251, 2019.

[19] F. Wu, T. Wu, and M. R. Yuce, "An internet-of-things (IoT) network system for connected safety and health monitoring applications," *Sensors (Switzerland)*, vol. 19, no. 1, 2019.

Journal of Physics: Conference Series **2327** (2022) 012075 doi:10.1088/1742-6596/2327/1/012075

A Critical Review on Hand Gesture Recognition using sEMG: Challenges, Application, Process and Techniques

Davinder Kumar[1], Aman Ganesh[2]
School of Electronics and Electrical Engg., Lovely Professional University, Punjab

[1]dav_thakur83@yahoo.co.in, [2]aman.23332@lpu.co.in ORCID:0000-0002-7950-1064.

Abstract. Hand gesture recognition systems are gaining popularity these days due to the ease with which humans and machines can communicate. The goal of hand gesture development is to improve interactions between humans and computers for the purpose of transmitting ideas. In a typical HGR systems, the main steps followed are, data collection, pre-processing, feature extraction and classification. For every stage, a significant number of techniques are available with various other sub steps. This study gives an overview of modern hand gesture recognition techniques, its Physiological and Anatomical Background, working and challenges faced by these systems. Moreover, the role of artificial intelligence in optimizing the performance of HGR systems is also delineated in this paper. Also, the precision and accuracy of the HGR approaches gets affected by the complexity and diversity of various hand movements, therefore, the need for implementing AI based ML and DL methods keeps on rising. Keeping this in mind, the performance of various ML algorithms in recognizing the visual and sensor-based hand gestures is investigated. Moreover, the commonly utilized framework in detecting hand gestures has been explored in numerous standard datasets.

1. Introduction

In today's world, with the development in the field of computer technology, PC's and technological devices have become an indisputable part of our life [1]. This is clearly evident not just from the computers in our hands but practically in every perspective of life, businesses, and the community. However, the biggest issue raised by the use of computers is interfacing. In recent times, the discipline of human and computer interaction or HCI have evolved not just in terms of interaction quality but also in terms of branching in its history [2]. Different researchers have continuously been working on the development of effective an efficient human computer interfaces by utilizing the basic communicative and manipulative skills of humans. In the year 1970, the first method of computer interaction was introduced by Myron W. Krueger via hand gestures. Human-Computer Interaction (HCI) research has exploded in recent years as a result of technological improvements that ultimately leads to designing of better sensors with enhanced processing abilities. Nowadays, humans can easily interact with a significant number of the applications and interfaces that have been designed by using different HCI systems. Moreover, it is also possible to interact with such applications when they are developed naturally through voice or gestures. Among the various human and computer interaction modes, hand gesture recognition (HGR) is one of the most prevalent and commonly used method that is briefly explored in this paper. HGR can be defined as the critical component of the HCI, that basically analyzes the various computer technologies developed for interpreting inputs given by the humans. Vision based methods and non-vision-based methods are the two types of the hand gesture recognition systems. Without the presence of any wearable gadgets, vision-based approaches utilize one or more cameras to gather the image or video of user's hands through image sensors. After that, an image-based feature extraction technique is used to extract the features for recognizing hand gestures. In most cases, a high level of identification accuracy can be predicted. On the other hand, in non-vision based HGR systems, a data glove is generally used for recording the input. A number of sensors can eb deployed on the gloves which include IMUs for sensing the finger movements, flex sensor for sensing the joint bends, EMGs for sensing muscle movements and so on. The non-vision based HGR methods are quite simple and effective when compared with the vision based HGR methods[5].Electroencephalogram (EEG), Electrocardiogram (ECG)and Electromyogram (EMG)are some of the various bio-signals that are used to detect the biometric features in brain, heart and muscles.

The electromyogram signals oftenly known as the EMG signal is a common biological signal that measures electrical impulses in muscles. The central nervous system of human beings is controlled by the brain which is responsible for delivering impulses to the forearm causing arm motions. Motor units (MUs) are the locations wherein the muscle fibers are triggered by the nerves, generating electrical signals that causes muscles to contract and relax. The voltage produce by the MU's is termed as the MUAP (motor unit action potential). Hence, an EMG signal may be thought of as the sum of MUAP along with noise that can be split into MU's that is the smallest entity in Human muscles. Currently, EMG can be utilized in medical tests globally in order to monitor the condition of muscles and nerve cells. Issues impacting the link among muscles and nerves, for example, can be easily detected with EMG. In addition to this, the EMG signals can also be used as the control signals for prosthetic instruments. On the basis of the position of the electrodes, the EMG signal can be categorized into two categories, those are, sEMG (surface EMG) and iEMG (intramuscular EMG). Surface EMG or sEMG are the sensors that have the ability to detect the muscle movements over the skin and proffer information from the nearby MU's (Motor unit). The sEMG sensors typically have two electrodes, one for referencing and the other for measuring the potential difference with regard to the prior sensor. While as, a thin wire is inserted in the muscle in intramuscular EMG or iEMG for detecting the muscle movement in motor units. Hand gesture categorization, silent speech recognition, stroke rehabilitation, robot control, and other applications have all benefited from the usage of sEMG, owing to its low cost and ease of collection, as well as the fact that it allows for more natural human-machine interaction [6-7].

1.1 Physiological and Anatomical Background

EMG or Electromyography is basically a study of muscular electrical impulses. In other words, EMG can also be described as the Myoelectric signals. During the last thirsty years, the myoelectric control signals have got a lot of notoriety because of its use in rehab and HCI. It is mechanism that acknowledges body motions by collecting and analyzing EMG data in order to control peripherals like prosthetics, rehabilitation aids, and other devices. Moreover, by implementing myoelectric signals the quality of people with physical disabilities or senility can be enhanced [8-9]. Muscle tissue, like nerve tissue, transmits electric signals, and the term assigned to such electric impulses denotes the power of muscle activity. This information is recorded in these muscular motion potentials by using the surface Electromyogram (sEMG) technique. However, signal distortion and signal to noise ratio (SNR) are the two problems that are faced while sensing and recording EMG signals. The SNR can be defined as the ratio of energy in a noisy signal. Usually, electric impulses which are not comprised of the targeted EMG signal are referred to as noise. Whereas, signal distortion is defined as the change in the relative contribution of any frequency component in an EMG signal. In order to obtain the muscle signals, two types of electrodes were utilized, those were invasive electrodes and non-invasive electrodes [10]. An electrode on the surface of skin (non-invasive) or a needle electrode (invasive) is used to determine a mixture of fibers of muscle action potentials from each motor unit fibers of muscle. The simplest model for EMG signal is given in equation 1.

$$Y(n) = \sum_{r=0}^{n-1} H(r)E(n-r) + W(n) \text{--------(1)}$$

Where, Y(n) depicts the modeled EMG signal, E(n) depicts the processed point that shows pitch trajectory, H(r), MUAP, W(n) illustrates the null mean additive white Gaussian noise and n represents the number of motor unit firings.

In the coming up section of this paper, the working of Myoelectric control systems using surface EMGs and various challenges faced in HGR systems is discussed. The main motive of conducting this review is to have a better understanding of current HGR systems and how the performance of these systems can be improved in the future. Moreover, it has been analyzed by incorporating the concept of Artificial Intelligence in modern HGR systems, there performance can be enhanced significantly. Therefore, in this manuscript, the role of AI in optimizing the performance of HGR systems is highlighted in order to have a better understanding of current HGR systems in the light of ML and DL. A number of recently published articles related to HGR have been reviewed and some findings have been observed after studying them that are mentioned at the end of paper.

1.2 How myoelectric control system works

In a typical myoelectric control system, hand movements are frequently employed to operate additional devices. Surface EMGs uses sensors to assess the muscle activities and is used to elicit hand motions. The intrinsic muscle movements are recorded by sEMG sensor from the skin surface whenever skeletal muscles are stimulated by the physiological brain activity and hence represents the formation and dissemination of compound nerve impulses, that are the coordinated action of motor units [11]. The data gathered from the sEMG sensors is then used to interpret the motion in limbs and detect hand gestures. A myoelectric technology with great spatiotemporal granularity will be able to decipher exclusionary and repeating patterns that could be used in myo-control and various identification jobs [12]. Figure. 1 shows the typical EMG sensor in which different components like electrodes, conditioning circuit, analog to digital converter and communication module are integrated.

Figure 1. Surface EMG sensor with different components

The raw EMG electric signals generated in the muscles is collected by the electrodes which are later on amplified, normalized and transferred by the conditioning circuit. The block diagram of the conditioning circuit that is deployed in the sEMG sensor is illustrated in Figure 2. on the basis of the instrumental amplifier, a differential amplifier i.e. A1 is developed that amplifies the amplitude of signals. Also, a high pass filter and the low pass filters are used to eliminate the DC from the signals and high frequency noises respectively. Moreover, the power line interferences are removed in the sEMG system by using the notch filter. Later on, the output generated by the conditioning circuit is digitalized by the analog to digital converter via the communication module [13].

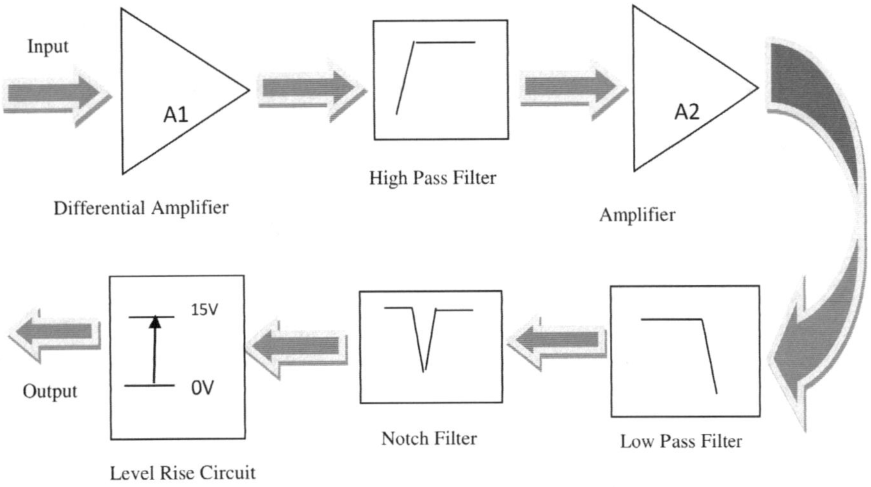

Figure 2. Block diagram of conditioning circuit

1.3 Stochastic process of EMG signals

Muscle fiber contractions can be divided into two categories when it comes to hand movements, one is dynamic and the other is static. The EMG electrical signals are described as the stochastic process which is influenced by the above mentioned two types of muscle contractions [14].

- **Dynamic muscle contraction:** The length of muscle fibers alters while the joints remain in motion during a dynamic contraction. A typical example of the dynamic contraction is when waving a hand to say hello to someone. In order to compute the activity of muscles during the dynamic contraction, the equation given in 2 can be utilized,

$$x(t) = a(t)w(t) + n(t) \text{-------------}(2)$$

where w(t) represents the stochastic aspect of the signal, a(t) represents the modulating signal that illustrates the intensity of the EMG and n(t) depicts different instrumentation and biological noises. Some of the problems faced in the dynamic contraction is calculating the time varying amplitude a(t) and predicting when value of a(t) is larger than zero.

- **Static muscle contractions:** In static muscle contractions, the invariance of muscular fibers and upper limb joints are involved. These upper limb muscles have more than 20 degree of freedom, encompassing the upper arm, elbow, wrist, forearm and hand as well. The amplitude of the surface EMG is minimum because of the narrowness and low muscle strength of upper limbs, thus making sEMG more challenging. The aggregate of a set of MUAPs is perhaps the common fundamental of the SEMG during a constant isometric contraction and can be given by the equation 3

$$x(t) = \sum_{i=1}^{N} S_i(t) * m_i(t) \text{--------}(3)$$

where N depicts the total no. of active MUs, $S_i(t)$represents impulses which represent the firing times of the ith MU, $m_i(t)$represents the MUAP of the ith MU respectively.

Nevertheless, given the challenges of the estimation methods in any non-stationary system, the mathematical models of the hand gesture recognition systems are not applied. Machine learning (ML) approaches, on the other hand, are popular since they can extrapolate a remedy for non-stationary processes utilizing a variety of strategies, such as covariate shift approaches, class-balance change, and segmentation in brief static periods.

1.4 Challenges faced in HGR systems
With the advancements in ML and deep neural networks, a new and innovative way has been developed for controlling the myoelectric protheses. In such contexts, surface EMG [15-17], are commonly used as inputs for the protheses controlling system. Whilst considerable academic study has been undertaken on using enhanced machine intelligence for hand gesture identification. However, the application of such new approaches in healthcare situations and in individuals' everyday lives has been constrained. As a result, there is indeed a notable disparity [18-19] among practical situations and new theoretical approaches in myoelectric prosthesis control, owing to a number of key problems.

One of the major problems faced is the change in surface EMG signals [20].The changes might occur due to the shifting of electrodes, variation in limb positions, non-stationary nature of surface EMG electric signals, muscular exhaustion, skin perspiration, or clinical characteristics connected to amputation surgery reducing the capability to capture accurately. This may result in difference between the likelihood distribution of sEMG signals from one person to another. Another biggest challenge faced in the sEMG control signals is the non-availability of large and labelled databases. In this sense, it is worth noting that DNNs operate admirably whenever a big training data is provided however, they often underperform when knowledge is constrained. In a laboratory context, big data collecting is conceivable, but this is not possible in real life. The usefulness of the myoelectric prosthesis can be considerably increased by tackling the foregoing difficulties. Therefore, it is necessary to develop flexible architectures that can transfer the information from source to destination even during the non-stationary discrepancies between them. Moreover, such models can also opt for impressive performance and that too without using any huge database.

2. Artificial Intelligence and HGR systems

Artificial intelligence (AI) can be used to correlate sEMG signals to prosthesis behavior. The purpose of feature extraction in sEMG-based gesture identification is to characterizes EMG impulses in an exclusionary manner. Extensive data processing and pattern recognition algorithms which can characterize and interpret large datasets are required to convert the complicated and potentially changeable data of the sEMG signal into meaningful control signals for prosthetic hands. Machine learning (ML) and deep learning (DL) are the two basic processing methods used in the sEMG pattern identification methodologies. The ML based approaches used for learning and performing operations is based on feature engineering. In traditional ML based HGR systems, several hurdles are faced which include processing raw EMG data as well [21]. Furthermore, a slew of research suggest that heuristic hand-crafted features play a big role in sEMG pattern recognition rate. Deep learning (DL) based approaches on the other hand relies on the feature learning rather than feature extraction. The characteristics of the DL methods relies on the hierarchical architectures which could easily extract information from various representative and hidden layers [22]. The abundance of training information available is one of the most important criteria for precise prediction, particularly when dealing with deep learning models. Hand gesture recognition provides an unusual situation in which a solitary person cannot reasonably be assumed to produce thousands and thousands of instances in a single session. Through pooling the observations of several users and massive amounts of data can be acquired. Such computational approaches permit feature extraction and model development to happen at the same time, allowing characteristics to be learnt dynamically rather which is better for complicated gesture recognition. Deep learning (DL) methods have lately been implemented in variety of applications which include, image identification, speech detection, machine translation and autonomous driving. The investigation of DL in pattern classification of sEMG was prompted by the increasing use of DL in physiological signal analysis and processing in current history, DL has showed bright possibilities in the field of sEMG pattern recognition for prosthetic hands, thanks to the introduction of massive sEMG huge datasets and the most current occurrence of optimization algorithms.

2.1 Hand gesture recognition systems

Although the techniques to gesture recognition models are relatively similar but the methods used at every level varies. Also, not all researchers employ all phases of the general framework. The general steps followed in any HGR system are, data collection, pre-processing, feature extraction, classification and performance analysis.

a) Data collection: The first step in any HGR system is to collect the necessary data from the available datasets. Some of the commonly utilized publicly databases are Ninapro2, CapMyo [23] and cls-hdemg (CLS) [24]. Out of the given databases, the ninapro2 is the largest publicly accessible HGR dataset in which ten more sub databases are added. MeganePro is another dataset for prosthetics control that has been used by various researchers in their approaches. Additionally, there are some private datasets that have also been used for evaluating the performance of HGR systems. Gadgets like Myo armband3, the Delsys Trigno4 and others are some of the commonly used sensors for collecting the EMG signals. The sEMG signals are acquired using sEMG sensors that can use one of two collection standards, sparse multi-channel sEMG or high-density sEMG (HD-sEMG), based on the density of the electrodes. In sparse multi-channelsEMG, a precise and anatomic position of the muscles is obtained. Ninapro is an example of this configuration. Whereas, in HD-sEMG a high quality of spatially and densely distributed are recorded. The example of this configuration is CapgMyo and csl-hdemgdatabases [25, 26 and 27].

b) Pre-processing: Since the surface EMG is a multi-channel signal which means that it gets influenced by the various environmental factors, tissue inferencing, fatigue in muscles, misplacement of electrodes and so on. these factors a huge noise and doesn't remain stationary which ultimately can conceal the critical information regarding muscle activities. Therefore, it is extremely important to handle these surface EMG electric impulses properly when dealing with pattern identification systems. For this, preprocessing must be implemented for retrieving usable data from the input signals. Pre-processing can be done at two stages i.e. filtering and segmenting [28]. Background interference is frequently rejected using the bandpass Butterworth filter using various boundaries, like 5–500 Hz [29–30], 10–450 Hz [31] or 10–500 Hz [32]. Regarding noise reduction, several investigations use a combination of high-pass and low-pass filters [33-34]. Following filtration, the EMG signals are frequently segmented into a sequence of packets using the sliding window approach. Windowing is by far the most common strategy in EMG pattern identification methods based on machine learning or deep learning.

c) Feature Extraction: Feature extraction is the most important and critical stage in detecting the different hand gestures. An informative and useful set of features is selected from the available processed characteristics, rather than a raw sEMG signal. these features are then passed to the classifier used in the system in order to enhance the effectiveness and precision of identifying Hand gesture. When dealing with the sEMG systems, the features sets can be categorized into three sections of time domain or TD, frequency domain or FD and combination of both i.e. time-frequency domain (TFD).

d) Classification: Classification is yet another important and critical step in determining the movement in muscles. During classification, a number of ML and DL algorithms are used by the experts in order to classify and categorize the various hand gestures. Some of the commonly used ML algorithms for classification are; SVM, NB, KNN, DT and random forest etc.

e) Performance Evaluation: Accuracy rate is typically among the most important criteria used to assess the effectiveness of hand gesture recognition systems. Nonetheless, there are various other parameters that must be taken into consideration while examining the functioning of Surface EMG based HGR systems. For offline evaluation, accuracy, recall, precision, standard deviation are some of the commonly used performance metrics. In addition to this, mean square error (MSE), RMSE (root mean square error), correlation coefficient (R) are some of the other metrics that depict the error measures.

3. Machine Learning based Hand Gestures Recognition Systems

Before demonstrating the application of various ML algorithms in HGR systems, it is necessary to firstly introduce the underlying ideas that underpin these methods. Generally, a ML algorithm can be divided into three categories, those are, supervised ML algorithms, unsupervised ML algorithms and semi-supervised ML algorithms. Each one of these strategies require data to uncover specific patterns or groupings. Over the years, a significant number of scholars have proposed and implemented various techniques for detecting the hand gestures. For example, the authors in [35], utilized YoBu data glove gesture-based database and also implemented (ELM) extreme learning machine to identify and recognize different hand gestures. Similarly, in [36], a ML-based hand gesture recognition system was developed using Computer Vision to construct various types of gesture programs which in turn resulted in highest accuracy rate. The authors in [37], presented an EMG armband-based wearing independent HGR system in which standard space was suggested for removing the impact of wearing positions. The authors anticipated the wearing position in armband by analyzing the sequential features at different levels which aids in unifying the basic features to the standard space. Moreover, random forest (RF) was utilized by the authors to classify and identify different hand gestures which provided an accuracy of 91.47%. also, the experts in [38], suggested a three-phase hand gesture recognition system, those are; pre-processing, feature extraction and classification. The preprocessing phase encompasses contrast enhancement, segmentation, rotational and filtration operations. In this research, a unique approach was provided for obtaining a rotation invariant motion picture by aligning the 1st principal component of segmented hand motions with vertical axes. In the second stage of feature extraction, various localized contour sequences (LCS) and block-based characteristic were extracted which were then concatenated in order to demonstrate stationary hand gestures. The final stage in the model was classification which was done using Support vector machine (SVM).Three separate hand alphabet datasets were used to develop and evaluate the present framework whose sensitivity rate was recorded 99.50%, 93.58% and 98.33% respectively. The authors utilized an unlabeled EMG database which was attained from Myo armband and proposed a new HGR system that was based on k-means clustering technique in [39]. The comparative analysis of these ML based hand gesture recognition systems is analyzed in table 1.

Table 1. Comparison between ML techniques in [35, 36, 37 and 38]

Phases	Details
In [35],	
Dataset used	Created database in which ten different types of static hand gestures were included.
Pre-processing	YoBu data glove
Feature extraction	A total of 54 dimensional features were extracted
Classification	Kernel-based SVM and ELM
Performance evaluation	ELM-Kernel performs better than SVM and ELM
In [36],	
Dataset used	Utilized Sabastian MarcelHand Gesture dataset.
Pre-processing	Distinguishing the primary colors precisely
Feature extraction	Haar-based features
Classification	ML
Performance evaluation	Accuracy rate 96%.
In [37],	
Dataset used	Armband data
Pre-processing	sequential features
Feature extraction	NA
Classification	Random forest (RF)
Performance evaluation	Accuracy achieved 91.47%

Journal of Physics: Conference Series **2327** (2022) 012075 doi:10.1088/1742-6596/2327/1/012075

In [38],	
Dataset used	Utilized 3 distinct hand alphabet datasets.
Pre-processing	image enhancement, segmentation, rotation and filtering process.
Feature extraction	localized contour sequences (LCS) and block-based features
Classification	SVM
Performance evaluation	sensitivity achieved 99.50%, 93.58% and 98.33% for three datasets.

Over the past few decades, a significant number of scholars have utilized support vector machine (SVM) classifier for categorizing various types of hand gestures. Some of the recently developed SVM based hand gesture recognition systems are discussed here; The authors in [40], utilized a Myo armband surface EMG dataset signals which were later on processed and normalized in order to minimize the noise effect in these signals. After this, the processed signal was segmented by using the sliding sub window technique for creating segments in every signal. moreover, the authors used the SVM with five sub-classes (i.e. fist, wave in and out, spreading of fingers and double pinch). The current scheme was able to achieve an accuracy rate of 89% only. Moreover, the authors in [41], utilized the two channeled sEMG dataset and developed an innovative method for detecting the various HG movements. The model used the SVM method, wherein, cuckoo search optimization algorithm was used so that the kernel function and parameters can be optimized. The accuracy rate of 98.12% was achieved by the system. Another hand gesture recognition system based on the SVM was developed by the experts in [42]. The authors utilized the raw sEMG signals which were processed and various statistical features were extracted from it. later on, these features were served to the SVM classifier which was trained on the basis of this data and produced results in the form of confusion matrix. Furthermore, A. I. Maqueda and others in [43] designed a robust vision based HGR system, whose performance was validated by creating a private database in which different batch of SVM classifiers were delivered to execute gesture characterization. The suggested model showed good results when a dynamic filter was used for enhancing the classification rate. On the basis of the average threshold crossing (ATC), the authors in [44] developed a low complexity approach for detecting the hand gestures. The amount of threshold crossing occurrences generated by a completely tailored collection panel were used to identify four separate wrist movements, those were; flexion, extension, abduction, and grip by leveraging EMG surface impulses from three forearm muscles. M. Tavakoli et al. in [45], developed a SVM and 2 channel EMG model with high dimensional feature sets. The model developed proved out to be resistant while dealing with uninvited body motions. Furthermore, the proposed approach was developed by including two methods, firstly when SVM thresholding was utilized for detecting gestures and secondly, in which locking gestures were used for detecting movements. A total of 5 hand movements was recognized by the system which were, opening and closing of hands, wrist flexion and extension and finally, double wrist flexion. In [46], G. Poon and others proposed SVM based model for detecting the bi-manual gestures. In order to overcome the problem of self-occlusion the authors utilized large number of cameras with varied viewing angles for acquiring hand gesture data. also, P. Parvathy et al. in [47], proposed a model in which data was collected from the Sebastian Marcel static hand posture dataset. This data was processed and important features were extracted from it and finally passed to the SVM classifier that recognized various hand gestures. Another paper was published [48,]in which the authors focused on detecting different hand gestures via ML and PCA methods and developed a bio-medical amplifier. Initially, the authors gathered the EMG data through surface electrodes. These signals were amplified by using the Bio-medical signal amplifier. The outcome produced by this amplifier was then given as input to the Band-pass filter which in turn was passed to analog to digital converter (ADC). The final output generated by the ADC was then passed to the classifier that provides 100% classification rate for the three hand gestures. The comparative table for the SVM based HGR system is given in table 2.

Table 2. Comparison between the systems based SVM based in [40 to 48]

Phases	Details
In [40],	
Dataset used	Myo armband
Pre-processing	Raw surface EMG electric impulses were processed to minimize noise effectand identify muscle activity areas.
Feature extraction	sliding sub-window
Classification	SVM
Performance evaluation	Accuracy achieved 89.0%.
In [42],	
Dataset used	EMG repository
Pre-processing	4th order Butterworth bandpass filter
Feature extraction	statistical features
Classification	SVM
Performance evaluation	confusion matrix
In [43],	
Dataset used	ASL and NTU Databases
Pre-processing	Finding the possible hand in any frame of a video stream.
Feature extraction	global spatial features
Classification	SVM
Performance evaluation	accuracy in global system was higher.
In [44],	
Dataset used	Dataset with 10 subjects where every movement was repeated ten times
Pre-processing	Reduced noise and analyzed various features.
Classification	SVM
Performance evaluation	Accuracy achieved 92.87%
In [45],	
Dataset used	Multi- levelled based EMG dataset
Pre-processing	Noise reduction and filtration
Feature extraction	High-dimensional space features
Classification	SVM
Performance evaluation	Accuracy lied in between 95% to 100% for a trained user
In [46],	
Dataset used	Dataset containing six bi-manual gestures
Pre-processing	Developed a multi-camera model for collecting multi-views bimanual gestures
Feature extraction	Segment-based, LDA, Histogram of Gradients (HOG), PCA, RF, Multichannel and LeNet
Classification	SVM
Performance evaluation	Overall accuracy of 70%
In [47],	
Dataset used	Sebastian Marcel static hand posture database
Pre-processing	Eliminating noise factor by using; Kalman filter, morphological filter and Gaussian and median filters

Journal of Physics: Conference Series **2327** (2022) 012075 doi:10.1088/1742-6596/2327/1/012075

Feature extraction	DWT with updated Speed Up Robust
Classification	SVM
Performance evaluation	Accuracy achieved 96.50%.
In [48],	
Dataset used	Data collected through sEMG signals from hands.
Pre-processing	Filtration using Band-Pass Filter
Classification	ML andPCA
Performance evaluation	Classification accuracy rate 100%

4. Deep learning and neural network based HGR methods

In addition to the ML algorithms, a number of Deep learning (DL) based hand gesture recognition systems were developed. The performance of such methods is effective and stable. Some of them are discussed here, A novel gesture identification model was developed by authors in [49] that was based on the Artificial neural network and surface EMG electric impulses in order to accomplish the real time response. The authors utilized the MYO armband device to record sEMG signals wherein five specific gestures were recorded. Moreover, a low pass filter was also utilized for removing the noise from the raw sEMG signals in the pre-processing stage. A sliding window technique was used along with the 5 specific time domain features to build a feature vector. Finally, the classification of the system, the ANN was applied to label every observation in sliding window. The performance of the suggested scheme was analyzed in terms of identification rate and response time whose values were 98.7% and 227.76ms respectively. Likewise, the authors in [50], presented a classification technique that relied on surface EMG (sEMG) data and hence facilitating for user mobility. Those surface EMG signals were gathered from the upper portion of forearm for providing a way to identify Indian sign language (ISL). The information was analyzed and different movements were classified with an accuracy rate of 97.5 percent, using an ANN that was based on Scaled Conjugate Gradient (SCG) training. A total of 120 records were used in the learning, which corresponded to four different wrist movements. In [51], the authors utilized just the two channels of sEMG, for proposing a new HGR system in which 6 hand movements were detected. The authors processed the raw sEMG signals in number of stages like filtration, detection of end points, feature extraction ad classification. The filtration was done by using the 1st order digital low pass infinite impulse response (IIR) with a frequency of 500Hz. The endpoints of the gesture were identified by the energy. The characteristics selected included short-time energy, zero-crossing rate and linear predictive coefficient (LPC) having 12 levels. The classification was done by utilizing the back propagation (BP) neural network. The suggested scheme yielded a recognition accuracy of 96.41 percent to 99.70 percent. Moreover, the authors in [52], developed a deep multi parallel based CNN HGR model that resolved the issue of losing critical information while extracting features in ML approaches. CNNs were proven to be effective to restrict the complications of feedforward NN through weight sharing and restricted local connections. The authors removed the concept of feature extraction and classified various hand movements directly by using multi levelled CNN. The approach achieved superior accuracy when compared to conventional ML methods. In [53], a Myo armband device was used to capture the readings for two databases of 18 and 17 subjects. The authors augmented the CNN with the help of transfer learning method to transfer data from 1st database and ease the data creation burden on particular person. The experimental results performed showcased that the suggested scheme is enough stable and precise for guiding a 6 degree of freedom robotic arm. Interestingly, an accuracy rate of 97.81 percent was achieved in the second database containing 17 individuals for seven hand/wrist movements in suggested CNN. Also, the experts in [54], suggested an improve attention based HGR system that was based on CNN and RNN for collecting the spatial features more effectively and efficiently. Furthermore, a novel sEMG image representation approach was also developed by using the conventional feature vectors that allowed DL models to identify tacit associations across channels for sparse multi-channel EMG signals. Rigorous testing was conducted on five sEMG standard datasets which revealed that the suggested method surpasses all previously published state-of-the-art techniques for both sparse multi-channel and high-density sEMG datasets. Researchers fixed the window size for NinaProDB1 and NinaProDB2 to 200ms, and 150 ms for BioPatRec sub-database, CapgMyo sub-database, and csl-hdemg datasets. The abovementioned standard datasets have recognition accuracies of 87.0 percent, 82.2 percent, 94.1 percent, 99.7 percent, and 94.5 percent accordingly, that was 9.2percent, 3.5 percent, 1.2 percent, 0.2 percent, and 5.2 percent greater than traditional systems. In order to overcome the issue of wearing too many sensors because of the amputee's restricted residual muscle, the authors in [55] developed a two-channel based surface EMG model for identifying six hand gestures. A number of important features were extracted by using the time and frequency domain feature approaches. Moreover, the authors also observed the performance of back propagation neural network (BPNN) over SVM in terms of mean and variance. An accuracy rate of 94.66 was achieved by using time domain feature technique and BPNN classification method. In [56], the experts proposed a model based on LSTM and RNN for detecting the hand gestures through surface EMG. The accuracy of the suggested model doesn't get altered even when the position of sensors was changed. In [57], a multi channeled surface EMG based amplification module with enhanced feature and optimization system was developed to achieve higher accuracy rate in DL methods. The developed feature set was formed by utilizing the various multi-dimensional surface EMG pictures by mixing CNN and LSTM models. The recommended scheme achieved a identification rate of 91.40% when FFT and RMS were utilized as feature data processing techniques.

In addition to the above-mentioned approaches, Park and Lee in [58], introduced a user adaptive CNN framework. This model was implemented on the data present in the NInaProdataset which resulted in the optimum classification and accuracy rate than Standard SVM models. Furthermore, Because the CNN architecture has many variables, Chen L. et al. in [59], created a compressed network model that not only minimized the intricacy of the model but also improves its accuracy rate.Pre-processing, hyperparameters, and system structure are some of the elements that were taken into consideration in this system.in addition to this, Asif et al. in [60] and Triwiyanto et al. [61], analyzed a number of methods that can be implemented on the CNNs to reduce the overall system complexity and enhance its performance. Tong and others in [62], proposed a dual-flow network design in which upper layer was made up of numerous CNN layers, while the lower layer was comprised of various LSTMs. The system extracts spatiotemporal features effectively through understanding the fusion info of the 2 layers. The comparison table for the given DL based systems is given in table 3 below.

Table 3. Comparison of various DL systems [49-62]

Phases	Details
In [49],	
Dataset used	Myo armband
Pre-processing	Improved quality of signal and reduce noise effect by using low pass filter.
Feature extraction	Build feature vector by using 5 time domain features via sliding window method.
Classification	ANN
Performance evaluation	Accuracy rate of 98.7%
In [51],	
Dataset used	Private
Pre-processing	Minimized noise effect by implementing Bandpass filter
Feature extraction	short-time energy, zero-crossing rate and linear predictive coefficient (LPC)
Classification	BPNN
Performance evaluation	96.41% to 99.70% accuracy rate

In [53],

Dataset used	two datasets of 18 and 17 able-bodied participants via Myo armband.
Classification	CNN
Performance evaluation	Accuracy rate of 97.81% on second dataset with 17 individuals.

In [54],

Dataset used	NinaProDB1, NinaProDB2, BioPatRec, CapgMyo and csl-hdemg databases.
Feature extraction	sEMG image representation technique
Classification	CNN-RNN
Performance evaluation	recognition accuracies for above mentioned databases are 87.0%, 82.2%, 94.1%, 99.7% and 94.5%,

In [55],

Dataset used	EMG dataset
Feature extraction	time and frequency domain features.
Classification	BPNN and SVM
Performance evaluation	accuracy 94.66%

In [57],

Dataset used	EMG dataset
Feature extraction	multidimensional sequential sEMG images
Classification	CNN-LSTM
Performance evaluation	Accuracy rate of 91.40%,

In [58],

Dataset used	NinaPro DB1
Time window	2,000 ms
Feature extraction	RMS $time \times ch$.
Classification	CNN
Performance evaluation	Accuracy achieved 94%

In [59],

Dataset used	Private
Time window	260m/s
Feature extraction	CWT
Classification	CNN
Performance evaluation	98.81%

In [60],

Dataset used	Private
Time window	150m/s
Feature extraction	Raw sEMG
Classification	CNN
Performance evaluation	Accuracy rate of 92%

In [61],

Dataset used	Private
Time window	200m/s
Feature extraction	Raw sEMG
Classification	CNN

Performance evaluation	Accuracy rate 93%
In [62],	
Dataset used	Private
Time window	NA
Feature extraction	SampEn, ZC, MAV
Classification	Dual-Flow Network
Performance evaluation	Accuracy rate 78.31

5. Other Deep learning methods

Aside from the approaches listed previously, there are numerous additional DL approaches that can be used to recognize and analyzes EMG patterns. Due to the simpler structure of Fully connected neural networks [63], BP neural networks, and artificial neural networks (ANN) are some of the commonly utilized techniques for offline and real-time recognition of sEMG. The time delay neural network (TDNN) is a well-known time series computing perceptron which processes the temporal range of significant characteristics in the original signal using tapped-delay lines so that a short-term memory could be sent into the classic feed-forward network design [64]. On the other hand, deep neural network provides two major benefits. Firstly, it could be used to determine immediate muscular contraction derived from natural motion and does not require repetitive isometric contracts. Secondly, Second, it can distinguish movement in two degrees of freedom simultaneously. In the estimation of simultaneous movement of the shoulder and elbow joints, TDNN analysis of combination with sEMG and kinematics datasets may demonstrate good performance [65]. Because the amount of training data samples in DL influences the effectiveness of the algorithm, the restricted databases gathered from many subjects must first be extended using data-augmentation techniques which can improve the model's resilience. Furthermore, because of the architecture of neural network training, the procedure of transfer learning becomes quite simple and effective.

6. Conclusion

In current world, hand gesture recognition systems are considered as one of the effective an efficient way of communication with the computers. These systems are applicable in number of other domains like, robotics, sign language, human-computer interaction, 3D games and so on. therefore, the need for researching HGR systems keep on rising because of its growing application in real world. In this paper, a number of HGR methods, based on ML and DL techniques are reviewed and compared with each other. After analyzing the literatures closely, it was observed that majority of the researchers utilized ML algorithms for detecting different hand gestures. However, the problem in the ML HGR system was that they undergo overfit problem when dealing with the least samples. Moreover, these systems must perform the pre-processing and feature extraction in order to provide optimum results, but they often tend to lose some critical information during these phases. Furthermore, it was also observed that among all the ML algorithms, most of the experts used support vector machine (SVM) in their work and were able to provide good results. In addition to this, we have also reviewed the performance of recently proposed Deep learning (DL) based HGR systems and compared the most common and prominent ones. Also, it was analyzed that identification of dynamic based hand gestures is more complicated and difficult than static gestures. This paper demonstrates the most important techniques, implementations, and obstacles of hand gesture recognition. This included various gesture acquisition methods, feature extracting processes their classification in the areas of sign language, robotics, and other fields, environmental problems, and datasets.

Acknowledgement

I would like to express my very great appreciation to **Dr. Anuj Jain** (Professor, School of Electronics and Electrical Engineering, Lovely Professional University, Phagwara, Punjab, India.) for his valuable and constructive suggestions during the planning and development of this research work. His willingness to give his time so generously has been very much appreciated.

References

[1] S. M. SaliShajideen and V. H. Preetha, "Human-Computer Interaction System Using 2D and 3D Hand Gestures," *2018 International Conference on Emerging Trends and Innovations In Engineering And Technological Research (ICETIETR),* 2018, pp. 1-4,

[2] G. Sinha, R. Shahi and M. Shankar, "Human Computer Interaction," *2010 3rd International Conference on Emerging Trends in Engineering and Technology*, 2010, pp. 1-4,

[3] H. A. Jalab and H. K. Omer, "Human computer interface using hand gesture recognition based on neural network," *2015 5th National Symposium on Information Technology: Towards New Smart World (NSITNSW),* 2015, pp. 1-6.

[4] I. Dhall, S. Vashisth and G. Aggarwal, "Automated Hand Gesture Recognition using a Deep Convolutional Neural Network model," *2020 10th International Conference on Cloud Computing, Data Science & Engineering (Confluence),* 2020, pp. 811-816.

[5] Song, Wei & Wang, et al., (2019). Design of a Wearable Smart sEMG Recorder Integrated Gradient Boosting Decision Tree based Hand Gesture Recognition. *IEEE transactions on biomedical circuits and systems.* PP. 10.1109/TBCAS.2019.2953998.

[6] Xiong, Dezhen, et al. "Deep learning for EMG-based human-machine interaction: A review." *IEEE/CAA Journal of Automatic aSinica* 8.3 (2021): 512-533.

[7] Ke, Weijie, et al. "Intersected EMG heatmaps and deep learning-based gesture recognition." *Proceedings of the 2020 12th International Conference on Machine Learning and Computing.* 2020.

[8] X. Chen, X. Zhang, Z. Zhao, J. Yang, V. Lantz and K. Wang, "Multiple Hand Gesture Recognition Based on Surface EMG Signal," *2007 1st International Conference on Bioinformatics and Biomedical Engineering, 2007*, pp. 506-509.

[9] X. Chen, Y. Li, R. Hu, X. Zhang and X. Chen, "Hand Gesture Recognition based on Surface Electromyography using Convolutional Neural Network with Transfer Learning Method," *in IEEE Journal of Biomedical and Health Informatics*, vol. 25, no. 4, pp. 1292-1304, April 2021.

[10] Sadikoglu, Fahreddin, CemalKavalcioglu, and Berk Dagman. "Electromyogram (EMG) signal detection, classification of EMG signals and diagnosis of neuropathy muscle disease." *Procedia computer science* 120 (2017): 422-429.

[11] Li, Wei, Ping Shi, and Hongliu Yu. "Gesture Recognition Using Surface Electromyography and Deep Learning for Prostheses Hand: State-of-the-Art, Challenges, and Future." *Frontiers in Neuroscience* 15 (2021).

[12] Rahimian, Elahe, et al. "Surface EMG-based hand gesture recognition via hybrid and dilated deep neural network architectures for neurorobotic prostheses." *Journal of Medical Robotics Research* 5.01n02 (2020): 2041001.

[13] C. Wu, Y. Yan, Q. Cao, F. Fei, D. Yang and A. Song, "A Low Cost Surface EMG Sensor Network for Hand Motion Recognition," *2018 IEEE 1st International Conference on Micro/Nano Sensors for AI, Healthcare, and Robotics (NSENS),* 2018, pp. 35-39.

[14] McGill, K.C. Surface electromyogram signal modelling. *Med. Biol. Eng. Comput.* **42,** 446–454 (2004).

[15] E. Rahimian, S. Zabihi, F. Atashzar, A. Asif, A. Mohammadi, "XceptionTime: Independent Time-Window XceptionTime Architecture for Hand Gesture Classification," *International Conference on Acoustics, Speech, and Signal Processing (ICASSP)*, pp. 1304-1308, 2020.

[16] P. Tsinganos, B. Cornelis, J. Cornelis, B, Jansen, and A. Skodras, "Improved Gesture Recognition Based on sEMG Signals and TCN," *International Conference on Acoustics, Speech, and Signal Processing (ICASSP),* pp. 1169-1173, 2019.

[17] E. Rahimian, S. Zabihi, S. F. Atashzar, A. Asif, and A. Mohammadi, "Surface EMG-Based Hand Gesture Recognition via Hybrid and Dilated Deep Neural Network Architectures for Neurorobotic Prostheses, *" Journal of Medical Robotics Research*, pp. 1-12, 2020.

[18] C. Castellini, et al., "Proceedings of the First Workshop on Peripheral Machine Interfaces: Going Beyond Traditional Surface Electromyography," *Frontiers in neurorobotics,* 8, p. 22, 2014.

[19] D. Farina, et al., "The Extraction of Neural Information from the Surface EMG for the Control of Upper-limb Prostheses: Emerging Avenues and Challenges," *IEEE Trans. Neural Syst. Rehabil. Eng.,* vol 22, no.4, pp. 797-809, 2014.

[20] R.Z. German, A.W. Crompton, and A.J. Thexton, "Variation in EMG Activity: A Hierarchical Approach," *American Zoologist,* vol. 48, no.2, pp. 283-293, 2008.

[21] Wang, J., Chen, Y., Hao, S., Peng, X., and Hu, L. (2019). Deep learning for sensor-based activity recognition: a survey. *Pattern Recognit. Lett.* 119, 3–11. doi: 10.1016/j.patrec.2018.02.010

[22] Côté-Allard, U., Fall, C. L., Drouin, A., Campeau-Lecours, A., Gosselin, C., Glette, K., et al. (2019). Deep learning for electromyographic hand gesture signal classification using transfer learning. *IEEE Trans. Neural Syst. Rehabil. Eng*. 27, 760–771.

[23] Y. Du, W. G. Jin, W. T. Wei, Y. Hu, and W. D. Geng, "Surface EMGbased inter-session gesture recognition enhanced by deep domain adaptation," *Sensors,* vol.17, no.3, pp.458, Mar. 2017.

[24] C. Amma, T. Krings, J. Böer, and T. Schultz, "Advancing musclecomputer interfaces with high-density electromyography," in Proc. *33rd Annu. ACM Conf. Human Factors in Computing Systems,* Seoul, Korea (South), 2015, pp. 929−938

[25] Atzori M, Gijsberts A, Castellini C, Caputo B, Hager AG, Elsig S, Giatsidis G, Bassetto F, Müller H, "Electromyography data for non-invasive naturally-controlled robotic hand prostheses", *Sci Data*. 2014; 1():140053.

[26] Amma C., Krings T., Boer J., Schultz T. (2015). Advancing muscle-computer interfaces with high-density electromyography, *in Proceedings of the 33rd Annual ACM Conference on Human Factors in Computing Systems* (Seoul: Association for Computing Machinery;), 929–938. 10.1145/2702123.2702501

[27] Geng W, Du Y, Jin W, Wei W, Hu Y, Li J, "Gesture recognition by instantaneous surface EMG images." *Sci Rep*. 2016 Nov 15; 6():36571.

[28] Ma Y., Chen B., Ren P., Zheng N., Indiveri G., Donati E. (2020). EMG-based gestures classification using a mixed-signal neuromorphic processing system. *IEEE J. Emerg. Top. Circuits Syst*. 10, 578–587. 10.1109/JETCAS.2020.3037951

[29] A. Ameri, M. A. Akhaee, E. Scheme, and K. Englehart, "Regression convolutional neural network for improved simultaneous EMG control," *J. Neural Eng*., vol.16, no.3, pp.036015, Jun. 2019.

[30] A. Ameri, M. A. Akhaee, E. Scheme, and K. Englehart, "A deep transfer learning approach to reducing the effect of electrode shift in EMG pattern recognition-based control," *IEEE Trans. Neural Syst. Rehabil. Eng.,* vol.28, no.2, pp.370–379, Feb. 2020.

[31] S. Tam, M. Boukadoum, A. Campeau-Lecours, and B. Gosselin, "A fully embedded adaptive real-time hand gesture classifier leveraging HD-sEMG and deep learning," *IEEE Trans. Biomed. Circuits Syst.,* vol.14, no.2, pp.232–243, Apr. 2020.

[32] Y. Chen, S. Yu, K. Ma, S. Y. Huang, G. F. Li, S. Q. Cai, and L. H. Xie, "A continuous estimation model of upper limb joint angles by using surface electromyography and deep learning method," *IEEE Acc*., vol.7, pp.174940–174950, Dec. 2019.

[33] W. T. Sun, H. X. Liu, R. Y. Tang, Y. R. Lang, J. P. He, and Q. Huang, "sEMG-based hand-gesture classification using a generative flow model," *Sensors*, vol.19, no.8, pp.1952, Apr. 2019.

[34] T. Z. Bao, A. Zaidi, S. G. Xie, and Z. Q. Zhang, "Surface-EMG based wrist kinematics estimation using convolutional neural network," in Proc. *IEEE 16th Int. Conf. Wearable and Implantable Body Sensor Networks*, Chicago, USA, 2019, pp. 1−4.

[35] D. Lu, Y. Yu, and H. Liu, "Gesture Recognition Using Data Glove: An Extreme Learning Machine Method," pp. 1349–1354, 2016.

[36] V. Gajjar, “Hand Gesture Real Time Paint Tool – Box : Machine Learning Approach,” 2017 IEEE Int. Conf. Power, *Control. Signals Instrum. Eng.*, pp. 856–860, 2017.

[37] Yingwei Zhang, Yiqiang Chen, Hanchao Yu, Xiaodong Yang, Wang Lu, and Hong Liu. 2018. Wearing-independent hand gesture recognition method based on EMG armband. *Personal Ubiquitous Comput*. 22, 3 (June 2018), 511–524.

[38] D. K. Ghosh and S. Ari, "Static Hand Gesture Recognition Using Mixture of Features and SVM Classifier," *2015 Fifth International Conference on Communication Systems and Network Technologies*, 2015, pp. 1094-1099, doi: 10.1109/CSNT.2015.18.

[39] Birur, Sushmalekha Shankar. Hand gesture recognition using unsupervised learning. *Diss.* San Francisco State University, 2018.

[40] W. Chen and Z. Zhang, "Hand Gesture Recognition using sEMG Signals Based on Support Vector Machine," *2019 IEEE 8th Joint International Information Technology and Artificial Intelligence Conference (ITAIC),* 2019, pp. 230-234,

[41] Aleksander Palkowski, Grzegorz Redlarski, "Basic Hand Gestures Classification Based on Surface Electromyography", *Computational and Mathematical Methods in Medicine*, vol. 2016, ArticleID 6481282, 7 pages, 2016.

[42] Mrs. Vijayalakshmi G V1 , Ajay J2 , Pavithra S3 , Pooja Eronisha A4 , Vanijayam K, “An Emg Based Hand Gesture Recognition System Using Svm”, issn (print): 2393-8374, (online): 2394-0697, volume-6, issue-6, 2019.

[43] A. I. Maqueda, C. R. Del-Blanco, F. Jaureguizar, and N. García, “Human-computer interaction based on visual hand-gesture recognition using volumetric spatiograms of local binary patterns,” *Comput. Vis. Image Underst.*, vol. 141, pp. 126–137, 2015, doi: 10.1016/j.cviu.2015.07.009.

[44] S. Sapienza et al., “On-Line Event-Driven Hand Gesture Recognition Based on Surface Electromyographic Signals,” Proc. - *IEEE Int. Symp. Circuits Syst.*, vol. 2018-May, 2018, doi: 10.1109/ISCAS.2018.8351065.

[45] M. Tavakoli, C. Benussi, P. Alhais Lopes, L. B. Osorio, and A. T. de Almeida, “Robust hand gesture recognition with a double channel surface EMG wearable armband and SVM classifier,” *Biomed. Signal Process. Control,* vol. 46, pp. 121–130, 2018, doi: 10.1016/j.bspc.2018.07.010.

[46] G. Poon, K. C. Kwan, and W. M. Pang, “Occlusion-robust bimanual gesture recognition by fusing multi-views,” *Multimed. Tools Appl.*, vol. 78, no. 16, pp. 23469–23488, 2019, doi: 10.1007/s11042-019-7660-y.

[47] P. Parvathy, K. Subramaniam, G. K. D. P. Venkatesan, P. Karthikaikumar, J. Varghese, and T. Jayasankar, “Development of hand gesture recognition system using machine learning,” *J. Ambient Intell. Humaniz. Comput.*, 2020.

[48] Gupta, Sauvik Das, et al. "Hand Gesture recognition and classification by discriminant and principal component analysis using machine learning techniques." Int J Adv Res Art Intel 1.9 (2012): 46-51.

[49] Zhang Z, Yang K, Qian J, Zhang L. Real-Time Surface EMG Pattern Recognition for Hand Gestures Based on an Artificial Neural Network. Sensors (Basel). 2019;19(14):3170. Published 2019 Jul 18. doi:10.3390/s19143170

[50] Kaginalkar A., Agrawal A. (2015) Towards EMG Based Gesture Recognition for Indian Sign Language Interpretation Using Artificial Neural Networks. In: *Stephanidis C. (eds) HCI International 2015 - Posters' Extended Abstracts. HCI 2015. Communications in Computer and Information Science,* vol 528. Springer, Cham.

[51] Yu, Hailong, et al. "A novel hand gesture recognition method based on 2-channel sEMG." *Technology and Health Care* 26.S1 (2018): 205-214.

[52] Su, Ziyi, et al. "Hand Gesture Recognition Based on sEMG Signal and Convolutional Neural Network." *International Journal of Pattern Recognition and Artificial Intelligence* 35.11 (2021): 2151012.

[53] Côté-Allard, Ulysse, et al. "Transfer learning for sEMG hand gestures recognition using convolutional neural networks." *2017 IEEE International Conference on Systems, Man, and Cybernetics (SMC)*. IEEE, 2017.

[54] Hu, Yu, et al. "A novel attention-based hybrid CNN-RNN architecture for sEMG-based gesture recognition." *PloS one* 13.10 (2018): e0206049.

[55] Yang, Yikang, et al. "A Time-Domain Hand Gesture Recognition System Based on Two-Channel sEMG Signals." *Proceedings of the 12th International Convention on Rehabilitation Engineering and Assistive Technology*. 2018.

[56] Hitoshi Tamura a ,Kazuki Itou and Yasushi Kambayashi, "A User Independent Method for Identifying Hand Gestures with sEMG*", Department of Robotics,* Nippon Institute of Technology, Saitama, Japan.

[57] Dianchun Bai, Tie Liu, Xinghua Han, Hongyu Yi, "Application Research on Optimization Algorithm of sEMG Gesture Recognition Based on Light CNN+LSTM Model", *Cyborg and Bionic Systems*, vol. 2021, Article ID 9794610, 12 pages, 2021.

[58] Park K., Lee S. (2016). Movement intention decoding based on deep learning for multiuser myoelectric interfaces, *in 2016 4th International Winter Conference on Brain-Computer Interface (BCI) (Gangwon:)*, 1–2. 10.1109/IWW-BCI.2016.7457459

[59] Chen L., Fu J., Wu Y., Li H., Zheng B. (2020). Hand gesture recognition using compact CNN via surface electromyography signals. *Sensors* 20:672. 10.3390/s20030672

[60] Asif AR, Waris A, Gilani SO, Jamil M, Ashraf H, Shafique M, Niazi IK, "Performance Evaluation of Convolutional Neural Network for Hand Gesture Recognition Using EMG.", *Sensors* (Basel). 2020 Mar 15; 20(6):.

[61] Triwiyanto T, Pawana IPA, Purnomo MH, "An Improved Performance of Deep Learning Based on Convolution Neural Network to Classify the Hand Motion by Evaluating Hyper Parameter", *IEEE Trans Neural Syst Rehabil Eng.* 2020 Jul; 28(7):1678-1688.

[62] Tong R., Zhang Y., Chen H., Liu H. (2019). Learn the temporal-spatial feature of sEMG via dual-flow network. *Int. J. Humanoid Robot.* 16:1941004. 10.1142/S0219843619410044

[63] Neacsu A. A., Cioroiu G., Radoi A., Burileanu C. (2019). Automatic EMG-based hand gesture recognition system using time-domain descriptors and fully-connected neural networks, *in 42nd International Conference on Telecommunications and Signal Processing (TSP) (Budapest: IEEE),* 232–235. 10.1109/TSP.2019.8768831

[64] Waibel A., Hanazawa T., Hinton G., Shikano K., Lang K. J. (1989). Phoneme recognition using time-delay neural networks. *IEEE Trans. Acoust. Speech Signal Process*. 37, 328–339. 10.1109/29.21701

[65] Day C. R., Chadwick E. K., Blana D. (2020). A comparative evaluation of time-delay, deep learning and echo state neural networks when used as simulated transhumeral prosthesis controllers, in *2020 International Joint Conference on Neural Networks* (IJCNN), 1–7. 10.1109/IJCNN48605.2020.9206772

Journal of Physics: Conference Series **2327** (2022) 012076 doi:10.1088/1742-6596/2327/1/012076

Past, Present and Future of Automated Mammographic Density Measurement for Breast Cancer Risk Prediction

Shivaji D. Pawar

Research Scholar, Department of Computer Science and Engineering, Lovely Professional University, Jalandhar (PB), India and Assistant Professor , SIES Graduate School of technology

https://orcid.org/0000-0002-6817-3026

Pratibha T. Joshi

Research Scholar, School of Electronics and Electrical Engineering Lovely Professional University, Jalandhar (PB), India and Assistant Professor ,SIES Graduate School of technology

https://orcid.org/0000-0001-5103-2481

Vishwayogita A. Savalkar

Research Scholar, Department of Computer Science and Engineering, Lovely Professional University, Jalandhar (PB), India and Assistant Professor, SIES Graduate School of technology

https://orcid.org/0000-0001-9043-3675

Kamal Kr. Sharma

Professor, School of Electronics and Electrical Engineering, Lovely Professional University, Jalandhar (PB), India.

https://orcid.org/0000-0002-6294-0829

Suhas G. Sapate

Professor, Department of Computer Science and Engineering, Annasaheb Dange College of Engineering and Technology, Ashta, Sangli (MH), India
https://orchid.org/0000-0002-4755-695X

Corresponding Author: pratibha.t.joshi@gmail.com

Abstract: Mammography is one of the essential screening technologies which is helpful to save the lives of women against breast cancer. Prediction of breast cancer from mammograms is not reached on its optimal level; hence there is a constant enhancement in clinical applications for mammographic breast density measurement. Optimal results in breast density measurement can be helpful to provide better care for women who have dense breasts. The sensitivity of digital mammograms reduces significantly in case dense breast, which may lead further to hide the cancerous lesions and may be converted into high stage breast cancer. Many research innovations and clinical applications are developed to support radiologists for the second opinion and predict breast cancer risk in advance. But still, there is an unsolved research question: which one is "dense breast" and which screening modularity is suitable for the dense breast to avoid the risk of breast cancer. Hence, currently, radiologists measure mammographic breast density with the help of BI-RADS classification, which is subjective.

The basic inclination of this paper is to provide an overview of different clinical applications used for breast density measurement in the past and present with their potential technologies, advantages, disadvantages and current status of all various clinical applications in the market and provides thoughts on the future of automated mammographic breast density measurement.

Keywords: *Breast cancer, mammographic breast density, Area based density measurement, volumetric breast density, and clinical applications.*

1. Introduction

In 1950, Robert Egan developed the concept of mammography by fusion technique of low kVp with high mA and single emulsion films for screening. After nine years, he published his mammography method results and became the first physician to detect breast cancer by using mammography [1]. After this success story of Robert Egan, mammography is the most commonly used screening technology for breast cancer worldwide. Until 2000, mammography was performed with screen-film cassettes. Screen-film mammography consisted of certain demerits such as low spatial resolution significantly higher in cost; hence in 2000, mammography transitioned from screen-film to digital detectors. This transition was known as Full Field Digital Mammography (FFDM) or Digital mammography, and by 2010, it was the most popular screening tool for breast cancer. Figure-1 illustrates the mammogram processing unit and comparison between digital and Film mammograms.

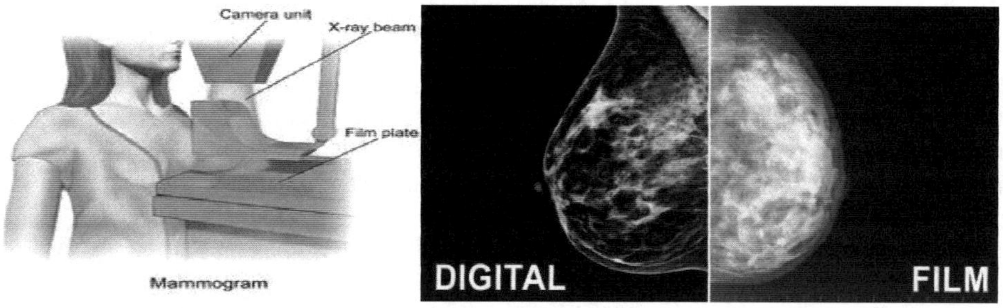

Figure.1- Mammogram processing unit and view of digital and Film mammogram.

Digital mammography uses computers and digital receptors in place of x-ray film. This enables electrical signals for reading and manipulating the image, which is helpful for radiologists' insights into the results more precisely than film mammography. Digital mammography is also fast in diagnosis, inexpensive, acts as the gold standard for detecting masses and helps calculate mammographic breast density, which is strongly associated with the risk of breast cancer [2-3].

Mammographic breast density term is used to measure and calibrate breast tissue appearing on the digital mammograms. Mammographic breast density varies from woman to woman and depends not on unique parameters but mostly on genetic and environmental factors. The breast is composed of different components such as fat, glandular tissue and connective tissues [4]. The fatty part appears dark on the mammogram. The remaining parts act as white, which is further calibrated as a mammographic breast density. Comparison between dense and fatty breasts is illustrated in Figure-2. The most commonly used standard for mammographic breast density measurement is "American

College of Radiology's Breast Imaging Reporting and Data System (BI-RADS) [5]", which is used by radiologists to classify mammograms as A, B, C or D, which is shown in figure-3.

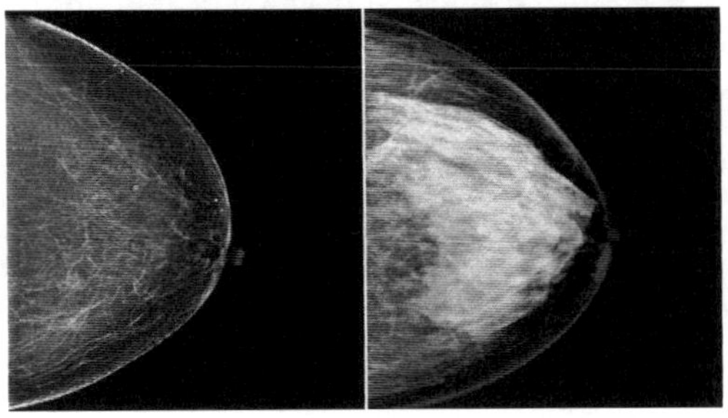

Figure.2- Fatty and dense breast

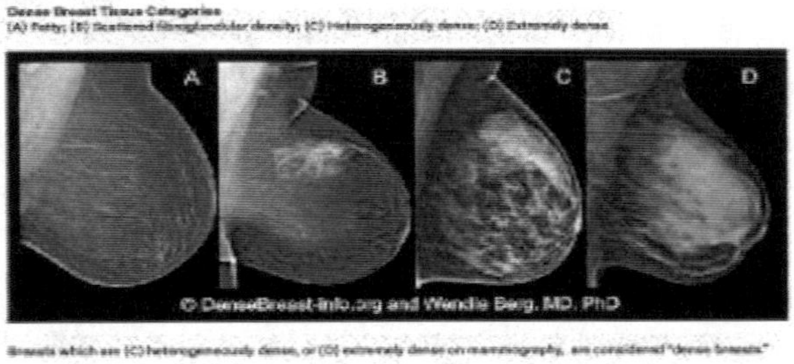

Figure.3- BIRADs Classification

Type A is the almost fatty breast which is lower density, and around 10% of women who undergoes a mammogram come under this class. Type B is known as scattered density; roughly 40% of women undergoing mammograms are under this category. Heterogeneously dense come under categories of type c, and almost 40% of women will have this category and finally Extremely dense, which is type D and, in this class, also there are 10% of women who undergo the mammograms where the sensitivity of mammogram reduces and hence this class is under risk of misprediction. Therefore, breast density measurement from a mammogram has been challenging for decades.

There are specific reasons why breast density measurement does not provide better results. Mammography creates 2D image anatomy. It does not have an absolute or external mammographic breast density measurement scale. Different manufacturers use different hardware; hence, there is variation in mammograms' quality, which is one significant cause for image processing algorithms not producing a consistent result of mammographic breast density. Breast compression during mammography is one reason for errors in breast density calculations—figure-4 shows how compression can change the breast density in a digital mammogram [6].

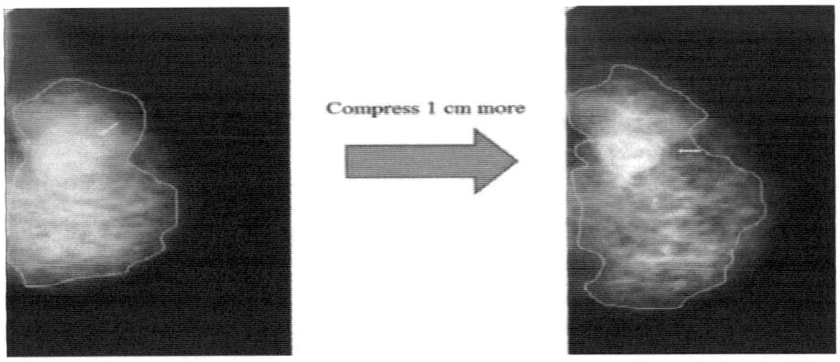

Figure.4- Breast compression changes the breast density

These demerits of digital mammography act as research challenges for mammographic breast density calculations.

Even though there is a lot of innovation in screening mammography, from film-based mammography to full-field digital mammography and breast density measurement clinical application from 2D to 3D, a measurement of breast density still does not reach its optimal solutions. The basic inclination behind this chapter is to study different 2D and 3D clinical applications deployed to measure breast density from the last four decades with their merits and demerits. This chapter is undoubtedly helpful for research scholars to get some insight into clinical application innovation in breast density measurement. And find out research direction for the future.

This paper hereafter organized as section 2, labelled as expectations of radiologists, section 3 gives an overview of clinical applications developed in the past, section 4 springs clinical application in the present, section 5 highlights future scope of research for mammographic breast density. Section 6 portages our findings during this study, and section 7 concludes the paper.

2. Expectations of Radiologists

As per radiologist's opinion, basic role of computer aided detection system (CAD) for screenings of mammographic breast density are

1. CAD system should be act as support system for radiologists to avoid misprediction of breast cancer.
2. Designed CAD system should consist all the variations of imaging parameters such as pixels intensity and contrast etc.so that it acts as reliable source of breast density measurement.
3. Sensitivity, accuracy, specificity and Reproducibility of CAD system should be on optimal level.
4. CAD system helpful to provide information to the women of their breast density and the need for additional tests for those with dense breasts.
5. CAD system should classify breast density as per BI-RADs classification.
6. CAD system should learn from its experience such that accuracy will improve over the period.

3. Systems in the past

Mammographic breast density has long research history from 1976 to still date. This research area has undergone different shades of thoughts, like pattern, area and volumetric breast density. This section will focus on the clinical application between the years 1990 to 2009.These clinical applications are divided into two subsections known as area-based and volumetric breast density

3.1 Area-based density Measurement

Wolfe has first noted in his research the association of high-density patterns and the risk of breast cancer. Wolfe classified breast density into four density patterns, N1, P1, P2 and DY and concluded in his research that density class DY was 37% more probability of developing breast cancer than N1 classes [7]. Tabar provided pattern-based breast density into five categories from I to V. Wolfe's results. Tabar was not widely accepted due to difficulty in assessment and more towards subjective nature. In 1980, film technology became more standardized and valuable to better breast density attenuation. Boyd visualized area-based mammographic breast density measurement for the first time with a good co-relation with breast cancer risk prediction [8].

In the initial phase of computerized clinical applications, all applications were based on the concept of semiautomatic area-based density measurement. Boyd, Byng and Yaffe developed the first semiautomatic automated clinical application on area-based density measurement in 1994, known as "cumulus".

This application remains the gold standard for 20 years for breast density measurement. This method uses interactive thresholding, brightness variations in average skewness, and the fractal dimensions of grayscale images. The fundamental limitations of this method were reader has to identify the threshold and boundaries of breast tissue to calculate the breast density [9-15]. The user interface of Cumulus is shown in figure-5, where edges are shown with red colour and fibro glandular tissue with green colour.

In 1998, an area-based computerized Clinical application known as "Madena", which Astrahan coded for measuring density patterns on a mammogram, was further suitable for tracking the changes in density patterns with medical treatments or over the period. This application works on histogram-based thresholding. Results obtained by this method are moderately correlated (r > 0.85) as compared with the results from the expert outlining approach [16].

In 2000, "ABDM" was another clinical application developed by Heine et al. [17]. This application was based on a statistical model that uses the concept of Chi-Square probability to separate the fatty breast from fibro glandular tissues on a digital mammogram with the support of digital filtering. This model uses multiresolution wavelet expansion and 1/f power law analysis with parametric modelling of a mammogram. Results obtained by the ABSM is shown in figure-5 in which the narrow upper curve is related to the fat area and the lower spread out curve corresponds to the denser area.

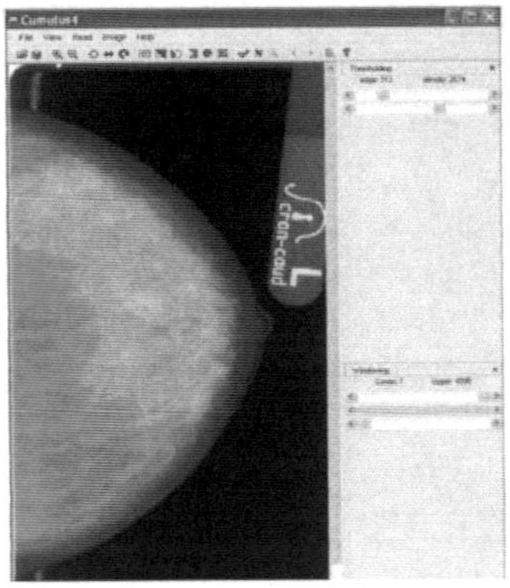

Figure.-5 Interface of Cumulus

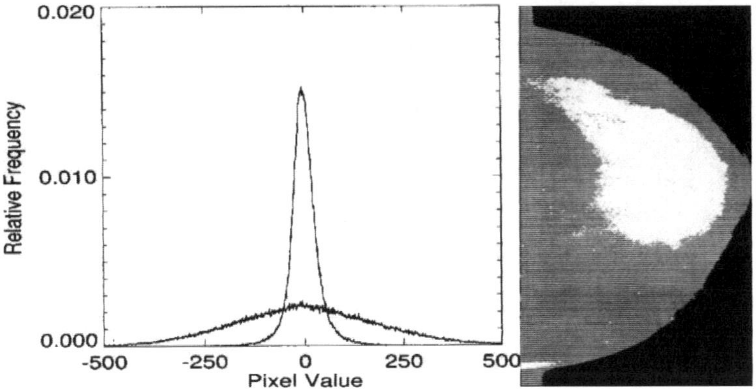

Figure.-6 The histograms associated with the ABDM.

In 2009, an Automatic application based on the concept of maximum Entropy and spatial information known as "MedDensity" was developed by Tagliafico et al. [18]. This clinical application was available in semiautomatic and automatic mode. In semiautomatic mode, the quantitative assessment of breast density was performed with the support of the thresholding and edge detection methods. In an automatic way, thresholds are adjusted automatically, shown on the left side of the computer interface in figure-7. There is a mask of the breast tissue and the value of mammographic density on the right side. The primary limitation of this method was that the radiologists manually identified the edges of the breast. Then the percentage of the dense breast is read into the black circle.

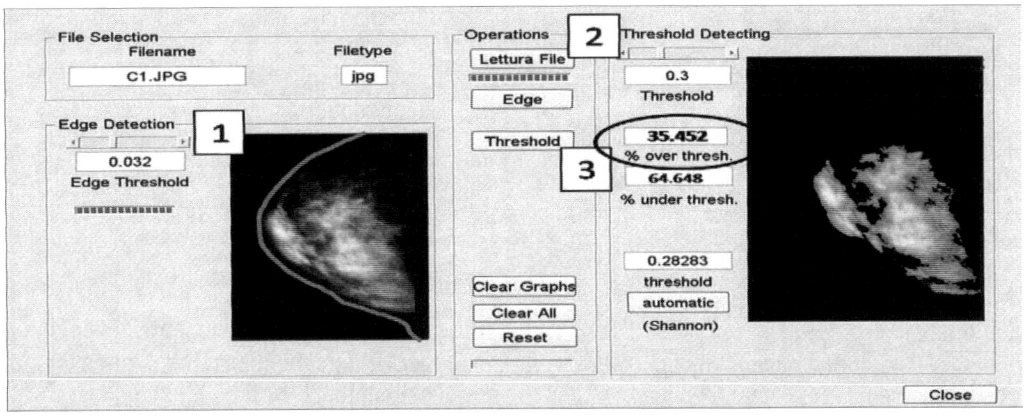

Figure. -7 Computer interface of MedDensity

3.2 Volumetric Breast Density

In 1992, another research paradigm was developed to calculate breast density as a three-dimensional measurement instead of two-dimensional breast density known as volumetric breast density. This concept was first presented in Higham's PhD thesis, and after 16 years, it became the backbone for some of the commercial software that was further approved by the FDA [19].

In the year 2003, Cumulus clinical application was proposed to measure volumetric breast density. In this method, the tissue-equivalent plastic object is calibrated with an imaging system with specific correction in variations in image parameters by keeping images of aluminium step-wedge placed during imaging and with the help of this information, volumetric breast density was calculated [20].

In the year 2008, Quantra another commercial clinical application that provides volumetric breast density analysis based on the physical model of the breast. In this application, the thickness of fibro glandular breast tissue associated with each pixel is aggregated to calculate the entire breast volume. And with the help of difference in attenuation of fibro glandular tissue pixels and fatty pixels, the volume of fibro glandular tissue was calculated to get the value of volumetric breast density [21]. Quantified analyses of all studied clinical applications from 1990 to 2009 are highlighted in Table-1.

Table 1. Quantified analysis of all clinical application in past (1990-2009).

Basic Approach	Implementation year	Name of clinical application	Software Algorithm used	Merits/demerit if any
Area-based Density measurement.	1994	Cumulus	Interactive thresholding	Semiautomatic
	1998	Madena	Histogram-based thresholding	Results obtained by this method are moderately corelate (r > 0.85)
	2000	ADBM	Multiresolution wavelet expansion and 1/f power law analysis with parametric modelling of mammogram	Frist statistical model used to calculate breast density
	2009	MedDensity	concept of maximum Entropy and spatial information	Available in semiautomatic and automatic mode
Volumetric Breast density	2003	Cumulus	Tissue equivalent plastic object is calibrated with imaging system	Frist volumetric application of breast density
	2008	Quantra	Works on physical model of the breast	Thickness of fibro glandular breast tissue was calculated with physical model.

4. Systems in the present

In this section, we have provided an overview of different clinical applications for breast density measurement from 2010 to 2019. These applications are divided into two subsections one is area while another is a volumetric breast density measurement.

4.1 Area-based density measurement

After 2010, advancement in computer vision algorithms provide good platform to process digital mammograms to provide more insights about the images. In 2012, two clinical applications, one is "M-Vu Breast Density" which work on Computer vision algorithms to process digital mammograms. In this algorithm structural and texture difference in fatty and fibro glandular tissue is used to classify area-based mammographic breast density according to the BI-RADS standard [22] and another one is "ImageJ" which is fully automated area-based breast density measurement tool which similar to the cumulus by using penalised regression with some additional features of mammograms to predict risk of breast cancer associated with breast density [23].

Optimal thresholding having ability to select optimal threshold which will classify object into two class. This concept is used in clinical applications "Auto density' in year 2013.In this application, an optimal threshold method is used to segment fibro glandular tissue from whole breast [24]. After 2013, there is no clinical application in clinical practise as research paradigm get shifted towards volumetric breast density.

4.2 Volumetric density measurement

In year 2010, Volpara is another FDA approved and validated by many Clinical trials and most commonly used clinical application which is based on relative physics model. This model used "hint" [25] and standard mammogram form "SMF" [26]. "hint" method gives normalise mammogram of interesting tissue by measuring anatomical information from the mammogram where each pixel represents thickness of fibro glandular tissue during compression. This information helps get volume of the whole breast. "SMF" provides estimation of fibro glandular tissue at each location of mammogram which is helpful to estimate the volume of fibro glandular tissue. This application also consists of set of calibrated datasets such as mAs and kVp which will provide realistic model of breast [27].

In 2014, 'Philips' Spectral Breast Density Measurement FDA approved clinical tool [28-29] which use concept of difference between energy spectrum of fatty and fibro glandular tissue. Thickness and density of each pixel is measured independently which is further used for estimation of volumetric breast density. This method does not use information of compressed breast like Volpara but results of this method also co-relate with BI-RADs classification. This is first clinical application in which non-contrast spectral imaging was used to measure photons counting technology which was divided into low and high energy categories.

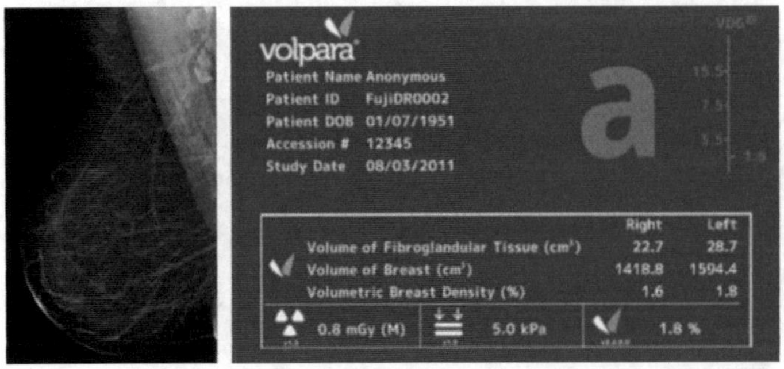

Figure. -8 Interface of Volpara

In 2018, Siemens Healthineers had cleared FDA approval for its clinical application named "MAMMOMAT". This application performs tomosynthesis scans at different angles with a range of 500 to provide a detailed 3D view, which is helpful to identify overlapping breast tissue. This application also has an additional integrated imaging tool known as Inspect, which will speed up the biopsies process and make it more comfortable by one click [30-31].

From the year 2010, mammographic breast density measurement technique is diverted more towards volumetric than the areas-based density measurement. Quantified analysis of systems in present illustrated in Table-2.

Table 2. Quantified analysis of systems in present (2010 to 2019)

Basic Approach	Implementation year	Name of clinical application	Software Algorithm used	Merits/demerit if any
Area-Based densitymeasurement	2012	M-Vu Breast Density	Structural and texture difference in fatty and fibro glandular tissue	Breast density according as per BI-RADS standard
	2012	ImageJ	Penalised regression	Use some additional features of mammograms to predict risk of breast cancer
	2013	Auto density	Optimal thresholding	Provide better segmentation results
Volumetric Breast density	2010	Volpara	Relative physics model	Use calibrated parameters such as such as mAs and kVp for measurement of breast thickness.
	2014	Philips	Difference between energy spectrum of fatty and fibro glandular tissue	First clinical application in which non-contrast spectral imaging was used to measure

2018	MAMMOMAT	Performs tomosynthesis scans at different angles with range of 50^0	photons counting technology. This technology is helpful to identify overlapping breast tissue.

5. A scope in the Future

In the USA, breast density notification to the individual patient after mammography screening is mandatory in 30 states [32]. In the future, this becomes a social need of all the countries to know about breast density and its correlation with breast cancer risk required supplementary screenings. Hence, future research needs to reduce inter reader variations and provide the most reliable and accurate clinical application to predict breast cancer in advance.

The research paradigm of breast density changes from area to volume and is now gaining more attention towards the machine and deep learning from the last few years. Artificial intelligence can learn complex features of objects [33-34] and is helpful for both volumetric and area-based density measurement. Hence, much deep learning and machine learning-based innovations are under clinical trials. These models use two or four class classification by using convolutional neural networks models. These models consist of five to six convolutional layers, three connected layers and SoftMax function for the category. Specific modification exists in terms of RELU hyperbolic tangent function [35]. Simultaneously, another path for research innovations is under development in terms of feature extraction, segmentation, and dimension reduction for machine learning classifications [36]. These modifications surely help make breast density measurement more objective than subjective and enable the use of such models in the future as a breast cancer risk prediction screening tool.

Past present and future research direction of clinical applications for breast density measurement are illustrated in Figure-9.

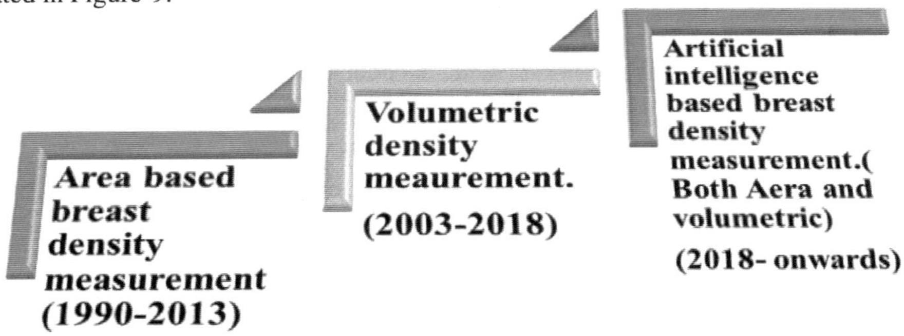

Figure. -9 Research directions for past, present and future of clinical applications for breast density measurement.

6. Our Observations & Findings

Even though there are many automated clinical applications in the market for mammographic breast density, it has advantages like faster, cheaper, and more objective but not providing the optimal results [37]. Following are some observations recorded during this study, which may be helpful to decide future research paths in breast density measurement.

1. We come across three essential breast density measurement approaches area-based, volumetric-based, and now deep learning. [40].

2. All the measurement of mammographic breast density's primary focus is to analyses breast's attenuation and absorption characteristics compared to physical breast tissue [38].
3. Different manufacturers perform digital mammograms at different clinical environments with the help of different proprietary algorithms and different imaging parameters that will change the contrast of each mammogram, which may be one of the causes of error in breast density measurement.
4. Assessment of three-dimensional breast density from two-dimension mammograms is a challenging task.
5. Compression of the breast during a mammography is one of the fundamental causes that hamper area-based density results.
6. To increase measurement reliability, volumetric breast density is another good option used by different manufacturers like Volpara, Quantra etc. Still, the method used to calculate variable breast thickness is other; hence validation of such clinical applications on the same platform are different.
7. There is no external or absolute ground truth for validations and standard datasets of different countries for confirmations [39].
8. In the future, Artificial intelligence can learn complex mammogram features. Hence, the future hope of research will provide a more reliable breast density measurement tool for predicting breast cancer.

7. Conclusion

Mammography is an essential tool to save women's lives against breast cancer. This tool is undoubtedly beneficial in educating women about their breast density, associated risk factors, and additional screening tests. Still, there is a constant need to improve the mammographic breast density measurement tool, which can be helpful to provide better care for a woman with dense breasts. In this paper, we have provided an overview of different clinical applications with potential technologies with some merits and demerits. This article is undoubtedly helpful to decide the future path of breast density measurement. This research area still requires certain modifications in screening technology, computer vision and artificial intelligence algorithms. In future, due to fast development in artificial intelligence models, there is new hope to have a reliable, more accurate clinical application for breast density measurement.

References

[1] Medich DC, Martel C. Medical Health Physics. Health Physics Society 2006 Summer School. Medical Physics Publishing. ISBN 1930524315 pp.25.
[2] Digital Mammography - Mammography - Imagines - The Women's Health & Wellness Resource Network". www.imaginis.com.
[3] (ACR), Radiological Society of North America (RSNA) and American College of Radiology. "Mammography (Mammogram)". radiologyinfo.org.
[4] Highnam, R., Brady, S. M., Yaffe, M. J., Karssemeijer, N., & Harvey, J. (2010). Robust Breast Composition Measurement - VolparaTM. Lecture Notes in Computer Science, 342–349. doi:10.1007/978-3-642-13666-5_46.
[5] https://www.bcna.org.au/breast-health-awareness/mammographic-density-and-screening.
[6] Ng, K.-H., & Lau, S. (2015). Vision 20/20: Mammographic breast density and its clinical applications. Medical Physics, 42(12), 7059–7077. doi:10.1118/1.4935141
[7] J. Wolfe, "Breast patterns as an index of risk for developing breast cancer," AJR, Am. J. Roentgenol. 126, 1130–1139 (1976).

[8] J. N. Wolfe, "Risk for breast cancer development determined by mammographic parenchymal pattern," Cancer 37(5), 2486–2492 (1976).

[9] I. T. Gram, E. Funkhouser, and L. Tabár, "The Tabár classification of mammographic parenchymal patterns," Eur. J. Radiol. 24(2), 131–136(1997).

[10] N. Boyd, J. Byng, R. Jong, E. Fishell, L. Little, A. Miller, G. Lockwood, D. Tritchler, and M. Yaffe, "Quantitative classification of mammographic densities and breast cancer risk: Results from the Canadian national breast screening study," J. Natl. Cancer Inst. 87, 670–675 (1995).

[11] N. F. Boyd, B. O'Sullivan, J. E. Campbell, E. Fishell, I. Simor, G. Cooke, and T. Germanson, "Mammographic signs as risk factors for breast cancer," Br. J. Cancer 45(2), 185–193 (1982).

[12] H. Lee-Han, G. Cooke, and N. F. Boyd, "Quantitative evaluation of mammographic densities: A comparison of methods of assessment," Eur. J. Cancer Prev. 4(4), 285–292 (1995).

[13] W. Byng, N. F. Boyd, E. Fishell, R. A. Jong, and M. J. Yaffe, "The quantitative analysis of mammographic densities," Phys. Med. Biol. 39(10), 1629–1638 (1994).

[14] W. Byng, M. J. Yaffe, R. A. Jong, R. S. Shumak, G. A. Lockwood, D. L. Tritchler, and N. F. Boyd, "Analysis of mammographic density and breast cancer risk from digitized mammograms," Radiographics 18(6), 1587–1598 (1998).

[15] M. J. Yaffe, N. F. Boyd, J. W. Byng, R. A. Jong, E. Fishell, G. A. Lockwood, L. E. Little, and D. L. Tritchler, "Breast cancer risk and measured mammographic density," Eur. J. Cancer Prev. 7(Suppl. 1), S47–S55 (1998).

[16] G. Ursin, M. A. Astrahan, M. Salane, Y. R. Parisky, J. G. Pearce, J.R. Daniels, M. C. Pike, and D. V. Spicer, "The detection of changes in mammographic densities," Cancer Epidemiol., Biomarkers Prev. 7(1),43–47 (1998

[17] J. J. Heine and R. P. Velthuizen, "A statistical methodology for mammographic density detection," Med. Phys. 27(12), 2644–2651 (2000).

[18] A. Tagliafico, G. Tagliafico, S. Tosto, F. Chiesa, C. Martinoli, L. E. Derchi, and M. Calabrese, "Mammographic density estimation: Comparison among BI-RADS categories, a semi-automated software and a fully automated one," Breast 18(1), 35–40 (2009).

[19] Hartman, R. Highnam, R. Warren, and V. Jackson, "Volumetric assessment of breast tissue composition from FFDM images," in Digital Mammography, edited by E. Krupinski (Springer, Berlin, Heidelberg, 2008), Vol. 5116, pp. 33–39.

[20] O. Pawluczyk, B. Augustine, M. Yaffe, D. Rico, J. Yang, G. Maudsley, and. Boyd, "A volumetric method for estimation of breast density on digitized screen-film mammograms," Med. Phys. 30, 352–364 (2003).

[21] K. Hartman, R. Highnam, R. Warren, and V. Jackson, "Volumetric assessment of breast tissue composition from FFDM images," in Digi- tal Mammography, edited by E. Krupinski (Springer, Berlin, Heidelberg, 2008), Vol. 5116, pp. 33–39.

[22] Vucomp, Inc., Vucomp—Breast density. http://www.vucomp.com/products/breast-density.

[23] J. Li, L. Szekely, L. Eriksson, B. Heddson, A. Sundbom, K. Czene, P. Hall, and K. Humphreys, "High-throughput mammographic-density measurement: A tool for risk prediction of breast cancer," Breast Cancer Res. 14(4), R114 (2012).

[24] C. Nickson, Y. Arzhaeva, Z. Aitken, T. Elgindy, M. Buckley, M. Li, D. English, and A. Kavanagh, "AutoDensity: An automated method to measure mammographic breast density that predicts breast cancer risk and screening outcomes," Breast Cancer Res. 15(5), R80 (2013).

[25] R. Highnam and M. Brady, Mammographic Image Analysis (Kluwer Academic, Dordrecht, The Netherlands, 1999).

[26] M. Brady, D. Gavaghan, A. Simpson, M. M. Parada, and R. Highnam, "diamond: A grid-enabled federated database of annotated mammograms," in Grid Computing: Making the Global Infrastructure a Reality, edited by F. Berman, G. Fox, and A. G. Hey (John Wiley &

Sons, Ltd., Chichester, 2003), pp. 923–943.

[27] K. Hartman, R. Highnam, R. Warren, and V. Jackson, "Volumetric assessment of breast tissue composition from FFDM images," in Digital Mammography, edited by E. Krupinski (Springer, Berlin, Heidelberg, 2008), Vol. 5116, pp. 33–39.

[28] J. L. Ducote and S. Molloi, "Quantification of breast density with dual energy mammography: An experimental feasibility study," Med. Phys.37(2), 793–801 (2010).

[29] H. Ding and S. Molloi, "Quantification of breast density with spectral mammography based on a scanned multi-slit photon-counting detector: A feasibility study," Phys. Med. Biol. 57(15), 4719–4738 (2012).

[30] https://www.medgadget.com/2018/03/mammomat-revelation-mammography-system-with-automatic-breast-density-measurement.

[31] https://www.siemens-healthineers.com/mammography/digital-mammography/mammomat-revelation#FEATURES_BENEFITS

[32] Breast density notification legislation, available online: http://densebreast-info.org/legislation.aspx.

[33] K.K. Sharma, S.D. Pawar, B. Bali, Proactive preventive and evidence-based artificial intelligence models: future healthcare, in International Conference on Intelligent Computing and Smart Communication 2019 (Springer, Singapore, 2020), pp. 463–472.Google Scholar

[34] Pawar, S. D., Sharma, K. K., & Sapate, S. G. (2021). Advances in Machine Learning and Deep Learning Approaches for Mammographic Breast Density Measurement for Breast Cancer Risk Prediction: An Overview. Design of Intelligent Applications Using Machine Learning and Deep Learning Techniques, 125-143.

[35] Mohamed, A. A., Luo, Y., Peng, H., Jankowitz, R. C., & Wu, S. (2017). Understanding Clinical Mammographic Breast Density Assessment: A Deep Learning Perspective. Journal of Digital Imaging, 31(4), 387–392. doi:10.1007/s10278-017-0022-2.

[36] Pawar, Shivaji and Sapate, Suhas and Sharma, Kamal, Machine Learning Approach towards Mammographic Breast Density Measurement for Breast Cancer Risk Prediction: An Overview SSRN: https://ssrn.com/abstract=3599187

[37] Sapate, S. G., Mahajan, A., Talbar, S. N., Sable, N., Desai, S., & Thakur, M. (2018). Radiomics based detection and characterization of suspicious lesions on full field digital mammograms. Computer Methods and Programs in Biomedicine, 163, 1–20. https://doi.org/10.1016/j.cmpb.2018.05.017

[38] Mark A. Sak, Peter J. ittrup Current and Future Methods for Measuring Breast Density: A Brief Comparative Review.

[39] Sapate, S., Talbar, S., Mahajan, A., Sable, N., Desai, S., & Thakur, M. (2020). Breast cancer diagnosis using abnormalities on ipsilateral views of digital mammograms. Biocybernetics and Biomedical Engineering, 40(1), 290–305. https://doi.org/10.1016/j.bbe.2019.04.008.

[40] Pawar,S. D., Sharma, K. K., Sapate, S. G., & Yadav, G. Y. (2021). Segmentation of pectoral muscle from digital mammograms with depth-first search algorithm towards breast density classification. Biocybernetics and Biomedical Engineering, 41(3), 1224-1241.

Brain Tumor Detection Using Artificial Convolutional Neural Networks

B Ramu[1], Sandeep Bansal[2], Navjot Rathour[3]

[1]Research Scholar, School of Electronics & Electrical Engineering, Lovely Professional University, Jalandhar, Punjab, India
[2]Associate Professor, School of Electronics & Electrical Engineering, Lovely Professional University, Jalandhar, Punjab, India
[3]Associate Professor, Uttaranchal Institute of Technology, Uttaranchal University, Dehradun, Uttarakhand, India
Corresponding author: sandeep.15732@lpu.co.in[2]

[ORCiD Id's- 0000-0002-0871-1582 [1], 000-003-0975-7934 [2], 0000-0002-7003-2257 [3]]

***Abstract*:** Automatic flaw detection in pictorial representation associate degreed CT brain photos is crucial in an exceptional type of diagnostic and therapeutic applications. Due to the large quantity of data in pictorial representation scans and additionally the crooked boundaries, growth segmentation and identification established tough. Associate degree automatic growth detection methodology was used during this investigation that improved accuracy, yield, and diagnostic time. The goal is to kind the tissue into three groups. There as three sorts of cancer: ancient, early, and malignant. The number of data in pictorial representation and CT scans is simply too nice for human interpretation and analysis. Tumor segmentation in resonance imaging (MRI) has emerged as a rising study subject inside the sphere of medical imaging in recent years. The flexibility to accurately notice the size and size of a growth is important among the identification of the sickness. Pre-processing of imaging footage, feature extraction, classification supported ANN and a cluster of the tumor half because the four phases of the diagnostic approach. The choices as retrieved victimization wave transformation once the image has been pre-processed (DWT). The photos as classified into ancient and pathological brain footage victimization an artificial Neural Network (ANN). With a spotlight on the special Fuzzy c-means cluster, the simplest formula for tumor detection is projected. The goal of this paper is to improve the ANFIS architecture so that it can achieve high classification accuracy and quantify tumor thickness and volume
Keywords— ANN, SFCM, Clustering, DWT

1. Introduction

The brain tumor is one of the common and the deadliest brain disease that has affected and ruined several lives in the world. According to a new case study on cancer, more than one lack people are diagnosed with brain tumors every year around the globe. There are complications in the conventional treatment of brain tumors. So, to overcome this, the advancement of technology gives accurate results in the detection of brain tumors. The chances of survival of a tumor-infected patient can be increased if the tumor has been detected accurately in its early stages. The introduction of e-health care systems into the medical field helps clinical experts to provide better care to patients. The advancement in technology increases efficiency and productivity. It made the work easier for the doctors.

A brain tumor could be an assortment of abnormal brain cells. Your brain is protected by a extremely robust. Any sort of development in such a touch location may be venturous. Brain tumors will either be malignant or non-cancerous. The term "malignant" refers to cancerous tumors, whereas "benign" refers to non-cancerous tumors. We have a tendency to area unit initial characteristic MRI brain scans as traditional or pathological exploitation artificial neural networks and extracting GLCM characteristics from the photographs. If the result's abnormal, we have a tendency to classify it as benign or malignant to ascertain cancer's stage. To classify the info as benign or malignant, we have a tendency to use special Fuzzy C-Means cluster. Tumors whether or not benign and malignant area unit each fatal. A malignant neoplasm brain tumor could be a tumor that grows within the brain and is cancerous. It differs from a benign brain tumor that is not malignant and develops additional slowly. Benign tumors don't perpetually need treatment. However, if symptoms area unit a problem, medical care could also be needed.

For benign tumors, surgery could be a well-liked therapeutic possibility. Resonance Imaging (MRI) could be a sort of imaging that involves the utilization of radio waves. It is a non-invasive imaging approach that produces three-dimensional anatomy footage with pinpoint accuracy. it is also wont to forestall illness, diagnose issues, and keep track of patients. It's supported with-it technology that stimulates and monitors changes within the movement axis of protons within the water that produces up biological tissues.

A little light emission beams is accumulated at a patient and immediately spun round the body, making signals that are dissected by the machine's PC to think of cross-sectional pictures or "cuts" of the body. These cuts are expressed as geological pictures, and that they offer extra exact subtleties than old x-beams. When a progression of the machine's standard produces sequent cuts, which would then be able to be carefully "stacked" along to make a three-dimensional picture of the patient, giving further developed distinguishing proof and situation of rudimentary constructions besides as any sores or irregularities. Counterfeit neural organizations are made inside the very technique that the human cerebrum is, with physical cell hubs interconnected in a really network-like design. A large number of neurons summon the human mind. Each substantial cell fuses a cell body that measures data via conveying information to and from the mind (information sources and yields). Counterfeit neural organizations are made inside the very strategy that the human cerebrum is, with physical cell hubs joined along in an incredibly network-like style. There are a huge number of neurons inside the human mind. Each physical cell consolidates a phone body that measures data by moving information towards (inputs) and reserved from (yields) the cerebrum.

2. Literature Survey

[1] E. Ben George, Jeba Rosline. The elemental ideas of the cuckoo search calculation are analyzed during this work, even as its application to the division of neural structure tumors utilizing engaging reverberation imaging (MRI). The human neural structure is that the most fascinating organ, and recognizing tumor-like ailments as a result of tough} of characteristic the cerebrum's segments is improbably difficult. The neoplasm could periodically arise with the same strength because of the remainder of the body. The radiologist's trip is created a lot of difficulty by the manner that the neoplasm, edema, blood cluster, and a few mind tissues all appear one thing similar. Neural structure tumors are usually recognized by a radiotherapist once an Associate in Nursing completes an audit of man footage that consumes most of the day. The key innovative endeavor is to foster a demonstrative strategy enthusiastic about the most effective streamlining system called the cuckoo search, which can allow the radiotherapist to accumulate precise outcomes.

[2] Kun-Huang Chen, Kung-Jeng Wang. Quite probably the foremost difficult components of utilizing microarray data are deciding a way to decide few valuable qualities among an excellent several qualities connected to malignancy. Various analysts utilize a scope of machine insight innovations to analyze quality articulation data. This investigation plans to foster a rare system that consolidates molecule swarm streamlining with an alternative tree because the classifier to attain fruitful quality determination from several mortal qualities that may facilitate distinguish malignancies. This investigation is concerned with the proficiency of our projected technique thereto of alternative notable benchmark characterization ways (support vector machine, self-sorting out the map, back propagation neural organization, C4.5 alternative tree, Naive Bayes, CART alternative tree, and versatile neuro-fluffy deduction framework) and behaviors probes eleven quality articulation malignant growth datasets.

[3] Joel St. George R, Anitha Jeba Kumari D. a perfect system for fragmenting cellular breakdown within the lungs footage is projected during this examination. The image is created utilizing Otsu's thresholding approach, and therefore the bar chart leveling of the still up within the air. It is a factual technique for gathering photos relying upon however comparative they're. The Fuzzy K-implies bunching calculation and therefore the Particle Swarm improvement technique are used for image division. Totally different qualities are calculable and thought of, together with PSNR and MSE. The perfect arrangement is discovered utilizing the planned improvement technique.

[4] Stelios Krinidis and Vassilios Chatzis. This work presents an image grouping variation of the flossy c-implies (FCM) strategy. In one among a sort flossy strategy, the projected technique coordinates close spatial knowledge and dark level knowledge. Flossy close knowledge C-Means is that the name of the new calculation (FLICM). FLICM will tackle the downsides of existing flossy c-implies calculations whereas in addition additional developing execution. Additional develops bunching execution the employment of a flossy close (both spatial and dim level) likeness life could be a very important component of FLICM, that expects to ensure clamor cold-heartedness and movie detail maintenance. Moreover, not in the slightest degree like all past flossy c-implies calculations declared within the writing, the projected approach is total while not tentatively adjusted boundaries.

[5] A.Sivarama Krishna anticipated a productive and inventive revelation of the brain tumor region from a picture that turned into wrapped up utilizing the Fluffy approach gathering calculation and histogram equalization. The

deterioration of pictures is accomplished by the utilization of the foremost figure assessment is done to decrease the degree of the wavelet coefficient. The results of the expected FCM clustering precisely pulled back tumor zone from the MR pictures.

[6] M. Sufyan discovered using a better lateral approach to segmentation of brain tumors that specifically relies on Sobel's feature detection. Provided them pals plots binary threshold activity with Sobel's technique and dig different ranges using a stable contour process. After reaching the top of this glory, most of the cancer cells are extracted from the received image using the depth values.

3. Proposed Methodology

3.1. Preprocessing

Resizing the pixel, diverting it from shading to grayscale (MRI and CT pictures are frequently grayscale), and afterward sifting it is all-important for the preprocessing interaction. Sifting might be done in an assortment of ways, including middle, mean, Gaussian, etc. To eliminate commotion from an image, we utilize channels. The middle methodology is utilized to channel the picture. The middle channel is a honing channel that permits the edges of the picture to be honed seen particularly to follow the odd segment or tumor.

(A). Reducing the picture size
(B). Grayscale image conversion
(C). Filtering by the median

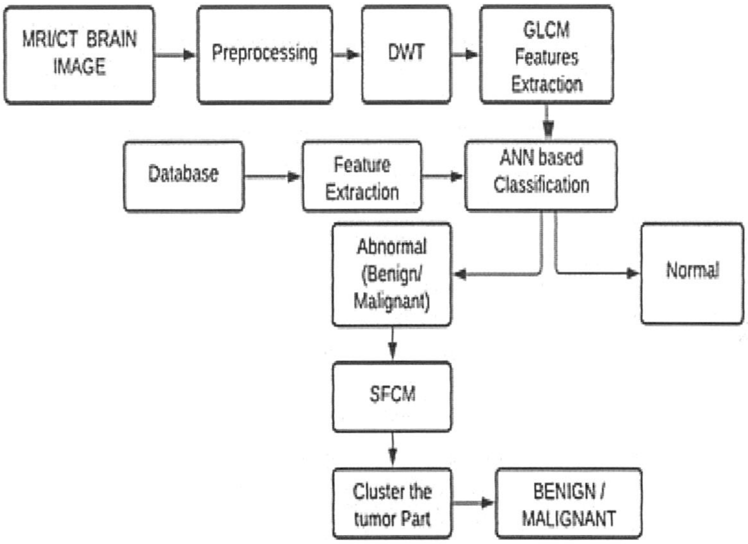

Figure 1. Block diagram of brain tumor detection

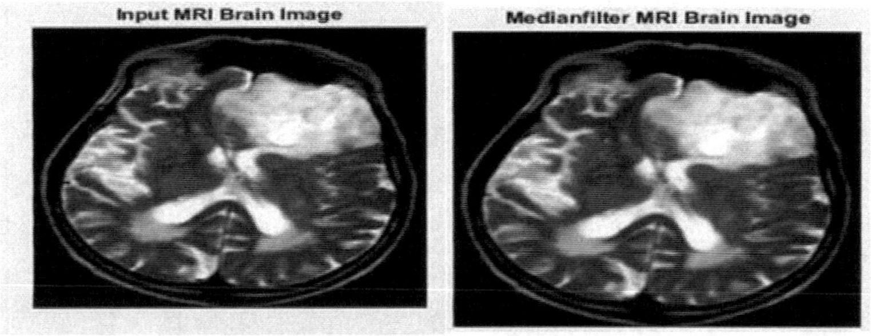

Figure 2. Input and Median filter MRI Image

3.2. DWT

The wavelets are tested discretely in DWT, which is a change wavelet. DWT gives spatial and recurrence area picture data simultaneously. The image will be breaking down in DWT utilizing a blend of the scientific channel bank and the annihilation activity. The channel bank is contained a couple of channels with high and low passes to coordinate with every deterioration stage. You are separated utilizing a low-pass channel. The 2D change is made by two separate 1D changes. Low-recurrence data, like estimation coefficients, and high-recurrence data, like point-by-point coefficients, are both remembered for 1D DWT. 2D DWT isolates the picture into four sub-groups: low-recurrence evenly and in an upward direction coordinated segments (CA), low-recurrence on a level plane engaged, and high-recurrence evenly engaged. High-recurrence evenly every now and again given even (CH) and in an upward direction directed segments (CV) (CD). The amount of circle space required increments as the picture goal increments. DWT is a procedure for contracting a picture's size without forfeiting quality, bringing about the more prominent goal. We're utilizing four layers of DWT to acquire a superior image of the picture's edges. The Contribution for a higher level of the DWT strategy will be the past DWT estimate coefficients framework. The two photographs with the most obvious edges have browsed the assortment of photographs in the four levels of the DWT approach. Those two pictures' GLCM attributes have been taken.

3.3 GLCM Features

The co-event lattice is (otherwise called a dim level co-event grids or GLCMs) is a network that portraying the appropriation of co-happening pixel esteems (grayscale values) at a specific offset all through a picture. The GLCM capacities process the number of sets of pixels with novel qualities that are available in an image in a specific spatial relationship, produce a GLCM, etc. The surface of an image may then be portrayed utilizing factual estimations separated from this lattice. You can utilize gray corps to recover measurements from the GLCMs whenever you've created them utilizing the gray comatrix. These measurements depict the energy, relationship, differentiation, and homogeneity of an image. In the wake of picking the best two photos from the four DWT levels, to get GLCM highlights, we utilize a dim lattice and concentrate four attributes from GLCM. Since entropy is definitely not an underlying capacity in GLCM, we figure it was utilizing an equation after removing four elements.

Contrast: It returns a measure of the intensity contrast between a pixel and its neighborover the entire picture. There is no contrast in a steady image. Contrast $\sum p,q|p-q|^2\ m(p,q)$ Where m (p, q) pixel at location (p, q)

Homogeneity: Returns a value indicating how close the element distribution of theGLCM is to the diagonal.

Energy: The sum of squared components of the GLCM is returned. The energy of aconstant image is 1

$$\text{Energy}=\sum(p,q)m(p,q)^2 \quad (1)$$

Correlation: The sum of squared components of the GLCM is returned. The energy of a constant image is 1. The correlation of a totally positively or negatively linked image is 1 or -1. Correlation is Nan for a constant image.

$$\text{Correlation}=\sum{}^{(p,q)m(p,q)}\big/_{(1+|p-q|)} \quad (2)$$

Entropy: The term "entropy" refers to a measure of randomness.

$$\sum b=0toL-1m(p,q)*\log 2\{m(p,q)\} \quad (3)$$

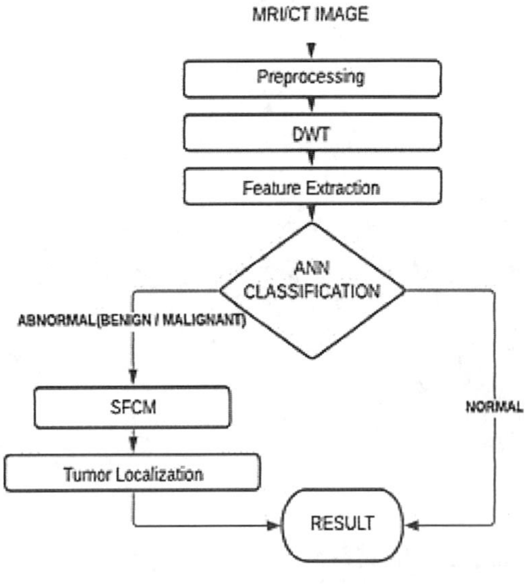

Figure 3. Flow chart of the project

3.4 Artificial Neural Networks

An artificial neural network (ANN) is created of a whole lot or thousands of process units that area unit connected via nodes. These processing units' area unit made of input and output units. The input units acquire multiple sources and patterns of knowledge supported an interior weight system, and also the human brain tries to find out regarding the data bestowed so as to supply one output report. ANNs would like a group of learning rules referred to as back propagation, Associate in Nursing word form for backward propagation of error, to good their output outcomes, very much like individuals use rules and directions to get a result or output. It starts with a learning method that teaches it to acknowledge patterns in recordings, whether or not visually, vocally, or textually. The network compares its actual output to what it absolutely was meant to produce the expected output during this observation activity. The distinction between the 2 outcomes is adjusted via back propagation. This suggests the network works backward, from the output unit to the input units, fixing the weights of the interactions between the units till the distinction between the $64000 and expected result emerges. When finishing the ANN coaching, the classifier examines the provided image within the info to examine whether or not the tumor will be found by examination every constituent. If a tumor is detected by examination every constituent, the classifier shows a message box indicating that the tumor is compact. The method of ANN includes:

- The Convolutional Neural Network takes knowledge within the style of patterns and vector photos from the skin setting. These inputs are selected by the notation x (n) for n range of inputs.
- The weights appointed to every input are increased by it. Weights are the items of knowledge that the neural network utilizes to unravel a drag. Weight could be common thanks to indicate the intensity of connections between neurons within a Neural Network.
- All of the input nodes are combined along within the physical laptop. If the summation is zero, the output is biased to create it non-zero or to extend the device reaction. Bias' weight and input are invariably set to at least one.
- The total relates to any numerical price between zero and eternity. The response is restricted by the brink price so it achieves the required price. During this scenario, the number is transmitted on via associate degree activation mechanism.

We may classify the image as regular or abnormal based on the characteristics collected by the ANN. If the image is correct, the procedure will be completed at this point. If the image is skewed, we cluster the tumor component and locate the tumor's position using Spatial Fuzzy C-means. This is the project's most essential stage.

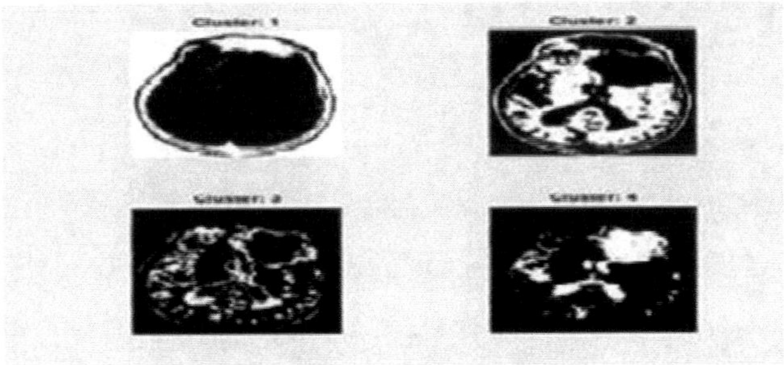

Figure 4. Segmented images

3.5 SFCM

The shorthand for spatial Fuzzy C-Means is spatial Fuzzy C-Means. This approach is finished by developing a chance perform thatcomeswith spatial info from every pixel's encompassing region.Through areplacement membr ershipfeature,thisgeographical data isleveraged tocomeupwith native membershipvalues forevery constituent.Finall y,new agglomeration centers canbe created lookingon the native and international membershipfunctions.Byaddressi ng theproblems of condition tonoise and intensity irregularity in MRI information, the SFCM technique enhances segmentation outcomes. The experimental findings on various generated and real-patient MRI brain footage reveal that the SFCM formula performs higher on image segmentation compared to existing FCM-based ways. J. C. Bezdek, [2] projected the fuzzy c-means (FCM) technique. The concept of FCM is to use the weights that minimize the entire weighted mean-square error.

$$J(wqk, z(k)) = S(k=1,K)S(k=1,K)(wqk)\|X(q)-z(k)\|2S(k=1,K)(wqk)=1 \quad (4)$$

$$wqk = (1/(Dqk)2)1/(p-1)/S(k=1,K)(1/(Dqk)2)1/(p-1), P>1 \quad (5)$$

With the FCM, every feature vector will belong to any cluster with a fuzzy truth worth (between zero and 1), that is calculated victimization the equation. The approach allocates a feature vector to a cluster supported the feature the vector's greatest night across all clusters. This is often the growth agglomeration and a growth localization methodology that was utilized during this experiment. With the FCM, every feature vector will belong to any cluster with a fuzzy truth worth (between zero and 1), that is calculated victimization AN equation. The approach allocates a feature vector to a cluster supported the feature vector's the best night across all clusters. This is often the growth agglomeration and a growth localization methodology that was utilized during this experiment.

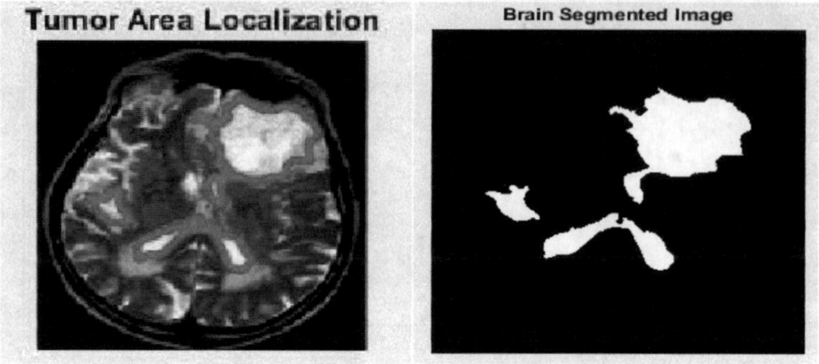

Figure 5. Tumor area Localization and segmented image

4. RESULTS

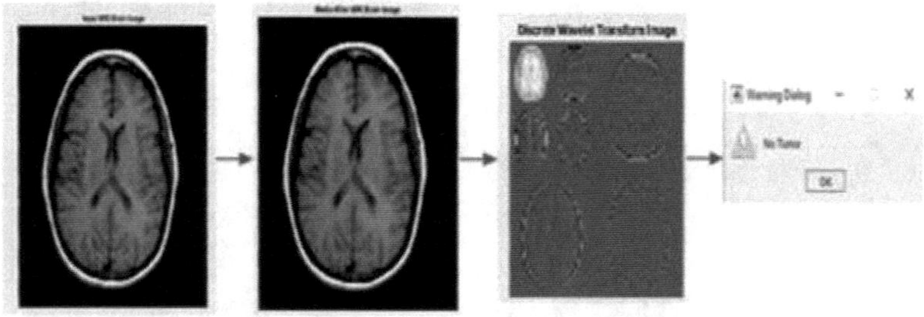

Figure 6. Result of a normal MRI brain image

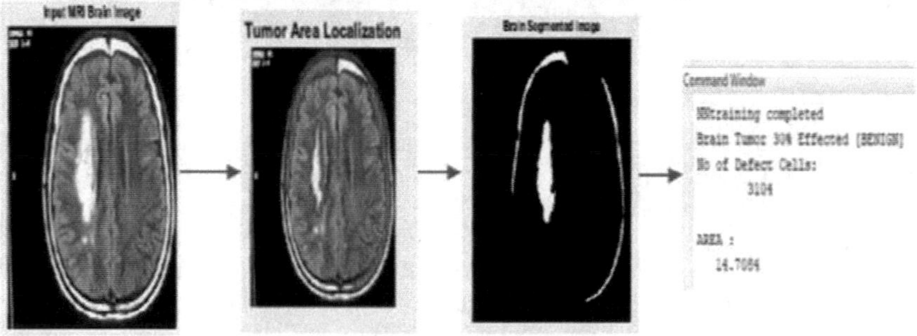

Figure 7. Result of a Benign MRI brain image

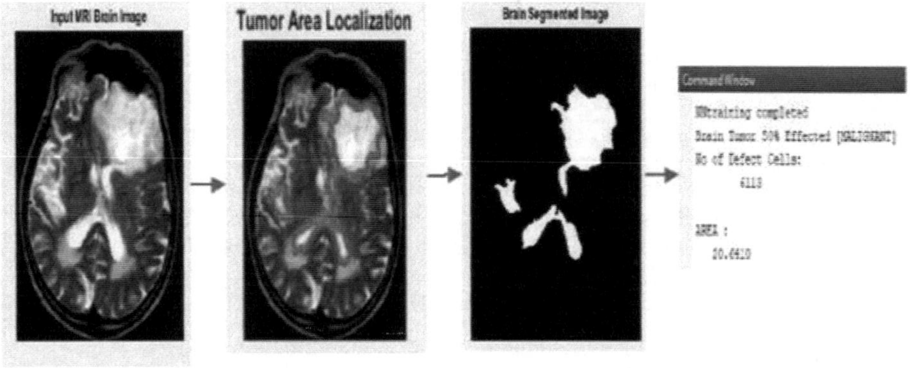

Figure 8. Result of a malignant brain image

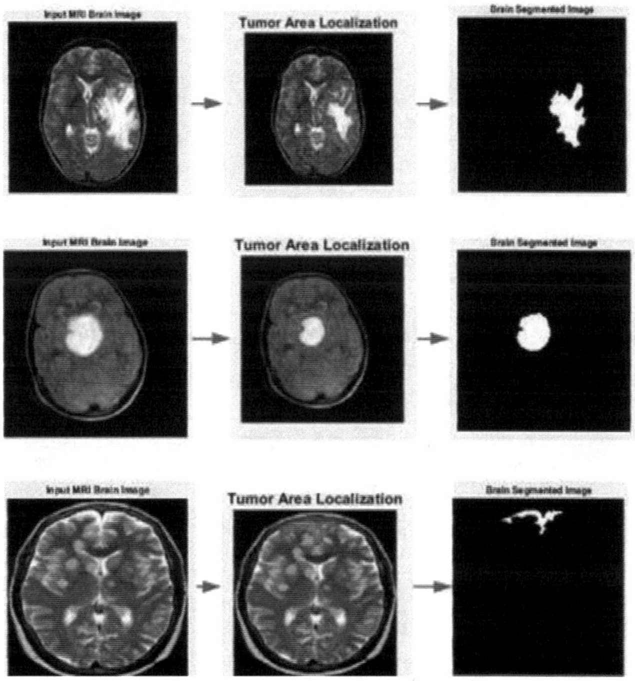

Figure 9. Result of different MRI brain images

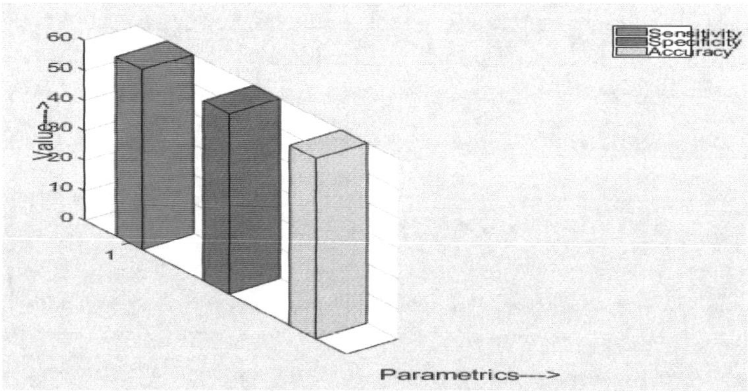

Figure 10. Performance Metrics

5. Conclusion

The use of ANFIS to the image differentiation of MRI brain tumors is described in this work. DWT, ANN, and SFCM are used in the brain tumor detection and classification system. The suggested approach employs several layers of wavelets, with DWT being used to achieve high accuracy. In segmentation, the DWT reduces edge loss. The results reveal that ANNs with the right sets of training data can discriminate between aberrant and normal tumor areas and properly categories them as benign, malignant, or normal brain tumors. Experiments using ANFIS as an image classifier provide promising results. Early detection of a brain tumor is possible, and the results are accurate. The goal of this project is to improve the ANFIS architecture so that it can achieve high classification accuracy and quantify tumor thickness and volume. in future research to improve accuracy by integrating more effective segmentation and extraction techniques with real-time images and clinical cases using a wider data set covering various scenarios.

6. References

[1] Asra Aslam, Ekram Khan, M.M. Sufyan Beg, Improved Edge Detection Algorithm for Brain Tumor Segmentation, Procedia Computer Science, Volume 58,2015, Pp 430-437, ISSN 1877-0509.

[2] M.Sharma and S.Mukherjee, "Artificial Neural Network Fuzzy Inference System (ANFIS) For Brain Tumor Detection ", November 2012.

[3] Lina Chato and Shahram Latifi, "Machine Learning and Deep Learning Techniques to Predict Overall Survival of Brain Tumor Patients using MRI Images", 2017 IEEE, 17th International Conference on Bioinformatics and Bioengineering.

[4] 2020 fourth international conference on 'Computing methodologies and communication'.

[5] H. Rahoumaa , Rabab Hamed M. Alyb , Hesham F.A. Hameda.16th International Learning & Technology Conference 2019, "Brain Cancer Diagnosis and Prediction Based on Neural Gas Network and Adaptive Neuro Fuzzy Kamel A Robust Fuzzy Local Information C-Means Clustering Algorithm", IEEE transactions on image processing, vol. 19, no. 5, may 2019.

[6] Jiasheng HAO, Yi SHEN, Hongbing XU and Jianxiao ZOU, "A Region Entropy Based Objective Evaluation Method for Image Segmentation", International Instrumentation and Measurement Technology Conference Singapore, 5-7 May 2009.

[7] Jianzhong Wang, Lili Dou, Na Che, Di Liu, Baoxue Zhang and Jun Kong. "Local-based Fuzzy Clustering for Segmentation of MR Brain Images".

[8] Johannes R.Sveinsson and Jon Atli Benediktsson."Combined Wavelet and Curvelet Denoising of SAR Images using TV segmentation".

[9] Asra Aslam, Ekram Khan, M.M. Sufyan Beg, Improved Edge Detection Algorithm for Brain Tumor Segmentation, Procedia Computer Science, Volume 58,2015, Pp 430-437, ISSN 1877-0509.

[10] Asra Aslam, Ekram Khan, M.M. Sufyan Beg, Improved Edge Detection Algorithm for Brain Tumor Segmentation, Procedia Computer Science, Volume 58,2015, Pp 430-437, ISSN 1877-0509.

[11] Muhammad Arif, F. Ajesh, Shermin Shamsudheen , Oana Geman ,Diana Izdrui , and Dragos Vicoveanu, Brain Tumor Detection and Classification by MRI Using Biologically Inspired Orthogonal Wavelet Transform and Deep Learning Techniques, Journal of Healthcare Engineering Volume 2022, Article ID 2693621, https://doi.org/10.1155/2022/2693621.

Detection of Lung cancer using Digital image Processing techniques and Artificial Neural Networks

Badireddygari Anurag Reddy[1] and Danvir Mandal[2]
[1]Research Scholar, Lovely Professional University, Phagwara, Punjab.
[2]Associate Professor, Lovely Professional University, Phagwara Punjab.
https://orcid.org/0000-0003-0065-5050

Abstract - In today's life digital image processing plays keys role. Digital Image processing using in medical field changes the medical world drastically. Medical imaging is growing speedily due to developments in image processing techniques including image recognition, analysis and enhancement. Now a day's different types diseases that have facing humans in daily life. In this Lung cancer is one of the major diseases that show devastating in humans heavily. Lung cancer is one of the major causes of death in humans. Lung cancer is normally carried out by trained professionals and these are majorly helpful in early stage of detection. This detection method introduces the possibility of human errors which consequences an automated process. In this paper it is mainly aims to detect the lung cancer at an early stage through an automated process and reduces the human errors and gets the accurate results. There are different types of processing techniques are currently using. Artificial neural network processing plays important role to identify the cancer tissues which are in lungs. Early stage of detection will save the human life.

Index Terms – lung cancer, image processing techniques, histogram, segmentation, artificial neural network.

1. INTRODUCTION

Lung cancer is one of the diseases and we can also call it as lung carcinoma and it's a malign lung tumor. It is mainly distinguished by the unconstrained cell growth in lung tissues. These are collected of cancer cells that are classified into two types. One is malignant tumor and second one is benign tumor. If the identification takes time then it causes death of human life. There are mainly two types of lung cancers are namely
a). SCLC
b). NSCLC
In this SCLC is referred as small cell lung carcinoma and NSCLC is referred as non small cell lung carcinoma. In general NSCLC contributes more than 84% of overall lung cancers and SCLC contributes nearly 16% of lung cancers. The growth rate of tumor is mainly categorized into grades format. In Grade 1: cells which are in lungs are looks like normal cells. In grade 1 cell is commonly slow spreading. Where as in Grade 2: cells are looks like in abnormal position and these cells are very fast spreading in human body. Grade 2 cells spreading four times than grade 1 cell. Another two grades are also present and these spreading abnormally in human body compare to other grades. The morality rate of lung cancer is lower and it never been higher. World research fund international (WCRF) gave first rank to the Lung cancer or respiratory system cancer. This research published result of humans how they are affecting by cancer. In the year 2017 nearly 22,65,600 new cases are identified in human bodies. It is almost being 20% of all cancers in world wide. It indicates more threat to the world.
In the year 2017 total 17,65,600 men were diagnosed worldwide. This is nearly 21 percent of all cancers in male population. It is most dangerous disease in male. While 5,00,000 women were diagnosed in that year which is 10 percent of women population in the world. According to WCRF it is third most dangerous disease in women. Finding or tackling the feral disease is always a difficult process but by using advanced technologies it is easy to detect cells and remission is also takes place. The cost of advanced technologies usage is more expensive.

Journal of Physics: Conference Series **2327** (2022) 012078 doi:10.1088/1742-6596/2327/1/012078

One of the important points to note that in growing countries large population of people are affected by different diseases. In that cancer is one of the diseases commonly identifying in people. Diagnosing the disease will be the most expensive and consultation charges are also heavy. Here our design is also aims to identifying cancer tissues in human body with affordable with a demographic scenario. From cancer image archive we have to use the Lung CT diagnosis data base. For various cancers large number of medical images is archived publicly. In this our database consists of Diagnostic enhanced CT scan images which are in Dicom format. Our database can handle nearly 6252 images.

In general CT images are available in JPEG format. So we have to convert that JPEG format images into Dicom format images of lungs. These images are scan and check if any abnormality is present in images through Image processing techniques. After completing the scanning process system checks is there any abnormality is present in the image. If anything found then it separates that image. These types of system are called as back propagation artificial neural networks.
Image processing techniques are used to convert the JPEG images into gray scale, Histogram equalization, thresholding and future extraction. The BPN (Back propagation Network) is trained using 80 images and it consists of total six input neurons, twelve hidden neurons and two output neurons. The output decides whether tumor is benign or malignant. The accuracy of this design method is nearly 80%.

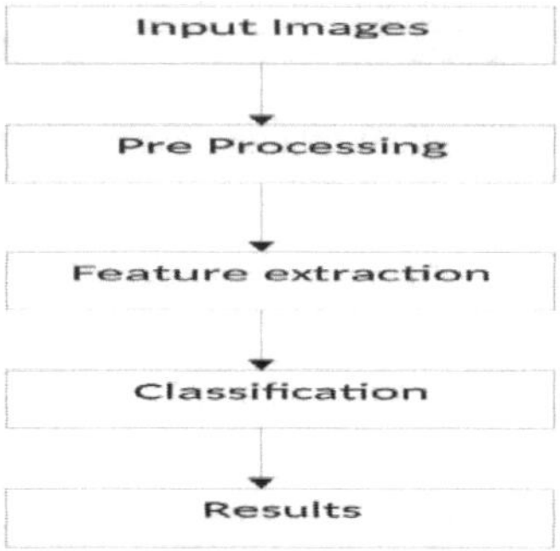

Figure 1. Flow chart of proposed method of cancer detection.

Figure 2. Flow chart of preprocessing Techniques.

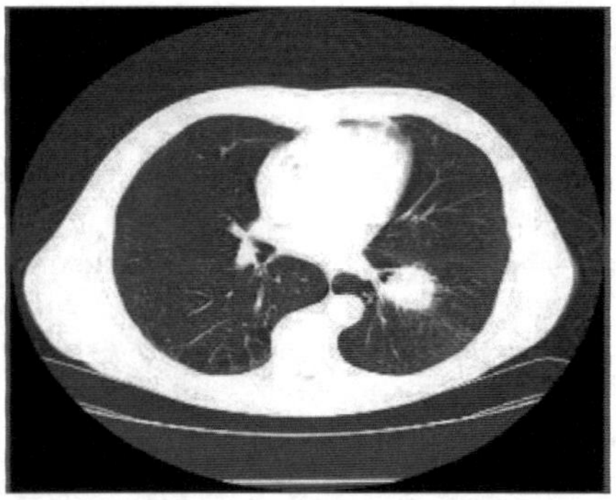

Figure 3. Image converted in Grayscale [1]

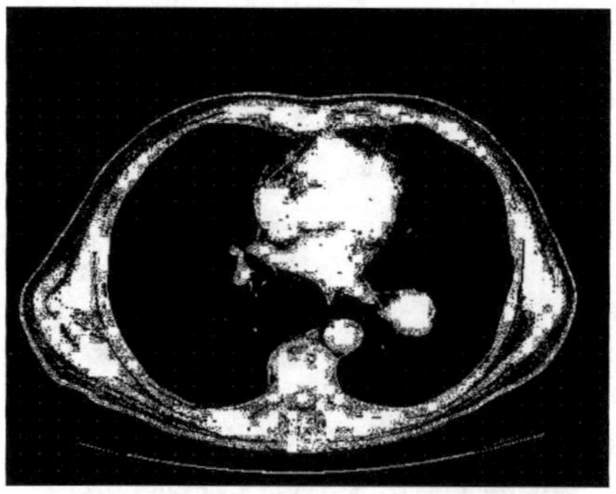

Figure 4. Binary version image of Fig3[1]

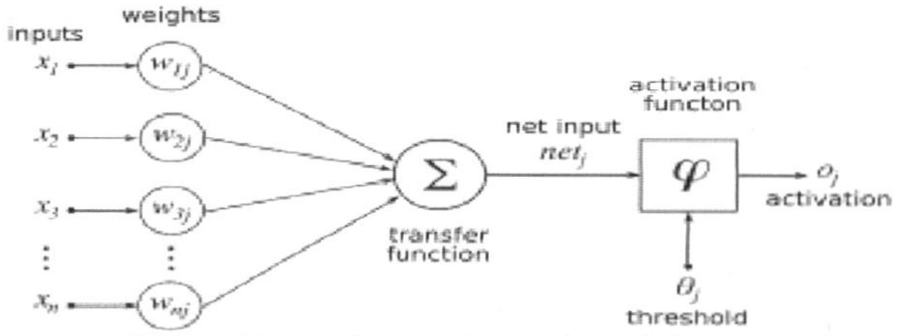

Figure 5. Neural network training method

2. PROPOSED METHOD

The proposed method begins with Image pre processing, Image segmentation & Histogram equalization, Feature extraction and lung cancer cell identification and result.

A. Image Pre processing

If the images in binary form then extraction of images is executed correctly. The binary forms of image are black and white images. In general images are in Red, Green and Blue form. Images are in logical representation format that is combination of three matrices. The dimension of each pixel is 1*n form, here n is called number of pixels. This type of data is complex in format and it is converted into grayscale image. Gray scale images are logical representation and each pixel represented in discrete format. The discrete level representation is out of 0 to 255. The lowest level represented as darkest color and highest level represented as brighter color. The dimensions are represented in pixels format. Generally medical images are default in grayscale but it checks if any other components are in images. To remove these type of components convert these into grayscale. It helps to detach hardware level errors and it will be occurred during image scanning process.

B. Image Segmentation & Histogram equalization

Image segmentation is plot between the tones and these are represented in the images and tones are shared by number of pixels. Histogram at high level indicates thick color and histogram at low level indicates dark images.

Image segmentation leads a clear and image and print image as compared to primary image. This image includes sharp boarder and objects are highlighted. By this the dynamic range is and contrasts are increased. Image segmentation will provide better contrast result.

C. Thresholding

It is one of the processes used for separation of an image by detaching the data pixels. In grayscale the pixels are above or below the constant level. This unchanging level is called as threshold. Images are retaining the required information and if any information is not related then it neglects that information. For segmentation purpose we use thresholding and section of interest. This thresholding also transforms color images into binary image or black and white image. In general pixels of binary image are described in binary 1 and binary 0. Binary 1 is represented as white image or data presence and binary 0 is represented as absence of data. This is mainly helps the data extracting from normal images.

It is clearly observed and noticed that tumors in the picture images are lie in the top above side of the spectrum in gray scale image. Due to this we checked the pixels range in between 160 to 230. In this range all types of pixels are turned into white.

Table 1: Extracted features in terms of pixels for a given object.

Area: 16
Minor axis Length: 2.53
Major axis Length: 15.23
Eccentricity: 0.924
Solidity: 0.596
Perimeter: 23.698

D. Feature extraction

The steps involved in future extraction are it calculates the certain properties of the object which is right side of the image. The above mentioned ranges are used to train the neural networks and it is used for conclusion statement. This process is called as Future extraction. Here we are using MATLAB and in this regionprops function is used. This type of function is used to calculate nearly 25 types of futures of different identified images. If the numbers of future objects are increased then the efficiency is reduced. So limit the future objects then increases the efficiency and accuracy. In this observation there are total 6 future objects are analyzed i.e. Area, Minor axis Length, Major axis Length, Eccentricity, Solidity, Perimeter. For every image initially calculating mean value and these images are in binary format. By running loop insignificant futures have been removed and objects were having greater area than 3 percent of the largest object. All the futures of mean value were computed and fed into 7*70 matrixes and this matrix is fed with the neural network.

3. BACK PROPAGATION NETWORK

Neural networks can handle or capable of large number of simultaneous calculations. Our detection system generally uses back propagation network. Our proposed network consists of total three layers. The three layers are input layer, hidden layer and final output layer. In that input layer having 6 neurons and it is one for one extracted future, hidden layer having 5 neurons and while the output layer having only 2 neurons. The outputs are may be malignant and benign. At each stage after every epoch it minimizes the errors. Back propagation network is mainly working on this principle only. It checks variation between the outputs and their corresponding inputs of a given layer.

Binarization is one of the methods used to convert the color pixels into two different classes. In that black and white images are used. The matrix that gives information and checks whether given image depicts cancerous lungs. A matrix of 2*70 is used for this purpose. Given image is represented by every column. In the first line row it is cancerous images are present and non cancerous images are present in the second line row. One important point to note is both row and column cannot be 1 or 0 at a time. This type of matrix is basically used for training purpose and it's for verification. A MATLAB tools are used to design and for pattern purpose. A Sigmoid training function network is used for all the layers. Based on mean square error we evaluate the performance of the network. After every epoch our network trains itself updates weights. In the final network stops its updates of weights whenever it reaches minimum squared error.

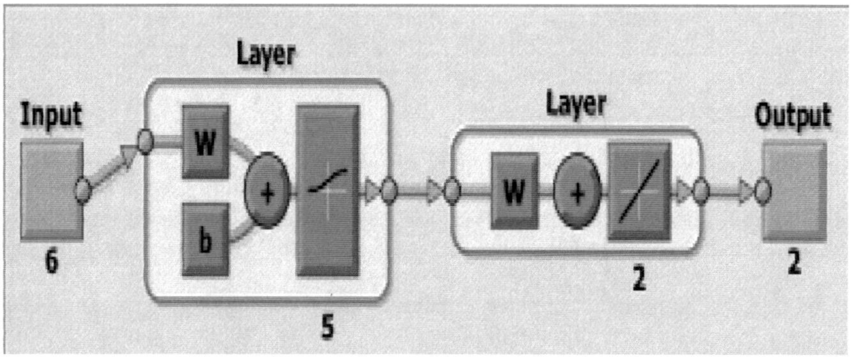

Figure 6. Neural network using MALAB

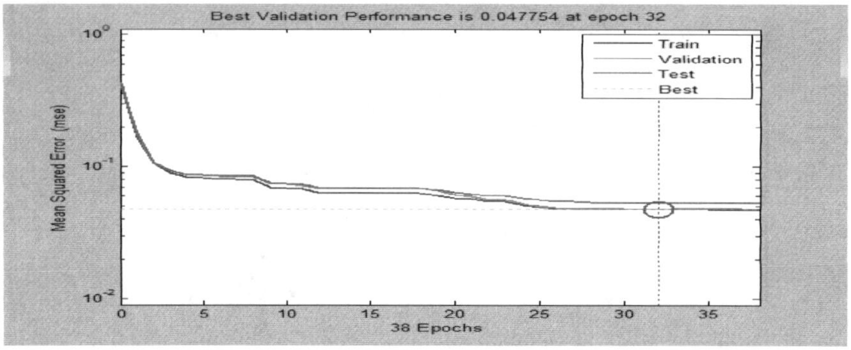

Figure 7. A graph of best validation performance of the system

4. OBSERVATION

In this mean squared error it is observed that at 32 epoch getting best validation performance i.e 0.04754 and error in the validation is minimum of 0.2. Error validation is also is reduces if the performance is increases. In this graph it is having total 38 epochs. Then the system automatically stops its updating weights after these 38 epochs. By using these functions we can achieve efficiency of 80%.

5. CONCLUSION

In the detection of lung cancer computer Aided Detection system was used. By using this performance of this system is accurate and gives better results. This system helps to detect the incorrect diagnosis and remove those results. The system will run effectively if the large data base used for detecting lung cancer disease. Without presence of the radiology the system gives better result by using large data base. In the future proposed work can be evaluated by well known algorithm about this lung cancer detection. Deep learning algorithm and machine learning algorithm methods are used to improve the performance of the system and accuracy.

6. REFERENCES

[1]. S.Kalaivani, Pramit Chatterjee, Rishi Gupta (2017) "Lung cancer detection using digital image processing and artificial neural networks." 2017 International conference of Electronics, Communication and Aerospace Technology (ICECA). DOI: 10.1109/ICECA.2017.821277.

[2]. Shamala B. Terdale, K.V.Kulhalli, "CAD System for Lung Cancer Detection using ANN", IOSR Journal of Electronics & Communication Engineering (IOSR-JECE), ISSN: 2278-2834, ISBN: 2278-8735, PP: 11-15, 2012.

[3]. S. Logesh kumar, M. Swathy, S. Sathish, J. Sivaraman and M. Rajasekar, "Identification of Lung Cancer Cell using Watershed Segmentation on CT Images", Indian Journal of Science and Technology, Vol 9(1), DOI: 10.17485/ijst/2016/v9i1/85765, January 2016.

[4]. I. Ali, G. R. Hart, G. Gunabushanam, Y. Liang, W. Muhammad, B. Nartowt, M. Kane, X. Ma, and J. Deng, "Lung nodule detection via deep reinforcement learning" Frontiers Oncol., vol. 8, p. 108, Apr. 2018. doi: 10.3389/fonc.2018.00108.

[5]. Grove, O., Berglund, A. E., Schabath, M. B., Aerts, H. J. W. L., Dekker, A., Wang, H., Gillies, R. J. (2015, March 4). Quantitative Computed Tomography Descriptors Associate Tumor Shape Complexity and Intra tumor Heterogeneity with Prognosis in Lung Aden carcinoma. (A. Muñoz -Barrutia, Ed.)PLOS ONE. Public Library of Science (PLoS). http://doi.org/10.1371/journal.pone.0118261.

[6]. Sangamithraa, P., & Govindaraju, S. (2016) "Lung tumour detection and classification using EK-Mean clustering." 2016 International Conference on Wireless Communications, Signal Processing and Networking (Wispnet). DOI: 10.1109/WiSPNET.2016.7566533.

AUTHOR INDEX